AF677115

Journal of
Physical and Chemical Reference Data

Monograph No. 7

Atomic Transition Probabilities of Carbon, Nitrogen, and Oxygen
A Critical Data Compilation

W. L. Wiese
J. R. Fuhr
and
T. M. Deters

National Institute of Standards and Technology
Gaithersburg, Maryland 20899-0001

Published by the **American Chemical Society** and the **American Institute of Physics** for the **National Institute of Standards and Technology**

Copyright © 1996 by the U.S. Secretary of Commerce on behalf of the United States. All rights reserved.
This copyright is assigned to the American Institute of Physics and the American Chemical Society.

International Standard Book Number
1-56396-602-6

American Institute of Physics
500 Sunnyside Boulevard
Woodbury, New York 11797-2999

Printed in the United States of America

Foreword

The *Journal of Physical and Chemical Reference Data* is published jointly by the American Institute of Physics and the American Chemical Society for the National Institute of Standards and Technology (NIST). Its objective is to provide critically evaluated physical and chemical property data, fully documented as to the original sources and the criteria used for evaluation. One of the principal sources of material for the journal is the NIST Standard Reference Data Program, a program promoting the compilation and critical evaluation of property data.

The regular issues of the *Journal of Physical and Chemical Reference Data* are published bimonthly and contain compilations and critical data reviews of moderate length. Longer works, volumes of collected tables, and other material unsuited to a periodical format have previously been published as *Supplements to the Journal*. Beginning in 1989 the generic title of these works has been changed to *Monograph*, which reflects their character as independent publications. This volume, ''Atomic Transition Probabilities of Carbon, Nitrogen, and Oxygen — A Critical Data Compilation'' by W. L. Wiese, J. R. Fuhr, and T. M. Deters is presented as *Monograph No. 7* of the *Journal of Physical and Chemical Reference Data.*

Jean W. Gallagher, Editor
Journal of Physical and Chemical Reference Data

Prologue

In our previous tabulation of atomic transition probabilities for these elements,[1] which was published about 30 years ago, we made extensive use of the simple semi-empirical Coulomb Approximation by Bates and Damgaard[2] to supplement rather sparse data available at that time. We greatly appreciated having this valuable source available to us, and we expressed our gratitude in the following way:

> If there is no other data source
> Use the Coulomb Approximation, of course.
> The results should be certainly fine
> For any moderately or highly excited line.*

For this second compilation, the much more sophisticated theoretical approach of the Opacity Project[3–9] has provided the very large majority of data. It is thus only fitting to express our appreciation similarly:

> We have used a near limitless data source,
> It is the Opacity Project, of course.
> Its results are certainly fine
> For most any *LS*-coupled line.

*Presented in the introduction to F III of Ref. 1.

W. L. Wiese
J. R. Fuhr
T. M. Deters

Atomic Transition Probabilities of Carbon, Nitrogen, and Oxygen

A Critical Data Compilation

W. L. Wiese, J. R. Fuhr, and T. M. Deters

National Institute of Standards and Technology, Gaithersburg, Maryland 20899–0001

Atomic transition probabilities have been critically compiled for about 13 000 spectral lines of the three elements carbon, nitrogen, and oxygen (nuclear charges $Z = 6-8$), based on all available theoretical and experimental literature sources. All stages of ionization are covered, and the data are presented in separate tables for each element and ion. Separate listings are given for allowed (electric dipole) and forbidden (magnetic dipole and quadrupole, electric quadrupole) transitions. Each data table is arranged as a multiplet table, i.e., the spectral lines are grouped in multiplets, and these are arranged in terms of ascending lower and upper excitation energies. This arrangement is essentially equivalent to a grouping of multiplets into transition arrays and their ordering in terms of ascending quantum numbers. For each line, the transition probability for spontaneous emission A, the oscillator strength (or f-value), and the line strength S are given, along with the spectroscopic designation, the wavelength, the statistical weights, and the energy levels of the upper and lower states. For allowed lines the absorption oscillator strength is listed, while for forbidden transitions the type of transition is identified. In addition, the estimated uncertainty and the source are indicated. In short introductions which precede the tables for each ion, the main justifications for the choice of the adopted data and for the accuracy ratings are discussed. A general introduction contains a discussion of our method of evaluation and the principal criteria for our judgements.

Key words: allowed and forbidden transitions; atomic ions; carbon; f-values; line strengths; nitrogen; oscillator strengths; oxygen; stage of ionization; transition probabilities.

Contents

List of Tables

List of Figures

1. Background

This is the first part of a new effort to update, revise and expand the reference data tables on atomic transition probabilities[a] for the light elements hydrogen through neon which were published by the National Bureau of Standards (NBS) almost 30 years ago.[1] This new tabulation has been undertaken mainly because a vast amount of new, high-quality material has become available during the last few years, primarily from sophisticated atomic structure calculations. Since this material is so extensive, the new tables will have to be published in two parts. This first part contains all spectra of the elements carbon, nitrogen, and oxygen. A second part containing the spectra of hydrogen, helium, lithium, beryllium, boron, fluorine and neon will follow soon.

A key event has been the large-scale production of new data by members of the Opacity Project,[3–9] an international collaboration of about 20 atomic structure theoreticians under the leadership of M. Seaton, during the late 1980s and early 1990s. This project has produced about 10^5 multiplet f-values for the spectra of the light elements hydrogen through neon. The calculations were done with an advanced quantum mechanical approach in which the critical issue of the mutual interaction of the atomic electrons—the "electron correlation" problem—has been given detailed treatment. Another important development has been the similarly sophisticated, but less extensive, calculational work of A. Hibbert and coworkers with the CIV 3 code, addressed to some neutral, singly and doubly ionized spectra of carbon, nitrogen and oxygen.[10–15] These calculations are more detailed than those of the Opacity Project insofar as data have been calculated not only for multiplets but for individual spectral lines. Again, the critical problem of electron correlation was addressed with a detailed configuration interaction treatment.

Based mainly on comparisons with experimental data, it has been estimated that the uncertainties of these two theoretical approaches should be typically of the order of $\pm 10\%$. But there is a lack of experimental data which are accurate enough to sensitively test the calculated results. Some emission measurements of relative transition probabilities exist, with uncertainties estimated to be in the range from 5 to 10%. There are also some lifetime measurements available—with uncertainties in the 3 to 10% range—which have either been utilized directly for comparisons of resonance lines, or have been combined with the emission data to provide absolute f-values that are entirely based on experiments. Nevertheless, the issue of the accuracy of theoretical data is still not satisfactorily settled, especially for the neutral spectra of carbon and nitrogen.

[a]Throughout these tables, we will use the terms atomic transition probability, oscillator strength (f-value) and line strength on an interchangeable basis, since these are equivalent quantities.

2. Brief Discussion of the Principal Data Sources

2.1. General Remarks

Since the publication of our earlier volume,[1] our knowledge of atomic transition probabilities for light elements has broadly advanced, due mainly to the above-mentioned important theoretical developments. As a result of this work, the volume of reliable data has increased more than tenfold, and the data quality has typically increased by factors of about 2–5.

The sources selected for these tables are totally different from those utilized in our earlier tabulation. It is therefore appropriate to briefly review the principal contributions and to provide references to papers where they are more extensively described and reviewed. But first some general remarks are in order on the theoretical approaches, which provide almost all of the tabulated data.

It has long been recognized that in many-electron atoms and ions, the mutual interactions between the atomic electrons—also known as electron correlation—is the most critical factor for the accurate calculation of transition probabilities.[16,17] Because of this interaction, the wavefunction of an atomic level usually cannot be accurately described by that of a single configuration. Thus, recent atomic structure calculations—including those from our two main data sources (OP and CIV 3)— have usually been carried out in a multiconfigurational framework. These calculations approximate the wavefunction of an atomic state by a linear combination of wavefunctions of this and related states of the same total quantum numbers L and S and of the same parity. In essence, each state of a complex atom or ion is no longer thought of as resulting from a specific single configuration, but as consisting of a mixture of configurations — usually a dominant one plus admixtures of related configurations. For example, the ground term of a Be-like ion, usually designated as $2s^2\ {}^1\mathrm{S}$, is actually better described as

$$a_1 2s^2\ {}^1\mathrm{S} + a_2 2p^2\ {}^1\mathrm{S} + a_3 2s3s\ {}^1\mathrm{S} + \text{other configurations of even parity}$$

which form a ${}^1\mathrm{S}$ state, with the a_i's being the mixing coefficients. This multiconfiguration treatment has been shown to be very successful in reproducing accurate experimental level energies if it is sufficiently detailed, i.e., if it includes a large number of possible interacting configurations.

Multiconfiguration or configuration-interaction approaches do not provide estimates of the uncertainties of the calculated transition probabilities. The quality of the calculations may, however, be realistically judged (a) from comparisons with accurate experimental lifetime and emission data, (b) from convergence studies, in which the change in the A-value is studied as more configurations are gradually added, and (c) from the degree of agreement between the results in the "dipole length" and "dipole velocity" formulations, which are often both presented. (Good agreement is a necessary, but not sufficient, condition.)

Data comparisons show that extensive configuration interaction calculations, which include at least 10, but preferably 100 and more interacting configurations, are needed for neutral C, N, and O and their lower, also relatively complex ions to produce accurate theoretical f-value data. Consequently, we have based our tabulated data only on calculations with extensive configuration interaction treatments and have not considered other, less detailed theoretical material.

Nevertheless, for some multiplets of neutral carbon and nitrogen, sizeable differences remain even between these multiconfiguration results and accurate experimental data. More experimental comparison material, especially on transitions between higher levels, is needed to assess the theoretical data in a more systematic manner.

2.2. The Opacity Project

The large majority of the tabulated data for allowed or electric dipole (E1) transitions—we estimate 90%—is either based exclusively on or involves the results of the Opacity Project (OP). This project is an international theoretical collaboration which was formed in 1984 under the leadership of M. Seaton and is now completed. It has involved about 20 participating atomic structure theoreticians from research groups in the United Kingdom, France, Germany, the United States and Venezuela.

The principal work of this project has been the mass production of atomic data by sophisticated *ab initio* atomic structure calculations. Detailed descriptions of this approach as well as selected results have been presented in a series of papers entitled "Atomic Data for Opacity Calculations" in the Journal of Physics B starting in late 1987.[3–9] These references [3–9] represent papers which are specifically concerned with the spectra of C, N and O in various stages of ionization. In addition to atomic transition probabilities, energy levels and photoionization and collision cross sections have been calculated, too. We were fortunate to obtain a complete set of the OP multiplet data for atomic line strengths from A. Pradhan of Ohio State University, a member of the Opacity team, who keeps a comprehensive depository of Opacity data for North America. At this point, we would like to acknowledge our deep gratitude to him for his invaluable help. The timely appearance of this book owes indeed much to his kindness. We also thank the other members of the Opacity team for their cooperation and especially M. Seaton for his numerous suggestions.

We should also note that the Opacity team has published a book[18] which contains their transition probability data plus selected results on photo-ionization cross-sections, etc. This book includes some data on very weak as well as on some far-infrared transitions, mostly between fairly high quantum numbers, which we have omitted.

The Opacity approach differs from the normal configuration-interaction type atomic structure calculations insofar as it is based on an approximation which is usually applied to calculate electron-ion or electron-atom collision data—the close coupling (CC) approximation. For the calculation of oscillator strengths of discrete transitions, this method has been extended to the case of electrons with negative energies, i.e., to captured electrons which undergo bound-bound transitions in the field of a target ion with n electrons. Thus the wavefunction for some state of an atomic ion (or atom) with $n + 1$ electrons is constructed from that of a system consisting of a "target" ion with n electrons plus an additional electron in its field bound to the ion. Considerable effort has been devoted to obtaining accurate target ion representations using the configuration interaction codes CIV 3[19] or Superstructure (SS).[20] The numerical approach used to solve the close-coupling integro-differential equations is based on an R-matrix method developed by some members of the OP team.[3] It is important to note that in the OP calculations only multiplet data were obtained, and it was not attempted to produce data for individual spectral lines. Since LS-coupling is normally a good approximation for the light atoms and ions which have been calculated by the OP team, line data may be obtained from the well-known LS-coupling line strength fractions.[17,21]

2.3. The CIV 3 Code

Extensive configuration interaction calculations with the CIV 3 code (CIV 3 = Configuration Interaction code Version 3) performed by Hibbert *et al.*[10–15] for C I, N I, and O I, as well as for N II, O II, and N III, have also yielded fairly large sets of data. While these are appreciably smaller than the OP calculations, they still comprise typically a few hundred lines per spectrum. These calculations are more detailed than the OP work insofar as not only multiplet but also individual line data are calculated. This has been done by including in the Hamiltonian the Breit-Pauli type relativistic terms in addition to the usual nonrelativistic electrostatic interactions. Thus the line data are produced in intermediate coupling and intersystem line strengths are also obtained. The normally rather weak intersystem lines are more difficult to calculate, as comparisons with the few available experimental data show (Table 1). Large differences are observed especially between the experiments and the CIV 3 calculations for C I and N I lines. In those cases, we have selected the experimental data since these have been measured at levels of accuracy similar to LS-allowed lines.

2.4. Other Multiconfiguration Calculations and Calculations of Forbidden Lines

Much more limited data sets resulting from other multiconfiguration calculations of approximately equal – or in some cases even greater – sophistication have also been used when available. We have especially utilized works by Froese Fischer, Nussbaumer, Weiss, and their colleagues, which are referenced in the introductory comments to the various spectra. We have also used calculations of this type for forbidden lines, i.e. magnetic dipole (M1) and electric and magnetic quadrupole (E2 and M2) transitions.

Under LS-coupling conditions, magnetic dipole radiation takes place only between the fine structure levels of the same spectroscopic term. These non-relativistic M1 line strengths S_{M1} may be calculated explicitly, and Pasternak[22] and Shortley[23] provided the following formulas:

TABLE 1. Comparisons of experimental and theoretical results for the line strengths of intersystem lines.

Spectrum	Transition	Wavelength (Å)	Experiment	Estim. Uncert.[1]	Theory	Selected Data[2]
C I	$2p^2\,{}^3\mathrm{P} - 2p3d\,{}^3\mathrm{F}^\circ$	1279.23	7.96−02[a]	±25%	6.71−02[b]	Expt.
		1279.06	9.29−03[a]	±25%	1.63−02[b]	Expt.
		1279.50	4.19−03[a]	±28%	9.59−03[b]	Expt.
	${}^1\mathrm{D} - {}^3\mathrm{F}^\circ$	1470.09	1.65−02[a]	±25%	1.36−02[b]	Aver.
	${}^1\mathrm{D} - {}^3\mathrm{D}^\circ$	1467.88	1.42−02[a]	±25%	5.04−03[b]	Expt.
		1468.41	1.83−02[a]	±25%	1.53−02[b]	Expt.
N I	$2s^22p^3\,{}^4\mathrm{S}^\circ - 2s^22p^2({}^3\mathrm{P})3d\,{}^2\mathrm{F}$	954.104	8.48−02[c]	±50%	1.96−03[f]	Expt.
	${}^4\mathrm{S}^\circ - {}^4\mathrm{D}$	952.303	2.35−02[c]	±14%	5.03−02[f]	Expt.
		952.415	2.13−02[c]	±12%	5.38−02[f]	Expt.
		952.523	7.53−03[c]	±20%	2.32−02[f]	Expt.
	$2s^22p^3\,{}^2\mathrm{P}^\circ - 2s^22p^2({}^3\mathrm{P})3d\,{}^2\mathrm{F}$	1316.29	9.62−03[d]	±40%	1.09−02[f]	Expt.
	$2s^22p^2({}^3\mathrm{P})3p\,{}^4\mathrm{D}^\circ - 2s^22p^2({}^3\mathrm{P})3d\,{}^2\mathrm{F}$	9997.73	3.63+00[e]	±14%	1.63+00[f]	Expt.
		10017.8	6.73+00[e]	±14%	1.25+00[f]	Expt.
		9947.07	4.20+00[e]	±14%	2.05+00[f]	Expt.
		9980.42	2.39+00[e]	±14%	1.63+00[f]	Expt.
	$2s^22p^2({}^3\mathrm{P})3p\,{}^4\mathrm{P}^\circ - 2s^22p^2({}^3\mathrm{P})3d\,{}^2\mathrm{F}$	10730.5	3.77+00[e]	±14%	5.87−01[f]	Expt.
C II	$2s^22p\,{}^2\mathrm{P}^\circ - 2s2p^2\,{}^4\mathrm{P}$	2325.40	1.91−06[g]	±14%	2.01−06[h]	Aver.
C III	$2s^2\,{}^1\mathrm{S} - 2s2p\,{}^3\mathrm{P}^\circ$	1908.73	1.25−06[i]	±6%	1.10−06[j]	Aver.
N II	$2s^22p^2\,{}^3\mathrm{P} - 2s2p^3\,{}^5\mathrm{S}^\circ$	2139.01	1.32−06[k,l,m]	±10%	1.13−06[n,o]	Expt.
		2142.77	3.05−06[k,l,m]	±10%	2.67−06[n,o]	Expt.
	$2p3s\,{}^3\mathrm{P}^\circ - 2p3p\,{}^1\mathrm{P}$	6379.62	2.35+00[k]	±15%	1.54+00[p]	Expt.
	${}^3\mathrm{P}^\circ - {}^1\mathrm{D}$	3955.85	2.08+00[k]	±12%	1.97+00[p]	Aver.
	${}^3\mathrm{P}^\circ - {}^1\mathrm{S}$	3408.13	4.71−01[k]	±15%	3.93−01[p]	Aver.
	${}^1\mathrm{P}^\circ - {}^3\mathrm{D}$	5747.30	1.73+00[k]	±12%	1.63+00[p]	Aver.
		5767.45	6.79−01[k]	±15%	7.28−01[p]	Aver.
	${}^1\mathrm{P}^\circ - {}^3\mathrm{S}$	5073.59	5.11−01[k]	±15%	4.90−01[p]	Aver.
	${}^1\mathrm{P}^\circ - {}^3\mathrm{P}$	4654.53	5.93−01[k]	±12%	5.33−01[p]	Aver.
		4667.21	4.52−01[k]	±15%	3.75−01[p]	Aver.
		4674.91	5.10−01[k]	±15%	4.72−01[p]	Aver.
N III	$2s^22p\,{}^2\mathrm{P}^\circ - 2s2p^2\,{}^4\mathrm{P}$	1749.67	4.89−06[q]	±14%	2.80−06[r]	Expt.

[1]Expanded uncertainties, i.e., twice the combined standard uncertainties from both random and systematic effects, are given for the experiments whenever it was possible to derive these from the literature data.
[2]"Aver." indicates that the average of the cited experimental and theoretical data has been selected for the tables.

References

[a]C. Goldbach, M. Martin and G. Nollez, Astron. Astrophys. **221**, 155 (1989).
[b]A. Hibbert, E. Biemont, M. Godefroid and N. Vaeck, Astron. Astrophys., Suppl. Ser. **99**, 179 (1993).
[c]C. Goldbach, T. Lüdtke, M. Martin and G. Nollez, Astron. Astrophys. **266**, 605 (1992).
[d]C. Goldbach and G. Nollez, Astron. Astrophys. **201**, 189 (1988).
[e]J. Musielok, W. L. Wiese and G. Veres, Phys. Rev. A 51,3588 (1995).
[f]A. Hibbert, E. Biemont, M. Godefroid and N. Vaeck, Astron. Astrophys. Suppl. Ser. **88**, 505 (1991).
[g]Z. Fang, V. H. S. Kwong, J. Wang and W. H. Parkinson, Phys. Rev. A **48**, 1114 (1993).
[h]D. J. Lennon, P. L. Dufton, A. Hibbert and A. E. Kingston, Astrophys. J. **294**, 200 (1985).
[i]V. H. S. Kwong, Z. Fang, T. T. Gibbons, W. H. Parkinson and P. L. Smith, Astrophys. J. **411**, 431 (1993).
[j]A. Weiss, unpublished.
[k]J. Musielok, J. M. Bridges, S. Djurović and W. L. Wiese, to be published.
[l]A. G. Calamai and C. E. Johnson, Phys. Rev. A **44**, 218 (1991).
[m]C. E. Johnson, unpublished.
[n]A. Hibbert and D. R. Bates, Planet Space Sci. **29**, 263 (1981).
[o]C. Froese Fischer and H. P. Saha, Phys. Scr. **32**, 181 (1985).
[p]A. Weiss, to be published.
[q]Z. Fang, V. H. S. Kwong and W. H. Parkinson, Astrophys. J. **413**, L141 (1993).
[r]R. P. Stafford, A. Hibbert and K. L. Bell, Mon. Not. R. Astron. Soc. **260**, L11 (1993).

$$S_{M1}(SLJ, SLJ) = (2J+1)\,[S(S+1) - L(L+1) + 3J(J+1)]^2\,[4J(J+1)]^{-1},$$

and

$$S_{M1}(SLJ, SLJ-1) = [J^2 - (L-S)^2]\,[(S+L+1)^2 - J^2]\,(4J)^{-1}.$$

(The latter formula also applies to the case of S_{M1} ($SLJ+1$, SLJ) = S_{M1} (SLJ, $SLJ+1$) since S_{M1} is a symmetric quantity.)

Since the spectra of carbon, nitrogen and oxygen are always quite close to LS-coupling (as is evident from the line strengths of intersystem lines which are often smaller by several orders of magnitude (see Table 1)), we have used these formulas to obtain the strengths of M1 lines when they were not explicitly given. We have also calculated the A-values from S_{M1}, using experimental wavelengths. Froese Fischer and Saha[24] have tested these formulas and published a comparison of line strengths for the $^3P_1 - {}^3P_2$ M1 fine structure transitions of carbon-like ions. They found that relativistic effects causing departures from the LS-coupling value of 2.50 start to become noticeable only for ions of charge states of about ten and higher.

2.5. Emission Experiments

The emission experiments utilized here were mostly performed with wall-stabilized arcs. These are moderate-to-high density sources operating in a steady-state mode under highly reproducible, well controlled conditions[25–28]. In optically thin conditions, the observed local emission intensities of the spectral lines are directly proportional to their atomic transition probabilities. However, the line intensities are also proportional to the number densities (state populations) of the radiating atoms or ions, and this density determination represents the principal problem of this method. Actually, in many recent emission experiments—and in all of those utilized here—the determination of excited-state densities is partially avoided by measuring transition probabilities on a relative scale only.

Direct measurements of the ratios of transition probabilities without any plasma analysis could be carried out for cases where two or more spectral lines are emitted from the same upper atomic state since the state density cancels out. These so-called "branching ratio" measurements have been done accurately with photoelectric techniques and advanced radiometric standards—the latter being necessary to compare line signals from different positions in the spectrum. The relative emission data were then normalized with lifetime measurements to obtain absolute values.

Emission measurements could be extended to groups of lines[26–28] with similar excitation energies when the plasma source was of moderate or high density and when the plasma temperatures could be determined accurately. In such plasmas, the populations of excited states follow a Boltzmann distribution, and these Boltzmann factors depend on the temperature only. The plasma temperature measurement thus connects all line intensity data—which are still on a relative scale. As in the branching-ratio method, normalization of the data to an absolute scale has been done via some other method, usually using lifetime results. For several spectra of this tabulation such normalizations have been carried out (see, e.g., Refs. 26 and 27).

Even as relative values, emission data have been quite valuable for this compilation. They have produced information on a variety of transitions on a uniform scale, have yielded accurate results for individual spectral lines and thus have provided reliable experimental information on the spectroscopic coupling scheme. This has been very important here because the primary source of data, the Opacity Project, provides only multiplet data. In order to obtain the always tabulated individual line data, we have utilized LS-coupling fractions,[21] since LS-coupling is normally an excellent assumption for the spectra of light elements.[17] The emission measurements have provided valuable guidance for assessing its validity—or sometimes the degree of departure from it—for principal transitions in the various spectra (see, e.g., Ref. 28).

Relative emission measurements can be quite accurate, with uncertainties typically being ±5% or less for lines that are emitted from an optically thin plasma layer (i.e., without self-absorption). Somewhat larger uncertainties are encountered for weak lines which do not rise far above the continuum background or for cases where lines partially overlap. Also, in rare cases, unrecognized impurity lines may produce some distortion, leading to an increase in a measured line intensity. However, this possibility is remote here, since the C, N, and O spectra are among the best known, and the spectra of gas admixtures, i.e., usually noble gases used as buffer gases, are equally well known.

2.6. Lifetime Measurements

A number of measurements of mean radiative lifetimes of excited atomic states have been performed for neutral C, N, and O as well as for many of their ions. The two main approaches for lifetime measurements are delay-time and beam measurements.[29] In a typical delay-time measurement, atoms or atomic ions are generated in a gas cell or discharge and are excited by electrons or photons. The delayed fluorescence signals are then counted on a time-resolved basis and are analyzed in terms of an exponential decay which yields a mean decay- or life-time. In the second approach, fast ion beams pass through a foil (foil excitation of beam-foil spectroscopy experiments) or are crossed with tunable lasers for excitation. In both cases the subsequent spontaneous emission of the fast ion beam is observed and analyzed as above. In earlier lifetime measurements, the experimental technology was not yet available to excite the atomic or ionic levels in a selective manner. Thus, numerous levels were excited simultaneously, and electrons cascaded from higher levels to the level to be studied, thereby repopulating it at the same time as it was depopulated.

The advent of tunable lasers has made possible the selective excitation of specific atomic levels, so that lifetime measurements can be made cascade-free. For the spectra of this tabulation, both the beam-laser and the delay-time approaches have been applied by several research teams.

3. Data Assessment

The central issue of a critical data compilation is the uniform critical assessment of the data, since this provides the basis for the data selection and the assignment of the numerical accuracies.

3.1. Main Criteria

All data have been reviewed by us with respect to the following four main criteria:

1. the author's evaluation and numerical estimate of his/her uncertainties,
2. the degree of agreement on his/her results with other reliable data,
3. the author's consideration of the critical factors affecting his/her results,
4. the degree of fit of his/her results into established systematic trends, or the reasons for possible deviations.

As indicated earlier, uncertainty estimates for calculated data are generally provided only on the basis of comparisons with experimental results from lifetime and emission measurements. Also, the agreement of results between the "dipole length" and "dipole velocity" representations has been utilized as an indicator of accuracy. Uncertainty estimates are usually given in a global fashion, lumping together all treated transitions, or at least certain classes of transitions. These estimates therefore have the character of a typical uncertainty for a reasonably strong transition.

The authors of the selected experimental papers have usually made detailed evaluations of the various sources of uncertainties in their measurements. Experimental uncertainty estimates have been given individually for each measured line strength or lifetime, and as a rule represent assessments of both measurement and systematic uncertainties. Measurement errors typically have been obtained by performing measurements repeatedly under the same conditions and by applying error statistics. Systematic errors have been estimated by considering the effects of approximations and limitations in the applied techniques, models and assumptions. Individual standard uncertainties have been combined using the usual method of "root-sum-of-squares." Finally, we established, whenever possible, "expanded uncertainties" of twice the standard deviations in order to obtain a uniform level of confidence of approximately 95% for the interval in which the experimental data should lie.

We found on occasion that discussions of experimental uncertainties by the authors have been incomplete. For example, sometimes not all sources of systematic error were addressed, or it was not clarified whether the stated measurement uncertainties represent a one-standard or multiple-standard deviations. Also, we noticed a tendency to underestimate experimental uncertainties, since on several occasions the differences between the results of various measurements were appreciably outside the mutually estimated uncertainties.

We have checked the second criterion—the degree of agreement among different data sets—for all lines. When larger disagreements are encountered we analyze such situations in detail. Especially interesting cases of agreements or disagreements are mentioned in the introductory comments of the pertinent spectra. In some instances we present graphical data comparisons.

3.2. The Critical Factors for the Determination of Atomic Transition Probabilities

The third point we have listed among our criteria is the author's consideration and treatment of the "critical factors" in his/her method. These are the factors that strongly affect the results. We require that these critical factors are adequately addressed and taken into account for any paper to be included in this compilation of reference data. This is the most important criterion by which we judge each contribution.

1. *Theoretical Methods*: As noted earlier, theoretical approaches have provided the large majority of the data for this compilation. The most important critical factor for calculated data is the electron correlation problem which must be taken into account by a detailed treatment of configuration interaction. It has been shown many times (a) by comparisons with experimental results, (b) by convergence studies in the calculations, i.e., by the inclusion of more and more interacting configurations, and (c) by the agreement, or lack thereof, of results in the dipole-length and dipole-velocity representations, that extensive treatments of configuration interaction are necessary in order to obtain reliable results for most atomic systems compiled here. Especially for the neutral atoms of this compilation, the number of interacting configurations to be considered for the lower atomic states must be in the tens, even in the hundreds, in order to obtain reliable results. We have therefore utilized only calculations which are based on extensive configuration interaction treatments. For a number of transitions in C I and N I, even these approximations do not appear to be adequate. In N I, because of close coincidences of the energies for the levels nd ^{2}P, ^{4}P and $(n+1)s$ ^{2}P, ^{4}P for $n = 3, 4, 5, \ldots$ and also for the levels $2s2p^4$ ^{4}P and $3s$ ^{4}P, configuration interaction effects are very pronounced so that even the most sophisticated calculations available exhibit strong disagreements for transitions starting or ending in these levels. Similar coincidences in energy level positions—and thus similar problems—occur for the $(n+1)s$ ^{1}P, ^{3}P and nd ^{1}P, ^{3}P levels of C I ($n = 3, 4, 5\ldots$).

For the determination of the strengths of individual lines, another critical factor for calculated data is the detailed consideration and treatment of the spectroscopic coupling. Again the two neutral spectra of C I and N I are most sensitively affected. Recent experiments have shown considerable departures from LS-coupling, which is usually considered to be an excellent approximation for light-element spectra. Some theorists have addressed the coupling problem and have calculated individual line strengths in intermediate coupling. However, these calculations often yield greater deviations from LS-coupling than emission experiments indicate, e.g., as discussed in the introduction to N I. The Opacity Project is restricted to multiplet data, however, and where it is the only data source we have applied LS coupling fractions to obtain individual line data. The experiments show that the stronger lines in multiplets depart much less from LS coupling than the weaker lines.

Before leaving the subject of calculated data, we would like to note that we have always used *line strengths* as the primary data and have converted these to A- and f-values by applying experimental wavelengths (energy differences), since these are much more precise than calculated ones. Some authors have used calculated wavelengths to obtain fully theoretical A-values. In these cases, we have reconverted their A-values to line strengths using their calculated wavelengths and have then, as usual, done all our conversions with experimental wavelength data.

2. *Emission Measurements*: For accurate measurements of *branching ratios* with emission sources, two critical factors must be considered:

(a) The lines must be emitted from an optically thin layer, i.e. self-absorption must be absent. For approximately homogeneous plasma layers, small amounts of self-absorption are also acceptable, provided the optical depth of the observed layer can be reliably determined, so that an accurate correction may be made.

(b) Radiometric calibrations of the line signals at various wavelengths must be done with accurate standards such as tungsten strip lamps, to take into account variations in sensitivity of the spectroscopic instrumentation with wavelength.

In emission measurements of *relative* oscillator strengths within a spectrum, the relative populations of ions or atoms in various excited states must also be accurately known (except, of course, when the upper level of the different transitions is the same). For emission sources (plasmas) in partial local thermodynamic equilibrium (LTE), the populations of excited states are distributed according to the Boltzmann population factors.[24] According to well-established validity criteria, partial LTE is readily attained in moderate and high density plasmas, i.e. for electron densities above a certain minimum value. The density of free electrons thus needs to be determined. In addition, the plasma temperature enters into the Boltzmann factors and must be reliably measured.

The emission experiments utilized for these tables were mostly carried out with wall-stabilized arcs, and the measured plasma conditions were such that partial LTE existed. Often, the electron densities were higher than the required minimum value by several orders of magnitude. Many of the emission results were given on absolute scales, which were provided by accurate lifetime data. Emission data are available mainly for the spectra of neutral and singly ionized atoms.

3. *Lifetime Measurements*: In lifetime experiments, critical factors include radiative cascading, radiative imprisonment, and collisional changes of level populations. Unrecognized line blending, e.g. with impurity lines, may also be a critical factor. Radiative imprisonment and collisional effects have been readily taken into account by varying, and especially by reducing, the number density of atoms in the measurement cell and by extrapolating to zero density. Radiative cascading could usually not be avoided in earlier lifetime measurement techniques such as phase-shift, electron beam, and beam-foil experiments, because the excitation of atomic levels was nonselective. Thus, numerous levels got excited simultaneously, and electrons cascaded from higher levels to the level to be studied, thereby repopulating it at the same time as it was being depopulated. These cascading effects tend to lengthen the radiative lifetime and are often apparent in the resulting radiative decay curves, which do not appear as single exponential decays, but as superpositions of several decays representing the primary and principal cascading levels. Therefore, the radiative decay curves often have been analyzed in terms of multiple-exponential decays, with the observed decays being fitted to 2 or 3 exponential decays plus a background. However, it has been shown[30] that this fitting procedure is seriously deficient because the decay curves may be fitted equally well to 2, 3, or 4 exponential decays, but the *primary* decay rates are appreciably different in each case. Many comparisons in the literature show that lifetime data thus corrected for cascading effects often differ systematically from the results of other accurate methods by as much as 20 to 50%, with these lifetimes being too long.[31] We have therefore not utilized such "multi-exponentially analyzed" lifetimes in this compilation. A cascade-correction procedure that has consistently provided reliable results is the ANDC method[32] (ANDC = arbitrarily normalized decay curves). In this technique, not only the decay of the primary level but also the decays of the main feeder levels are experimentally determined. (There are also relatively rare cases where the decay curve shows only a single dominant decay and all other cascades are of very different time scales, so that no correction procedure is required).[33]

Many recent lifetime experiments have been done with different versions of tunable-laser-based selective excitation techniques. This eliminates cascading as a critical factor, and most of these experiments have indeed provided very accurate results, though they are not entirely free of systematic errors either. Polarization effects due to the laser excitation and interference by external magnetic fields need to be given special attention, but often have not been discussed. Several groups in this field have described these phenomena in detail.[34–36]

3.3. Utilization Of Systematic Trends

We have occasionally used systematic trends of oscillator strengths along an isoelectronic sequence,[37] i.e., the scaling of f-values for a specific transition with nuclear charge Z, to test the consistency of data for successive isoelectronic ions, especially when they were obtained from different sources. This has been especially helpful for some forbidden lines where this analysis has provided valuable assistance when different but usually reliable methods disagreed.

4. Selection Procedure and Assignment of Uncertainties

For each spectrum we grouped the literature references assembled in our NIST data center according to the various theoretical and experimental approaches. As an example, Table 2 shows the distribution of data sources available for O I since 1966, the publication date of our previous tables. We then discarded work based on those theoretical or experimental approaches that are superseded by more advanced ones;

TABLE 2. Literature References for O I since 1966

		Number of References
A.	Calculations:	
	Semi-empirical (incl. Coulomb Approximation)	12
	Single configuration Hartree-Fock (HF)	3
	Limited multi-configuration HF	17
	Extensive multi-configuration HF	13
	Forbidden Transitions	4
B.	Experiments:	
	Emission	10
	Lifetimes, non-selective excitation	20
	Lifetimes, state-selective excitation	7
	Lifetimes of "metastable" states	4
	Absorption	7
	Electron Scattering	1
	TOTAL	98

e.g., we used only theoretical methods based on extensive configuration interaction treatments and discarded all work donewith semi-empirical, single-configuration Hartree-Fock or limited multi-configuration Hartree-Fock methods.

Also, with respect to lifetime results, we have only made use of data which are not affected by or are properly corrected for cascading effects, which includes beam-foil lifetime data corrected with the ANDC method. For the spectra included in these tables, sufficient numbers of such experimental results exist so that we did not need to include earlier lifetime results that were deficient on the critical factor of cascading effects. Likewise, for emission experiments, we did not use earlier measurements based on photographic techniques, but only recent photoelectric measurements.

Preliminary tables were set up containing all data from advanced methods. We analyzed these data in detail, searched for likely sources of numerical errors, and in several instances, made inquiries to the respective authors to try to determine the cause of major discrepancies.

Because the final evaluation and selection of sets of best data for each spectrum are highly dependent on the particular material available, the main justifications for the selections are given in the introductions to the individual spectra. However, we always adhere to two general rules on the final selection process:

(a) If one approach, theoretical or experimental, appears to be clearly better than all others, we have utilized only those results.

(b) When data from several different methods of comparable quality are available, either theoretical and/or experimental, we have averaged these to obtain the tabulated values. In the averaging procedure we have utilized each particular approach only once so that a fair balance between comparable-quality approaches is maintained. For example, when the CIV 3 code was applied to the same transition several times, usually with similar results, we used only the results of that version which we deemed to be most accurate. However, when data existed from the OP project as well as from the CIV 3 code and from other calculations with extensive multi-configuration treatments, we used the results of all these independent approaches and tabulated the arithmetic averages. Similarly, we have averaged lifetime results obtained with different techniques and/or with significantly different instrumentation. However, we have used only those lifetime techniques involving selective excitation or the ANDC correction. If the same experimental group obtained different results for a specific transition, we considered the latest result as the most accurate one, which superseded all earlier ones.

The final step in the evaluation is an estimate of the uncertainties of the selected data. At the present state of our knowledge, we still find it impossible to assign specific numerical uncertainties to each transition. We therefore adhere again to our earlier devised classification scheme wherein the data are grouped in several levels of accuracy which differ by steps of about factors of three. We use the following notation:

AA	...... uncertainties within	1%
A	...... uncertainties within	3%
B	.... uncertainties within	10%
C	.... uncertainties within	25%
D	...... uncertainties within	50%
E	.... uncertainties larger than 50% (but typically within factors of 2–3)	

We regard this compilation as a table of *reference* data and we therefore generally limit the multiplet entries to material we estimate to be uncertain by no more than ±25%. But in order to present all or most of the fairly prominent transitions of a spectrum, as well as to present all components of multiplets including the weak lines, we include some transitions of lesser quality.

As in our earlier work,[1] the word uncertainty is used with the meaning "extent of possible error" or "possible deviation from the true value." While this is not a precise definition of uncertainty, we have found it impossible to find a better common denominator when we consider the range of error discussions in the various papers and the fact that even sophisticated calculations only provide error estimates by comparisons with experimental or other advanced theoretical data.

For the prominent lines of the various spectra, normally transitions between lower excited states, we have obtained good estimates of the numerical uncertainties from comparisons among various experimental and theoretical results, combined with our evaluations of the capabilities and limitations of the various methods. But for most transitions between higher excited levels, the only data sources are the Opacity Project and sometimes CIV 3 calculations. As noted before, no intrinsic error estimates exist for these theoretical data. Thus, we developed a general scheme for assigning uncertainties for these higher transitions which is based on and extrapolated from our uncertainty estimates for the lower transitions:

(1) All multiplets of at least moderate strength, i.e., with line strengths $S \geqslant 3 \times 10^{-2}$, are normally assigned an accuracy rating of B, i.e., within ±10%. For weaker multiplets, the accuracy ratings are gradually decreased: first to B− for strengths in the range of $3 \times 10^{-2} > S \geqslant 1 \times 10^{-2}$; then assignments of C are given to multiplets with strengths in the range of $1 \times 10^{-2} > S \geqslant 1 \times 10^{-3}$; and finally, D for multiplets of lower line strengths. We have made further differentiations with plus and minus signs for the stronger and weaker lines within the C range.

(2) For the Li, Be and B-like spectra and for O I which are known to adhere very closely to *LS* coupling, the same uncertainties have been assigned to both multiplet values and to the individual lines in multiplets when they were obtained from *LS* coupling data.

(3) For the C- and N-like spectra, especially the neutral atoms, which are of more complex atomic structure and show appreciable deviations from *LS*-coupling, the uncertainty estimates for the lines have been lowered by one step from the multiplet estimates, when *LS* coupling fractions have been used. Thus, when multiplet values are assigned a B rating, the stronger lines ($S \geq 3 \times 10^{-2}$) have been assigned C+ accuracy ratings and for weaker lines the accuracies have been reduced further, to C−, D+, etc., since we have observed that weaker lines in multiplets usually show larger deviations from *LS* coupling than the stronger lines.

(4) We have taken any strong discrepancies between the length and velocity forms of the line-strength data into consideration in our accuracy estimates.

Data with uncertainties of Class AA exist only for the hydrogen-like species C VI, N VII and O VIII and for many transitions of the helium-like species C V, N VI, and O VII, where the listed data are essentially correct to the number of significant figures presented.

On the other end of the scale—i.e., transitions that are rather uncertain and thus belong to Class E—we have included only those lines that complete a set of multiplet data.

We have made further differentiations in the uncertainty classification scheme by applying plus or minus signs to some transitions to indicate that these lines are significantly better or worse, respectively, than the average line in this class, but that they do not quite belong to the next higher or lower class. If applications allow a choice of several lines of the same class, these should therefore be the first or last to use, respectively.

The new, much more advanced material compiled here allows extensive comparisons with the earlier data[1] compiled at NBS in 1966 which were taken from totally different sources. We have made many such comparisons between the old and new data and could thus perform checks on our earlier assessments of uncertainties. Three examples are given in Figs. 1, 2 and 3, in which the old and new data from C I, N I, and O I are compared. The comparisons show that our earlier error estimates are indeed realistic, i.e., for the majority of the data the differences between the new and the old data do not exceed the previous estimates (the new, usually higher quality data should be much closer to the "true" data, but, of course, they still contain significant uncertainties.)

In summary: While data obtained from theoretical approaches generally do not contain error estimates, many authors have compared their data with other experimental or theoretical results, have derived their data in both the dipole length and dipole velocity forms, and—based on the degree of agreement—have provided rough estimates of their uncertainties. Some authors have also performed studies of the convergence of their calculated data, as more and more interacting configurations are added. We find these estimates for the theoretical data generally to be quite realistic, and our own extensive comparisons of data from the most advanced sources have produced similar, often slightly more conservative conclusions.

Also, we sometimes have been more conservative with error estimates for experimental results when some systematic errors were apparently not considered, as in some laser-induced fluorescence lifetime measurements.

5. Arrangement of the Tables

We have essentially maintained the setup of our earlier critical compilations of atomic transition probabilities, since a sampling of a large number of users of those tables indicated that this was the preferred format. Thus, we present again detailed basic information for each spectral line: the spectroscopic notation, the wavelength, lower and upper energy levels and their respective statistical weights, the transition probability for spontaneous emission *A* and several equivalent expressions.

The first part of the tables contains data important for the identification of spectral lines, specifically the spectroscopic notation, the wavelengths, the energy levels of lower and upper states, and the respective statistical weights $g = 2J + 1$, where J is the total angular momentum quantum number. We present two wavelength columns, one giving the wavelength in vacuum or, for infrared lines above 1μ, the vacuum wave number; the other, for lines in the near ultraviolet, visible and near infrared spectrum ($2000 \text{ Å} < \lambda < 50{,}000 \text{ Å}$) giving the wavelength in air. Except for the hydrogen-like spectra, all wavelength data are basically experimental and the energy level data are derived from them. Many of the data are from compilations by C. E. Moore[38] which were recently re-issued in book form. These also contain new spectral data for O II which were critically evaluated and compiled by W. C. Martin *et al.*[39] Other wavelength and energy level data were obtained from our colleagues of the Atomic Energy Levels Data Center at NIST and the tables of Kelly.[40]

The tables are grouped according to multiplets, and they are arranged in order of increasing lower and upper energy levels. We have always listed individual lines within multiplets unless line data were unavailable or the wavelength positions of the lines coincided, which is the case for multiplets involving lower and upper levels with high n and ℓ values. Our arrangement follows exactly that of the multiplet tables by C. E. Moore. Because of its arrangement in multiplets and because of its comprehensive nature, this compilation may also serve as a general multiplet table.

Finally, in order to find lines quickly by wavelength in each spectrum, we have provided finding lists ordered in increasing wavelengths or wavenumbers at the beginning of each spectrum.

Most of our final tabulations were done during 1993, and 1994, especially during the second half of 1993 and early 1994.

We have expressed the atomic transition probabilities by four different quantities because different user communities have different preferences. Thus besides the transition probability for spontaneous emission A, we present the (absorption) oscillator strength f as well as the line strength S and log gf.

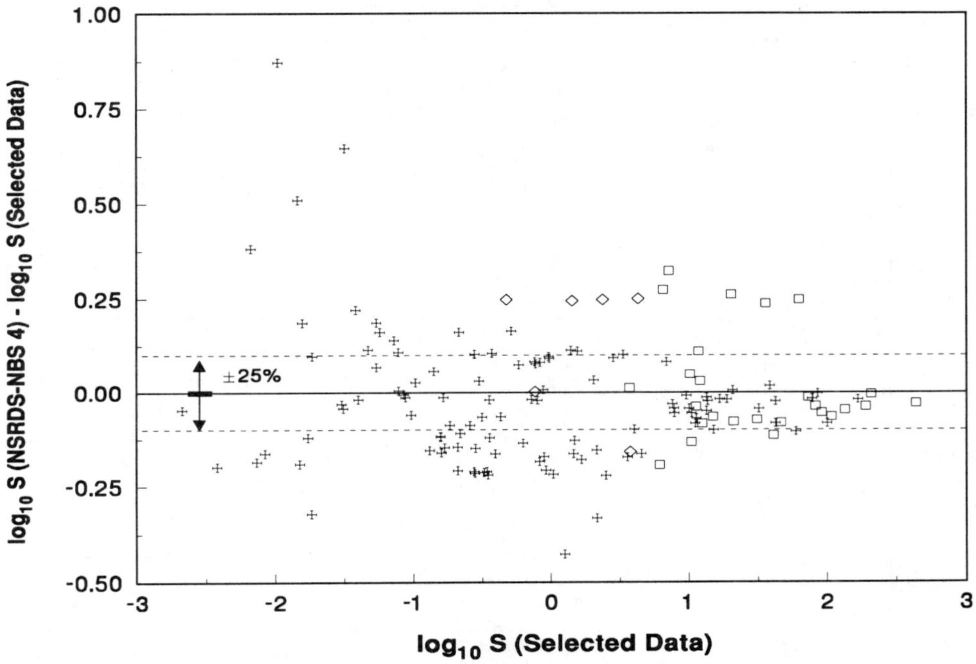

FIG. 1. Comparison of the C I data selected for this volume with those critically compiled in the NBS tables of 1966[1]. The ratio of old to new data is plotted on a logarithmic scale versus the logarithm of the line strength S of the new data. The broken lines indicate a band of deviations of ±25% around a perfect ratio of 1.00. Data from the earlier tables which were estimated to be uncertain by less than ±25% are shown as squares (coded "C" in the 1966 compilation); crosses denote earlier data with uncertainties within ±50% ("D" class); and diamonds indicate earlier data uncertain by more than ±50% (class "E"). The uncertainties of the newly compiled data are not considered in this comparison since they are estimated to be usually much smaller (but not negligible). It is seen that for the large majority of the 1966 data the estimated uncertainties are quite realistic.

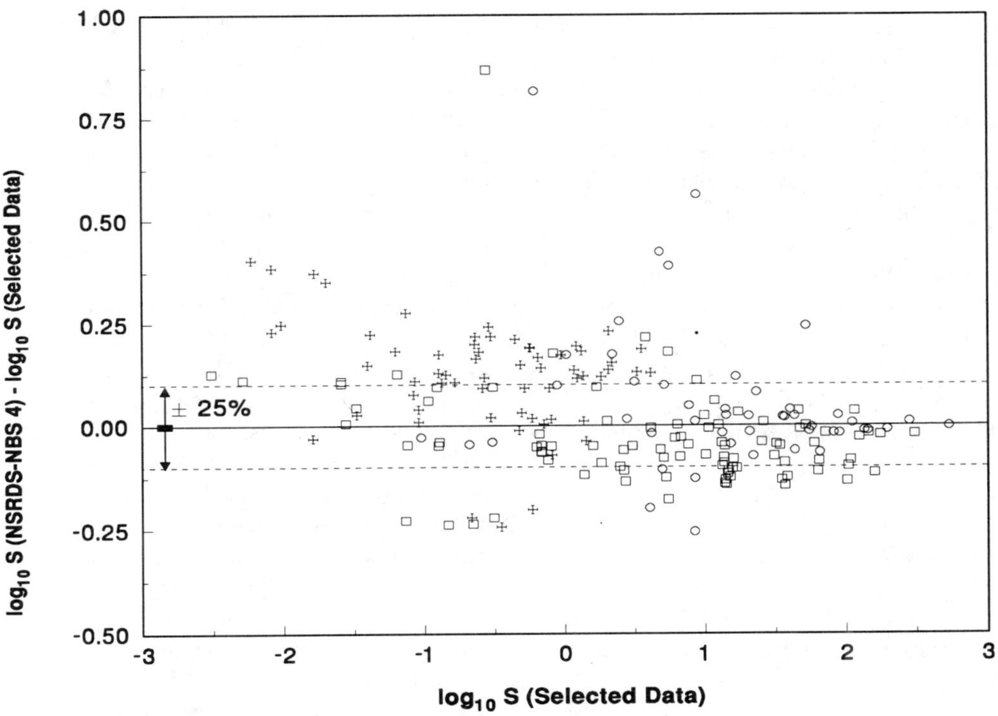

FIG. 2. Comparison of the N I data selected for this volume with those critically compiled in the earlier NBS tables of 1966[1]. The ratio of old to new data is plotted on a logarithmic scale versus the log S of the new data. For the explanation of the symbols see Figure 1. This figure contains an additional symbol, a circle, which denotes data in the earlier tables estimated to be uncertain by less than ±10%. The comparisons show that for the large majority of the 1966 data the estimated uncertainties are realistic, considering that the data tabulated in this volume also contain uncertainties, albeit much smaller, which are not included.

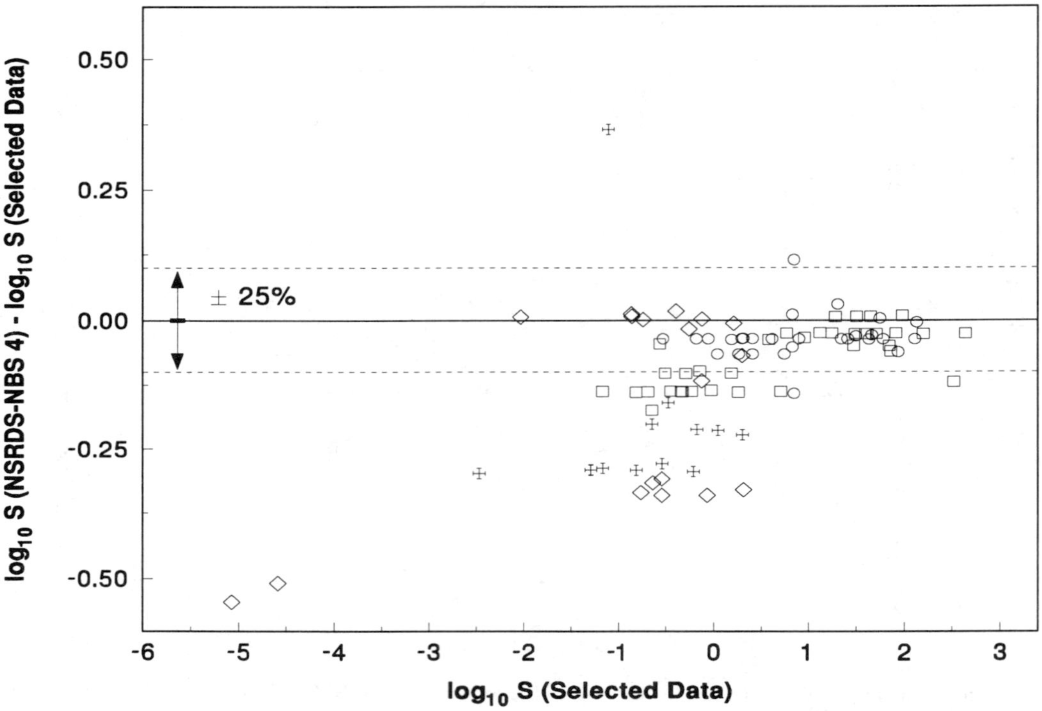

FIG. 3. Comparison of the O I data selected for this volume with those critically compiled in the earlier NBS tables of 1966[1]. The ratio of old to new data is plotted on a logarithmic scale versus the log S of the new data. For the explanation of the symbols see Figures 1 and 2. The comparisons show that for the large majority of the 1966 data the estimated uncertainties are realistic, considering that the data tabulated in this volume also contain uncertainties, albeit much smaller, which are not included.

The conversion factors between the tabulated quantities A, f, and S are listed in Table 3, based on the latest set of fundamental constants.[41] For the numerical conversions, we have used the wavelengths listed in the tables, which are experimentally derived. In some calculations theoretical wavelengths were used, but we have replaced these with experimental data (which are much more accurate) by regenerating the line strengths first and reconverting them with experimental data.

Occasionally, units other than A, f and S are found in the literature, for example the emission oscillator strength or the transition probability for absorption. These are related by the following formulas to the quantities we have given:

(a) The emission oscillator strength f_{ki} (i.e., i = lower, k = upper state) is related to the absorption oscillator strength f_{ik} (which we use) by

$$f_{ki} = -\ (g_i/g_k)\ f_{ik}.$$

(b) The transition probability of absorption B_{ik} is related to A_{ki} by

$$B_{ik} = 6.01\ \lambda^3\ (g_k/g_i)\ A_{ki}.$$

(c) The transition probability of induced emission B_{ki} is related to A_{ki} by

$$B_{ki} = 6.01\ \lambda^3\ A_{ki}.$$

The wavelength λ is in Angstrom units.

The material for the individual spectra is subdivided into a main table for allowed (electric dipole) transitions and a separate small table for forbidden (magnetic dipole magnetic quadrupole, and electric quadrupole) transitions. Intercombination or intersystem lines, while forbidden in LS coupling, are listed under allowed transitions since they are electric dipole lines.

In the source column, we have indicated by the letters "LS" that the line data have been obtained from LS coupling fractions.[21] We have always applied LS data when only multiplet values were available. Sometimes the letter "n" appears when line data are renormalized to an averaged multiplet value, since these are different from the ones originally obtained by the authors.

For forbidden lines, a few changes in the tabular arrangements have been made. First, we have indicated the type of transitions by listing "M1" for magnetic dipole and "E2" for electric quadrupole lines. (Higher-order forbidden lines are rarely available and only a few M2 lines are included.) Furthermore, the column containing f is omitted since this quantity is rarely utilized for forbidden lines. The line strength "S" has different atomic units from "S" for E1 transitions as given in Table 3, and is also used infrequently. When both M1 and E2 transitions occur at the same wavelength, the total transition probabilities are obtained by adding the magnetic dipole and electric quadrupole probabilities.

Most authors who have carried out recent calculations for S and A for E2 transitions follow a definition for S(E2) given by Cowan[17] and others. Since this now appears to be the preferred definition, we follow this, too. (This is reflected in the change of the conversion factor from that given in our earlier volume[1].)

Table 3. Conversion factors
The factor in each box converts by multiplication the quantity above it into the one at its left for the type of transition indicated.

	A_{ki}	f_{ik}	S
A_{ki} =	1	E1 $\frac{6.670_3 \times 10^{15} g_i}{g_k \lambda^2}$	E1 $\frac{2.026_1 \times 10^{18}}{g_k \lambda^3}$ E2 $\frac{1.119_9 \times 10^{18}}{g_k \lambda^5}$ M1 $\frac{2.697_4 \times 10^{13}}{g_k \lambda^3}$ M2 $\frac{1.491_0 \times 10^{13}}{g_k \lambda^5}$
f_{ik} =	E1 $\frac{1.499_2 \times 10^{-16} \lambda^2 g_k}{g_i}$	1	E1 $\frac{303.7_6}{g_i \lambda}$
S =	E1 $4.935_5 \times 10^{-19} g_k \lambda^3$ E2 $8.929_4 \times 10^{-19} g_k \lambda^5$ M1 $3.707_3 \times 10^{-14} g_k \lambda^3$ M2 $6.707_0 \times 10^{-14} g_k \lambda^5$	E1 $3.292_1 \times 10^{-3} g_i \lambda$	1

The line strength (S) is given in atomic units; formulas and values for these quantities in SI units are as follows:

For E1 transitions, $a_0^2 e^2 = 7.188_3 \times 10^{-59}$ m^2 C^2.
For E2 transitions, $a_0^4 e^2 = 2.012_9 \times 10^{-79}$ m^4 C^2.
For M1 transitions, $\mu_B^2 = (eh/4\pi m_e)^2 = 8.600_7 \times 10^{-47}$ J^2 T^{-2}.
For M2 transitions, $\mu_B^2 a_0^2 = 2.408_5 \times 10^{-67}$ J^2 m^2 T^{-2},

where a_0, e, m_e, and h are the Bohr radius, electron charge, electron mass, and Planck constant, respectively, and μ_B is the Bohr magneton.

The transition probability (A_{ki}) is in units of s^{-1}, and the f-value is dimensionless. The wavelength (λ) is given in Ångström units, and g_i and g_k are the statistical weights of the lower and upper level, respectively.

For the atomic constants entering into the relations given in this table, we have used the recommendations of the CODATA Task Group on Fundamental Constants (E. R. Cohen and B. N. Taylor, Rev. Mod. Phys. **59**, 1121 (1987)).

We should note that the definition of the line strength S for higher-order forbidden transitions (E2, M2, etc.) is not uniform in the literature. We follow the definitions given, e.g., by B. W. Shore and D. H. Menzel, in *Principles of Atomic Spectra*, (Wiley, New York, 1968), p. 440 and I. I. Sobelman in *Atomic Spectra and Radiative Transitions*, 2nd ed. (Springer-Verlag, Berlin, 1992), sec. 9.3. which are now widely used. In earlier NIST compilations, we used a different definition, both for the E2 line strength, which yields S-values that are 2/3 of the present values, and for the M2 line strength, which yields S-values that are 9/4 of the present values.

6. Key to Abbreviations and Symbols Used in the Tables

1. Symbols for indication of accuracy:

AA. . . . uncertainties within 1 percent,
A . . . . . uncertainties within 3 percent,
B . . . . . uncertainties within 10 percent,
C . . . . . uncertainties within 25 percent,
D . . . . . uncertainties within 50 percent,
E . . . . . uncertainties greater than 50 percent.

2. Abbreviations appearing in the source column of allowed transitions:

LS = *LS* coupling rules applied
n = normalized to a scale different from that of the author

3. Special symbols used in the wavelength and energy level columns:

Numbers in italics indicate multiplet values, i.e., weighted averages of line values.

4. Notation for exponents: In all tables, we have shown the power of ten by the customary notation, i.e., the numerical entries are followed by a plus or minus sign plus two digits indicating the exponent. For example, 5.89−03 stands for 5.89×10^{-3}.

Useful Relations

(A) Statistical weights:

The statistical weights are related to the total angular momentum or quantum number J_L (for one-electron spectra: j_ℓ) of a level (i.e., initial or final state of a line) by

$$g_L = 2J_L + 1,$$

and to the quantum numbers of a term (initial or final state of a multiplet) by

$$g_M = (2L + 1)(2S + 1).$$

(The "multiplet" values g_M may also be obtained by summing over all possible "line" values g_L. S is the resultant spin.)

(B) Relations between the strengths of allowed lines and the total multiplet strength:

1. Line strength S:

$$S(i,k) = \sum_{J_i, J_k} S(J_i, J_k)$$

or

$$S\,(\text{Multiplet}) = \Sigma\, S\,(\text{line})$$

(k denotes the upper and i the lower term).

2. Absorption oscillator strength f_{ik}:

$$f_{ik}^{\text{multiplet}} = \frac{1}{\bar{\lambda}_{ik} \sum_{J_i} (2J_i + 1)} \sum_{J_i J_k} (2J_i + 1) \times \lambda\,(J_i, J_k) \times f\,(J_i, J_k).$$

The mean wavelength for the multiplet, $\bar{\lambda}_{ik}$, may be obtained from the *weighted* energy levels. Often the wavelength differences for the lines within a multiplet are small, in which case the wavelength factors may be neglected.

3. Transition probability A_{ki}:

$$A_{ki}^{\text{multiplet}} = \frac{1}{\bar{\lambda}_{ik}^{\,3} \sum_{J_k} (2J_k + 1)} \sum_{J_i J_k} (2J_k + 1) \times \lambda\,(J_i, J_k)^3 \times A\,(J_i, J_k).$$

Relative strengths $S(J_i, J_k)$ of the components of a multiplet are listed for the case of *LS* coupling in C. W. Allen, *Astrophysical Quantitites*, 3rd ed. (The Athlone Press, London, 1973); H. E. White and A. Y. Eliason, Phys. Rev. **44**, 753 (1933); B. W. Shore and D. H. Menzel, *Principles of Atomic Structure*, p. 447 (John Wiley & Sons, Inc., New York, 1968); L. Goldberg, Astrophys. J. **82**, 1 (1935) and **84**, 11 (1936).

7. Future Plans and Acknowledgements

We plan to continue this critical compilation work with analogous tables for the elements hydrogen, helium, lithium, beryllium, boron, fluorine, and neon. We anticipate that these data will also be mainly based on the results of the Opacity Project.

It is a pleasure to acknowledge the assistance and cooperation of many colleagues in this field.

We are especially grateful to A. Pradhan who transferred to us electronically all Opacity Project data, and to M. Seaton and other members of the OP project who gave us valuable advice and clarifications on these data. We also owe special thanks to A. Hibbert and his colleagues for providing us with results of their CIV 3 calculations prior to publication and for giving us valuable comments.

We would also like to thank several colleagues who, in response to our urging, undertook special experiments or calculations to improve on some of the still inadequate results. We would like to mention especially the calculations by Weiss and Suskin on Be-like and C-like ions; the *LS*-coupling tests for B-like and C-like spectra by S. Glenzer, H. -J. Kunze, Y.-K. Kim, and J. Musielok, and the emission studies on N I, N II, and O II by J. M. Bridges, S. Djurović, J. Musielok and G. Veres.

We also acknowledge helpful communications from W. R. Johnson on the definitions and conversion formulae for magnetic quadrupole transitions.

Our colleagues from the NIST Atomic Energy Levels Data Center, W.C. Martin and A. Musgrove, have generously furnished us new data and advice on energy levels and wavelengths.

One of us (W. L. Wiese) performed part of the critical compilation work during extended visits to the Ruhr University, Bochum, Germany and the University of Rostock, Germany, as a recipient of the Humboldt Senior Scientist Award. He would like to thank H.-J. Kunze, G. Roepke and S. Guenter for their hospitality, and the Alexander von Humboldt Foundation for providing him this opportunity.

Finally, we would like to acknowledge the valuable contributions of C. Riland, R. Clark, and G. Dalton for assistance in setting up the tables and for advice and improvements in electronic programming, preparation of macros, etc. We also gratefully acknowledge the efficient typing of the text material by H. Felrice, M. King, and E. Schuster, as well as the editing of text material by E. Schuster.

8. References

[1]W. L. Wiese, M. W. Smith and B. M. Glennon, "Atomic Transition Probabilities—Hydrogen through Neon," Vol. I, NSRDS-NBS 4 (1966).

[2]D. R. Bates and A. Damgaard, Phil. Trans. Roy. Soc. London **A242**, 101 (1949).

[3]M. J. Seaton, J. Phys. B **20**, 6363 (1987).

[4]K. A. Berrington, P. G. Burke, U. Butler, M. J. Seaton, P. J. Storey, K. T. Taylor and Yu Yan, J. Phys. B **20** 6379 (1987).

[5]Yu Yan, K. T. Taylor and M. J. Seaton, J. Phys. B **20** 6399 (1987).

[6]J. A. Fernley, K. T. Taylor and M. J. Seaton, J. Phys. B **20**, 6457 (1987).

[7]D. Luo, A. K. Pradhan, H. E. Saraph, P. J. Storey and Yan Yu, J. Phys. B **22**, 389 (1989).

[8]D. Luo and A. K. Pradhan, J. Phys. B **22**, 3377 (1989).

[9]J. A. Tully, M. J. Seaton and K. A. Berrington, J. Phys. B **23**, 3811 (1990).

[10]A. Hibbert, E. Biemont, M. Godefroid and N. Vaeck, Astron. Astrophys., Suppl. Ser. **99**, 179 (1993).

[11]A. Hibbert, E. Biemont, M. Godefroid and N. Vaeck, Astron. Astrophys., Suppl. Ser. **88**, 505 (1991).

[12]A. Hibbert, E. Biemont, M. Godefroid, and N. Vaeck, J. Phys. B **24**, 3943 (1991).

[13]K. L. Bell, C. A. Ramsbottom and A. Hibbert, J. Phys. B **25**, 1735 (1992).

[14]K. L. Bell, A. Hibbert, R. P. Stafford, and B. M. McLaughlin, Phys. Scr. **50**, 343 (1994).

[15]K. L. Bell, A. Hibbert, R. P. Stafford and T. Brage, Mon. Not. R. Astron. Soc. **272**, 909 (1995).

[16]C. Froese Fischer, "The Hartree-Fock Method for Atoms," John Wiley and Sons, New York (1977).

[17]R. D. Cowan, "The Theory of Atomic Structure and Spectra," Univ. of Calif. Press,Berkeley, CA (1981).

[18]The Opacity Project Team, "The Opacity Project—Vol. I," Institute of Physics, Bristol, England (1995).

[19]A. Hibbert, Comput. Phys. Commun. **9**, 141 (1975).

[20]W. Eissner, M. Jones and H. Nussbaumer, Comput. Phys. Commun. **8**, 270 (1974).

[21]C. W. Allen, "Astrophysical Quantities," 3rd Ed. The Athlone Press, London (1973).

[22]S. Pasternak, Astrophys. J. **92**, 129 (1940).

[23]G. H. Shortley, Phys. Rev. **57**, 225 (1940).

[24]C. Froese Fischer and H. P. Saha, Phys. Scr. **32**, 181 (1985).

[25]W. L. Wiese, in "Methods of Experimental Physics," B. Bederson and W. Fite,Eds., Vol. **7B**, 307, Academic Press, New York (1968).

[26]C. Goldbach, and G. Nollez, Astron. Astrophys. **181**, 203 (1987).

[27]Q. Zhu, J. M. Bridges, T. Hahn and W. L. Wiese, Phys. Rev. A **40**, 3721 (1989).

[28]S. Glenzer, H.-J. Kunze, J. Musielok, Y.-K. Kim and W. L. Wiese, Phys. Rev. A **49**, 221 (1994).

[29]W. L. Wiese, in "Progress in Atomic Spectroscopy," Part B, W. Hanle and H. Kleinpoppen, Eds. Plenum Press, New York (1979).

[30]C. Lanczos, "Applied Analysis" (Prentice Hall, New York, 1956).

[31]For example, see N. Reistad, R. Hutton, A. E. Nilsson, I. Martinson and S. Mannervik, Phys. Scr. **34**, 151 (1986).

[32]L. J. Curtis, H. G. Berry, and J. Bromander, Phys. Lett. A **34**. 169 (1971).

[33]R. R. Haar, L. J. Curtis, T. J. Kvale, D. J. Beideck, I. Martinson, and R. Hellborg, Astron. Astrophys. **241**, 321 (1991).

[34]T. Fujimoto, C. Goto, and K. Fukuda, Phys. Scr. **26**, 443 (1982).

[35]P. Hannaford and R. M. Lowe, Opt. Eng. **22**, 532 (1983).

[36]W. Schade, L. Wolejko, and V. Helbig, Phys. Rev. A **47**, 2099 (1993).

[37]W. L. Wiese and A. W. Weiss, Phys. Rev. **175**, 50 (1968).

[38]C. E. Moore, "Tables of Spectra of Hydrogen, Carbon, Nitrogen and Oxygen Atoms and Ions," Edited by J. W. Gallagher, CRC Press, Boca Raton, FL (1993).

[39]W. C. Martin, V. Kaufman, and A. Musgrove, J. Phys. Chem. Ref. Data **22**, 1179(1993).

[40]R. L. Kelly, "Atomic and Ionic Spectrum Lines below 2000 Angstroms: Hydrogen through Krypton," J. Phys. Chem. Ref. Data **16**, Suppl. 1 (1987).

[41]E. R. Cohen and B. N. Taylor, Rev. Mod. Phys. **59**, 1121 (1987).

Note Added in Proof

Just before going to press, we learned about new high-quality data resulting from the SAM project (SAM = systematic and accurate multiconfiguration calculations of atomic properties). These calculations are done in the Breit-Pauli approximation which includes spin-orbit and other relativistic corrections. The calculations thus yield data for intersystem lines in addition to the usual allowed transitions. The new results would have been included in our selected data if they had been available to us earlier.

In the following, we present these results in a table (in the column labelled A_{ki} (SAM)) together with our selected data. Except for a few intersystem lines for C II and N III, the agreement is very good and supports our estimated accuracy ratings.

References for SAM Sources

[1]M. Godefroid, P. Jönsson, and C. Froese Fischer, private communication (1995).
[2]C. Froese Fischer, Phys. Scr. **49**, 323 (1994).
[3]C. Froese Fischer, M. Godefroid, and J. Olsen, private communication (1995).
[4]T. Brage and C. Froese Fischer, private communication (1995).
[5]C. Froese Fischer, private communication (1995).
[6]C. Froese Fischer, to be published.
[7]T. Brage and A. Hibbert, to be published.
[8]T. Brage, C. Froese Fischer, and P. G. Judge, Astrophys. J. **445**, 457 (1995).
[9]J. Fleming, T. Brage, K. L. Bell, N. Vaeck, A. Hibbert, M. R. Godefroid, and C. Froese Fischer, Astrophys. J., in press (1995).
[10]T. Brage, P. G. Judge, and P. Brekke, Astrophys. J., in press (1995).
[11]A. Ynnerman and C. Froese Fischer, Phys. Rev. A **51**, 2020 (1995).

Table for Note Added in Proof

Spectrum	Transition	Wavelength (Å)	g_i	g_k	A_{ki} (tabulated) $10^8\ s^{-1}$	A_{ki} (SAM) $10^8\ s^{-1}$	SAM Source
C I	$2s^22p^2\ ^3P - 2s2p^3\ ^3D^\circ$	1561.1	9	15	1.18+00	1.21+00	1
C II	$2s^22p\ ^2P^\circ - 2s2p^2\ ^2S$	1036.8	6	2	2.28+01	2.20+01	1
	$2s^22p\ ^2P^\circ - 2s2p^2\ ^2P$	904.08	6	6	4.10+01	4.07+01	1
	$2s^22p\ ^2P^\circ - 2s2p^2\ ^2D$	1335.3	6	10	2.85+00	2.88+00	1
	$2s^22p\ ^2P^\circ - 2s2p^2\ ^4P$	2325.40	4	6	5.26−07	4.61−07	2
		2323.50	2	4	1.53−08	1.44−08	2
		2326.93	4	4	1.12−07	9.43−08	2
		2324.69	2	2	7.30−07	6.26−07	2
		2328.12	4	2	7.63−07	6.96−07	2
C III	$2s^2\ ^1S - 2s3p\ ^1P^\circ$	386.203	1	3	3.46+01	3.59+01	3
	$2s^2\ ^1S - 2s2p\ ^3P^\circ$	1908.73	1	3	1.14−06	1.05−06	4
	$2s2p\ ^1P^\circ - 2p^2\ ^1S$	1247.38	3	1	2.08+01	2.01+01	4
	$2s2p\ ^3P^\circ - 2p^2\ ^3P$	1175.71	5	5	9.86+00	9.84+00	4
		1175.59	3	3	3.29+00	3.28+00	4
		1176.37	5	3	5.47+00	5.46+00	4
		1175.99	3	1	1.31+01	1.31+01	4
		1174.93	3	5	3.29+00	3.29+00	4
		1175.26	1	3	4.39+00	4.38+00	4
	$2s^2\ ^1S - 2s2p\ ^1P^\circ$	977.020	1	3	1.77+01	1.76+01	4
	$2s^2\ ^1S - 2s3p\ ^3P^\circ$	385.043	1	3	5.20−03	5.09−03	5
	$2s^2\ ^1S - 2s2p\ ^3P^\circ$	1906.68	1	5	5.19−11	5.12−11	4
N II	$2s^22p^2\ ^3P - 2s2p^3\ ^3D^\circ$	1085.1	9	15	3.88+00	3.75+00	6
	$2s^22p^2\ ^3P - 2s2p^3\ ^3P^\circ$	916.35	9	9	1.32+01	1.29+01	6
	$2s^22p^2\ ^3P - 2s2p^3\ ^3S^\circ$	645.00	9	3	1.09+02	1.10+02	6
	$2s^22p^2\ ^3P - 2s2p3s\ ^3P^\circ$	671.49	9	9	9.64+00	1.16+01	6
	$2s^22p^2\ ^3P - 2s2p3d\ ^3D^\circ$	533.67	9	15	4.13+01	4.42+01	6
	$2s^22p^2\ ^3P - 2s2p3d\ ^3P^\circ$	529.68	9	9	2.43+01	2.58+01	6
	$2s^22p^2\ ^3P - 2s2p4s\ ^3P^\circ$	508.74	9	9	2.54+00	2.81+00	6
	$2s2p^3\ ^3D^\circ - 2s^22p3p\ ^3P$	1275.6	15	9	6.67−01	5.90−01	6
	$2s2p^3\ ^3P^\circ - 2s^22p3p\ ^3P$	1628.2	9	9	2.53−02	2.60−02	6
	$2s2p^3\ ^3S^\circ - 2s^22p3p\ ^3P$	6445.8	3	9	–	1.34−04	6
	$2p3s\ ^3P^\circ - 2p3p\ ^3P$	4623.2	9	9	1.00+00	1.00+00	6
	$2p3p\ ^3P - 2p3d\ ^3D^\circ$	5938.5	9	15	5.56−01	5.52−01	6
	$2p3p\ ^3P - 2p3d\ ^3P^\circ$	5478.8	9	9	3.03−01	3.05−01	6
	$2s2p^3\ ^3D^\circ - 2s2p^23s\ ^3P$	836.41	15	9	1.46+01	1.77+01	6
	$2s2p^3\ ^3P^\circ - 2s2p^23s\ ^3P$	974.80	9	9	1.77+00	1.29+00	6
	$2s2p^3\ ^3S^\circ - 2s2p^23s\ ^3P$	1764.4	3	9	5.25−01	5.48−01	6
	$2s^22p3s\ ^3P^\circ - 2s2p^23s\ ^3P$	1592.6	9	9	1.56−01	1.55−01	6
	$2s^22p4s\ ^3P^\circ - 2s2p^23s\ ^3P$	6598.7	9	9	2.11−01	1.46−01	6
	$2p3p\ ^3P - 2p4s\ ^3P^\circ$	3842.7	9	9	9.19−01	9.07−01	6
	$2s^22p^2\ ^3P - 2s2p^3\ ^5S^\circ$	2139.01	3	5	5.49−07	5.26−07	7
		2142.77	5	5	1.27−06	1.28−06	7

Table for Note Added in Proof—Continued

Spectrum	Transition	Wavelength (Å)	g_i	g_k	A_{ki} (tabulated) 10^8 s^{-1}	A_{ki} (SAM) 10^8 s^{-1}	SAM Source
N III	$2s^22p\ ^2P^o - 2s2p^2\ ^4P$	1749.67	4	6	3.08−06	2.82−06	8
		1746.82	2	4	9.49−08	9.11−08	8
		1752.16	4	4	6.50−07	6.51−07	8
		1748.65	2	2	4.97−06	3.61−06	8
		1754.00	4	2	5.22−06	3.72−06	8
N IV	$2s^2\ ^1S - 2s2p\ ^1P^o$	765.147	1	3	2.32+01	2.31+01	9
	$2s^2\ ^1S - 2s2p\ ^3P^o$	1486.50	1	3	5.95−06	5.80−06	9
	$2s2p\ ^1P^o - 2p^2\ ^1S$	955.334	3	1	2.92+01	2.89+01	9
	$2s2p\ ^3P^o - 2p^2\ ^3P$	923.220	5	5	1.32+01	1.32+01	9
		923.056	3	3	4.40+00	4.39+00	9
		924.284	5	3	7.30+00	7.29+00	9
		923.676	3	1	1.76+01	1.75+01	9
		921.994	3	5	4.42+00	4.41+00	9
		922.519	1	3	5.88+00	5.87+00	9
	$2s2p\ ^3P^o - 2p^2\ ^1S$	594.895	3	1	5.81−05	9.03−05	9
O IV	$2s^22p\ ^2P^o - 2s2p^2\ ^4P$	1407.39	4	2	–	1.43−05	10
		1404.81	4	4	–	2.94−06	10
		1401.16	4	6	–	1.17−05	10
		1399.77	2	2	–	1.47−05	10
		1397.20	2	4	–	3.84−07	10
	$2s^22p\ ^2P^o - 2s2p^2\ ^2D$	790.199	4	6	7.08+00	7.18+00	10
		787.710	2	4	5.95+00	6.08+00	10
		790.112	4	4	1.18+00	1.17+00	10
	$2s^22p\ ^2P^o - 2s2p^2\ ^2S$	609.829	4	2	2.40+01	2.33+01	10
		608.397	2	2	1.21+01	1.25+01	10
	$2s^22p\ ^2P^o - 2s2p^2\ ^2P$	554.513	4	4	6.06+01	6.04+01	10
		554.076	2	2	4.86+01	4.76+01	10
		555.263	4	2	2.41+01	2.47+01	10
		553.329	2	4	1.22+01	1.21+01	10
	$2s2p^2\ ^4P - 2p^3\ ^4S^o$	625.853	6	4	3.19+01	3.17+01	10
		625.127	4	4	2.13+01	2.12+01	10
		624.619	2	4	1.07+01	1.06+01	10
O V	$2s^2\ ^1S - 2s2p\ ^3P^o$	1218.34	1	3	2.34−05	2.21−05	11
	$2s^2\ ^1S - 2s2p\ ^1P^o$	629.732	1	3	2.87+01	2.85+01	11

Carbon

C I

Ground State: $1s^2 2s^2 2p^2\ {}^3P_0$

Ionization Energy: 11.260 eV = 90820.42 cm^{-1}

Allowed Transitions

List of tabulated lines

Wavelength (Å)	No.	Wavelength (Å)	No.	Wavelength (Å)	No.	Wavelength (Å)	No.
in vacuum		1188.83	18	1280.60	5	1992.01	28
		1188.99	18	1280.85	5	1993.62	28
945.191	27	1189.07	18	1288.42	45		
945.338	27	1189.25	18	1311.36	44	in air	
945.579	27	1189.45	18	1328.83	4		
1122.00	26	1189.63	18	1329.09	4	2478.56	49
1122.33	26	1190.02	17	1329.10	4	2582.90	48
1122.34	26	1190.25	17	1329.12	4	2902.23	153
1122.45	26	1191.84	16	1329.58	4	2903.27	153
1122.65	26	1192.22	15	1329.60	4	2905.00	153
1122.99	26	1192.45	15	1354.29	43	4371.37	76
1128.82	25	1192.83	15	1355.84	42	4734.26	86
1129.13	25	1193.01	14	1357.13	41	4734.92	86
1129.16	25	1193.03	14	1357.66	40	4735.16	86
1129.20	25	1193.24	14	1359.28	39	4738.21	86
1129.40	25	1193.26	14	1359.44	38	4738.46	86
1129.75	25	1193.39	14	1364.16	37	4742.56	86
1138.38	24	1193.65	14	1431.60	58	4762.31	66
1138.56	24	1193.68	12	1432.10	58	4762.53	66
1138.60	24	1194.00	12	1432.53	58	4766.67	66
1138.74	24	1194.06	12	1459.03	36	4770.03	66
1138.95	24	1194.23	12	1463.34	35	4771.74	66
1139.09	24	1194.30	13	1467.40	34	4775.90	66
1139.51	23	1194.41	12	1467.88	33	4812.92	65
1139.77	22	1194.49	13	1468.41	33	4817.37	65
1139.79	23	1194.61	12	1470.09	32	4826.80	65
1139.81	23	1260.74	11	1472.23	31	4932.05	75
1139.86	23	1260.93	11	1481.76	30	5011.26	85
1140.01	23	1261.00	11	1542.18	57	5012.03	85
1140.01	22	1261.12	11	1560.31	3	5012.28	85
1140.12	22	1261.43	11	1560.68	3	5017.09	84
1140.22	22	1261.55	11	1560.71	3	5017.79	84
1140.32	22	1266.41	47	1561.34	3	5018.06	84
1140.36	23	1270.14	10	1561.37	3	5023.84	83
1140.57	22	1274.11	9	1561.44	3	5024.92	83
1155.81	21	1274.98	46	1600.82	56	5039.06	82
1155.98	21	1276.48	8	1602.97	55	5040.12	82
1156.03	21	1276.75	8	1606.96	54	5041.48	81
1156.20	21	1277.19	8	1608.44	53	5041.76	81
1156.39	21	1277.25	7	1613.38	2	5041.79	81
1156.56	21	1277.28	7	1613.80	2	5052.17	74
1157.77	20	1277.51	7	1614.51	2	5053.52	98
1157.91	20	1277.55	7	1656.27	1	5268.95	97
1158.02	20	1277.72	7	1656.93	1	5290.99	118
1158.03	19	1277.95	7	1657.01	1	5296.93	118
1158.13	20	1279.06	6	1657.38	1	5300.87	118
1158.32	19	1279.23	6	1657.91	1	5302.36	118
1158.40	19	1279.50	6	1658.12	1	5306.31	118
1158.49	20	1279.89	5	1751.83	52	5306.83	118
1158.54	19	1280.14	5	1763.91	51	5380.34	73
1158.67	19	1280.33	5	1765.37	50	5534.26	117
1158.91	19	1280.40	5	1930.90	29	5540.76	117

List of tabulated lines — Continued

Wavelength (Å)	No.	Wavelength (Å)	No.	Wavelength (Å)	No.	Wavelength (Å)	No.
5545.05	117	6417.54	130	8015.00	143	10183.0	268
5547.27	117	6568.73	171	8018.57	143	10198.6	268
5551.02	117	6586.27	170	8021.25	143	10202.8	268
5551.58	117	6587.61	95	8026.55	143	10213.4	268
5603.73	132	6591.46	148	8028.18	143	10217.6	268
5614.81	132	6595.24	148	8045.34	141	10224.5	268
5616.48	132	6596.85	148	8058.62	141	10225.8	178
5668.94	96	6602.41	148	8062.36	141	10273.9	280
5780.72	152	6605.77	148	8070.43	141	10275.0	190
5784.03	152	6611.35	148	8078.01	142	10284.2	247
5784.86	152	6654.61	169	8078.48	141	10310.3	190
5788.15	152	6655.52	94	8083.79	141	10311.7	190
5791.73	152	6662.74	147	8155.64	63	10326.3	190
5793.12	80	6671.85	147	8182.68	63	10327.7	190
5794.10	80	6674.12	147	8335.15	72	10328.6	190
5794.47	80	6679.64	147	8753.07	165	10382.1	107
5795.02	152	6683.97	147	8873.36	164	10405.0	107
5800.23	80	6688.79	147	8918.63	163	10413.6	107
5800.60	80	6711.32	93	8972.84	162	10449.9	107
5805.20	80	6828.12	92	9061.44	62	10481.0	203
5805.81	151	6970.28	113	9062.49	62	10495.5	303
5812.72	151	7022.13	112	9078.29	62	10541.2	90
5813.51	151	7056.89	111	9088.51	62	10541.8	303
5817.70	151	7065.89	110	9094.83	62	10545.6	303
5819.49	151	7071.10	110	9111.81	62	10550.6	303
5824.64	151	7074.86	110	9182.83	161	10554.5	303
5877.34	131	7076.48	110	9398.52	231	10559.3	303
5889.52	131	7085.48	110	9405.73	71	10577.7	89
5892.02	131	7087.83	110	9413.07	231	10593.5	89
5969.32	79	7089.46	109	9413.29	251	10631.4	70
5986.40	116	7093.24	110	9435.62	231	10654.6	70
5990.98	116	7100.12	109	9438.69	231	10668.6	70
5993.50	115	7108.93	109	9443.69	231	10683.1	60
5994.00	116	7111.47	108	9450.28	231	10685.3	60
5996.05	116	7113.18	108	9455.21	250	10691.2	60
6001.12	115	7115.17	108	9603.03	61	10707.3	60
6002.98	116	7115.18	109	9620.78	61	10729.5	60
6003.67	116	7116.99	109	9627.62	281	10754.0	60
6006.02	116	7119.66	109	9629.92	204	10759.3	88
6007.18	115	7122.20	108	9658.43	61	10795.7	88
6010.68	115	7132.11	108	9941.35	180	10827.7	202
6012.23	114	7139.18	108	10003.6	249	11056.7	106
6013.17	115	7202.27	168	10047.8	191	11107.1	246
6013.21	114	7216.01	182	10049.1	191	11122.1	279
6014.83	115	7241.32	167	10060.9	230	11260.7	245
6016.45	114	7338.35	64	10062.7	191	11328.4	302
6019.90	114	7364.74	166	10064.0	191	11330.3	87
6028.56	114	7473.31	129	10064.9	191	11335.1	267
6032.02	114	7476.18	129	10077.6	230	11336.1	105
6062.09	150	7483.45	129	10077.8	248	11348.1	267
6065.55	150	7612.10	128	10082.3	191	11353.3	267
6066.65	150	7634.91	127	10093.4	269	11356.4	229
6070.83	150	7662.44	126	10096.8	179	11366.4	267
6074.20	150	7685.19	126	10099.5	269	11367.0	267
6078.39	150	7692.49	126	10101.3	230	11377.7	229
6100.03	149	7832.64	146	10105.4	230	11379.7	302
6107.66	149	7837.10	146	10107.8	269	11381.4	267
6108.52	149	7840.25	146	10112.7	230	11383.3	189
6113.15	149	7848.24	146	10113.9	269	11383.4	189
6115.84	149	7852.86	146	10118.2	230	11384.9	302
6120.81	149	7860.88	146	10118.7	269	11384.9	189
6292.36	172	7892.20	145	10123.9	91	11385.1	189
6397.97	130	7981.96	144	10132.7	269	11386.2	189
6413.55	130	7987.90	181	10172.0	268	11390.0	302

List of tabulated lines — Continued

Wavelength (Å)	No.	Wavelength (Å)	No.	Wavelength (Å)	No.	Wavelength (Å)	No.
11395.2	302	12072.8	338	13251.8	352	14378.8	332
11400.9	302	12076.0	338	13256.2	352	14379.5	331
11404.1	229	12121.6	338	13273.5	352	14399.6	136
11409.6	69	12124.8	338	13366.7	351	14400.0	332
11413.1	188	12378.7	316	13409.9	121	14403.3	136
11414.8	188	12393.6	353	13429.9	351	14408.4	332
11415.0	188	12395.9	353	13439.5	351	14413.5	331
11416.2	229	12407.9	353	13502.3	120	14417.7	332
11422.4	229	12511.7	100	13519.0	393	14420.1	136
11425.5	229	12516.4	124	13548.7	139	14428.4	331
11498.2	266	12544.9	100	13559.7	120	14429.0	136
11512.3	266	12549.5	140	13571.5	139	14434.4	330
11528.3	266	12562.1	140	13581.3	120	14435.0	330
11536.5	187	12569.0	140	13609.3	139	14442.2	136
11538.2	187	12581.6	140	13628.4	370	14448.8	331
11538.3	187	12601.5	140	13651.2	370	14452.6	331
11540.6	266	12614.1	140	13655.9	370	14463.4	330
11547.1	266	12708.5	337	13674.2	370	14467.5	265
11556.5	104	12712.6	337	13690.0	370	14471.8	136
11559.5	266	12720.8	372	13697.8	78	14474.2	330
11561.3	266	12728.4	337	13703.3	78	14497.2	265
11584.8	104	12742.3	372	13705.3	78	14498.3	265
11600.2	103	12743.0	337	13708.4	370	14519.5	265
11614.5	103	12744.7	372	13723.3	243	14528.1	265
11619.3	103	12757.4	372	13724.3	392	14533.5	330
11622.2	317	12758.9	337	13734.4	369	14539.1	265
11628.8	103	12774.4	372	13741.9	78	14542.5	67
11648.0	103	12780.6	336	13743.9	78	14544.4	330
11652.9	125	12787.2	372	13756.5	315	14551.7	278
11658.8	125	12789.3	371	13765.0	369	14553.7	391
11659.7	103	12807.9	336	13765.3	78	14604.7	390
11661.9	244	12811.3	336	13768.9	369	14622.6	215
11669.6	125	12813.4	337	13773.2	368	14628.4	135
11674.1	103	12814.1	335	13793.9	369	14629.5	215
11729.9	59	12817.2	371	13807.9	368	14636.5	215
11748.2	102	12819.2	371	13812.1	368	14637.0	135
11753.3	102	12845.0	335	13822.0	369	14661.8	215
11754.8	102	12849.8	371	13840.3	368	14672.3	135
11756.4	59	12856.4	336	13850.5	368	14697.0	215
11777.5	102	12862.9	336	13857.0	369	14729.5	215
11801.1	102	12866.3	336	13875.3	368	14732.4	228
11812.6	59	12872.2	335	13939.9	200	14738.3	134
11819.0	101	12874.2	371	13993.5	138	14768.2	228
11824.0	102	12879.9	334	14106.7	216	14779.6	134
11848.7	101	12881.0	335	14112.5	216	14782.9	134
11863.0	101	12884.9	334	14136.0	216	14806.7	134
11879.6	101	12887.5	336	14138.6	314	14807.9	228
11892.9	101	12900.3	335	14151.6	333	14828.6	350
11895.8	101	12903.4	335	14157.6	333	14832.6	134
11917.1	340	12904.5	371	14161.4	242	14835.2	350
11919.3	340	12910.3	334	14169.4	216	14841.0	228
11930.4	340	12916.1	334	14174.8	216	14843.7	228
11946.0	340	12949.8	201	14180.5	333	14844.1	228
11957.1	340	12966.1	334	14195.3	333	14851.8	134
11958.2	339	12970.6	68	14198.9	216	14857.4	389
11983.3	339	12972.0	334	14218.3	333	14860.3	350
11985.1	339	12998.0	301	14286.0	333	14971.9	264
12005.0	340	13060.5	301	14301.1	332	14995.8	264
12031.4	339	13074.0	301	14306.8	137	15018.0	264
12032.4	339	13076.3	301	14331.2	332	15036.4	349
12033.2	339	13089.9	301	14332.2	137	15049.9	264
12041.6	338	13097.4	301	14339.5	332	15051.9	264
12048.7	338	13160.7	123	14340.8	331	15074.0	264
12059.7	339	13242.2	122	14374.4	137	15084.0	264

List of tabulated lines — Continued

Wavelength (Å)	No.	Wavelength (Å)	No.	Wavelength (Å)	No.	Wavelength (Å)	No.
15116.4	349	17227.7	211	17813.5	299	18844.9	289
15132.8	349	17234.5	225	17813.7	258	18926.6	288
15303.0	367	17242.9	226	17814.0	258	18972.5	409
15330.6	367	17243.4	226	17826.4	259	18989.7	347
15337.6	367	17274.9	224	17827.0	259	19127.7	347
15353.8	241	17302.0	212	17830.4	328	19163.2	347
15364.4	367	17321.3	211	17837.0	156	19194.3	364
15380.6	367	17321.6	211	17847.5	258	19235.1	364
15407.6	367	17323.5	224	17847.8	258	19248.8	364
15506.1	366	17324.2	224	17853.4	77	19268.2	287
15540.2	366	17337.4	225	17857.0	77	19296.2	364
15550.0	366	17338.6	225	17863.3	328	19316.6	364
15552.8	365	17346.4	226	17889.5	156	19364.4	364
15561.7	366	17408.0	263	17890.3	328	19645.9	414
15597.0	365	17408.4	263	17893.5	327	19699.6	414
15597.5	366	17414.9	329	17916.2	328	19722.0	154
15602.3	365	17426.1	329	17917.9	238	19743.3	363
15638.3	365	17427.4	224	17918.4	184	19781.7	363
15642.0	366	17428.1	224	17921.2	401	19790.7	363
15651.8	402	17448.6	158	17953.8	327	19808.7	414
15655.9	365	17451.0	263	17959.3	99	19814.7	363
15683.0	365	17451.3	263	17962.0	99	19820.8	362
15727.4	199	17456.0	223	17966.2	99	19822.2	414
15784.5	198	17465.0	329	17970.4	328	19845.5	414
15784.9	198	17475.6	222	17976.7	328	19848.6	363
15852.6	197	17476.0	222	17997.6	328	19863.3	414
16004.9	196	17478.0	262	18016.0	327	19892.7	362
16021.4	195	17483.2	329	18021.9	184	19912.4	362
16021.7	195	17483.4	263	18026.1	184	19920.7	363
16333.9	160	17505.7	223	18030.6	99	19971.1	362
16415.8	186	17506.2	223	18034.8	99	20002.4	421
16419.3	186	17515.9	157	18047.5	327	20009.8	362
16465.1	186	17521.3	262	18054.2	155	20044.1	362
16468.6	186	17521.8	262	18061.5	326	20045.5	421
16470.9	186	17522.4	329	18061.9	327	20098.4	421
16496.1	388	17526.0	222	18062.4	326	20112.3	421
16505.2	186	17526.3	222	18077.1	327	20130.9	427
16559.6	387	17545.2	261	18083.0	184	20171.2	421
16607.2	386	17551.0	240	18087.3	184	20185.2	421
16855.1	214	17551.4	240	18090.1	184	20215.0	421
16862.9	313	17554.0	262	18091.7	326	20990.5	408
16888.2	214	17554.5	262	18122.9	326	21023.2	176
16888.6	214	17592.2	157	18139.8	99	21122.4	442
16890.4	159	17605.2	260	18154.5	277	21155.8	298
16954.1	213	17611.9	223	18157.0	155	21156.3	298
16965.3	300	17612.4	223	18201.5	326	21173.4	384
16978.2	214	17619.1	299	18221.1	275	21191.3	298
16978.5	214	17622.2	239	18230.6	276	21191.8	298
16988.1	300	17625.4	329	18231.1	276	21211.5	298
17007.7	300	17632.8	222	18233.1	326	21259.3	297
17030.6	300	17637.5	260	18320.7	274	21259.9	297
17043.3	300	17638.2	260	18321.4	274	21295.2	297
17044.7	213	17672.0	77	18451.5	348	21310.7	442
17045.2	213	17684.6	77	18459.0	410	21326.4	442
17048.7	385	17727.2	299	18464.1	348	21364.4	383
17078.5	312	17752.2	299	18507.7	348	21369.9	442
17086.3	185	17768.9	258	18524.6	273	21385.6	442
17086.9	300	17774.5	299	18525.2	273	21409.1	442
17090.1	185	17776.9	77	18547.3	272	21514.5	382
17092.5	185	17781.3	311	18547.7	272	21691.2	296
17126.4	227	17786.2	77	18711.4	177	21691.7	296
17126.8	227	17789.7	77	18831.2	183	21728.6	296
17174.8	227	17793.3	259	18838.9	183	21906.3	407
17194.2	226	17799.6	299	18844.5	289	22156.2	119

List of tabulated lines — Continued

Wavelength (Å)	No.	Wavelength (Å)	No.	Wavelength (Å)	No.	Wavelength (Å)	No.
22160.4	119	29989.6	325	33145.4	346	37834.6	342
22166.8	119	30106.9	325	33167.0	321	37932.5	218
22641.8	381	30159.3	325	33201.8	346	37939.9	341
22781.0	413	30162.6	308	33212.3	439	37948.7	206
22834.8	420	30213.4	433	33229.8	320	38170.6	218
22853.3	413	30278.0	325	33283.1	235	38400.2	359
22865.7	420	30358.4	418	33285.7	321	38418.5	218
22906.6	175	30451.7	418	33323.7	207	38532.7	285
22928.8	420	30523.9	307	33324.0	439	38662.8	218
22960.0	420	30524.8	418	33345.7	346	38679.1	218
22985.5	420	30564.9	425	33356.3	439	38697.5	218
22989.4	413	30586.8	325	33438.3	320	38956.0	358
23010.7	413	30619.1	418	33454.8	207	39181.2	358
23036.6	420	30625.3	418	33468.9	439	39463.1	358
23049.8	413	30692.1	307	33486.3	235	39562.1	253
23063.0	413	30692.6	418	33526.5	439	39729.1	253
23154.8	174	30714.8	412	33652.6	235	39793.0	357
23179.7	426	30764.2	324	33662.0	320	39843.7	357
23239.6	435	30846.3	412	33719.0	319	39865.0	253
23261.7	419	30854.2	271	33729.4	319	39911.8	357
23320.1	419	30942.9	324	33760.1	439	40079.4	357
23380.5	419	30968.7	305	33774.7	319	40083.9	357
23402.5	419	30977.7	208	33802.6	320	40091.1	253
23479.1	419	31039.9	208	33809.5	207	40165.5	253
23501.3	419	31066.1	412	33815.4	320	40262.6	253
23538.5	419	31153.7	208	33876.0	320	40374.4	357
23580.1	400	31156.1	412	33933.7	319	40395.0	253
23644.2	434	31200.6	412	34064.9	304	40552.3	354
24004.0	399	31205.5	412	34086.3	295	40623.0	233
24143.2	173	31281.6	208	34154.0	345	40656.8	294
24177.4	210	31297.5	208	34159.4	319	40674.5	205
24412.6	441	31343.7	305	34322.1	319	40788.4	294
24443.4	194	31377.7	432	34402.7	219	40854.4	354
24448.0	441	31393.1	306	34483.3	438	40870.1	205
24490.3	441	31461.0	208	34676.3	192	40935.0	233
24525.9	441	31465.6	305	34704.2	431	40948.9	354
24556.8	441	31595.1	417	34926.8	438	41197.8	354
24722.6	441	31689.2	193	35040.5	438	41283.2	356
24790.5	440	31702.9	417	35086.0	438	41357.7	356
25037.4	440	31776.1	323	35200.8	438	41408.6	354
25062.7	440	31784.6	417	35264.5	438	41509.7	354
25119.2	440	31878.8	417	35558.0	254	41596.4	356
25144.6	440	31967.0	417	35639.9	361	42115.8	284
25177.0	440	31974.0	220	35762.5	361	42293.6	217
25257.0	257	32027.7	193	35828.3	361	42608.7	380
25559.1	237	32062.3	417	35995.9	361	42845.9	380
25624.4	133	32077.3	417	36061.6	344	44008.9	270
25691.9	133	32116.5	323	36063.9	361	44053.3	252
25697.5	133	32221.3	193	36159.8	234	44189.6	379
25706.1	133	32303.5	322	36233.7	361	44980.7	232
25833.5	133	32500.5	322	36739.3	360	45908.8	378
25842.2	133	32718.2	321	36820.9	318	46420.6	355
26869.8	406	32772.6	255	36821.0	343	47436.2	377
27366.9	209	32804.2	321	36939.5	360	48086.0	376
27997.9	221	32901.8	286	37077.1	318	48335.9	376
28127.4	221	32911.5	321	37190.0	360	48737.7	283
28135.7	310	32920.3	321	37212.7	206	48759.1	376
28758.4	256	32925.2	255	37219.8	341	49171.6	293
28816.4	309	32969.6	255	37240.3	206	49364.1	293
28964.7	256	33040.8	255	37337.7	206	49460.8	374
29150.7	236	33116.1	321	37502.4	206	49616.5	293
29272.5	405	33124.1	255	37680.3	206	49812.5	293
29943.5	325	33130.8	255	37762.9	341	49921.5	293

List of tabulated lines — Continued

Wavenumber (cm^{-1})	No.	Wavenumber (cm^{-1})	No.	Wavenumber (cm^{-1})	No.	Wavenumber (cm^{-1})	No.
21.06	423	1100.54	437	1353.91	403	1625.60	396
21.23	423	1101.53	415	1354.81	416	1727.83	373
27.78	422	1112.28	415	1365.56	416	1780.44	397
814.12	436	1115.21	415	1367.66	429	1781.87	292
819.25	436	1130.23	415	1368.82	416	1786.25	292
819.38	428	1136.14	415	1383.51	416	1794.18	292
832.24	436	1146.89	415	1452.38	424	1831.58	282
834.27	436	1186.40	411	1459.35	291	1864.38	398
847.26	436	1187.95	411	1463.73	291	1902.19	404
881.87	436	1189.10	411	1471.66	291	1977.60	375
1053.52	437	1202.97	411	1482.49	291	1982.62	374
1067.73	437	1223.71	411	1490.42	291	1991.47	375
1072.86	437	1237.58	411	1500.70	394	1992.23	293
1085.85	437	1239.69	290	1523.76	291		
1087.55	437	1349.48	416	1568.42	395		
1097.26	415	1350.87	416	1593.03	430		

This is one of the few tabulated spectra for which the data situation is not yet satisfactory despite numerous theoretical and experimental efforts.

As principal data sources we have selected two recent calculations which are based on extensive configuration interaction treatments. The most comprehensive source for multiplet data is the Opacity Project (OP),[1] which is reviewed in the general introduction. Hibbert *et al.*[2] have also calculated the oscillator strengths of a large number of individual lines with the configuration interaction code CIV 3 (see general introduction).

In addition, we have utilized the results of two smaller calculations: by Nussbaumer and Storey[3] who used the Superstructure Code as well as the close-coupling approach in conjunction with the R-matrix technique, and by Weiss,[4] who applied his superposition-of-configurations (SOC) code. While these calculations are of similar levels of sophistication to those above, Nussbaumer and Storey applied a smaller configuration basis set than the others. Therefore, especially for weak intersystem lines, the data of Nussbaumer and Storey are expected to be less accurate, and we have not included their results in those cases.

For many strong transitions, the results of these four calculations differ by no more than a few percent, but for weaker transitions, the differences become much larger, reaching 50% and more. Also, larger discrepancies and appreciable differences with experimental data occur for transitions starting or ending in ^{1}P° and ^{3}P° terms of configurations with excited *s* and *d* electrons. These ^{1}P° and ^{3}P° terms are energetically very close for pairs *nd* ^{1}P° and $(n+1)s$ ^{1}P°, or *nd* ^{3}P° and $(n+1)s$ ^{3}P°, with $n = 3,4,5$ . . , so that these interact strongly with each other. Figures 1, 2 and 3 illustrate this situation. Figure 1 shows, on a greatly expanded scale, the positions of the $2p3d$ ^{1}P°, ^{3}P° and the $2p4s$ ^{1}P°, ^{3}P° energy levels, indicating their closeness to each other. Fig. 2 shows a comparison between the OP and CIV 3 results for all multiplet data, including those involving the ^{1}P° and ^{3}P° terms. In Fig. 3 the multiplets with ^{1}P° or ^{3}P° as initial or final terms are omitted. Almost all discrepanies occur for multiplets with these terms. In these cases, we have estimated the uncertainties in a conservative manner, especially when only one data source was available.

Some emission measurements with wall-stabilized arcs[5–8] provide additional data and valuable comparisons. Results by Goldbach *et al.*[5,6] have been utilized for several VUV multiplets and intersystem lines. The experimental data are usually in excellent agreement with the *selected* theoretical results, within or close to the estimated experimental uncertainties of ± (12–30)%. But the experimental data for individual line strengths in multiplets have been used only when they were available for the complete set of lines. The emission data by Jones and Wiese[7] for some $3s$-$4p$ multiplets were not applied because their absolute scale, based on older literature data involving both other emission and lifetime results, is fairly uncertain. However, on a relative scale their work closely supports the data of Hibbert *et al.*[2]

The intermediate coupling calculations of Hibbert *et al.*,[2] Nussbaumer and Storey,[3] and Weiss,[4] as well as all emission experiments—especially new measurements by Musielok *et al.*[8]—show that considerable departures from *LS* coupling occur in this spectrum, especially for the weaker lines in *np*-*md* multiplets, where the differences reach factors of two or more. Therefore, we have used conservative error estimates when we had to rely on *LS* coupling fractions for individual lines in multiplets. For C I, we have encountered one of the few instances where the *f*-values of several intersystem lines—six to be exact—have been measured[6] as well as calculated.[2] The comparisons reveal a still unsatisfactory situation: The results for three lines agree within 20%, but the data for the other three lines disagree by factors of 1.75, 2.33 and 2.85. We have in all cases selected the experimental data, and have applied conservative error estimates for other intersystem lines where only theoretical data were available.

Finally, several lifetime measurements by Haar *et al.*[9] with the beam-foil spectroscopy method are in close agreement with the selected data[1–5] and thus provide another strong indication of support. In this experiment, the normally critical cascading effects were found to make only small contributions for the levels that were studied.

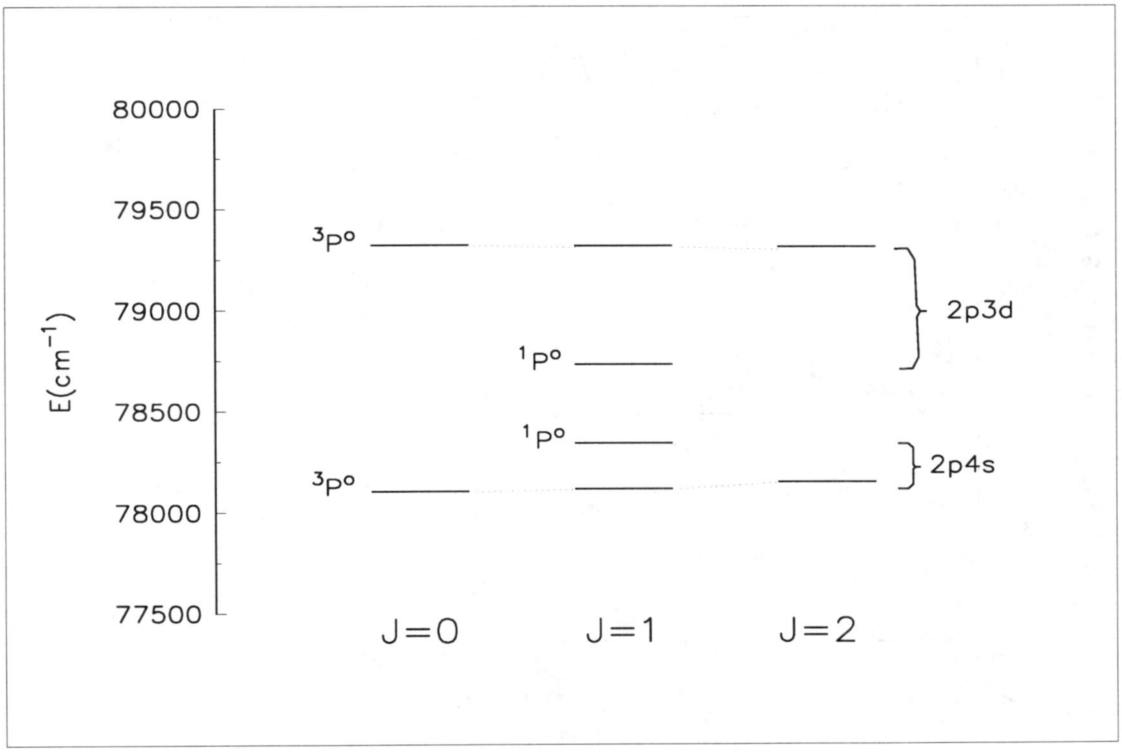

FIG. 1. Partial energy level diagram of C I. The $2p3d$ $^1P^o$, $^3P^o$ and the $2p4s$ $^1P^o$, $^3P^o$ levels are shown on a greatly expanded energy scale.

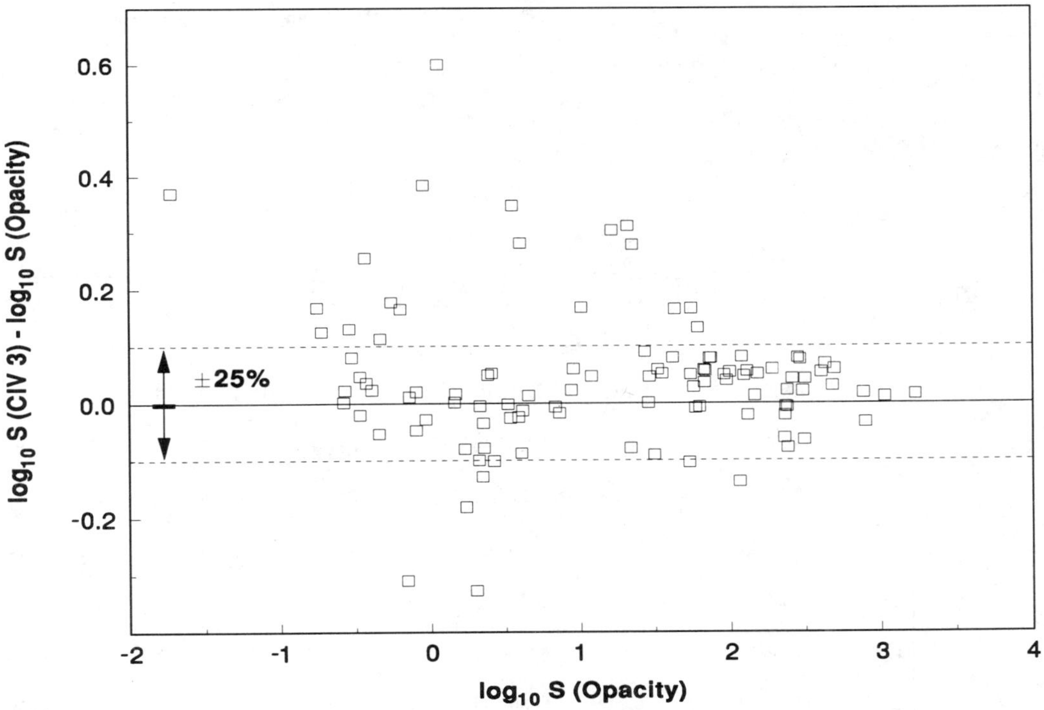

FIG. 2. Comparison of the line strengths from the Opacity Project (OP, by Luo and Pradhan[1]) and from the CIV 3 code calculations (Hibbert *et al.*[2]) for all *multiplet* values of C I where the two calculations overlap. The ratio of the CIV 3 line strength data to the OP data is plotted on a logarithmic scale versus log S of OP. The broken lines indicate a band of deviations of ±25% from a perfect ratio of 1.00.

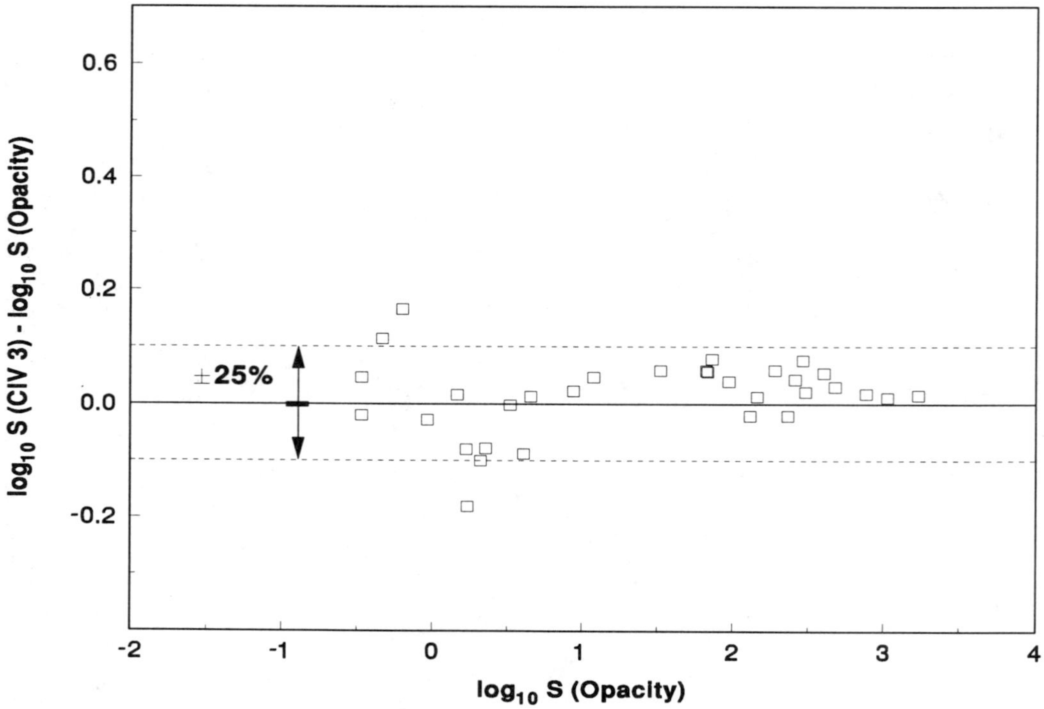

FIG. 3. Same comparison as in Figure 2, but omitting all multiplets starting or ending in $^1P^o$ or $^3P^o$ terms of configurations with excited s ($n \geq 4$) and d electrons.

References

[1]D. Luo and A. K. Pradhan, J. Phys. B **22**, 3377 (1989).
[2]A. Hibbert, E. Biemont, M. Godefroid, and N. Vaeck, Astron. Astrophys., Suppl. Ser. **99**, 179 (1993).
[3]H. Nussbaumer and P. J. Storey, Astron. Astrophys. **140**, 383 (1984).
[4]A. W. Weiss, private communication.
[5]C. Goldbach and G. Nollez, Astron. Astrophys. **181**, 203 (1987).
[6]C. Goldbach, M. Martin and G. Nollez, Astron. Astrophys. **221**, 155 (1989).
[7]D. W. Jones and W. L. Wiese, Phys. Rev. A **29**, 2597 (1984).
[8]J. Musielok, G. Veres, and W. L. Wiese, to be published.
[9]R. R. Haar, L. J. Curtis, T. J. Kvale, D. J. Beideck, I. Martinson, and R. Hellborg, Astron. Astrophys. **241**, 321 (1991).

C I: Allowed Transitions

No.	Transition Array	Mult.	λ_{air} (Å)	λ_{vac} (Å) or σ (cm^{-1})*	E_i (cm^{-1})	E_k (cm^{-1})	g_i–g_k	A_{ki} (10^8 s^{-1})	f_{ik}	S (at. u.)	log gf	Acc.	Source
1.	$2p^2$– $2p3s$	^3P–$^3P^o$		*1657.2*	*29.6*	– *60373.0*	9–9	3.40+00†	1.40–01	6.88+00	0.101	A	1,2,3,4,5
				1657.01	43.40	– 60393.14	5–5	2.52+00	1.04–01	2.83+00	–0.285	A	2n,3n,4n, 5n
				1657.38	16.40	– 60352.63	3–3	8.64–01	3.56–02	5.83–01	–0.971	A	2n,3n,4n, 5n
				1658.12	43.40	– 60352.63	5–3	1.44+00	3.56–02	9.73–01	–0.749	A	2n,3n,4n, 5n
				1657.91	16.40	– 60333.43	3–1	3.43+00	4.72–02	7.72–01	–0.849	A	2n,3n,4n, 5n
				1656.27	16.40	– 60393.14	3–5	8.58–01	5.88–02	9.62–01	–0.754	A	2n,3n,4n, 5n
				1656.93	0.00	– 60352.63	1–3	1.13+00	1.39–01	7.61–01	–0.855	A	2n,3n,4n, 5n
2.		^3P–$^1P^o$											
				1613.38	0.00	– 61981.82	1–3	3.01–04	3.53–05	1.87–04	–4.452	C+	2,4
				1613.80	16.40	– 61981.82	3–3	2.21–04	8.63–06	1.38–04	–4.587	C+	2,4
				1614.51	43.40	– 61981.82	5–3	2.64–04	6.18–06	1.64–04	–4.510	C+	2,4

C I: Allowed Transitions — Continued

No.	Transition Array	Mult.	λ_{air} (Å)	λ_{vac} (Å) or σ (cm^{-1})*	E_i (cm^{-1})	E_k (cm^{-1})	g_i–g_k	A_{ki} (10^8 s^{-1})	f_{ik}	S (at. u.)	log *gf*	Acc.	Source
3.	$2s^22p^2$– $2s2p^3$	^3P–$^3D°$		*1561.1*	*29.6*	*64088.8*	9–15	1.18+00	7.18–02	3.32+00	–0.189	A	1,2,5
				1561.44	43.40	64086.92	5–7	1.18+00	6.03–02	1.55+00	–0.521	A	2*n*
				1560.68	16.40	64090.95	3–5	8.86–01	5.39–02	8.31–01	–0.791	A	2*n*
				1560.31	0.00	64089.85	1–3	6.57–01	7.19–02	3.70–01	–1.143	A	2*n*
				1561.34	43.40	64090.95	5–5	2.94–01	1.08–02	2.76–01	–1.269	A	2*n*
				1560.71	16.40	64089.85	3–3	4.92–01	1.80–02	2.77–01	–1.269	A	2*n*
				1561.37	43.40	64089.85	5–3	3.26–02	7.16–04	1.84–02	–2.446	A	2*n*
4.		^3P–$^3P°$		*1329.3*	*29.6*	*75254.9*	9–9	2.39+00	6.34–02	2.50+00	–0.244	B	1,2,5
				1329.58	43.40	75255.27	5–5	1.79+00	4.76–02	1.04+00	–0.624	B	2*n*
				1329.12	16.40	75253.97	3–3	6.04–01	1.60–02	2.10–01	–1.319	B	2*n*
				1329.60	43.40	75253.97	5–3	1.00+00	1.59–02	3.48–01	–1.100	B	2*n*
				1329.09	16.40	75256.12	3–1	2.41+00	2.13–02	2.79–01	–1.195	B	2*n*
				1329.10	16.40	75255.27	3–5	5.89–01	2.60–02	3.41–01	–1.108	B	2*n*
				1328.83	0.00	75253.97	1–3	7.95–01	6.31–02	2.76–01	–1.200	B	2*n*
5.	$2p^2$– $2p4s$	^3P–$^3P°$		*1280.4*	*29.6*	*78132.9*	9–9	8.55–01	2.10–02	7.98–01	–0.723	B	1,2,3,6
				1280.33	43.40	78148.09	5–5	5.77–01	1.42–02	2.99–01	–1.149	B	2*n*,3*n*,6*n*
				1280.40	16.40	78116.74	3–3	1.73–01	4.26–03	5.39–02	–1.893	B	2*n*,3*n*,6*n*
				1280.85	43.40	78116.74	5–3	3.33–01	4.91–03	1.04–01	–1.610	B	2*n*,3*n*,6*n*
				1280.60	16.40	78104.98	3–1	8.22–01	6.74–03	8.52–02	–1.694	B	2*n*,3*n*,6*n*
				1279.89	16.40	78148.09	3–5	3.08–01	1.26–02	1.59–01	–1.422	B	2*n*,3*n*,6*n*
				1280.14	0.00	78116.74	1–3	3.11–01	2.29–02	9.66–02	–1.640	B	2*n*,3*n*,6*n*
6.	$2p^2$– $2p3d$	^3P–$^3F°$											
				1279.23	43.40	78215.51	5–7	1.10–01	3.78–03	7.96–02	–1.724	C	6
				1279.06	16.40	78199.07	3–5	1.80–02	7.36–04	9.29–03	–2.656	C	6
				1279.50	43.40	78199.07	5–5	8.10–03	1.99–04	4.19–03	–3.003	D+	6
7.		^3P–$^3D°$		*1277.5*	*29.6*	*78309.8*	9–15	2.32+00	9.44–02	3.57+00	–0.071	B–	1,2,3,5
				1277.55	43.40	78318.25	5–7	2.31+00	7.91–02	1.66+00	–0.403	B–	2*n*,3*n*
				1277.28	16.40	78307.63	3–5	1.73+00	7.06–02	8.91–01	–0.674	B–	2*n*,3*n*
				1277.25	0.00	78293.49	1–3	1.27+00	9.35–02	3.93–01	–1.029	B–	2*n*,3*n*
				1277.72	43.40	78307.63	5–5	6.35–01	1.55–02	3.27–01	–1.110	B–	2*n*,3*n*
				1277.51	16.40	78293.49	3–3	9.12–01	2.23–02	2.82–01	–1.174	B–	2*n*,3*n*
				1277.95	43.40	78293.49	5–3	5.56–02	8.17–04	1.72–02	–2.389	C+	2*n*,3*n*
8.	$2p^2$– $2p4s$	^3P–$^1P°$											
				1277.19	43.40	78340.28	5–3	1.86–02	2.73–04	5.74–03	–2.865	C	2
				1276.75	16.40	78340.28	3–3	1.03–01	2.52–03	3.18–02	–2.121	C	2
				1276.48	0.00	78340.28	1–3	6.13–02	4.49–03	1.89–02	–2.347	C	2
9.	$2p^2$– $2p3d$	^3P–$^1F°$											
				1274.11	43.40	78529.62	5–7	1.03–02	3.51–04	7.35–03	–2.756	C	2
10.		^3P–$^1P°$											
				1270.14	0.00	78731.27	1–3	3.81–03	2.76–04	1.15–03	–3.559	C	2
11.		^3P–$^3P°$		*1261.3*	*29.6*	*79314.9*	9–9	1.66+00	3.96–02	1.48+00	–0.448	B+	1,2,3,5
				1261.55	43.40	79310.85	5–5	1.27+00	3.03–02	6.29–01	–0.820	B+	2*n*,3*n*
				1261.00	16.40	79318.78	3–3	4.42–01	1.05–02	1.31–01	–1.500	B+	2*n*,3*n*
				1261.43	43.40	79318.78	5–3	7.06–01	1.01–02	2.10–01	–1.297	B+	2*n*,3*n*
				1260.93	16.40	79323.16	3–1	1.70+00	1.35–02	1.68–01	–1.392	B+	2*n*,3*n*
				1261.12	16.40	79310.85	3–5	3.71–01	1.47–02	1.84–01	–1.354	B+	2*n*,3*n*
				1260.74	0.00	79318.78	1–3	5.32–01	3.80–02	1.58–01	–1.420	B+	2*n*,3*n*

C I: Allowed Transitions — Continued

No.	Transition Array	Mult.	λ_{air} (Å)	λ_{vac} (Å) or σ (cm^{-1})*	E_i (cm^{-1})	E_k (cm^{-1})	g_i–g_k	A_{ki} (10^8 s^{-1})	f_{ik}	S (at. u.)	log gf	Acc.	Source
12.	$2p^2$–$2p5s$	^{3}P–^{3}P°		*1194.1*	*29.6*	*83772.5*	9–9	5.14−01	1.10−02	3.89−01	−1.005	B−	1,2
				1194.06	43.40	83791.04	5–5	2.73−01	5.85−03	1.15−01	−1.534	B−	2*n*
				1194.23	16.40	83752.41	3–3	8.23−02	1.76−03	2.08−02	−2.277	C+	2*n*
				1194.61	43.40	83752.41	5–3	1.79−01	2.29−03	4.51−02	−1.940	C+	2*n*
				1194.41	16.40	83740.06	3–1	4.47−01	3.19−03	3.76−02	−2.019	C+	2*n*
				1193.68	16.40	83791.04	3–5	2.86−01	1.02−02	1.20−01	−1.515	B−	2*n*
				1194.00	0.00	83752.41	1–3	2.00−01	1.28−02	5.03−02	−1.892	B−	2*n*
13.	$2p^2$–$2p4d$	^{3}P–^{3}F°											
				1194.49	43.40	83761.26	5–7	7.07−02	2.12−03	4.16−02	−1.975	C	2
				1194.30	16.40	83747.39	3–5	4.59−02	1.63−03	1.93−02	−2.309	C	2
14.		^{3}P–^{3}D°		*1193.2*	*29.6*	*83839.5*	9–15	1.22+00	4.35−02	1.54+00	−0.407	B−	1,2
				1193.24	43.40	83848.83	5–7	1.22+00	3.65−02	7.18−01	−0.738	B−	2*n*
				1193.01	16.40	83838.08	3–5	8.17−01	2.91−02	3.42−01	−1.060	B−	2*n*
				1193.03	0.00	83820.13	1–3	7.05−01	4.51−02	1.77−01	−1.345	B−	2*n*
				1193.39	43.40	83838.08	5–5	4.09−01	8.73−03	1.71−01	−1.360	B−	2*n*
				1193.26	16.40	83820.13	3–3	4.84−01	1.03−02	1.22−01	−1.508	B−	2*n*
				1193.65	43.40	83820.13	5–3	2.68−02	3.44−04	6.76−03	−2.764	C	2*n*
15.	$2p^2$–$2p5s$	^{3}P–^{1}P°											
				1192.83	43.40	83877.31	5–3	2.19−02	2.80−04	5.49−03	−2.854	C	2
				1192.45	16.40	83877.31	3–3	6.63−02	1.41−03	1.67−02	−2.372	C	2
				1192.22	0.00	83877.31	1–3	1.30−02	8.32−04	3.26−03	−3.080	C	2
16.	$2p^2$–$2p4d$	^{3}P–^{1}F°											
				1191.84	43.40	83947.43	5–7	2.18−02	6.49−04	1.27−02	−2.489	C	2
17.		^{3}P–^{1}P°											
				1190.02	0.00	84032.15	1–3	7.35−03	4.68−04	1.83−03	−3.330	C	2
				1190.25	16.40	84032.15	3–3	5.87−03	1.25−04	1.47−03	−3.427	C	2
18.		^{3}P–^{3}P°		*1189.3*	*29.6*	*84109.4*	9–9	5.72−01	1.21−02	4.28−01	−0.962	B	1,2
				1189.63	43.40	84103.10	5–5	4.51−01	9.57−03	1.87−01	−1.320	B	2*n*
				1189.07	16.40	84116.09	3–3	1.83−01	3.88−03	4.56−02	−1.934	B−	2*n*
				1189.45	43.40	84116.09	5–3	2.64−01	3.36−03	6.59−02	−1.774	B	2*n*
				1188.99	16.40	84121.22	3–1	6.48−01	4.58−03	5.38−02	−1.862	B	2*n*
				1189.25	16.40	84103.10	3–5	8.03−02	2.84−03	3.33−02	−2.070	B−	2*n*
				1188.83	0.00	84116.09	1–3	1.68−01	1.07−02	4.18−02	−1.972	B−	2*n*
19.	$2p^2$–$2p6s$	^{3}P–^{3}P°		*1158.5*	*29.6*	*86351.6*	9–9	2.77−01	5.57−03	1.91−01	−1.300	C+	1
				1158.40	43.40	86369.60	5–5	2.08−01	4.18−03	7.97−02	−1.680	C−	*LS*
				1158.54	16.40	86331.63	3–3	6.92−02	1.39−03	1.59−02	−2.379	C−	*LS*
				1158.91	43.40	86331.63	5–3	1.15−01	1.39−03	2.66−02	−2.157	C−	*LS*
				1158.67	16.40	86321.94	3–1	2.77−01	1.86−03	2.12−02	−2.254	C−	*LS*
				1158.03	16.40	86369.60	3–5	6.93−02	2.32−03	2.66−02	−2.157	C−	*LS*
				1158.32	0.00	86331.63	1–3	9.23−02	5.57−03	2.12−02	−2.254	C−	*LS*
20.	$2p^2$–$2p5d$	^{3}P–^{3}D°		*1158.0*	*29.6*	*86387.9*	9–15	7.28−01	2.44−02	8.37−01	−0.658	B	1
				1158.02	43.40	86397.80	5–7	7.28−01	2.05−02	3.91−01	−0.989	C+	*LS*
				1157.77	16.40	86389.38	3–5	5.46−01	1.83−02	2.09−01	−1.260	C+	*LS*
				1157.91	0.00	86362.52	1–3	4.05−01	2.44−02	9.30−02	−1.613	C+	*LS*
				1158.13	43.40	86389.38	5–5	1.82−01	3.66−03	6.98−02	−1.738	C+	*LS*
				1158.13	16.40	86362.52	3–3	3.03−01	6.10−03	6.98−02	−1.738	C+	*LS*
				1158.49	43.40	86362.52	5–3	2.02−02	2.44−04	4.65−03	−2.914	C−	*LS*

C I: Allowed Transitions — Continued

No.	Transition Array	Mult.	λ_{air} (Å)	λ_{vac} (Å) or σ (cm^{-1})*	E_i (cm^{-1})	E_k (cm^{-1})	g_i–g_k	A_{ki} (10^8 s^{-1})	f_{ik}	S (at. u.)	log gf	Acc.	Source
21.		^{3}P–^{3}P°		*1156.3*	*29.6*	*86512.8*	9–9	3.02−01	6.05−03	2.07−01	−1.264	C+	1
				1156.56	43.40	86506.70	5–5	2.26−01	4.54−03	8.64−02	−1.644	C−	*LS*
				1156.03	16.40	86519.47	3–3	7.55−02	1.51−03	1.73−02	−2.343	C−	*LS*
				1156.39	43.40	86519.47	5–3	1.26−01	1.51−03	2.88−02	−2.121	C−	*LS*
				1155.98	16.40	86523.16	3–1	3.02−01	2.02−03	2.30−02	−2.218	C−	*LS*
				1156.20	16.40	86506.70	3–5	7.55−02	2.52−03	2.88−02	−2.121	C−	*LS*
				1155.81	0.00	86519.47	1–3	1.01−01	6.05−03	2.30−02	−2.218	C−	*LS*
22.	$2p^2$– $2p7s$	^{3}P–^{3}P°		*1140.2*	*29.6*	*87737.3*	9–9	1.68−01	3.28−03	1.11−01	−1.530	C+	1
				1140.12	43.40	87753.73	5–5	1.26−01	2.46−03	4.61−02	−1.910	C−	*LS*
				1140.22	16.40	87718.56	3–3	4.20−02	8.19−04	9.23−03	−2.609	D	*LS*
				1140.57	43.40	87718.56	5–3	7.00−02	8.19−04	1.54−02	−2.388	C−	*LS*
				1140.32	16.40	87711.37	3–1	1.68−01	1.09−03	1.23−02	−2.485	C−	*LS*
				1139.77	16.40	87753.73	3–5	4.21−02	1.37−03	1.54−02	−2.387	C−	*LS*
				1140.01	0.00	87718.56	1–3	5.61−02	3.28−03	1.23−02	−2.484	C−	*LS*
23.	$2p^2$– $2p6d$	^{3}P–^{3}D°		*1139.8*	*29.6*	*87767.4*	9–15	4.34−01	1.41−02	4.75−01	−0.897	B	1
				1139.81	43.40	87777.17	5–7	4.34−01	1.18−02	2.22−01	−1.228	C+	*LS*
				1139.51	16.40	87773.09	3–5	3.25−01	1.06−02	1.19−01	−1.499	C+	*LS*
				1139.79	0.00	87735.3	1–3	2.41−01	1.41−02	5.28−02	−1.852	C+	*LS*
				1139.86	43.40	87773.09	5–5	1.08−01	2.11−03	3.96−02	−1.977	C+	*LS*
				1140.01	16.40	87735.3	3–3	1.81−01	3.52−03	3.96−02	−1.977	C+	*LS*
				1140.36	43.40	87735.3	5–3	1.20−02	1.41−04	2.64−03	−3.153	D	*LS*
24.		^{3}P–^{3}P°		*1138.8*	*29.6*	*87837.9*	9–9	1.72−01	3.34−03	1.13−01	−1.522	C+	1
				1139.09	43.40	87832.54	5–5	1.29−01	2.51−03	4.70−02	−1.902	C−	*LS*
				1138.60	16.40	87843.91	3–3	4.30−02	8.36−04	9.40−03	−2.601	D	*LS*
				1138.95	43.40	87843.91	5–3	7.16−02	8.36−04	1.57−02	−2.379	C−	*LS*
				1138.56	16.40	87846.9	3–1	1.72−01	1.11−03	1.25−02	−2.476	C−	*LS*
				1138.74	16.40	87832.54	3–5	4.30−02	1.39−03	1.57−02	−2.379	C−	*LS*
				1138.38	0.00	87843.91	1–3	5.74−02	3.34−03	1.25−02	−2.476	C−	*LS*
25.	$2p^2$– $2p7d$	^{3}P–^{3}D°		*1129.1*	*29.6*	*88596.5*	9–15	2.77−01	8.82−03	2.95−01	−1.100	B	1
				1129.13	43.40	88606.8	5–7	2.77−01	7.41−03	1.38−01	−1.431	C+	*LS*
				1128.82	16.40	88604.75	3–5	2.08−01	6.62−03	7.38−02	−1.702	C+	*LS*
				1129.20	0.00	88558.6	1–3	1.54−01	8.82−03	3.28−02	−2.054	C+	*LS*
				1129.16	43.40	88604.75	5–5	6.92−02	1.32−03	2.46−02	−2.179	C	*LS*
				1129.40	16.40	88558.6	3–3	1.15−01	2.21−03	2.46−02	−2.179	C	*LS*
				1129.75	43.40	88558.6	5–3	7.68−03	8.82−05	1.64−03	−3.356	D	*LS*
26.	$2p^2$– $2p8d$	^{3}P–^{3}D°		*1122.3*	*29.6*	*89132.9*	9–15	1.87−01	5.89−03	1.96−01	−1.276	B	1
				1122.33	43.40	89143.9	5–7	1.87−01	4.95−03	9.14−02	−1.607	C+	*LS*
				1122.00	16.40	89142.66	3–5	1.41−01	4.42−03	4.90−02	−1.877	C+	*LS*
				1122.45	0.00	89091.1	1–3	1.04−01	5.89−03	2.18−02	−2.230	C	*LS*
				1122.34	43.40	89142.66	5–5	4.68−02	8.84−04	1.63−02	−2.355	C	*LS*
				1122.65	16.40	89091.1	3–3	7.79−02	1.47−03	1.63−02	−2.355	C	*LS*
				1122.99	43.40	89091.1	5–3	5.19−03	5.89−05	1.09−03	−3.531	D	*LS*
27.	$2s^22p^2$– $2s2p^3$	^{3}P–^{3}S°		*945.46*	*29.6*	105798.7	9–3	3.41+01	1.52−01	4.27+00	0.137	B	1
				945.579	43.40	105798.7	5–3	1.89+01	1.52−01	2.37+00	−0.118	C+	*LS*
				945.338	16.40	105798.7	3–3	1.14+01	1.52−01	1.42+00	−0.340	C+	*LS*
				945.191	0.00	105798.7	1–3	3.79+00	1.52−01	4.74−01	−0.817	C+	*LS*

C I: Allowed Transitions — Continued

No.	Transition Array	Mult.	λ_{air} (Å)	λ_{vac} (Å) or σ (cm^{-1})*	E_i (cm^{-1})	E_k (cm^{-1})	g_i–g_k	A_{ki} (10^8 s^{-1})	f_{ik}	S (at. u.)	log gf	Acc.	Source
28.	$2p^2$– $2p3s$	$^1D-^3P^o$											
				1993.62	10192.63	60352.63	5–3	7.82–04	2.79–05	9.17–04	−3.855	C	4
				1992.01	10192.63	60393.14	5–5	4.01–06	2.39–07	7.83–06	−5.923	D	4
29.		$^1D-^1P^o$		1930.90	10192.63	61981.82	5–3	3.51+00	1.18–01	3.74+00	−0.230	B+	1,2,3,4,6
30.	$2p^2$– $2p3d$	$^1D-^1D^o$		1481.76	10192.63	77679.82	5–5	3.92–01	1.29–02	3.15–01	−1.191	B+	1,2,3,6
31.	$2p^2$– $2p4s$	$^1D-^3P^o$											
				1472.23	10192.63	78116.74	5–3	8.01–03	1.56–04	3.79–03	−3.107	C	2
32.	$2p^2$– $2p3d$	$^1D-^3F^o$											
				1470.09	10192.63	78215.51	5–7	1.37–02	6.22–04	1.50–02	−2.507	C+	2,6
33.		$^1D-^3D^o$											
				1467.88	10192.63	78318.25	5–7	1.30–02	5.88–04	1.42–02	−2.532	C	6
				1468.41	10192.63	78293.49	5–3	3.90–02	7.56–04	1.83–02	−2.422	C	6
34.	$2p^2$– $2p4s$	$^1D-^1P^o$		1467.40	10192.63	78340.28	5–3	5.49–01	1.06–02	2.57–01	−1.274	B+	1,2,3,6
35.	$2p^2$– $2p3d$	$^1D-^1F^o$		1463.34	10192.63	78529.62	5–7	1.88+00	8.47–02	2.04+00	−0.373	B	1,2,3,5
36.		$^1D-^1P^o$		1459.03	10192.63	78731.27	5–3	4.76–01	9.12–03	2.19–01	−1.341	B	1,2,3,6
37.	$2p^2$– $2p4d$	$^1D-^1D^o$		1364.16	10192.63	83497.62	5–5	1.57–01	4.39–03	9.86–02	−1.658	C	2
38.	$2p^2$– $2p5s$	$^1D-^3P^o$											
				1359.44	10192.63	83752.41	5–3	9.74–03	1.62–04	3.62–03	−3.092	C	2
39.	$2p^2$– $2p4d$	$^1D-^3F^o$											
				1359.28	10192.63	83761.26	5–7	2.16–02	8.36–04	1.87–02	−2.379	C	2
40.		$^1D-^3D^o$											
				1357.66	10192.63	83848.83	5–7	1.08–02	4.17–04	9.33–03	−2.680	C	2
41.	$2p^2$– $2p5s$	$^1D-^1P^o$		1357.13	10192.63	83877.31	5–3	1.35–01	2.23–03	4.99–02	−1.952	C	2
42.	$2p^2$– $2p4d$	$^1D-^1F^o$		1355.84	10192.63	83947.43	5–7	1.04+00	4.03–02	9.00–01	−0.696	B	1,2
43.		$^1D-^1P^o$		1354.29	10192.63	84032.15	5–3	2.95–01	4.87–03	1.09–01	−1.613	C	2
44.	$2p^2$– $2p5d$	$^1D-^1F^o$		1311.36	10192.63	86449.19	5–7	5.89–01	2.12–02	4.59–01	−0.974	B	1
45.	$2p^2$– $2p6d$	$^1D-^1F^o$		1288.42	10192.63	87806.93	5–7	3.51–01	1.22–02	2.59–01	−1.214	B	1

C I: Allowed Transitions — Continued

No.	Transition Array	Mult.	λ_{air} (Å)	λ_{vac} (Å) or σ (cm^{-1})*	E_i (cm^{-1})	E_k (cm^{-1})	g_i–g_k	A_{ki} (10^8 s^{-1})	f_{ik}	S (at. u.)	log gf	Acc.	Source
46.	$2p^2$– $2p7d$	^{1}D–^{1}F°		1274.98	10192.63	– 88625.00	5–7	2.25–01	7.67–03	1.61–01	–1.416	B–	1
47.	$2p^2$– $2p8d$	^{1}D–^{1}F°		1266.41	10192.63	– 89155.70	5–7	1.52–01	5.12–03	1.07–01	–1.592	B–	1
48.	$2p^2$– $2p3s$	^{1}S–^{3}P°											
			2582.90	2583.67	21648.01	– 60352.63	1–3	7.09–05	2.13–05	1.81–04	–4.672	C	4
49.		^{1}S–^{1}P°	2478.56	2479.31	21648.01	– 61981.82	1–3	3.40–01	9.39–02	7.67–01	–1.027	B+	1,2,3,4
50.	$2p^2$– $2p3d$	^{1}S–^{3}D°											
				1765.37	21648.01	– 78293.49	1–3	1.04–02	1.45–03	8.43–03	–2.838	C	2
51.	$2p^2$– $2p4s$	^{1}S–^{1}P°		1763.91	21648.01	– 78340.28	1–3	3.59–02	5.03–03	2.92–02	–2.298	C	2
52.	$2p^2$– $2p3d$	^{1}S–^{1}P°		1751.83	21648.01	– 78731.27	1–3	9.07–01	1.25–01	7.22–01	–0.903	B	1,2,3,6
53.	$2p^2$– $2p4d$	^{1}S–^{3}D°											
				1608.44	21648.01	– 83820.13	1–3	1.74–02	2.03–03	1.07–02	–2.693	C	2
54.	$2p^2$– $2p5s$	^{1}S–^{1}P°		1606.96	21648.01	– 83877.31	1–3	5.30–02	6.15–03	3.26–02	–2.211	C	2
55.	$2p^2$– $2p4d$	^{1}S–^{1}P°		1602.97	21648.01	– 84032.15	1–3	4.51–01	5.22–02	2.75–01	–1.283	B	1,2
56.		^{1}S–^{3}P°											
				1600.82	21648.01	– 84116.09	1–3	3.57–03	4.12–04	2.17–03	–3.385	C	2
57.	$2p^2$– $2p5d$	^{1}S–^{1}P°		1542.18	21648.01	– 86491.41	1–3	2.22–01	2.37–02	1.21–01	–1.625	C+	1
58.	$2s2p^3$– $2s2p^2(^4\mathrm{P})3s$	^{5}S°–^{5}P		*1432.0*	33735.20	– *103570*	5–15	2.08+00	1.91–01	4.51+00	–0.019	B	1,6
				1431.60	33735.20	– 103587.3	5–7	2.11+00	9.07–02	2.14+00	–0.343	C+	6n
				1432.10	33735.20	– 103562.5	5–5	2.01+00	6.18–02	1.46+00	–0.510	C+	6n
				1432.53	33735.20	– 103541.8	5–3	2.11+00	3.89–02	9.18–01	–0.711	C+	6n
59.	$2p3s$– $2p3p$	^{3}P°–^{1}P											
			11756.4	8503.70 cm^{-1}	60352.63	– 68856.33	3–3	2.50–04	5.19–04	6.03–02	–2.808	D	2,4
			11729.9	8522.90 cm^{-1}	60333.43	– 68856.33	1–3	2.19–05	1.35–04	5.23–03	–3.868	D	2,4
			11812.6	8463.19 cm^{-1}	60393.14	– 68856.33	5–3	2.63–09	3.30–09	6.42–07	–7.782	D	4
60.		^{3}P°–^{3}D	*10693*	*9349.0* cm^{-1}	*60373.0*	– *69722.0*	9–15	1.84–01	5.27–01	1.67+02	0.676	B	1,2,3,4
			10691.2	9350.89 cm^{-1}	60393.14	– 69744.03	5–7	1.84–01	4.43–01	7.79+01	0.345	B	2n,3n,4n
			10683.1	9358.03 cm^{-1}	60352.63	– 69710.66	3–5	1.40–01	4.00–01	4.23+01	0.080	B	2n,3n,4n
			10685.3	9356.05 cm^{-1}	60333.43	– 69689.48	1–3	1.04–01	5.34–01	1.88+01	–0.273	B	2n,3n,4n
			10729.5	9317.52 cm^{-1}	60393.14	– 69710.66	5–5	4.40–02	7.60–02	1.34+01	–0.420	B	2n,3n,4n
			10707.3	9336.85 cm^{-1}	60352.63	– 69689.48	3–3	7.53–02	1.30–01	1.37+01	–0.410	B	2n,3n,4n
			10754.0	9296.34 cm^{-1}	60393.14	– 69689.48	5–3	4.76–03	4.95–03	8.77–01	–1.606	B	2n,3n,4n

C I: Allowed Transitions — Continued

No.	Transition Array	Mult.	λ_{air} (Å)	λ_{vac} (Å) or σ (cm^{-1})*	E_i (cm^{-1})		E_k (cm^{-1})	g_i-g_k	A_{ki} (10^8 s^{-1})	f_{ik}	S (at. u.)	log gf	Acc.	Source
61.		$^3P^o-^3S$	*9639.7*	*9642.3*	*60373.0*	–	70743.95	9–3	2.42–01	1.12–01	3.21+01	0.005	B	1,2,3,4
			9658.43	9661.08	60393.14	–	70743.95	5–3	1.25–01	1.05–01	1.67+01	–0.280	B	2n,3n,4n
			9620.78	9623.42	60352.63	–	70743.95	3–3	8.62–02	1.20–01	1.14+01	–0.445	B	2n,3n,4n
			9603.03	9605.67	60333.43	–	70743.95	1–3	3.06–02	1.27–01	4.01+00	–0.896	B	2n,3n,4n
62.		$^3P^o-^3P$	*9086.8*	*9089.3*	*60373.0*	–	*71374.9*	9–9	3.01–01	3.72–01	1.00+02	0.525	B	1,2,3,4
			9094.83	9097.33	60393.14	–	71385.38	5–5	2.28–01	2.82–01	4.23+01	0.150	B	2n,3n,4n
			9078.29	9080.78	60352.63	–	71364.90	3–3	7.07–02	8.75–02	7.84+00	–0.581	B	2n,3n,4n
			9111.81	9114.31	60393.14	–	71364.90	5–3	1.35–01	1.01–01	1.51+01	–0.298	B	2n,3n,4n
			9088.51	9091.01	60352.63	–	71352.51	3–1	3.00–01	1.24–01	1.11+01	–0.429	B	2n,3n,4n
			9061.44	9063.92	60352.63	–	71385.38	3–5	7.31–02	1.50–01	1.34+01	–0.346	B	2n,3n,4n
			9062.49	9064.98	60333.43	–	71364.90	1–3	9.48–02	3.50–01	1.05+01	–0.456	B	2n,3n,4n
63.		$^3P^o-^1D$												
			8155.64	8157.88	60352.63	–	72610.72	3–5	1.29–05	2.15–05	1.73–03	–4.191	C	4
			8182.68	8184.93	60393.14	–	72610.72	5–5	8.20–06	8.24–06	1.11–03	–4.385	C	4
64.		$^3P^o-^1S$												
			7338.35	7340.38	60352.63	–	73975.91	3–1	2.97–05	7.99–06	5.79–04	–4.621	C	4
65.	$2p\,3s-$ $2p\,4p$	$^3P^o-^3S$	*4822.1*	*4823.5*	*60373.0*	–	81105.03	9–3	1.90–03	2.21–04	3.16–02	–2.701	C	1,2
			4826.80	4828.14	60393.14	–	81105.03	5–3	6.28–04	1.32–04	1.05–02	–3.182	C	2n
			4817.37	4818.72	60352.63	–	81105.03	3–3	8.76–04	3.05–04	1.45–02	–3.039	C	2n
			4812.92	4814.27	60333.43	–	81105.03	1–3	4.03–04	4.20–04	6.66–03	–3.377	C	2n
66.		$^3P^o-^3P$	*4769.4*	*4770.7*	*60373.0*	–	*81334.2*	9–9	1.06–02	3.63–03	5.13–01	–1.486	C	1,2
			4771.74	4773.08	60393.14	–	81343.99	5–5	7.97–03	2.72–03	2.14–01	–1.866	C	2n
			4766.67	4768.01	60352.63	–	81325.76	3–3	2.36–03	8.06–04	3.79–02	–2.617	C	2n
			4775.90	4777.23	60393.14	–	81325.76	5–3	4.84–03	9.92–04	7.80–02	–2.304	C	2n
			4770.03	4771.36	60352.63	–	81311.01	3–1	1.07–02	1.21–03	5.72–02	–2.439	C	2n
			4762.53	4763.86	60352.63	–	81343.99	3–5	2.72–03	1.54–03	7.25–02	–2.335	C	2n
			4762.31	4763.64	60333.43	–	81325.76	1–3	3.37–03	3.44–03	5.39–02	–2.463	C	2n
67.	$2p\,3s-$ $2p\,3p$	$^1P^o-^1P$	14542.5	6874.51 cm^{-1}	61981.82	–	68856.33	3–3	8.55–02	2.71–01	3.90+01	–0.090	B	1,2,3,4
68.		$^1P^o-^3D$												
			12970.6	7707.66 cm^{-1}	61981.82	–	69689.48	3–3	1.47–04	3.71–04	4.76–02	–2.953	C	2,4
69.		$^1P^o-^3S$												
			11409.6	8762.13 cm^{-1}	61981.82	–	70743.95	3–3	6.41–05	1.25–04	1.41–02	–3.425	C	2,4
70.		$^1P^o-^3P$												
			10668.6	9370.69 cm^{-1}	61981.82	–	71352.51	3–1	1.25–05	7.13–06	7.51–04	–4.670	C	4
			10654.6	9383.08 cm^{-1}	61981.82	–	71364.90	3–3	3.44–05	5.85–05	6.16–03	–3.756	C	4
			10631.4	9403.56 cm^{-1}	61981.82	–	71385.38	3–5	4.18–07	1.18–06	1.24–04	–5.451	C	4
71.		$^1P^o-^1D$	9405.73	9408.31	61981.82	–	72610.72	3–5	2.91–01	6.43–01	5.97+01	0.285	B+	1,2,3,4
72.		$^1P^o-^1S$	8335.15	8337.44	61981.82	–	73975.91	3–1	3.51–01	1.22–01	1.00+01	–0.437	B+	1,2,3,4

C I: Allowed Transitions — Continued

No.	Transition Array	Mult.	λ_{air} (Å)	λ_{vac} (Å) or σ (cm^{-1})*	E_i (cm^{-1})	E_k (cm^{-1})	g_i-g_k	A_{ki} (10^8 s^{-1})	f_{ik}	S (at. u.)	log *gf*	Acc.	Source
73.	$2p3s-$ $2p4p$	$^1P^o-^1P$	5380.34	5381.83	61981.82	80562.85	3−3	1.86−02	8.08−03	4.30−01	−1.615	B	2
74.		$^1P^o-^1D$	5052.17	5053.58	61981.82	81769.79	3−5	2.60−02	1.66−02	8.26−01	−1.304	B	2
75.		$^1P^o-^1S$	4932.05	4933.43	61981.82	82251.71	3−1	6.02−02	7.32−03	3.57−01	−1.658	B	2
76.	$2p3s-$ $2p5p$	$^1P^o-^1P$	4371.37	4372.60	61981.82	84851.53	3−3	1.27−02	3.64−03	1.57−01	−1.962	B−	1
77.	$2s2p^3-$ $2s^22p3p$	$^3D^o-^3D$	*17747*	*5633.1* cm^{-1}	*64088.8*	*69722.0*	15−15	3.62−04	1.71−03	1.50+00	−1.591	B	1,2
			17672.0	5657.11 cm^{-1}	64086.92	69744.03	7−7	3.24−04	1.52−03	6.19−01	−1.973	B	2n
			17789.7	5619.71 cm^{-1}	64090.95	69710.66	5−5	2.39−04	1.13−03	3.32−01	−2.247	B	2n
			17853.4	5599.63 cm^{-1}	64089.85	69689.48	3−3	2.56−04	1.22−03	2.16−01	−2.435	B	2n
			17776.9	5623.74 cm^{-1}	64086.92	69710.66	7−5	6.26−05	2.12−04	8.68−02	−2.829	B	2n
			17857.0	5598.53 cm^{-1}	64090.95	69689.48	5−3	9.39−05	2.69−04	7.92−02	−2.871	B	2n
			17684.6	5653.08 cm^{-1}	64090.95	69744.03	5−7	4.49−05	2.95−04	8.58−02	−2.832	B	2n
			17786.2	5620.81 cm^{-1}	64089.85	69710.66	3−5	5.64−05	4.46−04	7.84−02	−2.873	B	2n
78.		$^3D^o-^3P$	*13721*	*7286.1* cm^{-1}	*64088.8*	*71374.9*	15−9	4.00−03	6.77−03	4.59+00	−0.993	B	1,2
			13697.8	7298.46 cm^{-1}	64086.92	71385.38	7−5	3.36−03	6.75−03	2.13+00	−1.325	B	2n
			13743.9	7273.95 cm^{-1}	64090.95	71364.90	5−3	3.02−03	5.13−03	1.16+00	−1.591	B	2n
			13765.3	7262.66 cm^{-1}	64089.85	71352.51	3−1	4.03−03	3.81−03	5.19−01	−1.941	B	2n
			13705.3	7294.43 cm^{-1}	64090.95	71385.38	5−5	5.86−04	1.65−03	3.72−01	−2.083	B	2n
			13741.9	7275.05 cm^{-1}	64089.85	71364.90	3−3	9.93−04	2.81−03	3.81−01	−2.074	B	2n
			13703.3	7295.53 cm^{-1}	64089.85	71385.38	3−5	3.86−05	1.81−04	2.45−02	−3.265	B−	2n
79.	$2s2p^3-$ $2s^22p4p$	$^3D^o-^3D$											
			5969.32	5970.97	64086.92	80834.61	7−7	2.51−04	1.34−04	1.85−02	−3.027	C	2
80.		$^3D^o-^3P$	*5797.0*	*5798.6*	*64088.8*	*81334.2*	15−9	4.11−03	1.24−03	3.56−01	−1.729	B−	1,2
			5793.12	5794.73	64086.92	81343.99	7−5	3.44−03	1.24−03	1.65−01	−2.062	B−	2n
			5800.60	5802.21	64090.95	81325.76	5−3	3.04−03	9.19−04	8.77−02	−2.338	B−	2n
			5805.20	5806.81	64089.85	81311.01	3−1	4.12−03	6.93−04	3.97−02	−2.682	C+	2n
			5794.47	5796.08	64090.95	81343.99	5−5	6.44−04	3.24−04	3.09−02	−2.790	C+	2n
			5800.23	5801.84	64089.85	81325.76	3−3	1.04−03	5.26−04	3.01−02	−2.802	C+	2n
			5794.10	5795.71	64089.85	81343.99	3−5	4.42−05	3.70−05	2.12−03	−3.954	D	2n
81.	$2s2p^3-$ $2s^22p(^2P^o_{1/2})4f$	$^3D^o-[^5/_2]$											
			5041.79	5043.19	64090.95	83919.65	5−7	3.28−03	1.75−03	1.45−01	−2.058	C	2
			5041.48	5042.89	64089.85	83919.76	3−5	5.25−03	3.33−03	1.66−01	−2.000	C	2
			5041.76	5043.17	64090.95	83919.76	5−5	5.70−04	2.17−04	1.80−02	−2.964	C	2
82.		$^3D^o-[^7/_2]$											
			5039.06	5040.46	64086.92	83926.37	7−9	4.73−03	2.31−03	2.69−01	−1.791	C	2
			5040.12	5041.53	64090.95	83926.20	5−7	1.87−03	9.96−04	8.27−02	−2.303	C	2
83.	$2s2p^3-$ $2s^22p(^2P^o_{3/2})4f$	$^3D^o-[^7/_2]$											
			5023.84	5025.24	64086.92	83986.45	7−9	1.81−03	8.83−04	1.02−01	−2.209	C	2
			5024.92	5026.32	64090.95	83986.22	5−7	7.06−04	3.74−04	3.10−02	−2.728	C	2

C I: Allowed Transitions — Continued

No.	Transition Array	Mult.	λ_{air} (Å)	λ_{vac} (Å) or σ (cm^{-1})*	E_i (cm^{-1})	E_k (cm^{-1})	g_i-g_k	A_{ki} (10^8 s^{-1})	f_{ik}	S (at. u.)	log gf	Acc.	Source
84.		$^3D^\circ-[^5/_2]$											
			5017.09	5018.49	64086.92	84013.25	7–7	1.32–03	5.00–04	5.78–02	–2.456	C	2
			5018.06	5019.46	64090.95	84013.40	5–5	7.67–04	2.90–04	2.39–02	–2.839	C	2
			5017.79	5019.19	64089.85	84013.40	3–5	3.12–04	1.96–04	9.74–03	–3.230	C	2
85.		$^3D^\circ-[^3/_2]$											
			5012.28	5013.67	64090.95	84036.40	5–5	4.36–04	1.64–04	1.35–02	–3.086	C	2
			5012.03	5013.43	64089.85	84036.29	3–3	8.18–04	3.08–04	1.53–02	–3.034	C	2
			5011.26	5012.66	64086.92	84036.40	7–5	9.91–05	2.67–05	3.08–03	–3.729	C	2
86.	$2s2p^3$– $2s^22p5p$	$^3D^\circ-^3P$	*4736.6*	*4738.0*	*64088.8*	*85195.0*	15–9	3.04–03	6.14–04	1.44–01	–2.035	B–	1
			4734.26	4735.58	64086.92	85203.64	7–5	2.56–03	6.15–04	6.71–02	–2.366	C	*LS*
			4738.46	4739.79	64090.95	85188.95	5–3	2.28–03	4.61–04	3.59–02	–2.638	C	*LS*
			4742.56	4743.89	64089.85	85169.61	3–1	3.03–03	3.41–04	1.60–02	–2.990	C	*LS*
			4735.16	4736.49	64090.95	85203.64	5–5	4.57–04	1.54–04	1.20–02	–3.115	C	*LS*
			4738.21	4739.54	64089.85	85188.95	3–3	7.60–04	2.56–04	1.20–02	–3.115	C	*LS*
			4734.92	4736.24	64089.85	85203.64	3–5	3.05–05	1.71–05	7.98–04	–4.291	D–	*LS*
87.	$2p3p$– $2p3d$	$^1P-^1D^\circ$	11330.3	8823.49 cm^{-1}	68856.33	77679.82	3–5	2.03–01	6.53–01	7.31+01	0.292	B	1,2,3
88.	$2p3p$– $2p4s$	$^1P-^3P^\circ$											
			10759.3	9291.76 cm^{-1}	68856.33	78148.09	3–5	5.52–05	1.60–04	1.70–02	–3.319	C	2
			10795.7	9260.41 cm^{-1}	68856.33	78116.74	3–3	7.16–04	1.25–03	1.34–01	–2.425	C	2
89.	$2p3p$– $2p3d$	$^1P-^3D^\circ$											
			10577.7	9451.30 cm^{-1}	68856.33	78307.63	3–5	4.87–05	1.36–04	1.42–02	–3.388	C	2
			10593.5	9437.16 cm^{-1}	68856.33	78293.49	3–3	7.23–05	1.22–04	1.27–02	–3.438	C	2
90.	$2p3p$– $2p4s$	$^1P-^1P^\circ$	10541.2	9483.95 cm^{-1}	68856.33	78340.28	3–3	8.00–03	1.33–02	1.39+00	–1.398	D	1,2,3
91.	$2p3p$– $2p3d$	$^1P-^1P^\circ$	10123.9	9874.94 cm^{-1}	68856.33	78731.27	3–3	2.02–01	3.10–01	3.10+01	–0.032	C+	1,2,3
92.	$2p3p$– $2p4d$	$^1P-^1D^\circ$	6828.12	6830.00	68856.33	83497.62	3–5	9.89–03	1.15–02	7.77–01	–1.461	C	1,2
93.	$2p3p$– $2p5s$	$^1P-^3P^\circ$											
			6711.32	6713.18	68856.33	83752.41	3–3	9.99–04	6.75–04	4.48–02	–2.693	C	2
94.		$^1P-^1P^\circ$	6655.52	6657.36	68856.33	83877.31	3–3	5.03–03	3.34–03	2.19–01	–1.999	C	1,2
95.	$2p3p$– $2p4d$	$^1P-^1P^\circ$	6587.61	6589.43	68856.33	84032.15	3–3	5.09–02	3.31–02	2.15+00	–1.003	B	1,2
96.	$2p3p$– $2p5d$	$^1P-^1P^\circ$	5668.94	5670.52	68856.33	86491.41	3–3	2.35–02	1.13–02	6.34–01	–1.469	C	1
97.	$2p3p$– $2p6d$	$^1P-^1P^\circ$	5268.95	5270.41	68856.33	87830.17	3–3	1.27–02	5.27–03	2.74–01	–1.801	C	1
98.	$2p3p$– $2p7d$	$^1P-^1P^\circ$	5053.52	5054.92	68856.33	88639.02	3–3	7.66–03	2.93–03	1.46–01	–2.056	C	1

C I: Allowed Transitions — Continued

No.	Transition Array	Mult.	λ_{air} (Å)	λ_{vac} (Å) or σ (cm^{-1})*	E_i (cm^{-1})	E_k (cm^{-1})	g_i-g_k	A_{ki} (10^8 s^{-1})	f_{ik}	S (at. u.)	log gf	Acc.	Source
99.	$2s^22p3p-$ $2s2p^3$	$^3D-^3P^o$	*18069*	*5532.9* cm^{-1}	*69722.0*	*75254.9*	15–9	4.81–03	1.41–02	1.26+01	–0.674	C	1,2
			18139.8	5511.24 cm^{-1}	69744.03	75255.27	7–5	3.96–03	1.40–02	5.84+00	–1.010	C	2n
			18034.8	5543.31 cm^{-1}	69710.66	75253.97	5–3	3.78–03	1.11–02	3.28+00	–1.258	C	2n
			17959.3	5566.64 cm^{-1}	69689.48	75256.12	3–1	5.14–03	8.29–03	1.47+00	–1.605	C	2n
			18030.6	5544.61 cm^{-1}	69710.66	75255.27	5–5	6.48–04	3.16–03	9.38–01	–1.801	C	2n
			17966.2	5564.49 cm^{-1}	69689.48	75253.97	3–3	1.19–03	5.76–03	1.02+00	–1.763	C	2n
			17962.0	5565.79 cm^{-1}	69689.48	75255.27	3–5	4.01–05	3.23–04	5.74–02	–3.013	C	2n
100.	$2p3p-$ $2p3d$	$^3D-^1D^o$											
			12544.9	7969.16 cm^{-1}	69710.66	77679.82	5–5	1.93–04	4.54–04	9.39–02	–2.644	C	2
			12511.7	7990.34 cm^{-1}	69689.48	77679.82	3–5	2.70–05	1.06–04	1.30–02	–3.499	C	2
101.	$2p3p-$ $2p4s$	$^3D-^3P^o$	*11886*	*8410.9* cm^{-1}	*69722.0*	*78132.9*	15–9	1.14–01	1.46–01	8.54+01	0.339	B–	1,2,3
			11895.8	8404.06 cm^{-1}	69744.03	78148.09	7–5	9.24–02	1.40–01	3.84+01	–0.008	B–	2n,3n
			11892.9	8406.08 cm^{-1}	69710.66	78116.74	5–3	8.31–02	1.06–01	2.07+01	–0.276	B–	2n,3n
			11879.6	8415.50 cm^{-1}	69689.48	78104.98	3–1	1.16–01	8.18–02	9.60+00	–0.610	B–	2n,3n
			11848.7	8437.43 cm^{-1}	69710.66	78148.09	5–5	1.91–02	4.01–02	7.83+00	–0.698	B–	2n,3n
			11863.0	8427.26 cm^{-1}	69689.48	78116.74	3–3	3.08–02	6.51–02	7.62+00	–0.710	B–	2n,3n
			11819.0	8458.61 cm^{-1}	69689.48	78148.09	3–5	3.08–03	1.08–02	1.26+00	–1.491	B–	2n,3n
102.	$2p3p-$ $2p3d$	$^3D-^3F^o$	*11755*	*8504.4* cm^{-1}	*69722.0*	*78226.4*	15–21	2.58–01	7.49–01	4.35+02	1.050	B	1,2,3
			11753.3	8505.91 cm^{-1}	69744.03	78249.94	7–9	2.63–01	7.00–01	1.90+02	0.690	B	2n,3n
			11754.8	8504.85 cm^{-1}	69710.66	78215.51	5–7	2.40–01	6.95–01	1.35+02	0.541	B	2n,3n
			11748.2	8509.59 cm^{-1}	69689.48	78199.07	3–5	2.29–01	7.89–01	9.16+01	0.374	B	2n,3n
			11801.1	8471.48 cm^{-1}	69744.03	78215.51	7–7	1.26–02	2.63–02	7.16+00	–0.734	B	2n,3n
			11777.5	8488.41 cm^{-1}	69710.66	78199.07	5–5	2.90–02	6.04–02	1.17+01	–0.520	B	2n,3n
			11824.0	8455.04 cm^{-1}	69744.03	78199.07	7–5	2.94–05	4.40–05	1.20–02	–3.511	B–	2n,3n
103.		$^3D-^3D^o$	*11641*	*8587.8* cm^{-1}	*69722.0*	*78309.8*	15–15	7.06–02	1.44–01	8.26+01	0.333	B–	1,2,3
			11659.7	8574.22 cm^{-1}	69744.03	78318.25	7–7	7.48–02	1.52–01	4.10+01	0.028	B–	2n,3n
			11628.8	8596.97 cm^{-1}	69710.66	78307.63	5–5	5.42–02	1.10–01	2.10+01	–0.260	B–	2n,3n
			11619.3	8604.01 cm^{-1}	69689.48	78293.49	3–3	4.39–02	8.88–02	1.02+01	–0.574	B–	2n,3n
			11674.1	8563.60 cm^{-1}	69744.03	78307.63	7–5	1.57–02	2.29–02	6.16+00	–0.795	B–	2n,3n
			11648.0	8582.83 cm^{-1}	69710.66	78293.49	5–3	1.58–02	1.93–02	3.70+00	–1.016	B–	2n,3n
			11614.5	8607.59 cm^{-1}	69710.66	78318.25	5–7	5.13–04	1.45–03	2.78–01	–2.139	D	2n,3n
			11600.2	8618.15 cm^{-1}	69689.48	78307.63	3–5	5.36–04	1.80–03	2.07–01	–2.267	D	2n,3n
104.	$2p3p-$ $2p4s$	$^3D-^1P^o$											
			11584.8	8629.62 cm^{-1}	69710.66	78340.28	5–3	4.03–03	4.86–03	9.27–01	–1.614	C	2
			11556.5	8650.80 cm^{-1}	69689.48	78340.28	3–3	2.00–03	4.01–03	4.57–01	–1.920	C	2
105.	$2p3p-$ $2p3d$	$^3D-^1F^o$											
			11336.1	8818.96 cm^{-1}	69710.66	78529.62	5–7	1.52–03	4.09–03	7.64–01	–1.689	C	2
106.		$^3D-^1P^o$											
			11056.7	9041.79 cm^{-1}	69689.48	78731.27	3–3	7.13–04	1.31–03	1.43–01	–2.407	C	2
107.		$^3D-^3P^o$											
			10449.9	9566.82 cm^{-1}	69744.03	79310.85	7–5	2.12–04	2.48–04	5.98–02	–2.760	C	2
			10405.0	9608.12 cm^{-1}	69710.66	79318.78	5–3	1.12–04	1.09–04	1.87–02	–3.263	C	2
			10413.6	9600.19 cm^{-1}	69710.66	79310.85	5–5	7.41–05	1.21–04	2.07–02	–3.220	C	2
			10382.1	9629.30 cm^{-1}	69689.48	79318.78	3–3	6.62–05	1.07–04	1.10–02	–3.493	C	2

C I: Allowed Transitions — Continued

No.	Transition Array	Mult.	λ_{air} (Å)	λ_{vac} (Å) or σ (cm^{-1})*	E_i (cm^{-1})		E_k (cm^{-1})	g_i-g_k	A_{ki} (10^8 s^{-1})	f_{ik}	S (at. u.)	log gf	Acc.	Source
108.	$2p3p-$ $2p4d$	^{3}D–^{3}F°	*7114.5*	*7116.4*	*69722.0*	–	*83773.9*	15–21	2.39–02	2.54–02	8.93+00	–0.419	B–	1,2
			7113.18	7115.14	69744.03	–	83798.57	7–9	2.47–02	2.41–02	3.94+00	–0.774	B–	2*n*
			7115.17	7117.13	69710.66	–	83761.26	5–7	2.19–02	2.32–02	2.72+00	–0.935	B–	2*n*
			7111.47	7113.43	69689.48	–	83747.39	3–5	2.17–02	2.74–02	1.92+00	–1.086	B–	2*n*
			7132.11	7134.08	69744.03	–	83761.26	7–7	1.17–03	8.94–04	1.47–01	–2.204	B–	2*n*
			7122.20	7124.17	69710.66	–	83747.39	5–5	2.01–03	1.53–03	1.80–01	–2.116	B–	2*n*
			7139.18	7141.14	69744.03	–	83747.39	7–5	2.06–04	1.13–04	1.85–02	–3.103	C+	2*n*
109.	$2p3p-$ $2p5s$	^{3}D–^{3}P°	*7115.2*	*7117.2*	*69722.0*	–	*83772.5*	15–9	4.22–02	1.92–02	6.76+00	–0.540	B	1,2
			7116.99	7118.95	69744.03	–	83791.04	7–5	3.26–02	1.77–02	2.90+00	–0.907	B	2*n*
			7119.66	7121.62	69710.66	–	83752.41	5–3	3.12–02	1.42–02	1.67+00	–1.149	B	2*n*
			7115.18	7117.14	69689.48	–	83740.06	3–1	4.43–02	1.12–02	7.88–01	–1.473	B	2*n*
			7100.12	7102.08	69710.66	–	83791.04	5–5	8.96–03	6.77–03	7.91–01	–1.470	B	2*n*
			7108.93	7110.89	69689.48	–	83752.41	3–3	1.12–02	8.52–03	5.98–01	–1.592	B	2*n*
			7089.46	7091.41	69689.48	–	83791.04	3–5	1.49–04	1.87–04	1.31–02	–3.252	B–	2*n*
110.	$2p3p-$ $2p4d$	^{3}D–^{3}D°	*7081.4*	*7083.4*	*69722.0*	–	*83839.5*	15–15	7.16–03	5.38–03	1.88+00	–1.093	C	1,2
			7087.83	7089.79	69744.03	–	83848.83	7–7	6.85–03	5.16–03	8.42–01	–1.443	C	2*n*
			7076.48	7078.43	69710.66	–	83838.08	5–5	2.98–03	2.23–03	2.60–01	–1.952	D+	2*n*
			7074.86	7076.82	69689.48	–	83820.13	3–3	3.40–03	2.55–03	1.78–01	–2.116	D+	2*n*
			7093.24	7095.19	69744.03	–	83838.08	7–5	5.32–03	2.87–03	4.69–01	–1.697	D+	2*n*
			7085.48	7087.44	69710.66	–	83820.13	5–3	1.87–03	8.44–04	9.85–02	–2.375	D+	2*n*
			7071.10	7073.05	69710.66	–	83848.83	5–7	2.72–04	2.86–04	3.33–02	–2.845	D	2*n*
			7065.89	7067.84	69689.48	–	83838.08	3–5	2.26–05	2.81–05	1.96–03	–4.073	E	2*n*
111.	$2p3p-$ $2p5s$	^{3}D–^{1}P°												
			7056.89	7058.83	69710.66	–	83877.31	5–3	1.58–03	7.06–04	8.21–02	–2.452	C	2
112.	$2p3p-$ $2p4d$	^{3}D–^{1}F°												
			7022.13	7024.07	69710.66	–	83947.43	5–7	4.10–04	4.24–04	4.91–02	–2.673	C	2
113.		^{3}D–^{1}P°												
			6970.28	6972.20	69689.48	–	84032.15	3–3	3.30–04	2.40–04	1.65–02	–3.142	C	2
114.	$2p3p-$ $2p5d$	^{3}D–^{3}F°	*6014.8*	*6016.5*	*69722.0*	–	*86343.0*	15–21	4.35–03	3.30–03	9.80–01	–1.305	C	1
			6013.21	6014.88	69744.03	–	86369.47	7–9	4.35–03	3.03–03	4.20–01	–1.673	D	*LS*
			6016.45	6018.11	69710.66	–	86327.16	5–7	3.86–03	2.93–03	2.90–01	–1.834	D–	*LS*
			6012.23	6013.89	69689.48	–	86317.64	3–5	3.65–03	3.30–03	1.96–01	–2.004	D–	*LS*
			6028.56	6030.22	69744.03	–	86327.16	7–7	4.81–04	2.62–04	3.64–02	–2.737	D–	*LS*
			6019.90	6021.56	69710.66	–	86317.64	5–5	6.76–04	3.67–04	3.64–02	–2.736	D–	*LS*
			6032.02	6033.69	69744.03	–	86317.64	7–5	1.90–05	7.39–06	1.03–03	–4.286	D–	*LS*
115.	$2p3p-$ $2p6s$	^{3}D–^{3}P°	*6011.7*	*6013.4*	*69722.0*	–	*86351.6*	15–9	2.13–02	6.93–03	2.06+00	–0.983	C	1
			6013.17	6014.83	69744.03	–	86369.60	7–5	1.79–02	6.93–03	9.61–01	–1.314	D	*LS*
			6014.83	6016.50	69710.66	–	86331.63	5–3	1.60–02	5.20–03	5.15–01	–1.585	D	*LS*
			6010.68	6012.34	69689.48	–	86321.94	3–1	2.13–02	3.85–03	2.29–01	–1.937	D	*LS*
			6001.12	6002.78	69710.66	–	86369.60	5–5	3.22–03	1.74–03	1.72–01	–2.061	D	*LS*
			6007.18	6008.84	69689.48	–	86331.63	3–3	5.34–03	2.89–03	1.72–01	–2.062	D	*LS*
			5993.50	5995.16	69689.48	–	86369.60	3–5	2.15–04	1.93–04	1.14–02	–3.237	D–	*LS*

C I: Allowed Transitions — Continued

No.	Transition Array	Mult.	λ_{air} (Å)	λ_{vac} (Å) or σ (cm^{-1})*	E_i (cm^{-1})	E_k (cm^{-1})	g_i–g_k	A_{ki} (10^8 s^{-1})	f_{ik}	S (at. u.)	log gf	Acc.	Source
116.	$2p3p$–$2p5d$	^{3}D–^{3}D°	*5998.6*	*6000.3*	*69722.0*	*86387.9*	15–15	2.02–03	1.09–03	3.23–01	–1.786	C	1
			6002.98	6004.65	69744.03	86397.80	7–7	1.79–03	9.68–04	1.34–01	–2.169	D	*LS*
			5994.00	5995.66	69710.66	86389.38	5–5	1.41–03	7.60–04	7.49–02	–2.420	D	*LS*
			5996.05	5997.71	69689.48	86362.52	3–3	1.52–03	8.18–04	4.85–02	–2.610	D	*LS*
			6006.02	6007.68	69744.03	86389.38	7–5	3.14–04	1.21–04	1.68–02	–3.071	D–	*LS*
			6003.67	6005.34	69710.66	86362.52	5–3	5.04–04	1.63–04	1.62–02	–3.088	D–	*LS*
			5990.98	5992.64	69710.66	86397.80	5–7	2.26–04	1.70–04	1.68–02	–3.070	D–	*LS*
			5986.40	5988.06	69689.48	86389.38	3–5	3.05–04	2.73–04	1.62–02	–3.086	D–	*LS*
117.	$2p3p$–$2p7s$	^{3}D–^{3}P°	*5549.3*	*5550.8*	*69722.0*	*87737.3*	15–9	1.21–02	3.36–03	9.21–01	–1.297	C	1
			5551.02	5552.56	69744.03	87753.73	7–5	1.02–02	3.36–03	4.30–01	–1.629	D	*LS*
			5551.58	5553.12	69710.66	87718.56	5–3	9.09–03	2.52–03	2.30–01	–1.900	D–	*LS*
			5547.27	5548.81	69689.48	87711.37	3–1	1.21–02	1.87–03	1.02–01	–2.252	D–	*LS*
			5540.76	5542.29	69710.66	87753.73	5–5	1.83–03	8.41–04	7.67–02	–2.376	D–	*LS*
			5545.05	5546.59	69689.48	87718.56	3–3	3.04–03	1.40–03	7.67–02	–2.376	D–	*LS*
			5534.26	5535.80	69689.48	87753.73	3–5	1.22–04	9.36–05	5.12–03	–3.552	D–	*LS*
118.	$2p3p$–$2p8s$	^{3}D–^{3}P°	*5304.7*	*5306.2*	*69722.0*	*88568.0*	15–9	7.55–03	1.91–03	5.01–01	–1.543	C	1
			5306.31	5307.79	69744.03	88584.26	7–5	6.33–03	1.91–03	2.34–01	–1.874	D	*LS*
			5306.83	5308.31	69710.66	88549.06	5–3	5.65–03	1.43–03	1.25–01	–2.145	D	*LS*
			5302.36	5303.84	69689.48	88543.76	3–1	7.56–03	1.06–03	5.56–02	–2.497	D	*LS*
			5296.93	5298.41	69710.66	88584.26	5–5	1.14–03	4.78–04	4.17–02	–2.621	D	*LS*
			5300.87	5302.35	69689.48	88549.06	3–3	1.89–03	7.97–04	4.17–02	–2.622	D	*LS*
			5290.99	5292.47	69689.48	88584.26	3–5	7.61–05	5.32–05	2.78–03	–3.797	D–	*LS*
119.	$2s^22p3p$–$2s2p^3$	^{3}S–^{3}P°	*22162*	*4511.0* cm^{-1}	70743.95	*75254.9*	3–9	5.60–04	1.24–02	2.71+00	–1.430	C+	1,2
			22160.4	4511.32 cm^{-1}	70743.95	75255.27	3–5	6.65–04	8.17–03	1.79+00	–1.611	C	2*n*
			22166.8	4510.02 cm^{-1}	70743.95	75253.97	3–3	4.51–04	3.33–03	7.29–01	–2.001	C	2*n*
			22156.2	4512.17 cm^{-1}	70743.95	75256.12	3–1	3.63–04	8.90–04	1.95–01	–2.573	C	2*n*
120.	$2p3p$–$2p4s$	^{3}S–^{3}P°	*13530*	*7388.9* cm^{-1}	70743.95	*78132.9*	3–9	6.58–02	5.42–01	7.25+01	0.211	C+	1,2,3
			13502.3	7404.14 cm^{-1}	70743.95	78148.09	3–5	6.62–02	3.02–01	4.03+01	–0.043	C+	2*n*,3*n*
			13559.7	7372.79 cm^{-1}	70743.95	78116.74	3–3	6.52–02	1.80–01	2.41+01	–0.268	C+	2*n*,3*n*
			13581.3	7361.03 cm^{-1}	70743.95	78104.98	3–1	6.60–02	6.08–02	8.16+00	–0.739	C+	2*n*,3*n*
121.	$2p3p$–$2p3d$	^{3}S–^{3}F°											
			13409.9	7455.12 cm^{-1}	70743.95	78199.07	3–5	5.64–05	2.53–04	3.36–02	–3.119	C	2
122.		^{3}S–^{3}D°											
			13242.2	7549.54 cm^{-1}	70743.95	78293.49	3–3	9.74–05	2.56–04	3.35–02	–3.114	C	2
123.	$2p3p$–$2p4s$	^{3}S–^{1}P°											
			13160.7	7596.33 cm^{-1}	70743.95	78340.28	3–3	7.57–04	1.97–03	2.56–01	–2.229	C	2
124.	$2p3p$–$2p3d$	^{3}S–^{1}P°											
			12516.4	7987.32 cm^{-1}	70743.95	78731.27	3–3	8.82–05	2.07–04	2.56–02	–3.207	C	2
125.		^{3}S–^{3}P°	*11664*	*8570.9* cm^{-1}	70743.95	*79314.9*	3–9	8.88–02	5.43–01	6.26+01	0.212	B+	1,2,3
			11669.6	8566.90 cm^{-1}	70743.95	79310.85	3–5	9.13–02	3.11–01	3.59+01	–0.030	B	2*n*,3*n*
			11658.8	8574.83 cm^{-1}	70743.95	79318.78	3–3	8.62–02	1.76–01	2.02+01	–0.278	B	2*n*,3*n*
			11652.9	8579.21 cm^{-1}	70743.95	79323.16	3–1	8.36–02	5.68–02	6.54+00	–0.769	B	2*n*,3*n*

C I: Allowed Transitions — Continued

No.	Transition Array	Mult.	λ_{air} (Å)	λ_{vac} (Å) or σ (cm^{-1})*	E_i (cm^{-1})	E_k (cm^{-1})	g_i–g_k	A_{ki} (10^8 s^{-1})	f_{ik}	S (at. u.)	log gf	Acc.	Source
126.	2p3p– 2p5s	^3S–$^3P^o$	*7673.3*	*7675.5*	70743.95	*83772.5*	3–9	1.18–02	3.12–02	2.37+00	–1.029	B–	1,2
			7662.44	7664.54	70743.95	83791.04	3–5	1.20–02	1.76–02	1.33+00	–1.277	B–	2n
			7685.19	7687.31	70743.95	83752.41	3–3	1.14–02	1.01–02	7.66–01	–1.519	B–	2n
			7692.49	7694.61	70743.95	83740.06	3–1	1.18–02	3.50–03	2.66–01	–1.979	B–	2n
127.	2p3p– 2p4d	^3S–$^3D^o$											
			7634.91	7637.01	70743.95	83838.08	3–5	1.56–04	2.27–04	1.71–02	–3.168	C	2
128.	2p3p– 2p5s	^3S–$^1P^o$											
			7612.10	7614.20	70743.95	83877.31	3–3	5.20–04	4.52–04	3.40–02	–2.868	C	2
129.	2p3p– 2p4d	^3S–$^3P^o$	*7479.9*	*7482.0*	70743.95	*84109.4*	3–9	1.04–02	2.61–02	1.93+00	–1.107	C+	1,2
			7483.45	7485.51	70743.95	84103.10	3–5	1.01–02	1.42–02	1.05+00	–1.371	C+	2n
			7476.18	7478.23	70743.95	84116.09	3–3	1.06–02	8.88–03	6.56–01	–1.574	C+	2n
			7473.31	7475.37	70743.95	84121.22	3–1	1.08–02	3.00–03	2.22–01	–2.045	C+	2n
130.	2p3p– 2p6s	^3S–$^3P^o$	*6405.3*	*6407.1*	70743.95	*86351.6*	3–9	5.41–03	9.98–03	6.32–01	–1.524	C	1
			6397.97	6399.73	70743.95	86369.60	3–5	5.43–03	5.55–03	3.51–01	–1.778	D	LS
			6413.55	6415.32	70743.95	86331.63	3–3	5.39–03	3.32–03	2.11–01	–2.001	D	LS
			6417.54	6419.31	70743.95	86321.94	3–1	5.38–03	1.11–03	7.02–02	–2.479	D	LS
131.	2p3p– 2p7s	^3S–$^3P^o$	*5883.0*	*5884.7*	70743.95	*87737.3*	3–9	2.77–03	4.31–03	2.51–01	–1.888	C	1
			5877.34	5878.97	70743.95	87753.73	3–5	2.78–03	2.40–03	1.39–01	–2.143	D	LS
			5889.52	5891.15	70743.95	87718.56	3–3	2.76–03	1.44–03	8.36–02	–2.366	D	LS
			5892.02	5893.65	70743.95	87711.37	3–1	2.76–03	4.79–04	2.79–02	–2.843	D–	LS
132.	2p3p– 2p8s	^3S–$^3P^o$	*5608.8*	*5610.4*	70743.95	*88568.0*	3–9	1.62–03	2.29–03	1.27–01	–2.163	C	1
			5603.73	5605.28	70743.95	88584.26	3–5	1.62–03	1.27–03	7.04–02	–2.418	D	LS
			5614.81	5616.37	70743.95	88549.06	3–3	1.61–03	7.62–04	4.23–02	–2.641	D	LS
			5616.48	5618.04	70743.95	88543.76	3–1	1.61–03	2.54–04	1.41–02	–3.118	D–	LS
133.	$2s^22p3p$– $2s2p^3$	^3P–$^3P^o$	*25766*	*3880.0* cm^{-1}	*71374.9*	*75254.9*	9–9	6.94–03	6.92–02	5.28+01	–0.206	C+	1,2
			25833.5	3869.89 cm^{-1}	71385.38	75255.27	5–5	5.17–03	5.18–02	2.20+01	–0.587	C+	2n
			25706.1	3889.07 cm^{-1}	71364.90	75253.97	3–3	1.80–03	1.79–02	4.54+00	–1.271	C+	2n
			25842.2	3868.59 cm^{-1}	71385.38	75253.97	5–3	2.83–03	1.70–02	7.25+00	–1.070	C+	2n
			25691.9	3891.22 cm^{-1}	71364.90	75256.12	3–1	7.07–03	2.33–02	5.92+00	–1.155	C+	2n
			25697.5	3890.37 cm^{-1}	71364.90	75255.27	3–5	1.71–03	2.83–02	7.17+00	–1.072	C+	2n
			25624.4	3901.46 cm^{-1}	71352.51	75253.97	1–3	2.37–03	6.99–02	5.90+00	–1.155	C+	2n
134.	2p3p– 2p4s	^3P–$^3P^o$	*14793*	*6757.9* cm^{-1}	*71374.9*	*78132.9*	9–9	1.51–02	4.96–02	2.18+01	–0.350	B–	1,2,3
			14782.9	6762.71 cm^{-1}	71385.38	78148.09	5–5	1.38–02	4.51–02	1.10+01	–0.647	B–	2n,3n
			14806.7	6751.84 cm^{-1}	71364.90	78116.74	3–3	3.91–03	1.29–02	1.88+00	–1.413	B–	2n,3n
			14851.8	6731.36 cm^{-1}	71385.38	78116.74	5–3	6.41–03	1.27–02	3.11+00	–1.196	B–	2n,3n
			14832.6	6740.08 cm^{-1}	71364.90	78104.98	3–1	1.33–02	1.46–02	2.14+00	–1.359	B–	2n,3n
			14738.3	6783.19 cm^{-1}	71364.90	78148.09	3–5	2.31–03	1.25–02	1.82+00	–1.425	B–	2n,3n
			14779.6	6764.23 cm^{-1}	71352.51	78116.74	1–3	3.84–03	3.77–02	1.84+00	–1.424	B–	2n,3n
135.	2p3p– 2p3d	^3P–$^3F^o$											
			14637.0	6830.13 cm^{-1}	71385.38	78215.51	5–7	4.84–03	2.18–02	5.25+00	–0.963	C	2
			14628.4	6834.17 cm^{-1}	71364.90	78199.07	3–5	2.11–03	1.13–02	1.63+00	–1.471	C	2
			14672.3	6813.69 cm^{-1}	71385.38	78199.07	5–5	4.39–04	1.42–03	3.43–01	–2.149	C	2

C I: Allowed Transitions — Continued

No.	Transition Array	Mult.	λ_{air} (Å)	λ_{vac} (Å) or σ (cm^{-1})*	E_i (cm^{-1})	E_k (cm^{-1})	g_i–g_k	A_{ki} (10^8 s^{-1})	f_{ik}	S (at. u.)	log gf	Acc.	Source
136.		^{3}P–^{3}D°	*14416*	*6934.9* cm^{-1}	*71374.9*	*78309.8*	9–15	1.24–01	6.42–01	2.74+02	0.762	B	1,2,3
			14420.1	6932.87 cm^{-1}	71385.38	78318.25	5–7	1.23–01	5.37–01	1.28+02	0.429	B	2*n*,3*n*
			14399.6	6942.73 cm^{-1}	71364.90	78307.63	3–5	9.89–02	5.12–01	7.29+01	0.187	B	2*n*,3*n*
			14403.3	6940.98 cm^{-1}	71352.51	78293.49	1–3	6.88–02	6.42–01	3.05+01	–0.192	B	2*n*,3*n*
			14442.2	6922.25 cm^{-1}	71385.38	78307.63	5–5	2.78–02	8.70–02	2.07+01	–0.361	B	2*n*,3*n*
			14429.0	6928.59 cm^{-1}	71364.90	78293.49	3–3	4.84–02	1.51–01	2.16+01	–0.343	B	2*n*,3*n*
			14471.8	6908.11 cm^{-1}	71385.38	78293.49	5–3	2.78–03	5.24–03	1.25+00	–1.581	B	2*n*,3*n*
137.	2*p*3*p*– 2*p*4*s*	^{3}P–^{1}P°											
			14374.4	6954.90 cm^{-1}	71385.38	78340.28	5–3	2.45–04	4.56–04	1.08–01	–2.642	C	2
			14332.2	6975.38 cm^{-1}	71364.90	78340.28	3–3	3.88–03	1.19–02	1.69+00	–1.446	C	2
			14306.8	6987.77 cm^{-1}	71352.51	78340.28	1–3	6.13–03	5.65–02	2.66+00	–1.248	C	2
138.	2*p*3*p*– 2*p*3*d*	^{3}P–^{1}F°											
			13993.5	7144.24 cm^{-1}	71385.38	78529.62	5–7	4.28–04	1.76–03	4.05–01	–2.056	C	2
139.		^{3}P–^{1}P°											
			13609.3	7345.89 cm^{-1}	71385.38	78731.27	5–3	3.83–05	6.38–05	1.43–02	–3.496	C	2
			13571.5	7366.37 cm^{-1}	71364.90	78731.27	3–3	2.99–04	8.25–04	1.11–01	–2.606	C	2
			13548.7	7378.76 cm^{-1}	71352.51	78731.27	1–3	6.95–05	5.74–04	2.56–02	–3.241	C	2
140.		^{3}P–^{3}P°	*12591*	*7940.0* cm^{-1}	*71374.9*	*79314.9*	9–9	1.21–01	2.89–01	1.08+02	0.415	B–	1,2,3
			12614.1	7925.47 cm^{-1}	71385.38	79310.85	5–5	9.39–02	2.24–01	4.66+01	0.050	B–	2*n*,3*n*
			12569.0	7953.88 cm^{-1}	71364.90	79318.78	3–3	3.55–02	8.41–02	1.04+01	–0.598	B–	2*n*,3*n*
			12601.5	7933.40 cm^{-1}	71385.38	79318.78	5–3	5.05–02	7.22–02	1.50+01	–0.443	B–	2*n*,3*n*
			12562.1	7958.26 cm^{-1}	71364.90	79323.16	3–1	1.27–01	1.01–01	1.25+01	–0.521	B–	2*n*,3*n*
			12581.6	7945.95 cm^{-1}	71364.90	79310.85	3–5	2.45–02	9.68–02	1.20+01	–0.537	B–	2*n*,3*n*
			12549.5	7966.27 cm^{-1}	71352.51	79318.78	1–3	3.84–02	2.72–01	1.12+01	–0.565	B–	2*n*,3*n*
141.	2*p*3*p*– 2*p*5*s*	^{3}P–^{3}P°	*8063.9*	*8066.1*	*71374.9*	*83772.5*	9–9	1.43–02	1.39–02	3.32+00	–0.903	B	1,2
			8058.62	8060.84	71385.38	83791.04	5–5	1.09–02	1.06–02	1.41+00	–1.275	B	2*n*
			8070.43	8072.65	71364.90	83752.41	3–3	3.81–03	3.72–03	2.97–01	–1.952	B	2*n*
			8083.79	8086.02	71385.38	83752.41	5–3	6.31–03	3.71–03	4.94–01	–1.732	B	2*n*
			8078.48	8080.70	71364.90	83740.06	3–1	1.56–02	5.08–03	4.05–01	–1.817	B	2*n*
			8045.34	8047.55	71364.90	83791.04	3–5	2.71–03	4.39–03	3.48–01	–1.881	B	2*n*
			8062.36	8064.58	71352.51	83752.41	1–3	4.77–03	1.39–02	3.70–01	–1.856	B	2*n*
142.	2*p*3*p*– 2*p*4*d*	^{3}P–^{3}F°											
			8078.01	8080.23	71385.38	83761.26	5–7	1.24–04	1.69–04	2.25–02	–3.072	C	2
143.		^{3}P–^{3}D°	*8020.5*	*8022.7*	*71374.9*	*83839.5*	9–15	3.73–03	6.00–03	1.43+00	–1.267	D	1,2
			8021.25	8023.46	71385.38	83848.83	5–7	3.18–03	4.30–03	5.67–01	–1.668	D	2*n*
			8015.00	8017.20	71364.90	83838.08	3–5	4.69–03	7.53–03	5.96–01	–1.646	D	2*n*
			8018.57	8020.78	71352.51	83820.13	1–3	2.56–03	7.41–03	1.96–01	–2.130	D	2*n*
			8028.18	8030.39	71385.38	83838.08	5–5	2.27–05	2.19–05	2.90–03	–3.960	D–	2*n*
			8026.55	8028.76	71364.90	83820.13	3–3	8.36–04	8.07–04	6.40–02	–2.616	D	2*n*
144.	2*p*3*p*– 2*p*5*s*	^{3}P–^{1}P°											
			7981.96	7984.16	71352.51	83877.31	1–3	3.62–04	1.04–03	2.73–02	–2.984	C	2
145.	2*p*3*p*– 2*p*4*d*	^{3}P–^{1}P°											
			7892.20	7894.37	71364.90	84032.15	3–3	1.60–04	1.49–04	1.16–02	–3.349	C	2

C I: Allowed Transitions — Continued

No.	Transition Array	Mult.	λ_{air} (Å)	λ_{vac} (Å) or σ (cm^{-1})*	E_i (cm^{-1})	E_k (cm^{-1})	g_i-g_k	A_{ki} (10^8 s^{-1})	f_{ik}	S (at. u.)	log gf	Acc.	Source
146.		$^3P-^3P^o$	*7850.5*	*7852.7*	*71374.9*	*84109.4*	9–9	1.89–02	1.74–02	4.06+00	–0.804	B–	1,2
			7860.88	7863.04	71385.38	84103.10	5–5	1.53–02	1.42–02	1.84+00	–1.148	B–	2*n*
			7840.25	7842.41	71364.90	84116.09	3–3	5.17–03	4.76–03	3.69–01	–1.845	B–	2*n*
			7852.86	7855.02	71385.38	84116.09	5–3	7.49–03	4.15–03	5.37–01	–1.683	B–	2*n*
			7837.10	7839.25	71364.90	84121.22	3–1	1.81–02	5.55–03	4.30–01	–1.778	B–	2*n*
			7848.24	7850.40	71364.90	84103.10	3–5	4.02–03	6.19–03	4.80–01	–1.731	B–	2*n*
			7832.64	7834.79	71352.51	84116.09	1–3	5.65–03	1.56–02	4.02–01	–1.807	B–	2*n*
147.	2*p*3*p*– 2*p*6*s*	$^3P-^3P^o$	*6675.2*	*6677.0*	*71374.9*	*86351.6*	9–9	8.91–03	5.95–03	1.18+00	–1.271	C+	1
			6671.85	6673.69	71385.38	86369.60	5–5	6.69–03	4.47–03	4.91–01	–1.651	C–	*LS*
			6679.64	6681.49	71364.90	86331.63	3–3	2.22–03	1.49–03	9.81–02	–2.351	C–	*LS*
			6688.79	6690.64	71385.38	86331.63	5–3	3.69–03	1.49–03	1.64–01	–2.129	C–	*LS*
			6683.97	6685.81	71364.90	86321.94	3–1	8.87–03	1.98–03	1.31–01	–2.226	C–	*LS*
			6662.74	6664.58	71364.90	86369.60	3–5	2.24–03	2.48–03	1.64–01	–2.128	C–	*LS*
			6674.12	6675.96	71352.51	86331.63	1–3	2.97–03	5.95–03	1.31–01	–2.225	C–	*LS*
148.	2*p*3*p*– 2*p*5*d*	$^3P-^3P^o$	*6604.1*	*6605.9*	*71374.9*	*86512.8*	9–9	5.94–03	3.88–03	7.60–01	–1.457	C	1
			6611.35	6613.18	71385.38	86506.70	5–5	4.44–03	2.91–03	3.17–01	–1.837	D	*LS*
			6596.85	6598.67	71364.90	86519.47	3–3	1.49–03	9.72–04	6.33–02	–2.535	D	*LS*
			6605.77	6607.60	71385.38	86519.47	5–3	2.47–03	9.70–04	1.06–01	–2.314	D	*LS*
			6595.24	6597.06	71364.90	86523.16	3–1	5.96–03	1.30–03	8.44–02	–2.410	D	*LS*
			6602.41	6604.23	71364.90	86506.70	3–5	1.27–02	1.38–02	8.98–01	–1.384	D	*LS*
			6591.46	6593.28	71352.51	86519.47	1–3	1.99–03	3.89–03	8.44–02	–2.410	D	*LS*
149.	2*p*3*p*– 2*p*7*s*	$^3P-^3P^o$	*6109.9*	*6111.6*	*71374.9*	*87737.3*	9–9	5.55–03	3.11–03	5.63–01	–1.553	C	1
			6107.66	6109.35	71385.38	87753.73	5–5	4.17–03	2.33–03	2.34–01	–1.933	D	*LS*
			6113.15	6114.84	71364.90	87718.56	3–3	1.39–03	7.76–04	4.69–02	–2.633	D	*LS*
			6120.81	6122.51	71385.38	87718.56	5–3	2.30–03	7.75–04	7.81–02	–2.411	D	*LS*
			6115.84	6117.53	71364.90	87711.37	3–1	5.54–03	1.03–03	6.25–02	–2.508	D	*LS*
			6100.03	6101.72	71364.90	87753.73	3–5	1.39–03	1.30–03	7.81–02	–2.410	D	*LS*
			6108.52	6110.21	71352.51	87718.56	1–3	1.85–03	3.11–03	6.25–02	–2.508	D	*LS*
150.	2*p*3*p*– 2*p*6*d*	$^3P-^3P^o$	*6072.5*	*6074.2*	*71374.9*	*87837.9*	9–9	2.57–03	1.42–03	2.56–01	–1.893	C	1
			6078.39	6080.08	71385.38	87832.54	5–5	1.92–03	1.06–03	1.06–01	–2.274	D	*LS*
			6066.65	6068.33	71364.90	87843.91	3–3	6.44–04	3.55–04	2.13–02	–2.972	D–	*LS*
			6074.20	6075.88	71385.38	87843.91	5–3	1.07–03	3.55–04	3.55–02	–2.751	D	*LS*
			6065.55	6067.22	71364.90	87846.9	3–1	2.58–03	4.74–04	2.84–02	–2.847	D–	*LS*
			6070.83	6072.52	71364.90	87832.54	3–5	6.43–04	5.92–04	3.55–02	–2.751	D	*LS*
			6062.09	6063.77	71352.51	87843.91	1–3	8.61–04	1.42–03	2.84–02	–2.847	D–	*LS*
151.	2*p*3*p*– 2*p*8*s*	$^3P-^3P^o$	*5814.7*	*5816.3*	*71374.9*	*88568.0*	9–9	3.64–03	1.85–03	3.18–01	–1.780	C	1
			5812.72	5814.33	71385.38	88584.26	5–5	2.73–03	1.38–03	1.33–01	–2.160	D	*LS*
			5817.70	5819.31	71364.90	88549.06	3–3	9.09–04	4.61–04	2.65–02	–2.859	D–	*LS*
			5824.64	5826.26	71385.38	88549.06	5–3	1.51–03	4.61–04	4.42–02	–2.638	D	*LS*
			5819.49	5821.11	71364.90	88543.76	3–1	3.63–03	6.15–04	3.53–02	–2.734	D	*LS*
			5805.81	5807.42	71364.90	88584.26	3–5	9.15–04	7.70–04	4.42–02	–2.636	D	*LS*
			5813.51	5815.12	71352.51	88549.06	1–3	1.21–03	1.85–03	3.53–02	–2.734	D	*LS*
152.	2*p*3*p*– 2*p*7*d*	$^3P-^3P^o$	*5789.9*	*5791.6*	*71374.9*	*88641.4*	9–9	1.35–03	6.79–04	1.16–01	–2.214	C	1
			5795.02	5796.62	71385.38	88636.8	5–5	1.01–03	5.08–04	4.85–02	–2.595	D	*LS*
			5784.86	5786.47	71364.90	88646.6	3–3	3.38–04	1.70–04	9.70–03	–3.293	D	*LS*
			5791.73	5793.33	71385.38	88646.6	5–3	5.62–04	1.70–04	1.62–02	–3.072	D–	*LS*
			5784.03	5785.63	71364.90	88649.1	3–1	1.35–03	2.26–04	1.29–02	–3.168	D–	*LS*
			5788.15	5789.75	71364.90	88636.8	3–5	3.38–04	2.83–04	1.62–02	–3.071	D–	*LS*
			5780.72	5782.32	71352.51	88646.6	1–3	4.52–04	6.80–04	1.29–02	–3.168	D–	*LS*

C I: Allowed Transitions — Continued

No.	Transition Array	Mult.	λ_{air} (Å)	λ_{vac} (Å) or σ (cm^{-1})*	E_i (cm^{-1})	E_k (cm^{-1})	g_i-g_k	A_{ki} (10^8 s^{-1})	f_{ik}	S (at. u.)	log gf	Acc.	Source
153.	$2s^22p3p-$ $2s2p^3$	$^3P-^3S^\circ$	*2904.1*	*2905.0*	*71374.9*	– 105798.7	9–3	3.88–02	1.64–03	1.41–01	–1.832	C	1
			2905.00	2905.85	71385.38	– 105798.7	5–3	2.15–02	1.63–03	7.82–02	–2.088	D	*LS*
			2903.27	2904.12	71364.90	– 105798.7	3–3	1.29–02	1.64–03	4.69–02	–2.309	D	*LS*
			2902.23	2903.08	71352.51	– 105798.7	1–3	4.32–03	1.64–03	1.56–02	–2.786	D–	*LS*
154.	$2p3p-$ $2p3d$	$^1D-^1D^\circ$	19722.0	5069.10 cm^{-1}	72610.72	– 77679.82	5–5	1.88–02	1.10–01	3.56+01	–0.261	B+	1,2,3
155.	$2p3p-$ $2p4s$	$^1D-^3P^\circ$											
			18157.0	5506.02 cm^{-1}	72610.72	– 78116.74	5–3	6.20–04	1.84–03	5.50–01	–2.037	C	2
			18054.2	5537.37 cm^{-1}	72610.72	– 78148.09	5–5	6.32–06	3.09–05	9.18–03	–3.811	C	2
156.	$2p3p-$ $2p3d$	$^1D-^3F^\circ$											
			17889.5	5588.35 cm^{-1}	72610.72	– 78199.07	5–5	4.31–05	2.07–04	6.09–02	–2.985	C	2
			17837.0	5604.79 cm^{-1}	72610.72	– 78215.51	5–7	6.57–04	4.39–03	1.29+00	–1.659	C	2
157.		$^1D-^3D^\circ$											
			17515.9	5707.53 cm^{-1}	72610.72	– 78318.25	5–7	2.32–04	1.50–03	4.31–01	–2.126	C	2
			17592.2	5682.77 cm^{-1}	72610.72	– 78293.49	5–3	5.84–03	1.63–02	4.71+00	–1.090	C	2
158.	$2p3p-$ $2p4s$	$^1D-^1P^\circ$	17448.6	5729.56 cm^{-1}	72610.72	– 78340.28	5–3	7.50–02	2.06–01	5.91+01	0.012	B+	1,2,3
159.	$2p3p-$ $2p3d$	$^1D-^1F^\circ$	16890.4	5918.90 cm^{-1}	72610.72	– 78529.62	5–7	1.24–01	7.44–01	2.07+02	0.571	B	1,2,3
160.		$^1D-^1P^\circ$	16333.9	6120.55 cm^{-1}	72610.72	– 78731.27	5–3	6.13–03	1.47–02	3.96+00	–1.133	C	2,3
161.	$2p3p-$ $2p4d$	$^1D-^1D^\circ$	9182.83	9185.35	72610.72	– 83497.62	5–5	2.78–03	3.51–03	5.30–01	–1.756	C	1,2
162.	$2p3p-$ $2p5s$	$^1D-^3P^\circ$											
			8972.84	8975.30	72610.72	– 83752.41	5–3	6.67–04	4.84–04	7.14–02	–2.617	C	2
163.	$2p3p-$ $2p4d$	$^1D-^3D^\circ$											
			8918.63	8921.08	72610.72	– 83820.13	5–3	1.05–03	7.50–04	1.10–01	–2.426	C	2
164.	$2p3p-$ $2p5s$	$^1D-^1P^\circ$	8873.36	8875.80	72610.72	– 83877.31	5–3	2.05–02	1.45–02	2.12+00	–1.140	B–	1,2
165.	$2p3p-$ $2p4d$	$^1D-^1P^\circ$	8753.07	8755.47	72610.72	– 84032.15	5–3	5.23–03	3.60–03	5.19–01	–1.744	D	1,2
166.	$2p3p-$ $2p5d$	$^1D-^1D^\circ$	7364.74	7366.76	72610.72	– 86185.20	5–5	3.58–03	2.91–03	3.53–01	–1.837	C	1
167.	$2p3p-$ $2p6s$	$^1D-^1P^\circ$	7241.32	7243.32	72610.72	– 86416.55	5–3	8.83–03	4.16–03	4.96–01	–1.682	D	1
168.	$2p3p-$ $2p5d$	$^1D-^1P^\circ$	7202.27	7204.25	72610.72	– 86491.41	5–3	5.19–03	2.42–03	2.87–01	–1.917	D	1
169.	$2p3p-$ $2p6d$	$^1D-^1D^\circ$	6654.61	6656.45	72610.72	– 87633.75	5–5	2.85–03	1.90–03	2.08–01	–2.023	D	1

C I: Allowed Transitions — Continued

No.	Transition Array	Mult.	λ_{air} (Å)	λ_{vac} (Å) or σ (cm^{-1})*	E_i (cm^{-1})	E_k (cm^{-1})	g_i–g_k	A_{ki} (10^8 s^{-1})	f_{ik}	S (at. u.)	log gf	Acc.	Source
170.	$2p3p$– $2p7s$	^{1}D–^{1}P°	6586.27	6588.09	72610.72	87789.63	5–3	4.64–03	1.81–03	1.97–01	–2.043	D–	1
171.	$2p3p$– $2p6d$	^{1}D–^{1}P°	6568.73	6570.54	72610.72	87830.17	5–3	3.50–03	1.36–03	1.47–01	–2.168	D–	1
172.	$2p3p$– $2p7d$	^{1}D–^{1}D°	6292.36	6294.10	72610.72	88498.62	5–5	2.08–03	1.24–03	1.28–01	–2.209	D	1
173.	$2p3p$– $2p4s$	^{1}S–^{3}P°											
			24143.2	4140.83 cm^{-1}	73975.91	78116.74	1–3	1.23–04	3.23–03	2.57–01	–2.491	C	2
174.	$2p3p$– $2p3d$	^{1}S–^{3}D°											
			23154.8	4317.58 cm^{-1}	73975.91	78293.49	1–3	2.76–03	6.65–02	5.07+00	–1.177	C	2
175.	$2p3p$– $2p4s$	^{1}S–^{1}P°	22906.6	4364.37 cm^{-1}	73975.91	78340.28	1–3	2.57–02	6.07–01	4.58+01	–0.217	B	1,2,3
176.	$2p3p$– $2p3d$	^{1}S–^{1}P°	21023.2	4755.36 cm^{-1}	73975.91	78731.27	1–3	2.01–02	3.99–01	2.76+01	–0.399	B–	1,2,3
177.		^{1}S–^{3}P°											
			18711.4	5342.87 cm^{-1}	73975.91	79318.78	1–3	3.63–05	5.72–04	3.52–02	–3.243	C	2
178.	$2p3p$– $2p5s$	^{1}S–^{3}P°											
			10225.8	9776.50 cm^{-1}	73975.91	83752.41	1–3	2.39–05	1.12–04	3.78–03	–3.949	C	2
179.		^{1}S–^{1}P°	10096.8	9901.40 cm^{-1}	73975.91	83877.31	1–3	1.76–04	8.08–04	2.69–02	–3.092	C	2
180.	$2p3p$– $2p4d$	^{1}S–^{1}P°	9941.35	9944.07	73975.91	84032.15	1–3	2.32–03	1.03–02	3.37–01	–1.987	C	1,2
181.	$2p3p$– $2p5d$	^{1}S–^{1}P°	7987.90	7990.09	73975.91	86491.41	1–3	2.97–03	8.52–03	2.24–01	–2.070	D	1
182.	$2p3p$– $2p6d$	^{1}S–^{1}P°	7216.01	7218.00	73975.91	87830.17	1–3	2.36–03	5.52–03	1.31–01	–2.258	D	1
183.	$2s2p^3$– $2s^22p4p$	^{3}P°–^{1}P											
			18831.2	5308.88 cm^{-1}	75253.97	80562.85	3–3	5.85–05	3.11–04	5.79–02	–3.030	C	2
			18838.9	5306.73 cm^{-1}	75256.12	80562.85	1–3	1.89–05	3.02–04	1.87–02	–3.520	C	2
184.		^{3}P°–^{3}D	*17987*	*5558.1* cm^{-1}	*75254.9*	*80813.1*	9–15	7.35–03	5.94–02	3.17+01	–0.272	D	1,2
			17918.4	5579.34 cm^{-1}	75255.27	80834.61	5–7	7.32–03	4.93–02	1.46+01	–0.608	D	2n
			18021.9	5547.30 cm^{-1}	75253.97	80801.27	3–5	5.82–03	4.72–02	8.41+00	–0.849	D	2n
			18090.1	5526.39 cm^{-1}	75256.12	80782.51	1–3	4.33–03	6.37–02	3.80+00	–1.196	D	2n
			18026.1	5546.00 cm^{-1}	75255.27	80801.27	5–5	1.56–03	7.61–03	2.26+00	–1.420	D	2n
			18083.0	5528.54 cm^{-1}	75253.97	80782.51	3–3	2.88–03	1.41–02	2.52+00	–1.373	D	2n
			18087.3	5527.24 cm^{-1}	75255.27	80782.51	5–3	1.55–04	4.55–04	1.36–01	–2.643	D	2n
185.		^{3}P°–^{3}S	*17089*	*5850.1* cm^{-1}	*75254.9*	81105.03	9–3	2.09–03	3.04–03	1.54+00	–1.562	D–	1,2
			17090.1	5849.76 cm^{-1}	75255.27	81105.03	5–3	6.25–04	1.64–03	4.62–01	–2.085	D–	2n
			17086.3	5851.06 cm^{-1}	75253.97	81105.03	3–3	1.01–03	4.43–03	7.47–01	–1.877	D–	2n
			17092.5	5848.91 cm^{-1}	75256.12	81105.03	1–3	4.48–04	5.89–03	3.32–01	–2.230	D–	2n

C I: Allowed Transitions — Continued

No.	Transition Array	Mult.	λ_{air} (Å)	λ_{vac} (Å) or σ (cm^{-1})*	E_i (cm^{-1})	E_k (cm^{-1})	g_i-g_k	A_{ki} (10^8 s^{-1})	f_{ik}	S (at. u.)	log gf	Acc.	Source
186.		$^3P^o-^3P$	*16445*	*6079.3* cm^{-1}	*75254.9*	– *81334.2*	9−9	1.24−02	5.03−02	2.45+01	−0.344	D	1,2
			16419.3	6088.72 cm^{-1}	75255.27	– 81343.99	5−5	9.58−03	3.87−02	1.05+01	−0.713	D	2*n*
			16465.1	6071.79 cm^{-1}	75253.97	– 81325.76	3−3	2.91−03	1.18−02	1.92+00	−1.450	D	2*n*
			16468.6	6070.49 cm^{-1}	75255.27	– 81325.76	5−3	5.72−03	1.39−02	3.78+00	−1.156	D	2*n*
			16505.2	6057.04 cm^{-1}	75253.97	– 81311.01	3−1	1.25−02	1.70−02	2.77+00	−1.292	D	2*n*
			16415.8	6090.02 cm^{-1}	75253.97	– 81343.99	3−5	2.84−03	1.91−02	3.10+00	−1.241	D	2*n*
			16470.9	6069.64 cm^{-1}	75256.12	– 81325.76	1−3	3.73−03	4.55−02	2.47+00	−1.342	D	2*n*
187.	$2s2p^3-$ $2s^22p(^2P^o_{1/2})4f$	$^3P^o-[^5/_2]$											
			11538.3	8664.38 cm^{-1}	75255.27	– 83919.65	5−7	2.96−03	8.26−03	1.57+00	−1.384	C	2
			11536.5	8665.79 cm^{-1}	75253.97	– 83919.76	3−5	8.52−04	2.83−03	3.23−01	−2.071	C	2
			11538.2	8664.49 cm^{-1}	75255.27	– 83919.76	5−5	2.93−04	5.85−04	1.11−01	−2.534	C	2
188.	$2s2p^3-$ $2s^22p(^2P^o_{3/2})4f$	$^3P^o-[^5/_2]$											
			11415.0	8757.98 cm^{-1}	75255.27	– 84013.25	5−7	2.05−02	5.62−02	1.06+01	−0.551	C	2
			11413.1	8759.43 cm^{-1}	75253.97	– 84013.40	3−5	6.35−03	2.07−02	2.33+00	−1.207	C	2
			11414.8	8758.13 cm^{-1}	75255.27	– 84013.40	5−5	2.18−03	4.27−03	8.02−01	−1.671	C	2
189.		$^3P^o-[^3/_2]$											
			11383.3	8782.43 cm^{-1}	75253.97	– 84036.40	3−5	9.91−03	3.21−02	3.61+00	−1.016	C	2
			11386.2	8780.17 cm^{-1}	75256.12	– 84036.29	1−3	1.26−02	7.35−02	2.75+00	−1.134	C	2
			11384.9	8781.13 cm^{-1}	75255.27	– 84036.40	5−5	3.37−03	6.54−03	1.23+00	−1.485	C	2
			11383.4	8782.32 cm^{-1}	75253.97	– 84036.29	3−3	9.47−03	1.84−02	2.07+00	−1.258	C	2
			11385.1	8781.02 cm^{-1}	75255.27	– 84036.29	5−3	6.47−04	7.55−04	1.41−01	−2.423	C	2
190.	$2s2p^3-$ $2s^22p5p$	$^3P^o-^3D$	*10297*	*9708.6* cm^{-1}	*75254.9*	– *84963.5*	9−15	1.59−03	4.21−03	1.29+00	−1.421	C	1
			10275.0	9729.70 cm^{-1}	75255.27	– 84984.97	5−7	1.60−03	3.55−03	6.00−01	−1.751	D	*LS*
			10310.3	9696.39 cm^{-1}	75253.97	– 84950.36	3−5	1.19−03	3.16−03	3.21−01	−2.024	D	*LS*
			10328.6	9679.22 cm^{-1}	75256.12	– 84935.34	1−3	8.75−04	4.20−03	1.43−01	−2.377	D	*LS*
			10311.7	9695.09 cm^{-1}	75255.27	– 84950.36	5−5	3.96−04	6.31−04	1.07−01	−2.501	D	*LS*
			10326.3	9681.37 cm^{-1}	75253.97	– 84935.34	3−3	6.57−04	1.05−03	1.07−01	−2.502	D	*LS*
			10327.7	9680.07 cm^{-1}	75255.27	– 84935.34	5−3	4.38−05	4.20−05	7.14−03	−3.678	D−	*LS*
191.		$^3P^o-^3P$	*10058*	*9940.0* cm^{-1}	*75254.9*	– *85195.0*	9−9	2.78−03	4.22−03	1.26+00	−1.420	C	1
			10049.1	9948.37 cm^{-1}	75255.27	– 85203.64	5−5	2.09−03	3.17−03	5.24−01	−1.800	D	*LS*
			10062.7	9934.98 cm^{-1}	75253.97	– 85188.95	3−3	6.95−04	1.06−03	1.05−01	−2.500	D	*LS*
			10064.0	9933.68 cm^{-1}	75255.27	– 85188.95	5−3	1.16−03	1.05−03	1.75−01	−2.278	D	*LS*
			10082.3	9915.64 cm^{-1}	75253.97	– 85169.61	3−1	2.76−03	1.40−03	1.40−01	−2.375	D	*LS*
			10047.8	9949.67 cm^{-1}	75253.97	– 85203.64	3−5	6.98−04	1.76−03	1.75−01	−2.277	D	*LS*
			10064.9	9932.83 cm^{-1}	75256.12	– 85188.95	1−3	9.26−04	4.22−03	1.40−01	−2.375	D	*LS*
192.	$2p3d-$ $2p4p$	$^1D^o-^1P$	34676.3	2883.03 cm^{-1}	77679.82	– 80562.85	5−3	1.18−02	1.28−01	7.30+01	−0.194	B−	1,2
193.		$^1D^o-^3D$											
			31689.2	3154.79 cm^{-1}	77679.82	– 80834.61	5−7	5.57−06	1.17−04	6.13−02	−3.231	C	2
			32027.7	3121.45 cm^{-1}	77679.82	– 80801.27	5−5	2.23−05	3.43−04	1.81−01	−2.766	C	2
			32221.3	3102.69 cm^{-1}	77679.82	– 80782.51	5−3	3.65−05	3.41−04	1.81−01	−2.768	C	2
194.		$^1D^o-^1D$	24443.4	4089.97 cm^{-1}	77679.82	– 81769.79	5−5	3.49−03	3.13−02	1.26+01	−0.806	B	1,2
195.	$2p3d-$ $2p(^2P^o_{1/2})4f$	$^1D^o-[^5/_2]$											
			16021.7	6239.83 cm^{-1}	77679.82	– 83919.65	5−7	6.19−02	3.34−01	8.80+01	0.222	C	2
			16021.4	6239.94 cm^{-1}	77679.82	– 83919.76	5−5	2.54−03	9.77−03	2.58+00	−1.311	C	2

C I: Allowed Transitions — Continued

No.	Transition Array	Mult.	λ_{air} (Å)	λ_{vac} (Å) or σ (cm^{-1})*	E_i (cm^{-1})	E_k (cm^{-1})	g_i-g_k	A_{ki} (10^8 s^{-1})	f_{ik}	S (at. u.)	log gf	Acc.	Source
196.		$^1D^o-[^7/_2]$											
			16004.9	6246.38 cm^{-1}	77679.82	83926.20	5–7	6.37–02	3.43–01	9.03+01	0.234	C	2
197.	$2p3d-$ $2p(^2P^o_{3/2})4f$	$^1D^o-[^7/_2]$											
			15852.6	6306.40 cm^{-1}	77679.82	83986.22	5–7	2.09–02	1.10–01	2.88+01	−0.259	C	2
198.		$^1D^o-[^5/_2]$											
			15784.9	6333.43 cm^{-1}	77679.82	84013.25	5–7	7.14–03	3.73–02	9.70+00	−0.729	C	2
			15784.5	6333.58 cm^{-1}	77679.82	84013.40	5–5	1.37–02	5.10–02	1.33+01	−0.593	C	2
199.		$^1D^o-[^3/_2]$											
			15727.4	6356.58 cm^{-1}	77679.82	84036.40	5–5	1.12–02	4.15–02	1.08+01	−0.683	C	2
200.	$2p3d-$ $2p5p$	$^1D^o-^1P$	13939.9	7171.71 cm^{-1}	77679.82	84851.53	5–3	3.04–03	5.32–03	1.22+00	−1.575	B−	1
201.		$^1D^o-^1D$	12949.8	7719.99 cm^{-1}	77679.82	85399.81	5–5	1.10–03	2.78–03	5.92–01	−1.858	B−	1
202.	$2p3d-$ $2p6p$	$^1D^o-^1P$	10827.7	9233.04 cm^{-1}	77679.82	86912.86	5–3	1.56–03	1.64–03	2.93–01	−2.086	B−	1
203.		$^1D^o-^1D$	10481.0	9538.44 cm^{-1}	77679.82	87218.26	5–5	5.48–04	9.02–04	1.56–01	−2.346	B−	1
204.	$2p3d-$ $2p7p$	$^1D^o-^1P$	9629.92	9632.56	77679.82	88061.28	5–3	9.16–04	7.64–04	1.21–01	−2.418	C+	1
205.	$2p4s-$ $2p4p$	$^3P^o-^1P$											
			40870.1	2446.11 cm^{-1}	78116.74	80562.85	3–3	4.81–04	1.21–02	4.87+00	−1.442	C	2
			40674.5	2457.87 cm^{-1}	78104.98	80562.85	1–3	2.58–05	1.92–03	2.57–01	−2.717	C	2
206.		$^3P^o-^3D$	*37300*	*2680.2* cm^{-1}	*78132.9*	*80813.1*	9–15	1.98–02	6.88–01	7.60+02	0.792	B	1,2
			37212.7	2686.52 cm^{-1}	78148.09	80834.61	5–7	1.96–02	5.70–01	3.49+02	0.455	B	2*n*
			37240.3	2684.53 cm^{-1}	78116.74	80801.27	3–5	1.53–02	5.30–01	1.95+02	0.201	B	2*n*
			37337.7	2677.53 cm^{-1}	78104.98	80782.51	1–3	1.15–02	7.21–01	8.86+01	−0.142	B	2*n*
			37680.3	2653.18 cm^{-1}	78148.09	80801.27	5–5	4.60–03	9.79–02	6.07+01	−0.310	B	2*n*
			37502.4	2665.77 cm^{-1}	78116.74	80782.51	3–3	7.80–03	1.65–01	6.09+01	−0.307	B	2*n*
			37948.7	2634.42 cm^{-1}	78148.09	80782.51	5–3	6.97–04	9.03–03	5.64+00	−1.345	B	2*n*
207.		$^3P^o-^3S$	*33636*	*2972.2* cm^{-1}	*78132.9*	81105.03	9–3	4.22–02	2.39–01	2.38+02	0.332	B	1,2
			33809.5	2956.94 cm^{-1}	78148.09	81105.03	5–3	2.09–02	2.15–01	1.20+02	0.031	B	2*n*
			33454.8	2988.29 cm^{-1}	78116.74	81105.03	3–3	1.57–02	2.63–01	8.70+01	−0.103	B	2*n*
			33323.7	3000.05 cm^{-1}	78104.98	81105.03	1–3	5.76–03	2.88–01	3.15+01	−0.541	B	2*n*
208.		$^3P^o-^3P$	*31228*	*3201.4* cm^{-1}	*78132.9*	*81334.2*	9–9	2.14–02	3.13–01	2.89+02	0.449	B	1,2
			31281.6	3195.90 cm^{-1}	78148.09	81343.99	5–5	1.72–02	2.53–01	1.30+02	0.102	B	2*n*
			31153.7	3209.02 cm^{-1}	78116.74	81325.76	3–3	4.16–03	6.05–02	1.86+01	−0.741	B	2*n*
			31461.0	3177.67 cm^{-1}	78148.09	81325.76	5–3	1.07–02	9.50–02	4.92+01	−0.324	B	2*n*
			31297.5	3194.27 cm^{-1}	78116.74	81311.01	3–1	2.02–02	9.89–02	3.06+01	−0.528	B	2*n*
			30977.7	3227.25 cm^{-1}	78116.74	81343.99	3–5	4.92–03	1.18–01	3.61+01	−0.451	B	2*n*
			31039.9	3220.78 cm^{-1}	78104.98	81325.76	1–3	5.59–03	2.42–01	2.48+01	−0.616	B	2*n*
209.		$^3P^o-^1D$											
			27366.9	3653.05 cm^{-1}	78116.74	81769.79	3–5	1.92–04	3.59–03	9.69–01	−1.968	C	2

C I: Allowed Transitions — Continued

No.	Transition Array	Mult.	λ_{air} (Å)	λ_{vac} (Å) or σ (cm^{-1})*	E_i (cm^{-1})	E_k (cm^{-1})	g_i-g_k	A_{ki} (10^8 s^{-1})	f_{ik}	S (at. u.)	log gf	Acc.	Source
210.		$^{3}P^{o}-{}^{1}S$											
			24177.4	4134.97 cm^{-1}	78116.74	82251.71	3–1	3.76–04	1.10–03	2.62–01	–2.482	C	2
211.	$2p4s-$ $2p(^2P^o_{1/2})4f$	$^{3}P^{o}-[^5/_2]$											
			17321.6	5771.56 cm^{-1}	78148.09	83919.65	5–7	4.43–03	2.79–02	7.95+00	–0.856	C	2
			17227.7	5803.02 cm^{-1}	78116.74	83919.76	3–5	1.53–03	1.14–02	1.93+00	–1.467	C	2
			17321.3	5771.67 cm^{-1}	78148.09	83919.76	5–5	3.99–04	1.80–03	5.13–01	–2.046	C	2
212.		$^{3}P^{o}-[^7/_2]$											
			17302.0	5778.11 cm^{-1}	78148.09	83926.20	5–7	5.89–04	3.71–03	1.06+00	–1.732	C	2
213.	$2p4s-$ $2p(^2P^o_{3/2})4f$	$^{3}P^{o}-[^5/_2]$											
			17045.2	5865.16 cm^{-1}	78148.09	84013.25	5–7	2.05–02	1.25–01	3.51+01	–0.204	C	2
			16954.1	5896.66 cm^{-1}	78116.74	84013.40	3–5	5.05–03	3.63–02	6.08+00	–0.963	C	2
			17044.7	5865.31 cm^{-1}	78148.09	84013.40	5–5	1.38–03	6.00–03	1.68+00	–1.523	C	2
214.		$^{3}P^{o}-[^3/_2]$											
			16888.2	5919.66 cm^{-1}	78116.74	84036.40	3–5	1.08–02	7.72–02	1.29+01	–0.636	C	2
			16855.1	5931.31 cm^{-1}	78104.98	84036.29	1–3	1.22–02	1.55–01	8.63+00	–0.808	C	2
			16978.2	5888.31 cm^{-1}	78148.09	84036.40	5–5	2.93–03	1.27–02	3.54+00	–1.198	C	2
			16888.6	5919.55 cm^{-1}	78116.74	84036.29	3–3	8.29–03	3.55–02	5.92+00	–0.973	C	2
			16978.5	5888.20 cm^{-1}	78148.09	84036.29	5–3	4.58–04	1.19–03	3.32–01	–2.226	C	2
215.	$2p4s-$ $2p5p$	$^{3}P^{o}-{}^{3}D$	*14636*	*6830.7* cm^{-1}	*78132.9*	*84963.5*	9–15	5.39–05	2.88–04	1.25–01	–2.586	C	1
			14622.6	6836.88 cm^{-1}	78148.09	84984.97	5–7	5.40–05	2.43–04	5.84–02	–2.916	D	*LS*
			14629.5	6833.62 cm^{-1}	78116.74	84950.36	3–5	4.05–05	2.16–04	3.13–02	–3.188	D	*LS*
			14636.5	6830.36 cm^{-1}	78104.98	84935.34	1–3	2.99–05	2.88–04	1.39–02	–3.540	D	*LS*
			14697.0	6802.27 cm^{-1}	78148.09	84950.36	5–5	1.33–05	4.31–05	1.04–02	–3.667	D	*LS*
			14661.8	6818.60 cm^{-1}	78116.74	84935.34	3–3	2.23–05	7.20–05	1.04–02	–3.666	D	*LS*
			14729.5	6787.25 cm^{-1}	78148.09	84935.34	5–3	1.47–06	2.87–06	6.95–04	–4.844	D–	*LS*
216.		$^{3}P^{o}-{}^{3}P$	*14156*	*7062.1* cm^{-1}	*78132.9*	*85195.0*	9–9	9.17–04	2.75–03	1.16+00	–1.606	C	1
			14169.4	7055.55 cm^{-1}	78148.09	85203.64	5–5	6.86–04	2.06–03	4.81–01	–1.986	D	*LS*
			14136.0	7072.21 cm^{-1}	78116.74	85188.95	3–3	2.30–04	6.89–04	9.63–02	–2.684	D	*LS*
			14198.9	7040.86 cm^{-1}	78148.09	85188.95	5–3	3.78–04	6.86–04	1.60–01	–2.464	D	*LS*
			14174.8	7052.87 cm^{-1}	78116.74	85169.61	3–1	9.13–04	9.17–04	1.28–01	–2.561	D	*LS*
			14106.7	7086.90 cm^{-1}	78116.74	85203.64	3–5	2.32–04	1.15–03	1.60–01	–2.462	D	*LS*
			14112.5	7083.97 cm^{-1}	78104.98	85188.95	1–3	3.08–04	2.76–03	1.28–01	–2.559	D	*LS*
217.	$2p3d-$ $2p4p$	$^{3}F^{o}-{}^{1}P$											
			42293.6	2363.78 cm^{-1}	78199.07	80562.85	5–3	2.66–05	4.29–04	2.99–01	–2.669	C	2
218.		$^{3}F^{o}-{}^{3}D$	*38648*	*2586.7* cm^{-1}	*78226.4*	*80813.1*	21–15	1.16–02	1.85–01	4.94+02	0.589	B	1,2
			38679.1	2584.67 cm^{-1}	78249.94	80834.61	9–7	1.07–02	1.87–01	2.14+02	0.226	B	2*n*
			38662.8	2585.76 cm^{-1}	78215.51	80801.27	7–5	1.08–02	1.72–01	1.54+02	0.082	B	2*n*
			38697.5	2583.44 cm^{-1}	78199.07	80782.51	5–3	1.20–02	1.61–01	1.03+02	–0.093	B	2*n*
			38170.6	2619.10 cm^{-1}	78215.51	80834.61	7–7	4.50–04	9.83–03	8.65+00	–1.162	B	2*n*
			38418.5	2602.20 cm^{-1}	78199.07	80801.27	5–5	1.06–03	2.34–02	1.48+01	–0.932	B	2*n*
			37932.5	2635.54 cm^{-1}	78199.07	80834.61	5–7	3.64–06	1.10–04	6.87–02	–3.260	B	2*n*
219.		$^{3}F^{o}-{}^{3}S$											
			34402.7	2905.96 cm^{-1}	78199.07	81105.03	5–3	1.72–05	1.83–04	1.04–01	–3.038	C	2

C I: Allowed Transitions — Continued

No.	Transition Array	Mult.	λ_{air} (Å)	λ_{vac} (Å) or σ (cm^{-1})*	E_i (cm^{-1})	E_k (cm^{-1})	g_i-g_k	A_{ki} (10^8 s^{-1})	f_{ik}	S (at. u.)	log gf	Acc.	Source
220.		$^3F^o-^3P$											
			31974.0	3126.69 cm^{-1}	78199.07	81325.76	5−3	2.65−04	2.44−03	1.28+00	−1.914	C	2
221.		$^3F^o-^1D$											
			28127.4	3554.28 cm^{-1}	78215.51	81769.79	7−5	5.96−05	5.05−04	3.27−01	−2.452	C	2
			27997.9	3570.72 cm^{-1}	78199.07	81769.79	5−5	9.28−06	1.09−04	5.03−02	−3.263	C	2
222.	$2p3d-$ $2p(^2P^o_{1/2})4f$	$^3F^o-[^5/_2]$											
			17476.0	5720.58 cm^{-1}	78199.07	83919.65	5−7	4.73−03	3.03−02	8.73+00	−0.819	C	2
			17526.3	5704.14 cm^{-1}	78215.51	83919.65	7−7	7.39−03	3.41−02	1.38+01	−0.623	C	2
			17475.6	5720.69 cm^{-1}	78199.07	83919.76	5−5	1.01−02	4.64−02	1.34+01	−0.634	C	2
			17632.8	5669.71 cm^{-1}	78249.94	83919.65	9−7	2.55−04	9.26−04	4.84−01	−2.079	C	2
			17526.0	5704.25 cm^{-1}	78215.51	83919.76	7−5	1.07−03	3.53−03	1.43+00	−1.607	C	2
223.		$^3F^o-[^7/_2]$											
			17505.7	5710.86 cm^{-1}	78215.51	83926.37	7−9	6.41−02	3.79−01	1.53+02	0.423	C	2
			17456.0	5727.13 cm^{-1}	78199.07	83926.20	5−7	5.95−02	3.80−01	1.09+02	0.279	C	2
			17611.9	5676.43 cm^{-1}	78249.94	83926.37	9−9	4.70−03	2.19−02	1.14+01	−0.706	C	2
			17506.2	5710.69 cm^{-1}	78215.51	83926.20	7−7	1.70−04	7.81−04	3.15−01	−2.262	C	2
			17612.4	5676.26 cm^{-1}	78249.94	83926.20	9−7	1.78−04	6.43−04	3.36−01	−2.237	C	2
224.	$2p3d-$ $2p(^2P^o_{3/2})4f$	$^3F^o-[^7/_2]$											
			17323.5	5770.94 cm^{-1}	78215.51	83986.45	7−9	4.38−02	2.53−01	1.01+02	0.249	C	2
			17274.9	5787.15 cm^{-1}	78199.07	83986.22	5−7	8.83−02	5.53−01	1.57+02	0.442	C	2
			17427.4	5736.51 cm^{-1}	78249.94	83986.45	9−9	1.77−02	8.06−02	4.16+01	−0.139	C	2
			17324.2	5770.71 cm^{-1}	78215.51	83986.22	7−7	1.50−02	6.75−02	2.69+01	−0.326	C	2
			17428.1	5736.28 cm^{-1}	78249.94	83986.22	9−7	6.34−04	2.25−03	1.16+00	−1.694	C	2
225.		$^3F^o-[^9/_2]$											
			17338.6	5765.92 cm^{-1}	78249.94	84015.86	9−11	1.62−01	8.91−01	4.58+02	0.904	C	2
			17234.5	5800.74 cm^{-1}	78215.51	84016.25	7−9	4.90−02	2.81−01	1.11+02	0.293	C	2
			17337.4	5766.31 cm^{-1}	78249.94	84016.25	9−9	4.04−03	1.82−02	9.35+00	−0.786	C	2
226.		$^3F^o-[^5/_2]$											
			17346.4	5763.31 cm^{-1}	78249.94	84013.25	9−7	1.42−03	5.00−03	2.57+00	−1.347	C	2
			17242.9	5797.89 cm^{-1}	78215.51	84013.40	7−5	4.45−04	1.42−03	5.64−01	−2.003	C	2
			17243.4	5797.74 cm^{-1}	78215.51	84013.25	7−7	8.11−05	3.62−04	1.44−01	−2.596	C	2
			17194.2	5814.33 cm^{-1}	78199.07	84013.40	5−5	3.99−04	1.77−03	5.01−01	−2.053	C	2
			17194.2	5814.33 cm^{-1}	78199.07	84013.40	5−7	4.77−04	2.96−03	8.38−01	−1.830	C	2
227.		$^3F^o-[^3/_2]$											
			17174.8	5820.89 cm^{-1}	78215.51	84036.40	7−5	1.89−04	5.96−04	2.36−01	−2.379	C	2
			17126.8	5837.22 cm^{-1}	78199.07	84036.29	5−3	3.40−04	8.97−04	2.53−01	−2.348	C	2
			17126.4	5837.33 cm^{-1}	78199.07	84036.40	5−5	8.11−04	3.57−03	1.01+00	−1.749	C	2
228.	$2p3d-$ $2p5p$	$^3F^o-^3D$	*14839*	*6737.2* cm^{-1}	*78226.4*	*84963.5*	21−15	2.64−03	6.22−03	6.38+00	−0.884	C	1
			14843.7	6735.03 cm^{-1}	78249.94	84984.97	9−7	2.42−03	6.21−03	2.73+00	−1.252	D	*LS*
			14844.1	6734.85 cm^{-1}	78215.51	84950.36	7−5	2.34−03	5.52−03	1.89+00	−1.413	D	*LS*
			14841.0	6736.27 cm^{-1}	78199.07	84935.34	5−3	2.63−03	5.22−03	1.28+00	−1.583	D	*LS*
			14768.2	6769.46 cm^{-1}	78215.51	84984.97	7−7	2.13−04	6.96−04	2.37−01	−2.312	D	*LS*
			14807.9	6751.29 cm^{-1}	78199.07	84950.36	5−5	2.96−04	9.72−04	2.37−01	−2.314	D	*LS*
			14732.4	6785.90 cm^{-1}	78199.07	84984.97	5−7	6.05−06	2.75−05	6.68−03	−3.861	D−	*LS*

C I: Allowed Transitions — Continued

No.	Transition Array	Mult.	λ_{air} (Å)	λ_{vac} (Å) or σ (cm^{-1})*	E_i (cm^{-1})	E_k (cm^{-1})	g_i–g_k	A_{ki} (10^8 s^{-1})	f_{ik}	S (at. u.)	log gf	Acc.	Source
229.	2p3d– 2p6p	^{3}F°–^{3}D	*11420*	*8754.4* cm^{-1}	*78226.4*	*86980.8*	21–15	1.36–03	1.90–03	1.50+00	–1.399	C	1
			11422.4	8752.32 cm^{-1}	78249.94	87002.26	9–7	1.25–03	1.90–03	6.43–01	–1.767	D	*LS*
			11425.5	8749.94 cm^{-1}	78215.51	86965.45	7–5	1.21–03	1.69–03	4.44–01	–1.928	D	*LS*
			11416.2	8757.09 cm^{-1}	78199.07	86956.16	5–3	1.36–03	1.60–03	3.00–01	–2.098	D	*LS*
			11377.7	8786.75 cm^{-1}	78215.51	87002.26	7–7	1.09–04	2.12–04	5.57–02	–2.828	D	*LS*
			11404.1	8766.38 cm^{-1}	78199.07	86965.45	5–5	1.52–04	2.97–04	5.57–02	–2.829	D	*LS*
			11356.4	8803.19 cm^{-1}	78199.07	87002.26	5–7	3.11–06	8.41–06	1.57–03	–4.376	D–	*LS*
230.	2p3d– 2p7p	^{3}F°–^{3}D	*10111*	*9887.4* cm^{-1}	*78226.4*	*88113.8*	21–15	8.08–04	8.84–04	6.18–01	–1.731	C	1
			10112.7	9885.87 cm^{-1}	78249.94	88135.81	9–7	7.42–04	8.84–04	2.65–01	–2.099	D	*LS*
			10118.2	9880.52 cm^{-1}	78215.51	88096.03	7–5	7.16–04	7.85–04	1.83–01	–2.260	D	*LS*
			10105.4	9892.94 cm^{-1}	78199.07	88092.01	5–3	8.09–04	7.43–04	1.24–01	–2.430	D	*LS*
			10077.6	9920.30 cm^{-1}	78215.51	88135.81	7–7	6.49–05	9.89–05	2.30–02	–3.160	D–	*LS*
			10101.3	9896.96 cm^{-1}	78199.07	88096.03	5–5	9.03–05	1.38–04	2.30–02	–3.161	D–	*LS*
			10060.9	9936.74 cm^{-1}	78199.07	88135.81	5–7	1.84–06	3.91–06	6.48–04	–4.709	D–	*LS*
231.	2p3d– 2p8p	^{3}F°–^{3}D	*9443.2*	*9445.7*	*78226.4*	*88813.1*	21–15	5.20–04	4.96–04	3.24–01	–1.982	C	1
			9443.69	9446.28	78249.94	88836.12	9–7	4.77–04	4.96–04	1.39–01	–2.350	D	*LS*
			9450.28	9452.88	78215.51	88794.3	7–5	4.61–04	4.41–04	9.60–02	–2.511	D	*LS*
			9438.69	9441.28	78199.07	88790.85	5–3	5.21–04	4.17–04	6.48–02	–2.681	D	*LS*
			9413.07	9415.66	78215.51	88836.12	7–7	4.18–05	5.55–05	1.20–02	–3.411	D–	*LS*
			9435.62	9438.21	78199.07	88794.3	5–5	5.81–05	7.75–05	1.20–02	–3.412	D–	*LS*
			9398.52	9401.10	78199.07	88836.12	5–7	1.18–06	2.19–06	3.40–04	–4.960	D–	*LS*
232.	2p4s– 2p4p	^{1}P°–^{1}P	44980.7	2222.57 cm^{-1}	78340.28	80562.85	3–3	7.38–03	2.24–01	9.94+01	–0.173	C+	1,2
233.		^{1}P°–^{3}D											
			40623.0	2460.99 cm^{-1}	78340.28	80801.27	3–5	1.45–04	5.99–03	2.41+00	–1.745	C	2
			40935.0	2442.23 cm^{-1}	78340.28	80782.51	3–3	2.15–05	5.39–04	2.18–01	–2.791	C	2
234.		^{1}P°–^{3}S											
			36159.8	2764.75 cm^{-1}	78340.28	81105.03	3–3	1.75–04	3.44–03	1.23+00	–1.987	C	2
235.		^{1}P°–^{3}P											
			33283.1	3003.71 cm^{-1}	78340.28	81343.99	3–5	9.01–06	2.49–04	8.20–02	–3.126	C	2
			33486.3	2985.48 cm^{-1}	78340.28	81325.76	3–3	9.76–05	1.64–03	5.43–01	–2.308	C	2
			33652.6	2970.73 cm^{-1}	78340.28	81311.01	3–1	1.45–03	8.22–03	2.73+00	–1.608	C	2
236.		^{1}P°–^{1}D	29150.7	3429.51 cm^{-1}	78340.28	81769.79	3–5	3.64–02	7.72–01	2.22+02	0.365	B–	1,2
237.		^{1}P°–^{1}S	25559.1	3911.43 cm^{-1}	78340.28	82251.71	3–1	7.04–02	2.30–01	5.80+01	–0.161	B	1,2
238.	2p4s– 2p(^{2}P°$_{1/2}$)4f	^{1}P°–[$^5/_2$]											
			17917.9	5579.48 cm^{-1}	78340.28	83919.76	3–5	2.22–03	1.78–02	3.15+00	–1.273	C	2
239.	2p4s– 2p(^{2}P°$_{3/2}$)4f	^{1}P°–[$^5/_2$]											
			17622.2	5673.12 cm^{-1}	78340.28	84013.40	3–5	2.00–02	1.55–01	2.71+01	–0.331	C	2
240.		^{1}P°–[$^3/_2$]											
			17551.0	5696.12 cm^{-1}	78340.28	84036.40	3–5	1.04–02	8.02–02	1.39+01	–0.619	C	2
			17551.4	5696.01 cm^{-1}	78340.28	84036.29	3–3	2.31–03	1.07–02	1.85+00	–1.494	C	2

C I: Allowed Transitions — Continued

No.	Transition Array	Mult.	λ_{air} (Å)	λ_{vac} (Å) or σ (cm^{-1})*	E_i (cm^{-1})	E_k (cm^{-1})	g_i–g_k	A_{ki} (10^8 s^{-1})	f_{ik}	S (at. u.)	log gf	Acc.	Source
241.	2p4s– 2p5p	^{1}P°–^{1}P	15353.8	6511.25 cm^{-1}	78340.28	84851.53	3–3	1.05–03	3.69–03	5.60–01	–1.955	C	1
242.		^{1}P°–^{1}D	14161.4	7059.53 cm^{-1}	78340.28	85399.81	3–5	5.14–03	2.58–02	3.61+00	–1.112	C	1
243.		^{1}P°–^{1}S	13723.3	7284.90 cm^{-1}	78340.28	85625.18	3–1	1.78–02	1.68–02	2.27+00	–1.298	C	1
244.	2p4s– 2p6p	^{1}P°–^{1}P	11661.9	8572.58 cm^{-1}	78340.28	86912.86	3–3	1.15–03	2.35–03	2.71–01	–2.152	C	1
245.		^{1}P°–^{1}D	11260.7	8877.98 cm^{-1}	78340.28	87218.26	3–5	1.83–03	5.79–03	6.44–01	–1.760	C	1
246.		^{1}P°–^{1}S	11107.1	9000.76 cm^{-1}	78340.28	87341.04	3–1	8.05–03	4.97–03	5.45–01	–1.827	C	1
247.	2p4s– 2p7p	^{1}P°–^{1}P	10284.2	9721.00 cm^{-1}	78340.28	88061.28	3–3	8.66–04	1.37–03	1.39–01	–2.385	C	1
248.		^{1}P°–^{1}D	10077.8	9920.09 cm^{-1}	78340.28	88260.37	3–5	8.88–04	2.25–03	2.24–01	–2.170	C	1
249.		^{1}P°–^{1}S	10003.6	9993.70 cm^{-1}	78340.28	88333.98	3–1	4.45–03	2.23–03	2.20–01	–2.175	C	1
250.	2p4s– 2p8p	^{1}P°–^{1}D	9455.21	9457.80	78340.28	88913.56	3–5	5.07–04	1.13–03	1.06–01	–2.469	D	1
251.		^{1}P°–^{1}S	9413.29	9415.88	78340.28	88960.64	3–1	2.75–03	1.22–03	1.13–01	–2.437	D	1
252.	2p3d– 2p4p	^{3}D°–^{1}P											
			44053.3	2269.36 cm^{-1}	78293.49	80562.85	3–3	2.30–04	6.69–03	2.91+00	–1.697	C	2
253.		^{3}D°–^{3}D	*39936*	*2503.3* cm^{-1}	*78309.8*	*80813.1*	15–15	2.09–03	5.00–02	9.86+01	–0.125	B–	1,2
			39729.1	2516.36 cm^{-1}	78318.25	80834.61	7–7	2.20–03	5.22–02	4.77+01	–0.438	B–	2n
			40091.1	2493.64 cm^{-1}	78307.63	80801.27	5–5	1.31–03	3.15–02	2.08+01	–0.803	B–	2n
			40165.5	2489.02 cm^{-1}	78293.49	80782.51	3–3	1.26–03	3.04–02	1.21+01	–1.040	B–	2n
			40262.6	2483.02 cm^{-1}	78318.25	80801.27	7–5	2.20–07	3.82–06	3.54–03	–4.573	C–	2n
			40395.0	2474.88 cm^{-1}	78307.63	80782.51	5–3	4.68–05	6.86–04	4.56–01	–2.464	B–	2n
			39562.1	2526.98 cm^{-1}	78307.63	80834.61	5–7	5.64–04	1.85–02	1.21+01	–1.033	B–	2n
			39865.0	2507.78 cm^{-1}	78293.49	80801.27	3–5	3.53–04	1.40–02	5.53+00	–1.376	B–	2n
254.		^{3}D°–^{3}S											
			35558.0	2811.54 cm^{-1}	78293.49	81105.03	3–3	3.13–05	5.93–04	2.08–01	–2.750	C	2
255.		^{3}D°–^{3}P	*33054*	*3024.5* cm^{-1}	*78309.8*	*81334.2*	15–9	1.41–02	1.39–01	2.26+02	0.318	B	1,2
			33040.8	3025.74 cm^{-1}	78318.25	81343.99	7–5	1.17–02	1.37–01	1.04+02	–0.019	B	2n
			33124.1	3018.13 cm^{-1}	78307.63	81325.76	5–3	1.20–02	1.19–01	6.47+01	–0.226	B	2n
			33130.8	3017.52 cm^{-1}	78293.49	81311.01	3–1	1.46–02	8.02–02	2.62+01	–0.619	B	2n
			32925.2	3036.36 cm^{-1}	78307.63	81343.99	5–5	1.49–03	2.42–02	1.31+01	–0.917	B	2n
			32969.6	3032.27 cm^{-1}	78293.49	81325.76	3–3	3.28–03	5.34–02	1.74+01	–0.796	B	2n
			32772.6	3050.50 cm^{-1}	78293.49	81343.99	3–5	8.21–05	2.20–03	7.13–01	–2.180	B	2n
256.		^{3}D°–^{1}D											
			28964.7	3451.54 cm^{-1}	78318.25	81769.79	7–5	1.25–05	1.13–04	7.52–02	–3.103	C	2
			28758.4	3476.30 cm^{-1}	78293.49	81769.79	3–5	2.44–03	5.05–02	1.43+01	–0.820	C	2

C I: Allowed Transitions — Continued

No.	Transition Array	Mult.	λ_{air} (Å)	λ_{vac} (Å) or σ (cm^{-1})*	E_i (cm^{-1})	E_k (cm^{-1})	g_i-g_k	A_{ki} (10^8 s^{-1})	f_{ik}	S (at. u.)	log gf	Acc.	Source
257.		^{3}D°–^{1}S											
			25257.0	3958.22 cm^{-1}	78293.49 –	82251.71	3–1	5.62–03	1.79–02	4.47+00	–1.270	C	2
258.	$2p3d$– $2p(^2P^\circ_{1/2})4f$	^{3}D°–[$^5/_2$]											
			17814.0	5612.02 cm^{-1}	78307.63 –	83919.65	5–7	6.17–02	4.11–01	1.21+02	0.313	C	2
			17768.9	5626.27 cm^{-1}	78293.49 –	83919.76	3–5	1.11–01	8.78–01	1.54+02	0.420	C	2
			17847.8	5601.40 cm^{-1}	78318.25 –	83919.65	7–7	2.33–03	1.11–02	4.58+00	–1.109	C	2
			17813.7	5612.13 cm^{-1}	78307.63 –	83919.76	5–5	1.51–02	7.20–02	2.11+01	–0.444	C	2
			17847.5	5601.51 cm^{-1}	78318.25 –	83919.76	7–5	2.61–04	8.90–04	3.66–01	–2.205	C	2
259.		^{3}D°–[$^7/_2$]											
			17826.4	5608.12 cm^{-1}	78318.25 –	83926.37	7–9	7.57–02	4.64–01	1.91+02	0.511	C	2
			17793.3	5618.57 cm^{-1}	78307.63 –	83926.20	5–7	2.71–02	1.80–01	5.28+01	–0.045	C	2
			17827.0	5607.95 cm^{-1}	78318.25 –	83926.20	7–7	5.31–03	2.53–02	1.04+01	–0.751	C	2
260.	$2p3d$– $2p(^2P^\circ_{3/2})4f$	^{3}D°–[$^7/_2$]											
			17637.5	5668.20 cm^{-1}	78318.25 –	83986.45	7–9	5.19–02	3.12–01	1.27+02	0.339	C	2
			17605.2	5678.59 cm^{-1}	78307.63 –	83986.22	5–7	2.87–02	1.86–01	5.41+01	–0.030	C	2
			17638.2	5667.97 cm^{-1}	78318.25 –	83986.22	7–7	5.37–03	2.51–02	1.02+01	–0.756	C	2
261.		^{3}D°–[$^9/_2$]											
			17545.2	5698.00 cm^{-1}	78318.25 –	84016.25	7–9	4.13–03	2.45–02	9.92+00	–0.765	C	2
262.		^{3}D°–[$^5/_2$]											
			17521.8	5705.62 cm^{-1}	78307.63 –	84013.25	5–7	1.88–03	1.21–02	3.50+00	–1.218	C	2
			17478.0	5719.91 cm^{-1}	78293.49 –	84013.40	3–5	1.01–03	7.68–03	1.33+00	–1.637	C	2
			17554.5	5695.00 cm^{-1}	78318.25 –	84013.25	7–7	2.88–02	1.33–01	5.38+01	–0.031	C	2
			17521.3	5705.77 cm^{-1}	78307.63 –	84013.40	5–5	1.85–02	8.54–02	2.46+01	–0.370	C	2
			17554.0	5695.15 cm^{-1}	78318.25 –	84013.40	7–5	2.78–03	9.17–03	3.71+00	–1.192	C	2
263.		^{3}D°–[$^3/_2$]											
			17483.4	5718.15 cm^{-1}	78318.25 –	84036.40	7–5	2.44–03	7.97–03	3.21+00	–1.253	C	2
			17451.3	5728.66 cm^{-1}	78307.63 –	84036.29	5–3	6.77–03	1.86–02	5.33+00	–1.033	C	2
			17451.0	5728.77 cm^{-1}	78307.63 –	84036.40	5–5	9.18–03	4.19–02	1.20+01	–0.678	C	2
			17408.4	5742.80 cm^{-1}	78293.49 –	84036.29	3–3	1.63–02	7.41–02	1.27+01	–0.653	C	2
			17408.0	5742.91 cm^{-1}	78293.49 –	84036.40	3–5	6.99–05	5.29–04	9.10–02	–2.799	C	2
264.	$2p3d$– $2p5p$	^{3}D°–^{3}D	*15025*	*6653.7* cm^{-1}	*78309.8* –	*84963.5*	15–15	4.05–04	1.37–03	1.02+00	–1.687	C	1
			14995.8	6666.72 cm^{-1}	78318.25 –	84984.97	7–7	3.62–04	1.22–03	4.22–01	–2.069	D	*LS*
			15049.9	6642.73 cm^{-1}	78307.63 –	84950.36	5–5	2.80–04	9.52–04	2.36–01	–2.322	D	*LS*
			15051.9	6641.85 cm^{-1}	78293.49 –	84935.34	3–3	3.02–04	1.03–03	1.53–01	–2.512	D	*LS*
			15074.0	6632.11 cm^{-1}	78318.25 –	84950.36	7–5	6.25–05	1.52–04	5.29–02	–2.973	D	*LS*
			15084.0	6627.71 cm^{-1}	78307.63 –	84935.34	5–3	1.00–04	2.05–04	5.08–02	–2.990	D	*LS*
			14971.9	6677.34 cm^{-1}	78307.63 –	84984.97	5–7	4.56–05	2.15–04	5.29–02	–2.970	D	*LS*
			15018.0	6656.87 cm^{-1}	78293.49 –	84950.36	3–5	6.08–05	3.43–04	5.08–02	–2.988	D	*LS*
265.		^{3}D°–^{3}P	*14520*	*6885.2* cm^{-1}	*78309.8* –	*85195.0*	15–9	3.87–03	7.34–03	5.26+00	–0.958	C	1
			14519.5	6885.39 cm^{-1}	78318.25 –	85203.64	7–5	3.25–03	7.34–03	2.45+00	–1.289	D	*LS*
			14528.1	6881.32 cm^{-1}	78307.63 –	85188.95	5–3	2.90–03	5.50–03	1.32+00	–1.561	D	*LS*
			14539.1	6876.12 cm^{-1}	78293.49 –	85169.61	3–1	3.85–03	4.07–03	5.85–01	–1.913	D	*LS*
			14497.2	6896.01 cm^{-1}	78307.63 –	85203.64	5–5	5.83–04	1.84–03	4.38–01	–2.037	D	*LS*
			14498.3	6895.46 cm^{-1}	78293.49 –	85188.95	3–3	9.71–04	3.06–03	4.38–01	–2.037	D	*LS*
			14467.5	6910.15 cm^{-1}	78293.49 –	85203.64	3–5	3.91–05	2.05–04	2.92–02	–3.212	D–	*LS*

C I: Allowed Transitions — Continued

No.	Transition Array	Mult.	λ_{air} (Å)	λ_{vac} (Å) or σ (cm^{-1})*	E_i (cm^{-1})	E_k (cm^{-1})	g_i–g_k	A_{ki} (10^8 s^{-1})	f_{ik}	S (at. u.)	log gf	Acc.	Source
266.	$2p3d-2p6p$	$^3D^o-^3D$	*11530*	*8671.0* cm^{-1}	*78309.8*	*86980.8*	15–15	2.19–04	4.37–04	2.49–01	–2.184	C	1
			11512.3	8684.01 cm^{-1}	78318.25	87002.26	7–7	1.96–04	3.89–04	1.03–01	–2.565	D	*LS*
			11547.1	8657.82 cm^{-1}	78307.63	86965.45	5–5	1.52–04	3.03–04	5.77–02	–2.819	D	*LS*
			11540.6	8662.67 cm^{-1}	78293.49	86956.16	3–3	1.64–04	3.27–04	3.73–02	–3.008	D	*LS*
			11561.3	8647.20 cm^{-1}	78318.25	86965.45	7–5	3.39–05	4.85–05	1.29–02	–3.469	D–	*LS*
			11559.5	8648.53 cm^{-1}	78307.63	86956.16	5–3	5.43–05	6.53–05	1.24–02	–3.486	D–	*LS*
			11498.2	8694.63 cm^{-1}	78307.63	87002.26	5–7	2.46–05	6.83–05	1.29–02	–3.467	D–	*LS*
			11528.3	8671.96 cm^{-1}	78293.49	86965.45	3–5	3.29–05	1.09–04	1.24–02	–3.485	D–	*LS*
267.		$^3D^o-^3P$	*11366*	*8796.1* cm^{-1}	*78309.8*	*87105.9*	15–9	1.98–03	2.30–03	1.29+00	–1.463	C	1
			11367.0	8794.96 cm^{-1}	78318.25	87113.21	7–5	1.66–03	2.30–03	6.02–01	–1.794	D	*LS*
			11366.4	8795.49 cm^{-1}	78307.63	87103.12	5–3	1.48–03	1.72–03	3.22–01	–2.065	D	*LS*
			11381.4	8783.87 cm^{-1}	78293.49	87077.36	3–1	1.97–03	1.27–03	1.43–01	–2.418	D	*LS*
			11353.3	8805.58 cm^{-1}	78307.63	87113.21	5–5	2.97–04	5.75–04	1.07–01	–2.542	D	*LS*
			11348.1	8809.63 cm^{-1}	78293.49	87103.12	3–3	4.96–04	9.58–04	1.07–01	–2.541	D	*LS*
			11335.1	8819.72 cm^{-1}	78293.49	87113.21	3–5	1.99–05	6.40–05	7.16–03	–3.717	D–	*LS*
268.	$2p3d-2p7p$	$^3D^o-^3D$	*10197*	*9804.0* cm^{-1}	*78309.8*	*88113.8*	15–15	1.34–04	2.09–04	1.05–01	–2.503	C	1
			10183.0	9817.56 cm^{-1}	78318.25	88135.81	7–7	1.20–04	1.86–04	4.37–02	–2.885	D	*LS*
			10213.4	9788.40 cm^{-1}	78307.63	88096.03	5–5	9.30–05	1.45–04	2.45–02	–3.138	D–	*LS*
			10202.8	9798.52 cm^{-1}	78293.49	88092.01	3–3	1.01–04	1.57–04	1.58–02	–3.327	D–	*LS*
			10224.5	9777.78 cm^{-1}	78318.25	88096.03	7–5	2.08–05	2.33–05	5.48–03	–3.788	D–	*LS*
			10217.6	9784.38 cm^{-1}	78307.63	88092.01	5–3	3.34–05	3.13–05	5.27–03	–3.805	D–	*LS*
			10172.0	9828.18 cm^{-1}	78307.63	88135.81	5–7	1.51–05	3.27–05	5.48–03	–3.786	D–	*LS*
			10198.6	9802.54 cm^{-1}	78293.49	88096.03	3–5	2.01–05	5.23–05	5.27–03	–3.804	D–	*LS*
269.		$^3D^o-^3P$	*10116*	*9882.2* cm^{-1}	*78309.8*	*88192.0*	15–9	1.17–03	1.08–03	5.38–01	–1.791	C	1
			10118.7	9879.97 cm^{-1}	78318.25	88198.22	7–5	9.83–04	1.08–03	2.51–01	–2.123	D	*LS*
			10113.9	9884.67 cm^{-1}	78307.63	88192.30	5–3	8.79–04	8.08–04	1.35–01	–2.393	D	*LS*
			10132.7	9866.38 cm^{-1}	78293.49	88159.87	3–1	1.16–03	5.98–04	5.98–02	–2.746	D	*LS*
			10107.8	9890.59 cm^{-1}	78307.63	88198.22	5–5	1.76–04	2.70–04	4.49–02	–2.870	D	*LS*
			10099.5	9898.81 cm^{-1}	78293.49	88192.30	3–3	2.94–04	4.50–04	4.49–02	–2.870	D	*LS*
			10093.4	9904.73 cm^{-1}	78293.49	88198.22	3–5	1.18–05	3.00–05	2.99–03	–4.046	D–	*LS*
270.	$2p3d-2p4p$	$^1F^o-^3D$											
			44008.9	2271.65 cm^{-1}	78529.62	80801.27	7–5	2.99–05	6.20–04	6.29–01	–2.362	C	2
271.		$^1F^o-^1D$	30854.2	3240.17 cm^{-1}	78529.62	81769.79	7–5	1.75–02	1.78–01	1.27+02	0.096	B	1,2
272.	$2p3d-2p(^2P^o_{1/2})4f$	$^1F^o-[5/2]$											
			18547.7	5390.03 cm^{-1}	78529.62	83919.65	7–7	5.66–03	2.92–02	1.25+01	–0.689	C	2
			18547.3	5390.14 cm^{-1}	78529.62	83919.76	7–5	3.55–05	1.31–04	5.59–02	–3.039	C	2
273.		$^1F^o-[7/2]$											
			18525.2	5396.58 cm^{-1}	78529.62	83926.20	7–7	5.71–03	2.94–02	1.25+01	–0.687	C	2
			18524.6	5396.75 cm^{-1}	78529.62	83926.37	7–9	9.87–03	6.53–02	2.79+01	–0.340	C	2
274.	$2p3d-2p(^2P^o_{3/2})4f$	$^1F^o-[7/2]$											
			18320.7	5456.83 cm^{-1}	78529.62	83986.45	7–9	4.28–02	2.77–01	1.17+02	0.287	C	2
			18321.4	5456.60 cm^{-1}	78529.62	83986.22	7–7	3.35–03	1.69–02	7.12+00	–0.928	C	2
275.		$^1F^o-[9/2]$											
			18221.1	5486.63 cm^{-1}	78529.62	84016.25	7–9	9.82–02	6.29–01	2.64+02	0.644	C	2

C I: Allowed Transitions — Continued

No.	Transition Array	Mult.	λ_{air} (Å)	λ_{vac} (Å) or σ (cm^{-1})*	E_i (cm^{-1})	E_k (cm^{-1})	g_i–g_k	A_{ki} (10^8 s^{-1})	f_{ik}	S (at. u.)	log gf	Acc.	Source
276.		^{1}F^o–[$^5/_2$]											
			18231.1	5483.63 cm^{-1}	78529.62	84013.25	7–7	1.51–03	7.51–03	3.16+00	–1.279	C	2
			18230.6	5483.78 cm^{-1}	78529.62	84013.40	7–5	4.04–04	1.44–03	6.04–01	–1.997	C	2
277.		^{1}F^o–[$^3/_2$]											
			18154.5	5506.78 cm^{-1}	78529.62	84036.40	7–5	4.24–04	1.50–03	6.27–01	–1.980	C	2
278.	$2p3d$–$2p5p$	^{1}F^o–^{1}D	14551.7	6870.19 cm^{-1}	78529.62	85399.81	7–5	5.18–03	1.18–02	3.94+00	–1.085	B	1
279.	$2p3d$–$2p6p$	^{1}F^o–^{1}D	11122.1	8988.64 cm^{-1}	78529.62	87518.26	7–5	2.89–03	3.83–03	9.81–01	–1.572	B	1
280.	$2p3d$–$2p7p$	^{1}F^o–^{1}D	10273.9	9730.75 cm^{-1}	78529.62	88260.37	7–5	1.53–03	1.73–03	4.09–01	–1.917	B	1
281.	$2p3d$–$2p8p$	^{1}F^o–^{1}D	9627.62	9630.26	78529.62	88913.56	7–5	9.78–04	9.71–04	2.15–01	–2.168	B	1
282.	$2p3d$–$2p4p$	^{1}P^o–^{1}P		1831.58 cm^{-1}	78731.27	80562.85	3–3	5.43–03	2.43–01	1.31+02	–0.138	C+	1,2
283.		^{1}P^o–^{3}D											
			48737.7	2051.24 cm^{-1}	78731.27	80782.51	3–3	1.16–04	4.12–03	1.98+00	–1.908	C	2
284.		^{1}P^o–^{3}S											
			42115.8	2373.76 cm^{-1}	78731.27	81105.03	3–3	3.70–05	9.85–04	4.10–01	–2.529	C	2
285.		^{1}P^o–^{3}P											
			38532.7	2594.49 cm^{-1}	78731.27	81325.76	3–3	3.86–05	8.60–04	3.27–01	–2.589	C	2
286.		^{1}P^o–^{1}D	32901.8	3038.52 cm^{-1}	78731.27	81769.79	3–5	3.67–03	9.92–02	3.23+01	–0.526	D	1,2
287.	$2p3d$–$2p(^2P^o_{1/2})4f$	^{1}P^o–[$^5/_2$]											
			19268.2	5188.49 cm^{-1}	78731.27	83919.76	3–5	4.40–03	4.08–02	7.77+00	–0.912	C	2
288.	$2p3d$–$2p(^2P^o_{3/2})4f$	^{1}P^o–[$^5/_2$]											
			18926.6	5282.13 cm^{-1}	78731.27	84013.40	3–5	5.11–02	4.57–01	8.55+01	0.137	C	2
289.		^{1}P^o–[$^3/_2$]											
			18844.5	5305.13 cm^{-1}	78731.27	84036.40	3–5	4.17–02	3.70–01	6.89+01	0.046	C	2
			18844.9	5305.02 cm^{-1}	78731.27	84036.29	3–3	3.34–05	1.78–04	3.31–02	–3.273	C	2
290.	$2p3d$–$2p4p$	^{3}P^o–^{1}P											
				1239.69 cm^{-1}	79323.16	80562.85	1–3	5.27–07	1.54–04	4.10–02	–3.812	C	2
291.		^{3}P^o–^{3}D		*1498.2* cm^{-1}	*79314.9*	*80813.1*	9–15	3.09–04	3.43–02	6.79+01	–0.510	C	1,2
				1523.76 cm^{-1}	79310.85	80834.61	5–7	3.37–04	3.05–02	3.29+01	–0.817	C	2n
				1482.49 cm^{-1}	79318.78	80801.27	3–5	2.31–04	2.63–02	1.75+01	–1.104	C	2n
				1459.35 cm^{-1}	79323.16	80782.51	1–3	1.57–04	3.32–02	7.49+00	–1.479	C	2n
				1490.42 cm^{-1}	79310.85	80801.27	5–5	6.32–05	4.26–03	4.71+00	–1.671	C	2n
				1463.73 cm^{-1}	79318.78	80782.51	3–3	1.08–04	7.53–03	5.08+00	–1.646	C	2n
				1471.66 cm^{-1}	79310.85	80782.51	5–3	5.04–06	2.09–04	2.34–01	–2.980	C	2n

C I: Allowed Transitions — Continued

No.	Transition Array	Mult.	λ_{air} (Å)	λ_{vac} (Å) or σ (cm^{-1})*	E_i (cm^{-1})	E_k (cm^{-1})	g_i-g_k	A_{ki} ($10^8 s^{-1}$)	f_{ik}	S (at. u.)	log gf	Acc.	Source
292.		$^3P^o-^3S$		*1790.2* cm^{-1}	*79314.9*	81105.03	9–3	1.85–03	2.88–02	4.76+01	–0.587	B–	1,2
				1794.18 cm^{-1}	79310.85	81105.03	5–3	1.28–03	3.58–02	3.29+01	–0.747	B–	2*n*
				1786.25 cm^{-1}	79318.78	81105.03	3–3	4.54–04	2.13–02	1.18+01	–1.194	B–	2*n*
				1781.87 cm^{-1}	79323.16	81105.03	1–3	1.13–04	1.60–02	2.96+00	–1.795	B–	2*n*
293.		$^3P^o-^3P$	*49506*	*2019.4* cm^{-1}	*79314.9*	*81334.2*	9–9	6.14–03	2.26–01	3.31+02	0.308	B	1,2
			49171.6	2033.14 cm^{-1}	79310.85	81343.99	5–5	4.90–03	1.78–01	1.44+02	–0.052	B	2*n*
			49812.5	2006.98 cm^{-1}	79318.78	81325.76	3–3	1.77–03	6.58–02	3.24+01	–0.705	B	2*n*
			49616.5	2014.91 cm^{-1}	79310.85	81325.76	5–3	2.13–03	4.71–02	3.84+01	–0.628	B	2*n*
				1992.23 cm^{-1}	79318.78	81311.01	3–1	5.68–03	7.15–02	3.55+01	–0.668	B	2*n*
			49364.1	2025.21 cm^{-1}	79318.78	81343.99	3–5	1.47–03	8.95–02	4.36+01	–0.571	B	2*n*
			49921.5	2002.60 cm^{-1}	79323.16	81325.76	1–3	2.03–03	2.27–01	3.74+01	–0.643	B	2*n*
294.		$^3P^o-^1D$											
			40656.8	2458.94 cm^{-1}	79310.85	81769.79	5–5	7.56–06	1.87–04	1.25–01	–3.028	C	2
			40788.4	2451.01 cm^{-1}	79318.78	81769.79	3–5	1.25–05	5.18–04	2.09–01	–2.808	C	2
295.		$^3P^o-^1S$											
			34086.3	2932.93 cm^{-1}	79318.78	82251.71	3–1	4.44–05	2.58–04	8.69–02	–3.111	C	2
296.	$2p3d-2p(^2P^o_{1/2})4f$	$^3P^o-[5/2]$											
			21691.7	4608.80 cm^{-1}	79310.85	83919.65	5–7	7.51–03	7.42–02	2.65+01	–0.430	C	2
			21728.6	4600.98 cm^{-1}	79318.78	83919.76	3–5	2.11–03	2.49–02	5.35+00	–1.127	C	2
			21691.2	4608.91 cm^{-1}	79310.85	83919.76	5–5	7.40–04	5.22–03	1.86+00	–1.583	C	2
297.	$2p3d-2p(^2P^o_{3/2})4f$	$^3P^o-[5/2]$											
			21259.9	4702.40 cm^{-1}	79310.85	84013.25	5–7	6.56–02	6.22–01	2.18+02	0.493	C	2
			21295.2	4694.62 cm^{-1}	79318.78	84013.40	3–5	2.00–02	2.27–01	4.78+01	–0.166	C	2
			21259.3	4702.55 cm^{-1}	79310.85	84013.40	5–5	7.03–03	4.77–02	1.67+01	–0.623	C	2
298.		$^3P^o-[3/2]$											
			21191.3	4717.62 cm^{-1}	79318.78	84036.40	3–5	3.41–02	3.82–01	8.01+01	0.060	C	2
			21211.5	4713.13 cm^{-1}	79323.16	84036.29	1–3	4.23–02	8.56–01	5.98+01	–0.067	C	2
			21155.8	4725.55 cm^{-1}	79310.85	84036.40	5–5	1.14–02	7.63–02	2.66+01	–0.419	C	2
			21191.8	4717.51 cm^{-1}	79318.78	84036.29	3–3	3.22–02	2.17–01	4.53+01	–0.187	C	2
			21156.3	4725.44 cm^{-1}	79310.85	84036.29	5–3	2.19–03	8.82–03	3.07+00	–1.355	C	2
299.	$2p3d-2p5p$	$^3P^o-^3D$	*17699*	*5648.6* cm^{-1}	*79314.9*	*84963.5*	9–15	4.99–04	3.91–03	2.05+00	–1.454	C	1
			17619.1	5674.12 cm^{-1}	79310.85	84984.97	5–7	5.06–04	3.30–03	9.56–01	–1.783	D	*LS*
			17752.2	5631.58 cm^{-1}	79318.78	84950.36	3–5	3.71–04	2.92–03	5.12–01	–2.057	D	*LS*
			17813.5	5612.18 cm^{-1}	79323.16	84935.34	1–3	2.72–04	3.88–03	2.28–01	–2.411	D	*LS*
			17727.2	5639.51 cm^{-1}	79310.85	84950.36	5–5	1.24–04	5.85–04	1.71–01	–2.534	D	*LS*
			17799.6	5616.56 cm^{-1}	79318.78	84935.34	3–3	2.04–04	9.71–04	1.71–01	–2.535	D	*LS*
			17774.5	5624.49 cm^{-1}	79310.85	84935.34	5–3	1.37–05	3.89–05	1.14–02	–3.711	D	*LS*
300.		$^3P^o-^3P$	*17002*	*5880.1* cm^{-1}	*79314.9*	*85195.0*	9–9	1.53–04	6.64–04	3.34–01	–2.224	C	1
			16965.3	5892.79 cm^{-1}	79310.85	85203.64	5–5	1.16–04	4.99–04	1.39–01	–2.603	D	*LS*
			17030.6	5870.17 cm^{-1}	79318.78	85188.95	3–3	3.81–05	1.66–04	2.79–02	–3.304	D	*LS*
			17007.7	5878.10 cm^{-1}	79310.85	85188.95	5–3	6.37–05	1.66–04	4.64–02	–3.081	D	*LS*
			17086.9	5850.83 cm^{-1}	79318.78	85169.61	3–1	1.51–04	2.20–04	3.71–02	–3.180	D	*LS*
			16988.1	5884.86 cm^{-1}	79318.78	85203.64	3–5	3.84–05	2.77–04	4.64–02	–3.081	D	*LS*
			17043.3	5865.79 cm^{-1}	79323.16	85188.95	1–3	5.07–05	6.62–04	3.71–02	–3.179	D	*LS*

C I: Allowed Transitions — Continued

No.	Transition Array	Mult.	λ_{air} (Å)	λ_{vac} (Å) or σ (cm^{-1})*	E_i (cm^{-1})	E_k (cm^{-1})	g_i-g_k	A_{ki} ($10^8 s^{-1}$)	f_{ik}	S (at. u.)	log gf	Acc.	Source
301.	$2p3d-$ $2p6p$	$^3P^o-^3D$	*13041*	*7665.9* cm^{-1}	*79314.9*	*86980.8*	9–15	3.56–04	1.51–03	5.84–01	–1.866	C	1
			12998.0	7691.41 cm^{-1}	79310.85	87002.26	5–7	3.59–04	1.27–03	2.73–01	–2.196	D	*LS*
			13074.0	7646.67 cm^{-1}	79318.78	86965.45	3–5	2.65–04	1.13–03	1.46–01	–2.469	D	*LS*
			13097.4	7633.00 cm^{-1}	79323.16	86956.16	1–3	1.95–04	1.51–03	6.49–02	–2.822	D	*LS*
			13060.5	7654.60 cm^{-1}	79310.85	86965.45	5–5	8.86–05	2.26–04	4.87–02	–2.946	D	*LS*
			13089.9	7637.38 cm^{-1}	79318.78	86956.16	3–3	1.47–04	3.77–04	4.87–02	–2.947	D	*LS*
			13076.3	7645.31 cm^{-1}	79310.85	86956.16	5–3	9.80–06	1.51–05	3.25–03	–4.123	D–	*LS*
302.	$2p3d-$ $2p7p$	$^3P^o-^3D$	*11362*	*8798.9* cm^{-1}	*79314.9*	*88113.8*	9–15	2.33–04	7.53–04	2.54–01	–2.169	C	1
			11328.4	8824.96 cm^{-1}	79310.85	88135.81	5–7	2.36–04	6.34–04	1.18–01	–2.499	D	*LS*
			11390.0	8777.25 cm^{-1}	79318.78	88096.03	3–5	1.74–04	5.63–04	6.34–02	–2.772	D	*LS*
			11400.9	8768.85 cm^{-1}	79323.16	88092.01	1–3	1.28–04	7.50–04	2.82–02	–3.125	D–	*LS*
			11379.7	8785.18 cm^{-1}	79310.85	88096.03	5–5	5.81–05	1.13–04	2.11–02	–3.249	D–	*LS*
			11395.2	8773.23 cm^{-1}	79318.78	88092.01	3–3	9.64–05	1.88–04	2.11–02	–3.249	D–	*LS*
			11384.9	8781.16 cm^{-1}	79310.85	88092.01	5–3	6.45–06	7.52–06	1.41–03	–4.425	D–	*LS*
303.	$2p3d-$ $2p8p$	$^3P^o-^3D$	*10525*	*9498.3* cm^{-1}	*79314.9*	*88813.1*	9–15	1.57–04	4.36–04	1.36–01	–2.406	C	1
			10495.5	9525.27 cm^{-1}	79310.85	88836.12	5–7	1.59–04	3.67–04	6.34–02	–2.736	D	*LS*
			10550.6	9475.52 cm^{-1}	79318.78	88794.3	3–5	1.17–04	3.26–04	3.40–02	–3.010	D	*LS*
			10559.3	9467.69 cm^{-1}	79323.16	88790.85	1–3	8.66–05	4.34–04	1.51–02	–3.362	D–	*LS*
			10541.8	9483.45 cm^{-1}	79310.85	88794.3	5–5	3.92–05	6.53–05	1.13–02	–3.486	D–	*LS*
			10554.5	9472.07 cm^{-1}	79318.78	88790.85	3–3	6.51–05	1.09–04	1.13–02	–3.487	D–	*LS*
			10545.6	9480.00 cm^{-1}	79310.85	88790.85	5–3	4.35–06	4.35–06	7.55–04	–4.663	D–	*LS*
304.	$2p4p-$ $2p4d$	$^1P-^1D^o$	34064.9	2934.77 cm^{-1}	80562.85	83497.62	3–5	3.19–02	9.26–01	3.11+02	0.444	B	1,2
305.	$2p4p-$ $2p5s$	$^1P-^3P^o$											
			30968.7	3228.19 cm^{-1}	80562.85	83791.04	3–5	1.31–05	3.15–04	9.64–02	–3.024	C	2
			31343.7	3189.56 cm^{-1}	80562.85	83752.41	3–3	1.08–03	1.59–02	4.93+00	–1.321	C	2
			31465.6	3177.21 cm^{-1}	80562.85	83740.06	3–1	9.32–05	4.61–04	1.43–01	–2.859	C	2
306.	$2p4p-$ $2p4d$	$^1P-^3F^o$											
			31393.1	3184.54 cm^{-1}	80562.85	83747.39	3–5	1.52–05	3.75–04	1.16–01	–2.949	C	2
307.		$^1P-^3D^o$											
			30692.1	3257.28 cm^{-1}	80562.85	83820.13	3–3	3.01–04	4.25–03	1.29+00	–1.894	C	2
			30523.9	3275.23 cm^{-1}	80562.85	83838.08	3–5	1.61–05	3.74–04	1.13–01	–2.950	C	2
308.	$2p4p-$ $2p5s$	$^1P-^1P^o$	30162.6	3314.46 cm^{-1}	80562.85	83877.31	3–3	1.41–03	1.92–02	5.73+00	–1.239	D–	1,2
309.	$2p4p-$ $2p4d$	$^1P-^1P^o$	28816.4	3469.30 cm^{-1}	80562.85	84032.15	3–3	3.67–02	4.57–01	1.30+02	0.137	B	1,2
310.		$^1P-^3P^o$											
			28135.7	3553.24 cm^{-1}	80562.85	84116.09	3–3	8.45–05	1.00–03	2.79–01	–2.521	C	2
311.	$2p4p-$ $2p5d$	$^1P-^1D^o$	17781.3	5622.35 cm^{-1}	80562.85	86185.20	3–5	2.86–03	2.26–02	3.97+00	–1.169	B	1
312.	$2p4p-$ $2p6s$	$^1P-^1P^o$	17078.5	5853.70 cm^{-1}	80562.85	86416.55	3–3	6.43–04	2.81–03	4.74–01	–2.074	D	1
313.	$2p4p-$ $2p5d$	$^1P-^1P^o$	16862.9	5928.56 cm^{-1}	80562.85	86491.41	3–3	1.31–02	5.60–02	9.32+00	–0.775	D	1

C I: Allowed Transitions — Continued

No.	Transition Array	Mult.	λ_{air} (Å)	λ_{vac} (Å) or σ (cm^{-1})*	E_i (cm^{-1})	E_k (cm^{-1})	g_i–g_k	A_{ki} (10^8 s^{-1})	f_{ik}	S (at. u.)	log gf	Acc.	Source
314.	$2p4p$– $2p6d$	^{1}P–^{1}D°	14138.6	7070.90 cm^{-1}	80562.85	87633.75	3–5	4.45–04	2.22–03	3.10–01	–2.176	D	1
315.		^{1}P–^{1}P°	13756.5	7267.32 cm^{-1}	80562.85	87830.17	3–3	6.66–03	1.89–02	2.57+00	–1.246	D	1
316.	$2p4p$– $2p7d$	^{1}P–^{1}P°	12378.7	8076.17 cm^{-1}	80562.85	88639.02	3–3	3.91–03	8.98–03	1.10+00	–1.570	D	1
317.	$2p4p$– $2p8d$	^{1}P–^{1}P°	11622.2	8601.89 cm^{-1}	80562.85	89164.74	3–3	2.51–03	5.07–03	5.82–01	–1.818	D	1
318.	$2p4p$– $2p4d$	^{3}D–^{1}D°											
			37077.1	2696.35 cm^{-1}	80801.27	83497.62	5–5	1.23–04	2.54–03	1.55+00	–1.896	C	2
			36820.9	2715.11 cm^{-1}	80782.51	83497.62	3–5	3.81–06	1.29–04	4.69–02	–3.412	C	2
319.		^{3}D–^{3}F°	*33765*	*2960.9* cm^{-1}	*80813.1*	*83773.9*	15–21	4.31–02	1.03+00	1.72+03	1.189	B	1,2
			33729.4	2963.96 cm^{-1}	80834.61	83798.57	7–9	4.41–02	9.67–01	7.52+02	0.831	B	2n
			33774.7	2959.99 cm^{-1}	80801.27	83761.26	5–7	4.01–02	9.60–01	5.34+02	0.681	B	2n
			33719.0	2964.88 cm^{-1}	80782.51	83747.39	3–5	3.92–02	1.11+00	3.71+02	0.524	B	2n
			34159.4	2926.65 cm^{-1}	80834.61	83761.26	7–7	1.82–03	3.18–02	2.50+01	–0.653	B	2n
			33933.7	2946.12 cm^{-1}	80801.27	83747.39	5–5	3.62–03	6.24–02	3.49+01	–0.506	B	2n
			34322.1	2912.78 cm^{-1}	80834.61	83747.39	7–5	1.71–04	2.15–03	1.70+00	–1.822	B	2n
320.	$2p4p$– $2p5s$	^{3}D–^{3}P°	*33781*	*2959.4* cm^{-1}	*80813.1*	*83772.5*	15–9	2.70–02	2.77–01	4.62+02	0.618	B–	1,2
			33815.4	2956.43 cm^{-1}	80834.61	83791.04	7–5	1.95–02	2.39–01	1.86+02	0.223	B–	2n
			33876.0	2951.14 cm^{-1}	80801.27	83752.41	5–3	1.98–02	2.05–01	1.14+02	0.010	B–	2n
			33802.6	2957.55 cm^{-1}	80782.51	83740.06	3–1	2.89–02	1.65–01	5.51+01	–0.305	B–	2n
			33438.3	2989.77 cm^{-1}	80801.27	83791.04	5–5	7.06–03	1.18–01	6.52+01	–0.228	B–	2n
			33662.0	2969.90 cm^{-1}	80782.51	83752.41	3–3	7.24–03	1.23–01	4.09+01	–0.433	B–	2n
			33229.8	3008.53 cm^{-1}	80782.51	83791.04	3–5	5.24–05	1.45–03	4.75–01	–2.362	B–	2n
321.	$2p4p$– $2p4d$	^{3}D–^{3}D°	*33033*	*3026.4* cm^{-1}	*80813.1*	*83839.5*	15–15	1.19–02	1.95–01	3.19+02	0.467	C+	1,2
			33167.0	3014.22 cm^{-1}	80834.61	83848.83	7–7	1.24–02	2.05–01	1.57+02	0.157	C+	2n
			32920.3	3036.81 cm^{-1}	80801.27	83838.08	5–5	6.74–03	1.10–01	5.94+01	–0.261	C+	2n
			32911.5	3037.62 cm^{-1}	80782.51	83820.13	3–3	6.32–03	1.03–01	3.33+01	–0.512	C+	2n
			33285.7	3003.47 cm^{-1}	80834.61	83838.08	7–5	5.56–03	6.60–02	5.06+01	–0.335	C+	2n
			33116.1	3018.86 cm^{-1}	80801.27	83820.13	5–3	3.14–03	3.09–02	1.69+01	–0.810	C+	2n
			32804.2	3047.56 cm^{-1}	80801.27	83848.83	5–7	1.23–04	2.79–03	1.51+00	–1.856	C+	2n
			32718.2	3055.57 cm^{-1}	80782.51	83838.08	3–5	1.39–05	3.71–04	1.20–01	–2.953	C+	2n
322.	$2p4p$– $2p5s$	^{3}D–^{1}P°											
			32500.5	3076.04 cm^{-1}	80801.27	83877.31	5–3	1.31–03	1.25–02	6.68+00	–1.204	C	2
			32303.5	3094.80 cm^{-1}	80782.51	83877.31	3–3	2.04–05	3.19–04	1.02–01	–3.018	C	2
323.	$2p4p$– $2p4d$	^{3}D–^{1}F°											
			31776.1	3146.16 cm^{-1}	80801.27	83947.43	5–7	5.96–04	1.26–02	6.61+00	–1.200	C	2
			32116.5	3112.82 cm^{-1}	80834.61	83947.43	7–7	3.54–05	5.48–04	4.06–01	–2.416	C	2
324.		^{3}D–^{1}P°											
			30764.2	3249.64 cm^{-1}	80782.51	84032.15	3–3	1.03–03	1.46–02	4.44+00	–1.359	C	2
			30942.9	3230.88 cm^{-1}	80801.27	84032.15	5–3	1.36–05	1.17–04	5.97–02	–3.232	C	2

C I: Allowed Transitions — Continued

No.	Transition Array	Mult.	λ_{air} (Å)	λ_{vac} (Å) or σ (cm^{-1})*	E_i (cm^{-1})	E_k (cm^{-1})	g_i-g_k	A_{ki} ($10^8 s^{-1}$)	f_{ik}	S (at. u.)	log gf	Acc.	Source
325.		$^3D-^3P^o$	*30328*	*3296.4* cm^{-1}	*80813.1*	*84109.4*	15–9	4.69–04	3.88–03	5.81+00	−1.235	D	1,2
			30586.8	3268.49 cm^{-1}	80834.61	84103.10	7–5	4.29–04	4.30–03	3.03+00	−1.521	D	2*n*
			30159.3	3314.82 cm^{-1}	80801.27	84116.09	5–3	2.26–04	1.85–03	9.19–01	−2.034	D	2*n*
			29943.5	3338.71 cm^{-1}	80782.51	84121.22	3–1	2.54–04	1.14–03	3.36–01	−2.467	D	2*n*
			30278.0	3301.83 cm^{-1}	80801.27	84103.10	5–5	1.50–04	2.06–03	1.02+00	−1.988	D	2*n*
			29989.6	3333.58 cm^{-1}	80782.51	84116.09	3–3	1.07–04	1.44–03	4.27–01	−2.364	D	2*n*
			30106.9	3320.59 cm^{-1}	80782.51	84103.10	3–5	1.02–05	2.32–04	6.90–02	−3.158	D	2*n*
326.	2*p*4*p*– 2*p*5*d*	$^3D-^3F^o$	*18078*	*5529.9* cm^{-1}	*80813.1*	*86343.0*	15–21	7.17–03	4.92–02	4.39+01	−0.132	C	1
			18062.4	5534.86 cm^{-1}	80834.61	86369.47	7–9	7.19–03	4.52–02	1.88+01	−0.500	D	*LS*
			18091.7	5525.89 cm^{-1}	80801.27	86327.16	5–7	6.36–03	4.37–02	1.30+01	−0.661	D	*LS*
			18061.5	5535.13 cm^{-1}	80782.51	86317.64	3–5	6.04–03	4.92–02	8.78+00	−0.831	D	*LS*
			18201.5	5492.55 cm^{-1}	80834.61	86327.16	7–7	7.83–04	3.89–03	1.63+00	−1.565	D	*LS*
			18122.9	5516.37 cm^{-1}	80801.27	86317.64	5–5	1.11–03	5.47–03	1.63+00	−1.563	D	*LS*
			18233.1	5483.03 cm^{-1}	80834.61	86317.64	7–5	3.07–05	1.09–04	4.60–02	−3.116	D	*LS*
327.	2*p*4*p*– 2*p*6*s*	$^3D-^3P^o$	*18050*	*5538.6* cm^{-1}	*80813.1*	*86351.6*	15–9	1.02–02	2.98–02	2.66+01	−0.349	C	1
			18061.9	5534.99 cm^{-1}	80834.61	86369.60	7–5	8.53–03	2.98–02	1.24+01	−0.681	D	*LS*
			18077.1	5530.36 cm^{-1}	80801.27	86331.63	5–3	7.60–03	2.23–02	6.65+00	−0.952	D	*LS*
			18047.5	5539.43 cm^{-1}	80782.51	86321.94	3–1	1.02–02	1.66–02	2.95+00	−1.303	D	*LS*
			17953.8	5568.33 cm^{-1}	80801.27	86369.60	5–5	1.55–03	7.50–03	2.22+00	−1.426	D	*LS*
			18016.0	5549.12 cm^{-1}	80782.51	86331.63	3–3	2.56–03	1.25–02	2.22+00	−1.428	D	*LS*
			17893.5	5587.09 cm^{-1}	80782.51	86369.60	3–5	1.04–04	8.36–04	1.48–01	−2.601	D	*LS*
328.	2*p*4*p*– 2*p*5*d*	$^3D-^3D^o$	*17933*	*5574.9* cm^{-1}	*80813.1*	*86387.9*	15–15	2.31–03	1.11–02	9.84+00	−0.778	C	1
			17970.4	5563.19 cm^{-1}	80834.61	86397.80	7–7	2.04–03	9.86–03	4.08+00	−1.161	D	*LS*
			17890.3	5588.11 cm^{-1}	80801.27	86389.38	5–5	1.62–03	7.76–03	2.28+00	−1.411	D	*LS*
			17916.2	5580.01 cm^{-1}	80782.51	86362.52	3–3	1.73–03	8.34–03	1.48+00	−1.601	D	*LS*
			17997.6	5554.77 cm^{-1}	80834.61	86389.38	7–5	3.56–04	1.23–03	5.12–01	−2.064	D	*LS*
			17976.7	5561.25 cm^{-1}	80801.27	86362.52	5–3	5.72–04	1.66–03	4.92–01	−2.080	D	*LS*
			17863.3	5596.53 cm^{-1}	80801.27	86397.80	5–7	2.60–04	1.74–03	5.12–01	−2.060	D	*LS*
			17830.4	5606.87 cm^{-1}	80782.51	86389.38	3–5	3.52–04	2.80–03	4.92–01	−2.076	D	*LS*
329.		$^3D-^3P^o$	*17540*	*5699.7* cm^{-1}	*80813.1*	*86512.8*	15–9	3.94–04	1.09–03	9.44–01	−1.787	C	1
			17625.4	5672.09 cm^{-1}	80834.61	86506.70	7–5	3.26–04	1.08–03	4.41–01	−2.120	D	*LS*
			17483.2	5718.20 cm^{-1}	80801.27	86519.47	5–3	2.98–04	8.20–04	2.36–01	−2.387	D	*LS*
			17414.9	5740.65 cm^{-1}	80782.51	86523.16	3–1	4.02–04	6.10–04	1.05–01	−2.738	D	*LS*
			17522.4	5705.43 cm^{-1}	80801.27	86506.70	5–5	5.93–05	2.73–04	7.87–02	−2.865	D	*LS*
			17426.1	5736.96 cm^{-1}	80782.51	86519.47	3–3	1.00–04	4.57–04	7.87–02	−2.863	D	*LS*
			17465.0	5724.19 cm^{-1}	80782.51	86506.70	3–5	3.99–06	3.04–05	5.24–03	−4.040	D−	*LS*
330.	2*p*4*p*– 2*p*6*d*	$^3D-^3F^o$	*14448*	*6919.3* cm^{-1}	*80813.1*	*87732.4*	15–21	2.27–03	9.94–03	7.10+00	−0.826	C	1
			14434.4	6926.00 cm^{-1}	80834.61	87760.61	7–9	2.28–03	9.14–03	3.04+00	−1.194	D	*LS*
			14463.4	6912.11 cm^{-1}	80801.27	87713.38	5–7	2.01–03	8.83–03	2.10+00	−1.355	D	*LS*
			14435.0	6925.70 cm^{-1}	80782.51	87708.21	3–5	1.91–03	9.95–03	1.42+00	−1.525	D	*LS*
			14533.5	6878.77 cm^{-1}	80834.61	87713.38	7–7	2.48–04	7.87–04	2.64–01	−2.259	D	*LS*
			14474.2	6906.94 cm^{-1}	80801.27	87708.21	5–5	3.52–04	1.11–03	2.64–01	−2.257	D	*LS*
			14544.4	6873.60 cm^{-1}	80834.61	87708.21	7–5	9.79–06	2.22–05	7.43–03	−3.809	D−	*LS*
331.	2*p*4*p*– 2*p*7*s*	$^3D-^3P^o$	*14438*	*6924.2* cm^{-1}	*80813.1*	*87737.3*	15–9	5.43–03	1.02–02	7.25+00	−0.816	C	1
			14448.8	6919.12 cm^{-1}	80834.61	87753.73	7–5	4.55–03	1.02–02	3.38+00	−1.148	D	*LS*
			14452.6	6917.29 cm^{-1}	80801.27	87718.56	5–3	4.06–03	7.62–03	1.81+00	−1.419	D	*LS*
			14428.4	6928.86 cm^{-1}	80782.51	87711.37	3–1	5.44–03	5.66–03	8.06–01	−1.770	D	*LS*
			14379.5	6952.46 cm^{-1}	80801.27	87753.73	5–5	8.24–04	2.55–03	6.04–01	−1.894	D	*LS*
			14413.5	6936.05 cm^{-1}	80782.51	87718.56	3–3	1.36–03	4.25–03	6.04–01	−1.895	D	*LS*
			14340.8	6971.22 cm^{-1}	80782.51	87753.73	3–5	5.54–05	2.85–04	4.03–02	−3.069	D	*LS*

C I: Allowed Transitions — Continued

No.	Transition Array	Mult.	λ_{air} (Å)	λ_{vac} (Å) or σ (cm^{-1})*	E_i (cm^{-1})	E_k (cm^{-1})	g_i-g_k	A_{ki} (10^8 s^{-1})	f_{ik}	S (at. u.)	log gf	Acc.	Source
332.	2p4p–2p6d	^{3}D–^{3}D°	*14376*	*6954.4* cm^{-1}	*80813.1*	*87767.4*	15–15	8.42–04	2.61–03	1.85+00	–1.407	C	1
			14400.0	6942.56 cm^{-1}	80834.61	87777.17	7–7	7.45–04	2.32–03	7.68–01	–1.790	D	*LS*
			14339.5	6971.82 cm^{-1}	80801.27	87773.09	5–5	5.91–04	1.82–03	4.30–01	–2.041	D	*LS*
			14378.8	6952.79 cm^{-1}	80782.51	87735.3	3–3	6.31–04	1.96–03	2.78–01	–2.231	D	*LS*
			14408.4	6938.48 cm^{-1}	80834.61	87773.09	7–5	1.31–04	2.90–04	9.63–02	–2.692	D	*LS*
			14417.7	6934.03 cm^{-1}	80801.27	87735.3	5–3	2.09–04	3.90–04	9.26–02	–2.710	D	*LS*
			14331.2	6975.90 cm^{-1}	80801.27	87777.17	5–7	9.47–05	4.08–04	9.63–02	–2.690	D	*LS*
			14301.1	6990.58 cm^{-1}	80782.51	87773.09	3–5	1.28–04	6.56–04	9.26–02	–2.706	D	*LS*
333.		^{3}D–^{3}P°	*14231*	*7024.8* cm^{-1}	*80813.1*	*87837.9*	15–9	3.14–04	5.71–04	4.01–01	–2.067	C	1
			14286.0	6997.93 cm^{-1}	80834.61	87832.54	7–5	2.60–04	5.69–04	1.87–01	–2.400	D	*LS*
			14195.3	7042.64 cm^{-1}	80801.27	87843.91	5–3	2.37–04	4.29–04	1.00–01	–2.668	D	*LS*
			14151.6	7064.39 cm^{-1}	80782.51	87846.9	3–1	3.19–04	3.19–04	4.46–02	–3.019	D	*LS*
			14218.3	7031.27 cm^{-1}	80801.27	87832.54	5–5	4.72–05	1.43–04	3.35–02	–3.146	D	*LS*
			14157.6	7061.40 cm^{-1}	80782.51	87843.91	3–3	7.96–05	2.39–04	3.35–02	–3.144	D	*LS*
			14180.5	7050.03 cm^{-1}	80782.51	87832.54	3–5	3.17–06	1.59–05	2.23–03	–4.321	D–	*LS*
334.	2p4p–2p7d	^{3}D–^{3}F°	*12895*	*7753.1* cm^{-1}	*80813.1*	*88566*	15–21	9.65–04	3.37–03	2.14+00	–1.297	C	1
			12879.9	7761.89 cm^{-1}	80834.61	88596.5	7–9	9.68–04	3.10–03	9.19–01	–1.664	D	*LS*
			12910.3	7743.63 cm^{-1}	80801.27	88544.9	5–7	8.54–04	2.99–03	6.35–01	–1.825	D	*LS*
			12884.9	7758.89 cm^{-1}	80782.51	88541.4	3–5	8.13–04	3.37–03	4.29–01	–1.995	D	*LS*
			12966.1	7710.29 cm^{-1}	80834.61	88544.9	7–7	1.06–04	2.67–04	7.97–02	–2.729	D	*LS*
			12916.1	7740.13 cm^{-1}	80801.27	88541.4	5–5	1.50–04	3.75–04	7.97–02	–2.727	D	*LS*
			12972.0	7706.79 cm^{-1}	80834.61	88541.4	7–5	4.17–06	7.52–06	2.25–03	–4.279	D–	*LS*
335.	2p4p–2p8s	^{3}D–^{3}P°	*12891*	*7755.0* cm^{-1}	*80813.1*	*88568.0*	15–9	3.27–03	4.89–03	3.11+00	–1.134	C	1
			12900.3	7749.65 cm^{-1}	80834.61	88584.26	7–5	2.74–03	4.89–03	1.45+00	–1.466	D	*LS*
			12903.4	7747.79 cm^{-1}	80801.27	88549.06	5–3	2.45–03	3.67–03	7.79–01	–1.737	D	*LS*
			12881.0	7761.25 cm^{-1}	80782.51	88543.76	3–1	3.28–03	2.72–03	3.46–01	–2.088	D	*LS*
			12845.0	7782.99 cm^{-1}	80801.27	88584.26	5–5	4.96–04	1.23–03	2.60–01	–2.212	D	*LS*
			12872.2	7766.55 cm^{-1}	80782.51	88549.06	3–3	8.22–04	2.04–03	2.60–01	–2.213	D	*LS*
			12814.1	7801.75 cm^{-1}	80782.51	88584.26	3–5	3.33–05	1.37–04	1.73–02	–3.387	D–	*LS*
336.	2p4p–2p7d	^{3}D–^{3}D°	*12844*	*7783.4* cm^{-1}	*80813.1*	*88596.5*	15–15	3.98–04	9.86–04	6.25–01	–1.830	C	1
			12862.9	7772.19 cm^{-1}	80834.61	88606.8	7–7	3.53–04	8.74–04	2.59–01	–2.213	D	*LS*
			12811.3	7803.48 cm^{-1}	80801.27	88604.75	5–5	2.79–04	6.88–04	1.45–01	–2.464	D	*LS*
			12856.4	7776.09 cm^{-1}	80782.51	88558.6	3–3	2.98–04	7.38–04	9.38–02	–2.655	D	*LS*
			12866.3	7770.14 cm^{-1}	80834.61	88604.75	7–5	6.18–05	1.10–04	3.25–02	–3.115	D	*LS*
			12887.5	7757.33 cm^{-1}	80801.27	88558.6	5–3	9.86–05	1.47–04	3.13–02	–3.133	D	*LS*
			12807.9	7805.53 cm^{-1}	80801.27	88606.8	5–7	4.48–05	1.54–04	3.25–02	–3.113	D	*LS*
			12780.6	7822.24 cm^{-1}	80782.51	88604.75	3–5	6.07–05	2.48–04	3.13–02	–3.129	D	*LS*
337.		^{3}D–^{3}P°	*12771*	*7828.4* cm^{-1}	*80813.1*	*88641*	15–9	2.33–04	3.42–04	2.15–01	–2.290	C	1
			12813.4	7802.19 cm^{-1}	80834.61	88636.8	7–5	1.94–04	3.40–04	1.01–01	–2.623	D	*LS*
			12743.0	7845.33 cm^{-1}	80801.27	88646.6	5–3	1.76–04	2.57–04	5.39–02	–2.892	D	*LS*
			12708.5	7866.59 cm^{-1}	80782.51	88649.1	3–1	2.36–04	1.91–04	2.39–02	–3.243	D–	*LS*
			12758.9	7835.53 cm^{-1}	80801.27	88636.8	5–5	3.50–05	8.55–05	1.80–02	–3.369	D–	*LS*
			12712.6	7864.09 cm^{-1}	80782.51	88646.6	3–3	5.90–05	1.43–04	1.80–02	–3.368	D–	*LS*
			12728.4	7854.29 cm^{-1}	80782.51	88636.8	3–5	2.35–06	9.52–06	1.20–03	–4.544	D–	*LS*
338.	2p4p–2p8d	^{3}D–^{3}F°	*12057*	*8292.0* cm^{-1}	*80813.1*	*89105*	15–21	4.92–04	1.50–03	8.93–01	–1.648	C	1
			12041.6	8302.29 cm^{-1}	80834.61	89136.9	7–9	4.94–04	1.38–03	3.83–01	–2.015	D	*LS*
			12072.8	8280.83 cm^{-1}	80801.27	89082.1	5–7	4.35–04	1.33–03	2.65–01	–2.177	D	*LS*
			12048.7	8297.39 cm^{-1}	80782.51	89079.9	3–5	4.14–04	1.50–03	1.79–01	–2.346	D	*LS*
			12121.6	8247.49 cm^{-1}	80834.61	89082.1	7–7	5.39–05	1.19–04	3.32–02	–3.080	D	*LS*
			12076.0	8278.63 cm^{-1}	80801.27	89079.9	5–5	7.63–05	1.67–04	3.32–02	–3.079	D	*LS*
			12124.8	8245.29 cm^{-1}	80834.61	89079.9	7–5	2.13–06	3.35–06	9.36–04	–4.630	D–	*LS*

C I: Allowed Transitions — Continued

No.	Transition Array	Mult.	λ_{air} (Å)	λ_{vac} (Å) or σ (cm^{-1})*	E_i (cm^{-1})	E_k (cm^{-1})	g_i–g_k	A_{ki} (10^8 s^{-1})	f_{ik}	S (at. u.)	log gf	Acc.	Source
339.		^{3}D–^{3}D°	*12016*	*8319.9* cm^{-1}	*80813.1*	*89132.9*	15–15	2.20–04	4.77–04	2.83–01	–2.146	C	1
			12031.4	8309.29 cm^{-1}	80834.61	89143.9	7–7	1.95–04	4.23–04	1.17–01	–2.528	D	*LS*
			11985.1	8341.39 cm^{-1}	80801.27	89142.66	5–5	1.54–04	3.33–04	6.56–02	–2.779	D	*LS*
			12032.4	8308.59 cm^{-1}	80782.51	89091.1	3–3	1.65–04	3.57–04	4.24–02	–2.970	D	*LS*
			12033.2	8308.05 cm^{-1}	80834.61	89142.66	7–5	3.42–05	5.31–05	1.47–02	–3.430	D–	*LS*
			12059.7	8289.83 cm^{-1}	80801.27	89091.1	5–3	5.45–05	7.13–05	1.41–02	–3.448	D–	*LS*
			11983.3	8342.63 cm^{-1}	80801.27	89143.9	5–7	2.47–05	7.46–05	1.47–02	–3.428	D–	*LS*
			11958.2	8360.15 cm^{-1}	80782.51	89142.66	3–5	3.35–05	1.20–04	1.41–02	–3.445	D–	*LS*
340.		^{3}D–^{3}P°	*11969*	*8352.8* cm^{-1}	*80813.1*	*89166*	15–9	1.72–04	2.22–04	1.31–01	–2.477	C	1
			12005.0	8327.59 cm^{-1}	80834.61	89162.2	7–5	1.44–04	2.21–04	6.13–02	–2.810	D	*LS*
			11946.0	8368.73 cm^{-1}	80801.27	89170.0	5–3	1.30–04	1.67–04	3.28–02	–3.078	D	*LS*
			11917.1	8388.99 cm^{-1}	80782.51	89171.5	3–1	1.75–04	1.24–04	1.46–02	–3.430	D–	*LS*
			11957.1	8360.93 cm^{-1}	80801.27	89162.2	5–5	2.59–05	5.56–05	1.09–02	–3.556	D–	*LS*
			11919.3	8387.49 cm^{-1}	80782.51	89170.0	3–3	4.36–05	9.29–05	1.09–02	–3.555	D–	*LS*
			11930.4	8379.69 cm^{-1}	80782.51	89162.2	3–5	1.74–06	6.19–06	7.29–04	–4.731	D–	*LS*
341.	2*p*4*p*– 2*p*5*s*	^{3}S–^{3}P°	*37479*	*2667.5* cm^{-1}	81105.03	*83772.5*	3–9	1.33–02	8.38–01	3.10+02	0.400	B–	1,2
			37219.8	2686.01 cm^{-1}	81105.03	83791.04	3–5	1.36–02	4.70–01	1.73+02	0.149	B–	2*n*
			37762.9	2647.38 cm^{-1}	81105.03	83752.41	3–3	1.28–02	2.73–01	1.02+02	–0.087	B–	2*n*
			37939.9	2635.03 cm^{-1}	81105.03	83740.06	3–1	1.32–02	9.48–02	3.55+01	–0.546	B–	2*n*
342.	2*p*4*p*– 2*p*4*d*	^{3}S–^{3}F°											
			37834.6	2642.36 cm^{-1}	81105.03	83747.39	3–5	8.80–05	3.15–03	1.18+00	–2.025	C	2
343.		^{3}S–^{3}D°											
			36821.0	2715.10 cm^{-1}	81105.03	83820.13	3–3	1.92–04	3.91–03	1.42+00	–1.931	C	2
344.	2*p*4*p*– 2*p*5*s*	^{3}S–^{1}P°											
			36061.6	2772.28 cm^{-1}	81105.03	83877.31	3–3	6.25–04	1.22–02	4.35+00	–1.437	C	2
345.	2*p*4*p*– 2*p*4*d*	^{3}S–^{1}P°											
			34154.0	2927.12 cm^{-1}	81105.03	84032.15	3–3	7.33–05	1.28–03	4.33–01	–2.415	C	2
346.		^{3}S–^{3}P°	*33275*	*3004.4* cm^{-1}	81105.03	*84109.4*	3–9	1.42–02	7.07–01	2.32+02	0.327	B	1,2
			33345.7	2998.07 cm^{-1}	81105.03	84103.10	3–5	1.52–02	4.23–01	1.39+02	0.104	B	2*n*
			33201.8	3011.06 cm^{-1}	81105.03	84116.09	3–3	1.31–02	2.17–01	7.11+01	–0.187	B	2*n*
			33145.4	3016.19 cm^{-1}	81105.03	84121.22	3–1	1.22–02	6.67–02	2.18+01	–0.699	B	2*n*
347.	2*p*4*p*– 2*p*6*s*	^{3}S–^{3}P°	*19055*	*5246.6* cm^{-1}	81105.03	*86351.6*	3–9	3.59–03	5.87–02	1.10+01	–0.754	C+	1
			18989.7	5264.57 cm^{-1}	81105.03	86369.60	3–5	3.63–03	3.27–02	6.13+00	–1.008	C–	*LS*
			19127.7	5226.60 cm^{-1}	81105.03	86331.63	3–3	3.55–03	1.95–02	3.68+00	–1.233	C–	*LS*
			19163.2	5216.91 cm^{-1}	81105.03	86321.94	3–1	3.53–03	6.48–03	1.23+00	–1.711	C–	*LS*
348.	2*p*4*p*– 2*p*5*d*	^{3}S–^{3}P°	*18487*	*5407.8* cm^{-1}	81105.03	*86512.8*	3–9	2.89–03	4.44–02	8.11+00	–0.875	C+	1
			18507.7	5401.67 cm^{-1}	81105.03	86506.70	3–5	2.88–03	2.47–02	4.51+00	–1.131	C–	*LS*
			18464.1	5414.44 cm^{-1}	81105.03	86519.47	3–3	2.90–03	1.48–02	2.70+00	–1.352	C–	*LS*
			18451.5	5418.13 cm^{-1}	81105.03	86523.16	3–1	2.91–03	4.95–03	9.02–01	–1.829	C–	*LS*
349.	2*p*4*p*– 2*p*7*s*	^{3}S–^{3}P°	*15074*	*6632.3* cm^{-1}	81105.03	*87737.3*	3–9	1.67–03	1.71–02	2.54+00	–1.290	C+	1
			15036.4	6648.70 cm^{-1}	81105.03	87753.73	3–5	1.68–03	9.52–03	1.41+00	–1.544	C–	*LS*
			15116.4	6613.53 cm^{-1}	81105.03	87718.56	3–3	1.66–03	5.68–03	8.48–01	–1.769	C–	*LS*
			15132.8	6606.34 cm^{-1}	81105.03	87711.37	3–1	1.65–03	1.89–03	2.83–01	–2.246	C–	*LS*

C I: Allowed Transitions — Continued

No.	Transition Array	Mult.	λ_{air} (Å)	λ_{vac} (Å) or σ (cm^{-1})*	E_i (cm^{-1})	E_k (cm^{-1})	g_i–g_k	A_{ki} (10^8 s^{-1})	f_{ik}	S (at. u.)	log gf	Acc.	Source
350.	2p4p– 2p6d	^{3}S–^{3}P°	*14848*	*6732.9* cm^{-1}	81105.03	*87837.9*	3–9	9.65–04	9.57–03	1.40+00	–1.542	C+	1
			14860.3	6727.51 cm^{-1}	81105.03	87832.54	3–5	9.62–04	5.31–03	7.79–01	–1.798	C–	*LS*
			14835.2	6738.88 cm^{-1}	81105.03	87843.91	3–3	9.67–04	3.19–03	4.68–01	–2.019	C–	*LS*
			14828.6	6741.87 cm^{-1}	81105.03	87846.9	3–1	9.69–04	1.06–03	1.56–01	–2.496	C–	*LS*
351.	2p4p– 2p8s	^{3}S–^{3}P°	*13396*	*7463.0* cm^{-1}	81105.03	*88568.0*	3–9	9.34–04	7.54–03	9.98–01	–1.646	C	1
			13366.7	7479.23 cm^{-1}	81105.03	88584.26	3–5	9.40–04	4.20–03	5.54–01	–1.900	D	*LS*
			13429.9	7444.03 cm^{-1}	81105.03	88549.06	3–3	9.27–04	2.51–03	3.33–01	–2.124	D	*LS*
			13439.5	7438.73 cm^{-1}	81105.03	88543.76	3–1	9.25–04	8.35–04	1.11–01	–2.601	D	*LS*
352.	2p4p– 2p7d	^{3}S–^{3}P°	*13265*	*7536.4* cm^{-1}	81105.03	*88641*	3–9	4.20–04	3.32–03	4.35–01	–2.001	C	1
			13273.5	7531.77 cm^{-1}	81105.03	88636.8	3–5	4.19–04	1.84–03	2.42–01	–2.257	D	*LS*
			13256.2	7541.57 cm^{-1}	81105.03	88646.6	3–3	4.21–04	1.11–03	1.45–01	–2.478	D	*LS*
			13251.8	7544.07 cm^{-1}	81105.03	88649.1	3–1	4.21–04	3.70–04	4.84–02	–2.955	D	*LS*
353.	2p4p– 2p8d	^{3}S–^{3}P°	*12402*	*8060.8* cm^{-1}	81105.03	*89166*	3–9	2.16–04	1.49–03	1.83–01	–2.349	C	1
			12407.9	8057.17 cm^{-1}	81105.03	89162.2	3–5	2.16–04	8.29–04	1.02–01	–2.604	D	*LS*
			12395.9	8064.97 cm^{-1}	81105.03	89170.0	3–3	2.16–04	4.98–04	6.10–02	–2.826	D	*LS*
			12393.6	8066.47 cm^{-1}	81105.03	89171.5	3–1	2.16–04	1.66–04	2.03–02	–3.303	D–	*LS*
354.	2p4p– 2p5s	^{3}P–^{3}P°	*41002*	*2438.3* cm^{-1}	*81334.2*	*83772.5*	9–9	4.50–03	1.13–01	1.38+02	0.009	C+	1,2
			40854.4	2447.05 cm^{-1}	81343.99	83791.04	5–5	4.64–03	1.16–01	7.81+01	–0.236	C+	2*n*
			41197.8	2426.65 cm^{-1}	81325.76	83752.41	3–3	1.06–03	2.71–02	1.10+01	–1.091	C+	2*n*
			41509.7	2408.42 cm^{-1}	81343.99	83752.41	5–3	1.90–03	2.94–02	2.01+01	–0.833	C+	2*n*
			41408.6	2414.30 cm^{-1}	81325.76	83740.06	3–1	3.59–03	3.07–02	1.26+01	–1.035	C+	2*n*
			40552.3	2465.28 cm^{-1}	81325.76	83791.04	3–5	2.95–04	1.21–02	4.85+00	–1.440	C+	2*n*
			40948.9	2441.40 cm^{-1}	81311.01	83752.41	1–3	1.09–03	8.22–02	1.11+01	–1.085	C+	2*n*
355.	2p4p– 2p4d	^{3}P–^{1}D°											
			46420.6	2153.63 cm^{-1}	81343.99	83497.62	5–5	5.33–06	1.72–04	1.32–01	–3.065	C	2
356.		^{3}P–^{3}F°											
			41357.7	2417.27 cm^{-1}	81343.99	83761.26	5–7	1.02–03	3.66–02	2.49+01	–0.737	C	2
			41283.2	2421.63 cm^{-1}	81325.76	83747.39	3–5	4.22–04	1.80–02	7.33+00	–1.268	C	2
			41596.4	2403.40 cm^{-1}	81343.99	83747.39	5–5	2.00–04	5.18–03	3.55+00	–1.587	C	2
357.		^{3}P–^{3}D°	*39905*	*2505.3* cm^{-1}	*81334.2*	*83839.5*	9–15	2.28–02	9.08–01	1.07+03	0.912	B	1,2
			39911.8	2504.84 cm^{-1}	81343.99	83848.83	5–7	2.25–02	7.54–01	4.95+02	0.576	B	2*n*
			39793.0	2512.32 cm^{-1}	81325.76	83838.08	3–5	2.03–02	8.04–01	3.16+02	0.383	B	2*n*
			39843.7	2509.12 cm^{-1}	81311.01	83820.13	1–3	1.43–02	1.02+00	1.34+02	0.008	B	2*n*
			40083.9	2494.09 cm^{-1}	81343.99	83838.08	5–5	3.05–03	7.34–02	4.84+01	–0.436	B	2*n*
			40079.4	2494.37 cm^{-1}	81325.76	83820.13	3–3	8.12–03	1.96–01	7.74+01	–0.232	B	2*n*
			40374.4	2476.14 cm^{-1}	81343.99	83820.13	5–3	2.86–04	4.20–03	2.79+00	–1.678	B	2*n*
358.	2p4p– 2p5s	^{3}P–^{1}P°											
			39463.1	2533.32 cm^{-1}	81343.99	83877.31	5–3	6.28–05	8.81–04	5.72–01	–2.356	C	2
			39181.2	2551.55 cm^{-1}	81325.76	83877.31	3–3	5.89–04	1.36–02	5.25+00	–1.390	C	2
			38956.0	2566.30 cm^{-1}	81311.01	83877.31	1–3	9.40–04	6.42–02	8.23+00	–1.193	C	2
359.	2p4p– 2p4d	^{3}P–^{1}F°											
			38400.2	2603.44 cm^{-1}	81343.99	83947.43	5–7	2.39–04	7.41–03	4.68+00	–1.431	C	2

C I: Allowed Transitions — Continued

No.	Transition Array	Mult.	λ_{air} (Å)	λ_{vac} (Å) or σ (cm^{-1})*	E_i (cm^{-1})	E_k (cm^{-1})	g_i–g_k	A_{ki} (10^8 s^{-1})	f_{ik}	S (at. u.)	log gf	Acc.	Source
360.		^{3}P–^{1}P°											
			36939.5	2706.39 cm^{-1}	81325.76	84032.15	3–3	4.58–04	9.38–03	3.42+00	–1.551	C	2
			37190.0	2688.16 cm^{-1}	81343.99	84032.15	5–3	1.04–04	1.30–03	7.93–01	–2.188	C	2
			36739.3	2721.14 cm^{-1}	81311.01	84032.15	1–3	7.50–06	4.55–04	5.51–02	–3.342	C	2
361.		^{3}P–^{3}P°	*36024*	*2775.2* cm^{-1}	*81334.2*	*84109.4*	9–9	2.54–02	4.94–01	5.27+02	0.648	C+	1,2
			36233.7	2759.11 cm^{-1}	81343.99	84103.10	5–5	2.10–02	4.13–01	2.46+02	0.315	C+	2*n*
			35828.3	2790.33 cm^{-1}	81325.76	84116.09	3–3	8.97–03	1.73–01	6.11+01	–0.286	C+	2*n*
			36063.9	2772.10 cm^{-1}	81343.99	84116.09	5–3	1.02–02	1.20–01	7.10+01	–0.223	C+	2*n*
			35762.5	2795.46 cm^{-1}	81325.76	84121.22	3–1	2.76–02	1.77–01	6.24+01	–0.276	C+	2*n*
			35995.9	2777.34 cm^{-1}	81325.76	84103.10	3–5	3.36–03	1.09–01	3.87+01	–0.486	C+	2*n*
			35639.9	2805.08 cm^{-1}	81311.01	84116.09	1–3	7.10–03	4.06–01	4.76+01	–0.392	C+	2*n*
362.	2*p*4*p*–2*p*6*s*	^{3}P–^{3}P°	*19925*	*5017.4* cm^{-1}	*81334.2*	*86351.6*	9–9	2.69–03	1.60–02	9.44+00	–0.842	C	1
			19892.7	5025.61 cm^{-1}	81343.99	86369.60	5–5	2.03–03	1.20–02	3.93+00	–1.221	D	*LS*
			19971.1	5005.87 cm^{-1}	81325.76	86331.63	3–3	6.67–04	3.99–03	7.87–01	–1.922	D	*LS*
			20044.1	4987.64 cm^{-1}	81343.99	86331.63	5–3	1.10–03	3.97–03	1.31+00	–1.702	D	*LS*
			20009.8	4996.18 cm^{-1}	81325.76	86321.94	3–1	2.65–03	5.31–03	1.05+00	–1.798	D	*LS*
			19820.8	5043.84 cm^{-1}	81325.76	86369.60	3–5	6.82–04	6.70–03	1.31+00	–1.697	D	*LS*
			19912.4	5020.62 cm^{-1}	81311.01	86331.63	1–3	8.97–04	1.60–02	1.05+00	–1.796	D	*LS*
363.	2*p*4*p*–2*p*5*d*	^{3}P–^{3}D°	*19782*	*5053.7* cm^{-1}	*81334.2*	*86387.9*	9–15	2.74–03	2.68–02	1.57+01	–0.617	C+	1
			19781.7	5053.81 cm^{-1}	81343.99	86397.80	5–7	2.74–03	2.25–02	7.34+00	–0.948	C–	*LS*
			19743.3	5063.62 cm^{-1}	81325.76	86389.38	3–5	2.07–03	2.02–02	3.93+00	–1.218	C–	*LS*
			19790.7	5051.51 cm^{-1}	81311.01	86362.52	1–3	1.52–03	2.68–02	1.75+00	–1.572	C–	*LS*
			19814.7	5045.39 cm^{-1}	81343.99	86389.38	5–5	6.83–04	4.02–03	1.31+00	–1.697	C–	*LS*
			19848.6	5036.76 cm^{-1}	81325.76	86362.52	3–3	1.13–03	6.68–03	1.31+00	–1.698	C–	*LS*
			19920.7	5018.53 cm^{-1}	81343.99	86362.52	5–3	7.46–05	2.66–04	8.74–02	–2.875	C–	*LS*
364.		^{3}P–^{3}P°	*19305*	*5178.5* cm^{-1}	*81334.2*	*86512.8*	9–9	6.23–03	3.48–02	1.99+01	–0.504	C	1
			19364.4	5162.71 cm^{-1}	81343.99	86506.70	5–5	4.63–03	2.60–02	8.30+00	–0.885	D	*LS*
			19248.8	5193.71 cm^{-1}	81325.76	86519.47	3–3	1.57–03	8.73–03	1.66+00	–1.582	D	*LS*
			19316.6	5175.48 cm^{-1}	81343.99	86519.47	5–3	2.59–03	8.70–03	2.77+00	–1.361	D	*LS*
			19235.1	5197.40 cm^{-1}	81325.76	86523.16	3–1	6.30–03	1.17–02	2.21+00	–1.457	D	*LS*
			19296.2	5180.94 cm^{-1}	81325.76	86506.70	3–5	1.56–03	1.45–02	2.77+00	–1.361	D	*LS*
			19194.3	5208.46 cm^{-1}	81311.01	86519.47	1–3	2.11–03	3.50–02	2.21+00	–1.456	D	*LS*
365.	2*p*4*p*–2*p*7*s*	^{3}P–^{3}P°	*15613*	*6403.1* cm^{-1}	*81334.2*	*87737.3*	9–9	1.69–03	6.17–03	2.85+00	–1.255	C	1
			15597.0	6409.74 cm^{-1}	81343.99	87753.73	5–5	1.27–03	4.63–03	1.19+00	–1.635	D	*LS*
			15638.3	6392.80 cm^{-1}	81325.76	87718.56	3–3	4.20–04	1.54–03	2.38–01	–2.335	D	*LS*
			15683.0	6374.57 cm^{-1}	81343.99	87718.56	5–3	6.94–04	1.54–03	3.97–01	–2.115	D	*LS*
			15655.9	6385.61 cm^{-1}	81325.76	87711.37	3–1	1.67–03	2.05–03	3.17–01	–2.211	D	*LS*
			15552.8	6427.97 cm^{-1}	81325.76	87753.73	3–5	4.27–04	2.58–03	3.97–01	–2.111	D	*LS*
			15602.3	6407.55 cm^{-1}	81311.01	87718.56	1–3	5.64–04	6.18–03	3.17–01	–2.209	D	*LS*
366.	2*p*4*p*–2*p*6*d*	^{3}P–^{3}D°	*15540*	*6433.2* cm^{-1}	*81334.2*	*87767.4*	9–15	6.57–04	3.97–03	1.83+00	–1.447	C	1
			15540.2	6433.18 cm^{-1}	81343.99	87777.17	5–7	6.57–04	3.33–03	8.52–01	–1.778	D	*LS*
			15506.1	6447.33 cm^{-1}	81325.76	87773.09	3–5	4.96–04	2.98–03	4.57–01	–2.049	D	*LS*
			15561.7	6424.29 cm^{-1}	81311.01	87735.3	1–3	3.64–04	3.96–03	2.03–01	–2.402	D	*LS*
			15550.0	6429.10 cm^{-1}	81343.99	87773.09	5–5	1.64–04	5.94–04	1.52–01	–2.527	D	*LS*
			15597.5	6409.54 cm^{-1}	81325.76	87735.3	3–3	2.71–04	9.88–04	1.52–01	–2.528	D	*LS*
			15642.0	6391.31 cm^{-1}	81343.99	87735.3	5–3	1.79–05	3.94–05	1.01–02	–3.706	D–	*LS*

C I: Allowed Transitions — Continued

No.	Transition Array	Mult.	λ_{air} (Å)	λ_{vac} (Å) or σ (cm^{-1})*	E_i (cm^{-1})	E_k (cm^{-1})	g_i-g_k	A_{ki} (10^8 s^{-1})	f_{ik}	S (at. u.)	log gf	Acc.	Source
367.		^{3}P–^{3}P°	*15372*	*6503.7* cm^{-1}	*81334.2*	*87837.9*	9–9	2.63–03	9.30–03	4.24+00	–1.077	C	1
			15407.6	6488.55 cm^{-1}	81343.99	87832.54	5–5	1.96–03	6.96–03	1.77+00	–1.458	D	*LS*
			15337.6	6518.15 cm^{-1}	81325.76	87843.91	3–3	6.61–04	2.33–03	3.53–01	–2.155	D	*LS*
			15380.6	6499.92 cm^{-1}	81343.99	87843.91	5–3	1.09–03	2.32–03	5.89–01	–1.935	D	*LS*
			15330.6	6521.14 cm^{-1}	81325.76	87846.9	3–1	2.65–03	3.11–03	4.71–01	–2.030	D	*LS*
			15364.4	6506.78 cm^{-1}	81325.76	87832.54	3–5	6.58–04	3.88–03	5.89–01	–1.934	D	*LS*
			15303.0	6532.90 cm^{-1}	81311.01	87843.91	1–3	8.87–04	9.35–03	4.71–01	–2.029	D	*LS*
368.	2*p*4*p*–2*p*8*s*	^{3}P–^{3}P°	*13820*	*7233.8* cm^{-1}	*81334.2*	*88568.0*	9–9	1.11–03	3.18–03	1.30+00	–1.544	C	1
			13807.9	7240.27 cm^{-1}	81343.99	88584.26	5–5	8.34–04	2.38–03	5.42–01	–1.924	D	*LS*
			13840.3	7223.30 cm^{-1}	81325.76	88549.06	3–3	2.76–04	7.93–04	1.08–01	–2.624	D	*LS*
			13875.3	7205.07 cm^{-1}	81343.99	88549.06	5–3	4.57–04	7.91–04	1.81–01	–2.403	D	*LS*
			13850.5	7218.00 cm^{-1}	81325.76	88543.76	3–1	1.10–03	1.06–03	1.44–01	–2.499	D	*LS*
			13773.2	7258.50 cm^{-1}	81325.76	88584.26	3–5	2.80–04	1.33–03	1.81–01	–2.400	D	*LS*
			13812.1	7238.05 cm^{-1}	81311.01	88549.06	1–3	3.70–04	3.18–03	1.44–01	–2.498	D	*LS*
369.	2*p*4*p*–2*p*7*d*	^{3}P–^{3}D°	*13766*	*7262.2* cm^{-1}	*81334.2*	*88596.5*	9–15	2.16–04	1.02–03	4.18–01	–2.035	C	1
			13765.0	7262.81 cm^{-1}	81343.99	88606.8	5–7	2.16–04	8.61–04	1.95–01	–2.366	D	*LS*
			13734.4	7278.99 cm^{-1}	81325.76	88604.75	3–5	1.63–04	7.70–04	1.04–01	–2.636	D	*LS*
			13793.9	7247.59 cm^{-1}	81311.01	88558.6	1–3	1.19–04	1.02–03	4.64–02	–2.990	D	*LS*
			13768.9	7260.76 cm^{-1}	81343.99	88604.75	5–5	5.41–05	1.54–04	3.48–02	–3.114	D	*LS*
			13822.0	7232.84 cm^{-1}	81325.76	88558.6	3–3	8.91–05	2.55–04	3.48–02	–3.116	D	*LS*
			13857.0	7214.61 cm^{-1}	81343.99	88558.6	5–3	5.89–06	1.02–05	2.32–03	–4.293	D–	*LS*
370.		^{3}P–^{3}P°	*13681*	*7307.2* cm^{-1}	*81334.2*	*88641*	9–9	1.38–03	3.86–03	1.57+00	–1.459	C	1
			13708.4	7292.81 cm^{-1}	81343.99	88636.8	5–5	1.03–03	2.89–03	6.52–01	–1.840	D	*LS*
			13655.9	7320.84 cm^{-1}	81325.76	88646.6	3–3	3.46–04	9.67–04	1.30–01	–2.537	D	*LS*
			13690.0	7302.61 cm^{-1}	81343.99	88646.6	5–3	5.72–04	9.65–04	2.17–01	–2.317	D	*LS*
			13651.2	7323.34 cm^{-1}	81325.76	88649.1	3–1	1.39–03	1.29–03	1.74–01	–2.412	D	*LS*
			13674.2	7311.04 cm^{-1}	81325.76	88636.8	3–5	3.45–04	1.61–03	2.17–01	–2.316	D	*LS*
			13628.4	7335.59 cm^{-1}	81311.01	88646.6	1–3	4.64–04	3.88–03	1.74–01	–2.412	D	*LS*
371.	2*p*4*p*–2*p*8*d*	^{3}P–^{3}D°	*12819*	*7798.7* cm^{-1}	*81334.2*	*89132.9*	9–15	8.72–05	3.58–04	1.36–01	–2.492	C	1
			12817.2	7799.91 cm^{-1}	81343.99	89143.9	5–7	8.72–05	3.01–04	6.35–02	–2.823	D	*LS*
			12789.3	7816.90 cm^{-1}	81325.76	89142.66	3–5	6.59–05	2.69–04	3.40–02	–3.093	D	*LS*
			12849.8	7780.09 cm^{-1}	81311.01	89091.1	1–3	4.81–05	3.57–04	1.51–02	–3.447	D–	*LS*
			12819.2	7798.67 cm^{-1}	81343.99	89142.66	5–5	2.18–05	5.37–05	1.13–02	–3.571	D–	*LS*
			12874.2	7765.34 cm^{-1}	81325.76	89091.1	3–3	3.59–05	8.91–05	1.13–02	–3.573	D–	*LS*
			12904.5	7747.11 cm^{-1}	81343.99	89091.1	5–3	2.37–06	3.56–06	7.56–04	–4.750	D–	*LS*
372.		^{3}P–^{3}P°	*12765*	*7831.6* cm^{-1}	*81334.2*	*89166*	9–9	8.20–04	2.00–03	7.58–01	–1.744	C	1
			12787.2	7818.21 cm^{-1}	81343.99	89162.2	5–5	6.12–04	1.50–03	3.16–01	–2.125	D	*LS*
			12744.7	7844.24 cm^{-1}	81325.76	89170.0	3–3	2.06–04	5.02–04	6.32–02	–2.822	D	*LS*
			12774.4	7826.01 cm^{-1}	81343.99	89170.0	5–3	3.41–04	5.01–04	1.05–01	–2.602	D	*LS*
			12742.3	7845.74 cm^{-1}	81325.76	89171.5	3–1	8.25–04	6.69–04	8.42–02	–2.697	D	*LS*
			12757.4	7836.44 cm^{-1}	81325.76	89162.2	3–5	2.05–04	8.35–04	1.05–01	–2.601	D	*LS*
			12720.8	7858.99 cm^{-1}	81311.01	89170.0	1–3	2.76–04	2.01–03	8.42–02	–2.697	D	*LS*
373.	2*p*4*p*–2*p*4*d*	^{1}D–^{1}D°		1727.83 cm^{-1}	81769.79	83497.62	5–5	3.08–03	1.54–01	1.47+02	–0.112	B+	1,2
374.	2*p*4*p*–2*p*5*s*	^{1}D–^{3}P°											
				1982.62 cm^{-1}	81769.79	83752.41	5–3	3.73–04	8.54–03	7.09+00	–1.370	C	2
			49460.8	2021.25 cm^{-1}	81769.79	83791.04	5–5	4.29–06	1.57–04	1.28–01	–3.104	C	2

C I: Allowed Transitions — Continued

No.	Transition Array	Mult.	λ_{air} (Å)	λ_{vac} (Å) or σ (cm^{-1})*	E_i (cm^{-1})	E_k (cm^{-1})	g_i-g_k	A_{ki} (10^8 s^{-1})	f_{ik}	S (at. u.)	log gf	Acc.	Source
375.	2p4p– 2p4d	$^1D-^3F^o$											
				1991.47 cm^{-1}	81769.79	83761.26	5–7	2.65–04	1.40–02	1.16+01	–1.154	C	2
				1977.60 cm^{-1}	81769.79	83747.39	5–5	3.26–05	1.25–03	1.04+00	–2.204	C	2
376.		$^1D-^3D^o$											
			48759.1	2050.34 cm^{-1}	81769.79	83820.13	5–3	8.16–04	1.75–02	1.40+01	–1.059	C	2
			48335.9	2068.29 cm^{-1}	81769.79	83838.08	5–5	4.79–06	1.68–04	1.34–01	–3.076	C	2
			48086.0	2079.04 cm^{-1}	81769.79	83848.83	5–7	1.35–04	6.55–03	5.19+00	–1.485	C	2
377.	2p4p– 2p5s	$^1D-^1P^o$	47436.2	2107.52 cm^{-1}	81769.79	83877.31	5–3	1.56–02	3.15–01	2.46+02	0.198	B+	1,2
378.	2p4p– 2p4d	$^1D-^1F^o$	45908.8	2177.64 cm^{-1}	81769.79	83947.43	5–7	2.34–02	1.04+00	7.83+02	0.714	B+	1,2
379.		$^1D-^1P^o$	44189.6	2262.36 cm^{-1}	81769.79	84032.15	5–3	2.23–03	3.92–02	2.85+01	–0.708	B+	1,2
380.		$^1D-^3P^o$											
			42845.9	2333.31 cm^{-1}	81769.79	84103.10	5–5	3.36–05	9.26–04	6.53–01	–2.334	C	2
			42608.7	2346.30 cm^{-1}	81769.79	84116.09	5–3	7.44–05	1.22–03	8.53–01	–2.216	C	2
381.	2p4p– 2p5d	$^1D-^1D^o$	22641.8	4415.41 cm^{-1}	81769.79	86185.20	5–5	2.00–04	1.54–03	5.74–01	–2.113	C	1
382.	2p4p– 2p6s	$^1D-^1P^o$	21514.5	4646.76 cm^{-1}	81769.79	86416.55	5–3	4.55–03	1.89–02	6.71+00	–1.024	C+	1
383.	2p4p– 2p5d	$^1D-^1F^o$	21364.4	4679.40 cm^{-1}	81769.79	86449.19	5–7	1.59–03	1.52–02	5.35+00	–1.119	B–	1
384.		$^1D-^1P^o$	21173.4	4721.62 cm^{-1}	81769.79	86491.41	5–3	1.79–03	7.22–03	2.52+00	–1.442	C+	1
385.	2p4p– 2p6d	$^1D-^1D^o$	17048.7	5863.96 cm^{-1}	81769.79	87633.75	5–5	4.43–04	1.93–03	5.42–01	–2.015	C	1
386.	2p4p– 2p7s	$^1D-^1P^o$	16607.2	6019.84 cm^{-1}	81769.79	87789.63	5–3	2.19–03	5.45–03	1.49+00	–1.565	C	1
387.	2p4p– 2p6d	$^1D-^1F^o$	16559.6	6037.14 cm^{-1}	81769.79	87806.93	5–7	2.03–04	1.17–03	3.18–01	–2.234	D	1
388.		$^1D-^1P^o$	16496.1	6060.38 cm^{-1}	81769.79	87830.17	5–3	1.22–03	2.98–03	8.11–01	–1.826	C	1
389.	2p4p– 2p7d	$^1D-^1D^o$	14857.4	6728.83 cm^{-1}	81769.79	88498.62	5–5	4.16–04	1.38–03	3.37–01	–2.162	C	1
390.	2p4p– 2p8s	$^1D-^1P^o$	14604.7	6845.22 cm^{-1}	81769.79	88615.01	5–3	1.26–03	2.41–03	5.80–01	–1.919	C	1
391.	2p4p– 2p7d	$^1D-^1P^o$	14553.7	6869.23 cm^{-1}	81769.79	88639.02	5–3	8.34–04	1.59–03	3.81–01	–2.100	C	1
392.	2p4p– 2p8d	$^1D-^1D^o$	13724.3	7284.37 cm^{-1}	81769.79	89054.16	5–5	3.38–04	9.53–04	2.15–01	–2.322	C	1

C I: Allowed Transitions — Continued

No.	Transition Array	Mult.	λ_{air} (Å)	λ_{vac} (Å) or σ (cm^{-1})*	E_i (cm^{-1})	E_k (cm^{-1})	g_i–g_k	A_{ki} (10^8 s^{-1})	f_{ik}	S (at. u.)	log gf	Acc.	Source
393.		^{1}D–^{1}P°	13519.0	7394.95 cm^{-1}	81769.79	89164.74	5–3	5.87–04	9.65–04	2.15–01	–2.317	C	1
394.	2p4p–2p5s	^{1}S–^{3}P°		1500.70 cm^{-1}	82251.71	83752.41	1–3	6.92–05	1.38–02	3.03+00	–1.860	C	2
395.	2p4p–2p4d	^{1}S–^{3}D°		1568.42 cm^{-1}	82251.71	83820.13	1–3	5.00–04	9.15–02	1.92+01	–1.039	C	2
396.	2p4p–2p5s	^{1}S–^{1}P°		1625.60 cm^{-1}	82251.71	83877.31	1–3	6.24–03	1.06+00	2.15+02	0.026	B–	1,2
397.	2p4p–2p4d	^{1}S–^{1}P°		1780.44 cm^{-1}	82251.71	84032.15	1–3	2.68–03	3.79–01	7.01+01	–0.421	B	1,2
398.		^{1}S–^{3}P°		1864.38 cm^{-1}	82251.71	84116.09	1–3	1.17–04	1.51–02	2.67+00	–1.821	C	2
399.	2p4p–2p6s	^{1}S–^{1}P°	24004.0	4164.84 cm^{-1}	82251.71	86416.55	1–3	2.54–04	6.58–03	5.20–01	–2.182	C	1
400.	2p4p–2p5d	^{1}S–^{1}P°	23580.1	4239.70 cm^{-1}	82251.71	86491.41	1–3	7.10–05	1.78–03	1.38–01	–2.751	D	1
401.	2p4p–2p6d	^{1}S–^{1}P°	17921.2	5578.46 cm^{-1}	82251.71	87830.17	1–3	2.49–04	3.59–03	2.12–01	–2.445	D	1
402.	2p4p–2p7d	^{1}S–^{1}P°	15651.8	6387.31 cm^{-1}	82251.71	88639.02	1–3	2.54–04	2.79–03	1.44–01	–2.554	D	1
403.	2p4d–2p5p	^{1}D°–^{1}P		1353.91 cm^{-1}	83497.62	84851.53	5–3	4.84–03	2.37–01	2.88+02	0.074	C+	1
404.		^{1}D°–^{1}D		1902.19 cm^{-1}	83497.62	85399.81	5–5	1.38–03	5.72–02	4.95+01	–0.543	C+	1
405.	2p4d–2p6p	^{1}D°–^{1}P	29272.5	3415.24 cm^{-1}	83497.62	86912.86	5–3	1.62–03	1.25–02	6.01+00	–1.205	C+	1
406.		^{1}D°–^{1}D	26869.8	3720.64 cm^{-1}	83497.62	87218.26	5–5	5.61–04	6.07–03	2.69+00	–1.518	C+	1
407.	2p4d–2p7p	^{1}D°–^{1}P	21906.3	4563.66 cm^{-1}	83497.62	88061.28	5–3	8.87–04	3.83–03	1.38+00	–1.718	C+	1
408.		^{1}D°–^{1}D	20990.5	4762.75 cm^{-1}	83497.62	88260.37	5–5	3.07–04	2.03–03	7.01–01	–1.994	D	1
409.	2p4d–2p8p	^{1}D°–^{1}P	18972.5	5269.36 cm^{-1}	83497.62	88766.98	5–3	5.52–04	1.79–03	5.58–01	–2.049	D	1
410.		^{1}D°–^{1}D	18459.0	5415.94 cm^{-1}	83497.62	88913.56	5–5	1.90–04	9.70–04	2.95–01	–2.314	D	1

C I: Allowed Transitions — Continued

No.	Transition Array	Mult.	λ_{air} (Å)	λ_{vac} (Å) or σ (cm^{-1})*	E_i (cm^{-1})	E_k (cm^{-1})	g_i–g_k	A_{ki} (10^8 s^{-1})	f_{ik}	S (at. u.)	log gf	Acc.	Source
411.	$2p4d$–$2p5p$	$^3F°$–^3D		*1189.6* cm^{-1}	*83773.9*	*84963.5*	21–15	4.63–03	3.50–01	2.04+03	0.867	B–	1
				1186.40 cm^{-1}	83798.57	84984.97	9–7	4.22–03	3.50–01	8.73+02	0.498	C	LS
				1189.10 cm^{-1}	83761.26	84950.36	7–5	4.11–03	3.11–01	6.03+02	0.338	C	LS
				1187.95 cm^{-1}	83747.39	84935.34	5–3	4.61–03	2.94–01	4.07+02	0.167	C	LS
				1223.71 cm^{-1}	83761.26	84984.97	7–7	4.01–04	4.02–02	7.56+01	–0.551	C	LS
				1202.97 cm^{-1}	83747.39	84950.36	5–5	5.34–04	5.53–02	7.56+01	–0.558	C	LS
				1237.58 cm^{-1}	83747.39	84984.97	5–7	1.17–05	1.60–03	2.13+00	–2.096	C	LS
412.	$2p4d$–$2p6p$	$^3F°$–^3D	*31175*	*3206.8* cm^{-1}	*83773.9*	*86980.8*	21–15	1.27–03	1.32–02	2.85+01	–0.556	B–	1
			31205.5	3203.69 cm^{-1}	83798.57	87002.26	9–7	1.16–03	1.32–02	1.22+01	–0.925	C	LS
			31200.6	3204.19 cm^{-1}	83761.26	86965.45	7–5	1.13–03	1.17–02	8.44+00	–1.085	C	LS
			31156.1	3208.77 cm^{-1}	83747.39	86956.16	5–3	1.27–03	1.11–02	5.70+00	–1.255	C	LS
			30846.3	3241.00 cm^{-1}	83761.26	87002.26	7–7	1.04–04	1.49–03	1.06+00	–1.982	C	LS
			31066.1	3218.06 cm^{-1}	83747.39	86965.45	5–5	1.43–04	2.07–03	1.06+00	–1.985	C	LS
			30714.8	3254.87 cm^{-1}	83747.39	87002.26	5–7	2.98–06	5.90–05	2.99–02	–3.530	C–	LS
413.	$2p4d$–$2p7p$	$^3F°$–^3D	*23036*	*4339.8* cm^{-1}	*83773.9*	*88113.8*	21–15	6.94–04	3.94–03	6.28+00	–1.082	C	1
			23049.8	4337.24 cm^{-1}	83798.57	88135.81	9–7	6.36–04	3.94–03	2.69+00	–1.450	C–	LS
			23063.0	4334.77 cm^{-1}	83761.26	88096.03	7–5	6.14–04	3.50–03	1.86+00	–1.611	C–	LS
			23010.7	4344.62 cm^{-1}	83747.39	88092.01	5–3	6.96–04	3.31–03	1.26+00	–1.781	C–	LS
			22853.3	4374.55 cm^{-1}	83761.26	88135.81	7–7	5.65–05	4.43–04	2.33–01	–2.509	C–	LS
			22989.4	4348.64 cm^{-1}	83747.39	88096.03	5–5	7.78–05	6.16–04	2.33–01	–2.511	C–	LS
			22781.0	4388.42 cm^{-1}	83747.39	88135.81	5–7	1.61–06	1.75–05	6.58–03	–4.057	C–	LS
414.	$2p4d$–$2p8p$	$^3F°$–^3D	*19839*	*5039.2* cm^{-1}	*83773.9*	*88813.1*	21–15	4.34–04	1.83–03	2.51+00	–1.416	C	1
			19845.5	5037.55 cm^{-1}	83798.57	88836.12	9–7	3.98–04	1.83–03	1.07+00	–1.784	D	LS
			19863.3	5033.04 cm^{-1}	83761.26	88794.3	7–5	3.84–04	1.62–03	7.42–01	–1.945	D	LS
			19822.2	5043.46 cm^{-1}	83747.39	88790.85	5–3	4.35–04	1.54–03	5.01–01	–2.115	D	LS
			19699.6	5074.86 cm^{-1}	83761.26	88836.12	7–7	3.52–05	2.05–04	9.31–02	–2.843	D	LS
			19808.7	5046.91 cm^{-1}	83747.39	88794.3	5–5	4.85–05	2.86–04	9.31–02	–2.845	D	LS
			19645.9	5088.73 cm^{-1}	83747.39	88836.12	5–7	1.00–06	8.12–06	2.63–03	–4.391	D	LS
415.	$2p4d$–$2p5p$	$^3D°$–^3D		*1124.0* cm^{-1}	*83839.5*	*84963.5*	15–15	7.74–04	9.18–02	4.03+02	0.139	B–	1
				1136.14 cm^{-1}	83848.83	84984.97	7–7	7.11–04	8.25–02	1.67+02	–0.239	C	LS
				1112.28 cm^{-1}	83838.08	84950.36	5–5	5.22–04	6.32–02	9.36+01	–0.500	C	LS
				1115.21 cm^{-1}	83820.13	84935.34	3–3	5.67–04	6.83–02	6.05+01	–0.688	C	LS
				1101.53 cm^{-1}	83848.83	84950.36	7–5	1.14–04	1.00–02	2.10+01	–1.154	C	LS
				1097.26 cm^{-1}	83838.08	84935.34	5–3	1.80–04	1.34–02	2.02+01	–1.172	C	LS
				1146.89 cm^{-1}	83838.08	84984.97	5–7	9.17–05	1.46–02	2.10+01	–1.136	C	LS
				1130.23 cm^{-1}	83820.13	84950.36	3–5	1.18–04	2.31–02	2.02+01	–1.160	C	LS
416.		$^3D°$–^3P		*1355.5* cm^{-1}	*83839.5*	*85195.0*	15–9	5.64–03	2.76–01	1.00+03	0.617	B–	1
				1354.81 cm^{-1}	83848.83	85203.64	7–5	4.73–03	2.76–01	4.69+02	0.285	C	LS
				1350.87 cm^{-1}	83838.08	85188.95	5–3	4.18–03	2.06–01	2.51+02	0.013	C	LS
				1349.48 cm^{-1}	83820.13	85169.61	3–1	5.56–03	1.53–01	1.12+02	–0.340	C	LS
				1365.56 cm^{-1}	83838.08	85203.64	5–5	8.64–04	6.95–02	8.37+01	–0.459	C	LS
				1368.82 cm^{-1}	83820.13	85188.95	3–3	1.45–03	1.16–01	8.37+01	–0.458	C	LS
				1383.51 cm^{-1}	83820.13	85203.64	3–5	5.99–05	7.82–03	5.58+00	–1.630	C	LS
417.	$2p4d$–$2p6p$	$^3D°$–^3D	*31826*	*3141.3* cm^{-1}	*83839.5*	*86980.8*	15–15	1.80–04	2.74–03	4.30+00	–1.387	C+	1
			31702.9	3153.43 cm^{-1}	83848.83	87002.26	7–7	1.62–04	2.44–03	1.78+00	–1.768	C–	LS
			31967.0	3127.37 cm^{-1}	83838.08	86965.45	5–5	1.24–04	1.90–03	9.97–01	–2.023	C–	LS
			31878.8	3136.03 cm^{-1}	83820.13	86956.16	3–3	1.34–04	2.05–03	6.45–01	–2.212	C–	LS
			32077.3	3116.62 cm^{-1}	83848.83	86965.45	7–5	2.74–05	3.02–04	2.24–01	–2.674	C–	LS
			32062.3	3118.08 cm^{-1}	83838.08	86956.16	5–3	4.40–05	4.07–04	2.15–01	–2.691	C–	LS
			31595.1	3164.18 cm^{-1}	83838.08	87002.26	5–7	2.05–05	4.30–04	2.24–01	–2.668	C–	LS
			31784.6	3145.32 cm^{-1}	83820.13	86965.45	3–5	2.71–05	6.85–04	2.15–01	–2.687	C–	LS

C I: Allowed Transitions — Continued

No.	Transition Array	Mult.	λ_{air} (Å)	λ_{vac} (Å) or σ (cm^{-1})*	E_i (cm^{-1})	E_k (cm^{-1})	g_i-g_k	A_{ki} (10^8 s^{-1})	f_{ik}	S (at. u.)	log *gf*	Acc.	Source
418.		$^3D^o-^3P$	*30607*	*3266.4* cm^{-1}	*83839.5*	*87105.9*	15–9	1.64–03	1.38–02	2.09+01	−0.683	C+	1
			30625.3	3264.38 cm^{-1}	83848.83	87113.21	7–5	1.38–03	1.38–02	9.75+00	−1.015	C−	*LS*
			30619.1	3265.04 cm^{-1}	83838.08	87103.12	5–3	1.23–03	1.04–02	5.22+00	−1.286	C−	*LS*
			30692.6	3257.23 cm^{-1}	83820.13	87077.36	3–1	1.63–03	7.66–03	2.32+00	−1.639	C−	*LS*
			30524.8	3275.13 cm^{-1}	83838.08	87113.21	5–5	2.48–04	3.46–03	1.74+00	−1.761	C−	*LS*
			30451.7	3282.99 cm^{-1}	83820.13	87103.12	3–3	4.16–04	5.79–03	1.74+00	−1.760	C−	*LS*
			30358.4	3293.08 cm^{-1}	83820.13	87113.21	3–5	1.68–05	3.87–04	1.16–01	−2.935	C−	*LS*
419.	2*p*4*d*– 2*p*7*p*	$^3D^o-^3D$	*23389*	*4274.3* cm^{-1}	*83839.5*	*88113.8*	15–15	9.97–05	8.18–04	9.45–01	−1.911	C	1
			23320.1	4286.98 cm^{-1}	83848.83	88135.81	7–7	8.94–05	7.29–04	3.92–01	−2.292	D	*LS*
			23479.1	4257.95 cm^{-1}	83838.08	88096.03	5–5	6.86–05	5.67–04	2.19–01	−2.547	D	*LS*
			23402.5	4271.88 cm^{-1}	83820.13	88092.01	3–3	7.47–05	6.13–04	1.42–01	−2.735	D	*LS*
			23538.5	4247.20 cm^{-1}	83848.83	88096.03	7–5	1.53–05	9.06–05	4.91–02	−3.198	D	*LS*
			23501.3	4253.93 cm^{-1}	83838.08	88092.01	5–3	2.46–05	1.22–04	4.72–02	−3.214	D	*LS*
			23261.7	4297.73 cm^{-1}	83838.08	88135.81	5–7	1.13–05	1.28–04	4.91–02	−3.193	D	*LS*
			23380.5	4275.90 cm^{-1}	83820.13	88096.03	3–5	1.50–05	2.05–04	4.72–02	−3.212	D	*LS*
420.		$^3D^o-^3P$	*22969*	*4352.5* cm^{-1}	*83839.5*	*88192.0*	15–9	8.86–04	4.20–03	4.77+00	−1.200	C+	1
			22985.5	4349.39 cm^{-1}	83848.83	88198.22	7–5	7.42–04	4.20–03	2.22+00	−1.532	C−	*LS*
			22960.0	4354.22 cm^{-1}	83838.08	88192.30	5–3	6.65–04	3.15–03	1.19+00	−1.802	C−	*LS*
			23036.6	4339.74 cm^{-1}	83820.13	88159.87	3–1	8.78–04	2.33–03	5.30–01	−2.156	C−	*LS*
			22928.8	4360.14 cm^{-1}	83838.08	88198.22	5–5	1.34–04	1.05–03	3.97–01	−2.279	C−	*LS*
			22865.7	4372.17 cm^{-1}	83820.13	88192.30	3–3	2.24–04	1.76–03	3.97–01	−2.278	C−	*LS*
			22834.8	4378.09 cm^{-1}	83820.13	88198.22	3–5	9.01–06	1.17–04	2.65–02	−3.453	D+	*LS*
421.	2*p*4*d*– 2*p*8*p*	$^3D^o-^3D$	*20101*	*4973.6* cm^{-1}	*83839.5*	*88813.1*	15–15	6.34–05	3.84–04	3.81–01	−2.240	D	1
			20045.5	4987.29 cm^{-1}	83848.83	88836.12	7–7	5.68–05	3.42–04	1.58–01	−2.621	D−	*LS*
			20171.2	4956.22 cm^{-1}	83838.08	88794.3	5–5	4.36–05	2.66–04	8.84–02	−2.876	D−	*LS*
			20112.3	4970.72 cm^{-1}	83820.13	88790.85	3–3	4.74–05	2.88–04	5.71–02	−3.064	D−	*LS*
			20215.0	4945.47 cm^{-1}	83848.83	88794.3	7–5	9.72–06	4.25–05	1.98–02	−3.526	D−	*LS*
			20185.2	4952.77 cm^{-1}	83838.08	88790.85	5–3	1.56–05	5.73–05	1.90–02	−3.543	D−	*LS*
			20002.4	4998.04 cm^{-1}	83838.08	88836.12	5–7	7.16–06	6.02–05	1.98–02	−3.522	D−	*LS*
			20098.4	4974.17 cm^{-1}	83820.13	88794.3	3–5	9.51–06	9.59–05	1.90–02	−3.541	D−	*LS*
422.	$2p(^2P^o_{1/2})4f$– 2*p*4*d*	$[^5/_2]-^1F^o$											
				27.78 cm^{-1}	83919.65	83947.43	7–7	1.68–09	3.25–04	2.70+01	−2.642	C	2
423.		$[^7/_2]-^1F^o$											
				21.23 cm^{-1}	83926.20	83947.43	7–7	6.92–10	2.30–04	2.50+01	−2.793	C	2
				21.06 cm^{-1}	83926.37	83947.43	9–7	8.19–10	2.15–04	3.03+01	−2.713	C	2
424.	2*p*4*d*– 2*p*5*p*	$^1F^o-^1D$		1452.38 cm^{-1}	83947.43	85399.81	7–5	6.96–03	3.53–01	5.61+02	0.393	B−	1
425.	2*p*4*d*– 2*p*6*p*	$^1F^o-^1D$	30564.9	3270.83 cm^{-1}	83947.43	87218.26	7–5	2.30–03	2.30–02	1.62+01	−0.792	B−	1
426.	2*p*4*d*– 2*p*7*p*	$^1F^o-^1D$	23179.7	4312.94 cm^{-1}	83947.43	88260.37	7–5	1.25–03	7.19–03	3.84+00	−1.298	C+	1
427.	2*p*4*d*– 2*p*8*p*	$^1F^o-^1D$	20130.9	4966.13 cm^{-1}	83947.43	88913.56	7–5	7.75–04	3.36–03	1.56+00	−1.628	C+	1
428.	2*p*4*d*– 2*p*5*p*	$^1P^o-^1P$		819.38 cm^{-1}	84032.15	84851.53	3–3	2.14–03	4.79–01	5.77+02	0.157	B−	1

C I: Allowed Transitions — Continued

No.	Transition Array	Mult.	λ_{air} (Å)	λ_{vac} (Å) or σ (cm^{-1})*	E_i (cm^{-1})	E_k (cm^{-1})	g_i–g_k	A_{ki} (10^8 s^{-1})	f_{ik}	S (at. u.)	log gf	Acc.	Source
429.		$^1P^o-^1D$		1367.66 cm^{-1}	84032.15	85399.81	3–5	1.68–03	2.25–01	1.62+02	–0.171	B–	1
430.		$^1P^o-^1S$		1593.03 cm^{-1}	84032.15	85625.18	3–1	6.27–04	1.23–02	7.65+00	–1.431	C+	1
431.	$2p4d-2p6p$	$^1P^o-^1P$	34704.2	2880.71 cm^{-1}	84032.15	86912.86	3–3	2.17–05	3.92–04	1.35–01	–2.929	D	1
432.		$^1P^o-^1D$	31377.7	3186.11 cm^{-1}	84032.15	87218.26	3–5	1.45–04	3.58–03	1.11+00	–1.969	C	1
433.		$^1P^o-^1S$	30213.4	3308.89 cm^{-1}	84032.15	87341.04	3–1	4.72–04	2.15–03	6.42–01	–2.190	D	1
434.	$2p4d-2p7p$	$^1P^o-^1D$	23644.2	4228.22 cm^{-1}	84032.15	88260.37	3–5	4.11–05	5.74–04	1.34–01	–2.764	D	1
435.		$^1P^o-^1S$	23239.6	4301.83 cm^{-1}	84032.15	88333.98	3–1	3.13–04	8.46–04	1.94–01	–2.595	D	1
436.	$2p4d-2p5p$	$^3P^o-^3D$		*854.1* cm^{-1}	*84109.4*	*84963.5*	9–15	1.53–04	5.23–02	1.81+02	–0.327	B–	1
				881.87 cm^{-1}	84103.10	84984.97	5–7	1.68–04	4.54–02	8.47+01	–0.644	C	*LS*
				834.27 cm^{-1}	84116.09	84950.36	3–5	1.07–04	3.83–02	4.54+01	–0.939	C	*LS*
				814.12 cm^{-1}	84121.22	84935.34	1–3	7.35–05	4.99–02	2.02+01	–1.302	C	*LS*
				847.26 cm^{-1}	84103.10	84950.36	5–5	3.73–05	7.79–03	1.51+01	–1.410	C	*LS*
				819.25 cm^{-1}	84116.09	84935.34	3–3	5.62–05	1.25–02	1.51+01	–1.424	C	*LS*
				832.24 cm^{-1}	84103.10	84935.34	5–3	3.93–06	5.10–04	1.01+00	–2.594	C	*LS*
437.		$^3P^o-^3P$		*1085.5* cm^{-1}	*84109.4*	*85195.0*	9–9	3.28–03	4.17–01	1.14+03	0.575	B–	1
				1100.54 cm^{-1}	84103.10	85203.64	5–5	2.57–03	3.17–01	4.75+02	0.201	C	*LS*
				1072.86 cm^{-1}	84116.09	85188.95	3–3	7.92–04	1.03–01	9.49+01	–0.509	C	*LS*
				1085.85 cm^{-1}	84103.10	85188.95	5–3	1.37–03	1.04–01	1.58+02	–0.282	C	*LS*
				1053.52 cm^{-1}	84116.09	85169.61	3–1	3.00–03	1.35–01	1.27+02	–0.392	C	*LS*
				1087.55 cm^{-1}	84116.09	85203.64	3–5	8.25–04	1.74–01	1.58+02	–0.282	C	*LS*
				1067.73 cm^{-1}	84121.22	85188.95	1–3	1.04–03	4.11–01	1.27+02	–0.387	C	*LS*
438.	$2p4d-2p6p$	$^3P^o-^3D$	*34818*	*2871.3* cm^{-1}	*84109.4*	*86980.8*	9–15	4.51–05	1.37–03	1.41+00	–1.910	C	1
			34483.3	2899.16 cm^{-1}	84103.10	87002.26	5–7	4.64–05	1.16–03	6.57–01	–2.237	D	*LS*
			35086.0	2849.36 cm^{-1}	84116.09	86965.45	3–5	3.30–05	1.02–03	3.52–01	–2.516	D	*LS*
			35264.5	2834.94 cm^{-1}	84121.22	86956.16	1–3	2.41–05	1.35–03	1.57–01	–2.870	D	*LS*
			34926.8	2862.35 cm^{-1}	84103.10	86965.45	5–5	1.12–05	2.04–04	1.17–01	–2.991	D	*LS*
			35200.8	2840.07 cm^{-1}	84116.09	86956.16	3–3	1.82–05	3.38–04	1.17–01	–2.994	D	*LS*
			35040.5	2853.06 cm^{-1}	84103.10	86956.16	5–3	1.23–06	1.36–05	7.83–03	–4.169	D–	*LS*
439.		$^3P^o-^3P$	*33364*	*2996.4* cm^{-1}	*84109.4*	*87105.9*	9–9	3.12–04	5.21–03	5.15+00	–1.329	C+	1
			33212.3	3010.11 cm^{-1}	84103.10	87113.21	5–5	2.37–04	3.93–03	2.15+00	–1.707	C–	*LS*
			33468.9	2987.03 cm^{-1}	84116.09	87103.12	3–3	7.73–05	1.30–03	4.29–01	–2.409	C–	*LS*
			33324.0	3000.02 cm^{-1}	84103.10	87103.12	5–3	1.31–04	1.30–03	7.16–01	–2.186	C–	*LS*
			33760.1	2961.27 cm^{-1}	84116.09	87077.36	3–1	3.01–04	1.72–03	5.73–01	–2.288	C–	*LS*
			33356.3	2997.12 cm^{-1}	84116.09	87113.21	3–5	7.81–05	2.17–03	7.16–01	–2.186	C–	*LS*
			33526.5	2981.90 cm^{-1}	84121.22	87103.12	1–3	1.03–04	5.19–03	5.73–01	–2.285	C–	*LS*

C I: Allowed Transitions — Continued

No.	Transition Array	Mult.	λ_{air} (Å)	λ_{vac} (Å) or σ (cm^{-1})*	E_i (cm^{-1})	E_k (cm^{-1})	g_i–g_k	A_{ki} ($10^8 s^{-1}$)	f_{ik}	S (at. u.)	log gf	Acc.	Source
440.	$2p4d$–$2p7p$	$^3P^o$–^3D	*24966*	*4004.3* cm^{-1}	*84109.4*	*88113.8*	9–15	4.84–05	7.54–04	5.58–01	−2.168	D	1
			24790.5	4032.71 cm^{-1}	84103.10	88135.81	5–7	4.95–05	6.38–04	2.60–01	−2.496	D–	*LS*
			25119.2	3979.94 cm^{-1}	84116.09	88096.03	3–5	3.57–05	5.62–04	1.39–01	−2.773	D–	*LS*
			25177.0	3970.79 cm^{-1}	84121.22	88092.01	1–3	2.62–05	7.48–04	6.20–02	−3.126	D–	*LS*
			25037.4	3992.93 cm^{-1}	84103.10	88096.03	5–5	1.20–05	1.13–04	4.65–02	−3.249	D–	*LS*
			25144.6	3975.92 cm^{-1}	84116.09	88092.01	3–3	1.97–05	1.87–04	4.65–02	−3.251	D–	*LS*
			25062.7	3988.91 cm^{-1}	84103.10	88092.01	5–3	1.33–06	7.51–06	3.10–03	−4.425	D–	*LS*
441.		$^3P^o$–^3P	*24488*	*4082.5* cm^{-1}	*84109.4*	*88192.0*	9–9	1.19–04	1.07–03	7.79–01	−2.015	D	1
			24412.6	4095.12 cm^{-1}	84103.10	88198.22	5–5	9.04–05	8.08–04	3.25–01	−2.394	D–	*LS*
			24525.9	4076.21 cm^{-1}	84116.09	88192.30	3–3	2.97–05	2.68–04	6.49–02	−3.095	D–	*LS*
			24448.0	4089.20 cm^{-1}	84103.10	88192.30	5–3	5.00–05	2.69–04	1.08–01	−2.871	D–	*LS*
			24722.6	4043.78 cm^{-1}	84116.09	88159.87	3–1	1.16–04	3.55–04	8.66–02	−2.973	D–	*LS*
			24490.3	4082.13 cm^{-1}	84116.09	88198.22	3–5	2.99–05	4.47–04	1.08–01	−2.872	D–	*LS*
			24556.8	4071.08 cm^{-1}	84121.22	88192.30	1–3	3.95–05	1.07–03	8.66–02	−2.970	D–	*LS*
442.	$2p4d$–$2p8p$	$^3P^o$–^3D	*21254*	*4703.7* cm^{-1}	*84109.4*	*88813.1*	9–15	3.78–05	4.27–04	2.69–01	−2.416	D	1
			21122.4	4733.02 cm^{-1}	84103.10	88836.12	5–7	3.85–05	3.61–04	1.25–01	−2.744	D–	*LS*
			21369.9	4678.21 cm^{-1}	84116.09	88794.3	3–5	2.79–05	3.18–04	6.72–02	−3.020	D–	*LS*
			21409.1	4669.63 cm^{-1}	84121.22	88790.85	1–3	2.05–05	4.23–04	2.98–02	−3.373	D–	*LS*
			21310.7	4691.20 cm^{-1}	84103.10	88794.3	5–5	9.37–06	6.38–05	2.24–02	−3.496	D–	*LS*
			21385.6	4674.76 cm^{-1}	84116.09	88790.85	3–3	1.55–05	1.06–04	2.24–02	−3.498	D–	*LS*
			21326.4	4687.75 cm^{-1}	84103.10	88790.85	5–3	1.04–06	4.25–06	1.49–03	−4.673	D–	*LS*

*Wavelengths (Å) are always given unless cm^{-1} is indicated.
†The entries in the A_{ki}, f_{ik}, and S columns are listed in scientific notation; e.g., 5.89–03 stands for 5.89×10^{-3}.

C I

Forbidden Transitions

The tabulated data are selected from two recent advanced calculations, which include extensive treatments of the electron correlation problem: the configuration-interaction calculations by Hibbert *et al.*[1] with the C IV 3 code and the multiconfiguration Hartree-Fock calculations by Froese Fischer and Saha.[2] The wavefunctions developed for these calculations include many configurations that are expected to mix with the levels under consideration, and the results of the two papers typically agree within ±10%. The line strengths of the M1 fine structure transitions within the ^3P term of the ground state configuration are the same as derived from general expansion formulas first developed by Shortley *et al.*[3] for the case of near *LS*-coupling conditions.

References

[1]A. Hibbert, E. Biemont, M. Godefroid, and N. Vaeck, Astron. Astrophys., Suppl. Ser. **99**, 179 (1993).
[2]C. Froese Fischer and H. P. Saha, Phys. Scr. **32**, 181 (1985).
[3]G. H. Shortley, L. H. Aller, J. G. Baker, and D. H. Menzel, Astrophys. J. **93**, 178 (1941).

C I: Forbidden Transitions

No.	Transition Array	Mult.	λ_{air} (Å)	λ_{vac} (Å) or σ (cm^{-1})*	E_i (cm^{-1})		E_k (cm^{-1})	g_i–g_k	Type	A_{ki} (s^{-1})	S (at. u.)	Acc.	Source
1.	$2p^2$–$2p^2$	^{3}P–^{3}P											
				16.40 cm^{-1}	0.00	–	16.40	1–3	M1	7.88–08	2.00+00	B	1,2
				43.40 cm^{-1}	0.00	–	43.40	1–5	E2	1.81–14	5.24+00	B	1,2
				27.00 cm^{-1}	16.40	–	43.40	3–5	M1	2.65–07	2.50+00	B	1,2
				27.00 cm^{-1}	16.40	–	43.40	3–5	E2	3.80–15	1.18+01	B+	1,2
2.		^{3}P–^{1}D											
			9808.32	9811.01	0.00	–	10192.63	1–5	E2	8.62–08	3.50–05	B	1,2
			9824.13	9826.82	16.40	–	10192.63	3–5	M1	6.05–05	1.06–05	B+	1,2
			9824.13	9826.82	16.40	–	10192.63	3–5	E2	9.46–08	3.87–05	C+	1,2
			9850.26	9852.96	43.40	–	10192.63	5–5	M1	1.80–04	3.19–05	B	1,2
			9850.26	9852.96	43.40	–	10192.63	5–5	E2	1.03–06	4.28–04	B	1,2
3.		^{3}P–^{1}S											
			4621.57	4622.86	16.40	–	21648.01	3–1	M1	2.10–03	7.68–06	B	1,2
			4627.35	4628.64	43.40	–	21648.01	5–1	E2	1.79–05	3.40–05	C+	1,2
4.		^{1}D–^{1}S											
			8727.13	8729.52	10192.63	–	21648.01	5–1	E2	6.34–01	2.87+01	B	1,2

*Wavelengths (Å) are always given unless cm^{-1} is indicated.

C II

Boron Isoelectronic Sequence

Ground State: $1s^22s^22p\ ^2P^\circ_{1/2}$

Ionization Energy: 24.383 eV = 196664.7 cm^{-1}

Allowed Transitions

List of tabulated lines

Wavelength (Å)	No.	Wavelength (Å)	No.	Wavelength (Å)	No.	Wavelength (Å)	No.
in vacuum		530.359	14	547.291	25	564.582	23
		530.454	14	549.320	12	564.608	23
438.774	16	531.679	26	549.379	12	564.635	23
438.825	16	531.721	26	549.511	12	564.652	23
438.896	16	531.742	26	549.570	12	564.663	23
438.947	16	531.784	26	551.681	11	564.698	23
466.353	15	531.837	26	551.874	11	564.725	23
466.408	15	531.864	26	560.239	10	576.875	9
466.491	15	531.917	26	560.437	10	577.086	9
466.546	15	543.258	13	560.439	10	594.800	8
524.342	27	543.443	13	562.298	24	595.022	8
524.411	27	543.445	13	562.338	24	595.024	8
524.419	27	547.068	25	562.367	24	600.251	22
524.472	27	547.140	25	562.408	24	600.337	22
524.473	27	547.153	25	562.473	24	600.353	22
524.533	27	547.206	25	562.498	24	600.416	22
524.550	27	547.211	25	562.562	24	600.424	22
530.275	14	547.277	25	564.565	23	600.503	22

List of tabulated lines — Continued

Wavelength (Å)	No.	Wavelength (Å)	No.	Wavelength (Å)	No.	Wavelength (Å)	No.
600.518	22	1066.13	32	1633.68	105	2025.94	104
635.994	7	1126.99	17	1633.69	105	2038.09	149
636.251	7	1127.13	17	1633.96	105	2040.06	149
641.537	21	1127.27	17	1634.31	105	2041.00	149
641.593	21	1127.41	17	1634.63	105	2048.50	103
641.627	21	1127.63	17	1635.16	105	2049.49	103
641.684	21	1138.94	51	1689.81	123	2051.38	103
641.771	21	1139.33	51	1690.35	123	2051.79	138
641.801	21	1139.47	51	1690.57	123	2051.80	138
641.888	21	1141.62	31	1691.11	123	2051.81	138
651.211	20	1141.66	31	1701.92	57	2052.18	138
651.234	20	1141.74	31	1702.14	57	2052.39	138
651.268	20	1156.08	43	1720.46	48	2052.86	138
651.304	20	1209.11	70	1721.01	48	2053.34	138
651.328	20	1209.27	70	1721.68	48	2053.45	138
651.345	20	1209.48	70	1722.24	48	2054.39	138
651.388	20	1209.65	70	1758.10	69	2091.14	102
651.424	20	1230.02	42	1758.80	69	2091.19	102
686.415	37	1230.30	42	1759.15	69	2091.65	102
686.427	37	1269.02	41	1760.40	28	2091.70	102
686.476	37	1269.14	41	1760.47	28	2092.17	102
686.488	37	1291.22	82	1760.82	28	2092.73	102
687.053	6	1291.31	82	1851.56	68	2093.16	102
687.345	6	1291.75	82	1851.94	68	2094.14	102
687.352	6	1323.86	30	1910.21	67	2113.63	157
806.384	19	1323.91	30	1910.59	67	2114.63	157
806.533	19	1323.95	30	1910.62	67	2115.02	157
806.568	19	1324.00	30	1915.32	56	2115.75	157
806.677	19	1334.53	2	1916.01	56	2116.02	157
806.687	19	1335.66	2	1921.98	139	2116.75	157
806.830	19	1335.71	2	1922.90	139	2137.41	64
806.861	19	1384.00	40	1923.20	139	2137.90	64
809.676	36	1384.36	40	1923.73	139	2137.92	64
809.692	36	1424.33	59	1924.03	139	2154.70	148
809.747	36	1469.64	107	1924.25	139	2155.39	148
809.764	36	1469.72	107	1924.56	139	2156.26	148
858.092	5	1469.92	107	1924.66	139	2173.85	55
858.559	5	1470.15	107	1926.33	97	2174.17	55
903.623	4	1470.43	107	1926.58	97	2187.48	122
903.962	4	1470.69	107	1926.77	47	2188.38	122
904.142	4	1471.12	107	1927.02	47	2188.71	122
904.480	4	1488.77	50	1927.29	97	2189.61	122
909.278	52	1489.19	50	1927.53	97	2195.51	39
909.513	52	1489.69	50	1928.31	47	2238.48	156
909.620	52	1490.11	50	1928.55	47	2239.23	156
972.163	35	1490.38	29	1987.32	66	2239.30	156
972.187	35	1490.44	29	1987.76	66	2240.05	156
972.366	35	1504.76	58	1988.09	66	2240.99	156
996.367	34	1529.29	98	1988.53	66	2241.17	156
996.392	34	1529.45	98	1994.69	130	2242.12	156
996.465	34	1529.91	98	1995.44	130	2266.52	137
1009.86	18	1530.07	98	1995.74	130	2267.81	137
1010.08	18	1546.30	49	1996.50	130	2268.15	137
1010.37	18	1546.47	49			2268.89	137
1036.34	3	1547.29	49	in air		2269.38	137
1037.02	3	1547.46	49			2269.68	137
1039.68	71	1563.27	81	2017.92	65	2270.12	137
1039.80	71	1563.40	81	2018.38	65	2270.18	137
1039.97	71	1564.02	81	2023.13	104	2321.32	78
1040.09	71	1621.02	106	2023.24	104	2321.39	78
1063.28	33	1621.64	106	2023.82	104	2323.14	78
1063.31	33	1622.82	106	2024.10	104	2323.22	78
1065.89	32	1633.33	105	2024.78	104	2323.50	1
1065.92	32	1633.42	105	2025.08	104	2324.69	1

List of tabulated lines — Continued

Wavelength (Å)	No.	Wavelength (Å)	No.	Wavelength (Å)	No.	Wavelength (Å)	No.
2325.40	1	2766.11	146	4306.34	95	5889.78	72
2326.93	1	2767.68	146	4307.58	95	5891.60	72
2328.12	1	2836.71	38	4309.31	73	6095.29	142
2401.76	63	2837.60	38	4309.58	73	6098.51	142
2402.40	63	2904.63	154	4312.80	73	6102.56	142
2426.65	77	2905.72	154	4313.11	152	6257.18	91
2426.73	77	2906.01	154	4317.27	152	6259.56	91
2430.77	147	2907.10	154	4318.61	152	6259.80	91
2433.49	147	2908.83	154	4321.66	152	6514.96	119
2434.90	147	2908.99	154	4323.11	152	6522.64	119
2509.13	46	2910.72	154	4325.83	152	6522.93	119
2511.74	46	2966.18	153	4326.16	152	6578.05	53
2512.06	46	2966.65	153	4637.63	94	6582.88	53
2538.98	155	2966.87	153	4638.92	94	6723.32	112
2540.42	155	2967.62	153	4639.07	94	6723.65	112
2540.89	155	2967.86	153	4735.46	44	6724.56	132
2541.95	155	2968.09	153	4737.97	44	6727.07	132
2542.43	155	2968.84	153	4744.77	44	6727.26	132
2543.39	155	2969.59	153	4747.28	44	6731.07	132
2543.49	155	3077.82	161	4862.58	86	6733.58	132
2543.70	162	3079.70	161	4867.07	86	6734.00	132
2544.99	162	3079.89	161	5032.13	121	6738.61	132
2545.16	162	3081.15	161	5035.94	121	6742.43	132
2546.02	162	3081.78	161	5040.71	121	6750.54	132
2546.46	162	3082.35	161	5120.08	93	6755.16	132
2546.80	162	3083.04	161	5121.83	93	6779.94	99
2547.30	162	3083.05	161	5125.21	93	6780.59	99
2547.31	162	3120.85	159	5126.96	93	6783.91	99
2547.32	162	3122.12	159	5132.95	101	6787.21	99
2591.40	143	3124.15	159	5133.28	101	6791.47	99
2591.84	143	3125.42	159	5137.26	101	6798.10	99
2592.71	143	3165.46	79	5139.17	101	6800.69	99
2601.07	136	3165.97	79	5143.49	101	6812.28	99
2601.43	136	3167.94	79	5145.16	101	7046.25	144
2601.47	136	3183.50	45	5151.08	101	7053.09	144
2602.04	136	3187.70	45	5244.05	160	7063.68	144
2602.41	136	3298.01	114	5249.53	160	7105.77	131
2603.16	136	3298.09	114	5249.90	160	7112.48	131
2603.74	136	3299.20	114	5253.56	160	7113.04	131
2604.88	136	3299.27	114	5256.09	160	7115.63	131
2605.63	136	3360.89	75	5257.24	160	7119.76	131
2620.21	80	3361.05	75	5259.06	160	7119.91	131
2620.56	80	3361.72	75	5259.66	160	7125.72	131
2622.54	80	3387.50	88	5259.76	160	7132.47	131
2622.89	80	3581.76	134	5332.89	92	7134.10	131
2640.53	135	3584.97	134	5334.79	92	7231.33	60
2640.58	135	3585.80	134	5368.45	113	7236.42	60
2640.89	135	3587.65	134	5368.66	113	7237.17	60
2641.40	135	3588.91	134	5535.35	85	7508.89	118
2641.53	135	3589.65	134	5537.61	85	7519.49	118
2642.33	135	3590.76	134	5640.55	100	7519.93	118
2643.28	135	3590.88	134	5648.07	100	7530.57	118
2643.43	135	3831.73	96	5662.46	100	7975.65	117
2644.87	135	3835.72	96	5818.31	133	7987.61	117
2669.86	76	3836.70	96	5822.98	133	8028.85	151
2669.97	76	3881.01	87	5823.18	133	8037.73	151
2727.31	129	3918.97	61	5827.85	133	8039.40	151
2728.72	129	3920.68	61	5835.08	133	8048.31	151
2729.21	129	3922.08	54	5836.37	133	8062.10	151
2730.63	129	4009.88	145	5843.62	133	8062.80	151
2746.49	62	4017.27	145	5856.06	133	8076.64	151
2747.28	62	4021.17	145	5879.55	120	8682.52	84
2747.33	62	4267.00	74	5886.04	120	8696.70	84
2765.12	146	4267.26	74	5889.28	72	9231.12	90

List of tabulated lines — Continued

Wavelength (Å)	No.	Wavelength (Å)	No.	Wavelength (Å)	No.	Wavenumber (cm^{-1})	No.
9236.82	90	10190.5	158	17833.6	89	605.79	108
9238.30	128	10237.2	116	17852.7	89	623.85	108
9251.01	128	10256.3	116	17854.9	89	624.56	108
9267.26	128	10257.0	116	18896.6	83	853.89	109
9863.06	150	10338.6	111	18920.5	83	854.60	109
9868.46	150	10339.4	111	21171.5	141	949.82	140
9870.78	150	10358.9	111	21259.5	141	969.37	140
9878.99	150	12562.0	127	21647.0	115		
9882.68	150	12592.0	127	21735.4	115		
9884.40	150	13941.2	110	23419.3	125		
9892.60	150	13942.6	110	23523.7	125		
9900.83	150	13955.5	110	23527.0	125		
10144.3	158	15868.7	126	23632.4	125		
10155.5	158	15914.8	126	28648.4	124		
10179.2	158	15916.6	126	28804.8	124		

For the absolute scale of the $2s^22p\ ^2P^\circ - 2s2p^2\ ^4P$ intersystem lines from the ground state, we have selected the arithmetic mean of the lifetime data by Fang *et al.*[1] obtained with an ion trap, and of the multi-configuration calculations by Lennon *et al.*[3] which were done with the C IV 3 atomic structure code. The lifetime data for these long-lived states and the theoretical results are in very good agreement. For the individual lines, we have utilized the results of Lennon *et al.*,[3] renormalized to this scale. For some other intercombination lines, Nussbaumer and Storey's[4] calculational results are available which were obtained with the SUPERSTRUCTURE code. These data, which are in fair agreement with Refs. 1 and 3 for the principal intersystem lines cited above, are estimated to be of moderate accuracy.

For almost all multiplet data we used the results of the Opacity Project (OP), which is reviewed in the general introduction. For C II, the OP calculations were carried out by Yu Yan *et al.*[2] Nussbaumer and Storey[4] also performed calculations for some multiplets with the same computational approach as OP, using slightly different target functions, and obtained agreement with the OP results within a few percent for all but some very weak transitions. Since the same approach was used, we present the more recent OP results, and utilize the work of Nussbaumer and Storey only for a few transitions for which no OP results are available.

The OP data may be compared with the radiative lifetime measurements by Reistad *et al.*,[5] carried out with the beam-foil technique. They subjected their lifetime measurements to an extensive cascade analysis with the ANDC technique (see the general introduction), which has proven to be an accurate method of cascade correction. Comparisons may be made for eight excited levels, and the worst disagreement is found to be 14%. For this particular level, Reistad *et al.* estimate an uncertainty of ±13% for their result. This comparison is shown graphically in Fig. 1. The mean value of the eight experiment/theory lifetime ratios is 0.964 with a two-standard deviation of ±4%.

The relative strengths of some C II lines have been measured in emission with wall-stabilized arcs by Goly and Weniger,[6] by Glenzer *et al.*,[7] and with a hollow cathode by Huber *et al.*[8] In the last two experiments, the measurements were limited to branching ratios or line ratios within multiplets. No deviations from *LS*-coupling were observed within the experimental error limits (estimated to be normally in the range from 5–10%) for the investigated *ns-mp* and *np-md* transitions ($n = 2,3$; $m = 2,3,4$), except for one possible slight violation. This concerns the two lines of the $3s\ ^2S_{1/2} - 3p\ ^2P^\circ_{1/2,3/2}$ multiplet, where Huber *et al.*[8] measured a line intensity ratio of 2.34:1, compared to an *LS*-coupling ratio of 2:1. However, their uncertainties are almost as large as this deviation and their data contain the approximation that the $3p\ ^3P^\circ_{1/2}$ and $3p\ ^3P^\circ_{3/2}$ levels have identical lifetimes. On the other hand, the recent stabilized-arc work by Glenzer *et al.* has produced a ratio of 2.00 ± 0.08 for these lines. Also, Nussbaumer and Storey, who calculated the transition probabilities for 11 multiplets in intermediate coupling, using the SUPERSTRUCTURE code, obtained a ratio of 2.00:1 for the lines of this multiplet. For the other 10 multiplets, they found line ratios within 1–2% of the *LS*-coupling values, except for the very weak multiplet $2s2p^2\ ^2P - 2s^23p\ ^2P^\circ$, where the deviations reach 40%. (We have tabulated their line data for this case.) Therefore, *LS*-coupling appears to be, on the whole, a very good approximation for the lines covered in this spectrum. We have used it extensively to obtain individual line strengths from the OP multiplet values.

References

[1] Z. Fang, V. H. S. Kwong, J. Wang, and W. H. Parkinson, Phys. Rev. A **48**, 1114 (1993).

[2] Yu Yan, K. T. Taylor, and M. J. Seaton, J. Phys. B **20**, 6399 (1987).

[3] D. J. Lennon, P. L. Dufton, A. Hibbert, and A. E. Kingston, Astrophys. J. **294**, 200 (1985).

[4] H. Nussbaumer and P. J. Storey, Astron. Astrophys. **96**, 91 (1981).

[5] N. Reistad, R. Hutton, A. E. Nilsson, I. Martinson, and S. Mannervik, Phys. Scr. **34**, 151 (1986).

[6] A. Goly and S. Weniger, J. Quant. Spectrosc. Radiat. Transfer **28**, 389 (1982).

[7] S. Glenzer, H.-J. Kunze, J. Musielok, Y.-K. Kim, and W. L. Wiese, Phys. Rev. A **49**, 221 (1994).

[8] M. C. E. Huber, R. J. Sandeman, and G. P. Tozzi, Phys. Scr. **T8**, 95 (1984).

[9] Y. Baudinet-Robinet, P.-D. Dumont, H.-P. Garnir, and A. El Himdy, Phys. Rev. A **40**, 6321 (1989).

[10] A. W. Weiss, to be published.

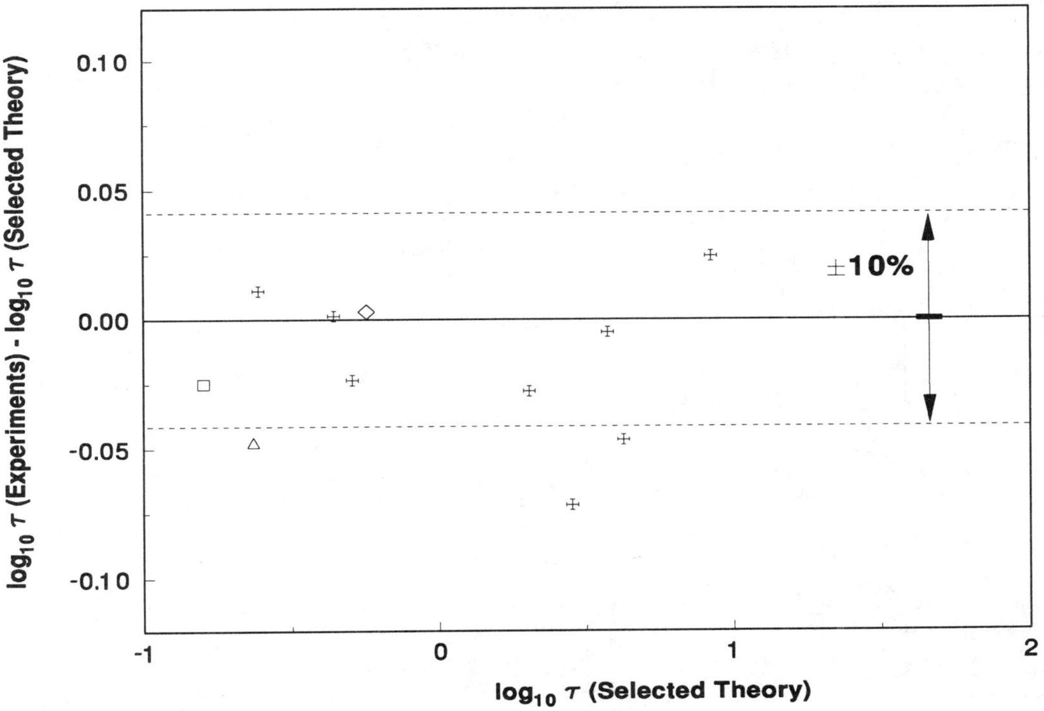

FIG. 1. The beam foil lifetimes τ of Reistad *et al.*,[5] which are ANDC corrected, are compared with lifetime data derived from the OP calculations of Yu Yan *et al.*[2] The experiment/theory ratios are plotted on a logarithmic scale versus the logarithm of the lifetimes (in ns) calculated from OP. Crosses denote the C II data. Also shown are, in an analogous manner, two available lifetime ratios for C III: Again, Reistad *et al.*[5] (diamond) and the beam-foil laser data of Baudinet-Robinet *et al.*[9] (square) are plotted versus our selected data, the SOC calculations by Weiss.[10] Finally, we show one comparison for C IV, Baudinet-Robinet *et al.*[9] (triangle) versus the OP calculations. Further details are given in the introductory comments to the two respective spectra.

C II: Allowed Transitions

No.	Transition Array	Mult.	λ_{air} (Å)	λ_{vac} (Å) or σ (cm^{-1})*	E_i (cm^{-1})		E_k (cm^{-1})	g_i–g_k	A_{ki} (10^8 s^{-1})	f_{ik}	S (at. u.)	log gf	Acc.	Source
1.	$2s^22p$ – $2s2p^2$	^{2}P°–^{4}P												
			2325.40	2326.11	63.42	–	43053.6	4–6	5.26–07	6.40–08	1.96–06	–6.592	B	1,3
			2323.50	2324.21	0	–	43025.3	2–4	1.53–08	2.48–09	3.79–08	–8.305	C	3*n*
			2326.93	2327.64	63.42	–	43025.3	4–4	1.12–07	9.10–09	2.79–07	–7.439	C	3*n*
			2324.69	2325.40	0	–	43003.3	2–2	7.30–07	5.92–08	9.06–07	–6.927	C+	3*n*
			2328.12	2328.84	63.42	–	43003.3	4–2	7.63–07	3.10–08	9.51–07	–6.906	C+	3*n*
2.		^{2}P°–^{2}D		*1335.3*	*42.3*	–	*74931.1*	6–10	2.85+00	1.27–01	3.35+00	–0.119	B	2
				1335.71	63.42	–	74930.10	4–6	2.84+00	1.14–01	2.01+00	–0.341	B	*LS*
				1334.53	0.00	–	74932.62	2–4	2.37+00	1.27–01	1.11+00	–0.596	B	*LS*
				1335.66	63.42	–	74932.62	4–4	4.74–01	1.27–02	2.23–01	–1.295	B	*LS*
3.		^{2}P°–^{2}S		*1036.8*	*42.3*	–	96493.74	6–2	2.28+01	1.22–01	2.51+00	–0.134	B	2
				1037.02	63.42	–	96493.74	4–2	1.52+01	1.22–01	1.67+00	–0.310	B	*LS*
				1036.34	0.00	–	96493.74	2–2	7.61+00	1.23–01	8.36–01	–0.611	B	*LS*
4.		^{2}P°–^{2}P		*904.08*	*42.3*	–	*110651.8*	6–6	4.10+01	5.03–01	8.98+00	0.480	B	2
				904.142	63.42	–	110665.56	4–4	3.42+01	4.19–01	4.99+00	0.224	B	*LS*
				903.962	0.00	–	110624.17	2–2	2.74+01	3.35–01	2.00+00	–0.174	B	*LS*
				904.480	63.42	–	110624.17	4–2	1.37+01	8.38–02	9.98–01	–0.475	B	*LS*
				903.623	0.00	–	110665.56	2–4	6.85+00	1.68–01	9.98–01	–0.474	B	*LS*

C II: Allowed Transitions — Continued

No.	Transition Array	Mult.	λ_{air} (Å)	λ_{vac} (Å) or σ (cm^{-1})*	E_i (cm^{-1})	E_k (cm^{-1})	g_i-g_k	A_{ki} (10^8 s^{-1})	f_{ik}	S (at. u.)	log gf	Acc.	Source
5.	$2s^22p-$ $2s^23s$	$^2P^o-^2S$		*858.40*	*42.3*	116537.65	6−2	3.53+00	1.30−02	2.21−01	−1.108	B	2
				858.559	63.42	116537.65	4−2	2.35+00	1.30−02	1.47−01	−1.284	B	*LS*
				858.092	0.00	116537.65	2−2	1.18+00	1.30−02	7.35−02	−1.585	B	*LS*
6.	$2s^22p-$ $2s^23d$	$^2P^o-^2D$		*687.25*	*42.3*	*145550.1*	6−10	2.84+01	3.36−01	4.56+00	0.304	B	2
				687.345	63.42	145550.70	4−6	2.84+01	3.02−01	2.73+00	0.082	B	*LS*
				687.053	0.00	145549.27	2−4	2.37+01	3.35−01	1.52+00	−0.173	B	*LS*
				687.352	63.42	145549.27	4−4	4.74+00	3.36−02	3.04−01	−0.872	B	*LS*
7.	$2s^22p-$ $2s^24s$	$^2P^o-^2S$		*636.17*	*42.3*	157234.07	6−2	3.03+00	6.13−03	7.70−02	−1.435	B	2
				636.251	63.42	157234.07	4−2	2.02+00	6.13−03	5.13−02	−1.611	B	*LS*
				635.994	0.00	157234.07	2−2	1.01+00	6.13−03	2.57−02	−1.912	B−	*LS*
8.	$2s^22p-$ $2s^24d$	$^2P^o-^2D$		*594.95*	*42.3*	*168124.2*	6−10	1.33+01	1.18−01	1.38+00	−0.151	B	2
				595.022	63.42	168124.45	4−6	1.33+01	1.06−01	8.29−01	−0.373	B	*LS*
				594.800	0.00	168123.74	2−4	1.11+01	1.18−01	4.60−01	−0.629	B	*LS*
				595.024	63.42	168123.74	4−4	2.22+00	1.18−02	9.22−02	−1.327	B	*LS*
9.	$2s^22p-$ $2s^25s$	$^2P^o-^2S$		*577.02*	*42.3*	173347.84	6−2	1.41+00	2.34−03	2.67−02	−1.852	B−	2
				577.086	63.42	173347.84	4−2	9.38−01	2.34−03	1.78−02	−2.028	B−	*LS*
				576.875	0.00	173347.84	2−2	4.70−01	2.34−03	8.90−03	−2.329	C+	*LS*
10.	$2s^22p-$ $2s^25d$	$^2P^o-^2D$		*560.37*	*42.3*	*178495.5*	6−10	7.28+00	5.71−02	6.32−01	−0.465	B	2
				560.437	63.42	178495.71	4−6	7.27+00	5.14−02	3.79−01	−0.687	B	*LS*
				560.239	0.00	178495.11	2−4	6.06+00	5.71−02	2.10−01	−0.943	B	*LS*
				560.439	63.42	178495.11	4−4	1.21+00	5.71−03	4.21−02	−1.641	B	*LS*
11.	$2s^22p-$ $2s^26s$	$^2P^o-^2S$		*551.81*	*42.3*	181264.24	6−2	7.36−01	1.12−03	1.22−02	−2.173	B−	2
				551.874	63.42	181264.24	4−2	4.90−01	1.12−03	8.13−03	−2.349	C+	*LS*
				551.681	0.00	181264.24	2−2	2.45−01	1.12−03	4.07−03	−2.650	C+	*LS*
12.	$2s^22p-$ $2s2p(^3P^o)3p$	$^2P^o-^2P$		*549.47*	*42.3*	*182036.9*	6−6	4.02+00	1.82−02	1.98−01	−0.962	B	2
				549.511	63.42	182043.41	4−4	3.35+00	1.52−02	1.10−01	−1.217	B	*LS*
				549.379	0.00	182023.86	2−2	2.68+00	1.21−02	4.39−02	−1.615	B	*LS*
				549.570	63.42	182023.86	4−2	1.34+00	3.03−03	2.20−02	−1.916	B−	*LS*
				549.320	0.00	182043.41	2−4	6.71−01	6.07−03	2.20−02	−1.916	B−	*LS*
13.	$2s^22p-$ $2s^26d$	$^2P^o-^2D$		*543.38*	*42.3*	*184075.0*	6−10	4.74+00	3.50−02	3.75−01	−0.678	B	2
				543.443	63.42	184075.28	4−6	4.74+00	3.15−02	2.25−01	−0.900	B	*LS*
				543.258	0.00	184074.59	2−4	3.95+00	3.49−02	1.25−01	−1.156	B	*LS*
				543.445	63.42	184074.59	4−4	7.90−01	3.50−03	2.50−02	−1.854	B−	*LS*
14.	$2s^22p-$ $2s2p(^3P^o)3p$	$^2P^o-^2D$		*530.34*	*42.3*	*188601.5*	6−10	5.14+00	3.61−02	3.78−01	−0.664	B	2
				530.359	63.42	188615.07	4−6	5.14+00	3.25−02	2.27−01	−0.886	B	*LS*
				530.275	0.00	188581.25	2−4	4.28+00	3.61−02	1.26−01	−1.142	B	*LS*
				530.454	63.42	188581.25	4−4	8.56−01	3.61−03	2.52−02	−1.840	B−	*LS*
15.	$2s^22p-$ $2s2p(^3P^o)4p$	$^2P^o-^2P$		*466.46*	*42.3*	*214421.4*	6−6	2.72+00	8.89−03	8.19−02	−1.273	B	2
				466.491	63.42	214429.95	4−4	2.27+00	7.41−03	4.55−02	−1.528	B	*LS*
				466.408	0.00	214404.33	2−2	1.82+00	5.93−03	1.82−02	−1.926	B−	*LS*
				466.546	63.42	214404.33	4−2	9.08−01	1.48−03	9.10−03	−2.227	C+	*LS*
				466.353	0.00	214429.95	2−4	4.54−01	2.96−03	9.10−03	−2.227	C+	*LS*

C II: Allowed Transitions — Continued

No.	Transition Array	Mult.	λ_{air} (Å)	λ_{vac} (Å) or σ (cm^{-1})*	E_i (cm^{-1})	E_k (cm^{-1})	g_i–g_k	A_{ki} (10^8 s^{-1})	f_{ik}	S (at. u.)	log gf	Acc.	Source
16.	$2s^22p$– $2s2p(^3P^o)5p$	$^2P^o$–^2P		*438.87*	*42.3*	*– 227898.9*	6–6	1.33+00	3.85–03	3.34–02	–1.636	B	2
				438.896	63.42	– 227907.71	4–4	1.11+00	3.21–03	1.86–02	–1.891	B–	*LS*
				438.825	0.00	– 227881.21	2–2	8.90–01	2.57–03	7.42–03	–2.289	C+	*LS*
				438.947	63.42	– 227881.21	4–2	4.45–01	6.42–04	3.71–03	–2.590	C+	*LS*
				438.774	0.00	– 227907.71	2–4	2.23–01	1.28–03	3.71–03	–2.590	C+	*LS*
17.	$2s2p^2$– $2s^23p$	^4P–$^2P^o$											
				1127.63	43053.6	– 131735.52	6–4	2.60–06	3.31–08	7.37–07	–6.702	D	4
				1127.41	43025.3	– 131724.37	4–2	2.03–07	1.93–09	2.87–08	–8.112	D	4
				1127.27	43025.3	– 131735.52	4–4	2.23–07	4.24–09	6.29–08	–7.771	D	4
				1127.13	43003.3	– 131724.37	2–2	5.57–07	1.06–08	7.87–08	–7.674	D	4
				1126.99	43003.3	– 131735.52	2–4	9.77–07	3.72–08	2.76–07	–7.128	D	4
18.	$2s2p^2$– $2p^3$	^4P–$^4S^o$		*1010.2*	*43035.8*	– 142027.1	12–4	3.43+01	1.75–01	6.97+00	0.321	B	2
				1010.37	43053.6	– 142027.1	6–4	1.71+01	1.75–01	3.49+00	0.020	B	*LS*
				1010.08	43025.3	– 142027.1	4–4	1.14+01	1.75–01	2.32+00	–0.156	B	*LS*
				1009.86	43003.3	– 142027.1	2–4	5.71+00	1.75–01	1.16+00	–0.457	B	*LS*
19.	$2s2p^2$– $2s2p(^3P^o)3s$	^4P–$^4P^o$		*806.62*	*43035.8*	*– 167009.3*	12–12	1.38+01	1.35–01	4.30+00	0.210	B	2
				806.568	43053.6	– 167035.71	6–6	9.69+00	9.45–02	1.51+00	–0.246	B	*LS*
				806.677	43025.3	– 166990.73	4–4	1.85+00	1.80–02	1.91–01	–1.143	B	*LS*
				806.687	43003.3	– 166967.13	2–2	2.31+00	2.25–02	1.20–01	–1.347	B	*LS*
				806.861	43053.6	– 166990.73	6–4	6.22+00	4.05–02	6.45–01	–0.614	B	*LS*
				806.830	43025.3	– 166967.13	4–2	1.15+01	5.62–02	5.98–01	–0.648	B	*LS*
				806.384	43025.3	– 167035.71	4–6	4.16+00	6.08–02	6.45–01	–0.614	B	*LS*
				806.533	43003.3	– 166990.73	2–4	5.77+00	1.13–01	5.98–01	–0.648	B	*LS*
20.	$2s2p^2$– $2s2p(^3P^o)3d$	^4P–$^4D^o$		*651.31*	*43035.8*	*– 196572.8*	12–20	4.19+01	4.44–01	1.14+01	0.727	B	2
				651.345	43053.6	– 196581.96	6–8	4.19+01	3.55–01	4.57+00	0.329	B	*LS*
				651.268	43025.3	– 196571.82	4–6	2.93+01	2.80–01	2.40+00	0.049	B	*LS*
				651.211	43003.3	– 196563.41	2–4	1.75+01	2.22–01	9.52–01	–0.352	B	*LS*
				651.388	43053.6	– 196571.82	6–6	1.26+01	7.99–02	1.03+00	–0.319	B	*LS*
				651.304	43025.3	– 196563.41	4–4	2.24+01	1.42–01	1.22+00	–0.245	B	*LS*
				651.234	43003.3	– 196557.87	2–2	3.49+01	2.22–01	9.52–01	–0.352	B	*LS*
				651.424	43053.6	– 196563.41	6–4	2.09+00	8.88–03	1.14–01	–1.273	B	*LS*
				651.328	43025.3	– 196557.87	4–2	6.98+00	2.22–02	1.90–01	–1.051	B	*LS*
21.		^4P–$^4P^o$		*641.76*	*43035.8*	*– 198856.9*	12–12	2.20+01	1.36–01	3.44+00	0.212	B	2
				641.888	43053.6	– 198844.00	6–6	1.54+01	9.49–02	1.20+00	–0.244	B	*LS*
				641.684	43025.3	– 198865.25	4–4	2.93+00	1.81–02	1.53–01	–1.141	B	*LS*
				641.537	43003.3	– 198879.01	2–2	3.66+00	2.26–02	9.55–02	–1.345	B	*LS*
				641.801	43053.6	– 198865.25	6–4	9.88+00	4.07–02	5.16–01	–0.612	B	*LS*
				641.627	43025.3	– 198879.01	4–2	1.83+01	5.65–02	4.78–01	–0.646	B	*LS*
				641.771	43025.3	– 198844.00	4–6	6.59+00	6.10–02	5.16–01	–0.612	B	*LS*
				641.593	43003.3	– 198865.25	2–4	9.16+00	1.13–01	4.78–01	–0.646	B	*LS*
22.	$2s2p^2$– $2s2p(^3P^o)4s$	^4P–$^4P^o$		*600.39*	*43035.8*	*– 209595.4*	12–12	3.63+00	1.96–02	4.66–01	–0.628	B	2
				600.353	43053.6	– 209622.32	6–6	2.54+00	1.37–02	1.63–01	–1.084	B	*LS*
				600.416	43025.3	– 209576.46	4–4	4.84–01	2.62–03	2.07–02	–1.980	B–	*LS*
				600.424	43003.3	– 209552.36	2–2	6.05–01	3.27–03	1.29–02	–2.184	B–	*LS*
				600.518	43053.6	– 209576.46	6–4	1.63+00	5.89–03	6.98–02	–1.452	B	*LS*
				600.503	43025.3	– 209552.36	4–2	3.02+00	8.18–03	6.47–02	–1.485	B	*LS*
				600.251	43025.3	– 209622.32	4–6	1.09+00	8.83–03	6.98–02	–1.452	B	*LS*
				600.337	43003.3	– 209576.46	2–4	1.51+00	1.64–02	6.47–02	–1.485	B	*LS*

C II: Allowed Transitions — Continued

No.	Transition Array	Mult.	λ_{air} (Å)	λ_{vac} (Å) or σ (cm^{-1})*	E_i (cm^{-1})	E_k (cm^{-1})	g_i–g_k	A_{ki} (10^8 s^{-1})	f_{ik}	S (at. u.)	log gf	Acc.	Source
23.	$2s2p^2$– $2s2p(^3P^\circ)4d$	^{4}P–^{4}D°		*564.64*	*43035.8*	*– 220140.7*	12–20	1.84+01	1.47–01	3.27+00	0.246	B	2
				564.663	43053.6	– 220150.49	6–8	1.84+01	1.17–01	1.31+00	–0.152	B	*LS*
				564.608	43025.3	– 220139.41	4–6	1.29+01	9.25–02	6.88–01	–0.432	B	*LS*
				564.565	43003.3	– 220130.86	2–4	7.68+00	7.34–02	2.73–01	–0.833	B	*LS*
				564.698	43053.6	– 220139.41	6–6	5.53+00	2.64–02	2.95–01	–0.800	B	*LS*
				564.635	43025.3	– 220130.86	4–4	9.83+00	4.70–02	3.49–01	–0.726	B	*LS*
				564.582	43003.3	– 220125.51	2–2	1.54+01	7.34–02	2.73–01	–0.833	B	*LS*
				564.725	43053.6	– 220130.86	6–4	9.21–01	2.94–03	3.27–02	–1.754	B	*LS*
				564.652	43025.3	– 220125.51	4–2	3.07+00	7.34–03	5.46–02	–1.532	B	*LS*
24.		^{4}P–^{4}P°		*562.47*	*43035.8*	*– 220824.1*	12–12	1.05+01	4.97–02	1.10+00	–0.225	B	2
				562.562	43053.6	– 220811.69	6–6	7.33+00	3.48–02	3.86–01	–0.681	B	*LS*
				562.408	43025.3	– 220832.15	4–4	1.40+00	6.62–03	4.90–02	–1.577	B	*LS*
				562.298	43003.3	– 220845.07	2–2	1.75+00	8.28–03	3.07–02	–1.781	B	*LS*
				562.498	43053.6	– 220832.15	6–4	4.71+00	1.49–02	1.66–01	–1.049	B	*LS*
				562.367	43025.3	– 220845.07	4–2	8.73+00	2.07–02	1.53–01	–1.082	B	*LS*
				562.473	43025.3	– 220811.69	4–6	3.14+00	2.23–02	1.66–01	–1.049	B	*LS*
				562.338	43003.3	– 220832.15	2–4	4.37+00	4.14–02	1.53–01	–1.082	B	*LS*
25.	$2s2p^2$– $2s2p(^3P^\circ)5s$	^{4}P–^{4}P°		*547.18*	*43035.8*	*– 225790.8*	12–12	1.44+00	6.47–03	1.40–01	–1.110	B	2
				547.153	43053.6	– 225817.8	6–6	1.01+00	4.53–03	4.90–02	–1.566	B	*LS*
				547.206	43025.3	– 225771.8	4–4	1.92–01	8.63–04	6.22–03	–2.462	C+	*LS*
				547.211	43003.3	– 225748.0	2–2	2.40–01	1.08–03	3.89–03	–2.666	C+	*LS*
				547.291	43053.6	– 225771.8	6–4	6.48–01	1.94–03	2.10–02	–1.934	B–	*LS*
				547.277	43025.3	– 225748.0	4–2	1.20+00	2.70–03	1.94–02	–1.967	B–	*LS*
				547.068	43025.3	– 225817.8	4–6	4.33–01	2.91–03	2.10–02	–1.934	B–	*LS*
				547.140	43003.3	– 225771.8	2–4	6.01–01	5.39–03	1.94–02	–1.967	B–	*LS*
26.	$2s2p^2$– $2s2p(^3P^\circ)5d$	^{4}P–^{4}P°		*531.83*	*43035.8*	*– 231064.8*	12–12	5.45+00	2.31–02	4.85–01	–0.557	B	2
				531.917	43053.6	– 231052.9	6–6	3.81+00	1.62–02	1.70–01	–1.013	B	*LS*
				531.784	43025.3	– 231071.7	4–4	7.26–01	3.08–03	2.16–02	–1.909	B–	*LS*
				531.679	43003.3	– 231086.6	2–2	9.09–01	3.85–03	1.35–02	–2.113	B–	*LS*
				531.864	43053.6	– 231071.7	6–4	2.45+00	6.93–03	7.28–02	–1.381	B	*LS*
				531.742	43025.3	– 231086.6	4–2	4.54+00	9.63–03	6.74–02	–1.414	B	*LS*
				531.837	43025.3	– 231052.9	4–6	1.63+00	1.04–02	7.28–02	–1.381	B	*LS*
				531.721	43003.3	– 231071.7	2–4	2.27+00	1.93–02	6.74–02	–1.414	B	*LS*
27.	$2s2p^2$– $2s2p(^3P^\circ)6s$	^{4}P–^{4}P°		*524.45*	*43035.8*	*– 233713.3*	12–12	7.54–01	3.11–03	6.44–02	–1.428	B	2
				524.419	43053.6	– 233740.7	6–6	5.28–01	2.18–03	2.25–02	–1.884	B–	*LS*
				524.472	43025.3	– 233693.3	4–4	1.00–01	4.14–04	2.86–03	–2.780	C	*LS*
				524.473	43003.3	– 233670.9	2–2	1.26–01	5.18–04	1.79–03	–2.985	C	*LS*
				524.550	43053.6	– 233693.3	6–4	3.39–01	9.32–04	9.66–03	–2.252	C+	*LS*
				524.533	43025.3	– 233670.9	4–2	6.28–01	1.29–03	8.94–03	–2.286	C+	*LS*
				524.342	43025.3	– 233740.7	4–6	2.26–01	1.40–03	9.66–03	–2.252	C+	*LS*
				524.411	43003.3	– 233693.3	2–4	3.14–01	2.59–03	8.94–03	–2.286	C+	*LS*
28.	$2s2p^2$– $2s^23p$	^{2}D–^{2}P°		*1760.5*	*74931.1*	*– 131731.8*	10–6	4.28–01	1.19–02	6.92–01	–0.923	B	2
				1760.40	74930.10	– 131735.52	6–4	3.86–01	1.19–02	4.15–01	–1.145	B	*LS*
				1760.82	74932.62	– 131724.37	4–2	4.28–01	9.94–03	2.31–01	–1.401	B	*LS*
				1760.47	74932.62	– 131735.52	4–4	4.28–02	1.99–03	4.61–02	–2.099	B	*LS*
29.	$2s2p^2$– $2p^3$	^{2}D–^{4}S°											
				1490.38	74930.10	– 142027.1	6–4	4.50–07	9.99–09	2.94–07	–7.222	D	4
				1490.44	74932.62	– 142027.1	4–4	1.20–07	3.98–09	7.81–08	–7.798	D	4

C II: Allowed Transitions — Continued

No.	Transition Array	Mult.	λ_{air} (Å)	λ_{vac} (Å) or σ (cm^{-1})*	E_i (cm^{-1})	E_k (cm^{-1})	g_i-g_k	A_{ki} (10^8 s^{-1})	f_{ik}	S (at. u.)	log gf	Acc.	Source
30.		$^2D-^2D^\circ$		*1323.9*	*74931.1*	*– 150463.6*	10–10	4.81+00	1.26–01	5.51+00	0.102	B	2
				1323.95	74930.10	– 150461.58	6–6	4.49+00	1.18–01	3.08+00	–0.150	B	*LS*
				1323.91	74932.62	– 150466.69	4–4	4.33+00	1.14–01	1.98+00	–0.342	B	*LS*
				1323.86	74930.10	– 150466.69	6–4	4.81–01	8.42–03	2.20–01	–1.296	B	*LS*
				1324.00	74932.62	– 150461.58	4–6	3.20–01	1.26–02	2.20–01	–1.296	B	*LS*
31.	$2s2p^2-$ $2s^24p$	$^2D-^2P^\circ$		*1141.7*	*74931.1*	*– 162522.3*	10–6	1.60+00	1.87–02	7.04–01	–0.727	B	2
				1141.62	74930.10	– 162524.57	6–4	1.44+00	1.87–02	4.23–01	–0.949	B	*LS*
				1141.74	74932.62	– 162517.89	4–2	1.60+00	1.56–02	2.35–01	–1.205	B	*LS*
				1141.66	74932.62	– 162524.57	4–4	1.60–01	3.12–03	4.70–02	–1.903	B	*LS*
32.	$2s2p^2-$ $2p^3$	$^2D-^2P^\circ$		*1066.0*	*74931.1*	*– 168742.0*	10–6	1.64+01	1.67–01	5.87+00	0.223	B	2
				1065.89	74930.10	– 168748.30	6–4	1.47+01	1.67–01	3.52+00	0.001	B	*LS*
				1066.13	74932.62	– 168729.53	4–2	1.63+01	1.39–01	1.95+00	–0.255	B	*LS*
				1065.92	74932.62	– 168748.30	4–4	1.64+00	2.79–02	3.91–01	–0.953	B	*LS*
33.	$2s2p^2-$ $2s^24f$	$^2D-^2F^\circ$		*1063.3*	*74931.1*	– 168978.34	10–14	1.54–01	3.66–03	1.28–01	–1.437	B	4
				1063.28	74930.10	– 168978.34	6–8	1.54–01	3.48–03	7.31–02	–1.680	C+	*LS*
				1063.31	74932.62	– 168978.34	4–6	1.44–01	3.66–03	5.12–02	–1.835	C+	*LS*
				1063.28	74930.10	– 168978.34	6–6	1.03–02	1.74–04	3.66–03	–2.981	C–	*LS*
34.	$2s2p^2-$ $2s^25p$	$^2D-^2P^\circ$		*996.40*	*74931.1*	*– 175292.3*	10–6	8.69–01	7.76–03	2.55–01	–1.110	B	2
				996.367	74930.10	– 175294.75	6–4	7.83–01	7.76–03	1.53–01	–1.332	B	*LS*
				996.465	74932.62	– 175287.39	4–2	8.68–01	6.46–03	8.48–02	–1.587	B	*LS*
				996.392	74932.62	– 175294.75	4–4	8.70–02	1.29–03	1.70–02	–2.286	B–	*LS*
35.	$2s2p^2-$ $2s2p(^3P^\circ)3s$	$^2D-^2P^\circ$		*972.23*	*74931.1*	*– 177787.2*	10–6	3.88–01	3.30–03	1.06–01	–1.482	B	2
				972.163	74930.10	– 177793.54	6–4	3.49–01	3.30–03	6.33–02	–1.704	B	*LS*
				972.366	74932.62	– 177774.59	4–2	3.87–01	2.74–03	3.51–02	–1.960	B	*LS*
				972.187	74932.62	– 177793.54	4–4	3.88–02	5.49–04	7.03–03	–2.658	C+	*LS*
36.	$2s2p^2-$ $2s2p(^3P^\circ)3d$	$^2D-^2D^\circ$		*809.71*	*74931.1*	*– 198432.0*	10–10	1.45+01	1.42–01	3.79+00	0.152	B	2
				809.676	74930.10	– 198436.31	6–6	1.35+01	1.33–01	2.12+00	–0.099	B	*LS*
				809.764	74932.62	– 198425.43	4–4	1.30+01	1.28–01	1.36+00	–0.291	B	*LS*
				809.747	74930.10	– 198425.43	6–4	1.44+00	9.47–03	1.51–01	–1.246	B	*LS*
				809.692	74932.62	– 198436.31	4–6	9.63–01	1.42–02	1.51–01	–1.246	B	*LS*
37.	$2s2p^2-$ $2s2p(^3P^\circ)4d$	$^2D-^2D^\circ$		*686.44*	*74931.1*	*– 220609.3*	10–10	6.88+00	4.86–02	1.10+00	–0.314	B	2
				686.415	74930.10	– 220614.51	6–6	6.42+00	4.53–02	6.15–01	–0.565	B	*LS*
				686.488	74932.62	– 220601.53	4–4	6.19+00	4.37–02	3.95–01	–0.757	B	*LS*
				686.476	74930.10	– 220601.53	6–4	6.88–01	3.24–03	4.39–02	–1.711	B	*LS*
				686.427	74932.62	– 220614.51	4–6	4.59–01	4.86–03	4.39–02	–1.711	B	*LS*
38.	$2s2p^2-$ $2s^23p$	$^2S-^2P^\circ$	*2837.0*	*2837.8*	96493.74	*– 131731.8*	2–6	3.98–01	1.44–01	2.69+00	–0.541	B	2
			2836.71	2837.54	96493.74	– 131735.52	2–4	3.98–01	9.60–02	1.79+00	–0.717	B	*LS*
			2837.60	2838.44	96493.74	– 131724.37	2–2	3.97–01	4.80–02	8.96–01	–1.018	B	*LS*
39.	$2s2p^2-$ $2p^3$	$^2S-^4S^\circ$											
			2195.51	2196.19	96493.74	– 142027.1	2–4	5.09–08	7.35–09	1.06–07	–7.833	D	4
40.		$^2S-^2P^\circ$		*1384.1*	96493.74	*– 168742.0*	2–6	2.73–02	2.35–03	2.14–02	–2.328	B–	4
				1384.00	96493.74	– 168748.30	2–4	2.73–02	1.57–03	1.43–02	–2.504	B–	*LS*
				1384.36	96493.74	– 168729.53	2–2	2.72–02	7.83–04	7.13–03	–2.805	C+	*LS*

C II: Allowed Transitions — Continued

No.	Transition Array	Mult.	λ_{air} (Å)	λ_{vac} (Å) or σ (cm^{-1})*	E_i (cm^{-1})	E_k (cm^{-1})	g_i–g_k	A_{ki} (10^8 s^{-1})	f_{ik}	S (at. u.)	log gf	Acc.	Source
41.	$2s2p^2$– $2s^25p$	^{2}S–^{2}P°		*1269.1*	96493.74	– *175292.3*	2–6	3.81–01	2.76–02	2.30–01	–1.258	B	2
				1269.02	96493.74	– 175294.75	2–4	3.81–01	1.84–02	1.54–01	–1.435	B	*LS*
				1269.14	96493.74	– 175287.39	2–2	3.81–01	9.19–03	7.68–02	–1.736	B	*LS*
42.	$2s2p^2$– $2s2p(^3P°)3s$	^{2}S–^{2}P°		*1230.1*	96493.74	– *177787.2*	2–6	1.80+00	1.22–01	9.92–01	–0.611	B	2
				1230.02	96493.74	– 177793.54	2–4	1.80+00	8.16–02	6.61–01	–0.787	B	*LS*
				1230.30	96493.74	– 177774.59	2–2	1.80+00	4.08–02	3.31–01	–1.088	B	*LS*
43.	$2s2p^2$– $2s^26p$	^{2}S–^{2}P°		1156.08	96493.74	– 182993.23	2–6	7.56–02	4.55–03	3.46–02	–2.041	B	2
44.	$2s2p^2$– $2s^23p$	^{2}P–^{2}P°	*4742.5*	*4743.8*	*110651.8*	– *131731.8*	6–6	6.90–04	2.33–04	2.18–02	–2.855	B–	2
			4744.77	4746.09	110665.56	– 131735.52	4–4	5.73–04	1.93–04	1.21–02	–3.111	B–	4*n*
			4737.97	4739.29	110624.17	– 131724.37	2–2	5.34–04	1.80–04	5.61–03	–3.444	C+	4*n*
			4747.28	4748.61	110665.56	– 131724.37	4–2	2.24–04	3.79–05	2.37–03	–3.820	C	4*n*
			4735.46	4736.79	110624.17	– 131735.52	2–4	8.28–05	5.57–05	1.74–03	–3.953	C	4*n*
45.	$2s2p^2$– $2p^3$	^{2}P–^{4}S°											
			3187.70	3188.62	110665.56	– 142027.1	4–4	7.68–07	1.17–07	4.91–06	–6.330	D	4
			3183.50	3184.42	110624.17	– 142027.1	2–4	1.49–07	4.54–08	9.52–07	–7.042	D	4
46.		^{2}P–^{2}D°	*2511.1*	*2511.8*	*110651.8*	– *150463.6*	6–10	5.43–01	8.55–02	4.24+00	–0.290	B	2
			2512.06	2512.81	110665.56	– 150461.58	4–6	5.42–01	7.69–02	2.54+00	–0.512	B	*LS*
			2509.13	2509.88	110624.17	– 150466.69	2–4	4.53–01	8.55–02	1.41+00	–0.767	B	*LS*
			2511.74	2512.49	110665.56	– 150466.69	4–4	9.04–02	8.55–03	2.83–01	–1.466	B	*LS*
47.	$2s2p^2$– $2s^24p$	^{2}P–^{2}P°		*1927.9*	*110651.8*	– *162522.3*	6–6	1.16–01	6.46–03	2.46–01	–1.412	B	2
				1928.31	110665.56	– 162524.57	4–4	9.65–02	5.38–03	1.37–01	–1.667	B	*LS*
				1927.02	110624.17	– 162517.89	2–2	7.74–02	4.31–03	5.46–02	–2.065	B	*LS*
				1928.55	110665.56	– 162517.89	4–2	3.86–02	1.08–03	2.73–02	–2.366	B–	*LS*
				1926.77	110624.17	– 162524.57	2–4	1.93–02	2.15–03	2.73–02	–2.366	B–	*LS*
48.	$2s2p^2$– $2p^3$	^{2}P–^{2}P°		*1721.5*	*110651.8*	– *168742.0*	6–6	3.30+00	1.47–01	4.99+00	–0.055	B	2
				1721.68	110665.56	– 168748.30	4–4	2.75+00	1.22–01	2.77+00	–0.310	B	*LS*
				1721.01	110624.17	– 168729.53	2–2	2.20+00	9.79–02	1.11+00	–0.708	B	*LS*
				1722.24	110665.56	– 168729.53	4–2	1.10+00	2.45–02	5.55–01	–1.009	B	*LS*
				1720.46	110624.17	– 168748.30	2–4	5.52–01	4.90–02	5.55–01	–1.009	B	*LS*
49.	$2s2p^2$– $2s^25p$	^{2}P–^{2}P°		*1547.0*	*110651.8*	– *175292.3*	6–6	5.35–01	1.92–02	5.87–01	–0.939	B	2
				1547.29	110665.56	– 175294.75	4–4	4.46–01	1.60–02	3.26–01	–1.194	B	*LS*
				1546.47	110624.17	– 175287.39	2–2	3.57–01	1.28–02	1.30–01	–1.592	B	*LS*
				1547.46	110665.56	– 175287.39	4–2	1.78–01	3.20–03	6.52–02	–1.893	B	*LS*
				1546.30	110624.17	– 175294.75	2–4	8.93–02	6.40–03	6.52–02	–1.893	B	*LS*
50.	$2s2p^2$– $2s2p(^3P°)3s$	^{2}P–^{2}P°		*1489.5*	*110651.8*	– *177787.2*	6–6	3.73–01	1.24–02	3.65–01	–1.128	B	2
				1489.69	110665.56	– 177793.54	4–4	3.11–01	1.03–02	2.03–01	–1.383	B	*LS*
				1489.19	110624.17	– 177774.59	2–2	2.49–01	8.28–03	8.12–02	–1.781	B	*LS*
				1490.11	110665.56	– 177774.59	4–2	1.24–01	2.07–03	4.06–02	–2.082	B	*LS*
				1488.77	110624.17	– 177793.54	2–4	6.23–02	4.14–03	4.06–02	–2.082	B	*LS*
51.	$2s2p^2$– $2s2p(^3P°)3d$	^{2}P–^{2}D°		*1139.2*	*110651.8*	– *198432.0*	6–10	5.04+00	1.63–01	3.68+00	–0.008	B	2
				1139.33	110665.56	– 198436.31	4–6	5.04+00	1.47–01	2.21+00	–0.230	B	*LS*
				1138.94	110624.17	– 198425.43	2–4	4.20+00	1.63–01	1.23+00	–0.486	B	*LS*
				1139.47	110665.56	– 198425.43	4–4	8.40–01	1.63–02	2.45–01	–1.185	B	*LS*

C II: Allowed Transitions — Continued

No.	Transition Array	Mult.	λ_{air} (Å)	λ_{vac} (Å) or σ (cm^{-1})*	E_i (cm^{-1})	E_k (cm^{-1})	g_i–g_k	A_{ki} (10^8 s^{-1})	f_{ik}	S (at. u.)	log gf	Acc.	Source
52.	$2s2p^2$– $2s2p(^3P^o)4d$	^{2}P–^{2}D°		*909.44*	*110651.8*	– *220609.3*	6–10	1.49+00	3.08–02	5.54–01	–0.733	B	2
				909.513	110665.56	– 220614.51	4–6	1.49+00	2.77–02	3.32–01	–0.955	B	*LS*
				909.278	110624.17	– 220601.53	2–4	1.24+00	3.08–02	1.84–01	–1.210	B	*LS*
				909.620	110665.56	– 220601.53	4–4	2.49–01	3.08–03	3.69–02	–1.909	B	*LS*
53.	$2s^23s$– $2s^23p$	^{2}S–^{2}P°	*6579.7*	*6581.5*	116537.65	– *131731.8*	2–6	3.62–01	7.05–01	3.06+01	0.149	B	2
			6578.05	6579.87	116537.65	– 131735.52	2–4	3.63–01	4.70–01	2.04+01	–0.026	B	*LS*
			6582.88	6584.70	116537.65	– 131724.37	2–2	3.62–01	2.35–01	1.02+01	–0.328	B	*LS*
54.	$2s^23s$– $2p^3$	^{2}S–^{4}S°											
			3922.08	3923.19	116537.65	– 142027.1	2–4	1.01–07	4.68–08	1.21–06	–7.029	D	4
55.	$2s^23s$– $2s^24p$	^{2}S–^{2}P°	*2174.0*	*2174.6*	116537.65	– *162522.3*	2–6	2.31–01	4.90–02	7.02–01	–1.009	B	2
			2173.85	2174.53	116537.65	– 162524.57	2–4	2.31–01	3.27–02	4.68–01	–1.185	B	*LS*
			2174.17	2174.85	116537.65	– 162517.89	2–2	2.31–01	1.63–02	2.34–01	–1.486	B	*LS*
56.	$2s^23s$– $2p^3$	^{2}S–^{2}P°		*1915.6*	116537.65	– *168742.0*	2–6	3.89–02	6.41–03	8.09–02	–1.892	B	2
				1915.32	116537.65	– 168748.30	2–4	3.89–02	4.28–03	5.39–02	–2.068	B	*LS*
				1916.01	116537.65	– 168729.53	2–2	3.88–02	2.14–03	2.70–02	–2.369	B–	*LS*
57.	$2s^23s$– $2s^25p$	^{2}S–^{2}P°		*1702.0*	116537.65	– *175292.3*	2–6	9.80–02	1.28–02	1.43–01	–1.593	B	2
				1701.92	116537.65	– 175294.75	2–4	9.80–02	8.51–03	9.54–02	–1.769	B	*LS*
				1702.14	116537.65	– 175287.39	2–2	9.80–02	4.26–03	4.77–02	–2.070	B	*LS*
58.	$2s^23s$– $2s^26p$	^{2}S–^{2}P°		1504.76	116537.65	– 182993.2	2–6	8.50–02	8.66–03	8.58–02	–1.761	B	2
59.	$2s^23s$– $2s^27p$	^{2}S–^{2}P°		1424.33	116537.65	– 186745.9	2–6	6.44–02	5.88–03	5.51–02	–1.930	B	2
60.	$2s^23p$– $2s^23d$	^{2}P°–^{2}D	*7234.8*	*7236.8*	*131731.8*	– *145550.1*	6–10	4.22–01	5.52–01	7.89+01	0.520	B	2
			7236.42	7238.41	131735.52	– 145550.70	4–6	4.22–01	4.97–01	4.73+01	0.298	B	*LS*
			7231.33	7233.33	131724.37	– 145549.27	2–4	3.52–01	5.52–01	2.63+01	0.043	B	*LS*
			7237.17	7239.16	131735.52	– 145549.27	4–4	7.03–02	5.52–02	5.26+00	–0.656	B	*LS*
61.	$2s^23p$– $2s^24s$	^{2}P°–^{2}S	*3920.1*	*3921.2*	*131731.8*	– 157234.07	6–2	1.91+00	1.46–01	1.13+01	–0.056	B	2
			3920.68	3921.79	131735.52	– 157234.07	4–2	1.27+00	1.46–01	7.56+00	–0.232	B	*LS*
			3918.97	3920.08	131724.37	– 157234.07	2–2	6.36–01	1.46–01	3.78+00	–0.533	B	*LS*
62.	$2s^23p$– $2s^24d$	^{2}P°–^{2}D	*2747.0*	*2747.8*	*131731.8*	– *168124.2*	6–10	5.23–01	9.86–02	5.35+00	–0.228	B	2
			2747.28	2748.09	131735.52	– 168124.45	4–6	5.23–01	8.87–02	3.21+00	–0.450	B	*LS*
			2746.49	2747.30	131724.37	– 168123.74	2–4	4.36–01	9.85–02	1.78+00	–0.705	B	*LS*
			2747.33	2748.14	131735.52	– 168123.74	4–4	8.71–02	9.86–03	3.57–01	–1.404	B	*LS*
63.	$2s^23p$– $2s^25s$	^{2}P°–^{2}S	*2402.2*	*2402.9*	*131731.8*	– 173347.84	6–2	7.69–01	2.22–02	1.05+00	–0.876	B	2
			2402.40	2403.13	131735.52	– 173347.84	4–2	5.13–01	2.22–02	7.02–01	–1.052	B	*LS*
			2401.76	2402.49	131724.37	– 173347.84	2–2	2.57–01	2.22–02	3.51–01	–1.353	B	*LS*
64.	$2s^23p$– $2s^25d$	^{2}P°–^{2}D	*2137.7*	*2138.4*	*131731.8*	– *178495.5*	6–10	4.47–01	5.11–02	2.16+00	–0.514	B	2
			2137.90	2138.57	131735.52	– 178495.71	4–6	4.47–01	4.60–02	1.29+00	–0.735	B	*LS*
			2137.41	2138.09	131724.37	– 178495.11	2–4	3.73–01	5.10–02	7.18–01	–0.991	B	*LS*
			2137.92	2138.60	131735.52	– 178495.11	4–4	7.45–02	5.11–03	1.44–01	–1.690	B	*LS*

C II: Allowed Transitions — Continued

No.	Transition Array	Mult.	λ_{air} (Å)	λ_{vac} (Å) or σ (cm^{-1})*	E_i (cm^{-1})	E_k (cm^{-1})	g_i-g_k	A_{ki} (10^8 s^{-1})	f_{ik}	S (at. u.)	log gf	Acc.	Source
65.	$2s^23p-$ $2s^26s$	$^2P^o-^2S$	*2018.2*	*2018.9*	*131731.8*	181264.24	6−2	3.96−01	8.06−03	3.21−01	−1.316	B	2
			2018.38	2019.03	131735.52	181264.24	4−2	2.64−01	8.06−03	2.14−01	−1.492	B	*LS*
			2017.92	2018.58	131724.37	181264.24	2−2	1.32−01	8.06−03	1.07−01	−1.793	B	*LS*
66.	$2s^23p-$ $2s2p(^3P^o)3p$	$^2P^o-^2P$		*1987.9*	*131731.8*	*182036.9*	6−6	1.91−01	1.13−02	4.44−01	−1.168	B	2
				1987.76	131735.52	182043.41	4−4	1.59−01	9.43−03	2.47−01	−1.424	B	*LS*
				1988.09	131724.37	182023.86	2−2	1.27−01	7.54−03	9.87−02	−1.822	B	*LS*
				1988.53	131735.52	182023.86	4−2	6.36−02	1.88−03	4.93−02	−2.123	B	*LS*
				1987.32	131724.37	182043.41	2−4	3.18−02	3.77−03	4.93−02	−2.122	B	*LS*
67.	$2s^23p-$ $2s^26d$	$^2P^o-^2D$		*1910.5*	*131731.8*	*184075.0*	6−10	3.36−01	3.07−02	1.16+00	−0.735	B	2
				1910.59	131735.52	184075.28	4−6	3.36−01	2.76−02	6.94−01	−0.957	B	*LS*
				1910.21	131724.37	184074.59	2−4	2.80−01	3.06−02	3.85−01	−1.213	B	*LS*
				1910.62	131735.52	184074.59	4−4	5.60−02	3.07−03	7.72−02	−1.911	B	*LS*
68.	$2s^23p-$ $2s^27s$	$^2P^o-^2S$		*1851.8*	*131731.8*	185732.93	6−2	2.27−01	3.88−03	1.42−01	−1.633	B	2
				1851.94	131735.52	185732.93	4−2	1.51−01	3.88−03	9.47−02	−1.809	B	*LS*
				1851.56	131724.37	185732.93	2−2	7.55−02	3.88−03	4.73−02	−2.110	B	*LS*
69.	$2s^23p-$ $2s2p(^3P^o)3p$	$^2P^o-^2D$		*1758.4*	*131731.8*	*188601.5*	6−10	3.94−01	3.04−02	1.06+00	−0.739	B	2
				1758.10	131735.52	188615.07	4−6	3.94−01	2.74−02	6.34−01	−0.961	B	*LS*
				1758.80	131724.37	188581.25	2−4	3.27−01	3.04−02	3.52−01	−1.216	B	*LS*
				1759.15	131735.52	188581.25	4−4	6.55−02	3.04−03	7.04−02	−1.915	B	*LS*
70.	$2s^23p-$ $2s2p(^3P^o)4p$	$^2P^o-^2P$		*1209.3*	*131731.8*	*214421.4*	6−6	1.03−01	2.25−03	5.37−02	−1.870	B	2
				1209.27	131735.52	214429.95	4−4	8.55−02	1.87−03	2.98−02	−2.125	B−	*LS*
				1209.48	131724.37	214404.33	2−2	6.83−02	1.50−03	1.19−02	−2.523	B−	*LS*
				1209.65	131735.52	214404.33	4−2	3.41−02	3.75−04	5.97−03	−2.824	C+	*LS*
				1209.11	131724.37	214429.95	2−4	1.71−02	7.49−04	5.97−03	−2.824	C+	*LS*
71.	$2s^23p-$ $2s2p(^3P^o)5p$	$^2P^o-^2P$		*1039.9*	*131731.8*	*227898.9*	6−6	2.10−01	3.41−03	7.00−02	−1.689	B	2
				1039.80	131735.52	227907.71	4−4	1.75−01	2.84−03	3.89−02	−1.945	B	*LS*
				1039.97	131724.37	227881.21	2−2	1.40−01	2.27−03	1.56−02	−2.343	B−	*LS*
				1040.09	131735.52	227881.21	4−2	7.00−02	5.68−04	7.78−03	−2.644	C+	*LS*
				1039.68	131724.37	227907.71	2−4	3.51−02	1.14−03	7.78−03	−2.644	C+	*LS*
72.	$2s^23d-$ $2s^24p$	$^2D-^2P^o$	*5890.4*	*5892.0*	*145550.1*	*162522.3*	10−6	3.50−01	1.09−01	2.12+01	0.038	B	2
			5889.78	5891.41	145550.70	162524.57	6−4	3.15−01	1.09−01	1.27+01	−0.184	B	*LS*
			5891.60	5893.23	145549.27	162517.89	4−2	3.49−01	9.09−02	7.05+00	−0.440	B	*LS*
			5889.28	5890.91	145549.27	162524.57	4−4	3.50−02	1.82−02	1.41+00	−1.138	B	*LS*
73.	$2s^23d-$ $2p^3$	$^2D-^2P^o$	*4310.6*	*4311.9*	*145550.1*	*168742.0*	10−6	1.26−03	2.11−04	3.00−02	−2.675	B	2
			4309.58	4310.79	145550.70	168748.30	6−4	1.14−03	2.11−04	1.80−02	−2.897	B−	*LS*
			4312.80	4314.02	145549.27	168729.53	4−2	1.26−03	1.76−04	9.99−03	−3.153	C+	*LS*
			4309.31	4310.53	145549.27	168748.30	4−4	1.27−04	3.52−05	2.00−03	−3.851	C	*LS*
74.	$2s^23d-$ $2s^24f$	$^2D-^2F^o$	*4267.2*	*4268.4*	*145550.1*	168978.34	10−14	2.38+00	9.11−01	1.28+02	0.960	B	4
			4267.26	4268.46	145550.70	168978.34	6−8	2.38+00	8.68−01	7.31+01	0.717	C+	*LS*
			4267.00	4268.20	145549.27	168978.34	4−6	2.23+00	9.11−01	5.12+01	0.562	C+	*LS*
			4267.26	4268.46	145550.70	168978.34	6−6	1.59−01	4.34−02	3.66+00	−0.584	C+	*LS*

C II: Allowed Transitions — Continued

No.	Transition Array	Mult.	λ_{air} (Å)	λ_{vac} (Å) or σ (cm^{-1})*	E_i (cm^{-1})	E_k (cm^{-1})	g_i-g_k	A_{ki} (10^8 s^{-1})	f_{ik}	S (at. u.)	log gf	Acc.	Source
75.	$2s^23d-$ $2s^25p$	^{2}D$-^2$P°	*3361.3*	*3362.2*	*145550.1*	*175292.3*	10−6	1.13−01	1.15−02	1.28+00	−0.938	B	2
			3361.05	3362.02	145550.70	175294.75	6−4	1.02−01	1.15−02	7.65−01	−1.160	B	*LS*
			3361.72	3362.69	145549.27	175287.39	4−2	1.13−01	9.59−03	4.25−01	−1.416	B	*LS*
			3360.89	3361.86	145549.27	175294.75	4−4	1.13−02	1.92−03	8.50−02	−2.114	B	*LS*
76.	$2s^23d-$ $2s^26p$	^{2}D$-^2$P°	*2669.9*	*2670.7*	*145550.1*	182993.23	10−6	5.08−02	3.26−03	2.87−01	−1.487	B	2
			2669.97	2670.76	145550.70	182993.23	6−4	4.57−02	3.26−03	1.72−01	−1.709	B	*LS*
			2669.86	2670.66	145549.27	182993.23	4−2	5.08−02	2.71−03	9.54−02	−1.964	B	*LS*
			2669.86	2670.66	145549.27	182993.23	4−4	5.08−03	5.43−04	1.91−02	−2.663	B−	*LS*
77.	$2s^23d-$ $2s^27p$	^{2}D$-^2$P°	*2426.7*	*2427.4*	*145550.1*	186745.9	10−6	3.46−02	1.83−03	1.46−01	−1.737	B	2
			2426.73	2427.47	145550.70	186745.9	6−4	3.11−02	1.83−03	8.78−02	−1.959	B	*LS*
			2426.65	2427.38	145549.27	186745.9	4−2	3.46−02	1.53−03	4.88−02	−2.214	B	*LS*
			2426.65	2427.38	145549.27	186745.9	4−4	3.46−03	3.05−04	9.76−03	−2.913	C+	*LS*
78.	$2s^23d-$ $2s2p(^3$P°$)3d$	^{2}D$-^2$D°	*2322.1*	*2322.8*	*145550.1*	*188601.5*	10−10	3.72−03	3.01−04	2.30−02	−2.522	B−	2
			2321.39	2322.11	145550.70	188615.07	6−6	3.48−03	2.81−04	1.29−02	−2.773	B−	*LS*
			2323.14	2323.85	145549.27	188581.25	4−4	3.35−03	2.71−04	8.28−03	−2.966	C+	*LS*
			2323.22	2323.93	145550.70	188581.25	6−4	3.72−04	2.00−05	9.20−04	−3.920	D	*LS*
			2321.32	2322.03	145549.27	188615.07	4−6	2.48−04	3.01−05	9.20−04	−3.919	D	*LS*
79.	$2p^3-$ $2s2p(^3$P°$)3p$	^{2}D°$-^2$P	*3166.3*	*3167.2*	*150463.6*	*182036.9*	10−6	2.39−01	2.15−02	2.24+00	−0.667	B	2
			3165.46	3166.38	150461.58	182043.41	6−4	2.15−01	2.15−02	1.35+00	−0.889	B	*LS*
			3167.94	3168.85	150466.69	182023.86	4−2	2.38−01	1.79−02	7.47−01	−1.145	B	*LS*
			3165.97	3166.89	150466.69	182043.41	4−4	2.39−02	3.59−03	1.50−01	−1.843	B	*LS*
80.		^{2}D°$-^2$D	*2621.3*	*2622.1*	*150463.6*	*188601.5*	10−10	5.39−02	5.56−03	4.80−01	−1.255	B	2
			2620.21	2620.99	150461.58	188615.07	6−6	5.04−02	5.19−03	2.69−01	−1.507	B	*LS*
			2622.89	2623.67	150466.69	188581.25	4−4	4.85−02	5.00−03	1.73−01	−1.699	B	*LS*
			2622.54	2623.32	150461.58	188581.25	6−4	5.39−03	3.70−04	1.92−02	−2.653	B−	*LS*
			2620.56	2621.34	150466.69	188615.07	4−6	3.60−03	5.56−04	1.92−02	−2.653	B−	*LS*
81.	$2p^3-$ $2s2p(^3$P°$)4p$	^{2}D°$-^2$P		*1563.5*	*150463.6*	*214421.4*	10−6	1.32−01	2.91−03	1.50−01	−1.536	B	2
				1563.27	150461.58	214429.95	6−4	1.19−01	2.91−03	8.98−02	−1.758	B	*LS*
				1564.02	150466.69	214404.33	4−2	1.32−01	2.42−03	4.99−02	−2.014	B	*LS*
				1563.40	150466.69	214429.95	4−4	1.32−02	4.85−04	9.98−03	−2.712	C+	*LS*
82.	$2p^3-$ $2s2p(^3$P°$)5p$	^{2}D°$-^2$P		*1291.4*	*150463.6*	*227898.9*	10−6	6.19−02	9.29−04	3.95−02	−2.032	B	2
				1291.22	150461.58	227907.71	6−4	5.58−02	9.29−04	2.37−02	−2.254	B−	*LS*
				1291.75	150466.69	227881.21	4−2	6.18−02	7.73−04	1.32−02	−2.510	B−	*LS*
				1291.31	150466.69	227907.71	4−4	6.20−03	1.55−04	2.63−03	−3.208	C	*LS*
83.	$2s^24s-$ $2s^24p$	^{2}S$-^2$P°	*18905*	*5288.2* cm^{-1}	157234.07	*162522.3*	2−6	6.94−02	1.12+00	1.39+02	0.348	B	2
			18896.6	5290.50 cm^{-1}	157234.07	162524.57	2−4	6.95−02	7.44−01	9.26+01	0.173	B	*LS*
			18920.5	5283.82 cm^{-1}	157234.07	162517.89	2−2	6.92−02	3.71−01	4.63+01	−0.129	B	*LS*
84.	$2s^24s-$ $2p^3$	^{2}S$-^2$P°	*8687.3*	*8689.7*	157234.07	*168742.0*	2−6	4.14−02	1.41−01	8.04+00	−0.551	B	2
			8682.52	8684.91	157234.07	168748.30	2−4	4.15−02	9.38−02	5.36+00	−0.727	B	*LS*
			8696.70	8699.09	157234.07	168729.53	2−2	4.13−02	4.68−02	2.68+00	−1.029	B	*LS*
85.	$2s^24s-$ $2s^25p$	^{2}S$-^2$P°	*5536.1*	*5537.7*	157234.07	*175292.3*	2−6	1.75−02	2.41−02	8.79−01	−1.317	B	2
			5535.35	5536.89	157234.07	175294.75	2−4	1.75−02	1.61−02	5.86−01	−1.493	B	*LS*
			5537.61	5539.15	157234.07	175287.39	2−2	1.75−02	8.04−03	2.93−01	−1.794	B	*LS*

C II: Allowed Transitions — Continued

No.	Transition Array	Mult.	λ_{air} (Å)	λ_{vac} (Å) or σ (cm^{-1})*	E_i (cm^{-1})	E_k (cm^{-1})	g_i-g_k	A_{ki} (10^8 s^{-1})	f_{ik}	S (at. u.)	log gf	Acc.	Source
86.	$2s^24s-$ $2s2p(^3P^\circ)3s$	$^2S-^2P^\circ$	*4864.1*	*4865.4*	157234.07	*177787.2*	2–6	2.34–02	2.49–02	7.97–01	–1.303	B	2
			4862.58	4863.94	157234.07	177793.54	2–4	2.34–02	1.66–02	5.31–01	–1.479	B	*LS*
			4867.07	4868.43	157234.07	177774.59	2–2	2.33–02	8.29–03	2.66–01	–1.781	B	*LS*
87.	$2s^24s-$ $2s^26p$	$^2S-^2P^\circ$	3881.01	3882.1	157234.07	182993.2	2–6	3.99–03	2.70–03	6.91–02	–2.267	B	2
88.	$2s^24s-$ $2s^27p$	$^2S-^2P^\circ$	3387.50	3388.5	157234.07	186745.9	2–6	6.72–03	3.47–03	7.73–02	–2.159	B	2
89.	$2s^24p-$ $2s^24d$	$^2P^\circ-^2D$	*17846*	*5601.9* cm^{-1}	*162522.3*	*168124.2*	6–10	1.15–01	9.19–01	3.24+02	0.741	B	2
			17852.7	5599.88 cm^{-1}	162524.57	168124.45	4–6	1.15–01	8.27–01	1.94+02	0.519	B	*LS*
			17833.6	5605.85 cm^{-1}	162517.89	168123.74	2–4	9.63–02	9.18–01	1.08+02	0.264	B	*LS*
			17854.9	5599.17 cm^{-1}	162524.57	168123.74	4–4	1.92–02	9.18–02	2.16+01	–0.435	B	*LS*
90.	$2s^24p-$ $2s^25s$	$^2P^\circ-^2S$	*9234.9*	*9237.4*	*162522.3*	173347.84	6–2	4.65–01	1.98–01	3.62+01	0.075	B	2
			9236.82	9239.35	162524.57	173347.84	4–2	3.10–01	1.98–01	2.41+01	–0.101	B	*LS*
			9231.12	9233.65	162517.89	173347.84	2–2	1.55–01	1.98–01	1.21+01	–0.402	B	*LS*
91.	$2s^24p-$ $2s^25d$	$^2P^\circ-^2D$	*6258.8*	*6260.5*	*162522.3*	*178495.5*	6–10	3.51–02	3.44–02	4.25+00	–0.685	B	2
			6259.56	6261.29	162524.57	178495.71	4–6	3.51–02	3.09–02	2.55+00	–0.907	B	*LS*
			6257.18	6258.91	162517.89	178495.11	2–4	2.93–02	3.44–02	1.42+00	–1.163	B	*LS*
			6259.80	6261.53	162524.57	178495.11	4–4	5.85–03	3.44–03	2.83–01	–1.862	B	*LS*
92.	$2s^24p-$ $2s^26s$	$^2P^\circ-^2S$	*5334.2*	*5335.6*	*162522.3*	181264.24	6–2	2.07–01	2.94–02	3.10+00	–0.753	B	2
			5334.79	5336.27	162524.57	181264.24	4–2	1.38–01	2.94–02	2.07+00	–0.929	B	*LS*
			5332.89	5334.37	162517.89	181264.24	2–2	6.90–02	2.94–02	1.03+00	–1.230	B	*LS*
93.	$2s^24p-$ $2s2p(^3P^\circ)3p$	$^2P^\circ-^2P$	*5122.9*	*5124.4*	*162522.3*	*182036.9*	6–6	4.82–02	1.90–02	1.92+00	–0.944	B	2
			5121.83	5123.26	162524.57	182043.41	4–4	4.02–02	1.58–02	1.07+00	–1.199	B	*LS*
			5125.21	5126.64	162517.89	182023.86	2–2	3.21–02	1.26–02	4.26–01	–1.597	B	*LS*
			5126.96	5128.39	162524.57	182023.86	4–2	1.60–02	3.16–03	2.13–01	–1.899	B	*LS*
			5120.08	5121.50	162517.89	182043.41	2–4	8.04–03	6.32–03	2.13–01	–1.898	B	*LS*
94.	$2s^24p-$ $2s^26d$	$^2P^\circ-^2D$	*4638.5*	*4639.8*	*162522.3*	*184075.0*	6–10	5.49–02	2.95–02	2.71+00	–0.752	B	2
			4638.92	4640.22	162524.57	184075.28	4–6	5.49–02	2.66–02	1.62+00	–0.974	B	*LS*
			4637.63	4638.93	162517.89	184074.59	2–4	4.57–02	2.95–02	9.01–01	–1.229	B	*LS*
			4639.07	4640.37	162524.57	184074.59	4–4	9.15–03	2.95–03	1.80–01	–1.928	B	*LS*
95.	$2s^24p-$ $2s^27s$	$^2P^\circ-^2S$	*4307.2*	*4308.4*	*162522.3*	185732.93	6–2	1.12–01	1.03–02	8.80–01	–1.207	B	2
			4307.58	4308.79	162524.57	185732.93	4–2	7.44–02	1.03–02	5.87–01	–1.383	B	*LS*
			4306.34	4307.55	162517.89	185732.93	2–2	3.72–02	1.03–02	2.93–01	–1.684	B	*LS*
96.	$2s^24p-$ $2s2p(^3P^\circ)3p$	$^2P^\circ-^2D$	*3833.3*	*3834.4*	*162522.3*	*188602.1*	6–10	2.04–01	7.49–02	5.67+00	–0.347	B	2
			3831.73	3832.81	162524.57	188615.07	4–6	2.04–01	6.74–02	3.40+00	–0.569	B	*LS*
			3835.72	3836.80	162517.89	188581.25	2–4	1.69–01	7.48–02	1.89+00	–0.825	B	*LS*
			3836.70	3837.79	162524.57	188581.25	4–4	3.39–02	7.48–03	3.78–01	–1.524	B	*LS*
97.	$2s^24p-$ $2s2p(^3P^\circ)4p$	$^2P^\circ-^2P$		*1926.8*	*162522.3*	*214421.4*	6–6	1.36–02	7.57–04	2.88–02	–2.343	B–	2
				1926.58	162524.57	214429.95	4–4	1.13–02	6.31–04	1.60–02	–2.598	B–	*LS*
				1927.29	162517.89	214404.33	2–2	9.06–03	5.04–04	6.40–03	–2.996	C+	*LS*
				1927.53	162524.57	214404.33	4–2	4.53–03	1.26–04	3.20–03	–3.297	C+	*LS*
				1926.33	162517.89	214429.95	2–4	2.27–03	2.52–04	3.20–03	–3.297	C+	*LS*

C II: Allowed Transitions — Continued

No.	Transition Array	Mult.	λ_{air} (Å)	λ_{vac} (Å) or σ (cm^{-1})*	E_i (cm^{-1})	E_k (cm^{-1})	g_i–g_k	A_{ki} (10^8 s^{-1})	f_{ik}	S (at. u.)	log gf	Acc.	Source
98.	$2s^24p$– $2s2p(^3P^o)5p$	$^2P^o$–^2P		*1529.6*	*162522.3*	– *227898.9*	6–6	1.54–02	5.39–04	1.63–02	–2.490	B–	2
				1529.45	162524.57	– 227907.71	4–4	1.28–02	4.50–04	9.06–03	–2.745	C+	*LS*
				1529.91	162517.89	– 227881.21	2–2	1.02–02	3.60–04	3.62–03	–3.143	C+	*LS*
				1530.07	162524.57	– 227881.21	4–2	5.12–03	8.99–05	1.81–03	–3.444	C	*LS*
				1529.29	162517.89	– 227907.71	2–4	2.56–03	1.80–04	1.81–03	–3.444	C	*LS*
99.	$2s2p(^3P^o)3s$– $2s2p(^3P^o)3p$	$^4P^o$–^4D	*6785.9*	*6787.8*	*167009.3*	– *181741.7*	12–20	3.64–01	4.19–01	1.12+02	0.702	B	2
			6783.91	6785.78	167035.71	– 181772.41	6–8	3.65–01	3.36–01	4.50+01	0.304	B	*LS*
			6779.94	6781.81	166990.73	– 181736.05	4–6	2.56–01	2.64–01	2.36+01	0.024	B	*LS*
			6780.59	6782.47	166967.13	– 181711.03	2–4	1.52–01	2.10–01	9.37+00	–0.377	B	*LS*
			6800.69	6802.56	167035.71	– 181736.05	6–6	1.09–01	7.53–02	1.01+01	–0.345	B	*LS*
			6791.47	6793.34	166990.73	– 181711.03	4–4	1.94–01	1.34–01	1.20+01	–0.271	B	*LS*
			6787.21	6789.08	166967.13	– 181696.66	2–2	3.04–01	2.10–01	9.37+00	–0.378	B	*LS*
			6812.28	6814.16	167035.71	– 181711.03	6–4	1.80–02	8.35–03	1.12+00	–1.300	B	*LS*
			6798.10	6799.98	166990.73	– 181696.66	4–2	6.04–02	2.09–02	1.87+00	–1.077	B	*LS*
100.		$^4P^o$–^4S	*5654.0*	*5655.6*	*167009.3*	– 184690.98	12–4	5.89–01	9.41–02	2.10+01	0.053	B	2
			5662.46	5664.03	167035.71	– 184690.98	6–4	2.93–01	9.40–02	1.05+01	–0.249	B	*LS*
			5648.07	5649.64	166990.73	– 184690.98	4–4	1.97–01	9.42–02	7.01+00	–0.424	B	*LS*
			5640.55	5642.12	166967.13	– 184690.98	2–4	9.89–02	9.44–02	3.50+00	–0.724	B	*LS*
101.		$^4P^o$–^4P	*5141.9*	*5143.3*	*167009.3*	– *186452.1*	12–12	9.29–01	3.68–01	7.48+01	0.645	B	2
			5145.16	5146.60	167035.71	– 186466.02	6–6	6.49–01	2.58–01	2.62+01	0.189	B	*LS*
			5139.17	5140.61	166990.73	– 186443.69	4–4	1.24–01	4.91–02	3.32+00	–0.707	B	*LS*
			5137.26	5138.69	166967.13	– 186427.35	2–2	1.55–01	6.14–02	2.08+00	–0.911	B	*LS*
			5151.08	5152.52	167035.71	– 186443.69	6–4	4.16–01	1.10–01	1.12+01	–0.179	B	*LS*
			5143.49	5144.93	166990.73	– 186427.35	4–2	7.73–01	1.53–01	1.04+01	–0.212	B	*LS*
			5133.28	5134.71	166990.73	– 186466.02	4–6	2.80–01	1.66–01	1.12+01	–0.178	B	*LS*
			5132.95	5134.38	166967.13	– 186443.69	2–4	3.89–01	3.07–01	1.04+01	–0.211	B	*LS*
102.	$2s2p(^3P^o)3s$– $2s2p(^3P^o)4p$	$^4P^o$–^4D	*2091.7*	*2092.4*	*167009.3*	– *214801.0*	12–20	2.41–01	2.63–02	2.17+00	–0.501	B	2
			2091.65	2092.31	167035.71	– 214829.77	6–8	2.41–01	2.10–02	8.70–01	–0.899	B	*LS*
			2091.19	2091.85	166990.73	– 214795.27	4–6	1.69–01	1.66–02	4.56–01	–1.178	B	*LS*
			2091.14	2091.80	166967.13	– 214772.84	2–4	1.00–01	1.32–02	1.81–01	–1.580	B	*LS*
			2093.16	2093.82	167035.71	– 214795.27	6–6	7.20–02	4.73–03	1.96–01	–1.547	B	*LS*
			2092.17	2092.83	166990.73	– 214772.84	4–4	1.28–01	8.42–03	2.32–01	–1.473	B	*LS*
			2091.70	2092.37	166967.13	– 214759.91	2–2	2.01–01	1.32–02	1.81–01	–1.580	B	*LS*
			2094.14	2094.81	167035.71	– 214772.84	6–4	1.20–02	5.26–04	2.17–02	–2.501	B–	*LS*
			2092.73	2093.40	166990.73	– 214759.91	4–2	4.00–02	1.31–03	3.62–02	–2.279	B	*LS*
103.		$^4P^o$–^4S	*2050.3*	*2050.9*	*167009.3*	– 215767.77	12–4	6.58–02	1.38–03	1.12–01	–1.780	B	2
			2051.38	2052.04	167035.71	– 215767.77	6–4	3.28–02	1.38–03	5.60–02	–2.082	B	*LS*
			2049.49	2050.14	166990.73	– 215767.77	4–4	2.19–02	1.38–03	3.73–02	–2.257	B	*LS*
			2048.50	2049.15	166967.13	– 215767.77	2–4	1.10–02	1.38–03	1.87–02	–2.558	B–	*LS*
104.		$^4P^o$–^4P	*2024.5*	*2025.2*	*167009.3*	– *216387.3*	12–12	1.54–02	9.46–04	7.57–02	–1.945	B	2
			2025.08	2025.73	167035.71	– 216400.57	6–6	1.08–02	6.62–04	2.65–02	–2.401	B–	*LS*
			2024.10	2024.75	166990.73	– 216379.59	4–4	2.06–03	1.26–04	3.36–03	–3.297	C+	*LS*
			2023.82	2024.47	166967.13	– 216362.84	2–2	2.57–03	1.58–04	2.10–03	–3.501	C	*LS*
			2025.94	2026.59	167035.71	– 216379.59	6–4	6.92–03	2.84–04	1.14–02	–2.769	B–	*LS*
			2024.78	2025.44	166990.73	– 216362.84	4–2	1.28–02	3.94–04	1.05–02	–2.802	B–	*LS*
			2023.24	2023.89	166990.73	– 216400.57	4–6	4.63–03	4.26–04	1.14–02	–2.768	B–	*LS*
			2023.13	2023.78	166967.13	– 216379.59	2–4	6.43–03	7.89–04	1.05–02	–2.802	B–	*LS*

C II: Allowed Transitions — Continued

No.	Transition Array	Mult.	λ_{air} (Å)	λ_{vac} (Å) or σ (cm^{-1})*	E_i (cm^{-1})	E_k (cm^{-1})	g_i–g_k	A_{ki} (10^8 s^{-1})	f_{ik}	S (at. u.)	log gf	Acc.	Source
105.	$2s2p(^3P^o)3s$– $2s2p(^3P^o)5p$	$^4P^o$–^4D		*1633.7*	*167009.3*	– *228218.6*	12–20	1.73–01	1.15–02	7.44–01	–0.859	B	2
				1633.69	167035.71	– 228246.97	6–8	1.73–01	9.22–03	2.98–01	–1.257	B	*LS*
				1633.42	166990.73	– 228211.8	4–6	1.21–01	7.27–03	1.56–01	–1.537	B	*LS*
				1633.33	166967.13	– 228191.9	2–4	7.21–02	5.77–03	6.20–02	–1.938	B	*LS*
				1634.63	167035.71	– 228211.8	6–6	5.18–02	2.07–03	6.70–02	–1.905	B	*LS*
				1633.96	166990.73	– 228191.9	4–4	9.22–02	3.69–03	7.94–02	–1.831	B	*LS*
				1633.68	166967.13	– 228178.6	2–2	1.44–01	5.77–03	6.20–02	–1.938	B	*LS*
				1635.16	167035.71	– 228191.9	6–4	8.62–03	2.30–04	7.44–03	–2.859	C+	*LS*
				1634.31	166990.73	– 228178.6	4–2	2.88–02	5.76–04	1.24–02	–2.637	B–	*LS*
106.		$^4P^o$–^4S		*1622.1*	*167009.3*	– 228656.84	12–4	6.54–02	8.60–04	5.51–02	–1.986	B	2
				1622.82	167035.71	– 228656.84	6–4	3.27–02	8.59–04	2.76–02	–2.288	B–	*LS*
				1621.64	166990.73	– 228656.84	4–4	2.18–02	8.60–04	1.84–02	–2.463	B–	*LS*
				1621.02	166967.13	– 228656.84	2–4	1.09–02	8.60–04	9.18–03	–2.764	C+	*LS*
107.	$2s2p(^3P^o)3s$– $2s2p(^3P^o)6p$	$^4P^o$–^4D		*1470.0*	*167009.3*	– *235038*	12–20	1.10–01	5.96–03	3.46–01	–1.146	B	2
				1469.92	167035.71	– 235066.8	6–8	1.10–01	4.77–03	1.38–01	–1.544	B	*LS*
				1469.72	166990.73	– 235031	4–6	7.73–02	3.75–03	7.27–02	–1.823	B	*LS*
				1469.64	166967.13	– 235011	2–4	4.60–02	2.98–03	2.88–02	–2.225	B–	*LS*
				1470.69	167035.71	– 235031	6–6	3.31–02	1.07–03	3.11–02	–2.192	B	*LS*
				1470.15	166990.73	– 235011	4–4	5.88–02	1.91–03	3.69–02	–2.118	B	*LS*
				1469.92	166967.13	– 234998	2–2	9.20–02	2.98–03	2.88–02	–2.225	B–	*LS*
				1471.12	167035.71	– 235011	6–4	5.50–03	1.19–04	3.46–03	–3.146	C+	*LS*
				1470.43	166990.73	– 234998	4–2	1.84–02	2.98–04	5.77–03	–2.924	C+	*LS*
108.	$2s^24d$– $2p^3$	^2D–$^2P^o$		*617.8* cm^{-1}	*168124.2*	– *168742.0*	10–6	3.68–05	8.67–03	4.62+01	–1.062	B	2
				623.85 cm^{-1}	168124.45	– 168748.30	6–4	3.41–05	8.76–03	2.77+01	–1.280	B	*LS*
				605.79 cm^{-1}	168123.74	– 168729.53	4–2	3.47–05	7.08–03	1.54+01	–1.548	B	*LS*
				624.56 cm^{-1}	168123.74	– 168748.30	4–4	3.80–06	1.46–03	3.08+00	–2.233	B	*LS*
109.	$2s^24d$– $2s^24f$	^2D–$^2F^o$		*854.2* cm^{-1}	*168124.2*	– 168978.34	10–14	3.49–04	1.00–01	3.87+02	0.002	B	4
				853.89 cm^{-1}	168124.45	– 168978.34	6–8	3.49–04	9.56–02	2.21+02	–0.241	C+	*LS*
				854.60 cm^{-1}	168123.74	– 168978.34	4–6	3.27–04	1.00–01	1.55+02	–0.396	C+	*LS*
				853.89 cm^{-1}	168124.45	– 168978.34	6–6	2.33–05	4.78–03	1.11+01	–1.542	C+	*LS*
110.	$2s^24d$– $2s^25p$	^2D–$^2P^o$	*13947*	*7168.1* cm^{-1}	*168124.2*	– *175292.3*	10–6	9.40–02	1.64–01	7.55+01	0.216	B	2
			13942.6	7170.30 cm^{-1}	168124.45	– 175294.75	6–4	8.46–02	1.64–01	4.53+01	–0.006	B	*LS*
			13955.5	7163.65 cm^{-1}	168123.74	– 175287.39	4–2	9.37–02	1.37–01	2.51+01	–0.262	B	*LS*
			13941.2	7171.01 cm^{-1}	168123.74	– 175294.75	4–4	9.41–03	2.74–02	5.03+00	–0.960	B	*LS*
111.	$2s^24d$– $2s2p(^3P^o)3s$	^2D–$^2P^o$	*10346*	*9663.0* cm^{-1}	*168124.2*	– *177787.2*	10–6	3.06–03	2.95–03	1.00+00	–1.531	B	2
			10339.4	9669.09 cm^{-1}	168124.45	– 177793.54	6–4	2.76–03	2.95–03	6.02–01	–1.752	B	*LS*
			10358.9	9650.85 cm^{-1}	168123.74	– 177774.59	4–2	3.05–03	2.45–03	3.34–01	–2.009	B	*LS*
			10338.6	9669.80 cm^{-1}	168123.74	– 177793.54	4–4	3.07–04	4.91–04	6.69–02	–2.706	B	*LS*
112.	$2s^24d$– $2s^26p$	^2D–$^2P^o$	*6723.5*	*6725.4*	*168124.2*	– 182993.23	10–6	5.16–02	2.10–02	4.64+00	–0.678	B	2
			6723.65	6725.50	168124.45	– 182993.23	6–4	4.64–02	2.10–02	2.79+00	–0.900	B	*LS*
			6723.32	6725.18	168123.74	– 182993.23	4–2	5.15–02	1.75–02	1.55+00	–1.156	B	*LS*
			6723.32	6725.18	168123.74	– 182993.23	4–4	5.16–03	3.50–03	3.09–01	–1.854	B	*LS*
113.	$2s^24d$– $2s^27p$	^2D–$^2P^o$	*5368.6*	*5370.1*	*168124.2*	– 186745.9	10–6	3.11–02	8.07–03	1.43+00	–1.093	B	2
			5368.66	5370.15	168124.45	– 186745.9	6–4	2.80–02	8.07–03	8.56–01	–1.315	B	*LS*
			5368.45	5369.95	168123.74	– 186745.9	4–2	3.11–02	6.72–03	4.75–01	–1.571	B	*LS*
			5368.45	5369.95	168123.74	– 186745.9	4–4	3.11–03	1.35–03	9.51–02	–2.269	B	*LS*

C II: Allowed Transitions — Continued

No.	Transition Array	Mult.	λ_{air} (Å)	λ_{vac} (Å) or σ (cm^{-1})*	E_i (cm^{-1})	E_k (cm^{-1})	g_i-g_k	A_{ki} (10^8 s^{-1})	f_{ik}	S (at. u.)	log gf	Acc.	Source
114.	$2s^24d-$ $2s2p(^3P^o)3d$	$^2D-^2D^o$	*3298.5*	*3299.5*	*168124.2*	*– 198432.0*	10–10	2.65–03	4.32–04	4.69–02	–2.365	B	2
			3298.09	3299.04	168124.45	– 198436.31	6–6	2.47–03	4.03–04	2.63–02	–2.616	B–	LS
			3299.20	3300.15	168123.74	– 198425.43	4–4	2.38–03	3.89–04	1.69–02	–2.808	B–	LS
			3299.27	3300.22	168124.45	– 198425.43	6–4	2.65–04	2.88–05	1.88–03	–3.763	C	LS
			3298.01	3298.96	168123.74	– 198436.31	4–6	1.77–04	4.32–05	1.88–03	–3.762	C	LS
115.	$2p^3-$ $2s^25s$	$^2P^o-^2S$	*21706*	*4605.8 cm^{-1}*	*168742.0*	– 173347.84	6–2	2.92–02	6.87–02	2.95+01	–0.385	B	2
			21735.4	4599.54 cm^{-1}	168748.30	– 173347.84	4–2	1.94–02	6.86–02	1.96+01	–0.562	B	LS
			21647.0	4618.31 cm^{-1}	168729.53	– 173347.84	2–2	9.80–03	6.89–02	9.82+00	–0.861	B	LS
116.	$2p^3-$ $2s^25d$	$^2P^o-^2D$	*10250*	*9753.5 cm^{-1}*	*168742.0*	*– 178495.5*	6–10	4.41–02	1.16–01	2.34+01	–0.159	B	2
			10256.3	9747.41 cm^{-1}	168748.30	– 178495.71	4–6	4.40–02	1.04–01	1.41+01	–0.381	B	LS
			10237.2	9765.58 cm^{-1}	168729.53	– 178495.11	2–4	3.68–02	1.16–01	7.80+00	–0.636	B	LS
			10257.0	9746.81 cm^{-1}	168748.30	– 178495.11	4–4	7.33–03	1.16–02	1.56+00	–1.335	B	LS
117.	$2p^3-$ $2s^26s$	$^2P^o-^2S$	*7983.6*	*7985.8*	*168742.0*	– 181264.24	6–2	2.16–03	6.89–04	1.09–01	–2.384	B	2
			7987.61	7989.81	168748.30	– 181264.24	4–2	1.44–03	6.88–04	7.24–02	–2.560	B	LS
			7975.65	7977.85	168729.53	– 181264.24	2–2	7.23–04	6.89–04	3.62–02	–2.861	B	LS
118.	$2p^3-$ $2s2p(^3P^o)3p$	$^2P^o-^2P$	*7519.6*	*7521.7*	*168742.0*	*– 182036.9*	6–6	9.13–02	7.74–02	1.15+01	–0.333	B	2
			7519.49	7521.56	168748.30	– 182043.41	4–4	7.61–02	6.45–02	6.39+00	–0.588	B	LS
			7519.93	7522.00	168729.53	– 182023.86	2–2	6.08–02	5.16–02	2.55+00	–0.986	B	LS
			7530.57	7532.64	168748.30	– 182023.86	4–2	3.03–02	1.29–02	1.28+00	–1.288	B	LS
			7508.89	7510.96	168729.53	– 182043.41	2–4	1.53–02	2.58–02	1.28+00	–1.287	B	LS
119.	$2p^3-$ $2s^26d$	$^2P^o-^2D$	*6520.1*	*6521.9*	*168742.0*	*– 184075.0*	6–10	8.16–03	8.66–03	1.12+00	–1.284	B	2
			6522.64	6524.44	168748.30	– 184075.28	4–6	8.15–03	7.79–03	6.69–01	–1.506	B	LS
			6514.96	6516.76	168729.53	– 184074.59	2–4	6.81–03	8.66–03	3.72–01	–1.761	B	LS
			6522.93	6524.74	168748.30	– 184074.59	4–4	1.36–03	8.66–04	7.44–02	–2.460	B	LS
120.	$2p^3-$ $2s^27s$	$^2P^o-^2S$	*5883.9*	*5885.5*	*168742.0*	– 185732.93	6–2	2.35–03	4.06–04	4.72–02	–2.613	B	2
			5886.04	5887.68	168748.30	– 185732.93	4–2	1.56–03	4.06–04	3.15–02	–2.789	B	LS
			5879.55	5881.18	168729.53	– 185732.93	2–2	7.84–04	4.06–04	1.57–02	–3.090	B–	LS
121.	$2p^3-$ $2s2p(^3P^o)3p$	$^2P^o-^2D$	*5034.0*	*5035.4*	*168742.0*	*– 188601.5*	6–10	3.15–01	2.00–01	1.99+01	0.078	B	2
			5032.13	5033.53	168748.30	– 188615.07	4–6	3.16–01	1.80–01	1.19+01	–0.143	B	LS
			5035.94	5037.35	168729.53	– 188581.25	2–4	2.62–01	1.99–01	6.61+00	–0.399	B	LS
			5040.71	5042.11	168748.30	– 188581.25	4–4	5.23–02	1.99–02	1.32+00	–1.098	B	LS
122.	$2p^3-$ $2s2p(^3P^o)4p$	$^2P^o-^2P$	*2188.5*	*2189.2*	*168742.0*	*– 214421.4*	6–6	4.05–01	2.91–02	1.26+00	–0.758	B	2
			2188.38	2189.06	168748.30	– 214429.95	4–4	3.37–01	2.42–02	6.98–01	–1.014	B	LS
			2188.71	2189.39	168729.53	– 214404.33	2–2	2.70–01	1.94–02	2.79–01	–1.412	B	LS
			2189.61	2190.29	168748.30	– 214404.33	4–2	1.35–01	4.84–03	1.40–01	–1.713	B	LS
			2187.48	2188.16	168729.53	– 214429.95	2–4	6.76–02	9.69–03	1.40–01	–1.712	B	LS
123.	$2p^3-$ $2s2p(^3P^o)5p$	$^2P^o-^2P$		*1690.4*	*168742.0*	*– 227898.9*	6–6	3.28–01	1.40–02	4.69–01	–1.074	B	2
				1690.35	168748.30	– 227907.71	4–4	2.73–01	1.17–02	2.60–01	–1.330	B	LS
				1690.57	168729.53	– 227881.21	2–2	2.18–01	9.36–03	1.04–01	–1.728	B	LS
				1691.11	168748.30	– 227881.21	4–2	1.09–01	2.34–03	5.21–02	–2.029	B	LS
				1689.81	168729.53	– 227907.71	2–4	5.47–02	4.68–03	5.21–02	–2.029	B	LS

C II: Allowed Transitions — Continued

No.	Transition Array	Mult.	λ_{air} (Å)	λ_{vac} (Å) or σ (cm^{-1})*	E_i (cm^{-1})	E_k (cm^{-1})	g_i-g_k	A_{ki} (10^8 s^{-1})	f_{ik}	S (at. u.)	log gf	Acc.	Source
124.	$2s2p(^3P^\circ)3s-2s^26s$	$^2P^\circ-^2S$	*28753*	*3477.0 cm^{-1}*	*177787.2*	181264.24	6–2	4.90–02	2.03–01	1.15+02	0.085	B	2
			28804.8	3470.70 cm^{-1}	177793.54	181264.24	4–2	3.25–02	2.02–01	7.67+01	–0.092	B	*LS*
			28648.4	3489.65 cm^{-1}	177774.59	181264.24	2–2	1.65–02	2.03–01	3.83+01	–0.391	B	*LS*
125.	$2s2p(^3P^\circ)3s-2s2p(^3P^\circ)3p$	$^2P^\circ-^2P$	*23525*	*4249.7 cm^{-1}*	*177787.2*	*182036.9*	6–6	5.83–03	4.84–02	2.25+01	–0.537	B	2
			23523.7	4249.87 cm^{-1}	177793.54	182043.41	4–4	4.86–03	4.03–02	1.25+01	–0.792	B	*LS*
			23527.0	4249.27 cm^{-1}	177774.59	182023.86	2–2	3.89–03	3.23–02	5.00+00	–1.190	B	*LS*
			23632.4	4230.32 cm^{-1}	177793.54	182023.86	4–2	1.92–03	8.03–03	2.50+00	–1.493	B	*LS*
			23419.3	4268.82 cm^{-1}	177774.59	182043.41	2–4	9.85–04	1.62–02	2.50+00	–1.489	B	*LS*
126.	$2s2p(^3P^\circ)3s-2s^26d$	$^2P^\circ-^2D$	*15899*	*6287.8 cm^{-1}*	*177787.2*	*184075.0*	6–10	2.94–02	1.86–01	5.83+01	0.047	B	2
			15914.8	6281.74 cm^{-1}	177793.54	184075.28	4–6	2.93–02	1.67–01	3.50+01	–0.175	B	*LS*
			15868.7	6300.00 cm^{-1}	177774.59	184074.59	2–4	2.46–02	1.86–01	1.94+01	–0.430	B	*LS*
			15916.6	6281.05 cm^{-1}	177793.54	184074.59	4–4	4.88–03	1.86–02	3.89+00	–1.130	B	*LS*
127.	$2s2p(^3P^\circ)3s-2s^27s$	$^2P^\circ-^2S$	*12582*	*7945.7 cm^{-1}*	*177787.2*	185732.93	6–2	7.40–03	5.86–03	1.46+00	–1.454	B	2
			12592.0	7939.39 cm^{-1}	177793.54	185732.93	4–2	4.92–03	5.85–03	9.70–01	–1.631	B	*LS*
			12562.0	7958.34 cm^{-1}	177774.59	185732.93	2–2	2.48–03	5.86–03	4.85–01	–1.931	B	*LS*
128.	$2s2p(^3P^\circ)3s-2s2p(^3P^\circ)3p$	$^2P^\circ-^2D$	*9244.5*	*9247.0*	*177787.2*	*188601.5*	6–10	3.33–02	7.12–02	1.30+01	–0.369	B	2
			9238.30	9240.84	177793.54	188615.07	4–6	3.34–02	6.41–02	7.80+00	–0.591	B	*LS*
			9251.01	9253.55	177774.59	188581.25	2–4	2.77–02	7.11–02	4.33+00	–0.847	B	*LS*
			9267.26	9269.81	177793.54	188581.25	4–4	5.52–03	7.10–03	8.67–01	–1.547	B	*LS*
129.	$2s2p(^3P^\circ)3s-2s2p(^3P^\circ)4p$	$^2P^\circ-^2P$	*2728.9*	*2729.7*	*177787.2*	*214421.4*	6–6	3.97–01	4.44–02	2.39+00	–0.575	B	2
			2728.72	2729.53	177793.54	214429.95	4–4	3.31–01	3.70–02	1.33+00	–0.830	B	*LS*
			2729.21	2730.02	177774.59	214404.33	2–2	2.65–01	2.96–02	5.31–01	–1.228	B	*LS*
			2730.63	2731.44	177793.54	214404.33	4–2	1.32–01	7.39–03	2.66–01	–1.529	B	*LS*
			2727.31	2728.11	177774.59	214429.95	2–4	6.63–02	1.48–02	2.66–01	–1.529	B	*LS*
130.	$2s2p(^3P^\circ)3s-2s2p(^3P^\circ)5p$	$^2P^\circ-^2P$		*1995.5*	*177787.2*	*227898.9*	6–6	2.21–01	1.32–02	5.20–01	–1.101	B	2
				1995.44	177793.54	227907.71	4–4	1.84–01	1.10–02	2.89–01	–1.356	B	*LS*
				1995.74	177774.59	227881.21	2–2	1.47–01	8.80–03	1.16–01	–1.754	B	*LS*
				1996.50	177793.54	227881.21	4–2	7.36–02	2.20–03	5.78–02	–2.056	B	*LS*
				1994.69	177774.59	227907.71	2–4	3.69–02	4.40–03	5.78–02	–2.055	B	*LS*
131.	$2s2p(^3P^\circ)3p-2s2p(^3P^\circ)3d$	$^4D-^4F^\circ$	*7118.0*	*7120.0*	*181741.7*	*195786.7*	20–28	4.19–01	4.46–01	2.09+02	0.950	B	2
			7119.91	7121.87	181772.41	195813.66	8–10	4.19–01	3.98–01	7.46+01	0.503	B	*LS*
			7115.63	7117.59	181736.05	195785.74	6–8	3.60–01	3.64–01	5.12+01	0.339	B	*LS*
			7113.04	7115.00	181711.03	195765.85	4–6	3.15–01	3.58–01	3.36+01	0.157	B	*LS*
			7112.48	7114.44	181696.66	195752.58	2–4	2.94–01	4.46–01	2.09+01	–0.049	B	*LS*
			7134.10	7136.06	181772.41	195785.74	8–8	5.93–02	4.53–02	8.51+00	–0.441	B	*LS*
			7125.72	7127.69	181736.05	195765.85	6–6	1.02–01	7.74–02	1.09+01	–0.333	B	*LS*
			7119.76	7121.72	181711.03	195752.58	4–4	1.17–01	8.91–02	8.36+00	–0.448	B	*LS*
			7105.77	7107.73	181696.66	195765.85	8–6	4.07–03	2.31–03	4.33–01	–1.733	B	*LS*
			7132.47	7134.43	181736.05	195752.58	6–4	8.33–03	4.24–03	5.97–01	–1.595	B	*LS*

C II: Allowed Transitions — Continued

No.	Transition Array	Mult.	λ_{air} (Å)	λ_{vac} (Å) or σ (cm^{-1})*	E_i (cm^{-1})	E_k (cm^{-1})	$g_i - g_k$	A_{ki} (10^8 s^{-1})	f_{ik}	S (at. u.)	log gf	Acc.	Source
132.		^{4}D–^{4}D°	*6740.7*	*6742.6*	*181741.7*	*– 196572.8*	20–20	1.26–01	8.59–02	3.81+01	0.235	B	2
			6750.54	6752.40	181772.41	– 196581.96	8–8	1.08–01	7.35–02	1.31+01	–0.230	B	*LS*
			6738.61	6740.47	181736.05	– 196571.82	6–6	7.23–02	4.93–02	6.56+00	–0.529	B	*LS*
			6731.07	6732.93	181711.03	– 196563.41	4–4	5.06–02	3.44–02	3.05+00	–0.861	B	*LS*
			6727.07	6728.93	181696.66	– 196557.87	2–2	6.34–02	4.30–02	1.91+00	–1.065	B	*LS*
			6755.16	6757.03	181772.41	– 196571.82	8–6	2.38–02	1.22–02	2.17+00	–1.010	B	*LS*
			6742.43	6744.29	181736.05	– 196563.41	6–4	4.41–02	2.00–02	2.67+00	–0.920	B	*LS*
			6733.58	6735.44	181711.03	– 196557.87	4–2	6.32–02	2.15–02	1.91+00	–1.066	B	*LS*
			6734.00	6735.86	181736.05	– 196581.96	6–8	1.80–02	1.63–02	2.17+00	–1.009	B	*LS*
			6727.26	6729.12	181711.03	– 196571.82	4–6	2.96–02	3.01–02	2.67+00	–0.919	B	*LS*
			6724.56	6726.42	181696.66	– 196563.41	2–4	3.17–02	4.30–02	1.91+00	–1.065	B	*LS*
133.		^{4}D–^{4}P°	*5841.1*	*5842.8*	*181741.7*	*– 198856.9*	20–12	6.69–02	2.05–02	7.90+00	–0.387	B	2
			5856.06	5857.69	181772.41	– 198844.00	8–6	5.31–02	2.05–02	3.16+00	–0.786	B	*LS*
			5836.37	5837.98	181736.05	– 198865.25	6–4	4.22–02	1.44–02	1.66+00	–1.064	B	*LS*
			5823.18	5824.80	181711.03	– 198879.01	4–2	3.38–02	8.58–03	6.58–01	–1.464	B	*LS*
			5843.62	5845.24	181736.05	– 198844.00	6–6	1.20–02	6.16–03	7.11–01	–1.433	B	*LS*
			5827.85	5829.47	181711.03	– 198865.25	4–4	2.16–02	1.10–02	8.42–01	–1.358	B	*LS*
			5818.31	5819.93	181696.66	– 198879.01	2–2	3.38–02	1.72–02	6.58–01	–1.464	B	*LS*
			5835.08	5836.70	181711.03	– 198844.00	4–6	1.34–03	1.03–03	7.90–02	–2.386	B	*LS*
			5822.98	5824.59	181696.66	– 198865.25	2–4	3.38–03	3.43–03	1.32–01	–2.163	B	*LS*
134.	2s2p(^{3}P°)3p – 2s2p(^{3}P°)4s	^{4}D–^{4}P°	*3589.2*	*3590.2*	*181741.7*	*– 209595.4*	20–12	9.77–01	1.13–01	2.68+01	0.355	B	2
			3589.65	3590.68	181772.41	– 209622.32	8–6	7.81–01	1.13–01	1.07+01	–0.043	B	*LS*
			3590.88	3591.90	181736.05	– 209576.46	6–4	6.15–01	7.92–02	5.62+00	–0.323	B	*LS*
			3590.76	3591.78	181711.03	– 209552.36	4–2	4.88–01	4.71–02	2.23+00	–0.724	B	*LS*
			3584.97	3585.99	181736.05	– 209622.32	6–6	1.76–01	3.40–02	2.41+00	–0.690	B	*LS*
			3587.65	3588.68	181711.03	– 209576.46	4–4	3.13–01	6.04–02	2.85+00	–0.617	B	*LS*
			3588.91	3589.93	181696.66	– 209552.36	2–2	4.89–01	9.43–02	2.23+00	–0.724	B	*LS*
			3581.76	3582.78	181711.03	– 209622.32	4–6	1.97–02	5.67–03	2.68–01	–1.644	B	*LS*
			3585.80	3586.83	181696.66	– 209576.46	2–4	4.90–02	1.89–02	4.46–01	–1.423	B	*LS*
135.	2s2p(^{3}P°)3p – 2s2p(^{3}P°)4d	^{4}D–^{4}F°	*2641.2*	*2642.0*	*181741.7*	*– 219591.9*	20–28	6.99–01	1.02–01	1.78+01	0.311	B	2
			2641.40	2642.18	181772.41	– 219619.88	8–10	6.99–01	9.14–02	6.36+00	–0.136	B	*LS*
			2640.89	2641.68	181736.05	– 219590.76	6–8	6.00–01	8.36–02	4.36+00	–0.299	B	*LS*
			2640.58	2641.37	181711.03	– 219570.15	4–6	5.25–01	8.23–02	2.86+00	–0.483	B	*LS*
			2640.53	2641.32	181696.66	– 219556.54	2–4	4.90–01	1.02–01	1.78+00	–0.689	B	*LS*
			2643.43	2644.22	181772.41	– 219590.76	8–8	9.94–02	1.04–02	7.25–01	–1.079	B	*LS*
			2642.33	2643.12	181736.05	– 219570.15	6–6	1.70–01	1.78–02	9.28–01	–0.972	B	*LS*
			2641.53	2642.32	181711.03	– 219556.54	4–4	1.96–01	2.05–02	7.12–01	–1.087	B	*LS*
			2644.87	2645.66	181772.41	– 219570.15	8–6	6.73–03	5.30–04	3.69–02	–2.373	B	*LS*
			2643.28	2644.07	181736.05	– 219556.54	6–4	1.40–02	9.74–04	5.09–02	–2.233	B	*LS*
136.		^{4}D–^{4}D°	*2603.5*	*2604.2*	*181741.7*	*– 220140.7*	20–20	1.33–01	1.35–02	2.31+00	–0.569	B	2
			2604.88	2605.65	181772.41	– 220150.49	8–8	1.14–01	1.16–02	7.93–01	–1.034	B	*LS*
			2603.16	2603.94	181736.05	– 220139.41	6–6	7.61–02	7.73–03	3.98–01	–1.334	B	*LS*
			2602.04	2602.82	181711.03	– 220130.86	4–4	5.32–02	5.40–03	1.85–01	–1.666	B	*LS*
			2601.43	2602.21	181696.66	– 220125.51	2–2	6.65–02	6.75–03	1.16–01	–1.870	B	*LS*
			2605.63	2606.41	181772.41	– 220139.41	8–6	2.51–02	1.92–03	1.32–01	–1.814	B	*LS*
			2603.74	2604.52	181736.05	– 220130.86	6–4	4.64–02	3.15–03	1.62–01	–1.724	B	*LS*
			2602.41	2603.19	181711.03	– 220125.51	4–2	6.64–02	3.37–03	1.16–01	–1.870	B	*LS*
			2602.41	2603.19	181736.05	– 220150.49	6–8	1.89–02	2.56–03	1.32–01	–1.813	B	*LS*
			2601.47	2602.24	181711.03	– 220139.41	4–6	3.10–02	4.72–03	1.62–01	–1.724	B	*LS*
			2601.07	2601.85	181696.66	– 220130.86	2–4	3.33–02	6.75–03	1.16–01	–1.870	B	*LS*

C II: Allowed Transitions — Continued

No.	Transition Array	Mult.	λ_{air} (Å)	λ_{vac} (Å) or σ (cm^{-1})*	E_i (cm^{-1})	E_k (cm^{-1})	g_i-g_k	A_{ki} (10^8 s^{-1})	f_{ik}	S (at. u.)	log gf	Acc.	Source
137.	$2s2p(^3P^o)3p-$ $2s2p(^3P^o)5s$	$^4D-^4P^o$	*2269.5*	*2270.2*	*181741.7*	*– 225790.8*	20–12	4.18–01	1.94–02	2.89+00	–0.412	B	2
			2269.68	2270.39	181772.41	– 225817.8	8–6	3.34–01	1.94–02	1.16+00	–0.810	B	*LS*
			2270.18	2270.88	181736.05	– 225771.8	6–4	2.63–01	1.36–02	6.08–01	–1.090	B	*LS*
			2270.12	2270.82	181711.03	– 225748.0	4–2	2.09–01	8.07–03	2.41–01	–1.491	B	*LS*
			2267.81	2268.51	181736.05	– 225817.8	6–6	7.54–02	5.81–03	2.60–01	–1.457	B	*LS*
			2268.89	2269.59	181711.03	– 225771.8	4–4	1.34–01	1.03–02	3.09–01	–1.384	B	*LS*
			2269.38	2270.08	181696.66	– 225748.0	2–2	2.09–01	1.61–02	2.41–01	–1.491	B	*LS*
			2266.52	2267.23	181711.03	– 225817.8	4–6	8.39–03	9.69–04	2.89–02	–2.411	B–	*LS*
			2268.15	2268.85	181696.66	– 225771.8	2–4	2.09–02	3.23–03	4.82–02	–2.190	B	*LS*
138.	$2s2p(^3P^o)3p-$ $2s2p(^3P^o)5d$	$^4D-^4F^o$	*2052.1*	*2052.7*	*181741.7*	*– 230456.9*	20–28	5.16–01	4.56–02	6.16+00	–0.040	B	2
			2052.18	2052.84	181772.41	– 230485.5	8–10	5.16–01	4.07–02	2.20+00	–0.487	B	*LS*
			2051.81	2052.46	181736.05	– 230458	6–8	4.42–01	3.72–02	1.51+00	–0.651	B	*LS*
			2051.80	2052.46	181711.03	– 230433	4–6	3.87–01	3.66–02	9.90–01	–0.834	B	*LS*
			2051.79	2052.45	181696.66	– 230419	2–4	3.61–01	4.56–02	6.16–01	–1.040	B	*LS*
			2053.34	2054.00	181772.41	– 230458	8–8	7.34–02	4.64–03	2.51–01	–1.431	B	*LS*
			2052.86	2053.52	181736.05	– 230433	6–6	1.25–01	7.92–03	3.21–01	–1.323	B	*LS*
			2052.39	2053.05	181711.03	– 230419	4–4	1.44–01	9.12–03	2.46–01	–1.438	B	*LS*
			2054.39	2055.05	181772.41	– 230433	8–6	4.97–03	2.36–04	1.28–02	–2.724	B–	*LS*
			2053.45	2054.11	181736.05	– 230419	6–4	1.03–02	4.34–04	1.76–02	–2.585	B–	*LS*
139.	$2s2p(^3P^o)3p-$ $2s2p(^3P^o)6s$	$^4D-^4P^o$		*1924.1*	*181741.7*	*– 233713.3*	20–12	2.25–01	7.48–03	9.48–01	–0.825	B	2
				1924.25	181772.41	– 233740.7	8–6	1.80–01	7.48–03	3.79–01	–1.223	B	*LS*
				1924.66	181736.05	– 233693.3	6–4	1.41–01	5.24–03	1.99–01	–1.503	B	*LS*
				1924.56	181711.03	– 233670.9	4–2	1.12–01	3.12–03	7.90–02	–1.904	B	*LS*
				1922.90	181736.05	– 233740.7	6–6	4.05–02	2.25–03	8.53–02	–1.870	B	*LS*
				1923.73	181711.03	– 233693.3	4–4	7.20–02	3.99–03	1.01–01	–1.797	B	*LS*
				1924.03	181696.66	– 233670.9	2–2	1.12–01	6.24–03	7.90–02	–1.904	B	*LS*
				1921.98	181711.03	– 233740.7	4–6	4.51–03	3.75–04	9.48–03	–2.824	C+	*LS*
				1923.20	181696.66	– 233693.3	2–4	1.13–02	1.25–03	1.58–02	–2.603	B–	*LS*
140.	$2s2p(^3P^o)3p-$ $2s^26p$	$^2P-^2P^o$		*956.3* cm^{-1}	*182036.9*	– 182993.23	6–6	4.56–06	7.48–04	1.54+00	–2.348	B	2
				949.82 cm^{-1}	182043.41	– 182993.23	4–4	3.73–06	6.19–04	8.58–01	–2.606	B	*LS*
				969.37 cm^{-1}	182023.86	– 182993.23	2–2	3.17–06	5.05–04	3.43–01	–2.995	B	*LS*
				949.82 cm^{-1}	182043.41	– 182993.23	4–2	1.49–06	1.24–04	1.72–01	–3.305	B	*LS*
				969.37 cm^{-1}	182023.86	– 182993.23	2–4	7.92–07	2.53–04	1.72–01	–3.297	B	*LS*
141.	$2s2p(^3P^o)3p-$ $2s^27p$	$^2P-^2P^o$	*21230*	*4709.0* cm^{-1}	*182036.9*	– 186745.9	6–6	1.50–04	1.01–03	4.25–01	–2.216	B	2
			21259.5	4702.5 cm^{-1}	182043.41	– 186745.9	4–4	1.24–04	8.43–04	2.36–01	–2.472	B	*LS*
			21171.5	4722.0 cm^{-1}	182023.86	– 186745.9	2–2	1.01–04	6.77–04	9.44–02	–2.868	B	*LS*
			21259.5	4702.5 cm^{-1}	182043.41	– 186745.9	4–2	4.97–05	1.69–04	4.72–02	–3.171	B	*LS*
			21171.5	4722.0 cm^{-1}	182023.86	– 186745.9	2–4	2.52–05	3.38–04	4.72–02	–3.169	B	*LS*
142.	$2s2p(^3P^o)3p-$ $2s2p(^3P^o)3d$	$^2P-^2D^o$	*6097.7*	*6099.4*	*182036.9*	*– 198432.0*	6–10	5.03–01	4.68–01	5.63+01	0.448	B	2
			6098.51	6100.20	182043.41	– 198436.31	4–6	5.03–01	4.21–01	3.38+01	0.226	B	*LS*
			6095.29	6096.98	182023.86	– 198425.43	2–4	4.20–01	4.67–01	1.88+01	–0.029	B	*LS*
			6102.56	6104.25	182043.41	– 198425.43	4–4	8.37–02	4.67–02	3.76+00	–0.728	B	*LS*
143.	$2s2p(^3P^o)3p-$ $2s2p(^3P^o)4d$	$^2P-^2D^o$	*2591.8*	*2592.5*	*182036.9*	*– 220609.3*	6–10	2.45–01	4.11–02	2.10+00	–0.608	B	2
			2591.84	2592.61	182043.41	– 220614.51	4–6	2.45–01	3.70–02	1.26+00	–0.830	B	*LS*
			2591.40	2592.17	182023.86	– 220601.53	2–4	2.04–01	4.11–02	7.00–01	–1.086	B	*LS*
			2592.71	2593.49	182043.41	– 220601.53	4–4	4.08–02	4.11–03	1.40–01	–1.784	B	*LS*

C II: Allowed Transitions — Continued

No.	Transition Array	Mult.	λ_{air} (Å)	λ_{vac} (Å) or σ (cm^{-1})*	E_i (cm^{-1})	E_k (cm^{-1})	g_i-g_k	A_{ki} (10^8 s^{-1})	f_{ik}	S (at. u.)	log gf	Acc.	Source
144.	$2s2p(^3P^\circ)3p-2s2p(^3P^\circ)3d$	$^4S-^4P^\circ$	*7057.3*	*7059.2*	184690.98	*198856.9*	4–12	3.18–01	7.13–01	6.62+01	0.455	B	2
			7063.68	7065.63	184690.98	198844.00	4–6	3.17–01	3.56–01	3.31+01	0.153	B	*LS*
			7053.09	7055.04	184690.98	198865.25	4–4	3.19–01	2.38–01	2.21+01	–0.022	B	*LS*
			7046.25	7048.19	184690.98	198879.01	4–2	3.20–01	1.19–01	1.10+01	–0.323	B	*LS*
145.	$2s2p(^3P^\circ)3p-2s2p(^3P^\circ)4s$	$^4S-^4P^\circ$	*4014.2*	*4015.4*	184690.98	*209595.4*	4–12	9.63–02	6.98–02	3.69+00	–0.554	B	2
			4009.88	4011.02	184690.98	209622.32	4–6	9.66–02	3.49–02	1.84+00	–0.855	B	*LS*
			4017.27	4018.41	184690.98	209576.46	4–4	9.61–02	2.32–02	1.23+00	–1.032	B	*LS*
			4021.17	4022.30	184690.98	209552.36	4–2	9.58–02	1.16–02	6.15–01	–1.333	B	*LS*
146.	$2s2p(^3P^\circ)3p-2s2p(^3P^\circ)4d$	$^4S-^4P^\circ$	*2766.7*	*2767.5*	184690.98	*220824.1*	4–12	2.14–01	7.37–02	2.69+00	–0.530	B	2
			2767.68	2768.49	184690.98	220811.69	4–6	2.14–01	3.68–02	1.34+00	–0.832	B	*LS*
			2766.11	2766.93	184690.98	220832.15	4–4	2.14–01	2.46–02	8.95–01	–1.007	B	*LS*
			2765.12	2765.94	184690.98	220845.07	4–2	2.14–01	1.23–02	4.48–01	–1.308	B	*LS*
147.	$2s2p(^3P^\circ)3p-2s2p(^3P^\circ)5s$	$^4S-^4P^\circ$	*2432.4*	*2433.1*	184690.98	*225791*	4–12	4.46–02	1.19–02	3.80–01	–1.324	B	2
			2430.77	2431.50	184690.98	225817.8	4–6	4.47–02	5.94–03	1.90–01	–1.624	B	*LS*
			2433.49	2434.23	184690.98	225771.8	4–4	4.45–02	3.95–03	1.27–01	–1.801	B	*LS*
			2434.90	2435.64	184690.98	225748.0	4–2	4.44–02	1.98–03	6.33–02	–2.102	B	*LS*
148.	$2s2p(^3P^\circ)3p-2s2p(^3P^\circ)5d$	$^4S-^4P^\circ$	*2155.7*	*2156.4*	184690.98	*231065*	4–12	1.87–01	3.91–02	1.11+00	–0.806	B	2
			2156.26	2156.94	184690.98	231052.9	4–6	1.87–01	1.96–02	5.55–01	–1.107	B	*LS*
			2155.39	2156.07	184690.98	231071.7	4–4	1.87–01	1.30–02	3.70–01	–1.283	B	*LS*
			2154.70	2155.38	184690.98	231086.6	4–2	1.87–01	6.52–03	1.85–01	–1.584	B	*LS*
149.	$2s2p(^3P^\circ)3p-2s2p(^3P^\circ)6s$	$^4S-^4P^\circ$	*2039.2*	*2039.9*	184690.98	*233713*	4–12	2.37–02	4.43–03	1.19–01	–1.751	B	2
			2038.09	2038.75	184690.98	233740.7	4–6	2.37–02	2.22–03	5.95–02	–2.052	B	*LS*
			2040.06	2040.72	184690.98	233693.3	4–4	2.37–02	1.48–03	3.97–02	–2.229	B	*LS*
			2041.00	2041.65	184690.98	233670.9	4–2	2.36–02	7.38–04	1.98–02	–2.530	B–	*LS*
150.	$2s2p(^3P^\circ)3p-2s2p(^3P^\circ)3d$	$^4P-^4D^\circ$	*9878.0*	*9880.7*	*186452.1*	*196572.8*	12–20	1.33–01	3.24–01	1.26+02	0.589	B	2
			9882.68	9885.39	186466.02	196581.96	6–8	1.33–01	2.59–01	5.05+01	0.191	B	*LS*
			9870.78	9873.49	186443.69	196571.82	4–6	9.31–02	2.04–01	2.65+01	–0.088	B	*LS*
			9863.06	9865.77	186427.35	196563.41	2–4	5.56–02	1.62–01	1.05+01	–0.489	B	*LS*
			9892.60	9895.31	186466.02	196571.82	6–6	3.96–02	5.82–02	1.14+01	–0.457	B	*LS*
			9878.99	9881.70	186443.69	196563.41	4–4	7.08–02	1.04–01	1.35+01	–0.383	B	*LS*
			9868.46	9871.16	186427.35	196557.87	2–2	1.11–01	1.62–01	1.05+01	–0.489	B	*LS*
			9900.83	9903.55	186466.02	196563.41	6–4	6.59–03	6.46–03	1.26+00	–1.412	B	*LS*
			9884.40	9887.11	186443.69	196557.87	4–2	2.21–02	1.62–02	2.11+00	–1.189	B	*LS*
151.		$^4P-^4P^\circ$	*8059.2*	*8061.4*	*186452.1*	*198856.9*	12–12	1.01–01	9.87–02	3.14+01	0.074	B	2
			8076.64	8078.86	186466.02	198844.00	6–6	7.05–02	6.90–02	1.10+01	–0.383	B	*LS*
			8048.31	8050.52	186443.69	198865.25	4–4	1.36–02	1.32–02	1.40+00	–1.278	B	*LS*
			8028.85	8031.06	186427.35	198879.01	2–2	1.71–02	1.65–02	8.73–01	–1.481	B	*LS*
			8062.80	8065.02	186466.02	198865.25	6–4	4.56–02	2.96–02	4.71+00	–0.751	B	*LS*
			8039.40	8041.61	186443.69	198879.01	4–2	8.51–02	4.12–02	4.37+00	–0.783	B	*LS*
			8062.10	8064.31	186443.69	198844.00	4–6	3.04–02	4.44–02	4.71+00	–0.750	B	*LS*
			8037.73	8039.94	186427.35	198865.25	2–4	4.26–02	8.25–02	4.37+00	–0.783	B	*LS*

C II: Allowed Transitions — Continued

No.	Transition Array	Mult.	λ_{air} (Å)	λ_{vac} (Å) or σ (cm^{-1})*	E_i (cm^{-1})	E_k (cm^{-1})	g_i-g_k	A_{ki} (10^8 s^{-1})	f_{ik}	S (at. u.)	log gf	Acc.	Source
152.	$2s2p(^3P^o)3p-2s2p(^3P^o)4s$	$^4P-^4P^o$	*4319.7*	*4320.9*	*186452.1*	*209595.4*	12–12	8.40–01	2.35–01	4.01+01	0.450	B	2
			4317.27	4318.48	186466.02	209622.32	6–6	5.89–01	1.65–01	1.40+01	–0.005	B	*LS*
			4321.66	4322.87	186443.69	209576.46	4–4	1.12–01	3.13–02	1.78+00	–0.902	B	*LS*
			4323.11	4324.32	186427.35	209552.36	2–2	1.40–01	3.91–02	1.11+00	–1.106	B	*LS*
			4325.83	4327.05	186466.02	209576.46	6–4	3.77–01	7.04–02	6.02+00	–0.374	B	*LS*
			4326.16	4327.38	186443.69	209552.36	4–2	6.97–01	9.78–02	5.57+00	–0.408	B	*LS*
			4313.11	4314.32	186443.69	209622.32	4–6	2.53–01	1.06–01	6.02+00	–0.373	B	*LS*
			4318.61	4319.82	186427.35	209576.46	2–4	3.50–01	1.96–01	5.57+00	–0.407	B	*LS*
153.	$2s2p(^3P^o)3p-2s2p(^3P^o)4d$	$^4P-^4D^o$	*2967.5*	*2968.4*	*186452.1*	*220140.7*	12–20	6.35–01	1.40–01	1.64+01	0.224	B	2
			2967.86	2968.73	186466.02	220150.49	6–8	6.34–01	1.12–01	6.55+00	–0.174	B	*LS*
			2966.87	2967.74	186443.69	220139.41	4–6	4.44–01	8.80–02	3.44+00	–0.454	B	*LS*
			2966.18	2967.05	186427.35	220130.86	2–4	2.65–01	6.98–02	1.36+00	–0.855	B	*LS*
			2968.84	2969.70	186466.02	220139.41	6–6	1.90–01	2.51–02	1.47+00	–0.822	B	*LS*
			2967.62	2968.49	186443.69	220130.86	4–4	3.38–01	4.47–02	1.75+00	–0.748	B	*LS*
			2966.65	2967.52	186427.35	220125.51	2–2	5.29–01	6.98–02	1.36+00	–0.855	B	*LS*
			2969.59	2970.46	186466.02	220130.86	6–4	3.17–02	2.79–03	1.64–01	–1.776	B	*LS*
			2968.09	2968.96	186443.69	220125.51	4–2	1.06–01	6.98–03	2.73–01	–1.554	B	*LS*
154.		$^4P-^4P^o$	*2908.5*	*2909.3*	*186452.1*	*220824.1*	12–12	2.55–01	3.24–02	3.72+00	–0.411	B	2
			2910.72	2911.58	186466.02	220811.69	6–6	1.78–01	2.27–02	1.30+00	–0.867	B	*LS*
			2907.10	2907.95	186443.69	220832.15	4–4	3.41–02	4.32–03	1.65–01	–1.762	B	*LS*
			2904.63	2905.48	186427.35	220845.07	2–2	4.27–02	5.40–03	1.03–01	–1.966	B	*LS*
			2908.99	2909.84	186466.02	220832.15	6–4	1.15–01	9.71–03	5.58–01	–1.234	B	*LS*
			2906.01	2906.86	186443.69	220845.07	4–2	2.13–01	1.35–02	5.17–01	–1.267	B	*LS*
			2908.83	2909.68	186443.69	220811.69	4–6	7.66–02	1.46–02	5.58–01	–1.234	B	*LS*
			2905.72	2906.57	186427.35	220832.15	2–4	1.07–01	2.70–02	5.17–01	–1.267	B	*LS*
155.	$2s2p(^3P^o)3p-2s2p(^3P^o)5s$	$^4P-^4P^o$	*2541.3*	*2542.0*	*186452.1*	*225791*	12–12	2.87–01	2.78–02	2.79+00	–0.476	B	2
			2540.42	2541.18	186466.02	225817.8	6–6	2.01–01	1.95–02	9.78–01	–0.932	B	*LS*
			2541.95	2542.71	186443.69	225771.8	4–4	3.83–02	3.71–03	1.24–01	–1.829	B	*LS*
			2542.43	2543.19	186427.35	225748.0	2–2	4.78–02	4.64–03	7.76–02	–2.033	B	*LS*
			2543.39	2544.16	186466.02	225771.8	6–4	1.29–01	8.34–03	4.19–01	–1.301	B	*LS*
			2543.49	2544.25	186443.69	225748.0	4–2	2.39–01	1.16–02	3.88–01	–1.334	B	*LS*
			2538.98	2539.74	186443.69	225817.8	4–6	8.65–02	1.25–02	4.19–01	–1.300	B	*LS*
			2540.89	2541.65	186427.35	225771.8	2–4	1.20–01	2.32–02	3.88–01	–1.334	B	*LS*
156.	$2s2p(^3P^o)3p-2s2p(^3P^o)5d$	$^4P-^4P^o$	*2240.8*	*2241.5*	*186452.1*	*231065*	12–12	1.90–01	1.43–02	1.26+00	–0.766	B	2
			2242.12	2242.81	186466.02	231052.9	6–6	1.33–01	1.00–02	4.43–01	–1.222	B	*LS*
			2240.05	2240.75	186443.69	231071.7	4–4	2.53–02	1.91–03	5.62–02	–2.118	B	*LS*
			2238.48	2239.18	186427.35	231086.6	2–2	3.17–02	2.38–03	3.51–02	–2.322	B	*LS*
			2241.17	2241.87	186466.02	231071.7	6–4	8.54–02	4.29–03	1.90–01	–1.590	B	*LS*
			2239.30	2240.00	186443.69	231086.6	4–2	1.58–01	5.96–03	1.76–01	–1.623	B	*LS*
			2240.99	2241.69	186443.69	231052.9	4–6	5.69–02	6.43–03	1.90–01	–1.590	B	*LS*
			2239.23	2239.93	186427.35	231071.7	2–4	7.92–02	1.19–02	1.76–01	–1.623	B	*LS*
157.	$2s2p(^3P^o)3p-2s2p(^3P^o)6s$	$^4P-^4P^o$	*2115.2*	*2115.9*	*186452.1*	*233713*	12–12	1.42–01	9.54–03	7.97–01	–0.941	B	2
			2114.63	2115.30	186466.02	233740.7	6–6	9.97–02	6.68–03	2.79–01	–1.397	B	*LS*
			2115.75	2116.42	186443.69	233693.3	4–4	1.90–02	1.27–03	3.54–02	–2.293	B	*LS*
			2116.02	2116.69	186427.35	233670.9	2–2	2.37–02	1.59–03	2.21–02	–2.498	B–	*LS*
			2116.75	2117.42	186466.02	233693.3	6–4	6.39–02	2.86–03	1.20–01	–1.765	B	*LS*
			2116.75	2117.42	186443.69	233670.9	4–2	1.18–01	3.97–03	1.11–01	–1.799	B	*LS*
			2113.63	2114.30	186443.69	233740.7	4–6	4.28–02	4.30–03	1.20–01	–1.765	B	*LS*
			2115.02	2115.69	186427.35	233693.3	2–4	5.93–02	7.95–03	1.11–01	–1.798	B	*LS*

C II: Allowed Transitions — Continued

No.	Transition Array	Mult.	λ_{air} (Å)	λ_{vac} (Å) or σ (cm^{-1})*	E_i (cm^{-1})	E_k (cm^{-1})	g_i-g_k	A_{ki} (10^8 s^{-1})	f_{ik}	S (at. u.)	log gf	Acc.	Source
158.	$2s2p(^3P^\circ)3p-2s2p(^3P^\circ)3d$	$^2D-^2D^\circ$	10170	9830.5 cm^{-1}	188601.5	198432.0	10–10	2.97–02	4.61–02	1.54+01	–0.336	B	2
			10179.2	9821.24 cm^{-1}	188615.07	198436.31	6–6	2.77–02	4.30–02	8.64+00	–0.589	B	LS
			10155.5	9844.18 cm^{-1}	188581.25	198425.43	4–4	2.69–02	4.15–02	5.55+00	–0.780	B	LS
			10190.5	9810.36 cm^{-1}	188615.07	198425.43	6–4	2.95–03	3.07–03	6.17–01	–1.735	B	LS
			10144.3	9855.06 cm^{-1}	188581.25	198436.31	4–6	2.00–03	4.62–03	6.17–01	–1.733	B	LS
159.	$2s2p(^3P^\circ)3p-2s2p(^3P^\circ)4d$	$^2D-^2D^\circ$	3123.3	3124.2	188601.5	220609.3	10–10	1.44–01	2.11–02	2.17+00	–0.676	B	2
			3124.15	3125.05	188615.07	220614.51	6–6	1.34–01	1.97–02	1.21+00	–0.928	B	LS
			3122.12	3123.02	188581.25	220601.53	4–4	1.30–01	1.90–02	7.80–01	–1.120	B	LS
			3125.42	3126.32	188615.07	220601.53	6–4	1.44–02	1.40–03	8.66–02	–2.075	B	LS
			3120.85	3121.76	188581.25	220614.51	4–6	9.63–03	2.11–03	8.66–02	–2.074	B	LS
160.	$2s2p(^3P^\circ)3d-2s2p(^3P^\circ)4p$	$^4F^\circ-^4D$	5257.7	5259.2	195786.7	214801.0	28–20	2.70–01	7.98–02	3.87+01	0.349	B	2
			5257.24	5258.70	195813.66	214829.77	10–8	2.41–01	7.98–02	1.38+01	–0.098	B	LS
			5259.06	5260.52	195785.74	214795.27	8–6	2.20–01	6.84–02	9.47+00	–0.262	B	LS
			5259.76	5261.22	195765.85	214772.84	6–4	2.16–01	5.98–02	6.21+00	–0.445	B	LS
			5259.66	5261.13	195752.58	214759.91	4–2	2.69–01	5.58–02	3.87+00	–0.651	B	LS
			5249.53	5250.99	195785.74	214829.77	8–8	2.76–02	1.14–02	1.57+00	–1.040	B	LS
			5253.56	5255.02	195765.85	214795.27	6–6	4.70–02	1.94–02	2.02+00	–0.933	B	LS
			5256.09	5257.55	195752.58	214772.84	4–4	5.40–02	2.23–02	1.55+00	–1.049	B	LS
			5244.05	5245.51	195765.85	214829.77	6–8	1.41–03	7.73–04	8.01–02	–2.334	B	LS
			5249.90	5251.36	195752.58	214795.27	4–6	2.58–03	1.60–03	1.10–01	–2.194	B	LS
161.	$2s2p(^3P^\circ)3d-2s2p(^3P^\circ)5p$	$^4F^\circ-^4D$	3082.5	3083.4	195786.7	228218.6	28–20	1.06–01	1.08–02	3.06+00	–0.520	B	2
			3082.35	3083.25	195813.66	228246.97	10–8	9.46–02	1.08–02	1.09+00	–0.967	B	LS
			3083.04	3083.94	195785.74	228211.8	8–6	8.65–02	9.24–03	7.50–01	–1.131	B	LS
			3083.04	3083.94	195765.85	228191.9	6–4	8.51–02	8.08–03	4.92–01	–1.314	B	LS
			3083.05	3083.94	195752.58	228178.6	4–2	1.06–01	7.54–03	3.06–01	–1.520	B	LS
			3079.70	3080.60	195785.74	228246.97	8–8	1.08–02	1.54–03	1.25–01	–1.910	B	LS
			3081.15	3082.05	195765.85	228211.8	6–6	1.84–02	2.62–03	1.60–01	–1.803	B	LS
			3081.78	3082.68	195752.58	228191.9	4–4	2.12–02	3.02–03	1.23–01	–1.918	B	LS
			3077.82	3078.71	195765.85	228246.97	6–8	5.51–04	1.04–04	6.34–03	–3.203	C+	LS
			3079.89	3080.79	195752.58	228211.8	4–6	1.01–03	2.16–04	8.75–03	–3.064	C+	LS
162.	$2s2p(^3P^\circ)3d-2s2p(^3P^\circ)6p$	$^4F^\circ-^4D$	2546.9	2547.7	195786.7	235038	28–20	5.51–02	3.83–03	8.98–01	–0.970	B	2
			2546.80	2547.57	195813.66	235066.8	10–8	4.92–02	3.83–03	3.21–01	–1.417	B	LS
			2547.31	2548.08	195785.74	235031	8–6	4.49–02	3.28–03	2.20–01	–1.581	B	LS
			2547.32	2548.09	195765.85	235011	6–4	4.42–02	2.87–03	1.44–01	–1.764	B	LS
			2547.30	2548.07	195752.58	234998	4–2	5.50–02	2.68–03	8.98–02	–1.970	B	LS
			2544.99	2545.76	195785.74	235066.8	8–8	5.62–03	5.45–04	3.66–02	–2.360	B	LS
			2546.02	2546.79	195765.85	235031	6–6	9.58–03	9.31–04	4.68–02	–2.253	B	LS
			2546.46	2547.22	195752.58	235011	4–4	1.10–02	1.07–03	3.59–02	–2.368	B	LS
			2543.70	2544.47	195765.85	235066.8	6–8	2.86–04	3.70–05	1.86–03	–3.653	C	LS
			2545.16	2545.93	195752.58	235031	4–6	5.25–04	7.66–05	2.57–03	–3.514	C	LS

*Wavelengths (Å) are always given unless cm^{-1} is indicated.

C II

Forbidden Transitions

We have selected the results of multi-configuration Hartree-Fock calculations with relativistic Breit-Pauli terms by Froese Fischer[1] for the M1 and E2 fine structure transitions in the $^2P^o$ term of the ground state configuration.

Reference

[1]C. Froese Fischer, J. Phys. B **16**, 157 (1983).

C II: Forbidden Transitions

No.	Transition Array	Mult.	λ_{air} (Å)	λ_{vac} (Å) or σ (cm^{-1})*	E_i (cm^{-1})		E_k (cm^{-1})	g_i-g_k	Type	A_{ki} (s^{-1})	S (at. u.)	Acc.	Source
1.	$2p-2p$	$^2P^o-^2P^o$											
				63.42 cm^{-1}	0.00	–	63.42	2–4	M1	2.30–06	1.34+00	A	1
				63.42 cm^{-1}	0.00	–	63.42	2–4	E2	1.38–13	4.81+00	B	1

*Wavelengths (Å) are always given unless cm^{-1} is indicated.

C III

Beryllium Isoelectronic Sequence

Ground State: $1s^22s^2\ ^1S_0$

Ionization Energy: 47.887 eV = 386241.0 cm^{-1}

Allowed Transitions

List of tabulated lines

Wavelength (Å)	No.	Wavelength (Å)	No.	Wavelength (Å)	No.	Wavelength (Å)	No.
in vacuum		347.773	22	386.203	3	473.410	60
		347.841	22	388.969	17	477.625	33
270.324	11	358.740	42	389.005	17	483.534	49
274.051	10	360.552	21	389.090	17	483.596	49
280.026	9	360.569	21	390.045	40	483.601	49
288.423	8	360.600	21	409.322	39	483.609	49
291.326	7	360.625	21	411.958	38	483.713	49
310.170	6	360.627	21	416.769	64	483.720	49
314.335	27	360.673	21	423.438	37	492.650	59
314.358	27	363.754	20	433.339	36	493.335	48
314.414	27	363.785	20	446.329	63	493.371	48
319.213	26	363.860	20	450.734	35	493.405	48
319.237	26	369.398	19	455.323	50	493.468	48
319.294	26	369.399	19	455.380	50	493.521	48
321.289	25	369.422	19	455.382	50	493.584	48
321.314	25	369.432	19	455.384	50	497.917	58
321.372	25	369.475	19	455.479	50	499.425	47
322.574	5	369.508	19	455.481	50	499.464	47
327.784	43	371.686	18	456.580	62	499.497	47
330.588	24	371.697	18	459.466	16	499.535	47
330.614	24	371.719	18	459.514	16	499.583	47
330.676	24	371.743	18	459.516	16	499.615	47
341.146	23	371.775	18	459.627	16	506.632	57
341.174	23	371.797	18	459.633	16	511.523	56
341.239	23	379.065	65	459.635	16	533.601	73
341.239	23	379.253	41	460.049	34	535.288	55
347.744	22	385.043	4	468.942	61	538.080	15

List of tabulated lines — Continued

Wavelength (Å)	No.	Wavelength (Å)	No.	Wavelength (Å)	No.	Wavelength (Å)	No.
538.149	15	1051.15	106	1427.84	75	1923.38	112
538.312	15	1069.69	13	1427.91	103	1923.49	112
548.317	72	1070.17	131	1428.18	103	1962.71	315
554.413	46	1070.33	13	1428.50	75	1987.23	311
554.487	46	1111.57	182	1428.55	103	1991.81	180
554.502	46	1125.63	119	1428.67	103		
554.509	46	1125.64	119	1428.94	103	in air	
554.633	46	1125.67	119	1429.10	103		
554.648	46	1125.68	119	1477.63	114	2009.34	99
565.528	54	1131.05	130	1477.69	114	2009.56	99
566.482	53	1139.90	95	1478.02	114	2010.09	99
572.771	71	1148.89	118	1478.04	114	2051.61	154
574.281	32	1148.91	118	1478.11	114	2061.40	195
585.265	45	1148.92	118	1478.30	114	2061.65	195
585.401	45	1148.94	118	1478.33	114	2062.00	195
585.428	45	1160.58	94	1479.89	102	2064.85	239
585.499	45	1165.62	105	1480.01	102	2089.47	172
585.610	45	1165.70	105	1480.30	102	2089.57	172
585.662	45	1165.87	105	1497.56	125	2089.78	172
609.040	70	1165.88	105	1512.93	181	2091.80	111
609.272	52	1169.99	228	1531.83	91	2091.85	111
622.130	69	1174.93	12	1541.12	124	2091.97	111
690.521	31	1175.26	12	1558.29	276	2092.01	111
714.876	68	1175.59	12	1576.48	113	2092.06	111
750.750	133	1175.71	12	1576.89	101	2092.11	111
768.467	14	1175.99	12	1577.30	113	2100.44	146
784.397	67	1176.10	204	1577.37	113	2100.48	146
814.626	87	1176.37	12	1577.53	101	2100.52	146
817.758	44	1210.08	129	1577.86	101	2143.34	141
817.863	44	1240.28	128	1577.88	113	2144.83	141
817.950	44	1246.43	104	1577.91	113	2147.98	141
817.988	44	1246.52	104	1577.98	113	2162.93	122
818.181	44	1246.72	104	1578.00	101	2176.95	121
818.269	44	1247.38	30	1578.14	101	2181.83	171
841.480	79	1256.47	76	1578.47	101	2181.94	171
841.488	79	1256.54	76	1586.68	295	2182.17	171
841.493	79	1256.55	93	1591.44	81	2186.85	226
849.432	86	1256.58	76	1620.07	100	2202.53	161
884.524	51	1296.32	117	1620.34	100	2212.89	179
902.580	109	1296.33	117	1620.59	100	2289.98	322
902.627	109	1296.34	117	1620.68	100	2291.07	194
902.733	109	1296.37	117	1620.74	100	2291.38	194
909.592	85	1296.39	117	1621.09	100	2291.62	202
914.626	132	1306.08	250	1623.25	227	2291.82	194
933.735	97	1308.71	66	1645.03	90	2296.87	29
943.168	78	1329.19	82	1646.95	299	2307.37	210
943.189	78	1337.98	240	1703.54	203	2457.25	225
943.218	78	1347.20	116	1718.08	163	2609.75	110
943.998	108	1347.25	116	1779.09	123	2609.82	110
944.048	108	1347.27	116	1785.02	155	2610.01	110
944.165	108	1347.31	116	1805.17	142	2614.41	110
962.395	107	1347.33	116	1806.22	142	2614.49	110
962.447	107	1347.38	116	1808.46	142	2616.63	110
962.569	107	1347.95	92	1880.60	162	2641.40	275
977.020	2	1381.65	127	1894.29	89	2670.24	190
981.077	77	1406.49	156	1897.36	173	2671.32	190
981.214	77	1423.90	126	1897.44	173	2671.34	190
981.462	77	1425.90	115	1897.57	249	2672.95	190
1001.99	96	1426.19	115	1897.62	173	2672.98	190
1004.60	84	1426.22	115	1908.73	1	2673.00	190
1026.43	157	1426.45	75	1922.96	112	2673.77	178
1040.72	83	1426.72	115	1923.16	112	2681.61	140
1050.94	106	1426.74	115	1923.27	112	2683.92	140
1051.01	106	1426.80	115	1923.34	112	2683.94	140

List of tabulated lines — Continued

Wavelength (Å)	No.	Wavelength (Å)	No.	Wavelength (Å)	No.	Wavelength (Å)	No.
2688.83	140	3609.05	169	4383.53	187	6154.16	185
2688.86	140	3609.07	169	4388.02	187	6155.12	185
2688.87	140	3609.62	169	4390.52	274	6156.69	185
2693.75	254	3609.68	169	4405.75	199	6160.01	185
2695.53	254	3609.70	169	4443.15	221	6163.96	185
2697.25	170	3653.29	235	4515.35	168	6167.55	320
2697.42	170	3655.97	235	4515.81	168	6205.56	198
2697.76	170	3656.11	235	4516.79	168	6350.77	244
2698.13	254	3656.21	235	4647.42	74	6460.29	273
2716.01	248	3660.93	235	4650.25	74	6727.48	136
2727.55	153	3661.08	235	4651.02	138	6731.04	136
2751.83	201	3703.70	177	4651.47	74	6742.15	136
2774.58	209	3789.43	258	4652.05	138	6744.39	136
2794.57	217	3806.12	281	4659.06	138	6762.17	136
2796.46	217	3806.60	281	4663.64	138	6773.39	136
2796.48	217	3806.95	281	4665.86	138	6774.95	264
2799.48	217	3838.91	238	4670.49	280	6786.76	264
2799.51	217	3844.51	246	4671.22	280	6793.36	264
2799.54	217	3883.82	188	4671.74	280	6851.18	212
2805.02	189	3885.94	188	4673.95	138	6853.68	212
2805.10	189	3886.15	188	4724.17	214	6857.24	212
2805.16	189	3889.14	188	4724.71	214	6862.69	212
2806.23	189	3889.46	188	4725.45	214	6868.78	212
2806.31	189	3889.67	188	4730.18	214	6870.53	234
2808.07	189	3914.91	266	4730.92	214	6872.04	212
2844.12	28	3918.85	266	4739.66	214	6879.37	234
2847.97	28	3921.05	266	5048.95	284	6880.50	234
2849.05	152	4026.16	285	5158.52	279	6881.10	212
2850.30	28	4054.49	270	5159.40	279	6882.07	234
2857.02	160	4056.06	222	5160.03	279	6898.11	234
2863.71	224	4067.94	193	5175.81	297	6899.69	234
2874.25	145	4068.92	193	5244.66	137	7037.25	149
2874.44	145	4070.26	193	5249.11	220	7157.73	237
2874.72	145	4070.31	193	5253.58	137	7353.88	243
2906.74	298	4121.84	200	5272.52	137	7433.11	307
2942.54	216	4152.51	215	5337.40	186	7433.54	307
2942.63	216	4156.50	215	5341.47	186	7434.13	307
2942.70	216	4156.74	215	5345.88	186	7486.56	288
2944.66	216	4162.88	215	5348.82	186	7486.64	288
2944.75	216	4163.25	215	5353.23	186	7486.72	288
2948.04	216	4163.49	215	5359.93	186	7486.80	288
2982.12	120	4166.24	310	5376.19	257	7486.96	288
3038.89	223	4166.97	245	5407.96	300	7576.70	135
3099.78	253	4173.09	208	5408.36	300	7578.15	252
3102.14	253	4186.90	207	5408.99	300	7586.41	135
3105.58	253	4247.31	176	5481.64	293	7592.27	252
3145.94	247	4252.10	265	5695.92	88	7595.31	135
3151.86	139	4255.30	192	5758.11	269	7612.45	135
3155.07	139	4255.38	192	5823.17	303	7612.75	252
3161.90	139	4255.41	192	5826.42	219	7612.91	252
3170.01	159	4256.37	192	5840.69	262	7625.99	135
3182.61	259	4256.45	192	5841.06	262	7634.97	135
3258.02	144	4256.75	265	5841.32	262	7707.43	175
3259.51	144	4257.89	192	5858.34	213	7771.76	229
3262.27	144	4259.35	265	5863.25	213	7780.41	229
3288.83	267	4281.98	301	5871.68	213	7796.00	229
3291.62	267	4282.23	301	5872.10	213	8021.15	292
3293.17	267	4282.63	301	5880.56	213	8255.27	251
3357.79	231	4296.60	294	5894.07	213	8255.57	251
3357.91	231	4325.56	150	5894.45	230	8255.67	251
3358.00	231	4379.48	187	5895.28	230	8272.04	251
3498.24	325	4379.93	187	5896.43	230	8272.33	251
3506.78	151	4380.57	187	6147.87	185	8296.54	251
3608.78	169	4382.90	187	6149.28	185	8317.04	296

List of tabulated lines — Continued

Wavelength (Å)	No.	Wavelength (Å)	No.	Wavelength (Å)	No.	Wavenumber (cm^{-1})	No.
8321.16	283	9699.57	211	18529.6	263	54.39	289
8332.99	184	9701.10	98	18531.0	263	167.12	260
8341.60	184	9705.41	98	18531.5	263	192.81	260
8347.95	184	9706.44	98	18619.7	263	207.11	260
8357.86	184	9715.09	98	18620.2	263	389.02	312
8358.72	184	9717.75	98	18669.9	263	426.93	236
8367.96	167	9718.79	98	18724.9	165	553.63	183
8371.32	167	9719.62	211	18845.5	165	559.53	183
8375.04	184	9733.27	211	18862.6	165	567.92	183
8392.44	167	9756.58	211	18907.7	268	576.94	183
8394.03	167	9859.40	242	18924.8	165	582.84	183
8397.40	167	10125.9	302	18932.9	165	592.37	183
8408.92	167	10292	148	18950.1	165	1073.20	191
8500.32	80	10541.2	166	21074.8	255	1091.78	191
8582.09	317	10543.7	166	21217.5	324	1097.68	191
8582.31	317	10549.0	166	23228.1	261	1122.36	191
8582.62	317	11209.8	309	23241.0	261	1130.75	191
8652.14	287	11640.1	306	23258.9	261	1132	319
8652.25	287	11641.2	306	25647.2	321	1136.65	191
8652.57	287	11642.6	306	27399.2	164	1139.88	134
8652.93	287	11790.9	218	27539.0	164	1208.26	134
8653.05	287	11981.2	143	27575.3	164	1240.59	134
8653.53	287	11988.1	143	27639.3	164	1257.89	277
8931.65	314	11991.3	143	27656.5	164	1260.28	277
8942.57	323	12113.9	241	27693.2	164	1260.43	277
9331.01	197	12194.2	256	29401.1	272	1263.59	277
9358.37	158	12542.7	232	37249.7	174	1263.74	277
9409.14	278	12555.6	232	45408	290	1264.17	277
9412.07	278	12576.0	232	45583.0	304	1510.47	205
9414.19	278	12583.9	232	45605.1	304	1617.30	282
9565.53	291	12614.3	232	45621.5	304	1623.25	308
9573.95	233	12634.9	232			1654.74	286
9580.18	233	14476.1	318			1655.17	286
9593.32	233	14727.9	316			1655.32	286
9597.81	206	14728.5	316			1656.11	286
9617.03	233	14729.5	316			1656.54	286
9627.60	233	15231.3	305			1658.26	286
9651.47	233	15233.2	305			1715.19	147
9696.48	211	15235.6	305			1808.29	196
9696.54	211	16221.2	313			1837.40	271

For five singlets and one triplet, fully correlated superposition-of-configurations (SOC) wavefunctions have been calculated by Weiss,[1] in which several thousand interacting configurations are included for describing both upper and lower states. The dipole-length and dipole-velocity results agree typically within a few tenths of a percent—indicating a level of electron-correlation treatment which puts this work in a class by itself.

For most other transitions, we have utilized the results of the Opacity Project (OP), which is reviewed in the general introduction. For this Be-like ion, the OP calculations were carried out by Tully *et al.*[2] For a few transitions, data of similarly advanced configuration-interaction calculations by Hibbert,[3] carried out with the CIV 3 code, are available and are averaged with the OP data with which they are in very close agreement.

Allard *et al.*[4] have undertaken an extensive compilation of C III transition probabilities, very similar in spirit to this one, and have also utilized the OP data as the main source of their tables. But they have averaged the OP results with less advanced earlier work, using the OP results with a weight factor of five and the earlier work with a weight of one, while our approach is to omit all less advanced work (weight of zero). Thus, our tabulated results often differ, but usually only slightly, from those of Allard *et al.*[4] However, we utilize their tables for a few multiplets for which they had special OP data at their disposal and for an intercombination line. For this latter type of transition, for which no OP results exist, we have otherwise utilized the new calculations by Weiss.[1]

For the $2s^2\,{}^1S_0 - 2s2p\,{}^3P_1$ intersystem line, which connects the lowest excited term to the ground state, Kwong *et al.*[5] determined the transition probability by measuring the lifetime of the upper level with a radiofrequency ion trap. Their result agrees within 13% with Weiss' calculation,[1] and the average has been tabulated.

For another Be-like ion, O v, Hibbert[6] calculated the transition probabilities of individual lines in triplets involving s, p and d electrons in intermediate coupling. His line ratios within multiplets are very close to those obtained from LS-coupling, typically within 1 to 3%. Somewhat larger differences, up to 25%, are encountered only for weak multiplet components. Therefore, we assume that LS-coupling is also a very good approximation for the analogous spectrum of C III and we estimate that for the stronger lines in multiplets the same accuracy ratings as for the multiplet value apply. However, we expect some deterioration in accuracy for weak lines and for triplet lines involving f and g electrons, and we have accordingly lowered the ratings in those cases.

Two advanced lifetime measurements are available as sensitive tests of the tabulated data. Reistad *et al.*[7] measured the lifetime of the $2s2p$ $^1P^0_1$ level with the beam foil method and applied the ANDC technique (see general introduction) for a reliable analysis and correction of the critical cascading effects. Their result, estimated by the authors to be accurate to ±3.5%, agrees with our tabulated value, which is from Weiss' work,[1] within 0.7%. Baudinet-Robinet *et al.*[8] measured the lifetime of the $2s3d$ $^1D_2$ level with the beam-foil-laser method, which provides selective level excitation. Their result, estimated to be accurate to ±6.7%, differs from our tabulated data (the principal transition is again from the work of Weiss[1]) by 6.1%. Graphical comparisons of these lifetimes with our selected theoretical data are shown in Fig. 1 of the C II introduction.

References

[1] A. W. Weiss, to be published.
[2] J. A. Tully, M. J. Seaton and K. A. Berrington, J. Phys. B **23**, 3811 (1990).
[3] A. Hibbert, J. Phys. B **9**, 2805 (1976).
[4] N. Allard, M.-C. Artru, T. Lanz, and M. Le Dourneuf, Astron. Astrophys., Suppl. Ser. **84**, 563 (1990) and **91**, 399 (1991).
[5] V. H. S. Kwong, Z. Fang, T. T. Gibbons, W. H. Parkinson and P. L. Smith, Astrosphys. J. **411**. 431 (1993).
[6] A. Hibbert, J. Phys. B **13**, 1721 (1980).
[7] N. Reistad, R. Hutton, A. E. Nilsson, I. Martinson and S. Mannervik, Phys. Scr. **34**, 151 (1986).
[8] Y. Baudinet-Robinet, P.-D. Dumont, H. P. Garnir, and A. El Himdy, Phys. Rev. A **40**, 6321 (1989).

C III: Allowed Transitions

No.	Transition Array	Mult.	λ_{air} (Å)	λ_{vac} (Å) or σ (cm^{-1})*	E_i (cm^{-1})	E_k (cm^{-1})	g_i-g_k	A_{ki} (10^8 s^{-1})	f_{ik}	S (at. u.)	log gf	Acc.	Source
1.	$2s^2-2s2p$	$^1S-^3P^o$											
				1908.73	0	– 52390.75	1–3	1.14–06	1.87–07	1.17–06	–6.729	B+	1,5
2.		$^1S-^1P^o$		977.020	0	– 102352.04	1–3	1.767+01	7.586–01	2.440+00	–0.1200	A+	1
3.	$2s^2-2s3p$	$^1S-^1P^o$		386.203	0	– 258931.29	1–3	3.46+01	2.32–01	2.96–01	–0.634	B	2
4.		$^1S-^3P^o$											
				385.043	0	– 259711.22	1–3	5.20–03	3.47–05	4.40–05	–4.460	C+	4
5.	$2s^2-2p3s$	$^1S-^1P^o$		322.574	0	– 310006.32	1–3	9.66+00	4.52–02	4.80–02	–1.345	B	2
6.	$2s^2-2s4p$	$^1S-^1P^o$		310.170	0	– 322404.20	1–3	6.56+00	2.84–02	2.90–02	–1.547	B–	2
7.	$2s^2-2s5p$	$^1S-^1P^o$		291.326	0	– 343258.03	1–3	1.18+01	4.50–02	4.32–02	–1.346	B	2
8.	$2s^2-2p3d$	$^1S-^1P^o$		288.423	0	– 346712.73	1–3	8.73–01	3.26–03	3.10–03	–2.486	C	2
9.	$2s^2-2s6p$	$^1S-^1P^o$		280.026	0	– 357109.68	1–3	4.67+00	1.65–02	1.52–02	–1.783	B–	2
10.	$2s^2-2s7p$	$^1S-^1P^o$		274.051	0	– 364896	1–3	3.35+00	1.13–02	1.02–02	–1.947	B–	2
11.	$2s^2-2s8p$	$^1S-^1P^o$		270.324	0	– 369926	1–3	2.36+00	7.75–03	6.90–03	–2.111	C+	2

C III: Allowed Transitions — Continued

No.	Transition Array	Mult.	λ_{air} (Å)	λ_{vac} (Å) or σ (cm^{-1})*	E_i (cm^{-1})	E_k (cm^{-1})	g_i–g_k	A_{ki} (10^8 s^{-1})	f_{ik}	S (at. u.)	log gf	Acc.	Source
12.	$2s2p-2p^2$	$^3P^o-^3P$		*1175.7*	*52419.4*	– *137477.7*	9–9	1.314+01	2.724–01	9.488+00	0.3894	A+	1
				1175.71	52447.11	– 137502.01	5–5	9.856+00	2.043–01	3.953+00	0.0092	A+	1
				1175.59	52390.75	– 137454.40	3–3	3.287+00	6.809–02	7.906–01	–0.6898	A+	1
				1176.37	52447.11	– 137454.40	5–3	5.468+00	6.807–02	1.318+00	–0.4681	A+	1
				1175.99	52390.75	– 137425.70	3–1	1.313+01	9.075–02	1.054+00	–0.5650	A+	1
				1174.93	52390.75	– 137502.01	3–5	3.293+00	1.136–01	1.318+00	–0.4676	A+	1
				1175.26	52367.06	– 137454.40	1–3	4.385+00	2.724–01	1.054+00	–0.5648	A+	1
13.		$^3P^o-^1D$											
				1070.33	52447.11	– 145876.13	5–5	2.53–04	4.34–06	7.65–05	–4.663	B	1
				1069.69	52390.75	– 145876.13	3–5	4.07–05	1.16–06	1.23–05	–5.457	B	1
14.		$^3P^o-^1S$											
				768.467	52390.75	– 182519.88	3–1	1.14–05	3.35–08	2.55–07	–6.997	B	1
15.	$2s2p-2s3s$	$^3P^o-^3S$		*538.23*	*52419.4*	– 238213	9–3	3.67+01	5.32–02	8.48–01	–0.320	A	2,3
				538.312	52447.11	– 238213	5–3	2.04+01	5.32–02	4.71–01	–0.575	B+	*LS*
				538.149	52390.75	– 238213	3–3	1.23+01	5.32–02	2.83–01	–0.797	B+	*LS*
				538.080	52367.06	– 238213	1–3	4.09+00	5.32–02	9.42–02	–1.274	B+	*LS*
16.	$2s2p-2s3d$	$^3P^o-^3D$		*459.57*	*52419.4*	– *270013.0*	9–15	1.06+02	5.61–01	7.64+00	0.703	A	2,3
				459.627	52447.11	– 270014.74	5–7	1.06+02	4.71–01	3.56+00	0.372	B+	*LS*
				459.514	52390.75	– 270011.93	3–5	7.97+01	4.21–01	1.91+00	0.101	B+	*LS*
				459.466	52367.06	– 270010.83	1–3	5.91+01	5.61–01	8.49–01	–0.251	B+	*LS*
				459.633	52447.11	– 270011.93	5–5	2.66+01	8.41–02	6.36–01	–0.376	B+	*LS*
				459.516	52390.75	– 270010.83	3–3	4.43+01	1.40–01	6.36–01	–0.376	B+	*LS*
				459.635	52447.11	– 270010.83	5–3	2.95+00	5.61–03	4.24–02	–1.552	B	*LS*
17.	$2s2p-2s4s$	$^3P^o-^3S$		*389.05*	*52419.4*	– 309457.17	9–3	1.72+01	1.30–02	1.50–01	–0.930	B	2
				389.090	52447.11	– 309457.17	5–3	9.58+00	1.30–02	8.36–02	–1.186	B	*LS*
				389.005	52390.75	– 309457.17	3–3	5.75+00	1.30–02	5.01–02	–1.407	B	*LS*
				388.969	52367.06	– 309457.17	1–3	1.92+00	1.31–02	1.67–02	–1.884	B–	*LS*
18.	$2s2p-2s4d$	$^3P^o-^3D$		*371.73*	*52419.4*	– *321434.5*	9–15	4.55+01	1.57–01	1.73+00	0.150	B	2
				371.743	52447.11	– 321450.05	5–7	4.55+01	1.32–01	8.08–01	–0.181	B	*LS*
				371.697	52390.75	– 321426.74	3–5	3.41+01	1.18–01	4.33–01	–0.452	B	*LS*
				371.686	52367.06	– 321411.31	1–3	2.53+01	1.57–01	1.92–01	–0.804	B	*LS*
				371.775	52447.11	– 321426.74	5–5	1.14+01	2.36–02	1.44–01	–0.929	B	*LS*
				371.719	52390.75	– 321411.31	3–3	1.90+01	3.93–02	1.44–01	–0.929	B	*LS*
				371.797	52447.11	– 321411.31	5–3	1.26+00	1.57–03	9.61–03	–2.105	C+	*LS*
19.	$2s2p-2p3p$	$^3P^o-^3D$		*369.42*	*52419.4*	– *323114.7*	9–15	1.99–02	6.78–05	7.42–04	–3.215	D	4
				369.422	52447.11	– 323140.33	5–7	1.99–02	5.69–05	3.46–04	–3.546	D	*LS*
				369.398	52390.75	– 323101.36	3–5	1.49–02	5.08–05	1.85–04	–3.817	D	*LS*
				369.399	52367.06	– 323076.88	1–3	1.10–02	6.78–05	8.24–05	–4.169	D	*LS*
				369.475	52447.11	– 323101.36	5–5	4.97–03	1.02–05	6.18–05	–4.294	D	*LS*
				369.432	52390.75	– 323076.88	3–3	8.28–03	1.69–05	6.18–05	–4.294	D	*LS*
				369.508	52447.11	– 323076.88	5–3	5.52–04	6.78–07	4.12–06	–5.470	D–	*LS*
20.		$^3P^o-^3S$		*363.82*	*52419.4*	– 327278.27	9–3	1.46+01	9.65–03	1.04–01	–1.061	B	2
				363.860	52447.11	– 327278.27	5–3	8.10+00	9.65–03	5.78–02	–1.317	B	*LS*
				363.785	52390.75	– 327278.27	3–3	4.86+00	9.65–03	3.47–02	–1.538	B	*LS*
				363.754	52367.06	– 327278.27	1–3	1.62+00	9.65–03	1.16–02	–2.015	B–	*LS*

C III: Allowed Transitions — Continued

No.	Transition Array	Mult.	λ_{air} (Å)	λ_{vac} (Å) or σ (cm^{-1})*	E_i (cm^{-1})	E_k (cm^{-1})	g_i-g_k	A_{ki} (10^8 s^{-1})	f_{ik}	S (at. u.)	log gf	Acc.	Source
21.		$^3P^o-^3P$		*360.61*	*52419.4*	*329724.7*	9–9	1.24+01	2.41–02	2.58–01	–0.664	B	2
				360.625	52447.11	329743.57	5–5	9.27+00	1.81–02	1.07–01	–1.044	B	*LS*
				360.600	52390.75	329706.47	3–3	3.09+00	6.02–03	2.15–02	–1.743	B–	*LS*
				360.673	52447.11	329706.47	5–3	5.15+00	6.02–03	3.58–02	–1.521	B	*LS*
				360.627	52390.75	329685.38	3–1	1.24+01	8.03–03	2.86–02	–1.618	B–	*LS*
				360.552	52390.75	329743.57	3–5	3.09+00	1.00–02	3.58–02	–1.521	B	*LS*
				360.569	52367.06	329706.47	1–3	4.12+00	2.41–02	2.86–02	–1.618	B–	*LS*
22.	$2s2p-$ $2s5s$	$^3P^o-^3S$		*347.81*	*52419.4*	339934.72	9–3	3.21+00	1.94–03	2.00–02	–1.758	B–	2
				347.841	52447.11	339934.72	5–3	1.78+00	1.94–03	1.11–02	–2.013	B–	*LS*
				347.773	52390.75	339934.72	3–3	1.07+00	1.94–03	6.67–03	–2.235	C+	*LS*
				347.744	52367.06	339934.72	1–3	3.57–01	1.94–03	2.22–03	–2.712	C	*LS*
23.	$2s2p-$ $2s5d$	$^3P^o-^3D$		*341.21*	*52419.4*	*345496.9*	9–15	1.78+01	5.17–02	5.23–01	–0.332	B	2
				341.239	52447.11	345497.15	5–7	1.78+01	4.34–02	2.44–01	–0.663	B	*LS*
				341.174	52390.75	345496.72	3–5	1.33+01	3.88–02	1.31–01	–0.934	B	*LS*
				341.146	52367.06	345496.57	1–3	9.88+00	5.17–02	5.81–02	–1.286	B	*LS*
				341.239	52447.11	345496.72	5–5	4.44+00	7.76–03	4.36–02	–1.411	B	*LS*
				341.174	52390.75	345496.57	3–3	7.41+00	1.29–02	4.36–02	–1.411	B	*LS*
				341.239	52447.11	345496.57	5–3	4.94–01	5.17–04	2.91–03	–2.587	C	*LS*
24.	$2s2p-$ $2s6s$	$^3P^o-^3S$		*330.65*	*52419.4*	354858.03	9–3	2.86+00	1.56–03	1.53–02	–1.852	B–	2
				330.676	52447.11	354858.03	5–3	1.59+00	1.56–03	8.50–03	–2.107	C+	*LS*
				330.614	52390.75	354858.03	3–3	9.53–01	1.56–03	5.10–03	–2.329	C+	*LS*
				330.588	52367.06	354858.03	1–3	3.18–01	1.56–03	1.70–03	–2.806	C	*LS*
25.	$2s2p-$ $2s7s$	$^3P^o-^3S$		*321.34*	*52419.4*	363613	9–3	1.99+00	1.03–03	9.80–03	–2.033	C+	2
				321.372	52447.11	363613	5–3	1.11+00	1.03–03	5.44–03	–2.289	C+	*LS*
				321.314	52390.75	363613	3–3	6.65–01	1.03–03	3.27–03	–2.510	C	*LS*
				321.289	52367.06	363613	1–3	2.22–01	1.03–03	1.09–03	–2.987	C	*LS*
26.	$2s2p-$ $2s7d$	$^3P^o-^3D$		*319.27*	*52419.4*	365638	9–15	6.26+00	1.59–02	1.51–01	–0.843	B	2
				319.294	52447.11	365638	5–7	6.26+00	1.34–02	7.04–02	–1.174	B	*LS*
				319.237	52390.75	365638	3–5	4.70+00	1.20–02	3.77–02	–1.445	B	*LS*
				319.213	52367.06	365638	1–3	3.48+00	1.59–02	1.68–02	–1.797	B–	*LS*
				319.294	52447.11	365638	5–5	1.56+00	2.39–03	1.26–02	–1.922	B–	*LS*
				319.237	52390.75	365638	3–3	2.61+00	3.99–03	1.26–02	–1.922	B–	*LS*
				319.294	52447.11	365638	5–3	1.74–01	1.59–04	8.38–04	–3.099	D	*LS*
27.	$2s2p-$ $2s8d$	$^3P^o-^3D$		*314.39*	*52419.4*	370499	9–15	4.15+00	1.02–02	9.54–02	–1.035	B	2
				314.414	52447.11	370499	5–7	4.15+00	8.60–03	4.45–02	–1.366	B	*LS*
				314.358	52390.75	370499	3–5	3.11+00	7.68–03	2.39–02	–1.637	B–	*LS*
				314.335	52367.06	370499	1–3	2.31+00	1.02–02	1.06–02	–1.990	B–	*LS*
				314.414	52447.11	370499	5–5	1.04+00	1.54–03	7.95–03	–2.115	C+	*LS*
				314.358	52390.75	370499	3–3	1.73+00	2.56–03	7.95–03	–2.115	C+	*LS*
				314.414	52447.11	370499	5–3	1.15–01	1.02–04	5.30–04	–3.291	D	*LS*
28.	$2s2p-$ $2p^2$	$^1P^o-^3P$											
			2844.12	2844.95	102352.04	137502.01	3–5	1.22–05	2.46–06	6.92–05	–5.131	B	1
			2847.97	2848.81	102352.04	137454.40	3–3	1.38–07	1.68–08	4.74–07	–7.297	B	1
			2850.30	2851.14	102352.04	137425.70	3–1	2.67–06	1.08–07	3.05–06	–6.488	B	1

C III: Allowed Transitions — Continued

No.	Transition Array	Mult.	λ_{air} (Å)	λ_{vac} (Å) or σ (cm^{-1})*	E_i (cm^{-1})	E_k (cm^{-1})	g_i-g_k	A_{ki} (10^8 s^{-1})	f_{ik}	S (at. u.)	log gf	Acc.	Source
29.		$^1P^o-^1D$	2296.87	2297.58	102352.04	145876.13	3–5	1.376+00	1.815–01	4.118+00	–0.2641	A+	1
30.		$^1P^o-^1S$		1247.38	102352.04	182519.88	3–1	2.082+01	1.619–01	1.994+00	–0.3137	A+	1
31.	$2s2p-2s3s$	$^1P^o-^1S$		690.521	102352.04	247170.26	3–1	8.30+00	1.98–02	1.35–01	–1.227	A	1
32.	$2s2p-2s3d$	$^1P^o-^1D$		574.281	102352.04	276482.86	3–5	6.24+01	5.14–01	2.92+00	0.188	A	1
33.	$2s2p-2s4s$	$^1P^o-^1S$		477.625	102352.04	311721.51	3–1	2.60+00	2.97–03	1.40–02	–2.050	B–	2
34.	$2s2p-2p3p$	$^1P^o-^1P$		460.049	102352.04	319720.35	3–3	2.28+01	7.23–02	3.29–01	–0.664	B	2
35.	$2s2p-2s4d$	$^1P^o-^1D$		450.734	102352.04	324212.49	3–5	3.23+01	1.64–01	7.29–01	–0.308	B	2
36.	$2s2p-2p3p$	$^1P^o-^1D$		433.339	102352.04	333118.21	3–5	2.40+01	1.12–01	4.81–01	–0.472	B	2
37.	$2s2p-2s5s$	$^1P^o-^1S$		423.438	102352.04	338514.33	3–1	3.64–02	3.26–05	1.37–04	–4.009	D	4
38.	$2s2p-2p3p$	$^1P^o-^1S$		411.958	102352.04	345095.43	3–1	4.06+00	3.44–03	1.40–02	–1.986	B–	2
39.	$2s2p-2s5d$	$^1P^o-^1D$		409.322	102352.04	346658.34	3–5	9.74+00	4.08–02	1.65–01	–0.913	B	2
40.	$2s2p-2s6d$	$^1P^o-^1D$		390.045	102352.04	358732.93	3–5	6.38+00	2.43–02	9.35–02	–1.138	B	2
41.	$2s2p-2s7d$	$^1P^o-^1D$		379.253	102352.04	366028	3–5	4.17+00	1.50–02	5.61–02	–1.347	B	2
42.	$2s2p-2p4p$	$^1P^o-^1P$		358.740	102352.04	381105.4	3–3	9.60+00	1.85–02	6.56–02	–1.255	B	2
43.	$2s2p-2p5p$	$^1P^o-^1P$		327.784	102352.04	407431	3–3	4.87+00	7.85–03	2.54–02	–1.628	B–	2
44.	$2p^2-2s3p$	$^3P-^3P^o$		*818.06*	*137477.7*	*259717.9*	9–9	1.93–02	1.94–04	4.70–03	–2.758	D	2
				818.181	137502.01	259724.30	5–5	1.45–02	1.45–04	1.96–03	–3.138	D	*LS*
				817.950	137454.40	259711.22	3–3	4.83–03	4.85–05	3.92–04	–3.837	D–	*LS*
				818.269	137502.01	259711.22	5–3	8.05–03	4.85–05	6.53–04	–3.616	D–	*LS*
				817.988	137454.40	259705.55	3–1	1.93–02	6.46–05	5.22–04	–3.712	D–	*LS*
				817.863	137454.40	259724.30	3–5	4.84–03	8.08–05	6.53–04	–3.615	D–	*LS*
				817.758	137425.70	259711.22	1–3	6.45–03	1.94–04	5.22–04	–3.712	D–	*LS*
45.	$2p^2-2p3s$	$^3P-^3P^o$		*585.46*	*137477.7*	*308283.3*	9–9	2.62+01	1.35–01	2.34+00	0.083	B	2
				585.428	137502.01	308317.29	5–5	1.97+01	1.01–01	9.73–01	–0.297	B	*LS*
				585.499	137454.40	308248.91	3–3	6.55+00	3.37–02	1.95–01	–0.996	B	*LS*
				585.662	137502.01	308248.91	5–3	1.09+01	3.37–02	3.24–01	–0.774	B	*LS*
				585.610	137454.40	308216.58	3–1	2.62+01	4.49–02	2.60–01	–0.871	B	*LS*
				585.265	137454.40	308317.29	3–5	6.56+00	5.61–02	3.24–01	–0.774	B	*LS*
				585.401	137425.70	308248.91	1–3	8.74+00	1.35–01	2.60–01	–0.871	B	*LS*

C III: Allowed Transitions — Continued

No.	Transition Array	Mult.	λ_{air} (Å)	λ_{vac} (Å) or σ (cm^{-1})*	E_i (cm^{-1})	E_k (cm^{-1})	g_i–g_k	A_{ki} (10^8 s^{-1})	f_{ik}	S (at. u.)	log gf	Acc.	Source
46.	$2p^2$–$2s4p$	^{3}P–^{3}P°		*554.57*	*137477.7*	*317798.9*	9–9	8.05−01	3.71−03	6.10−02	−1.476	B	2
				554.633	137502.01	317801.30	5–5	6.04−01	2.78−03	2.54−02	−1.856	B−	*LS*
				554.502	137454.40	317796.51	3–3	2.01−01	9.28−04	5.08−03	−2.555	C+	*LS*
				554.648	137502.01	317796.51	5–3	3.35−01	9.28−04	8.47−03	−2.333	C+	*LS*
				554.509	137454.40	317794.26	3–1	8.05−01	1.24−03	6.78−03	−2.430	C+	*LS*
				554.487	137454.40	317801.30	3–5	2.01−01	1.55−03	8.47−03	−2.333	C+	*LS*
				554.413	137425.70	317796.51	1–3	2.69−01	3.71−03	6.78−03	−2.430	C+	*LS*
47.	$2p^2$–$2p3d$	^{3}P–^{3}D°		*499.51*	*137477.7*	*337675.2*	9–15	1.19+02	7.39−01	1.09+01	0.823	B	2
				499.535	137502.01	337688.04	5–7	1.19+02	6.21−01	5.11+00	0.492	B	*LS*
				499.464	137454.40	337668.89	3–5	8.90+01	5.55−01	2.74+00	0.221	B	*LS*
				499.425	137425.70	337655.98	1–3	6.59+01	7.40−01	1.22+00	−0.131	B	*LS*
				499.583	137502.01	337668.89	5–5	2.96+01	1.11−01	9.12−01	−0.256	B	*LS*
				499.497	137454.40	337655.98	3–3	4.94+01	1.85−01	9.12−01	−0.256	B	*LS*
				499.615	137502.01	337655.98	5–3	3.29+00	7.39−03	6.08−02	−1.432	B	*LS*
48.		^{3}P–^{3}P°		*493.49*	*137477.7*	*340114.8*	9–9	4.69+01	1.71−01	2.51+00	0.188	B	2
				493.584	137502.01	340101.84	5–5	3.52+01	1.29−01	1.04+00	−0.192	B	*LS*
				493.405	137454.40	340127.53	3–3	1.17+01	4.29−02	2.09−01	−0.891	B	*LS*
				493.521	137502.01	340127.53	5–3	1.96+01	4.28−02	3.48−01	−0.669	B	*LS*
				493.371	137454.40	340141.83	3–1	4.70+01	5.71−02	2.78−01	−0.766	B	*LS*
				493.468	137454.40	340101.84	3–5	1.17+01	7.14−02	3.48−01	−0.669	B	*LS*
				493.335	137425.70	340127.53	1–3	1.57+01	1.71−01	2.78−01	−0.766	B	*LS*
49.	$2p^2$–$2s5p$	^{3}P–^{3}P°		*483.66*	*137477.7*	*344234.7*	9–9	1.52+01	5.33−02	7.64−01	−0.319	B	2
				483.720	137502.01	344232.98	5–5	1.14+01	4.00−02	3.18−01	−0.699	B	*LS*
				483.601	137454.40	344236.29	3–3	3.80+00	1.33−02	6.37−02	−1.398	B	*LS*
				483.713	137502.01	344236.29	5–3	6.33+00	1.33−02	1.06−01	−1.176	B	*LS*
				483.596	137454.40	344238.68	3–1	1.52+01	1.78−02	8.49−02	−1.273	B	*LS*
				483.609	137454.40	344232.98	3–5	3.80+00	2.22−02	1.06−01	−1.176	B	*LS*
				483.534	137425.70	344236.29	1–3	5.07+00	5.33−02	8.49−02	−1.273	B	*LS*
50.	$2p^2$–$2s6p$	^{3}P–^{3}P°		*455.43*	*137477.7*	*357050.7*	9–9	2.81−01	8.74−04	1.18−02	−2.104	B−	2
				455.479	137502.01	357051.23	5–5	2.11−01	6.56−04	4.92−03	−2.484	C+	*LS*
				455.382	137454.40	357050.17	3–3	7.03−02	2.19−04	9.83−04	−3.183	D	*LS*
				455.481	137502.01	357050.17	5–3	1.17−01	2.19−04	1.64−03	−2.961	C	*LS*
				455.384	137454.40	357049.38	3–1	2.81−01	2.92−04	1.31−03	−3.058	C	*LS*
				455.380	137454.40	357051.23	3–5	7.03−02	3.64−04	1.64−03	−2.961	C	*LS*
				455.323	137425.70	357050.17	1–3	9.38−02	8.75−04	1.31−03	−3.058	C	*LS*
51.	$2p^2$–$2s3p$	^{1}D–^{1}P°		884.524	145876.13	258931.29	5–3	3.93+00	2.77−02	4.03−01	−0.859	B	2
52.	$2p^2$–$2p3s$	^{1}D–^{1}P°		609.272	145876.13	310006.32	5–3	3.83+00	1.28−02	1.28−01	−1.194	B	2
53.	$2p^2$–$2s4p$	^{1}D–^{1}P°		566.482	145876.13	322404.20	5–3	2.07+01	5.98−02	5.58−01	−0.524	B	2
54.	$2p^2$–$2s4f$	^{1}D–^{1}F°		565.528	145876.13	322702.02	5–7	1.29+01	8.65−02	8.05−01	−0.364	B	2
55.	$2p^2$–$2p3d$	^{1}D–^{1}D°		535.288	145876.13	332691.28	5–5	5.17+01	2.22−01	1.96+00	0.045	B	2
56.		^{1}D–^{1}F°		511.523	145876.13	341370.94	5–7	7.39+01	4.06−01	3.42+00	0.307	B	2

C III: Allowed Transitions — Continued

No.	Transition Array	Mult.	λ_{air} (Å)	λ_{vac} (Å) or σ (cm^{-1})*	E_i (cm^{-1})	E_k (cm^{-1})	g_i–g_k	A_{ki} (10^8 s^{-1})	f_{ik}	S (at. u.)	log gf	Acc.	Source
57.	$2p^2$– $2s5p$	^1D–$^1P^\circ$		506.632	145876.13	343258.03	5–3	3.12–01	7.19–04	6.00–03	–2.444	C+	2
58.	$2p^2$– $2p3d$	^1D–$^1P^\circ$		497.917	145876.13	346712.73	5–3	1.94+00	4.33–03	3.55–02	–1.664	B	2
59.	$2p^2$– $2s5f$	^1D–$^1F^\circ$		492.650	145876.13	348859.99	5–7	2.93+01	1.49–01	1.21+00	–0.126	B	2
60.	$2p^2$– $2s6p$	^1D–$^1P^\circ$		473.410	145876.13	357109.68	5–3	3.84–02	7.75–05	6.04–04	–3.412	D	4
61.	$2p^2$– $2s6f$	^1D–$^1F^\circ$		468.942	145876.13	359121.95	5–7	1.38+00	6.35–03	4.90–02	–1.498	B	2
62.	$2p^2$– $2s7p$	^1D–$^1P^\circ$		456.580	145876.13	364896	5–3	1.63–01	3.06–04	2.30–03	–2.815	C	2
63.	$2p^2$– $2s8p$	^1D–$^1P^\circ$		446.329	145876.13	369926	5–3	2.73–01	4.90–04	3.60–03	–2.611	C+	2
64.	$2p^2$– $2p4d$	^1D–$^1D^\circ$		416.769	145876.13	385817.2	5–5	1.73+01	4.51–02	3.10–01	–0.647	B	2
65.	$2p^2$– $2p5d$	^1D–$^1D^\circ$		379.065	145876.13	409683	5–5	8.03+00	1.73–02	1.08–01	–1.063	B	2
66.	$2p^2$– $2s3p$	^1S–$^1P^\circ$		1308.71	182519.88	258931.29	1–3	3.27–01	2.52–02	1.09–01	–1.598	B	2
67.	$2p^2$– $2p3s$	^1S–$^1P^\circ$		784.397	182519.88	310006.32	1–3	2.33+00	6.46–02	1.67–01	–1.190	B	2
68.	$2p^2$– $2s4p$	^1S–$^1P^\circ$		714.876	182519.88	322404.20	1–3	7.01+00	1.61–01	3.79–01	–0.793	B	2
69.	$2p^2$– $2s5p$	^1S–$^1P^\circ$		622.130	182519.88	343258.03	1–3	4.23+00	7.36–02	1.51–01	–1.133	B	2
70.	$2p^2$– $2p3d$	^1S–$^1P^\circ$		609.040	182519.88	346712.73	1–3	5.09+01	8.49–01	1.70+00	–0.071	B	2
71.	$2p^2$– $2s6p$	^1S–$^1P^\circ$		572.771	182519.88	357109.68	1–3	8.88–01	1.31–02	2.47–02	–1.883	B–	2
72.	$2p^2$– $2s7p$	^1S–$^1P^\circ$		548.317	182519.88	364896	1–3	9.42–02	1.27–03	2.30–03	–2.895	C	2
73.	$2p^2$– $2s8p$	^1S–$^1P^\circ$		533.601	182519.88	369926	1–3	1.62–03	2.08–05	3.65–05	–4.682	D	4
74.	$2s3s$– $2s3p$	^3S–$^3P^\circ$	*4648.8*	*4650.1*	238213.00	*259717.9*	3–9	7.25–01	7.05–01	3.24+01	0.325	A	2,3
			4647.42	4648.72	238213.00	259724.30	3–5	7.26–01	3.92–01	1.80+01	0.070	B+	*LS*
			4650.25	4651.55	238213.00	259711.22	3–3	7.25–01	2.35–01	1.08+01	–0.152	B+	*LS*
			4651.47	4652.78	238213.00	259705.55	3–1	7.24–01	7.83–02	3.60+00	–0.629	B+	*LS*
75.	$2s3s$– $2p3s$	^3S–$^3P^\circ$		*1427.1*	238213.00	*308283.3*	3–9	3.21+00	2.94–01	4.14+00	–0.055	B	2
				1426.45	238213.00	308317.29	3–5	3.21+00	1.63–01	2.30+00	–0.310	B	*LS*
				1427.84	238213.00	308248.91	3–3	3.20+00	9.79–02	1.38+00	–0.532	B	*LS*
				1428.50	238213.00	308216.58	3–1	3.20+00	3.26–02	4.60–01	–1.009	B	*LS*

C III: Allowed Transitions — Continued

No.	Transition Array	Mult.	λ_{air} (Å)	λ_{vac} (Å) or σ (cm^{-1})*	E_i (cm^{-1})	E_k (cm^{-1})	$g_i - g_k$	A_{ki} (10^8 s^{-1})	f_{ik}	S (at. u.)	log gf	Acc.	Source
76.	$2s3s-2s4p$	^{3}S–^{3}P°		*1256.5*	238213.00	*317798.9*	3–9	1.08+00	7.67−02	9.52−01	−0.638	B	2
				1256.47	238213.00	317801.30	3–5	1.08+00	4.26−02	5.29−01	−0.893	B	*LS*
				1256.54	238213.00	317796.51	3–3	1.08+00	2.56−02	3.17−01	−1.115	B	*LS*
				1256.58	238213.00	317794.26	3–1	1.08+00	8.52−03	1.06−01	−1.592	B	*LS*
77.	$2s3s-2p3d$	^{3}S–^{3}P°		*981.34*	238213.00	*340114.8*	3–9	3.71−01	1.61−02	1.56−01	−1.316	B	2
				981.462	238213.00	340101.84	3–5	3.71−01	8.94−03	8.66−02	−1.572	B	*LS*
				981.214	238213.00	340127.53	3–3	3.72−01	5.36−03	5.20−02	−1.794	B	*LS*
				981.077	238213.00	340141.83	3–1	3.72−01	1.79−03	1.73−02	−2.271	B−	*LS*
78.	$2s3s-2s5p$	^{3}S–^{3}P°		*943.20*	238213.00	*344234.7*	3–9	4.50−01	1.80−02	1.68−01	−1.267	B	2
				943.218	238213.00	344232.98	3–5	4.50−01	1.00−02	9.33−02	−1.522	B	*LS*
				943.189	238213.00	344236.29	3–3	4.50−01	6.01−03	5.60−02	−1.744	B	*LS*
				943.168	238213.00	344238.68	3–1	4.51−01	2.00−03	1.87−02	−2.221	B−	*LS*
79.	$2s3s-2s6p$	^{3}S–^{3}P°		*841.48*	238213.00	*357050.7*	3–9	5.21−01	1.66−02	1.38−01	−1.303	B	2
				841.480	238213.00	357051.23	3–5	5.21−01	9.22−03	7.66−02	−1.558	B	*LS*
				841.488	238213.00	357050.17	3–3	5.21−01	5.53−03	4.60−02	−1.780	B	*LS*
				841.493	238213.00	357049.38	3–1	5.21−01	1.84−03	1.53−02	−2.257	B−	*LS*
80.	$2s3s-2s3p$	^{1}S–^{1}P°	8500.32	8502.66	247170.26	258931.29	1–3	1.01−01	3.29−01	9.20+00	−0.483	B	2
81.	$2s3s-2p3s$	^{1}S–^{1}P°		1591.44	247170.26	310006.32	1–3	1.88−01	2.14−02	1.12−01	−1.669	C+	2
82.	$2s3s-2s4p$	^{1}S–^{1}P°		1329.19	247170.26	322404.20	1–3	1.03+01	8.20−01	3.59+00	−0.086	B	2
83.	$2s3s-2s5p$	^{1}S–^{1}P°		1040.72	247170.26	343258.03	1–3	7.19−02	3.50−03	1.20−02	−2.456	C	2
84.	$2s3s-2p3d$	^{1}S–^{1}P°		1004.60	247170.26	346712.73	1–3	1.20+00	5.46−02	1.81−01	−1.263	B	2
85.	$2s3s-2s6p$	^{1}S–^{1}P°		909.592	247170.26	357109.68	1–3	5.21−01	1.94−02	5.80−02	−1.713	B	2
86.	$2s3s-2s7p$	^{1}S–^{1}P°		849.432	247170.26	364896	1–3	3.37−01	1.09−02	3.06−02	−1.961	B	2
87.	$2s3s-2s8p$	^{1}S–^{1}P°		814.626	247170.26	369926	1–3	2.39−01	7.12−03	1.91−02	−2.147	B−	2
88.	$2s3p-2s3d$	^{1}P°–^{1}D	5695.92	5697.50	258931.29	276482.86	3–5	4.27−01	3.46−01	1.95+01	0.016	B	2
89.	$2s3p-2s4s$	^{1}P°–^{1}S		1894.29	258931.29	311721.51	3–1	5.40+00	9.69−02	1.81+00	−0.537	B	2
90.	$2s3p-2p3p$	^{1}P°–^{1}P		1645.03	258931.29	319720.35	3–3	5.24+00	2.12−01	3.45+00	−0.196	B	2
91.	$2s3p-2s4d$	^{1}P°–^{1}D		1531.83	258931.29	324212.49	3–5	3.47+00	2.03−01	3.08+00	−0.215	B	2
92.	$2s3p-2p3p$	^{1}P°–^{1}D		1347.95	258931.29	333118.21	3–5	6.62−01	3.01−02	4.00−01	−1.045	B	2

C III: Allowed Transitions — Continued

No.	Transition Array	Mult.	λ_{air} (Å)	λ_{vac} (Å) or σ (cm^{-1})*	E_i (cm^{-1})	E_k (cm^{-1})	g_i-g_k	A_{ki} (10^8 s^{-1})	f_{ik}	S (at. u.)	log gf	Acc.	Source
93.	2s3p–2s5s	$^1P^o-^1S$		1256.55	258931.29	338514.33	3–1	3.23+00	2.55–02	3.16–01	–1.116	B	2
94.	2s3p–2p3p	$^1P^o-^1S$		1160.58	258931.29	345095.43	3–1	2.45–01	1.65–03	1.89–02	–2.306	B–	2
95.	2s3p–2s5d	$^1P^o-^1D$		1139.90	258931.29	346658.34	3–5	2.73+00	8.86–02	9.97–01	–0.575	B	2
96.	2s3p–2s6d	$^1P^o-^1D$		1001.99	258931.29	358732.93	3–5	1.67+00	4.18–02	4.14–01	–0.902	B	2
97.	2s3p–2s7d	$^1P^o-^1D$		933.735	258931.29	366028	3–5	1.09+00	2.36–02	2.18–01	–1.149	B	2
98.	2s3p–2s3d	$^3P^o-^3D$	*9710.6*	*9713.3*	*259717.9*	*270013.0*	9–15	7.89–02	1.86–01	5.35+01	0.224	A	2,3
			9715.09	9717.76	259724.30	270014.74	5–7	7.88–02	1.56–01	2.50+01	–0.108	B+	*LS*
			9705.41	9708.07	259711.22	270011.93	3–5	5.93–02	1.40–01	1.34+01	–0.378	B+	*LS*
			9701.10	9703.76	259705.55	270010.83	1–3	4.40–02	1.86–01	5.94+00	–0.730	B+	*LS*
			9717.75	9720.41	259724.30	270011.93	5–5	1.97–02	2.79–02	4.46+00	–0.856	B+	*LS*
			9706.44	9709.11	259711.22	270010.83	3–3	3.29–02	4.65–02	4.46+00	–0.855	B+	*LS*
			9718.79	9721.45	259724.30	270010.83	5–3	2.19–03	1.86–03	2.97–01	–2.032	B+	*LS*
99.	2s3p–2s4s	$^3P^o-^3S$	*2009.8*	*2010.5*	*259717.9*	309457.17	9–3	6.91+00	1.40–01	8.31+00	0.099	B	2
			2010.09	2010.74	259724.30	309457.17	5–3	3.84+00	1.40–01	4.62+00	–0.156	B	*LS*
			2009.56	2010.21	259711.22	309457.17	3–3	2.30+00	1.40–01	2.77+00	–0.378	B	*LS*
			2009.34	2009.98	259705.55	309457.17	1–3	7.69–01	1.40–01	9.23–01	–0.855	B	*LS*
100.	2s3p–2s4d	$^3P^o-^3D$		*1620.3*	*259717.9*	*321434.5*	9–15	8.31+00	5.45–01	2.62+01	0.691	B	2
				1620.07	259724.30	321450.05	5–7	8.32+00	4.58–01	1.22+01	0.360	B	*LS*
				1620.34	259711.22	321426.74	3–5	6.23+00	4.09–01	6.54+00	0.089	B	*LS*
				1620.59	259705.55	321411.31	1–3	4.61+00	5.45–01	2.91+00	–0.264	B	*LS*
				1620.68	259724.30	321426.74	5–5	2.08+00	8.18–02	2.18+00	–0.388	B	*LS*
				1620.74	259711.22	321411.31	3–3	3.46+00	1.36–01	2.18+00	–0.388	B	*LS*
				1621.09	259724.30	321411.31	5–3	2.31–01	5.45–03	1.45–01	–1.565	B	*LS*
101.	2s3p–2p3p	$^3P^o-^3D$		*1577.4*	*259717.9*	*323114.7*	9–15	2.05–01	1.27–02	5.94–01	–0.941	C+	2
				1576.89	259724.30	323140.33	5–7	2.05–01	1.07–02	2.77–01	–1.272	C+	*LS*
				1577.53	259711.22	323101.36	3–5	1.53–01	9.54–03	1.49–01	–1.543	C+	*LS*
				1578.00	259705.55	323076.88	1–3	1.14–01	1.27–02	6.60–02	–1.896	C+	*LS*
				1577.86	259724.30	323101.36	5–5	5.11–02	1.91–03	4.95–02	–2.021	C+	*LS*
				1578.14	259711.22	323076.88	3–3	8.51–02	3.18–03	4.95–02	–2.021	C+	*LS*
				1578.47	259724.30	323076.88	5–3	5.67–03	1.27–04	3.30–03	–3.197	D	*LS*
102.		$^3P^o-^3S$		*1480.2*	*259717.9*	327278.27	9–3	2.18+00	2.38–02	1.04+00	–0.669	B	2
				1480.30	259724.30	327278.27	5–3	1.21+00	2.38–02	5.81–01	–0.924	B	*LS*
				1480.01	259711.22	327278.27	3–3	7.26–01	2.38–02	3.48–01	–1.146	B	*LS*
				1479.89	259705.55	327278.27	1–3	2.42–01	2.38–02	1.16–01	–1.623	B	*LS*
103.		$^3P^o-^3P$		*1428.4*	*259717.9*	*329724.7*	9–9	3.54+00	1.08–01	4.58+00	–0.011	B	2
				1428.18	259724.30	329743.57	5–5	2.66+00	8.12–02	1.91+00	–0.391	B	*LS*
				1428.67	259711.22	329706.47	3–3	8.85–01	2.71–02	3.82–01	–1.090	B	*LS*
				1428.94	259724.30	329706.47	5–3	1.47+00	2.71–02	6.37–01	–0.869	B	*LS*
				1429.10	259711.22	329685.38	3–1	3.54+00	3.61–02	5.09–01	–0.966	B	*LS*
				1427.91	259711.22	329743.57	3–5	8.86–01	4.51–02	6.37–01	–0.868	B	*LS*
				1428.55	259705.55	329706.47	1–3	1.18+00	1.08–01	5.09–01	–0.965	B	*LS*

C III: Allowed Transitions — Continued

No.	Transition Array	Mult.	λ_{air} (Å)	λ_{vac} (Å) or σ (cm^{-1})*	E_i (cm^{-1})	E_k (cm^{-1})	g_i–g_k	A_{ki} (10^8 s^{-1})	f_{ik}	S (at. u.)	log gf	Acc.	Source
104.	$2s3p$–$2s5s$	$^3P^o$–^3S		*1246.6*	*259717.9*	– 339934.72	9–3	1.58+00	1.23–02	4.53–01	–0.957	B	2
				1246.72	259724.30	– 339934.72	5–3	8.77–01	1.23–02	2.52–01	–1.213	B	*LS*
				1246.52	259711.22	– 339934.72	3–3	5.26–01	1.23–02	1.51–01	–1.434	B	*LS*
				1246.43	259705.55	– 339934.72	1–3	1.75–01	1.23–02	5.03–02	–1.911	B	*LS*
105.	$2s3p$–$2s5d$	$^3P^o$–^3D		*1165.8*	*259717.9*	*– 345496.9*	9–15	3.11+00	1.06–01	3.65+00	–0.022	B	2
				1165.87	259724.30	– 345497.15	5–7	3.11+00	8.87–02	1.70+00	–0.353	B	*LS*
				1165.70	259711.22	– 345496.72	3–5	2.33+00	7.92–02	9.12–01	–0.624	B	*LS*
				1165.62	259705.55	– 345496.57	1–3	1.73+00	1.06–01	4.05–01	–0.976	B	*LS*
				1165.88	259724.30	– 345496.72	5–5	7.77–01	1.58–02	3.04–01	–1.101	B	*LS*
				1165.70	259711.22	– 345496.57	3–3	1.30+00	2.64–02	3.04–01	–1.101	B	*LS*
				1165.88	259724.30	– 345496.57	5–3	8.63–02	1.06–03	2.03–02	–2.278	B–	*LS*
106.	$2s3p$–$2s6s$	$^3P^o$–^3S		*1051.1*	*259717.9*	– 354858.03	9–3	1.07+00	5.90–03	1.84–01	–1.275	B	2
				1051.15	259724.30	– 354858.03	5–3	5.93–01	5.90–03	1.02–01	–1.531	B	*LS*
				1051.01	259711.22	– 354858.03	3–3	3.56–01	5.90–03	6.12–02	–1.752	B	*LS*
				1050.94	259705.55	– 354858.03	1–3	1.19–01	5.90–03	2.04–02	–2.229	B–	*LS*
107.	$2s3p$–$2s7s$	$^3P^o$–^3S		*962.51*	*259717.9*	– 363613	9–3	6.64–01	3.08–03	8.77–02	–1.558	B	2
				962.569	259724.30	– 363613	5–3	3.69–01	3.08–03	4.87–02	–1.813	B	*LS*
				962.447	259711.22	– 363613	3–3	2.21–01	3.08–03	2.92–02	–2.035	B–	*LS*
				962.395	259705.55	– 363613	1–3	7.38–02	3.08–03	9.74–03	–2.512	C+	*LS*
108.	$2s3p$–$2s7d$	$^3P^o$–^3D		*944.11*	*259717.9*	– 365638	9–15	1.16+00	2.58–02	7.23–01	–0.633	B	2
				944.165	259724.30	– 365638	5–7	1.16+00	2.17–02	3.37–01	–0.964	B	*LS*
				944.048	259711.22	– 365638	3–5	8.71–01	1.94–02	1.81–01	–1.235	B	*LS*
				943.998	259705.55	– 365638	1–3	6.45–01	2.59–02	8.03–02	–1.587	B	*LS*
				944.165	259724.30	– 365638	5–5	2.90–01	3.88–03	6.03–02	–1.713	B	*LS*
				944.048	259711.22	– 365638	3–3	4.84–01	6.46–03	6.03–02	–1.712	B	*LS*
				944.165	259724.30	– 365638	5–3	3.22–02	2.58–04	4.02–03	–2.889	C+	*LS*
109.	$2s3p$–$2s8d$	$^3P^o$–^3D		*902.68*	*259717.9*	– 370499	9–15	7.77–01	1.58–02	4.23–01	–0.847	B	2
				902.733	259724.30	– 370499	5–7	7.77–01	1.33–02	1.97–01	–1.178	B	*LS*
				902.627	259711.22	– 370499	3–5	5.83–01	1.19–02	1.06–01	–1.449	B	*LS*
				902.580	259705.55	– 370499	1–3	4.32–01	1.58–02	4.70–02	–1.801	B	*LS*
				902.733	259724.30	– 370499	5–5	1.94–01	2.37–03	3.53–02	–1.926	B	*LS*
				902.627	259711.22	– 370499	3–3	3.24–01	3.96–03	3.53–02	–1.926	B	*LS*
				902.733	259724.30	– 370499	5–3	2.16–02	1.58–04	2.35–03	–3.102	C	*LS*
110.	$2s3d$–$2p3s$	^3D–$^3P^o$	*2612.2*	*2613.0*	*270013.0*	*– 308283.3*	15–9	4.54–02	2.78–03	3.59–01	–1.379	B	2
			2610.01	2610.79	270014.74	– 308317.29	7–5	3.82–02	2.79–03	1.68–01	–1.710	B	*LS*
			2614.49	2615.27	270011.93	– 308248.91	5–3	3.39–02	2.09–03	8.98–02	–1.982	B	*LS*
			2616.63	2617.41	270010.83	– 308216.58	3–1	4.51–02	1.54–03	3.99–02	–2.334	B	*LS*
			2609.82	2610.60	270011.93	– 308317.29	5–5	6.82–03	6.97–04	2.99–02	–2.458	B–	*LS*
			2614.41	2615.19	270010.83	– 308248.91	3–3	1.13–02	1.16–03	2.99–02	–2.459	B–	*LS*
			2609.75	2610.53	270010.83	– 308317.29	3–5	4.55–04	7.74–05	2.00–03	–3.634	C	*LS*
111.	$2s3d$–$2s4p$	^3D–$^3P^o$	*2092.0*	*2092.7*	*270013.0*	*– 317798.9*	15–9	1.09+00	4.30–02	4.44+00	–0.191	B	2
			2091.97	2092.64	270014.74	– 317801.30	7–5	9.17–01	4.30–02	2.07+00	–0.522	B	*LS*
			2092.06	2092.73	270011.93	– 317796.51	5–3	8.19–01	3.22–02	1.11+00	–0.793	B	*LS*
			2092.11	2092.78	270010.83	– 317794.26	3–1	1.09+00	2.39–02	4.93–01	–1.145	B	*LS*
			2091.85	2092.52	270011.93	– 317801.30	5–5	1.64–01	1.07–02	3.70–01	–1.270	B	*LS*
			2092.01	2092.68	270010.83	– 317796.51	3–3	2.73–01	1.79–02	3.70–01	–1.270	B	*LS*
			2091.80	2092.47	270010.83	– 317801.30	3–5	1.09–02	1.19–03	2.47–02	–2.446	B–	*LS*

C III: Allowed Transitions — Continued

No.	Transition Array	Mult.	λ_{air} (Å)	λ_{vac} (Å) or σ (cm^{-1})*	E_i (cm^{-1})	E_k (cm^{-1})	g_i-g_k	A_{ki} (10^8 s^{-1})	f_{ik}	S (at. u.)	log gf	Acc.	Source
112.	$2s3d-2s4f$	$^3D-^3F^\circ$		*1923.1*	*270013.0*	*– 322011.8*	15–21	8.11+00	6.30–01	5.98+01	0.975	B	2
				1922.96	270014.74	– 322017.97	7–9	8.12+00	5.78–01	2.56+01	0.607	C+	*LS*
				1923.16	270011.93	– 322009.58	5–7	7.21+00	5.60–01	1.77+01	0.447	C+	*LS*
				1923.34	270010.83	– 322003.68	3–5	6.81+00	6.30–01	1.20+01	0.276	C+	*LS*
				1923.27	270014.74	– 322009.58	7–7	9.04–01	5.01–02	2.22+00	–0.455	C+	*LS*
				1923.38	270011.93	– 322003.68	5–5	1.27+00	7.02–02	2.22+00	–0.455	C+	*LS*
				1923.49	270014.74	– 322003.68	7–5	3.57–02	1.41–03	6.27–02	–2.005	C+	*LS*
113.	$2s3d-2p3d$	$^3D-^3F^\circ$		*1577.1*	*270013.0*	*– 333421.0*	15–21	4.74+00	2.47–01	1.93+01	0.569	B	2
				1576.48	270014.74	– 333447.24	7–9	4.74+00	2.27–01	8.25+00	0.201	B	*LS*
				1577.30	270011.93	– 333411.55	5–7	4.21+00	2.20–01	5.70+00	0.041	B	*LS*
				1577.88	270010.83	– 333387.01	3–5	3.97+00	2.47–01	3.85+00	–0.130	B	*LS*
				1577.37	270014.74	– 333411.55	7–7	5.28–01	1.97–02	7.15–01	–0.861	B	*LS*
				1577.91	270011.93	– 333387.01	5–5	7.38–01	2.75–02	7.15–01	–0.861	B	*LS*
				1577.98	270014.74	– 333387.01	7–5	2.08–02	5.55–04	2.02–02	–2.411	B–	*LS*
114.		$^3D-^3D^\circ$		*1477.9*	*270013.0*	*– 337675.2*	15–15	3.73+00	1.22–01	8.92+00	0.263	B	2
				1477.69	270014.74	– 337688.04	7–7	3.32+00	1.09–01	3.70+00	–0.119	B	*LS*
				1478.04	270011.93	– 337668.89	5–5	2.60+00	8.51–02	2.07+00	–0.371	B	*LS*
				1478.30	270010.83	– 337655.98	3–3	2.80+00	9.17–02	1.34+00	–0.561	B	*LS*
				1478.11	270014.74	– 337668.89	7–5	5.82–01	1.36–02	4.64–01	–1.021	B	*LS*
				1478.33	270011.93	– 337655.98	5–3	9.32–01	1.83–02	4.46–01	–1.038	B	*LS*
				1477.63	270011.93	– 337688.04	5–7	4.16–01	1.91–02	4.64–01	–1.021	B	*LS*
				1478.02	270010.83	– 337668.89	3–5	5.60–01	3.06–02	4.46–01	–1.038	B	*LS*
115.		$^3D-^3P^\circ$		*1426.5*	*270013.0*	*– 340114.8*	15–9	3.32+00	6.08–02	4.28+00	–0.040	B	2
				1426.80	270014.74	– 340101.84	7–5	2.79+00	6.08–02	2.00+00	–0.371	B	*LS*
				1426.22	270011.93	– 340127.53	5–3	2.49+00	4.56–02	1.07+00	–0.642	B	*LS*
				1425.90	270010.83	– 340141.83	3–1	3.32+00	3.38–02	4.76–01	–0.994	B	*LS*
				1426.74	270011.93	– 340101.84	5–5	4.98–01	1.52–02	3.57–01	–1.119	B	*LS*
				1426.19	270010.83	– 340127.53	3–3	8.31–01	2.53–02	3.57–01	–1.119	B	*LS*
				1426.72	270010.83	– 340101.84	3–5	3.32–02	1.69–03	2.38–02	–2.296	B–	*LS*
116.	$2s3d-2s5p$	$^3D-^3P^\circ$		*1347.3*	*270013.0*	*– 344234.7*	15–9	1.87–01	3.05–03	2.03–01	–1.339	B	2
				1347.38	270014.74	– 344232.98	7–5	1.57–01	3.05–03	9.48–02	–1.670	B	*LS*
				1347.27	270011.93	– 344236.29	5–3	1.40–01	2.29–03	5.08–02	–1.941	B	*LS*
				1347.20	270010.83	– 344238.68	3–1	1.87–01	1.70–03	2.26–02	–2.293	B–	*LS*
				1347.33	270011.93	– 344232.98	5–5	2.80–02	7.63–04	1.69–02	–2.418	B–	*LS*
				1347.25	270010.83	– 344236.29	3–3	4.67–02	1.27–03	1.69–02	–2.418	B–	*LS*
				1347.31	270010.83	– 344232.98	3–5	1.87–03	8.48–05	1.13–03	–3.595	C	*LS*
117.	$2s3d-2s5f$	$^3D-^3F^\circ$		*1296.3*	*270013.0*	*– 347153.9*	15–21	6.50+00	2.29–01	1.47+01	0.537	B	2
				1296.33	270014.74	– 347155.41	7–9	6.50+00	2.11–01	6.29+00	0.169	C+	*LS*
				1296.32	270011.93	– 347153.26	5–7	5.78+00	2.04–01	4.35+00	0.008	C+	*LS*
				1296.33	270010.83	– 347151.89	3–5	5.46+00	2.29–01	2.94+00	–0.162	C+	*LS*
				1296.37	270014.74	– 347153.26	7–7	7.25–01	1.83–02	5.45–01	–0.893	C+	*LS*
				1296.34	270011.93	– 347151.89	5–5	1.01+00	2.56–02	5.45–01	–0.893	C+	*LS*
				1296.39	270014.74	– 347151.89	7–5	2.86–02	5.15–04	1.54–02	–2.443	C	*LS*
118.	$2s3d-2s6p$	$^3D-^3P^\circ$		*1148.9*	*270013.0*	*– 357050.7*	15–9	7.08–02	8.41–04	4.77–02	–1.899	B–	2
				1148.94	270014.74	– 357051.23	7–5	5.95–02	8.41–04	2.23–02	–2.230	B–	*LS*
				1148.92	270011.93	– 357050.17	5–3	5.31–02	6.31–04	1.19–02	–2.501	B–	*LS*
				1148.92	270010.83	– 357049.38	3–1	7.08–02	4.67–04	5.30–03	–2.853	C+	*LS*
				1148.91	270011.93	– 357051.23	5–5	1.06–02	2.10–04	3.98–03	–2.978	C+	*LS*
				1148.91	270010.83	– 357050.17	3–3	1.77–02	3.50–04	3.98–03	–2.978	C+	*LS*
				1148.89	270010.83	– 357051.23	3–5	7.08–04	2.34–05	2.65–04	–4.155	D	*LS*

C III: Allowed Transitions — Continued

No.	Transition Array	Mult.	λ_{air} (Å)	λ_{vac} (Å) or σ (cm^{-1})*	E_i (cm^{-1})	E_k (cm^{-1})	g_i-g_k	A_{ki} ($10^8\ s^{-1}$)	f_{ik}	S (at. u.)	log gf	Acc.	Source
119.	2s3d– 2s6f	$^3D-^3F^o$		*1125.7*	*270013.0*	*– 358850.4*	15–21	3.06+00	8.14–02	4.53+00	0.087	B	2
				1125.67	270014.74	– 358850.74	7–9	3.06+00	7.48–02	1.94+00	–0.281	C+	*LS*
				1125.64	270011.93	– 358850.32	5–7	2.72+00	7.23–02	1.34+00	–0.442	C+	*LS*
				1125.63	270010.83	– 358850.02	3–5	2.57+00	8.14–02	9.05–01	–0.612	C+	*LS*
				1125.68	270014.74	– 358850.32	7–7	3.41–01	6.48–03	1.68–01	–1.343	C+	*LS*
				1125.64	270011.93	– 358850.02	5–5	4.78–01	9.07–03	1.68–01	–1.343	C+	*LS*
				1125.68	270014.74	– 358850.02	7–5	1.35–02	1.83–04	4.74–03	–2.893	C–	*LS*
120.	2s3d– 2p3s	$^1D-^1P^o$	2982.12	2982.99	276482.86	– 310006.32	5–3	1.52–01	1.22–02	5.97–01	–1.216	B	2
121.	2s3d– 2s4p	$^1D-^1P^o$	2176.95	2177.64	276482.86	– 322404.20	5–3	2.15+00	9.19–02	3.29+00	–0.338	B	2
122.	2s3d– 2s4f	$^1D-^1F^o$	2162.93	2163.60	276482.86	– 322702.02	5–7	8.18+00	8.03–01	2.86+01	0.604	B	2
123.	2s3d– 2p3d	$^1D-^1D^o$		1779.09	276482.86	– 332691.28	5–5	1.47+00	6.99–02	2.05+00	–0.456	B	2
124.		$^1D-^1F^o$		1541.12	276482.86	– 341370.94	5–7	1.38+00	6.86–02	1.74+00	–0.464	B	2
125.	2s3d– 2s5p	$^1D-^1P^o$		1497.56	276482.86	– 343258.03	5–3	1.48+00	2.98–02	7.36–01	–0.826	B	2
126.	2s3d– 2p3d	$^1D-^1P^o$		1423.90	276482.86	– 346712.73	5–3	2.27+00	4.15–02	9.72–01	–0.683	B	2
127.	2s3d– 2s5f	$^1D-^1F^o$		1381.65	276482.86	– 348859.99	5–7	8.65+00	3.47–01	7.88+00	0.239	B	2
128.	2s3d– 2s6p	$^1D-^1P^o$		1240.28	276482.86	– 357109.68	5–3	8.60–02	1.19–03	2.43–02	–2.225	B–	2
129.	2s3d– 2s6f	$^1D-^1F^o$		1210.08	276482.86	– 359121.95	5–7	3.29+00	1.01–01	2.02+00	–0.296	B	2
130.	2s3d– 2s7p	$^1D-^1P^o$		1131.05	276482.86	– 364896	5–3	9.48–02	1.09–03	2.03–02	–2.263	B–	2
131.	2s3d– 2s8p	$^1D-^1P^o$		1070.17	276482.86	– 369926	5–3	7.71–02	7.95–04	1.40–02	–2.401	B–	2
132.	2s3d– 2p4d	$^1D-^1D^o$		914.626	276482.86	– 385817.2	5–5	3.04–01	3.81–03	5.74–02	–1.720	B	2
133.	2s3d– 2p5d	$^1D-^1D^o$		750.750	276482.86	– 409683	5–5	1.21–01	1.02–03	1.26–02	–2.293	B	2
134.	2p3s– 2s4s	$^3P^o-^3S$		*1173.9* cm^{-1}	*308283.3*	– 309457.17	9–3	9.47–06	3.43–04	8.67–01	–2.510	B	2
				1139.88 cm^{-1}	308317.29	– 309457.17	5–3	4.82–06	3.33–04	4.81–01	–2.778	B	*LS*
				1208.26 cm^{-1}	308248.91	– 309457.17	3–3	3.44–06	3.53–04	2.89–01	–2.975	B	*LS*
				1240.59 cm^{-1}	308216.58	– 309457.17	1–3	1.24–06	3.63–04	9.63–02	–3.440	B	*LS*

C III: Allowed Transitions — Continued

No.	Transition Array	Mult.	λ_{air} (Å)	λ_{vac} (Å) or σ (cm^{-1})*	E_i (cm^{-1})	E_k (cm^{-1})	g_i-g_k	A_{ki} (10^8 s^{-1})	f_{ik}	S (at. u.)	log gf	Acc.	Source
135.	$2p\,3s-2s\,4d$	$^3P^o-^3D$	*7601.8*	*7603.9*	*308283.3*	*321434.5*	9–15	4.65–02	6.72–02	1.51+01	–0.219	B	2
			7612.45	7614.55	308317.29	321450.05	5–7	4.63–02	5.63–02	7.06+00	–0.550	B	*LS*
			7586.41	7588.50	308248.91	321426.74	3–5	3.51–02	5.05–02	3.78+00	–0.820	B	*LS*
			7576.70	7578.78	308216.58	321411.31	1–3	2.61–02	6.74–02	1.68+00	–1.171	B	*LS*
			7625.99	7628.09	308317.29	321426.74	5–5	1.15–02	1.00–02	1.26+00	–1.299	B	*LS*
			7595.31	7597.40	308248.91	321411.31	3–3	1.94–02	1.68–02	1.26+00	–1.297	B	*LS*
			7634.97	7637.07	308317.29	321411.31	5–3	1.28–03	6.69–04	8.40–02	–2.476	B	*LS*
136.	$2p\,3s-2p\,3p$	$^3P^o-^3D$	*6740.6*	*6742.5*	*308283.3*	*323114.7*	9–15	2.00–01	2.27–01	4.53+01	0.310	B	2
			6744.39	6746.25	308317.29	323140.33	5–7	1.99–01	1.90–01	2.11+01	–0.021	B	*LS*
			6731.04	6732.90	308248.91	323101.36	3–5	1.50–01	1.70–01	1.13+01	–0.292	B	*LS*
			6727.48	6729.34	308216.58	323076.88	1–3	1.12–01	2.27–01	5.03+00	–0.644	B	*LS*
			6762.17	6764.04	308317.29	323101.36	5–5	4.95–02	3.39–02	3.77+00	–0.771	B	*LS*
			6742.15	6744.01	308248.91	323076.88	3–3	8.32–02	5.67–02	3.77+00	–0.769	B	*LS*
			6773.39	6775.26	308317.29	323076.88	5–3	5.47–03	2.26–03	2.52–01	–1.948	B	*LS*
137.		$^3P^o-^3S$	*5263.1*	*5264.6*	*308283.3*	327278.27	9–3	4.72–01	6.53–02	1.02+01	–0.231	B	2
			5272.52	5273.99	308317.29	327278.27	5–3	2.61–01	6.52–02	5.66+00	–0.487	B	*LS*
			5253.58	5255.04	308248.91	327278.27	3–3	1.58–01	6.54–02	3.39+00	–0.707	B	*LS*
			5244.66	5246.12	308216.58	327278.27	1–3	5.30–02	6.55–02	1.13+00	–1.184	B	*LS*
138.		$^3P^o-^3P$	*4662.6*	*4663.9*	*308283.3*	*329724.7*	9–9	9.06–01	2.95–01	4.08+01	0.424	B	2
			4665.86	4667.17	308317.29	329743.57	5–5	6.78–01	2.21–01	1.70+01	0.044	B	*LS*
			4659.06	4660.36	308248.91	329706.47	3–3	2.27–01	7.39–02	3.40+00	–0.654	B	*LS*
			4673.95	4675.26	308317.29	329706.47	5–3	3.75–01	7.36–02	5.66+00	–0.434	B	*LS*
			4663.64	4664.95	308248.91	329685.38	3–1	9.05–01	9.84–02	4.53+00	–0.530	B	*LS*
			4651.02	4652.32	308248.91	329743.57	3–5	2.28–01	1.23–01	5.66+00	–0.432	B	*LS*
			4652.05	4653.35	308216.58	329706.47	1–3	3.04–01	2.96–01	4.53+00	–0.529	B	*LS*
139.	$2p\,3s-2s\,5s$	$^3P^o-^3S$	*3158.5*	*3159.4*	*308283.3*	339934.72	9–3	8.57–02	4.27–03	4.00–01	–1.415	B	2
			3161.90	3162.81	308317.29	339934.72	5–3	4.75–02	4.27–03	2.22–01	–1.671	B	*LS*
			3155.07	3155.99	308248.91	339934.72	3–3	2.87–02	4.28–03	1.33–01	–1.892	B	*LS*
			3151.86	3152.77	308216.58	339934.72	1–3	9.58–03	4.28–03	4.44–02	–2.368	B	*LS*
140.	$2p\,3s-2s\,5d$	$^3P^o-^3D$	*2686.4*	*2687.2*	*308283.3*	*345496.9*	9–15	1.87–03	3.38–04	2.69–02	–2.517	C+	2
			2688.83	2689.63	308317.29	345497.15	5–7	1.87–03	2.84–04	1.26–02	–2.848	C+	*LS*
			2683.92	2684.72	308248.91	345496.72	3–5	1.41–03	2.54–04	6.73–03	–3.119	C	*LS*
			2681.61	2682.40	308216.58	345496.57	1–3	1.05–03	3.38–04	2.99–03	–3.470	C–	*LS*
			2688.86	2689.66	308317.29	345496.72	5–5	4.67–04	5.06–05	2.24–03	–3.597	C–	*LS*
			2683.94	2684.73	308248.91	345496.57	3–3	7.82–04	8.45–05	2.24–03	–3.596	C–	*LS*
			2688.87	2689.67	308317.29	345496.57	5–3	5.19–05	3.38–06	1.49–04	–4.773	D–	*LS*
141.	$2p\,3s-2s\,6s$	$^3P^o-^3S$	*2146.4*	*2147.1*	*308283.3*	354858.03	9–3	3.71–02	8.54–04	5.43–02	–2.115	B	2
			2147.98	2148.66	308317.29	354858.03	5–3	2.05–02	8.53–04	3.02–02	–2.370	B	*LS*
			2144.83	2145.50	308248.91	354858.03	3–3	1.24–02	8.54–04	1.81–02	–2.591	B–	*LS*
			2143.34	2144.02	308216.58	354858.03	1–3	4.13–03	8.55–04	6.03–03	–3.068	C+	*LS*
142.	$2p\,3s-2s\,7s$	$^3P^o-^3S$		*1807.3*	*308283.3*	363613	9–3	2.60–02	4.24–04	2.27–02	–2.418	B–	2
				1808.46	308317.29	363613	5–3	1.44–02	4.24–04	1.26–02	–2.674	B–	*LS*
				1806.22	308248.91	363613	3–3	8.67–03	4.24–04	7.57–03	–2.895	C+	*LS*
				1805.17	308216.58	363613	1–3	2.90–03	4.24–04	2.52–03	–3.372	C	*LS*

C III: Allowed Transitions — Continued

No.	Transition Array	Mult.	λ_{air} (Å)	λ_{vac} (Å) or σ (cm^{-1})*	E_i (cm^{-1})	E_k (cm^{-1})	g_i–g_k	A_{ki} (10^8 s^{-1})	f_{ik}	S (at. u.)	log gf	Acc.	Source
143.	2s4s–2s4p	^{3}S–^{3}P°	*11985*	*8341.8* cm^{-1}	309457.17	– *317798.9*	3–9	1.61–01	1.04+00	1.23+02	0.493	B–	2
			11981.2	8344.13 cm^{-1}	309457.17	– 317801.30	3–5	1.61–01	5.77–01	6.82+01	0.238	B–	*LS*
			11988.1	8339.34 cm^{-1}	309457.17	– 317796.51	3–3	1.61–01	3.46–01	4.09+01	0.016	B–	*LS*
			11991.3	8337.09 cm^{-1}	309457.17	– 317794.26	3–1	1.60–01	1.15–01	1.36+01	–0.461	B–	*LS*
144.	2s4s–2p3d	^{3}S–^{3}P°	*3260.9*	*3261.8*	309457.17	– *340114.9*	3–9	4.89–02	2.34–02	7.53–01	–1.154	B	2
			3262.27	3263.21	309457.17	– 340101.84	3–5	4.88–02	1.30–02	4.18–01	–1.409	B	*LS*
			3259.51	3260.44	309457.17	– 340127.83	3–3	4.89–02	7.80–03	2.51–01	–1.631	B	*LS*
			3258.02	3258.96	309457.17	– 340141.83	3–1	4.90–02	2.60–03	8.37–02	–2.108	B	*LS*
145.	2s4s–2s5p	^{3}S–^{3}P°	*2874.6*	*2875.4*	309457.17	– *344234.7*	3–9	1.19–01	4.41–02	1.25+00	–0.878	B	2
			2874.72	2875.56	309457.17	– 344232.98	3–5	1.19–01	2.45–02	6.96–01	–1.133	B	*LS*
			2874.44	2875.29	309457.17	– 344236.29	3–3	1.19–01	1.47–02	4.18–01	–1.355	B	*LS*
			2874.25	2875.09	309457.17	– 344238.68	3–1	1.19–01	4.90–03	1.39–01	–1.832	B	*LS*
146.	2s4s–2s6p	^{3}S–^{3}P°	*2100.5*	*2101.1*	309457.17	– *357050.7*	3–9	1.64–01	3.25–02	6.74–01	–1.011	B	2
			2100.44	2101.10	309457.17	– 357051.23	3–5	1.64–01	1.81–02	3.75–01	–1.266	B	*LS*
			2100.48	2101.15	309457.17	– 357050.17	3–3	1.64–01	1.08–02	2.25–01	–1.488	B	*LS*
			2100.52	2101.18	309457.17	– 357049.38	3–1	1.64–01	3.61–03	7.49–02	–1.965	B	*LS*
147.	2p3s–2s4s	^{1}P°–^{1}S		1715.19 cm^{-1}	310006.32	– 311721.51	3–1	2.44–03	4.14–02	2.38+01	–0.906	B	2
148.	2p3s–2p3p	^{1}P°–^{1}P	10292	9714.03 cm^{-1}	310006.32	– 319720.35	3–3	2.80–02	4.44–02	4.52+00	–0.875	B	2
149.	2p3s–2s4d	^{1}P°–^{1}D	7037.25	7039.19	310006.32	– 324212.49	3–5	7.46–01	9.23–01	6.41+01	0.442	B	2
150.	2p3s–2p3p	^{1}P°–^{1}D	4325.56	4326.78	310006.32	– 333118.21	3–5	1.24–01	5.80–02	2.48+00	–0.759	B	2
151.	2p3s–2s5s	^{1}P°–^{1}S	3506.78	3507.79	310006.32	– 338514.33	3–1	7.21+00	4.43–01	1.53+01	0.124	B	2
152.	2p3s–2p3p	^{1}P°–^{1}S	2849.05	2849.89	310006.32	– 345095.43	3–1	1.95–01	7.89–03	2.22–01	–1.626	B	2
153.	2p3s–2s5d	^{1}P°–^{1}D	2727.55	2728.36	310006.32	– 346658.34	3–5	2.82+00	5.24–01	1.41+01	0.197	B	2
154.	2p3s–2s6d	^{1}P°–^{1}D	2051.61	2052.27	310006.32	– 358732.93	3–5	1.22+00	1.28–01	2.59+00	–0.416	B	2
155.	2p3s–2s7d	^{1}P°–^{1}D		1785.02	310006.32	– 366028	3–5	6.75–01	5.38–02	9.48–01	–0.792	B	2
156.	2p3s–2p4p	^{1}P°–^{1}P		1406.49	310006.32	– 381105.4	3–3	9.38+00	2.78–01	3.86+00	–0.079	B	2
157.	2p3s–2p5p	^{1}P°–^{1}P		1026.43	310006.32	– 407431	3–3	3.17–01	5.00–03	5.07–02	–1.824	B	2
158.	2s4s–2s4p	^{1}S–^{1}P°	9358.37	9360.94	311721.51	– 322404.20	1–3	1.59–01	6.28–01	1.93+01	–0.202	B	2

C III: Allowed Transitions — Continued

No.	Transition Array	Mult.	λ_{air} (Å)	λ_{vac} (Å) or σ (cm^{-1})*	E_i (cm^{-1})	E_k (cm^{-1})	g_i–g_k	A_{ki} (10^8 s^{-1})	f_{ik}	S (at. u.)	log gf	Acc.	Source
159.	2s4s–2s5p	^1S–$^1P^o$	3170.01	3170.93	311721.51	– 343258.03	1–3	2.26–01	1.02–01	1.07+00	–0.990	B	2
160.	2s4s–2p3d	^1S–$^1P^o$	2857.02	2857.86	311721.51	– 346712.73	1–3	3.77–02	1.39–02	1.30–01	–1.858	B	2
161.	2s4s–2s6p	^1S–$^1P^o$	2202.53	2203.22	311721.51	– 357109.68	1–3	3.40–01	7.42–02	5.38–01	–1.130	B	2
162.	2s4s–2s7p	^1S–$^1P^o$		1880.60	311721.51	– 364896	1–3	3.28–01	5.22–02	3.23–01	–1.283	B	2
163.	2s4s–2s8p	^1S–$^1P^o$		1718.08	311721.51	– 369926	1–3	3.34–01	4.43–02	2.51–01	–1.353	B	2
164.	2s4p–2s4d	$^3P^o$–^3D	*27498*	*3635.6* cm^{-1}	*317798.9*	*– 321434.5*	9–15	1.35–02	2.54–01	2.07+02	0.360	B	2
			27399.2	3648.75 cm^{-1}	317801.30	– 321450.05	5–7	1.36–02	2.14–01	9.68+01	0.030	B	*LS*
			27539.0	3630.23 cm^{-1}	317796.51	– 321426.74	3–5	1.00–02	1.91–01	5.18+01	–0.243	B	*LS*
			27639.3	3617.05 cm^{-1}	317794.26	– 321411.31	1–3	7.36–03	2.53–01	2.30+01	–0.597	B	*LS*
			27575.3	3625.44 cm^{-1}	317801.30	– 321426.74	5–5	3.34–03	3.81–02	1.73+01	–0.721	B	*LS*
			27656.5	3614.80 cm^{-1}	317796.51	– 321411.31	3–3	5.51–03	6.32–02	1.73+01	–0.722	B	*LS*
			27693.2	3610.01 cm^{-1}	317801.30	– 321411.31	5–3	3.66–04	2.53–03	1.15+00	–1.899	B	*LS*
165.	2s4p–2p3p	$^3P^o$–^3D	*18807*	*5315.7* cm^{-1}	*317798.9*	*– 323114.7*	9–15	1.05–02	9.27–02	5.16+01	–0.079	B	2
			18724.9	5339.03 cm^{-1}	317801.30	– 323140.33	5–7	1.06–02	7.82–02	2.41+01	–0.408	B	*LS*
			18845.5	5304.85 cm^{-1}	317796.51	– 323101.36	3–5	7.81–03	6.94–02	1.29+01	–0.682	B	*LS*
			18924.8	5282.62 cm^{-1}	317794.26	– 323076.88	1–3	5.71–03	9.21–02	5.74+00	–1.036	B	*LS*
			18862.6	5300.06 cm^{-1}	317801.30	– 323101.36	5–5	2.60–03	1.39–02	4.30+00	–1.159	B	*LS*
			18932.9	5280.37 cm^{-1}	317796.51	– 323076.88	3–3	4.28–03	2.30–02	4.30+00	–1.161	B	*LS*
			18950.1	5275.58 cm^{-1}	317801.30	– 323076.88	5–3	2.85–04	9.20–04	2.87–01	–2.337	B	*LS*
166.		$^3P^o$–^3S	*10546*	*9479.3* cm^{-1}	*317798.9*	– 327278.27	9–3	3.59–02	1.99–02	6.24+00	–0.746	B	2
			10549.0	9476.97 cm^{-1}	317801.30	– 327278.27	5–3	1.99–02	1.99–02	3.46+00	–1.001	B	*LS*
			10543.7	9481.76 cm^{-1}	317796.51	– 327278.27	3–3	1.20–02	2.00–02	2.08+00	–1.223	B	*LS*
			10541.2	9484.01 cm^{-1}	317794.26	– 327278.27	1–3	3.99–03	2.00–02	6.93–01	–1.700	B	*LS*
167.	–	$^3P^o$–^3P	*8382.9*	*8385.2*	*317798.9*	*– 329724.7*	9–9	1.26–03	1.32–03	3.29–01	–1.924	B	2
			8371.32	8373.62	317801.30	– 329743.57	5–5	9.46–04	9.95–04	1.37–01	–2.303	B	*LS*
			8394.03	8396.33	317796.51	– 329706.47	3–3	3.13–04	3.31–04	2.74–02	–3.003	B–	*LS*
			8397.40	8399.71	317801.30	– 329706.47	5–3	5.21–04	3.31–04	4.57–02	–2.782	B	*LS*
			8408.92	8411.23	317796.51	– 329685.38	3–1	1.25–03	4.40–04	3.66–02	–2.879	B	*LS*
			8367.96	8370.26	317796.51	– 329743.57	3–5	3.16–04	5.53–04	4.57–02	–2.780	B	*LS*
			8392.44	8394.75	317794.26	– 329706.47	1–3	4.17–04	1.32–03	3.66–02	–2.878	B	*LS*
168.	2s4p–2s5s	$^3P^o$–^3S	*4516.3*	*4517.6*	*317798.9*	– 339934.72	9–3	1.72+00	1.75–01	2.34+01	0.197	B	2
			4516.79	4518.05	317801.30	– 339934.72	5–3	9.54–01	1.75–01	1.30+01	–0.058	B	*LS*
			4515.81	4517.08	317796.51	– 339934.72	3–3	5.73–01	1.75–01	7.81+00	–0.280	B	*LS*
			4515.35	4516.62	317794.26	– 339934.72	1–3	1.91–01	1.75–01	2.60+00	–0.757	B	*LS*
169.	2s4p–2s5d	$^3P^o$–^3D	*3609.3*	*3610.4*	*317798.9*	*– 345496.9*	9–15	9.80–01	3.19–01	3.41+01	0.458	B	2
			3609.62	3610.65	317801.30	– 345497.15	5–7	9.80–01	2.68–01	1.59+01	0.127	B	*LS*
			3609.05	3610.08	317796.51	– 345496.72	3–5	7.36–01	2.39–01	8.53+00	–0.144	B	*LS*
			3608.78	3609.81	317794.26	– 345496.57	1–3	5.45–01	3.19–01	3.79+00	–0.496	B	*LS*
			3609.68	3610.71	317801.30	– 345496.72	5–5	2.45–01	4.79–02	2.84+00	–0.621	B	*LS*
			3609.07	3610.10	317796.51	– 345496.57	3–3	4.09–01	7.98–02	2.84+00	–0.621	B	*LS*
			3609.70	3610.72	317801.30	– 345496.57	5–3	2.72–02	3.19–03	1.90–01	–1.797	B	*LS*

C III: Allowed Transitions — Continued

No.	Transition Array	Mult.	λ_{air} (Å)	λ_{vac} (Å) or σ (cm^{-1})*	E_i (cm^{-1})	E_k (cm^{-1})	g_i-g_k	A_{ki} (10^8 s^{-1})	f_{ik}	S (at. u.)	log gf	Acc.	Source
170.	$2s4p-$ $2s6s$	^{3}P°$-^3$S	*2697.6*	*2698.4*	*317798.9*	– 354858.03	9−3	9.16−01	3.33−02	2.66+00	−0.523	B	2
			2697.76	2698.57	317801.30	– 354858.03	5−3	5.09−01	3.33−02	1.48+00	−0.778	B	*LS*
			2697.42	2698.22	317796.51	– 354858.03	3−3	3.06−01	3.33−02	8.88−01	−1.000	B	*LS*
			2697.25	2698.05	317794.26	– 354858.03	1−3	1.02−01	3.33−02	2.96−01	−1.477	B	*LS*
171.	$2s4p-$ $2s7s$	^{3}P°$-^3$S	*2182.1*	*2182.7*	*317798.9*	– 363613	9−3	5.57−01	1.33−02	8.57−01	−0.923	B	2
			2182.17	2182.85	317801.30	– 363613	5−3	3.09−01	1.33−02	4.76−01	−1.179	B	*LS*
			2181.94	2182.62	317796.51	– 363613	3−3	1.86−01	1.33−02	2.86−01	−1.400	B	*LS*
			2181.83	2182.51	317794.26	– 363613	1−3	6.19−02	1.33−02	9.53−02	−1.877	B	*LS*
172.	$2s4p-$ $2s7d$	^{3}P°$-^3$D	*2089.7*	*2090.3*	*317798.9*	– 365638	9−15	4.49−01	4.90−02	3.04+00	−0.355	B	2
			2089.78	2090.45	317801.30	– 365638	5−7	4.49−01	4.12−02	1.42+00	−0.686	B	*LS*
			2089.57	2090.24	317796.51	– 365638	3−5	3.37−01	3.68−02	7.59−01	−0.957	B	*LS*
			2089.47	2090.14	317794.26	– 365638	1−3	2.50−01	4.90−02	3.37−01	−1.310	B	*LS*
			2089.78	2090.45	317801.30	– 365638	5−5	1.12−01	7.35−03	2.53−01	−1.435	B	*LS*
			2089.57	2090.24	317796.51	– 365638	3−3	1.87−01	1.23−02	2.53−01	−1.435	B	*LS*
			2089.78	2090.45	317801.30	– 365638	5−3	1.25−02	4.90−04	1.69−02	−2.611	B−	*LS*
173.	$2s4p-$ $2s8d$	^{3}P°$-^3$D		*1897.5*	*317798.9*	– 370499	9−15	3.09−01	2.78−02	1.57+00	−0.601	B	2
				1897.62	317801.30	– 370499	5−7	3.09−01	2.34−02	7.31−01	−0.932	B	*LS*
				1897.44	317796.51	– 370499	3−5	2.32−01	2.09−02	3.91−01	−1.203	B	*LS*
				1897.36	317794.26	– 370499	1−3	1.72−01	2.78−02	1.74−01	−1.555	B	*LS*
				1897.62	317801.30	– 370499	5−5	7.74−02	4.18−03	1.30−01	−1.680	B	*LS*
				1897.44	317796.51	– 370499	3−3	1.29−01	6.96−03	1.30−01	−1.680	B	*LS*
				1897.62	317801.30	– 370499	5−3	8.60−03	2.78−04	8.70−03	−2.856	C+	*LS*
174.	$2p3p-$ $2s4p$	^{1}P$-^1$P°	37249.7	2683.85 cm^{-1}	319720.35	– 322404.20	3−3	7.51−04	1.56−02	5.75+00	−1.329	B	2
175.	$2p3p-$ $2p3d$	^{1}P$-^1$D°	7707.43	7709.55	319720.35	– 332691.28	3−5	1.30−01	1.93−01	1.47+01	−0.236	B	2
176.	$2p3p-$ $2s5p$	^{1}P$-^1$P°	4247.31	4248.51	319720.35	– 343258.03	3−3	1.58−01	4.28−02	1.79+00	−0.892	B	2
177.	$2p3p-$ $2p3d$	^{1}P$-^1$P°	3703.70	3704.75	319720.35	– 346712.73	3−3	5.90−01	1.21−01	4.44+00	−0.439	B	2
178.	$2p3p-$ $2s6p$	^{1}P$-^1$P°	2673.77	2674.56	319720.35	– 357109.68	3−3	1.70−03	1.82−04	4.80−03	−3.263	C	2
179.	$2p3p-$ $2s7p$	^{1}P$-^1$P°	2212.89	2213.58	319720.35	– 364896	3−3	1.36−02	9.97−04	2.18−02	−2.524	B−	2
180.	$2p3p-$ $2s8p$	^{1}P$-^1$P°		1991.81	319720.35	– 369926	3−3	2.62−02	1.56−03	3.07−02	−2.330	B	2
181.	$2p3p-$ $2p4d$	^{1}P$-^1$D°		1512.93	319720.35	– 385817.2	3−5	4.37+00	2.50−01	3.73+00	−0.125	B	2
182.	$2p3p-$ $2p5d$	^{1}P$-^1$D°		1111.57	319720.35	– 409683	3−5	2.51+00	7.75−02	8.51−01	−0.634	B	2

C III: Allowed Transitions — Continued

No.	Transition Array	Mult.	λ_{air} (Å)	λ_{vac} (Å) or σ (cm^{-1})*	E_i (cm^{-1})	E_k (cm^{-1})	g_i–g_k	A_{ki} (10^8 s^{-1})	f_{ik}	S (at. u.)	log gf	Acc.	Source
183.	2s4d–2s4f	$^3D-^3F^o$		*577.2* cm^{-1}	*321434.5*	*– 322011.8*	15–21	2.21–05	1.39–02	1.19+02	–0.680	B	2
				567.92 cm^{-1}	321450.05	– 322017.97	7–9	2.10–05	1.26–02	5.10+01	–1.055	C+	*LS*
				582.84 cm^{-1}	321426.74	– 322009.58	5–7	2.02–05	1.25–02	3.53+01	–1.205	C+	*LS*
				592.37 cm^{-1}	321411.31	– 322003.68	3–5	2.01–05	1.43–02	2.38+01	–1.368	C+	*LS*
				559.53 cm^{-1}	321450.05	– 322009.58	7–7	2.24–06	1.07–03	4.42+00	–2.124	C+	*LS*
				576.94 cm^{-1}	321426.74	– 322003.68	5–5	3.44–06	1.55–03	4.42+00	–2.111	C+	*LS*
				553.63 cm^{-1}	321450.05	– 322003.68	7–5	8.58–08	3.00–05	1.25–01	–3.678	C+	*LS*
184.	2s4d–2p3d	$^3D-^3F^o$	*8340.4*	*8342.7*	*321434.5*	*– 333421.0*	15–21	2.69–01	3.93–01	1.62+02	0.771	B	2
			8332.99	8335.29	321450.05	– 333447.24	7–9	2.70–01	3.61–01	6.94+01	0.403	B	*LS*
			8341.60	8343.90	321426.74	– 333411.55	5–7	2.39–01	3.49–01	4.80+01	0.242	B	*LS*
			8347.95	8350.24	321411.31	– 333387.01	3–5	2.26–01	3.93–01	3.24+01	0.071	B	*LS*
			8357.86	8360.16	321450.05	– 333411.55	7–7	2.98–02	3.12–02	6.01+00	–0.660	B	*LS*
			8358.72	8361.02	321426.74	– 333387.01	5–5	4.17–02	4.37–02	6.01+00	–0.660	B	*LS*
			8375.04	8377.34	321450.05	– 333387.01	7–5	1.17–03	8.79–04	1.70–01	–2.211	B	*LS*
185.		$^3D-^3D^o$	*6155.7*	*6157.4*	*321434.5*	*– 337675.2*	15–15	1.47–02	8.35–03	2.54+00	–0.902	B	2
			6156.69	6158.40	321450.05	– 337688.04	7–7	1.31–02	7.42–03	1.05+00	–1.284	B	*LS*
			6155.12	6156.82	321426.74	– 337668.89	5–5	1.02–02	5.81–03	5.89–01	–1.537	B	*LS*
			6154.16	6155.87	321411.31	– 337655.98	3–3	1.10–02	6.27–03	3.81–01	–1.726	B	*LS*
			6163.96	6165.67	321450.05	– 337668.89	7–5	2.28–03	9.30–04	1.32–01	–2.187	B	*LS*
			6160.01	6161.72	321426.74	– 337655.98	5–3	3.67–03	1.25–03	1.27–01	–2.203	B	*LS*
			6147.87	6149.57	321426.74	– 337688.04	5–7	1.64–03	1.30–03	1.32–01	–2.186	B	*LS*
			6149.28	6150.98	321411.31	– 337668.89	3–5	2.21–03	2.09–03	1.27–01	–2.203	B	*LS*
186.		$^3D-^3P^o$	*5351.7*	*5353.2*	*321434.5*	*– 340114.8*	15–9	6.57–02	1.69–02	4.48+00	–0.595	B	2
			5359.93	5361.42	321450.05	– 340101.84	7–5	5.50–02	1.69–02	2.09+00	–0.927	B	*LS*
			5345.88	5347.37	321426.74	– 340127.53	5–3	4.95–02	1.27–02	1.12+00	–1.197	B	*LS*
			5337.40	5338.88	321411.31	– 340141.83	3–1	6.63–02	9.43–03	4.97–01	–1.548	B	*LS*
			5353.23	5354.72	321426.74	– 340101.84	5–5	9.85–03	4.23–03	3.73–01	–1.674	B	*LS*
			5341.47	5342.96	321411.31	– 340127.53	3–3	1.65–02	7.07–03	3.73–01	–1.673	B	*LS*
			5348.82	5350.30	321411.31	– 340101.84	3–5	6.58–04	4.71–04	2.49–02	–2.850	B–	*LS*
187.	2s4d–2s5p	$^3D-^3P^o$	*4384.7*	*4385.9*	*321434.5*	*– 344234.7*	15–9	2.57–01	4.45–02	9.63+00	–0.176	B	2
			4388.02	4389.25	321450.05	– 344232.98	7–5	2.16–01	4.44–02	4.49+00	–0.507	B	*LS*
			4382.90	4384.13	321426.74	– 344236.29	5–3	1.93–01	3.34–02	2.41+00	–0.778	B	*LS*
			4379.48	4380.71	321411.31	– 344238.68	3–1	2.58–01	2.47–02	1.07+00	–1.130	B	*LS*
			4383.53	4384.76	321426.74	– 344232.98	5–5	3.86–02	1.11–02	8.02–01	–1.255	B	*LS*
			4379.93	4381.16	321411.31	– 344236.29	3–3	6.45–02	1.85–02	8.02–01	–1.255	B	*LS*
			4380.57	4381.80	321411.31	– 344232.98	3–5	2.58–03	1.24–03	5.35–02	–2.431	B	*LS*
188.	2s4d–2s5f	$^3D-^3F^o$	*3887.0*	*3888.1*	*321434.5*	*– 347153.9*	15–21	1.14+00	3.60–01	6.91+01	0.733	B	2
			3889.14	3890.24	321450.05	– 347155.41	7–9	1.13+00	3.31–01	2.96+01	0.364	C+	*LS*
			3885.94	3887.04	321426.74	– 347153.26	5–7	1.01+00	3.20–01	2.05+01	0.204	C+	*LS*
			3883.82	3884.92	321411.31	– 347151.89	3–5	9.57–01	3.61–01	1.38+01	0.034	C+	*LS*
			3889.46	3890.56	321450.05	– 347153.26	7–7	1.26–01	2.87–02	2.57+00	–0.698	C+	*LS*
			3886.15	3887.25	321426.74	– 347151.89	5–5	1.77–01	4.02–02	2.57+00	–0.697	C+	*LS*
			3889.67	3890.77	321450.05	– 347151.89	7–5	4.99–03	8.08–04	7.24–02	–2.247	C+	*LS*
189.	2s4d–2s6p	$^3D-^3P^o$	*2806.9*	*2807.7*	*321434.5*	*– 357050.7*	15–9	1.89–01	1.34–02	1.86+00	–0.697	B	2
			2808.07	2808.90	321450.05	– 357051.23	7–5	1.59–01	1.34–02	8.67–01	–1.028	B	*LS*
			2806.31	2807.14	321426.74	– 357050.17	5–3	1.42–01	1.01–02	4.64–01	–1.299	B	*LS*
			2805.16	2805.99	321411.31	– 357049.38	3–1	1.89–01	7.45–03	2.06–01	–1.651	B	*LS*
			2806.23	2807.06	321426.74	– 357051.23	5–5	2.84–02	3.35–03	1.55–01	–1.776	B	*LS*
			2805.10	2805.93	321411.31	– 357050.17	3–3	4.74–02	5.59–03	1.55–01	–1.776	B	*LS*
			2805.02	2805.84	321411.31	– 357051.23	3–5	1.90–03	3.73–04	1.03–02	–2.952	B–	*LS*

C III: Allowed Transitions — Continued

No.	Transition Array	Mult.	λ_{air} (Å)	λ_{vac} (Å) or σ (cm^{-1})*	E_i (cm^{-1})	E_k (cm^{-1})	g_i–g_k	A_{ki} ($10^8\ s^{-1}$)	f_{ik}	S (at. u.)	log gf	Acc.	Source
190.	$2s4d$–$2s6f$	^{3}D–^{3}F°	*2671.9*	*2672.7*	*321434.5*	*358850.4*	15–21	6.98–01	1.05–01	1.38+01	0.196	B	2
			2672.95	2673.75	321450.05	358850.74	7–9	6.97–01	9.60–02	5.91+00	–0.173	C+	*LS*
			2671.32	2672.11	321426.74	358850.32	5–7	6.21–01	9.30–02	4.09+00	–0.333	C+	*LS*
			2670.24	2671.03	321411.31	358850.02	3–5	5.87–01	1.05–01	2.76+00	–0.503	C+	*LS*
			2672.98	2673.78	321450.05	358850.32	7–7	7.77–02	8.32–03	5.13–01	–1.235	C+	*LS*
			2671.34	2672.13	321426.74	358850.02	5–5	1.09–01	1.17–02	5.13–01	–1.234	C+	*LS*
			2673.00	2673.80	321450.05	358850.02	7–5	3.07–03	2.35–04	1.45–02	–2.784	C	*LS*
191.	$2s4f$–$2p3p$	^{3}F°–^{3}D		*1102.9* cm^{-1}	*322011.8*	*323114.7*	21–15	1.41–04	1.25–02	7.81+01	–0.582	B	2
				1122.36 cm^{-1}	322017.97	323140.33	9–7	1.37–04	1.27–02	3.35+01	–0.943	C+	*LS*
				1091.78 cm^{-1}	322009.58	323101.36	7–5	1.22–04	1.10–02	2.31+01	–1.115	C+	*LS*
				1073.20 cm^{-1}	322003.68	323076.88	5–3	1.30–04	1.02–02	1.56+01	–1.293	C+	*LS*
				1130.75 cm^{-1}	322009.58	323140.33	7–7	1.21–05	1.42–03	2.90+00	–2.002	C+	*LS*
				1097.68 cm^{-1}	322003.68	323101.36	5–5	1.55–05	1.93–03	2.90+00	–2.015	C+	*LS*
				1136.65 cm^{-1}	322003.68	323140.33	5–7	3.48–07	5.65–05	8.18–02	–3.549	C+	*LS*
192.	$2s4f$–$2s5d$	^{3}F°–^{3}D	*4256.8*	*4258.0*	*322011.8*	*345496.9*	21–15	3.10–02	6.01–03	1.77+00	–0.899	B	2
			4257.89	4259.09	322017.97	345497.15	9–7	2.84–02	6.01–03	7.58–01	–1.267	C+	*LS*
			4256.45	4257.65	322009.58	345496.72	7–5	2.75–02	5.34–03	5.24–01	–1.427	C+	*LS*
			4255.41	4256.61	322003.68	345496.57	5–3	3.10–02	5.05–03	3.54–01	–1.598	C+	*LS*
			4256.37	4257.57	322009.58	345497.15	7–7	2.46–03	6.69–04	6.57–02	–2.329	C+	*LS*
			4255.38	4256.58	322003.68	345496.72	5–5	3.45–03	9.38–04	6.57–02	–2.329	C+	*LS*
			4255.30	4256.50	322003.68	345497.15	5–7	6.96–05	2.64–05	1.85–03	–3.879	D	*LS*
193.	$2s4f$–$2s5g$	^{3}F°–^{3}G	*4069.3*	*4070.4*	*322011.8*	*346579.3*	21–27	3.28+00	1.05+00	2.94+02	1.342	B	4*n*
			4070.26	4071.41	322017.97	346579.49	9–11	3.28+00	9.95–01	1.20+02	0.952	C+	4
			4068.92	4070.07	322009.58	346579.21	7–9	3.08+00	9.82–01	9.20+01	0.837	C+	4
			4067.94	4069.09	322003.68	346579.21	5–7	3.02+00	1.05+00	7.01+01	0.719	C+	4
			4070.31	4071.46	322017.97	346579.21	9–9	2.05–01	5.08–02	6.13+00	–0.340	C+	*LS*
			4068.92	4070.07	322009.58	346579.21	7–7	2.63–01	6.54–02	6.13+00	–0.340	C+	*LS*
			4070.31	4071.46	322017.97	346579.21	9–7	4.01–03	7.74–04	9.33–02	–2.157	C+	*LS*
194.	$2s4f$–$2s7d$	^{3}F°–^{3}D	*2291.5*	*2292.2*	*322011.8*	365638	21–15	6.36–03	3.58–04	5.67–02	–2.124	C+	2
			2291.82	2292.52	322017.97	365638	9–7	5.84–03	3.58–04	2.43–02	–2.492	C–	*LS*
			2291.38	2292.08	322009.58	365638	7–5	5.65–03	3.18–04	1.68–02	–2.653	C–	*LS*
			2291.07	2291.77	322003.68	365638	5–3	6.36–03	3.01–04	1.13–02	–2.823	C–	*LS*
			2291.38	2292.08	322009.58	365638	7–7	5.06–04	3.99–05	2.11–03	–3.554	D	*LS*
			2291.07	2291.77	322003.68	365638	5–5	7.09–04	5.58–05	2.11–03	–3.554	D	*LS*
			2291.07	2291.77	322003.68	365638	5–7	1.43–05	1.57–06	5.94–05	–5.104	D–	*LS*
195.	$2s4f$–$2s8d$	^{3}F°–^{3}D	*2061.7*	*2062.4*	*322011.8*	370499	21–15	3.68–03	1.68–04	2.39–02	–2.453	C+	2
			2062.00	2062.66	322017.97	370499	9–7	3.38–03	1.68–04	1.02–02	–2.821	C–	*LS*
			2061.65	2062.31	322009.58	370499	7–5	3.27–03	1.49–04	7.08–03	–2.982	D	*LS*
			2061.40	2062.05	322003.68	370499	5–3	3.68–03	1.41–04	4.78–03	–3.152	D	*LS*
			2061.65	2062.31	322009.58	370499	7–7	2.93–04	1.87–05	8.88–04	–3.884	D–	*LS*
			2061.40	2062.05	322003.68	370499	5–5	4.10–04	2.62–05	8.88–04	–3.884	D–	*LS*
			2061.40	2062.05	322003.68	370499	5–7	8.27–06	7.38–07	2.50–05	–5.433	D–	*LS*
196.	$2s4p$–$2s4d$	^{1}P°–^{1}D		1808.29 cm^{-1}	322404.20	324212.49	3–5	7.05–04	5.39–02	2.94+01	–0.791	B	2
197.	$2s4p$–$2p3p$	^{1}P°–^{1}D	9331.01	9333.57	322404.20	333118.21	3–5	1.05–01	2.28–01	2.10+01	–0.164	B	2
198.	$2s4p$–$2s5s$	^{1}P°–^{1}S	6205.56	6207.27	322404.20	338514.33	3–1	1.03–03	1.97–04	1.21–02	–3.227	D	2

C III: Allowed Transitions — Continued

No.	Transition Array	Mult.	λ_{air} (Å)	λ_{vac} (Å) or σ (cm^{-1})*	E_i (cm^{-1})	E_k (cm^{-1})	g_i–g_k	A_{ki} (10^8 s^{-1})	f_{ik}	S (at. u.)	log gf	Acc.	Source
199.	$2s4p$– $2p3p$	$^1P^o$–^1S	4405.75	4406.99	322404.20	– 345095.43	3–1	4.41–01	4.28–02	1.86+00	–0.892	B	2
200.	$2s4p$– $2s5d$	$^1P^o$–^1D	4121.84	4123.01	322404.20	– 346658.34	3–5	8.53–01	3.62–01	1.47+01	0.036	B	4
201.	$2s4p$– $2s6d$	$^1P^o$–^1D	2751.83	2752.64	322404.20	– 358732.93	3–5	4.71–03	8.90–04	2.42–02	–2.573	B–	2
202.	$2s4p$– $2s7d$	$^1P^o$–^1D	2291.62	2292.33	322404.20	– 366028	3–5	6.59–03	8.66–04	1.96–02	–2.585	B–	2
203.	$2s4p$– $2p4p$	$^1P^o$–^1P		1703.54	322404.20	– 381105.4	3–3	2.60–03	1.13–04	1.90–03	–3.470	D–	2
204.	$2s4p$– $2p5p$	$^1P^o$–^1P		1176.10	322404.20	– 407431	3–3	2.74–01	5.67–03	6.59–02	–1.769	B	2
205.	$2s4f$– $2s4d$	$^1F^o$–^1D		1510.47 cm^{-1}	322702.02	– 324212.49	7–5	9.67–04	4.54–02	6.92+01	–0.498	B	2
206.	$2s4f$– $2p3p$	$^1F^o$–^1D	9597.81	9600.44	322702.02	– 333118.21	7–5	3.07–04	3.02–04	6.69–02	–2.674	B	2
207.	$2s4f$– $2s5g$	$^1F^o$–^1G	4186.90	4188.08	322702.02	– 346579.31	7–9	3.50+00	1.18+00	1.14+02	0.918	B	4
208.	$2s4f$– $2s5d$	$^1F^o$–^1D	4173.09	4174.26	322702.02	– 346658.34	7–5	2.01–02	3.74–03	3.60–01	–1.582	B	2
209.	$2s4f$– $2s6d$	$^1F^o$–^1D	2774.58	2775.39	322702.02	– 358732.93	7–5	9.00–03	7.43–04	4.75–02	–2.284	B	2
210.	$2s4f$– $2s7d$	$^1F^o$–^1D	2307.37	2308.08	322702.02	– 366028	7–5	4.71–03	2.69–04	1.43–02	–2.725	B–	2
211.	$2p3p$– $2p3d$	^3D–$^3F^o$	*9700.1*	*9702.8*	*323114.7*	*– 333421.0*	15–21	8.47–03	1.67–02	8.01+00	–0.601	C+	2
			9699.57	9702.23	323140.33	– 333447.24	7–9	8.47–03	1.54–02	3.43+00	–0.969	C+	*LS*
			9696.48	9699.14	323101.36	– 333411.55	5–7	7.53–03	1.49–02	2.37+00	–1.129	C+	*LS*
			9696.54	9699.20	323076.88	– 333387.01	3–5	7.12–03	1.67–02	1.60+00	–1.299	C+	*LS*
			9733.27	9735.94	323140.33	– 333411.55	7–7	9.34–04	1.33–03	2.97–01	–2.032	C+	*LS*
			9719.62	9722.28	323101.36	– 333387.01	5–5	1.31–03	1.86–03	2.97–01	–2.032	C+	*LS*
			9756.58	9759.26	323140.33	– 333387.01	7–5	3.66–05	3.73–05	8.39–03	–3.583	C	*LS*
212.		^3D–$^3D^o$	*6866.0*	*6867.9*	*323114.7*	*– 337675.2*	15–15	5.03–02	3.56–02	1.21+01	–0.273	B	2
			6872.04	6873.93	323140.33	– 337688.04	7–7	4.46–02	3.16–02	5.00+00	–0.656	B	*LS*
			6862.69	6864.58	323101.36	– 337668.89	5–5	3.51–02	2.48–02	2.80+00	–0.907	B	*LS*
			6857.24	6859.13	323076.88	– 337655.98	3–3	3.79–02	2.67–02	1.81+00	–1.096	B	*LS*
			6881.10	6882.99	323140.33	– 337668.89	7–5	7.80–03	3.95–03	6.27–01	–1.558	B	*LS*
			6868.78	6870.67	323101.36	– 337655.98	5–3	1.26–02	5.33–03	6.03–01	–1.574	B	*LS*
			6853.68	6855.57	323101.36	– 337688.04	5–7	5.64–03	5.56–03	6.27–01	–1.556	B	*LS*
			6851.18	6853.07	323076.88	– 337668.89	3–5	7.60–03	8.91–03	6.03–01	–1.573	B	*LS*
213.		^3D–$^3P^o$	*5880.7*	*5882.3*	*323114.7*	*– 340114.8*	15–9	1.33–01	4.13–02	1.20+01	–0.208	B	2
			5894.07	5895.70	323140.33	– 340101.84	7–5	1.11–01	4.12–02	5.60+00	–0.540	B	*LS*
			5871.68	5873.31	323101.36	– 340127.53	5–3	1.00–01	3.10–02	3.00+00	–0.809	B	*LS*
			5858.34	5859.96	323076.88	– 340141.83	3–1	1.34–01	2.30–02	1.33+00	–1.161	B	*LS*
			5880.56	5882.19	323101.36	– 340101.84	5–5	1.99–02	1.03–02	9.99–01	–1.287	B	*LS*
			5863.25	5864.88	323076.88	– 340127.53	3–3	3.35–02	1.73–02	9.99–01	–1.286	B	*LS*
			5872.10	5873.73	323076.88	– 340101.84	3–5	1.33–03	1.15–03	6.66–02	–2.463	B	*LS*

C III: Allowed Transitions — Continued

No.	Transition Array	Mult.	λ_{air} (Å)	λ_{vac} (Å) or σ (cm^{-1})*	E_i (cm^{-1})	E_k (cm^{-1})	g_i–g_k	A_{ki} (10^8 s^{-1})	f_{ik}	S (at. u.)	log gf	Acc.	Source
214.	2p3p–2s5p	^{3}D–^{3}P°	*4733.5*	*4734.8*	*323114.7*	– *344234.7*	15–9	5.28–03	1.06–03	2.49–01	–1.797	B	2
			4739.66	4740.99	323140.33	– 344232.98	7–5	4.42–03	1.06–03	1.16–01	–2.128	B	*LS*
			4730.18	4731.50	323101.36	– 344236.29	5–3	3.97–03	7.99–04	6.22–02	–2.399	B	*LS*
			4724.17	4725.50	323076.88	– 344238.68	3–1	5.31–03	5.92–04	2.76–02	–2.750	B–	*LS*
			4730.92	4732.24	323101.36	– 344232.98	5–5	7.93–04	2.66–04	2.07–02	–2.876	B–	*LS*
			4724.71	4726.03	323076.88	– 344236.29	3–3	1.33–03	4.44–04	2.07–02	–2.875	B–	*LS*
			4725.45	4726.77	323076.88	– 344232.98	3–5	5.31–05	2.96–05	1.38–03	–4.051	C	*LS*
215.	2p3p–2s5f	^{3}D–^{3}F°	*4158.7*	*4159.9*	*323114.7*	– *347153.9*	15–21	7.08–01	2.57–01	5.28+01	0.586	B	2
			4162.88	4164.05	323140.33	– 347155.41	7–9	7.06–01	2.36–01	2.26+01	0.218	C+	*LS*
			4156.50	4157.68	323101.36	– 347153.26	5–7	6.31–01	2.29–01	1.56+01	0.058	C+	*LS*
			4152.51	4153.68	323076.88	– 347151.89	3–5	5.98–01	2.58–01	1.06+01	–0.112	C+	*LS*
			4163.25	4164.42	323140.33	– 347153.26	7–7	7.87–02	2.04–02	1.96+00	–0.844	C+	*LS*
			4156.74	4157.91	323101.36	– 347151.89	5–5	1.11–01	2.87–02	1.96+00	–0.844	C+	*LS*
			4163.49	4164.66	323140.33	– 347151.89	7–5	3.11–03	5.77–04	5.53–02	–2.394	C+	*LS*
216.	2p3p–2s6p	^{3}D–^{3}P°	*2945.9*	*2946.7*	*323114.7*	– *357050.7*	15–9	1.84–02	1.44–03	2.09–01	–1.666	B	2
			2948.04	2948.90	323140.33	– 357051.23	7–5	1.54–02	1.44–03	9.76–02	–1.998	B	*LS*
			2944.75	2945.61	323101.36	– 357050.17	5–3	1.38–02	1.08–03	5.23–02	–2.268	B	*LS*
			2942.70	2943.56	323076.88	– 357049.38	3–1	1.85–02	7.99–04	2.32–02	–2.620	B–	*LS*
			2944.66	2945.52	323101.36	– 357051.23	5–5	2.76–03	3.59–04	1.74–02	–2.745	B–	*LS*
			2942.63	2943.49	323076.88	– 357050.17	3–3	4.61–03	5.99–04	1.74–02	–2.745	B–	*LS*
			2942.54	2943.40	323076.88	– 357051.23	3–5	1.85–04	4.00–05	1.16–03	–3.921	C	*LS*
217.	2p3p–2s6f	^{3}D–^{3}F°	*2797.5*	*2798.3*	*323114.7*	– *358850.4*	15–21	4.42–01	7.25–02	1.00+01	0.037	B	2
			2799.48	2800.30	323140.33	– 358850.74	7–9	4.41–01	6.66–02	4.30+00	–0.332	C+	*LS*
			2796.46	2797.28	323101.36	– 358850.32	5–7	3.93–01	6.45–02	2.97+00	–0.492	C+	*LS*
			2794.57	2795.39	323076.88	– 358850.02	3–5	3.72–01	7.26–02	2.00+00	–0.662	C+	*LS*
			2799.51	2800.34	323140.33	– 358850.32	7–7	4.91–02	5.77–03	3.72–01	–1.394	C+	*LS*
			2796.48	2797.31	323101.36	– 358850.02	5–5	6.90–02	8.09–03	3.72–01	–1.393	C+	*LS*
			2799.54	2800.36	323140.33	– 358850.02	7–5	1.94–03	1.63–04	1.05–02	–2.943	C	*LS*
218.	2s4d–2p3d	^{1}D–^{1}D°	11790.9	8478.79 cm^{-1}	324212.49	– 332691.28	5–5	7.14–05	1.49–04	2.89–02	–3.128	D–	2
219.		^{1}D–^{1}F°	5826.42	5828.03	324212.49	– 341370.94	5–7	7.32–01	5.21–01	5.00+01	0.416	B	2
220.	2s4d–2s5p	^{1}D–^{1}P°	5249.11	5250.57	324212.49	– 343258.03	5–3	3.90–01	9.66–02	8.35+00	–0.316	B	2
221.	2s4d–2p3d	^{1}D–^{1}P°	4443.15	4444.40	324212.49	– 346712.73	5–3	8.09–03	1.44–03	1.05–01	–2.144	B	2
222.	2s4d–2s5f	^{1}D–^{1}F°	4056.06	4057.21	324212.49	– 348859.99	5–7	1.07+00	3.68–01	2.46+01	0.265	B	2
223.	2s4d–2s6p	^{1}D–^{1}P°	3038.89	3039.77	324212.49	– 357109.68	5–3	1.99–01	1.66–02	8.28–01	–1.082	B	2
224.	2s4d–2s6f	^{1}D–^{1}F°	2863.71	2864.55	324212.49	– 359121.95	5–7	1.05+00	1.80–01	8.51+00	–0.045	B	2
225.	2s4d–2s7p	^{1}D–^{1}P°	2457.25	2458.00	324212.49	– 364896	5–3	1.18–01	6.42–03	2.60–01	–1.493	B	2

C III: Allowed Transitions — Continued

No.	Transition Array	Mult.	λ_{air} (Å)	λ_{vac} (Å) or σ (cm^{-1})*	E_i (cm^{-1})	E_k (cm^{-1})	g_i-g_k	A_{ki} (10^8 s^{-1})	f_{ik}	S (at. u.)	log gf	Acc.	Source
226.	2s4d– 2s8p	^{1}D–^{1}P°	2186.85	2187.54	324212.49	– 369926	5–3	7.32–02	3.15–03	1.13–01	–1.803	B	2
227.	2s4d– 2p4d	^{1}D–^{1}D°		1623.25	324212.49	– 385817.2	5–5	2.79+00	1.10–01	2.94+00	–0.259	B	2
228.	2s4d– 2p5d	^{1}D–^{1}D°		1169.99	324212.49	– 409683	5–5	1.87–02	3.84–04	7.40–03	–2.716	C+	2
229.	2p3p– 2p3d	^{3}S–^{3}P°	*7788.1*	*7790.2*	327278.27	– *340114.8*	3–9	1.76–01	4.79–01	3.68+01	0.157	B	2
			7796.00	7798.14	327278.27	– 340101.84	3–5	1.75–01	2.66–01	2.05+01	–0.098	B	*LS*
			7780.41	7782.55	327278.27	– 340127.53	3–3	1.76–01	1.60–01	1.23+01	–0.319	B	*LS*
			7771.76	7773.90	327278.27	– 340141.83	3–1	1.77–01	5.33–02	4.09+00	–0.796	B	*LS*
230.	2p3p– 2s5p	^{3}S–^{3}P°	*5895.8*	*5897.5*	327278.27	– *344234.7*	3–9	1.86–03	2.91–03	1.70–01	–2.059	C+	2
			5896.43	5898.07	327278.27	– 344232.98	3–5	1.86–03	1.62–03	9.42–02	–2.314	C+	*LS*
			5895.28	5896.91	327278.27	– 344236.29	3–3	1.86–03	9.71–04	5.65–02	–2.536	C+	*LS*
			5894.45	5896.08	327278.27	– 344238.68	3–1	1.86–03	3.24–04	1.88–02	–3.013	C	*LS*
231.	2p3p– 2s6p	^{3}S–^{3}P°	*3357.9*	*3358.8*	327278.27	– *357050.7*	3–9	2.04–03	1.03–03	3.43–02	–2.508	B–	2
			3357.79	3358.75	327278.27	– 357051.23	3–5	2.04–03	5.74–04	1.91–02	–2.764	B–	*LS*
			3357.91	3358.87	327278.27	– 357050.17	3–3	2.04–03	3.45–04	1.14–02	–2.985	B–	*LS*
			3358.00	3358.96	327278.27	– 357049.38	3–1	2.04–03	1.15–04	3.81–03	–3.463	C	*LS*
232.	2p3p– 2p3d	^{3}P–^{3}D°	*12574*	*7950.5* cm^{-1}	*329724.7*	– *337675.2*	9–15	3.29–02	1.30–01	4.85+01	0.069	B	2
			12583.9	7944.47 cm^{-1}	329743.57	– 337688.04	5–7	3.29–02	1.09–01	2.26+01	–0.263	B	*LS*
			12555.6	7962.42 cm^{-1}	329706.47	– 337668.89	3–5	2.48–02	9.77–02	1.21+01	–0.533	B	*LS*
			12542.7	7970.60 cm^{-1}	329685.38	– 337655.98	1–3	1.84–02	1.30–01	5.39+00	–0.885	B	*LS*
			12614.3	7925.32 cm^{-1}	329743.57	– 337668.89	5–5	8.16–03	1.95–02	4.04+00	–1.012	B	*LS*
			12576.0	7949.51 cm^{-1}	329706.47	– 337655.98	3–3	1.37–02	3.25–02	4.04+00	–1.011	B	*LS*
			12634.9	7912.41 cm^{-1}	329743.57	– 337655.98	5–3	9.02–04	1.29–03	2.69–01	–2.189	B–	*LS*
233.		^{3}P–^{3}P°	*9621.9*	*9624.5*	*329724.7*	– *340114.8*	9–9	2.11–02	2.93–02	8.35+00	–0.579	C+	2
			9651.47	9654.12	329743.57	– 340101.84	5–5	1.57–02	2.19–02	3.48+00	–0.961	C+	*LS*
			9593.32	9595.95	329706.47	– 340127.53	3–3	5.32–03	7.35–03	6.96–01	–1.657	C+	*LS*
			9627.60	9630.24	329743.57	– 340127.53	5–3	8.78–03	7.32–03	1.16+00	–1.437	C+	*LS*
			9580.18	9582.80	329706.47	– 340141.83	3–1	2.14–02	9.81–03	9.28–01	–1.531	C+	*LS*
			9617.03	9619.67	329706.47	– 340101.84	3–5	5.28–03	1.22–02	1.16+00	–1.436	C+	*LS*
			9573.95	9576.57	329685.38	– 340127.53	1–3	7.14–03	2.94–02	9.28–01	–1.531	C+	*LS*
234.	2p3p– 2s5p	^{3}P–^{3}P°	*6889.9*	*6891.8*	*329724.7*	– *344234.7*	9–9	1.83–02	1.30–02	2.65+00	–0.932	B	2
			6899.69	6901.59	329743.57	– 344232.98	5–5	1.36–02	9.73–03	1.11+00	–1.313	B	*LS*
			6880.50	6882.40	329706.47	– 344236.29	3–3	4.58–03	3.25–03	2.21–01	–2.011	B	*LS*
			6898.11	6900.02	329743.57	– 344236.29	5–3	7.58–03	3.24–03	3.68–01	–1.790	B	*LS*
			6879.37	6881.27	329706.47	– 344238.68	3–1	1.83–02	4.34–03	2.95–01	–1.886	B	*LS*
			6882.07	6883.97	329706.47	– 344232.98	3–5	4.58–03	5.42–03	3.68–01	–1.789	B	*LS*
			6870.53	6872.42	329685.38	– 344236.29	1–3	6.14–03	1.30–02	2.95–01	–1.885	B	*LS*
235.	2p3p– 2s6p	^{3}P–^{3}P°	*3658.5*	*3659.5*	*329724.7*	– *357050.7*	9–9	2.86–03	5.75–04	6.23–02	–2.286	C+	2
			3660.93	3661.98	329743.57	– 357051.23	5–5	2.14–03	4.31–04	2.60–02	–2.667	C+	*LS*
			3656.11	3657.15	329706.47	– 357050.17	3–3	7.17–04	1.44–04	5.19–03	–3.365	D	*LS*
			3661.08	3662.12	329743.57	– 357050.17	5–3	1.19–03	1.44–04	8.65–03	–3.144	D	*LS*
			3656.21	3657.26	329706.47	– 357049.38	3–1	2.87–03	1.92–04	6.92–03	–3.240	D	*LS*
			3655.97	3657.01	329706.47	– 357051.23	3–5	7.17–04	2.40–04	8.65–03	–3.143	D	*LS*
			3653.29	3654.33	329685.38	– 357050.17	1–3	9.58–04	5.75–04	6.92–03	–3.240	D	*LS*

C III: Allowed Transitions — Continued

No.	Transition Array	Mult.	λ_{air} (Å)	λ_{vac} (Å) or σ (cm^{-1})*	E_i (cm^{-1})	E_k (cm^{-1})	g_i–g_k	A_{ki} (10^8 s^{-1})	f_{ik}	S (at. u.)	log gf	Acc.	Source
236.	$2p3d$–$2p3p$	$^1D^o$–^1D		426.93 cm^{-1}	332691.28	– 333118.21	5–5	1.43–06	1.18–03	4.54+00	–2.230	B	2
237.	$2p3d$–$2s5d$	$^1D^o$–^1D	7157.73	7159.70	332691.28	– 346658.34	5–5	5.32–04	4.09–04	4.82–02	–2.689	D	2
238.	$2p3d$–$2s6d$	$^1D^o$–^1D	3838.91	3840.00	332691.28	– 358732.93	5–5	7.16–04	1.58–04	1.00–02	–3.102	C	2
239.	$2p3d$–$2p4p$	$^1D^o$–^1P	2064.85	2065.51	332691.28	– 381105.4	5–3	6.88–01	2.64–02	8.98–01	–0.879	B	2
240.	$2p3d$–$2p5p$	$^1D^o$–^1P		1337.98	332691.28	– 407431	5–3	2.72–01	4.38–03	9.65–02	–1.659	B	2
241.	$2p3p$–$2p3d$	^1D–$^1F^o$	12113.9	8252.73 cm^{-1}	333118.21	– 341370.94	5–7	5.66–02	1.74–01	3.48+01	–0.059	B	2
242.	$2p3p$–$2s5p$	^1D–$^1P^o$	9859.40	9862.11	333118.21	– 343258.03	5–3	2.55–02	2.23–02	3.62+00	–0.952	B–	2
243.	$2p3p$–$2p3d$	^1D–$^1P^o$	7353.88	7355.91	333118.21	– 346712.73	5–3	3.09–02	1.51–02	1.82+00	–1.123	B	2
244.	$2p3p$–$2s5f$	^1D–$^1F^o$	6350.77	6352.52	333118.21	– 348859.99	5–7	3.10–02	2.62–02	2.74+00	–0.882	B	2
245.	$2p3p$–$2s6p$	^1D–$^1P^o$	4166.97	4168.15	333118.21	– 357109.68	5–3	7.31–02	1.14–02	7.83–01	–1.243	B	2
246.	$2p3p$–$2s6f$	^1D–$^1F^o$	3844.51	3845.60	333118.21	– 359121.95	5–7	4.76–02	1.48–02	9.35–01	–1.132	B	2
247.	$2p3p$–$2s7p$	^1D–$^1P^o$	3145.94	3146.85	333118.21	– 364896	5–3	8.62–02	7.68–03	3.98–01	–1.416	B	2
248.	$2p3p$–$2s8p$	^1D–$^1P^o$	2716.01	2716.82	333118.21	– 369926	5–3	1.07–01	7.11–03	3.18–01	–1.449	B	2
249.	$2p3p$–$2p4d$	^1D–$^1D^o$		1897.57	333118.21	– 385817.2	5–5	1.41+00	7.61–02	2.38+00	–0.420	B	2
250.	$2p3p$–$2p5d$	^1D–$^1D^o$		1306.08	333118.21	– 409683	5–5	1.24+00	3.18–02	6.84–01	–0.799	B	2
251.	$2p3d$–$2s5d$	$^3F^o$–^3D	*8278.7*	*8281.0*	*333421.0*	*– 345496.9*	21–15	7.47–02	5.48–02	3.14+01	0.061	B	2
			8296.54	8298.82	333447.24	– 345497.15	9–7	6.81–02	5.47–02	1.34+01	–0.308	B	*LS*
			8272.33	8274.60	333411.55	– 345496.72	7–5	6.65–02	4.87–02	9.29+00	–0.467	B	*LS*
			8255.67	8257.94	333387.01	– 345496.57	5–3	7.53–02	4.62–02	6.27+00	–0.637	B	*LS*
			8272.04	8274.31	333411.55	– 345497.15	7–7	5.96–03	6.11–03	1.16+00	–1.369	B	*LS*
			8255.57	8257.84	333387.01	– 345496.72	5–5	8.39–03	8.57–03	1.16+00	–1.368	B	*LS*
			8255.27	8257.54	333387.01	– 345497.15	5–7	1.69–04	2.42–04	3.29–02	–2.918	B	*LS*
252.	$2p3d$–$2s5g$	$^3F^o$–^3G	*7597.7*	*7599.8*	*333421.0*	*– 346579.3*	21–27	2.93–01	3.26–01	1.71+02	0.835	B	4*n*
			7612.75	7614.84	333447.24	– 346579.49	9–11	2.92–01	3.10–01	7.00+01	0.446	C+	4
			7592.27	7594.36	333411.55	– 346579.21	7–9	2.75–01	3.05–01	5.34+01	0.330	C+	4
			7578.15	7580.24	333387.01	– 346579.21	5–7	2.70–01	3.26–01	4.06+01	0.212	C+	4
			7612.91	7615.00	333447.24	– 346579.21	9–9	1.82–02	1.58–02	3.56+00	–0.847	C+	*LS*
			7592.27	7594.36	333411.55	– 346579.21	7–7	2.36–02	2.04–02	3.56+00	–0.846	C+	*LS*
			7612.91	7615.00	333447.24	– 346579.21	9–7	3.56–04	2.41–04	5.43–02	–2.664	C+	*LS*

C III: Allowed Transitions — Continued

No.	Transition Array	Mult.	λ_{air} (Å)	λ_{vac} (Å) or σ (cm^{-1})*	E_i (cm^{-1})	E_k (cm^{-1})	g_i-g_k	A_{ki} ($10^8 s^{-1}$)	f_{ik}	S (at. u.)	log gf	Acc.	Source
253.	$2p3d-$ $2s7d$	$^3F^o-^3D$	*3103.1*	*3104.0*	*333421.0*	365638	21–15	1.76–02	1.82–03	3.90–01	–1.419	B	2
			3105.58	3106.48	333447.24	365638	9–7	1.61–02	1.81–03	1.67–01	–1.787	B	*LS*
			3102.14	3103.04	333411.55	365638	7–5	1.57–02	1.61–03	1.15–01	–1.947	B	*LS*
			3099.78	3100.68	333387.01	365638	5–3	1.77–02	1.53–03	7.79–02	–2.117	B	*LS*
			3102.14	3103.04	333411.55	365638	7–7	1.40–03	2.02–04	1.45–02	–2.849	B–	*LS*
			3099.78	3100.68	333387.01	365638	5–5	1.97–03	2.84–04	1.45–02	–2.848	B–	*LS*
			3099.78	3100.68	333387.01	365638	5–7	3.96–05	8.00–06	4.08–04	–4.398	D	*LS*
254.	$2p3d-$ $2s8d$	$^3F^o-^3D$	*2696.2*	*2697.0*	*333421.0*	370499	21–15	1.13–02	8.82–04	1.64–01	–1.732	B	2
			2698.13	2698.93	333447.24	370499	9–7	1.04–02	8.81–04	7.05–02	–2.101	B	*LS*
			2695.53	2696.33	333411.55	370499	7–5	1.01–02	7.84–04	4.87–02	–2.261	B	*LS*
			2693.75	2694.55	333387.01	370499	5–3	1.14–02	7.41–04	3.29–02	–2.431	B–	*LS*
			2695.53	2696.33	333411.55	370499	7–7	9.02–04	9.83–05	6.11–03	–3.162	C+	*LS*
			2693.75	2694.55	333387.01	370499	5–5	1.26–03	1.38–04	6.11–03	–3.162	C+	*LS*
			2693.75	2694.55	333387.01	370499	5–7	2.55–05	3.88–06	1.72–04	–4.712	D	*LS*
255.	$2s5s-$ $2s5p$	$^1S-^1P^o$	21074.8	4743.70 cm^{-1}	338514.33	343258.03	1–3	5.76–02	1.15+00	7.99+01	0.061	B	2
256.	$2s5s-$ $2p3d$	$^1S-^1P^o$	12194.2	8198.40 cm^{-1}	338514.33	346712.73	1–3	1.16–02	7.74–02	3.11+00	–1.111	B	2
257.	$2s5s-$ $2s6p$	$^1S-^1P^o$	5376.19	5377.69	338514.33	357109.68	1–3	3.29–02	4.28–02	7.58–01	–1.368	B	2
258.	$2s5s-$ $2s7p$	$^1S-^1P^o$	3789.43	3790.51	338514.33	364896	1–3	5.32–02	3.44–02	4.29–01	–1.463	B	2
259.	$2s5s-$ $2s8p$	$^1S-^1P^o$	3182.61	3183.53	338514.33	369926	1–3	5.77–02	2.63–02	2.76–01	–1.580	B	2
260.	$2s5s-$ $2p3d$	$^3S-^3P^o$		*180.1* cm^{-1}	339934.72	*340114.8*	3–9	9.70–07	1.35–02	7.38+01	–1.394	B	2
				167.12 cm^{-1}	339934.72	340101.84	3–5	7.75–07	6.93–03	4.10+01	–1.682	B	*LS*
				192.81 cm^{-1}	339934.72	340127.53	3–3	1.19–06	4.80–03	2.46+01	–1.842	B	*LS*
				207.11 cm^{-1}	339934.72	340141.83	3–1	1.48–06	1.72–03	8.20+00	–2.288	B	*LS*
261.	$2s5s-$ $2s5p$	$^3S-^3P^o$	*23249*	*4300.0* cm^{-1}	339934.72	*344234.7*	3–9	4.79–02	1.17+00	2.68+02	0.544	B	2
			23258.9	4298.26 cm^{-1}	339934.72	344232.98	3–5	4.79–02	6.47–01	1.49+02	0.288	B	*LS*
			23241.0	4301.57 cm^{-1}	339934.72	344236.29	3–3	4.80–02	3.89–01	8.92+01	0.067	B	*LS*
			23228.1	4303.96 cm^{-1}	339934.72	344238.68	3–1	4.80–02	1.30–01	2.97+01	–0.410	B	*LS*
262.	$2s5s-$ $2s6p$	$^3S-^3P^o$	*5840.9*	*5842.5*	339934.72	*357050.7*	3–9	4.57–02	7.01–02	4.05+00	–0.677	B	2
			5840.69	5842.31	339934.72	357051.23	3–5	4.57–02	3.89–02	2.25+00	–0.932	B	*LS*
			5841.06	5842.67	339934.72	357050.17	3–3	4.57–02	2.34–02	1.35+00	–1.154	B	*LS*
			5841.32	5842.94	339934.72	357049.38	3–1	4.57–02	7.79–03	4.49–01	–1.631	B	*LS*
263.	$2p3d-$ $2s5d$	$^3P^o-^3D$	*18575*	*5382.0* cm^{-1}	*340114.8*	*345496.9*	9–15	3.86–02	3.33–01	1.83+02	0.476	B	2
			18529.6	5395.31 cm^{-1}	340101.84	345497.15	5–7	3.89–02	2.80–01	8.55+01	0.147	B	*LS*
			18619.7	5369.19 cm^{-1}	340127.53	345496.72	3–5	2.87–02	2.49–01	4.58+01	–0.127	B	*LS*
			18669.9	5354.74 cm^{-1}	340141.83	345496.57	1–3	2.11–02	3.31–01	2.04+01	–0.480	B	*LS*
			18531.0	5394.88 cm^{-1}	340101.84	345496.72	5–5	9.71–03	5.00–02	1.53+01	–0.602	B	*LS*
			18620.2	5369.04 cm^{-1}	340127.53	345496.57	3–3	1.60–02	8.30–02	1.53+01	–0.604	B	*LS*
			18531.5	5394.73 cm^{-1}	340101.84	345496.57	5–3	1.08–03	3.34–03	1.02+00	–1.778	B	*LS*

C III: Allowed Transitions — Continued

No.	Transition Array	Mult.	λ_{air} (Å)	λ_{vac} (Å) or σ (cm^{-1})*	E_i (cm^{-1})	E_k (cm^{-1})	g_i–g_k	A_{ki} (10^8 s^{-1})	f_{ik}	S (at. u.)	log gf	Acc.	Source
264.	$2p3d$–$2s6s$	^{3}P°–^{3}S	*6780.9*	*6782.8*	*340114.8*	354858.03	9–3	1.64−01	3.77−02	7.58+00	−0.469	B	2
			6774.95	6776.82	340101.84	354858.03	5–3	9.14−02	3.78−02	4.21+00	−0.724	B	*LS*
			6786.76	6788.64	340127.53	354858.03	3–3	5.46−02	3.77−02	2.53+00	−0.947	B	*LS*
			6793.36	6795.23	340141.83	354858.03	1–3	1.81−02	3.76−02	8.42−01	−1.424	B	*LS*
265.	$2p3d$–$2s7s$	^{3}P°–^{3}S	*4254.5*	*4255.7*	*340114.8*	363613	9–3	1.08−01	9.79−03	1.23+00	−1.055	B	2
			4252.10	4253.30	340101.84	363613	5–3	6.02−02	9.79−03	6.86−01	−1.310	B	*LS*
			4256.75	4257.95	340127.53	363613	3–3	3.60−02	9.78−03	4.11−01	−1.532	B	*LS*
			4259.35	4260.55	340141.83	363613	1–3	1.20−02	9.78−03	1.37−01	−2.010	B	*LS*
266.	$2p3d$–$2s7d$	^{3}P°–^{3}D	*3916.9*	*3918.0*	*340114.8*	365638	9–15	2.54−02	9.73−03	1.13+00	−1.057	B	2
			3914.91	3916.02	340101.84	365638	5–7	2.54−02	8.18−03	5.27−01	−1.388	B	*LS*
			3918.85	3919.96	340127.53	365638	3–5	1.90−02	7.30−03	2.83−01	−1.660	B	*LS*
			3921.05	3922.16	340141.83	365638	1–3	1.41−02	9.72−03	1.26−01	−2.012	B	*LS*
			3914.91	3916.02	340101.84	365638	5–5	6.35−03	1.46−03	9.42−02	−2.136	B	*LS*
			3918.85	3919.96	340127.53	365638	3–3	1.06−02	2.43−03	9.42−02	−2.137	B	*LS*
			3914.91	3916.02	340101.84	365638	5–3	7.06−04	9.74−05	6.28−03	−3.313	C+	*LS*
267.	$2p3d$–$2s8d$	^{3}P°–^{3}D	*3290.2*	*3291.2*	*340114.8*	370499	9–15	2.09−02	5.66−03	5.52−01	−1.293	B	2
			3288.83	3289.78	340101.84	370499	5–7	2.09−02	4.75−03	2.57−01	−1.624	B	*LS*
			3291.62	3292.56	340127.53	370499	3–5	1.57−02	4.24−03	1.38−01	−1.895	B	*LS*
			3293.17	3294.11	340141.83	370499	1–3	1.16−02	5.65−03	6.13−02	−2.248	B	*LS*
			3288.83	3289.78	340101.84	370499	5–5	5.23−03	8.49−04	4.60−02	−2.372	B	*LS*
			3291.62	3292.56	340127.53	370499	3–3	8.70−03	1.41−03	4.60−02	−2.373	B	*LS*
			3288.83	3289.78	340101.84	370499	5–3	5.81−04	5.66−05	3.06−03	−3.548	C	*LS*
268.	$2p3d$–$2s5d$	^{1}F°–^{1}D	18907.7	5287.40 cm^{-1}	341370.94	346658.34	7–5	3.76−02	1.44−01	6.28+01	0.003	B	2
269.	$2p3d$–$2s6d$	^{1}F°–^{1}D	5758.11	5759.71	341370.94	358732.93	7–5	3.98−03	1.41−03	1.88−01	−2.004	B	2
270.	$2p3d$–$2s7d$	^{1}F°–^{1}D	4054.49	4055.63	341370.94	366028	7–5	1.46−03	2.57−04	2.40−02	−2.745	B−	2
271.	$2s5p$–$2p3p$	^{1}P°–^{1}S		1837.40 cm^{-1}	343258.03	345095.43	3–1	3.24−03	4.80−02	2.58+01	−0.842	B	2
272.	$2s5p$–$2s5d$	^{1}P°–^{1}D	29401.1	3400.31 cm^{-1}	343258.03	346658.34	3–5	3.43−02	7.42−01	2.15+02	0.347	B	2
273.	$2s5p$–$2s6d$	^{1}P°–^{1}D	6460.29	6462.08	343258.03	358732.93	3–5	1.32−01	1.38−01	8.81+00	−0.383	B	2
274.	$2s5p$–$2s7d$	^{1}P°–^{1}D	4390.52	4391.75	343258.03	366028	3–5	1.16−01	5.60−02	2.43+00	−0.774	B	2
275.	$2s5p$–$2p4p$	^{1}P°–^{1}P	2641.40	2642.19	343258.03	381105.4	3–3	3.10−01	3.24−02	8.46−01	−1.012	B	2
276.	$2s5p$–$2p5p$	^{1}P°–^{1}P		1558.29	343258.03	407431	3–3	3.31+00	1.21−01	1.86+00	−0.442	B	2

C III: Allowed Transitions — Continued

No.	Transition Array	Mult.	λ_{air} (Å)	λ_{vac} (Å) or σ (cm^{-1})*	E_i (cm^{-1})	E_k (cm^{-1})	g_i–g_k	A_{ki} (10^8 s^{-1})	f_{ik}	S (at. u.)	log gf	Acc.	Source
277.	2s5p–2s5d	^{3}P°–^{3}D		*1262.2* cm^{-1}	*344234.7*	– *345496.9*	9–15	1.53–03	2.39–01	5.62+02	0.333	B	2
				1264.17 cm^{-1}	344232.98	– 345497.15	5–7	1.53–03	2.01–01	2.62+02	0.003	B	*LS*
				1260.43 cm^{-1}	344236.29	– 345496.72	3–5	1.14–03	1.79–01	1.40+02	–0.269	B	*LS*
				1257.89 cm^{-1}	344238.68	– 345496.57	1–3	8.39–04	2.39–01	6.24+01	–0.622	B	*LS*
				1263.74 cm^{-1}	344232.98	– 345496.72	5–5	3.83–04	3.59–02	4.68+01	–0.745	B	*LS*
				1260.28 cm^{-1}	344236.29	– 345496.57	3–3	6.33–04	5.97–02	4.68+01	–0.747	B	*LS*
				1263.59 cm^{-1}	344232.98	– 345496.57	5–3	4.25–05	2.40–03	3.12+00	–1.922	B	*LS*
278.	2s5p–2s6s	^{3}P°–^{3}S	*9410.7*	*9413.3*	*344234.7*	– 354858.03	9–3	5.39–01	2.39–01	6.65+01	0.332	B	2
			9409.14	9411.72	344232.98	– 354858.03	5–3	2.99–01	2.39–01	3.70+01	0.077	B	*LS*
			9412.07	9414.65	344236.29	– 354858.03	3–3	1.79–01	2.39–01	2.22+01	–0.145	B	*LS*
			9414.19	9416.77	344238.68	– 354858.03	1–3	5.98–02	2.38–01	7.39+00	–0.623	B	*LS*
279.	2s5p–2s7s	^{3}P°–^{3}S	*5159.0*	*5160.4*	*344234.7*	– 363613	9–3	2.52–01	3.35–02	5.13+00	–0.520	B	2
			5158.52	5159.95	344232.98	– 363613	5–3	1.40–01	3.35–02	2.85+00	–0.776	B	*LS*
			5159.40	5160.83	344236.29	– 363613	3–3	8.40–02	3.35–02	1.71+00	–0.998	B	*LS*
			5160.03	5161.47	344238.68	– 363613	1–3	2.80–02	3.35–02	5.70–01	–1.475	B	*LS*
280.	2s5p–2s7d	^{3}P°–^{3}D	*4670.9*	*4672.2*	*344234.7*	– 365638	9–15	1.79–01	9.78–02	1.35+01	–0.055	B	2
			4670.49	4671.80	344232.98	– 365638	5–7	1.79–01	8.22–02	6.32+00	–0.386	B	*LS*
			4671.22	4672.52	344236.29	– 365638	3–5	1.34–01	7.34–02	3.39+00	–0.657	B	*LS*
			4671.74	4673.05	344238.68	– 365638	1–3	9.96–02	9.78–02	1.50+00	–1.010	B	*LS*
			4670.49	4671.80	344232.98	– 365638	5–5	4.48–02	1.47–02	1.13+00	–1.135	B	*LS*
			4671.22	4672.52	344236.29	– 365638	3–3	7.47–02	2.45–02	1.13+00	–1.135	B	*LS*
			4670.49	4671.80	344232.98	– 365638	5–3	4.98–03	9.78–04	7.52–02	–2.311	B	*LS*
281.	2s5p–2s8d	^{3}P°–^{3}D	*3806.4*	*3807.5*	*344234.7*	– 370499	9–15	1.22–01	4.43–02	5.00+00	–0.399	B	2
			3806.12	3807.20	344232.98	– 370499	5–7	1.22–01	3.72–02	2.33+00	–0.730	B	*LS*
			3806.60	3807.68	344236.29	– 370499	3–5	9.17–02	3.32–02	1.25+00	–1.001	B	*LS*
			3806.95	3808.03	344238.68	– 370499	1–3	6.79–02	4.43–02	5.55–01	–1.354	B	*LS*
			3806.12	3807.20	344232.98	– 370499	5–5	3.06–02	6.65–03	4.16–01	–1.479	B	*LS*
			3806.60	3807.68	344236.29	– 370499	3–3	5.09–02	1.11–02	4.16–01	–1.479	B	*LS*
			3806.12	3807.20	344232.98	– 370499	5–3	3.40–03	4.43–04	2.78–02	–2.655	B–	*LS*
282.	2p3p–2p3d	^{1}S–^{1}P°		1617.30 cm^{-1}	345095.43	– 346712.73	1–3	7.16–04	1.23–01	2.51+01	–0.910	B	2
283.	2p3p–2s6p	^{1}S–^{1}P°	8321.16	8323.45	345095.43	– 357109.68	1–3	1.05–01	3.27–01	8.96+00	–0.486	B	2
284.	2p3p–2s7p	^{1}S–^{1}P°	5048.95	5050.36	345095.43	– 364896	1–3	3.23–02	3.70–02	6.15–01	–1.432	B	2
285.	2p3p–2s8p	^{1}S–^{1}P°	4026.16	4027.29	345095.43	– 369926	1–3	9.09–03	6.63–03	8.79–02	–2.178	B	2
286.	2s5d–2s5f	^{3}D–^{3}F°		*1657.0* cm^{-1}	*345496.9*	– *347153.9*	15–21	4.05–03	3.10–01	9.22+02	0.667	B	2
				1658.26 cm^{-1}	345497.15	– 347155.41	7–9	4.06–03	2.84–01	3.95+02	0.299	B	*LS*
				1656.54 cm^{-1}	345496.72	– 347153.26	5–7	3.59–03	2.75–01	2.73+02	0.138	B	*LS*
				1655.32 cm^{-1}	345496.57	– 347151.89	3–5	3.39–03	3.09–01	1.84+02	–0.033	B	*LS*
				1656.11 cm^{-1}	345497.15	– 347153.26	7–7	4.50–04	2.46–02	3.43+01	–0.764	B	*LS*
				1655.17 cm^{-1}	345496.72	– 347151.89	5–5	6.30–04	3.45–02	3.43+01	–0.764	B	*LS*
				1654.74 cm^{-1}	345497.15	– 347151.89	7–5	1.77–05	6.94–04	9.66–01	–2.314	B	*LS*

C III: Allowed Transitions — Continued

No.	Transition Array	Mult.	λ_{air} (Å)	λ_{vac} (Å) or σ (cm^{-1})*	E_i (cm^{-1})	E_k (cm^{-1})	g_i–g_k	A_{ki} (10^8 s^{-1})	f_{ik}	S (at. u.)	log gf	Acc.	Source
287.	2s5d–2s6p	^3D–$^3P^o$	*8652.8*	*8655.2*	*345496.9*	– *357050.7*	15–9	1.72–01	1.16–01	4.95+01	0.240	B	2
			8652.57	8654.95	345497.15	– 357051.23	7–5	1.45–01	1.16–01	2.31+01	–0.091	B	*LS*
			8653.05	8655.42	345496.72	– 357050.17	5–3	1.29–01	8.69–02	1.24+01	–0.362	B	*LS*
			8653.53	8655.90	345496.57	– 357049.38	3–1	1.72–01	6.44–02	5.50+00	–0.714	B	*LS*
			8652.25	8654.63	345496.72	– 357051.23	5–5	2.58–02	2.90–02	4.13+00	–0.839	B	*LS*
			8652.93	8655.31	345496.57	– 357050.17	3–3	4.30–02	4.83–02	4.13+00	–0.839	B	*LS*
			8652.14	8654.52	345496.57	– 357051.23	3–5	1.72–03	3.22–03	2.75–01	–2.015	B	*LS*
288.	2s5d–2s6f	^3D–$^3F^o$	*7486.6*	*7488.7*	*345496.9*	– *358850.4*	15–21	4.98–01	5.87–01	2.17+02	0.944	B	2
			7486.56	7488.62	345497.15	– 358850.74	7–9	4.98–01	5.39–01	9.30+01	0.576	C+	*LS*
			7486.56	7488.62	345496.72	– 358850.32	5–7	4.43–01	5.21–01	6.42+01	0.416	C+	*LS*
			7486.64	7488.70	345496.57	– 358850.02	3–5	4.19–01	5.87–01	4.34+01	0.245	C+	*LS*
			7486.80	7488.86	345497.15	– 358850.32	7–7	5.55–02	4.67–02	8.06+00	–0.486	C+	*LS*
			7486.72	7488.79	345496.72	– 358850.02	5–5	7.77–02	6.54–02	8.06+00	–0.486	C+	*LS*
			7486.96	7489.03	345497.15	– 358850.02	7–5	2.19–03	1.32–03	2.27–01	–2.035	C+	*LS*
289.	2s5d–2p3d	^1D–$^1P^o$		54.39 cm^{-1}	346658.34	– 346712.73	5–3	3.63–08	1.10–03	3.34+01	–2.258	B	2
290.	2s5d–2s5f	^1D–$^1F^o$	45408	2201.65 cm^{-1}	346658.34	– 348859.99	5–7	8.18–03	3.54–01	2.65+02	0.248	B	2
291.	2s5d–2s6p	^1D–$^1P^o$	9565.53	9568.15	346658.34	– 357109.68	5–3	2.25–01	1.85–01	2.92+01	–0.033	B	2
292.	2s5d–2s6f	^1D–$^1F^o$	8021.15	8023.36	346658.34	– 359121.95	5–7	5.01–01	6.77–01	8.94+01	0.529	B	2
293.	2s5d–2s7p	^1D–$^1P^o$	5481.64	5483.16	346658.34	– 364896	5–3	1.29–01	3.50–02	3.16+00	–0.757	B	2
294.	2s5d–2s8p	^1D–$^1P^o$	4296.60	4297.81	346658.34	– 369926	5–3	8.60–02	1.43–02	1.01+00	–1.146	B	2
295.	2s5d–2p5d	^1D–$^1D^o$		1586.68	346658.34	– 409683	5–5	2.22+00	8.37–02	2.18+00	–0.379	B	2
296.	2p3d–2s6d	$^1P^o$–^1D	8317.04	8319.33	346712.73	– 358732.93	3–5	7.61–02	1.32–01	1.08+01	–0.404	B	2
297.	2p3d–2s7d	$^1P^o$–^1D	5175.81	5177.25	346712.73	– 366028	3–5	4.43–02	2.97–02	1.52+00	–1.050	B	2
298.	2p3d–2p4p	$^1P^o$–^1P	2906.74	2907.60	346712.73	– 381105.4	3–3	1.30–01	1.65–02	4.75–01	–1.304	B	2
299.	2p3d–2p5p	$^1P^o$–^1P		1646.95	346712.73	– 407431	3–3	5.25–02	2.13–03	3.47–02	–2.194	B–	2
300.	2s5f–2s7d	$^3F^o$–^3D	*5408.5*	*5410.0*	*347153.9*	– 365638	21–15	3.57–02	1.12–02	4.18+00	–0.629	B	2
			5408.99	5410.50	347155.41	– 365638	9–7	3.27–02	1.12–02	1.79+00	–0.997	C+	*LS*
			5408.36	5409.87	347153.26	– 365638	7–5	3.17–02	9.93–03	1.24+00	–1.158	C+	*LS*
			5407.96	5409.47	347151.89	– 365638	5–3	3.57–02	9.39–03	8.36–01	–1.328	C+	*LS*
			5408.36	5409.87	347153.26	– 365638	7–7	2.84–03	1.25–03	1.55–01	–2.060	C+	*LS*
			5407.96	5409.47	347151.89	– 365638	5–5	3.98–03	1.74–03	1.55–01	–2.060	C+	*LS*
			5407.96	5409.47	347151.89	– 365638	5–7	8.01–05	4.92–05	4.38–03	–3.609	D–	*LS*

C III: Allowed Transitions — Continued

No.	Transition Array	Mult.	λ_{air} (Å)	λ_{vac} (Å) or σ (cm⁻¹)*	E_i (cm⁻¹)	E_k (cm⁻¹)	g_i–g_k	A_{ki} (10^8 s⁻¹)	f_{ik}	S (at. u.)	log gf	Acc.	Source
301.	2s5f–2s8d	$^3F^o$–^3D	*4282.3*	*4283.5*	*347153.9*	– 370499	21–15	2.08–02	4.08–03	1.21+00	–1.067	B	2
			4282.63	4283.83	347155.41	– 370499	9–7	1.91–02	4.08–03	5.18–01	–1.435	C+	*LS*
			4282.23	4283.44	347153.26	– 370499	7–5	1.85–02	3.63–03	3.58–01	–1.595	C+	*LS*
			4281.98	4283.19	347151.89	– 370499	5–3	2.08–02	3.43–03	2.42–01	–1.766	C+	*LS*
			4282.23	4283.44	347153.26	– 370499	7–7	1.65–03	4.55–04	4.49–02	–2.497	C+	*LS*
			4281.98	4283.19	347151.89	– 370499	5–5	2.32–03	6.37–04	4.49–02	–2.497	C+	*LS*
			4281.98	4283.19	347151.89	– 370499	5–7	4.67–05	1.80–05	1.27–03	–4.046	D	*LS*
302.	2s5f–2s6d	$^1F^o$–^1D	10125.9	9872.94 cm⁻¹	348859.99	– 358732.93	7–5	6.51–02	7.16–02	1.67+01	–0.300	B	2
303.	2s5f–2s7d	$^1F^o$–^1D	5823.17	5824.79	348859.99	– 366028	7–5	2.86–02	1.04–02	1.39+00	–1.138	B	2
304.	2s6s–2s6p	^3S–$^3P^o$	*45595*	*2192.6* cm⁻¹	354858.03	– *357050.7*	3–9	1.76–02	1.64+00	7.40+02	0.693	B	2
			45583.0	2193.20 cm⁻¹	354858.03	– 357051.23	3–5	1.76–02	9.13–01	4.11+02	0.438	B	*LS*
			45605.1	2192.14 cm⁻¹	354858.03	– 357050.17	3–3	1.76–02	5.48–01	2.47+02	0.216	B	*LS*
			45621.5	2191.35 cm⁻¹	354858.03	– 357049.38	3–1	1.75–02	1.82–01	8.22+01	–0.262	B	*LS*
305.	2s6p–2s7s	$^3P^o$–^3S	*15234*	*6562* cm⁻¹	*357050.7*	– 363613	9–3	2.95–01	3.42–01	1.55+02	0.489	B	2
			15235.6	6562 cm⁻¹	357051.23	– 363613	5–3	1.64–01	3.42–01	8.59+01	0.234	B	*LS*
			15233.2	6563 cm⁻¹	357050.17	– 363613	3–3	9.84–02	3.42–01	5.15+01	0.012	B	*LS*
			15231.3	6564 cm⁻¹	357049.38	– 363613	1–3	3.28–02	3.42–01	1.72+01	–0.465	B	*LS*
306.	2s6p–2s7d	$^3P^o$–^3D	*11642*	*8587* cm⁻¹	*357050.7*	– 365638	9–15	9.55–02	3.24–01	1.12+02	0.464	B	2
			11642.6	8587 cm⁻¹	357051.23	– 365638	5–7	9.55–02	2.72–01	5.21+01	0.133	B	*LS*
			11641.2	8588 cm⁻¹	357050.17	– 365638	3–5	7.16–02	2.43–01	2.79+01	–0.138	B	*LS*
			11640.1	8589 cm⁻¹	357049.38	– 365638	1–3	5.31–02	3.24–01	1.24+01	–0.490	B	*LS*
			11642.6	8587 cm⁻¹	357051.23	– 365638	5–5	2.39–02	4.85–02	9.30+00	–0.615	B	*LS*
			11641.2	8588 cm⁻¹	357050.17	– 365638	3–3	3.98–02	8.09–02	9.30+00	–0.615	B	*LS*
			11642.6	8587 cm⁻¹	357051.23	– 365638	5–3	2.65–03	3.24–03	6.20–01	–1.791	B	*LS*
307.	2s6p–2s8d	$^3P^o$–^3D	*7433.8*	*7435.9*	*357050.7*	– 370499	9–15	7.59–02	1.05–01	2.31+01	–0.025	B	2
			7434.13	7436.18	357051.23	– 370499	5–7	7.59–02	8.81–02	1.08+01	–0.356	B	*LS*
			7433.54	7435.59	357050.17	– 370499	3–5	5.69–02	7.87–02	5.78+00	–0.627	B	*LS*
			7433.11	7435.15	357049.38	– 370499	1–3	4.22–02	1.05–01	2.57+00	–0.979	B	*LS*
			7434.13	7436.18	357051.23	– 370499	5–5	1.90–02	1.57–02	1.93+00	–1.104	B	*LS*
			7433.54	7435.59	357050.17	– 370499	3–3	3.16–02	2.62–02	1.93+00	–1.104	B	*LS*
			7434.13	7436.18	357051.23	– 370499	5–3	2.11–03	1.05–03	1.28–01	–2.280	B	*LS*
308.	2s6p–2s6d	$^1P^o$–^1D		1623.25 cm⁻¹	357109.68	– 358732.93	3–5	9.37–03	8.88–01	5.41+02	0.426	B	2
309.	2s6p–2s7d	$^1P^o$–^1D	11209.8	8918 cm⁻¹	357109.68	– 366028	3–5	5.56–02	1.75–01	1.93+01	–0.281	B	2
310.	2s6p–2p4p	$^1P^o$–^1P	4166.24	4167.41	357109.68	– 381105.4	3–3	2.30–02	5.99–03	2.47–01	–1.745	B	2
311.	2s6p–2p5p	$^1P^o$–^1P		1987.23	357109.68	– 407431	3–3	1.10–02	6.52–04	1.28–02	–2.709	B–	2
312.	2s6d–2s6f	^1D–$^1F^o$		389.02 cm⁻¹	358732.93	– 359121.95	5–7	1.25–04	1.74–01	7.36+02	–0.060	B	2
313.	2s6d–2s7p	^1D–$^1P^o$	16221.2	6163 cm⁻¹	358732.93	– 364896	5–3	1.17–01	2.77–01	7.41+01	0.142	B	2

C III: Allowed Transitions — Continued

No.	Transition Array	Mult.	λ_{air} (Å)	λ_{vac} (Å) or σ (cm^{-1})*	E_i (cm^{-1})	E_k (cm^{-1})	$g_i - g_k$	A_{ki} ($10^8 s^{-1}$)	f_{ik}	S (at. u.)	log gf	Acc.	Source
314.	$2s6d$– $2s8p$	$^1D-^1P^o$	8931.65	8934.10	358732.93	369926	5–3	6.74–02	4.84–02	7.11+00	–0.617	B	2
315.	$2s6d$– $2p5d$	$^1D-^1D^o$		1962.71	358732.93	409683	5–5	6.70–03	3.87–04	1.25–02	–2.713	C+	2
316.	$2s6f$– $2s7d$	$^3F^o-^3D$	*14729*	*6788* cm^{-1}	*358850.4*	365638	21–15	3.92–02	9.11–02	9.27+01	0.282	B	2
			14729.5	6787 cm^{-1}	358850.74	365638	9–7	3.60–02	9.11–02	3.97+01	–0.086	C+	*LS*
			14728.5	6788 cm^{-1}	358850.32	365638	7–5	3.48–02	8.09–02	2.75+01	–0.247	C+	*LS*
			14727.9	6788 cm^{-1}	358850.02	365638	5–3	3.92–02	7.65–02	1.85+01	–0.417	C+	*LS*
			14728.5	6788 cm^{-1}	358850.32	365638	7–7	3.12–03	1.01–02	3.44+00	–1.149	C+	*LS*
			14727.9	6788 cm^{-1}	358850.02	365638	5–5	4.37–03	1.42–02	3.44+00	–1.149	C+	*LS*
			14727.9	6788 cm^{-1}	358850.02	365638	5–7	8.80–05	4.01–04	9.72–02	–2.698	C+	*LS*
317.	$2s6f$– $2s8d$	$^3F^o-^3D$	*8582.4*	*8584.7*	*358850.4*	370499	21–15	2.08–02	1.64–02	9.75+00	–0.462	B	2
			8582.62	8584.97	358850.74	370499	9–7	1.91–02	1.64–02	4.18+00	–0.830	C+	*LS*
			8582.31	8584.66	358850.32	370499	7–5	1.85–02	1.46–02	2.89+00	–0.991	C+	*LS*
			8582.09	8584.44	358850.02	370499	5–3	2.08–02	1.38–02	1.95+00	–1.161	C+	*LS*
			8582.31	8584.66	358850.32	370499	7–7	1.66–03	1.83–03	3.62–01	–1.892	C+	*LS*
			8582.09	8584.44	358850.02	370499	5–5	2.32–03	2.56–03	3.62–01	–1.892	C+	*LS*
			8582.09	8584.44	358850.02	370499	5–7	4.67–05	7.23–05	1.02–02	–3.442	C	*LS*
318.	$2s6f$– $2s7d$	$^1F^o-^1D$	14476.1	6906 cm^{-1}	359121.95	366028	7–5	2.78–02	6.23–02	2.08+01	–0.360	B	2
319.	$2s7p$– $2s7d$	$^1P^o-^1D$		1132 cm^{-1}	364896	366028	3–5	6.05–03	1.18+00	1.03+03	0.549	B	2
320.	$2s7p$– $2p4p$	$^1P^o-^1P$	6167.55	6169.26	364896	381105.4	3–3	1.17–02	6.66–03	4.06–01	–1.699	B	2
321.	$2s7d$– $2s8p$	$^1D-^1P^o$	25647.2	3898 cm^{-1}	366028	369926	5–3	6.44–02	3.81–01	1.61+02	0.280	B	2
322.	$2s7d$– $2p5d$	$^1D-^1D^o$	2289.98	2290.69	366028	409683	5–5	1.31–03	1.03–04	3.90–03	–3.286	C+	2
323.	$2s8p$– $2p4p$	$^1P^o-^1P$	8942.57	8945.02	369926	381105.4	3–3	6.34–03	7.60–03	6.72–01	–1.642	B	2
324.	$2p4p$– $2p4d$	$^1P-^1D^o$	21217.5	4711.8 cm^{-1}	381105.4	385817.2	3–5	2.78–02	3.13–01	6.55+01	–0.028	B	2
325.	$2p4p$– $2p5d$	$^1P-^1D^o$	3498.24	3499.24	381105.4	409683	3–5	7.07–01	2.16–01	7.47+00	–0.188	B	2

*Wavelengths (Å) are always given unless cm^{-1} is indicated.

C III

Forbidden Transitions

List of tabulated lines

Wavelength (Å)	No.	Wavelength (Å)	No.	Wavelength (Å)	No.	Wavenumber (cm^{-1})	No.
in vacuum		1176.37	6	2844.12	9	23.69 cm^{-1}	4
		1176.77	6	2847.97	9	28.70 cm^{-1}	11
685.513	3	1906.68	1	11830.5	12	47.61 cm^{-1}	11
727.262	2			11870.8	12	56.36 cm^{-1}	4
727.514	2	in air		11938.3	12	76.31 cm^{-1}	11
768.800	8					80.05 cm^{-1}	4
1069.41	7	1999.95	5				
1069.69	7	2000.90	5				
1070.33	7	2003.16	5				
1174.61	6	2218.30	13				
1174.93	6	2220.65	13				
1175.59	6	2296.87	10				
1175.71	6	2728.17	14				

We have selected the data of Glass,[1-3] who calculated transition probabilities utilizing the CIV 3 code (see general introduction), i.e., by constructing advanced wavefunctions from configuration interaction expansions. He also included some relativistic corrections. The agreement with earlier, less comprehensive (and somewhat less advanced) calculations by Nussbaumer and Storey[4] is usually within a few percent, except for two weak M1 lines, $^1S_0$ – $^3P_1$ in the $2s^2$–$2p^2$ transition array and $^3P^o_1$ – $^1P^o_1$ in the $2s2p$ configuration, where the discrepancies are large.

For the M1 fine structure transitions in the ^3P terms of the $2s2p$ and $2p^2$ configurations we have used the line strengths S obtained from general non-relativistic formulas for conditions of pure LS-coupling, given by Pasternak.[5] The A-values published by Glass, converted to S, are in close agreement, while the A-values of Nussbaumer and Storey[4] convert exactly to those of the formulas.

References

[1] R. Glass, Astrophys. Space Sci. **87**, 41 (1982).
[2] R. Glass, Astrophys. Space Sci. **91**, 417 (1983).
[3] R. Glass, Astrophys. Space Sci. **92**, 307 (1983).
[4] H. Nussbaumer and P. J. Storey, Astron. Astrophys. **64**, 139 (1978).
[5] S. Pasternak, Astrophys. J. **92**, 129 (1940).

C III: Forbidden Transitions

No.	Transition Array	Mult.	λ_{air} (Å)	λ_{vac} (Å) or σ (cm^{-1})*	E_i (cm^{-1})	E_k (cm^{-1})	g_i–g_k	Type	A_{ki} (s^{-1})	S (at. u.)	Acc.	Source
1.	$2s^2$–$2s2p$	^1S–$^3P^o$										
				1906.68	0.00	– 52447.11	1–5	M2	5.19–03	4.38+01	B+	1,4
2.	$2s^2$–$2p^2$	^1S–^3P										
				727.514	0.00	– 137454.40	1–3	M1	7.21–04	3.09–08	D	2,4
				727.262	0.00	– 137502.01	1–5	E2	3.92–02	3.56–05	B	3,4
3.		^1S–^1D										
				685.513	0.00	– 145876.13	1–5	E2	2.40+03	1.62+00	B+	3,4
4.	$2s2p$–$2s2p$	$^3P^o$–$^3P^o$										
				23.69 cm^{-1}	52367.06	– 52390.75	1–3	M1	2.39–07	2.00+00	A	4
				80.05 cm^{-1}	52367.06	– 52447.11	1–5	E2	1.50–13	2.03+00	C	3
				56.36 cm^{-1}	52390.75	– 52447.11	3–5	M1	2.41–06	2.50+00	A	4
				56.36 cm^{-1}	52390.75	– 52447.11	3–5	E2	5.09–14	3.99+00	C	3

C III: Forbidden Transitions — Continued

No.	Transition Array	Mult.	λ_{air} (Å)	λ_{vac} (Å) or σ (cm^{-1})*	E_i (cm^{-1})	E_k (cm^{-1})	g_i-g_k	Type	A_{ki} (s^{-1})	S (at. u.)	Acc.	Source
5.		$^3P^o-^1P^o$										
			1999.95	2000.60	52367.06	102352.04	1–3	M1	1.50–03	1.33–06	B	2,4
			2000.90	2001.55	52390.75	102352.04	3–3	M1	3.30–03	2.94–06	E	2,4
			2000.90	2001.55	52390.75	102352.04	3–3	E2	6.01–05	5.17–06	D	3
			2003.16	2003.81	52447.11	102352.04	5–3	M1	1.87–03	1.67–06	B	2,4
			2003.16	2003.81	52447.11	102352.04	5–3	E2	4.91–05	4.25–06	D	3
6.	$2s2p-2p^2$	$^3P^o-^3P$										
				1176.77	52447.11	137425.70	5–1	M2	3.92–02	5.93+00	B	1
				1175.59	52390.75	137454.40	3–3	M2	3.74–02	1.69+01	B	1
				1176.37	52447.11	137454.40	5–3	M2	8.04–05	3.64–02	C	1
				1174.61	52367.06	137502.01	1–5	M2	8.03–03	6.02+00	B	1
				1174.93	52390.75	137502.01	3–5	M2	3.37–05	2.53–02	C	1
				1175.71	52447.11	137502.01	5–5	M2	4.65–02	3.50+01	B	1
7.		$^3P^o-^1D$										
				1069.41	52367.06	145876.13	1–5	M2	3.25–02	1.52+01	B	1
				1069.69	52390.75	145876.13	3–5	M2	7.35–02	3.45+01	B	1
				1070.33	52447.11	145876.13	5–5	M2	5.76–02	2.71+01	B	1
8.		$^3P^o-^1S$										
				768.800	52447.11	182519.88	5–1	M2	1.98–01	3.56+00	B	1
9.		$^1P^o-^3P$										
			2847.97	2848.81	102352.04	137454.40	3–3	M2	2.94–04	1.11+01	B	1
			2844.12	2844.95	102352.04	137502.01	3–5	M2	5.34–04	3.34+01	B	1
10.		$^1P^o-^1D$										
			2296.87	2297.58	102352.04	145876.13	3–5	M2	5.43–05	1.17+00	C	1
11.	$2p^2-2p^2$	$^3P-^3P$										
				28.70 cm^{-1}	137425.70	137454.40	1–3	M1	5.31–07	2.50+00	B+	5
				76.31 cm^{-1}	137425.70	137502.01	1–5	E2	1.10–13	1.89+00	E	3
				47.61 cm^{-1}	137454.40	137502.01	3–5	M1	1.16–06	2.00+00	B+	5
				47.61 cm^{-1}	137454.40	137502.01	3–5	E2	1.43–14	2.61+00	E	3
12.		$^3P-^1D$										
			11830.5	8450.43 cm^{-1}	137425.70	145876.13	1–5	E2	8.56–09	8.86–06	D	3
			11870.8	8421.73 cm^{-1}	137454.40	145876.13	3–5	M1	1.98–04	6.14–05	C	2
			11870.8	8421.73 cm^{-1}	137454.40	145876.13	3–5	E2	8.88–08	9.36–05	D	3
			11938.3	8374.12 cm^{-1}	137502.01	145876.13	5–5	M1	6.28–04	1.98–04	C+	2
			11938.3	8374.12 cm^{-1}	137502.01	145876.13	5–5	E2	6.10–07	6.61–04	D	3
13.		$^3P-^1S$										
			2218.30	2218.99	137454.40	182519.88	3–1	M1	1.85–02	7.48–06	C	2
			2220.65	2221.34	137502.01	182519.88	5–1	E2	1.73–03	8.36–05	C	3
14.		$^1D-^1S$										
			2728.17	2728.98	145876.13	182519.88	5–1	E2	4.39+01	5.94+00	B	3,4

*Wavelengths (Å) are always given unless cm^{-1} is indicated.

C IV

Lithium Isoelectronic Sequence

Ground State: $1s^22s\ ^2S_{1/2}$

Ionization Energy: 64.492 eV = 520178.4 cm^{-1}

Allowed Transitions

List of tabulated lines

Wavelength (Å)	No.	Wavelength (Å)	No.	Wavelength (Å)	No.	Wavelength (Å)	No.
in vacuum		259.539	17	596.698	61	1086.1	95
		259.540	17	608.83	47	1097.32	70
196.89	11	260.1	221	614.795	46	1097.34	70
197.82	10	262.547	16	614.914	46	1107.59	41
199.04	9	262.621	16	628.650	60	1107.93	41
200.676	8	283.6	220	628.660	60	1107.98	41
203.054	7	289.141	15	628.692	60	1109.0	85
203.055	7	289.228	15	660.850	45	1118.25	84
205.0	230	289.231	15	660.983	45	1118.41	84
205.1	229	296.856	14	660.988	45	1128	250
205.2	228	296.951	14	672.342	44	1135.6	102
206.6	227	300.4	233	672.485	44	1136.63	94
206.633	6	304.6	232	684.866	31	1136.64	94
206.634	6	312.420	2	684.900	31	1136.68	94
208.5	226	312.451	2	685.405	59	1155	251
208.8	240	314.4	219	685.424	59	1168.9	56
208.9	239	337.9	231	685.455	59	1184.7	83
212.414	5	384.031	13	726.9	241	1198.40	55
212.416	5	384.174	13	770.190	43	1198.55	55
218.6	238	384.190	13	770.367	43	1198.59	55
218.9	237	419.525	12	770.377	43	1199.91	82
222.785	4	419.714	12	776.3	242	1200.10	82
222.789	4	487.67	38	797.967	42	1210.61	69
224.4	225	493.43	37	798.169	42	1210.65	69
225.50	28	501.09	36	799.66	58	1215.1	101
226.69	27	511.596	35	806.562	57	1217.39	93
228.29	26	511.599	35	806.608	57	1217.41	93
230.45	25	527.342	34	806.630	57	1217.45	93
230.812	24	527.346	34	938.46	74	1230.04	40
230.870	24	532.07	54	943.8	243	1230.52	40
231.0	224	538.72	53	948.090	30	1315.8	81
233.55	23	546.39	65	948.208	30	1325	218
234.101	22	547.86	52	960.04	73	1344.18	80
234.160	22	552.176	33	966.5	244	1344.41	80
238.23	21	552.183	33	989.47	72	1353.4	100
239.107	20	553.64	64	1003.0	88	1358.42	92
239.169	20	560.48	51	1007	245	1358.47	92
239.3	236	562.789	50	1013	246	1358.50	92
240.2	235	562.889	50	1024.8	105	1440.28	68
244.903	3	563.30	63	1024.9	97	1440.36	68
244.911	3	576.582	62	1026.9	87	1548.19	1
245.769	19	576.586	62	1029	247	1550.77	1
245.833	19	576.617	62	1031.27	71	1585.81	79
245.834	19	579.18	49	1031.28	71	1586.11	79
247.341	18	582.748	48	1049.7	104	1586.14	79
247.407	18	582.856	48	1050.7	96	1599.0	114
247.7	234	595.487	32	1060.7	86	1641.0	99
254.4	223	595.500	32	1073	248	1653.63	78
256.4	222	596.660	61	1083	249	1653.99	78
259.468	17	596.666	61	1085.0	103	1654.46	91

List of tabulated lines — Continued

Wavelength (Å)	No.	Wavelength (Å)	No.	Wavelength (Å)	No.	Wavenumber (cm^{-1})	No.
1654.56	91	2907.7	135	6875.2	174	112	217
1654.57	91	2928.3	162	7143.6	179	176	202
1662.7	113	2934.91	128	7213.4	186	186	214
1690.4	126	2935.06	128	7250.53	173	186	209
1721.8	140	2935.12	128	7251.79	173	250	188
1723.6	133	2953.33	118	7381.9	150	371	170
1753.0	112	2953.95	118	7563.8	192	577.9	198
1759.5	125	2972.8	145	7821.7	196	579.0	198
1793.5	139	3141.7	155	7946.62	158	591.4	148
1797.9	132	3220.6	161	7947.38	158	592.6	148
1860.9	124	3385.22	144	7948.14	158	595.3	148
1888.67	111	3385.35	144	8379.22	149	826.9	183
1888.71	111	3607.3	154	8381.96	149	828.5	183
1899.0	138	3707.91	153	9290.1	185	1025.3	115
1903.9	131	3708.45	153	9830.8	201	1027.0	115
		3710.15	160	9879.5	191	1032.4	115
in air		3710.30	160	10229.2	165	1242.4	164
		3710.32	160	10230.8	165	1244.8	164
2014.2	123	3934.28	108	10303	205	1991.0	141
2045.13	122	3934.89	108	10324	195	1994.9	141
2045.43	122	3967.1	169	10618	208		
2059.0	137	4153.7	177	11388	172		
2064.28	130	4219.06	143	12254	178		
2064.33	130	4219.34	143	12861	200		
2064.36	130	4238.7	182	12979.4	171		
2103.94	67	4384.0	168	12983.5	171		
2104.25	67	4440.34	117	13545	204		
2121.98	110	4441.50	117	14243	207		
2122.06	110	4441.74	117	14330.7	66		
2278.9	121	4554.3	152	14357.7	66		
2336.01	120	4597.3	176	14999.9	184		
2336.3	136	4664.8	134	15002.4	184		
2336.39	120	4718.0	181	16628	190		
2347.19	129	4735.99	159	17833	211		
2347.28	129	4736.26	159	17894	194		
2404.44	77	4736.35	159	18834	213		
2405.10	77	4785.48	127	19009.8	189		
2405.20	77	4785.87	127	19015.6	189		
2471	252	4786.37	127	20699.6	39		
2524.4	90	4788.90	151	20790.4	39		
2530.0	106	4789.79	151	20835.9	39		
2533.8	98	5016.62	116	21383	199		
2555.3	147	5018.40	116	23342	203		
2591.09	109	5073.2	167	25497	206		
2591.25	109	5361.0	175	28609.4	107		
2594.82	89	5526.0	180	28667.7	107		
2595.09	89	5801.31	29	30442	210		
2595.30	89	5811.97	29	34331	212		
2684.5	157	6150.2	187	39862	215		
2697.72	76	6404.50	166	45237	216		
2698.67	76	6404.95	166	49369.3	75		
2722.1	146	6433.5	193	49592.1	75		
2736.3	163	6586.9	197	49690.7	75		
2819.2	119	6591.53	142				
2863.0	156	6592.57	142				

Our tables are mainly based on the extensive results of the Opacity Project (OP), which is reviewed in the general introduction. For C IV, the OP calculations were carried out by Peach *et al.*[1]

A few other comprehensive, but less sophisticated calculations exist.[2,3] For C IV—an ion with only three electrons, two of which are normally in the closed $1s$ shell—these simpler calculational approaches, such as the single-configuration self-consistent field approximation (calculations by Biemont[2]) and the semi-empirical Coulomb approximation (calculations by Lindgard and Nielsen[3]) also produce fairly accurate results. They are indeed in very good agreement with the OP results, typically within ±10%, and have been used for transitions involving higher angular momentum quantum numbers for which no OP data are available.

A critical compilation[4] performed in 1976 contains data for about 40 multiplets, which were critically selected from the then available literature. The OP data agree with these earlier results normally within a few percent. But for some higher $ms-np$ transitions (m=2–4 and n=4–8), differences up to ±20% are encountered.

A lifetime measurement by Baudinet-Robinet *et al.*[5] for the $3s\ ^2S_{1/2}$ level, which was selectively excited using the beam-foil-laser technique, has yielded the value $\tau = 0.21 \pm 0.02\ ns$, close to the OP result of $0.234\ ns$. The ratio of these two data sources is shown in a graphical comparison of lifetime data and calculated results for the spectra of several carbon ions in Fig. 1 of the C II introduction.

Several calculations and measurements have also been carried out for transitions between doubly excited quartet levels. These levels have energies above the first ionization potential, but are metastable against autoionization. We have selected the results of Laughlin[6] which were calculated with the model-potential method. These calculations should be quite reliable, which is also indicated by the close agreement (< 8%) with beam-foil lifetime data.[7,8]

References

1. G. Peach, H. E. Saraph, and M. J. Seaton, J. Phys. B **21**, 3669 (1988).
2. E. Biémont, Astron. Astrophys., Suppl. Ser. **27**, 489 (1977).
3. A. Lindgard and S. E. Nielsen, At. Data Nucl. Data Tables **19**, 533 (1977).
4. G. A. Martin and W. L. Wiese, J. Phys. Chem. Ref. Data **5**, 537 (1976).
5. Y. Baudinet-Robinet, P.-D. Dumont, H. P. Garnir, and A. El Himdy, Phys. Rev. A **40**, 6321 (1989).
6. C. Laughlin, Z. Phys. D **9**, 273 (1988).
7. J. H. Blanke, P. H. Heckmann, and E. Träbert, Phys. Scr. **32**, 509 (1985).
8. B. F. Davis and K. T. Chung, Phys. Rev. A **37**, 111 (1988).

C IV: Allowed Transitions

No.	Transition Array	Mult.	λ_{air} (Å)	λ_{vac} (Å) or σ (cm^{-1})*	E_i (cm^{-1})	E_k (cm^{-1})	g_i-g_k	A_{ki} (10^8 s^{-1})	f_{ik}	S (at. u.)	log gf	Acc.	Source
1.	$1s^22s-1s^22p$	^{2}S–^{2}P°		*1549.0*	0.0	*64556*	2–6	2.65+00	2.85–01	2.91+00	–0.243	A	1
				1548.19	0.0	64591.7	2–4	2.65+00	1.90–01	1.94+00	–0.419	A	*LS*
				1550.77	0.0	64484.0	2–2	2.64+00	9.51–02	9.71–01	–0.721	A	*LS*
2.	$1s^22s-1s^23p$	^{2}S–^{2}P°		*312.43*	0.0	*320071*	2–6	4.63+01	2.03–01	4.18–01	–0.391	A–	1
				312.420	0.0	320081.7	2–4	4.63+01	1.35–01	2.79–01	–0.567	A–	*LS*
				312.451	0.0	320050.1	2–2	4.63+01	6.77–02	1.39–01	–0.868	A–	*LS*
3.	$1s^22s-1s^24p$	^{2}S–^{2}P°		*244.91*	0.0	*408320*	2–6	2.27+01	6.13–02	9.88–02	–0.912	B+	1
				244.903	0.0	408324.2	2–4	2.27+01	4.08–02	6.59–02	–1.088	B+	*LS*
				244.911	0.0	408311.1	2–2	2.27+01	2.04–02	3.29–02	–1.389	B+	*LS*
4.	$1s^22s-1s^25p$	^{2}S–^{2}P°		*222.79*	0.0	*448861*	2–6	1.21+01	2.71–02	3.97–02	–1.267	B	1
				222.785	0.0	448862.9	2–4	1.21+01	1.80–02	2.65–02	–1.443	B	*LS*
				222.789	0.0	448855.8	2–2	1.21+01	9.02–03	1.32–02	–1.744	B	*LS*
5.	$1s^22s-1s^26p$	^{2}S–^{2}P°		*212.41*	0.0	*470778*	2–6	7.15+00	1.45–02	2.03–02	–1.537	B	1
				212.414	0.0	470778.9	2–4	7.15+00	9.68–03	1.35–02	–1.713	B	*LS*
				212.416	0.0	470775.0	2–2	7.15+00	4.84–03	6.77–03	–2.014	B	*LS*
6.	$1s^22s-1s^27p$	^{2}S–^{2}P°		*206.63*	0.0	*483950*	2–6	4.55+00	8.75–03	1.19–02	–1.757	B	1
				206.633	0.0	483950.8	2–4	4.55+00	5.83–03	7.93–03	–1.933	B	*LS*
				206.634	0.0	483948.4	2–2	4.55+00	2.92–03	3.97–03	–2.234	B	*LS*

C IV: Allowed Transitions — Continued

No.	Transition Array	Mult.	λ_{air} (Å)	λ_{vac} (Å) or σ (cm^{-1})*	E_i (cm^{-1})	E_k (cm^{-1})	g_i–g_k	A_{ki} (10^8 s^{-1})	f_{ik}	S (at. u.)	log gf	Acc.	Source
7.	$1s^22s-1s^28p$	$^2S-^2P^o$		*203.05*	0.0	– *492479*	2–6	3.07+00	5.68–03	7.60–03	–1.944	B	1
				203.054	0.0	– 492479.3	2–4	3.07+00	3.79–03	5.07–03	–2.120	B	*LS*
				203.055	0.0	– 492477.7	2–2	3.07+00	1.89–03	2.53–03	–2.421	B	*LS*
8.	$1s^22s-1s^29p$	$^2S-^2P^o$		*200.68*	0.0	– *498315*	2–6	2.11+00	3.83–03	5.05–03	–2.116	B	3
				200.676	0.0	– 498315.7	2–4	2.11+00	2.55–03	3.37–03	–2.292	B	*LS*
				200.676	0.0	– 498314.6	2–2	2.11+00	1.27–03	1.68–03	–2.593	B	*LS*
9.	$1s^22s-1s^210p$	$^2S-^2P^o$		199.04	0.0	– 502412	2–6	1.66+00	2.96–03	3.88–03	–2.228	B	3
10.	$1s^22s-1s^211p$	$^2S-^2P^o$		197.82	0.0	– 505510	2–6	1.25+00	2.20–03	2.87–03	–2.356	B	3
11.	$1s^22s-1s^212p$	$^2S-^2P^o$		196.89	0.0	– 507906	2–6	8.96–01	1.56–03	2.03–03	–2.505	B	3
12.	$1s^22p-1s^23s$	$^2P^o-^2S$		*419.65*	*64556*	– 302849.0	6–2	4.27+01	3.76–02	3.12–01	–0.647	A	1
				419.714	64591.7	– 302849.0	4–2	2.85+01	3.76–02	2.08–01	–0.823	A	*LS*
				419.525	64484.0	– 302849.0	2–2	1.42+01	3.76–02	1.04–01	–1.124	A	*LS*
13.	$1s^22p-1s^23d$	$^2P^o-^2D$		*384.13*	*64556*	– *324886*	6–10	1.76+02	6.47–01	4.91+00	0.589	A	1
				384.174	64591.7	– 324890.3	4–6	1.75+02	5.82–01	2.95+00	0.367	A	*LS*
				384.031	64484.0	– 324879.8	2–4	1.46+02	6.47–01	1.64+00	0.112	A	*LS*
				384.190	64591.7	– 324879.8	4–4	2.92+01	6.47–02	3.27–01	–0.587	A	*LS*
14.	$1s^22p-1s^24s$	$^2P^o-^2S$		*296.92*	*64556*	– 401348.1	6–2	1.59+01	7.02–03	4.12–02	–1.375	A–	1
				296.951	64591.7	– 401348.1	4–2	1.06+01	7.02–03	2.75–02	–1.551	A–	*LS*
				296.856	64484.0	– 401348.1	2–2	5.32+00	7.03–03	1.37–02	–1.852	A–	*LS*
15.	$1s^22p-1s^24d$	$^2P^o-^2D$		*289.20*	*64556*	– *410339*	6–10	5.87+01	1.23–01	7.01–01	–0.133	A	1
				289.228	64591.7	– 410340.1	4–6	5.87+01	1.10–01	4.21–01	–0.355	A	*LS*
				289.141	64484.0	– 410336.1	2–4	4.90+01	1.23–01	2.34–01	–0.610	A	*LS*
				289.231	64591.7	– 410336.1	4–4	9.78+00	1.23–02	4.67–02	–1.309	A	*LS*
16.	$1s^22p-1s^25s$	$^2P^o-^2S$		*262.60*	*64556*	– 445368.5	6–2	7.72+00	2.66–03	1.38–02	–1.797	B	1
				262.621	64591.7	– 445368.5	4–2	5.15+00	2.66–03	9.20–03	–1.973	B	*LS*
				262.547	64484.0	– 445368.5	2–2	2.57+00	2.66–03	4.60–03	–2.274	B	*LS*
17.	$1s^22p-1s^25d$	$^2P^o-^2D$		*259.52*	*64556*	– *449889*	6–10	2.73+01	4.60–02	2.36–01	–0.559	A	1
				259.539	64591.7	– 449889.9	4–6	2.73+01	4.14–02	1.41–01	–0.781	A	*LS*
				259.468	64484.0	– 449888.2	2–4	2.28+01	4.60–02	7.85–02	–1.036	A	*LS*
				259.540	64591.7	– 449888.2	4–4	4.55+00	4.60–03	1.57–02	–1.736	A	*LS*
18.	$1s^22p-1s^26s$	$^2P^o-^2S$		*247.39*	*64556*	– 468784.0	6–2	4.28+00	1.31–03	6.40–03	–2.105	B	1
				247.407	64591.7	– 468784.0	4–2	2.85+00	1.31–03	4.27–03	–2.281	B	*LS*
				247.341	64484.0	– 468784.0	2–2	1.43+00	1.31–03	2.13–03	–2.582	B	*LS*
19.	$1s^22p-1s^26d$	$^2P^o-^2D$		*245.81*	*64556*	– *471371*	6–10	1.50+01	2.27–02	1.10–01	–0.866	B+	1
				245.833	64591.7	– 471371.5	4–6	1.50+01	2.04–02	6.61–02	–1.088	B+	*LS*
				245.769	64484.0	– 471370.3	2–4	1.25+01	2.27–02	3.67–02	–1.343	B+	*LS*
				245.834	64591.7	– 471370.3	4–4	2.50+00	2.27–03	7.35–03	–2.042	B+	*LS*

C IV: Allowed Transitions — Continued

No.	Transition Array	Mult.	λ_{air} (Å)	λ_{vac} (Å) or σ (cm^{-1})*	E_i (cm^{-1})	E_k (cm^{-1})	g_i-g_k	A_{ki} (10^8 s^{-1})	f_{ik}	S (at. u.)	log gf	Acc.	Source
20.	$1s^22p-$ $1s^27s$	$^2P^o-^2S$		*239.15*	*64556*	– 482706.0	6–2	2.60+00	7.42–04	3.51–03	–2.351	B	2
				239.169	64591.7	– 482706.0	4–2	1.73+00	7.42–04	2.34–03	–2.528	B	*LS*
				239.107	64484.0	– 482706.0	2–2	8.66–01	7.42–04	1.17–03	–2.828	B	*LS*
21.	$1s^22p-$ $1s^27d$	$^2P^o-^2D$		238.23	64556	– 484321	6–10	9.27+00	1.31–02	6.19–02	–1.103	B	2
22.	$1s^22p-$ $1s^28s$	$^2P^o-^2S$		*234.14*	*64556*	– 491650.8	6–2	4.84+01	1.33–02	6.13–02	–1.099	B	1
				234.160	64591.7	– 491650.8	4–2	3.22+01	1.33–02	4.09–02	–1.276	B	*LS*
				234.101	64484.0	– 491650.8	2–2	1.61+01	1.33–02	2.04–02	–1.577	B	*LS*
23.	$1s^22p-$ $1s^28d$	$^2P^o-^2D$		233.55	64556	– 492729	6–10	6.04+00	8.24–03	3.80–02	–1.306	B	1
24.	$1s^22p-$ $1s^29s$	$^2P^o-^2S$		*230.85*	*64556*	– 497736.7	6–2	1.17+00	3.12–04	1.42–03	–2.728	B	3
				230.870	64591.7	– 497736.7	4–2	7.80–01	3.12–04	9.48–04	–2.904	B	*LS*
				230.812	64484.0	– 497736.7	2–2	3.90–01	3.12–04	4.74–04	–3.205	B	*LS*
25.	$1s^22p-$ $1s^29d$	$^2P^o-^2D$		230.45	64556	– 498491	6–10	4.22+00	5.61–03	2.55–02	–1.473	B	3
26.	$1s^22p-$ $1s^210d$	$^2P^o-^2D$		228.29	64556	– 502598	6–10	3.03+00	3.95–03	1.78–02	–1.626	B	3
27.	$1s^22p-$ $1s^211d$	$^2P^o-^2D$		226.69	64556	– 505696	6–10	2.34+00	3.01–03	1.35–02	–1.744	B	3
28.	$1s^22p-$ $1s^212d$	$^2P^o-^2D$		225.50	64556	– 508018	6–10	1.80+00	2.29–03	1.02–02	–1.863	B	3
29.	$1s^23s-$ $1s^23p$	$^2S-^2P^o$	*5804.8*	*5806.5*	302849.0	– *320071*	2–6	3.17–01	4.80–01	1.83+01	–0.018	A	1
			5801.31	5802.92	302849.0	– 320081.7	2–4	3.17–01	3.20–01	1.22+01	–0.194	A	*LS*
			5811.97	5813.58	302849.0	– 320050.1	2–2	3.16–01	1.60–01	6.12+00	–0.495	A	*LS*
30.	$1s^23s-$ $1s^24p$	$^2S-^2P^o$		*948.13*	302849.0	– *408320*	2–6	5.03+00	2.04–01	1.27+00	–0.390	B	1
				948.090	302849.0	– 408324.2	2–4	5.04+00	1.36–01	8.47–01	–0.566	B	*LS*
				948.208	302849.0	– 408311.1	2–2	5.03+00	6.79–02	4.24–01	–0.867	B	*LS*
31.	$1s^23s-$ $1s^25p$	$^2S-^2P^o$		*684.88*	302849.0	– *448861*	2–6	3.11+00	6.56–02	2.96–01	–0.882	B	1
				684.866	302849.0	– 448862.9	2–4	3.11+00	4.37–02	1.97–01	–1.058	B	*LS*
				684.900	302849.0	– 448855.8	2–2	3.11+00	2.19–02	9.85–02	–1.360	B	*LS*
32.	$1s^23s-$ $1s^26p$	$^2S-^2P^o$		*595.49*	302849.0	– *470778*	2–6	1.91+00	3.05–02	1.19–01	–1.215	B	1
				595.487	302849.0	– 470778.9	2–4	1.91+00	2.03–02	7.96–02	–1.391	B	*LS*
				595.500	302849.0	– 470775.0	2–2	1.91+00	1.02–02	3.98–02	–1.692	B	*LS*
33.	$1s^23s-$ $1s^27p$	$^2S-^2P^o$		*552.18*	302849.0	– *483950*	2–6	1.24+00	1.69–02	6.16–02	–1.470	B	1
				552.176	302849.0	– 483950.8	2–4	1.24+00	1.13–02	4.11–02	–1.646	B	*LS*
				552.183	302849.0	– 483948.4	2–2	1.24+00	5.65–03	2.05–02	–1.947	B	*LS*
34.	$1s^23s-$ $1s^28p$	$^2S-^2P^o$		*527.34*	302849.0	– *492479*	2–6	8.40–01	1.05–02	3.65–02	–1.677	B	1
				527.342	302849.0	– 492479.3	2–4	8.40–01	7.01–03	2.43–02	–1.853	B	*LS*
				527.346	302849.0	– 492477.7	2–2	8.40–01	3.50–03	1.22–02	–2.154	B	*LS*

C IV: Allowed Transitions — Continued

No.	Transition Array	Mult.	λ_{air} (Å)	λ_{vac} (Å) or σ (cm^{-1})*	E_i (cm^{-1})	E_k (cm^{-1})	g_i-g_k	A_{ki} ($10^8 s^{-1}$)	f_{ik}	S (at. u.)	log gf	Acc.	Source
35.	$1s^23s-1s^29p$	$^2S-^2P^o$		*511.60*	302849.0	*498315*	2–6	5.94–01	7.00–03	2.36–02	–1.854	B	3
				511.596	302849.0	498315.7	2–4	5.94–01	4.66–03	1.57–02	–2.030	B	*LS*
				511.599	302849.0	498314.6	2–2	5.94–01	2.33–03	7.85–03	–2.331	B	*LS*
36.	$1s^23s-1s^210p$	$^2S-^2P^o$		501.09	302849.0	502412	2–6	4.78–01	5.40–03	1.78–02	–1.967	B	3
37.	$1s^23s-1s^211p$	$^2S-^2P^o$		493.43	302849.0	505510	2–6	3.61–01	3.95–03	1.28–02	–2.102	B	3
38.	$1s^23s-1s^212p$	$^2S-^2P^o$		487.67	302849.0	507906	2–6	2.54–01	2.72–03	8.74–03	–2.264	B	3
39.	$1s^23p-1s^23d$	$^2P^o-^2D$	*20763*	*4815* cm^{-1}	*320071*	*324886*	6–10	5.81–03	6.26–02	2.57+01	–0.425	A	1
			20790.4	4808.6 cm^{-1}	320081.7	324890.3	4–6	5.79–03	5.62–02	1.54+01	–0.648	A	*LS*
			20699.6	4829.7 cm^{-1}	320050.1	324879.8	2–4	4.89–03	6.28–02	8.56+00	–0.901	A	*LS*
			20835.9	4798.1 cm^{-1}	320081.7	324879.8	4–4	9.58–04	6.24–03	1.71+00	–1.603	A	*LS*
40.	$1s^23p-1s^24s$	$^2P^o-^2S$		*1230.4*	*320071*	401348.1	6–2	1.08+01	8.15–02	1.98+00	–0.311	A	1
				1230.52	320081.7	401348.1	4–2	7.18+00	8.14–02	1.32+00	–0.487	A	*LS*
				1230.04	320050.1	401348.1	2–2	3.59+00	8.15–02	6.60–01	–0.788	A	*LS*
41.	$1s^23p-1s^24d$	$^2P^o-^2D$		*1107.8*	*320071*	*410339*	6–10	1.76+01	5.41–01	1.18+01	0.511	A	1
				1107.93	320081.7	410340.1	4–6	1.76+01	4.86–01	7.10+00	0.289	A	*LS*
				1107.59	320050.1	410336.1	2–4	1.47+01	5.41–01	3.94+00	0.034	A	*LS*
				1107.98	320081.7	410336.1	4–4	2.94+00	5.40–02	7.89–01	–0.665	A	*LS*
42.	$1s^23p-1s^25s$	$^2P^o-^2S$		*798.10*	*320071*	445368.5	6–2	4.89+00	1.56–02	2.46–01	–1.029	A	1
				798.169	320081.7	445368.5	4–2	3.26+00	1.56–02	1.64–01	–1.206	A	*LS*
				797.967	320050.1	445368.5	2–2	1.63+00	1.56–02	8.18–02	–1.507	A	*LS*
43.	$1s^23p-1s^25d$	$^2P^o-^2D$		*770.31*	*320071*	*449889*	6–10	8.91+00	1.32–01	2.01+00	–0.101	A	1
				770.367	320081.7	449889.9	4–6	8.91+00	1.19–01	1.21+00	–0.323	A	*LS*
				770.190	320050.1	449888.2	2–4	7.43+00	1.32–01	6.70–01	–0.578	A	*LS*
				770.377	320081.7	449888.2	4–4	1.48+00	1.32–02	1.34–01	–1.277	A	*LS*
44.	$1s^23p-1s^26s$	$^2P^o-^2S$		*672.44*	*320071*	468784.0	6–2	2.66+00	6.02–03	7.99–02	–1.443	A–	1
				672.485	320081.7	468784.0	4–2	1.77+00	6.02–03	5.33–02	–1.619	A–	*LS*
				672.342	320050.1	468784.0	2–2	8.88–01	6.02–03	2.66–02	–1.920	A–	*LS*
45.	$1s^23p-1s^26d$	$^2P^o-^2D$		*660.94*	*320071*	*471371*	6–10	5.03+00	5.49–02	7.16–01	–0.483	A	1
				660.983	320081.7	471371.5	4–6	5.03+00	4.94–02	4.30–01	–0.704	A	*LS*
				660.850	320050.1	471370.3	2–4	4.19+00	5.49–02	2.39–01	–0.960	A	*LS*
				660.988	320081.7	471370.3	4–4	8.38–01	5.49–03	4.78–02	–1.659	A	*LS*
46.	$1s^23p-1s^27s$	$^2P^o-^2S$		*614.87*	*320071*	482706.0	6–2	1.62+00	3.05–03	3.71–02	–1.737	B+	1
				614.914	320081.7	482706.0	4–2	1.08+00	3.05–03	2.47–02	–1.913	B+	*LS*
				614.795	320050.1	482706.0	2–2	5.39–01	3.06–03	1.24–02	–2.214	B+	*LS*
47.	$1s^23p-1s^27d$	$^2P^o-^2D$		608.83	320071	484321	6–10	3.11+00	2.88–02	3.46–01	–0.763	A	1

C IV: Allowed Transitions — Continued

No.	Transition Array	Mult.	λ_{air} (Å)	λ_{vac} (Å) or σ (cm^{-1})*	E_i (cm^{-1})	E_k (cm^{-1})	g_i-g_k	A_{ki} ($10^8 s^{-1}$)	f_{ik}	S (at. u.)	log gf	Acc.	Source
48.	$1s^23p-1s^28s$	$^2P^o-^2S$		*582.82*	*320071*	– *491650.8*	6–2	1.05+00	1.79–03	2.06–02	–1.969	B+	1
				582.856	320081.7	– 491650.8	4–2	7.03–01	1.79–03	1.37–02	–2.145	B+	*LS*
				582.748	320050.1	– 491650.8	2–2	3.52–01	1.79–03	6.87–03	–2.446	B+	*LS*
49.	$1s^23p-1s^28d$	$^2P^o-^2D$		579.18	320071	– 492729	6–10	2.06+00	1.72–02	1.97–01	–0.986	A–	1
50.	$1s^23p-1s^29s$	$^2P^o-^2S$		*562.86*	*320071*	– *497736.7*	6–2	7.26–01	1.15–03	1.28–02	–2.162	B	3
				562.889	320081.7	– 497736.7	4–2	4.84–01	1.15–03	8.52–03	–2.338	B	*LS*
				562.789	320050.1	– 497736.7	2–2	2.42–01	1.15–03	4.26–03	–2.639	B	*LS*
51.	$1s^23p-1s^29d$	$^2P^o-^2D$		560.48	320071	– 498491	6–10	1.43+00	1.12–02	1.24–01	–1.172	B	3
52.	$1s^23p-1s^210d$	$^2P^o-^2D$		547.86	320071	– 502598	6–10	1.03+00	7.76–03	8.40–02	–1.332	B	3
53.	$1s^23p-1s^211d$	$^2P^o-^2D$		538.72	320071	– 505696	6–10	7.77–01	5.63–03	5.99–02	–1.471	B	3
54.	$1s^23p-1s^212d$	$^2P^o-^2D$		532.07	320071	– 508018	6–10	5.94–01	4.20–03	4.42–02	–1.598	B	3
55.	$1s^23d-1s^24p$	$^2D-^2P^o$		*1198.6*	*324886*	– *408320*	10–6	1.28+00	1.66–02	6.54–01	–0.780	B+	1
				1198.55	324890.3	– 408324.2	6–4	1.15+00	1.66–02	3.93–01	–1.002	B+	*LS*
				1198.59	324879.8	– 408311.1	4–2	1.28+00	1.38–02	2.18–01	–1.257	B+	*LS*
				1198.40	324879.8	– 408324.2	4–4	1.28–01	2.76–03	4.36–02	–1.956	B+	*LS*
56.	$1s^23d-1s^24f$	$^2D-^2F^o$		1168.9	324886	– 410434	10–14	3.54+01	1.02+00	3.91+01	1.007	B	2
57.	$1s^23d-1s^25p$	$^2D-^2P^o$		*806.62*	*324886*	– *448861*	10–6	5.47–01	3.20–03	8.50–02	–1.495	B+	1
				806.630	324890.3	– 448862.9	6–4	4.92–01	3.20–03	5.10–02	–1.717	B+	*LS*
				806.608	324879.8	– 448855.8	4–2	5.47–01	2.67–03	2.83–02	–1.972	B+	*LS*
				806.562	324879.8	– 448862.9	4–4	5.47–02	5.34–04	5.67–03	–2.671	B+	*LS*
58.	$1s^23d-1s^25f$	$^2D-^2F^o$		799.66	324886	– 449940	10–14	1.17+01	1.57–01	4.13+00	0.195	B	2
59.	$1s^23d-1s^26p$	$^2D-^2P^o$		*685.44*	*324886*	– *470778*	10–6	2.86–01	1.21–03	2.73–02	–1.917	B	1
				685.455	324890.3	– 470778.9	6–4	2.58–01	1.21–03	1.64–02	–2.139	B	*LS*
				685.424	324879.8	– 470775.0	4–2	2.86–01	1.01–03	9.10–03	–2.394	B	*LS*
				685.405	324879.8	– 470778.9	4–4	2.86–02	2.02–04	1.82–03	–3.093	B	*LS*
60.	$1s^23d-1s^27p$	$^2D-^2P^o$		*628.68*	*324886*	– *483950*	10–6	1.70–01	6.04–04	1.25–02	–2.219	B–	1
				628.692	324890.3	– 483950.8	6–4	1.53–01	6.04–04	7.50–03	–2.441	B–	*LS*
				628.660	324879.8	– 483948.4	4–2	1.70–01	5.03–04	4.17–03	–2.696	B–	*LS*
				628.650	324879.8	– 483950.8	4–4	1.70–02	1.01–04	8.33–04	–3.395	B–	*LS*
61.	$1s^23d-1s^28p$	$^2D-^2P^o$		*596.68*	*324886*	– *492479*	10–6	1.10–01	3.51–04	6.90–03	–2.454	B–	1
				596.698	324890.3	– 492479.3	6–4	9.87–02	3.51–04	4.14–03	–2.676	B–	*LS*
				596.666	324879.8	– 492477.7	4–2	1.10–01	2.93–04	2.30–03	–2.931	B–	*LS*
				596.660	324879.8	– 492479.3	4–4	1.10–02	5.85–05	4.60–04	–3.630	B–	*LS*

C IV: Allowed Transitions — Continued

No.	Transition Array	Mult.	λ_{air} (Å)	λ_{vac} (Å) or σ (cm^{-1})*	E_i (cm^{-1})	E_k (cm^{-1})	g_i-g_k	A_{ki} (10^8 s^{-1})	f_{ik}	S (at. u.)	log gf	Acc.	Source
62.	$1s^23d-1s^29p$	$^2D-^2P^o$		*576.60*	*324886*	– *498315*	10–6	7.58–02	2.27–04	4.31–03	–2.644	B	3
				576.617	324890.3	– 498315.7	6–4	6.82–02	2.27–04	2.58–03	–2.866	B	*LS*
				576.586	324879.8	– 498314.6	4–2	7.58–02	1.89–04	1.44–03	–3.121	B	*LS*
				576.582	324879.8	– 498315.7	4–4	7.58–03	3.78–05	2.87–04	–3.820	B	*LS*
63.	$1s^23d-1s^210p$	$^2D-^2P^o$		563.30	324886	– 502412	10–6	6.41–02	1.83–04	3.39–03	–2.738	B	3
64.	$1s^23d-1s^211p$	$^2D-^2P^o$		553.64	324886	– 505510	10–6	4.77–02	1.32–04	2.40–03	–2.881	B	3
65.	$1s^23d-1s^212p$	$^2D-^2P^o$		546.39	324886	– 507906	10–6	3.07–02	8.25–05	1.48–03	–3.084	B	3
66.	$1s^24s-1s^24p$	$^2S-^2P^o$	*14340*	*6972* cm^{-1}	401348.1	– *408320*	2–6	7.14–02	6.60–01	6.23+01	0.121	A–	1
			14330.7	6976.1 cm^{-1}	401348.1	– 408324.2	2–4	7.15–02	4.40–01	4.15+01	–0.055	A–	*LS*
			14357.7	6963.0 cm^{-1}	401348.1	– 408311.1	2–2	7.11–02	2.20–01	2.08+01	–0.357	A–	*LS*
67.	$1s^24s-1s^25p$	$^2S-^2P^o$	*2104.0*	*2104.7*	401348.1	– *448861*	2–6	1.08+00	2.15–01	2.97+00	–0.367	B+	1
			2103.94	2104.61	401348.1	– 448862.9	2–4	1.08+00	1.43–01	1.98+00	–0.543	B+	*LS*
			2104.25	2104.92	401348.1	– 448855.8	2–2	1.08+00	7.15–02	9.91–01	–0.845	B+	*LS*
68.	$1s^24s-1s^26p$	$^2S-^2P^o$		*1440.3*	401348.1	– *470778*	2–6	7.56–01	7.05–02	6.69–01	–0.850	B	1
				1440.28	401348.1	– 470778.9	2–4	7.56–01	4.70–02	4.46–01	–1.027	B	*LS*
				1440.36	401348.1	– 470775.0	2–2	7.56–01	2.35–02	2.23–01	–1.328	B	*LS*
69.	$1s^24s-1s^27p$	$^2S-^2P^o$		*1210.6*	401348.1	– *483950*	2–6	5.08–01	3.35–02	2.67–01	–1.174	B	1
				1210.61	401348.1	– 483950.8	2–4	5.08–01	2.23–02	1.78–01	–1.351	B	*LS*
				1210.65	401348.1	– 483948.4	2–2	5.08–01	1.12–02	8.89–02	–1.652	B	*LS*
70.	$1s^24s-1s^28p$	$^2S-^2P^o$		*1097.3*	401348.1	– *492479*	2–6	3.51–01	1.90–02	1.37–01	–1.420	B–	1
				1097.32	401348.1	– 492479.3	2–4	3.51–01	1.27–02	9.15–02	–1.597	B–	*LS*
				1097.34	401348.1	– 492477.7	2–2	3.51–01	6.33–03	4.57–02	–1.898	B–	*LS*
71.	$1s^24s-1s^29p$	$^2S-^2P^o$		*1031.3*	401348.1	– *498315*	2–6	2.46–01	1.18–02	7.99–02	–1.629	B	3
				1031.27	401348.1	– 498315.7	2–4	2.46–01	7.84–03	5.32–02	–1.805	B	*LS*
				1031.28	401348.1	– 498314.6	2–2	2.46–01	3.92–03	2.66–02	–2.106	B	*LS*
72.	$1s^24s-1s^210p$	$^2S-^2P^o$		989.47	401348.1	– 502412	2–6	2.01–01	8.84–03	5.76–02	–1.752	B	3
73.	$1s^24s-1s^211p$	$^2S-^2P^o$		960.04	401348.1	– 505510	2–6	1.52–01	6.30–03	3.98–02	–1.900	B	3
74.	$1s^24s-1s^212p$	$^2S-^2P^o$		938.46	401348.1	– 507906	2–6	1.06–01	4.19–03	2.59–02	–2.077	B	3
75.	$1s^24p-1s^24d$	$^2P^o-^2D$	*49523*	*2019* cm^{-1}	*408320*	– *410339*	6–10	1.82–03	1.11–01	1.09+02	–0.176	A	1
			49592.1	2015.9 cm^{-1}	408324.2	– 410340.1	4–6	1.81–03	1.00–01	6.53+01	–0.398	A	*LS*
			49369.3	2025.0 cm^{-1}	408311.1	– 410336.1	2–4	1.53–03	1.12–01	3.63+01	–0.651	A	*LS*
			49690.7	2011.9 cm^{-1}	408324.2	– 410336.1	4–4	3.00–04	1.11–02	7.26+00	–1.353	A	*LS*

C IV: Allowed Transitions — Continued

No.	Transition Array	Mult.	λ_{air} (Å)	λ_{vac} (Å) or σ (cm^{-1})*	E_i (cm^{-1})	E_k (cm^{-1})	g_i–g_k	A_{ki} (10^8 s^{-1})	f_{ik}	S (at. u.)	log *gf*	Acc.	Source
76.	$1s^24p-$ $1s^25s$	$^2P°-^2S$	*2698.3*	*2699.2*	*408320*	– 445368.5	6–2	3.51+00	1.28–01	6.81+00	–0.116	A	1
			2698.67	2699.47	408324.2	– 445368.5	4–2	2.34+00	1.28–01	4.54+00	–0.292	A	*LS*
			2697.72	2698.52	408311.1	– 445368.5	2–2	1.17+00	1.28–01	2.27+00	–0.593	A	*LS*
77.	$1s^24p-$ $1s^25d$	$^2P°-^2D$	*2404.9*	*2405.6*	*408320*	– *449889*	6–10	3.56+00	5.14–01	2.44+01	0.489	A	1
			2405.10	2405.83	408324.2	– 449889.9	4–6	3.56+00	4.63–01	1.47+01	0.268	A	*LS*
			2404.44	2405.17	408311.1	– 449888.2	2–4	2.97+00	5.14–01	8.14+00	0.012	A	*LS*
			2405.20	2405.93	408324.2	– 449888.2	4–4	5.93–01	5.14–02	1.63+00	–0.687	A	*LS*
78.	$1s^24p-$ $1s^26s$	$^2P°-^2S$		*1653.9*	*408320*	– 468784.0	6–2	1.80+00	2.46–02	8.03–01	–0.831	A–	1
				1653.99	408324.2	– 468784.0	4–2	1.20+00	2.46–02	5.35–01	–1.007	A–	*LS*
				1653.63	408311.1	– 468784.0	2–2	6.00–01	2.46–02	2.68–01	–1.308	A–	*LS*
79.	$1s^24p-$ $1s^26d$	$^2P°-^2D$		*1586.0*	*408320*	– *471371*	6–10	2.16+00	1.36–01	4.26+00	–0.089	A	1
				1586.11	408324.2	– 471371.5	4–6	2.16+00	1.22–01	2.55+00	–0.311	A	*LS*
				1585.81	408311.1	– 471370.3	2–4	1.80+00	1.36–01	1.42+00	–0.566	A	*LS*
				1586.14	408324.2	– 471370.3	4–4	3.60–01	1.36–02	2.84–01	–1.265	A	*LS*
80.	$1s^24p-$ $1s^27s$	$^2P°-^2S$		*1344.3*	*408320*	– 482706.0	6–2	1.06+00	9.59–03	2.55–01	–1.240	A–	1
				1344.41	408324.2	– 482706.0	4–2	7.08–01	9.59–03	1.70–01	–1.416	A–	*LS*
				1344.18	408311.1	– 482706.0	2–2	3.54–01	9.59–03	8.49–02	–1.717	A–	*LS*
81.	$1s^24p-$ $1s^27d$	$^2P°-^2D$		1315.8	408320	– 484321	6–10	1.36+00	5.90–02	1.53+00	–0.451	A–	1
82.	$1s^24p-$ $1s^28s$	$^2P°-^2S$		*1200.0*	*408320*	– 491650.8	6–2	6.83–01	4.91–03	1.17–01	–1.530	B+	1
				1200.10	408324.2	– 491650.8	4–2	4.55–01	4.91–03	7.77–02	–1.706	B+	*LS*
				1199.91	408311.1	– 491650.8	2–2	2.28–01	4.92–03	3.88–02	–2.007	B+	*LS*
83.	$1s^24p-$ $1s^28d$	$^2P°-^2D$		1184.7	408320	– 492729	6–10	9.11–01	3.19–02	7.47–01	–0.718	A–	1
84.	$1s^24p-$ $1s^29s$	$^2P°-^2S$		*1118.4*	*408320*	– 497736.7	6–2	4.55–01	2.84–03	6.28–02	–1.768	B	3
				1118.41	408324.2	– 497736.7	4–2	3.03–01	2.84–03	4.19–02	–1.944	B	*LS*
				1118.25	408311.1	– 497736.7	2–2	1.52–01	2.84–03	2.09–02	–2.245	B	*LS*
85.	$1s^24p-$ $1s^29d$	$^2P°-^2D$		1109.0	408320	– 498491	6–10	6.27–01	1.93–02	4.22–01	–0.937	B	3
86.	$1s^24p-$ $1s^210d$	$^2P°-^2D$		1060.7	408320	– 502598	6–10	4.56–01	1.28–02	2.69–01	–1.114	B	3
87.	$1s^24p-$ $1s^211d$	$^2P°-^2D$		1026.9	408320	– 505696	6–10	3.37–01	8.88–03	1.80–01	–1.274	B	3
88.	$1s^24p-$ $1s^212d$	$^2P°-^2D$		1003.0	408320	– 508018	6–10	2.57–01	6.45–03	1.28–01	–1.412	B	3
89.	$1s^24d-$ $1s^25p$	$^2D-^2P°$	*2595.1*	*2595.9*	*410339*	– *448861*	10–6	6.67–01	4.04–02	3.45+00	–0.393	A–	1
			2595.09	2595.87	410340.1	– 448862.9	6–4	6.01–01	4.04–02	2.07+00	–0.615	A–	*LS*
			2595.30	2596.07	410336.1	– 448855.8	4–2	6.67–01	3.37–02	1.15+00	–0.870	A–	*LS*
			2594.82	2595.60	410336.1	– 448862.9	4–4	6.68–02	6.74–03	2.30–01	–1.569	A–	*LS*

C IV: Allowed Transitions — Continued

No.	Transition Array	Mult.	λ_{air} (Å)	λ_{vac} (Å) or σ (cm^{-1})*	E_i (cm^{-1})	E_k (cm^{-1})	g_i–g_k	A_{ki} (10^8 s^{-1})	f_{ik}	S (at. u.)	log gf	Acc.	Source
90.	$1s^24d$– $1s^25f$	^{2}D–^{2}F°	2524.4	2525.2	410339	– 449940	10–14	6.63+00	8.88–01	7.38+01	0.948	B	2
91.	$1s^24d$– $1s^26p$	^{2}D–^{2}P°		*1654.6*	*410339*	– *470778*	10–6	3.29–01	8.11–03	4.42–01	–1.091	B+	1
				1654.57	410340.1	– 470778.9	6–4	2.96–01	8.11–03	2.65–01	–1.313	B+	*LS*
				1654.56	410336.1	– 470775.0	4–2	3.29–01	6.76–03	1.47–01	–1.568	B+	*LS*
				1654.46	410336.1	– 470778.9	4–4	3.29–02	1.35–03	2.94–02	–2.267	B+	*LS*
92.	$1s^24d$– $1s^27p$	^{2}D–^{2}P°		*1358.5*	*410339*	– *483950*	10–6	1.88–01	3.12–03	1.40–01	–1.505	B	1
				1358.50	410340.1	– 483950.8	6–4	1.69–01	3.12–03	8.38–02	–1.727	B	*LS*
				1358.47	410336.1	– 483948.4	4–2	1.88–01	2.60–03	4.66–02	–1.982	B	*LS*
				1358.42	410336.1	– 483950.8	4–4	1.88–02	5.21–04	9.31–03	–2.681	B	*LS*
93.	$1s^24d$– $1s^28p$	^{2}D–^{2}P°		*1217.4*	*410339*	– *492479*	10–6	1.18–01	1.58–03	6.33–02	–1.802	B–	1
				1217.45	410340.1	– 492479.3	6–4	1.07–01	1.58–03	3.80–02	–2.023	B–	*LS*
				1217.41	410336.1	– 492477.7	4–2	1.18–01	1.32–03	2.11–02	–2.279	B–	*LS*
				1217.39	410336.1	– 492479.3	4–4	1.18–02	2.63–04	4.22–03	–2.978	B–	*LS*
94.	$1s^24d$– $1s^29p$	^{2}D–^{2}P°		*1136.7*	*410339*	– *498315*	10–6	7.78–02	9.05–04	3.39–02	–2.044	B	3
				1136.68	410340.1	– 498315.7	6–4	7.01–02	9.05–04	2.03–02	–2.265	B	*LS*
				1136.64	410336.1	– 498314.6	4–2	7.78–02	7.54–04	1.13–02	–2.521	B	*LS*
				1136.63	410336.1	– 498315.7	4–4	7.78–03	1.51–04	2.26–03	–3.220	B	*LS*
95.	$1s^24d$– $1s^210p$	^{2}D–^{2}P°		1086.1	410339	– 502412	10–6	6.36–02	6.75–04	2.41–02	–2.171	B	3
96.	$1s^24d$– $1s^211p$	^{2}D–^{2}P°		1050.7	410339	– 505510	10–6	4.67–02	4.64–04	1.61–02	–2.333	B	3
97.	$1s^24d$– $1s^212p$	^{2}D–^{2}P°		1024.9	410339	– 507906	10–6	3.03–02	2.86–04	9.66–03	–2.543	B	3
98.	$1s^24f$– $1s^25d$	^{2}F°–^{2}D	2533.8	2534.5	410434	– 449889	14–10	1.31–01	9.01–03	1.05+00	–0.899	B	2
99.	$1s^24f$– $1s^26d$	^{2}F°–^{2}D		1641.0	410434	– 471371	14–10	5.57–02	1.61–03	1.22–01	–1.648	B	2
100.	$1s^24f$– $1s^27d$	^{2}F°–^{2}D		1353.4	410434	– 484321	14–10	2.92–02	5.73–04	3.57–02	–2.096	B	2
101.	$1s^24f$– $1s^28d$	^{2}F°–^{2}D		1215.1	410434	– 492729	14–10	1.74–02	2.76–04	1.54–02	–2.413	B	2
102.	$1s^24f$– $1s^29d$	^{2}F°–^{2}D		1135.6	410434	– 498491	14–10	1.09–02	1.50–04	7.85–03	–2.678	B	3
103.	$1s^24f$– $1s^210d$	^{2}F°–^{2}D		1085.0	410434	– 502598	14–10	7.83–03	9.87–05	4.94–03	–2.859	B	3
104.	$1s^24f$– $1s^211d$	^{2}F°–^{2}D		1049.7	410434	– 505696	14–10	4.45–03	5.25–05	2.54–03	–3.133	B	3
105.	$1s^24f$– $1s^212d$	^{2}F°–^{2}D		1024.8	410434	– 508018	14–10	3.04–03	3.42–05	1.62–03	–3.319	B	3

C IV: Allowed Transitions — Continued

No.	Transition Array	Mult.	λ_{air} (Å)	λ_{vac} (Å) or σ (cm^{-1})*	E_i (cm^{-1})	E_k (cm^{-1})	g_i–g_k	A_{ki} (10^8 s^{-1})	f_{ik}	S (at. u.)	log gf	Acc.	Source
106.	$1s^24f$–$1s^25g$	$^2F^o$–^2G	2530.0	2530.8	410434	– 449948	14–18	1.09+01	1.35+00	1.57+02	1.275	B	3
107.	$1s^25s$–$1s^25p$	^2S–$^2P^o$	*28629*	*3492* cm^{-1}	445368.5	– *448861*	2–6	2.26–02	8.35–01	1.57+02	0.223	A	1
			28609.4	3494.4 cm^{-1}	445368.5	– 448862.9	2–4	2.27–02	5.57–01	1.05+02	0.047	A	*LS*
			28667.7	3487.3 cm^{-1}	445368.5	– 448855.8	2–2	2.26–02	2.78–01	5.25+01	–0.255	A	*LS*
108.	$1s^25s$–$1s^26p$	^2S–$^2P^o$	*3934.5*	*3935.6*	445368.5	– *470778*	2–6	3.30–01	2.30–01	5.95+00	–0.338	A–	1
			3934.28	3935.40	445368.5	– 470778.9	2–4	3.30–01	1.53–01	3.97+00	–0.514	A–	*LS*
			3934.89	3936.00	445368.5	– 470775.0	2–2	3.30–01	7.66–02	1.98+00	–0.815	A–	*LS*
109.	$1s^25s$–$1s^27p$	^2S–$^2P^o$	*2591.1*	*2591.9*	445368.5	– *483950*	2–6	2.52–01	7.60–02	1.30+00	–0.818	B+	1
			2591.09	2591.86	445368.5	– 483950.8	2–4	2.52–01	5.06–02	8.64–01	–0.994	B+	*LS*
			2591.25	2592.02	445368.5	– 483948.4	2–2	2.52–01	2.53–02	4.32–01	–1.296	B+	*LS*
110.	$1s^25s$–$1s^28p$	^2S–$^2P^o$	*2122.0*	*2122.7*	445368.5	– *492479*	2–6	1.80–01	3.64–02	5.08–01	–1.138	B	1
			2121.98	2122.66	445368.5	– 492479.3	2–4	1.80–01	2.43–02	3.39–01	–1.314	B	*LS*
			2122.06	2122.73	445368.5	– 492477.7	2–2	1.80–01	1.21–02	1.69–01	–1.615	B	*LS*
111.	$1s^25s$–$1s^29p$	^2S–$^2P^o$		*1888.7*	445368.5	– *498315*	2–6	1.29–01	2.07–02	2.58–01	–1.382	B	3
				1888.67	445368.5	– 498315.7	2–4	1.29–01	1.38–02	1.72–01	–1.558	B	*LS*
				1888.71	445368.5	– 498314.6	2–2	1.29–01	6.91–03	8.60–02	–1.859	B	*LS*
112.	$1s^25s$–$1s^210p$	^2S–$^2P^o$		1753.0	445368.5	– 502412	2–6	1.09–01	1.50–02	1.73–01	–1.522	B	3
113.	$1s^25s$–$1s^211p$	^2S–$^2P^o$		1662.7	445368.5	– 505510	2–6	8.32–02	1.03–02	1.13–01	–1.684	B	3
114.	$1s^25s$–$1s^212p$	^2S–$^2P^o$		1599.0	445368.5	– 507906	2–6	5.77–02	6.63–03	6.98–02	–1.877	B	3
115.	$1s^25p$–$1s^25d$	$^2P^o$–^2D		*1029* cm^{-1}	*448861*	– *449889*	6–10	6.55–04	1.55–01	2.97+02	–0.032	A	1
				1027.0 cm^{-1}	448862.9	– 449889.9	4–6	6.52–04	1.39–01	1.78+02	–0.255	A	*LS*
				1032.4 cm^{-1}	448855.8	– 449888.2	2–4	5.52–04	1.55–01	9.90+01	–0.508	A	*LS*
				1025.3 cm^{-1}	448862.9	– 449888.2	4–4	1.08–04	1.54–02	1.98+01	–1.210	A	*LS*
116.	$1s^25p$–$1s^26s$	$^2P^o$–^2S	*5017.8*	*5019.2*	*448861*	– 468784.0	6–2	1.39+00	1.75–01	1.73+01	0.021	A	1
			5018.40	5019.80	448862.9	– 468784.0	4–2	9.27–01	1.75–01	1.16+01	–0.155	A	*LS*
			5016.62	5018.01	448855.8	– 468784.0	2–2	4.64–01	1.75–01	5.78+00	–0.456	A	*LS*
117.	$1s^25p$–$1s^26d$	$^2P^o$–^2D	*4441.1*	*4442.4*	*448861*	– *471371*	6–10	1.05+00	5.16–01	4.52+01	0.490	A	1
			4441.50	4442.75	448862.9	– 471371.5	4–6	1.05+00	4.64–01	2.71+01	0.269	A	*LS*
			4440.34	4441.58	448855.8	– 471370.3	2–4	8.72–01	5.16–01	1.51+01	0.013	A	*LS*
			4441.74	4442.98	448862.9	– 471370.3	4–4	1.74–01	5.16–02	3.02+00	–0.686	A	*LS*
118.	$1s^25p$–$1s^27s$	$^2P^o$–^2S	*2953.7*	*2954.6*	*448861*	– 482706.0	6–2	7.75–01	3.38–02	1.97+00	–0.693	A–	1
			2953.95	2954.81	448862.9	– 482706.0	4–2	5.17–01	3.38–02	1.31+00	–0.869	A–	*LS*
			2953.33	2954.19	448855.8	– 482706.0	2–2	2.58–01	3.38–02	6.57–01	–1.170	A–	*LS*
119.	$1s^25p$–$1s^27d$	$^2P^o$–^2D	2819.2	2820.1	448861	– 484321	6–10	7.07–01	1.40–01	7.82+00	–0.074	A	1

C IV: Allowed Transitions — Continued

No.	Transition Array	Mult.	λ_{air} (Å)	λ_{vac} (Å) or σ (cm^{-1})*	E_i (cm^{-1})	E_k (cm^{-1})	g_i-g_k	A_{ki} (10^8 s^{-1})	f_{ik}	S (at. u.)	log gf	Acc.	Source
120.	$1s^25p-$ $1s^28s$	$^2P^o-^2S$	*2336.3*	*2337.0*	*448861*	– 491650.8	6–2	4.85–01	1.32–02	6.11–01	–1.100	B+	1
			2336.39	2337.11	448862.9	– 491650.8	4–2	3.23–01	1.32–02	4.07–01	–1.276	B+	*LS*
			2336.01	2336.72	448855.8	– 491650.8	2–2	1.62–01	1.32–02	2.04–01	–1.577	B+	*LS*
121.	$1s^25p-$ $1s^28d$	$^2P^o-^2D$	2278.9	2279.6	448861	– 492729	6–10	4.80–01	6.23–02	2.80+00	–0.427	A–	1
122.	$1s^25p-$ $1s^29s$	$^2P^o-^2S$	*2045.3*	*2046.0*	*448861*	– 497736.7	6–2	3.25–01	6.80–03	2.75–01	–1.389	B	3
			2045.43	2046.09	448862.9	– 497736.7	4–2	2.17–01	6.80–03	1.83–01	–1.565	B	*LS*
			2045.13	2045.79	448855.8	– 497736.7	2–2	1.08–01	6.80–03	9.16–02	–1.866	B	*LS*
123.	$1s^25p-$ $1s^29d$	$^2P^o-^2D$	2014.2	2014.9	448861	– 498491	6–10	3.37–01	3.42–02	1.36+00	–0.688	B	3
124.	$1s^25p-$ $1s^210d$	$^2P^o-^2D$		1860.9	448861	– 502598	6–10	2.49–01	2.16–02	7.92–01	–0.888	B	3
125.	$1s^25p-$ $1s^211d$	$^2P^o-^2D$		1759.5	448861	– 505696	6–10	1.83–01	1.41–02	4.91–01	–1.072	B	3
126.	$1s^25p-$ $1s^212d$	$^2P^o-^2D$		1690.4	448861	– 508018	6–10	1.39–01	9.96–03	3.32–01	–1.224	B	3
127.	$1s^25d-$ $1s^26p$	$^2D-^2P^o$	*4786.0*	*4787.3*	*449889*	– *470778*	10–6	3.31–01	6.82–02	1.07+01	–0.166	A	1
			4785.87	4787.21	449889.9	– 470778.9	6–4	2.98–01	6.82–02	6.45+00	–0.388	A	*LS*
			4786.37	4787.71	449888.2	– 470775.0	4–2	3.31–01	5.68–02	3.58+00	–0.643	A	*LS*
			4785.48	4786.82	449888.2	– 470778.9	4–4	3.31–02	1.14–02	7.16–01	–1.342	A	*LS*
128.	$1s^25d-$ $1s^27p$	$^2D-^2P^o$	*2935.1*	*2935.9*	*449889*	– *483950*	10–6	1.81–01	1.40–02	1.35+00	–0.854	B+	1
			2935.06	2935.92	449889.9	– 483950.8	6–4	1.62–01	1.40–02	8.11–01	–1.076	B+	*LS*
			2935.12	2935.98	449888.2	– 483948.4	4–2	1.81–01	1.17–02	4.51–01	–1.331	B+	*LS*
			2934.91	2935.77	449888.2	– 483950.8	4–4	1.81–02	2.33–03	9.01–02	–2.030	B+	*LS*
129.	$1s^25d-$ $1s^28p$	$^2D-^2P^o$	*2347.3*	*2348.0*	*449889*	– *492479*	10–6	1.10–01	5.46–03	4.22–01	–1.263	B+	1
			2347.28	2348.00	449889.9	– 492479.3	6–4	9.91–02	5.46–03	2.53–01	–1.485	B+	*LS*
			2347.28	2348.00	449888.2	– 492477.7	4–2	1.10–01	4.55–03	1.41–01	–1.740	B+	*LS*
			2347.19	2347.91	449888.2	– 492479.3	4–4	1.10–02	9.10–04	2.81–02	–2.439	B+	*LS*
130.	$1s^25d-$ $1s^29p$	$^2D-^2P^o$	*2064.3*	*2065.0*	*449889*	– *498315*	10–6	7.28–02	2.79–03	1.90–01	–1.554	B	3
			2064.36	2065.01	449889.9	– 498315.7	6–4	6.55–02	2.79–03	1.14–01	–1.776	B	*LS*
			2064.33	2064.99	449888.2	– 498314.6	4–2	7.28–02	2.33–03	6.32–02	–2.031	B	*LS*
			2064.28	2064.94	449888.2	– 498315.7	4–4	7.28–03	4.65–04	1.26–02	–2.730	B	*LS*
131.	$1s^25d-$ $1s^210p$	$^2D-^2P^o$		1903.9	449889	– 502412	10–6	5.90–02	1.93–03	1.21–01	–1.716	B	3
132.	$1s^25d-$ $1s^211p$	$^2D-^2P^o$		1797.9	449889	– 505510	10–6	4.36–02	1.27–03	7.51–02	–1.897	B	3
133.	$1s^25d-$ $1s^212p$	$^2D-^2P^o$		1723.6	449889	– 507906	10–6	2.88–02	7.69–04	4.37–02	–2.114	B	3
134.	$1s^25f-$ $1s^26d$	$^2F^o-^2D$	4664.8	4666.1	449940	– 471371	14–10	1.01–01	2.36–02	5.08+00	–0.481	B	2

C IV: Allowed Transitions — Continued

No.	Transition Array	Mult.	λ_{air} (Å)	λ_{vac} (Å) or σ (cm^{-1})*	E_i (cm^{-1})	E_k (cm^{-1})	g_i–g_k	A_{ki} (10^8 s^{-1})	f_{ik}	S (at. u.)	log gf	Acc.	Source
135.	$1s^25f$– $1s^27d$	$^2F^o$–^2D	2907.7	2908.6	449940	– 484321	14–10	4.96–02	4.49–03	6.02–01	–1.201	B	2
136.	$1s^25f$– $1s^28d$	$^2F^o$–^2D	2336.3	2337.0	449940	– 492729	14–10	2.82–02	1.65–03	1.78–01	–1.636	B	2
137.	$1s^25f$– $1s^29d$	$^2F^o$–^2D	2059.0	2059.7	449940	– 498491	14–10	1.86–02	8.44–04	8.01–02	–1.928	B	3
138.	$1s^25f$– $1s^210d$	$^2F^o$–^2D		1899.0	449940	– 502598	14–10	1.34–02	5.19–04	4.55–02	–2.138	B	3
139.	$1s^25f$– $1s^211d$	$^2F^o$–^2D		1793.5	449940	– 505696	14–10	8.04–03	2.77–04	2.29–02	–2.411	B	3
140.	$1s^25f$– $1s^212d$	$^2F^o$–^2D		1721.8	449940	– 508018	14–10	5.65–03	1.79–04	1.42–02	–2.600	B	3
141.	$1s^26s$– $1s^26p$	^2S–$^2P^o$		*1994* cm^{-1}	468784.0	– *470778*	2–6	8.98–03	1.02+00	3.36+02	0.308	B	2
				1994.9 cm^{-1}	468784.0	– 470778.9	2–4	9.00–03	6.78–01	2.24+02	0.132	B	*LS*
				1991.0 cm^{-1}	468784.0	– 470775.0	2–2	8.94–03	3.38–01	1.12+02	–0.170	B	*LS*
142.	$1s^26s$– $1s^27p$	^2S–$^2P^o$	*6591.9*	*6593.7*	468784.0	– *483950*	2–6	1.26–01	2.47–01	1.07+01	–0.306	B+	1
			6591.53	6593.35	468784.0	– 483950.8	2–4	1.26–01	1.65–01	7.15+00	–0.482	B+	*LS*
			6592.57	6594.39	468784.0	– 483948.4	2–2	1.26–01	8.23–02	3.57+00	–0.783	B+	*LS*
143.	$1s^26s$– $1s^28p$	^2S–$^2P^o$	*4219.1*	*4220.3*	468784.0	– *492479*	2–6	1.02–01	8.17–02	2.27+00	–0.787	B	1
			4219.06	4220.25	468784.0	– 492479.3	2–4	1.02–01	5.44–02	1.51+00	–0.963	B	*LS*
			4219.34	4220.53	468784.0	– 492477.7	2–2	1.02–01	2.72–02	7.56–01	–1.264	B	*LS*
144.	$1s^26s$– $1s^29p$	^2S–$^2P^o$	*3385.3*	*3386.2*	468784.0	– *498315*	2–6	7.70–02	3.97–02	8.85–01	–1.100	B	3
			3385.22	3386.19	468784.0	– 498315.7	2–4	7.70–02	2.65–02	5.90–01	–1.276	B	*LS*
			3385.35	3386.32	468784.0	– 498314.6	2–2	7.70–02	1.32–02	2.95–01	–1.577	B	*LS*
145.	$1s^26s$– $1s^210p$	^2S–$^2P^o$	2972.8	2973.7	468784.0	– 502412	2–6	6.49–02	2.58–02	5.05–01	–1.287	B	3
146.	$1s^26s$– $1s^211p$	^2S–$^2P^o$	2722.1	2722.9	468784.0	– 505510	2–6	4.92–02	1.64–02	2.94–01	–1.484	B	3
147.	$1s^26s$– $1s^212p$	^2S–$^2P^o$	2555.3	2556.1	468784.0	– 507906	2–6	3.36–02	9.88–03	1.66–01	–1.704	B	3
148.	$1s^26p$– $1s^26d$	$^2P^o$–^2D		*593* cm^{-1}	*470778*	– *471371*	6–10	2.59–04	1.84–01	6.12+02	0.043	B	2
				592.6 cm^{-1}	470778.9	– 471371.5	4–6	2.58–04	1.65–01	3.67+02	–0.179	B	*LS*
				595.3 cm^{-1}	470775.0	– 471370.3	2–4	2.18–04	1.85–01	2.04+02	–0.433	B	*LS*
				591.4 cm^{-1}	470778.9	– 471370.3	4–4	4.28–05	1.83–02	4.08+01	–1.135	B	*LS*
149.	$1s^26p$– $1s^27s$	$^2P^o$–^2S	*8381.1*	*8383.4*	*470778*	– 482706.0	6–2	6.35–01	2.23–01	3.69+01	0.127	A	1
			8381.96	8384.27	470778.9	– 482706.0	4–2	4.23–01	2.23–01	2.46+01	–0.050	A	*LS*
			8379.22	8381.53	470775.0	– 482706.0	2–2	2.12–01	2.23–01	1.23+01	–0.351	A	*LS*
150.	$1s^26p$– $1s^27d$	$^2P^o$–^2D	7381.9	7383.9	470778	– 484321	6–10	3.89–01	5.29–01	7.72+01	0.502	A	1

C IV: Allowed Transitions — Continued

No.	Transition Array	Mult.	λ_{air} (Å)	λ_{vac} (Å) or σ (cm^{-1})*	E_i (cm^{-1})	E_k (cm^{-1})	g_i–g_k	A_{ki} (10^8 s^{-1})	f_{ik}	S (at. u.)	log gf	Acc.	Source
151.	$1s^26p$–$1s^28s$	$^2P^o$–^2S	*4789.5*	*4790.8*	*470778*	– 491650.8	6–2	3.76–01	4.31–02	4.07+00	–0.588	A	1
			4789.79	4791.13	470778.9	– 491650.8	4–2	2.50–01	4.31–02	2.72+00	–0.764	A	*LS*
			4788.90	4790.24	470775.0	– 491650.8	2–2	1.25–01	4.31–02	1.36+00	–1.065	A	*LS*
152.	$1s^26p$–$1s^28d$	$^2P^o$–^2D	4554.3	4555.6	470778	– 492729	6–10	2.82–01	1.46–01	1.31+01	–0.058	A	1
153.	$1s^26p$–$1s^29s$	$^2P^o$–^2S	*3708.3*	*3709.3*	*470778*	– 497736.7	6–2	2.47–01	1.70–02	1.24+00	–0.992	B	3
			3708.45	3709.50	470778.9	– 497736.7	4–2	1.65–01	1.70–02	8.29–01	–1.168	B	*LS*
			3707.91	3708.96	470775.0	– 497736.7	2–2	8.23–02	1.70–02	4.14–01	–1.469	B	*LS*
154.	$1s^26p$–$1s^29d$	$^2P^o$–^2D	3607.3	3608.4	470777.6	– 498491	6–10	2.01–01	6.55–02	4.67+00	–0.406	B	3
155.	$1s^26p$–$1s^210d$	$^2P^o$–^2D	3141.7	3142.6	470777.6	– 502598	6–10	1.48–01	3.65–02	2.27+00	–0.659	B	3
156.	$1s^26p$–$1s^211d$	$^2P^o$–^2D	2863.0	2863.8	470777.6	– 505696	6–10	1.07–01	2.19–02	1.24+00	–0.882	B	3
157.	$1s^26p$–$1s^212d$	$^2P^o$–^2D	2684.5	2685.3	470777.6	– 508018	6–10	8.07–02	1.45–02	7.71–01	–1.059	B	3
158.	$1s^26d$–$1s^27p$	^2D–$^2P^o$	*7947.6*	*7949.8*	*471371*	– *483950*	10–6	1.73–01	9.83–02	2.57+01	–0.007	A–	1
			7947.38	7949.57	471371.5	– 483950.8	6–4	1.56–01	9.83–02	1.54+01	–0.229	A–	*LS*
			7948.14	7950.33	471370.3	– 483948.4	4–2	1.73–01	8.19–02	8.57+00	–0.485	A–	*LS*
			7946.62	7948.81	471370.3	– 483950.8	4–4	1.73–02	1.64–02	1.71+00	–1.183	A–	*LS*
159.	$1s^26d$–$1s^28p$	^2D–$^2P^o$	*4736.3*	*4737.6*	*471371*	– *492479*	10–6	1.01–01	2.05–02	3.19+00	–0.689	B+	1
			4736.26	4737.59	471371.5	– 492479.3	6–4	9.13–02	2.05–02	1.92+00	–0.911	B+	*LS*
			4736.35	4737.67	471370.3	– 492477.7	4–2	1.01–01	1.71–02	1.06+00	–1.166	B+	*LS*
			4735.99	4737.32	471370.3	– 492479.3	4–4	1.01–02	3.41–03	2.13–01	–1.865	B+	*LS*
160.	$1s^26d$–$1s^29p$	^2D–$^2P^o$	*3710.3*	*3711.4*	*471371*	– *498315*	10–6	6.65–02	8.24–03	1.01+00	–1.084	B	3
			3710.32	3711.37	471371.5	– 498315.7	6–4	5.98–02	8.24–03	6.04–01	–1.306	B	*LS*
			3710.30	3711.36	471370.3	– 498314.6	4–2	6.65–02	6.86–03	3.35–01	–1.561	B	*LS*
			3710.15	3711.21	471370.3	– 498315.7	4–4	6.65–03	1.37–03	6.71–02	–2.260	B	*LS*
161.	$1s^26d$–$1s^210p$	^2D–$^2P^o$	3220.6	3221.5	471371	– 502412	10–6	5.08–02	4.74–03	5.03–01	–1.324	B	3
162.	$1s^26d$–$1s^211p$	^2D–$^2P^o$	2928.3	2929.2	471371	– 505510	10–6	3.62–02	2.79–03	2.69–01	–1.554	B	3
163.	$1s^26d$–$1s^212p$	^2D–$^2P^o$	2736.3	2737.1	471371	– 507906	10–6	2.35–02	1.59–03	1.43–01	–1.799	B	3
164.	$1s^27s$–$1s^27p$	^2S–$^2P^o$		*1244* cm^{-1}	482706.0	– *483950*	2–6	4.09–03	1.19+00	6.29+02	0.376	B	2
				1244.8 cm^{-1}	482706.0	– 483950.8	2–4	4.10–03	7.93–01	4.20+02	0.200	B	*LS*
				1242.4 cm^{-1}	482706.0	– 483948.4	2–2	4.08–03	3.96–01	2.10+02	–0.101	B	*LS*
165.	$1s^27s$–$1s^28p$	^2S–$^2P^o$	*10230*	*9773* cm^{-1}	482706.0	– *492479*	2–6	5.57–02	2.62–01	1.77+01	–0.280	B	2
			10229.2	9773.3 cm^{-1}	482706.0	– 492479.3	2–4	5.57–02	1.75–01	1.18+01	–0.456	B	*LS*
			10230.8	9771.7 cm^{-1}	482706.0	– 492477.7	2–2	5.57–02	8.74–02	5.89+00	–0.757	B	*LS*

C IV: Allowed Transitions — Continued

No.	Transition Array	Mult.	λ_{air} (Å)	λ_{vac} (Å) or σ (cm^{-1})*	E_i (cm^{-1})	E_k (cm^{-1})	g_i–g_k	A_{ki} (10^8 s^{-1})	f_{ik}	S (at. u.)	log gf	Acc.	Source
166.	$1s^27s$–$1s^29p$	^{2}S–^{2}P°	*6404.7*	*6406.4*	482706.0	– *498315*	2–6	4.70–02	8.67–02	3.66+00	–0.761	B	3
			6404.50	6406.27	482706.0	– 498315.7	2–4	4.70–02	5.78–02	2.44+00	–0.937	B	*LS*
			6404.95	6406.72	482706.0	– 498314.6	2–2	4.70–02	2.89–02	1.22+00	–1.238	B	*LS*
167.	$1s^27s$–$1s^210p$	^{2}S–^{2}P°	5073.2	5074.6	482706.0	– 502412	2–6	4.27–02	4.94–02	1.65+00	–1.005	B	3
168.	$1s^27s$–$1s^211p$	^{2}S–^{2}P°	4384.0	4385.2	482706.0	– 505510	2–6	3.34–02	2.89–02	8.35–01	–1.238	B	3
169.	$1s^27s$–$1s^212p$	^{2}S–^{2}P°	3967.1	3968.3	482706.0	– 507906	2–6	2.32–02	1.64–02	4.29–01	–1.484	B	3
170.	$1s^27p$–$1s^27d$	^{2}P°–^{2}D		371 cm^{-1}	483950	– 484321	6–10	1.22–04	2.21–01	1.18+03	0.123	B	2
171.	$1s^27p$–$1s^28s$	^{2}P°–^{2}S	*12982*	*7701* cm^{-1}	*483950*	– 491650.8	6–2	3.24–01	2.73–01	6.99+01	0.214	B	2
			12983.5	7700.0 cm^{-1}	483950.8	– 491650.8	4–2	2.16–01	2.73–01	4.66+01	0.038	B	*LS*
			12979.4	7702.4 cm^{-1}	483948.4	– 491650.8	2–2	1.08–01	2.73–01	2.33+01	–0.263	B	*LS*
172.	$1s^27p$–$1s^28d$	^{2}P°–^{2}D	11388	8779 cm^{-1}	483950	– 492729	6–10	1.71–01	5.54–01	1.25+02	0.521	B	2
173.	$1s^27p$–$1s^29s$	^{2}P°–^{2}S	*7251.4*	*7253.4*	*483950*	– 497736.7	6–2	1.98–01	5.21–02	7.46+00	–0.505	B	3
			7251.79	7253.79	483950.8	– 497736.7	4–2	1.32–01	5.21–02	4.97+00	–0.681	B	*LS*
			7250.53	7252.53	483948.4	– 497736.7	2–2	6.60–02	5.21–02	2.49+00	–0.982	B	*LS*
174.	$1s^27p$–$1s^29d$	^{2}P°–^{2}D	6875.2	6877.1	483950	– 498491	6–10	1.28–01	1.51–01	2.05+01	–0.044	B	3
175.	$1s^27p$–$1s^210d$	^{2}P°–^{2}D	5361.0	5362.5	483950	– 502598	6–10	9.75–02	7.00–02	7.42+00	–0.377	B	3
176.	$1s^27p$–$1s^211d$	^{2}P°–^{2}D	4597.3	4598.5	483950	– 505696	6–10	7.11–02	3.76–02	3.41+00	–0.647	B	3
177.	$1s^27p$–$1s^212d$	^{2}P°–^{2}D	4153.7	4154.9	483950	– 508018	6–10	5.46–02	2.35–02	1.93+00	–0.850	B	3
178.	$1s^27d$–$1s^28p$	^{2}D–^{2}P°	12254	8158 cm^{-1}	484321	– 492479	10–6	9.52–02	1.29–01	5.19+01	0.109	B	2
179.	$1s^27d$–$1s^29p$	^{2}D–^{2}P°	7143.6	7145.5	484321	– 498315	10–6	5.91–02	2.72–02	6.39+00	–0.566	B	3
180.	$1s^27d$–$1s^210p$	^{2}D–^{2}P°	5526.0	5527.5	484321	– 502412	10–6	4.56–02	1.25–02	2.28+00	–0.902	B	3
181.	$1s^27d$–$1s^211p$	^{2}D–^{2}P°	4718.0	4719.3	484321	– 505510	10–6	3.31–02	6.64–03	1.03+00	–1.178	B	3
182.	$1s^27d$–$1s^212p$	^{2}D–^{2}P°	4238.7	4239.9	484321	– 507906	10–6	2.20–02	3.56–03	4.97–01	–1.448	B	3
183.	$1s^28s$–$1s^28p$	^{2}S–^{2}P°		*828* cm^{-1}	491650.8	– *492479*	2–6	2.07–03	1.36+00	1.08+03	0.435	B	2
				828.5 cm^{-1}	491650.8	– 492479.3	2–4	2.08–03	9.08–01	7.22+02	0.259	B	*LS*
				826.9 cm^{-1}	491650.8	– 492477.7	2–2	2.07–03	4.53–01	3.61+02	–0.043	B	*LS*

C IV: Allowed Transitions — Continued

No.	Transition Array	Mult.	λ_{air} (Å)	λ_{vac} (Å) or σ (cm^{-1})*	E_i (cm^{-1})	E_k (cm^{-1})	g_i–g_k	A_{ki} (10^8 s^{-1})	f_{ik}	S (at. u.)	log gf	Acc.	Source
184.	$1s^28s$– $1s^29p$	^2S–$^2P^o$	*15001*	*6665* cm^{-1}	491650.8	*498315*	2–6	2.85–02	2.88–01	2.85+01	–0.239	B	3
			14999.9	6664.9 cm^{-1}	491650.8	498315.7	2–4	2.85–02	1.92–01	1.90+01	–0.415	B	*LS*
			15002.4	6663.8 cm^{-1}	491650.8	498314.6	2–2	2.85–02	9.61–02	9.49+00	–0.716	B	*LS*
185.	$1s^28s$– $1s^210p$	^2S–$^2P^o$	9290.1	9292.6	491650.8	502412	2–6	2.84–02	1.10–01	6.75+00	–0.656	B	3
186.	$1s^28s$– $1s^211p$	^2S–$^2P^o$	7213.4	7215.4	491650.8	505510	2–6	2.18–02	5.10–02	2.42+00	–0.992	B	3
187.	$1s^28s$– $1s^212p$	^2S–$^2P^o$	6150.2	6151.9	491650.8	507906	2–6	1.46–02	2.48–02	1.00+00	–1.305	B	3
188.	$1s^28p$– $1s^28d$	$^2P^o$–^2D		250 cm^{-1}	492479	492729	6–10	6.43–05	2.58–01	2.04+03	0.189	B	2
189.	$1s^28p$– $1s^29s$	$^2P^o$–^2S	*19014*	*5258* cm^{-1}	*492479*	497736.7	6–2	1.78–01	3.21–01	1.21+02	0.285	B	3
			19015.6	5257.4 cm^{-1}	492479.3	497736.7	4–2	1.18–01	3.21–01	8.04+01	0.109	B	*LS*
			19009.8	5259.0 cm^{-1}	492477.7	497736.7	2–2	5.93–02	3.21–01	4.02+01	–0.192	B	*LS*
190.	$1s^28p$– $1s^29d$	$^2P^o$–^2D	16628	6012 cm^{-1}	492479	498491	6–10	8.31–02	5.74–01	1.89+02	0.537	B	3
191.	$1s^28p$– $1s^210d$	$^2P^o$–^2D	9879.5	9882.2	492479	502598	6–10	6.58–02	1.61–01	3.13+01	–0.016	B	3
192.	$1s^28p$– $1s^211d$	$^2P^o$–^2D	7563.8	7565.9	492479	505696	6–10	4.66–02	6.66–02	9.96+00	–0.398	B	3
193.	$1s^28p$– $1s^212d$	$^2P^o$–^2D	6433.5	6435.3	492479	508018	6–10	3.51–02	3.63–02	4.61+00	–0.662	B	3
194.	$1s^28d$– $1s^29p$	^2D–$^2P^o$	17894	5587 cm^{-1}	492729	498315	10–6	5.74–02	1.65–01	9.75+01	0.219	B	3
195.	$1s^28d$– $1s^210p$	^2D–$^2P^o$	10324	9684 cm^{-1}	492729	502412	10–6	4.10–02	3.93–02	1.34+01	–0.406	B	3
196.	$1s^28d$– $1s^211p$	^2D–$^2P^o$	7821.7	7823.8	492729	505510	10–6	2.74–02	1.51–02	3.89+00	–0.821	B	3
197.	$1s^28d$– $1s^212p$	^2D–$^2P^o$	6586.9	6588.7	492729	507906	10–6	1.72–02	6.73–03	1.46+00	–1.172	B	3
198.	$1s^29s$– $1s^29p$	^2S–$^2P^o$		*579* cm^{-1}	497736.7	*498315*	2–6	1.13–03	1.52+00	1.73+03	0.483	B	3
				579.0 cm^{-1}	497736.7	498315.7	2–4	1.13–03	1.01+00	1.15+03	0.307	B	*LS*
				577.9 cm^{-1}	497736.7	498314.6	2–2	1.13–03	5.06–01	5.77+02	0.005	B	*LS*
199.	$1s^29s$– $1s^210p$	^2S–$^2P^o$	21383	4675 cm^{-1}	497736.7	502412	2–6	1.85–02	3.80–01	5.36+01	–0.119	B	3
200.	$1s^29s$– $1s^211p$	^2S–$^2P^o$	12861	7773 cm^{-1}	497736.7	505510	2–6	1.61–02	1.20–01	1.02+01	–0.619	B	3
201.	$1s^29s$– $1s^212p$	^2S–$^2P^o$	9830.8	9833.5	497736.7	507906	2–6	1.10–02	4.79–02	3.10+00	–1.019	B	3

C IV: Allowed Transitions — Continued

No.	Transition Array	Mult.	λ_{air} (Å)	λ_{vac} (Å) or σ (cm^{-1})*	E_i (cm^{-1})	E_k (cm^{-1})	g_i–g_k	A_{ki} (10^8 s^{-1})	f_{ik}	S (at. u.)	log gf	Acc.	Source
202.	$1s^29p$– $1s^29d$	$^2P^o$–^2D		176 cm^{-1}	498315	498491	6–10	3.86–05	3.11–01	3.49+03	0.271	B	3
203.	$1s^29p$– $1s^210d$	$^2P^o$–^2D	23342	4283 cm^{-1}	498315	502598	6–10	4.50–02	6.13–01	2.83+02	0.566	B	3
204.	$1s^29p$– $1s^211d$	$^2P^o$–^2D	13545	7381 cm^{-1}	498315	505696	6–10	3.40–02	1.56–01	4.17+01	−0.029	B	3
205.	$1s^29p$– $1s^212d$	$^2P^o$–^2D	10303	9703 cm^{-1}	498315	508018	6–10	2.60–02	6.89–02	1.40+01	−0.384	B	3
206.	$1s^29d$– $1s^210p$	^2D–$^2P^o$	25497	3921 cm^{-1}	498491	502412	10–6	3.95–02	2.31–01	1.94+02	0.364	B	3
207.	$1s^29d$– $1s^211p$	^2D–$^2P^o$	14243	7019 cm^{-1}	498491	505510	10–6	2.65–02	4.83–02	2.27+01	−0.316	B	3
208.	$1s^29d$– $1s^212p$	^2D–$^2P^o$	10618	9415 cm^{-1}	498491	507906	10–6	1.65–02	1.67–02	5.85+00	−0.777	B	3
209.	$1s^210p$– $1s^210d$	$^2P^o$–^2D		186 cm^{-1}	502412	502598	6–10	7.02–05	5.07–01	5.39+03	0.483	B	3
210.	$1s^210p$– $1s^211d$	$^2P^o$–^2D	30442	3284 cm^{-1}	502412	505696	6–10	2.05–02	4.75–01	2.86+02	0.455	B	3
211.	$1s^210p$– $1s^212d$	$^2P^o$–^2D	17833	5606 cm^{-1}	502412	508018	6–10	1.74–02	1.38–01	4.87+01	−0.081	B	3
212.	$1s^210d$– $1s^211p$	^2D–$^2P^o$	34331	2912 cm^{-1}	502598	505510	10–6	2.45–02	2.60–01	2.94+02	0.415	B	3
213.	$1s^210d$– $1s^212p$	^2D–$^2P^o$	18834	5308 cm^{-1}	502598	507906	10–6	1.51–02	4.81–02	2.99+01	−0.318	B	3
214.	$1s^211p$– $1s^211d$	$^2P^o$–^2D		186 cm^{-1}	505510	505696	6–10	1.04–04	7.51–01	7.97+03	0.654	B	3
215.	$1s^211p$– $1s^212d$	$^2P^o$–^2D	39862	2508 cm^{-1}	505510	508018	6–10	1.21–02	4.81–01	3.79+02	0.460	B	3
216.	$1s^211d$– $1s^212p$	^2D–$^2P^o$	45237	2210 cm^{-1}	505696	507906	10–6	1.62–02	2.98–01	4.44+02	0.475	B	3
217.	$1s^212p$– $1s^212d$	$^2P^o$–^2D		112 cm^{-1}	507906	508018	6–10	3.24–05	6.46–01	1.14+04	0.588	B	3
218.	$1s2s2p$– $1s2p^2$	$^4P^o$–^4P		1325			12–12	6.30+00	1.66–01	8.67+00	0.299	B	6
219.	$1s2s2p$– $1s2s3s$	$^4P^o$–^4S		314.4			12–4	8.49+01	4.19–02	5.21–01	−0.298	B	6
220.	$1s2s2p$– $1s2s3d$	$^4P^o$–^4D		283.6			12–20	3.36+02	6.75–01	7.57+00	0.909	B	6
221.	$1s2s2p$– $1s2p3p$	$^4P^o$–^4D		260.1			12–20	1.35+01	2.28–02	2.35–01	−0.562	B	6

C IV: Allowed Transitions — Continued

No.	Transition Array	Mult.	λ_{air} (Å)	λ_{vac} (Å) or σ (cm^{-1})*	E_i (cm^{-1})	E_k (cm^{-1})	g_i–g_k	A_{ki} (10^8 s^{-1})	f_{ik}	S (at. u.)	log gf	Acc.	Source
222.		^{4}P°–^{4}S		256.4			12–4	6.43+01	2.11–02	2.14–01	–0.596	B	6
223.		^{4}P°–^{4}P		254.4			12–12	5.91+01	5.73–02	5.76–01	–0.162	B	6
224.	1s2s2p–1s2s4s	^{4}P°–^{4}S		231.0			12–4	1.84+01	4.91–03	4.48–02	–1.230	B	6
225.	1s2s2p–1s2s4d	^{4}P°–^{4}D		224.4			12–20	1.07+02	1.34–01	1.19+00	0.207	B	6
226.	1s2s2p–1s2s5s	^{4}P°–^{4}S		208.5			12–4	2.78+01	6.04–03	4.98–02	–1.140	B	6
227.	1s2s2p–1s2p4p	^{4}P°–^{4}D		206.6			12–20	3.51+01	3.74–02	3.05–01	–0.348	B	6
228.		^{4}P°–^{4}P		205.2			12–12	3.02+01	1.91–02	1.54–01	–0.641	B	6
229.	1s2s2p–1s2s5d	^{4}P°–^{4}D		205.1			12–20	3.31+01	3.48–02	2.82–01	–0.379	B	6
230.	1s2s2p–1s2p4p	^{4}P°–^{4}S		205.0			12–4	1.56+01	3.28–03	2.65–02	–1.405	B	6
231.	1s2p^2–1s2p3s	^{4}P–^{4}P°		337.9			12–12	6.11+01	1.05–01	1.40+00	0.099	B	6
232.	1s2p^2–1s2p3d	^{4}P–^{4}D°		304.6			12–20	3.44+02	7.97–01	9.59+00	0.981	B	6
233.		^{4}P–^{4}P°		300.4			12–12	1.77+02	2.39–01	2.84+00	0.458	B	6
234.	1s2p^2–1s2p4s	^{4}P–^{4}P°		247.7			12–12	1.66+01	1.53–02	1.49–01	–0.737	B	6
235.	1s2p^2–1s2p4d	^{4}P–^{4}D°		240.2			12–20	1.29+02	1.86–01	1.77+00	0.349	B	6
236.		^{4}P–^{4}P°		239.3			12–12	6.56+01	5.63–02	5.32–01	–0.170	B	6
237.	1s2p^2–1s2p5d	^{4}P–^{4}D°		218.9			12–20	6.27+01	7.51–02	6.49–01	–0.045	B	6
238.		^{4}P–^{4}P°		218.6			12–12	2.95+01	2.11–02	1.82–01	–0.596	B	6
239.	1s2p^2–1s2p6d	^{4}P–^{4}D°		208.9			12–20	3.51+01	3.83–02	3.16–01	–0.338	B	6
240.		^{4}P–^{4}P°		208.8			12–12	1.82+01	1.19–02	9.81–02	–0.846	B	6
241.	1s2s3d–1s2s5f	^{4}D–^{4}F°		726.9			20–28	1.05+01	1.16–01	5.57+00	0.367	B	6

C IV: Allowed Transitions — Continued

No.	Transition Array	Mult.	λ_{air} (Å)	λ_{vac} (Å) or σ (cm^{-1})*	E_i (cm^{-1})	E_k (cm^{-1})	g_i-g_k	A_{ki} (10^8 s^{-1})	f_{ik}	S (at. u.)	log gf	Acc.	Source
242.	$1s2p3d-$ $1s2p5f$	$^4D°-^4F$		776.3			20–28	1.06+01	1.34−01	6.85+00	0.428	B	6
243.	$1s2s3p-$ $1s2s4d$	$^4P°-^4D$		943.8			12–20	1.84+01	4.09−01	1.53+01	0.691	B	6
244.	$1s2p3p-$ $1s2p4d$	$^4D-^4F°$		966.5			20–28	1.37+01	2.69−01	1.71+01	0.730	B	6
245.	$1s2p3d-$ $1s2p4f$	$^4F°-^4G$		1007			28–36	2.64+01	5.16−01	4.79+01	1.160	B	6
246.	$1s2p3p-$ $1s2p4d$	$^4P-^4D°$		1013			12–20	1.32+01	3.38−01	1.35+01	0.609	B	6
247.	$1s2s3d-$ $1s2s4f$	$^4D-^4F°$		1029			20–28	3.90+01	8.67−01	5.88+01	1.239	B	6
248.	$1s2s3p-$ $1s2s4s$	$^4P°-^4S$		1073			12–4	1.12+01	6.45−02	2.73+00	−0.112	B	6
249.	$1s2p3d-$ $1s2s5g$	$^4F°-^4G$		1083			28–36	1.12+01	2.53−01	2.53+01	0.851	B	6
250.	$1s2p3d-$ $1s2p4f$	$^4D°-^4F$		1128			20–28	3.27+01	8.73−01	6.48+01	1.242	B	6
251.		$^4P°-^4D$		1155			12–20	2.81+01	9.37−01	4.27+01	1.051	B	6
252.	$1s2p4f-$ $1s2p5g$	$^4F-^4G°$		2471			28–36	1.06+01	1.25+00	2.84+02	1.543	B	6

*Wavelengths (Å) are always given unless cm^{-1} is indicated.

C V

Helium Isoelectronic Sequence

Ground State: $1s^2\ ^1S_0$

Ionization Energy: 392.007 eV = 3162395 cm^{-1}

Allowed Transitions

List of tabulated lines

Wavelength (Å)	No.	Wavelength (Å)	No.	Wavelength (Å)	No.	Wavelength (Å)	No.
in vacuum		260.135	17	506.265	48	1651.26	84
		260.143	17	506.363	48	1651.30	76
33.4262	3	260.227	17	511.480	63	1753.82	99
34.9728	2	267.267	27	515.312	57	1817.26	105
40.2678	1	271.882	26	515.314	57	1856.27	113
141.252	10	357.589	38	515.345	57	1862.09	104
143.951	9	372.524	44	517.328	66	1873.86	116
148.329	8	372.975	55	672.041	34	1874.13	109
150.520	25	372.977	55	711.833	40		
150.523	25	373.031	55	717.375	47	in air	
150.551	25	374.211	54	717.383	47		
153.488	16	374.213	54	717.581	47	2270.89	4
153.555	24	374.267	54	737.029	74	2277.27	4
153.558	24	375.408	37	745.534	62	2277.92	4
153.587	24	381.797	60	756.779	46	2737.66	98
156.228	7	381.798	60	756.788	46	2842.90	120
157.194	32	381.815	60	757.009	46	2873.33	103
158.375	15	384.377	69	760.104	65	2934.21	125
158.479	23	391.602	43	762.509	83	2960.40	112
158.482	23	392.180	53	762.959	56	2991.46	130
158.513	23	392.183	53	762.964	56	3012.49	124
160.505	31	392.242	53	763.031	56	3014.90	115
165.885	30	394.228	52	767.691	82	3021.07	132
167.220	14	394.231	52	777.064	92	3023.07	127
167.382	22	394.291	52	778.685	87	3042.65	108
167.386	22	402.179	59	782.059	96	3057.50	102
167.420	22	402.181	59	816.953	73	3526.66	11
168.414	21	402.199	59	847.339	81	4566.77	119
168.417	21	404.795	68	856.959	80	4773.92	123
168.452	21	406.710	36	864.966	91	4914.86	129
173.279	6	425.064	42	868.448	86	5006.57	131
175.655	29	425.986	51	871.501	95	5050.24	126
186.346	13	425.989	51	981.310	72	5096.33	122
186.691	20	426.059	51	1022.69	79	8420.72	33
186.695	20	429.823	50	1045.09	78	8448.12	33
186.738	20	429.826	50	1047.02	90	8449.19	33
189.255	19	429.897	50	1056.56	85	12208.8	39
189.260	19	436.856	64	1057.78	94	16943.1	45
189.304	19	438.320	58	1422.63	100	16945.4	45
197.022	28	438.322	58	1467.19	107	16949.7	45
227.182	5	438.344	58	1474.53	71	17025.3	45
227.202	5	440.856	67	1486.50	106	17058.7	45
227.203	5	472.168	35	1493.66	114	17061.1	45
247.315	12	495.390	41	1500.78	110	20700.9	70
248.664	18	497.050	49	1504.01	117	41137.7	97
248.672	18	497.054	49	1557.18	77	41161.4	75
248.741	18	497.149	49	1610.00	89		
248.748	18	506.261	48	1639.16	93		

List of tabulated lines — Continued

Wavenumber (cm^{-1})	No.
238.6	128
371.9	111
706.0	121
732.9	88
1230.1	101
1389.7	118
1838.3	61

We have mainly utilized the results of Cann and Thakkar,[1] who calculated the oscillator strengths of about 100 nS − mP and nP − mD transitions ($n, m < 7$) with variationally determined, highly correlated wavefunctions. This work is an expansion of, and is similar in spirit to, the classic variational calculations on He-like ions by Schiff and Pekeris.[2] Cann and Thakkar improved this variational approach by the use of exponential correlation factors which yield more accurate wavefunctions with a similar number of term expansions. Comparisons of their results with the earlier data by Schiff and Pekeris[2] and with similar work by Kono and Hattori[3] in the 1980s show that the differences do not exceed 0.1%.

The presented data are the arithmetic mean values of the "dipole length" and "dipole velocity" results of Cann and Thakkar.[1] Their numerical values are given to at least three, but most often five or six significant figures, and the differences between their length and velocity data do not exceed seven units in the last digit shown. These differences, which they consider to be their main indicator of accuracy, are smaller than 0.1% for all but a few transitions.

For some transitions involving higher excited states, which their material does not cover, we have used the results of the Opacity Project (OP), which is reviewed in the general introduction. The calculations for He-like ions were carried out by Fernley *et al.*[4] and also are estimated to be very accurate, as comparisons with the above work of Cann and Thakkar show.

It should be noted that all variational calculations[1–3] have been done in a non-relativistic framework. This is an excellent approximation for neutral helium and the lower ions, but for highly charged He-like ions, relativistic effects are expected to become significant. For C V, N VI and O VII, some test calculations by Kim and Weiss[5] show that such effects are still negligible. They become noticeable only for the line strengths of P-D transitions, where deviations from non-relativistic values may sometimes reach 0.2%.

References

[1]N. M. Cann and A. J. Thakkar, Phys. Rev. A **46**, 5397 (1992).
[2]B. Schiff and C. L. Pekeris, Phys. Rev. **134**, A638 (1964).
[3]A. Kono and S. Hattori, Phys. Rev. A **30**, 2093 (1984).
[4]J. A. Fernley, K. T. Taylor and M. J. Seaton, J. Phys. B **20**, 6457 (1987).
[5]Y.-K. Kim and A. W. Weiss (private communication), 1993.

C V: Allowed Transitions

No.	Transition Array	Mult.	λ_{air} (Å)	λ_{vac} (Å) or σ (cm^{-1})*	E_i (cm^{-1})	E_k (cm^{-1})	g_i–g_k	A_{ki} (10^8 s^{-1})	f_{ik}	S (at. u.)	log gf	Acc.	Source
1.	1s^2– 1s2p	^{1}S–^{1}P°		40.2678	0.0	2483372.8	1–3	8.873+03	6.471−01	8.578−02	−0.1891	AA	1
2.	1s^2– 1s3p	^{1}S–^{1}P°		34.9728	0.0	2859368.6	1–3	2.554+03	1.405−01	1.617−02	−0.8524	AA	1
3.	1s^2– 1s4p	^{1}S–^{1}P°		33.4262	0.0	2991662.4	1–3	1.065+03	5.353−02	5.891−03	−1.2714	AA	1
4.	1s2s– 1s2p	^{3}S–^{3}P°	*2273.9*	*2274.6*	2411271.2	*2455234*	3–9	5.646−01	1.314−01	2.951+00	−0.4043	AA	1
			2270.89	2271.59	2411271.2	2455293.2	3–5	5.669−01	7.309−02	1.640+00	−0.6590	AA	*LS*
			2277.92	2278.63	2411271.2	2455157.3	3–3	5.616−01	4.372−02	9.838−01	−0.8822	AA	*LS*
			2277.27	2277.97	2411271.2	2455169.9	3–1	5.621−01	1.458−02	3.279−01	−1.3592	AA	*LS*
5.	1s2s– 1s3p	^{3}S–^{3}P°		*227.19*	2411271.2	*2851428*	3–9	1.363+02	3.165−01	7.101−01	−0.0225	AA	1
				227.182	2411271.2	2851446.0	3–5	1.363+02	1.758−01	3.945−01	−0.2778	AA	*LS*
				227.203	2411271.2	2851406.0	3–3	1.363+02	1.055−01	2.367−01	−0.4997	AA	*LS*
				227.202	2411271.2	2851407.5	3–1	1.363+02	3.516−02	7.890−02	−0.9768	AA	*LS*
6.	1s2s– 1s4p	^{3}S–^{3}P°		173.279	2411271.2	2988374.3	3–9	6.195+01	8.367−02	1.432−01	−0.6003	AA	1
7.	1s2s– 1s5p	^{3}S–^{3}P°		156.228	2411271.2	3051362.9	3–9	3.226+01	3.541−02	5.463−02	−0.9738	AA	1

C V: Allowed Transitions — Continued

No.	Transition Array	Mult.	λ_{air} (Å)	λ_{vac} (Å) or σ (cm^{-1})*	E_i (cm^{-1})	E_k (cm^{-1})	g_i-g_k	A_{ki} (10^8 s^{-1})	f_{ik}	S (at. u.)	log gf	Acc.	Source
8.	1s2s– 1s6p	^{3}S–^{3}P°		148.329	2411271.2	3085449.5	3–9	1.877+01	1.858–02	2.722–02	–1.2539	AA	1
9.	1s2s– 1s7p	^{3}S–^{3}P°		143.951	2411271.2	3105951.0	3–9	1.18+01	1.10–02	1.57–02	–1.480	A	4
10.	1s2s– 1s8p	^{3}S–^{3}P°		141.252	2411271.2	3119224.8	3–9	7.99+00	7.17–03	1.00–02	–1.667	A	4
11.	1s2s– 1s2p	^{1}S–^{1}P°	3526.66	3527.67	2455025.5	2483372.8	1–3	1.663–01	9.306–02	1.081+00	–1.0313	AA	1
12.	1s2s– 1s3p	^{1}S–^{1}P°		247.315	2455025.5	2859368.6	1–3	1.278+02	3.517–01	2.863–01	–0.4538	AA	1
13.	1s2s– 1s4p	^{1}S–^{1}P°		186.346	2455025.5	2991662.4	1–3	5.768+01	9.009–02	5.527–02	–1.0453	AA	1
14.	1s2s– 1s5p	^{1}S–^{1}P°		167.220	2455025.5	3053041.3	1–3	3.001+01	3.775–02	2.078–02	–1.4231	AA	1
15.	1s2s– 1s6p	^{1}S–^{1}P°		158.375	2455025.5	3086438.8	1–3	1.748+01	1.972–02	1.028–02	–1.7051	AA	1
16.	1s2s– 1s7p	^{1}S–^{1}P°		153.488	2455025.5	3106541	1–3	1.10+01	1.17–02	5.90–03	–1.933	A	4
17.	1s2p– 1s3s	^{3}P°–^{3}S		*260.19*	*2455234*	2839573.8	9–3	6.680+01	2.260–02	1.742–01	–0.6916	AA	1
				260.227	2455293.2	2839573.8	5–3	3.710+01	2.260–02	9.679–02	–0.9470	AA	*LS*
				260.135	2455157.3	2839573.8	3–3	2.228+01	2.260–02	5.807–02	–1.1687	AA	*LS*
				260.143	2455169.9	2839573.8	1–3	7.426+00	2.260–02	1.936–02	–1.6458	AA	*LS*
18.	1s2p– 1s3d	^{3}P°–^{3}D		*248.71*	*2455234*	*2857312*	9–15	4.247+02	6.565–01	4.838+00	0.7715	AA	1
				248.741	2455293.2	2857318.0	5–7	4.246+02	5.514–01	2.258+00	0.4404	AA	*LS*
				248.664	2455157.3	2857306.5	3–5	3.187+02	4.924–01	1.209+00	0.1695	AA	*LS*
				248.672	2455169.9	2857305.7	1–3	2.361+02	6.566–01	5.375–01	–0.1827	AA	*LS*
				248.748	2455293.2	2857306.5	5–5	1.061+02	9.846–02	4.031–01	–0.3078	AA	*LS*
				248.664	2455157.3	2857305.7	3–3	1.771+02	1.641–01	4.031–01	–0.3076	AA	*LS*
				248.748	2455293.2	2857305.7	5–3	1.179+01	6.564–03	2.688–02	–1.4839	AA	*LS*
19.	1s2p– 1s4s	^{3}P°–^{3}S		*189.28*	*2455234*	2983544.9	9–3	2.597+01	4.650–03	2.608–02	–1.3783	AA	1
				189.304	2455293.2	2983544.9	5–3	1.442+01	4.649–03	1.449–02	–1.6336	AA	*LS*
				189.255	2455157.3	2983544.9	3–3	8.661+00	4.651–03	8.693–03	–1.8554	AA	*LS*
				189.260	2455169.9	2983544.9	1–3	2.887+00	4.651–03	2.898–03	–2.3325	AA	*LS*
20.	1s2p– 1s4d	^{3}P°–^{3}D		*186.72*	*2455234*	2990803.1	9–15	1.413+02	1.231–01	6.810–01	0.0445	AA	1
				186.738	2455293.2	2990803.1	5–7	1.413+02	1.034–01	3.178–01	–0.2865	AA	*LS*
				186.691	2455157.3	2990803.1	3–5	1.060+02	9.234–02	1.703–01	–0.5575	AA	*LS*
				186.695	2455169.9	2990803.1	1–3	7.854+01	1.231–01	7.567–02	–0.9097	AA	*LS*
				186.738	2455293.2	2990803.1	5–5	3.532+01	1.846–02	5.675–02	–1.0347	AA	*LS*
				186.691	2455157.3	2990803.1	3–3	5.891+01	3.078–02	5.675–02	–1.0346	AA	*LS*
				186.738	2455293.2	2990803.1	5–3	3.924+00	1.231–03	3.783–03	–2.2108	AA	*LS*
21.	1s2p– 1s5s	^{3}P°–^{3}S		*168.44*	*2455234*	3048932.7	9–3	1.270+01	1.800–03	8.983–03	–1.7905	AA	1
				168.452	2455293.2	3048932.7	5–3	7.051+00	1.800–03	4.991–03	–2.0458	AA	*LS*
				168.414	2455157.3	3048932.7	3–3	4.234+00	1.800–03	2.994–03	–2.2676	AA	*LS*
				168.417	2455169.9	3048932.7	1–3	1.411+00	1.800–03	9.981–04	–2.7447	AA	*LS*

C V: Allowed Transitions — Continued

No.	Transition Array	Mult.	λ_{air} (Å)	λ_{vac} (Å) or σ (cm^{-1})*	E_i (cm^{-1})	E_k (cm^{-1})	g_i-g_k	A_{ki} (10^8 s^{-1})	f_{ik}	S (at. u.)	log gf	Acc.	Source
22.	1s2p– 1s5d	$^3P^o-^3D$		*167.40*	*2455234*	3052593.0	9–15	6.559+01	4.593–02	2.278–01	–0.3837	AA	1
				167.420	2455293.2	3052593.0	5–7	6.557+01	3.858–02	1.063–01	–0.7147	AA	*LS*
				167.382	2455157.3	3052593.0	3–5	4.921+01	3.445–02	5.695–02	–0.9857	AA	*LS*
				167.386	2455169.9	3052593.0	1–3	3.645+01	4.593–02	2.531–02	–1.3379	AA	*LS*
				167.420	2455293.2	3052593.0	5–5	1.639+01	6.889–03	1.898–02	–1.4629	AA	*LS*
				167.382	2455157.3	3052593.0	3–3	2.734+01	1.148–02	1.898–02	–1.4628	AA	*LS*
				167.420	2455293.2	3052593.0	5–3	1.822+00	4.593–04	1.266–03	–2.6390	AA	*LS*
23.	1s2p– 1s6d	$^3P^o-^3D$		*158.50*	*2455234*	3086155.5	9–15	3.608+01	2.265–02	1.064–01	–0.6907	AA	1
				158.513	2455293.2	3086155.5	5–7	3.607+01	1.902–02	4.963–02	–1.0218	AA	*LS*
				158.479	2455157.3	3086155.5	3–5	2.707+01	1.699–02	2.659–02	–1.2928	AA	*LS*
				158.482	2455169.9	3086155.5	1–3	2.005+01	2.265–02	1.182–02	–1.6449	AA	*LS*
				158.513	2455293.2	3086155.5	5–5	9.017+00	3.397–03	8.863–03	–1.7700	AA	*LS*
				158.479	2455157.3	3086155.5	3–3	1.504+01	5.662–03	8.863–03	–1.7699	AA	*LS*
				158.513	2455293.2	3086155.5	5–3	1.002+00	2.264–04	5.909–04	–2.9461	AA	*LS*
24.	1s2p– 1s7d	$^3P^o-^3D$		*153.57*	*2455234*	3106390.8	9–15	2.20+01	1.30–02	5.91–02	–0.932	A	4
				153.587	2455293.2	3106390.8	5–7	2.20+01	1.09–02	2.76–02	–1.263	A	*LS*
				153.555	2455157.3	3106390.8	3–5	1.65+01	9.74–03	1.48–02	–1.534	A	*LS*
				153.558	2455169.9	3106390.8	1–3	1.22+01	1.30–02	6.57–03	–1.886	A	*LS*
				153.587	2455293.2	3106390.8	5–5	5.51+00	1.95–03	4.93–03	–2.011	A	*LS*
				153.555	2455157.3	3106390.8	3–3	9.19+00	3.25–03	4.93–03	–2.011	A	*LS*
				153.587	2455293.2	3106390.8	5–3	6.12–01	1.30–04	3.28–04	–3.188	A	*LS*
25.	1s2p– 1s8d	$^3P^o-^3D$		*150.54*	*2455234*	3119520.3	9–15	1.45+01	8.21–03	3.66–02	–1.132	A	4
				150.551	2455293.2	3119520.3	5–7	1.45+01	6.89–03	1.71–02	–1.463	A	*LS*
				150.520	2455157.3	3119520.3	3–5	1.09+01	6.16–03	9.15–03	–1.734	A	*LS*
				150.523	2455169.9	3119520.3	1–3	8.05+00	8.21–03	4.07–03	–2.086	A	*LS*
				150.551	2455293.2	3119520.3	5–5	3.62+00	1.23–03	3.05–03	–2.211	A	*LS*
				150.520	2455157.3	3119520.3	3–3	6.04+00	2.05–03	3.05–03	–2.211	A	*LS*
				150.551	2455293.2	3119520.3	5–3	4.02–01	8.21–05	2.03–04	–3.387	A	*LS*
26.	1s2p– 1s3s	$^1P^o-^1S$		271.882	2483372.8	2851180	3–1	5.699+01	2.105–02	5.653–02	–1.1996	AA	1
27.	1s2p– 1s3d	$^1P^o-^1D$		267.267	2483372.8	2857530.3	3–5	3.947+02	7.045–01	1.860+00	0.3250	AA	1
28.	1s2p– 1s4d	$^1P^o-^1D$		197.022	2483372.8	2990929.5	3–5	1.232+02	1.195–01	2.325–01	–0.4455	AA	1
29.	1s2p– 1s5d	$^1P^o-^1D$		175.655	2483372.8	3052669.4	3–5	5.579+01	4.301–02	7.461–02	–0.8893	AA	1
30.	1s2p– 1s6d	$^1P^o-^1D$		165.885	2483372.8	3086200.2	3–5	3.030+01	2.083–02	3.413–02	–1.2041	AA	1
31.	1s2p– 1s7d	$^1P^o-^1D$		160.505	2483372.8	3106407	3–5	1.84+01	1.19–02	1.88–02	–1.449	A	4
32.	1s2p– 1s8d	$^1P^o-^1D$		157.194	2483372.8	3119530	3–5	1.20+01	7.41–03	1.15–02	–1.653	A	4
33.	1s3s– 1s3p	$^3S-^3P^o$	*8433.2*	*8435.5*	2839573.8	*2851428*	3–9	6.868–02	2.198–01	1.831+01	–0.1809	AA	1
			8420.72	8423.04	2839573.8	2851446.0	3–5	6.898–02	1.223–01	1.017+01	–0.4355	AA	*LS*
			8449.19	8451.51	2839573.8	2851406.0	3–3	6.829–02	7.313–02	6.104+00	–0.6588	AA	*LS*
			8448.12	8450.44	2839573.8	2851407.5	3–1	6.832–02	2.438–02	2.035+00	–1.1359	AA	*LS*

C v: Allowed Transitions — Continued

No.	Transition Array	Mult.	λ_{air} (Å)	λ_{vac} (Å) or σ (cm^{-1})*	E_i (cm^{-1})	E_k (cm^{-1})	g_i-g_k	A_{ki} (10^8 s^{-1})	f_{ik}	S (at. u.)	log gf	Acc.	Source
34.	1s3s– 1s4p	^{3}S–^{3}P°		672.041	2839573.8	2988374.3	3–9	1.677+01	3.407–01	2.261+00	0.0095	AA	1
35.	1s3s– 1s5p	^{3}S–^{3}P°		472.168	2839573.8	3051362.9	3–9	9.494+00	9.520–02	4.439–01	–0.5442	AA	1
36.	1s3s– 1s6p	^{3}S–^{3}P°		406.710	2839573.8	3085449.5	3–9	5.652+00	4.205–02	1.689–01	–0.8992	AA	1
37.	1s3s– 1s7p	^{3}S–^{3}P°		375.408	2839573.8	3105951.0	3–9	3.60+00	2.28–02	8.46–02	–1.165	A	4
38.	1s3s– 1s8p	^{3}S–^{3}P°		357.589	2839573.8	3119224.8	3–9	2.43+00	1.40–02	4.93–02	–1.378	A	4
39.	1s3s– 1s3p	^{1}S–^{1}P°	12208.8	8189 cm^{-1}	2851180	2859368.6	1–3	2.411–02	1.617–01	6.501+00	–0.7913	AA	1
40.	1s3s– 1s4p	^{1}S–^{1}P°		711.833	2851180	2991662.4	1–3	1.664+01	3.792–01	8.887–01	–0.4211	AA	1
41.	1s3s– 1s5p	^{1}S–^{1}P°		495.390	2851180	3053041.3	1–3	9.323+00	1.029–01	1.678–01	–0.9876	AA	1
42.	1s3s– 1s6p	^{1}S–^{1}P°		425.064	2851180	3086438.8	1–3	5.538+00	4.500–02	6.297–02	–1.3468	AA	1
43.	1s3s– 1s7p	^{1}S–^{1}P°		391.602	2851180	3106541	1–3	3.53+00	2.44–02	3.14–02	–1.613	A	4
44.	1s3s– 1s8p	^{1}S–^{1}P°		372.524	2851180	3119619	1–3	2.38+00	1.48–02	1.82–02	–1.829	A	4
45.	1s3p– 1s3d	^{3}P°–^{3}D	*16993*	*5883.3* cm^{-1}	*2851428*	*2857312*	9–15	6.719–03	4.850–02	2.443+01	–0.3600	AA	1
			17025.3	5872.0 cm^{-1}	2851446.0	2857318.0	5–7	6.680–03	4.066–02	1.140+01	–0.6918	AA	*LS*
			16943.1	5900.5 cm^{-1}	2851406.0	2857306.5	3–5	5.083–03	3.648–02	6.106+00	–0.9608	AA	*LS*
			16949.7	5898.2 cm^{-1}	2851407.5	2857305.7	1–3	3.761–03	4.862–02	2.714+00	–1.3132	AA	*LS*
			17058.7	5860.5 cm^{-1}	2851446.0	2857306.5	5–5	1.660–03	7.247–03	2.035+00	–1.4409	AA	*LS*
			16945.4	5899.7 cm^{-1}	2851406.0	2857305.7	3–3	2.823–03	1.216–02	2.035+00	–1.4380	AA	*LS*
			17061.1	5859.7 cm^{-1}	2851446.0	2857305.7	5–3	1.844–04	4.831–04	1.357–01	–2.6170	AA	*LS*
46.	1s3p– 1s4s	^{3}P°–^{3}S		*756.91*	*2851428*	2983544.9	9–3	1.797+01	5.146–02	1.154+00	–0.3343	AA	1
				757.009	2851446.0	2983544.9	5–3	9.980+00	5.144–02	6.410–01	–0.5897	AA	*LS*
				756.779	2851406.0	2983544.9	3–3	5.993+00	5.146–02	3.846–01	–0.8114	AA	*LS*
				756.788	2851407.5	2983544.9	1–3	1.998+00	5.146–02	1.282–01	–1.2885	AA	*LS*
47.	1s3p– 1s4d	^{3}P°–^{3}D		*717.49*	*2851428*	2990803.1	9–15	4.318+01	5.554–01	1.181+01	0.6989	AA	1
				717.581	2851446.0	2990803.1	5–7	4.316+01	4.665–01	5.510+00	0.3678	AA	*LS*
				717.375	2851406.0	2990803.1	3–5	3.240+01	4.166–01	2.952+00	0.0969	AA	*LS*
				717.383	2851407.5	2990803.1	1–3	2.400+01	5.555–01	1.312+00	–0.2553	AA	*LS*
				717.581	2851446.0	2990803.1	5–5	1.079+01	8.330–02	9.839–01	–0.3804	AA	*LS*
				717.375	2851406.0	2990803.1	3–3	1.800+01	1.389–01	9.839–01	–0.3802	AA	*LS*
				717.581	2851446.0	2990803.1	5–3	1.199+00	5.553–03	6.560–02	–1.5565	AA	*LS*
48.	1s3p– 1s5s	^{3}P°–^{3}S		*506.32*	*2851428*	3048932.7	9–3	8.501+00	1.089–02	1.634–01	–1.0087	AA	1
				506.363	2851446.0	3048932.7	5–3	4.721+00	1.089–02	9.076–02	–1.2640	AA	*LS*
				506.261	2851406.0	3048932.7	3–3	2.834+00	1.089–02	5.446–02	–1.4858	AA	*LS*
				506.265	2851407.5	3048932.7	1–3	9.448–01	1.089–02	1.815–02	–1.9629	AA	*LS*

C v: Allowed Transitions — Continued

No.	Transition Array	Mult.	λ_{air} (Å)	λ_{vac} (Å) or σ (cm^{-1})*	E_i (cm^{-1})	E_k (cm^{-1})	g_i-g_k	A_{ki} (10^8 s^{-1})	f_{ik}	S (at. u.)	log gf	Acc.	Source
49.	1s3p– 1s5d	$^3P^o-^3D$		*497.11*	*2851428*	3052593.0	9–15	2.164+01	1.336–01	1.968+00	0.0801	AA	1
				497.149	2851446.0	3052593.0	5–7	2.164+01	1.122–01	9.185–01	–0.2509	AA	*LS*
				497.050	2851406.0	3052593.0	3–5	1.624+01	1.002–01	4.921–01	–0.5219	AA	*LS*
				497.054	2851407.5	3052593.0	1–3	1.203+01	1.336–01	2.187–01	–0.8740	AA	*LS*
				497.149	2851446.0	3052593.0	5–5	5.409+00	2.004–02	1.640–01	–0.9991	AA	*LS*
				497.050	2851406.0	3052593.0	3–3	9.021+00	3.341–02	1.640–01	–0.9990	AA	*LS*
				497.149	2851446.0	3052593.0	5–3	6.010–01	1.336–03	1.093–02	–2.1752	AA	*LS*
50.	1s3p– 1s6s	$^3P^o-^3S$		*429.86*	*2851428*	3084059.8	9–3	4.689+00	4.330–03	5.515–02	–1.4093	AA	1
				429.897	2851446.0	3084059.8	5–3	2.604+00	4.329–03	3.064–02	–1.6646	AA	*LS*
				429.823	2851406.0	3084059.8	3–3	1.563+00	4.330–03	1.838–02	–1.8864	AA	*LS*
				429.826	2851407.5	3084059.8	1–3	5.211–01	4.330–03	6.127–03	–2.3635	AA	*LS*
51.	1s3p– 1s6d	$^3P^o-^3D$		*426.03*	*2851428*	3086155.5	9–15	1.217+01	5.520–02	6.967–01	–0.3038	AA	1
				426.059	2851446.0	3086155.5	5–7	1.217+01	4.636–02	3.251–01	–0.6349	AA	*LS*
				425.986	2851406.0	3086155.5	3–5	9.131+00	4.140–02	1.742–01	–0.9059	AA	*LS*
				425.989	2851407.5	3086155.5	1–3	6.764+00	5.520–02	7.741–02	–1.2581	AA	*LS*
				426.059	2851446.0	3086155.5	5–5	3.042+00	8.279–03	5.806–02	–1.3831	AA	*LS*
				425.986	2851406.0	3086155.5	3–3	5.073+00	1.380–02	5.806–02	–1.3830	AA	*LS*
				426.059	2851446.0	3086155.5	5–3	3.380–01	5.519–04	3.871–03	–2.5592	AA	*LS*
52.	1s3p– 1s7s	$^3P^o-^3S$		*394.26*	*2851428*	3105066	9–3	2.88+00	2.23–03	2.61–02	–1.697	A	4
				394.291	2851446.0	3105066	5–3	1.60+00	2.23–03	1.45–02	–1.952	A	*LS*
				394.228	2851406.0	3105066	3–3	9.59–01	2.23–03	8.70–03	–2.174	A	*LS*
				394.231	2851407.5	3105066	1–3	3.20–01	2.23–03	2.90–03	–2.651	A	*LS*
53.	1s3p– 1s7d	$^3P^o-^3D$		*392.21*	*2851428*	3106390.8	9–15	7.51+00	2.89–02	3.36–01	–0.585	A	4
				392.242	2851446.0	3106390.8	5–7	7.51+00	2.43–02	1.57–01	–0.916	A	*LS*
				392.180	2851406.0	3106390.8	3–5	5.64+00	2.17–02	8.39–02	–1.187	A	*LS*
				392.183	2851407.5	3106390.8	1–3	4.18+00	2.89–02	3.73–02	–1.539	A	*LS*
				392.242	2851446.0	3106390.8	5–5	1.88+00	4.33–03	2.80–02	–1.664	A	*LS*
				392.180	2851406.0	3106390.8	3–3	3.13+00	7.22–03	2.80–02	–1.664	A	*LS*
				392.242	2851446.0	3106390.8	5–3	2.09–01	2.89–04	1.86–03	–2.840	A	*LS*
54.	1s3p– 1s8s	$^3P^o-^3S$		*374.24*	*2851428*	3118635	9–3	1.88+00	1.32–03	1.46–02	–1.926	A	4
				374.267	2851446.0	3118635	5–3	1.04+00	1.32–03	8.11–03	–2.182	A	*LS*
				374.211	2851406.0	3118635	3–3	6.27–01	1.32–03	4.87–03	–2.403	A	*LS*
				374.213	2851407.5	3118635	1–3	2.09–01	1.32–03	1.62–03	–2.880	A	*LS*
55.	1s3p– 1s8d	$^3P^o-^3D$		*373.01*	*2851428*	3119520.3	9–15	4.97+00	1.73–02	1.91–01	–0.809	A	4
				373.031	2851446.0	3119520.3	5–7	4.97+00	1.45–02	8.90–02	–1.140	A	*LS*
				372.975	2851406.0	3119520.3	3–5	3.73+00	1.29–02	4.77–02	–1.411	A	*LS*
				372.977	2851407.5	3119520.3	1–3	2.76+00	1.73–02	2.12–02	–1.763	A	*LS*
				373.031	2851446.0	3119520.3	5–5	1.24+00	2.59–03	1.59–02	–1.888	A	*LS*
				372.975	2851406.0	3119520.3	3–3	2.07+00	4.32–03	1.59–02	–1.888	A	*LS*
				373.031	2851446.0	3119520.3	5–3	1.38–01	1.73–04	1.06–03	–3.064	A	*LS*
56.	1s3d– 1s4p	$^3D-^3P^o$		*762.99*	*2857312*	2988374.3	15–9	2.926+00	1.532–02	5.774–01	–0.6385	AA	1
				763.031	2857318.0	2988374.3	7–5	2.458+00	1.532–02	2.694–01	–0.9696	AA	*LS*
				762.964	2857306.5	2988374.3	5–3	2.195+00	1.149–02	1.443–01	–1.2406	AA	*LS*
				762.959	2857305.7	2988374.3	3–1	2.926+00	8.513–03	6.415–02	–1.5928	AA	*LS*
				762.964	2857306.5	2988374.3	5–5	4.390–01	3.831–03	4.811–02	–1.7177	AA	*LS*
				762.959	2857305.7	2988374.3	3–3	7.317–01	6.385–03	4.811–02	–1.7177	AA	*LS*
				762.959	2857305.7	2988374.3	3–5	2.927–02	4.257–04	3.208–03	–2.8938	AA	*LS*

C V: Allowed Transitions — Continued

No.	Transition Array	Mult.	λ_{air} (Å)	λ_{vac} (Å) or σ (cm^{-1})*	E_i (cm^{-1})	E_k (cm^{-1})	g_i–g_k	A_{ki} (10^8 s^{-1})	f_{ik}	S (at. u.)	log gf	Acc.	Source
57.	1s3d–1s5p	^{3}D–^{3}P°		*515.33*	*2857312*	3051362.9	15–9	1.248+00	2.980–03	7.584–02	–1.3497	AA	1
				515.345	2857318.0	3051362.9	7–5	1.048+00	2.980–03	3.539–02	–1.6807	AA	*LS*
				515.314	2857306.5	3051362.9	5–3	9.358–01	2.235–03	1.896–02	–1.9517	AA	*LS*
				515.312	2857305.7	3051362.9	3–1	1.248+00	1.656–03	8.426–03	–2.3039	AA	*LS*
				515.314	2857306.5	3051362.9	5–5	1.872–01	7.451–04	6.320–03	–2.4288	AA	*LS*
				515.312	2857305.7	3051362.9	3–3	3.119–01	1.242–03	6.320–03	–2.4288	AA	*LS*
				515.312	2857305.7	3051362.9	3–5	1.248–02	8.279–05	4.213–04	–3.6049	AA	*LS*
58.	1s3d–1s6p	^{3}D–^{3}P°		*438.33*	*2857312*	3085449.5	15–9	6.520–01	1.127–03	2.439–02	–1.7721	AA	1
				438.344	2857318.0	3085449.5	7–5	5.476–01	1.127–03	1.138–02	–2.1031	AA	*LS*
				438.322	2857306.5	3085449.5	5–3	4.890–01	8.451–04	6.098–03	–2.3741	AA	*LS*
				438.320	2857305.7	3085449.5	3–1	6.520–01	6.260–04	2.710–03	–2.7263	AA	*LS*
				438.322	2857306.5	3085449.5	5–5	9.780–02	2.817–04	2.033–03	–2.8512	AA	*LS*
				438.320	2857305.7	3085449.5	3–3	1.630–01	4.695–04	2.033–03	–2.8512	AA	*LS*
				438.320	2857305.7	3085449.5	3–5	6.520–03	3.130–05	1.355–04	–4.0273	AA	*LS*
59.	1s3d–1s7p	^{3}D–^{3}P°		*402.19*	*2857312*	3105951.0	15–9	3.84–01	5.59–04	1.11–02	–2.077	A	4
				402.199	2857318.0	3105951.0	7–5	3.23–01	5.59–04	5.18–03	–2.408	A	*LS*
				402.181	2857306.5	3105951.0	5–3	2.88–01	4.19–04	2.78–03	–2.679	A	*LS*
				402.179	2857305.7	3105951.0	3–1	3.84–01	3.11–04	1.23–03	–3.031	A	*LS*
				402.181	2857306.5	3105951.0	5–5	5.76–02	1.40–04	9.25–04	–3.156	A	*LS*
				402.179	2857305.7	3105951.0	3–3	9.60–02	2.33–04	9.25–04	–3.156	A	*LS*
				402.179	2857305.7	3105951.0	3–5	3.84–03	1.55–05	6.17–05	–4.332	A	*LS*
60.	1s3d–1s8p	^{3}D–^{3}P°		*381.81*	*2857312*	3119224.8	15–9	2.47–01	3.24–04	6.10–03	–2.314	A	4
				381.815	2857318.0	3119224.8	7–5	2.07–01	3.24–04	2.85–03	–2.645	A	*LS*
				381.798	2857306.5	3119224.8	5–3	1.85–01	2.43–04	1.53–03	–2.916	A	*LS*
				381.797	2857305.7	3119224.8	3–1	2.47–01	1.80–04	6.78–04	–3.268	A	*LS*
				381.798	2857306.5	3119224.8	5–5	3.70–02	8.09–05	5.08–04	–3.393	A	*LS*
				381.797	2857305.7	3119224.8	3–3	6.17–02	1.35–04	5.08–04	–3.393	A	*LS*
				381.797	2857305.7	3119224.8	3–5	2.47–03	8.99–06	3.39–05	–4.569	A	*LS*
61.	1s3d–1s3p	^{1}D–^{1}P°		1838.3 cm^{-1}	2857530.3	2859368.6	5–3	3.438–04	9.150–03	8.193+00	–1.3396	AA	1
62.	1s3d–1s4p	^{1}D–^{1}P°		745.534	2857530.3	2991662.4	5–3	1.956+00	9.779–03	1.200–01	–1.3107	AA	1
63.	1s3d–1s5p	^{1}D–^{1}P°		511.480	2857530.3	3053041.3	5–3	8.455–01	1.990–03	1.675–02	–2.0023	AA	1
64.	1s3d–1s6p	^{1}D–^{1}P°		436.856	2857530.3	3086438.8	5–3	4.428–01	7.602–04	5.467–03	–2.4201	AA	1
65.	1s3p–1s4d	^{1}P°–^{1}D		760.104	2859368.6	2990929.5	3–5	4.423+01	6.384–01	4.793+00	0.2822	AA	1
66.	1s3p–1s5d	^{1}P°–^{1}D		517.328	2859368.6	3052669.4	3–5	2.101+01	1.405–01	7.178–01	–0.3752	AA	1
67.	1s3p–1s6d	^{1}P°–^{1}D		440.856	2859368.6	3086200.2	3–5	1.156+01	5.615–02	2.445–01	–0.7735	AA	1
68.	1s3p–1s7d	^{1}P°–^{1}D		404.795	2859368.6	3106407	3–5	7.06+00	2.89–02	1.16–01	–1.062	A	4
69.	1s3p–1s8d	^{1}P°–^{1}D		384.377	2859368.6	3119530	3–5	4.63+00	1.71–02	6.49–02	–1.290	A	4

C V: Allowed Transitions — Continued

No.	Transition Array	Mult.	λ_{air} (Å)	λ_{vac} (Å) or σ (cm^{-1})*	E_i (cm^{-1})	E_k (cm^{-1})	g_i-g_k	A_{ki} (10^8 s^{-1})	f_{ik}	S (at. u.)	log gf	Acc.	Source
70.	1s4s– 1s4p	^{3}S–^{3}P°	20700.9	4829.4 cm^{-1}	2983544.9	2988374.3	3–9	1.569–02	3.025–01	6.186+01	–0.0422	AA	1
71.	1s4s– 1s5p	^{3}S–^{3}P°		1474.53	2983544.9	3051362.9	3–9	3.830+00	3.745–01	5.454+00	0.0506	AA	1
72.	1s4s– 1s6p	^{3}S–^{3}P°		981.310	2983544.9	3085449.5	3–9	2.457+00	1.064–01	1.031+00	–0.4959	AA	1
73.	1s4s– 1s7p	^{3}S–^{3}P°		816.953	2983544.9	3105951.0	3–9	1.59+00	4.79–02	3.86–01	–0.843	A	4
74.	1s4s– 1s8p	^{3}S–^{3}P°		737.029	2983544.9	3119224.8	3–9	1.08+00	2.64–02	1.92–01	–1.101	A	4
75.	1s4p– 1s4d	^{3}P°–^{3}D	41161.4	2428.8 cm^{-1}	2988374.3	2990803.1	9–15	2.008–03	8.505–02	1.038+02	–0.1161	AA	1
76.	1s4p– 1s5s	^{3}P°–^{3}S		1651.30	2988374.3	3048932.7	9–3	6.059+00	8.256–02	4.039+00	–0.1290	AA	1
77.	1s4p– 1s5d	^{3}P°–^{3}D		1557.18	2988374.3	3052593.0	9–15	8.805+00	5.335–01	2.461+01	0.6813	AA	1
78.	1s4p– 1s6s	^{3}P°–^{3}S		1045.09	2988374.3	3084059.8	9–3	3.237+00	1.767–02	5.470–01	–0.7986	AA	1
79.	1s4p– 1s6d	^{3}P°–^{3}D		1022.69	2988374.3	3086155.5	9–15	5.304+00	1.386–01	4.200+00	0.0960	AA	1
80.	1s4p– 1s7s	^{3}P°–^{3}S		856.959	2988374.3	3105066	9–3	1.94+00	7.11–03	1.81–01	–1.194	A	4
81.	1s4p– 1s7d	^{3}P°–^{3}D		847.339	2988374.3	3106390.8	9–15	3.34+00	5.99–02	1.50+00	–0.269	A	4
82.	1s4p– 1s8s	^{3}P°–^{3}S		767.691	2988374.3	3118635	9–3	1.26+00	3.70–03	8.41–02	–1.478	A	4
83.	1s4p– 1s8d	^{3}P°–^{3}D		762.509	2988374.3	3119520.3	9–15	2.22+00	3.23–02	7.29–01	–0.537	A	4
84.	1s4d– 1s5p	^{3}D–^{3}P°		1651.26	2990803.1	3051362.9	15–9	1.529+00	3.750–02	3.057+00	–0.2499	AA	1
85.	1s4d– 1s6p	^{3}D–^{3}P°		1056.56	2990803.1	3085449.5	15–9	7.557–01	7.589–03	3.959–01	–0.9437	AA	1
86.	1s4d– 1s7p	^{3}D–^{3}P°		868.448	2990803.1	3105951.0	15–9	4.31–01	2.93–03	1.26–01	–1.358	A	4
87.	1s4d– 1s8p	^{3}D–^{3}P°		778.685	2990803.1	3119224.8	15–9	2.72–01	1.48–03	5.70–02	–1.653	A	4
88.	1s4d– 1s4p	^{1}D–^{1}P°		732.9 cm^{-1}	2990929.5	2991662.4	5–3	9.373–05	1.570–02	3.525+01	–1.1052	AA	1
89.	1s4d– 1s5p	^{1}D–^{1}P°		1610.00	2990929.5	3053041.3	5–3	1.078+00	2.513–02	6.660–01	–0.9008	AA	1
90.	1s4d– 1s6p	^{1}D–^{1}P°		1047.02	2990929.5	3086438.8	5–3	5.413–01	5.338–03	9.199–02	–1.5737	AA	1

C V: Allowed Transitions — Continued

No.	Transition Array	Mult.	λ_{air} (Å)	λ_{vac} (Å) or σ (cm^{-1})*	E_i (cm^{-1})	E_k (cm^{-1})	g_i-g_k	A_{ki} (10^8 s^{-1})	f_{ik}	S (at. u.)	log gf	Acc.	Source
91.	$1s4d-1s7p$	$^1D-^1P^o$		864.966	2990929.5	3106541	5−3	3.09−01	2.08−03	2.96−02	−1.983	A	4
92.	$1s4d-1s8p$	$^1D-^1P^o$		777.064	2990929.5	3119619	5−3	1.94−01	1.06−03	1.35−02	−2.278	A	4
93.	$1s4p-1s5d$	$^1P^o-^1D$		1639.16	2991662.4	3052669.4	3−5	9.439+00	6.337−01	1.026+01	0.2790	AA	1
94.	$1s4p-1s6d$	$^1P^o-^1D$		1057.78	2991662.4	3086200.2	3−5	5.427+00	1.517−01	1.585+00	−0.3418	AA	1
95.	$1s4p-1s7d$	$^1P^o-^1D$		871.501	2991662.4	3106407	3−5	3.33+00	6.33−02	5.45−01	−0.722	A	4
96.	$1s4p-1s8d$	$^1P^o-^1D$		782.059	2991662.4	3119530	3−5	2.20+00	3.35−02	2.59−01	−0.997	A	4
97.	$1s5s-1s5p$	$^3S-^3P^o$	41137.7	2430.2 cm^{-1}	3048932.7	3051362.9	3−9	5.035−03	3.834−01	1.558+02	0.0608	AA	1
98.	$1s5s-1s6p$	$^3S-^3P^o$	2737.66	2738.47	3048932.7	3085449.5	3−9	1.223+00	4.125−01	1.116+01	0.0925	AA	1
99.	$1s5s-1s7p$	$^3S-^3P^o$		1753.82	3048932.7	3105951.0	3−9	8.50−01	1.18−01	2.04+00	−0.452	A	4
100.	$1s5s-1s8p$	$^3S-^3P^o$		1422.63	3048932.7	3119224.8	3−9	5.86−01	5.33−02	7.49−01	−0.796	A	4
101.	$1s5p-1s5d$	$^3P^o-^3D$		1230.1 cm^{-1}	3051362.9	3052593.0	9−15	7.119−04	1.176−01	2.832+02	0.0245	AA	1
102.	$1s5p-1s6s$	$^3P^o-^3S$	3057.50	3058.39	3051362.9	3084059.8	9−3	2.454+00	1.147−01	1.040+01	0.0139	AA	1
103.	$1s5p-1s6d$	$^3P^o-^3D$	2873.33	2874.17	3051362.9	3086155.5	9−15	2.607+00	5.381−01	4.583+01	0.6851	AA	1
104.	$1s5p-1s7s$	$^3P^o-^3S$		1862.09	3051362.9	3105066	9−3	1.42+00	2.47−02	1.36+00	−0.653	A	4
105.	$1s5p-1s7d$	$^3P^o-^3D$		1817.26	3051362.9	3106390.8	9−15	1.75+00	1.44−01	7.76+00	0.113	A	4
106.	$1s5p-1s8s$	$^3P^o-^3S$		1486.50	3051362.9	3118635	9−3	9.04−01	9.98−03	4.40−01	−1.047	A	4
107.	$1s5p-1s8d$	$^3P^o-^3D$		1467.19	3051362.9	3119520.3	9−15	1.18+00	6.36−02	2.76+00	−0.243	A	4
108.	$1s5d-1s6p$	$^3D-^3P^o$	3042.65	3043.54	3052593.0	3085449.5	15−9	7.612−01	6.342−02	9.532+00	−0.0217	AA	1
109.	$1s5d-1s7p$	$^3D-^3P^o$		1874.13	3052593.0	3105951.0	15−9	4.16−01	1.31−02	1.22+00	−0.706	A	4
110.	$1s5d-1s8p$	$^3D-^3P^o$		1500.78	3052593.0	3119224.8	15−9	2.54−01	5.14−03	3.81−01	−1.113	A	4
111.	$1s5d-1s5p$	$^1D-^1P^o$		371.9 cm^{-1}	3052669.4	3053041.3	5−3	3.300−05	2.146−02	9.499+01	−0.9694	AA	1

C V: Allowed Transitions — Continued

No.	Transition Array	Mult.	λ_{air} (Å)	λ_{vac} (Å) or σ (cm^{-1})*	E_i (cm^{-1})	E_k (cm^{-1})	g_i-g_k	A_{ki} (10^8 s^{-1})	f_{ik}	S (at. u.)	log gf	Acc.	Source
112.	$1s5d-1s6p$	$^1D-^1P^o$	2960.40	2961.26	3052669.4	3086438.8	5–3	5.541–01	4.370–02	2.130+00	−0.6605	AA	1
113.	$1s5d-1s7p$	$^1D-^1P^o$		1856.27	3052669.4	3106541	5–3	3.06–01	9.49–03	2.90–01	−1.324	A	4
114.	$1s5d-1s8p$	$^1D-^1P^o$		1493.66	3052669.4	3119619	5–3	1.87–01	3.76–03	9.25–02	−1.726	A	4
115.	$1s5p-1s6d$	$^1P^o-^1D$	3014.90	3015.78	3053041.3	3086200.2	3–5	2.871+00	6.524–01	1.943+01	0.2916	AA	1
116.	$1s5p-1s7d$	$^1P^o-^1D$		1873.86	3053041.3	3106407	3–5	1.83+00	1.61–01	2.98+00	−0.316	A	4
117.	$1s5p-1s8d$	$^1P^o-^1D$		1504.01	3053041.3	3119530	3–5	1.22+00	6.87–02	1.02+00	−0.686	A	4
118.	$1s6s-1s6p$	$^3S-^3P^o$		1389.7 cm^{-1}	3084059.8	3085449.5	3–9	1.989–03	4.633–01	3.293+02	0.1430	AA	1
119.	$1s6s-1s7p$	$^3S-^3P^o$	4566.77	4568.05	3084059.8	3105951.0	3–9	4.83–01	4.53–01	2.04+01	0.133	A	4
120.	$1s6s-1s8p$	$^3S-^3P^o$	2842.90	2843.74	3084059.8	3119224.8	3–9	3.55–01	1.29–01	3.62+00	−0.413	A	4
121.	$1s6p-1s6d$	$^3P^o-^3D$		706.0 cm^{-1}	3085449.5	3086155.5	9–15	2.956–04	1.482–01	6.219+02	0.1251	AA	1
122.	$1s6p-1s7s$	$^3P^o-^3S$	5096.33	5097.75	3085449.5	3105066	9–3	1.14+00	1.47–01	2.23+01	0.123	A	4
123.	$1s6p-1s7d$	$^3P^o-^3D$	4773.92	4775.25	3085449.5	3106390.8	9–15	9.76–01	5.56–01	7.86+01	0.699	A	4
124.	$1s6p-1s8s$	$^3P^o-^3S$	3012.49	3013.36	3085449.5	3118635	9–3	7.01–01	3.18–02	2.84+00	−0.543	A	4
125.	$1s6p-1s8d$	$^3P^o-^3D$	2934.21	2935.06	3085449.5	3119520.3	9–15	6.99–01	1.50–01	1.31+01	0.131	A	4
126.	$1s6d-1s7p$	$^3D-^3P^o$	5050.24	5051.65	3086155.5	3105951.0	15–9	3.99–01	9.16–02	2.28+01	0.138	A	4
127.	$1s6d-1s8p$	$^3D-^3P^o$	3023.07	3023.95	3086155.5	3119224.8	15–9	2.34–01	1.93–02	2.88+00	−0.539	A	4
128.	$1s6d-1s6p$	$^1D-^1P^o$		238.6 cm^{-1}	3086200.2	3086438.8	5–3	1.704–05	2.692–02	1.857+02	−0.8709	AA	1
129.	$1s6d-1s7p$	$^1D-^1P^o$	4914.86	4916.23	3086200.2	3106541	5–3	2.946–01	6.402–02	5.180+00	−0.4947	AA	1
130.	$1s6d-1s8p$	$^1D-^1P^o$	2991.46	2992.33	3086200.2	3119619	5–3	1.766–01	1.421–02	6.998–01	−1.1484	AA	1
131.	$1s6p-1s7d$	$^1P^o-^1D$	5006.57	5007.96	3086438.8	3106407	3–5	1.090+00	6.825–01	3.375+01	0.3113	AA	1
132.	$1s6p-1s8d$	$^1P^o-^1D$	3021.07	3021.95	3086438.8	3119530	3–5	7.472–01	1.704–01	5.084+00	−0.2914	AA	1

*Wavelengths (Å) are always given unless cm^{-1} is indicated.

C V

Forbidden Transitions

Most tabulated data are from calculations which are based on correlated, but nonrelativistic, wavefunctions.[1,3–5] Drake has used either a nuclear charge expansion calculation[1] with the perturbation coefficient evaluated through ninth order or he has used 50-term variational wavefunctions.[3] For the helium-like carbon and nitrogen spectra, where high precision experimental data for the M1 line are available[2] (see below and the N VI introduction on forbidden lines), the agreement between the experiments and Drake's results is almost perfect. Godefroid and Verhaegen[5] use an extensive multiconfiguration approach and have calculated their data both in the length and velocity form. These two forms usually agree within four significant figures.

For the magnetic dipole transition from $1s2s\ ^3S_1$ to the ground state, a high-precision measurement of the decay rate has been made by Schmidt *et al.*[2] on an ion storage ring. Their result, estimated to be accurate within 0.2% (one standard deviation), agrees perfectly with Drake's work.

References

[1]G. W. F. Drake, Phys. Rev. A **3**, 908 (1971).

[2]H. T. Schmidt, P. Forck, M. Grieser, D. Habs, J. Kenntner, G. Miersch, R. Repnow, U. Schramm, T. Schüssler, D. Schwalm, and A. Wolf, Phys. Rev. Lett. **72**, 1616(1994).

[3]G. W. F. Drake, Astrophys. J. **158**, 1199 (1969).

[4]B. Kundu, P. K. Mukherjee, and H. P. Roy, Phys. Scr. **39**, 722 (1989).

[5]M. Godefroid and G. Verhaegen, J. Phys. B. **13**, 3081 (1980).

C V: Forbidden Transitions

No.	Transition Array	Mult.	λ_{air} (Å)	λ_{vac} (Å) or σ (cm⁻¹)*	E_i (cm⁻¹)	E_k (cm⁻¹)	g_i–g_k	Type	A_{ki} (s⁻¹)	S (at. u.)	Acc.	Source
1.	$1s^2$–$1s2s$	^{1}S–^{3}S										
				41.4721	0	2411262	1–3	M1	4.857+01	3.853–07	AA	1,2
2.	$1s^2$–$1s2p$	^{1}S–^{3}P°										
				40.7285	0	2455284	1–5	M2	2.62+04	9.85–01	A	3
	$1s^2$–$1s3p$	^{1}S–^{3}P°										
				35.0703	0	2851418	1–5	M2	9.30+03	1.65–01	C	4
	s^2–$s3d$	^{1}S–^{1}D										
				34.9953	0	2857529	1–5	E2	1.296+07	3.037–03	AA	5
		^{1}S–^{3}P°										
				33.4632	0	2988359	1–5	M2	4.16+03	5.85–02	C	4
		^{1}S–^{3}P°										
				32.7726	0	3051332	1–5	M2	2.15+03	2.73–02	C	4
		^{1}S–^{3}P°										
				32.4103	0	3085435	1–5	M2	1.27+03	1.52–02	C	4
		^{1}S–^{3}P°										
				32.1964	0	3105933	1–5	M2	8.48+02	9.84–03	C	4
		^{1}S–^{1}D										
				248.444	2455024	2857529	1–5	E2	6.183+05	2.613+00	AA	5
		^{1}P°–^{1}F°										
				197.024	2483371	2990923	3–7	E2	6.428+05	1.193+00	AA	5

's given unless cm⁻¹ is indicated.

C VI

Hydrogen Isoelectronic Sequence

Ground State: $1s\ ^2S_{1/2}$

Ionization Energy: 489.981 eV = 3952061.3 cm^{-1}

Allowed Transitions

We have tabulated numerical data for hydrogen-like ions only for the spectrum of N VII (nuclear charge $Z = 7$). Data for C VI ($Z = 6$) may be obtained by scaling the data for the *corresponding transitions* of N VII according to the following relationships:[1]

$$f(\text{C VI}) = f(\text{N VII}),$$

$$A(\text{C VI}) = \left(\frac{6}{7}\right)^4 A(\text{N VII}) = 0.5398\, A(\text{N VII}),$$

$$S(\text{C VI}) = \left(\frac{6}{7}\right)^{-2} S(\text{N VII}) = 1.3611\, S(\text{N VII}).$$

The wavelength and energy level data for C VI may be obtained from the calculations of Erickson,[2] and Johnson and Soff.[3]

References

[1]A. Corney, Atomic and Laser Spectroscopy, Oxford Univ. Press, Oxford (1977).
[2]G. W. Erickson, J. Phys. Chem. Ref. Data **6**, 831 (1977).
[3]W. R. Johnson and G. Soff, At. Data Nucl. Data Tables **33**, 405 (1985).

Nitrogen

N I

Ground State: $1s^22s^22p^3\ {}^4S^{\circ}_{3/2}$

Ionization Energy: 14.534 eV = 117225.7 cm^{-1}

Allowed Transitions

List of tabulated lines

Wavelength (Å)	No.	Wavelength (Å)	No.	Wavelength (Å)	No.	Wavelength (Å)	No.
in vacuum		1090.20	44	1319.68	32	4660.46	54
		1093.23	43	1326.56	31	4662.91	119
865.227	14	1093.25	43	1326.57	31	4669.89	54
865.835	14	1095.09	42	1327.92	31	4670.68	119
866.180	14	1095.16	42	1411.93	30	4678.59	54
869.404	13	1097.24	23	1411.94	30	4914.94	53
870.026	13	1098.15	23	1411.95	30	4935.12	53
870.367	13	1098.26	23	1492.63	15	5199.84	67
875.277	12	1100.36	22	1492.82	15	5201.61	67
875.657	12	1100.47	22	1494.68	15	5281.20	63
875.721	12	1101.29	22	1742.72	29	5310.39	95
876.067	11	1106.55	41	1742.73	29	5328.62	62
876.646	11	1134.17	2	1745.25	29	5333.37	95
876.988	11	1134.41	2	1745.26	29	5344.05	62
885.973	10	1134.98	2			5346.49	95
886.226	10	1163.88	21	in air		5354.39	62
886.333	10	1164.00	21			5356.62	62
887.458	9	1164.21	21	3818.27	58	5367.01	62
888.024	9	1164.32	21	3822.03	58	5372.61	62
888.372	9	1167.45	19	3830.43	58	5372.67	62
906.207	8	1168.33	20	3834.22	58	5378.27	62
906.432	8	1168.42	19	3887.57	57	5471.80	94
906.619	8	1168.54	19	3892.21	57	5496.52	94
909.697	7	1170.67	18	3900.18	57	5510.19	94
910.278	7	1171.08	18	4011.13	56	5511.70	77
910.645	7	1171.20	18	4024.56	56	5513.09	77
952.303	6	1176.51	17	4095.05	110	5518.10	77
952.415	6	1176.63	17	4099.94	55	5518.58	77
952.523	6	1177.69	17	4107.90	110	5519.97	77
953.415	5	1188.97	39	4109.95	55	5529.38	77
953.655	5	1189.23	39	4113.97	55	5529.52	77
953.970	5	1190.92	38	4137.64	49	5529.99	77
954.104	4	1190.93	38	4143.43	49	5545.00	77
963.990	3	1191.01	38	4151.48	49	5545.14	77
964.626	3	1191.02	38	4249.87	109	5557.38	76
965.041	3	1199.55	1	4250.16	109	5560.34	76
980.623	28	1200.22	1	4264.00	109	5564.27	76
980.706	28	1200.71	1	4273.28	123	5564.38	76
989.265	27	1225.03	37	4278.13	108	5575.87	76
989.335	27	1225.37	37	4279.17	108	5575.98	76
989.350	27	1228.41	36	4292.16	108	5580.08	76
990.773	26	1228.79	36	4293.21	108	5591.76	76
990.829	26	1243.17	16	4356.29	107	5713.14	93
990.858	26	1243.18	16	4385.54	122	5729.35	93
990.914	26	1243.31	16	4392.41	122	5732.11	93
1000.14	25	1310.54	35	4435.11	121	5734.40	85
1000.23	25	1310.94	35	4435.43	121	5735.91	85
1067.61	24	1310.95	35	4442.14	121	5739.94	85
1068.51	24	1316.04	34	4442.45	121	5740.46	85
1068.61	24	1316.29	33	4465.90	120	5741.97	85
1082.71	45	1319.00	32	4474.17	120	5746.96	92
1090.19	44	1319.67	32	4651.82	54	5752.50	85

List of tabulated lines — Continued

Wavelength (Å)	No.	Wavelength (Å)	No.	Wavelength (Å)	No.	Wavelength (Å)	No.
5752.65	85	6741.67	83	8683.40	46	10507.0	81
5753.16	85	6752.03	83	8686.15	46	10513.4	81
5771.97	92	6759.20	83	8703.25	46	10520.6	81
5786.83	92	6769.61	83	8711.70	46	10533.8	81
5999.43	66	6926.67	82	8718.84	46	10539.6	81
6008.47	66	6945.18	82	8728.90	46	10549.6	81
6187.12	106	6951.60	82	8747.37	46	10563.3	81
6194.31	106	6960.50	82	9028.92	65	10584.6	236
6201.98	91	6973.07	82	9045.88	127	10591.0	236
6214.41	91	6979.18	82	9049.49	127	10596.0	236
6216.51	106	6982.03	82	9049.89	127	10623.2	80
6219.65	91	7306.56	104	9060.48	65	10644.0	80
6223.76	106	7318.98	104	9187.45	126	10653.0	80
6237.67	105	7347.57	104	9187.86	126	10672.7	245
6238.59	105	7351.36	89	9207.59	126	10673.9	80
6268.46	105	7360.13	104	9208.00	126	10689.1	245
6275.51	90	7366.15	89	9361.21	233	10713.5	80
6303.92	90	7378.51	89	9382.12	233	10718.0	80
6321.51	90	7406.12	103	9386.81	51	10730.5	79
6398.40	75	7406.24	103	9392.79	51	10757.9	80
6402.36	75	7423.64	48	9395.85	64	10786.8	230
6405.26	75	7428.75	102	9460.68	51	10983.4	235
6407.67	75	7442.30	48	9464.17	64	10988.7	235
6411.65	75	7442.62	102	9530.49	237	11180.1	68
6419.87	75	7448.26	103	9534.50	237	11227.1	68
6420.64	75	7468.31	48	9776.90	72	11237.6	68
6423.07	75	7485.18	102	9786.78	72	11257.3	264
6440.94	75	7507.61	101	9788.29	72	11266.2	68
6441.71	75	7546.21	101	9798.56	72	11291.7	68
6481.71	74	7550.91	101	9810.01	72	11294.3	68
6482.70	74	7587.57	88	9814.02	72	11313.9	68
6483.75	74	7628.18	88	9822.75	72	11323.2	68
6484.81	74	7654.04	88	9834.61	72	11505.7	229
6491.22	74	7703.51	201	9863.33	72	11537.3	229
6499.52	74	7727.21	201	9872.15	72	11566.1	61
6506.30	74	7886.22	116	9883.38	71	11625.2	61
6507.02	74	7893.98	116	9905.52	71	11651.5	61
6521.11	74	7898.98	130	9909.22	71	11762.5	234
6595.84	118	7899.28	130	9931.47	71	11768.6	234
6603.21	118	7908.47	116	9947.07	70	11964.1	100
6606.18	73	7915.42	130	9965.75	71	11998.3	100
6611.40	118	8105.67	115	9968.51	71	12074.5	100
6622.54	73	8129.17	115	9980.42	70	12085.7	263
6626.99	73	8150.68	115	9997.73	70	12109.3	100
6636.94	73	8166.23	129	10003.0	71	12123.5	263
6644.96	73	8166.56	129	10017.8	70	12130.0	78
6646.50	73	8174.45	115	10054.3	71	12186.8	78
6648.26	117	8184.86	47	10105.1	69	12203.9	78
6651.15	117	8188.01	47	10108.9	69	12229.1	144
6653.46	73	8200.36	47	10112.5	69	12231.4	78
6656.51	73	8201.44	129	10114.6	69	12267.8	144
6664.06	117	8201.76	129	10128.3	69	12270.8	78
6666.96	117	8210.72	47	10147.3	69	12288.8	87
6700.49	84	8216.34	47	10164.8	69	12289.2	78
6704.84	84	8223.13	47	10166.8	69	12295.3	196
6706.11	84	8242.39	47	10200.0	69	12298.5	78
6708.76	84	8567.74	52	10211.5	232	12328.8	87
6713.11	84	8594.00	52	10213.2	232	12345.1	144
6720.97	83	8629.24	52	10238.1	232	12381.6	87
6722.61	84	8655.75	128	10376.3	231	12384.5	144
6723.45	84	8655.88	52	10382.4	231	12419.2	196
6726.12	84	8656.11	128	10401.9	231	12461.3	99
6729.28	83	8666.94	128	10408.1	231	12469.6	99
6733.32	83	8680.28	46	10500.3	81	12469.8	262

List of tabulated lines — Continued

Wavelength (Å)	No.	Wavelength (Å)	No.	Wavelength (Å)	No.	Wavelength (Å)	No.
12472.3	262	14090.1	155	15822.8	163	17982.6	187
12490.5	196	14143.5	142	15845.5	163	18019.8	186
12510.0	262	14172.4	142	15865.5	163	18428.2	166
12512.5	262	14195.2	155	15870.1	140	18429.3	192
12551.7	125	14232.5	155	15895.5	163	18509.5	166
12552.5	125	14298.9	142	15900.9	163	18509.6	166
12581.0	99	14313.2	86	15926.6	163	18539.6	192
12615.6	125	14328.4	142	15931.1	163	18586.3	192
12688.7	136	14454.6	86	16254.8	176	18726.4	200
12708.9	98	14455.1	141	16266.9	176	18800.6	200
12716.3	261	14455.3	260	16302.7	176	18867.1	200
12758.1	261	14509.4	260	16306.8	176	19094.2	191
12761.0	136	14522.8	113	16319.0	176	19359.6	191
12767.4	261	14548.5	86	16358.0	175	19526.5	191
12771.5	98	14598.4	113	16398.1	176	19680.8	244
12873.3	136	14604.6	113	16399.3	176	19736.8	244
12897.3	98	14617.4	141	16403.4	176	24531.8	253
12969.6	143	14681.1	113	16406.7	167	25029.9	253
12976.2	151	14693.5	194	16410.7	175	25090.1	259
13010.5	143	14757.1	59	16444.8	175	25253.6	259
13019.8	151	14801.2	194	16445.7	167	25321.5	253
13041.4	151	14819.6	194	16471.2	167	26149.6	133
13045.9	135	14868.9	59	16498.0	175	26458.6	133
13051.9	135	14919.3	193	16535.0	175	26536.4	168
13069.9	135	14952.0	59	16545.1	176	26714.2	168
13085.5	151	14966.6	59	16546.3	176	26945.8	133
13100.1	143	14974.4	160	16596.9	175	27489.7	173
13122.4	135	15050.9	59	16634.4	175	27563.0	173
13146.6	135	15052.4	160	16658.7	174	27593.8	173
13169.3	135	15089.0	193	16682.4	174	27638.7	173
13172.2	124	15095.0	59	16708.7	174	27712.8	173
13173.0	124	15102.3	59	16712.3	174	27837.6	172
13196.5	195	15112.4	160	16713.3	174	27857.2	173
13241.1	135	15146.7	59	16785.7	175	27871.6	173
13254.3	134	15152.4	149	16813.8	174	27917.5	173
13264.5	134	15185.6	149	16814.8	174	27990.4	172
13284.7	134	15191.0	193	16860.6	148	28050.8	172
13292.4	134	15198.5	153	16861.4	174	28206.0	172
13295.4	124	15241.5	149	16862.8	181	28284.1	173
13296.2	124	15275.1	149	16875.8	181	28299.1	173
13333.2	134	15277.2	153	16902.7	181	28388.1	172
13341.3	195	15286.6	153	16907.1	181	28401.9	252
13371.8	134	15331.0	153	16920.2	181	28496.4	172
13385.8	134	15376.4	159	16928.5	148	28682.3	172
13422.1	195	15378.0	153	16968.4	174	28699.2	139
13429.6	50	15392.0	153	16973.9	180	29031.0	139
13448.1	97	15411.1	153	17018.8	180	29081.0	252
13455.9	134	15422.9	153	17019.5	181	29096.3	171
13464.5	60	15429.2	153	17020.8	181	29107.5	171
13534.6	60	15458.7	159	17025.3	181	29112.1	171
13544.6	60	15474.4	159	17039.8	148	29122.5	171
13581.3	50	15496.2	112	17067.4	180	29135.2	172
13587.7	97	15557.7	159	17112.8	180	29263.3	171
13588.5	97	15582.3	112	17152.6	180	29274.7	179
13602.3	114	15591.8	159	17233.8	180	29346.3	139
13615.6	60	15621.8	159	17274.2	180	29357.8	179
13624.2	114	15656.2	159	17347.3	188	29357.9	179
13649.7	60	15679.0	140	17403.9	188	29408.7	179
13651.6	60	15682.9	112	17561.6	188	29421.6	171
13668.6	114	15771.1	112	17619.6	188	29467.6	252
13686.0	60	15792.2	163	17731.5	187	29492.5	179
13713.4	150	15794.1	169	17758.0	187	29576.0	171
13813.0	150	15810.4	169	17773.6	186	29589.5	171
13887.0	150	15810.4	140	17794.3	186	29634.5	132

List of tabulated lines — Continued

Wavelength (Å)	No.	Wavelength (Å)	No.	Wavelength (Å)	No.	Wavenumber (cm^{-1})	No.
29666.3	132	35177.4	199	47394.0	164	182.388	216
29669.6	178	35304.2	177	47541.2	152	234.745	216
29693.4	139	35458.2	177	47612.9	215	258.545	216
29699.4	179	35467.1	199	47662.6	152	275.867	211
29715.9	179	35499.1	177	47713.4	152	300.943	211
29768.0	179	35706.3	96	47728.8	152	324.743	211
29771.8	132	35712.5	96	47733.8	157	417.216	205
29807.2	178	36378.4	251	48188.4	215	441.016	205
29898.3	171	37045.7	251	48205.2	215	481.226	257
29911.9	178	37161.4	251	48462.0	164	485.330	205
30032.0	132	37630.2	198	48795.2	215	487.762	257
30051.7	178	37794.6	250	48921.7	156	507.024	257
30173.0	132	37902.4	146	48987.1	162	513.560	257
30258.6	178	37988.8	198	49028.4	164	607.250	258
30280.2	132	38150.1	158	49283.6	162	626.004	258
30410.4	185	38202.7	198	49504.5	162	633.048	258
30427.1	178	38572.3	198	49696.6	162	743.027	223
30626.6	185	38660.8	158			748.104	221
30639.2	178	38903.2	250			751.816	223
30661.3	132	38940.5	146			757.772	221
31075.1	185	39059.1	158			763.802	221
31300.9	185	39534.3	146			766.239	223
31678.8	131	39588.7	250			769.695	221
31754.3	170	39744.8	197			779.363	221
31764.6	131	39855.1	165			793.553	218
31876.8	131	40237.4	165			800.796	221
31908.0	131	40294.7	189			802.660	221
32122.8	170	40384.0	197			808.553	221
32133.4	131	40497.9	165			815.144	218
32152.2	184	40851.0	197			815.722	212
32212.7	184	40926.4	154			820.846	218
32272.3	170	41230.0	228			840.798	212
32337.2	131	41467.0	189			842.437	218
32469.4	131	41526.6	197			857.653	221
32477.8	170	41551.3	228			859.517	221
32645.3	183	41638.7	228			865.184	218
32691.9	170	41966.5	228			876.133	222
32740.4	170	42139.4	154			881.295	218
32854.9	131	42151.4	154			884.288	222
32863.5	170	42243.0	189			885.801	222
32907.8	138	42607.1	227			890.181	222
32914.8	183	42851.9	137			895.422	254
32951.9	170	43628.3	227			899.849	222
32959.5	184	43722.7	226			904.042	218
32978.2	138	43769.3	209			906.909	255
33182.5	147	43848.1	226			921.582	219
33612.7	147	44086.1	227			923.578	222
33626.9	147	44269.3	209			925.442	222
33694.9	183	44310.6	226			931.335	222
33761.4	138	44310.8	137			935.630	219
34068.8	147	45114.6	209			947.970	254
34113.1	177	45196.5	157			948.875	219
34225.0	182	45646.0	209			948.913	210
34421.2	190	45742.1	152			952.762	210
34597.5	177	45915.1	157			953.665	213
34747.8	190	46050.8	157			954.506	254
34758.2	96	46462.5	152			956.728	210
34760.6	177	46491.8	152			957.071	206
34764.1	96	46797.1	157			957.640	210
34949.5	177	46864.6	164			959.457	255
34973.1	199	46904.8	152			960.899	218
35024.7	190	47140.3	157			962.923	219
35042.1	182	47286.2	152			968.088	213
35069.2	182	47382.0	157			970.504	210

List of tabulated lines — Continued

Wavenumber (cm^{-1})	No.	Wavenumber (cm^{-1})	No.	Wavenumber (cm^{-1})	No.	Wavenumber (cm^{-1})	No.
978.741	213	1368.90	249	1607.878	214	1765.509	241
985.629	255	1382.817	224	1608.719	242	1769.79	161
985.670	219	1400.137	203	1616.399	239	1774.36	161
991.620	210	1406.029	224	1618.9	247	1782.07	161
993.164	213	1415.076	203	1619.304	217	1793.97	161
1004.077	219	1429.124	203	1620.443	240	1796.859	161
1009.362	210	1435.365	224	1623.432	247	1801.775	208
1013.585	210	1458.577	224	1624.986	247	1805.91	161
1021.446	256	1459.363	203	1632.9	247	1817.341	208
1025.185	206	1470.278	203	1632.954	214	1818.19	161
1026.824	219	1473.411	203	1634.00	240	1823.770	145
1040.200	256	1497.661	246	1637.480	247	1846.975	225
1048.477	210	1499.452	239	1644.848	204	1849.561	225
1064.102	248	1511.22	246	1649.757	241	1851.035	161
1073.994	256	1536.852	239	1650.206	241	1857.587	111
1092.748	256	1538.815	246	1659.301	240	1858.072	111
1095.014	207	1543.147	217	1672.86	240	1860.09	161
1108.244	248	1552.38	246	1685.502	214	1862.330	145
1109.437	207	1555.950	239	1689.135	204	1872.63	156
1151.354	220	1556.309	239	1694.45	240	1872.773	225
1163.128	207	1563.586	240	1705.060	241	1892.24	156
1177.551	207	1566.42	246	1705.11	240	1893.732	111
1178.647	220	1566.756	217	1706.614	241	1899.00	156
1180.021	248	1574.707	204	1707.063	241	1918.61	156
1201.394	220	1582.278	247	1713.784	243	1953.227	156
1238.497	202	1583.832	247	1726.70	240	1954.73	156
1277.355	202	1584.281	247	1728.207	243	1984.21	162
1287.047	202	1584.63	246	1732.538	243	1987.48	162
1308.638	202	1587.760	242	1739.3	241	1989.349	156
1323.091	249	1593.709	239	1740.126	238	1996.11	162
1331.334	202	1594.296	242	1743.918	241	1999.76	162
1334.212	202	1594.808	239	1745.472	241		
1336.65	249	1595.167	239	1746.662	238		
1347.496	202	1598.67	246	1749.227	208		
1352.925	202	1600.475	239	1760.9	241		
1358.983	203	1602.183	242	1762.46	161		

For the spectrum of neutral nitrogen the status of atomic transition probability data is still not satisfactory, despite numerous calculations and experiments.

As usual, our principal data sources are advanced atomic structure calculations, but we have utilized some experimental emission data, too. The main source of multiplet data is the Opacity Project (OP),[1] which is reviewed in the general introduction. Bell and Berrington[2] undertook additional calculations for selected transitions with the same R-matrix approach used by OP, paying special attention to the description of the wavefunctions of some N$^+$ target states. Hibbert *et al.*[3] have calculated the oscillator strengths of a large number of individual lines in intermediate coupling with the configuration-interaction code CIV 3 (see the general introduction) and have included some intersystem lines. Weiss and Suskin[4] have carried out calculations with the superposition-of-configurations (SOC) code and have mainly addressed multiplets and lines that were the subject of recent experimental studies.

The results of these calculations agree well with each other for many transitions. However, a number of large discrepancies are also encountered, sometimes surpassing factors of two. These occur especially between OP and the other calculations. For some transitions, the differences between the calculations and the experimental results are equally large, again primarily involving the OP data. These discrepancies involve especially the ns ^{2}P and ^{4}P, the nd ^{2}P and ^{4}P, and the $2s2p^4$ ^{4}P terms. Pairs of nd and $(n{+}1)s$ ^{2}P or ^{4}P terms are energetically very close and thus strongly interact with each other, presenting difficult calculational problems. Also, the $2s2p^4$ ^{4}P term is energetically close to the $2s^22p^23s$ ^{4}P term, causing the mixing to be quite strong. An analogous situation has been found for C I and was discussed in the introductory comments for that spectrum.

Hibbert *et al.*,[3] Weiss and Suskin,[4] and Bell and Berrington[2] have given special attention to the optimization of the orbitals for the spectroscopic terms noted above. Their results for the $2p$ ^{4}S° – $3s$ ^{4}P, $2s^22p^3$ ^{4}S° – $2s2p^4$ ^{4}P, and the $3s$ ^{4}P – $3p$ ^{4}S°, ^{4}P° and ^{4}D° multiplets are in close agreement with each other and with the results of emission experiments of Goldbach *et al.*,[5,7] Zhu *et al.*[8] and Musielok *et al.*,[9] which also have been utilized for these tables. The agreement is typically within ten percent, while the differences with the OP results are much larger. On the other hand, for the $2s^22p^3\ ^2P^o - 2s^22p^23d\ ^2P$, ^2D multiplets, the OP results[1] and the emission experiment of Goldbach *et al.*[6] agree within 2% in each case, while the experiment disagrees with the calculated data of Hibbert *et al.*[3] by 28% and 66%, respectively. *A* graphical comparison of *all* emission data with the CIV 3 calculations of Hibbert *et al.*[3] is shown in Fig. 1.

All emission experiments[5–9] show that *LS*-coupling is closely approximated for multiplets involving *s–p* or *p–s* transitions, while for *p–d* multiplets some large departures from this coupling scheme are observed, surpassing factors of two. On the other hand, the agreement between recent emission results[9] and intermediate coupling calculations[3] for *p-d* multiplets is much better. Because of these observed deviations from *LS* coupling, we have assigned conservative uncertainty estimates for lines in pertinent multiplets when *LS* coupling data are the only available source. This especially applies to weak lines.

The intermediate coupling calculations of Hibbert *et al.*[3] have also yielded data for intersystem lines, but the results differ appreciably from the above cited emission experiments[6,7,9], as seen in Fig 1. Since the experimental data for the intersystem lines are practically as accurate as for the *LS*-allowed lines, we have given preference to these data when available.

We should note that all emission experiments derive their absolute scales from lifetime data, usually from a combination of several results. Thus, the data of Goldbach *et al.*[5] for the resonance multiplet (multiplet no. 1), which is the reference transition for their absolute scale, is actually the average of six earlier lifetime measurements, and only the relative data for the three component lines are the results of their own measurements. In this case, the lifetimes were obtained with non-selective excitation techniques. However, the radiative decay rate for the resonance transition is so much faster than those of possible cascading transitions that cascading effects are expected to introduce uncertainties only of the order of a few percent.

The absolute scales of the emission data of Zhu *et al.*[8] and Musielok *et al.*[9] are based on averages of several lifetimes obtained with the state-selective laser-induced-fluorescence (LIF) excitation technique.

References

[1]V. M. Burke and D. J. Lennon, to be published.

[2]K. L. Bell and K. A. Berrington, J. Phys. B **24**, 933 (1991).

[3]A. Hibbert, E. Biemont, M. Godefroid and N. Vaeck, Astron. Astrophys., Suppl. Ser. **88**, 505 (1991).

[4]A. W. Weiss and M. Suskin (unpublished).

[5]C. Goldbach, M. Martin, G. Nollez, P. Plomdeur, J.-P. Zimmermann, D. Babic, Astron. Astrophys. **161**, 47 (1986).

[6]C. Goldbach, T. Lüdtke, M. Martin and G. Nollez, Astron. Astrophys. **266**, 605 (1992).

[7]C. Goldbach and G. Nollez, Astron. Astrophys. **201**, 189 (1988).

[8]Q. Zhu, J. M. Bridges, T. Hahn and W. L. Wiese, Phys. Rev. A **40**, 3721 (1989).

[9]J. Musielok, W. L. Wiese, and G. Veres, Phys. Rev. A **51**, 3588 (1995).

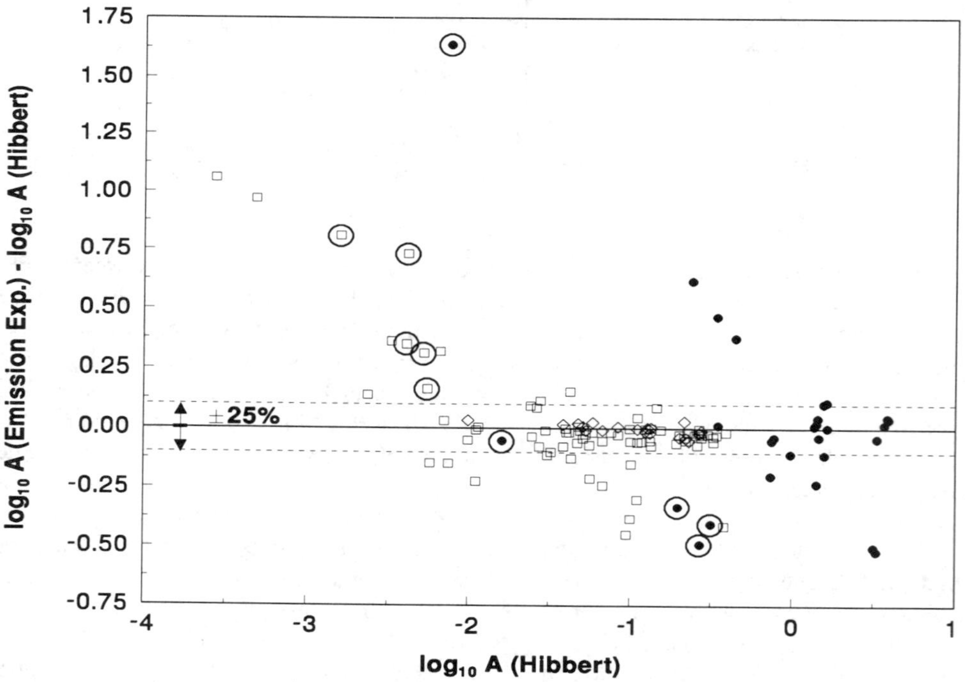

FIG. 1. Comparison of emission data[5–9] with the CIV 3 calculations of Hibbert *et al.*[3] The experiment/theory ratios are plotted on a logarithmic scale versus the logarithm of the calculated transition probabilities *A* (in $10^8\ s^{-1}$). Ratios involving the emission data of Goldbach *et al.*[5–7] are denoted by filled circles, those of Zhu *et al.*[8] by diamonds, and those of Musielok *et al.*[9] by squares. Intersystem lines are indicated by circles around the respective symbols. The broken lines indicate a band of deviations of ±25% around a perfect ratio of 1.00.

N I: Allowed Transitions

No.	Transition Array	Mult.	λ_{air} (Å)	λ_{vac} (Å) or σ (cm^{-1})*	E_i (cm^{-1})	E_k (cm^{-1})	g_i–g_k	A_{ki} (10^8 s^{-1})	f_{ik}	S (at. u.)	log gf	Acc.	Source
1.	$2s^22p^3-$ $2s^22p^2(^3P)3s$	$^4S^o-^4P$		*1200.0*	0.000 –	*83335.60*	4–12	4.00+00	2.59−01	4.09+00	0.015	B+	2,3,4,5
				1199.55	0.000 –	83364.620	4–6	4.01+00	1.30−01	2.05+00	−0.285	B+	3*n*,4*n*,5*n*
				1200.22	0.000 –	83317.830	4–4	3.99+00	8.62−02	1.36+00	−0.462	B+	3*n*,4*n*,5*n*
				1200.71	0.000 –	83284.070	4–2	3.98+00	4.30−02	6.80−01	−0.764	B+	3*n*,4*n*,5*n*
2.	$2s^22p^3-$ $2s2p^4$	$^4S^o-^4P$		*1134.7*	0.000 –	*88132.45*	4–12	1.52+00	8.82−02	1.32+00	−0.452	B	2,3,4,7
				1134.98	0.000 –	88107.260	4–6	1.50+00	4.33−02	6.48−01	−0.761	B	3*n*,4*n*,7*n*
				1134.41	0.000 –	88151.170	4–4	1.54+00	2.96−02	4.43−01	−0.926	B	3*n*,4*n*,7*n*
				1134.17	0.000 –	88170.570	4–2	1.58+00	1.52−02	2.28−01	−1.215	B	3*n*,4*n*,7*n*
3.	$2s^22p^3-$ $2s^22p^2(^3P)4s$	$^4S^o-^4P$		*964.38*	0.000 –	*103693.9*	4–12	6.75−01	2.82−02	3.58−01	−0.947	B	2,3,7
				963.990	0.000 –	103735.48	4–6	7.07−01	1.48−02	1.88−01	−1.228	B	3*n*,7*n*
				964.626	0.000 –	103667.16	4–4	6.76−01	9.43−03	1.20−01	−1.424	B	3*n*,7*n*
				965.041	0.000 –	103622.51	4–2	5.76−01	4.02−03	5.11−02	−1.794	B	3*n*,7*n*
4.	$2s^22p^3-$ $2s^22p^2(^3P)3d$	$^4S^o-^2F$											
				954.104	0.000 –	104810.360	4–6	3.30−01	6.75−03	8.48−02	−1.569	D	7
5.		$^4S^o-^4P$		*953.77*	0.000 –	*104846.8*	4–12	1.78+00	7.30−02	9.17−01	−0.534	B+	2,3,7
				953.970	0.000 –	104825.110	4–6	1.70+00	3.49−02	4.38−01	−0.855	B	3*n*,7*n*
				953.655	0.000 –	104859.73	4–4	1.83+00	2.50−02	3.14−01	−1.000	B	3*n*,7*n*
				953.415	0.000 –	104886.10	4–2	1.93+00	1.31−02	1.65−01	−1.280	B	3*n*,7*n*
6.		$^4S^o-^4D$											
				952.303	0.000 –	105008.55	4–6	9.19−02	1.88−03	2.35−02	−2.125	C+	7
				952.415	0.000 –	104996.27	4–4	1.25−01	1.70−03	2.13−02	−2.167	C+	7
				952.523	0.000 –	104984.37	4–2	8.82−02	6.00−04	7.53−03	−2.620	C	7
7.	$2s^22p^3-$ $2s^22p^2(^3P)5s$	$^4S^o-^4P$		*910.05*	0.000 –	*109884.21*	4–12	2.93−01	1.09−02	1.31−01	−1.361	C	2
				909.697	0.000 –	109926.661	4–6	2.93−01	5.45−03	6.53−02	−1.661	D+	*LS*
				910.278	0.000 –	109856.520	4–4	2.92−01	3.63−03	4.35−02	−1.838	D+	*LS*
				910.645	0.000 –	109812.233	4–2	2.92−01	1.82−03	2.18−02	−2.139	D+	*LS*
8.	$2s^22p^3-$ $2s^22p^2(^3P)4d$	$^4S^o-^4P$		*906.49*	0.000 –	*110315.90*	4–12	9.98−01	3.69−02	4.40−01	−0.831	C	2
				906.619	0.000 –	110299.974	4–6	9.98−01	1.84−02	2.20−01	−1.132	D+	*LS*
				906.432	0.000 –	110322.721	4–4	9.99−01	1.23−02	1.47−01	−1.308	D+	*LS*
				906.207	0.000 –	110350.014	4–2	9.99−01	6.15−03	7.34−02	−1.609	D+	*LS*
9.	$2s^22p^3-$ $2s^22p^2(^3P)6s$	$^4S^o-^4P$		*887.80*	0.000 –	*112638.14*	4–12	1.51−01	5.35−03	6.25−02	−1.670	C	2
				887.458	0.000 –	112681.389	4–6	1.51−01	2.68−03	3.13−02	−1.970	D+	*LS*
				888.024	0.000 –	112609.612	4–4	1.51−01	1.78−03	2.08−02	−2.147	D+	*LS*
				888.372	0.000 –	112565.470	4–2	1.51−01	8.91−04	1.04−02	−2.448	D+	*LS*
10.	$2s^22p^3-$ $2s^22p^2(^3P)5d$	$^4S^o-^4P$		*886.24*	0.000 –	*112836.6*	4–12	5.58−01	1.97−02	2.30−01	−1.103	C	2
				886.333	0.000 –	112824.459	4–6	5.58−01	9.85−03	1.15−01	−1.405	D+	*LS*
				886.226	0.000 –	112838.02	4–4	5.58−01	6.57−03	7.66−02	−1.581	D+	*LS*
				885.973	0.000 –	112870.27	4–2	5.58−01	3.28−03	3.83−02	−1.881	D+	*LS*
11.	$2s^22p^3-$ $2s^22p^2(^3P)7s$	$^4S^o-^4P$		*876.41*	0.000 –	*114101.42*	4–12	8.80−02	3.04−03	3.51−02	−1.915	C	2
				876.067	0.000 –	114146.525	4–6	8.81−02	1.52−03	1.75−02	−2.216	D+	*LS*
				876.646	0.000 –	114071.153	4–4	8.79−02	1.01−03	1.17−02	−2.392	D+	*LS*
				876.988	0.000 –	114026.654	4–2	8.78−02	5.06−04	5.85−03	−2.694	D	*LS*

N I: Allowed Transitions — Continued

No.	Transition Array	Mult.	λ_{air} (Å)	λ_{vac} (Å) or σ (cm^{-1})*	E_i (cm^{-1})	E_k (cm^{-1})	g_i-g_k	A_{ki} (10^8 s^{-1})	f_{ik}	S (at. u.)	log gf	Acc.	Source
12.	$2s^22p^3-$ $2s^22p^2(^3P)6d$	$^4S^o-^4P$		*875.63*	0.000	*114204*	4–12	3.39–01	1.17–02	1.35–01	–1.330	C	2
				875.721	0.000	114191.6	4–6	3.39–01	5.85–03	6.75–02	–1.631	D+	*LS*
				875.657	0.000	114200.0	4–4	3.39–01	3.90–03	4.50–02	–1.807	D+	*LS*
				875.277	0.000	114249.5	4–2	3.40–01	1.95–03	2.25–02	–2.108	D+	*LS*
13.	$2s^22p^3-$ $2s^22p^2(^3P)8s$	$^4S^o-^4P$		*869.77*	0.000	*114973*	4–12	5.55–02	1.89–03	2.16–02	–2.121	C	2
				869.404	0.000	115021.3	4–6	5.56–02	9.45–04	1.08–02	–2.422	D+	*LS*
				870.026	0.000	114939.1	4–4	5.55–02	6.30–04	7.22–03	–2.599	D	*LS*
				870.367	0.000	114894.0	4–2	5.54–02	3.15–04	3.61–03	–2.900	D	*LS*
14.	$2s^22p^3-$ $2s^22p^2(^3P)9s$	$^4S^o-^4P$		*865.59*	0.000	*115528*	4–12	3.74–02	1.26–03	1.44–02	–2.298	C	2
				865.227	0.000	115576.6	4–6	3.74–02	6.30–04	7.18–03	–2.598	D	*LS*
				865.835	0.000	115495.5	4–4	3.74–02	4.20–04	4.79–03	–2.775	D	*LS*
				866.180	0.000	115449.5	4–2	3.73–02	2.10–04	2.39–03	–3.076	E	*LS*
15.	$2s^22p^3-$ $2s^22p^2(^3P)3s$	$^2D^o-^2P$		*1493.3*	*19227.95*	*86192.79*	10–6	3.56+00	7.15–02	3.52+00	–0.146	B+	1,3,5
				1492.63	19224.464	86220.510	6–4	3.13+00	6.98–02	2.06+00	–0.378	B+	3*n*,5*n*
				1494.68	19233.177	86137.350	4–2	3.72+00	6.23–02	1.23+00	–0.603	B+	3*n*,5*n*
				1492.82	19233.177	86220.510	4–4	3.51–01	1.17–02	2.30–01	–1.329	B+	3*n*,5*n*
16.	$2s^22p^3-$ $2s^22p^2(^1D)3s$	$^2D^o-^2D$		*1243.2*	*19227.95*	*99663.62*	10–10	3.43+00	7.96–02	3.26+00	–0.099	B+	1,3,5
				1243.18	19224.464	99663.427	6–6	3.21+00	7.43–02	1.82+00	–0.351	B	3*n*
				1243.31	19233.177	99663.912	4–4	3.09+00	7.16–02	1.17+00	–0.543	B	3*n*
				1243.17	19224.464	99663.912	6–4	3.34–01	5.15–03	1.27–01	–1.510	B	3*n*
				1243.31	19233.177	99663.427	4–6	2.35–01	8.16–03	1.34–01	–1.486	B	3*n*
17.	$2s^22p^3-$ $2s^22p^2(^3P)4s$	$^2D^o-^2P$		*1176.9*	*19227.95*	*104196.03*	10–6	1.02+00	1.28–02	4.94–01	–0.894	B	1,7
				1176.51	19224.464	104221.630	6–4	9.22–01	1.28–02	2.97–01	–1.116	B	*LS*
				1177.69	19233.177	104144.820	4–2	1.02+00	1.06–02	1.65–01	–1.372	B	*LS*
				1176.63	19233.177	104221.630	4–4	1.02–01	2.13–03	3.30–02	–2.070	B	*LS*
18.	$2s^22p^3-$ $2s^22p^2(^3P)3d$	$^2D^o-^2P$		*1171.0*	*19227.95*	*104628.32*	10–6	5.69–01	7.02–03	2.71–01	–1.154	C	3
				1171.08	19224.464	104615.470	6–4	5.60–01	7.68–03	1.78–01	–1.337	C	3
				1170.67	19233.177	104654.030	4–2	5.31–01	5.46–03	8.41–02	–1.661	C	3
				1171.20	19233.177	104615.470	4–4	2.84–02	5.84–04	9.01–03	–2.631	C–	3
19.		$^2D^o-^2F$		*1167.9*	*19227.95*	*104850.93*	10–14	1.29+00	3.69–02	1.42+00	–0.433	B	1,3,7
				1167.45	19224.464	104881.350	6–8	1.29+00	3.52–02	8.13–01	–0.675	B	3*n*
				1168.54	19233.177	104810.360	4–6	1.24+00	3.81–02	5.86–01	–0.817	B	3*n*
				1168.42	19224.464	104810.360	6–6	4.24–02	8.67–04	2.00–02	–2.284	B	3*n*
20.		$^2D^o-^4P$											
				1168.33	19233.177	104825.110	4–6	2.36–01	7.25–03	1.12–01	–1.538	B	7
21.		$^2D^o-^2D$		*1164.1*	*19227.95*	*105134.18*	10–10	7.57–01	1.54–02	5.89–01	–0.813	C+	3
				1163.88	19224.464	105143.710	6–6	7.52–01	1.53–02	3.51–01	–1.038	C+	3
				1164.32	19233.177	105119.880	4–4	6.94–01	1.41–02	2.16–01	–1.249	C+	3
				1164.21	19224.464	105119.880	6–4	5.17–02	7.01–04	1.61–02	–2.376	C+	3
				1164.00	19233.177	105143.710	4–6	1.27–02	3.86–04	5.92–03	–2.811	C	3

N I: Allowed Transitions — Continued

No.	Transition Array	Mult.	λ_{air} (Å)	λ_{vac} (Å) or σ (cm^{-1})*	E_i (cm^{-1})	E_k (cm^{-1})	g_i-g_k	A_{ki} (10^8 s^{-1})	f_{ik}	S (at. u.)	log gf	Acc.	Source
22.	$2s^22p^3-$ $2s^22p^2(^3P)5s$	$^2D^o-^2P$		*1100.7*	*19227.95*	*110081.13*	10–6	3.99–01	4.35–03	1.58–01	–1.361	C	1
				1100.36	19224.464	110103.834	6–4	3.60–01	4.35–03	9.46–02	–1.583	D+	*LS*
				1101.29	19233.177	110035.720	4–2	3.99–01	3.62–03	5.26–02	–1.839	D+	*LS*
				1100.47	19233.177	110103.834	4–4	4.00–02	7.25–04	1.05–02	–2.537	D+	*LS*
23.	$2s^22p^3-$ $2s^22p^2(^3P)4d$	$^2D^o-^2F$		*1097.7*	*19227.95*	*110329.82*	10–14	6.35–01	1.61–02	5.80–01	–0.795	C+	1
				1097.24	19224.464	110362.462	6–8	6.35–01	1.53–02	3.31–01	–1.037	C–	*LS*
				1098.26	19233.177	110286.305	4–6	5.91–01	1.60–02	2.32–01	–1.193	C–	*LS*
				1098.15	19224.464	110286.305	6–6	4.23–02	7.64–04	1.66–02	–2.339	C–	*LS*
24.	$2s^22p^3-$ $2s^22p^2(^3P)5d$	$^2D^o-^2F$		*1068.0*	*19227.95*	*112857.50*	10–14	3.52–01	8.43–03	2.97–01	–1.074	C+	1
				1067.61	19224.464	112891.238	6–8	3.53–01	8.03–03	1.69–01	–1.317	C–	*LS*
				1068.61	19233.177	112812.518	4–6	3.28–01	8.43–03	1.19–01	–1.472	C–	*LS*
				1068.51	19224.464	112812.518	6–6	2.34–02	4.01–04	8.47–03	–2.618	D+	*LS*
25.	$2s^22p^3-$ $2s^22p^2(^1D)4s$	$^2D^o-^2D$		*1000.2*	*19227.95*	*119210.0*	10–10	4.00–01	6.00–03	1.98–01	–1.222	C+	3
				1000.14	19224.464	119210.0	6–6	3.72–01	5.58–03	1.10–01	–1.475	C+	3
				1000.23	19233.177	119210.0	4–4	3.62–01	5.44–03	7.16–02	–1.663	C+	3
				1000.14	19224.464	119210.0	6–4	3.92–02	3.92–04	7.74–03	–2.629	C	3
				1000.23	19233.177	119210.0	4–6	2.72–02	6.11–04	8.05–03	–2.612	C	3
26.	$2s^22p^3-$ $2s^22p^2(^1D)3d$	$^2D^o-^2D$		*990.84*	*19227.95*	*120152*	10–10	2.51+00	3.70–02	1.21+00	–0.432	C+	3
				990.829	19224.464	120150.1	6–6	2.35+00	3.46–02	6.77–01	–0.683	C+	3
				990.858	19233.177	120155.8	4–4	2.26+00	3.33–02	4.35–01	–0.875	C+	3
				990.773	19224.464	120155.8	6–4	2.49–01	2.44–03	4.78–02	–1.834	C+	3
				990.914	19233.177	120150.1	4–6	1.65–01	3.65–03	4.76–02	–1.836	C+	3
27.		$^2D^o-^2P$		*989.29*	*19227.95*	*120310*	10–6	1.11+00	9.75–03	3.18–01	–1.011	C	3
				989.265	19224.464	120309.6	6–4	9.95–01	9.74–03	1.90–01	–1.233	C	3
				989.335	19233.177	120311.2	4–2	1.12+00	8.19–03	1.07–01	–1.485	C	3
				989.350	19233.177	120309.6	4–4	1.09–01	1.60–03	2.08–02	–2.195	C	3
28.	$2s^22p^3-$ $2s2p^4$	$^2D^o-^2D$											
				980.623	19224.464	121200.5	6–6	3.33+00	4.81–02	9.31–01	–0.540	C+	3
				980.706	19233.177	121200.5	4–6	2.29–01	4.96–03	6.40–02	–1.703	C+	3
29.	$2s^22p^3-$ $2s^22p^2(^3P)3s$	$^2P^o-^2P$		*1743.6*	*28839.18*	*86192.79*	6–6	1.39+00	6.32–02	2.18+00	–0.421	B+	1,3,5
				1742.73	28839.306	86220.510	4–4	1.16+00	5.29–02	1.21+00	–0.675	B	3*n*
				1745.25	28838.920	86137.350	2–2	9.22–01	4.21–02	4.84–01	–1.075	B	3*n*
				1745.26	28839.306	86137.350	4–2	4.45–01	1.02–02	2.33–01	–1.391	B	3*n*
				1742.72	28838.920	86220.510	2–4	2.34–01	2.13–02	2.44–01	–1.371	B	3*n*
30.	$2s^22p^3-$ $2s^22p^2(^1D)3s$	$^2P^o-^2D$		*1411.9*	*28839.18*	*99663.62*	6–10	5.14–01	2.56–02	7.14–01	–0.813	C+	1,3,6
				1411.95	28839.306	99663.427	4–6	5.10–01	2.29–02	4.25–01	–1.039	C	3*n*
				1411.93	28838.920	99663.912	2–4	4.25–01	2.54–02	2.36–01	–1.294	C	3*n*
				1411.94	28839.306	99663.912	4–4	9.59–02	2.87–03	5.33–02	–1.941	C	3*n*
31.	$2s^22p^3-$ $2s^22p^2(^3P)4s$	$^2P^o-^2P$		*1327.0*	*28839.18*	*104196.03*	6–6	1.07–01	2.83–03	7.43–02	–1.770	C	6
				1326.57	28839.306	104221.630	4–4	8.95–02	2.36–03	4.13–02	–2.025	C	*LS*
				1327.92	28838.920	104144.820	2–2	7.14–02	1.89–03	1.65–02	–2.423	C	*LS*
				1327.92	28839.306	104144.820	4–2	3.57–02	4.72–04	8.25–03	–2.724	C	*LS*
				1326.56	28838.920	104221.630	2–4	1.79–02	9.45–04	8.25–03	–2.724	C	*LS*

N I: Allowed Transitions — Continued

No.	Transition Array	Mult.	λ_{air} (Å)	λ_{vac} (Å) or σ (cm^{-1})*	E_i (cm^{-1})	E_k (cm^{-1})	g_i-g_k	A_{ki} (10^8 s^{-1})	f_{ik}	S (at. u.)	log gf	Acc.	Source
32.	$2s^22p^3-$ $2s^22p^2(^3P)3d$	$^2P^o-^2P$		*1319.5*	*28839.18*	*104628.32*	6–6	8.35−01	2.18−02	5.68−01	−0.883	B	1,3,6
				1319.68	28839.306	104615.470	4–4	6.33−01	1.65−02	2.87−01	−1.180	B	3*n*
				1319.00	28838.920	104654.030	2–2	5.62−01	1.47−02	1.27−01	−1.533	B	3*n*
				1319.00	28839.306	104654.030	4–2	2.76−01	3.60−03	6.25−02	−1.842	B	3*n*
				1319.67	28838.920	104615.470	2–4	2.02−01	1.05−02	9.14−02	−1.677	B	3*n*
33.		$^2P^o-^2F$											
				1316.29	28839.306	104810.360	4–6	1.42−02	5.55−04	9.62−03	−2.654	D	6
34.		$^2P^o-^4P$											
				1316.04	28839.306	104825.110	4–6	4.04−04	1.58−05	2.73−04	−4.201	D+	6
35.		$^2P^o-^2D$		*1310.7*	*28839.18*	*105134.18*	6–10	8.48−01	3.64−02	9.43−01	−0.661	B+	1,6
				1310.54	28839.306	105143.710	4–6	8.42−01	3.25−02	5.61−01	−0.886	B	3*n*
				1310.94	28838.920	105119.880	2–4	6.68−01	3.44−02	2.97−01	−1.162	B	3*n*
				1310.95	28839.306	105119.880	4–4	1.89−01	4.87−03	8.41−02	−1.710	C+	3*n*
36.	$2s^22p^3-$ $2s^22p^2(^3P)4d$	$^2P^o-^2P$		*1228.7*	*28839.18*	*110228.47*	6–6	4.05−01	9.18−03	2.23−01	−1.259	C	1
				1228.79	28839.306	110220.107	4–4	3.38−01	7.65−03	1.24−01	−1.515	D+	*LS*
				1228.41	28838.920	110245.183	2–2	2.70−01	6.12−03	4.95−02	−1.912	D+	*LS*
				1228.41	28839.306	110245.183	4–2	1.35−01	1.53−03	2.47−02	−2.213	D+	*LS*
				1228.79	28838.920	110220.107	2–4	6.76−02	3.06−03	2.47−02	−2.213	D+	*LS*
37.		$^2P^o-^2D$		*1225.2*	*28839.18*	*110460.96*	6–10	4.41−01	1.65−02	4.00−01	−1.004	C+	1
				1225.03	28839.306	110470.244	4–6	4.41−01	1.49−02	2.40−01	−1.225	C−	*LS*
				1225.37	28838.920	110447.032	2–4	3.67−01	1.65−02	1.33−01	−1.481	C−	*LS*
				1225.37	28839.306	110447.032	4–4	7.34−02	1.65−03	2.67−02	−2.180	C−	*LS*
38.	$2s^22p^3-$ $2s^22p^2(^3P)5d$	$^2P^o-^2P$		*1191.0*	*28839.18*	*112803.21*	6–6	2.19−01	4.65−03	1.10−01	−1.554	C	1
				1191.02	28839.306	112801.031	4–4	1.82−01	3.88−03	6.08−02	−1.809	D+	*LS*
				1190.92	28838.920	112807.567	2–2	1.46−01	3.10−03	2.43−02	−2.207	D+	*LS*
				1190.93	28839.306	112807.567	4–2	7.30−02	7.76−04	1.22−02	−2.508	D+	*LS*
				1191.01	28838.920	112801.031	2–4	3.65−02	1.55−03	1.22−02	−2.508	D+	*LS*
39.		$^2P^o-^2D$		*1189.1*	*28839.18*	*112938.31*	6–10	2.48−01	8.76−03	2.06−01	−1.279	C+	1
				1188.97	28839.306	112945.809	4–6	2.48−01	7.89−03	1.23−01	−1.501	C−	*LS*
				1189.23	28838.920	112927.055	2–4	2.07−01	8.76−03	6.86−02	−1.756	C−	*LS*
				1189.23	28839.306	112927.055	4–4	4.13−02	8.76−04	1.37−02	−2.455	C−	*LS*
40.	$2s^22p^3-$ $2s^22p^2(^1S)3s$	$^2P^o-^2S$		1143.6	28839.18	116278.558	6–2	8.92+00	5.83−02	1.32+00	−0.456	B	7
41.	$2s^22p^3-$ $2s^22p^2(^1D)4s$	$^2P^o-^2D$		*1106.6*	*28839.18*	119210.0	6–10	3.62−02	1.11−03	2.42−02	−2.177	C+	3
				1106.55	28839.306	119210.0	4–6	3.52−02	9.69−04	1.41−02	−2.412	C+	3
				1106.55	28838.920	119210.0	2–4	3.04−02	1.11−03	8.12−03	−2.652	C	3
				1106.55	28839.306	119210.0	4–4	7.39−03	1.36−04	1.98−03	−3.265	C−	3
42.	$2s^22p^3-$ $2s^22p^2(^1D)3d$	$^2P^o-^2D$											
				1095.16	28839.306	120150.1	4–6	1.40−02	3.77−04	5.43−03	−2.822	C	3
				1095.09	28838.920	120155.8	2–4	1.15−02	4.14−04	2.98−03	−3.082	C−	3

N I: Allowed Transitions — Continued

No.	Transition Array	Mult.	λ_{air} (Å)	λ_{vac} (Å) or σ (cm^{-1})*	E_i (cm^{-1})	E_k (cm^{-1})	g_i-g_k	A_{ki} (10^8 s^{-1})	f_{ik}	S (at. u.)	log gf	Acc.	Source
43.		$^2P^o-^2P$		*1093.2*	*28839.18*	*120310*	6–6	1.19+00	2.13−02	4.60−01	−0.894	C	3
				1093.25	28839.306	120309.6	4–4	9.96−01	1.79−02	2.57−01	−1.146	C	3
				1093.23	28838.920	120311.2	2–2	7.88−01	1.41−02	1.02−01	−1.549	C	3
				1093.23	28839.306	120311.2	4–2	3.85−01	3.45−03	4.96−02	−1.861	C	3
				1093.25	28838.920	120309.6	2–4	1.99−01	7.13−03	5.13−02	−1.846	C	3
44.		$^2P^o-^2S$		*1090.2*	*28839.18*	120566.0	6–2	1.96+00	1.17−02	2.51−01	−1.155	C+	1
				1090.20	28839.306	120566.0	4–2	1.31+00	1.17−02	1.67−01	−1.331	C−	*LS*
				1090.19	28838.920	120566.0	2–2	6.55−01	1.17−02	8.37−02	−1.632	C−	*LS*
45.	$2s^22p^3-2s2p^4$	$^2P^o-^2D$											
				1082.71	28839.306	121200.5	4–6	1.26+00	3.31−02	4.72−01	−0.878	C+	3
46.	$2s^22p^2(^3P)3s-2s^22p^2(^3P)3p$	$^4P-^4D^o$	*8691.6*	*8694.0*	*83335.60*	*94837.78*	12–20	2.47−01	4.66−01	1.60+02	0.747	B+	3,4,9
			8680.28	8682.67	83364.620	94881.820	6–8	2.46−01	3.70−01	6.35+01	0.346	B+	3,4,9
			8683.40	8685.79	83317.830	94830.890	4–6	1.80−01	3.05−01	3.48+01	0.086	B+	3,4,9
			8686.15	8688.54	83284.070	94793.490	2–4	1.09−01	2.48−01	1.42+01	−0.305	B+	3,4,9
			8718.84	8721.23	83364.620	94830.890	6–6	6.75−02	7.70−02	1.33+01	−0.335	B+	3,4,9
			8711.70	8714.10	83317.830	94793.490	4–4	1.28−01	1.46−01	1.67+01	−0.234	B+	3,4,9
			8703.25	8705.64	83284.070	94770.880	2–2	2.10−01	2.38−01	1.37+01	−0.322	B+	3,4,9
			8747.37	8749.77	83364.620	94793.490	6–4	1.04−02	7.98−03	1.38+00	−1.320	B+	3,4,9
			8728.90	8731.30	83317.830	94770.880	4–2	3.76−02	2.15−02	2.47+00	−1.065	B+	3,4,9
47.		$^4P-^4P^o$	*8211.8*	*8214.1*	*83335.60*	*95509.86*	12–12	3.11−01	3.14−01	1.02+02	0.576	A	3,4,9
			8216.34	8218.59	83364.620	95532.150	6–6	2.23−01	2.26−01	3.67+01	0.133	B+	3,4,9
			8210.72	8212.97	83317.830	95493.690	4–4	4.84−02	4.89−02	5.29+00	−0.709	B+	3,4,9
			8200.36	8202.61	83284.070	95475.310	2–2	4.95−02	5.00−02	2.70+00	−1.000	B+	3,4,9
			8242.39	8244.66	83364.620	95493.690	6–4	1.36−01	9.24−02	1.50+01	−0.256	B+	3,4,9
			8223.13	8225.39	83317.830	95475.310	4–2	2.64−01	1.34−01	1.45+01	−0.271	B+	3,4,9
			8184.86	8187.11	83317.830	95532.150	4–6	8.58−02	1.29−01	1.39+01	−0.286	B+	3,4,9
			8188.01	8190.26	83284.070	95493.690	2–4	1.27−01	2.55−01	1.38+01	−0.292	B+	3,4,9
48.		$^4P-^4S^o$	*7452.2*	*7454.2*	*83335.60*	96750.840	12–4	3.77−01	1.05−01	3.09+01	0.099	A	3,4,9
			7468.31	7470.37	83364.620	96750.840	6–4	1.93−01	1.08−01	1.59+01	−0.189	B+	3,4,9
			7442.30	7444.35	83317.830	96750.840	4–4	1.24−01	1.03−01	1.01+01	−0.384	B+	3,4,9
			7423.64	7425.69	83284.070	96750.840	2–4	5.95−02	9.84−02	4.81+00	−0.706	B+	3,4,9
49.	$2s^22p^2(^3P)3s-2s^22p^2(^3P)4p$	$^4P-^4S^o$	*4146.5*	*4147.7*	*83335.60*	107445.622	12–4	1.90−02	1.63−03	2.67−01	−1.708	B	8
			4151.48	4152.65	83364.620	107445.622	6–4	1.01−02	1.74−03	1.43−01	−1.981	C+	8
			4143.43	4144.60	83317.830	107445.622	4–4	6.09−03	1.57−03	8.57−02	−2.202	C+	8
			4137.64	4138.81	83284.070	107445.622	2–4	2.80−03	1.44−03	3.92−02	−2.542	C+	8
50.	$2s^22p^2(^3P)3s-2s^22p^2(^3P)3p$	$^2P-^2S^o$	*13530*	*7388.76* cm^{-1}	*86192.79*	93581.550	6–2	9.83−02	9.00−02	2.41+01	−0.268	B	1,3
			13581.3	7361.040 cm^{-1}	86220.510	93581.550	4–2	6.39−02	8.84−02	1.58+01	−0.452	B	3*n*
			13429.6	7444.200 cm^{-1}	86137.350	93581.550	2–2	3.44−02	9.32−02	8.24+00	−0.730	B	3*n*
51.		$^2P-^2D^o$	*9395.3*	*9397.9*	*86192.79*	*96833.50*	6–10	2.63−01	5.81−01	1.08+02	0.542	B+	1,3,9
			9392.79	9395.37	86220.510	96864.050	4–6	2.63−01	5.22−01	6.46+01	0.320	B	3*n*,9*n*
			9386.81	9389.38	86137.350	96787.680	2–4	2.24−01	5.92−01	3.66+01	0.073	B	3*n*,9*n*
			9460.68	9463.27	86220.510	96787.680	4–4	3.98−02	5.35−02	6.66+00	−0.670	B	3*n*,9*n*

N I: Allowed Transitions — Continued

No.	Transition Array	Mult.	λ_{air} (Å)	λ_{vac} (Å) or σ (cm^{-1})*	E_i (cm^{-1})	E_k (cm^{-1})	g_i–g_k	A_{ki} (10^8 s^{-1})	f_{ik}	S (at. u.)	log gf	Acc.	Source
52.		$^2P-^2P^o$	*8617.5*	*8619.8*	*86192.79* –	*97793.95*	6–6	3.15–01	3.51–01	5.97+01	0.323	B	1,3,9
			8629.24	8631.61	86220.510 –	97805.840	4–4	2.66–01	2.97–01	3.38+01	0.075	B	3*n*,9*n*
			8594.00	8596.36	86137.350 –	97770.180	2–2	2.09–01	2.31–01	1.31+01	−0.334	B	3*n*,9*n*
			8655.88	8658.26	86220.510 –	97770.180	4–2	1.05–01	5.90–02	6.73+00	−0.627	B	3*n*,9*n*
			8567.74	8570.09	86137.350 –	97805.840	2–4	4.92–02	1.08–01	6.12+00	−0.664	B	3*n*,9*n*
53.	$2s^22p^2(^3P)3s-2s^22p^2(^3P)4p$	$^2P-^2S^o$	*4928.4*	*4929.7*	*86192.79* –	106477.800	6–2	2.57–02	3.12–03	3.04–01	−1.728	B	1,8
			4935.12	4936.49	86220.510 –	106477.800	4–2	1.76–02	3.21–03	2.09–01	−1.891	B	8*n*
			4914.94	4916.31	86137.350 –	106477.800	2–2	8.08–03	2.93–03	9.48–02	−2.232	B	8*n*
54.		$^2P-^2P^o$	*4666.7*	*4668.0*	*86192.79* –	*107615.01*	6–6	9.72–03	3.18–03	2.93–01	−1.720	C+	9
			4669.89	4671.20	86220.510 –	107628.283	4–4	7.49–03	2.45–03	1.51–01	−2.009	C+	9
			4660.46	4661.76	86137.350 –	107588.469	2–2	6.72–03	2.19–03	6.72–02	−2.359	C+	9
			4678.59	4679.90	86220.510 –	107588.469	4–2	2.69–03	4.41–04	2.72–02	−2.753	C+	9
			4651.82	4653.13	86137.350 –	107628.283	2–4	2.40–03	1.56–03	4.77–02	−2.506	C+	9
55.	$2s^22p^2(^3P)3s-2s^22p^2(^1D)3p$	$^2P-^2D^o$	*4106.9*	*4108.0*	*86192.79* –	*110535.33*	6–10	3.99–02	1.68–02	1.37+00	−0.996	C+	9
			4109.95	4111.11	86220.510 –	110544.850	4–6	3.90–02	1.48–02	8.02–01	−1.227	C+	9
			4099.94	4101.10	86137.350 –	110521.050	2–4	3.48–02	1.75–02	4.73–01	−1.455	C+	9
			4113.97	4115.13	86220.510 –	110521.050	4–4	6.62–03	1.68–03	9.11–02	−2.172	C+	9
56.	$2s^22p^2(^3P)3s-2s^22p^2(^3P)5p$	$^2P-^2S^o$	*4020.1*	*4021.2*	*86192.79* –	111060.905	6–2	1.71–02	1.38–03	1.10–01	−2.081	C+	1
			4024.56	4025.70	86220.510 –	111060.905	4–2	1.14–02	1.38–03	7.33–02	−2.257	C–	*LS*
			4011.13	4012.27	86137.350 –	111060.905	2–2	5.75–03	1.39–03	3.66–02	−2.557	C–	*LS*
57.		$^2P-^2D^o$	*3891.2*	*3892.3*	*86192.79* –	*111884.59*	6–10	2.50–02	9.46–03	7.27–01	−1.246	C+	1
			3892.21	3893.31	86220.510 –	111905.609	4–6	2.50–02	8.51–03	4.36–01	−1.468	C–	*LS*
			3887.57	3888.67	86137.350 –	111853.061	2–4	2.09–02	9.47–03	2.42–01	−1.723	C–	*LS*
			3900.18	3901.29	86220.510 –	111853.061	4–4	4.14–03	9.44–04	4.85–02	−2.423	C–	*LS*
58.	$2s^22p^2(^3P)3s-2s^22p^2(^1D)3p$	$^2P-^2P^o$	*3827.6*	*3828.7*	*86192.79* –	*112311.21*	6–6	5.58–02	1.23–02	9.27–01	−1.134	C+	3
			3830.43	3831.52	86220.510 –	112319.805	4–4	4.67–02	1.03–02	5.19–01	−1.386	C+	3
			3822.03	3823.12	86137.350 –	112294.007	2–2	3.70–02	8.10–03	2.04–01	−1.791	C+	3
			3834.22	3835.31	86220.510 –	112294.007	4–2	1.89–02	2.09–03	1.05–01	−2.079	C+	3
			3818.27	3819.35	86137.350 –	112319.805	2–4	8.95–03	3.91–03	9.84–02	−2.107	C+	3
59.	$2s2p^4-2s^22p^2(^3P)3p$	$^4P-^4D^o$	*14909*	*6705.33* cm^{-1}	*88132.45* –	*94837.78*	12–20	9.55–03	5.31–02	3.13+01	−0.196	B	3,4
			14757.1	6774.560 cm^{-1}	88107.260 –	94881.820	6–8	9.96–03	4.34–02	1.26+01	−0.585	B	3,4
			14966.6	6679.720 cm^{-1}	88151.170 –	94830.890	4–6	6.67–03	3.36–02	6.63+00	−0.871	B	3,4
			15095.0	6622.920 cm^{-1}	88170.570 –	94793.490	2–4	3.87–03	2.64–02	2.63+00	−1.277	B	3,4
			14868.9	6723.630 cm^{-1}	88107.260 –	94830.890	6–6	2.80–03	9.29–03	2.73+00	−1.254	B	3,4
			15050.9	6642.320 cm^{-1}	88151.170 –	94793.490	4–4	4.85–03	1.65–02	3.26+00	−1.181	B	3,4
			15146.7	6600.310 cm^{-1}	88170.570 –	94770.880	2–2	7.51–03	2.58–02	2.58+00	−1.287	B	3,4
			14952.0	6686.230 cm^{-1}	88107.260 –	94793.490	6–4	4.46–04	9.98–04	2.95–01	−2.223	C+	3,4
			15102.3	6619.710 cm^{-1}	88151.170 –	94770.880	4–2	1.47–03	2.52–03	5.01–01	−1.997	C+	3,4
60.		$^4P-^4P^o$	*13551*	*7377.41* cm^{-1}	*88132.45* –	*95509.86*	12–12	7.19–03	1.98–02	1.06+01	−0.624	C	3,4
			13464.5	7424.890 cm^{-1}	88107.260 –	95532.150	6–6	5.36–03	1.46–02	3.87+00	−1.059	C+	3,4
			13615.6	7342.520 cm^{-1}	88151.170 –	95493.690	4–4	1.12–03	3.11–03	5.58–01	−1.905	C+	3,4
			13686.0	7304.740 cm^{-1}	88170.570 –	95475.310	2–2	1.08–03	3.02–03	2.72–01	−2.219	C	3,4
			13534.6	7386.430 cm^{-1}	88107.260 –	95493.690	6–4	3.14–03	5.75–03	1.54+00	−1.462	C	3,4
			13649.7	7324.140 cm^{-1}	88151.170 –	95475.310	4–2	5.87–03	8.20–03	1.47+00	−1.484	C	3,4
			13544.6	7380.980 cm^{-1}	88151.170 –	95532.150	4–6	1.99–03	8.22–03	1.47+00	−1.483	D	3,4
			13651.6	7323.120 cm^{-1}	88170.570 –	95493.690	2–4	2.84–03	1.59–02	1.43+00	−1.499	D	3,4

N I: Allowed Transitions — Continued

No.	Transition Array	Mult.	λ_{air} (Å)	λ_{vac} (Å) or σ (cm^{-1})*	E_i (cm^{-1})	E_k (cm^{-1})	g_i–g_k	A_{ki} (10^8 s^{-1})	f_{ik}	S (at. u.)	log gf	Acc.	Source
61.		^{4}P–^{4}S°	*11600*	*8618.39* cm^{-1}	*88132.45*	96750.840	12–4	4.04–02	2.72–02	1.25+01	–0.486	C+	3,4
			11566.1	8643.580 cm^{-1}	88107.260	96750.840	6–4	2.08–02	2.79–02	6.37+00	–0.776	C+	3,4
			11625.2	8599.670 cm^{-1}	88151.170	96750.840	4–4	1.32–02	2.67–02	4.09+00	–0.971	C+	3,4
			11651.5	8580.270 cm^{-1}	88170.570	96750.840	2–4	6.39–03	2.60–02	2.00+00	–1.283	C+	3,4
62.	$2s2p^4$– $2s^22p^2(^3P)4p$	^{4}P–^{4}D°	*5348.7*	*5350.2*	*88132.45*	*106823.33*	12–20	2.03–03	1.45–03	3.06–01	–1.760	C	1
			5328.62	5330.10	88107.260	106868.635	6–8	2.05–03	1.16–03	1.23–01	–2.156	D+	*LS*
			5356.62	5358.11	88151.170	106814.459	4–6	1.41–03	9.12–04	6.43–02	–2.438	D+	*LS*
			5372.61	5374.10	88170.570	106778.337	2–4	8.34–04	7.22–04	2.55–02	–2.841	D+	*LS*
			5344.05	5345.54	88107.260	106814.459	6–6	6.10–04	2.61–04	2.76–02	–2.805	D+	*LS*
			5367.01	5368.50	88151.170	106778.337	4–4	1.07–03	4.62–04	3.27–02	–2.733	D+	*LS*
			5378.27	5379.77	88170.570	106758.731	2–2	1.66–03	7.21–04	2.55–02	–2.841	D+	*LS*
			5354.39	5355.88	88107.260	106778.337	6–4	1.01–04	2.90–05	3.06–03	–3.760	D	*LS*
			5372.67	5374.16	88151.170	106758.731	4–2	3.33–04	7.22–05	5.11–03	–3.540	D	*LS*
63.		^{4}P–^{4}P°											
			5281.20	5282.67	88107.260	107037.069	6–6	2.45–03	1.02–03	1.07–01	–2.211	C	8
64.	$2s^22p^2(^3P)3p$– $2s^22p^2(^3P)4s$	^{2}S°–^{2}P	*9418.5*	*9421.1*	93581.550	*104196.03*	2–6	1.66–03	6.63–03	4.11–01	–1.877	C	3
			9395.85	9398.43	93581.550	104221.630	2–4	2.82–04	7.47–04	4.62–02	–2.826	C	3
			9464.17	9466.77	93581.550	104144.820	2–2	4.36–03	5.86–03	3.65–01	–1.931	C	3
65.	$2s^22p^2(^3P)3p$– $2s^22p^2(^3P)3d$	^{2}S°–^{2}P	*9049.9*	*9052.4*	93581.550	*104628.32*	2–6	2.98–01	1.10+00	6.54+01	0.341	B	1,3,9
			9060.48	9062.96	93581.550	104615.470	2–4	2.95–01	7.28–01	4.34+01	0.163	B	3*n*,9*n*
			9028.92	9031.40	93581.550	104654.030	2–2	3.02–01	3.69–01	2.19+01	–0.132	B	3*n*,9*n*
66.	$2s^22p^2(^3P)3p$– $2s^22p^2(^3P)4d$	^{2}S°–^{2}P	*6005.5*	*6007.1*	93581.550	*110228.47*	2–6	3.60–02	5.84–02	2.31+00	–0.932	C+	9
			6008.47	6010.14	93581.550	110220.107	2–4	3.58–02	3.88–02	1.53+00	–1.110	C+	9
			5999.43	6001.09	93581.550	110245.183	2–2	3.64–02	1.96–02	7.76–01	–1.406	C+	9
67.	$2s^22p^2(^3P)3p$– $2s^22p^2(^3P)5d$	^{2}S°–^{2}P	*5201.0*	*5202.5*	93581.550	*112803.21*	2–6	1.87–02	2.27–02	7.79–01	–1.342	C	1
			5201.61	5203.05	93581.550	112801.031	2–4	1.87–02	1.52–02	5.19–01	–1.518	D+	*LS*
			5199.84	5201.29	93581.550	112807.567	2–2	1.87–02	7.58–03	2.60–01	–1.819	D+	*LS*
68.	$2s^22p^2(^3P)3p$– $2s^22p^2(^3P)4s$	^{4}D°–^{4}P	*11289*	*8856.1* cm^{-1}	*94837.78*	*103693.9*	20–12	1.35–01	1.55–01	1.15+02	0.491	B	1,3
			11291.7	8853.66 cm^{-1}	94881.820	103735.48	8–6	1.08–01	1.55–01	4.61+01	0.093	B	3*n*
			11313.9	8836.27 cm^{-1}	94830.890	103667.16	6–4	9.02–02	1.15–01	2.58+01	–0.160	B	3*n*
			11323.2	8829.02 cm^{-1}	94793.490	103622.51	4–2	7.36–02	7.08–02	1.06+01	–0.548	B	3*n*
			11227.1	8904.59 cm^{-1}	94830.890	103735.48	6–6	2.09–02	3.95–02	8.76+00	–0.626	B	3*n*
			11266.2	8873.67 cm^{-1}	94793.490	103667.16	4–4	4.13–02	7.86–02	1.17+01	–0.502	B	3*n*
			11294.3	8851.63 cm^{-1}	94770.880	103622.51	2–2	6.92–02	1.32–01	9.85+00	–0.577	B	3*n*
			11180.1	8941.99 cm^{-1}	94793.490	103735.48	4–6	2.01–03	5.65–03	8.32–01	–1.646	C+	3*n*
			11237.6	8896.28 cm^{-1}	94770.880	103667.16	2–4	6.00–03	2.27–02	1.68+00	–1.342	B	3*n*
69.	$2s^22p^2(^3P)3p$– $2s^22p^2(^3P)3d$	^{4}D°–^{4}F	*10117*	*9881.80* cm^{-1}	*94837.78*	*104719.58*	20–28	3.75–01	8.05–01	5.37+02	1.207	B+	1,3,9
			10114.6	9883.950 cm^{-1}	94881.820	104765.770	8–10	3.82–01	7.33–01	1.95+02	0.768	B+	3*n*,9*n*
			10112.5	9886.060 cm^{-1}	94830.890	104716.950	6–8	3.30–01	6.74–01	1.35+02	0.607	B+	3*n*,9*n*
			10108.9	9889.570 cm^{-1}	94793.490	104683.060	4–6	2.93–01	6.73–01	8.97+01	0.430	B+	3*n*,9*n*
			10105.1	9893.250 cm^{-1}	94770.880	104664.130	2–4	2.70–01	8.28–01	5.51+01	0.219	B+	3*n*,9*n*
			10164.8	9835.130 cm^{-1}	94881.820	104716.950	8–8	3.98–02	6.16–02	1.65+01	–0.307	B+	3*n*,9*n*
			10147.3	9852.170 cm^{-1}	94830.890	104683.060	6–6	7.44–02	1.15–01	2.30+01	–0.161	B+	3*n*,9*n*
			10128.3	9870.640 cm^{-1}	94793.490	104664.130	4–4	9.89–02	1.52–01	2.03+01	–0.215	B+	3*n*,9*n*
			10200.0	9801.240 cm^{-1}	94881.820	104683.060	8–6	2.81–03	3.29–03	8.85–01	–1.579	C+	3*n*,9*n*
			10166.8	9833.240 cm^{-1}	94830.890	104664.130	6–4	4.95–03	5.11–03	1.03+00	–1.513	C+	3*n*,9*n*

N I: Allowed Transitions — Continued

No.	Transition Array	Mult.	λ_{air} (Å)	λ_{vac} (Å) or σ (cm^{-1})*	E_i (cm^{-1})	E_k (cm^{-1})	g_i-g_k	A_{ki} (10^8 s^{-1})	f_{ik}	S (at. u.)	log gf	Acc.	Source
70.		$^4D^o-^2F$											
			9997.73	9999.530 cm^{-1}	94881.820	104881.350	8–8	9.20–03	1.38–02	3.63+00	–0.957	C+	9
			10017.8	9979.470 cm^{-1}	94830.890	104810.360	6–6	2.26–02	3.40–02	6.73+00	–0.690	C+	9
			9947.07	9949.79	94830.890	104881.350	6–8	1.08–02	2.14–02	4.20+00	–0.892	C+	9
			9980.42	9983.16	94793.490	104810.360	4–6	8.10–03	1.82–02	2.39+00	–1.139	C+	9
71.		$^4D^o-^4P$	*9988.2*	*9991.0*	*94837.78*	*104846.8*	20–12	4.34–02	3.90–02	2.56+01	–0.108	C+	9
			10054.3	9943.290 cm^{-1}	94881.820	104825.110	8–6	1.39–02	1.58–02	4.19+00	–0.897	C+	9
			9968.51	9971.24	94830.890	104859.73	6–4	4.50–03	4.47–03	8.81–01	–1.571	C+	9
			9905.52	9908.24	94793.490	104886.10	4–2	3.11–03	2.29–03	2.99–01	–2.038	C+	9
			10003.0	9994.220 cm^{-1}	94830.890	104825.110	6–6	2.28–02	3.41–02	6.75+00	–0.688	C+	9
			9931.47	9934.20	94793.490	104859.73	4–4	3.64–02	5.38–02	7.04+00	–0.667	C+	9
			9883.38	9886.09	94770.880	104886.10	2–2	2.93–02	4.29–02	2.79+00	–1.066	C+	9
			9965.75	9968.48	94793.490	104825.110	4–6	7.60–03	1.70–02	2.23+00	–1.168	C+	9
			9909.22	9911.93	94770.880	104859.73	2–4	7.58–03	2.23–02	1.46+00	–1.350	C+	9
72.		$^4D^o-^4D$	*9830.6*	*9833.3*	*94837.78*	*105007.3*	20–20	9.37–02	1.36–01	8.79+01	0.434	B	1,3,9
			9863.33	9866.04	94881.820	105017.600	8–8	9.62–02	1.40–01	3.65+01	0.051	B	*3n*,*9n*
			9822.75	9825.44	94830.890	105008.55	6–6	4.95–02	7.16–02	1.39+01	–0.367	B	*3n*,*9n*
			9798.56	9801.25	94793.490	104996.27	4–4	2.75–02	3.97–02	5.12+00	–0.799	C+	*3n*,*9n*
			9788.29	9790.97	94770.880	104984.37	2–2	2.99–02	4.30–02	2.77+00	–1.066	B	*3n*,*9n*
			9872.15	9874.86	94881.820	105008.55	8–6	2.97–02	3.26–02	8.48+00	–0.583	C+	*3n*,*9n*
			9834.61	9837.31	94830.890	104996.27	6–4	4.50–02	4.35–02	8.46+00	–0.583	B	*3n*,*9n*
			9810.01	9812.70	94793.490	104984.37	4–2	5.30–02	3.82–02	4.94+00	–0.815	B	*3n*,*9n*
			9814.02	9816.71	94830.890	105017.600	6–8	6.56–03	1.26–02	2.45+00	–1.120	C+	*3n*,*9n*
			9786.78	9789.47	94793.490	105008.55	4–6	1.13–02	2.43–02	3.13+00	–1.012	B	*3n*,*9n*
			9776.90	9779.58	94770.880	104996.27	2–4	1.18–02	3.39–02	2.18+00	–1.169	B	*3n*,*9n*
73.	$2s^22p^2(^3P)3p-$ $2s^22p^2(^3P)5s$	$^4D^o-^4P$	*6644.3*	*6646.1*	*94837.78*	*109884.21*	20–12	4.36–02	1.73–02	7.58+00	–0.461	C	1
			6644.96	6646.80	94881.820	109926.661	8–6	3.49–02	1.73–02	3.03+00	–0.859	D+	*LS*
			6653.46	6655.29	94830.890	109856.520	6–4	2.74–02	1.21–02	1.59+00	–1.139	D+	*LS*
			6656.51	6658.35	94793.490	109812.233	4–2	2.17–02	7.20–03	6.31–01	–1.541	D+	*LS*
			6622.54	6624.37	94830.890	109926.661	6–6	7.93–03	5.21–03	6.82–01	–1.505	D+	*LS*
			6636.94	6638.77	94793.490	109856.520	4–4	1.40–02	9.25–03	8.08–01	–1.432	D+	*LS*
			6646.50	6648.34	94770.880	109812.233	2–2	2.18–02	1.44–02	6.31–01	–1.540	D+	*LS*
			6606.18	6608.00	94793.490	109926.661	4–6	8.87–04	8.71–04	7.58–02	–2.458	D+	*LS*
			6626.99	6628.82	94770.880	109856.520	2–4	2.20–03	2.89–03	1.26–01	–2.238	D+	*LS*
74.	$2s^22p^2(^3P)3p-$ $2s^22p^2(^3P)4d$	$^4D^o-^4F$	*6485.6*	*6487.4*	*94837.78*	*110252.27*	20–28	4.90–02	4.32–02	1.85+01	–0.063	C+	1
			6482.70	6484.49	94881.820	110303.233	8–10	4.90–02	3.86–02	6.59+00	–0.510	C–	*LS*
			6484.81	6486.60	94830.890	110247.288	6–8	4.20–02	3.53–02	4.52+00	–0.674	C–	*LS*
			6483.75	6485.54	94793.490	110212.396	4–6	3.67–02	3.47–02	2.97+00	–0.857	C–	*LS*
			6481.71	6483.50	94770.880	110194.654	2–4	3.43–02	4.32–02	1.85+00	–1.063	C–	*LS*
			6506.30	6508.10	94881.820	110247.288	8–8	6.91–03	4.39–03	7.51–01	–1.455	C–	*LS*
			6499.52	6501.31	94830.890	110212.396	6–6	1.18–02	7.50–03	9.62–01	–1.347	C–	*LS*
			6491.22	6493.02	94793.490	110194.654	4–4	1.37–02	8.64–03	7.38–01	–1.462	C–	*LS*
			6521.11	6522.91	94881.820	110212.396	8–6	4.66–04	2.23–04	3.82–02	–2.749	C–	*LS*
			6507.02	6508.82	94830.890	110194.654	6–4	9.70–04	4.10–04	5.27–02	–2.609	C–	*LS*

N I: Allowed Transitions — Continued

No.	Transition Array	Mult.	λ_{air} (Å)	λ_{vac} (Å) or σ (cm^{-1})*	E_i (cm^{-1})	E_k (cm^{-1})	g_i-g_k	A_{ki} (10^8 s^{-1})	f_{ik}	S (at. u.)	log *gf*	Acc.	Source
75.		$^4D^o-^4D$	*6424.3*	*6426.1*	*94837.78*	*110399.37*	20–20	1.56–02	9.66–03	4.09+00	–0.714	C+	1
			6440.94	6442.72	94881.820	110403.220	8–8	1.33–02	8.26–03	1.40+00	–1.180	C–	*LS*
			6420.64	6422.42	94830.890	110401.356	6–6	8.97–03	5.54–03	7.03–01	–1.478	C–	*LS*
			6407.67	6409.45	94793.490	110395.463	4–4	6.30–03	3.88–03	3.27–01	–1.810	C–	*LS*
			6402.36	6404.13	94770.880	110385.795	2–2	7.89–03	4.85–03	2.04–01	–2.013	C–	*LS*
			6441.71	6443.49	94881.820	110401.356	8–6	2.94–03	1.37–03	2.33–01	–1.959	C–	*LS*
			6423.07	6424.85	94830.890	110395.463	6–4	5.47–03	2.26–03	2.86–01	–1.869	C–	*LS*
			6411.65	6413.42	94793.490	110385.795	4–2	7.86–03	2.42–03	2.04–01	–2.014	C–	*LS*
			6419.87	6421.65	94830.890	110403.220	6–8	2.23–03	1.84–03	2.33–01	–1.958	C–	*LS*
			6405.26	6407.03	94793.490	110401.356	4–6	3.68–03	3.39–03	2.86–01	–1.867	C–	*LS*
			6398.40	6400.17	94770.880	110395.463	2–4	3.95–03	4.85–03	2.04–01	–2.013	C–	*LS*
76.	$2s^22p^2(^3P)3p-$ $2s^22p^2(^3P)5d$	$^4D^o-^4F$	*5563.5*	*5565.1*	*94837.78*	*112807.04*	20–28	1.42–02	9.20–03	3.37+00	–0.735	C+	1
			5560.34	5561.88	94881.820	112861.348	8–10	1.42–02	8.22–03	1.20+00	–1.182	C–	*LS*
			5564.27	5565.81	94830.890	112797.725	6–8	1.21–02	7.51–03	8.25–01	–1.346	C–	*LS*
			5564.27	5565.81	94793.490	112760.325	4–6	1.06–02	7.39–03	5.41–01	–1.529	C–	*LS*
			5557.38	5558.93	94770.880	112759.966	2–4	9.94–03	9.21–03	3.37–01	–1.735	C–	*LS*
			5580.08	5581.63	94881.820	112797.725	8–8	2.00–03	9.33–04	1.37–01	–2.127	C–	*LS*
			5575.87	5577.42	94830.890	112760.325	6–6	3.42–03	1.59–03	1.76–01	–2.019	C–	*LS*
			5564.38	5565.92	94793.490	112759.966	4–4	3.96–03	1.84–03	1.35–01	–2.133	C–	*LS*
			5591.76	5593.31	94881.820	112760.325	8–6	1.35–04	4.74–05	6.98–03	–3.421	D+	*LS*
			5575.98	5577.53	94830.890	112759.966	6–4	2.81–04	8.74–05	9.62–03	–3.280	D+	*LS*
77.		$^4D^o-^4D$	*5531.8*	*5533.4*	*94837.78*	*112910*	20–20	5.09–03	2.33–03	8.50–01	–1.331	C+	1
			5545.00	5546.54	94881.820	112911.079	8–8	4.33–03	2.00–03	2.91–01	–1.797	C–	*LS*
			5529.52	5531.05	94830.890	112910.630	6–6	2.92–03	1.34–03	1.46–01	–2.095	C–	*LS*
			5518.58	5520.11	94793.490	112909.076	4–4	2.05–03	9.36–04	6.80–02	–2.427	C–	*LS*
			5513.09	5514.62	94770.880	112904.5	2–2	2.57–03	1.17–03	4.25–02	–2.631	C–	*LS*
			5545.14	5546.68	94881.820	112910.630	8–6	9.59–04	3.32–04	4.84–02	–2.576	C–	*LS*
			5529.99	5531.53	94830.890	112909.076	6–4	1.78–03	5.45–04	5.95–02	–2.486	C–	*LS*
			5519.97	5521.50	94793.490	112904.5	4–2	2.56–03	5.85–04	4.25–02	–2.631	C–	*LS*
			5529.38	5530.92	94830.890	112911.079	6–8	7.26–04	4.43–04	4.84–02	–2.575	C–	*LS*
			5518.10	5519.63	94793.490	112910.630	4–6	1.20–03	8.19–04	5.95–02	–2.485	C–	*LS*
			5511.70	5513.23	94770.880	112909.076	2–4	1.29–03	1.17–03	4.25–02	–2.630	C–	*LS*
78.	$2s^22p^2(^3P)3p-$ $2s^22p^2(^3P)4s$	$^4P^o-^4P$	*12216*	*8184.0* cm^{-1}	*95509.86*	*103693.9*	12–12	9.62–02	2.15–01	1.04+02	0.412	B	1,3
			12186.8	8203.33 cm^{-1}	95532.150	103735.48	6–6	7.11–02	1.58–01	3.82+01	–0.022	B	3*n*
			12231.4	8173.47 cm^{-1}	95493.690	103667.16	4–4	1.52–02	3.40–02	5.48+00	–0.866	B	3*n*
			12270.8	8147.20 cm^{-1}	95475.310	103622.51	2–2	1.43–02	3.24–02	2.61+00	–1.189	B	3*n*
			12289.2	8135.01 cm^{-1}	95532.150	103667.16	6–4	3.76–02	5.69–02	1.38+01	–0.467	B	3*n*
			12298.5	8128.82 cm^{-1}	95493.690	103622.51	4–2	7.41–02	8.40–02	1.36+01	–0.474	B	3*n*
			12130.0	8241.79 cm^{-1}	95493.690	103735.48	4–6	2.95–02	9.76–02	1.56+01	–0.408	B	3*n*
			12203.9	8191.85 cm^{-1}	95475.310	103667.16	2–4	4.08–02	1.82–01	1.47+01	–0.438	B	3*n*
79.	$2s^22p^2(^3P)3p-$ $2s^22p^2(^3P)3d$	$^4P^o-^2F$											
			10730.5	9316.670 cm^{-1}	95493.690	104810.360	4–6	1.03–02	2.67–02	3.77+00	–0.972	C+	9
80.		$^4P^o-^4P$	*10707*	*9337.0* cm^{-1}	*95509.86*	*104846.8*	12–12	7.18–02	1.23–01	5.22+01	0.171	B	9†
			10757.9	9292.960 cm^{-1}	95532.150	104825.110	6–6	2.37–02	4.12–02	8.75+00	–0.607	C+	9
			10673.9	9366.04 cm^{-1}	95493.690	104859.73	4–4	2.57–03	4.39–03	6.18–01	–1.755	D	3
			10623.2	9410.79 cm^{-1}	95475.310	104886.10	2–2	3.39–02	5.74–02	4.01+00	–0.940	C+	9
			10718.0	9327.58 cm^{-1}	95532.150	104859.73	6–4	2.29–02	2.64–02	5.58+00	–0.801	C+	9
			10644.0	9392.41 cm^{-1}	95493.690	104886.10	4–2	4.01–02	3.41–02	4.78+00	–0.865	C+	9
			10713.5	9331.420 cm^{-1}	95493.690	104825.110	4–6	4.18–02	1.08–01	1.52+01	–0.365	C+	9
			10653.0	9384.42 cm^{-1}	95475.310	104859.73	2–4	5.55–02	1.89–01	1.33+01	–0.423	C+	9

N I: Allowed Transitions — Continued

No.	Transition Array	Mult.	λ_{air} (Å)	λ_{vac} (Å) or σ (cm^{-1})*	E_i (cm^{-1})	E_k (cm^{-1})	g_i-g_k	A_{ki} (10^8 s^{-1})	f_{ik}	S (at. u.)	log gf	Acc.	Source
81.		$^4P^o-^4D$	*10526*	*9497.4* cm^{-1}	*95509.86*	– *105007.3*	12–20	2.43–01	6.73–01	2.80+02	0.907	B	3,9
			10539.6	9485.450 cm^{-1}	95532.150	– 105017.600	6–8	2.39–01	5.31–01	1.11+02	0.503	B	9*n*
			10507.0	9514.86 cm^{-1}	95493.690	– 105008.55	4–6	1.25–01	3.10–01	4.29+01	0.094	B	9*n*
			10500.3	9520.96 cm^{-1}	95475.310	– 104996.27	2–4	6.17–02	2.04–01	1.41+01	–0.389	B	9*n*
			10549.6	9476.40 cm^{-1}	95532.150	– 105008.55	6–6	1.15–01	1.92–01	4.01+01	0.062	B	9*n*
			10520.6	9502.58 cm^{-1}	95493.690	– 104996.27	4–4	1.54–01	2.56–01	3.55+01	0.010	B	9*n*
			10513.4	9509.06 cm^{-1}	95475.310	– 104984.37	2–2	1.81–01	2.99–01	2.07+01	–0.223	B	9*n*
			10563.3	9464.12 cm^{-1}	95532.150	– 104996.27	6–4	3.30–02	3.68–02	7.68+00	–0.656	B	9*n*
			10533.8	9490.68 cm^{-1}	95493.690	– 104984.37	4–2	7.28–02	6.06–02	8.40+00	–0.616	B	9*n*
82.	$2s^22p^2(^3P)3p-2s^22p^2(^3P)5s$	$^4P^o-^4P$	*6954.9*	*6956.8*	*95509.86*	– *109884.21*	12–12	2.53–02	1.84–02	5.05+00	–0.657	B	1,9
			6945.18	6947.09	95532.150	– 109926.661	6–6	1.83–02	1.33–02	1.82+00	–1.099	B	9*n*
			6960.50	6962.42	95493.690	– 109856.520	4–4	4.67–03	3.40–03	3.11–01	–1.867	B	9*n*
			6973.07	6975.00	95475.310	– 109812.233	2–2	3.83–03	2.79–03	1.28–01	–2.253	B	9*n*
			6979.18	6981.11	95532.150	– 109856.520	6–4	9.83–03	4.79–03	6.60–01	–1.542	B	9*n*
			6982.03	6983.95	95493.690	– 109812.233	4–2	2.04–02	7.44–03	6.84–01	–1.526	B	9*n*
			6926.67	6928.58	95493.690	– 109926.661	4–6	7.75–03	8.36–03	7.63–01	–1.476	B	9*n*
			6951.60	6953.52	95475.310	– 109856.520	2–4	1.03–02	1.49–02	6.82–01	–1.526	B	9*n*
83.	$2s^22p^2(^3P)3p-2s^22p^2(^3P)4d$	$^4P^o-^4P$	*6752.1*	*6754.0*	*95509.86*	– *110315.90*	12–12	1.16–02	7.90–03	2.11+00	–1.023	C	1
			6769.61	6771.48	95532.150	– 110299.974	6–6	8.02–03	5.51–03	7.37–01	–1.480	D+	*LS*
			6741.67	6743.53	95493.690	– 110322.721	4–4	1.55–03	1.05–03	9.36–02	–2.375	D+	*LS*
			6720.97	6722.82	95475.310	– 110350.014	2–2	1.95–03	1.32–03	5.85–02	–2.578	D+	*LS*
			6759.20	6761.06	95532.150	– 110322.721	6–4	5.18–03	2.37–03	3.16–01	–1.848	D+	*LS*
			6729.28	6731.14	95493.690	– 110350.014	4–2	9.73–03	3.30–03	2.93–01	–1.879	D+	*LS*
			6752.03	6753.89	95493.690	– 110299.974	4–6	3.47–03	3.55–03	3.16–01	–1.847	D+	*LS*
			6733.32	6735.18	95475.310	– 110322.721	2–4	4.85–03	6.60–03	2.93–01	–1.880	D+	*LS*
84.		$^4P^o-^4D$	*6714.3*	*6716.1*	*95509.86*	– *110399.37*	12–20	3.57–02	4.03–02	1.07+01	–0.316	C+	1
			6722.61	6724.47	95532.150	– 110403.220	6–8	3.56–02	3.22–02	4.27+00	–0.714	C–	*LS*
			6706.11	6707.96	95493.690	– 110401.356	4–6	2.51–02	2.54–02	2.24+00	–0.993	C–	*LS*
			6700.49	6702.34	95475.310	– 110395.463	2–4	1.50–02	2.02–02	8.90–01	–1.394	C–	*LS*
			6723.45	6725.31	95532.150	– 110401.356	6–6	1.07–02	7.24–03	9.61–01	–1.362	C–	*LS*
			6708.76	6710.61	95493.690	– 110395.463	4–4	1.91–02	1.29–02	1.14+00	–1.288	C–	*LS*
			6704.84	6706.69	95475.310	– 110385.795	2–2	2.99–02	2.02–02	8.90–01	–1.395	C–	*LS*
			6726.12	6727.98	95532.150	– 110395.463	6–4	1.78–03	8.04–04	1.07–01	–2.317	C–	*LS*
			6713.11	6714.97	95493.690	– 110385.795	4–2	5.96–03	2.01–03	1.78–01	–2.094	C–	*LS*
85.	$2s^22p^2(^3P)3p-2s^22p^2(^3P)5d$	$^4P^o-^4D$	*5745.5*	*5747.1*	*95509.86*	– *112910*	12–20	1.07–02	8.80–03	2.00+00	–0.976	C+	1
			5752.50	5754.09	95532.150	– 112911.079	6–8	1.06–02	7.03–03	7.99–01	–1.375	C–	*LS*
			5739.94	5741.54	95493.690	– 112910.630	4–6	7.49–03	5.55–03	4.20–01	–1.654	C–	*LS*
			5734.40	5736.00	95475.310	– 112909.076	2–4	4.47–03	4.41–03	1.67–01	–2.055	C–	*LS*
			5752.65	5754.24	95532.150	– 112910.630	6–6	3.19–03	1.58–03	1.80–01	–2.023	C–	*LS*
			5740.46	5742.05	95493.690	– 112909.076	4–4	5.71–03	2.82–03	2.13–01	–1.948	C–	*LS*
			5735.91	5737.50	95475.310	– 112904.5	2–2	8.94–03	4.41–03	1.67–01	–2.055	C–	*LS*
			5753.16	5754.76	95532.150	– 112909.076	6–4	5.31–04	1.76–04	2.00–02	–2.977	C–	*LS*
			5741.97	5743.56	95493.690	– 112904.5	4–2	1.78–03	4.40–04	3.33–02	–2.754	C–	*LS*
86.	$2s^22p^2(^3P)3p-2s^22p^2(^3P)4s$	$^4S^o-^4P$	*14399*	*6943.0* cm^{-1}	96750.840	– *103693.9*	4–12	1.24–02	1.16–01	2.19+01	–0.335	C	2,3
			14313.2	6984.64 cm^{-1}	96750.840	– 103735.48	4–6	1.33–02	6.12–02	1.15+01	–0.611	C	3*n*
			14454.6	6916.32 cm^{-1}	96750.840	– 103667.16	4–4	1.18–02	3.69–02	7.02+00	–0.831	C	3*n*
			14548.5	6871.67 cm^{-1}	96750.840	– 103622.51	4–2	1.10–02	1.75–02	3.36+00	–1.154	C	3*n*
87.	$2s^22p^2(^3P)3p-2s^22p^2(^3P)3d$	$^4S^o-^4P$	*12348*	*8096.0* cm^{-1}	96750.840	– *104846.8*	4–12	1.29–01	8.85–01	1.44+02	0.549	C+	2,3
			12381.6	8074.270 cm^{-1}	96750.840	– 104825.110	4–6	1.28–01	4.41–01	7.19+01	0.246	C+	3*n*
			12328.8	8108.89 cm^{-1}	96750.840	– 104859.73	4–4	1.27–01	2.89–01	4.69+01	0.062	C+	3*n*
			12288.8	8135.26 cm^{-1}	96750.840	– 104886.10	4–2	1.38–01	1.56–01	2.52+01	–0.205	C+	3*n*

N I: Allowed Transitions — Continued

No.	Transition Array	Mult.	λ_{air} (Å)	λ_{vac} (Å) or σ (cm^{-1})*	E_i (cm^{-1})	E_k (cm^{-1})	g_i-g_k	A_{ki} (10^8 s^{-1})	f_{ik}	S (at. u.)	log gf	Acc.	Source
88.	$2s^22p^2(^3P)3p-$ $2s^22p^2(^3P)5s$	$^4S^o-^4P$	*7612.1*	*7614.2*	96750.840	*109884.21*	4–12	6.94–03	1.81–02	1.81+00	–1.140	C	2
			7587.57	7589.66	96750.840	109926.661	4–6	7.01–03	9.08–03	9.07–01	–1.440	D+	*LS*
			7628.18	7630.28	96750.840	109856.520	4–4	6.90–03	6.02–03	6.05–01	–1.618	D+	*LS*
			7654.04	7656.15	96750.840	109812.233	4–2	6.83–03	3.00–03	3.02–01	–1.921	D+	*LS*
89.	$2s^22p^2(^3P)3p-$ $2s^22p^2(^3P)4d$	$^4S^o-^4P$	*7369.9*	*7371.9*	96750.840	*110315.90*	4–12	1.31–02	3.20–02	3.11+00	–0.893	C	2
			7378.51	7380.55	96750.840	110299.974	4–6	1.30–02	1.60–02	1.55+00	–1.194	D+	*LS*
			7366.15	7368.18	96750.840	110322.721	4–4	1.31–02	1.07–02	1.04+00	–1.370	D+	*LS*
			7351.36	7353.39	96750.840	110350.014	4–2	1.32–02	5.35–03	5.18–01	–1.670	D+	*LS*
90.	$2s^22p^2(^3P)3p-$ $2s^22p^2(^3P)6s$	$^4S^o-^4P$	*6292.6*	*6294.3*	96750.840	*112638.14*	4–12	3.72–03	6.63–03	5.50–01	–1.576	C	2
			6275.51	6277.25	96750.840	112681.389	4–6	3.75–03	3.32–03	2.75–01	–1.876	D+	*LS*
			6303.92	6305.66	96750.840	112609.612	4–4	3.70–03	2.21–03	1.83–01	–2.054	D+	*LS*
			6321.51	6323.26	96750.840	112565.470	4–2	3.67–03	1.10–03	9.16–02	–2.357	D+	*LS*
91.	$2s^22p^2(^3P)3p-$ $2s^22p^2(^3P)5d$	$^4S^o-^4P$	*6215.0*	*6216.7*	96750.840	*112836.6*	4–12	2.78–03	4.83–03	3.95–01	–1.714	C	2
			6219.65	6221.37	96750.840	112824.459	4–6	2.77–03	2.41–03	1.98–01	–2.015	D+	*LS*
			6214.41	6216.13	96750.840	112838.02	4–4	2.78–03	1.61–03	1.32–01	–2.191	D+	*LS*
			6201.98	6203.69	96750.840	112870.27	4–2	2.80–03	8.07–04	6.59–02	–2.491	D+	*LS*
92.	$2s^22p^2(^3P)3p-$ $2s^22p^2(^3P)7s$	$^4S^o-^4P$	*5761.9*	*5763.5*	96750.840	*114101.42*	4–12	2.19–03	3.27–03	2.48–01	–1.883	C	2
			5746.96	5748.55	96750.840	114146.525	4–6	2.21–03	1.64–03	1.24–01	–2.183	D+	*LS*
			5771.97	5773.57	96750.840	114071.153	4–4	2.18–03	1.09–03	8.27–02	–2.361	D+	*LS*
			5786.83	5788.44	96750.840	114026.654	4–2	2.16–03	5.43–04	4.14–02	–2.663	D+	*LS*
93.	$2s^22p^2(^3P)3p-$ $2s^22p^2(^3P)6d$	$^4S^o-^4P$	*5728.0*	*5729.6*	96750.840	*114204*	4–12	8.87–04	1.31–03	9.88–02	–2.281	C	2
			5732.11	5733.70	96750.840	114191.6	4–6	8.85–04	6.55–04	4.94–02	–2.582	D+	*LS*
			5729.35	5730.93	96750.840	114200.0	4–4	8.87–04	4.37–04	3.29–02	–2.758	D+	*LS*
			5713.14	5714.72	96750.840	114249.5	4–2	8.94–04	2.19–04	1.65–02	–3.058	D+	*LS*
94.	$2s^22p^2(^3P)3p-$ $2s^22p^2(^3P)8s$	$^4S^o-^4P$	*5486.4*	*5487.9*	96750.840	*114973*	4–12	1.38–03	1.87–03	1.35–01	–2.126	C	2
			5471.80	5473.32	96750.840	115021.3	4–6	1.39–03	9.37–04	6.76–02	–2.426	D+	*LS*
			5496.52	5498.05	96750.840	114939.1	4–4	1.37–03	6.22–04	4.50–02	–2.604	D+	*LS*
			5510.19	5511.72	96750.840	114894.0	4–2	1.36–03	3.10–04	2.25–02	–2.906	D+	*LS*
95.	$2s^22p^2(^3P)3p-$ $2s^22p^2(^3P)9s$	$^4S^o-^4P$	*5324.0*	*5325.5*	96750.840	*115528*	4–12	9.33–04	1.19–03	8.35–02	–2.322	C	2
			5310.39	5311.87	96750.840	115576.6	4–6	9.40–04	5.97–04	4.17–02	–2.622	D+	*LS*
			5333.37	5334.85	96750.840	115495.5	4–4	9.28–04	3.96–04	2.78–02	–2.800	D+	*LS*
			5346.49	5347.98	96750.840	115449.5	4–2	9.21–04	1.98–04	1.39–02	–3.102	D+	*LS*
96.	$2s^22p^2(^3P)3p-$ $2s^22p^2(^1D)3s$	$^2D^o-^2D$	*35325*	*2830.12* cm^{-1}	*96833.50*	*99663.62*	10–10	1.31–04	2.46–03	2.86+00	–1.610	B	1,3
			35712.5	2799.377 cm^{-1}	96864.050	99663.427	6–6	1.19–04	2.27–03	1.60+00	–1.866	B	3*n*
			34758.2	2876.232 cm^{-1}	96787.680	99663.912	4–4	1.22–04	2.21–03	1.01+00	–2.054	B	3*n*
			35706.3	2799.862 cm^{-1}	96864.050	99663.912	6–4	1.36–05	1.73–04	1.22–01	–2.984	B	3*n*
			34764.1	2875.747 cm^{-1}	96787.680	99663.427	4–6	9.95–06	2.70–04	1.24–01	–2.966	B	3*n*
97.	$2s^22p^2(^3P)3p-$ $2s^22p^2(^3P)4s$	$^2D^o-^2P$	*13579*	*7362.53* cm^{-1}	*96833.50*	*104196.03*	10–6	6.11–02	1.01–01	4.53+01	0.006	C	3
			13587.7	7357.580 cm^{-1}	96864.050	104221.630	6–4	5.36–02	9.89–02	2.66+01	–0.227	C	3
			13588.5	7357.140 cm^{-1}	96787.680	104144.820	4–2	6.63–02	9.18–02	1.64+01	–0.435	C	3
			13448.1	7433.950 cm^{-1}	96787.680	104221.630	4–4	4.91–03	1.33–02	2.36+00	–1.273	C	3

N I: Allowed Transitions — Continued

No.	Transition Array	Mult.	λ_{air} (Å)	λ_{vac} (Å) or σ (cm^{-1})*	E_i (cm^{-1})	E_k (cm^{-1})	g_i-g_k	A_{ki} (10^8 s^{-1})	f_{ik}	S (at. u.)	log gf	Acc.	Source
98.	$2s^22p^2(^3P)3p-$ $2s^22p^2(^3P)3d$	$^2D^o-^2P$	*12826*	*7794.82* cm^{-1}	*96833.50*	*104628.32*	10−6	2.92−02	4.32−02	1.82+01	−0.365	B	1,3
			12897.3	7751.420 cm^{-1}	96864.050	104615.470	6−4	2.59−02	4.32−02	1.10+01	−0.587	B	3*n*
			12708.9	7866.350 cm^{-1}	96787.680	104654.030	4−2	2.45−02	2.97−02	4.98+00	−0.925	B	3*n*
			12771.5	7827.790 cm^{-1}	96787.680	104615.470	4−4	5.47−03	1.34−02	2.25+00	−1.272	B	3*n*
99.		$^2D^o-^2F$	*12469*	*8017.43* cm^{-1}	*96833.50*	*104850.93*	10−14	2.27−01	7.41−01	3.04+02	0.870	B	1,3
			12469.6	8017.300 cm^{-1}	96864.050	104881.350	6−8	2.28−01	7.10−01	1.75+02	0.629	B	3*n*
			12461.3	8022.680 cm^{-1}	96787.680	104810.360	4−6	2.16−01	7.55−01	1.24+02	0.480	B	3*n*
			12581.0	7946.310 cm^{-1}	96864.050	104810.360	6−6	9.29−03	2.20−02	5.48+00	−0.878	C+	3*n*
100.		$^2D^o-^2D$	*12044*	*8300.68* cm^{-1}	*96833.50*	*105134.18*	10−10	6.03−02	1.31−01	5.21+01	0.118	B	1,3
			12074.5	8279.660 cm^{-1}	96864.050	105143.710	6−6	6.12−02	1.34−01	3.19+01	−0.095	B	3*n*
			11998.3	8332.200 cm^{-1}	96787.680	105119.880	4−4	5.04−02	1.09−01	1.72+01	−0.361	B	3*n*
			12109.3	8255.830 cm^{-1}	96864.050	105119.880	6−4	7.53−03	1.10−02	2.64+00	−1.179	B	3*n*
			11964.1	8356.030 cm^{-1}	96787.680	105143.710	4−6	5.49−04	1.77−03	2.79−01	−2.150	C+	3*n*
101.	$2s^22p^2(^3P)3p-$ $2s^22p^2(^3P)5s$	$^2D^o-^2P$	*7546.4*	*7548.5*	*96833.50*	*110081.13*	10−6	2.29−02	1.18−02	2.92+00	−0.930	B	9
			7550.91	7552.99	96864.050	110103.834	6−4	1.79−02	1.02−02	1.52+00	−1.214	B	9
			7546.21	7548.29	96787.680	110035.720	4−2	2.85−02	1.22−02	1.21+00	−1.312	B	9
			7507.61	7509.68	96787.680	110103.834	4−4	2.30−03	1.95−03	1.93−01	−2.108	C+	9
102.	$2s^22p^2(^3P)3p-$ $2s^22p^2(^3P)4d$	$^2D^o-^2P$	*7463.4*	*7465.5*	*96833.50*	*110228.47*	10−6	8.94−03	4.48−03	1.10+00	−1.349	C	1
			7485.18	7487.24	96864.050	110220.107	6−4	7.97−03	4.47−03	6.60−01	−1.572	D+	*LS*
			7428.75	7430.80	96787.680	110245.183	4−2	9.06−03	3.75−03	3.67−01	−1.824	D+	*LS*
			7442.62	7444.67	96787.680	110220.107	4−4	9.01−04	7.48−04	7.34−02	−2.524	D+	*LS*
103.		$^2D^o-^2F$	*7407.4*	*7409.4*	*96833.50*	*110329.82*	10−14	1.63−02	1.88−02	4.58+00	−0.726	C+	1
			7406.24	7408.28	96864.050	110362.462	6−8	1.63−02	1.79−02	2.62+00	−0.969	C−	*LS*
			7406.12	7408.16	96787.680	110286.305	4−6	1.52−02	1.88−02	1.83+00	−1.124	C−	*LS*
			7448.26	7450.31	96864.050	110286.305	6−6	1.07−03	8.89−04	1.31−01	−2.273	C−	*LS*
104.		$^2D^o-^2D$	*7336.1*	*7338.1*	*96833.50*	*110460.96*	10−10	5.97−03	4.82−03	1.16+00	−1.317	C+	1
			7347.57	7349.59	96864.050	110470.244	6−6	5.55−03	4.49−03	6.52−01	−1.570	C−	*LS*
			7318.98	7320.99	96787.680	110447.032	4−4	5.41−03	4.35−03	4.19−01	−1.760	C−	*LS*
			7360.13	7362.15	96864.050	110447.032	6−4	5.91−04	3.20−04	4.65−02	−2.716	C−	*LS*
			7306.56	7308.57	96787.680	110470.244	4−6	4.03−04	4.84−04	4.65−02	−2.713	C−	*LS*
105.	$2s^22p^2(^3P)3p-$ $2s^22p^2(^3P)5d$	$^2D^o-^2F$	*6238.9*	*6240.6*	*96833.50*	*112857.50*	10−14	2.81−03	2.29−03	4.71−01	−1.639	C+	1
			6237.67	6239.40	96864.050	112891.238	6−8	2.81−03	2.19−03	2.69−01	−1.882	C−	*LS*
			6238.59	6240.31	96787.680	112812.518	4−6	2.62−03	2.29−03	1.89−01	−2.037	C−	*LS*
			6268.46	6270.19	96864.050	112812.518	6−6	1.85−04	1.09−04	1.35−02	−3.185	C−	*LS*
106.		$^2D^o-^2D$	*6207.6*	*6209.3*	*96833.50*	*112938.31*	10−10	1.34−03	7.74−04	1.58−01	−2.111	C+	1
			6216.51	6218.23	96864.050	112945.809	6−6	1.25−03	7.21−04	8.86−02	−2.364	C−	*LS*
			6194.31	6196.03	96787.680	112927.055	4−4	1.21−03	6.98−04	5.70−02	−2.554	C−	*LS*
			6223.76	6225.49	96864.050	112927.055	6−4	1.33−04	5.15−05	6.33−03	−3.510	D+	*LS*
			6187.12	6188.84	96787.680	112945.809	4−6	9.02−05	7.77−05	6.33−03	−3.508	D+	*LS*
107.	$2s^22p^2(^3P)3p-$ $2s^22p^2(^1D)3d$	$^2D^o-^2F$											
			4356.29	4357.52	96864.050	119812.9	6−8	5.10−02	1.93−02	1.66+00	−0.935	C+	3

N I: Allowed Transitions — Continued

No.	Transition Array	Mult.	λ_{air} (Å)	λ_{vac} (Å) or σ (cm^{-1})*	E_i (cm^{-1})	E_k (cm^{-1})	g_i-g_k	A_{ki} (10^8 s^{-1})	f_{ik}	S (at. u.)	log gf	Acc.	Source
108.		$^2D^o-^2D$	4287.2	4288.4	96833.50	120152	10–10	5.57–02	1.53–02	2.17+00	–0.814	C+	3
			4293.21	4294.42	96864.050	120150.1	6–6	5.21–02	1.44–02	1.22+00	–1.064	C+	3
			4278.13	4279.33	96787.680	120155.8	4–4	5.01–02	1.37–02	7.74–01	–1.260	C+	3
			4292.16	4293.37	96864.050	120155.8	6–4	5.57–03	1.03–03	8.70–02	–2.211	C+	3
			4279.17	4280.38	96787.680	120150.1	4–6	3.61–03	1.49–03	8.38–02	–2.226	C+	3
109.		$^2D^o-^2P$	4258.4	4259.6	96833.50	120310	10–6	2.52–02	4.11–03	5.77–01	–1.386	C	3
			4264.00	4265.20	96864.050	120309.6	6–4	2.26–02	4.12–03	3.47–01	–1.607	C	3
			4249.87	4251.06	96787.680	120311.2	4–2	2.59–02	3.50–03	1.96–01	–1.853	C	3
			4250.16	4251.35	96787.680	120309.6	4–4	2.24–03	6.07–04	3.40–02	–2.615	C	3
110.	$2s^22p^2(^3P)3p-2s2p^4$	$^2D^o-^2D$											
			4107.90	4109.06	96864.050	121200.5	6–6	1.23–02	3.12–03	2.53–01	–1.728	C+	3
			4095.05	4096.21	96787.680	121200.5	4–6	8.05–04	3.04–04	1.64–02	–2.916	C+	3
111.	$2s^22p^2(^3P)3p-2s^22p^2(^1D)3s$	$^2P^o-^2D$		1869.67 cm^{-1}	97793.95	99663.62	6–10	2.67–05	1.91–03	2.02+00	–1.941	B	1,3
				1857.587 cm^{-1}	97805.840	99663.427	4–6	2.57–05	1.67–03	1.19+00	–2.174	B	3n
				1893.732 cm^{-1}	97770.180	99663.912	2–4	2.34–05	1.96–03	6.81–01	–2.407	B	3n
				1858.072 cm^{-1}	97805.840	99663.912	4–4	4.92–06	2.14–04	1.51–01	–3.068	B	3n
112.	$2s^22p^2(^3P)3p-2s^22p^2(^3P)4s$	$^2P^o-^2P$	15616	6402.08 cm^{-1}	97793.95	104196.03	6–6	5.86–02	2.14–01	6.62+01	0.109	C	3
			15582.3	6415.790 cm^{-1}	97805.840	104221.630	4–4	5.05–02	1.84–01	3.77+01	–0.134	C	3
			15682.9	6374.640 cm^{-1}	97770.180	104144.820	2–2	3.59–02	1.33–01	1.37+01	–0.577	C	3
			15771.1	6338.980 cm^{-1}	97805.840	104144.820	4–2	1.68–02	3.13–02	6.51+00	–0.902	C	3
			15496.2	6451.450 cm^{-1}	97770.180	104221.630	2–4	1.12–02	8.08–02	8.24+00	–0.792	C	3
113.	$2s^22p^2(^3P)3p-2s^22p^2(^3P)3d$	$^2P^o-^2P$	14628	6834.37 cm^{-1}	97793.95	104628.32	6–6	2.78–02	8.92–02	2.58+01	–0.272	C	3
			14681.1	6809.630 cm^{-1}	97805.840	104615.470	4–4	1.75–02	5.65–02	1.09+01	–0.646	C	3
			14522.8	6883.850 cm^{-1}	97770.180	104654.030	2–2	2.31–02	7.32–02	7.00+00	–0.835	C	3
			14598.4	6848.190 cm^{-1}	97805.840	104654.030	4–2	1.14–02	1.82–02	3.49+00	–1.139	C	3
			14604.6	6845.290 cm^{-1}	97770.180	104615.470	2–4	7.06–03	4.52–02	4.35+00	–1.044	C	3
114.		$^2P^o-^2D$	13620	7340.23 cm^{-1}	97793.95	105134.18	6–10	1.40–01	6.49–01	1.75+02	0.591	B	1,3
			13624.2	7337.870 cm^{-1}	97805.840	105143.710	4–6	1.39–01	5.80–01	1.04+02	0.365	B	3n
			13602.3	7349.700 cm^{-1}	97770.180	105119.880	2–4	1.12–01	6.19–01	5.55+01	0.093	B	3n
			13668.6	7314.040 cm^{-1}	97805.840	105119.880	4–4	3.02–02	8.46–02	1.52+01	–0.471	B	3n
115.	$2s^22p^2(^3P)3p-2s^22p^2(^3P)5s$	$^2P^o-^2P$	8136.3	8138.6	97793.95	110081.13	6–6	1.40–02	1.39–02	2.23+00	–1.079	C	1
			8129.17	8131.41	97805.840	110103.834	4–4	1.17–02	1.16–02	1.24+00	–1.334	D+	LS
			8150.68	8152.92	97770.180	110035.720	2–2	9.29–03	9.25–03	4.96–01	–1.733	D+	LS
			8174.45	8176.70	97805.840	110035.720	4–2	4.60–03	2.31–03	2.48–01	–2.035	D+	LS
			8105.67	8107.90	97770.180	110103.834	2–4	2.36–03	4.65–03	2.48–01	–2.031	D+	LS
116.	$2s^22p^2(^3P)3p-2s^22p^2(^3P)4d$	$^2P^o-^2D$	7892.4	7894.5	97793.95	110460.96	6–10	8.22–03	1.28–02	1.99+00	–1.115	C+	1
			7893.98	7896.15	97805.840	110470.244	4–6	8.21–03	1.15–02	1.20+00	–1.337	C–	LS
			7886.22	7888.39	97770.180	110447.032	2–4	6.86–03	1.28–02	6.65–01	–1.592	C–	LS
			7908.47	7910.65	97805.840	110447.032	4–4	1.36–03	1.28–03	1.33–01	–2.292	C–	LS
117.	$2s^22p^2(^3P)3p-2s^22p^2(^3P)5d$	$^2P^o-^2P$	6660.7	6662.6	97793.95	112803.21	6–6	2.04–03	1.36–03	1.79–01	–2.089	C	1
			6666.96	6668.80	97805.840	112801.031	4–4	1.70–03	1.13–03	9.92–02	–2.345	D+	LS
			6648.26	6650.09	97770.180	112807.567	2–2	1.37–03	9.07–04	3.97–02	–2.742	D+	LS
			6664.06	6665.90	97805.840	112807.567	4–2	6.79–04	2.26–04	1.98–02	–3.044	D+	LS
			6651.15	6652.98	97770.180	112801.031	2–4	3.42–04	4.53–04	1.98–02	–3.043	D+	LS

N I: Allowed Transitions — Continued

No.	Transition Array	Mult.	λ_{air} (Å)	λ_{vac} (Å) or σ (cm^{-1})*	E_i (cm^{-1})	E_k (cm^{-1})	$g_i - g_k$	A_{ki} (10^8 s^{-1})	f_{ik}	S (at. u.)	log gf	Acc.	Source
118.		$^2P^o - ^2D$	*6601.3*	*6603.1*	*97793.95*	*112938.31*	6–10	1.09–03	1.18–03	1.54–01	–2.148	C+	1
			6603.21	6605.03	97805.840	112945.809	4–6	1.09–03	1.07–03	9.26–02	–2.370	C–	*LS*
			6595.84	6597.67	97770.180	112927.055	2–4	9.08–04	1.19–03	5.15–02	–2.625	C–	*LS*
			6611.40	6613.23	97805.840	112927.055	4–4	1.80–04	1.18–04	1.03–02	–3.325	C–	*LS*
119.	$2s^22p^2(^3P)3p - 2s^22p^2(^1D)4s$	$^2P^o - ^2D$	*4668.1*	*4669.4*	*97793.95*	119210.0	6–10	1.81–02	9.85–03	9.08–01	–1.229	C+	3
			4670.68	4671.99	97805.840	119210.0	4–6	1.80–02	8.86–03	5.45–01	–1.451	C+	3
			4662.91	4664.22	97770.180	119210.0	2–4	1.51–02	9.85–03	3.02–01	–1.706	C+	3
			4670.68	4671.99	97805.840	119210.0	4–4	3.02–03	9.89–04	6.08–02	–2.403	C+	3
120.	$2s^22p^2(^3P)3p - 2s^22p^2(^1D)3d$	$^2P^o - ^2D$											
			4474.17	4475.42	97805.840	120150.1	4–6	2.76–04	1.24–04	7.33–03	–3.303	C	3
			4465.90	4467.15	97770.180	120155.8	2–4	2.04–04	1.22–04	3.60–03	–3.612	C	3
121.		$^2P^o - ^2P$	*4440.0*	*4441.3*	*97793.95*	*120310*	6–6	4.50–02	1.33–02	1.17+00	–1.097	C	3
			4442.45	4443.70	97805.840	120309.6	4–4	3.81–02	1.13–02	6.60–01	–1.346	C	3
			4435.11	4436.36	97770.180	120311.2	2–2	2.99–02	8.81–03	2.57–01	–1.754	C	3
			4442.14	4443.39	97805.840	120311.2	4–2	1.41–02	2.09–03	1.22–01	–2.079	C	3
			4435.43	4436.67	97770.180	120309.6	2–4	7.51–03	4.43–03	1.29–01	–2.053	C	3
122.		$^2P^o - ^2S$	*4390.1*	*4391.3*	*97793.95*	120566.0	6–2	2.64–02	2.55–03	2.21–01	–1.816	C+	1
			4392.41	4393.64	97805.840	120566.0	4–2	1.76–02	2.54–03	1.47–01	–1.992	C–	*LS*
			4385.54	4386.77	97770.180	120566.0	2–2	8.84–03	2.55–03	7.36–02	–2.293	C–	*LS*
123.	$2s^22p^2(^3P)3p - 2s2p^4$	$^2P^o - ^2D$											
			4273.28	4274.48	97805.840	121200.5	4–6	8.28–03	3.40–03	1.91–01	–1.866	C+	3
124.	$2s^22p^2(^1D)3s - 2s^22p^2(^3P)4p$	$^2D - ^2D^o$	*13222*	*7561.36* cm^{-1}	*99663.62*	*107224.98*	10–10	4.51–03	1.18–02	5.15+00	–0.927	C+	1
			13172.2	7589.679 cm^{-1}	99663.427	107253.106	6–6	4.26–03	1.11–02	2.88+00	–1.177	C–	*LS*
			13296.2	7518.876 cm^{-1}	99663.912	107182.788	4–4	3.99–03	1.06–02	1.85+00	–1.373	C–	*LS*
			13295.4	7519.361 cm^{-1}	99663.427	107182.788	6–4	4.44–04	7.84–04	2.06–01	–2.327	C–	*LS*
			13173.0	7589.194 cm^{-1}	99663.912	107253.106	4–6	3.04–04	1.19–03	2.06–01	–2.323	C–	*LS*
125.		$^2D - ^2P^o$	*12573*	*7951.39* cm^{-1}	*99663.62*	*107615.01*	10–6	1.81–03	2.57–03	1.07+00	–1.589	C+	1
			12551.7	7964.856 cm^{-1}	99663.427	107628.283	6–4	1.64–03	2.58–03	6.39–01	–1.810	C–	*LS*
			12615.6	7924.557 cm^{-1}	99663.912	107588.469	4–2	1.79–03	2.14–03	3.55–01	–2.068	C–	*LS*
			12552.5	7964.371 cm^{-1}	99663.912	107628.283	4–4	1.82–04	4.30–04	7.10–02	–2.765	C–	*LS*
126.	$2s^22p^2(^1D)3s - 2s^22p^2(^1D)3p$	$^2D - ^2D^o$	*9195.7*	*9198.2*	*99663.62*	*110535.33*	10–10	2.61–01	3.31–01	1.00+02	0.520	C+	3
			9187.45	9189.97	99663.427	110544.850	6–6	2.44–01	3.09–01	5.61+01	0.268	C+	3
			9208.00	9210.53	99663.912	110521.050	4–4	2.33–01	2.96–01	3.59+01	0.074	C+	3
			9207.59	9210.12	99663.427	110521.050	6–4	2.70–02	2.29–02	4.17+00	–0.862	C+	3
			9187.86	9190.38	99663.912	110544.850	4–6	1.76–02	3.34–02	4.05+00	–0.874	C+	3
127.		$^2D - ^2F^o$	*9047.6*	*9050.1*	*99663.62*	*110713.26*	10–14	2.80–01	4.80–01	1.43+02	0.682	B	1,3
			9045.88	9048.36	99663.427	110715.152	6–8	2.80–01	4.58–01	8.18+01	0.439	B	3*n*
			9049.89	9052.37	99663.912	110710.739	4–6	2.60–01	4.80–01	5.72+01	0.283	B	3*n*
			9049.49	9051.98	99663.427	110710.739	6–6	1.88–02	2.31–02	4.14+00	–0.857	B	3*n*
128.	$2s^22p^2(^1D)3s - 2s^22p^2(^3P)5p$	$^2D - ^2P^o$	*8659.5*	*8661.9*	*99663.62*	*111208.46*	10–6	6.05–02	4.08–02	1.16+01	–0.389	C+	1
			8655.75	8658.13	99663.427	111213.271	6–4	5.46–02	4.08–02	6.98+00	–0.611	C–	*LS*
			8666.94	8669.32	99663.912	111198.848	4–2	6.04–02	3.40–02	3.88+00	–0.867	C–	*LS*
			8656.11	8658.49	99663.912	111213.271	4–4	6.06–03	6.81–03	7.76–01	–1.565	C–	*LS*

N I: Allowed Transitions — Continued

No.	Transition Array	Mult.	λ_{air} (Å)	λ_{vac} (Å) or σ (cm^{-1})*	E_i (cm^{-1})	E_k (cm^{-1})	g_i-g_k	A_{ki} (10^8 s^{-1})	f_{ik}	S (at. u.)	log gf	Acc.	Source
129.		$^2D-^2D^o$	*8180.4*	*8182.7*	*99663.62*	*111884.59*	10−10	1.40−01	1.41−01	3.79+01	0.148	C+	1
			8166.23	8168.48	99663.427	111905.609	6−6	1.32−01	1.32−01	2.12+01	−0.103	C−	*LS*
			8201.76	8204.02	99663.912	111853.061	4−4	1.25−01	1.26−01	1.36+01	−0.297	C−	*LS*
			8201.44	8203.69	99663.427	111853.061	6−4	1.39−02	9.35−03	1.52+00	−1.251	C−	*LS*
			8166.56	8168.80	99663.912	111905.609	4−6	9.39−03	1.41−02	1.52+00	−1.249	C−	*LS*
130.	$2s^22p^2(^1D)3s-2s^22p^2(^1D)3p$	$^2D-^2P^o$	*7904.5*	*7906.6*	*99663.62*	*112311.21*	10−6	3.14−01	1.77−01	4.60+01	0.247	C	1,3
			7898.98	7901.15	99663.427	112319.805	6−4	2.82−01	1.76−01	2.75+01	0.024	C	3*n*
			7915.42	7917.60	99663.912	112294.007	4−2	3.13−01	1.47−01	1.53+01	−0.231	C	3*n*
			7899.28	7901.46	99663.912	112319.805	4−4	3.28−02	3.07−02	3.19+00	−0.911	C	3*n*
131.	$2s^22p^2(^3P)4s-2s^22p^2(^3P)4p$	$^4P-^4D^o$	*31946*	*3129.5* cm^{-1}	*103693.9*	*106823.33*	12−20	3.06−02	7.80−01	9.85+02	0.972	C	1
			31908.0	3133.16 cm^{-1}	103735.48	106868.635	6−8	3.07−02	6.25−01	3.94+02	0.574	D+	*LS*
			31764.6	3147.30 cm^{-1}	103667.16	106814.459	4−6	2.18−02	4.94−01	2.07+02	0.296	D+	*LS*
			31678.8	3155.83 cm^{-1}	103622.51	106778.337	2−4	1.31−02	3.94−01	8.21+01	−0.104	D+	*LS*
			32469.4	3078.98 cm^{-1}	103735.48	106814.459	6−6	8.74−03	1.38−01	8.86+01	−0.081	D+	*LS*
			32133.4	3111.18 cm^{-1}	103667.16	106778.337	4−4	1.60−02	2.48−01	1.05+02	−0.003	D+	*LS*
			31876.8	3136.22 cm^{-1}	103622.51	106758.731	2−2	2.57−02	3.91−01	8.21+01	−0.107	D+	*LS*
			32854.9	3042.86 cm^{-1}	103735.48	106778.337	6−4	1.41−03	1.52−02	9.85+00	−1.041	D+	*LS*
			32337.2	3091.57 cm^{-1}	103667.16	106758.731	4−2	4.92−03	3.86−02	1.64+01	−0.812	D+	*LS*
132.		$^4P-^4P^o$	*30112*	*3320.1* cm^{-1}	*103693.9*	*107013.96*	12−12	3.94−02	5.35−01	6.37+02	0.808	C	1
			30280.2	3301.59 cm^{-1}	103735.48	107037.069	6−6	2.71−02	3.73−01	2.23+02	0.349	D+	*LS*
			30032.0	3328.87 cm^{-1}	103667.16	106996.032	4−4	5.29−03	7.15−02	2.83+01	−0.543	D+	*LS*
			29771.8	3357.97 cm^{-1}	103622.51	106980.480	2−2	6.79−03	9.02−02	1.77+01	−0.744	D+	*LS*
			30661.3	3260.55 cm^{-1}	103735.48	106996.032	6−4	1.68−02	1.58−01	9.55+01	−0.024	D+	*LS*
			30173.0	3313.32 cm^{-1}	103667.16	106980.480	4−2	3.26−02	2.23−01	8.84+01	−0.051	D+	*LS*
			29666.3	3369.91 cm^{-1}	103667.16	107037.069	4−6	1.24−02	2.44−01	9.55+01	−0.010	D+	*LS*
			29634.5	3373.52 cm^{-1}	103622.51	106996.032	2−4	1.72−02	4.53−01	8.84+01	−0.043	D+	*LS*
133.		$^4P-^4S^o$	*26647*	*3751.7* cm^{-1}	*103693.9*	107445.622	12−4	4.13−02	1.47−01	1.54+02	0.245	C	1
			26945.8	3710.14 cm^{-1}	103735.48	107445.622	6−4	2.00−02	1.45−01	7.72+01	−0.060	D+	*LS*
			26458.6	3778.46 cm^{-1}	103667.16	107445.622	4−4	1.41−02	1.48−01	5.15+01	−0.229	D+	*LS*
			26149.6	3823.11 cm^{-1}	103622.51	107445.622	2−4	7.29−03	1.49−01	2.57+01	−0.525	D+	*LS*
134.	$2s^22p^2(^3P)4s-2s^22p^2(^3P)5p$	$^4P-^4D^o$	*13296*	*7519.1* cm^{-1}	*103693.9*	*111212.94*	12−20	6.03−04	2.66−03	1.40+00	−1.495	C	1
			13284.7	7525.39 cm^{-1}	103735.48	111260.873	6−8	6.05−04	2.13−03	5.60−01	−1.893	D+	*LS*
			13264.5	7536.86 cm^{-1}	103667.16	111204.016	4−6	4.25−04	1.68−03	2.94−01	−2.172	D+	*LS*
			13254.3	7542.65 cm^{-1}	103622.51	111165.158	2−4	2.54−04	1.34−03	1.17−01	−2.573	D+	*LS*
			13385.8	7468.54 cm^{-1}	103735.48	111204.016	6−6	1.77−04	4.76−04	1.26−01	−2.544	D+	*LS*
			13333.2	7498.00 cm^{-1}	103667.16	111165.158	4−4	3.19−04	8.50−04	1.49−01	−2.469	D+	*LS*
			13292.4	7521.06 cm^{-1}	103622.51	111143.567	2−2	5.03−04	1.33−03	1.17−01	−2.574	D+	*LS*
			13455.9	7429.68 cm^{-1}	103735.48	111165.158	6−4	2.91−05	5.26−05	1.40−02	−3.501	D+	*LS*
			13371.8	7476.41 cm^{-1}	103667.16	111143.567	4−2	9.88−05	1.32−04	2.33−02	−3.276	D+	*LS*
135.		$^4P-^4P^o$	*13137*	*7610.0* cm^{-1}	*103693.9*	*111303.88*	12−12	2.16−03	5.58−03	2.89+00	−1.174	C	1
			13169.3	7591.32 cm^{-1}	103735.48	111326.798	6−6	1.50−03	3.89−03	1.01+00	−1.631	D+	*LS*
			13122.4	7618.48 cm^{-1}	103667.16	111285.644	4−4	2.88−04	7.44−04	1.29−01	−2.526	D+	*LS*
			13069.9	7649.09 cm^{-1}	103622.51	111271.596	2−2	3.65−04	9.34−04	8.04−02	−2.729	D+	*LS*
			13241.1	7550.16 cm^{-1}	103735.48	111285.644	6−4	9.47−04	1.66−03	4.34−01	−2.002	D+	*LS*
			13146.6	7604.44 cm^{-1}	103667.16	111271.596	4−2	1.79−03	2.32−03	4.02−01	−2.032	D+	*LS*
			13051.9	7659.64 cm^{-1}	103667.16	111326.798	4−6	6.59−04	2.53−03	4.34−01	−1.996	D+	*LS*
			13045.9	7663.13 cm^{-1}	103622.51	111285.644	2−4	9.17−04	4.68−03	4.02−01	−2.029	D+	*LS*

N I: Allowed Transitions — Continued

No.	Transition Array	Mult.	λ_{air} (Å)	λ_{vac} (Å) or σ (cm^{-1})*	E_i (cm^{-1})	E_k (cm^{-1})	g_i-g_k	A_{ki} ($10^8 s^{-1}$)	f_{ik}	S (at. u.)	log *gf*	Acc.	Source
136.		$^4P-^4S^o$	*12805*	*7807.5* cm^{-1}	*103693.9*	– 111501.368	12–4	4.84–03	3.96–03	2.00+00	–1.323	C	1
			12873.3	7765.89 cm^{-1}	103735.48	– 111501.368	6–4	2.38–03	3.94–03	1.00+00	–1.626	D+	*LS*
			12761.0	7834.21 cm^{-1}	103667.16	– 111501.368	4–4	1.63–03	3.98–03	6.68–01	–1.798	D+	*LS*
			12688.7	7878.86 cm^{-1}	103622.51	– 111501.368	2–4	8.28–04	4.00–03	3.34–01	–2.097	D+	*LS*
137.	$2s^22p^2(^3P)4s-2s^22p^2(^3P)4p$	$^2P-^2S^o$	*43814*	*2281.77* cm^{-1}	*104196.03*	– 106477.800	6–2	7.52–03	7.21–02	6.24+01	–0.364	C	1
			44310.8	2256.170 cm^{-1}	104221.630	– 106477.800	4–2	4.85–03	7.13–02	4.16+01	–0.545	D+	*LS*
			42851.9	2332.980 cm^{-1}	104144.820	– 106477.800	2–2	2.68–03	7.38–02	2.08+01	–0.831	D+	*LS*
138.		$^2P-^2D^o$	*33006*	*3028.95* cm^{-1}	*104196.03*	– *107224.98*	6–10	2.52–02	6.86–01	4.47+02	0.615	C	1
			32978.2	3031.476 cm^{-1}	104221.630	– 107253.106	4–6	2.53–02	6.18–01	2.68+02	0.393	D+	*LS*
			32907.8	3037.968 cm^{-1}	104144.820	– 107182.788	2–4	2.12–02	6.88–01	1.49+02	0.139	D+	*LS*
			33761.4	2961.158 cm^{-1}	104221.630	– 107182.788	4–4	3.93–03	6.71–02	2.98+01	–0.571	D+	*LS*
139.		$^2P-^2P^o$	*29240*	*3418.98* cm^{-1}	*104196.03*	– *107615.01*	6–6	4.36–02	5.59–01	3.23+02	0.526	C	1
			29346.3	3406.653 cm^{-1}	104221.630	– 107628.283	4–4	3.60–02	4.64–01	1.79+02	0.269	D+	*LS*
			29031.0	3443.649 cm^{-1}	104144.820	– 107588.469	2–2	2.97–02	3.75–01	7.18+01	–0.124	D+	*LS*
			29693.4	3366.839 cm^{-1}	104221.630	– 107588.469	4–2	1.39–02	9.18–02	3.59+01	–0.435	D+	*LS*
			28699.2	3483.463 cm^{-1}	104144.820	– 107628.283	2–4	7.69–03	1.90–01	3.59+01	–0.420	D+	*LS*
140.	$2s^22p^2(^3P)4s-2s^22p^2(^1D)3p$	$^2P-^2D^o$	*15770*	*6339.30* cm^{-1}	*104196.03*	– *110535.33*	6–10	9.49–04	5.90–03	1.84+00	–1.451	C	3
			15810.4	6323.220 cm^{-1}	104221.630	– 110544.850	4–6	9.17–04	5.16–03	1.07+00	–1.686	C	3
			15679.0	6376.230 cm^{-1}	104144.820	– 110521.050	2–4	8.66–04	6.39–03	6.59–01	–1.894	C	3
			15870.1	6299.420 cm^{-1}	104221.630	– 110521.050	4–4	1.32–04	5.00–04	1.05–01	–2.699	C	3
141.	$2s^22p^2(^3P)4s-2s^22p^2(^3P)5p$	$^2P-^2S^o$	*14563*	*6864.88* cm^{-1}	*104196.03*	– 111060.905	6–2	1.91–03	2.02–03	5.81–01	–1.916	C	1
			14617.4	6839.275 cm^{-1}	104221.630	– 111060.905	4–2	1.26–03	2.01–03	3.87–01	–2.094	D+	*LS*
			14455.1	6916.085 cm^{-1}	104144.820	– 111060.905	2–2	6.50–04	2.04–03	1.94–01	–2.390	D+	*LS*
142.		$^2P-^2P^o$	*14256*	*7012.43* cm^{-1}	*104196.03*	– *111208.46*	6–6	8.98–04	2.74–03	7.71–01	–1.785	C	1
			14298.9	6991.641 cm^{-1}	104221.630	– 111213.271	4–4	7.42–04	2.27–03	4.28–01	–2.041	D+	*LS*
			14172.4	7054.028 cm^{-1}	104144.820	– 111198.848	2–2	6.10–04	1.84–03	1.71–01	–2.435	D+	*LS*
			14328.4	6977.218 cm^{-1}	104221.630	– 111198.848	4–2	2.95–04	4.54–04	8.56–02	–2.741	D+	*LS*
			14143.5	7068.451 cm^{-1}	104144.820	– 111213.271	2–4	1.53–04	9.20–04	8.56–02	–2.735	D+	*LS*
143.		$^2P-^2D^o$	*13003*	*7688.56* cm^{-1}	*104196.03*	– *111884.59*	6–10	1.02–02	4.30–02	1.10+01	–0.589	C	1
			13010.5	7683.979 cm^{-1}	104221.630	– 111905.609	4–6	1.02–02	3.87–02	6.62+00	–0.811	D+	*LS*
			12969.6	7708.241 cm^{-1}	104144.820	– 111853.061	2–4	8.54–03	4.31–02	3.68+00	–1.065	D+	*LS*
			13100.1	7631.431 cm^{-1}	104221.630	– 111853.061	4–4	1.66–03	4.27–03	7.36–01	–1.768	D+	*LS*
144.	$2s^22p^2(^3P)4s-2s^22p^2(^1D)3p$	$^2P-^2P^o$	*12319*	*8115.18* cm^{-1}	*104196.03*	– *112311.21*	6–6	3.46–03	7.88–03	1.92+00	–1.325	C	3
			12345.1	8098.175 cm^{-1}	104221.630	– 112319.805	4–4	2.90–03	6.63–03	1.08+00	–1.576	C	3
			12267.8	8149.187 cm^{-1}	104144.820	– 112294.007	2–2	2.19–03	4.95–03	4.00–01	–2.005	C	3
			12384.5	8072.377 cm^{-1}	104221.630	– 112294.007	4–2	1.27–03	1.47–03	2.39–01	–2.232	C	3
			12229.1	8174.985 cm^{-1}	104144.820	– 112319.805	2–4	5.54–04	2.48–03	2.00–01	–2.304	C	3
145.	$2s^22p^2(^3P)3d-2s^22p^2(^3P)4p$	$^2P-^2S^o$		*1849.48* cm^{-1}	*104628.32*	– 106477.800	6–2	1.01–02	1.48–01	1.58+02	–0.053	C	1
				1862.330 cm^{-1}	104615.470	– 106477.800	4–2	6.88–03	1.49–01	1.05+02	–0.226	D+	*LS*
				1823.770 cm^{-1}	104654.030	– 106477.800	2–2	3.23–03	1.46–01	5.26+01	–0.536	D+	*LS*

N I: Allowed Transitions — Continued

No.	Transition Array	Mult.	λ_{air} (Å)	λ_{vac} (Å) or σ (cm^{-1})*	E_i (cm^{-1})	E_k (cm^{-1})	g_i-g_k	A_{ki} ($10^8 s^{-1}$)	f_{ik}	S (at. u.)	log gf	Acc.	Source
146.		$^2P-^2D^\circ$	*38501*	*2596.66* cm^{-1}	*104628.32*	*107224.98*	6–10	2.45–03	9.08–02	6.90+01	−0.264	C	1
			37902.4	2637.636 cm^{-1}	104615.470	107253.106	4–6	2.57–03	8.30–02	4.14+01	−0.479	D+	*LS*
			39534.3	2528.758 cm^{-1}	104654.030	107182.788	2–4	1.89–03	8.84–02	2.30+01	−0.753	D+	*LS*
			38940.5	2567.318 cm^{-1}	104615.470	107182.788	4–4	3.95–04	8.97–03	4.60+00	−1.445	D+	*LS*
147.		$^2P-^2P^\circ$	*33473*	*2986.69* cm^{-1}	*104628.32*	*107615.01*	6–6	2.36–04	3.96–03	2.62+00	−1.624	C	1
			33182.5	3012.813 cm^{-1}	104615.470	107628.283	4–4	2.02–04	3.33–03	1.46+00	−1.875	D+	*LS*
			34068.8	2934.439 cm^{-1}	104654.030	107588.469	2–2	1.49–04	2.59–03	5.82–01	−2.285	D+	*LS*
			33626.9	2972.999 cm^{-1}	104615.470	107588.469	4–2	7.75–05	6.57–04	2.91–01	−2.580	D+	*LS*
			33612.7	2974.253 cm^{-1}	104654.030	107628.283	2–4	3.88–05	1.32–03	2.91–01	−2.580	D+	*LS*
148.	$2s^22p^2(^3P)3d-2s^22p^2(^1D)3p$	$^2P-^2D^\circ$	*16924*	*5907.01* cm^{-1}	*104628.32*	*110535.33*	6–10	4.12–04	2.95–03	9.86–01	−1.752	B	1,3
			16860.6	5929.380 cm^{-1}	104615.470	110544.850	4–6	4.16–04	2.66–03	5.91–01	−1.973	B	3*n*
			17039.8	5867.020 cm^{-1}	104654.030	110521.050	2–4	2.93–04	2.55–03	2.87–01	−2.292	B	3*n*
			16928.5	5905.580 cm^{-1}	104615.470	110521.050	4–4	1.13–04	4.86–04	1.08–01	−2.711	B	3*n*
149.	$2s^22p^2(^3P)3d-2s^22p^2(^3P)5p$	$^2P-^2P^\circ$	*15193*	*6580.14* cm^{-1}	*104628.32*	*111208.46*	6–6	4.97–04	1.72–03	5.16–01	−1.986	C	1
			15152.4	6597.801 cm^{-1}	104615.470	111213.271	4–4	4.18–04	1.44–03	2.87–01	−2.240	D+	*LS*
			15275.1	6544.818 cm^{-1}	104654.030	111198.848	2–2	3.26–04	1.14–03	1.15–01	−2.642	D+	*LS*
			15185.6	6583.378 cm^{-1}	104615.470	111198.848	4–2	1.66–04	2.87–04	5.74–02	−2.940	D+	*LS*
			15241.5	6559.241 cm^{-1}	104654.030	111213.271	2–4	8.21–05	5.72–04	5.74–02	−2.942	D+	*LS*
150.		$^2P-^2D^\circ$	*13777*	*7256.27* cm^{-1}	*104628.32*	*111884.59*	6–10	9.70–04	4.60–03	1.25+00	−1.559	C	1
			13713.4	7290.139 cm^{-1}	104615.470	111905.609	4–6	9.84–04	4.16–03	7.51–01	−1.779	D+	*LS*
			13887.0	7199.031 cm^{-1}	104654.030	111853.061	2–4	7.90–04	4.57–03	4.17–01	−2.039	D+	*LS*
			13813.0	7237.591 cm^{-1}	104615.470	111853.061	4–4	1.60–04	4.59–04	8.35–02	−2.736	D+	*LS*
151.	$2s^22p^2(^3P)3d-2s^22p^2(^1D)3p$	$^2P-^2P^\circ$	*13012*	*7682.88* cm^{-1}	*104628.32*	*112311.21*	6–6	6.08–04	1.54–03	3.97–01	−2.033	C	3
			12976.2	7704.335 cm^{-1}	104615.470	112319.805	4–4	3.57–04	9.01–04	1.54–01	−2.443	C	3
			13085.5	7639.977 cm^{-1}	104654.030	112294.007	2–2	4.71–04	1.21–03	1.04–01	−2.616	C	3
			13019.8	7678.537 cm^{-1}	104615.470	112294.007	4–2	3.53–04	4.49–04	7.71–02	−2.745	C	3
			13041.4	7665.775 cm^{-1}	104654.030	112319.805	2–4	1.41–04	7.17–04	6.16–02	−2.843	C	3
152.	$2s^22p^2(^3P)3d-2s^22p^2(^3P)4p$	$^4F-^4D^\circ$	*47521*	*2103.8* cm^{-1}	*104719.6*	*106823.33*	28–20	7.25–03	1.75–01	7.68+02	0.691	C+	1
			47541.2	2102.87 cm^{-1}	104765.77	106868.635	10–8	6.47–03	1.75–01	2.74+02	0.244	C−	*LS*
			47662.6	2097.509 cm^{-1}	104716.950	106814.459	8–6	5.87–03	1.50–01	1.88+02	0.079	C−	*LS*
			47713.4	2095.277 cm^{-1}	104683.060	106778.337	6–4	5.76–03	1.31–01	1.23+02	−0.105	C−	*LS*
			47728.8	2094.601 cm^{-1}	104664.130	106758.731	4–2	7.16–03	1.22–01	7.68+01	−0.311	C−	*LS*
			46462.5	2151.685 cm^{-1}	104716.950	106868.635	8–8	7.90–04	2.56–02	3.13+01	−0.689	C−	*LS*
			46904.8	2131.399 cm^{-1}	104683.060	106814.459	6–6	1.31–03	4.32–02	4.01+01	−0.586	C−	*LS*
			47286.2	2114.207 cm^{-1}	104664.130	106778.337	4–4	1.47–03	4.93–02	3.07+01	−0.705	C−	*LS*
			45742.1	2185.575 cm^{-1}	104683.060	106868.635	6–8	4.21–05	1.76–03	1.59+00	−1.976	C−	*LS*
			46491.8	2150.329 cm^{-1}	104664.130	106814.459	4–6	7.37–05	3.58–03	2.19+00	−1.844	C−	*LS*
153.	$2s^22p^2(^3P)3d-2s^22p^2(^3P)5p$	$^4F-^4D^\circ$	*15396*	*6493.4* cm^{-1}	*104719.6*	*111212.94*	28–20	1.16–03	2.94–03	4.17+00	−1.085	C+	1
			15392.0	6495.10 cm^{-1}	104765.77	111260.873	10–8	1.03–03	2.94–03	1.49+00	−1.532	C−	*LS*
			15411.1	6487.066 cm^{-1}	104716.950	111204.016	8–6	9.42–04	2.52–03	1.02+00	−1.696	C−	*LS*
			15422.9	6482.098 cm^{-1}	104683.060	111165.158	6–4	9.25–04	2.20–03	6.70–01	−1.880	C−	*LS*
			15429.2	6479.437 cm^{-1}	104664.130	111143.567	4–2	1.15–03	2.05–03	4.17–01	−2.086	C−	*LS*
			15277.2	6543.923 cm^{-1}	104716.950	111260.873	8–8	1.21–04	4.22–04	1.70–01	−2.472	C−	*LS*
			15331.0	6520.956 cm^{-1}	104683.060	111204.016	6–6	2.04–04	7.18–04	2.17–01	−2.366	C−	*LS*
			15378.0	6501.028 cm^{-1}	104664.130	111165.158	4–4	2.32–04	8.24–04	1.67–01	−2.482	C−	*LS*
			15198.5	6577.813 cm^{-1}	104683.060	111260.873	6–8	6.23–06	2.88–05	8.64–03	−3.763	D+	*LS*
			15286.6	6539.886 cm^{-1}	104664.130	111204.016	4–6	1.13–05	5.92–05	1.19–02	−3.626	C−	*LS*

N I: Allowed Transitions — Continued

No.	Transition Array	Mult.	λ_{air} (Å)	λ_{vac} (Å) or σ (cm^{-1})*	E_i (cm^{-1})	E_k (cm^{-1})	g_i–g_k	A_{ki} (10^8 s^{-1})	f_{ik}	S (at. u.)	log gf	Acc.	Source
154.	$2s^22p^2(^3P)3d$–$2s^22p^2(^3P)4p$	^2F–$^2D^o$	*42111*	*2374.05* cm^{-1}	*104850.93*	*107224.98*	14–10	9.03–03	1.72–01	3.33+02	0.381	C+	1
			42151.4	2371.756 cm^{-1}	104881.350	107253.106	8–6	8.58–03	1.71–01	1.90+02	0.137	C–	*LS*
			42139.4	2372.428 cm^{-1}	104810.360	107182.788	6–4	9.02–03	1.60–01	1.33+02	–0.018	C–	*LS*
			40926.4	2442.746 cm^{-1}	104810.360	107253.106	6–6	4.69–04	1.18–02	9.51+00	–1.151	C–	*LS*
155.	$2s^22p^2(^3P)3d$–$2s^22p^2(^3P)5p$	^2F–$^2D^o$	*14213*	*7033.66* cm^{-1}	*104850.93*	*111884.59*	14–10	4.14–03	8.96–03	5.87+00	–0.902	C+	1
			14232.5	7024.259 cm^{-1}	104881.350	111905.609	8–6	3.93–03	8.95–03	3.35+00	–1.145	C–	*LS*
			14195.2	7042.701 cm^{-1}	104810.360	111853.061	6–4	4.16–03	8.37–03	2.35+00	–1.299	C–	*LS*
			14090.1	7095.249 cm^{-1}	104810.360	111905.609	6–6	2.02–04	6.03–04	1.68–01	–2.442	C–	*LS*
156.	$2s^22p^2(^3P)3d$–$2s^22p^2(^3P)4p$	^4P–$^4D^o$		*1976.5* cm^{-1}	*104846.8*	*106823.33*	12–20	4.42–04	2.83–02	5.65+01	–0.470	C	1
			48921.7	2043.525 cm^{-1}	104825.110	106868.635	6–8	4.88–04	2.34–02	2.26+01	–0.853	D+	*LS*
				1954.73 cm^{-1}	104859.73	106814.459	4–6	2.99–04	1.76–02	1.19+01	–1.152	D+	*LS*
				1892.24 cm^{-1}	104886.10	106778.337	2–4	1.62–04	1.35–02	4.70+00	–1.568	D+	*LS*
				1989.349 cm^{-1}	104825.110	106814.459	6–6	1.35–04	5.12–03	5.08+00	–1.513	D+	*LS*
				1918.61 cm^{-1}	104859.73	106778.337	4–4	2.16–04	8.78–03	6.02+00	–1.455	D+	*LS*
				1872.63 cm^{-1}	104886.10	106758.731	2–2	3.13–04	1.34–02	4.70+00	–1.572	D+	*LS*
				1953.227 cm^{-1}	104825.110	106778.337	6–4	2.13–05	5.58–04	5.65–01	–2.475	D+	*LS*
				1899.00 cm^{-1}	104859.73	106758.731	4–2	6.53–05	1.36–03	9.41–01	–2.265	D+	*LS*
157.		^4P–$^4P^o$	*46131*	*2167.1* cm^{-1}	*104846.8*	*107013.96*	12–12	1.19–03	3.79–02	6.91+01	–0.342	C	1
			45196.5	2211.959 cm^{-1}	104825.110	107037.069	6–6	8.85–04	2.71–02	2.42+01	–0.789	D+	*LS*
			46797.1	2136.30 cm^{-1}	104859.73	106996.032	4–4	1.52–04	4.98–03	3.07+00	–1.700	D+	*LS*
			47733.8	2094.38 cm^{-1}	104886.10	106980.480	2–2	1.79–04	6.11–03	1.92+00	–1.913	D+	*LS*
			46050.8	2170.922 cm^{-1}	104825.110	106996.032	6–4	5.38–04	1.14–02	1.04+01	–1.165	D+	*LS*
			47140.3	2120.75 cm^{-1}	104859.73	106980.480	4–2	9.28–04	1.55–02	9.60+00	–1.209	D+	*LS*
			45915.1	2177.34 cm^{-1}	104859.73	107037.069	4–6	3.62–04	1.71–02	1.04+01	–1.164	D+	*LS*
			47382.0	2109.93 cm^{-1}	104886.10	106996.032	2–4	4.57–04	3.08–02	9.60+00	–1.211	D+	*LS*
158.		^4P–$^4S^o$	*38469*	*2598.8* cm^{-1}	*104846.8*	107445.622	12–4	1.85–02	1.37–01	2.08+02	0.216	C	1
			38150.1	2620.512 cm^{-1}	104825.110	107445.622	6–4	9.49–03	1.38–01	1.04+02	–0.082	D+	*LS*
			38660.8	2585.89 cm^{-1}	104859.73	107445.622	4–4	6.08–03	1.36–01	6.94+01	–0.264	D+	*LS*
			39059.1	2559.52 cm^{-1}	104886.10	107445.622	2–4	2.95–03	1.35–01	3.47+01	–0.569	D+	*LS*
159.	$2s^22p^2(^3P)3d$–$2s^22p^2(^3P)5p$	^4P–$^4P^o$	*15483*	*6457.1* cm^{-1}	*104846.8*	*111303.88*	12–12	6.02–04	2.16–03	1.32+00	–1.586	C	1
			15376.4	6501.688 cm^{-1}	104825.110	111326.798	6–6	4.30–04	1.52–03	4.63–01	–2.039	D+	*LS*
			15557.7	6425.91 cm^{-1}	104859.73	111285.644	4–4	7.91–05	2.87–04	5.88–02	–2.940	D+	*LS*
			15656.2	6385.50 cm^{-1}	104886.10	111271.596	2–2	9.70–05	3.56–04	3.67–02	–3.147	D+	*LS*
			15474.4	6460.534 cm^{-1}	104825.110	111285.644	6–4	2.71–04	6.49–04	1.98–01	–2.410	D+	*LS*
			15591.8	6411.87 cm^{-1}	104859.73	111271.596	4–2	4.91–04	8.94–04	1.84–01	–2.446	D+	*LS*
			15458.7	6467.07 cm^{-1}	104859.73	111326.798	4–6	1.81–04	9.74–04	1.98–01	–2.409	D+	*LS*
			15621.8	6399.54 cm^{-1}	104886.10	111285.644	2–4	2.44–04	1.79–03	1.84–01	–2.447	D+	*LS*
160.		^4P–$^4S^o$	*15023*	*6654.5* cm^{-1}	*104846.8*	111501.368	12–4	3.60–03	4.06–03	2.41+00	–1.312	C	1
			14974.4	6676.258 cm^{-1}	104825.110	111501.368	6–4	1.82–03	4.08–03	1.21+00	–1.611	D+	*LS*
			15052.4	6641.64 cm^{-1}	104859.73	111501.368	4–4	1.19–03	4.06–03	8.04–01	–1.790	D+	*LS*
			15112.4	6615.27 cm^{-1}	104886.10	111501.368	2–4	5.90–04	4.04–03	4.02–01	–2.093	D+	*LS*

N I: Allowed Transitions — Continued

No.	Transition Array	Mult.	λ_{air} (Å)	λ_{vac} (Å) or σ (cm^{-1})*	E_i (cm^{-1})	E_k (cm^{-1})	g_i-g_k	A_{ki} (10^8 s^{-1})	f_{ik}	S (at. u.)	log gf	Acc.	Source
161.	$2s^22p^2(^3P)3d-2s^22p^2(^3P)4p$	$^4D-^4D^o$		*1816.0* cm^{-1}	*105007.3*	*106823.33*	20–20	9.28–04	4.22–02	1.53+02	–0.074	C+	1
				1851.035 cm^{-1}	105017.600	106868.635	8–8	8.43–04	3.69–02	5.24+01	–0.530	C–	*LS*
				1805.91 cm^{-1}	105008.55	106814.459	6–6	5.23–04	2.40–02	2.63+01	–0.841	C–	*LS*
				1782.07 cm^{-1}	104996.27	106778.337	4–4	3.51–04	1.65–02	1.22+01	–1.179	C–	*LS*
				1774.36 cm^{-1}	104984.37	106758.731	2–2	4.33–04	2.06–02	7.64+00	–1.385	C–	*LS*
				1796.859 cm^{-1}	105017.600	106814.459	8–6	1.71–04	5.94–03	8.71+00	–1.323	C–	*LS*
				1769.79 cm^{-1}	105008.55	106778.337	6–4	3.01–04	9.59–03	1.07+01	–1.240	C–	*LS*
				1762.46 cm^{-1}	104996.27	106758.731	4–2	4.24–04	1.02–02	7.64+00	–1.388	C–	*LS*
				1860.09 cm^{-1}	105008.55	106868.635	6–8	1.42–04	8.21–03	8.71+00	–1.308	C–	*LS*
				1818.19 cm^{-1}	104996.27	106814.459	4–6	2.17–04	1.48–02	1.07+01	–1.228	C–	*LS*
				1793.97 cm^{-1}	104984.37	106778.337	2–4	2.24–04	2.08–02	7.64+00	–1.380	C–	*LS*
162.		$^4D-^4P^o$	*49820*	*2006.7* cm^{-1}	*105007.3*	*107013.96*	20–12	5.94–03	1.33–01	4.35+02	0.424	C+	1
			49504.5	2019.469 cm^{-1}	105017.600	107037.069	8–6	4.85–03	1.34–01	1.74+02	0.029	C–	*LS*
				1987.48 cm^{-1}	105008.55	106996.032	6–4	3.64–03	9.20–02	9.14+01	–0.258	C–	*LS*
				1984.21 cm^{-1}	104996.27	106980.480	4–2	2.87–03	5.47–02	3.63+01	–0.660	C–	*LS*
			49283.6	2028.52 cm^{-1}	105008.55	107037.069	6–6	1.11–03	4.02–02	3.92+01	–0.617	C–	*LS*
				1999.76 cm^{-1}	104996.27	106996.032	4–4	1.88–03	7.05–02	4.64+01	–0.550	C–	*LS*
				1996.11 cm^{-1}	104984.37	106980.480	2–2	2.93–03	1.10–01	3.63+01	–0.658	C–	*LS*
			48987.1	2040.80 cm^{-1}	104996.27	107037.069	4–6	1.25–04	6.75–03	4.35+00	–1.569	C–	*LS*
			49696.6	2011.66 cm^{-1}	104984.37	106996.032	2–4	2.99–04	2.22–02	7.26+00	–1.353	C–	*LS*
163.	$2s^22p^2(^3P)3d-2s^22p^2(^3P)5p$	$^4D-^4P^o$	*15877*	*6296.6* cm^{-1}	*105007.3*	*111303.88*	20–12	9.54–04	2.16–03	2.26+00	–1.364	C+	1
			15845.5	6309.198 cm^{-1}	105017.600	111326.798	8–6	7.68–04	2.17–03	9.04–01	–1.761	C–	*LS*
			15926.6	6277.09 cm^{-1}	105008.55	111285.644	6–4	5.95–04	1.51–03	4.75–01	–2.043	C–	*LS*
			15931.1	6275.33 cm^{-1}	104996.27	111271.596	4–2	4.72–04	8.98–04	1.88–01	–2.445	C–	*LS*
			15822.8	6318.25 cm^{-1}	105008.55	111326.798	6–6	1.73–04	6.51–04	2.03–01	–2.408	C–	*LS*
			15895.5	6289.37 cm^{-1}	104996.27	111285.644	4–4	3.04–04	1.15–03	2.41–01	–2.336	C–	*LS*
			15900.9	6287.23 cm^{-1}	104984.37	111271.596	2–2	4.75–04	1.80–03	1.88–01	–2.444	C–	*LS*
			15792.2	6330.53 cm^{-1}	104996.27	111326.798	4–6	1.94–05	1.09–04	2.26–02	–3.362	C–	*LS*
			15865.5	6301.27 cm^{-1}	104984.37	111285.644	2–4	4.78–05	3.61–04	3.77–02	–3.142	C–	*LS*
164.	$2s^22p^2(^3P)3d-2s^22p^2(^3P)4p$	$^2D-^2D^o$	*47816*	*2090.80* cm^{-1}	*105134.18*	*107224.98*	10–10	1.30–03	4.45–02	7.01+01	–0.351	C+	1
			47394.0	2109.396 cm^{-1}	105143.710	107253.106	6–6	1.25–03	4.19–02	3.93+01	–0.599	C–	*LS*
			48462.0	2062.908 cm^{-1}	105119.880	107182.788	4–4	1.12–03	3.95–02	2.52+01	–0.801	C–	*LS*
			49028.4	2039.078 cm^{-1}	105143.710	107182.788	6–4	1.21–04	2.90–03	2.80+00	–1.760	C–	*LS*
			46864.6	2133.226 cm^{-1}	105119.880	107253.106	4–6	9.20–05	4.54–03	2.80+00	–1.741	C–	*LS*
165.		$^2D-^2P^o$	*40298*	*2480.83* cm^{-1}	*105134.18*	*107615.01*	10–6	9.43–03	1.38–01	1.83+02	0.139	C+	1
			40237.4	2484.573 cm^{-1}	105143.710	107628.283	6–4	8.52–03	1.38–01	1.10+02	–0.082	C–	*LS*
			40497.9	2468.589 cm^{-1}	105119.880	107588.469	4–2	9.29–03	1.14–01	6.09+01	–0.340	C–	*LS*
			39855.1	2508.403 cm^{-1}	105119.880	107628.283	4–4	9.74–04	2.32–02	1.22+01	–1.032	C–	*LS*
166.	$2s^22p^2(^3P)3d-2s^22p^2(^1D)3p$	$^2D-^2D^o$											
			18509.6	5401.140 cm^{-1}	105143.710	110544.850	6–6	6.92–04	3.55–03	1.30+00	–1.671	C+	3
			18509.5	5401.170 cm^{-1}	105119.880	110521.050	4–4	5.94–04	3.05–03	7.44–01	–1.913	C+	3
			18428.2	5424.970 cm^{-1}	105119.880	110544.850	4–6	6.22–05	4.75–04	1.15–01	–2.721	C+	3
167.	$2s^22p^2(^3P)3d-2s^22p^2(^3P)5p$	$^2D-^2P^o$	*16458*	*6074.28* cm^{-1}	*105134.18*	*111208.46*	10–6	1.08–03	2.64–03	1.43+00	–1.579	C+	1
			16471.2	6069.561 cm^{-1}	105143.710	111213.271	6–4	9.72–04	2.64–03	8.58–01	–1.801	C–	*LS*
			16445.7	6078.968 cm^{-1}	105119.880	111198.848	4–2	1.09–03	2.20–03	4.76–01	–2.056	C–	*LS*
			16406.7	6093.391 cm^{-1}	105119.880	111213.271	4–4	1.09–04	4.41–04	9.53–02	–2.753	C–	*LS*
168.	$2s^22p^2(^3P)4p-2s^22p^2(^3P)4d$	$^2S^o-^2P$	*26655*	*3750.67* cm^{-1}	106477.800	*110228.47*	2–6	4.07–02	1.30+00	2.28+02	0.415	C	1
			26714.2	3742.307 cm^{-1}	106477.800	110220.107	2–4	4.04–02	8.64–01	1.52+02	0.238	D+	*LS*
			26536.4	3767.383 cm^{-1}	106477.800	110245.183	2–2	4.12–02	4.35–01	7.60+01	–0.060	D+	*LS*

N I: Allowed Transitions — Continued

No.	Transition Array	Mult.	λ_{air} (Å)	λ_{vac} (Å) or σ (cm^{-1})*	E_i (cm^{-1})	E_k (cm^{-1})	g_i–g_k	A_{ki} (10^8 s^{-1})	f_{ik}	S (at. u.)	log gf	Acc.	Source
169.	$2s^22p^2(^3P)4p-$ $2s^22p^2(^3P)5d$	$^2S°-^2P$	*15805*	*6325.41* cm^{-1}	106477.800 –	*112803.21*	2–6	1.29–02	1.44–01	1.50+01	−0.539	C	1
			15810.4	6323.231 cm^{-1}	106477.800 –	112801.031	2–4	1.28–02	9.63–02	1.00+01	−0.715	D+	*LS*
			15794.1	6329.767 cm^{-1}	106477.800 –	112807.567	2–2	1.29–02	4.82–02	5.01+00	−1.016	D+	*LS*
170.	$2s^22p^2(^3P)4p-$ $2s^22p^2(^3P)5s$	$^4D°-^4P$	*32661*	*3060.88* cm^{-1}	*106823.33* –	*109884.21*	20–12	2.71–02	2.60–01	5.58+02	0.715	C	1
			32691.9	3058.026 cm^{-1}	106868.635 –	109926.661	8–6	2.16–02	2.59–01	2.23+02	0.317	D+	*LS*
			32863.5	3042.061 cm^{-1}	106814.459 –	109856.520	6–4	1.67–02	1.81–01	1.17+02	0.035	D+	*LS*
			32951.9	3033.896 cm^{-1}	106778.337 –	109812.233	4–2	1.32–02	1.07–01	4.65+01	−0.368	D+	*LS*
			32122.8	3112.202 cm^{-1}	106814.459 –	109926.661	6–6	5.12–03	7.92–02	5.03+01	−0.323	D+	*LS*
			32477.8	3078.183 cm^{-1}	106778.337 –	109856.520	4–4	8.81–03	1.39–01	5.96+01	−0.254	D+	*LS*
			32740.4	3053.502 cm^{-1}	106758.731 –	109812.233	2–2	1.34–02	2.16–01	4.65+01	−0.365	D+	*LS*
			31754.3	3148.324 cm^{-1}	106778.337 –	109926.661	4–6	5.89–04	1.34–02	5.58+00	−1.272	D+	*LS*
			32272.3	3097.789 cm^{-1}	106758.731 –	109856.520	2–4	1.40–03	4.38–02	9.31+00	−1.057	D+	*LS*
171.	$2s^22p^2(^3P)4p-$ $2s^22p^2(^3P)4d$	$^4D°-^4F$	*29156*	*3428.94* cm^{-1}	*106823.33* –	*110252.27*	20–28	5.67–02	1.01+00	1.94+03	1.306	C+	1
			29107.5	3434.598 cm^{-1}	106868.635 –	110303.233	8–10	5.70–02	9.05–01	6.94+02	0.860	C−	*LS*
			29122.5	3432.829 cm^{-1}	106814.459 –	110247.288	6–8	4.88–02	8.27–01	4.76+02	0.696	C−	*LS*
			29112.1	3434.059 cm^{-1}	106778.337 –	110212.396	4–6	4.27–02	8.14–01	3.12+02	0.513	C−	*LS*
			29096.3	3435.923 cm^{-1}	106758.731 –	110194.654	2–4	3.99–02	1.01+00	1.94+02	0.307	C−	*LS*
			29589.5	3378.653 cm^{-1}	106868.635 –	110247.288	8–8	7.73–03	1.01–01	7.91+01	−0.090	C−	*LS*
			29421.6	3397.937 cm^{-1}	106814.459 –	110212.396	6–6	1.34–02	1.74–01	1.01+02	0.019	C−	*LS*
			29263.3	3416.317 cm^{-1}	106778.337 –	110194.654	4–4	1.57–02	2.02–01	7.77+01	−0.093	C−	*LS*
			29898.3	3343.761 cm^{-1}	106868.635 –	110212.396	8–6	5.08–04	5.11–03	4.02+00	−1.388	C−	*LS*
			29576.0	3380.195 cm^{-1}	106814.459 –	110194.654	6–4	1.09–03	9.50–03	5.55+00	−1.244	C−	*LS*
172.		$^4D°-^4P$	*28624*	*3492.56* cm^{-1}	*106823.33* –	*110315.90*	20–12	4.31–03	3.18–02	5.99+01	−0.197	C	1
			29135.2	3431.339 cm^{-1}	106868.635 –	110299.974	8–6	3.27–03	3.12–02	2.39+01	−0.603	D+	*LS*
			28496.4	3508.262 cm^{-1}	106814.459 –	110322.721	6–4	2.75–03	2.23–02	1.26+01	−0.873	D+	*LS*
			27990.4	3571.677 cm^{-1}	106778.337 –	110350.014	4–2	2.30–03	1.35–02	4.99+00	−1.267	D+	*LS*
			28682.3	3485.515 cm^{-1}	106814.459 –	110299.974	6–6	7.71–04	9.51–03	5.39+00	−1.244	D+	*LS*
			28206.0	3544.384 cm^{-1}	106778.337 –	110322.721	4–4	1.44–03	1.72–02	6.39+00	−1.163	D+	*LS*
			27837.6	3591.283 cm^{-1}	106758.731 –	110350.014	2–2	2.34–03	2.72–02	4.99+00	−1.264	D+	*LS*
			28388.1	3521.637 cm^{-1}	106778.337 –	110299.974	4–6	8.84–05	1.60–03	5.99–01	−2.193	D+	*LS*
			28050.8	3563.990 cm^{-1}	106758.731 –	110322.721	2–4	2.29–04	5.40–03	9.98–01	−1.966	D+	*LS*
173.		$^4D°-^4D$	*27956*	*3576.04* cm^{-1}	*106823.33* –	*110399.37*	20–20	1.50–02	1.76–01	3.24+02	0.546	C+	1
			28284.1	3534.585 cm^{-1}	106868.635 –	110403.220	8–8	1.24–02	1.49–01	1.11+02	0.077	C−	*LS*
			27871.6	3586.897 cm^{-1}	106814.459 –	110401.356	6–6	8.69–03	1.01–01	5.57+01	−0.217	C−	*LS*
			27638.7	3617.126 cm^{-1}	106778.337 –	110395.463	4–4	6.22–03	7.12–02	2.59+01	−0.546	C−	*LS*
			27563.0	3627.064 cm^{-1}	106758.731 –	110385.795	2–2	7.83–03	8.92–02	1.62+01	−0.749	C−	*LS*
			28299.1	3532.721 cm^{-1}	106868.635 –	110401.356	8–6	2.75–03	2.48–02	1.85+01	−0.703	C−	*LS*
			27917.5	3581.004 cm^{-1}	106814.459 –	110395.463	6–4	5.28–03	4.11–02	2.27+01	−0.608	C−	*LS*
			27712.8	3607.458 cm^{-1}	106778.337 –	110385.795	4–2	7.71–03	4.44–02	1.62+01	−0.751	C−	*LS*
			27857.2	3588.761 cm^{-1}	106814.459 –	110403.220	6–8	2.16–03	3.35–02	1.85+01	−0.696	C−	*LS*
			27593.8	3623.019 cm^{-1}	106778.337 –	110401.356	4–6	3.64–03	6.24–02	2.27+01	−0.603	C−	*LS*
			27489.7	3636.732 cm^{-1}	106758.731 –	110395.463	2–4	3.95–03	8.95–02	1.62+01	−0.747	C−	*LS*
174.	$2s^22p^2(^3P)4p-$ $2s^22p^2(^3P)5d$	$^4D°-^4F$	*16707*	*5983.71* cm^{-1}	*106823.33* –	*112807.04*	20–28	1.40–02	8.17–02	8.99+01	0.214	C+	1
			16682.4	5992.713 cm^{-1}	106868.635 –	112861.348	8–10	1.40–02	7.31–02	3.21+01	−0.233	C−	*LS*
			16708.7	5983.266 cm^{-1}	106814.459 –	112797.725	6–8	1.20–02	6.68–02	2.20+01	−0.397	C−	*LS*
			16712.3	5981.988 cm^{-1}	106778.337 –	112760.325	4–6	1.05–02	6.57–02	1.45+01	−0.581	C−	*LS*
			16658.7	6001.235 cm^{-1}	106758.731 –	112759.966	2–4	9.85–03	8.20–02	8.99+00	−0.785	C−	*LS*
			16861.4	5929.090 cm^{-1}	106868.635 –	112797.725	8–8	1.93–03	8.24–03	3.66+00	−1.181	C−	*LS*
			16813.8	5945.866 cm^{-1}	106814.459 –	112760.325	6–6	3.33–03	1.41–02	4.69+00	−1.072	C−	*LS*
			16713.3	5981.629 cm^{-1}	106778.337 –	112759.966	4–4	3.90–03	1.63–02	3.60+00	−1.185	C−	*LS*
			16968.4	5891.690 cm^{-1}	106868.635 –	112760.325	8–6	1.29–04	4.17–04	1.86–01	−2.477	C−	*LS*
			16814.8	5945.507 cm^{-1}	106814.459 –	112759.966	6–4	2.74–04	7.74–04	2.57–01	−2.333	C−	*LS*

N I: Allowed Transitions — Continued

No.	Transition Array	Mult.	λ_{air} (Å)	λ_{vac} (Å) or σ (cm^{-1})*	E_i (cm^{-1})	E_k (cm^{-1})	g_i-g_k	A_{ki} ($10^8\ s^{-1}$)	f_{ik}	S (at. u.)	log gf	Acc.	Source
175.		$^4D^o-^4P$	*16625*	*6013.3* cm^{-1}	*106823.33*	*112836.6*	20−12	1.46−03	3.64−03	3.98+00	−1.138	C	1
			16785.7	5955.824 cm^{-1}	106868.635	112824.459	8−6	1.14−03	3.60−03	1.59+00	−1.540	D+	*LS*
			16596.9	6023.56 cm^{-1}	106814.459	112838.02	6−4	9.26−04	2.55−03	8.36−01	−1.815	D+	*LS*
			16410.7	6091.93 cm^{-1}	106778.337	112870.27	4−2	7.60−04	1.54−03	3.32−01	−2.212	D+	*LS*
			16634.4	6010.000 cm^{-1}	106814.459	112824.459	6−6	2.63−04	1.09−03	3.58−01	−2.184	D+	*LS*
			16498.0	6059.68 cm^{-1}	106778.337	112838.02	4−4	4.79−04	1.95−03	4.25−01	−2.107	D+	*LS*
			16358.0	6111.54 cm^{-1}	106758.731	112870.27	2−2	7.68−04	3.08−03	3.32−01	−2.210	D+	*LS*
			16535.0	6046.122 cm^{-1}	106778.337	112824.459	4−6	2.97−05	1.83−04	3.98−02	−3.136	D+	*LS*
			16444.8	6079.29 cm^{-1}	106758.731	112838.02	2−4	7.56−05	6.13−04	6.63−02	−2.912	D+	*LS*
176.		$^4D^o-^4D$	*16425*	*6087* cm^{-1}	*106823.33*	*112910*	20−20	4.17−03	1.69−02	1.83+01	−0.472	C+	1
			16545.1	6042.444 cm^{-1}	106868.635	112911.079	8−8	3.50−03	1.44−02	6.26+00	−0.939	C−	*LS*
			16399.3	6096.171 cm^{-1}	106814.459	112910.630	6−6	2.40−03	9.69−03	3.14+00	−1.235	C−	*LS*
			16306.8	6130.739 cm^{-1}	106778.337	112909.076	4−4	1.71−03	6.80−03	1.46+00	−1.565	C−	*LS*
			16266.9	6145.8 cm^{-1}	106758.731	112904.5	2−2	2.15−03	8.52−03	9.13−01	−1.768	C−	*LS*
			16546.3	6042.00 cm^{-1}	106868.635	112910.63	8−6	7.76−04	2.39−03	1.04+00	−1.719	C−	*LS*
			16403.4	6094.617 cm^{-1}	106814.459	112909.076	6−4	1.47−03	3.94−03	1.28+00	−1.626	C−	*LS*
			16319.0	6126.2 cm^{-1}	106778.337	112904.5	4−2	2.13−03	4.25−03	9.13−01	−1.770	C−	*LS*
			16398.1	6096.620 cm^{-1}	106814.459	112911.079	6−8	5.98−04	3.21−03	1.04+00	−1.715	C−	*LS*
			16302.7	6132.293 cm^{-1}	106778.337	112910.630	4−6	9.96−04	5.95−03	1.28+00	−1.623	C−	*LS*
			16254.8	6150.345 cm^{-1}	106758.731	112909.076	2−4	1.08−03	8.53−03	9.13−01	−1.768	C−	*LS*
177.	$2s^22p^2(^3P)4p-$ $2s^22p^2(^3P)5s$	$^4P^o-^4P$	*34831*	*2870.25* cm^{-1}	*107013.96*	*109884.21*	12−12	2.19−02	3.98−01	5.48+02	0.679	C	1
			34597.5	2889.592 cm^{-1}	107037.069	109926.661	6−6	1.56−02	2.81−01	1.92+02	0.226	D+	*LS*
			34949.5	2860.488 cm^{-1}	106996.032	109856.520	4−4	2.89−03	5.29−02	2.43+01	−0.675	D+	*LS*
			35304.2	2831.753 cm^{-1}	106980.480	109812.233	2−2	3.50−03	6.55−02	1.52+01	−0.883	D+	*LS*
			35458.2	2819.451 cm^{-1}	107037.069	109856.520	6−4	9.34−03	1.17−01	8.22+01	−0.153	D+	*LS*
			35499.1	2816.201 cm^{-1}	106996.032	109812.233	4−2	1.72−02	1.63−01	7.61+01	−0.186	D+	*LS*
			34113.1	2930.629 cm^{-1}	106996.032	109926.661	4−6	6.99−03	1.83−01	8.22+01	−0.136	D+	*LS*
			34760.6	2876.040 cm^{-1}	106980.480	109856.520	2−4	9.17−03	3.32−01	7.61+01	−0.177	D+	*LS*
178.	$2s^22p^2(^3P)4p-$ $2s^22p^2(^3P)4d$	$^4P^o-^4P$	*30277*	*3301.94* cm^{-1}	*107013.96*	*110315.90*	12−12	1.63−02	2.24−01	2.68+02	0.430	C	1
			30639.2	3262.905 cm^{-1}	107037.069	110299.974	6−6	1.10−02	1.55−01	9.39+01	−0.031	D+	*LS*
			30051.7	3326.689 cm^{-1}	106996.032	110322.721	4−4	2.22−03	3.01−02	1.19+01	−0.919	D+	*LS*
			29669.6	3369.534 cm^{-1}	106980.480	110350.014	2−2	2.89−03	3.81−02	7.45+00	−1.118	D+	*LS*
			30427.1	3285.652 cm^{-1}	107037.069	110322.721	6−4	7.23−03	6.69−02	4.02+01	−0.396	D+	*LS*
			29807.2	3353.982 cm^{-1}	106996.032	110350.014	4−2	1.42−02	9.49−02	3.72+01	−0.421	D+	*LS*
			30258.6	3303.942 cm^{-1}	106996.032	110299.974	4−6	4.90−03	1.01−01	4.02+01	−0.394	D+	*LS*
			29911.9	3342.241 cm^{-1}	106980.480	110322.721	2−4	7.05−03	1.89−01	3.72+01	−0.422	D+	*LS*
179.		$^4P^o-^4D$	*29530*	*3385.41* cm^{-1}	*107013.96*	*110399.37*	12−20	4.22−02	9.20−01	1.07+03	1.043	C+	1
			29699.4	3366.151 cm^{-1}	107037.069	110403.220	6−8	4.15−02	7.32−01	4.29+02	0.643	C−	*LS*
			29357.8	3405.324 cm^{-1}	106996.032	110401.356	4−6	3.01−02	5.83−01	2.25+02	0.368	C−	*LS*
			29274.7	3414.983 cm^{-1}	106980.480	110395.463	2−4	1.81−02	4.64−01	8.95+01	−0.032	C−	*LS*
			29715.9	3364.287 cm^{-1}	107037.069	110401.356	6−6	1.24−02	1.65−01	9.66+01	−0.005	C−	*LS*
			29408.7	3399.431 cm^{-1}	106996.032	110395.463	4−4	2.28−02	2.96−01	1.15+02	0.073	C−	*LS*
			29357.9	3405.315 cm^{-1}	106980.480	110385.795	2−2	3.58−02	4.63−01	8.95+01	−0.034	C−	*LS*
			29768.0	3358.394 cm^{-1}	107037.069	110395.463	6−4	2.06−03	1.83−02	1.07+01	−0.960	C−	*LS*
			29492.5	3389.763 cm^{-1}	106996.032	110385.795	4−2	7.07−03	4.61−02	1.79+01	−0.735	C−	*LS*
180.	$2s^22p^2(^3P)4p-$ $2s^22p^2(^3P)5d$	$^4P^o-^4P$	*17170*	*5822.7* cm^{-1}	*107013.96*	*112836.6*	12−12	3.48−03	1.54−02	1.04+01	−0.733	C	1
			17274.2	5787.390 cm^{-1}	107037.069	112824.459	6−6	2.39−03	1.07−02	3.66+00	−1.192	D+	*LS*
			17112.8	5841.99 cm^{-1}	106996.032	112838.02	4−4	4.69−04	2.06−03	4.64−01	−2.084	D+	*LS*
			16973.9	5889.79 cm^{-1}	106980.480	112870.27	2−2	6.01−04	2.60−03	2.90−01	−2.285	D+	*LS*
			17233.8	5800.95 cm^{-1}	107037.069	112838.02	6−4	1.55−03	4.60−03	1.57+00	−1.559	D+	*LS*
			17018.8	5874.24 cm^{-1}	106996.032	112870.27	4−2	2.98−03	6.47−03	1.45+00	−1.587	D+	*LS*
			17152.6	5828.427 cm^{-1}	106996.032	112824.459	4−6	1.05−03	6.94−03	1.57+00	−1.557	D+	*LS*
			17067.4	5857.54 cm^{-1}	106980.480	112838.02	2−4	1.48−03	1.29−02	1.45+00	−1.588	D+	*LS*

N I: Allowed Transitions — Continued

No.	Transition Array	Mult.	λ_{air} (Å)	λ_{vac} (Å) or σ (cm^{-1})*	E_i (cm^{-1})	E_k (cm^{-1})	g_i-g_k	A_{ki} (10^8 s^{-1})	f_{ik}	S (at. u.)	log gf	Acc.	Source
181.		$^4P^o-^4D$	*16956*	*5896* cm^{-1}	*107013.96*	– *112910*	12–20	1.05–02	7.53–02	5.04+01	–0.044	C+	1
			17019.5	5874.010 cm^{-1}	107037.069	– 112911.079	6–8	1.04–02	6.00–02	2.02+01	–0.444	C–	*LS*
			16902.7	5914.598 cm^{-1}	106996.032	– 112910.630	4–6	7.41–03	4.76–02	1.06+01	–0.720	C–	*LS*
			16862.8	5928.596 cm^{-1}	106980.480	– 112909.076	2–4	4.44–03	3.79–02	4.20+00	–1.121	C–	*LS*
			17020.8	5873.561 cm^{-1}	107037.069	– 112910.630	6–6	3.11–03	1.35–02	4.54+00	–1.091	C–	*LS*
			16907.1	5913.044 cm^{-1}	106996.032	– 112909.076	4–4	5.64–03	2.42–02	5.38+00	–1.015	C–	*LS*
			16875.8	5924.0 cm^{-1}	106980.480	– 112904.5	2–2	8.86–03	3.78–02	4.20+00	–1.121	C–	*LS*
			17025.3	5872.007 cm^{-1}	107037.069	– 112909.076	6–4	5.18–04	1.50–03	5.04–01	–2.046	C–	*LS*
			16920.2	5908.5 cm^{-1}	106996.032	– 112904.5	4–2	1.76–03	3.77–03	8.41–01	–1.821	C–	*LS*
182.	$2s^22p^2(^3P)4p-$ $2s^22p^2(^3P)5s$	$^2D^o-^2P$	*35003*	*2856.15* cm^{-1}	*107224.98*	– *110081.13*	10–6	1.90–02	2.09–01	2.41+02	0.321	C	1
			35069.2	2850.728 cm^{-1}	107253.106	– 110103.834	6–4	1.70–02	2.09–01	1.45+02	0.098	D+	*LS*
			35042.1	2852.932 cm^{-1}	107182.788	– 110035.720	4–2	1.89–02	1.74–01	8.04+01	–0.157	D+	*LS*
			34225.0	2921.046 cm^{-1}	107182.788	– 110103.834	4–4	2.03–03	3.57–02	1.61+01	–0.845	D+	*LS*
183.	$2s^22p^2(^3P)4p-$ $2s^22p^2(^3P)4d$	$^2D^o-^2P$	*33286*	*3003.49* cm^{-1}	*107224.98*	– *110228.47*	10–6	9.46–03	9.43–02	1.03+02	–0.025	C	1
			33694.9	2967.001 cm^{-1}	107253.106	– 110220.107	6–4	8.21–03	9.32–02	6.20+01	–0.253	D+	*LS*
			32645.3	3062.395 cm^{-1}	107182.788	– 110245.183	4–2	1.00–02	8.01–02	3.44+01	–0.494	D+	*LS*
			32914.8	3037.319 cm^{-1}	107182.788	– 110220.107	4–4	9.79–04	1.59–02	6.89+00	–1.197	D+	*LS*
184.		$^2D^o-^2F$	*32199*	*3104.84* cm^{-1}	*107224.98*	– *110329.82*	10–14	4.70–02	1.02+00	1.08+03	1.010	C+	1
			32152.2	3109.356 cm^{-1}	107253.106	– 110362.462	6–8	4.72–02	9.76–01	6.20+02	0.768	C–	*LS*
			32212.7	3103.517 cm^{-1}	107182.788	– 110286.305	4–6	4.38–02	1.02+00	4.34+02	0.612	C–	*LS*
			32959.5	3033.199 cm^{-1}	107253.106	– 110286.305	6–6	2.92–03	4.76–02	3.10+01	–0.544	C–	*LS*
185.		$^2D^o-^2D$	*30894*	*3235.98* cm^{-1}	*107224.98*	– *110460.96*	10–10	1.24–02	1.78–01	1.81+02	0.249	C+	1
			31075.1	3217.138 cm^{-1}	107253.106	– 110470.244	6–6	1.14–02	1.65–01	1.01+02	–0.005	C–	*LS*
			30626.6	3264.244 cm^{-1}	107182.788	– 110447.032	4–4	1.15–02	1.61–01	6.50+01	–0.190	C–	*LS*
			31300.9	3193.926 cm^{-1}	107253.106	– 110447.032	6–4	1.19–03	1.17–02	7.23+00	–1.154	C–	*LS*
			30410.4	3287.456 cm^{-1}	107182.788	– 110470.244	4–6	8.68–04	1.80–02	7.23+00	–1.142	C–	*LS*
186.	$2s^22p^2(^3P)4p-$ $2s^22p^2(^3P)5d$	$^2D^o-^2P$	*17922*	*5578.23* cm^{-1}	*107224.98*	– *112803.21*	10–6	3.44–03	9.93–03	5.86+00	–1.003	C	1
			18019.8	5547.925 cm^{-1}	107253.106	– 112801.031	6–4	3.04–03	9.87–03	3.51+00	–1.227	D+	*LS*
			17773.6	5624.779 cm^{-1}	107182.788	– 112807.567	4–2	3.52–03	8.34–03	1.95+00	–1.477	D+	*LS*
			17794.3	5618.243 cm^{-1}	107182.788	– 112801.031	4–4	3.51–04	1.67–03	3.90–01	–2.176	D+	*LS*
187.		$^2D^o-^2F$	*17749*	*5632.52* cm^{-1}	*107224.98*	– *112857.50*	10–14	9.21–03	6.09–02	3.56+01	–0.216	C+	1
			17731.5	5638.132 cm^{-1}	107253.106	– 112891.238	6–8	9.23–03	5.80–02	2.03+01	–0.458	C–	*LS*
			17758.0	5629.730 cm^{-1}	107182.788	– 112812.518	4–6	8.58–03	6.08–02	1.42+01	–0.614	C–	*LS*
			17982.6	5559.412 cm^{-1}	107253.106	– 112812.518	6–6	5.90–04	2.86–03	1.02+00	–1.765	C–	*LS*
188.		$^2D^o-^2D$	*17498*	*5713.33* cm^{-1}	*107224.98*	– *112938.31*	10–10	2.79–03	1.28–02	7.37+00	–0.893	C+	1
			17561.6	5692.703 cm^{-1}	107253.106	– 112945.809	6–6	2.57–03	1.19–02	4.13+00	–1.146	C–	*LS*
			17403.9	5744.267 cm^{-1}	107182.788	– 112927.055	4–4	2.55–03	1.16–02	2.65+00	–1.334	C–	*LS*
			17619.6	5673.949 cm^{-1}	107253.106	– 112927.055	6–4	2.73–04	8.47–04	2.95–01	–2.294	C–	*LS*
			17347.3	5763.021 cm^{-1}	107182.788	– 112945.809	4–6	1.91–04	1.29–03	2.95–01	–2.287	C–	*LS*
189.	$2s^22p^2(^3P)4p-$ $2s^22p^2(^3P)5s$	$^4S^o-^4P$	*40996*	*2438.59* cm^{-1}	107445.622	– *109884.21*	4–12	2.87–03	2.17–01	1.17+02	–0.061	C	2
			40294.7	2481.039 cm^{-1}	107445.622	– 109926.661	4–6	3.02–03	1.10–01	5.86+01	–0.355	D+	*LS*
			41467.0	2410.898 cm^{-1}	107445.622	– 109856.520	4–4	2.77–03	7.15–02	3.91+01	–0.544	D+	*LS*
			42243.0	2366.611 cm^{-1}	107445.622	– 109812.233	4–2	2.62–03	3.51–02	1.95+01	–0.853	D+	*LS*

N I: Allowed Transitions — Continued

No.	Transition Array	Mult.	λ_{air} (Å)	λ_{vac} (Å) or σ (cm^{-1})*	E_i (cm^{-1})	E_k (cm^{-1})	g_i-g_k	A_{ki} (10^8 s^{-1})	f_{ik}	S (at. u.)	log gf	Acc.	Source
190.	$2s^22p^2(^3P)4p-2s^22p^2(^3P)4d$	$^4S°-^4P$	*34830*	*2870.27* cm^{-1}	107445.622	*110315.90*	4−12	2.55−02	1.39+00	6.38+02	0.745	C	2
			35024.7	2854.352 cm^{-1}	107445.622	110299.974	4−6	2.50−02	6.91−01	3.19+02	0.442	D+	*LS*
			34747.8	2877.099 cm^{-1}	107445.622	110322.721	4−4	2.56−02	4.64−01	2.13+02	0.269	D+	*LS*
			34421.2	2904.392 cm^{-1}	107445.622	110350.014	4−2	2.64−02	2.34−01	1.06+02	−0.028	D+	*LS*
191.	$2s^22p^2(^3P)4p-2s^22p^2(^3P)6s$	$^4S°-^4P$	*19253*	*5192.52* cm^{-1}	107445.622	*112638.14*	4−12	1.34−03	2.23−02	5.66+00	−1.050	C	2
			19094.2	5235.767 cm^{-1}	107445.622	112681.389	4−6	1.37−03	1.12−02	2.83+00	−1.347	D+	*LS*
			19359.6	5163.990 cm^{-1}	107445.622	112609.612	4−4	1.31−03	7.39−03	1.89+00	−1.529	D+	*LS*
			19526.5	5119.848 cm^{-1}	107445.622	112565.470	4−2	1.28−03	3.66−03	9.43−01	−1.834	D+	*LS*
192.	$2s^22p^2(^3P)4p-2s^22p^2(^3P)5d$	$^4S°-^4P$	*18544*	*5391.0* cm^{-1}	107445.622	*112836.6*	4−12	4.94−03	7.65−02	1.87+01	−0.514	C	2
			18586.3	5378.837 cm^{-1}	107445.622	112824.459	4−6	4.91−03	3.82−02	9.34+00	−0.816	D+	*LS*
			18539.6	5392.40 cm^{-1}	107445.622	112838.02	4−4	4.95−03	2.55−02	6.23+00	−0.991	D+	*LS*
			18429.3	5424.65 cm^{-1}	107445.622	112870.27	4−2	5.04−03	1.28−02	3.11+00	−1.290	D+	*LS*
193.	$2s^22p^2(^3P)4p-2s^22p^2(^3P)7s$	$^4S°-^4P$	*15020*	*6655.80* cm^{-1}	107445.622	*114101.42*	4−12	7.67−04	7.79−03	1.54+00	−1.506	C	2
			14919.3	6700.903 cm^{-1}	107445.622	114146.525	4−6	7.83−04	3.92−03	7.71−01	−1.804	D+	*LS*
			15089.0	6625.531 cm^{-1}	107445.622	114071.153	4−4	7.57−04	2.58−03	5.14−01	−1.986	D+	*LS*
			15191.0	6581.032 cm^{-1}	107445.622	114026.654	4−2	7.42−04	1.28−03	2.57−01	−2.289	D+	*LS*
194.	$2s^22p^2(^3P)4p-2s^22p^2(^3P)6d$	$^4S°-^4P$	*14792*	*6758* cm^{-1}	107445.622	*114204*	4−12	1.77−03	1.74−02	3.39+00	−1.157	C	2
			14819.6	6746.0 cm^{-1}	107445.622	114191.6	4−6	1.76−03	8.68−03	1.70+00	−1.459	D+	*LS*
			14801.2	6754.4 cm^{-1}	107445.622	114200.0	4−4	1.76−03	5.80−03	1.13+00	−1.635	D+	*LS*
			14693.5	6803.9 cm^{-1}	107445.622	114249.5	4−2	1.80−03	2.92−03	5.65−01	−1.933	D+	*LS*
195.	$2s^22p^2(^3P)4p-2s^22p^2(^3P)8s$	$^4S°-^4P$	*13282*	*7527* cm^{-1}	107445.622	*114973*	4−12	4.82−04	3.83−03	6.70−01	−1.815	C	2
			13196.5	7575.7 cm^{-1}	107445.622	115021.3	4−6	4.92−04	1.93−03	3.35−01	−2.113	D+	*LS*
			13341.3	7493.5 cm^{-1}	107445.622	114939.1	4−4	4.76−04	1.27−03	2.23−01	−2.294	D+	*LS*
			13422.1	7448.4 cm^{-1}	107445.622	114894.0	4−2	4.67−04	6.32−04	1.12−01	−2.597	D+	*LS*
196.	$2s^22p^2(^3P)4p-2s^22p^2(^3P)9s$	$^4S°-^4P$	*12369*	*8083* cm^{-1}	107445.622	*115528*	4−12	3.22−04	2.22−03	3.62−01	−2.052	C	2
			12295.3	8131.0 cm^{-1}	107445.622	115576.6	4−6	3.28−04	1.12−03	1.81−01	−2.350	D+	*LS*
			12419.2	8049.9 cm^{-1}	107445.622	115495.5	4−4	3.19−04	7.37−04	1.21−01	−2.530	D+	*LS*
			12490.5	8003.9 cm^{-1}	107445.622	115449.5	4−2	3.13−04	3.66−04	6.03−02	−2.834	D+	*LS*
197.	$2s^22p^2(^3P)4p-2s^22p^2(^3P)5s$	$^2P°-^2P$	*40539*	*2466.12* cm^{-1}	*107615.01*	*110081.13*	6−6	2.34−02	5.76−01	4.61+02	0.539	C	1
			40384.0	2475.551 cm^{-1}	107628.283	110103.834	4−4	1.97−02	4.82−01	2.56+02	0.285	D+	*LS*
			40851.0	2447.251 cm^{-1}	107588.469	110035.720	2−2	1.52−02	3.81−01	1.03+02	−0.118	D+	*LS*
			41526.6	2407.437 cm^{-1}	107628.283	110035.720	4−2	7.25−03	9.37−02	5.13+01	−0.426	D+	*LS*
			39744.8	2515.365 cm^{-1}	107588.469	110103.834	2−4	4.14−03	1.96−01	5.13+01	−0.407	D+	*LS*
198.	$2s^22p^2(^3P)4p-2s^22p^2(^3P)4d$	$^2P°-^2P$	*38253*	*2613.45* cm^{-1}	*107615.01*	*110228.47*	6−6	3.91−03	8.57−02	6.47+01	−0.289	C	1
			38572.3	2591.824 cm^{-1}	107628.283	110220.107	4−4	3.17−03	7.08−02	3.60+01	−0.548	D+	*LS*
			37630.2	2656.714 cm^{-1}	107588.469	110245.183	2−2	2.73−03	5.81−02	1.44+01	−0.935	D+	*LS*
			38202.7	2616.900 cm^{-1}	107628.283	110245.183	4−2	1.31−03	1.43−02	7.19+00	−1.243	D+	*LS*
			37988.8	2631.638 cm^{-1}	107588.469	110220.107	2−4	6.65−04	2.88−02	7.19+00	−1.240	D+	*LS*
199.		$^2P°-^2D$	*35128*	*2845.95* cm^{-1}	*107615.01*	*110460.96*	6−10	3.14−02	9.69−01	6.72+02	0.764	C+	1
			35177.4	2841.961 cm^{-1}	107628.283	110470.244	4−6	3.13−02	8.70−01	4.03+02	0.542	C−	*LS*
			34973.1	2858.563 cm^{-1}	107588.469	110447.032	2−4	2.65−02	9.73−01	2.24+02	0.289	C−	*LS*
			35467.1	2818.749 cm^{-1}	107628.283	110447.032	4−4	5.09−03	9.59−02	4.48+01	−0.416	C−	*LS*

N I: Allowed Transitions — Continued

No.	Transition Array	Mult.	λ_{air} (Å)	λ_{vac} (Å) or σ (cm^{-1})*	E_i (cm^{-1})	E_k (cm^{-1})	g_i-g_k	A_{ki} (10^8 s^{-1})	f_{ik}	S (at. u.)	log gf	Acc.	Source
200.	$2s^22p^2(^3P)4p-$ $2s^22p^2(^3P)5d$	$^2P^o-^2D$	*18780*	*5323.30* cm^{-1}	*107615.01*	*112938.31*	6–10	5.12–03	4.51–02	1.67+01	−0.567	C+	1
			18800.6	5317.526 cm^{-1}	107628.283	112945.809	4–6	5.11–03	4.06–02	1.00+01	−0.790	C−	*LS*
			18726.4	5338.586 cm^{-1}	107588.469	112927.055	2–4	4.31–03	4.53–02	5.58+00	−1.043	C−	*LS*
			18867.1	5298.772 cm^{-1}	107628.283	112927.055	4–4	8.42–04	4.49–03	1.12+00	−1.745	C−	*LS*
201.	$2s^22p^2(^3P)4p-$ $2s^22p^2(^1D)3d$	$^2P^o-^2S$	*7719.3*	*7721.4*	*107615.01*	120566.0	6–2	2.07–02	6.16–03	9.39–01	−1.433	C+	1
			7727.21	7729.34	107628.283	120566.0	4–2	1.37–02	6.15–03	6.26–01	−1.609	C−	*LS*
			7703.51	7705.63	107588.469	120566.0	2–2	6.93–03	6.17–03	3.13–01	−1.909	C−	*LS*
202.	$2s^22p^2(^3P)5s-$ $2s^22p^2(^3P)5p$	$^4P-^4D^o$		*1328.73* cm^{-1}	*109884.21*	*111212.94*	12–20	7.06–03	9.99–01	2.97+03	1.079	C	1
				1334.212 cm^{-1}	109926.661	111260.873	6–8	7.15–03	8.03–01	1.19+03	0.683	D+	*LS*
				1347.496 cm^{-1}	109856.520	111204.016	4–6	5.16–03	6.38–01	6.24+02	0.407	D+	*LS*
				1352.925 cm^{-1}	109812.233	111165.158	2–4	3.11–03	5.09–01	2.48+02	0.008	D+	*LS*
				1277.355 cm^{-1}	109926.661	111204.016	6–6	1.88–03	1.73–01	2.67+02	0.016	D+	*LS*
				1308.638 cm^{-1}	109856.520	111165.158	4–4	3.60–03	3.15–01	3.17+02	0.100	D+	*LS*
				1331.334 cm^{-1}	109812.233	111143.567	2–2	5.92–03	5.01–01	2.48+02	0.001	D+	*LS*
				1238.497 cm^{-1}	109926.661	111165.158	6–4	2.86–04	1.86–02	2.97+01	−0.952	D+	*LS*
				1287.047 cm^{-1}	109856.520	111143.567	4–2	1.07–03	4.84–02	4.95+01	−0.713	D+	*LS*
203.		$^4P-^4P^o$		*1419.67* cm^{-1}	*109884.21*	*111303.88*	12–12	9.82–03	7.30–01	2.03+03	0.942	C	1
				1400.137 cm^{-1}	109926.661	111326.798	6–6	6.59–03	5.04–01	7.11+02	0.480	D+	*LS*
				1429.124 cm^{-1}	109856.520	111285.644	4–4	1.34–03	9.80–02	9.02+01	−0.407	D+	*LS*
				1459.363 cm^{-1}	109812.233	111271.596	2–2	1.78–03	1.25–01	5.64+01	−0.602	D+	*LS*
				1358.983 cm^{-1}	109926.661	111285.644	6–4	3.87–03	2.10–01	3.05+02	0.099	D+	*LS*
				1415.076 cm^{-1}	109856.520	111271.596	4–2	8.10–03	3.03–01	2.82+02	0.084	D+	*LS*
				1470.278 cm^{-1}	109856.520	111326.798	4–6	3.27–03	3.40–01	3.05+02	0.134	D+	*LS*
				1473.411 cm^{-1}	109812.233	111285.644	2–4	4.57–03	6.31–01	2.82+02	0.101	D+	*LS*
204.		$^4P-^4S^o$		*1617.16* cm^{-1}	*109884.21*	111501.368	12–4	8.62–03	1.65–01	4.02+02	0.296	C	1
				1574.707 cm^{-1}	109926.661	111501.368	6–4	3.98–03	1.60–01	2.01+02	−0.017	D+	*LS*
				1644.848 cm^{-1}	109856.520	111501.368	4–4	3.02–03	1.67–01	1.34+02	−0.174	D+	*LS*
				1689.135 cm^{-1}	109812.233	111501.368	2–4	1.64–03	1.72–01	6.70+01	−0.464	D+	*LS*
205.	$2s^22p^2(^3P)5s-$ $2s^22p^2(^1D)3p$	$^2P-^2D^o$		*454.20* cm^{-1}	*110081.13*	*110535.33*	6–10	1.35–04	1.63–01	7.10+02	−0.009	C	1
				441.016 cm^{-1}	110103.834	110544.850	4–6	1.24–04	1.43–01	4.26+02	−0.243	D+	*LS*
				485.330 cm^{-1}	110035.720	110521.050	2–4	1.37–04	1.75–01	2.37+02	−0.457	D+	*LS*
				417.216 cm^{-1}	110103.834	110521.050	4–4	1.74–05	1.50–02	4.73+01	−1.222	D+	*LS*
206.	$2s^22p^2(^3P)5s-$ $2s^22p^2(^3P)5p$	$^2P-^2S^o$		*979.78* cm^{-1}	*110081.13*	111060.905	6–2	9.29–04	4.83–02	9.74+01	−0.538	C	1
				957.071 cm^{-1}	110103.834	111060.905	4–2	5.77–04	4.72–02	6.49+01	−0.724	D+	*LS*
				1025.185 cm^{-1}	110035.720	111060.905	2–2	3.55–04	5.06–02	3.25+01	−0.995	D+	*LS*
207.		$^2P-^2P^o$		*1127.33* cm^{-1}	*110081.13*	*111208.46*	6–6	4.64–03	5.48–01	9.59+02	0.517	C	1
				1109.437 cm^{-1}	110103.834	111213.271	4–4	3.69–03	4.49–01	5.33+02	0.254	D+	*LS*
				1163.128 cm^{-1}	110035.720	111198.848	2–2	3.40–03	3.77–01	2.13+02	−0.123	D+	*LS*
				1095.014 cm^{-1}	110103.834	111198.848	4–2	1.42–03	8.86–02	1.07+02	−0.450	D+	*LS*
				1177.551 cm^{-1}	110035.720	111213.271	2–4	8.82–04	1.91–01	1.07+02	−0.419	D+	*LS*
208.		$^2P-^2D^o$		*1803.46* cm^{-1}	*110081.13*	*111884.59*	6–10	6.47–03	4.97–01	5.44+02	0.474	C	1
				1801.775 cm^{-1}	110103.834	111905.609	4–6	6.45–03	4.47–01	3.26+02	0.252	D+	*LS*
				1817.341 cm^{-1}	110035.720	111853.061	2–4	5.52–03	5.01–01	1.81+02	0.000	D+	*LS*
				1749.227 cm^{-1}	110103.834	111853.061	4–4	9.84–04	4.82–02	3.63+01	−0.715	D+	*LS*

N I: Allowed Transitions — Continued

No.	Transition Array	Mult.	λ_{air} (Å)	λ_{vac} (Å) or σ (cm^{-1})*	E_i (cm^{-1})	E_k (cm^{-1})	g_i–g_k	A_{ki} (10^8 s^{-1})	f_{ik}	S (at. u.)	log gf	Acc.	Source
209.	$2s^22p^2(^3P)5s-2s^22p^2(^1D)3p$	^{2}P–^{2}P°	*44829*	*2230.08* cm^{-1}	*110081.13*	*112311.21*	6–6	3.54–03	1.07–01	9.45+01	–0.194	C	1
			45114.6	2215.971 cm^{-1}	110103.834	112319.805	4–4	2.90–03	8.83–02	5.25+01	–0.452	D+	*LS*
			44269.3	2258.287 cm^{-1}	110035.720	112294.007	2–2	2.45–03	7.20–02	2.10+01	–0.842	D+	*LS*
			45646.0	2190.173 cm^{-1}	110103.834	112294.007	4–2	1.12–03	1.75–02	1.05+01	–1.156	D+	*LS*
			43769.3	2284.085 cm^{-1}	110035.720	112319.805	2–4	6.34–04	3.64–02	1.05+01	–1.138	D+	*LS*
210.	$2s^22p^2(^3P)4d-2s^22p^2(^3P)5p$	^{4}F–^{4}D°		*960.67* cm^{-1}	*110252.27*	*111212.94*	28–20	3.03–03	3.51–01	3.37+03	0.992	C+	1
				957.640 cm^{-1}	110303.233	111260.873	10–8	2.68–03	3.50–01	1.20+03	0.544	C–	*LS*
				956.728 cm^{-1}	110247.288	111204.016	8–6	2.44–03	3.00–01	8.25+02	0.380	C–	*LS*
				952.762 cm^{-1}	110212.396	111165.158	6–4	2.37–03	2.61–01	5.41+02	0.195	C–	*LS*
				948.913 cm^{-1}	110194.654	111143.567	4–2	2.92–03	2.43–01	3.37+02	–0.013	C–	*LS*
				1013.585 cm^{-1}	110247.288	111260.873	8–8	3.62–04	5.28–02	1.37+02	–0.375	C–	*LS*
				991.620 cm^{-1}	110212.396	111204.016	6–6	5.79–04	8.82–02	1.76+02	–0.277	C–	*LS*
				970.504 cm^{-1}	110194.654	111165.158	4–4	6.24–04	9.93–02	1.35+02	–0.401	C–	*LS*
				1048.477 cm^{-1}	110212.396	111260.873	6–8	2.04–05	3.70–03	6.97+00	–1.653	C–	*LS*
				1009.362 cm^{-1}	110194.654	111204.016	4–6	3.34–05	7.38–03	9.62+00	–1.530	C–	*LS*
211.	$2s^22p^2(^3P)4d-2s^22p^2(^1D)3p$	^{2}P–^{2}D°		*306.86* cm^{-1}	*110228.47*	*110535.33*	6–10	1.40–05	3.72–02	2.40+02	–0.651	C	1
				324.743 cm^{-1}	110220.107	110544.850	4–6	1.66–05	3.55–02	1.44+02	–0.848	D+	*LS*
				275.867 cm^{-1}	110245.183	110521.050	2–4	8.50–06	3.35–02	7.99+01	–1.174	D+	*LS*
				300.943 cm^{-1}	110220.107	110521.050	4–4	2.21–06	3.65–03	1.60+01	–1.836	D+	*LS*
212.	$2s^22p^2(^3P)4d-2s^22p^2(^3P)5p$	^{2}P–^{2}S°		*832.44* cm^{-1}	*110228.47*	111060.905	6–2	4.21–03	3.03–01	7.19+02	0.260	C	1
				840.798 cm^{-1}	110220.107	111060.905	4–2	2.89–03	3.06–01	4.80+02	0.088	D+	*LS*
				815.722 cm^{-1}	110245.183	111060.905	2–2	1.32–03	2.97–01	2.40+02	–0.226	D+	*LS*
213.		^{2}P–^{2}P°		*980.00* cm^{-1}	*110228.47*	*111208.46*	6–6	9.52–06	1.48–03	2.99+00	–2.050	C	1
				993.164 cm^{-1}	110220.107	111213.271	4–4	8.25–06	1.25–03	1.66+00	–2.300	D+	*LS*
				953.665 cm^{-1}	110245.183	111198.848	2–2	5.85–06	9.63–04	6.65–01	–2.715	D+	*LS*
				978.741 cm^{-1}	110220.107	111198.848	4–2	3.16–06	2.47–04	3.32–01	–3.005	D+	*LS*
				968.088 cm^{-1}	110245.183	111213.271	2–4	1.53–06	4.89–04	3.32–01	–3.010	D+	*LS*
214.		^{2}P–^{2}D°		*1656.12* cm^{-1}	*110228.47*	*111884.59*	6–10	1.60–03	1.46–01	1.74+02	–0.058	C	1
				1685.502 cm^{-1}	110220.107	111905.609	4–6	1.69–03	1.34–01	1.04+02	–0.272	D+	*LS*
				1607.878 cm^{-1}	110245.183	111853.061	2–4	1.22–03	1.42–01	5.80+01	–0.548	D+	*LS*
				1632.954 cm^{-1}	110220.107	111853.061	4–4	2.56–04	1.44–02	1.16+01	–1.240	D+	*LS*
215.	$2s^22p^2(^3P)4d-2s^22p^2(^1D)3p$	^{2}P–^{2}P°	*48001*	*2082.74* cm^{-1}	*110228.47*	*112311.21*	6–6	3.78–05	1.31–03	1.24+00	–2.106	C	1
			47612.9	2099.698 cm^{-1}	110220.107	112319.805	4–4	3.23–05	1.10–03	6.88–01	–2.358	D+	*LS*
			48795.2	2048.824 cm^{-1}	110245.183	112294.007	2–2	2.40–05	8.57–04	2.75–01	–2.766	D+	*LS*
			48205.2	2073.900 cm^{-1}	110220.107	112294.007	4–2	1.24–05	2.17–04	1.38–01	–3.062	D+	*LS*
			48188.4	2074.622 cm^{-1}	110245.183	112319.805	2–4	6.23–06	4.34–04	1.38–01	–3.062	D+	*LS*
216.		^{2}F–^{2}D°		*205.51* cm^{-1}	*110329.82*	*110535.33*	14–10	1.90–05	4.82–02	1.08+03	–0.171	C+	1
				182.388 cm^{-1}	110362.462	110544.850	8–6	1.27–05	4.28–02	6.18+02	–0.466	C–	*LS*
				234.745 cm^{-1}	110286.305	110521.050	6–4	2.84–05	5.14–02	4.32+02	–0.511	C–	*LS*
				258.545 cm^{-1}	110286.305	110544.850	6–6	1.80–06	4.04–03	3.09+01	–1.615	C–	*LS*
217.	$2s^22p^2(^3P)4d-2s^22p^2(^3P)5p$	^{2}F–^{2}D°		*1554.77* cm^{-1}	*110329.82*	*111884.59*	14–10	4.05–03	1.79–01	5.32+02	0.400	C+	1
				1543.147 cm^{-1}	110362.462	111905.609	8–6	3.77–03	1.78–01	3.04+02	0.154	C–	*LS*
				1566.756 cm^{-1}	110286.305	111853.061	6–4	4.15–03	1.69–01	2.13+02	0.005	C–	*LS*
				1619.304 cm^{-1}	110286.305	111905.609	6–6	2.18–04	1.25–02	1.52+01	–1.126	C–	*LS*

N I: Allowed Transitions — Continued

No.	Transition Array	Mult.	λ_{air} (Å)	λ_{vac} (Å) or σ (cm^{-1})*	E_i (cm^{-1})	E_k (cm^{-1})	g_i-g_k	A_{ki} (10^8 s^{-1})	f_{ik}	S (at. u.)	log gf	Acc.	Source
218.		^{4}P–^{4}D°		*897.05* cm^{-1}	*110315.90*	– *111212.94*	12–20	1.82–04	5.64–02	2.48+02	–0.169	C	1
				960.899 cm^{-1}	110299.974	– 111260.873	6–8	2.24–04	4.84–02	9.94+01	–0.537	D+	*LS*
				881.295 cm^{-1}	110322.721	– 111204.016	4–6	1.21–04	3.49–02	5.22+01	–0.855	D+	*LS*
				815.144 cm^{-1}	110350.014	– 111165.158	2–4	5.69–05	2.56–02	2.07+01	–1.290	D+	*LS*
				904.042 cm^{-1}	110299.974	– 111204.016	6–6	5.58–05	1.02–02	2.24+01	–1.212	D+	*LS*
				842.437 cm^{-1}	110322.721	– 111165.158	4–4	8.03–05	1.70–02	2.65+01	–1.168	D+	*LS*
				793.553 cm^{-1}	110350.014	– 111143.567	2–2	1.05–04	2.50–02	2.07+01	–1.302	D+	*LS*
				865.184 cm^{-1}	110299.974	– 111165.158	6–4	8.16–06	1.09–03	2.48+00	–2.185	D+	*LS*
				820.846 cm^{-1}	110322.721	– 111143.567	4–2	2.32–05	2.58–03	4.14+00	–1.986	D+	*LS*
219.		^{4}P–^{4}P°		*987.98* cm^{-1}	*110315.90*	– *111303.88*	12–12	4.74–04	7.27–02	2.91+02	–0.059	C	1
				1026.824 cm^{-1}	110299.974	– 111326.798	6–6	3.72–04	5.29–02	1.02+02	–0.498	D+	*LS*
				962.923 cm^{-1}	110322.721	– 111285.644	4–4	5.85–05	9.45–03	1.29+01	–1.423	D+	*LS*
				921.582 cm^{-1}	110350.014	– 111271.596	2–2	6.41–05	1.13–02	8.07+00	–1.646	D+	*LS*
				985.670 cm^{-1}	110299.974	– 111285.644	6–4	2.12–04	2.18–02	4.36+01	–0.884	D+	*LS*
				948.875 cm^{-1}	110322.721	– 111271.596	4–2	3.50–04	2.91–02	4.04+01	–0.934	D+	*LS*
				1004.077 cm^{-1}	110322.721	– 111326.798	4–6	1.49–04	3.33–02	4.36+01	–0.876	D+	*LS*
				935.630 cm^{-1}	110350.014	– 111285.644	2–4	1.68–04	5.74–02	4.04+01	–0.940	D+	*LS*
220.		^{4}P–^{4}S°		*1185.47* cm^{-1}	*110315.90*	– 111501.368	12–4	7.77–03	2.76–01	9.20+02	0.520	C	1
				1201.394 cm^{-1}	110299.974	– 111501.368	6–4	4.04–03	2.80–01	4.60+02	0.225	D+	*LS*
				1178.647 cm^{-1}	110322.721	– 111501.368	4–4	2.55–03	2.75–01	3.07+02	0.041	D+	*LS*
				1151.354 cm^{-1}	110350.014	– 111501.368	2–4	1.19–03	2.68–01	1.53+02	–0.270	D+	*LS*
221.		^{4}D–^{4}D°		*813.58* cm^{-1}	*110399.37*	– *111212.94*	20–20	3.63–04	8.22–02	6.65+02	0.216	C+	1
				857.653 cm^{-1}	110403.220	– 111260.873	8–8	3.65–04	7.43–02	2.28+02	–0.226	C–	*LS*
				802.660 cm^{-1}	110401.356	– 111204.016	6–6	2.00–04	4.65–02	1.14+02	–0.554	C–	*LS*
				769.695 cm^{-1}	110395.463	– 111165.158	4–4	1.23–04	3.11–02	5.32+01	–0.905	C–	*LS*
				757.772 cm^{-1}	110385.795	– 111143.567	2–2	1.47–04	3.83–02	3.33+01	–1.116	C–	*LS*
				800.796 cm^{-1}	110403.220	– 111204.016	8–6	6.58–05	1.15–02	3.79+01	–1.035	C–	*LS*
				763.802 cm^{-1}	110401.356	– 111165.158	6–4	1.05–04	1.80–02	4.66+01	–0.966	C–	*LS*
				748.104 cm^{-1}	110395.463	– 111143.567	4–2	1.41–04	1.89–02	3.33+01	–1.121	C–	*LS*
				859.517 cm^{-1}	110401.356	– 111260.873	6–8	6.10–05	1.65–02	3.79+01	–1.004	C–	*LS*
				808.553 cm^{-1}	110395.463	– 111204.016	4–6	8.32–05	2.86–02	4.66+01	–0.942	C–	*LS*
				779.363 cm^{-1}	110385.795	– 111165.158	2–4	7.98–05	3.94–02	3.33+01	–1.104	C–	*LS*
222.		^{4}D–^{4}P°		*904.51* cm^{-1}	*110399.37*	– *111303.88*	20–12	2.35–03	2.58–01	1.88+03	0.713	C+	1
				923.578 cm^{-1}	110403.220	– 111326.798	8–6	2.00–03	2.64–01	7.52+02	0.324	C–	*LS*
				884.288 cm^{-1}	110401.356	– 111285.644	6–4	1.38–03	1.77–01	3.95+02	0.026	C–	*LS*
				876.133 cm^{-1}	110395.463	– 111271.596	4–2	1.07–03	1.04–01	1.57+02	–0.380	C–	*LS*
				925.442 cm^{-1}	110401.356	– 111326.798	6–6	4.53–04	7.93–02	1.69+02	–0.323	C–	*LS*
				890.181 cm^{-1}	110395.463	– 111285.644	4–4	7.17–04	1.36–01	2.01+02	–0.266	C–	*LS*
				885.801 cm^{-1}	110385.795	– 111271.596	2–2	1.10–03	2.11–01	1.57+02	–0.375	C–	*LS*
				931.335 cm^{-1}	110395.463	– 111326.798	4–6	5.13–05	1.33–02	1.88+01	–1.274	C–	*LS*
				899.849 cm^{-1}	110385.795	– 111285.644	2–4	1.16–04	4.28–02	3.13+01	–1.067	C–	*LS*
223.		^{2}D–^{2}P°		*747.50* cm^{-1}	*110460.96*	– *111208.46*	10–6	1.09–03	1.76–01	7.74+02	0.245	C+	1
				743.027 cm^{-1}	110470.244	– 111213.271	6–4	9.66–04	1.75–01	4.64+02	0.021	C–	*LS*
				751.816 cm^{-1}	110447.032	– 111198.848	4–2	1.11–03	1.47–01	2.58+02	–0.230	C–	*LS*
				766.239 cm^{-1}	110447.032	– 111213.271	4–4	1.18–04	3.00–02	5.16+01	–0.920	C–	*LS*
224.		^{2}D–^{2}D°		*1423.63* cm^{-1}	*110460.96*	– *111884.59*	10–10	6.45–04	4.77–02	1.10+02	–0.322	C+	1
				1435.365 cm^{-1}	110470.244	– 111905.609	6–6	6.17–04	4.48–02	6.17+01	–0.570	C–	*LS*
				1406.029 cm^{-1}	110447.032	– 111853.061	4–4	5.59–04	4.24–02	3.97+01	–0.771	C–	*LS*
				1382.817 cm^{-1}	110470.244	– 111853.061	6–4	5.91–05	3.09–03	4.41+00	–1.732	C–	*LS*
				1458.577 cm^{-1}	110447.032	– 111905.609	4–6	4.62–05	4.88–03	4.41+00	–1.709	C–	*LS*

N I: Allowed Transitions — Continued

No.	Transition Array	Mult.	λ_{air} (Å)	λ_{vac} (Å) or σ (cm^{-1})*	E_i (cm^{-1})	E_k (cm^{-1})	g_i-g_k	A_{ki} (10^8 s^{-1})	f_{ik}	S (at. u.)	log gf	Acc.	Source
225.	$2s^22p^2(^3P)4d-2s^22p^2(^1D)3p$	$^2D-^2P^o$		1850.25 cm^{-1}	110460.96	112311.21	10−6	1.20−03	3.14−02	5.59+01	−0.502	C+	1
				1849.561 cm^{-1}	110470.244	112319.805	6−4	1.08−03	3.14−02	3.36+01	−0.724	C−	LS
				1846.975 cm^{-1}	110447.032	112294.007	4−2	1.19−03	2.62−02	1.86+01	−0.980	C−	LS
				1872.773 cm^{-1}	110447.032	112319.805	4−4	1.24−04	5.30−03	3.73+00	−1.673	C−	LS
226.	$2s^22p^2(^1D)3p-2s^22p^2(^3P)5d$	$^2D^o-^2P$	44082	2267.88 cm^{-1}	110535.33	112803.21	10−6	2.58−03	4.51−02	6.55+01	−0.346	C	1
			44310.6	2256.181 cm^{-1}	110544.850	112801.031	6−4	2.29−03	4.49−02	3.93+01	−0.570	D+	LS
			43722.7	2286.517 cm^{-1}	110521.050	112807.567	4−2	2.65−03	3.79−02	2.18+01	−0.819	D+	LS
			43848.1	2279.981 cm^{-1}	110521.050	112801.031	4−4	2.62−04	7.56−03	4.37+00	−1.519	D+	LS
227.		$^2D^o-^2F$	43051	2322.17 cm^{-1}	110535.33	112857.50	10−14	1.27−02	4.95−01	7.02+02	0.695	C+	1
			42607.1	2346.388 cm^{-1}	110544.850	112891.238	6−8	1.31−02	4.77−01	4.01+02	0.456	C−	LS
			43628.3	2291.468 cm^{-1}	110521.050	112812.518	4−6	1.14−02	4.89−01	2.81+02	0.291	C−	LS
			44086.1	2267.668 cm^{-1}	110544.850	112812.518	6−6	7.90−04	2.30−02	2.01+01	−0.860	C−	LS
228.		$^2D^o-^2D$	41604	2402.98 cm^{-1}	110535.33	112938.31	10−10	3.00−03	7.77−02	1.06+02	−0.109	C+	1
			41638.7	2400.959 cm^{-1}	110544.850	112945.809	6−6	2.79−03	7.25−02	5.96+01	−0.361	C−	LS
			41551.3	2406.005 cm^{-1}	110521.050	112927.055	4−4	2.71−03	7.01−02	3.83+01	−0.552	C−	LS
			41966.5	2382.205 cm^{-1}	110544.850	112927.055	6−4	2.92−04	5.14−03	4.26+00	−1.511	C−	LS
			41230.0	2424.759 cm^{-1}	110521.050	112945.809	4−6	2.05−04	7.85−03	4.26+00	−1.503	C−	LS
229.	$2s^22p^2(^1D)3p-2s^22p^2(^1D)4s$	$^2D^o-^2D$	11525	8675 cm^{-1}	110535.33	119210.0	10−10	1.27−03	2.52−03	9.57−01	−1.598	C+	3
			11537.3	8665.2 cm^{-1}	110544.850	119210.0	6−6	1.16−03	2.32−03	5.30−01	−1.856	C+	3
			11505.7	8689.0 cm^{-1}	110521.050	119210.0	4−4	1.09−03	2.16−03	3.27−01	−2.064	C+	3
			11537.3	8665.2 cm^{-1}	110544.850	119210.0	6−4	1.37−04	1.83−04	4.16−02	−2.960	C+	3
			11505.7	8689.0 cm^{-1}	110521.050	119210.0	4−6	1.31−04	3.89−04	5.89−02	−2.808	C+	3
230.	$2s^22p^2(^1D)3p-2s^22p^2(^1D)3d$	$^2D^o-^2F$											
			10786.8	9268.1 cm^{-1}	110544.850	119812.9	6−8	2.16−01	5.02−01	1.07+02	0.479	C+	3
231.		$^2D^o-^2D$	10395	9617 cm^{-1}	110535.33	120152	10−10	2.22−01	3.59−01	1.23+02	0.555	C+	3
			10408.1	9605.3 cm^{-1}	110544.850	120150.1	6−6	2.06−01	3.35−01	6.88+01	0.303	C+	3
			10376.3	9634.8 cm^{-1}	110521.050	120155.8	4−4	2.01−01	3.24−01	4.43+01	0.113	C+	3
			10401.9	9611.0 cm^{-1}	110544.850	120155.8	6−4	2.22−02	2.40−02	4.93+00	−0.842	C+	3
			10382.4	9629.1 cm^{-1}	110521.050	120150.1	4−6	1.47−02	3.57−02	4.89+00	−0.845	C+	3
232.		$^2D^o-^2P$	10228	9775 cm^{-1}	110535.33	120310	10−6	8.93−02	8.41−02	2.83+01	−0.075	C	3
			10238.1	9764.8 cm^{-1}	110544.850	120309.6	6−4	8.00−02	8.39−02	1.70+01	−0.298	C	3
			10211.5	9790.2 cm^{-1}	110521.050	120311.2	4−2	9.10−02	7.11−02	9.57+00	−0.546	C	3
			10213.2	9788.6 cm^{-1}	110521.050	120309.6	4−4	8.48−03	1.33−02	1.79+00	−1.275	C	3
233.	$2s^22p^2(^1D)3p-2s2p^4$	$^2D^o-^2D$											
			9382.12	9384.69	110544.850	121200.5	6−6	1.36−01	1.80−01	3.33+01	0.032	C+	3
			9361.21	9363.78	110521.050	121200.5	4−6	9.51−03	1.88−02	2.31+00	−1.125	C+	3
234.	$2s^22p^2(^1D)3p-2s^22p^2(^1D)4s$	$^2F^o-^2D$	11766	8497 cm^{-1}	110713.26	119210.0	14−10	1.16−01	1.71−01	9.30+01	0.380	C+	3
			11768.6	8494.8 cm^{-1}	110715.152	119210.0	8−6	1.10−01	1.71−01	5.31+01	0.137	C+	3
			11762.5	8499.3 cm^{-1}	110710.739	119210.0	6−4	1.16−01	1.60−01	3.72+01	−0.018	C+	3
			11762.5	8499.3 cm^{-1}	110710.739	119210.0	6−6	5.52−03	1.14−02	2.66+00	−1.163	C+	3
235.	$2s^22p^2(^1D)3p-2s^22p^2(^1D)3d$	$^2F^o-^2F$											
			10988.7	9097.7 cm^{-1}	110715.152	119812.9	8−8	1.02−01	1.85−01	5.37+01	0.171	C+	3
			10983.4	9102.2 cm^{-1}	110710.739	119812.9	6−8	3.71−03	8.94−03	1.94+00	−1.270	C+	3

N I: Allowed Transitions — Continued

No.	Transition Array	Mult.	λ_{air} (Å)	λ_{vac} (Å) or σ (cm^{-1})*	E_i (cm^{-1})	E_k (cm^{-1})	g_i-g_k	A_{ki} (10^8 s^{-1})	f_{ik}	S (at. u.)	log gf	Acc.	Source
236.		^{2}F°–^{2}D	*10591*	*9439* cm^{-1}	*110713.26*	*120152*	14–10	2.07–02	2.49–02	1.22+01	–0.457	C+	3
			10596.0	9434.9 cm^{-1}	110715.152	120150.1	8–6	1.98–02	2.49–02	6.96+00	–0.700	C+	3
			10584.6	9445.1 cm^{-1}	110710.739	120155.8	6–4	2.07–02	2.32–02	4.85+00	–0.857	C+	3
			10591.0	9439.4 cm^{-1}	110710.739	120150.1	6–6	1.04–03	1.74–03	3.65–01	–1.980	C+	3
237.	$2s^22p^2(^1D)3p-2s2p^4$	^{2}F°–^{2}D											
			9534.50	9537.12	110715.152	121200.5	8–6	2.89–03	2.95–03	7.42–01	–1.627	C+	3
			9530.49	9533.11	110710.739	121200.5	6–6	1.31–04	1.79–04	3.37–02	–2.970	C+	3
238.	$2s^22p^2(^3P)5p-2s^22p^2(^3P)5d$	^{2}S°–^{2}P		*1742.31* cm^{-1}	111060.905	*112803.21*	2–6	1.08–02	1.59+00	6.03+02	0.504	C	1
				1740.126 cm^{-1}	111060.905	112801.031	2–4	1.07–02	1.06+00	4.02+02	0.327	D+	*LS*
				1746.662 cm^{-1}	111060.905	112807.567	2–2	1.09–02	5.33–01	2.01+02	0.028	D+	*LS*
239.		^{4}D°–^{4}F		*1594.10* cm^{-1}	*111212.94*	*112807.04*	20–28	1.45–02	1.20+00	4.96+03	1.380	C+	1
				1600.475 cm^{-1}	111260.873	112861.348	8–10	1.47–02	1.08+00	1.77+03	0.935	C–	*LS*
				1593.709 cm^{-1}	111204.016	112797.725	6–8	1.25–02	9.80–01	1.21+03	0.769	C–	*LS*
				1595.167 cm^{-1}	111165.158	112760.325	4–6	1.09–02	9.65–01	7.97+02	0.587	C–	*LS*
				1616.399 cm^{-1}	111143.567	112759.966	2–4	1.06–02	1.22+00	4.96+02	0.386	C–	*LS*
				1536.852 cm^{-1}	111260.873	112797.725	8–8	1.86–03	1.18–01	2.02+02	–0.026	C–	*LS*
				1556.309 cm^{-1}	111204.016	112760.325	6–6	3.29–03	2.04–01	2.58+02	0.087	C–	*LS*
				1594.808 cm^{-1}	111165.158	112759.966	4–4	4.08–03	2.40–01	1.98+02	–0.017	C–	*LS*
				1499.452 cm^{-1}	111260.873	112760.325	8–6	1.17–04	5.85–03	1.03+01	–1.330	C–	*LS*
				1555.950 cm^{-1}	111204.016	112759.966	6–4	2.70–04	1.12–02	1.42+01	–1.174	C–	*LS*
240.		^{4}D°–^{4}P		*1623.7* cm^{-1}	*111212.94*	*112836.6*	20–12	1.18–03	4.02–02	1.63+02	–0.095	C	1
				1563.586 cm^{-1}	111260.873	112824.459	8–6	8.42–04	3.87–02	6.52+01	–0.509	D+	*LS*
				1634.00 cm^{-1}	111204.016	112838.02	6–4	7.57–04	2.83–02	3.42+01	–0.770	D+	*LS*
				1705.11 cm^{-1}	111165.158	112870.27	4–2	6.83–04	1.76–02	1.36+01	–1.153	D+	*LS*
				1620.443 cm^{-1}	111204.016	112824.459	6–6	2.11–04	1.20–02	1.47+01	–1.141	D+	*LS*
				1672.86 cm^{-1}	111165.158	112838.02	4–4	4.13–04	2.21–02	1.74+01	–1.054	D+	*LS*
				1726.70 cm^{-1}	111143.567	112870.27	2–2	7.09–04	3.56–02	1.36+01	–1.147	D+	*LS*
				1659.301 cm^{-1}	111165.158	112824.459	4–6	2.52–05	2.05–03	1.63+00	–2.085	D+	*LS*
				1694.45 cm^{-1}	111143.567	112838.02	2–4	6.70–05	6.99–03	2.72+00	–1.854	D+	*LS*
241.		^{4}D°–^{4}D		*1697* cm^{-1}	*111212.94*	*112910*	20–20	4.05–03	2.11–01	8.18+02	0.625	C+	1
				1650.206 cm^{-1}	111260.873	112911.079	8–8	3.19–03	1.76–01	2.80+02	0.148	C–	*LS*
				1706.614 cm^{-1}	111204.016	112910.630	6–6	2.36–03	1.22–01	1.41+02	–0.137	C–	*LS*
				1743.918 cm^{-1}	111165.158	112909.076	4–4	1.76–03	8.66–02	6.54+01	–0.460	C–	*LS*
				1760.9 cm^{-1}	111143.567	112904.5	2–2	2.26–03	1.09–01	4.09+01	–0.660	C–	*LS*
				1649.757 cm^{-1}	111260.873	112910.630	8–6	7.07–04	2.92–02	4.66+01	–0.632	C–	*LS*
				1705.060 cm^{-1}	111204.016	112909.076	6–4	1.44–03	4.94–02	5.72+01	–0.528	C–	*LS*
				1739.3 cm^{-1}	111165.158	112904.5	4–2	2.18–03	5.40–02	4.09+01	–0.666	C–	*LS*
				1707.063 cm^{-1}	111204.016	112911.079	6–8	5.88–04	4.03–02	4.66+01	–0.617	C–	*LS*
				1745.472 cm^{-1}	111165.158	112910.630	4–6	1.03–03	7.59–02	5.72+01	–0.518	C–	*LS*
				1765.509 cm^{-1}	111143.567	112909.076	2–4	1.14–03	1.10–01	4.09+01	–0.659	C–	*LS*
242.		^{2}P°–^{2}P		*1594.75* cm^{-1}	*111208.46*	*112803.21*	6–6	1.14–03	6.74–02	8.34+01	–0.393	C	1
				1587.760 cm^{-1}	111213.271	112801.031	4–4	9.40–04	5.59–02	4.63+01	–0.651	D+	*LS*
				1608.719 cm^{-1}	111198.848	112807.567	2–2	7.82–04	4.53–02	1.85+01	–1.043	D+	*LS*
				1594.296 cm^{-1}	111213.271	112807.567	4–2	3.81–04	1.12–02	9.27+00	–1.348	D+	*LS*
				1602.183 cm^{-1}	111198.848	112801.031	2–4	1.93–04	2.26–02	9.27+00	–1.346	D+	*LS*
243.		^{2}P°–^{2}D		*1729.84* cm^{-1}	*111208.46*	*112938.31*	6–10	1.14–02	9.56–01	1.09+03	0.758	C+	1
				1732.538 cm^{-1}	111213.271	112945.809	4–6	1.15–02	8.61–01	6.54+02	0.537	C–	*LS*
				1728.207 cm^{-1}	111198.848	112927.055	2–4	9.51–03	9.55–01	3.64+02	0.281	C–	*LS*
				1713.784 cm^{-1}	111213.271	112927.055	4–4	1.86–03	9.47–02	7.27+01	–0.422	C–	*LS*

N I: Allowed Transitions — Continued

No.	Transition Array	Mult.	λ_{air} (Å)	λ_{vac} (Å) or σ (cm^{-1})*	E_i (cm^{-1})	E_k (cm^{-1})	g_i-g_k	A_{ki} (10^8 s^{-1})	f_{ik}	S (at. u.)	log gf	Acc.	Source
244.	$2s^22p^2(^3P)5p-2s^22p^2(^1S)3s$	$^2P^o-^2S$	*19718*	*5070.10* cm^{-1}	*111208.46*	116278.558	6–2	2.27–04	4.41–04	1.72–01	–2.577	C+	1
			19736.8	5065.287 cm^{-1}	111213.271	116278.558	4–2	1.51–04	4.41–04	1.15–01	–2.754	C–	*LS*
			19680.8	5079.710 cm^{-1}	111198.848	116278.558	2–2	7.61–05	4.42–04	5.73–02	–3.053	C–	*LS*
245.	$2s^22p^2(^3P)5p-2s^22p^2(^1D)3d$	$^2P^o-^2S$	*10684*	*9357.5* cm^{-1}	*111208.46*	120566.0	6–2	6.75–02	3.85–02	8.12+00	–0.636	C+	1
			10689.1	9352.7 cm^{-1}	111213.271	120566.0	4–2	4.49–02	3.85–02	5.42+00	–0.813	C–	*LS*
			10672.7	9367.2 cm^{-1}	111198.848	120566.0	2–2	2.26–02	3.85–02	2.71+00	–1.113	C–	*LS*
246.	$2s^22p^2(^3P)5p-2s^22p^2(^3P)5d$	$^4P^o-^4P$		*1532.7* cm^{-1}	*111303.88*	*112836.6*	12–12	3.93–03	2.50–01	6.45+02	0.478	C	1
				1497.661 cm^{-1}	111326.798	112824.459	6–6	2.56–03	1.71–01	2.26+02	0.012	D+	*LS*
				1552.38 cm^{-1}	111285.644	112838.02	4–4	5.44–04	3.38–02	2.87+01	–0.869	D+	*LS*
				1598.67 cm^{-1}	111271.596	112870.27	2–2	7.42–04	4.35–02	1.79+01	–1.060	D+	*LS*
				1511.22 cm^{-1}	111326.798	112838.02	6–4	1.69–03	7.41–02	9.68+01	–0.352	D+	*LS*
				1584.63 cm^{-1}	111285.644	112870.27	4–2	3.62–03	1.08–01	8.96+01	–0.365	D+	*LS*
				1538.815 cm^{-1}	111285.644	112824.459	4–6	1.19–03	1.13–01	9.68+01	–0.344	D+	*LS*
				1566.42 cm^{-1}	111271.596	112838.02	2–4	1.75–03	2.13–01	8.96+01	–0.370	D+	*LS*
247.		$^4P^o-^4D$		*1606* cm^{-1}	*111303.88*	*112910*	12–20	1.13–02	1.10+00	2.70+03	1.119	C+	1
				1584.281 cm^{-1}	111326.798	112911.079	6–8	1.09–02	8.66–01	1.08+03	0.716	C–	*LS*
				1624.986 cm^{-1}	111285.644	112910.630	4–6	8.22–03	6.99–01	5.67+02	0.447	C–	*LS*
				1637.480 cm^{-1}	111271.596	112909.076	2–4	5.00–03	5.59–01	2.25+02	0.049	C–	*LS*
				1583.832 cm^{-1}	111326.798	112910.630	6–6	3.26–03	1.95–01	2.43+02	0.068	C–	*LS*
				1623.432 cm^{-1}	111285.644	112909.076	4–4	6.24–03	3.55–01	2.88+02	0.152	C–	*LS*
				1632.9 cm^{-1}	111271.596	112904.5	2–2	9.93–03	5.58–01	2.25+02	0.047	C–	*LS*
				1582.278 cm^{-1}	111326.798	112909.076	6–4	5.42–04	2.16–02	2.70+01	–0.887	C–	*LS*
				1618.9 cm^{-1}	111285.644	112904.5	4–2	1.93–03	5.53–02	4.50+01	–0.655	C–	*LS*
248.	$2s^22p^2(^3P)5p-2s^22p^2(^3P)6s$	$^4S^o-^4P$		*1136.78* cm^{-1}	111501.368	*112638.14*	4–12	8.88–04	3.09–01	3.58+02	0.092	C	2
				1180.021 cm^{-1}	111501.368	112681.389	4–6	9.93–04	1.60–01	1.79+02	–0.193	D+	*LS*
				1108.244 cm^{-1}	111501.368	112609.612	4–4	8.23–04	1.00–01	1.19+02	–0.396	D+	*LS*
				1064.102 cm^{-1}	111501.368	112565.470	4–2	7.28–04	4.82–02	5.97+01	–0.715	D+	*LS*
249.	$2s^22p^2(^3P)5p-2s^22p^2(^3P)5d$	$^4S^o-^4P$		*1335.2* cm^{-1}	111501.368	*112836.6*	4–12	6.74–03	1.70+00	1.68+03	0.833	C	2
				1323.091 cm^{-1}	111501.368	112824.459	4–6	6.56–03	8.42–01	8.38+02	0.528	D+	*LS*
				1336.65 cm^{-1}	111501.368	112838.02	4–4	6.76–03	5.67–01	5.59+02	0.356	D+	*LS*
				1368.90 cm^{-1}	111501.368	112870.27	4–2	7.26–03	2.90–01	2.79+02	0.065	D+	*LS*
250.	$2s^22p^2(^3P)5p-2s^22p^2(^3P)7s$	$^4S^o-^4P$	*38450*	*2600.05* cm^{-1}	111501.368	*114101.42*	4–12	4.04–04	2.69–02	1.36+01	–0.968	C	2
			37794.6	2645.157 cm^{-1}	111501.368	114146.525	4–6	4.26–04	1.37–02	6.81+00	–1.262	D+	*LS*
			38903.2	2569.785 cm^{-1}	111501.368	114071.153	4–4	3.90–04	8.86–03	4.54+00	–1.450	D+	*LS*
			39588.7	2525.286 cm^{-1}	111501.368	114026.654	4–2	3.70–04	4.35–03	2.27+00	–1.759	D+	*LS*
251.	$2s^22p^2(^3P)5p-2s^22p^2(^3P)6d$	$^4S^o-^4P$	*36990*	*2703* cm^{-1}	111501.368	*114204*	4–12	1.85–03	1.14–01	5.55+01	–0.341	C	2
			37161.4	2690.2 cm^{-1}	111501.368	114191.6	4–6	1.83–03	5.67–02	2.78+01	–0.644	D+	*LS*
			37045.7	2698.6 cm^{-1}	111501.368	114200.0	4–4	1.84–03	3.79–02	1.85+01	–0.819	D+	*LS*
			36378.4	2748.1 cm^{-1}	111501.368	114249.5	4–2	1.95–03	1.93–02	9.26+00	–1.112	D+	*LS*
252.	$2s^22p^2(^3P)5p-2s^22p^2(^3P)8s$	$^4S^o-^4P$	*28800*	*3471* cm^{-1}	111501.368	*114973*	4–12	2.44–04	9.10–03	3.45+00	–1.439	C	2
			28401.9	3519.9 cm^{-1}	111501.368	115021.3	4–6	2.54–04	4.61–03	1.73+00	–1.734	D+	*LS*
			29081.0	3437.7 cm^{-1}	111501.368	114939.1	4–4	2.37–04	3.00–03	1.15+00	–1.920	D+	*LS*
			29467.6	3392.6 cm^{-1}	111501.368	114894.0	4–2	2.28–04	1.48–03	5.75–01	–2.227	D+	*LS*

N I: Allowed Transitions — Continued

No.	Transition Array	Mult.	λ_{air} (Å)	λ_{vac} (Å) or σ (cm^{-1})*	E_i (cm^{-1})	E_k (cm^{-1})	g_i-g_k	A_{ki} (10^8 s^{-1})	f_{ik}	S (at. u.)	log gf	Acc.	Source
253.	$2s^22p^2(^3P)5p-$ $2s^22p^2(^3P)9s$	$^4S°-^4P$	*24826*	*4027* cm⁻¹	111501.368	*115528*	4–12	1.60–04	4.45–03	1.46+00	–1.750	C	2
			24531.8	4075.2 cm⁻¹	111501.368	115576.6	4–6	1.66–04	2.25–03	7.28–01	–2.045	D+	*LS*
			25029.9	3994.1 cm⁻¹	111501.368	115495.5	4–4	1.57–04	1.47–03	4.85–01	–2.230	D+	*LS*
			25321.5	3948.1 cm⁻¹	111501.368	115449.5	4–2	1.51–04	7.27–04	2.43–01	–2.536	D+	*LS*
254.	$2s^22p^2(^3P)5p-$ $2s^22p^2(^3P)5d$	$^2D°-^2P$		*918.62* cm⁻¹	*111884.59*	*112803.21*	10–6	1.34–03	1.42–01	5.10+02	0.153	C	1
				895.422 cm⁻¹	111905.609	112801.031	6–4	1.11–03	1.39–01	3.06+02	–0.080	D+	*LS*
				954.506 cm⁻¹	111853.061	112807.567	4–2	1.50–03	1.23–01	1.70+02	–0.307	D+	*LS*
				947.970 cm⁻¹	111853.061	112801.031	4–4	1.47–04	2.45–02	3.40+01	–1.009	D+	*LS*
255.		$^2D°-^2F$		*972.91* cm⁻¹	*111884.59*	*112857.50*	10–14	4.35–03	9.64–01	3.26+03	0.984	C+	1
				985.629 cm⁻¹	111905.609	112891.238	6–8	4.52–03	9.30–01	1.86+03	0.747	C–	*LS*
				959.457 cm⁻¹	111853.061	112812.518	4–6	3.89–03	9.51–01	1.30+03	0.580	C–	*LS*
				906.909 cm⁻¹	111905.609	112812.518	6–6	2.35–04	4.28–02	9.32+01	–0.590	C–	*LS*
256.		$^2D°-^2D$		*1053.72* cm⁻¹	*111884.59*	*112938.31*	10–10	1.37–03	1.85–01	5.77+02	0.266	C+	1
				1040.200 cm⁻¹	111905.609	112945.809	6–6	1.23–03	1.70–01	3.23+02	0.009	C–	*LS*
				1073.994 cm⁻¹	111853.061	112927.055	4–4	1.30–03	1.69–01	2.08+02	–0.169	C–	*LS*
				1021.446 cm⁻¹	111905.609	112927.055	6–4	1.25–04	1.19–02	2.31+01	–1.145	C–	*LS*
				1092.748 cm⁻¹	111853.061	112945.809	4–6	1.02–04	1.91–02	2.31+01	–1.116	C–	*LS*
257.	$2s^22p^2(^1D)3p-$ $2s^22p^2(^3P)5d$	$^2P°-^2P$		*492.00* cm⁻¹	*112311.21*	*112803.21*	6–6	1.19–05	7.34–03	2.94+01	–1.356	C	1
				481.226 cm⁻¹	112319.805	112801.031	4–4	9.24–06	5.98–03	1.64+01	–1.621	D+	*LS*
				513.560 cm⁻¹	112294.007	112807.567	2–2	8.99–06	5.10–03	6.54+00	–1.991	D+	*LS*
				487.762 cm⁻¹	112319.805	112807.567	4–2	3.85–06	1.21–03	3.27+00	–2.314	D+	*LS*
				507.024 cm⁻¹	112294.007	112801.031	2–4	2.16–06	2.52–03	3.27+00	–2.298	D+	*LS*
258.		$^2P°-^2D$		*627.10* cm⁻¹	*112311.21*	*112938.31*	6–10	7.07–04	4.49–01	1.41+03	0.430	C+	1
				626.004 cm⁻¹	112319.805	112945.809	4–6	7.03–04	4.03–01	8.48+02	0.208	C–	*LS*
				633.048 cm⁻¹	112294.007	112927.055	2–4	6.06–04	4.53–01	4.71+02	–0.043	C–	*LS*
				607.250 cm⁻¹	112319.805	112927.055	4–4	1.07–04	4.35–02	9.42+01	–0.760	C–	*LS*
259.	$2s^22p^2(^1D)3p-$ $2s^22p^2(^1S)3s$	$^2P°-^2S$	*25199*	*3967.35* cm⁻¹	*112311.21*	116278.558	6–2	8.84–04	2.81–03	1.40+00	–1.774	C+	1
			25253.6	3958.753 cm⁻¹	112319.805	116278.558	4–2	5.86–04	2.80–03	9.31–01	–1.951	C–	*LS*
			25090.1	3984.551 cm⁻¹	112294.007	116278.558	2–2	2.99–04	2.82–03	4.66–01	–2.249	C–	*LS*
260.	$2s^22p^2(^1D)3p-$ $2s^22p^2(^1D)4s$	$^2P°-^2D$	*14491*	*6899* cm⁻¹	*112311.21*	119210.0	6–10	6.23–02	3.27–01	9.36+01	0.293	C+	3
			14509.4	6890.2 cm⁻¹	112319.805	119210.0	4–6	6.20–02	2.94–01	5.61+01	0.070	C+	3
			14455.3	6916.0 cm⁻¹	112294.007	119210.0	2–4	5.23–02	3.28–01	3.12+01	–0.184	C+	3
			14509.4	6890.2 cm⁻¹	112319.805	119210.0	4–4	1.04–02	3.28–02	6.26+00	–0.882	C+	3
261.	$2s^22p^2(^1D)3p-$ $2s^22p^2(^1D)3d$	$^2P°-^2D$	*12750*	*7841* cm⁻¹	*112311.21*	*120152*	6–10	1.66–03	6.74–03	1.70+00	–1.393	C+	3
			12767.4	7830.3 cm⁻¹	112319.805	120150.1	4–6	1.69–03	6.21–03	1.04+00	–1.605	C+	3
			12716.3	7861.8 cm⁻¹	112294.007	120155.8	2–4	1.40–03	6.78–03	5.68–01	–1.868	C+	3
			12758.1	7836.0 cm⁻¹	112319.805	120155.8	4–4	2.06–04	5.03–04	8.45–02	–2.697	C+	3
262.		$^2P°-^2P$	*12498*	*7999* cm⁻¹	*112311.21*	*120310*	6–6	1.59–01	3.72–01	9.19+01	0.349	C	3
			12512.5	7989.8 cm⁻¹	112319.805	120309.6	4–4	1.32–01	3.10–01	5.11+01	0.094	C	3
			12469.8	8017.2 cm⁻¹	112294.007	120311.2	2–2	1.07–01	2.49–01	2.04+01	–0.303	C	3
			12510.0	7991.4 cm⁻¹	112319.805	120311.2	4–2	5.21–02	6.12–02	1.01+01	–0.611	C	3
			12472.3	8015.6 cm⁻¹	112294.007	120309.6	2–4	2.67–02	1.25–01	1.02+01	–0.603	C	3

N I: Allowed Transitions — Continued

No.	Transition Array	Mult.	λ_{air} (Å)	λ_{vac} (Å) or σ (cm^{-1})*	E_i (cm^{-1})	E_k (cm^{-1})	g_i-g_k	A_{ki} (10^8 s^{-1})	f_{ik}	S (at. u.)	log gf	Acc.	Source
263.		$^2P^o-^2S$	*12111*	*8254.8* cm^{-1}	*112311.21*	– 120566.0	6–2	1.56–01	1.14–01	2.73+01	−0.165	C+	1
			12123.5	8246.2 cm^{-1}	112319.805	– 120566.0	4–2	1.03–01	1.14–01	1.82+01	−0.341	C−	*LS*
			12085.7	8272.0 cm^{-1}	112294.007	– 120566.0	2–2	5.22–02	1.14–01	9.10+00	−0.641	C−	*LS*
264.	$2s^22p^2(^1D)3p-$ $2s2p^4$	$^2P^o-^2D$											
			11257.3	8880.7 cm^{-1}	112319.805	– 121200.5	4–6	2.92–02	8.34–02	1.24+01	−0.477	C+	3

*Wavelengths (Å) are always given unless cm^{-1} is indicated.
†For multiplet No. 80, the data for the weak line at 10673.9 Å are from Ref. 3.

N I

Forbidden Transitions

We have selected the results of two recent calculations and present the arithmetic averages. Butler and Zeippen[1] applied the well-known SUPERSTRUCTURE code, and Godefroid and Froese-Fischer[2] used a multi-configuration Hartree-Fock program to obtain their atomic wavefunctions. Both calculations contain extensive treatments of configuration interaction, but Ref. 2 does not include relativistic terms in the M1 transition operator which especially affect the $^4S-{}^2D$ transitions in the p^3 configuration. We have replaced the wavelength data used in Ref. 2 with slightly more accurate values based on the latest experimental term energies. The two calculations agree typically within 10–20%, except for the very weak E2 line at 8.71 cm^{-1}, where they disagree by a factor of 8.

For the M1 $^2D^o_{5/2}-{}^2D^o_{3/2}$ transition, we have applied the line strength obtained from general nonrelativistic expansion formulas first developed by Shortley *et al.*[3] For near *LS*-coupling conditions, S=2.40 is obtained. We used this value rather than reconverting to S from the published A-values of Ref. 1 and 2 with an inaccurate wavelength. For some neighboring lower ions of the nitrogen sequence for which—in contrast to N I—more accurate wavelength data are available, this line strength is indeed derived from the published A_{ki} values.

References

[1]K. Butler and C. J. Zeippen, Astron. Astrophys. **141**, 274 (1984).
[2]M. Godefroid and C. Froese Fischer, J. Phys. B **17**, 681 (1984).
[3]G. H. Shortley, L. H. Aller, J. G. Baker, and D. H. Menzel, Astrophys. J **93**, 178 (1941).

N I: Forbidden Transitions

No.	Transition Array	Mult.	λ_{air} (Å)	λ_{vac} (Å) or σ (cm^{-1})*	E_i (cm^{-1})	E_k (cm^{-1})	g_i-g_k	Type	A_{ki} (s^{-1})	S (at. u.)	Acc.	Source
1.	$2p^3-$ $2p^3$	$^4S^o-^2D^o$										
			5200.26	5201.71	0.00	– 19224.46	4–6	M1	2.45–07	7.65–09	C	1
			5200.26	5201.71	0.00	– 19224.46	4–6	E2	5.52–06	1.13–04	B	1,2
			5197.90	5199.35	0.00	– 19233.18	4–4	M1	1.90–05	3.95–07	C	1
			5197.90	5199.35	0.00	– 19233.18	4–4	E2	3.60–06	4.88–05	B	1,2
2.		$^4S^o-^2P^o$										
			3466.50	3467.49	0.00	– 28839.31	4–4	M1	6.21–03	3.84–05	B	1,2
			3466.50	3467.49	0.00	– 28839.31	4–4	E2	1.66–07	2.97–07	C+	1,2
			3466.54	3467.54	0.00	– 28838.92	4–2	M1	2.52–03	7.79–06	B	1,2
			3466.54	3467.54	0.00	– 28838.92	4–2	E2	1.14–08	1.02–08	C+	1,2

N I: Forbidden Transitions

No.	Transition Array	Mult.	λ_{air} (Å)	λ_{vac} (Å) or σ (cm^{-1})*	E_i (cm^{-1})	E_k (cm^{-1})	g_i–g_k	Type	A_{ki} (s^{-1})	S (at. u.)	Acc.	Source
3.		^{2}D°–^{2}D°										
				8.71 cm^{-1}	19224.46	– 19233.18	6–4	M1	1.07–08	2.40+00	B+	3
				8.71 cm^{-1}	19224.46	– 19233.18	6–4	E2	3.20–21	2.28–03	E	1,2
4.		^{2}D°–^{2}P°										
			10397.7	9614.84 cm^{-1}	19224.46	– 28839.31	6–4	M1	9.78–04	1.63–04	B	1,2
			10397.7	9614.84 cm^{-1}	19224.46	– 28839.31	6–4	E2	5.29–02	2.30+01	B	1,2
			10398.2	9614.46 cm^{-1}	19224.46	– 28838.92	6–2	E2	3.03–02	6.58+00	B	1,2
			10407.2	9606.13 cm^{-1}	19233.18	– 28839.31	4–4	M1	1.75–03	2.92–04	B	1,2
			10407.2	9606.13 cm^{-1}	19233.18	– 28839.31	4–4	E2	2.26–02	9.85+00	B	1,2
			10407.6	9605.74 cm^{-1}	19233.18	– 28838.92	4–2	M1	1.09–03	9.13–05	B	1,2
			10407.6	9605.74 cm^{-1}	19233.18	– 28838.92	4–2	E2	4.52–02	9.88+00	B	1,2

*Wavelengths (Å) are always given unless cm^{-1} is indicated.

N II

Carbon Isoelectronic Sequence

Ground State: $1s^22s^22p^2\ ^3P_0$

Ionization Energy: 29.601 eV = 238750.50 cm^{-1}

Allowed Transitions

List of tabulated lines

Wavelength (Å)	No.	Wavelength (Å)	No.	Wavelength (Å)	No.	Wavelength (Å)	No.
in vacuum		508.668	15	547.818	33	670.515	6
		508.697	15	559.762	49	670.884	6
437.265	18	508.794	15	572.069	32	671.016	5
437.286	18	508.928	15	574.650	31	671.386	5
437.329	18	509.006	15	576.232	30	671.411	5
437.358	18	510.758	35	580.802	29	671.630	5
437.443	18	513.849	34	580.904	29	671.773	5
437.515	18	525.983	14	582.156	28	672.001	5
474.493	17	528.529	13	583.925	27	693.774	4
474.546	17	529.355	12	584.128	27	694.169	4
474.602	17	529.413	12	599.644	8	715.254	25
474.706	17	529.491	12	599.819	8	745.841	43
474.787	17	529.637	12	600.115	8	746.984	24
474.891	17	529.722	12	605.437	48	747.606	23
475.647	16	529.867	12	610.116	47	748.369	23
475.698	16	530.343	50	629.167	51	775.965	22
475.757	16	533.511	11	629.447	51	816.740	42
475.803	16	533.581	11	629.670	51	836.187	59
475.884	16	533.650	11	635.197	46	836.279	59
475.943	16	533.729	11	640.121	45	836.289	59
485.849	37	533.815	11	644.634	7	836.616	59
487.379	36	533.884	11	644.837	7	836.627	59
505.986	52	534.637	10	645.178	7	836.837	59
506.054	52	536.300	9	646.209	44	858.376	41
506.153	52	536.365	9	660.286	26	860.205	40
508.484	15	536.536	9	670.296	6	898.575	58

List of tabulated lines — Continued

Wavelength (Å)	No.	Wavelength (Å)	No.	Wavelength (Å)	No.	Wavelength (Å)	No.
898.587	58	1345.34	53	1866.46	92	2522.24	147
900.724	57	1346.41	53	1868.24	92	2522.46	147
900.830	57	1346.44	53	1877.51	78	2524.51	147
900.843	57	1590.14	84	1878.62	78	2526.18	147
901.398	57	1590.56	84	1887.40	91	2590.94	138
901.411	57	1591.36	84	1904.89	150	2709.84	161
901.626	57	1592.12	84	1906.17	150	2734.73	172
903.962	56	1593.60	84	1908.30	150	2744.68	137
904.069	56	1594.82	84	1991.30	111	2780.94	136
904.748	56	1596.43	97			2799.22	160
904.855	56	1597.66	97	in air		2823.63	135
904.867	56	1598.43	97			3006.83	109
905.274	56	1627.35	62	2044.76	165	3023.67	171
905.286	56	1627.38	62	2064.99	100	3125.92	108
909.762	55	1628.90	62	2076.95	99	3176.87	19
915.612	3	1628.92	62	2079.97	99	3190.19	121
915.962	3	1629.08	62	2081.12	99	3196.39	121
916.012	3	1629.83	62	2088.66	124	3311.42	120
916.020	3	1635.00	71	2091.31	124	3318.10	120
916.701	3	1675.73	61	2094.19	124	3324.58	120
916.710	3	1675.75	61	2095.53	124	3328.72	120
974.554	66	1675.92	61	2096.20	124	3329.70	68
974.563	66	1678.89	38	2096.86	124	3330.32	120
975.012	66	1708.64	70	2130.18	164	3331.31	120
975.021	66	1709.66	113	2139.01	1	3408.13	77
975.078	66	1720.41	151	2142.77	1	3437.14	90
975.298	66	1720.46	151	2159.93	163	3451.28	130
1060.21	65	1720.53	151	2197.51	132	3593.60	129
1060.22	65	1721.57	151	2203.63	132	3609.10	129
1060.29	65	1722.19	151	2205.86	132	3615.86	129
1063.35	64	1723.30	151	2206.09	110	3663.83	146
1063.36	64	1724.65	69	2283.64	149	3668.57	146
1064.14	64	1725.03	83	2285.30	123	3676.44	146
1064.15	64	1732.43	96	2286.62	123	3829.80	145
1064.22	64	1740.31	60	2286.67	123	3838.37	145
1064.44	64	1743.20	60	2286.69	149	3841.14	224
1064.95	21	1743.23	60	2288.41	149	3842.19	145
1064.96	21	1745.05	60	2288.48	123	3842.45	224
1067.88	63	1745.08	60	2289.86	123	3847.40	145
1068.96	63	1745.26	60	2290.26	149	3848.00	224
1068.97	63	1763.64	101	2291.66	123	3850.02	224
1069.55	63	1765.14	101	2292.66	149	3855.10	145
1069.56	63	1766.08	101	2293.32	149	3855.37	224
1069.63	63	1772.74	82	2293.53	123	3855.59	224
1083.99	2	1780.55	95	2316.49	122	3856.06	145
1084.56	2	1789.40	112	2316.68	122	3864.32	224
1084.58	2	1830.53	81	2317.04	122	3913.89	229
1085.53	2	1831.59	81	2319.95	122	3915.41	229
1085.55	2	1836.17	81	2321.66	122	3915.57	229
1085.70	2	1839.93	94	2325.14	122	3919.00	107
1275.04	54	1840.98	80	2326.34	173	3919.65	229
1275.25	54	1842.28	80	2386.81	131	3924.64	229
1275.28	54	1843.36	80	2388.23	131	3930.43	229
1276.20	54	1844.26	80	2390.87	131	3955.85	76
1276.22	54	1845.62	80	2461.27	162	3977.31	76
1276.80	54	1848.00	80	2488.12	148	3995.00	89
1299.79	20	1851.81	93	2488.76	148	4098.87	187
1299.81	20	1852.72	93	2490.24	139	4102.90	187
1300.04	20	1857.87	79	2490.31	148	4106.99	187
1306.71	39	1858.55	79	2493.18	148	4108.02	187
1343.34	53	1859.26	79	2493.93	148	4111.04	187
1343.57	53	1859.64	79	2496.81	148	4112.07	187
1345.08	53	1862.59	79	2520.21	147	4114.33	106
1345.31	53	1864.36	79	2520.79	147	4123.12	106

List of tabulated lines — Continued

Wavelength (Å)	No.	Wavelength (Å)	No.	Wavelength (Å)	No.	Wavelength (Å)	No.
4124.08	234	4997.22	233	5535.35	232	6504.61	183
4133.67	234	5001.13	114	5535.38	232	6522.33	183
4145.77	234	5001.47	114	5540.06	232	6532.56	183
4227.74	159	5002.70	74	5543.47	232	6544.17	183
4236.36	119	5005.15	114	5551.92	232	6545.54	183
4352.22	193	5005.30	233	5552.68	232	6554.47	183
4374.99	105	5007.33	127	5565.26	232	6560.20	207
4376.63	193	5010.62	74	5588.59	186	6561.76	207
4379.58	105	5011.31	233	5631.72	241	6564.20	199
4393.85	118	5012.04	233	5666.63	73	6582.60	207
4447.03	104	5016.38	114	5676.02	73	6595.67	207
4459.94	117	5022.07	228	5679.56	73	6600.94	239
4465.53	117	5023.05	233	5686.21	73	6610.56	156
4467.01	158	5025.66	114	5710.77	73	6613.62	207
4475.89	117	5028.78	228	5730.66	73	6629.79	178
4477.68	117	5040.71	114	5747.30	86	6634.79	207
4488.09	117	5045.10	74	5767.45	86	6785.58	182
4507.56	117	5046.54	228	5893.15	250	6796.65	182
4564.76	103	5073.59	87	5897.25	250	6809.98	191
4601.48	75	5104.45	170	5899.83	250	6826.23	155
4607.15	75	5114.28	144	5910.33	250	6834.09	191
4613.87	75	5168.05	255	5914.07	250	6847.23	191
4621.39	75	5170.16	255	5927.22	250	6857.03	256
4630.54	75	5171.27	255	5927.81	142	6869.58	256
4643.09	75	5171.47	255	5931.78	142	6887.83	256
4654.53	88	5172.34	251	5940.24	142	6941.75	190
4667.21	88	5172.97	251	5941.65	142	6966.81	190
4674.91	88	5173.39	251	5952.39	142	6973.56	244
4694.27	253	5174.46	255	5954.28	200	6975.63	190
4695.90	253	5175.89	251	5960.91	142	7000.94	190
4697.64	253	5176.57	255	5990.67	176	7013.98	190
4698.55	253	5177.06	251	6061.57	192	7014.73	190
4700.03	253	5179.34	255	6065.00	141	7138.85	189
4702.50	253	5179.52	251	6114.60	175	7156.75	214
4704.25	253	5180.36	251	6136.89	175	7181.54	214
4706.40	253	5183.20	255	6150.75	175	7188.18	189
4709.44	128	5184.96	251	6167.75	175	7197.10	214
4709.58	253	5185.09	251	6170.16	175	7214.71	189
4712.07	253	5186.21	255	6173.31	175	7215.05	189
4718.38	253	5190.38	251	6174.34	245	7241.78	189
4721.58	253	5191.96	251	6209.42	185	7256.54	189
4774.24	116	5199.50	251	6210.36	180	7450.95	198
4776.22	219	5240.86	218	6218.69	185	7528.12	154
4779.72	116	5306.77	177	6242.41	194	7538.57	188
4781.19	116	5313.42	254	6284.32	157	7545.36	154
4788.14	116	5320.20	254	6285.69	140	7554.56	188
4793.65	116	5320.96	254	6286.11	140	7649.71	197
4803.29	116	5327.76	254	6309.25	140	7762.24	153
4810.30	116	5338.73	254	6318.80	184	7897.63	217
4812.08	252	5340.21	254	6328.40	184	7906.08	196
4815.62	252	5351.23	254	6330.80	240	7938.18	196
4821.80	252	5383.72	126	6340.58	184	8089.08	152
4822.56	252	5390.69	126	6346.86	184	8128.14	152
4828.77	252	5452.07	143	6356.54	184	8296.21	195
4837.78	252	5454.22	143	6357.58	184	8438.74	168
4838.85	252	5457.39	169	6366.79	72	8566.62	223
4847.90	252	5462.58	143	6379.62	72	8604.33	223
4860.17	115	5475.29	181	6384.32	174	8660.45	223
4860.61	252	5478.09	143	6399.15	179	8676.09	223
4895.12	67	5480.05	143	6433.44	98	8687.43	134
4987.38	127	5493.23	125	6435.61	72	8694.90	223
4991.24	233	5495.65	143	6457.68	98	8698.99	223
4994.36	233	5526.23	232	6482.05	85	8831.75	249
4994.37	127	5530.24	232	6491.80	183	8855.30	249

List of tabulated lines — Continued

Wavelength (Å)	No.	Wavelength (Å)	No.	Wavelength (Å)	No.	Wavenumber (cm^{-1})	No.
8893.29	249	13284.1	221	15154.4	203	997.92	247
8930.03	227	13300.9	226	15215.6	203	1003.00	247
8986.18	227	13328.3	221	15269.0	203	1028.02	247
9010.42	227	13343.1	226	15433.9	203	1074.19	247
9032.04	227	13344.8	212	16192.8	202	1076.24	247
9069.51	227	13375.0	221	16194.7	202	1122.41	247
9089.48	227	13425.7	133	16256.2	202	1509.52	246
9217.06	231	13425.9	226	16329.9	202	1513.41	246
9325.77	231	13436.6	221	16514.3	202	1525.99	246
9365.73	231	13501.3	221	16655.0	202	1539.62	246
9399.64	167	13810.7	230	16701.1	237	1561.63	246
9688.98	206	13858.6	230	17376.1	236	1587.84	246
10546.7	238	13947.9	230	17604.0	211		
10907.5	216	14125.7	225	17755.9	201		
10909.1	166	14178.9	225	17920.9	201		
11044.8	213	14195.2	225	18755.4	210		
11414.9	205	14287.1	204	18849.3	210		
12400.0	222	14321.0	225	20375.3	209		
12438.6	222	14328.9	225	20589.9	209		
12510.4	222	14337.7	220	21849.2	235		
12518.2	222	14358.6	220	22588.0	242		
12591.0	222	14364.9	220	23185.1	208		
12624.5	243	14393.7	204	38382.0	248		
12745.3	222	14471.2	220	39105.9	248		
12891.2	215	14474.0	225	39856.1	248		
13173.5	226	14538.1	220	40340.2	248		
13176.1	226	14645.6	204	41098.6	248		
13219.7	226	14675.3	220	41140.7	248		
13238.0	221	14995.4	203	46393.2	102		
13265.3	221	15036.2	203				

As for so many of these tabulated spectra, we have selected as our main source the Opacity Project (OP), which is reviewed in the general introduction. The numerical calculations for this carbon-like ion have been performed by Luo and Pradhan.[1]

Another fairly large calculation which includes an extensive treatment of configuration interaction was carried out by Bell *et al.*[2] with the CIV 3 code—an approach fundamentally different from OP. Comparisons could be made for about 75 multiplets which show that the results of the two theoretical methods agree within ±10% for about 80% of the transitions; however, for eight multiplets the disagreements exceed 40%. Figure 1 shows a graphical comparison of these data. For a few prominent multiplets, additional sophisticated calculations are available, and their results are also included in this table: (a) new calculations with the OP package by Bell and Berrington,[3] where the number of interacting configurations for the target ion has been enlarged compared to the original OP calculations, usually improving the agreement between "length" and "velocity" data; (b) multi-configuration Hartree Fock (MCHF) calculations in intermediate coupling by Fischer and Saha[4] for individual lines of the $2s^22p^2 - 2s2p^3$ transition array and by Ellis[9] for lines of the $2s^22p^2 - 2s^22p3s$ transition array; and (c) superposition-of- configurations calculations by Weiss.[10] When the results overlap, the spread in the data is typically about ±10%, and sometimes much better, as shown in Table 1.

For the two intercombination lines from the quasi-metastable $2s2p^3$ $^5S°$ level to the ground state, accurate lifetime measurements with ion traps have been carried out by Calamai and Johnson,[5] and Johnson *et al.*,[6] and their uncertainties were estimated to be ±6% and ±10%, respectively. We used the mean value of the two results, which agree within 6%, to re-normalize the individual line strengths calculated by Fischer and Saha[4] and Hibbert and Bates,[7] and measured in emission by Musielok *et al.*[8] with a wall-stabilized arc. The two calculations and the measurement agree within 10%, and again we have applied the mean value.

The emission experiment by Musielok *et al.*[8] has produced many relative line strengths within multiplets, as well as A-values for a number of unusually strong intersystem lines. Their intersystem line results and Weiss' calculated data[10] are in very good agreement and the mean values have been tabulated. The unusual strength of the intersystem lines arises from a near coincidence for the energy levels of the $2p3s$ ^1P and ^3P states, which occurs only for N II, but not for the neighboring isoelectronic species. The relative strength data by Musielok *et al.*[8] for the lines of some $3s$-$3p$ multiplets show only small deviations from LS-coupling, i.e., less than ±10%, which is also borne out by Weiss' intermediate coupling calculations. For some $3p$-$3d$ multiplets, Musielok *et al.*[11] did not find any significant deviations from LS-coupling either.

The selected calculations[1–4,9,10] may be compared to several advanced lifetime experiments, performed with the beam-foil laser technique, which allows selective excitation of the desired atomic levels. Lifetimes of eight N II levels have been precisely measured by Baudinet-Robinet *et al.*,[12] Dumont *et al.*[13] and Bastin *et al.*,[14] with estimated errors ranging from 1.6% to 4.9%. The agreement with the theoretical results is indeed very good, as shown in Fig. 2, yielding an average value of 0.965 for the ratio $\tau_{expt}-\tau_{calc}$.

References

[1]D. Luo and A. K. Pradhan, J. Phys. B **22**, 3377 (1989).
[2]K. L. Bell, A. Hibbert and R. P. Stafford, Phys. Scr. **52**, 240 (1995).
[3]K. L. Bell and K. A. Berrington, J. Phys. B **24**, 933 (1991).
[4]C. Froese Fischer and H. P. Saha, Phys. Scr. **32**, 181 (1985).
[5]A. G. Calamai and C. E. Johnson, Phys. Rev. A **44**, 218 (1991).
[6]B. C. Johnson, P. L. Smith, and W. H. Parkinson (private communication, cited in Ref. 5).
[7]A. Hibbert and D. R. Bates, Planet. Space Sci. **29**, 263 (1981).
[8]J. Musielok, J. M. Bridges, S. Djurović and W. L. Wiese (to be published).
[9]D. G. Ellis, Phys. Rev. A **47**, 161 (1993).
[10]A. W. Weiss (to be published).
[11]J. Musielok, W. L. Wiese, S. Glenzer and H.-J. Kunze (to be published).
[12]Y. Baudinet-Robinet, H. P. Garnir, P.-D. Dumont, and J. Résimont, Phys. Rev. A **42**, 1080 (1990).
[13]P.-D. Dumont, H.-P. Garnir, and Y. Baudinet-Robinet, Z. Phys. D **21**, 209 (1991).
[14]T. Bastin, Y. Baudinet-Robinet, H. P. Garnir, P. D. Dumont, Z. Phys. D **24**, 343 (1992).

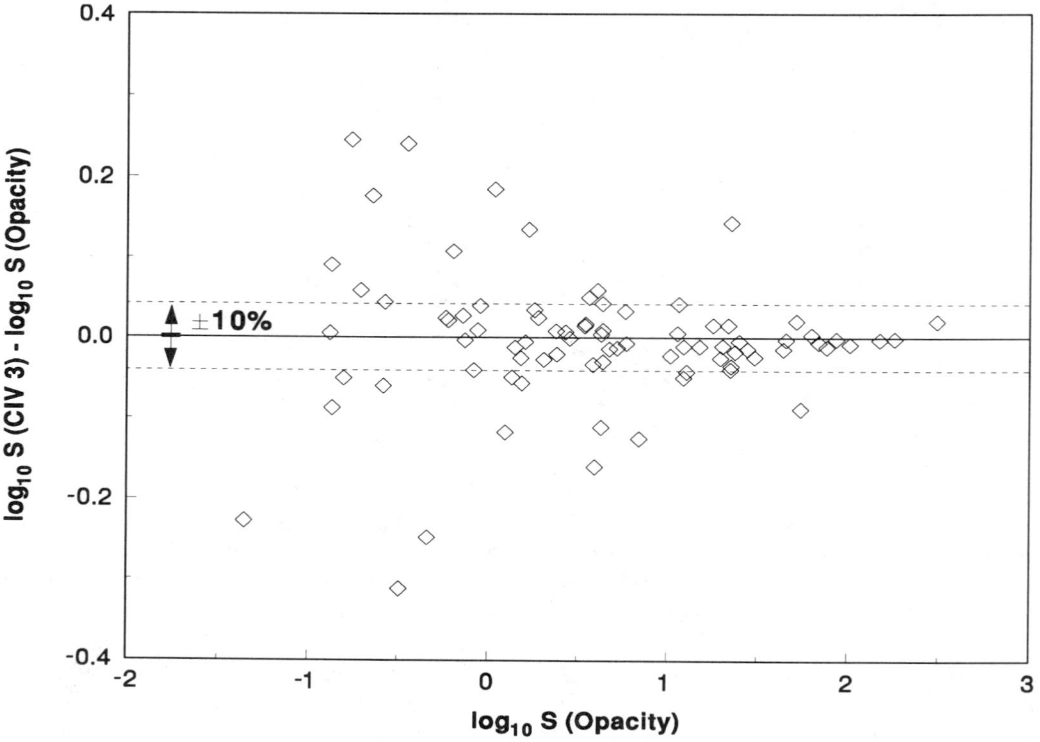

FIG. 1. Comparison of the OP multiplet data by Luo and Pradhan[1] with the results of the CIV 3 calculations by Bell *et al.*[2] The OP/CIV 3 multiplet ratios are plotted on a logarithmic scale versus log *S* of the OP data. The broken lines indicate a band of deviations of ±10% around a perfect ratio of 1.00. The figure shows that the scatter in the ratios increases significantly for the weaker multiplets.

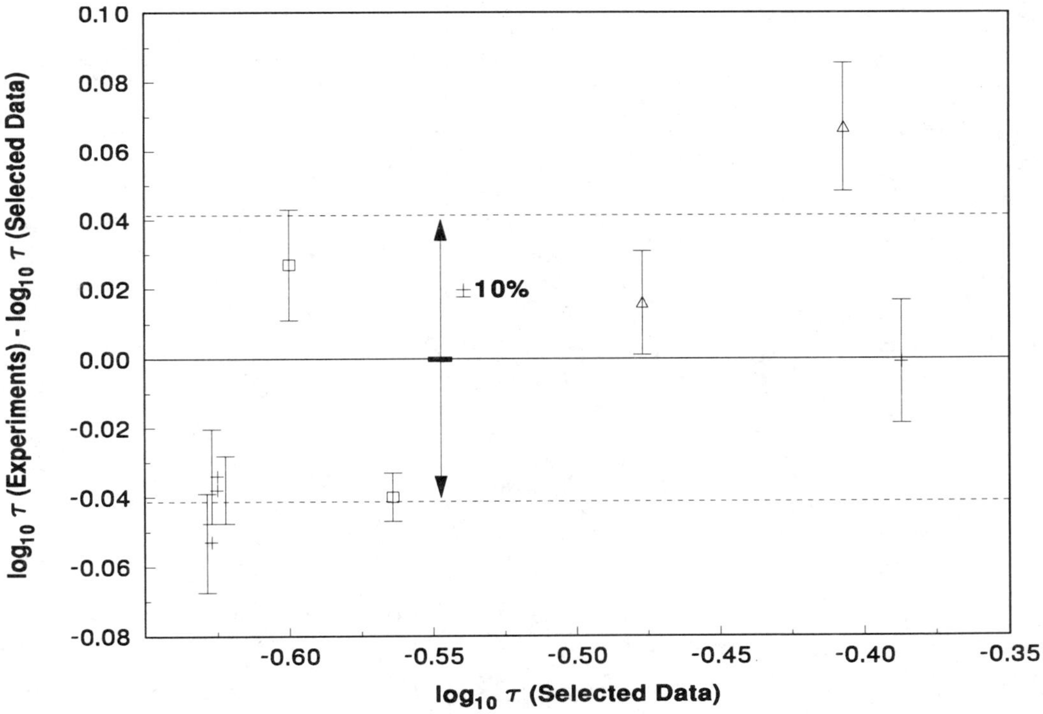

FIG. 2. Comparison of the lifetime data by Baudinet-Robinet *et al.*[12] (squares), Dumont *et al.*[13] (crosses), and Bastin *et al.*[14] (triangles) with the selected theoretical data (all in *ns*). The experiment/theory ratios are plotted on a logarithmic scale in the same manner as in Fig. 1. The uncertainty estimates given in the literature for the experimental lifetime data are indicated by the error flags.

TABLE 1. Comparison of theoretical data for N II multiplets when more than two data sources are available. Given are the multiplet line strengths (in atomic units).

Vacuum Wavelength Multiplet	Luo & Pradhan[1] (Å)	Bell *et al.*[2]	Bell & Berrington[3]	Froese Fischer & Saha[4]	Ellis[9]	Weiss[10]	
$2s^22p^2\ ^3P - 2s2p^3\ ^3D^o$	1085.1	3.49	3.62	3.86	3.52	—	—
$2s^22p^2\ ^3P - 2s2p^3\ ^3P^o$	916.35	4.35	4.44	4.67	4.39	—	—
$2p^2\ ^3P - 2p3s\ ^3P^o$	671.49	1.52	1.43	1.19	—	1.41	1.27
$2s^22p^2\ ^3P - 2s2p^3\ ^3S^o$	645.00	4.27	4.31	4.30	4.41	—	—
$2s^22p^2\ ^1D - 2s2p^3\ ^1D^o$	775.97	3.46	3.58	—	3.63	—	—
$2p^2\ ^1D - 2p3s\ ^1P^o$	746.98	2.39	2.43	—	—	2.19	2.30
$2s^22p^2\ ^1D - 2s2p^3\ ^1P^o$	660.29	1.55	1.35	—	1.82	—	—
$2s^22p^2\ ^1S - 2s2p^3\ ^1P^o$	745.84	0.723	0.769	—	0.814	—	—
$2p3s\ ^3P^o - 2p3p\ ^3D$	5680.9	69.1	68.1	70.8	—	—	69.3
$2p3s\ ^3P^o - 2p3p\ ^3S$	5030.2	12.3	11.9	—	—	—	12.3
$2p3s\ ^3P^o - 2p3p\ ^3P$	4624.5	45.4	45.1	41.0	—	—	45.8
$2p3s\ ^1P^o - 2p3p\ ^1P$	6483.8	12.8	11.6	—	—	—	12.1

N II: Allowed Transitions

No.	Transition Array	Mult.	λ_{air} (Å)	λ_{vac} (Å) or σ (cm^{-1})*	E_i (cm^{-1})	E_k (cm^{-1})	g_i–g_k	A_{ki} (10^8 s^{-1})	f_{ik}	S (at. u.)	log gf	Acc.	Source
1.	$2s^22p^2$– $2s2p^3$	^{3}P–^{5}S°											
			2139.01	2139.68	48.7	46784.6	3–5	5.49–07	6.28–08	1.33–06	–6.725	B–	4,5,6,7,8
			2142.77	2143.45	130.8	46784.6	5–5	1.27–06	8.75–08	3.09–06	–6.359	B–	4,5,6,7,8
2.		^{3}P–^{3}D°		*1085.1*	*89*	*92244*	9–15	3.88+00	1.14–01	3.67+00	0.011	B+	2,3,4
				1085.70	130.8	92237.2	5–7	3.87+00	9.57–02	1.71+00	–0.320	B+	2*n*,4*n*
				1084.58	48.7	92250.3	3–5	2.93+00	8.63–02	9.24–01	–0.587	B+	2*n*,4*n*
				1083.99	0.0	92251.8	1–3	2.18+00	1.15–01	4.11–01	–0.939	B+	2*n*,4*n*
				1085.55	130.8	92250.3	5–5	9.47–01	1.67–02	2.99–01	–1.078	B+	2*n*,4*n*
				1084.56	48.7	92251.8	3–3	1.61+00	2.83–02	3.03–01	–1.071	B+	2*n*,4*n*
				1085.53	130.8	92251.8	5–3	1.04–01	1.10–03	1.96–02	–2.260	B	2*n*,4*n*
3.		^{3}P–^{3}P°		*916.35*	*89*	*109218*	9–9	1.32+01	1.66–01	4.50+00	0.174	A–	2,3,4
				916.701	130.8	109217.6	5–5	9.90+00	1.25–01	1.88+00	–0.205	B+	2*n*,4*n*
				916.020	48.7	109216.6	3–3	3.33+00	4.19–02	3.79–01	–0.901	B+	2*n*,4*n*
				916.710	130.8	109216.6	5–3	5.46+00	4.13–02	6.23–01	–0.685	B+	2*n*,4*n*
				915.962	48.7	109223.5	3–1	1.32+01	5.53–02	5.00–01	–0.780	B+	2*n*,4*n*
				916.012	48.7	109217.6	3–5	3.26+00	6.83–02	6.18–01	–0.689	B+	2*n*,4*n*
				915.612	0.0	109216.6	1–3	4.38+00	1.65–01	4.98–01	–0.782	B+	2*n*,4*n*
4.		^{3}P–^{1}D°											
				694.169	130.8	144187.94	5–5	2.96–04	2.14–06	2.44–05	–4.972	C	4
				693.774	48.7	144187.94	3–5	5.71–05	6.87–07	4.71–06	–5.686	C	4
5.	$2p^2$– $2p3s$	^{3}P–^{3}P°		*671.49*	*89*	*149012.4*	9–9	9.64+00	6.52–02	1.30+00	–0.232	B	2,3,10
				671.386	130.8	149076.52	5–5	7.40+00	5.00–02	5.53–01	–0.602	B	2*n*,10*n*
				671.630	48.7	148940.17	3–3	2.27+00	1.53–02	1.02–01	–1.337	B	2*n*,10*n*
				672.001	130.8	148940.17	5–3	3.87+00	1.57–02	1.74–01	–1.104	B	2*n*,10*n*
				671.773	48.7	148908.59	3–1	9.85+00	2.22–02	1.47–01	–1.176	B	2*n*,10*n*
				671.016	48.7	149076.52	3–5	2.47+00	2.78–02	1.84–01	–1.079	B	2*n*,10*n*
				671.411	0.0	148940.17	1–3	3.04+00	6.17–02	1.36–01	–1.210	B	2*n*,10*n*
6.		^{3}P–^{1}P°											
				670.884	130.8	149187.80	5–3	3.31–01	1.34–03	1.48–02	–2.174	D	10
				670.515	48.7	149187.80	3–3	2.49–01	1.68–03	1.11–02	–2.299	D	10
				670.296	0.0	149187.80	1–3	3.18–01	6.43–03	1.42–02	–2.191	D	10
7.	$2s^22p^2$– $2s2p^3$	^{3}P–^{3}S°		*645.00*	*89*	155126.73	9–3	1.09+02	2.27–01	4.34+00	0.310	B+	2,3,4
				645.178	130.8	155126.73	5–3	6.07+01	2.27–01	2.41+00	0.055	B+	2*n*,4*n*
				644.837	48.7	155126.73	3–3	3.64+01	2.27–01	1.45+00	–0.167	B+	2*n*,4*n*
				644.634	0.0	155126.73	1–3	1.21+01	2.27–01	4.82–01	–0.644	B+	2*n*,4*n*
8.		^{3}P–^{1}P°											
				600.115	130.8	166765.66	5–3	9.96–05	3.23–07	3.19–06	–5.792	C	4
				599.819	48.7	166765.66	3–3	1.12–03	6.03–06	3.57–05	–4.743	C	4
				599.644	0	166765.66	1–3	2.11–05	3.41–07	6.74–07	–6.467	C	4
9.	$2p^2$– $2p3d$	^{3}P–^{3}F°											
				536.365	130.8	186570.98	5–7	1.25–01	7.53–04	6.65–03	–2.424	C	2
				536.300	48.7	186511.58	3–5	5.38–02	3.86–04	2.05–03	–2.936	C	2
				536.536	130.8	186511.58	5–5	1.96–02	8.46–05	7.47–04	–3.374	D	2
10.		^{3}P–^{1}D°											
				534.637	48.7	187091.37	3–5	2.18–01	1.56–03	8.22–03	–2.331	D	2

N II: Allowed Transitions — Continued

No.	Transition Array	Mult.	λ_{air} (Å)	λ_{vac} (Å) or σ (cm^{-1})*	E_i (cm^{-1})	E_k (cm^{-1})	g_i–g_k	A_{ki} (10^8 s^{-1})	f_{ik}	S (at. u.)	log gf	Acc.	Source
11.		^{3}P–^{3}D°		*533.67*	*89*	– *187470.9*	9–15	4.13+01	2.94–01	4.65+00	0.423	B+	1,2
				533.729	130.8	– 187491.90	5–7	4.13+01	2.47–01	2.17+00	0.092	B+	2*n*
				533.581	48.7	– 187461.56	3–5	3.20+01	2.28–01	1.20+00	–0.166	B+	2*n*
				533.511	0.0	– 187437.56	1–3	2.39+01	3.06–01	5.37–01	–0.514	B+	2*n*
				533.815	130.8	– 187461.56	5–5	9.19+00	3.93–02	3.45–01	–0.707	B+	2*n*
				533.650	48.7	– 187437.56	3–3	1.66+01	7.09–02	3.74–01	–0.672	B+	2*n*
				533.884	130.8	– 187437.56	5–3	9.43–01	2.42–03	2.12–02	–1.918	B	2*n*
12.		^{3}P–^{3}P°		*529.68*	*89*	– *188883.5*	9–9	2.43+01	1.02–01	1.60+00	–0.036	B+	1,2
				529.867	130.8	– 188857.37	5–5	1.94+01	8.15–02	7.11–01	–0.390	B+	2*n*
				529.491	48.7	– 188909.17	3–3	6.75+00	2.84–02	1.48–01	–1.070	B+	2*n*
				529.722	130.8	– 188909.17	5–3	1.03+01	2.61–02	2.28–01	–0.884	B+	2*n*
				529.413	48.7	– 188937.24	3–1	2.43+01	3.41–02	1.78–01	–0.990	B+	2*n*
				529.637	48.7	– 188857.37	3–5	4.92+00	3.45–02	1.80–01	–0.986	B+	2*n*
				529.355	0.0	– 188909.17	1–3	7.23+00	9.11–02	1.59–01	–1.041	B+	2*n*
13.		^{3}P–^{1}F°											
				528.529	130.8	– 189335.16	5–7	1.52–02	8.90–05	7.74–04	–3.352	D	2
14.		^{3}P–^{1}P°											
				525.983	0.0	– 190120.24	1–3	3.28–02	4.08–04	7.07–04	–3.389	D	2
15.	$2p^2$–$2p4s$	^{3}P–^{3}P°		*508.74*	*89*	– *196652.7*	9–9	2.54+00	9.87–03	1.49–01	–1.052	B	1,2
				508.697	130.8	– 196711.54	5–5	1.91+00	7.43–03	6.22–02	–1.430	B	2*n*
				508.794	48.7	– 196592.07	3–3	6.26–01	2.43–03	1.22–02	–2.137	B–	2*n*
				509.006	130.8	– 196592.07	5–3	1.05+00	2.45–03	2.05–02	–1.912	B–	2*n*
				508.928	48.7	– 196540.23	3–1	2.51+00	3.25–03	1.64–02	–2.011	B–	2*n*
				508.484	48.7	– 196711.54	3–5	6.47–01	4.18–03	2.10–02	–1.902	B–	2*n*
				508.668	0.0	– 196592.07	1–3	8.45–01	9.83–03	1.65–02	–2.007	B–	2*n*
16.	$2p^2$–$2p4d$	^{3}P–^{3}D°		*475.76*	*89*	– *210277.4*	9–15	2.10+01	1.19–01	1.67+00	0.029	B	1
				475.803	130.8	– 210301.68	5–7	2.10+01	9.98–02	7.81–01	–0.302	C+	*LS*
				475.698	48.7	– 210266.04	3–5	1.58+01	8.91–02	4.19–01	–0.573	C+	*LS*
				475.647	0.0	– 210239.83	1–3	1.17+01	1.19–01	1.86–01	–0.925	C+	*LS*
				475.884	130.8	– 210266.04	5–5	5.25+00	1.78–02	1.40–01	–1.050	C+	*LS*
				475.757	48.7	– 210239.83	3–3	8.75+00	2.97–02	1.40–01	–1.050	C+	*LS*
				475.943	130.8	– 210239.83	5–3	5.83–01	1.19–03	9.30–03	–2.226	C–	*LS*
17.		^{3}P–^{3}P°		*474.74*	*89*	– *210728.6*	9–9	1.29+01	4.35–02	6.13–01	–0.407	B	1
				474.891	130.8	– 210705.26	5–5	9.66+00	3.26–02	2.55–01	–0.787	C+	*LS*
				474.602	48.7	– 210751.43	3–3	3.22+00	1.09–02	5.10–02	–1.486	C+	*LS*
				474.787	130.8	– 210751.43	5–3	5.37+00	1.09–02	8.51–02	–1.264	C+	*LS*
				474.546	48.7	– 210776.45	3–1	1.29+01	1.45–02	6.81–02	–1.361	C+	*LS*
				474.706	48.7	– 210705.26	3–5	3.22+00	1.81–02	8.51–02	–1.264	C+	*LS*
				474.493	0.0	– 210751.43	1–3	4.30+00	4.36–02	6.81–02	–1.361	C+	*LS*
18.	$2s^22p^2$–$2s2p^23p$	^{3}P–^{3}D°		*437.32*	*89*	– *228752.3*	9–15	7.97+00	3.81–02	4.93–01	–0.465	B	1
				437.329	130.8	– 228791.83	5–7	7.97+00	3.20–02	2.30–01	–0.796	C+	*LS*
				437.286	48.7	– 228731.81	3–5	5.98+00	2.86–02	1.23–01	–1.067	C+	*LS*
				437.265	0.0	– 228694.30	1–3	4.43+00	3.81–02	5.48–02	–1.419	C+	*LS*
				437.443	130.8	– 228731.81	5–5	1.99+00	5.71–03	4.11–02	–1.544	C+	*LS*
				437.358	48.7	– 228694.30	3–3	3.32+00	9.52–03	4.11–02	–1.544	C+	*LS*
				437.515	130.8	– 228694.30	5–3	2.21–01	3.81–04	2.74–03	–2.721	D	*LS*

N II: Allowed Transitions — Continued

No.	Transition Array	Mult.	λ_{air} (Å)	λ_{vac} (Å) or σ (cm^{-1})*	E_i (cm^{-1})	E_k (cm^{-1})	g_i-g_k	A_{ki} (10^8 s^{-1})	f_{ik}	S (at. u.)	log gf	Acc.	Source
19.	$2s^22p^2-$ $2s2p^3$	$^1D-^5S^o$											
			3176.87	3177.79	15316.2	– 46784.6	5–5	9.33−12	1.41−12	7.38−11	−11.151	D	4
20.		$^1D-^3D^o$											
				1300.04	15316.2	– 92237.2	5–7	3.17−05	1.13−06	2.41−05	−5.250	C	4
				1299.81	15316.2	– 92250.3	5–5	6.23−06	1.58−07	3.38−06	−6.103	C	4
				1299.79	15316.2	– 92251.8	5–3	5.20−06	7.91−08	1.69−06	−6.403	C	4
21.		$^1D-^3P^o$											
				1064.95	15316.2	– 109217.6	5–5	1.90−05	3.23−07	5.67−06	−5.792	C	4
				1064.96	15316.2	– 109216.6	5–3	4.98−05	5.08−07	8.91−06	−5.595	C	4
22.		$^1D-^1D^o$		775.965	15316.2	– 144187.94	5–5	3.08+01	2.78−01	3.56+00	0.144	B+	1,2,4
23.	$2p^2-$ $2p3s$	$^1D-^3P^o$											
				747.606	15316.2	– 149076.52	5–5	2.52−04	2.11−06	2.60−05	−4.976	E	10
				748.369	15316.2	– 148940.17	5–3	3.83+00	1.93−02	2.38−01	−1.016	C	10
24.		$^1D-^1P^o$		746.984	15316.2	– 149187.80	5–3	3.85+01	1.93−01	2.38+00	−0.015	B+	1,2,10
25.	$2s^22p^2-$ $2s2p^3$	$^1D-^3S^o$											
				715.254	15316.2	– 155126.73	5–3	6.40−04	2.95−06	3.47−05	−4.832	C	4
26.		$^1D-^1P^o$		660.286	15316.2	– 166765.66	5–3	3.69+01	1.45−01	1.57+00	−0.140	C+	1,2,4
27.	$2p^2-$ $2p3d$	$^1D-^3F^o$											
				583.925	15316.2	– 186570.98	5–7	1.79−02	1.28−04	1.23−03	−3.193	D	2
				584.128	15316.2	– 186511.58	5–5	2.06−01	1.05−03	1.01−02	−2.278	D	2
28.		$^1D-^1D^o$		582.156	15316.2	– 187091.37	5–5	2.85+01	1.45−01	1.39+00	−0.140	B	1,2
29.		$^1D-^3D^o$											
				580.802	15316.2	– 187491.90	5–7	1.09−02	7.70−05	7.36−04	−3.415	D	2
				580.904	15316.2	– 187461.56	5–5	1.24−01	6.27−04	6.00−03	−2.504	D	2
30.		$^1D-^3P^o$											
				576.232	15316.2	– 188857.37	5–5	2.86−02	1.42−04	1.35−03	−3.147	D	2
31.		$^1D-^1F^o$		574.650	15316.2	– 189335.16	5–7	3.60+01	2.49−01	2.36+00	0.096	B	1,2
32.		$^1D-^1P^o$		572.069	15316.2	– 190120.24	5–3	4.01−02	1.18−04	1.11−03	−3.229	C	2
33.	$2p^2-$ $2p4s$	$^1D-^1P^o$		547.818	15316.2	– 197858.69	5–3	2.16+00	5.83−03	5.26−02	−1.535	C	2
34.	$2p^2-$ $2p4d$	$^1D-^1D^o$		513.849	15316.2	– 209925.76	5–5	1.24+01	4.90−02	4.14−01	−0.611	B	1

N II: Allowed Transitions — Continued

No.	Transition Array	Mult.	λ_{air} (Å)	λ_{vac} (Å) or σ (cm^{-1})*	E_i (cm^{-1})	E_k (cm^{-1})	g_i-g_k	A_{ki} (10^8 s^{-1})	f_{ik}	S (at. u.)	log gf	Acc.	Source
35.		^{1}D–^{1}F°		510.758	15316.2	– 211103.63	5–7	1.87+01	1.03–01	8.63–01	–0.290	B	1
36.	$2p^2$– $2p5d$	^{1}D–^{1}D°		487.379	15316.2	– 220495.36	5–5	6.32+00	2.25–02	1.81–01	–0.949	B	1
37.		^{1}D–^{1}F°		485.849	15316.2	– 221141.61	5–7	9.79+00	4.85–02	3.88–01	–0.615	B	1
38.	$2s^22p^2$– $2s2p^3$	^{1}S–^{3}D°											
				1678.89	32688.8	– 92251.8	1–3	3.36–06	4.26–07	2.35–06	–6.371	C	4
39.		^{1}S–^{3}P°											
				1306.71	32688.8	– 109216.6	1–3	1.98–05	1.52–06	6.54–06	–5.818	C	4
40.	$2p^2$– $2p3s$	^{1}S–^{3}P°											
				860.205	32688.8	– 148940.17	1–3	5.07–02	1.69–03	4.78–03	–2.773	D	10
41.		^{1}S–^{1}P°		858.376	32688.8	– 149187.80	1–3	4.15–01	1.38–02	3.89–02	–1.861	D	1,2,10
42.	$2s^22p^2$– $2s2p^3$	^{1}S–^{3}S°											
				816.740	32688.8	– 155126.73	1–3	9.87–05	2.96–06	7.96–06	–5.529	C	4
43.		^{1}S–^{1}P°		745.841	32688.8	– 166765.66	1–3	1.25+01	3.13–01	7.68–01	–0.505	B	1,2,4
44.	$2p^2$– $2p3d$	^{1}S–^{3}D°											
				646.209	32688.8	– 187437.56	1–3	1.12–02	2.10–04	4.46–04	–3.679	D	2
45.		^{1}S–^{3}P°											
				640.121	32688.8	– 188909.17	1–3	1.75–02	3.22–04	6.78–04	–3.492	D	2
46.		^{1}S–^{1}P°		635.197	32688.8	– 190120.24	1–3	2.33+01	4.23–01	8.85–01	–0.374	B	1,2
47.	$2p^2$– $2p4s$	^{1}S–^{3}P°											
				610.116	32688.8	– 196592.07	1–3	1.34–02	2.24–04	4.49–04	–3.650	D	2
48.		^{1}S–^{1}P°		605.437	32688.8	– 197858.69	1–3	3.77+00	6.21–02	1.24–01	–1.207	C+	1,2
49.	$2p^2$– $2p4d$	^{1}S–^{1}P°		559.762	32688.8	– 211336.16	1–3	1.14+01	1.60–01	2.95–01	–0.795	B	1
50.	$2p^2$– $2p5d$	^{1}S–^{1}P°		530.343	32688.8	– 221246.17	1–3	6.12+00	7.74–02	1.35–01	–1.111	B	1
51.	$2s2p^3$– $2s2p^2(^4$P$)3s$	^{5}S°–^{5}P		*629.36*	46784.6	– *205675.9*	5–15	1.07+01	1.91–01	1.98+00	–0.019	C+	1,2
				629.167	46784.6	– 205724.81	5–7	1.08+01	8.93–02	9.25–01	–0.350	C+	2n
				629.447	46784.6	– 205654.22	5–5	1.07+01	6.38–02	6.61–01	–0.496	C+	2n
				629.670	46784.6	– 205597.97	5–3	1.07+01	3.82–02	3.96–01	–0.718	C+	2n

N II: Allowed Transitions — Continued

No.	Transition Array	Mult.	λ_{air} (Å)	λ_{vac} (Å) or σ (cm^{-1})*	E_i (cm^{-1})	E_k (cm^{-1})	g_i-g_k	A_{ki} (10^8 s^{-1})	f_{ik}	S (at. u.)	log gf	Acc.	Source
52.	$2s2p^3-$ $2s2p^23d$	$^5S^o-^5P$		*506.09*	46784.6	– *244379.2*	5–15	4.51+01	5.19–01	4.33+00	0.414	B	1
				506.153	46784.6	– 244353.31	5–7	4.51+01	2.42–01	2.02+00	0.083	C+	*LS*
				506.054	46784.6	– 244391.88	5–5	4.51+01	1.73–01	1.44+00	–0.063	C+	*LS*
				505.986	46784.6	– 244418.52	5–3	4.51+01	1.04–01	8.65–01	–0.284	C+	*LS*
53.	$2s2p^3-$ $2s^22p3p$	$^3D^o-^3D$		*1344.6*	*92244*	– *166615.2*	15–15	7.38–02	2.00–03	1.33–01	–1.523	B	1,2
				1343.34	92237.2	– 166678.64	7–7	6.57–02	1.78–03	5.50–02	–1.905	B	2*n*
				1345.31	92250.3	– 166582.45	5–5	5.04–02	1.37–03	3.03–02	–2.165	B	2*n*
				1346.44	92251.8	– 166521.69	3–3	5.43–02	1.48–03	1.96–02	–2.354	B	2*n*
				1345.08	92237.2	– 166582.45	7–5	1.31–02	2.54–04	7.87–03	–2.750	B–	2*n*
				1346.41	92250.3	– 166521.69	5–3	2.00–02	3.26–04	7.23–03	–2.788	B–	2*n*
				1343.57	92250.3	– 166678.64	5–7	7.76–03	2.94–04	6.50–03	–2.833	B–	2*n*
				1345.34	92251.8	– 166582.45	3–5	1.06–02	4.78–04	6.34–03	–2.844	B–	2*n*
54.		$^3D^o-^3P$		*1275.6*	*92244*	– *170636.4*	15–9	6.67–01	9.77–03	6.15–01	–0.834	B	1,2
				1275.04	92237.2	– 170666.23	7–5	5.60–01	9.74–03	2.86–01	–1.166	B	2*n*
				1276.20	92250.3	– 170607.89	5–3	4.98–01	7.29–03	1.53–01	–1.438	B	2*n*
				1276.80	92251.8	– 170572.61	3–1	6.66–01	5.43–03	6.84–02	–1.788	B	2*n*
				1275.25	92250.3	– 170666.23	5–5	1.02–01	2.49–03	5.22–02	–1.905	B	2*n*
				1276.22	92251.8	– 170607.89	3–3	1.68–01	4.11–03	5.18–02	–1.909	B	2*n*
				1275.28	92251.8	– 170666.23	3–5	6.90–03	2.80–04	3.53–03	–3.075	C+	2*n*
55.	$2s2p^3-$ $2s^22p4p$	$^3D^o-^1P$											
				909.762	92251.8	– 202170.63	3–3	3.82–03	4.74–05	4.26–04	–3.847	D–	2
56.		$^3D^o-^3D$		*904.52*	*92244*	– *202799.9*	15–15	1.66–01	2.03–03	9.07–02	–1.516	D	2
				903.962	92237.2	– 202861.36	7–7	1.35–01	1.66–03	3.45–02	–1.935	D	2
				904.855	92250.3	– 202765.26	5–5	6.62–02	8.12–04	1.21–02	–2.391	D	2
				905.286	92251.8	– 202714.12	3–3	7.00–02	8.60–04	7.69–03	–2.588	D	2
				904.748	92237.2	– 202765.26	7–5	9.78–02	8.58–04	1.79–02	–2.222	D	2
				905.274	92250.3	– 202714.12	5–3	1.01–01	7.46–04	1.11–02	–2.428	D	2
				904.069	92250.3	– 202861.36	5–7	1.65–02	2.83–04	4.21–03	–2.849	D	2
				904.867	92251.8	– 202765.26	3–5	1.75–02	3.57–04	3.19–03	–2.970	D	2
57.		$^3D^o-^3P$		*901.06*	*92244*	– *203224.9*	15–9	4.22+00	3.08–02	1.37+00	–0.335	D	1,2
				900.724	92237.2	– 203258.98	7–5	3.54+00	3.07–02	6.38–01	–0.667	D	2*n*
				901.398	92250.3	– 203189.03	5–3	3.07+00	2.24–02	3.33–01	–0.950	D	2*n*
				901.626	92251.8	– 203162.48	3–1	4.29+00	1.74–02	1.55–01	–1.282	D	2*n*
				900.830	92250.3	– 203258.98	5–5	6.72–01	8.17–03	1.21–01	–1.389	D	2*n*
				901.411	92251.8	– 203189.03	3–3	1.07+00	1.30–02	1.16–01	–1.408	D	2*n*
				900.843	92251.8	– 203258.98	3–5	4.66–02	9.44–04	8.40–03	–2.548	D	2*n*
58.		$^3D^o-^3S$											
				898.575	92250.3	– 203537.66	5–3	1.16–01	8.39–04	1.24–02	–2.377	D	2
				898.587	92251.8	– 203537.66	3–3	3.91–02	4.73–04	4.20–03	–2.848	D	2
59.	$2s2p^3-$ $2s2p^2(^4P)3s$	$^3D^o-^3P$		*836.41*	*92244*	– *211802.9*	15–9	1.46+01	9.17–02	3.79+00	0.139	D+	1,2
				836.187	92237.2	– 211827.67	7–5	1.23+01	9.22–02	1.78+00	–0.190	D+	2*n*
				836.616	92250.3	– 211779.45	5–3	1.09+01	6.88–02	9.47–01	–0.464	D+	2*n*
				836.837	92251.8	– 211749.35	3–1	1.45+01	5.07–02	4.19–01	–0.818	D+	2*n*
				836.279	92250.3	– 211827.67	5–5	2.17+00	2.28–02	3.13–01	–0.944	D+	2*n*
				836.627	92251.8	– 211779.45	3–3	3.61+00	3.79–02	3.13–01	–0.944	D+	2*n*
				836.289	92251.8	– 211827.67	3–5	1.44–01	2.51–03	2.07–02	–2.124	D	2*n*

N II: Allowed Transitions — Continued

No.	Transition Array	Mult.	λ_{air} (Å)	λ_{vac} (Å) or σ (cm^{-1})*	E_i (cm^{-1})	E_k (cm^{-1})	g_i–g_k	A_{ki} ($10^8 s^{-1}$)	f_{ik}	S (at. u.)	log gf	Acc.	Source
60.	$2s2p^3$– $2s^22p3p$	$^3P^o$–^3D		*1742.2*	*109218*	– *166615.2*	9–15	2.42–01	1.83–02	9.46–01	–0.783	B	1,2
				1740.31	109217.6	– 166678.64	5–7	2.42–01	1.54–02	4.41–01	–1.113	B	2n
				1743.20	109216.6	– 166582.45	3–5	1.82–01	1.38–02	2.38–01	–1.382	B	2n
				1745.26	109223.5	– 166521.69	1–3	1.35–01	1.85–02	1.06–01	–1.734	B	2n
				1743.23	109217.6	– 166582.45	5–5	5.93–02	2.70–03	7.75–02	–1.870	B	2n
				1745.05	109216.6	– 166521.69	3–3	9.94–02	4.54–03	7.82–02	–1.866	B	2n
				1745.08	109217.6	– 166521.69	5–3	6.56–03	1.80–04	5.16–03	–3.047	C+	2n
61.		$^3P^o$–^3S		*1675.8*	*109218*	– 168892.21	9–3	8.59–01	1.20–02	5.98–01	–0.965	B	1,2
				1675.75	109217.6	– 168892.21	5–3	4.72–01	1.19–02	3.29–01	–1.224	B	2n
				1675.73	109216.6	– 168892.21	3–3	2.89–01	1.22–02	2.01–01	–1.437	B	2n
				1675.92	109223.5	– 168892.21	1–3	9.71–02	1.23–02	6.77–02	–1.911	B	2n
62.		$^3P^o$–^3P		*1628.2*	*109218*	– *170636.4*	9–9	2.53–02	1.01–03	4.85–02	–2.043	C	2
				1627.38	109217.6	– 170666.23	5–5	1.91–02	7.59–04	2.03–02	–2.421	C	2
				1628.90	109216.6	– 170607.89	3–3	3.89–03	1.55–04	2.49–03	–3.333	C–	2
				1628.92	109217.6	– 170607.89	5–3	1.64–02	3.92–04	1.05–02	–2.708	C	2
				1629.83	109216.6	– 170572.61	3–1	2.56–02	3.40–04	5.47–03	–2.991	C–	2
				1627.35	109216.6	– 170666.23	3–5	5.49–03	3.63–04	5.84–03	–2.962	C–	2
				1629.08	109223.5	– 170607.89	1–3	6.06–03	7.24–04	3.88–03	–3.141	C–	2
63.	$2s2p^3$– $2s^22p4p$	$^3P^o$–^3D		*1068.6*	*109218*	– *202799.9*	9–15	7.53–02	2.15–03	6.80–02	–1.713	C	2
				1067.88	109217.6	– 202861.36	5–7	7.46–02	1.79–03	3.14–02	–2.049	C	2
				1068.96	109216.6	– 202765.26	3–5	4.28–02	1.22–03	1.29–02	–2.436	C	2
				1069.63	109223.5	– 202714.12	1–3	3.14–02	1.62–03	5.70–03	–2.791	C–	2
				1068.97	109217.6	– 202765.26	5–5	3.38–02	5.79–04	1.02–02	–2.538	C	2
				1069.55	109216.6	– 202714.12	3–3	3.69–02	6.32–04	6.68–03	–2.722	C–	2
				1069.56	109217.6	– 202714.12	5–3	6.53–03	6.72–05	1.18–03	–3.474	D+	2
64.		$^3P^o$–^3P		*1063.8*	*109218*	– *203224.9*	9–9	4.45–01	7.56–03	2.38–01	–1.167	D+	1,2
				1063.36	109217.6	– 203258.98	5–5	3.24–01	5.49–03	9.60–02	–1.562	D+	2n
				1064.14	109216.6	– 203189.03	3–3	5.64–02	9.57–04	1.01–02	–2.542	D+	2n
				1064.15	109217.6	– 203189.03	5–3	2.74–01	2.79–03	4.88–02	–1.856	D+	2n
				1064.44	109216.6	– 203162.48	3–1	4.40–01	2.49–03	2.62–02	–2.126	D+	2n
				1063.35	109216.6	– 203258.98	3–5	1.22–01	3.45–03	3.62–02	–1.985	D+	2n
				1064.22	109223.5	– 203189.03	1–3	1.17–01	5.95–03	2.08–02	–2.226	D+	2n
65.		$^3P^o$–^3S		*1060.2*	*109218*	– 203537.66	9–3	9.33–01	5.24–03	1.65–01	–1.326	C	2
				1060.22	109217.6	– 203537.66	5–3	3.95–01	3.99–03	6.96–02	–1.700	C	2
				1060.21	109216.6	– 203537.66	3–3	3.82–01	6.43–03	6.73–02	–1.715	C	2
				1060.29	109223.5	– 203537.66	1–3	1.57–01	7.93–03	2.77–02	–2.101	C	2
66.	$2s2p^3$– $2s2p^2(^4P)3s$	$^3P^o$–^3P		*974.80*	*109218*	– *211802.9*	9–9	1.77+00	2.52–02	7.28–01	–0.644	C	1,2
				974.563	109217.6	– 211827.67	5–5	1.33+00	1.89–02	3.04–01	–1.024	C	2n
				975.012	109216.6	– 211779.45	3–3	4.50–01	6.41–03	6.17–02	–1.716	C	2n
				975.021	109217.6	– 211779.45	5–3	7.27–01	6.22–03	9.98–02	–1.507	C	2n
				975.298	109216.6	– 211749.35	3–1	1.77+00	8.40–03	8.09–02	–1.599	C	2n
				974.554	109216.6	– 211827.67	3–5	4.41–01	1.05–02	1.01–01	–1.503	C	2n
				975.078	109223.5	– 211779.45	1–3	5.92–01	2.53–02	8.13–02	–1.596	C	2n
67.	$2s2p^3$– $2s^22p3p$	$^1D^o$–^1P	4895.12	4896.48	144187.94	– 164610.76	5–3	4.26–02	9.19–03	7.41–01	–1.338	B+	1,2
68.		$^1D^o$–^1D	3329.70	3330.66	144187.94	– 174212.03	5–5	2.69–02	4.47–03	2.45–01	–1.651	B	1,2

N II: Allowed Transitions — Continued

No.	Transition Array	Mult.	λ_{air} (Å)	λ_{vac} (Å) or σ (cm^{-1})*	E_i (cm^{-1})	E_k (cm^{-1})	g_i–g_k	A_{ki} (10^8 s^{-1})	f_{ik}	S (at. u.)	log gf	Acc.	Source
69.	$2s2p^3$–$2s^22p4p$	$^1D^o$–^1P		1724.65	144187.94	– 202170.63	5–3	2.88–01	7.70–03	2.19–01	–1.415	B	2
70.		$^1D^o$–^3D											
				1708.64	144187.94	– 202714.12	5–3	2.24–03	5.89–05	1.66–03	–3.531	D	2
71.		$^1D^o$–^1D		1635.00	144187.94	– 205350.18	5–5	1.57–01	6.27–03	1.69–01	–1.504	B	2
72.	$2p3s$–$2p3p$	$^3P^o$–^1P											
			6435.61	6437.39	149076.52	– 164610.76	5–3	9.62–08	3.59–08	3.80–06	–6.746	E	10
			6379.62	6381.38	148940.17	– 164610.76	3–3	6.11–02	3.73–02	2.35+00	–0.951	C+	8
			6366.79	6368.55	148908.59	– 164610.76	1–3	1.01–04	1.84–04	3.86–03	–3.735	D	10
73.		$^3P^o$–^3D	*5679.3*	*5680.9*	*149012.4*	*– 166615.2*	9–15	5.12–01	4.12–01	6.94+01	0.570	A	2,3,10
			5679.56	5681.13	149076.52	– 166678.64	5–7	5.25–01	3.55–01	3.32+01	0.250	A	2*n*,10*n*
			5666.63	5668.20	148940.17	– 166582.45	3–5	3.74–01	3.00–01	1.68+01	–0.045	A	2*n*,10*n*
			5676.02	5677.59	148908.59	– 166521.69	1–3	2.96–01	4.29–01	8.01+00	–0.368	A	2*n*,10*n*
			5710.77	5712.35	149076.52	– 166582.45	5–5	1.24–01	6.07–02	5.71+00	–0.518	A	2*n*,10*n*
			5686.21	5687.79	148940.17	– 166521.69	3–3	1.94–01	9.43–02	5.29+00	–0.549	A	2*n*,10*n*
			5730.66	5732.25	149076.52	– 166521.69	5–3	1.34–02	3.95–03	3.73–01	–1.704	A	2*n*,10*n*
74.		$^3P^o$–^3S	*5028.8*	*5030.2*	*149012.4*	– 168892.21	9–3	6.45–01	8.16–02	1.22+01	–0.134	A	1,2,10
			5045.10	5046.51	149076.52	– 168892.21	5–3	3.42–01	7.84–02	6.51+00	–0.407	B+	2*n*,10*n*
			5010.62	5012.02	148940.17	– 168892.21	3–3	2.19–01	8.25–02	4.08+00	–0.606	B+	2*n*,10*n*
			5002.70	5004.10	148908.59	– 168892.21	1–3	8.45–02	9.52–02	1.57+00	–1.021	B+	2*n*,10*n*
75.		$^3P^o$–^3P	*4623.2*	*4624.5*	*149012.4*	*– 170636.4*	9–9	1.00+00	3.21–01	4.39+01	0.460	B+	2,3,10
			4630.54	4631.84	149076.52	– 170666.23	5–5	7.72–01	2.48–01	1.89+01	0.094	B+	2*n*,10*n*
			4613.87	4615.16	148940.17	– 170607.89	3–3	2.26–01	7.21–02	3.29+00	–0.665	B+	2*n*,10*n*
			4643.09	4644.39	149076.52	– 170607.89	5–3	4.51–01	8.75–02	6.69+00	–0.359	B+	2*n*,10*n*
			4621.39	4622.69	148940.17	– 170572.61	3–1	9.55–01	1.02–01	4.66+00	–0.514	B+	2*n*,10*n*
			4601.48	4602.77	148940.17	– 170666.23	3–5	2.35–01	1.24–01	5.66+00	–0.428	B+	2*n*,10*n*
			4607.15	4608.44	148908.59	– 170607.89	1–3	3.26–01	3.11–01	4.72+00	–0.507	B+	2*n*,10*n*
76.		$^3P^o$–^1D											
			3977.31	3978.44	149076.52	– 174212.03	5–5	5.53–05	1.31–05	8.60–04	–4.183	D	10
			3955.85	3956.97	148940.17	– 174212.03	3–5	1.31–01	5.13–02	2.00+00	–0.813	B	8,10
77.		$^3P^o$–^1S											
			3408.13	3409.11	148940.17	– 178273.38	3–1	2.19–01	1.27–02	4.28–01	–1.418	B	8,10
78.	$2p3s$–$2p4p$	$^3P^o$–^1P											
				1878.62	148940.17	– 202170.63	3–3	2.37–02	1.26–03	2.33–02	–2.424	D	2
				1877.51	148908.59	– 202170.63	1–3	4.77–04	7.57–05	4.68–04	–4.121	D	2
79.		$^3P^o$–^3D		*1859.2*	*149012.4*	*– 202799.9*	9–15	2.32–01	2.00–02	1.10+00	–0.745	C+	1,2
				1859.26	149076.52	– 202861.36	5–7	2.38–01	1.72–02	5.28–01	–1.065	C+	2*n*
				1857.87	148940.17	– 202765.26	3–5	1.76–01	1.51–02	2.78–01	–1.343	C+	2*n*
				1858.55	148908.59	– 202714.12	1–3	1.39–01	2.15–02	1.32–01	–1.667	C+	2*n*
				1862.59	149076.52	– 202765.26	5–5	5.13–02	2.67–03	8.18–02	–1.875	C+	2*n*
				1859.64	148940.17	– 202714.12	3–3	8.11–02	4.20–03	7.72–02	–1.899	C+	2*n*
				1864.36	149076.52	– 202714.12	5–3	5.45–03	1.70–04	5.23–03	–3.070	C+	2*n*

N II: Allowed Transitions — Continued

No.	Transition Array	Mult.	λ_{air} (Å)	λ_{vac} (Å) or σ (cm^{-1})*	E_i (cm^{-1})	E_k (cm^{-1})	g_i–g_k	A_{ki} (10^8 s^{-1})	f_{ik}	S (at. u.)	log gf	Acc.	Source
80.		$^3P^o-^3P$		*1844.6*	*149012.4*	– *203224.9*	9–9	8.60–02	4.39–03	2.40–01	–1.404	D	1,2
				1845.62	149076.52	– 203258.98	5–5	7.56–02	3.86–03	1.17–01	–1.714	D	2*n*
				1843.36	148940.17	– 203189.03	3–3	4.28–02	2.18–03	3.97–02	–2.184	D	2*n*
				1848.00	149076.52	– 203189.03	5–3	1.55–02	4.75–04	1.45–02	–2.624	D	2*n*
				1844.26	148940.17	– 203162.48	3–1	7.98–02	1.36–03	2.47–02	–2.390	D	2*n*
				1840.98	148940.17	– 203258.98	3–5	1.09–02	9.20–04	1.67–02	–2.559	D	2*n*
				1842.28	148908.59	– 203189.03	1–3	2.90–02	4.42–03	2.68–02	–2.355	D	2*n*
81.		$^3P^o-^3S$		*1834.0*	*149012.4*	– 203537.66	9–3	1.64–01	2.76–03	1.50–01	–1.605	C	1,2
				1836.17	149076.52	– 203537.66	5–3	1.08–01	3.29–03	9.94–02	–1.784	C	2*n*
				1831.59	148940.17	– 203537.66	3–3	4.24–02	2.13–03	3.86–02	–2.194	C	2*n*
				1830.53	148908.59	– 203537.66	1–3	1.31–02	1.97–03	1.19–02	–2.705	C	2*n*
82.		$^3P^o-^1D$											
				1772.74	148940.17	– 205350.18	3–5	3.22–03	2.53–04	4.43–03	–3.119	D	2
83.		$^3P^o-^1S$											
				1725.03	148940.17	– 206910.24	3–1	2.11–02	3.13–04	5.33–03	–3.027	D	2
84.	$2s^22p3s-$ $2s2p^2(^4P)3s$	$^3P^o-^3P$		*1592.6*	*149012.4*	– *211802.9*	9–9	1.56–01	5.94–03	2.80–01	–1.272	B	1,2
				1593.60	149076.52	– 211827.67	5–5	1.19–01	4.51–03	1.18–01	–1.646	B	2*n*
				1591.36	148940.17	– 211779.45	3–3	3.78–02	1.43–03	2.25–02	–2.366	B	2*n*
				1594.82	149076.52	– 211779.45	5–3	6.52–02	1.49–03	3.91–02	–2.128	B	2*n*
				1592.12	148940.17	– 211749.35	3–1	1.49–01	1.89–03	2.97–02	–2.247	B	2*n*
				1590.14	148940.17	– 211827.67	3–5	3.89–02	2.45–03	3.86–02	–2.133	B	2*n*
				1590.56	148908.59	– 211779.45	1–3	5.39–02	6.13–03	3.21–02	–2.213	B	2*n*
85.	$2p3s-$ $2p3p$	$^1P^o-^1P$	6482.05	6483.84	149187.80	– 164610.76	3–3	3.01–01	1.90–01	1.21+01	–0.245	B+	1,2,10
86.		$^1P^o-^3D$											
			5747.30	5748.89	149187.80	– 166582.45	3–5	3.40–02	2.31–02	1.59+00	–1.075	B	8,10
			5767.45	5769.05	149187.80	– 166521.69	3–3	2.44–02	1.22–02	6.94–01	–1.437	B	8,10
87.		$^1P^o-^3S$											
			5073.59	5075.01	149187.80	– 168892.21	3–3	2.59–02	1.00–02	5.01–01	–1.523	B	8,10
88.		$^1P^o-^3P$											
			4654.53	4655.83	149187.80	– 170666.23	3–5	2.43–02	1.32–02	6.05–01	–1.404	C+	8,10
			4667.21	4668.51	149187.80	– 170607.89	3–3	2.99–02	9.77–03	4.50–01	–1.533	C+	8,10
			4674.91	4676.22	149187.80	– 170572.61	3–1	1.05–01	1.014–02	5.28–01	–1.464	C+	8,10
89.		$^1P^o-^1D$	3995.00	3996.13	149187.80	– 174212.03	3–5	1.35+00	5.38–01	2.12+01	0.208	B	1,2,10
90.		$^1P^o-^1S$	3437.14	3438.13	149187.80	– 178273.38	3–1	2.07+00	1.22–01	4.14+00	–0.436	B	1,2,10
91.	$2p3s-$ $2p4p$	$^1P^o-^1P$		1887.40	149187.80	– 202170.63	3–3	3.63–01	1.94–02	3.61–01	–1.236	D	1,2
92.		$^1P^o-^3D$											
				1866.46	149187.80	– 202765.26	3–5	8.64–03	7.52–04	1.39–02	–2.646	D	2
				1868.24	149187.80	– 202714.12	3–3	1.14–02	5.95–04	1.10–02	–2.748	D	2

N II: Allowed Transitions — Continued

No.	Transition Array	Mult.	λ_{air} (Å)	λ_{vac} (Å) or σ (cm^{-1})*	E_i (cm^{-1})	E_k (cm^{-1})	$g_i - g_k$	A_{ki} (10^8 s^{-1})	f_{ik}	S (at. u.)	log gf	Acc.	Source
93.		$^1P^\circ - {}^3P$											
				1851.81	149187.80	203189.03	3–3	3.81–03	1.96–04	3.58–03	–3.231	D	2
				1852.72	149187.80	203162.48	3–1	4.34–03	7.45–05	1.36–03	–3.651	D	2
94.		$^1P^\circ - {}^3S$											
				1839.93	149187.80	203537.66	3–3	3.97–03	2.01–04	3.66–03	–3.219	D	2
95.		$^1P^\circ - {}^1D$		1780.55	149187.80	205350.18	3–5	4.03–02	3.19–03	5.61–02	–2.019	C	2
96.		$^1P^\circ - {}^1S$		1732.43	149187.80	206910.24	3–1	3.91–01	5.86–03	1.00–01	–1.755	C	2
97.	$2p3s - 2s2p^2(^4P)3s$	$^1P^\circ - {}^3P$											
				1596.43	149187.80	211827.67	3–5	1.78–03	1.13–04	1.79–03	–3.469	D	2
				1597.66	149187.80	211779.45	3–3	2.25–03	8.61–05	1.36–03	–3.588	D	2
				1598.43	149187.80	211749.35	3–1	1.02–02	1.30–04	2.05–03	–3.409	D	2
98.	$2s2p^3 - 2s^22p3p$	$^3S^\circ - {}^3P$											
			6433.44	6435.21	155126.73	170666.23	3–5	1.65–04	1.70–04	1.08–02	–3.292	C	2
			6457.68	6459.46	155126.73	170607.89	3–3	1.59–04	9.96–05	6.35–03	–3.525	C	2
99.	$2s2p^3 - 2s^22p4p$	$^3S^\circ - {}^3P$	*2078.4*	*2079.1*	155126.73	*203224.9*	3–9	5.24–02	1.02–02	2.09–01	–1.515	C+	1,2
			2076.95	2077.61	155126.73	203258.98	3–5	5.31–02	5.73–03	1.18–01	–1.765	C+	2*n*
			2079.97	2080.63	155126.73	203189.03	3–3	5.09–02	3.31–03	6.79–02	–2.004	C+	2*n*
			2081.12	2081.78	155126.73	203162.48	3–1	5.27–02	1.14–03	2.35–02	–2.466	C+	2*n*
100.		$^3S^\circ - {}^3S$	2064.99	2065.65	155126.73	203537.66	3–3	1.79–03	1.14–04	2.33–03	–3.465	D	2
101.	$2s2p^3 - 2s2p^2(^4P)3s$	$^3S^\circ - {}^3P$		*1764.4*	155126.73	*211802.9*	3–9	5.25–01	7.36–02	1.28+00	–0.656	C+	1,2
				1763.64	155126.73	211827.67	3–5	5.29–01	4.11–02	7.16–01	–0.909	C+	2*n*
				1765.14	155126.73	211779.45	3–3	5.22–01	2.44–02	4.25–01	–1.136	C+	2*n*
				1766.08	155126.73	211749.35	3–1	5.18–01	8.08–03	1.41–01	–1.616	C+	2*n*
102.	$2s^22p3p - 2s2p^3$	$^1P - {}^1P^\circ$	46393.2	2154.90 cm^{-1}	164610.76	166765.66	3–3	2.62–04	8.47–03	3.88+00	–1.595	C+	1,2
103.	$2p3p - 2p3d$	$^1P - {}^3F^\circ$											
			4564.76	4566.04	164610.76	186511.58	3–5	1.41–02	7.36–03	3.32–01	–1.656	D	2
104.		$^1P - {}^1D^\circ$	4447.03	4448.28	164610.76	187091.37	3–5	1.14+00	5.66–01	2.48+01	0.230	B	1,2
105.		$^1P - {}^3D^\circ$											
			4374.99	4376.21	164610.76	187461.56	3–5	5.55–03	2.66–03	1.15–01	–2.099	D	2
			4379.58	4380.82	164610.76	187437.56	3–3	1.45–03	4.17–04	1.81–02	–2.902	D	2
106.		$^1P - {}^3P^\circ$											
			4123.12	4124.29	164610.76	188857.37	3–5	3.92–04	1.67–04	6.78–03	–3.301	D	2
			4114.33	4115.50	164610.76	188909.17	3–3	1.42–03	3.60–04	1.46–02	–2.967	D	2

N II: Allowed Transitions — Continued

No.	Transition Array	Mult.	λ_{air} (Å)	λ_{vac} (Å) or σ (cm^{-1})*	E_i (cm^{-1})	E_k (cm^{-1})	g_i-g_k	A_{ki} (10^8 s^{-1})	f_{ik}	S (at. u.)	log gf	Acc.	Source
107.		$^1P-^1P^o$	3919.00	3920.11	164610.76	– 190120.24	3–3	6.76–01	1.56–01	6.03+00	−0.331	B	1,2
108.	$2p3p-2p4s$	$^1P-^3P^o$											
			3125.92	3126.83	164610.76	– 196592.07	3–3	6.11–03	8.96–04	2.77–02	−2.571	D	2
109.		$^1P-^1P^o$	3006.83	3007.71	164610.76	– 197858.69	3–3	9.19–01	1.25–01	3.70+00	−0.427	B	1,2
110.	$2p3p-2p4d$	$^1P-^1D^o$	2206.09	2206.77	164610.76	– 209925.76	3–5	7.04–02	8.56–03	1.86–01	−1.591	B	1
111.	$2p3p-2p5s$	$^1P-^1P^o$		1991.30	164610.76	– 214829.18	3–3	3.69–01	2.19–02	4.31–01	−1.182	B	1
112.	$2p3p-2p5d$	$^1P-^1D^o$		1789.40	164610.76	– 220495.36	3–5	1.26–01	1.01–02	1.79–01	−1.518	B	1
113.	$2p3p-2p6s$	$^1P-^1P^o$		1709.66	164610.76	– 223101.82	3–3	1.86–01	8.14–03	1.38–01	−1.612	B	1
114.	$2p3p-2p3d$	$^3D-^3F^o$	*5004.5*	*5005.9*	*166615.2*	*– 186591.8*	15–21	1.15+00	6.06–01	1.50+02	0.958	B	1,2
			5005.15	5006.55	166678.64	– 186652.49	7–9	1.16+00	5.58–01	6.44+01	0.592	B	2n
			5001.47	5002.87	166582.45	– 186570.98	5–7	1.05+00	5.50–01	4.53+01	0.439	B	2n
			5001.13	5002.53	166521.69	– 186511.58	3–5	9.76–01	6.10–01	3.02+01	0.263	B	2n
			5025.66	5027.06	166678.64	– 186570.98	7–7	1.07–01	4.06–02	4.71+00	−0.546	B	2n
			5016.38	5017.78	166582.45	– 186511.58	5–5	1.62–01	6.11–02	5.05+00	−0.515	B	2n
			5040.71	5042.12	166678.64	– 186511.58	7–5	3.78–03	1.03–03	1.20–01	−2.142	B	2n
115.		$^3D-^1D^o$											
			4860.17	4861.52	166521.69	– 187091.37	3–5	1.61–02	9.49–03	4.56–01	−1.546	D	2
116.		$^3D-^3D^o$	*4793.5*	*4794.8*	*166615.2*	*– 187470.9*	15–15	3.35–01	1.15–01	2.73+01	0.238	B	1,2
			4803.29	4804.63	166678.64	– 187491.90	7–7	3.18–01	1.10–01	1.22+01	−0.113	B	2n
			4788.14	4789.48	166582.45	– 187461.56	5–5	2.52–01	8.67–02	6.83+00	−0.363	B	2n
			4779.72	4781.06	166521.69	– 187437.56	3–3	2.52–01	8.63–02	4.07+00	−0.587	B	2n
			4810.30	4811.64	166678.64	– 187461.56	7–5	4.75–02	1.18–02	1.31+00	−1.084	B	2n
			4793.65	4794.99	166582.45	– 187437.56	5–3	7,77–02	1.61–02	1.27+00	−1.095	B	2n
			4781.19	4782.53	166582.45	– 187491.90	5–7	2.05–02	9.86–03	7.76–01	−1.307	B	2n
			4774.24	4775.58	166521.69	– 187461.56	3–5	3.24–02	1.84–02	8.70–01	−1.257	B	2n
117.		$^3D-^3P^o$	*4489.4*	*4490.7*	*166615.2*	*– 188883.5*	15–9	1.13–01	2.05–02	4.55+00	−0.512	B	1,2
			4507.56	4508.82	166678.64	– 188857.37	7–5	1.00–01	2.18–02	2.26+00	−0.817	B	2n
			4477.68	4478.94	166582.45	– 188909.17	5–3	8.85–02	1.60–02	1.18+00	−1.098	B	2n
			4459.94	4461.19	166521.69	– 188937.24	3–1	1.12–01	1.12–02	4.93–01	−1.474	B	2n
			4488.09	4489.35	166582.45	– 188857.37	5–5	1.30–02	3.93–03	2.91–01	−1.706	B	2n
			4465.53	4466.78	166521.69	– 188909.17	3–3	2.36–02	7.06–03	3.12–01	−1.674	B	2n
			4475.89	4477.14	166521.69	– 188857.37	3–5	6.68–04	3.35–04	1.48–02	−2.998	B−	2n
118.		$^3D-^1F^o$											
			4393.85	4395.08	166582.45	– 189335.16	5–7	4.13–04	1.67–04	1.21–02	−3.077	D	2
119.		$^3D-^1P^o$											
			4236.36	4237.55	166521.69	– 190120.24	3–3	1.66–03	4.48–04	1.87–02	−2.872	D	2

N II: Allowed Transitions — Continued

No.	Transition Array	Mult.	λ_{air} (Å)	λ_{vac} (Å) or σ (cm^{-1})*	E_i (cm^{-1})	E_k (cm^{-1})	g_i–g_k	A_{ki} (10^8 s^{-1})	f_{ik}	S (at. u.)	log gf	Acc.	Source
120.	2p3p–2p4s	^{3}D–^{3}P°	*3328.2*	*3329.2*	*166615.2*	– *196652.7*	15–9	1.16+00	1.16–01	1.91+01	0.241	B	1,2
			3328.72	3329.68	166678.64	– 196711.54	7–5	9.79–01	1.16–01	8.92+00	–0.090	B	2n
			3331.31	3332.26	166582.45	– 196592.07	5–3	8.83–01	8.82–02	4.84+00	–0.355	B	2n
			3330.32	3331.27	166521.69	– 196540.23	3–1	1.18+00	6.55–02	2.15+00	–0.707	B	2n
			3318.10	3319.05	166582.45	– 196711.54	5–5	1.68–01	2.77–02	1.51+00	–0.858	B	2n
			3324.58	3325.53	166521.69	– 196592.07	3–3	2.84–01	4.71–02	1.55+00	–0.850	B	2n
			3311.42	3312.37	166521.69	– 196711.54	3–5	1.09–02	2.99–03	9.79–02	–2.047	B	2n
121.		^{3}D–^{1}P°											
			3196.39	3197.32	166582.45	– 197858.69	5–3	2.09–03	1.92–04	1.01–02	–3.017	D	2
			3190.19	3191.12	166521.69	– 197858.69	3–3	3.30–03	5.04–04	1.59–02	–2.820	D	2
122.	2p3p–2p4d	^{3}D–^{3}F°	*2317.1*	*2317.8*	*166615.2*	– *209759.5*	15–21	1.97–01	2.22–02	2.54+00	–0.478	B	1
			2317.04	2317.75	166678.64	– 209823.91	7–9	1.97–01	2.04–02	1.09+00	–0.846	C+	LS
			2316.49	2317.20	166582.45	– 209737.97	5–7	1.75–01	1.97–02	7.52–01	–1.006	C+	LS
			2316.68	2317.39	166521.69	– 209673.65	3–5	1.65–01	2.22–02	5.08–01	–1.177	C+	LS
			2321.66	2322.38	166678.64	– 209737.97	7–7	2.18–02	1.76–03	9.43–02	–1.909	C+	LS
			2319.95	2320.66	166582.45	– 209673.65	5–5	3.06–02	2.47–03	9.43–02	–1.909	C+	LS
			2325.14	2325.85	166678.64	– 209673.65	7–5	8.57–04	4.96–05	2.66–03	–3.459	D	LS
123.		^{3}D–^{3}D°	*2289.6*	*2290.3*	*166615.2*	– *210277.4*	15–15	2.36–02	1.85–03	2.09–01	–1.556	B	1
			2291.66	2292.37	166678.64	– 210301.68	7–7	2.09–02	1.64–03	8.68–02	–1.939	C+	LS
			2288.48	2289.19	166582.45	– 210266.04	5–5	1.64–02	1.29–03	4.86–02	–2.191	C+	LS
			2286.67	2287.38	166521.69	– 210239.83	3–3	1.77–02	1.39–03	3.14–02	–2.380	C+	LS
			2293.53	2294.24	166678.64	– 210266.04	7–5	3.66–03	2.06–04	1.09–02	–2.841	C	LS
			2289.86	2290.56	166582.45	– 210239.83	5–3	5.89–03	2.78–04	1.05–02	–2.857	C	LS
			2286.62	2287.32	166582.45	– 210301.68	5–7	2.64–03	2.89–04	1.09–02	–2.840	C	LS
			2285.30	2286.01	166521.69	– 210266.04	3–5	3.55–03	4.64–04	1.05–02	–2.856	C	LS
124.	2p3p–2p5s	^{3}D–^{3}P°	*2095.4*	*2096.1*	*166615.2*	– *214322.9*	15–9	5.00–01	1.97–02	2.04+00	–0.528	B	1
			2095.53	2096.19	166678.64	– 214384.13	7–5	4.20–01	1.97–02	9.53–01	–0.859	C+	LS
			2096.86	2097.52	166582.45	– 214257.69	5–3	3.74–01	1.48–02	5.11–01	–1.131	C+	LS
			2096.20	2096.86	166521.69	– 214211.96	3–1	4.99–01	1.10–02	2.27–01	–1.483	C+	LS
			2091.31	2091.98	166582.45	– 214384.13	5–5	7.54–02	4.95–03	1.70–01	–1.607	C+	LS
			2094.19	2094.86	166521.69	– 214257.69	3–3	1.25–01	8.23–03	1.70–01	–1.607	C+	LS
			2088.66	2089.32	166521.69	– 214384.13	3–5	5.05–03	5.50–04	1.14–02	–2.782	C	LS
125.	2p3p–2p3d	^{3}S–^{1}D°											
			5493.23	5494.76	168892.21	– 187091.37	3–5	2.78–04	2.10–04	1.14–02	–3.201	D	2
126.		^{3}S–^{3}D°											
			5383.72	5385.22	168892.21	– 187461.56	3–5	3.31–03	2.40–03	1.28–01	–2.143	D	2
			5390.69	5392.19	168892.21	– 187437.56	3–3	1.83–03	7.97–04	4.24–02	–2.622	D	2
127.		^{3}S–^{3}P°	*5000.8*	*5002.2*	168892.21	– *188883.5*	3–9	7.75–01	8.72–01	4.31+01	0.418	B	1,2
			5007.33	5008.73	168892.21	– 188857.37	3–5	7.89–01	4.94–01	2.45+01	0.171	B	2n
			4994.37	4995.76	168892.21	– 188909.17	3–3	7.60–01	2.84–01	1.40+01	–0.069	B	2n
			4987.38	4988.77	168892.21	– 188937.24	3–1	7.48–01	9.30–02	4.58+00	–0.555	B	2n
128.		^{3}S–^{1}P°											
			4709.44	4710.75	168892.21	– 190120.24	3–3	1.27–03	4.24–04	1.97–02	–2.896	D	2

N II: Allowed Transitions — Continued

No.	Transition Array	Mult.	λ_{air} (Å)	λ_{vac} (Å) or σ (cm^{-1})*	E_i (cm^{-1})	E_k (cm^{-1})	g_i–g_k	A_{ki} (10^8 s^{-1})	f_{ik}	S (at. u.)	log gf	Acc.	Source
129.	$2p3p$– $2p4s$	^{3}S–^{3}P°	*3601.2*	*3602.2*	168892.21	– *196652.7*	3–9	1.31–01	7.65–02	2.72+00	–0.639	B	1,2
			3593.60	3594.62	168892.21	– 196711.54	3–5	1.21–01	3.89–02	1.38+00	–0.933	B	2*n*
			3609.10	3610.13	168892.21	– 196592.07	3–3	1.41–01	2.76–02	9.84–01	–1.082	B	2*n*
			3615.86	3616.90	168892.21	– 196540.23	3–1	1.53–01	1.00–02	3.57–01	–1.523	B	2*n*
130.		^{3}S–^{1}P°											
			3451.28	3452.27	168892.21	– 197858.69	3–3	1.14–03	2.03–04	6.92–03	–3.215	D	2
131.	$2p3p$– $2p4d$	^{3}S–^{3}P°	*2389.5*	*2390.3*	168892.21	– *210728.6*	3–9	3.29–02	8.44–03	1.99–01	–1.597	B	1
			2390.87	2391.60	168892.21	– 210705.26	3–5	3.28–02	4.68–03	1.11–01	–1.852	C+	*LS*
			2388.23	2388.96	168892.21	– 210751.43	3–3	3.29–02	2.81–03	6.64–02	–2.074	C+	*LS*
			2386.81	2387.53	168892.21	– 210776.45	3–1	3.30–02	9.38–04	2.21–02	–2.550	C	*LS*
132.	$2p3p$– $2p5s$	^{3}S–^{3}P°	*2200.5*	*2201.2*	168892.21	– *214322.9*	3–9	6.97–02	1.52–02	3.30–01	–1.342	B	1
			2197.51	2198.19	168892.21	– 214384.13	3–5	7.00–02	8.44–03	1.83–01	–1.596	C+	*LS*
			2203.63	2204.32	168892.21	– 214257.69	3–3	6.94–02	5.05–03	1.10–01	–1.819	C+	*LS*
			2205.86	2206.54	168892.21	– 214211.96	3–1	6.92–02	1.68–03	3.66–02	–2.297	C+	*LS*
133.	$2s2p^3$– $2s^22p3p$	^{1}P°–^{1}D	13425.7	7446.37 cm^{-1}	166765.66	– 174212.03	3–5	7.30–03	3.29–02	4.36+00	–1.006	C+	1,2
134.		^{1}P°–^{1}S	8687.43	8689.82	166765.66	– 178273.38	3–1	8.73–03	3.30–03	2.83–01	–2.005	C	1,2
135.	$2s2p^3$– $2s^22p4p$	^{1}P°–^{1}P	2823.63	2824.46	166765.66	– 202170.63	3–3	2.39–01	2.86–02	7.97–01	–1.067	C+	1,2
136.		^{1}P°–^{3}D											
			2780.94	2781.76	166765.66	– 202714.12	3–3	1.76–03	2.04–04	5.59–03	–3.214	D	2
137.		^{1}P°–^{3}P											
			2744.68	2745.49	166765.66	– 203189.03	3–3	4.28–04	4.84–05	1.31–03	–3.838	D	2
138.		^{1}P°–^{1}D	2590.94	2591.71	166765.66	– 205350.18	3–5	1.13–01	1.90–02	4.86–01	–1.244	D–	1,2
139.		^{1}P°–^{1}S	2490.24	2491.00	166765.66	– 206910.24	3–1	2.61–01	8.10–03	1.99–01	–1.614	C	2
140.	$2p3p$– $2p3d$	^{3}P–^{3}F°											
			6285.69	6287.43	170666.23	– 186570.98	5–7	9.37–04	7.77–04	8.05–02	–2.410	D	2
			6286.11	6287.85	170607.89	– 186511.58	3–5	3.93–04	3.88–04	2.41–02	–2.934	D	2
			6309.25	6311.00	170666.23	– 186511.58	5–5	1.47–04	8.80–05	9.14–03	–3.357	D	2
141.		^{3}P–^{1}D°											
			6065.00	6066.68	170607.89	– 187091.37	3–5	2.21–03	2.04–03	1.22–01	–2.214	D	2
142.		^{3}P–^{3}D°	*5938.5*	*5940.2*	*170636.4*	– *187470.9*	9–15	5.56–01	4.90–01	8.62+01	0.644	B	1,2
			5941.65	5943.30	170666.23	– 187491.90	5–7	5.54–01	4.11–01	4.02+01	0.313	B	2*n*
			5931.78	5933.43	170607.89	– 187461.56	3–5	4.27–01	3.76–01	2.20+01	0.052	B	2*n*
			5927.81	5929.46	170572.61	– 187437.56	1–3	3.22–01	5.09–01	9.93+00	–0.293	B	2*n*
			5952.39	5954.04	170666.23	– 187461.56	5–5	1.27–01	6.77–02	6.63+00	–0.471	B	2*n*
			5940.24	5941.89	170607.89	– 187437.56	3–3	2.26–01	1.20–01	7.02+00	–0.445	B	2*n*
			5960.91	5962.56	170666.23	– 187437.56	5–3	1.34–02	4.27–03	4.19–01	–1.671	B	2*n*

N II: Allowed Transitions — Continued

No.	Transition Array	Mult.	λ_{air} (Å)	λ_{vac} (Å) or σ (cm^{-1})*	E_i (cm^{-1})	E_k (cm^{-1})	g_i–g_k	A_{ki} (10^8 s^{-1})	f_{ik}	S (at. u.)	log gf	Acc.	Source
143.		^{3}P–^{3}P°	*5478.8*	*5480.3*	*170636.4*	– *188883.5*	9–9	3.03−01	1.36−01	2.21+01	0.089	B	1,2
			5495.65	5497.18	170666.23	– 188857.37	5–5	2.40−01	1.09−01	9.82+00	−0.266	B	2*n*
			5462.58	5464.10	170607.89	– 188909.17	3–3	1.00−01	4.47−02	2.41+00	−0.872	B	2*n*
			5480.05	5481.57	170666.23	– 188909.17	5–3	1.30−01	3.50−02	3.16+00	−0.756	B	2*n*
			5454.22	5455.73	170607.89	– 188937.24	3–1	3.34−01	4.96−02	2.67+00	−0.827	B	2*n*
			5478.09	5479.61	170607.89	– 188857.37	3–5	4.75−02	3.56−02	1.93+00	−0.971	B	2*n*
			5452.07	5453.59	170572.61	– 188909.17	1–3	8.89−02	1.19−01	2.14+00	−0.925	B	2*n*
144.		^{3}P–^{1}P°											
			5114.28	5115.71	170572.61	– 190120.24	1–3	2.60−04	3.06−04	5.16−03	−3.514	D	2
145.	2*p*3*p*– 2*p*4*s*	^{3}P–^{3}P°	*3842.7*	*3843.7*	*170636.4*	– *196652.7*	9–9	9.19−01	2.04−01	2.32+01	0.263	B	1,2
			3838.37	3839.46	170666.23	– 196711.54	5–5	6.98−01	1.54−01	9.74+00	−0.113	B	2*n*
			3847.40	3848.50	170607.89	– 196592.07	3–3	2.22−01	4.92−02	1.87+00	−0.831	B	2*n*
			3856.06	3857.16	170666.23	– 196592.07	5–3	3.71−01	4.97−02	3.15+00	−0.605	B	2*n*
			3855.10	3856.19	170607.89	– 196540.23	3–1	8.82−01	6.55−02	2.49+00	−0.707	B	2*n*
			3829.80	3830.88	170607.89	– 196711.54	3–5	2.42−01	8.88−02	3.36+00	−0.574	B	2*n*
			3842.19	3843.28	170572.61	– 196592.07	1–3	3.06−01	2.03−01	2.57+00	−0.692	B	2*n*
146.		^{3}P–^{1}P°											
			3676.44	3677.49	170666.23	– 197858.69	5–3	5.15−04	6.27−05	3.79−03	−3.504	D	2
			3668.57	3669.62	170607.89	– 197858.69	3–3	1.06−03	2.14−04	7.76−03	−3.192	D	2
			3663.83	3664.87	170572.61	– 197858.69	1–3	7.77−04	4.70−04	5.67−03	−3.328	D	2
147.	2*p*3*p*– 2*p*4*d*	^{3}P–^{3}D°	*2521.9*	*2522.6*	*170636.4*	– *210277.4*	9–15	2.26−01	3.60−02	2.69+00	−0.490	B	1
			2522.24	2522.99	170666.23	– 210301.68	5–7	2.26−01	3.02−02	1.25+00	−0.821	C+	*LS*
			2520.79	2521.55	170607.89	– 210266.04	3–5	1.70−01	2.70−02	6.72−01	−1.092	C+	*LS*
			2520.21	2520.97	170572.61	– 210239.83	1–3	1.26−01	3.60−02	2.99−01	−1.444	C+	*LS*
			2524.51	2525.26	170666.23	– 210266.04	5–5	5.64−02	5.39−03	2.24−01	−1.570	C+	*LS*
			2522.46	2523.22	170607.89	– 210239.83	3–3	9.42−02	8.99−03	2.24−01	−1.569	C+	*LS*
			2526.18	2526.94	170666.23	– 210239.83	5–3	6.25−03	3.59−04	1.49−02	−2.746	C	*LS*
148.		^{3}P–^{3}P°	*2493.5*	*2494.3*	*170636.4*	– *210728.6*	9–9	9.91−02	9.24−03	6.83−01	−1.080	B	1
			2496.81	2497.56	170666.23	– 210705.26	5–5	7.41−02	6.92−03	2.84−01	−1.461	C+	*LS*
			2490.31	2491.06	170607.89	– 210751.43	3–3	2.49−02	2.31−03	5.69−02	−2.159	C+	*LS*
			2493.93	2494.69	170666.23	– 210751.43	5–3	4.13−02	2.31−03	9.48−02	−1.937	C+	*LS*
			2488.76	2489.51	170607.89	– 210776.45	3–1	9.97−02	3.09−03	7.59−02	−2.033	C+	*LS*
			2493.18	2493.93	170607.89	– 210705.26	3–5	2.48−02	3.85−03	9.48−02	−1.937	C+	*LS*
			2488.12	2488.87	170572.61	– 210751.43	1–3	3.33−02	9.26−03	7.59−02	−2.033	C+	*LS*
149.	2*p*3*p*– 2*p*5*s*	^{3}P–^{3}P°	*2288.3*	*2289.0*	*170636.4*	– *214322.9*	9–9	3.04−01	2.39−02	1.62+00	−0.668	B	1
			2286.69	2287.39	170666.23	– 214384.13	5–5	2.29−01	1.79−02	6.75−01	−1.047	C+	*LS*
			2290.26	2290.96	170607.89	– 214257.69	3–3	7.59−02	5.97−03	1.35−01	−1.747	C+	*LS*
			2293.32	2294.03	170666.23	– 214257.69	5–3	1.26−01	5.96−03	2.25−01	−1.526	C+	*LS*
			2292.66	2293.36	170607.89	– 214211.96	3–1	3.03−01	7.95−03	1.80−01	−1.623	C+	*LS*
			2283.64	2284.34	170607.89	– 214384.13	3–5	7.66−02	9.98−03	2.25−01	−1.524	C+	*LS*
			2288.41	2289.11	170572.61	– 214257.69	1–3	1.01−01	2.39−02	1.80−01	−1.622	C+	*LS*
150.	2*s*^{2}2*p*3*p*– 2*s*2*p*^{2}3*p*	^{3}P–^{3}S°		*1907.2*	*170636.4*	– 223069.02	9–3	2.61−01	4.75−03	2.68−01	−1.369	B	1
				1908.30	170666.23	– 223069.02	5–3	1.45−01	4.75−03	1.49−01	−1.625	C+	*LS*
				1906.17	170607.89	– 223069.02	3–3	8.72−02	4.75−03	8.95−02	−1.846	C+	*LS*
				1904.89	170572.61	– 223069.02	1–3	2.91−02	4.76−03	2.98−02	−2.323	C	*LS*

N II: Allowed Transitions — Continued

No.	Transition Array	Mult.	λ_{air} (Å)	λ_{vac} (Å) or σ (cm^{-1})*	E_i (cm^{-1})	E_k (cm^{-1})	g_i-g_k	A_{ki} (10^8 s^{-1})	f_{ik}	S (at. u.)	log gf	Acc.	Source
151.		^{3}P–^{3}D°		*1720.7*	*170636.4*	*– 228752.3*	9–15	5.23–01	3.87–02	1.97+00	–0.458	B	1
				1720.41	170666.23	– 228791.83	5–7	5.24–01	3.25–02	9.21–01	–0.789	C+	*LS*
				1720.46	170607.89	– 228731.81	3–5	3.93–01	2.90–02	4.94–01	–1.060	C+	*LS*
				1720.53	170572.61	– 228694.30	1–3	2.91–01	3.87–02	2.19–01	–1.412	C+	*LS*
				1722.19	170666.23	– 228731.81	5–5	1.31–01	5.80–03	1.65–01	–1.537	C+	*LS*
				1721.57	170607.89	– 228694.30	3–3	2.18–01	9.68–03	1.65–01	–1.537	C+	*LS*
				1723.30	170666.23	– 228694.30	5–3	1.45–02	3.87–04	1.10–02	–2.714	C	*LS*
152.	2*p*3*p*– 2*p*3*d*	^{1}D–^{3}F°											
			8089.08	8091.30	174212.03	– 186570.98	5–7	8.82–05	1.21–04	1.61–02	–3.217	D	2
			8128.14	8130.38	174212.03	– 186511.58	5–5	4.72–04	4.67–04	6.25–02	–2.631	D	2
153.		^{1}D–^{1}D°	7762.24	7764.37	174212.03	– 187091.37	5–5	8.74–02	7.90–02	1.01+01	–0.403	B	1,2
154.		^{1}D–^{3}D°											
			7528.12	7530.19	174212.03	– 187491.90	5–7	6.10–05	7.26–05	9.00–03	–3.440	D	2
			7545.36	7547.44	174212.03	– 187461.56	5–5	3.95–04	3.38–04	4.20–02	–2.773	D	2
155.		^{1}D–^{3}P°											
			6826.23	6828.11	174212.03	– 188857.37	5–5	2.13–04	1.49–04	1.67–02	–3.128	D	2
156.		^{1}D–^{1}F°	6610.56	6612.39	174212.03	– 189335.16	5–7	6.34–01	5.81–01	6.33+01	0.464	B	1,2
157.		^{1}D–^{1}P°	6284.32	6286.06	174212.03	– 190120.24	5–3	7.74–02	2.75–02	2.85+00	–0.861	B	1,2
158.	2*p*3*p*– 2*p*4*s*	^{1}D–^{3}P°											
			4467.01	4468.27	174212.03	– 196592.07	5–3	2.25–03	4.04–04	2.97–02	–2.694	D	2
159.		^{1}D–^{1}P°	4227.74	4228.93	174212.03	– 197858.69	5–3	1.08+00	1.74–01	1.21+01	–0.060	B	1,2
160.	2*p*3*p*– 2*p*4*d*	^{1}D–^{1}D°	2799.22	2800.04	174212.03	– 209925.76	5–5	1.62–01	1.91–02	8.80–01	–1.020	B	1
161.		^{1}D–^{1}F°	2709.84	2710.64	174212.03	– 211103.63	5–7	2.64–01	4.08–02	1.82+00	–0.691	B	1
162.	2*p*3*p*– 2*p*5*s*	^{1}D–^{1}P°	2461.27	2462.01	174212.03	– 214829.18	5–3	4.11–01	2.24–02	9.07–01	–0.951	B	1
163.	2*p*3*p*– 2*p*5*d*	^{1}D–^{1}D°	2159.93	2160.61	174212.03	– 220495.36	5–5	1.23–01	8.60–03	3.06–01	–1.366	B	1
164.		^{1}D–^{1}F°	2130.18	2130.85	174212.03	– 221141.61	5–7	2.42–01	2.31–02	8.09–01	–0.938	B	1
165.	2*p*3*p*– 2*p*6*s*	^{1}D–^{1}P°	2044.76	2045.42	174212.03	– 223101.82	5–3	2.11–01	7.95–03	2.68–01	–1.401	B	1
166.	2*p*3*p*– 2*p*3*d*	^{1}S–^{3}D°											
			10909.1	9164.18 cm^{-1}	178273.38	– 187437.56	1–3	3.09–05	1.65–04	5.94–03	–3.781	D	2

N II: Allowed Transitions — Continued

No.	Transition Array	Mult.	λ_{air} (Å)	λ_{vac} (Å) or σ (cm^{-1})*	E_i (cm^{-1})	E_k (cm^{-1})	g_i-g_k	A_{ki} (10^8 s^{-1})	f_{ik}	S (at. u.)	log gf	Acc.	Source
167.		$^1S-^3P^o$											
			9399.64	9402.22	178273.38	188909.17	1–3	9.79–05	3.89–04	1.21–02	–3.410	D	2
168.		$^1S-^1P^o$	8438.74	8441.06	178273.38	190120.24	1–3	2.24–01	7.19–01	2.00+01	–0.143	B	1,2
169.	2p3p–2p4s	$^1S-^3P^o$											
			5457.39	5458.91	178273.38	196592.07	1–3	1.83–04	2.45–04	4.41–03	–3.610	D	2
170.		$^1S-^1P^o$	5104.45	5105.87	178273.38	197858.69	1–3	1.01–01	1.18–01	1.99+00	–0.927	C	1,2
171.	2p3p–2p4d	$^1S-^1P^o$	3023.67	3024.55	178273.38	211336.16	1–3	2.27–01	9.33–02	9.29–01	–1.030	B	1
172.	2p3p–2p5s	$^1S-^1P^o$	2734.73	2735.54	178273.38	214829.18	1–3	5.03–02	1.69–02	1.52–01	–1.772	B	1
173.	2p3p–2p5d	$^1S-^1P^o$	2326.34	2327.05	178273.38	221246.17	1–3	1.83–01	4.46–02	3.42–01	–1.351	B	1
174.	2p3d–2p4p	$^3F^o-^1P$											
			6384.32	6386.08	186511.58	202170.63	5–3	7.86–03	2.88–03	3.03–01	–1.841	D	2
175.		$^3F^o-^3D$	*6168.0*	*6169.8*	*186591.8*	*202799.9*	21–15	2.87–01	1.17–01	4.98+01	0.390	C+	1,2
			6167.75	6169.46	186652.49	202861.36	9–7	2.65–01	1.18–01	2.15+01	0.025	C+	2n
			6173.31	6175.02	186570.98	202765.26	7–5	2.62–01	1.07–01	1.52+01	–0.126	C+	2n
			6170.16	6171.87	186511.58	202714.12	5–3	2.85–01	9.77–02	9.92+00	–0.311	C+	2n
			6136.89	6138.59	186570.98	202861.36	7–7	1.90–02	1.08–02	1.52+00	–1.123	C+	2n
			6150.75	6152.45	186511.58	202765.26	5–5	2.89–02	1.64–02	1.66+00	–1.086	C+	2n
			6114.60	6116.29	186511.58	202861.36	5–7	4.76–04	3.74–04	3.76–02	–2.728	C+	2n
176.		$^3F^o-^3P$											
			5990.67	5992.33	186570.98	203258.98	7–5	1.56–04	6.01–05	8.29–03	–3.376	D	2
177.		$^3F^o-^1D$											
			5306.77	5308.25	186511.58	205350.18	5–5	2.20–04	9.27–05	8.10–03	–3.334	D	2
178.		$^1D^o-^1P$	6629.79	6631.63	187091.37	202170.63	5–3	2.69–01	1.06–01	1.16+01	–0.275	C+	1,2
179.		$^1D^o-^3D$											
			6399.15	6400.92	187091.37	202714.12	5–3	9.48–03	3.49–03	3.68–01	–1.758	D	2
180.		$^1D^o-^3P$											
			6210.36	6212.08	187091.37	203189.03	5–3	1.66–04	5.77–05	5.90–03	–3.540	D	2
181.		$^1D^o-^1D$	5475.29	5476.81	187091.37	205350.18	5–5	4.66–02	2.10–02	1.89+00	–0.979	B	1,2
182.		$^3D^o-^1P$											
			6796.65	6798.53	187461.56	202170.63	5–3	8.37–04	3.48–04	3.89–02	–2.760	D	2
			6785.58	6787.45	187437.56	202170.63	3–3	9.41–04	6.50–04	4.36–02	–2.710	D	2

N II: Allowed Transitions — Continued

No.	Transition Array	Mult.	λ_{air} (Å)	λ_{vac} (Å) or σ (cm^{-1})*	E_i (cm^{-1})	E_k (cm^{-1})	g_i–g_k	A_{ki} (10^8 s^{-1})	f_{ik}	S (at. u.)	log gf	Acc.	Source
183.		$^3D^o-^3D$	*6521.8*	*6523.6*	*187470.9*	*− 202799.9*	15−15	5.55−02	3.54−02	1.14+01	−0.274	B	1,2
			6504.61	6506.41	187491.90	− 202861.36	7−7	5.33−02	3.38−02	5.07+00	−0.626	B	2n
			6532.56	6534.37	187461.56	− 202765.26	5−5	3.66−02	2.34−02	2.52+00	−0.932	B	2n
			6544.17	6545.98	187437.56	− 202714.12	3−3	3.59−02	2.31−02	1.49+00	−1.160	B	2n
			6545.54	6547.35	187491.90	− 202765.26	7−5	1.07−02	4.89−03	7.38−01	−1.465	B	2n
			6554.47	6556.28	187461.56	− 202714.12	5−3	1.43−02	5.54−03	5.97−01	−1.558	B	2n
			6491.80	6493.59	187461.56	− 202861.36	5−7	5.49−03	4.86−03	5.19−01	−1.614	B	2n
			6522.33	6524.14	187437.56	− 202765.26	3−5	7.03−03	7.47−03	4.82−01	−1.649	B	2n
184.		$^3D^o-^3P$	*6345.8*	*6347.6*	*187470.9*	*− 203224.9*	15−9	2.61−01	9.45−02	2.96+01	0.151	B	1,2
			6340.58	6342.33	187491.90	− 203258.98	7−5	2.13−01	9.18−02	1.34+01	−0.192	B	2n
			6356.54	6358.30	187461.56	− 203189.03	5−3	1.92−01	6.97−02	7.29+00	−0.458	B	2n
			6357.58	6359.33	187437.56	− 203162.48	3−1	2.68−01	5.42−02	3.40+00	−0.789	B	2n
			6328.40	6330.15	187461.56	− 203258.98	5−5	4.28−02	2.57−02	2.68+00	−0.891	B	2n
			6346.86	6348.61	187437.56	− 203189.03	3−3	6.93−02	4.19−02	2.63+00	−0.901	B	2n
			6318.80	6320.55	187437.56	− 203258.98	3−5	3.02−03	3.01−03	1.88−01	−2.044	B	2n
185.		$^3D^o-^3S$											
			6218.69	6220.41	187461.56	− 203537.66	5−3	2.83−03	9.86−04	1.01−01	−2.307	D	2
			6209.42	6211.14	187437.56	− 203537.66	3−3	9.84−04	5.69−04	3.49−02	−2.768	D	2
186.		$^3D^o-^1D$											
			5588.59	5590.15	187461.56	− 205350.18	5−5	2.90−04	1.36−04	1.25−02	−3.167	D	2
187.	$2p3d-$ $2s2p^2(^4P)3s$	$^3D^o-^3P$	*4108.7*	*4109.8*	*187470.9*	*− 211802.9*	15−9	1.82−02	2.77−03	5.61−01	−1.382	C	2
			4108.02	4109.18	187491.90	− 211827.67	7−5	1.52−02	2.75−03	2.61−01	−1.715	C	2
			4111.04	4112.20	187461.56	− 211779.45	5−3	1.37−02	2.09−03	1.41−01	−1.982	C	2
			4112.07	4113.23	187437.56	− 211749.35	3−1	1.85−02	1.57−03	6.37−02	−2.328	C	2
			4102.90	4104.06	187461.56	− 211827.67	5−5	2.68−03	6.77−04	4.58−02	−2.470	C	2
			4106.99	4108.14	187437.56	− 211779.45	3−3	4.57−03	1.16−03	4.69−02	−2.460	C	2
			4098.87	4100.02	187437.56	− 211827.67	3−5	1.80−04	7.54−05	3.05−03	−3.646	C−	2
188.	$2p3d-$ $2p4p$	$^3P^o-^1P$											
			7538.57	7540.65	188909.17	− 202170.63	3−3	4.66−04	3.97−04	2.96−02	−2.924	D	2
			7554.56	7556.64	188937.24	− 202170.63	1−3	2.52−05	6.46−05	1.61−03	−4.189	D	2
189.		$^3P^o-^3D$	*7183.8*	*7185.8*	*188883.5*	*− 202799.9*	9−15	2.21−02	2.86−02	6.08+00	−0.590	C	1,2
			7138.85	7140.82	188857.37	− 202861.36	5−7	2.33−02	2.50−02	2.94+00	−0.903	C	2n
			7215.05	7217.04	188909.17	− 202765.26	3−5	1.52−02	1.98−02	1.41+00	−1.225	C	2n
			7256.54	7258.54	188937.24	− 202714.12	1−3	1.10−02	2.62−02	6.25−01	−1.582	C	2n
			7188.18	7190.16	188857.37	− 202765.26	5−5	6.34−03	4.91−03	5.81−01	−1.610	C	2n
			7241.78	7243.78	188909.17	− 202714.12	3−3	8.38−03	6.59−03	4.71−01	−1.704	C	2n
			7214.71	7216.70	188857.37	− 202714.12	5−3	9.03−04	4.23−04	5.02−02	−2.675	C	2n
190.		$^3P^o-^3P$	*6970.9*	*6972.8*	*188883.5*	*− 203224.9*	9−9	3.88−02	2.83−02	5.84+00	−0.595	B	1,2
			6941.75	6943.67	188857.37	− 203258.98	5−5	2.81−02	2.03−02	2.32+00	−0.994	B	2n
			7000.94	7002.87	188909.17	− 203189.03	3−3	9.82−04	7.22−04	4.99−02	−2.665	B	2n
			6975.63	6977.56	188857.37	− 203189.03	5−3	3.42−02	1.50−02	1.72+00	−1.126	B	2n
			7013.98	7015.91	188909.17	− 203162.48	3−1	3.00−02	7.38−03	5.11−01	−1.655	B	2n
			6966.81	6968.73	188909.17	− 203258.98	3−5	1.12−02	1.36−02	9.39−01	−1.388	B	2n
			7014.73	7016.66	188937.24	− 203189.03	1−3	5.83−03	1.29−02	2.98−01	−1.889	B	2n

N II: Allowed Transitions — Continued

No.	Transition Array	Mult.	λ_{air} (Å)	λ_{vac} (Å) or σ (cm^{-1})*	E_i (cm^{-1})	E_k (cm^{-1})	g_i-g_k	A_{ki} (10^8 s^{-1})	f_{ik}	S (at. u.)	log gf	Acc.	Source
191.		$^3P^o-^3S$	*6822.1*	*6824.0*	*188883.5*	– 203537.66	9–3	4.53–01	1.06–01	2.13+01	–0.022	B	1,2
			6809.98	6811.85	188857.37	– 203537.66	5–3	2.31–01	9.66–02	1.08+01	–0.316	B	2n
			6834.09	6835.98	188909.17	– 203537.66	3–3	1.63–01	1.14–01	7.70+00	–0.466	B	2n
			6847.23	6849.12	188937.24	– 203537.66	1–3	5.89–02	1.24–01	2.80+00	–0.906	B	2n
192.		$^3P^o-^1D$											
			6061.57	6063.25	188857.37	– 205350.18	5–5	1.14–04	6.29–05	6.28–03	–3.502	D	2
193.	$2p3d-$ $2s2p^2(^4P)3s$	$^3P^o-^3P$											
			4352.22	4353.45	188857.37	– 211827.67	5–5	2.59–04	7.36–05	5.27–03	–3.434	D	2
			4376.63	4377.86	188937.24	– 211779.45	1–3	9.27–05	7.99–05	1.15–03	–4.097	D	2
194.	$2p3d-$ $2p4p$	$^1F^o-^1D$	6242.41	6244.14	189335.16	– 205350.18	7–5	3.03–01	1.27–01	1.82+01	–0.052	B	1,2
195.		$^1P^o-^1P$	8296.21	8298.49	190120.24	– 202170.63	3–3	2.32–02	2.39–02	1.96+00	–1.144	B	1,2
196.		$^1P^o-^3D$											
			7906.08	7908.25	190120.24	– 202765.26	3–5	4.22–05	6.60–05	5.16–03	–3.703	D	2
			7938.18	7940.36	190120.24	– 202714.12	3–3	4.45–04	4.20–04	3.29–02	–2.899	D	2
197.		$^1P^o-^3P$											
			7649.71	7651.82	190120.24	– 203189.03	3–3	1.53–04	1.34–04	1.01–02	–3.396	D	2
198.		$^1P^o-^3S$											
			7450.95	7453.00	190120.24	– 203537.66	3–3	2.29–04	1.91–04	1.40–02	–3.243	D	2
199.		$^1P^o-^1D$	6564.20	6566.01	190120.24	– 205350.18	3–5	4.77–02	5.13–02	3.33+00	–0.812	D	1,2
200.		$^1P^o-^1S$	5954.28	5955.93	190120.24	– 206910.24	3–1	4.94–01	8.75–02	5.15+00	–0.581	B	1,2
201.	$2p4s-$ $2p4p$	$^3P^o-^1P$											
			17920.9	5578.56 cm^{-1}	196592.07	– 202170.63	3–3	1.79–03	8.64–03	1.53+00	–1.586	D	2
			17755.9	5630.40 cm^{-1}	196540.23	– 202170.63	1–3	1.21–04	1.72–03	1.01–01	–2.764	D	2
202.		$^3P^o-^3D$	*16263*	*6147.2* cm^{-1}	*196652.7*	*– 202799.9*	9–15	9.98–02	6.60–01	3.18+02	0.774	B	1,2
			16256.2	6149.82 cm^{-1}	196711.54	– 202861.36	5–7	1.00–01	5.56–01	1.49+02	0.444	B	2n
			16194.7	6173.19 cm^{-1}	196592.07	– 202765.26	3–5	8.23–02	5.40–01	8.64+01	0.209	B	2n
			16192.8	6173.89 cm^{-1}	196540.23	– 202714.12	1–3	6.17–02	7.28–01	3.88+01	–0.138	B	2n
			16514.3	6053.72 cm^{-1}	196711.54	– 202765.26	5–5	1.78–02	7.29–02	1.98+01	–0.438	B	2n
			16329.9	6122.05 cm^{-1}	196592.07	– 202714.12	3–3	3.54–02	1.42–01	2.29+01	–0.371	B	2n
			16655.0	6002.58 cm^{-1}	196711.54	– 202714.12	5–3	1.60–03	4.00–03	1.10+00	–1.699	B	2n
203.		$^3P^o-^3P$	*15211*	*6572.3* cm^{-1}	*196652.7*	*– 203224.9*	9–9	1.16–01	4.01–01	1.81+02	0.558	B	1,2
			15269.0	6547.44 cm^{-1}	196711.54	– 203258.98	5–5	9.39–02	3.28–01	8.25+01	0.215	B	2n
			15154.4	6596.96 cm^{-1}	196592.07	– 203189.03	3–3	4.54–02	1.56–01	2.34+01	–0.329	B	2n
			15433.9	6477.49 cm^{-1}	196711.54	– 203189.03	5–3	3.07–02	6.57–02	1.67+01	–0.483	B	2n
			15215.6	6570.41 cm^{-1}	196592.07	– 203162.48	3–1	1.16–01	1.34–01	2.01+01	–0.396	B	2n
			14995.4	6666.91 cm^{-1}	196592.07	– 203258.98	3–5	2.21–02	1.24–01	1.84+01	–0.429	B	2n
			15036.2	6648.80 cm^{-1}	196540.23	– 203189.03	1–3	3.94–02	4.01–01	1.98+01	–0.397	B	2n

N II: Allowed Transitions — Continued

No.	Transition Array	Mult.	λ_{air} (Å)	λ_{vac} (Å) or σ (cm^{-1})*	E_i (cm^{-1})	E_k (cm^{-1})	g_i-g_k	A_{ki} (10^8 s^{-1})	f_{ik}	S (at. u.)	log gf	Acc.	Source
204.		$^3P^o-^3S$	*14520*	*6885.0* cm^{-1}	*196652.7*	– 203537.66	9–3	1.17–01	1.23–01	5.30+01	0.045	B	1,2
			14645.6	6826.12 cm^{-1}	196711.54	– 203537.66	5–3	8.24–02	1.59–01	3.84+01	–0.099	B	2*n*
			14393.7	6945.59 cm^{-1}	196592.07	– 203537.66	3–3	2.69–02	8.35–02	1.19+01	–0.601	B	2*n*
			14287.1	6997.43 cm^{-1}	196540.23	– 203537.66	1–3	6.42–03	5.90–02	2.77+00	–1.229	B	2*n*
205.		$^3P^o-^1D$											
			11414.9	8758.11 cm^{-1}	196592.07	– 205350.18	3–5	2.30–04	7.49–04	8.44–02	–2.649	D	2
206.		$^3P^o-^1S$											
			9688.98	9691.64	196592.07	– 206910.24	3–1	5.12–04	2.41–04	2.30–02	–3.142	D	2
207.	$2s^22p\,4s-$ $2s\,2p^2(^4P)3s$	$^3P^o-^3P$	*6598.7*	*6600.6*	*196652.7*	*– 211802.9*	9–9	2.11–01	1.38–01	2.69+01	0.093	D	1,2
			6613.62	6615.45	196711.54	– 211827.67	5–5	1.58–01	1.03–01	1.13+01	–0.286	D	2*n*
			6582.60	6584.41	196592.07	– 211779.45	3–3	5.26–02	3.42–02	2.22+00	–0.989	D	2*n*
			6634.79	6636.62	196711.54	– 211779.45	5–3	8.80–02	3.49–02	3.81+00	–0.759	D	2*n*
			6595.67	6597.49	196592.07	– 211749.35	3–1	2.11–01	4.59–02	2.99+00	–0.861	D	2*n*
			6561.76	6563.57	196592.07	– 211827.67	3–5	5.25–02	5.65–02	3.66+00	–0.771	D	2*n*
			6560.20	6562.02	196540.23	– 211779.45	1–3	7.05–02	1.36–01	2.95+00	–0.865	D	2*n*
208.	$2p\,4s-$ $2p\,4p$	$^1P^o-^1P$	23185.1	4311.94 cm^{-1}	197858.69	– 202170.63	3–3	4.09–02	3.30–01	7.55+01	–0.005	B	1,2
209.		$^1P^o-^3D$											
			20375.3	4906.57 cm^{-1}	197858.69	– 202765.26	3–5	4.38–05	4.54–04	9.14–02	–2.866	D	2
			20589.9	4855.43 cm^{-1}	197858.69	– 202714.12	3–3	8.43–04	5.36–03	1.09+00	–1.794	D	2
210.		$^1P^o-^3P$											
			18755.4	5330.34 cm^{-1}	197858.69	– 203189.03	3–3	4.22–04	2.23–03	4.13–01	–2.175	D	2
			18849.3	5303.79 cm^{-1}	197858.69	– 203162.48	3–1	1.11–04	1.97–04	3.67–02	–3.228	D	2
211.		$^1P^o-^3S$											
			17604.0	5678.97 cm^{-1}	197858.69	– 203537.66	3–3	2.50–04	1.16–03	2.02–01	–2.457	D	2
212.		$^1P^o-^1D$	13344.8	7491.49 cm^{-1}	197858.69	– 205350.18	3–5	1.73–01	7.71–01	1.02+02	0.364	B	1,2
213.		$^1P^o-^1S$	11044.8	9051.55 cm^{-1}	197858.69	– 206910.24	3–1	2.24–01	1.36–01	1.49+01	–0.388	B	1,2
214.	$2p\,4s-$ $2s\,2p^2(^4P)3s$	$^1P^o-^3P$											
			7156.75	7158.72	197858.69	– 211827.67	3–5	2.27–04	2.91–04	2.06–02	–3.059	D	2
			7181.54	7183.52	197858.69	– 211779.45	3–3	2.06–04	1.59–04	1.13–02	–3.321	D	2
			7197.10	7199.08	197858.69	– 211749.35	3–1	9.98–04	2.59–04	1.84–02	–3.110	D	2
215.	$2p\,4p-$ $2p\,4d$	$^1P-^1D^o$	12891.2	7755.13 cm^{-1}	202170.63	– 209925.76	3–5	2.11–01	8.76–01	1.11+02	0.419	B	1
216.		$^1P-^1P^o$	10907.5	9165.53 cm^{-1}	202170.63	– 211336.16	3–3	1.32–01	2.35–01	2.53+01	–0.151	B	1
217.	$2p\,4p-$ $2p\,5s$	$^1P-^1P^o$	7897.63	7899.80	202170.63	– 214829.18	3–3	2.27–01	2.12–01	1.65+01	–0.196	B	1

N II: Allowed Transitions — Continued

No.	Transition Array	Mult.	λ_{air} (Å)	λ_{vac} (Å) or σ (cm^{-1})*	E_i (cm^{-1})	E_k (cm^{-1})	g_i-g_k	A_{ki} (10^8 s^{-1})	f_{ik}	S (at. u.)	log gf	Acc.	Source
218.	$2p4p-$ $2p5d$	^{1}P–^{1}P°	5240.86	5242.32	202170.63	– 221246.17	3–3	6.20–03	2.55–03	1.32–01	–2.116	B	1
219.	$2p4p-$ $2p6s$	^{1}P–^{1}P°	4776.22	4777.56	202170.63	– 223101.82	3–3	1.03–01	3.52–02	1.66+00	–0.976	B	1
220.	$2p4p-$ $2p4d$	^{3}D–^{3}F°	*14365*	*6959.6* cm^{-1}	*202799.9*	*– 209759.5*	15–21	2.13–01	9.23–01	6.55+02	1.141	B	1
			14358.6	6962.55 cm^{-1}	202861.36	– 209823.91	7–9	2.13–01	8.48–01	2.81+02	0.774	C+	*LS*
			14337.7	6972.71 cm^{-1}	202765.26	– 209737.97	5–7	1.91–01	8.22–01	1.94+02	0.614	C+	*LS*
			14364.9	6959.53 cm^{-1}	202714.12	– 209673.65	3–5	1.79–01	9.23–01	1.31+02	0.443	C+	*LS*
			14538.1	6876.61 cm^{-1}	202861.36	– 209737.97	7–7	2.29–02	7.26–02	2.43+01	–0.294	C+	*LS*
			14471.2	6908.39 cm^{-1}	202765.26	– 209673.65	5–5	3.25–02	1.02–01	2.43+01	–0.292	C+	*LS*
			14675.3	6812.29 cm^{-1}	202861.36	– 209673.65	7–5	8.80–04	2.03–03	6.86–01	–1.848	C+	*LS*
221.		^{3}D–^{3}D°	*13370*	*7477.6* cm^{-1}	*202799.9*	*– 210277.4*	15–15	6.46–02	1.73–01	1.14+02	0.415	B	1
			13436.6	7440.32 cm^{-1}	202861.36	– 210301.68	7–7	5.66–02	1.53–01	4.74+01	0.030	C+	*LS*
			13328.3	7500.78 cm^{-1}	202765.26	– 210266.04	5–5	4.54–02	1.21–01	2.65+01	–0.219	C+	*LS*
			13284.1	7525.71 cm^{-1}	202714.12	– 210239.83	3–3	4.94–02	1.31–01	1.72+01	–0.407	C+	*LS*
			13501.3	7404.68 cm^{-1}	202861.36	– 210266.04	7–5	9.79–03	1.91–02	5.95+00	–0.874	C+	*LS*
			13375.0	7474.57 cm^{-1}	202765.26	– 210239.83	5–3	1.61–02	2.60–02	5.72+00	–0.887	C+	*LS*
			13265.3	7536.42 cm^{-1}	202765.26	– 210301.68	5–7	7.37–03	2.72–02	5.95+00	–0.866	C+	*LS*
			13238.0	7551.92 cm^{-1}	202714.12	– 210266.04	3–5	9.99–03	4.37–02	5.72+00	–0.882	C+	*LS*
222.		^{3}D–^{3}P°	*12609*	*7928.7* cm^{-1}	*202799.9*	*– 210728.6*	15–9	2.02–02	2.89–02	1.80+01	–0.363	B	1
			12745.3	7843.90 cm^{-1}	202861.36	– 210705.26	7–5	1.64–02	2.86–02	8.39+00	–0.699	C+	*LS*
			12518.2	7986.17 cm^{-1}	202765.26	– 210751.43	5–3	1.55–02	2.18–02	4.50+00	–0.962	C+	*LS*
			12400.0	8062.33 cm^{-1}	202714.12	– 210776.45	3–1	2.12–02	1.63–02	2.00+00	–1.310	C+	*LS*
			12591.0	7940.00 cm^{-1}	202765.26	– 210705.26	5–5	3.04–03	7.23–03	1.50+00	–1.442	C+	*LS*
			12438.6	8037.31 cm^{-1}	202714.12	– 210751.43	3–3	5.26–03	1.22–02	1.50+00	–1.437	C+	*LS*
			12510.4	7991.14 cm^{-1}	202714.12	– 210705.26	3–5	2.07–04	8.08–04	9.99–02	–2.615	C+	*LS*
223.	$2p4p-$ $2p5s$	^{3}D–^{3}P°	*8675.9*	*8678.3*	*202799.9*	*– 214322.9*	15–9	3.04–01	2.06–01	8.82+01	0.490	B	1
			8676.09	8678.47	202861.36	– 214384.13	7–5	2.55–01	2.06–01	4.11+01	0.159	C+	*LS*
			8698.99	8701.38	202765.26	– 214257.69	5–3	2.26–01	1.54–01	2.20+01	–0.114	C+	*LS*
			8694.90	8697.29	202714.12	– 214211.96	3–1	3.02–01	1.14–01	9.80+00	–0.466	C+	*LS*
			8604.33	8606.69	202765.26	– 214384.13	5–5	4.67–02	5.19–02	7.35+00	–0.586	C+	*LS*
			8660.45	8662.83	202714.12	– 214257.69	3–3	7.64–02	8.59–02	7.35+00	–0.589	C+	*LS*
			8566.62	8568.97	202714.12	– 214384.13	3–5	3.16–03	5.79–03	4.90–01	–1.760	C+	*LS*
224.	$2s^22p4p-$ $2s2p^23p$	^{3}D–^{3}D°	*3852.1*	*3853.2*	*202799.9*	*– 228752.3*	15–15	5.33–03	1.18–03	2.25–01	–1.750	B	1
			3855.37	3856.47	202861.36	– 228791.83	7–7	4.72–03	1.05–03	9.35–02	–2.133	C+	*LS*
			3850.02	3851.11	202765.26	– 228731.81	5–5	3.71–03	8.25–04	5.23–02	–2.384	C+	*LS*
			3848.00	3849.09	202714.12	– 228694.30	3–3	4.01–03	8.90–04	3.38–02	–2.574	C+	*LS*
			3864.32	3865.41	202861.36	– 228731.81	7–5	8.23–04	1.32–04	1.17–02	–3.036	C	*LS*
			3855.59	3856.68	202765.26	– 228694.30	5–3	1.33–03	1.78–04	1.13–02	–3.052	C	*LS*
			3841.14	3842.23	202765.26	– 228791.83	5–7	5.99–04	1.85–04	1.17–02	–3.033	C	*LS*
			3842.45	3843.54	202714.12	– 228731.81	3–5	8.05–04	2.97–04	1.13–02	–3.050	C	*LS*
225.	$2p4p-$ $2p4d$	^{3}P–^{3}D°	*14243*	*7019.2* cm^{-1}	*203224.9*	*– 210244.1*	9–15	1.40–01	7.08–01	2.99+02	0.804	B	1
			14195.2	7042.70 cm^{-1}	203258.98	– 210301.68	5–7	1.41–01	5.97–01	1.39+02	0.475	C+	*LS*
			14328.9	6977.01 cm^{-1}	203189.03	– 210166.04	3–5	1.03–01	5.28–01	7.47+01	0.200	C+	*LS*
			14125.7	7077.35 cm^{-1}	203162.48	– 210239.83	1–3	7.95–02	7.14–01	3.32+01	–0.146	C+	*LS*
			14474.0	6907.06 cm^{-1}	203258.98	– 210166.04	5–5	3.33–02	1.04–01	2.49+01	–0.282	C+	*LS*
			14178.9	7050.80 cm^{-1}	203189.03	– 210239.83	3–3	5.90–02	1.78–01	2.49+01	–0.273	C+	*LS*
			14321.0	6980.85 cm^{-1}	203258.98	– 210239.83	5–3	3.82–03	7.04–03	1.66+00	–1.453	C+	*LS*

N II: Allowed Transitions — Continued

No.	Transition Array	Mult.	λ_{air} (Å)	λ_{vac} (Å) or σ (cm^{-1})*	E_i (cm^{-1})	E_k (cm^{-1})	g_i–g_k	A_{ki} (10^8 s^{-1})	f_{ik}	S (at. u.)	log gf	Acc.	Source
226.		^{3}P–^{3}P°	*13323*	*7503.6* cm^{-1}	*203224.9*	– *210728.6*	9–9	7.45–02	1.98–01	7.83+01	0.251	B	1
			13425.9	7446.28 cm^{-1}	203258.98	– 210705.26	5–5	5.46–02	1.48–01	3.26+01	–0.132	C+	*LS*
			13219.7	7562.40 cm^{-1}	203189.03	– 210751.43	3–3	1.91–02	5.00–02	6.52+00	–0.824	C+	*LS*
			13343.1	7492.45 cm^{-1}	203258.98	– 210751.43	5–3	3.09–02	4.95–02	1.09+01	–0.607	C+	*LS*
			13176.1	7587.42 cm^{-1}	203189.03	– 210776.45	3–1	7.70–02	6.68–02	8.70+00	–0.698	C+	*LS*
			13300.9	7516.23 cm^{-1}	203189.03	– 210705.26	3–5	1.87–02	8.27–02	1.09+01	–0.605	C+	*LS*
			13173.5	7588.95 cm^{-1}	203162.48	– 210751.43	1–3	2.57–02	2.01–01	8.70+00	–0.698	C+	*LS*
227.	2p4p– 2p5s	^{3}P–^{3}P°	*9008.2*	*9010.7*	*203224.9*	– *214322.9*	9–9	1.88–01	2.29–01	6.12+01	0.314	B	1
			8986.18	8988.64	203258.98	– 214384.13	5–5	1.42–01	1.72–01	2.55+01	–0.065	C+	*LS*
			9032.04	9034.52	203189.03	– 214257.69	3–3	4.67–02	5.71–02	5.10+00	–0.766	C+	*LS*
			9089.48	9091.98	203258.98	– 214257.69	5–3	7.64–02	5.68–02	8.49+00	–0.547	C+	*LS*
			9069.51	9072.00	203189.03	– 214211.96	3–1	1.85–01	7.59–02	6.80+00	–0.643	C+	*LS*
			8930.03	8932.48	203189.03	– 214384.13	3–5	4.83–02	9.63–02	8.49+00	–0.539	C+	*LS*
			9010.42	9012.90	203162.48	– 214257.69	1–3	6.27–02	2.29–01	6.80+00	–0.640	C+	*LS*
228.	2$s^2$2p4p– 2s2$p^2$3p	^{3}P–^{3}S°	*5037.9*	*5039.3*	*203224.9*	– 223069.02	9–3	9.75–02	1.24–02	1.85+00	–0.954	B	1
			5046.54	5047.95	203258.98	– 223069.02	5–3	5.39–02	1.23–02	1.03+00	–1.210	C+	*LS*
			5028.78	5030.18	203189.03	– 223069.02	3–3	3.27–02	1.24–02	6.15–01	–1.430	C+	*LS*
			5022.07	5023.47	203162.48	– 223069.02	1–3	1.09–02	1.24–02	2.05–01	–1.907	C+	*LS*
229.		^{3}P–^{3}D°	*3916.3*	*3917.4*	*203224.9*	– *228752.3*	9–15	3.22–01	1.23–01	1.43+01	0.045	B	1
			3915.41	3916.52	203258.98	– 228791.83	5–7	3.22–01	1.04–01	6.68+00	–0.286	C+	*LS*
			3913.89	3915.00	203189.03	– 228731.81	3–5	2.42–01	9.25–02	3.58+00	–0.557	C+	*LS*
			3915.57	3916.68	203162.48	– 228694.30	1–3	1.79–01	1.23–01	1.59+00	–0.909	C+	*LS*
			3924.64	3925.75	203258.98	– 228731.81	5–5	7.99–02	1.85–02	1.19+00	–1.035	C+	*LS*
			3919.65	3920.76	203189.03	– 228694.30	3–3	1.34–01	3.08–02	1.19+00	–1.034	C+	*LS*
			3930.43	3931.54	203258.98	– 228694.30	5–3	8.84–03	1.23–03	7.95–02	–2.212	C+	*LS*
230.	2p4p– 2p4d	^{3}S–^{3}P°	*13903*	*7190.9* cm^{-1}	203537.66	– *210728.6*	3–9	1.46–01	1.27+00	1.75+02	0.582	B	1
			13947.9	7167.60 cm^{-1}	203537.66	– 210705.26	3–5	1.45–01	7.05–01	9.71+01	0.325	C+	*LS*
			13858.6	7213.77 cm^{-1}	203537.66	– 210751.43	3–3	1.48–01	4.26–01	5.83+01	0.106	C+	*LS*
			13810.7	7238.79 cm^{-1}	203537.66	– 210776.45	3–1	1.49–01	1.42–01	1.94+01	–0.369	C+	*LS*
231.	2p4p– 2p5s	^{3}S–^{3}P°	*9269.4*	*9272.0*	203537.66	– *214322.9*	3–9	4.00–02	1.55–01	1.42+01	–0.333	B	1
			9217.06	9219.59	203537.66	– 214384.13	3–5	4.07–02	8.64–02	7.87+00	–0.586	C+	*LS*
			9325.77	9328.33	203537.66	– 214257.69	3–3	3.93–02	5.13–02	4.72+00	–0.813	C+	*LS*
			9365.73	9368.30	203537.66	– 214211.96	3–1	3.88–02	1.70–02	1.57+00	–1.292	C+	*LS*
232.	2s2p^2(^{4}P)3s– 2s2$p^2$3p	^{5}P–^{5}D°	*5537.4*	*5538.9*	*205675.9*	– *223730.1*	15–25	6.04–01	4.63–01	1.26+02	0.841	B	1
			5535.35	5536.88	205724.81	– 223785.51	7–9	6.04–01	3.57–01	4.55+01	0.398	C+	*LS*
			5530.24	5531.78	205654.22	– 223731.59	5–7	4.04–01	2.59–01	2.36+01	0.113	C+	*LS*
			5526.23	5527.77	205597.97	– 223688.45	3–5	2.13–01	1.62–01	8.85+00	–0.313	C+	*LS*
			5551.92	5553.46	205724.81	– 223731.59	7–7	2.00–01	9.23–02	1.18+01	–0.190	C+	*LS*
			5543.47	5545.01	205654.22	– 223688.45	5–5	3.51–01	1.62–01	1.48+01	–0.092	C+	*LS*
			5535.38	5536.92	205597.97	– 223658.55	3–3	4.53–01	2.08–01	1.14+01	–0.204	C+	*LS*
			5565.26	5566.80	205724.81	– 223688.45	7–5	3.97–02	1.32–02	1.69+00	–1.036	C+	*LS*
			5552.68	5554.22	205654.22	– 223658.55	5–3	1.50–01	4.15–02	3.79+00	–0.683	C+	*LS*
			5540.06	5541.60	205597.97	– 223643.30	3–1	6.03–01	9.25–02	5.06+00	–0.557	C+	*LS*

N II: Allowed Transitions — Continued

No.	Transition Array	Mult.	λ_{air} (Å)	λ_{vac} (Å) or σ (cm^{-1})*	E_i (cm^{-1})	E_k (cm^{-1})	g_i–g_k	A_{ki} (10^8 s^{-1})	f_{ik}	S (at. u.)	log gf	Acc.	Source
233.		^{5}P–^{5}P°	*5006.8*	*5008.2*	*205675.9*	*– 225643.1*	15–15	7.80–01	2.93–01	7.25+01	0.643	B	1
			5012.04	5013.43	205724.81	– 225671.22	7–7	5.19–01	1.95–01	2.26+01	0.136	C+	*LS*
			5005.30	5006.70	205654.22	– 225627.47	5–5	6.51–02	2.44–02	2.01+00	–0.913	C+	*LS*
			4997.22	4998.62	205597.97	– 225603.50	3–3	1.96–01	7.35–02	3.63+00	–0.657	C+	*LS*
			5023.05	5024.45	205724.81	– 225627.47	7–5	3.61–01	9.74–02	1.13+01	–0.166	C+	*LS*
			5011.31	5012.71	205654.22	– 225603.50	5–3	5.84–01	1.32–01	1.09+01	–0.181	C+	*LS*
			4994.36	4995.75	205654.22	– 225671.22	5–7	2.62–01	1.37–01	1.13+01	–0.164	C+	*LS*
			4991.24	4992.64	205597.97	– 225627.47	3–5	3.54–01	2.21–01	1.09+01	–0.179	C+	*LS*
234.		^{5}P–^{5}S°	*4137.4*	*4138.6*	*205675.9*	– 229838.96	15–5	1.59+00	1.36–01	2.77+01	0.309	B	1
			4145.77	4146.94	205724.81	– 229838.96	7–5	7.36–01	1.35–01	1.29+01	–0.023	C+	*LS*
			4133.67	4134.84	205654.22	– 229838.96	5–5	5.30–01	1.36–01	9.24+00	–0.168	C+	*LS*
			4124.08	4125.24	205597.97	– 229838.96	3–5	3.20–01	1.36–01	5.55+00	–0.389	C+	*LS*
235.	$2p4p$– $2p4d$	^{1}D–^{1}D°	21849.2	4575.58 cm^{-1}	205350.18	– 209925.76	5–5	1.70–02	1.22–01	4.38+01	–0.215	B	1
236.		^{1}D–^{1}F°	17376.1	5753.45 cm^{-1}	205350.18	– 211103.63	5–7	1.42–01	8.99–01	2.57+02	0.653	B	1
237.		^{1}D–^{1}P°	16701.1	5985.98 cm^{-1}	205350.18	– 211336.16	5–3	1.58–02	3.96–02	1.09+01	–0.704	B	1
238.	$2p4p$– $2p5s$	^{1}D–^{1}P°	10546.7	9479.00 cm^{-1}	205350.18	– 214829.18	5–3	2.77–01	2.78–01	4.82+01	0.142	B	1
239.	$2p4p$– $2p5d$	^{1}D–^{1}D°	6600.94	6602.76	205350.18	– 220495.36	5–5	1.86–02	1.21–02	1.32+00	–1.217	B	1
240.		^{1}D–^{1}F°	6330.80	6332.55	205350.18	– 221141.61	5–7	1.39–02	1.17–02	1.22+00	–1.234	B	1
241.	$2p4p$– $2p6s$	^{1}D–^{1}P°	5631.72	5633.28	205350.18	– 223101.82	5–3	1.18–01	3.36–02	3.11+00	–0.775	B	1
242.	$2p4p$– $2p4d$	^{1}S–^{1}P°	22588.0	4425.92 cm^{-1}	206910.24	– 211336.16	1–3	4.58–02	1.05+00	7.81+01	0.021	B	1
243.	$2p4p$– $2p5s$	^{1}S–^{1}P°	12624.5	7918.94 cm^{-1}	206910.24	– 214829.18	1–3	2.84–02	2.04–01	8.47+00	–0.691	B	1
244.	$2p4p$– $2p5d$	^{1}S–^{1}P°	6973.56	6975.48	206910.24	– 221246.17	1–3	2.62–02	5.73–02	1.31+00	–1.242	B	1
245.	$2p4p$– $2p6s$	^{1}S–^{1}P°	6174.34	6176.05	206910.24	– 223101.82	1–3	1.40–02	2.39–02	4.87–01	–1.621	B	1
246.	$2s^22p4d$– $2s2p^2(^4P)3s$	^{3}D°–^{3}P		*1525.5* cm^{-1}	*210277.4*	*– 211802.9*	15–9	9.69–04	3.74–02	1.21+02	–0.251	B	1
				1525.99 cm^{-1}	210301.68	– 211827.67	7–5	8.15–04	3.74–02	5.65+01	–0.582	C+	*LS*
				1513.41 cm^{-1}	210266.04	– 211779.45	5–3	7.09–04	2.78–02	3.03+01	–0.856	C+	*LS*
				1509.52 cm^{-1}	210239.83	– 211749.35	3–1	9.39–04	2.06–02	1.35+01	–1.210	C+	*LS*
				1561.63 cm^{-1}	210266.04	– 211827.67	5–5	1.56–04	9.58–03	1.01+01	–1.320	C+	*LS*
				1539.62 cm^{-1}	210239.83	– 211779.45	3–3	2.49–04	1.57–02	1.01+01	–1.326	C+	*LS*
				1587.84 cm^{-1}	210239.83	– 211827.67	3–5	1.09–05	1.08–03	6.73–01	–2.489	C+	*LS*

N II: Allowed Transitions — Continued

No.	Transition Array	Mult.	λ_{air} (Å)	λ_{vac} (Å) or σ (cm^{-1})*	E_i (cm^{-1})	E_k (cm^{-1})	g_i–g_k	A_{ki} (10^8 s^{-1})	f_{ik}	S (at. u.)	log gf	Acc.	Source
247.		$^3P^o$–^3P		*1074.3* cm^{-1}	*210728.6*	*– 211802.9*	9–9	8.72–05	1.13–02	3.12+01	–0.992	B	1
				1122.41 cm^{-1}	210705.26	– 211827.67	5–5	7.46–05	8.87–03	1.30+01	–1.353	C+	*LS*
				1028.02 cm^{-1}	210751.43	– 211779.45	3–3	1.91–05	2.71–03	2.60+00	–2.090	C+	*LS*
				1074.19 cm^{-1}	210705.26	– 211779.45	5–3	3.63–05	2.83–03	4.34+00	–1.849	C+	*LS*
				997.92 cm^{-1}	210751.43	– 211749.35	3–1	6.99–05	3.51–03	3.47+00	–1.978	C+	*LS*
				1076.24 cm^{-1}	210751.43	– 211827.67	3–5	2.19–05	4.73–03	4.34+00	–1.848	C+	*LS*
				1003.00 cm^{-1}	210776.45	– 211779.45	1–3	2.37–05	1.06–02	3.47+00	–1.976	C+	*LS*
248.	$2s2p^2(^4P)3s$– $2s^22p5s$	^3P–$^3P^o$	*39672*	*2520.0* cm^{-1}	*211802.9*	*– 214322.9*	9–9	3.87–03	9.12–02	1.07+02	–0.086	B	1
			39105.9	2556.46 cm^{-1}	211827.67	– 214384.13	5–5	3.03–03	6.94–02	4.47+01	–0.460	C+	*LS*
			40340.2	2478.24 cm^{-1}	211779.45	– 214257.69	3–3	9.19–04	2.24–02	8.94+00	–1.172	C+	*LS*
			41140.7	2430.02 cm^{-1}	211827.67	– 214257.69	5–3	1.44–03	2.20–02	1.49+01	–0.959	C+	*LS*
			41098.6	2432.51 cm^{-1}	211779.45	– 214211.96	3–1	3.48–03	2.94–02	1.19+01	–1.055	C+	*LS*
			38382.0	2604.68 cm^{-1}	211779.45	– 214384.13	3–5	1.07–03	3.93–02	1.49+01	–0.929	C+	*LS*
			39856.1	2508.34 cm^{-1}	211749.35	– 214257.69	1–3	1.27–03	9.08–02	1.19+01	–1.042	C+	*LS*
249.	$2s2p^2(^4P)3s$– $2s2p^23p$	^3P–$^3S^o$	*8873.7*	*8876.2*	*211802.9*	– 223069.02	9–3	7.47–02	2.94–02	7.73+00	–0.577	B	1
			8893.29	8895.73	211827.67	– 223069.02	5–3	4.12–02	2.93–02	4.30+00	–0.834	C+	*LS*
			8855.30	8857.73	211779.45	– 223069.02	3–3	2.51–02	2.95–02	2.58+00	–1.054	C+	*LS*
			8831.75	8834.18	211749.35	– 223069.02	1–3	8.42–03	2.95–02	8.59–01	–1.530	C+	*LS*
250.		^3P–$^3D^o$	*5898.3*	*5899.9*	*211802.9*	*– 228752.3*	9–15	2.88–01	2.50–01	4.37+01	0.352	B	1
			5893.15	5894.78	211827.67	– 228791.83	5–7	2.88–01	2.10–01	2.04+01	0.022	C+	*LS*
			5897.25	5898.88	211779.45	– 228731.81	3–5	2.16–01	1.88–01	1.09+01	–0.250	C+	*LS*
			5899.83	5901.46	211749.35	– 228694.30	1–3	1.60–01	2.50–01	4.86+00	–0.602	C+	*LS*
			5914.07	5915.71	211827.67	– 228731.81	5–5	7.13–02	3.74–02	3.64+00	–0.728	C+	*LS*
			5910.33	5911.96	211779.45	– 228694.30	3–3	1.19–01	6.24–02	3.64+00	–0.728	C+	*LS*
			5927.22	5928.87	211827.67	– 228694.30	5–3	7.87–03	2.49–03	2.43–01	–1.905	C+	*LS*
251.	$2s2p^23p$– $2s2p^23d$	$^5D^o$–^5F	*5178.0*	*5179.4*	*223730.1*	*– 243037.3*	25–35	1.07+00	6.02–01	2.57+02	1.178	B	1
			5179.52	5180.96	223785.51	– 243086.94	9–11	1.07+00	5.26–01	8.07+01	0.675	C+	*LS*
			5175.89	5177.33	223731.59	– 243046.56	7–9	8.93–01	4.61–01	5.50+01	0.509	C+	*LS*
			5173.39	5174.83	223688.45	– 243012.77	5–7	7.36–01	4.13–01	3.52+01	0.315	C+	*LS*
			5172.34	5173.78	223658.55	– 242986.76	3–5	6.01–01	4.02–01	2.05+01	0.081	C+	*LS*
			5172.97	5174.41	223643.30	– 242969.16	1–3	5.01–01	6.03–01	1.03+01	–0.220	C+	*LS*
			5190.38	5191.82	223785.51	– 243046.56	9–9	1.77–01	7.16–02	1.10+01	–0.191	C+	*LS*
			5184.96	5186.40	223731.59	– 243012.77	7–7	3.20–01	1.29–01	1.54+01	–0.045	C+	*LS*
			5180.36	5181.80	223688.45	– 242986.76	5–5	4.28–01	1.72–01	1.47+01	–0.065	C+	*LS*
			5177.06	5178.50	223658.55	– 242969.16	3–3	5.00–01	2.01–01	1.03+01	–0.220	C+	*LS*
			5199.50	5200.95	223785.51	– 243012.77	9–7	1.51–02	4.76–03	7.34–01	–1.368	C+	*LS*
			5191.96	5193.41	223731.59	– 242986.76	7–5	4.25–02	1.23–02	1.47+00	–1.066	C+	*LS*
			5185.09	5186.53	223688.45	– 242969.16	5–3	7.11–02	1.72–02	1.47+00	–1.066	C+	*LS*
252.		$^5D^o$–^5P	*4841.5*	*4842.8*	*223730.1*	*– 244379.2*	25–15	5.71–03	1.20–03	4.80–01	–1.522	B	1
			4860.61	4861.97	223785.51	– 244353.31	9–7	4.35–03	1.20–03	1.73–01	–1.967	C+	*LS*
			4838.85	4840.20	223731.59	– 244391.88	7–5	3.20–03	8.03–04	8.95–02	–2.250	C+	*LS*
			4822.56	4823.91	223688.45	– 244418.52	5–3	2.02–03	4.23–04	3.36–02	–2.675	C+	*LS*
			4847.90	4849.26	223731.59	– 244353.31	7–7	1.14–03	4.01–04	4.48–02	–2.552	C+	*LS*
			4828.77	4830.12	223688.45	– 244391.88	5–5	2.01–03	7.04–04	5.59–02	–2.454	C+	*LS*
			4815.62	4816.96	223658.55	– 244418.52	3–3	2.61–03	9.07–04	4.32–02	–2.565	C+	*LS*
			4837.78	4839.13	223688.45	– 244353.31	5–7	1.63–04	8.03–05	6.39–03	–3.396	C–	*LS*
			4821.80	4823.15	223658.55	– 244391.88	3–5	5.20–04	3.02–04	1.44–02	–3.043	C	*LS*
			4812.08	4813.43	223643.30	– 244418.52	1–3	1.16–03	1.21–03	1.92–02	–2.917	C	*LS*

N II: Allowed Transitions — Continued

No.	Transition Array	Mult.	λ_{air} (Å)	λ_{vac} (Å) or σ (cm^{-1})*	E_i (cm^{-1})	E_k (cm^{-1})	g_i-g_k	A_{ki} (10^8 s^{-1})	f_{ik}	S (at. u.)	log gf	Acc.	Source
253.		$^5D^o-^5D$	*4709.3*	*4710.6*	*223730.1*	*– 244958.7*	25–25	3.65–01	1.21–01	4.70+01	0.481	B	1
			4718.38	4719.70	223785.51	– 244973.31	9–9	3.02–01	1.01–01	1.41+01	–0.042	C+	*LS*
			4709.58	4710.90	223731.59	– 244958.95	7–7	1.82–01	6.06–02	6.58+00	–0.372	C+	*LS*
			4702.50	4703.82	223688.45	– 244947.77	5–5	9.15–02	3.03–02	2.35+00	–0.819	C+	*LS*
			4697.64	4698.95	223658.55	– 244939.89	3–3	3.06–02	1.01–02	4.70–01	–1.517	C+	*LS*
			4721.58	4722.90	223785.51	– 244958.95	9–7	7.75–02	2.01–02	2.82+00	–0.742	C+	*LS*
			4712.07	4713.38	223731.59	– 244947.77	7–5	1.46–01	3.46–02	3.76+00	–0.616	C+	*LS*
			4704.25	4705.56	223688.45	– 244939.89	5–3	2.13–01	4.25–02	3.29+00	–0.673	C+	*LS*
			4698.55	4699.87	223658.55	– 244935.74	3–1	3.67–01	4.05–02	1.88+00	–0.916	C+	*LS*
			4706.40	4707.72	223731.59	– 244973.31	7–9	6.09–02	2.60–02	2.82+00	–0.740	C+	*LS*
			4700.03	4701.35	223688.45	– 244958.95	5–7	1.05–01	4.86–02	3.76+00	–0.615	C+	*LS*
			4695.90	4697.21	223658.55	– 244947.77	3–5	1.29–01	7.09–02	3.29+00	–0.672	C+	*LS*
			4694.27	4695.59	223643.30	– 244939.89	1–3	1.23–01	1.22–01	1.88+00	–0.915	C+	*LS*
254.		$^5P^o-^5P$	*5335.8*	*5337.3*	*225643.1*	*– 244379.2*	15–15	5.56–01	2.37–01	6.25+01	0.551	B	1
			5351.23	5352.72	225671.22	– 244353.31	7–7	3.67–01	1.58–01	1.94+01	0.043	C+	*LS*
			5327.76	5329.24	225627.47	– 244391.88	5–5	4.65–02	1.98–02	1.74+00	–1.004	C+	*LS*
			5313.42	5314.90	225603.50	– 244418.52	3–3	1.41–01	5.95–02	3.12+00	–0.748	C+	*LS*
			5340.21	5341.69	225671.22	– 244391.88	7–5	2.59–01	7.90–02	9.72+00	–0.257	C+	*LS*
			5320.20	5321.68	225627.47	– 244418.52	5–3	4.20–01	1.07–01	9.37+00	–0.271	C+	*LS*
			5338.73	5340.21	225627.47	– 244353.31	5–7	1.85–01	1.11–01	9.72+00	–0.257	C+	*LS*
			5320.96	5322.44	225603.50	– 244391.88	3–5	2.52–01	1.78–01	9.37+00	–0.272	C+	*LS*
255.		$^5P^o-^5D$	*5175.7*	*5177.2*	*225643.1*	*– 244958.7*	15–25	8.69–01	5.82–01	1.49+02	0.941	B	1
			5179.34	5180.79	225671.22	– 244973.31	7–9	8.67–01	4.48–01	5.35+01	0.497	C+	*LS*
			5171.47	5172.91	225627.47	– 244958.95	5–7	5.81–01	3.26–01	2.78+01	0.212	C+	*LS*
			5168.05	5169.49	225603.50	– 244947.77	3–5	3.06–01	2.04–01	1.04+01	–0.213	C+	*LS*
			5183.20	5184.64	225671.22	– 244958.95	7–7	2.88–01	1.16–01	1.39+01	–0.090	C+	*LS*
			5174.46	5175.90	225627.47	– 244947.77	5–5	5.07–01	2.04–01	1.73+01	0.008	C+	*LS*
			5170.16	5171.60	225603.50	– 244939.89	3–3	6.54–01	2.62–01	1.34+01	–0.104	C+	*LS*
			5186.21	5187.65	225671.22	– 244947.77	7–5	5.76–02	1.66–02	1.98+00	–0.935	C+	*LS*
			5176.57	5178.01	225627.47	– 244939.89	5–3	2.17–01	5.23–02	4.46+00	–0.582	C+	*LS*
			5171.27	5172.71	225603.50	– 244935.74	3–1	8.71–01	1.16–01	5.95+00	–0.457	C+	*LS*
256.		$^5S^o-^5P$	*6875.6*	*6877.5*	229838.96	*– 244379.2*	5–15	2.51–01	5.33–01	6.03+01	0.425	B	1
			6887.83	6889.73	229838.96	– 244353.31	5–7	2.49–01	2.48–01	2.81+01	0.094	C+	*LS*
			6869.58	6871.47	229838.96	– 244391.88	5–5	2.51–01	1.78–01	2.01+01	–0.051	C+	*LS*
			6857.03	6858.92	229838.96	– 244418.52	5–3	2.53–01	1.07–01	1.21+01	–0.272	C+	*LS*

*Wavelengths (Å) are always given unless cm^{-1} is indicated.

N II

Forbidden Transitions

The tabulated data are selected from two advanced calculations, which include detailed treatments of the electron correlation problem: the multi-configuration calculations by Nussbaumer and Rusca[1] with the SUPERSTRUCTURE code and the multi-configuration Hartree-Fock calculations by Froese Fischer and Saha.[2] In these calculations, at least 11 configurations are included in the wavefunction expansions, and the results of the two papers typically agree within ± 10–20%. The line strengths of the M1 fine structure transitions within the ^{3}P term of the ground state configuration are the same as derived from general formulas first developed by Shortley *et al*.[3] for the case of near *LS*-coupling conditions.

References

[1]H. Nussbaumer and C. Rusca, Astron. Astrophys. **72**, 129 (1979).
[2]C. Froese Fischer and H. P. Saha, Phys. Scr. **32**, 181 (1985).
[3]G. H. Shortley, L. H. Aller, J. G. Baker, and D. H. Menzel, Astrophys. J **93**, 178 (1941).

N II: Forbidden Transitions

No.	Transition Array	Mult.	λ_{air} (Å)	λ_{vac} (Å) or σ (cm^{-1})*	E_i (cm^{-1})	E_k (cm^{-1})	g_i-g_k	Type	A_{ki} (s^{-1})	S (at. u.)	Acc.	Source
1.	$2p^2-$ $2p^2$	^{3}P–^{3}P										
				48.7 cm^{-1}	0.0	– 48.7	1–3	M1	2.07–06	2.00+00	B	1,2
				130.8 cm^{-1}	0.0	– 130.8	1–5	E2	9.69–13	1.13+00	C	1,2
				82.1 cm^{-1}	48.7	– 130.8	3–5	M1	7.40–06	2.50+00	B	1,2
				82.1 cm^{-1}	48.7	– 130.8	3–5	E2	2.59–13	3.10+00	C	2
2.		^{3}P–^{1}D										
			6527.23	6529.03	0.0	– 15316.2	1–5	E2	5.45–07	2.89–05	B	1,2
			6548.05	6549.86	48.7	– 15316.2	3–5	M1	9.19–04	4.79–05	B	1,2
			6548.05	6549.86	48.7	– 15316.2	3–5	E2	9.07–07	4.88–05	C+	2
			6583.45	6585.27	130.8	– 15316.2	5–5	M1	2.72–03	1.44–04	B	1,2
			6583.45	6585.27	130.8	– 15316.2	5–5	E2	7.53–06	4.16–04	C+	2
3.		^{3}P–^{1}S										
			3062.83	3063.72	48.7	– 32688.8	3–1	M1	3.15–02	3.36–05	B	1,2
			3070.55	3071.44	130.8	– 32688.8	5–1	E2	1.40–04	3.41–05	B	1,2
4.		^{1}D–^{1}S										
			5754.59	5756.19	15316.2	– 32688.8	5–1	E2	1.17+00	6.59+00	B	1,2

*Wavelengths (Å) are always given unless cm^{-1} is indicated.

N III

Boron Isoelectronic Sequence

Ground State: $1s^22s^22p\ ^2P^\circ_{1/2}$

Ionization Energy: 47.448 eV = 382703.8 cm^{-1}

Allowed Transitions

List of tabulated lines

Wavelength (Å)	No.	Wavelength (Å)	No.	Wavelength (Å)	No.	Wavelength (Å)	No.
in vacuum		265.232	40	274.231	35	276.326	16
		265.271	39	274.258	35	282.063	15
237.994	21	266.806	18	274.268	34	282.202	15
238.035	21	266.847	18	274.276	35	283.850	32
238.093	21	266.930	18	274.313	34	283.882	32
238.134	21	266.971	18	274.313	35	283.898	32
246.206	20	268.212	38	274.337	35	283.930	32
246.312	20	268.255	38	274.374	34	283.974	32
248.320	20	268.314	38	274.374	35	283.996	32
248.428	20	268.601	37	275.821	33	284.040	32
264.823	19	268.644	37	275.824	33	285.857	14
264.846	19	268.654	36	275.829	33	286.000	14
264.945	19	268.697	36	275.852	33	290.861	31
264.968	19	268.703	37	275.869	33	290.912	31
265.133	40	268.756	36	275.883	33	290.930	31
265.172	39	272.524	17	275.897	33	290.962	31
265.175	40	272.654	17	275.931	33	290.972	31
265.214	39	274.213	35	276.193	16	291.023	31

List of tabulated lines — Continued

Wavelength (Å)	No.	Wavelength (Å)	No.	Wavelength (Å)	No.	Wavelength (Å)	No.
291.031	31	374.198	6	543.767	84	774.357	115
292.442	13	374.434	6	543.836	84	774.572	115
292.591	13	374.442	6	543.879	84	774.707	115
292.592	13	382.706	51	553.370	60	774.923	115
299.660	12	382.716	51	553.446	60	802.278	171
299.816	12	393.809	50	572.945	83	802.508	171
303.825	30	393.819	50	572.992	83	802.508	171
303.855	30	398.884	71	573.102	83	802.905	171
303.880	30	399.044	71	577.296	24	803.527	171
303.910	30	399.060	71	577.310	24	812.828	149
303.960	30	413.681	49	577.495	24	813.306	149
303.985	30	413.792	49	577.509	24	813.577	149
304.035	30	413.803	49	577.765	24	814.056	149
305.756	11	418.705	48	580.379	100	822.062	188
305.919	11	418.900	48	580.393	100	822.230	188
311.548	10	418.911	48	580.652	100	822.329	188
311.631	10	428.184	47	580.666	100	822.497	188
311.717	10	428.196	47	582.309	82	823.314	92
314.715	9	428.236	47	582.357	82	823.796	92
314.856	9	428.248	47	582.555	82	824.575	114
314.888	9	433.915	26	601.467	59	824.819	114
317.808	54	434.019	26	601.877	59	832.367	196
317.809	54	434.067	26	602.993	68	834.515	170
317.814	54	434.132	26	603.083	68	834.763	170
317.816	54	434.137	26	603.393	68	835.026	169
321.079	29	434.249	26	603.484	68	835.193	170
321.139	29	434.285	26	660.551	67	835.275	169
321.163	29	449.563	62	661.032	67	835.705	169
321.201	29	449.567	62	661.046	67	835.865	170
321.205	29	451.871	5	661.527	67	836.378	169
321.267	29	452.227	5	679.316	99	838.021	113
321.285	29	456.070	46	679.335	99	838.273	113
323.431	8	456.073	46	679.470	99	847.514	81
323.487	8	456.087	46	679.490	99	847.616	81
323.613	8	460.609	86	684.998	4	848.151	81
323.670	8	460.640	86	685.393	23	848.253	81
332.135	7	460.793	86	685.515	4	848.266	126
332.327	7	467.431	76	685.673	23	848.306	126
338.340	53	467.649	76	685.740	23	854.241	209
338.347	53	467.794	76	685.817	4	854.552	209
338.351	53	472.237	61	686.055	23	854.645	208
338.359	53	472.395	61	686.122	23	854.957	208
338.810	73	500.924	85	686.336	4	854.983	209
338.936	73	500.960	85	691.193	44	855.388	208
351.986	52	501.163	85	691.225	44	855.726	91
351.994	52	507.906	70	691.397	44	855.786	91
358.282	28	508.089	70	757.886	116	859.102	112
358.327	28	508.190	70	758.078	116	859.177	187
358.359	28	508.374	70	758.093	116	859.360	187
358.404	28	509.579	45	758.177	159	859.367	112
358.474	28	509.596	45	758.285	116	859.437	187
358.508	28	509.890	45	758.487	159	859.621	187
358.578	28	526.942	101	763.334	3	867.536	195
362.822	27	526.954	101	764.351	3	871.802	80
362.839	27	527.143	101	771.545	22	871.862	58
362.872	27	527.155	101	771.901	22	871.911	80
362.901	27	530.041	69	772.384	22	872.044	80
362.918	27	530.272	69	772.889	43	872.135	58
362.942	27	530.351	69	772.929	43	872.153	80
362.979	27	530.464	25	772.955	43	881.192	111
363.007	27	530.565	25	773.118	127	881.470	111
370.639	72	530.632	25	773.151	127	883.855	135
370.791	72	530.733	25	773.177	127	884.146	135
370.793	72	530.861	25	773.210	127	884.632	135

List of tabulated lines — Continued

Wavelength (Å)	No.	Wavelength (Å)	No.	Wavelength (Å)	No.	Wavelength (Å)	No.
884.668	135	979.768	42	1090.41	133	1243.37	147
884.758	135	979.832	42	1090.62	133	1244.60	147
885.154	135	979.905	42	1091.25	133	1245.12	147
885.536	135	979.969	42	1094.55	165	1251.64	283
887.393	207	989.799	2	1095.29	165	1251.67	283
887.857	207	991.509	134	1095.55	165	1252.26	283
891.324	206	991.511	2	1095.98	165	1252.29	283
891.662	206	991.577	2	1096.42	165	1268.44	220
891.758	168	991.794	134	1096.45	165	1268.60	220
891.907	205	992.120	134	1096.72	165	1268.95	220
892.042	168	992.405	134	1096.84	165	1269.11	220
892.131	206	992.873	134	1098.72	185	1269.34	94
892.246	205	993.543	134	1099.02	185	1269.41	94
892.533	168	994.012	134	1099.70	185	1270.21	94
892.716	205	994.373	134	1100.00	185	1270.28	94
893.301	168	1001.70	166	1100.07	132	1287.77	146
908.115	167	1001.75	66	1100.82	132	1288.74	146
908.192	167	1002.06	166	1102.22	132	1289.65	146
908.419	167	1002.11	66	1103.04	96	1290.54	273
908.701	167	1002.11	166	1104.07	96	1290.62	146
908.713	167	1002.47	166	1104.12	96	1291.72	273
908.727	167	1002.85	66	1106.04	90	1291.75	273
908.910	167	1003.01	166	1106.34	90	1292.93	273
909.021	167	1003.09	166	1120.24	131	1304.32	164
909.222	167	1003.21	66	1120.26	131	1304.88	164
909.497	167	1003.64	166	1120.60	131	1304.93	164
909.762	320	1004.61	166	1120.81	131	1305.48	164
910.132	320	1005.99	57	1121.02	131	1306.41	164
910.569	320	1006.04	57	1121.71	131	1306.54	164
918.568	110	1012.19	255	1122.48	131	1307.46	164
918.871	110	1012.48	255	1135.76	107	1309.11	164
921.760	186	1012.60	255	1136.23	107	1315.88	158
921.972	186	1021.12	221	1140.05	123	1316.44	89
922.375	186	1021.13	221	1140.12	123	1316.82	158
922.587	186	1021.45	221	1144.55	95	1318.41	89
928.937	194	1021.47	221	1144.91	95	1322.64	236
929.144	194	1034.62	97	1144.96	95	1322.81	236
929.568	194	1034.67	97	1151.59	192	1324.19	236
942.413	79	1038.61	108	1153.18	192	1324.31	122
942.540	79	1038.99	108	1154.13	192	1324.36	236
943.020	79	1039.00	108	1160.04	124	1324.40	122
947.379	148	1049.66	193	1160.06	124	1324.41	122
948.106	148	1050.11	193	1160.11	124	1339.76	282
948.396	148	1050.71	193	1165.87	41	1339.79	282
949.125	148	1051.72	125	1165.96	41	1340.39	282
956.264	204	1051.78	125	1167.70	253	1340.43	282
956.484	204	1055.95	237	1168.09	253	1377.19	121
956.654	204	1055.97	237	1168.69	253	1379.19	121
956.874	204	1056.94	237	1182.97	65	1379.30	121
957.323	204	1056.96	237	1183.03	65	1386.83	191
957.414	204	1069.05	254	1184.51	65	1387.26	56
957.864	204	1069.38	254	1184.57	65	1387.30	105
960.067	109	1069.46	254	1194.50	201	1387.38	105
960.398	109	1084.69	202	1195.34	201	1387.45	191
967.982	98	1085.16	202	1195.59	201	1387.99	105
968.022	98	1085.19	202	1196.20	201	1388.50	191
975.833	203	1085.66	202	1196.62	201	1409.97	120
975.948	203	1086.31	202	1197.05	201	1410.08	120
976.143	203	1086.36	202	1197.23	201	1410.13	120
976.499	203	1087.00	202	1228.61	106	1448.63	200
976.510	203	1089.14	133	1229.16	106	1449.31	200
976.549	203	1089.25	133	1242.85	147	1449.53	200
976.905	203	1089.67	133	1243.06	75	1450.21	200
977.112	203	1089.88	133	1243.14	75	1451.35	200

List of tabulated lines — Continued

Wavelength (Å)	No.	Wavelength (Å)	No.	Wavelength (Å)	No.	Wavelength (Å)	No.
1451.45	200	1721.46	244	1841.72	190	2122.57	265
1452.59	200	1721.85	244	1845.69	247	2141.63	241
1467.02	252	1722.87	244	1845.75	247	2142.82	226
1467.62	252	1723.95	244	1845.87	247	2142.87	226
1468.77	252	1724.41	244	1846.14	247	2144.02	241
1472.15	249	1725.50	244	1846.41	247	2144.11	241
1472.62	249	1727.63	244	1846.49	247	2145.56	246
1473.23	249	1729.83	243	1846.83	247	2145.69	241
1474.02	249	1729.90	243	1846.87	247	2146.56	246
1474.31	249	1729.95	243	1847.45	247	2146.92	241
1474.78	249	1730.00	243	1876.04	181	2147.31	241
1475.47	249	1730.86	243	1876.91	181	2147.39	246
1475.76	249	1731.56	243	1885.06	119	2147.85	246
1484.16	219	1732.08	243	1885.22	119	2147.99	246
1484.86	219	1732.42	243	1885.26	119	2148.09	241
1498.40	281	1733.70	243	1896.24	299	2148.51	241
1498.44	281	1737.60	300	1897.87	299	2148.99	246
1500.03	281	1738.96	300	1898.48	143	2149.02	241
1500.07	281	1740.81	300	1898.88	298	2149.45	246
1508.75	291	1742.79	144	1900.08	299	2150.06	246
1510.08	291	1743.50	300	1900.52	298	2215.78	297
1516.97	184	1746.23	144	1902.57	143	2216.96	297
1517.54	184	1746.82	1	1902.73	298	2218.01	297
1517.74	184	1747.85	64	1903.27	299	2219.19	297
1518.31	184	1748.65	1	1905.93	298	2219.37	296
1549.78	260	1749.67	1	1907.92	251	2221.60	296
1551.09	260	1751.22	64	1908.08	251	2221.60	297
1552.17	260	1751.66	64	1908.94	251	2222.20	297
1553.49	260	1752.16	1	1917.55	259	2224.63	296
1554.33	260	1754.00	1	1918.49	259	2224.63	297
1555.10	260	1762.09	242	1919.30	259	2229.01	296
1555.95	260	1762.44	242	1919.56	259	2229.01	297
1558.38	78	1762.60	242	1919.78	259	2231.62	142
1558.51	78	1763.16	242	1920.64	259	2237.22	142
1558.72	78	1763.51	242	1920.85	259	2237.28	142
1558.86	78	1763.84	242	1921.31	259	2242.80	218
1558.90	235	1764.22	242	1947.04	199	2244.41	218
1561.06	235	1764.78	242	1949.28	199	2247.67	118
1574.31	104	1766.06	242	1949.90	199	2247.95	118
1575.20	104	1766.44	242	1951.52	199	2248.93	118
1575.65	104	1788.15	306	1952.33	199	2256.06	157
1576.08	309	1792.35	55	1953.77	199	2256.16	157
1576.54	104	1798.74	248	1953.96	199	2258.81	157
1576.91	309	1799.16	248	1958.73	305	2258.91	157
1584.42	183	1799.68	248	1985.08	180	2265.89	245
1585.05	183	1800.10	248	1986.07	180	2267.33	245
1585.89	183	1800.80	248	1992.58	317	2269.32	245
1586.52	183	1801.64	248	1994.36	317	2270.44	245
1612.98	145	1802.34	248	1995.50	316	2271.80	245
1615.93	145	1803.25	248	1996.46	317	2271.88	245
1666.28	272	1803.53	248	1997.28	316	2272.47	245
1668.25	272	1804.44	248	1999.39	316	2273.59	245
1668.53	272	1804.49	103			2274.14	245
1670.51	272	1805.66	103	in air		2274.82	245
1694.79	163	1809.96	182			2301.61	304
1696.56	163	1810.78	182	2029.19	280	2302.89	304
1697.16	163	1816.32	319	2029.26	280	2305.49	304
1698.18	163	1817.79	319	2032.53	280	2314.60	258
1699.00	163	1818.15	318	2032.61	280	2317.44	258
1699.34	163	1819.54	319	2106.47	179	2317.54	258
1699.96	163	1819.62	318	2107.57	179	2320.38	258
1700.03	163	1821.37	318	2116.52	265	2322.26	258
1719.58	244	1835.57	190	2117.58	265	2322.79	258
1720.44	244	1839.55	190	2121.50	265	2324.06	295

List of tabulated lines — Continued

Wavelength (Å)	No.	Wavelength (Å)	No.	Wavelength (Å)	No.	Wavelength (Å)	No.
2324.60	295	2800.85	217	3450.04	155	4523.56	128
2324.68	258	2803.20	217	3451.01	155	4528.54	213
2325.56	295	2862.86	271	3676.05	264	4530.86	128
2327.58	295	2865.61	271	3691.10	264	4534.58	128
2327.90	295	2868.70	271	3745.95	129	4539.70	175
2328.01	295	2871.45	271	3752.63	162	4544.82	213
2328.85	295	2897.27	74	3754.69	129	4544.84	175
2330.04	295	2919.42	224	3757.57	162	4547.30	128
2331.33	295	2919.52	224	3757.65	162	4551.40	213
2332.71	295	2942.40	233	3762.60	162	4579.88	290
2333.54	178	2950.10	233	3770.36	162	4589.18	231
2334.90	178	2972.55	156	3771.03	129	4590.08	290
2348.51	315	2977.33	156	3771.36	162	4591.98	230
2349.84	315	2978.84	156	3779.16	162	4592.17	290
2350.98	315	2983.64	156	3789.03	240	4604.18	231
2352.31	315	3053.87	294	3789.30	240	4605.16	311
2352.55	314	3057.65	294	3792.97	162	4610.55	230
2355.03	314	3058.10	294	3934.50	154	4610.74	230
2355.03	315	3061.89	294	3938.51	154	4615.17	223
2355.24	315	3067.01	294	3942.88	154	4615.41	223
2357.96	314	3067.63	294	3996.50	214	4616.39	311
2357.96	315	3072.77	294	3998.63	215	4621.04	311
2367.49	257	3079.41	232	4003.58	215	4623.05	231
2370.56	257	3079.59	232	4003.72	215	4630.61	311
2372.53	257	3081.14	294	4007.88	214	4634.13	102
2392.83	225	3087.84	232	4013.00	214	4636.40	311
2392.89	225	3109.14	288	4097.36	88	4640.64	102
2398.51	152	3112.57	288	4103.39	88	4641.85	102
2402.60	152	3113.07	270	4121.22	293	4641.96	311
2418.07	234	3118.74	270	4122.80	269	4646.04	311
2423.27	234	3118.77	288	4131.63	293	4679.42	278
2442.97	63	3119.05	288	4133.83	293	4679.83	278
2449.56	63	3119.97	270	4134.91	269	4692.24	278
2453.95	256	3120.59	288	4136.07	279	4692.66	278
2459.29	256	3125.28	288	4136.40	279	4853.84	138
2462.60	256	3125.67	270	4141.59	293	4858.70	160
2462.99	256	3130.30	288	4141.80	279	4858.98	160
2466.31	256	3173.84	303	4142.12	279	4861.27	160
2468.44	256	3219.33	303	4146.13	293	4867.12	160
2469.07	256	3223.52	303	4146.78	293	4867.17	160
2469.97	313	3304.03	176	4152.10	293	4873.10	138
2471.08	313	3306.63	176	4153.93	293	4873.60	160
2471.21	256	3306.75	176	4195.74	139	4880.66	138
2472.20	313	3311.83	312	4200.07	139	4881.78	160
2474.31	313	3316.27	312	4215.77	139	4884.14	160
2474.49	313	3316.74	312	4318.78	161	4896.58	160
2474.94	313	3321.19	312	4321.22	161	4897.14	212
2477.23	313	3327.03	312	4321.39	161	4899.10	212
2478.18	313	3327.22	312	4325.43	161	4904.78	212
2482.82	250	3329.49	117	4327.69	161	5030.96	287
2484.56	250	3330.11	117	4327.88	161	5038.31	287
2486.43	250	3333.07	312	4332.95	161	5047.19	287
2621.52	177	3342.09	117	4337.01	161	5054.59	287
2623.23	177	3342.76	140	4345.81	161	5066.75	287
2686.91	216	3353.98	130	4351.11	161	5084.92	287
2689.21	216	3354.32	130	4419.49	302	5097.24	287
2696.11	151	3355.46	140	4442.92	302	5106.74	287
2696.70	151	3358.78	130	4457.12	302	5159.99	268
2713.98	93	3360.98	130	4478.75	263	5178.97	268
2714.07	93	3365.80	130	4501.11	263	5260.86	198
2714.39	93	3367.36	130	4510.88	128	5262.52	308
2734.36	141	3374.07	130	4510.96	128	5269.04	308
2742.85	141	3443.62	155	4514.85	128	5270.57	198
2800.70	217	3444.59	155	4518.14	128	5271.79	308

List of tabulated lines — Continued

Wavelength (Å)	No.	Wavelength (Å)	No.	Wavelength (Å)	No.	Wavenumber (cm^{-1})	No.
5272.68	198	8654.36	276	12343	284	691.0	222
5282.43	198	9003.21	238	12668	301	778.5	222
5297.75	198	9009.86	238	12721	301	779.6	222
5298.95	198	9230.74	286	12809	301		
5314.36	198	9240.21	286	13240	307		
5320.87	229	9268.99	286	13300	307		
5327.19	229	9285.52	286	14015	266		
5352.46	229	9324.23	286	14156	266		
5488.97	77	9342.10	286	14233	310		
5817.79	228	9388.42	286	14299	310		
5820.57	228	9402.46	150	14324	310		
5847.94	228	9424.53	150	14391	310		
6365.84	239	9631.64	289	14501	310		
6394.75	239	9686.14	289	14504	310		
6445.34	197	10080	285	14616	310		
6450.79	197	10145	285	16485	172		
6454.08	197	10268	285	16504	172		
6463.09	197	10442	292	16572	172		
6467.02	197	10478	292	17516	274		
6468.57	197	10492	292	17522	274		
6478.76	197	10528	292	25429	136		
6487.84	197	10588	292	26183	136		
6505.36	174	10596	292	30120	211		
6515.91	174	10657	292	30412	211		
6761.33	262	10690	173	30947	211		
6812.43	262	10758	292	31254	211		
7356.27	227	10764	173	31459	189		
7371.51	227	10793	173	32054	189		
7404.54	227	10978	87	32498	189		
7405.70	277	10983	87	38689	153		
7406.74	277	11590	275	39066	153		
7419.99	227	11590	275	39367	210		
8307.51	137	12072	284	39515	153		
8344.95	137	12076	284	39849	210		
8386.39	137	12120	284	39866	210		
8424.56	137	12138	284	39908	153		
8444.34	267	12166	284	45356	261		
8495.29	267	12233	284	45378	261		
8652.93	276	12251	284	47758	261		

We have utilized the results of two sophisticated calculations for this boron-like spectrum: the configuration-interaction calculations for about 20 multiplets and 30 intersystem lines with the CIV 3 code by Bell *et al.*[1] and the large-scale close-coupling calculations with the R-matrix code by Yu Yan *et al.*[2] of the Opacity Project (OP) team (see general introduction). The results are usually in excellent agreement, differing by only a few percent, the largest disagreement being 16%.

While the OP calculations yield multiplet values only, the CIV 3 calculations by Bell *et al.*[1] extend to the fine-structure levels and thus provide data for individual lines. These line strengths within multiplets have been found to be very close to *LS*-coupling. This is also supported by an emission experiment of Glenzer *et al.*[3] in which the line ratios for several $3s$-$3p$ and $3p$-$3d$ doublets were measured to be within 1% of *LS*-coupling, with estimated uncertainties of 5–6%. We have therefore combined the OP multiplet data with *LS* line strengths and estimate that the assumption of *LS*-coupling causes no significant loss of accuracy for this spectrum. Exceptions may be some weak lines and transitions involving *f*-electrons for which we have lowered the accuracy ratings accordingly.

The theoretical results may be compared with three advanced beam-foil lifetime experiments, in which the influence of electron cascades on the population of the primary level was analyzed and accurately taken into account with the well-established ANDC technique (see general introduction). These are the experiments by Pinnington *et al.*[4] for the $3s\ ^2S$ and $2s2p^2\ ^2S$, ^2P and ^2D terms; by Westhuizen *et al.*[5] for the $3p\ ^4P_{1/2,3/2,5/2}$ levels, and by Kotze *et al.*[6] for the $3p\ ^4D_{3/2,5/2,7/2}$ levels. For the OP calculations, lifetimes may be compared for 11 levels, and the mean value for the ratio expt/OP is 1.03, with a two-standard deviation of ±10%. This comparison contains, however, two ratios which differ by about 30% from unity. For the data of Bell *et al.*,[1] comparisons with four lifetime results by Pinnington *et al.*[4] may be made, and the ratios are almost identical to those involving the OP data.

For the quasi-metastable $2s2p^2\ ^4P_{1/2,3/2,5/2}$ levels, which connect with the ground state only by intersystem lines, accurate lifetime measurements with an ion trap technique have been reported by Fang *et al.*[7], and the two-standard deviations of their uncertainties are approximately 14%. These data differ from the decay rates calculated by Bell *et al.*[1] by 8%, 16% and 52%, which is surprising since very similar measurements and calculations for the analogous levels of the isoelectronic ion of C II are in much better agreement, differing only by 4, 5, and 22% (see the C II introduction). We have applied the experimental results of Fang *et al.*[7] to establish the absolute scale and have thus renormalized the line strength data of Bell *et al.*[1]

References

[1] K. L. Bell, A. Hibbert, R. P. Stafford and T. Brage, Mon. Not. R. Astron. Soc. **272**, 909 (1995).

[2] Yu Yan, K. T. Taylor, and M. J. Seaton, J. Phys. B **20**, 6399 (1987) and to be published.

[3] S. Glenzer, H.-J. Kunze, J. Musielok, Y.-K. Kim, and W. L. Wiese, Phys Rev. A **49**, 221 (1994).

[4] E. H. Pinnington, W. Ansbacher, R. N. Gosselin, and J. A. Kernahan, Phys. Lett. A **114**, 373 (1986).

[5] P. van der Westhuizen, F. J. Coetzer, and T. C. Kotze in Atomic Spectra and Oscillator Strengths for Astrophysics and Fusion Research, p. 182 (J. E. Hansen, Ed.), North Holland, Amsterdam, 1990.

[6] T. C. Kotze, P. van der Westhuizen, and K. Visser, Phys. Rev. A **44**, 4180 (1991).

[7] Z. Fang, V. H. S. Kwong and W. H. Parkinson, Astrophys. J. **413**, L141 (1993).

N III: Allowed Transitions

No.	Transition Array	Mult.	λ_{air} (Å)	λ_{vac} (Å) or σ (cm^{-1})*	E_i (cm^{-1})	E_k (cm^{-1})	g_i-g_k	A_{ki} (10^8 s^{-1})	f_{ik}	S (at. u.)	log gf	Acc.	Source
1.	$2s^22p-$ $2s2p^2$	$^2P^o-^4P$											
				1749.67	174.4	57327.9	4−6	3.08−06	2.12−07	4.89−06	−6.072	C+	7
				1746.82	0.0	57246.8	2−4	9.49−08	8.68−09	9.99−08	−7.760	C	1*n*,7
				1752.16	174.4	57246.8	4−4	6.50−07	2.99−08	6.90−07	−6.922	C	1*n*,7
				1748.65	0.0	57187.1	2−2	4.97−06	2.28−07	2.62−06	−6.341	C	1*n*,7
				1754.00	174.4	57187.1	4−2	5.22−06	1.20−07	2.78−06	−6.317	C	1*n*,7
2.		$^2P^o-^2D$		*990.97*	*116*	*101027*	6−10	4.98+00	1.22−01	2.39+00	−0.135	B	1,2
				991.577	174.4	101023.9	4−6	4.97+00	1.10−01	1.43+00	−0.357	B	1*n*
				989.799	0.0	101030.6	2−4	4.18+00	1.23−01	8.00−01	−0.610	B	1*n*
				991.511	174.4	101030.6	4−4	8.17−01	1.20−02	1.57−01	−1.317	B	1*n*
3.		$^2P^o-^2S$		*764.01*	*116*	131004.3	6−2	2.81+01	8.19−02	1.24+00	−0.309	B	1,2
				764.351	174.4	131004.3	4−2	1.85+01	8.10−02	8.15−01	−0.489	B	1*n*
				763.334	0.0	131004.3	2−2	9.58+00	8.37−02	4.21−01	−0.776	B	1*n*
4.		$^2P^o-^2P$		*685.72*	*116*	*145949*	6−6	5.60+01	3.95−01	5.34+00	0.374	B	1,2
				685.817	174.4	145985.8	4−4	4.54+01	3.20−01	2.89+00	0.108	B	1*n*
				685.515	0.0	145875.7	2−2	3.83+01	2.70−01	1.22+00	−0.268	B	1*n*
				686.336	174.4	145875.7	4−2	1.95+01	6.88−02	6.22−01	−0.560	B	1*n*
				684.998	0.0	145985.8	2−4	9.63+00	1.36−01	6.11−01	−0.567	B	1*n*
5.	$2s^22p-$ $2s^23s$	$^2P^o-^2S$		*452.11*	*116*	221302.2	6−2	3.08+01	3.15−02	2.81−01	−0.724	B	1,2
				452.227	174.4	221302.2	4−2	2.05+01	3.15−02	1.87−01	−0.900	B	1*n*
				451.871	0.0	221302.2	2−2	1.03+01	3.14−02	9.34−02	−1.202	B	1*n*
6.	$2s^22p-$ $2s^23d$	$^2P^o-^2D$		*374.36*	*116*	*267242*	6−10	1.19+02	4.16−01	3.07+00	0.397	B	1,2
				374.434	174.4	267244.0	4−6	1.19+02	3.74−01	1.84+00	0.175	B	1*n*
				374.198	0.0	267238.4	2−4	9.89+01	4.15−01	1.02+00	−0.080	B	1*n*
				374.442	174.4	267238.4	4−4	1.98+01	4.16−02	2.05−01	−0.779	B	1*n*
7.	$2s^22p-$ $2s^24s$	$^2P^o-^2S$		*332.26*	*116*	301082.6	6−2	1.11+01	6.14−03	4.03−02	−1.434	B	2
				332.327	174.4	301082.6	4−2	7.42+00	6.14−03	2.69−02	−1.610	B−	*LS*
				332.135	0.0	301082.6	2−2	3.71+00	6.14−03	1.34−02	−1.911	B−	*LS*

N III: Allowed Transitions — Continued

No.	Transition Array	Mult.	λ_{air} (Å)	λ_{vac} (Å) or σ (cm^{-1})*	E_i (cm^{-1})	E_k (cm^{-1})	g_i–g_k	A_{ki} (10^8 s^{-1})	f_{ik}	S (at. u.)	log gf	Acc.	Source
8.	$2s^22p$ – $2s2p(^3P^o)3p$	$^2P^o$–^2P		*323.57*	*116*	– *309167*	6–6	3.31+01	5.20–02	3.33–01	–0.506	B	2
				323.613	174.4	– 309185.2	4–4	2.76+01	4.33–02	1.85–01	–0.761	B	*LS*
				323.487	0.0	– 309131.2	2–2	2.21+01	3.47–02	7.39–02	–1.159	B	*LS*
				323.670	174.4	– 309131.2	4–2	1.10+01	8.67–03	3.69–02	–1.460	B	*LS*
				323.431	0.0	– 309185.2	2–4	5.53+00	1.73–02	3.69–02	–1.460	B	*LS*
9.	$2s^22p$ – $2s^24d$	$^2P^o$–^2D		*314.81*	*116*	– *317767*	6–10	7.85+01	1.94–01	1.21+00	0.067	B	2
				314.856	174.4	– 317779.5	4–6	7.84+01	1.75–01	7.25–01	–0.155	B	*LS*
				314.715	0.0	– 317747.7	2–4	6.54+01	1.94–01	4.03–01	–0.410	B	*LS*
				314.888	174.4	– 317747.7	4–4	1.31+01	1.94–02	8.05–02	–1.110	B	*LS*
10.	$2s^22p$ – $2s2p(^3P^o)3p$	$^2P^o$–^2D		*311.61*	*116*	– *321031*	6–10	6.58+00	1.60–02	9.82–02	–1.019	B	2
				311.631	174.4	– 321066.8	4–6	6.57+00	1.44–02	5.89–02	–1.241	B	*LS*
				311.548	0.0	– 320978.2	2–4	5.48+00	1.60–02	3.27–02	–1.496	B	*LS*
				311.717	174.4	– 320978.2	4–4	1.09+00	1.59–03	6.55–03	–2.195	C+	*LS*
11.		$^2P^o$–^2S		*305.86*	*116*	– 327058.0	6–2	1.45+01	6.79–03	4.10–02	–1.390	B	2
				305.919	174.4	– 327058.0	4–2	9.67+00	6.79–03	2.73–02	–1.566	B–	*LS*
				305.756	0.0	– 327058.0	2–2	4.84+00	6.79–03	1.37–02	–1.867	B–	*LS*
12.	$2s^22p$ – $2s^25s$	$^2P^o$–^2S		*299.76*	*116*	– 333712.0	6–2	1.21+01	5.46–03	3.23–02	–1.485	B	2
				299.816	174.4	– 333712.0	4–2	8.09+00	5.45–03	2.15–02	–1.661	B–	*LS*
				299.660	0.0	– 333712.0	2–2	4.05+00	5.46–03	1.08–02	–1.962	B–	*LS*
13.	$2s^22p$ – $2s^25d$	$^2P^o$–^2D		*292.54*	*116*	– *341948*	6–10	2.26+01	4.83–02	2.79–01	–0.538	B	2
				292.591	174.4	– 341948.8	4–6	2.26+01	4.35–02	1.68–01	–0.760	B	*LS*
				292.442	0.0	– 341947.7	2–4	1.89+01	4.84–02	9.31–02	–1.015	B	*LS*
				292.592	174.4	– 341947.7	4–4	3.77+00	4.83–03	1.86–02	–1.714	B–	*LS*
14.	$2s^22p$ – $2s^26s$	$^2P^o$–^2S		*285.95*	*116*	– 349824.8	6–2	4.68+00	1.91–03	1.08–02	–1.940	B–	2
				286.000	174.4	– 349824.8	4–2	3.12+00	1.91–03	7.20–03	–2.117	C+	*LS*
				285.857	0.0	– 349824.8	2–2	1.56+00	1.91–03	3.60–03	–2.417	C+	*LS*
15.	$2s^22p$ – $2s^26d$	$^2P^o$–^2D		*282.16*	*116*	– 354530.5	6–10	1.36+01	2.70–02	1.51–01	–0.790	B	2
				282.202	174.4	– 354530.5	4–6	1.36+01	2.43–02	9.04–02	–1.012	B	*LS*
				282.063	0.0	– 354530.5	2–4	1.13+01	2.70–02	5.02–02	–1.267	B	*LS*
				282.202	174.4	– 354530.5	4–4	2.26+00	2.70–03	1.00–02	–1.966	B–	*LS*
16.	$2s^22p$ – $2s^27d$	$^2P^o$–^2D		*276.28*	*116*	– 362066	6–10	8.70+00	1.66–02	9.06–02	–1.002	B	2
				276.326	174.4	– 362066	4–6	8.70+00	1.49–02	5.44–02	–1.224	B	*LS*
				276.193	0.0	– 362066	2–4	7.26+00	1.66–02	3.02–02	–1.479	B	*LS*
				276.326	174.4	– 362066	4–4	1.45+00	1.66–03	6.04–03	–2.178	C+	*LS*
17.	$2s^22p$ – $2s^28d$	$^2P^o$–^2D		*272.61*	*116*	– 366940	6–10	5.93+00	1.10–02	5.93–02	–1.180	B	2
				272.654	174.4	– 366940	4–6	5.93+00	9.91–03	3.56–02	–1.402	B	*LS*
				272.524	0.0	– 366940	2–4	4.95+00	1.10–02	1.98–02	–1.657	B–	*LS*
				272.654	174.4	– 366940	4–4	9.88–01	1.10–03	3.95–03	–2.356	C+	*LS*
18.	$2s^22p$ – $2s2p(^3P^o)4p$	$^2P^o$–^2P		*266.90*	*116*	– *374785*	6–6	7.97+00	8.52–03	4.49–02	–1.292	B	2
				266.930	174.4	– 374804.8	4–4	6.64+00	7.10–03	2.49–02	–1.547	B–	*LS*
				266.847	0.0	– 374746.4	2–2	5.32+00	5.68–03	9.98–03	–1.945	C+	*LS*
				266.971	174.4	– 374746.4	4–2	2.66+00	1.42–03	4.99–03	–2.246	C+	*LS*
				266.806	0.0	– 374804.8	2–4	1.33+00	2.84–03	4.99–03	–2.246	C+	*LS*

N III: Allowed Transitions — Continued

No.	Transition Array	Mult.	λ_{air} (Å)	λ_{vac} (Å) or σ (cm^{-1})*	E_i (cm^{-1})	E_k (cm^{-1})	g_i-g_k	A_{ki} (10^8 s^{-1})	f_{ik}	S (at. u.)	log gf	Acc.	Source
19.	$2s^22p-$ $2s2p(^1P^o)3p$	$^2P^o-^2P$		*264.91*	*116*	– *377600*	6–6	1.87+01	1.96–02	1.03–01	–0.929	B	2
				264.945	174.4	– 377611.3	4–4	1.56+01	1.64–02	5.71–02	–1.184	B	*LS*
				264.846	0.0	– 377577.9	2–2	1.25+01	1.31–02	2.28–02	–1.582	B–	*LS*
				264.968	174.4	– 377577.9	4–2	6.22+00	3.27–03	1.14–02	–1.883	B–	*LS*
				264.823	0.0	– 377611.3	2–4	3.12+00	6.55–03	1.14–02	–1.883	B–	*LS*
20.	$2s^22p-$ $2s2p(^3P^o)5p$	$^2P^o-^2P$		*247.68*	*116*	– *403858*	6–6	8.96+00	8.24–03	4.03–02	–1.306	B	2
				248.428	174.4	– 402705.6	4–4	7.40+00	6.84–03	2.24–02	–1.563	B–	*LS*
				246.206	0.0	– 406163.9	2–2	6.08+00	5.52–03	8.96–03	–1.957	C+	*LS*
				246.312	174.4	– 406163.9	4–2	3.04+00	1.38–03	4.48–03	–2.258	C+	*LS*
				248.320	0.0	– 402705.6	2–4	1.48+00	2.74–03	4.48–03	–2.261	C+	*LS*
21.	$2s^22p-$ $2s2p(^3P^o)6p$	$^2P^o-^2P$		*238.07*	*116*	– *420154*	6–6	3.03+00	2.57–03	1.21–02	–1.811	B–	2
				238.093	174.4	– 420178.4	4–4	2.52+00	2.14–03	6.72–03	–2.067	C+	*LS*
				238.035	0.0	– 420106.1	2–2	2.02+00	1.72–03	2.69–03	–2.465	C	*LS*
				238.134	174.4	– 420106.1	4–2	1.01+00	4.29–04	1.34–03	–2.766	C	*LS*
				237.994	0.0	– 420178.4	2–4	5.05–01	8.58–04	1.34–03	–2.765	C	*LS*
22.	$2s2p^2-$ $2p^3$	$^4P-^4S^o$		*772.08*	*57277*	– 186797.1	12–4	4.90+01	1.46–01	4.45+00	0.244	B	1,2
				772.384	57327.9	– 186797.1	6–4	2.45+01	1.46–01	2.23+00	–0.058	B	1*n*
				771.901	57246.8	– 186797.1	4–4	1.64+01	1.46–01	1.48+00	–0.233	B	1*n*
				771.545	57187.1	– 186797.1	2–4	8.19+00	1.46–01	7.42–01	–0.534	B	1*n*
23.		$^4P-^2D^o$											
				686.122	57327.9	– 203074.6	6–6	1.06–04	7.45–07	1.01–05	–5.350	D	1
				685.673	57246.8	– 203088.9	4–4	3.22–05	2.27–07	2.05–06	–6.042	D	1
				686.055	57327.9	– 203088.9	6–4	4.63–06	2.18–08	2.95–07	–6.884	D	1
				685.740	57246.8	– 203074.6	4–6	2.19–06	2.32–08	2.09–07	–7.033	D	1
				685.393	57187.1	– 203088.9	2–4	7.68–07	1.08–08	4.88–08	–7.665	D	1
24.		$^4P-^2P^o$											
				577.765	57327.9	– 230408.6	6–4	3.43–05	1.14–07	1.31–06	–6.163	D	1
				577.509	57246.8	– 230404.3	4–2	7.86–06	1.96–08	1.49–07	–7.105	D	1
				577.495	57246.8	– 230408.6	4–4	1.09–04	5.46–07	4.16–06	–5.660	D	1
				577.310	57187.1	– 230404.3	2–2	2.46–05	1.23–07	4.67–07	–6.609	D	1
				577.296	57187.1	– 230408.6	2–4	7.82–06	7.81–08	2.97–07	–6.806	D	1
25.	$2s2p^2-$ $2s^23p$	$^4P-^2P^o$											
				530.861	57327.9	– 245701.3	6–4	7.53–06	2.12–08	2.22–07	–6.895	D	1
				530.733	57246.8	– 245665.4	4–2	7.96–07	1.68–09	1.17–08	–8.173	D	1
				530.632	57246.8	– 245701.3	4–4	2.12–07	8.94–10	6.25–09	–8.447	D	1
				530.565	57187.1	– 245665.4	2–2	2.60–06	1.10–08	3.83–08	–7.659	D	1
				530.464	57187.1	– 245701.3	2–4	3.02–06	2.55–08	8.89–08	–7.293	D	1
26.	$2s2p^2-$ $2s2p(^3P^o)3s$	$^4P-^4P^o$		*434.10*	*57277*	– *287639*	12–12	4.38+01	1.24–01	2.12+00	0.172	B	2
				434.067	57327.9	– 287706.9	6–6	3.07+01	8.67–02	7.43–01	–0.284	B	*LS*
				434.132	57246.8	– 287591.5	4–4	5.84+00	1.65–02	9.44–02	–1.180	B	*LS*
				434.137	57187.1	– 287529.4	2–2	7.30+00	2.06–02	5.90–02	–1.384	B	*LS*
				434.285	57327.9	– 287591.5	6–4	1.97+01	3.71–02	3.18–01	–0.652	B	*LS*
				434.249	57246.8	– 287529.4	4–2	3.65+01	5.16–02	2.95–01	–0.686	B	*LS*
				433.915	57246.8	– 287706.9	4–6	1.32+01	5.57–02	3.18–01	–0.652	B	*LS*
				434.019	57187.1	– 287591.5	2–4	1.83+01	1.03–01	2.95–01	–0.685	B	*LS*

N III: Allowed Transitions — Continued

No.	Transition Array	Mult.	λair (Å)	λvac (Å) or σ (cm⁻¹)*	E_i (cm⁻¹)	E_k (cm⁻¹)	g_i–g_k	A_{ki} (10^8 s⁻¹)	f_{ik}	S (at. u.)	log gf	Acc.	Source
27.	$2s2p^2$– $2s2p(^3P°)3d$	^4P–$^4D°$		*362.91*	*57277*	– *332829*	12–20	1.86+02	6.13–01	8.79+00	0.867	B	2
				362.942	57327.9	– 332854.0	6–8	1.86+02	4.90–01	3.51+00	0.469	B	*LS*
				362.872	57246.8	– 332826.0	4–6	1.30+02	3.86–01	1.84+00	0.189	B	*LS*
				362.822	57187.1	– 332804.4	2–4	7.76+01	3.06–01	7.32–01	–0.213	B	*LS*
				362.979	57327.9	– 332826.0	6–6	5.58+01	1.10–01	7.91–01	–0.179	B	*LS*
				362.901	57246.8	– 332804.4	4–4	9.93+01	1.96–01	9.37–01	–0.105	B	*LS*
				362.839	57187.1	– 332791.3	2–2	1.55+02	3.06–01	7.32–01	–0.213	B	*LS*
				363.007	57327.9	– 332804.4	6–4	9.30+00	1.23–02	8.79–02	–1.134	B	*LS*
				362.918	57246.8	– 332791.3	4–2	3.10+01	3.06–02	1.46–01	–0.912	B	*LS*
28.		^4P–$^4P°$		*358.47*	*57277*	– *336240*	12–12	1.01+02	1.94–01	2.75+00	0.367	B	2
				358.578	57327.9	– 336206.9	6–6	7.05+01	1.36–01	9.62–01	–0.089	B	*LS*
				358.404	57246.8	– 336261.6	4–4	1.34+01	2.59–02	1.22–01	–0.985	B	*LS*
				358.282	57187.1	– 336296.6	2–2	1.68+01	3.24–02	7.64–02	–1.189	B	*LS*
				358.508	57327.9	– 336261.6	6–4	4.53+01	5.82–02	4.12–01	–0.457	B	*LS*
				358.359	57246.8	– 336296.6	4–2	8.41+01	8.09–02	3.82–01	–0.490	B	*LS*
				358.474	57246.8	– 336206.9	4–6	3.02+01	8.74–02	4.12–01	–0.457	B	*LS*
				358.327	57187.1	– 336261.6	2–4	4.20+01	1.62–01	3.82–01	–0.490	B	*LS*
29.	$2s2p^2$– $2s2p(^3P°)4s$	^4P–$^4P°$		*321.18*	*57277*	– *368627*	12–12	1.32+01	2.04–02	2.59–01	–0.611	B	2
				321.163	57327.9	– 368696.3	6–6	9.23+00	1.43–02	9.05–02	–1.067	B	*LS*
				321.201	57246.8	– 368578.3	4–4	1.76+00	2.72–03	1.15–02	–1.964	B–	*LS*
				321.205	57187.1	– 368514.4	2–2	2.20+00	3.40–03	7.19–03	–2.168	C+	*LS*
				321.285	57327.9	– 368578.3	6–4	5.93+00	6.11–03	3.88–02	–1.435	B	*LS*
				321.267	57246.8	– 368514.4	4–2	1.10+01	8.49–03	3.59–02	–1.469	B	*LS*
				321.079	57246.8	– 368696.3	4–6	3.96+00	9.18–03	3.88–02	–1.435	B	*LS*
				321.139	57187.1	– 368578.3	2–4	5.50+00	1.70–02	3.59–02	–1.469	B	*LS*
30.	$2s2p^2$– $2s2p(^3P°)4d$	^4P–$^4P°$		*303.96*	*57277*	– *386270*	12–12	4.29+01	5.95–02	7.14–01	–0.147	B	2
				304.035	57327.9	– 386237.7	6–6	3.00+01	4.16–02	2.50–01	–0.603	B	*LS*
				303.910	57246.8	– 386291.8	4–4	5.73+00	7.93–03	3.17–02	–1.499	B	*LS*
				303.825	57187.1	– 386324.3	2–2	7.17+00	9.92–03	1.98–02	–1.703	B–	*LS*
				303.985	57327.9	– 386291.8	6–4	1.93+01	1.78–02	1.07–01	–0.970	B	*LS*
				303.880	57246.8	– 386324.3	4–2	3.58+01	2.48–02	9.92–02	–1.004	B	*LS*
				303.960	57246.8	– 386237.7	4–6	1.29+01	2.68–02	1.07–01	–0.970	B	*LS*
				303.855	57187.1	– 386291.8	2–4	1.79+01	4.96–02	9.92–02	–1.004	B	*LS*
31.	$2s2p^2$– $2s2p(^3P°)5s$	^4P–$^4P°$		*290.95*	*57277*	– *400982*	12–12	6.31+00	8.00–03	9.20–02	–1.017	B	2
				290.930	57327.9	– 401053.5	6–6	4.42+00	5.60–03	3.22–02	–1.473	B	*LS*
				290.962	57246.8	– 400934.2	4–4	8.41–01	1.07–03	4.09–03	–2.370	C+	*LS*
				290.972	57187.1	– 400862.5	2–2	1.05+00	1.33–03	2.56–03	–2.574	C	*LS*
				291.031	57327.9	– 400934.2	6–4	2.84+00	2.40–03	1.38–02	–1.842	B–	*LS*
				291.023	57246.8	– 400862.5	4–2	5.25+00	3.33–03	1.28–02	–1.875	B–	*LS*
				290.861	57246.8	– 401053.5	4–6	1.89+00	3.60–03	1.38–02	–1.841	B–	*LS*
				290.912	57187.1	– 400934.2	2–4	2.63+00	6.67–03	1.28–02	–1.875	B–	*LS*
32.	$2s2p^2$– $2s2p(^3P°)5d$	^4P–$^4P°$		*283.97*	*57277*	– *409425*	12–12	2.21+01	2.67–02	2.99–01	–0.494	B	2
				284.040	57327.9	– 409391.3	6–6	1.54+01	1.87–02	1.05–01	–0.951	B	*LS*
				283.930	57246.8	– 409445.8	4–4	2.94+00	3.56–03	1.33–02	–1.847	B–	*LS*
				283.850	57187.1	– 409486.2	2–2	3.68+00	4.45–03	8.32–03	–2.051	C+	*LS*
				283.996	57327.9	– 409445.8	6–4	9.93+00	8.01–03	4.49–02	–1.318	B	*LS*
				283.898	57246.8	– 409486.2	4–2	1.84+01	1.11–02	4.16–02	–1.352	B	*LS*
				283.974	57246.8	– 409391.3	4–6	6.62+00	1.20–02	4.49–02	–1.318	B	*LS*
				283.882	57187.1	– 409445.8	2–4	9.21+00	2.22–02	4.16–02	–1.352	B	*LS*

N III: Allowed Transitions — Continued

No.	Transition Array	Mult.	λ_{air} (Å)	λ_{vac} (Å) or σ (cm^{-1})*	E_i (cm^{-1})	E_k (cm^{-1})	g_i-g_k	A_{ki} (10^8 s^{-1})	f_{ik}	S (at. u.)	log gf	Acc.	Source
33.	$2s2p^2-$ $2p^2(^3P)3p$	$^4P-^4D^o$		*275.84*	*57277*	*- 419806*	12-20	2.02+01	3.84-02	4.18-01	-0.337	B	2
				275.829	57327.9	- 419871.7	6-8	2.02+01	3.07-02	1.67-01	-0.734	B	*LS*
				275.821	57246.8	- 419800.7	4-6	1.41+01	2.42-02	8.79-02	-1.014	B	*LS*
				275.824	57187.1	- 419737.6	2-4	8.42+00	1.92-02	3.49-02	-1.416	B	*LS*
				275.883	57327.9	- 419800.7	6-6	6.06+00	6.91-03	3.77-02	-1.382	B	*LS*
				275.869	57246.8	- 419737.6	4-4	1.08+01	1.23-02	4.46-02	-1.309	B	*LS*
				275.852	57187.1	- 419700.3	2-2	1.68+01	1.92-02	3.49-02	-1.416	B	*LS*
				275.931	57327.9	- 419737.6	6-4	1.01+00	7.68-04	4.18-03	-2.337	C+	*LS*
				275.897	57246.8	- 419700.3	4-2	3.36+00	1.92-03	6.97-03	-2.115	C+	*LS*
34.	$2s2p^2-$ $2s2p(^3P^o)6d$	$^4P-^4D^o$		*274.34*	*57277*	- 421794.2	12-20	6.60+00	1.24-02	1.35-01	-0.827	B	2
				274.374	57327.9	- 421794.2	6-8	6.60+00	9.93-03	5.38-02	-1.225	B	*LS*
				274.313	57246.8	- 421794.2	4-6	4.62+00	7.83-03	2.83-02	-1.504	B-	*LS*
				274.268	57187.1	- 421794.2	2-4	2.75+00	6.21-03	1.12-02	-1.906	B-	*LS*
				274.374	57327.9	- 421794.2	6-6	1.98+00	2.24-03	1.21-02	-1.873	B-	*LS*
				274.313	57246.8	- 421794.2	4-4	3.52+00	3.97-03	1.44-02	-1.799	B-	*LS*
				274.268	57187.1	- 421794.2	2-2	5.51+00	6.21-03	1.12-02	-1.906	B-	*LS*
				274.374	57327.9	- 421794.2	6-4	3.30-01	2.48-04	1.35-03	-2.827	C	*LS*
				274.313	57246.8	- 421794.2	4-2	1.10+00	6.21-04	2.24-03	-2.605	C	*LS*
35.		$^4P-^4P^o$		*274.31*	*57277*	*- 421823*	12-12	1.44+01	1.63-02	1.76-01	-0.709	B	2
				274.374	57327.9	- 421794.2	6-6	1.01+01	1.14-02	6.17-02	-1.165	B	*LS*
				274.276	57246.8	- 421843.2	4-4	1.92+00	2.17-03	7.84-03	-2.061	C+	*LS*
				274.213	57187.1	- 421867.2	2-2	2.41+00	2.71-03	4.90-03	-2.265	C+	*LS*
				274.337	57327.9	- 421843.2	6-4	6.49+00	4.88-03	2.65-02	-1.533	B-	*LS*
				274.258	57246.8	- 421867.2	4-2	1.20+01	6.78-03	2.45-02	-1.566	B-	*LS*
				274.313	57246.8	- 421794.2	4-6	4.33+00	7.33-03	2.65-02	-1.533	B-	*LS*
				274.231	57187.1	- 421843.2	2-4	6.02+00	1.36-02	2.45-02	-1.566	B-	*LS*
36.	$2s2p^2-$ $2s2p(^3P^o)7d$	$^4P-^4D^o$		*268.72*	*57277*	- 429412.9	12-20	1.22+01	2.21-02	2.34-01	-0.577	B	2
				268.756	57327.9	- 429412.9	6-8	1.22+01	1.76-02	9.36-02	-0.975	B	*LS*
				268.697	57246.8	- 429412.9	4-6	8.56+00	1.39-02	4.92-02	-1.255	B	*LS*
				268.654	57187.1	- 429412.9	2-4	5.10+00	1.10-02	1.95-02	-1.656	B-	*LS*
				268.756	57327.9	- 429412.9	6-6	3.67+00	3.97-03	2.11-02	-1.623	B-	*LS*
				268.697	57246.8	- 429412.9	4-4	6.52+00	7.06-03	2.50-02	-1.549	B-	*LS*
				268.654	57187.1	- 429412.9	2-2	1.02+01	1.10-02	1.95-02	-1.656	B-	*LS*
				268.756	57327.9	- 429412.9	6-4	6.11-01	4.41-04	2.34-03	-2.577	C	*LS*
				268.697	57246.8	- 429412.9	4-2	2.04+00	1.10-03	3.90-03	-2.355	C+	*LS*
37.		$^4P-^4P^o$		*268.67*	*57277*	- 429486.3	12-12	6.80+00	7.36-03	7.81-02	-1.054	B	2
				268.703	57327.9	- 429486.3	6-6	4.76+00	5.15-03	2.73-02	-1.510	B-	*LS*
				268.644	57246.8	- 429486.3	4-4	9.07-01	9.81-04	3.47-03	-2.406	C+	*LS*
				268.601	57187.1	- 429486.3	2-2	1.13+00	1.23-03	2.17-03	-2.610	C	*LS*
				268.703	57327.9	- 429486.3	6-4	3.06+00	2.21-03	1.17-02	-1.878	B-	*LS*
				268.644	57246.8	- 429486.3	4-2	5.67+00	3.07-03	1.08-02	-1.911	B-	*LS*
				268.644	57246.8	- 429486.3	4-6	2.04+00	3.31-03	1.17-02	-1.878	B-	*LS*
				268.601	57187.1	- 429486.3	2-4	2.84+00	6.13-03	1.08-02	-1.911	B-	*LS*
38.	$2s2p^2-$ $2p^2(^3P)3p$	$^4P-^4S^o$		*268.28*	*57277*	- 430025.9	12-4	1.16+01	4.19-03	4.44-02	-1.299	B	2
				268.314	57327.9	- 430025.9	6-4	5.82+00	4.19-03	2.22-02	-1.600	B-	*LS*
				268.255	57246.8	- 430025.9	4-4	3.88+00	4.19-03	1.48-02	-1.776	B-	*LS*
				268.212	57187.1	- 430025.9	2-4	1.94+00	4.19-03	7.40-03	-2.077	C+	*LS*

N III: Allowed Transitions — Continued

No.	Transition Array	Mult.	λ_{air} (Å)	λ_{vac} (Å) or σ (cm^{-1})*	E_i (cm^{-1})	E_k (cm^{-1})	g_i–g_k	A_{ki} (10^8 s^{-1})	f_{ik}	S (at. u.)	log gf	Acc.	Source
39.	$2s2p^2$– $2s2p(^3P^\circ)8d$	^{4}P–^{4}D°		*265.24*	*57277*	– *434301.2*	12–20	8.37+00	1.47–02	1.54–01	–0.753	B	2
				265.271	57327.9	– 434301.2	6–8	8.36+00	1.18–02	6.16–02	–1.151	B	*LS*
				265.214	57246.8	– 434301.2	4–6	5.86+00	9.27–03	3.24–02	–1.431	B	*LS*
				265.172	57187.1	– 434301.2	2–4	3.49+00	7.36–03	1.28–02	–1.832	B–	*LS*
				265.271	57327.9	– 434301.2	6–6	2.51+00	2.65–03	1.39–02	–1.799	B–	*LS*
				265.214	57246.8	– 434301.2	4–4	4.46+00	4.71–03	1.64–02	–1.725	B–	*LS*
				265.172	57187.1	– 434301.2	2–2	6.98+00	7.36–03	1.28–02	–1.832	B–	*LS*
				265.271	57327.9	– 434301.2	6–4	4.18–01	2.94–04	1.54–03	–2.753	C	*LS*
				265.214	57246.8	– 434301.2	4–2	1.39+00	7.35–04	2.57–03	–2.531	C	*LS*
40.		^{4}P–^{4}P°		*265.20*	*57277*	– *434356.6*	12–12	4.75+00	5.01–03	5.25–02	–1.221	B	2
				265.232	57327.9	– 434356.6	6–6	3.33+00	3.51–03	1.84–02	–1.677	B–	*LS*
				265.175	57246.8	– 434356.6	4–4	6.34–01	6.68–04	2.33–03	–2.573	C	*LS*
				265.133	57187.1	– 434356.6	2–2	7.93–01	8.35–04	1.46–03	–2.777	C	*LS*
				265.232	57327.9	– 434356.6	6–4	2.14+00	1.50–03	7.88–03	–2.045	C+	*LS*
				265.175	57246.8	– 434356.6	4–2	3.96+00	2.09–03	7.29–03	–2.078	C+	*LS*
				265.175	57246.8	– 434356.6	4–6	1.43+00	2.26–03	7.88–03	–2.045	C+	*LS*
				265.133	57187.1	– 434356.6	2–4	1.98+00	4.18–03	7.29–03	–2.078	C+	*LS*
41.	$2s2p^2$– $2p^3$	^{2}D–^{4}S°											
				1165.87	101023.9	– 186797.1	6–4	1.87–06	2.54–08	5.84–07	–6.818	D	1
				1165.96	101030.6	– 186797.1	4–4	1.50–08	3.05–10	4.68–09	–8.914	D	1
42.		^{2}D–^{2}D°		*979.88*	*101027*	– *203080*	10–10	9.87+00	1.42–01	4.58+00	0.153	B	1,2
				979.905	101023.9	– 203074.6	6–6	9.21+00	1.33–01	2.57+00	–0.099	B	1*n*
				979.832	101030.6	– 203088.9	4–4	8.84+00	1.27–01	1.64+00	–0.293	B	1*n*
				979.768	101023.9	– 203088.9	6–4	1.02+00	9.74–03	1.89–01	–1.233	B	1*n*
				979.969	101030.6	– 203074.6	4–6	6.66–01	1.44–02	1.86–01	–1.240	B	1*n*
43.		^{2}D–^{2}P°		*772.92*	*101027*	– *230407*	10–6	2.33+01	1.25–01	3.19+00	0.098	B	1,2
				772.889	101023.9	– 230408.6	6–4	2.09+01	1.25–01	1.91+00	–0.125	B	1*n*
				772.955	101030.6	– 230404.3	4–2	2.34+01	1.05–01	1.06+00	–0.378	B	1*n*
				772.929	101030.6	– 230408.6	4–4	2.36+00	2.11–02	2.15–01	–1.073	B	1*n*
44.	$2s2p^2$– $2s^23p$	^{2}D–^{2}P°		*691.27*	*101027*	– *245689*	10–6	1.02+00	4.40–03	1.00–01	–1.356	B	1,2
				691.193	101023.9	– 245701.3	6–4	9.22–01	4.40–03	6.01–02	–1.578	B	1*n*
				691.397	101030.6	– 245665.4	4–2	1.02+00	3.66–03	3.34–02	–1.834	B	1*n*
				691.225	101030.6	– 245701.3	4–4	1.02–01	7.32–04	6.66–03	–2.534	C+	1*n*
45.	$2s2p^2$– $2s2p(^3P^\circ)3s$	^{2}D–^{2}P°		*509.68*	*101027*	– *297227*	10–6	1.78+01	4.17–02	7.00–01	–0.380	B	2
				509.579	101023.9	– 297264.4	6–4	1.61+01	4.17–02	4.20–01	–0.602	B	*LS*
				509.890	101030.6	– 297151.2	4–2	1.78+01	3.47–02	2.33–01	–0.857	B	*LS*
				509.596	101030.6	– 297264.4	4–4	1.79+00	6.95–03	4.66–02	–1.556	B	*LS*
46.	$2s2p^2$– $2s^24f$	^{2}D–^{2}F°		*456.08*	*101027*	– *320288*	10–14	1.21+00	5.30–03	7.96–02	–1.276	B	2
				456.070	101023.9	– 320288.3	6–8	1.21+00	5.05–03	4.55–02	–1.519	C+	*LS*
				456.087	101030.6	– 320287.2	4–6	1.13+00	5.30–03	3.18–02	–1.674	C+	*LS*
				456.073	101023.9	– 320287.2	6–6	8.10–02	2.52–04	2.27–03	–2.820	D	*LS*
47.	$2s2p^2$– $2s2p(^3P^\circ)3d$	^{2}D–^{2}D°		*428.21*	*101027*	– *334557*	10–10	5.58+01	1.53–01	2.16+00	0.186	B	2
				428.184	101023.9	– 334568.3	6–6	5.21+01	1.43–01	1.21+00	–0.066	B	*LS*
				428.248	101030.6	– 334540.2	4–4	5.02+01	1.38–01	7.78–01	–0.258	B	*LS*
				428.236	101023.9	– 334540.2	6–4	5.58+00	1.02–02	8.65–02	–1.212	B	*LS*
				428.196	101030.6	– 334568.3	4–6	3.72+00	1.53–02	8.65–02	–1.212	B	*LS*

N III: Allowed Transitions — Continued

No.	Transition Array	Mult.	λ_{air} (Å)	λ_{vac} (Å) or σ (cm^{-1})*	E_i (cm^{-1})	E_k (cm^{-1})	g_i-g_k	A_{ki} (10^8 s^{-1})	f_{ik}	S (at. u.)	log gf	Acc.	Source
48.		^{2}D−^{2}F°		*418.79*	*101027*	*339808*	10−14	1.05+02	3.87−01	5.33+00	0.587	B	2
				418.705	101023.9	339855.5	6−8	1.05+02	3.68−01	3.05+00	0.344	B	*LS*
				418.911	101030.6	339744.6	4−6	9.79+01	3.87−01	2.13+00	0.189	B	*LS*
				418.900	101023.9	339744.6	6−6	7.00+00	1.84−02	1.52−01	−0.957	B	*LS*
49.		^{2}D−^{2}P°		*413.76*	*101027*	*342715*	10−6	1.93+00	2.97−03	4.05−02	−1.527	B	2
				413.792	101023.9	342691.5	6−4	1.74+00	2.97−03	2.43−02	−1.749	B−	*LS*
				413.681	101030.6	342762.5	4−2	1.93+00	2.48−03	1.35−02	−2.004	B−	*LS*
				413.803	101030.6	342691.5	4−4	1.93−01	4.95−04	2.70−03	−2.703	C	*LS*
50.	$2s2p^2-$ $2s^26f$	^{2}D−^{2}F°		*393.81*	*101027*	354954.1	10−14	6.82−01	2.22−03	2.88−02	−1.653	B−	2
				393.809	101023.9	354954.1	6−8	6.82−01	2.12−03	1.65−02	−1.896	C	*LS*
				393.819	101030.6	354954.1	4−6	6.37−01	2.22−03	1.15−02	−2.051	C	*LS*
				393.809	101023.9	354954.1	6−6	4.55−02	1.06−04	8.23−04	−3.197	D−	*LS*
51.	$2s2p^2-$ $2s^27f$	^{2}D−^{2}F°		*382.71*	*101027*	362320.9	10−14	6.48−01	1.99−03	2.51−02	−1.701	B−	2
				382.706	101023.9	362320.9	6−8	6.48−01	1.90−03	1.43−02	−1.944	C	*LS*
				382.716	101030.6	362320.9	4−6	6.05−01	1.99−03	1.00−02	−2.099	C	*LS*
				382.706	101023.9	362320.9	6−6	4.32−02	9.49−05	7.17−04	−3.245	D−	*LS*
52.	$2s2p^2-$ $2s2p(^3P°)4d$	^{2}D−^{2}D°		*351.99*	*101027*	385126	10−10	3.15+01	5.85−02	6.78−01	−0.233	B	2
				351.986	101023.9	385126	6−6	2.94+01	5.46−02	3.79−01	−0.485	B	*LS*
				351.994	101030.6	385126	4−4	2.83+01	5.26−02	2.44−01	−0.677	B	*LS*
				351.986	101023.9	385126	6−4	3.15+00	3.90−03	2.71−02	−1.631	B−	*LS*
				351.994	101030.6	385126	4−6	2.10+00	5.85−03	2.71−02	−1.631	B−	*LS*
53.	$2s2p^2-$ $2s2p(^1P°)3d$	^{2}D−^{2}D°		*338.35*	*101027*	*396581*	10−10	1.74+01	2.99−02	3.33−01	−0.525	B	2
				338.340	101023.9	396584.8	6−6	1.62+01	2.79−02	1.86−01	−0.776	B	*LS*
				338.359	101030.6	396574.9	4−4	1.57+01	2.69−02	1.20−01	−0.968	B	*LS*
				338.351	101023.9	396574.9	6−4	1.74+00	1.99−03	1.33−02	−1.923	B−	*LS*
				338.347	101030.6	396584.8	4−6	1.16+00	2.99−03	1.33−02	−1.923	B−	*LS*
54.	$2s2p^2-$ $2p^2(^3P)3p$	^{2}D−^{2}D°		*317.81*	*101027*	*415679*	10−10	1.54+00	2.33−03	2.44−02	−1.632	B−	2
				317.809	101023.9	415678.0	6−6	1.44+00	2.18−03	1.37−02	−1.884	B−	*LS*
				317.814	101030.6	415679.7	4−4	1.39+00	2.10−03	8.78−03	−2.076	C+	*LS*
				317.808	101023.9	415679.7	6−4	1.54−01	1.55−04	9.76−04	−3.030	D	*LS*
				317.816	101030.6	415678.0	4−6	1.03−01	2.33−04	9.76−04	−3.030	D	*LS*
55.	$2s2p^2-$ $2p^3$	^{2}S−^{4}S°											
				1792.35	131004.3	186797.1	2−4	6.90−08	6.65−09	7.84−08	−7.876	D	1
56.		^{2}S−^{2}D°											
				1387.26	131004.3	203088.9	2−4	2.13−04	1.23−05	1.12−04	−4.609	C	1
57.		^{2}S−^{2}P°		*1006.0*	131004.3	*230407*	2−6	2.54+00	1.16−01	7.65−01	−0.636	B	1,2
				1005.99	131004.3	230408.6	2−4	2.57+00	7.80−02	5.16−01	−0.807	B	1*n*
				1006.04	131004.3	230404.3	2−2	2.48+00	3.76−02	2.49−01	−1.124	B	1*n*
58.	$2s2p^2-$ $2s^23p$	^{2}S−^{2}P°		*871.96*	131004.3	*245689*	2−6	3.84−01	1.31−02	7.54−02	−1.580	B	1,2
				871.862	131004.3	245701.3	2−4	3.84−01	8.75−03	5.02−02	−1.757	B	1*n*
				872.135	131004.3	245665.4	2−2	3.85−01	4.39−03	2.52−02	−2.056	B	1*n*

N III: Allowed Transitions — Continued

No.	Transition Array	Mult.	λ_{air} (Å)	λ_{vac} (Å) or σ (cm^{-1})*	E_i (cm^{-1})	E_k (cm^{-1})	g_i–g_k	A_{ki} (10^8 s^{-1})	f_{ik}	S (at. u.)	log gf	Acc.	Source
59.	$2s2p^2$– $2s2p(^3P^\circ)3s$	^2S–$^2P^\circ$		*601.60*	131004.3	– *297227*	2–6	5.05+00	8.21–02	3.25–01	–0.784	B	2
				601.467	131004.3	– 297264.4	2–4	5.05+00	5.48–02	2.17–01	–0.960	B	*LS*
				601.877	131004.3	– 297151.2	2–2	5.04+00	2.74–02	1.08–01	–1.262	B	*LS*
60.	$2s2p^2$– $2s^24p$	^2S–$^2P^\circ$		*553.40*	131004.3	– *311707*	2–6	8.41–01	1.16–02	4.22–02	–1.635	B	2
				553.370	131004.3	– 311715.2	2–4	8.41–01	7.72–03	2.81–02	–1.811	B–	*LS*
				553.446	131004.3	– 311690.3	2–2	8.41–01	3.86–03	1.41–02	–2.112	B–	*LS*
61.	$2s2p^2$– $2s2p(^3P^\circ)3d$	^2S–$^2P^\circ$		*472.34*	131004.3	– *342715*	2–6	6.11+01	6.13–01	1.91+00	0.089	B	2
				472.395	131004.3	– 342691.5	2–4	6.11+01	4.09–01	1.27+00	–0.088	B	*LS*
				472.237	131004.3	– 342762.5	2–2	6.11+01	2.04–01	6.36–01	–0.388	B	*LS*
62.	$2s2p^2$– $2s2p(^1P^\circ)3s$	^2S–$^2P^\circ$		*449.56*	131004.3	– *353442*	2–6	5.19+00	4.72–02	1.40–01	–1.025	B	2
				449.563	131004.3	– 353442.5	2–4	5.19+00	3.14–02	9.31–02	–1.201	B	*LS*
				449.567	131004.3	– 353440.6	2–2	5.19+00	1.57–02	4.65–02	–1.502	B–	*LS*
63.	$2s2p^2$– $2p^3$	^2P–$^4S^\circ$											
			2449.56	2450.30	145985.8	– 186797.1	4–4	3.69–06	3.32–07	1.07–05	–5.877	D	1
			2442.97	2443.71	145875.7	– 186797.1	2–4	1.03–06	1.84–07	2.96–06	–6.434	D	1
64.		^2P–$^2D^\circ$		*1750.4*	*145949*	– *203080*	6–10	1.52+00	1.16–01	4.01+00	–0.157	B	1,2
				1751.66	145985.8	– 203074.6	4–6	1.51+00	1.04–01	2.40+00	–0.380	B	1*n*
				1747.85	145875.7	– 203088.9	2–4	1.28+00	1.17–01	1.34+00	–0.631	B	1*n*
				1751.22	145985.8	– 203088.9	4–4	2.48–01	1.14–02	2.62–01	–1.342	B	1*n*
65.		^2P–$^2P^\circ$		*1184.0*	*145949*	– *230407*	6–6	8.73+00	1.83–01	4.29+00	0.042	B	1,2
				1184.51	145985.8	– 230408.6	4–4	7.29+00	1.53–01	2.39+00	–0.212	B	1*n*
				1183.03	145875.7	– 230404.3	2–2	5.87+00	1.23–01	9.59–01	–0.608	B	1*n*
				1184.57	145985.8	– 230404.3	4–2	2.89+00	3.04–02	4.74–01	–0.915	B	1*n*
				1182.97	145875.7	– 230408.6	2–4	1.42+00	5.95–02	4.63–01	–0.925	B	1*n*
66.	$2s2p^2$– $2s^23p$	^2P–$^2P^\circ$		*1002.6*	*145949*	– *245689*	6–6	2.67–02	4.02–04	7.96–03	–2.617	C	1
				1002.85	145985.8	– 245701.3	4–4	2.22–02	3.34–04	4.41–03	–2.874	C	1
				1002.11	145875.7	– 245665.4	2–2	1.69–02	2.54–04	1.68–03	–3.293	C	1
				1003.21	145985.8	– 245665.4	4–2	9.16–03	6.91–05	9.13–04	–3.558	C	1
				1001.75	145875.7	– 245701.3	2–4	4.85–03	1.46–04	9.62–04	–3.535	C	1
67.	$2s2p^2$– $2s2p(^3P^\circ)3s$	^2P–$^2P^\circ$		*661.03*	*145949*	– *297227*	6–6	1.34+00	8.80–03	1.15–01	–1.277	B	2
				661.032	145985.8	– 297264.4	4–4	1.12+00	7.33–03	6.38–02	–1.533	B	*LS*
				661.046	145875.7	– 297151.2	2–2	8.95–01	5.87–03	2.55–02	–1.931	B–	*LS*
				661.527	145985.8	– 297151.2	4–2	4.47–01	1.47–03	1.28–02	–2.232	B–	*LS*
				660.551	145875.7	– 297264.4	2–4	2.24–01	2.94–03	1.28–02	–2.231	B–	*LS*
68.	$2s2p^2$– $2s^24p$	^2P–$^2P^\circ$		*603.29*	*145949*	– *311707*	6–6	1.92–01	1.05–03	1.25–02	–2.201	B–	2
				603.393	145985.8	– 311715.2	4–4	1.60–01	8.74–04	6.94–03	–2.456	C+	*LS*
				603.083	145875.7	– 311690.3	2–2	1.28–01	7.00–04	2.78–03	–2.854	C	*LS*
				603.484	145985.8	– 311690.3	4–2	6.40–02	1.75–04	1.39–03	–3.155	C	*LS*
				602.993	145875.7	– 311715.2	2–4	3.21–02	3.50–04	1.39–03	–3.155	C	*LS*
69.	$2s2p^2$– $2s2p(^3P^\circ)3d$	^2P–$^2D^\circ$		*530.20*	*145949*	– *334557*	6–10	1.88+01	1.32–01	1.38+00	–0.102	B	2
				530.272	145985.8	– 334568.3	4–6	1.88+01	1.19–01	8.28–01	–0.324	B	*LS*
				530.041	145875.7	– 334540.2	2–4	1.57+01	1.32–01	4.60–01	–0.579	B	*LS*
				530.351	145985.8	– 334540.2	4–4	3.13+00	1.32–02	9.21–02	–1.278	B	*LS*

N III: Allowed Transitions — Continued

No.	Transition Array	Mult.	λ_{air} (Å)	λ_{vac} (Å) or σ (cm^{-1})*	E_i (cm^{-1})	E_k (cm^{-1})	g_i–g_k	A_{ki} (10^8 s^{-1})	f_{ik}	S (at. u.)	log gf	Acc.	Source
70.		^{2}P–^{2}P°		*508.22*	*145949*	– *342715*	6–6	1.64+01	6.34–02	6.36–01	–0.420	B	2
				508.374	145985.8	– 342691.5	4–4	1.36+01	5.28–02	3.54–01	–0.675	B	*LS*
				507.906	145875.7	– 342762.5	2–2	1.09+01	4.23–02	1.41–01	–1.073	B	*LS*
				508.190	145985.8	– 342762.5	4–2	5.46+00	1.06–02	7.07–02	–1.374	B	*LS*
				508.089	145875.7	– 342691.5	2–4	2.73+00	2.11–02	7.07–02	–1.374	B	*LS*
71.	$2s2p^2$– $2s2p(^1P^o)3d$	^{2}P–^{2}D°		*398.99*	*145949*	– *396581*	6–10	1.19+02	4.75–01	3.74+00	0.455	B	2
				399.044	145985.8	– 396584.8	4–6	1.19+02	4.27–01	2.25+00	0.233	B	*LS*
				398.884	145875.7	– 396574.9	2–4	9.96+01	4.75–01	1.25+00	–0.022	B	*LS*
				399.060	145985.8	– 396574.9	4–4	1.99+01	4.75–02	2.50–01	–0.721	B	*LS*
72.	$2s2p^2$– $2p^2(^3P)3p$	^{2}P–^{2}D°		*370.74*	*145949*	– *415679*	6–10	2.08+01	7.15–02	5.23–01	–0.368	B	2
				370.793	145985.8	– 415678.0	4–6	2.08+01	6.43–02	3.14–01	–0.590	B	*LS*
				370.639	145875.7	– 415679.7	2–4	1.74+01	7.15–02	1.74–01	–0.845	B	*LS*
				370.791	145985.8	– 415679.7	4–4	3.47+00	7.14–03	3.49–02	–1.544	B	*LS*
73.		^{2}P–^{2}S°		*338.89*	*145949*	– 441026.6	6–2	3.65+01	2.09–02	1.40–01	–0.901	B	2
				338.936	145985.8	– 441026.6	4–2	2.43+01	2.09–02	9.34–02	–1.077	B	*LS*
				338.810	145875.7	– 441026.6	2–2	1.22+01	2.09–02	4.67–02	–1.378	B	*LS*
74.	$2p^3$– $2s^23s$	^{4}S°–^{2}S											
			2897.27	2898.12	186797.1	– 221302.2	4–2	1.77–08	1.11–09	4.25–08	–8.351	D+	1
75.	$2p^3$– $2s^23d$	^{4}S°–^{2}D											
				1243.14	186797.1	– 267238.4	4–4	3.16–08	7.31–10	1.20–08	–8.534	C	1
				1243.06	186797.1	– 267244.0	4–6	1.72–07	5.97–09	9.77–08	–7.622	C	1
76.	$2p^3$– $2p^2(^3P)3s$	^{4}S°–^{4}P		*467.57*	186797.1	– *400671*	4–12	2.12+01	2.08–01	1.28+00	–0.079	B	2
				467.431	186797.1	– 400732.2	4–6	2.12+01	1.04–01	6.42–01	–0.380	B	*LS*
				467.649	186797.1	– 400632.9	4–4	2.12+01	6.95–02	4.28–01	–0.556	B	*LS*
				467.794	186797.1	– 400566.2	4–2	2.12+01	3.47–02	2.14–01	–0.857	B	*LS*
77.	$2p^3$– $2s^23s$	^{2}D°–^{2}S											
			5488.97	5490.49	203088.9	– 221302.2	4–2	1.03–08	2.32–09	1.67–07	–8.033	D	1
78.	$2p^3$– $2s^23d$	^{2}D°–^{2}D		*1558.6*	*203080*	– *267242*	10–10	5.17–04	1.88–05	9.67–04	–3.725	D	1
				1558.38	203074.6	– 267244.0	6–6	4.83–04	1.76–05	5.41–04	–3.977	D	1
				1558.86	203088.9	– 267238.4	4–4	4.75–04	1.73–05	3.55–04	–4.160	D	1
				1558.51	203074.6	– 267238.4	6–4	4.92–05	1.20–06	3.68–05	–5.144	D	1
				1558.72	203088.9	– 267244.0	4–6	3.01–05	1.65–06	3.38–05	–5.182	D	1
79.	$2p^3$– $2s2p(^3P^o)3p$	^{2}D°–^{2}P		*942.62*	*203080*	– *309167*	10–6	1.57+00	1.26–02	3.90–01	–0.901	B	2
				942.413	203074.6	– 309185.2	6–4	1.42+00	1.26–02	2.34–01	–1.122	B	*LS*
				943.020	203088.9	– 309131.2	4–2	1.57+00	1.05–02	1.30–01	–1.378	B	*LS*
				942.540	203088.9	– 309185.2	4–4	1.57–01	2.09–03	2.60–02	–2.077	B–	*LS*
80.	$2p^3$– $2s^24d$	^{2}D°–^{2}D		*871.94*	*203080*	– *317767*	10–10	5.35–02	6.10–04	1.75–02	–2.215	B–	2
				871.802	203074.6	– 317779.5	6–6	4.99–02	5.69–04	9.80–03	–2.467	C+	*LS*
				872.153	203088.9	– 317747.7	4–4	4.81–02	5.49–04	6.30–03	–2.659	C+	*LS*
				872.044	203074.6	– 317747.7	6–4	5.35–03	4.06–05	7.00–04	–3.613	D	*LS*
				871.911	203088.9	– 317779.5	4–6	3.57–03	6.10–05	7.00–04	–3.613	D	*LS*

N III: Allowed Transitions — Continued

No.	Transition Array	Mult.	λ_{air} (Å)	λ_{vac} (Å) or σ (cm^{-1})*	E_i (cm^{-1})	E_k (cm^{-1})	g_i-g_k	A_{ki} ($10^8\ s^{-1}$)	f_{ik}	S (at. u.)	log gf	Acc.	Source
81.	$2p^3$– $2s2p(^3P^o)3p$	$^2D^o$–^2D		*847.81*	*203080*	– *321031*	10–10	4.44–01	4.79–03	1.34–01	–1.320	B	2
				847.514	203074.6	– 321066.8	6–6	4.15–01	4.47–03	7.48–02	–1.572	B	*LS*
				848.253	203088.9	– 320978.2	4–4	3.99–01	4.31–03	4.81–02	–1.764	B	*LS*
				848.151	203074.6	– 320978.2	6–4	4.44–02	3.19–04	5.34–03	–2.718	C+	*LS*
				847.616	203088.9	– 321066.8	4–6	2.96–02	4.79–04	5.34–03	–2.718	C+	*LS*
82.	$2p^3$– $2s2p(^3P^o)4p$	$^2D^o$–^2P		*582.39*	*203080*	– *374785*	10–6	3.03–01	9.23–04	1.77–02	–2.035	B–	2
				582.309	203074.6	– 374804.8	6–4	2.72–01	9.23–04	1.06–02	–2.256	B–	*LS*
				582.555	203088.9	– 374746.4	4–2	3.02–01	7.69–04	5.90–03	–2.512	C+	*LS*
				582.357	203088.9	– 374804.8	4–4	3.03–02	1.54–04	1.18–03	–3.211	C	*LS*
83.	$2p^3$– $2s2p(^1P^o)3p$	$^2D^o$–^2P		*573.00*	*203080*	– *377600*	10–6	5.87+00	1.73–02	3.27–01	–0.761	B	2
				572.945	203074.6	– 377611.3	6–4	5.28+00	1.73–02	1.96–01	–0.983	B	*LS*
				573.102	203088.9	– 377577.9	4–2	5.86+00	1.44–02	1.09–01	–1.238	B	*LS*
				572.992	203088.9	– 377611.3	4–4	5.87–01	2.89–03	2.18–02	–1.937	B–	*LS*
84.	$2p^3$– $2s2p(^3P^o)4f$	$^2D^o$–^2F		*543.81*	*203080*	– *386967*	10–14	2.65+00	1.65–02	2.95–01	–0.784	B	2
				543.767	203074.6	– 386977.0	6–8	2.65+00	1.57–02	1.68–01	–1.027	C+	*LS*
				543.879	203088.9	– 386953.4	4–6	2.47+00	1.65–02	1.18–01	–1.182	C+	*LS*
				543.836	203074.6	– 386953.4	6–6	1.77–01	7.84–04	8.42–03	–2.328	C–	*LS*
85.	$2p^3$– $2s2p(^3P^o)5p$	$^2D^o$–^2P		*501.00*	*203080*	– *402679*	10–6	5.43+00	1.23–02	2.02–01	–0.912	B	2
				500.924	203074.6	– 402705.6	6–4	4.89+00	1.23–02	1.21–01	–1.134	B	*LS*
				501.163	203088.9	– 402624.6	4–2	5.42+00	1.02–02	6.74–02	–1.389	B	*LS*
				500.960	203088.9	– 402705.6	4–4	5.43–01	2.04–03	1.35–02	–2.088	B–	*LS*
86.	$2p^3$– $2s2p(^3P^o)6p$	$^2D^o$–^2P		*460.67*	*203080*	– *420154*	10–6	2.14+00	4.09–03	6.20–02	–1.388	B	2
				460.609	203074.6	– 420178.4	6–4	1.93+00	4.09–03	3.72–02	–1.610	B	*LS*
				460.793	203088.9	– 420106.1	4–2	2.14+00	3.41–03	2.07–02	–1.866	B–	*LS*
				460.640	203088.9	– 420178.4	4–4	2.14–01	6.81–04	4.13–03	–2.565	C	*LS*
87.	$2s^23s$– $2p^3$	^2S–$^2P^o$	*10980*	*9105* cm^{-1}	221302.2	– *230407*	2–6	1.52–04	8.25–04	5.97–02	–2.783	B	1,2
			10978	9106.4 cm^{-1}	221302.2	– 230408.6	2–4	1.52–04	5.50–04	3.98–02	–2.959	B	1*n*
			10983	9102.1 cm^{-1}	221302.2	– 230404.3	2–2	1.52–04	2.75–04	1.99–02	–3.260	B	1*n*
88.	$2s^23s$– $2s^23p$	^2S–$^2P^o$	*4099.4*	*4100.6*	221302.2	– *245689*	2–6	8.69–01	6.57–01	1.77+01	0.119	B	1,2
			4097.36	4098.51	221302.2	– 245701.3	2–4	8.70–01	4.38–01	1.18+01	–0.057	B	1*n*
			4103.39	4104.55	221302.2	– 245665.4	2–2	8.67–01	2.19–01	5.91+00	–0.359	B	1*n*
89.	$2s^23s$– $2s2p(^3P^o)3s$	^2S–$^2P^o$		*1317.1*	221302.2	– *297227*	2–6	7.52–01	5.87–02	5.09–01	–0.931	B	2
				1316.44	221302.2	– 297264.4	2–4	7.53–01	3.91–02	3.39–01	–1.107	B	*LS*
				1318.41	221302.2	– 297151.2	2–2	7.50–01	1.95–02	1.70–01	–1.408	B	*LS*
90.	$2s^23s$– $2s^24p$	^2S–$^2P^o$		*1106.1*	221302.2	– *311707*	2–6	1.09+00	5.98–02	4.36–01	–0.922	B	2
				1106.04	221302.2	– 311715.2	2–4	1.09+00	3.99–02	2.90–01	–1.098	B	*LS*
				1106.34	221302.2	– 311690.3	2–2	1.09+00	1.99–02	1.45–01	–1.399	B	*LS*
91.	$2s^23s$– $2s^25p$	^2S–$^2P^o$		*855.75*	221302.2	– *338159*	2–6	1.55+00	5.12–02	2.89–01	–0.990	B	2
				855.726	221302.2	– 338162.1	2–4	1.55+00	3.41–02	1.92–01	–1.166	B	*LS*
				855.786	221302.2	– 338153.9	2–2	1.55+00	1.71–02	9.62–02	–1.467	B	*LS*

N III: Allowed Transitions — Continued

No.	Transition Array	Mult.	λ_{air} (Å)	λ_{vac} (Å) or σ (cm^{-1})*	E_i (cm^{-1})	E_k (cm^{-1})	g_i-g_k	A_{ki} ($10^8 s^{-1}$)	f_{ik}	S (at. u.)	log gf	Acc.	Source
92.	$2s^23s-$ $2s2p(^3P^\circ)3d$	$^2S-^2P^\circ$		*823.64*	221302.2	– *342715*	2–6	2.38–01	7.27–03	3.94–02	–1.838	B	2
				823.796	221302.2	– 342691.5	2–4	2.38–01	4.84–03	2.63–02	–2.014	B–	*LS*
				823.314	221302.2	– 342762.5	2–2	2.38–01	2.42–03	1.31–02	–2.315	B–	*LS*
93.	$2p^3-$ $2s^23d$	$^2P^\circ-^2D$	*2714.0*	*2714.8*	*230407*	– *267242*	6–10	1.27–02	2.34–03	1.25–01	–1.853	B	1,2
			2713.98	2714.78	230408.6	– 267244.0	4–6	1.27–02	2.11–03	7.53–02	–2.074	B	1*n*
			2714.07	2714.88	230404.3	– 267238.4	2–4	1.06–02	2.34–03	4.18–02	–2.330	B	1*n*
			2714.39	2715.19	230408.6	– 267238.4	4–4	2.11–03	2.33–04	8.34–03	–3.030	B	1*n*
94.	$2p^3-$ $2s2p(^3P^\circ)3p$	$^2P^\circ-^2P$		*1269.7*	*230407*	– *309167*	6–6	4.07–02	9.85–04	2.47–02	–2.228	B–	2
				1269.41	230408.6	– 309185.2	4–4	3.40–02	8.21–04	1.37–02	–2.484	B–	*LS*
				1270.21	230404.3	– 309131.2	2–2	2.71–02	6.56–04	5.49–03	–2.882	C+	*LS*
				1270.28	230408.6	– 309131.2	4–2	1.36–02	1.64–04	2.74–03	–3.183	C	*LS*
				1269.34	230404.3	– 309185.2	2–4	6.80–03	3.28–04	2.74–03	–3.183	C	*LS*
95.	$2p^3-$ $2s^24d$	$^2P^\circ-^2D$		*1144.7*	*230407*	– *317767*	6–10	1.99–01	6.53–03	1.48–01	–1.407	B	2
				1144.55	230408.6	– 317779.5	4–6	1.99–01	5.88–03	8.86–02	–1.629	B	*LS*
				1144.91	230404.3	– 317747.7	2–4	1.66–01	6.53–03	4.92–02	–1.884	B	*LS*
				1144.96	230408.6	– 317747.7	4–4	3.32–02	6.53–04	9.84–03	–2.583	C+	*LS*
96.	$2p^3-$ $2s2p(^3P^\circ)3p$	$^2P^\circ-^2D$		*1103.5*	*230407*	– *321031*	6–10	5.61–01	1.71–02	3.72–01	–0.989	B	2
				1103.04	230408.6	– 321066.8	4–6	5.62–01	1.54–02	2.23–01	–1.211	B	*LS*
				1104.07	230404.3	– 320978.2	2–4	4.67–01	1.71–02	1.24–01	–1.467	B	*LS*
				1104.12	230408.6	– 320978.2	4–4	9.34–02	1.71–03	2.48–02	–2.166	B–	*LS*
97.		$^2P^\circ-^2S$		*1034.7*	*230407*	– 327058.0	6–2	2.87+00	1.54–02	3.14–01	–1.035	B	2
				1034.67	230408.6	– 327058.0	4–2	1.92+00	1.54–02	2.09–01	–1.211	B	*LS*
				1034.62	230404.3	– 327058.0	2–2	9.58–01	1.54–02	1.05–01	–1.512	B	*LS*
98.	$2p^3-$ $2s^25s$	$^2P^\circ-^2S$		*968.01*	*230407*	– 333712.0	6–2	3.45–01	1.62–03	3.09–02	–2.013	B	2
				968.022	230408.6	– 333712.0	4–2	2.30–01	1.62–03	2.06–02	–2.189	B–	*LS*
				967.982	230404.3	– 333712.0	2–2	1.15–01	1.62–03	1.03–02	–2.490	B–	*LS*
99.	$2p^3-$ $2s2p(^1P^\circ)3p$	$^2P^\circ-^2P$		*679.38*	*230407*	– *377600*	6–6	2.46+00	1.70–02	2.29–01	–0.991	B	2
				679.335	230408.6	– 377611.3	4–4	2.05+00	1.42–02	1.27–01	–1.246	B	*LS*
				679.470	230404.3	– 377577.9	2–2	1.64+00	1.14–02	5.08–02	–1.644	B	*LS*
				679.490	230408.6	– 377577.9	4–2	8.20–01	2.84–03	2.54–02	–1.945	B–	*LS*
				679.316	230404.3	– 377611.3	2–4	4.10–01	5.68–03	2.54–02	–1.945	B–	*LS*
100.	$2p^3-$ $2s2p(^3P^\circ)5p$	$^2P^\circ-^2P$		*580.48*	*230407*	– *402679*	6–6	4.20+00	2.12–02	2.43–01	–0.895	B	2
				580.393	230408.6	– 402705.6	4–4	3.50+00	1.77–02	1.35–01	–1.150	B	*LS*
				580.652	230404.3	– 402624.6	2–2	2.80+00	1.41–02	5.41–02	–1.548	B	*LS*
				580.666	230408.6	– 402624.6	4–2	1.40+00	3.54–03	2.70–02	–1.849	B–	*LS*
				580.379	230404.3	– 402705.6	2–4	7.01–01	7.08–03	2.70–02	–1.849	B–	*LS*
101.	$2p^3-$ $2s2p(^3P^\circ)6p$	$^2P^\circ-^2P$		*527.02*	*230407*	– *420154*	6–6	1.52+00	6.35–03	6.61–02	–1.419	B	2
				526.954	230408.6	– 420178.4	4–4	1.27+00	5.29–03	3.67–02	–1.674	B	*LS*
				527.143	230404.3	– 420106.1	2–2	1.02+00	4.23–03	1.47–02	–2.072	B–	*LS*
				527.155	230408.6	– 420106.1	4–2	5.08–01	1.06–03	7.34–03	–2.373	C+	*LS*
				526.942	230404.3	– 420178.4	2–4	2.54–01	2.12–03	7.34–03	–2.373	C+	*LS*

N III: Allowed Transitions — Continued

No.	Transition Array	Mult.	λ_{air} (Å)	λ_{vac} (Å) or σ (cm^{-1})*	E_i (cm^{-1})	E_k (cm^{-1})	g_i-g_k	A_{ki} (10^8 s^{-1})	f_{ik}	S (at. u.)	log gf	Acc.	Source
102.	$2s^23p-$ $2s^23d$	$^2P^o-^2D$	*4638.4*	*4639.7*	*245689*	– *267242*	6–10	7.61–01	4.09–01	3.75+01	0.390	B	1,2
			4640.64	4641.94	245701.3	– 267244.0	4–6	7.60–01	3.68–01	2.25+01	0.168	B	1*n*
			4634.13	4635.42	245665.4	– 267238.4	2–4	6.36–01	4.09–01	1.25+01	–0.087	B	1*n*
			4641.85	4643.15	245701.3	– 267238.4	4–4	1.26–01	4.09–02	2.50+00	–0.786	B	1*n*
103.	$2s^23p-$ $2s^24s$	$^2P^o-^2S$		*1805.3*	*245689*	– 301082.6	6–2	7.45+00	1.21–01	4.33+00	–0.138	B	2
				1805.66	245701.3	– 301082.6	4–2	4.96+00	1.21–01	2.88+00	–0.314	B	*LS*
				1804.49	245665.4	– 301082.6	2–2	2.49+00	1.21–01	1.44+00	–0.615	B	*LS*
104.	$2s^23p-$ $2s2p(^3P^o)3p$	$^2P^o-^2P$		*1575.3*	*245689*	– *309167*	6–6	3.15–01	1.17–02	3.65–01	–1.153	B	2
				1575.20	245701.3	– 309185.2	4–4	2.63–01	9.77–03	2.03–01	–1.408	B	*LS*
				1575.65	245665.4	– 309131.2	2–2	2.10–01	7.82–03	8.11–02	–1.806	B	*LS*
				1576.54	245701.3	– 309131.2	4–2	1.05–01	1.95–03	4.05–02	–2.107	B	*LS*
				1574.31	245665.4	– 309185.2	2–4	5.26–02	3.91–03	4.05–02	–2.107	B	*LS*
105.	$2s^23p-$ $2s^24d$	$^2P^o-^2D$		*1387.4*	*245689*	– *317767*	6–10	4.96+00	2.39–01	6.54+00	0.156	B	2
				1387.38	245701.3	– 317779.5	4–6	4.96+00	2.15–01	3.92+00	–0.066	B	*LS*
				1387.30	245665.4	– 317747.7	2–4	4.14+00	2.39–01	2.18+00	–0.321	B	*LS*
				1387.99	245701.3	– 317747.7	4–4	8.26–01	2.39–02	4.36–01	–1.020	B	*LS*
106.	$2s^23p-$ $2s2p(^3P^o)3p$	$^2P^o-^2S$		*1229.0*	*245689*	– 327058.0	6–2	1.10–01	8.32–04	2.02–02	–2.302	B–	2
				1229.16	245701.3	– 327058.0	4–2	7.35–02	8.32–04	1.35–02	–2.478	B–	*LS*
				1228.61	245665.4	– 327058.0	2–2	3.68–02	8.32–04	6.73–03	–2.779	C+	*LS*
107.	$2s^23p-$ $2s^25s$	$^2P^o-^2S$		*1136.1*	*245689*	– 333712.0	6–2	4.04+00	2.61–02	5.85–01	–0.806	B	2
				1136.23	245701.3	– 333712.0	4–2	2.69+00	2.61–02	3.90–01	–0.982	B	*LS*
				1135.76	245665.4	– 333712.0	2–2	1.35+00	2.61–02	1.95–01	–1.283	B	*LS*
108.	$2s^23p-$ $2s^25d$	$^2P^o-^2D$		*1038.9*	*245689*	– *341948*	6–10	2.57+00	6.94–02	1.42+00	–0.381	B	2
				1038.99	245701.3	– 341948.8	4–6	2.57+00	6.24–02	8.54–01	–0.603	B	*LS*
				1038.61	245665.4	– 341947.7	2–4	2.14+00	6.94–02	4.74–01	–0.858	B	*LS*
				1039.00	245701.3	– 341947.7	4–4	4.29–01	6.94–03	9.49–02	–1.557	B	*LS*
109.	$2s^23p-$ $2s^26s$	$^2P^o-^2S$		*960.28*	*245689*	– 349824.8	6–2	2.26+00	1.04–02	1.97–01	–1.205	B	2
				960.398	245701.3	– 349824.8	4–2	1.50+00	1.04–02	1.32–01	–1.381	B	*LS*
				960.067	245665.4	– 349824.8	2–2	7.53–01	1.04–02	6.58–02	–1.682	B	*LS*
110.	$2s^23p-$ $2s^26d$	$^2P^o-^2D$		*918.76*	*245689*	– 354530.5	6–10	1.65+00	3.48–02	6.32–01	–0.680	B	2
				918.871	245701.3	– 354530.5	4–6	1.65+00	3.13–02	3.79–01	–0.902	B	*LS*
				918.568	245665.4	– 354530.5	2–4	1.38+00	3.48–02	2.11–01	–1.157	B	*LS*
				918.871	245701.3	– 354530.5	4–4	2.75–01	3.48–03	4.21–02	–1.856	B	*LS*
111.	$2s^23p-$ $2s^27s$	$^2P^o-^2S$		*881.37*	*245689*	– 359148.1	6–2	1.54+00	5.97–03	1.04–01	–1.446	B	2
				881.470	245701.3	– 359148.1	4–2	1.02+00	5.97–03	6.93–02	–1.622	B	*LS*
				881.192	245665.4	– 359148.1	2–2	5.13–01	5.97–03	3.46–02	–1.923	B	*LS*
112.	$2s^23p-$ $2s^27d$	$^2P^o-^2D$		*859.28*	*245689*	– 362066	6–10	1.09+00	2.01–02	3.41–01	–0.919	B	2
				859.367	245701.3	– 362066	4–6	1.09+00	1.81–02	2.04–01	–1.141	B	*LS*
				859.102	245665.4	– 362066	2–4	9.07–01	2.01–02	1.14–01	–1.396	B	*LS*
				859.367	245701.3	– 362066	4–4	1.81–01	2.01–03	2.27–02	–2.095	B–	*LS*

N III: Allowed Transitions — Continued

No.	Transition Array	Mult.	λ_{air} (Å)	λ_{vac} (Å) or σ (cm^{-1})*	E_i (cm^{-1})	E_k (cm^{-1})	g_i-g_k	A_{ki} (10^8 s^{-1})	f_{ik}	S (at. u.)	log gf	Acc.	Source
113.	$2s^23p-$ $2s^28s$	$^2P^o-^2S$		*838.19*	*245689*	364994.2	6–2	1.16+00	4.09–03	6.77–02	–1.610	B	2
				838.273	245701.3	364994.2	4–2	7.76–01	4.09–03	4.51–02	–1.786	B	*LS*
				838.021	245665.4	364994.2	2–2	3.88–01	4.09–03	2.26–02	–2.087	B–	*LS*
114.	$2s^23p-$ $2s^28d$	$^2P^o-^2D$		*824.74*	*245689*	366940	6–10	7.55–01	1.28–02	2.09–01	–1.113	B	2
				824.819	245701.3	366940	4–6	7.55–01	1.16–02	1.25–01	–1.335	B	*LS*
				824.575	245665.4	366940	2–4	6.30–01	1.28–02	6.97–02	–1.590	B	*LS*
				824.819	245701.3	366940	4–4	1.26–01	1.28–03	1.39–02	–2.290	B–	*LS*
115.	$2s^23p-$ $2s2p(^3P^o)4p$	$^2P^o-^2P$		*774.62*	*245689*	*374785*	6–6	3.21+00	2.89–02	4.43–01	–0.761	B	2
				774.572	245701.3	374804.8	4–4	2.68+00	2.41–02	2.46–01	–1.016	B	*LS*
				774.707	245665.4	374746.4	2–2	2.14+00	1.93–02	9.83–02	–1.414	B	*LS*
				774.923	245701.3	374746.4	4–2	1.07+00	4.82–03	4.92–02	–1.715	B	*LS*
				774.357	245665.4	374804.8	2–4	5.36–01	9.64–03	4.92–02	–1.715	B	*LS*
116.	$2s^23p-$ $2s2p(^1P^o)3p$	$^2P^o-^2P$		*758.09*	*245689*	*377600*	6–6	2.14+01	1.84–01	2.76+00	0.044	B	2
				758.093	245701.3	377611.3	4–4	1.78+01	1.54–01	1.53+00	–0.211	B	*LS*
				758.078	245665.4	377577.9	2–2	1.43+01	1.23–01	6.14–01	–0.609	B	*LS*
				758.285	245701.3	377577.9	4–2	7.13+00	3.07–02	3.07–01	–0.910	B	*LS*
				757.886	245665.4	377611.3	2–4	3.57+00	6.15–02	3.07–01	–0.910	B	*LS*
117.	$2s^23d-$ $2s2p(^3P^o)3s$	$^2D-^2P^o$	*3334.0*	*3335.0*	*267242*	*297227*	10–6	2.58–01	2.58–02	2.83+00	–0.588	B	2
			3330.11	3331.07	267244.0	297264.4	6–4	2.33–01	2.58–02	1.70+00	–0.810	B	*LS*
			3342.09	3343.05	267238.4	297151.2	4–2	2.56–01	2.15–02	9.44–01	–1.066	B	*LS*
			3329.49	3330.45	267238.4	297264.4	4–4	2.59–02	4.31–03	1.89–01	–1.764	B	*LS*
118.	$2s^23d-$ $2s^24p$	$^2D-^2P^o$	*2248.3*	*2249.0*	*267242*	*311707*	10–6	1.09+00	4.97–02	3.68+00	–0.304	B	2
			2247.95	2248.65	267244.0	311715.2	6–4	9.84–01	4.97–02	2.21+00	–0.525	B	*LS*
			2248.93	2249.62	267238.4	311690.3	4–2	1.09+00	4.14–02	1.23+00	–0.781	B	*LS*
			2247.67	2248.36	267238.4	311715.2	4–4	1.09–01	8.29–03	2.45–01	–1.480	B	*LS*
119.	$2s^23d-$ $2s^24f$	$^2D-^2F^o$		*1885.2*	*267242*	*320288*	10–14	1.17+01	8.76–01	5.44+01	0.942	B	2
				1885.22	267244.0	320288.3	6–8	1.17+01	8.34–01	3.11+01	0.699	C+	*LS*
				1885.06	267238.4	320287.2	4–6	1.10+01	8.76–01	2.17+01	0.545	C+	*LS*
				1885.26	267244.0	320287.2	6–6	7.83–01	4.17–02	1.55+00	–0.602	C+	*LS*
120.	$2s^23d-$ $2s^25p$	$^2D-^2P^o$		*1410.1*	*267242*	*338159*	10–6	5.73–01	1.02–02	4.76–01	–0.989	B	2
				1410.08	267244.0	338162.1	6–4	5.16–01	1.02–02	2.85–01	–1.211	B	*LS*
				1410.13	267238.4	338153.9	4–2	5.73–01	8.54–03	1.59–01	–1.466	B	*LS*
				1409.97	267238.4	338162.1	4–4	5.73–02	1.71–03	3.17–02	–2.165	B	*LS*
121.	$2s^23d-$ $2s2p(^3P^o)3d$	$^2D-^2F^o$		*1378.1*	*267242*	*339808*	10–14	8.46–03	3.37–04	1.53–02	–2.472	B–	2
				1377.19	267244.0	339855.5	6–8	8.48–03	3.21–04	8.74–03	–2.715	C+	*LS*
				1379.19	267238.4	339744.6	4–6	7.88–03	3.37–04	6.12–03	–2.870	C+	*LS*
				1379.30	267244.0	339744.6	6–6	5.63–04	1.60–05	4.37–04	–4.017	D	*LS*
122.	$2s^23d-$ $2s^25f$	$^2D-^2F^o$		*1324.4*	*267242*	*342750*	10–14	3.77+00	1.39–01	6.05+00	0.142	B	2
				1324.40	267244.0	342750.1	6–8	3.77+00	1.32–01	3.46+00	–0.101	C+	*LS*
				1324.31	267238.4	342749.2	4–6	3.52+00	1.39–01	2.42+00	–0.256	C+	*LS*
				1324.41	267244.0	342749.2	6–6	2.51–01	6.61–03	1.73–01	–1.402	C+	*LS*

N III: Allowed Transitions — Continued

No.	Transition Array	Mult.	λ_{air} (Å)	λ_{vac} (Å) or σ (cm^{-1})*	E_i (cm^{-1})	E_k (cm^{-1})	g_i-g_k	A_{ki} (10^8 s^{-1})	f_{ik}	S (at. u.)	log *gf*	Acc.	Source
123.	$2s^23d-$ $2s^26f$	$^2D-^2F^o$		*1140.1*	*267242*	– 354954.1	10–14	1.57+00	4.28–02	1.61+00	–0.369	B	2
				1140.12	267244.0	– 354954.1	6–8	1.57+00	4.07–02	9.17–01	–0.612	C+	*LS*
				1140.05	267238.4	– 354954.1	4–6	1.46+00	4.28–02	6.42–01	–0.767	C+	*LS*
				1140.12	267244.0	– 354954.1	6–6	1.04–01	2.04–03	4.59–02	–1.913	C+	*LS*
124.	$2s^23d-$ $2s2p(^1P^o)3s$	$^2D-^2P^o$		*1160.1*	*267242*	– *353442*	10–6	1.04–01	1.26–03	4.82–02	–1.899	B	2
				1160.11	267244.0	– 353442.5	6–4	9.38–02	1.26–03	2.89–02	–2.121	B–	*LS*
				1160.06	267238.4	– 353440.6	4–2	1.04–01	1.05–03	1.61–02	–2.376	B–	*LS*
				1160.04	267238.4	– 353442.5	4–4	1.04–02	2.10–04	3.21–03	–3.075	C+	*LS*
125.	$2s^23d-$ $2s^27f$	$^2D-^2F^o$		*1051.8*	*267242*	– 362320.9	10–14	7.66–01	1.78–02	6.16–01	–0.750	B	2
				1051.78	267244.0	– 362320.9	6–8	7.66–01	1.69–02	3.52–01	–0.993	C+	*LS*
				1051.72	267238.4	– 362320.9	4–6	7.15–01	1.78–02	2.46–01	–1.148	C+	*LS*
				1051.78	267244.0	– 362320.9	6–6	5.11–02	8.47–04	1.76–02	–2.294	C	*LS*
126.	$2s^23d-$ $2s2p(^3P^o)4d$	$^2D-^2D^o$		*848.29*	*267242*	– 385126	10–10	7.57–01	8.16–03	2.28–01	–1.088	B	2
				848.306	267244.0	– 385126	6–6	7.06–01	7.62–03	1.28–01	–1.340	B	*LS*
				848.266	267238.4	– 385126	4–4	6.81–01	7.35–03	8.21–02	–1.532	B	*LS*
				848.306	267244.0	– 385126	6–4	7.57–02	5.44–04	9.12–03	–2.486	C+	*LS*
				848.266	267238.4	– 385126	4–6	5.05–02	8.16–04	9.12–03	–2.486	C+	*LS*
127.	$2s^23d-$ $2s2p(^1P^o)3d$	$^2D-^2D^o$		*773.16*	*267242*	– *396581*	10–10	2.13+01	1.91–01	4.87+00	0.281	B	2
				773.151	267244.0	– 396584.8	6–6	1.99+01	1.78–01	2.72+00	0.030	B	*LS*
				773.177	267238.4	– 396574.9	4–4	1.92+01	1.72–01	1.75+00	–0.162	B	*LS*
				773.210	267244.0	– 396574.9	6–4	2.13+00	1.27–02	1.95–01	–1.117	B	*LS*
				773.118	267238.4	– 396584.8	4–6	1.42+00	1.91–02	1.95–01	–1.117	B	*LS*
128.	$2s2p(^3P^o)3s-$ $2s2p(^3P^o)3p$	$^4P^o-^4D$	*4517.3*	*4518.5*	*287639*	– *309770*	12–20	6.79–01	3.46–01	6.18+01	0.618	B	2
			4514.85	4516.12	287706.9	– 309849.8	6–8	6.80–01	2.77–01	2.47+01	0.221	B	*LS*
			4510.96	4512.23	287591.5	– 309753.5	4–6	4.77–01	2.18–01	1.30+01	–0.059	B	*LS*
			4510.88	4512.15	287529.4	– 309691.8	2–4	2.84–01	1.73–01	5.15+00	–0.460	B	*LS*
			4534.58	4535.85	287706.9	– 309753.5	6–6	2.01–01	6.21–02	5.56+00	–0.429	B	*LS*
			4523.56	4524.83	287591.5	– 309691.8	4–4	3.61–01	1.11–01	6.59+00	–0.354	B	*LS*
			4518.14	4519.41	287529.4	– 309656.2	2–2	5.65–01	1.73–01	5.15+00	–0.461	B	*LS*
			4547.30	4548.58	287706.9	– 309691.8	6–4	3.33–02	6.88–03	6.18–01	–1.384	B	*LS*
			4530.86	4532.13	287591.5	– 309656.2	4–2	1.12–01	1.73–02	1.03+00	–1.161	B	*LS*
129.		$^4P^o-^4S$	*3761.4*	*3762.5*	*287639*	– 314217.3	12–4	1.13+00	7.97–02	1.18+01	–0.020	B	2
			3771.03	3772.10	287706.9	– 314217.3	6–4	5.59–01	7.95–02	5.92+00	–0.322	B	*LS*
			3754.69	3755.76	287591.5	– 314217.3	4–4	3.78–01	7.98–02	3.95+00	–0.496	B	*LS*
			3745.95	3747.02	287529.4	– 314217.3	2–4	1.90–01	8.00–02	1.97+00	–0.796	B	*LS*
130.		$^4P^o-^4P$	*3363.8*	*3364.7*	*287639*	– *317359*	12–12	1.82+00	3.09–01	4.11+01	0.569	B	2
			3367.36	3368.33	287706.9	– 317395.2	6–6	1.27+00	2.16–01	1.44+01	0.113	B	*LS*
			3360.98	3361.94	287591.5	– 317336.2	4–4	2.44–01	4.13–02	1.83+00	–0.782	B	*LS*
			3358.78	3359.74	287529.4	– 317293.6	2–2	3.05–01	5.16–02	1.14+00	–0.986	B	*LS*
			3374.07	3375.04	287706.9	– 317336.2	6–4	8.13–01	9.25–02	6.16+00	–0.256	B	*LS*
			3365.80	3366.77	287591.5	– 317293.6	4–2	1.52+00	1.29–01	5.71+00	–0.288	B	*LS*
			3354.32	3355.29	287591.5	– 317395.2	4–6	5.51–01	1.40–01	6.16+00	–0.253	B	*LS*
			3353.98	3354.94	287529.4	– 317336.2	2–4	7.66–01	2.58–01	5.71+00	–0.287	B	*LS*

N III: Allowed Transitions — Continued

No.	Transition Array	Mult.	λ_{air} (Å)	λ_{vac} (Å) or σ (cm^{-1})*	E_i (cm^{-1})	E_k (cm^{-1})	g_i-g_k	A_{ki} (10^8 s^{-1})	f_{ik}	S (at. u.)	log gf	Acc.	Source
131.	$2s2p(^3P^o)3s-2s2p(^3P^o)4p$	$^4P^o-^4D$		*1120.6*	*287639*	– *376880*	12–20	1.74+00	5.45–02	2.41+00	–0.184	B	2
				1120.60	287706.9	– 376945.2	6–8	1.74+00	4.36–02	9.65–01	–0.582	B	*LS*
				1120.26	287591.5	– 376856.8	4–6	1.22+00	3.44–02	5.07–01	–0.862	B	*LS*
				1120.24	287529.4	– 376795.7	2–4	7.25–01	2.73–02	2.01–01	–1.263	B	*LS*
				1121.71	287706.9	– 376856.8	6–6	5.20–01	9.80–03	2.17–01	–1.231	B	*LS*
				1121.02	287591.5	– 376795.7	4–4	9.25–01	1.74–02	2.57–01	–1.156	B	*LS*
				1120.81	287529.4	– 376750.4	2–2	1.45+00	2.72–02	2.01–01	–1.264	B	*LS*
				1122.48	287706.9	– 376795.7	6–4	8.64–02	1.09–03	2.41–02	–2.185	B–	*LS*
				1120.26	287591.5	– 376856.8	4–2	2.90–01	2.73–03	4.02–02	–1.962	B	*LS*
132.		$^4P^o-^4S$		*1101.4*	*287639*	– 378432.8	12–4	8.91–01	5.40–03	2.35–01	–1.188	B	2
				1102.22	287706.9	– 378432.8	6–4	4.45–01	5.40–03	1.18–01	–1.490	B	*LS*
				1100.82	287591.5	– 378432.8	4–4	2.98–01	5.41–03	7.84–02	–1.665	B	*LS*
				1100.07	287529.4	– 378432.8	2–4	1.49–01	5.41–03	3.92–02	–1.966	B	*LS*
133.		$^4P^o-^4P$		*1090.2*	*287639*	– *379364*	12–12	7.94–01	1.41–02	6.09–01	–0.770	B	2
				1090.62	287706.9	– 379397.6	6–6	5.55–01	9.90–03	2.13–01	–1.226	B	*LS*
				1089.88	287591.5	– 379344.8	4–4	1.06–01	1.89–03	2.71–02	–2.122	B–	*LS*
				1089.67	287529.4	– 379300.1	2–2	1.33–01	2.36–03	1.69–02	–2.326	B–	*LS*
				1091.25	287706.9	– 379344.8	6–4	3.56–01	4.24–03	9.14–02	–1.594	B	*LS*
				1090.41	287591.5	– 379300.1	4–2	6.61–01	5.89–03	8.46–02	–1.627	B	*LS*
				1089.25	287591.5	– 379397.6	4–6	2.39–01	6.37–03	9.14–02	–1.594	B	*LS*
				1089.14	287529.4	– 379344.8	2–4	3.32–01	1.18–02	8.46–02	–1.627	B	*LS*
134.	$2s2p(^3P^o)3s-2s2p(^3P^o)4f$	$^4P^o-^4D$		*993.31*	*287639*	– *388312*	12–20	1.02–01	2.51–03	9.84–02	–1.522	B	2
				994.373	287706.9	– 388272.8	6–8	1.01–01	2.00–03	3.94–02	–1.920	C+	*LS*
				992.873	287591.5	– 388309.3	4–6	7.13–02	1.58–03	2.07–02	–2.199	C	*LS*
				991.794	287529.4	– 388356.8	2–4	4.26–02	1.26–03	8.20–03	–2.600	C–	*LS*
				994.012	287706.9	– 388309.3	6–6	3.04–02	4.51–04	8.86–03	–2.568	C–	*LS*
				992.405	287591.5	– 388356.8	4–4	5.44–02	8.03–04	1.05–02	–2.493	C	*LS*
				991.509	287529.4	– 388385.8	2–2	8.52–02	1.26–03	8.20–03	–2.600	C–	*LS*
				993.543	287706.9	– 388356.8	6–4	5.08–03	5.01–05	9.84–04	–3.522	D–	*LS*
				992.120	287591.5	– 388385.8	4–2	1.70–02	1.26–04	1.64–03	–3.299	D	*LS*
135.	$2s2p(^3P^o)3s-2p^2(^3P)3s$	$^4P^o-^4P$		*884.71*	*287639*	– *400671*	12–12	1.97+01	2.32–01	8.09+00	0.444	B	2
				884.758	287706.9	– 400732.2	6–6	1.38+01	1.62–01	2.83+00	–0.012	B	*LS*
				884.632	287591.5	– 400632.9	4–4	2.63+00	3.09–02	3.60–01	–0.908	B	*LS*
				884.668	287529.4	– 400566.2	2–2	3.29+00	3.86–02	2.25–01	–1.112	B	*LS*
				885.536	287706.9	– 400632.9	6–4	8.85+00	6.94–02	1.21+00	–0.380	B	*LS*
				885.154	287591.5	– 400566.2	4–2	1.64+01	9.64–02	1.12+00	–0.414	B	*LS*
				883.855	287591.5	– 400732.2	4–6	5.94+00	1.04–01	1.21+00	–0.380	B	*LS*
				884.146	287529.4	– 400632.9	2–4	8.24+00	1.93–01	1.12+00	–0.413	B	*LS*
136.	$2s2p(^3P^o)3s-2s^24s$	$^2P^o-^2S$	*25930*	*3856* cm^{-1}	*297227*	– 301082.6	6–2	4.60–03	1.55–02	7.91+00	–1.033	B	2
			26183	3818.2 cm^{-1}	297264.4	– 301082.6	4–2	2.98–03	1.53–02	5.28+00	–1.213	B	*LS*
			25429	3931.4 cm^{-1}	297151.2	– 301082.6	2–2	1.63–03	1.58–02	2.64+00	–1.502	B	*LS*
137.	$2s2p(^3P^o)3s-2s2p(^3P^o)3p$	$^2P^o-^2P$	*8372.9*	*8375.2*	*297227*	– *309167*	6–6	9.69–02	1.02–01	1.68+01	–0.214	B	2
			8386.39	8388.70	297264.4	– 309185.2	4–4	8.03–02	8.47–02	9.36+00	–0.470	B	*LS*
			8344.95	8347.25	297151.2	– 309131.2	2–2	6.52–02	6.81–02	3.74+00	–0.866	B	*LS*
			8424.56	8426.87	297264.4	– 309131.2	4–2	3.17–02	1.69–02	1.87+00	–1.171	B	*LS*
			8307.51	8309.79	297151.2	– 309185.2	2–4	1.65–02	3.42–02	1.87+00	–1.165	B	*LS*

N III: Allowed Transitions — Continued

No.	Transition Array	Mult.	λ_{air} (Å)	λ_{vac} (Å) or σ (cm^{-1})*	E_i (cm^{-1})	E_k (cm^{-1})	g_i-g_k	A_{ki} (10^8 s^{-1})	f_{ik}	S (at. u.)	log gf	Acc.	Source
138.	$2s2p(^3P^o)3s-2s^24d$	$^2P^o-^2D$	*4867.2*	*4868.5*	*297227*	– *317767*	6–10	2.38−02	1.41−02	1.36+00	−1.072	B	2
			4873.10	4874.46	297264.4	– 317779.5	4–6	2.38−02	1.27−02	8.14−01	−1.295	B	*LS*
			4853.84	4855.19	297151.2	– 317747.7	2–4	2.00−02	1.42−02	4.52−01	−1.548	B	*LS*
			4880.66	4882.03	297264.4	– 317747.7	4–4	3.94−03	1.41−03	9.05−02	−2.249	B	*LS*
139.	$2s2p(^3P^o)3s-2s2p(^3P^o)3p$	$^2P^o-^2D$	*4199.8*	*4201.0*	*297227*	– *321031*	6–10	1.12+00	4.94−01	4.10+01	0.472	B	2
			4200.07	4201.26	297264.4	– 321066.8	4–6	1.12+00	4.45−01	2.46+01	0.250	B	*LS*
			4195.74	4196.92	297151.2	– 320978.2	2–4	9.37−01	4.95−01	1.37+01	−0.004	B	*LS*
			4215.77	4216.95	297264.4	– 320978.2	4–4	1.85−01	4.93−02	2.73+00	−0.706	B	*LS*
140.		$^2P^o-^2S$	*3351.3*	*3352.2*	*297227*	– 327058.0	6–2	1.13+00	6.34−02	4.20+00	−0.420	B	2
			3355.46	3356.43	297264.4	– 327058.0	4–2	7.51−01	6.33−02	2.80+00	−0.596	B	*LS*
			3342.76	3343.72	297151.2	– 327058.0	2–2	3.80−01	6.36−02	1.40+00	−0.896	B	*LS*
141.	$2s2p(^3P^o)3s-2s^25s$	$^2P^o-^2S$	*2740.0*	*2740.9*	*297227*	– 333712.0	6–2	3.25−01	1.22−02	6.61−01	−1.135	B	2
			2742.85	2743.66	297264.4	– 333712.0	4–2	2.16−01	1.22−02	4.40−01	−1.312	B	*LS*
			2734.36	2735.17	297151.2	– 333712.0	2–2	1.09−01	1.22−02	2.20−01	−1.612	B	*LS*
142.	$2s2p(^3P^o)3s-2s^25d$	$^2P^o-^2D$	*2235.4*	*2236.1*	*297227*	– *341948*	6–10	3.92−02	4.89−03	2.16−01	−1.533	B	2
			2237.22	2237.92	297264.4	– 341948.8	4–6	3.91−02	4.40−03	1.30−01	−1.755	B	*LS*
			2231.62	2232.32	297151.2	– 341947.7	2–4	3.28−02	4.90−03	7.20−02	−2.009	B	*LS*
			2237.28	2237.97	297264.4	– 341947.7	4–4	6.51−03	4.89−04	1.44−02	−2.709	B−	*LS*
143.	$2s2p(^3P^o)3s-2s^26s$	$^2P^o-^2S$		*1901.2*	*297227*	– 349824.8	6–2	7.03−02	1.27−03	4.77−02	−2.118	B	2
				1902.57	297264.4	– 349824.8	4–2	4.68−02	1.27−03	3.18−02	−2.294	B	*LS*
				1898.48	297151.2	– 349824.8	2–2	2.35−02	1.27−03	1.59−02	−2.594	B−	*LS*
144.	$2s2p(^3P^o)3s-2s^26d$	$^2P^o-^2D$		*1745.1*	*297227*	– 354530.5	6–10	7.36−03	5.60−04	1.93−02	−2.474	B−	2
				1746.23	297264.4	– 354530.5	4–6	7.34−03	5.04−04	1.16−02	−2.696	B−	*LS*
				1742.79	297151.2	– 354530.5	2–4	6.16−03	5.61−04	6.43−03	−2.950	C+	*LS*
				1746.23	297264.4	– 354530.5	4–4	1.22−03	5.60−05	1.29−03	−3.650	C	*LS*
145.	$2s2p(^3P^o)3s-2s^27s$	$^2P^o-^2S$		*1615.0*	*297227*	– 359148.1	6–2	3.03−02	3.95−04	1.26−02	−2.625	B−	2
				1615.93	297264.4	– 359148.1	4–2	2.02−02	3.95−04	8.40−03	−2.802	C+	*LS*
				1612.98	297151.2	– 359148.1	2–2	1.01−02	3.95−04	4.20−03	−3.102	C+	*LS*
146.	$2s2p(^3P^o)3s-2s2p(^3P^o)4p$	$^2P^o-^2P$		*1289.4*	*297227*	– *374785*	6–6	3.78+00	9.43−02	2.40+00	−0.247	B	2
				1289.65	297264.4	– 374804.8	4–4	3.15+00	7.86−02	1.33+00	−0.503	B	*LS*
				1288.74	297151.2	– 374746.4	2–2	2.53+00	6.29−02	5.34−01	−0.900	B	*LS*
				1290.62	297264.4	– 374746.4	4–2	1.26+00	1.57−02	2.67−01	−1.202	B	*LS*
				1287.77	297151.2	– 374804.8	2–4	6.33−01	3.15−02	2.67−01	−1.201	B	*LS*
147.	$2s2p(^3P^o)3s-2s2p(^1P^o)3p$	$^2P^o-^2P$		*1244.2*	*297227*	– *377600*	6–6	7.29−01	1.69−02	4.16−01	−0.993	B	2
				1244.60	297264.4	– 377611.3	4–4	6.07−01	1.41−02	2.31−01	−1.249	B	*LS*
				1243.37	297151.2	– 377577.9	2–2	4.87−01	1.13−02	9.24−02	−1.646	B	*LS*
				1245.12	297264.4	– 377577.9	4–2	2.42−01	2.82−03	4.62−02	−1.948	B	*LS*
				1242.85	297151.2	– 377611.3	2–4	1.22−01	5.65−03	4.62−02	−1.947	B	*LS*
148.	$2s2p(^3P^o)3s-2s2p(^3P^o)5p$	$^2P^o-^2P$		*948.30*	*297227*	– *402679*	6–6	7.47+00	1.01−01	1.89+00	−0.219	B	2
				948.396	297264.4	– 402705.6	4–4	6.22+00	8.39−02	1.05+00	−0.474	B	*LS*
				948.106	297151.2	– 402624.6	2–2	4.98+00	6.72−02	4.19−01	−0.872	B	*LS*
				949.125	297264.4	– 402624.6	4–2	2.48+00	1.68−02	2.10−01	−1.173	B	*LS*
				947.379	297151.2	– 402705.6	2–4	1.25+00	3.36−02	2.10−01	−1.172	B	*LS*

N III: Allowed Transitions — Continued

No.	Transition Array	Mult.	λ_{air} (Å)	λ_{vac} (Å) or σ (cm^{-1})*	E_i (cm^{-1})	E_k (cm^{-1})	g_i-g_k	A_{ki} (10^8 s^{-1})	f_{ik}	S (at. u.)	log gf	Acc.	Source
149.	$2s2p(^3P^o)3s-2s2p(^3P^o)6p$	$^2P^o-^2P$		*813.49*	*297227*	– *420154*	6–6	6.81–01	6.75–03	1.09–01	–1.392	B	2
				813.577	297264.4	– 420178.4	4–4	5.67–01	5.63–03	6.03–02	–1.648	B	*LS*
				813.306	297151.2	– 420106.1	2–2	4.54–01	4.50–03	2.41–02	–2.046	B–	*LS*
				814.056	297264.4	– 420106.1	4–2	2.26–01	1.12–03	1.21–02	–2.347	B–	*LS*
				812.828	297151.2	– 420178.4	2–4	1.14–01	2.25–03	1.21–02	–2.346	B–	*LS*
150.	$2s^24s-2s^24p$	$^2S-^2P^o$	*9409.7*	*9412.3*	301082.6	– *311707*	2–6	2.73–01	1.09+00	6.74+01	0.338	B	2
			9402.46	9405.04	301082.6	– 311715.2	2–4	2.74–01	7.26–01	4.49+01	0.162	B	*LS*
			9424.53	9427.11	301082.6	– 311690.3	2–2	2.72–01	3.62–01	2.25+01	–0.140	B	*LS*
151.	$2s^24s-2s^25p$	$^2S-^2P^o$	*2696.3*	*2697.1*	301082.6	– *338159*	2–6	1.59–01	5.19–02	9.21–01	–0.984	B	2
			2696.11	2696.91	301082.6	– 338162.1	2–4	1.59–01	3.46–02	6.14–01	–1.160	B	*LS*
			2696.70	2697.50	301082.6	– 338153.9	2–2	1.59–01	1.73–02	3.07–01	–1.461	B	*LS*
152.	$2s^24s-2s2p(^3P^o)3d$	$^2S-^2P^o$	*2401.2*	*2402.0*	301082.6	– *342715*	2–6	2.12–02	5.49–03	8.68–02	–1.959	B	2
			2402.60	2403.33	301082.6	– 342691.5	2–4	2.11–02	3.66–03	5.79–02	–2.136	B	*LS*
			2398.51	2399.24	301082.6	– 342762.5	2–2	2.12–02	1.83–03	2.89–02	–2.436	B–	*LS*
153.	$2s2p(^3P^o)3p-2s^24p$	$^2P-^2P^o$	*39400*	*2540* cm^{-1}	*309167*	– *311707*	6–6	9.35–05	2.17–03	1.69+00	–1.885	B	2
			39515	2530.0 cm^{-1}	309185.2	– 311715.2	4–4	7.70–05	1.80–03	9.38–01	–2.142	B	*LS*
			39066	2559.1 cm^{-1}	309131.2	– 311690.3	2–2	6.38–05	1.46–03	3.75–01	–2.535	B	*LS*
			39908	2505.1 cm^{-1}	309185.2	– 311690.3	4–2	2.99–05	3.57–04	1.88–01	–2.845	B	*LS*
			38689	2584.0 cm^{-1}	309131.2	– 311715.2	2–4	1.64–05	7.37–04	1.88–01	–2.832	B	*LS*
154.	$2s2p(^3P^o)3p-2s2p(^3P^o)3d$	$^2P-^2D^o$	*3937.4*	*3938.6*	*309167*	– *334557*	6–10	8.96–01	3.47–01	2.70+01	0.319	B	2
			3938.51	3939.63	309185.2	– 334568.3	4–6	8.96–01	3.12–01	1.62+01	0.097	B	*LS*
			3934.50	3935.61	309131.2	– 334540.2	2–4	7.49–01	3.47–01	9.00+00	–0.158	B	*LS*
			3942.88	3944.00	309185.2	– 334540.2	4–4	1.49–01	3.47–02	1.80+00	–0.858	B	*LS*
155.	$2s2p(^3P^o)3p-2s^25p$	$^2P-^2P^o$	*3448.2*	*3449.2*	*309167*	– *338159*	6–6	1.36–02	2.43–03	1.65–01	–1.837	B	2
			3450.04	3451.02	309185.2	– 338162.1	4–4	1.13–02	2.02–03	9.18–02	–2.093	B	*LS*
			3444.59	3445.58	309131.2	– 338153.9	2–2	9.10–03	1.62–03	3.67–02	–2.490	B–	*LS*
			3451.01	3452.00	309185.2	– 338153.9	4–2	4.52–03	4.04–04	1.84–02	–2.792	B–	*LS*
			3443.62	3444.61	309131.2	– 338162.1	2–4	2.28–03	8.10–04	1.84–02	–2.791	B–	*LS*
156.	$2s2p(^3P^o)3p-2s2p(^3P^o)3d$	$^2P-^2P^o$	*2979.9*	*2980.8*	*309167*	– *342715*	6–6	9.93–01	1.32–01	7.78+00	–0.101	B	2
			2983.64	2984.51	309185.2	– 342691.5	4–4	8.24–01	1.10–01	4.32+00	–0.356	B	*LS*
			2972.55	2973.42	309131.2	– 342762.5	2–2	6.67–01	8.84–02	1.73+00	–0.753	B	*LS*
			2977.33	2978.20	309185.2	– 342762.5	4–2	3.32–01	2.21–02	8.65–01	–1.054	B	*LS*
			2978.84	2979.71	309131.2	– 342691.5	2–4	1.66–01	4.41–02	8.65–01	–1.055	B	*LS*
157.	$2s2p(^3P^o)3p-2s2p(^1P^o)3s$	$^2P-^2P^o$	*2257.9*	*2258.6*	*309167*	– *353442*	6–6	3.75–02	2.87–03	1.28–01	–1.765	B	2
			2258.81	2259.51	309185.2	– 353442.5	4–4	3.12–02	2.39–03	7.10–02	–2.020	B	*LS*
			2256.16	2256.86	309131.2	– 353440.6	2–2	2.51–02	1.91–03	2.84–02	–2.418	B–	*LS*
			2258.91	2259.61	309185.2	– 353440.6	4–2	1.25–02	4.77–04	1.42–02	–2.719	B–	*LS*
			2256.06	2256.76	309131.2	– 353442.5	2–4	6.26–03	9.56–04	1.42–02	–2.719	B–	*LS*
158.	$2s2p(^3P^o)3p-2s2p(^3P^o)4d$	$^2P-^2D^o$		*1316.5*	*309167*	– 385126	6–10	2.54+00	1.10–01	2.86+00	–0.180	B	2
				1316.82	309185.2	– 385126	4–6	2.54+00	9.90–02	1.72+00	–0.402	B	*LS*
				1315.88	309131.2	– 385126	2–4	2.12+00	1.10–01	9.54–01	–0.657	B	*LS*
				1316.82	309185.2	– 385126	4–4	4.23–01	1.10–02	1.91–01	–1.357	B	*LS*

N III: Allowed Transitions — Continued

No.	Transition Array	Mult.	λ_{air} (Å)	λ_{vac} (Å) or σ (cm^{-1})*	E_i (cm^{-1})	E_k (cm^{-1})	g_i-g_k	A_{ki} (10^8 s^{-1})	f_{ik}	S (at. u.)	log gf	Acc.	Source
159.	$2s2p(^3P^o)3p-$ $2p^2(^3P)3p$	$^2P-^2S^o$		*758.38*	*309167*	- 441026.6	6−2	4.76+01	1.37−01	2.05+00	−0.086	B	2
				758.487	309185.2	- 441026.6	4−2	3.17+01	1.37−01	1.37+00	−0.262	B	*LS*
				758.177	309131.2	- 441026.6	2−2	1.59+01	1.37−01	6.83−01	−0.563	B	*LS*
160.	$2s2p(^3P^o)3p-$ $2s2p(^3P^o)3d$	$^4D-^4F^o$	*4864.6*	*4865.9*	*309770*	- *330321*	20−28	6.19−01	3.07−01	9.84+01	0.789	B	2
			4867.17	4868.53	309849.8	- 330389.9	8−10	6.18−01	2.74−01	3.52+01	0.341	B	*LS*
			4861.27	4862.63	309753.5	- 330318.5	6−8	5.32−01	2.51−01	2.41+01	0.178	B	*LS*
			4858.98	4860.34	309691.8	- 330266.5	4−6	4.66−01	2.47−01	1.58+01	−0.005	B	*LS*
			4858.70	4860.05	309656.2	- 330232.1	2−4	4.35−01	3.08−01	9.84+00	−0.211	B	*LS*
			4884.14	4885.51	309849.8	- 330318.5	8−8	8.71−02	3.12−02	4.01+00	−0.603	B	*LS*
			4873.60	4874.96	309753.5	- 330266.5	6−6	1.50−01	5.33−02	5.13+00	−0.495	B	*LS*
			4867.12	4868.48	309691.8	- 330232.1	4−4	1.73−01	6.14−02	3.94+00	−0.610	B	*LS*
			4896.58	4897.95	309849.8	- 330266.5	8−6	5.86−03	1.58−03	2.04−01	−1.898	B	*LS*
			4881.78	4883.15	309753.5	- 330232.1	6−4	1.22−02	2.92−03	2.81−01	−1.757	B	*LS*
161.		$^4D-^4D^o$	*4335.5*	*4336.7*	*309770*	- *332829*	20−20	2.14−01	6.02−02	1.72+01	0.080	B	2
			4345.81	4347.03	309849.8	- 332854.0	8−8	1.82−01	5.15−02	5.89+00	−0.385	B	*LS*
			4332.95	4334.16	309753.5	- 332826.0	6−6	1.23−01	3.45−02	2.95+00	−0.684	B	*LS*
			4325.43	4326.64	309691.8	- 332804.4	4−4	8.60−02	2.41−02	1.37+00	−1.015	B	*LS*
			4321.22	4322.44	309656.2	- 332791.3	2−2	1.08−01	3.02−02	8.59−01	−1.219	B	*LS*
			4351.11	4352.33	309849.8	- 332826.0	8−6	4.01−02	8.54−03	9.79−01	−1.165	B	*LS*
			4337.01	4338.23	309753.5	- 332804.4	6−4	7.47−02	1.40−02	1.20+00	−1.075	B	*LS*
			4327.88	4329.10	309691.8	- 332791.3	4−2	1.07−01	1.51−02	8.59−01	−1.220	B	*LS*
			4327.69	4328.91	309753.5	- 332854.0	6−8	3.06−02	1.15−02	9.79−01	−1.163	B	*LS*
			4321.39	4322.60	309691.8	- 332826.0	4−6	5.03−02	2.11−02	1.20+00	−1.073	B	*LS*
			4318.78	4319.99	309656.2	- 332804.4	2−4	5.40−02	3.02−02	8.59−01	−1.219	B	*LS*
162.		$^4D-^4P^o$	*3776.8*	*3777.9*	*309770*	- *336240*	20−12	1.31−01	1.68−02	4.18+00	−0.474	B	2
			3792.97	3794.04	309849.8	- 336206.9	8−6	1.03−01	1.67−02	1.67+00	−0.874	B	*LS*
			3771.36	3772.43	309753.5	- 336261.6	6−4	8.28−02	1.18−02	8.77−01	−1.151	B	*LS*
			3757.65	3758.72	309691.8	- 336296.6	4−2	6.64−02	7.03−03	3.48−01	−1.551	B	*LS*
			3779.16	3780.23	309753.5	- 336206.9	6−6	2.35−02	5.03−03	3.76−01	−1.520	B	*LS*
			3762.60	3763.67	309691.8	- 336261.6	4−4	4.24−02	8.99−03	4.45−01	−1.444	B	*LS*
			3752.63	3753.70	309656.2	- 336296.6	2−2	6.67−02	1.41−02	3.48−01	−1.550	B	*LS*
			3770.36	3771.44	309691.8	- 336206.9	4−6	2.63−03	8.41−04	4.18−02	−2.473	B	*LS*
			3757.57	3758.64	309656.2	- 336261.6	2−4	6.64−03	2.81−03	6.96−02	−2.250	B	*LS*
163.	$2s2p(^3P^o)3p-$ $2s2p(^3P^o)4s$	$^4D-^4P^o$		*1699.0*	*309770*	- *368627*	20−12	3.76+00	9.78−02	1.09+01	0.291	B	2
				1699.34	309849.8	- 368696.3	8−6	3.01+00	9.77−02	4.37+00	−0.107	B	*LS*
				1699.96	309753.5	- 368578.3	6−4	2.37+00	6.84−02	2.30+00	−0.387	B	*LS*
				1700.03	309691.8	- 368514.4	4−2	1.88+00	4.07−02	9.11−01	−0.788	B	*LS*
				1696.56	309753.5	- 368696.3	6−6	6.81−01	2.94−02	9.84−01	−0.754	B	*LS*
				1698.18	309691.8	- 368578.3	4−4	1.21+00	5.22−02	1.17+00	−0.681	B	*LS*
				1699.00	309656.2	- 368514.4	2−2	1.88+00	8.15−02	9.11−01	−0.788	B	*LS*
				1694.79	309691.8	- 368696.3	4−6	7.59−02	4.90−03	1.09−01	−1.708	B	*LS*
				1697.16	309656.2	- 368578.3	2−4	1.89−01	1.63−02	1.82−01	−1.486	B	*LS*
164.	$2s2p(^3P^o)3p-$ $2s2p(^3P^o)4d$	$^4D-^4P^o$		*1307.2*	*309770*	- *386270*	20−12	3.73−02	5.74−04	4.94−02	−1.940	B	2
				1309.11	309849.8	- 386237.7	8−6	2.97−02	5.73−04	1.98−02	−2.339	B−	*LS*
				1306.54	309753.5	- 386291.8	6−4	2.36−02	4.02−04	1.04−02	−2.618	B−	*LS*
				1304.93	309691.8	- 386324.3	4−2	1.88−02	2.40−04	4.12−03	−3.019	C+	*LS*
				1307.46	309753.5	- 386237.7	6−6	6.72−03	1.72−04	4.45−03	−2.986	C+	*LS*
				1305.48	309691.8	- 386291.8	4−4	1.20−02	3.07−04	5.27−03	−2.911	C+	*LS*
				1304.32	309656.2	- 386324.3	2−2	1.88−02	4.79−04	4.12−03	−3.018	C+	*LS*
				1306.41	309691.8	- 386237.7	4−6	7.48−04	2.87−05	4.94−04	−3.940	D	*LS*
				1304.88	309656.2	- 386291.8	2−4	1.88−03	9.58−05	8.23−04	−3.717	D	*LS*

N III: Allowed Transitions — Continued

No.	Transition Array	Mult.	λ_{air} (Å)	λ_{vac} (Å) or σ (cm^{-1})*	E_i (cm^{-1})	E_k (cm^{-1})	g_i-g_k	A_{ki} ($10^8 s^{-1}$)	f_{ik}	S (at. u.)	log gf	Acc.	Source
165.	$2s2p(^3P^\circ)3p-2s2p(^3P^\circ)5s$	$^4D-^4P^\circ$		*1096.3*	*309770*	– *400982*	20-12	1.90+00	2.05-02	1.48+00	-0.387	B	2
				1096.45	309849.8	– 401053.5	8-6	1.52+00	2.05-02	5.93-01	-0.785	B	*LS*
				1096.72	309753.5	– 400934.2	6-4	1.19+00	1.44-02	3.11-01	-1.065	B	*LS*
				1096.84	309691.8	– 400862.5	4-2	9.48-01	8.55-03	1.23-01	-1.466	B	*LS*
				1095.29	309753.5	– 401053.5	6-6	3.43-01	6.16-03	1.33-01	-1.432	B	*LS*
				1095.98	309691.8	– 400934.2	4-4	6.08-01	1.10-02	1.58-01	-1.358	B	*LS*
				1096.42	309656.2	– 400862.5	2-2	9.49-01	1.71-02	1.23-01	-1.466	B	*LS*
				1094.55	309691.8	– 401053.5	4-6	3.82-02	1.03-03	1.48-02	-2.386	B-	*LS*
				1095.55	309656.2	– 400934.2	2-4	9.51-02	3.42-03	2.47-02	-2.164	B-	*LS*
166.	$2s2p(^3P^\circ)3p-2s2p(^3P^\circ)5d$	$^4D-^4P^\circ$		*1003.5*	*309770*	– *409425*	20-12	1.72-02	1.56-04	1.03-02	-2.506	B-	2
				1004.61	309849.8	– 409391.3	8-6	1.37-02	1.56-04	4.12-03	-2.905	C+	*LS*
				1003.09	309753.5	– 409445.8	6-4	1.09-02	1.09-04	2.16-03	-3.184	C	*LS*
				1002.06	309691.8	– 409486.2	4-2	8.64-03	6.50-05	8.58-04	-3.585	D	*LS*
				1003.64	309753.5	– 409391.3	6-6	3.10-03	4.68-05	9.27-04	-3.552	D	*LS*
				1002.47	309691.8	– 409445.8	4-4	5.52-03	8.32-05	1.10-03	-3.478	C	*LS*
				1001.70	309656.2	– 409486.2	2-2	8.65-03	1.30-04	8.58-04	-3.585	D	*LS*
				1003.01	309691.8	– 409391.3	4-6	3.45-04	7.80-06	1.03-04	-4.506	D	*LS*
				1002.11	309656.2	– 409445.8	2-4	8.64-04	2.60-05	1.72-04	-4.284	D	*LS*
167.	$2s2p(^3P^\circ)3p-2p^2(^3P)3p$	$^4D-^4D^\circ$		*908.79*	*309770*	– *419806*	20-20	1.28+01	1.58-01	9.46+00	0.500	B	2
				908.910	309849.8	– 419871.7	8-8	1.09+01	1.35-01	3.24+00	0.035	B	*LS*
				908.701	309753.5	– 419800.7	6-6	7.32+00	9.06-02	1.63+00	-0.265	B	*LS*
				908.713	309691.8	– 419737.6	4-4	5.11+00	6.32-02	7.56-01	-0.597	B	*LS*
				908.727	309656.2	– 419700.3	2-2	6.38+00	7.90-02	4.73-01	-0.801	B	*LS*
				909.497	309849.8	– 419800.7	8-6	2.42+00	2.25-02	5.39-01	-0.745	B	*LS*
				909.222	309753.5	– 419737.6	6-4	4.46+00	3.69-02	6.62-01	-0.655	B	*LS*
				909.021	309691.8	– 419700.3	4-2	6.38+00	3.95-02	4.73-01	-0.801	B	*LS*
				908.115	309753.5	– 419871.7	6-8	1.82+00	3.00-02	5.39-01	-0.744	B	*LS*
				908.192	309691.8	– 419800.7	4-6	2.98+00	5.53-02	6.62-01	-0.655	B	*LS*
				908.419	309656.2	– 419737.6	2-4	3.19+00	7.90-02	4.73-01	-0.801	B	*LS*
168.	$2s2p(^3P^\circ)3p-2s2p(^3P^\circ)6d$	$^4D-^4D^\circ$		*892.67*	*309770*	– 421794.2	20-20	2.02+00	2.41-02	1.42+00	-0.316	B	2
				893.301	309849.8	– 421794.2	8-8	1.73+00	2.07-02	4.87-01	-0.781	B	*LS*
				892.533	309753.5	– 421794.2	6-6	1.16+00	1.38-02	2.44-01	-1.081	B	*LS*
				892.042	309691.8	– 421794.2	4-4	8.10-01	9.66-03	1.14-01	-1.413	B	*LS*
				891.758	309656.2	– 421794.2	2-2	1.01+00	1.21-02	7.10-02	-1.617	B	*LS*
				893.301	309849.8	– 421794.2	8-6	3.83-01	3.44-03	8.09-02	-1.561	B	*LS*
				892.533	309753.5	– 421794.2	6-4	7.08-01	5.63-03	9.93-02	-1.471	B	*LS*
				892.042	309691.8	– 421794.2	4-2	1.01+00	6.04-03	7.10-02	-1.617	B	*LS*
				892.533	309753.5	– 421794.2	6-8	2.88-01	4.59-03	8.09-02	-1.560	B	*LS*
				892.042	309691.8	– 421794.2	4-6	4.73-01	8.46-03	9.93-02	-1.471	B	*LS*
				891.758	309656.2	– 421794.2	2-4	5.07-01	1.21-02	7.10-02	-1.617	B	*LS*
169.	$2s2p(^3P^\circ)3p-2s2p(^3P^\circ)7d$	$^4D-^4D^\circ$		*835.82*	*309770*	– 429412.9	20-20	1.16-01	1.22-03	6.69-02	-1.614	B	2
				836.378	309849.8	– 429412.9	8-8	9.93-02	1.04-03	2.29-02	-2.079	B-	*LS*
				835.705	309753.5	– 429412.9	6-6	6.66-02	6.97-04	1.15-02	-2.379	B-	*LS*
				835.275	309691.8	– 429412.9	4-4	4.65-02	4.87-04	5.35-03	-2.711	C+	*LS*
				835.026	309656.2	– 429412.9	2-2	5.82-02	6.08-04	3.35-03	-2.915	C+	*LS*
				836.378	309849.8	– 429412.9	8-6	2.20-02	1.73-04	3.81-03	-2.859	C+	*LS*
				835.705	309753.5	– 429412.9	6-4	4.06-02	2.84-04	4.68-03	-2.769	C+	*LS*
				835.275	309691.8	– 429412.9	4-2	5.81-02	3.04-04	3.35-03	-2.915	C+	*LS*
				835.705	309753.5	– 429412.9	6-8	1.65-02	2.31-04	3.81-03	-2.858	C+	*LS*
				835.275	309691.8	– 429412.9	4-6	2.71-02	4.26-04	4.68-03	-2.769	C+	*LS*
				835.026	309656.2	– 429412.9	2-4	2.91-02	6.08-04	3.35-03	-2.915	C+	*LS*

N III: Allowed Transitions — Continued

No.	Transition Array	Mult.	λ_{air} (Å)	λ_{vac} (Å) or σ (cm^{-1})*	E_i (cm^{-1})	E_k (cm^{-1})	g_i-g_k	A_{ki} (10^8 s^{-1})	f_{ik}	S (at. u.)	log gf	Acc.	Source
170.		^{4}D–^{4}P°		*835.31*	*309770*	– 429486.3	20–12	6.32–02	3.96–04	2.18–02	–2.101	B–	2
				835.865	309849.8	– 429486.3	8–6	5.04–02	3.96–04	8.72–03	–2.499	C+	*LS*
				835.193	309753.5	– 429486.3	6–4	3.98–02	2.78–04	4.58–03	–2.779	C+	*LS*
				834.763	309691.8	– 429486.3	4–2	3.16–02	1.65–04	1.82–03	–3.180	C	*LS*
				835.193	309753.5	– 429486.3	6–6	1.14–02	1.19–04	1.96–03	–3.147	C	*LS*
				834.763	309691.8	– 429486.3	4–4	2.02–02	2.12–04	2.33–03	–3.073	C	*LS*
				834.515	309656.2	– 429486.3	2–2	3.17–02	3.31–04	1.82–03	–3.180	C	*LS*
				834.763	309691.8	– 429486.3	4–6	1.27–03	1.98–05	2.18–04	–4.101	D	*LS*
				834.515	309656.2	– 429486.3	2–4	3.17–03	6.61–05	3.63–04	–3.879	D	*LS*
171.	$2s2p(^3P°)3p-$ $2s2p(^3P°)8d$	^{4}D–^{4}D°		*803.01*	*309770*	– 434301.2	20–20	1.20–01	1.16–03	6.12–02	–1.635	B	2
				803.527	309849.8	– 434301.2	8–8	1.02–01	9.92–04	2.10–02	–2.100	B–	*LS*
				802.905	309753.5	– 434301.2	6–6	6.87–02	6.64–04	1.05–02	–2.400	B–	*LS*
				802.508	309691.8	– 434301.2	4–4	4.80–02	4.63–04	4.90–03	–2.732	C+	*LS*
				802.278	309656.2	– 434301.2	2–2	6.00–02	5.79–04	3.06–03	–2.936	C+	*LS*
				803.527	309849.8	– 434301.2	8–6	2.27–02	1.65–04	3.49–03	–2.880	C+	*LS*
				802.905	309753.5	– 434301.2	6–4	4.19–02	2.70–04	4.28–03	–2.790	C+	*LS*
				802.508	309691.8	– 434301.2	4–2	6.00–02	2.90–04	3.06–03	–2.936	C+	*LS*
				802.905	309753.5	– 434301.2	6–8	1.71–02	2.20–04	3.49–03	–2.880	C+	*LS*
				802.508	309691.8	– 434301.2	4–6	2.80–02	4.05–04	4.28–03	–2.790	C+	*LS*
				802.278	309656.2	– 434301.2	2–4	3.00–02	5.79–04	3.06–03	–2.936	C+	*LS*
172.	$2s^24p-$ $2s^24d$	^{2}P°–^{2}D	*16500*	*6060* cm^{-1}	*311707*	– *317767*	6–10	6.70–02	4.56–01	1.49+02	0.437	B	2
			16485	6064.3 cm^{-1}	311715.2	– 317779.5	4–6	6.72–02	4.10–01	8.91+01	0.215	B	*LS*
			16504	6057.4 cm^{-1}	311690.3	– 317747.7	2–4	5.58–02	4.56–01	4.95+01	–0.040	B	*LS*
			16572	6032.5 cm^{-1}	311715.2	– 317747.7	4–4	1.10–02	4.54–02	9.90+00	–0.741	B	*LS*
173.	$2s^24p-$ $2s2p(^3P°)3p$	^{2}P°–^{2}D	*10720*	*9324* cm^{-1}	*311707*	– *321031*	6–10	2.24–02	6.44–02	1.36+01	–0.413	B	2
			10690	9351.6 cm^{-1}	311715.2	– 321066.8	4–6	2.26–02	5.81–02	8.18+00	–0.634	B	*LS*
			10764	9287.9 cm^{-1}	311690.3	– 320978.2	2–4	1.85–02	6.41–02	4.55+00	–0.892	B	*LS*
			10793	9263.0 cm^{-1}	311715.2	– 320978.2	4–4	3.66–03	6.40–03	9.09–01	–1.592	B	*LS*
174.		^{2}P°–^{2}S	*6512.4*	*6514.2*	*311707*	– 327058.0	6–2	1.96–01	4.15–02	5.34+00	–0.604	B	2
			6515.91	6517.72	311715.2	– 327058.0	4–2	1.30–01	4.15–02	3.56+00	–0.780	B	*LS*
			6505.36	6507.15	311690.3	– 327058.0	2–2	6.55–02	4.16–02	1.78+00	–1.080	B	*LS*
175.	$2s^24p-$ $2s^25s$	^{2}P°–^{2}S	*4543.1*	*4544.4*	*311707*	– 333712.0	6–2	1.71+00	1.76–01	1.58+01	0.024	B	2
			4544.84	4546.12	311715.2	– 333712.0	4–2	1.14+00	1.76–01	1.05+01	–0.152	B	*LS*
			4539.70	4540.98	311690.3	– 333712.0	2–2	5.71–01	1.76–01	5.27+00	–0.453	B	*LS*
176.	$2s^24p-$ $2s^25d$	^{2}P°–^{2}D	*3305.8*	*3306.8*	*311707*	– *341948*	6–10	6.60–01	1.80–01	1.18+01	0.034	B	2
			3306.63	3307.58	311715.2	– 341948.8	4–6	6.59–01	1.62–01	7.06+00	–0.188	B	*LS*
			3304.03	3304.98	311690.3	– 341947.7	2–4	5.51–01	1.80–01	3.92+00	–0.443	B	*LS*
			3306.75	3307.70	311715.2	– 341947.7	4–4	1.10–01	1.80–02	7.84–01	–1.142	B	*LS*
177.	$2s^24p-$ $2s^26s$	^{2}P°–^{2}S	*2622.7*	*2623.4*	*311707*	– 349824.8	6–2	8.94–01	3.07–02	1.59+00	–0.734	B	2
			2623.23	2624.01	311715.2	– 349824.8	4–2	5.96–01	3.07–02	1.06+00	–0.910	B	*LS*
			2621.52	2622.30	311690.3	– 349824.8	2–2	2.98–01	3.08–02	5.31–01	–1.211	B	*LS*
178.	$2s^24p-$ $2s^26d$	^{2}P°–^{2}D	*2334.4*	*2335.1*	*311707*	– 354530.5	6–10	5.16–01	7.02–02	3.24+00	–0.375	B	2
			2334.90	2335.61	311715.2	– 354530.5	4–6	5.15–01	6.32–02	1.94+00	–0.597	B	*LS*
			2333.54	2334.26	311690.3	– 354530.5	2–4	4.30–01	7.02–02	1.08+00	–0.852	B	*LS*
			2334.90	2335.61	311715.2	– 354530.5	4–4	8.59–02	7.02–03	2.16–01	–1.552	B	*LS*

N III: Allowed Transitions — Continued

No.	Transition Array	Mult.	λ_{air} (Å)	λ_{vac} (Å) or σ (cm^{-1})*	E_i (cm^{-1})	E_k (cm^{-1})	g_i-g_k	A_{ki} (10^8 s^{-1})	f_{ik}	S (at. u.)	log gf	Acc.	Source
179.	$2s^24p-$ $2s^27s$	$^2P^o-^2S$	*2107.2*	*2107.9*	*311707*	– 359148.1	6–2	5.18–01	1.15–02	4.78–01	–1.162	B	2
			2107.57	2108.24	311715.2	– 359148.1	4–2	3.45–01	1.15–02	3.19–01	–1.338	B	*LS*
			2106.47	2107.14	311690.3	– 359148.1	2–2	1.73–01	1.15–02	1.59–01	–1.639	B	*LS*
180.	$2s^24p-$ $2s^27d$	$^2P^o-^2D$		*1985.7*	*311707*	– 362066	6–10	3.60–01	3.55–02	1.39+00	–0.672	B	2
				1986.07	311715.2	– 362066	4–6	3.60–01	3.19–02	8.34–01	–0.894	B	*LS*
				1985.08	311690.3	– 362066	2–4	3.00–01	3.55–02	4.64–01	–1.149	B	*LS*
				1986.07	311715.2	– 362066	4–4	5.99–02	3.54–03	9.27–02	–1.848	B	*LS*
181.	$2s^24p-$ $2s^28s$	$^2P^o-^2S$		*1876.6*	*311707*	– 364994.2	6–2	3.30–01	5.81–03	2.16–01	–1.457	B	2
				1876.91	311715.2	– 364994.2	4–2	2.20–01	5.81–03	1.44–01	–1.634	B	*LS*
				1876.04	311690.3	– 364994.2	2–2	1.10–01	5.82–03	7.18–02	–1.934	B	*LS*
182.	$2s^24p-$ $2s^28d$	$^2P^o-^2D$		*1810.5*	*311707*	– 366940	6–10	2.55–01	2.09–02	7.47–01	–0.902	B	2
				1810.78	311715.2	– 366940	4–6	2.55–01	1.88–02	4.48–01	–1.124	B	*LS*
				1809.96	311690.3	– 366940	2–4	2.13–01	2.09–02	2.49–01	–1.379	B	*LS*
				1810.78	311715.2	– 366940	4–4	4.25–02	2.09–03	4.98–02	–2.078	B–	*LS*
183.	$2s^24p-$ $2s2p(^3P^o)4p$	$^2P^o-^2P$		*1585.3*	*311707*	*– 374785*	6–6	5.25–01	1.98–02	6.20–01	–0.925	B	2
				1585.05	311715.2	– 374804.8	4–4	4.38–01	1.65–02	3.44–01	–1.180	B	*LS*
				1585.89	311690.3	– 374746.4	2–2	3.50–01	1.32–02	1.38–01	–1.579	B	*LS*
				1586.52	311715.2	– 374746.4	4–2	1.75–01	3.30–03	6.89–02	–1.880	B	*LS*
				1584.42	311690.3	– 374804.8	2–4	8.77–02	6.60–03	6.89–02	–1.879	B	*LS*
184.	$2s^24p-$ $2s2p(^1P^o)3p$	$^2P^o-^2P$		*1517.6*	*311707*	*– 377600*	6–6	3.59–02	1.24–03	3.72–02	–2.128	B	2
				1517.54	311715.2	– 377611.3	4–4	3.00–02	1.03–03	2.07–02	–2.383	B–	*LS*
				1517.74	311690.3	– 377577.9	2–2	2.40–02	8.27–04	8.27–03	–2.781	C+	*LS*
				1518.31	311715.2	– 377577.9	4–2	1.20–02	2.07–04	4.13–03	–3.083	C+	*LS*
				1516.97	311690.3	– 377611.3	2–4	6.00–03	4.14–04	4.13–03	–3.082	C+	*LS*
185.	$2s^24p-$ $2s2p(^3P^o)5p$	$^2P^o-^2P$		*1099.2*	*311707*	*– 402679*	6–6	5.11–01	9.26–03	2.01–01	–1.255	B	2
				1099.02	311715.2	– 402705.6	4–4	4.26–01	7.72–03	1.12–01	–1.510	B	*LS*
				1099.70	311690.3	– 402624.6	2–2	3.40–01	6.17–03	4.47–02	–1.909	B	*LS*
				1100.00	311715.2	– 402624.6	4–2	1.70–01	1.54–03	2.23–02	–2.210	B–	*LS*
				1098.72	311690.3	– 402705.6	2–4	8.53–02	3.09–03	2.23–02	–2.209	B–	*LS*
186.	$2s^24p-$ $2s2p(^3P^o)6p$	$^2P^o-^2P$		*922.11*	*311707*	*– 420154*	6–6	4.44–02	5.66–04	1.03–02	–2.469	B–	2
				921.972	311715.2	– 420178.4	4–4	3.70–02	4.71–04	5.72–03	–2.725	C+	*LS*
				922.375	311690.3	– 420106.1	2–2	2.95–02	3.77–04	2.29–03	–3.123	C	*LS*
				922.587	311715.2	– 420106.1	4–2	1.48–02	9.42–05	1.14–03	–3.424	C	*LS*
				921.760	311690.3	– 420178.4	2–4	7.40–03	1.89–04	1.14–03	–3.423	C	*LS*
187.	$2s^24p-$ $2s2p(^3P^o)7p$	$^2P^o-^2P$		*859.39*	*311707*	*– 428069*	6–6	1.28–01	1.42–03	2.41–02	–2.070	B–	2
				859.360	311715.2	– 428080.8	4–4	1.07–01	1.18–03	1.34–02	–2.325	B–	*LS*
				859.437	311690.3	– 428045.5	2–2	8.55–02	9.46–04	5.36–03	–2.723	C+	*LS*
				859.621	311715.2	– 428045.5	4–2	4.27–02	2.37–04	2.68–03	–3.024	C	*LS*
				859.177	311690.3	– 428080.8	2–4	2.14–02	4.73–04	2.68–03	–3.024	C	*LS*
188.	$2s^24p-$ $2s2p(^3P^o)8p$	$^2P^o-^2P$		*822.27*	*311707*	*– 433322*	6–6	1.76–01	1.79–03	2.90–02	–1.970	B–	2
				822.230	311715.2	– 433335.6	4–4	1.47–01	1.49–03	1.61–02	–2.225	B–	*LS*
				822.329	311690.3	– 433296.2	2–2	1.17–01	1.19–03	6.44–03	–2.623	C+	*LS*
				822.497	311715.2	– 433296.2	4–2	5.87–02	2.98–04	3.22–03	–2.924	C+	*LS*
				822.062	311690.3	– 433335.6	2–4	2.94–02	5.95–04	3.22–03	–2.924	C+	*LS*

N III: Allowed Transitions — Continued

No.	Transition Array	Mult.	λ_{air} (Å)	λ_{vac} (Å) or σ (cm⁻¹)*	E_i (cm⁻¹)	E_k (cm⁻¹)	g_i–g_k	A_{ki} (10^8 s⁻¹)	f_{ik}	S (at. u.)	log gf	Acc.	Source
189.	$2s2p(^3P^\circ)3p$ – $2s2p(^3P^\circ)3d$	$^4S-^4P^\circ$	*31820*	*3142* cm⁻¹	314217.3	– *317359*	4–12	1.59−03	7.22−02	3.02+01	−0.539	B	2
			31459	3177.9 cm⁻¹	314217.3	– 317395.2	4–6	1.64−03	3.65−02	1.51+01	−0.836	B	*LS*
			32054	3118.9 cm⁻¹	314217.3	– 317336.2	4–4	1.55−03	2.39−02	1.01+01	−1.020	B	*LS*
			32498	3076.3 cm⁻¹	314217.3	– 317293.6	4–2	1.49−03	1.18−02	5.04+00	−1.327	B	*LS*
190.	$2s2p(^3P^\circ)3p$ – $2s2p(^3P^\circ)4s$	$^4S-^4P^\circ$		*1837.9*	314217.3	– *368627*	4–12	3.53−01	5.37−02	1.30+00	−0.668	B	2
				1835.57	314217.3	– 368696.3	4–6	3.55−01	2.69−02	6.50−01	−0.969	B	*LS*
				1839.55	314217.3	– 368578.3	4–4	3.52−01	1.79−02	4.33−01	−1.146	B	*LS*
				1841.72	314217.3	– 368514.4	4–2	3.51−01	8.93−03	2.17−01	−1.447	B	*LS*
191.	$2s2p(^3P^\circ)3p$ – $2s2p(^3P^\circ)4d$	$^4S-^4P^\circ$		*1387.9*	314217.3	– *386270*	4–12	2.18+00	1.89−01	3.45+00	−0.122	B	2
				1388.50	314217.3	– 386237.7	4–6	2.18+00	9.44−02	1.73+00	−0.423	B	*LS*
				1387.45	314217.3	– 386291.8	4–4	2.18+00	6.30−02	1.15+00	−0.599	B	*LS*
				1386.83	314217.3	– 386324.3	4–2	2.18+00	3.15−02	5.75−01	−0.900	B	*LS*
192.	$2s2p(^3P^\circ)3p$ – $2s2p(^3P^\circ)5s$	$^4S-^4P^\circ$		*1152.5*	314217.3	– *400982*	4–12	1.08−01	6.42−03	9.75−02	−1.590	B	2
				1151.59	314217.3	– 401053.5	4–6	1.08−01	3.21−03	4.88−02	−1.891	B	*LS*
				1153.18	314217.3	– 400934.2	4–4	1.07−01	2.14−03	3.25−02	−2.067	B	*LS*
				1154.13	314217.3	– 400862.5	4–2	1.07−01	1.07−03	1.63−02	−2.369	B−	*LS*
193.	$2s2p(^3P^\circ)3p$ – $2s2p(^3P^\circ)5d$	$^4S-^4P^\circ$		*1050.3*	314217.3	– *409425*	4–12	1.57+00	7.80−02	1.08+00	−0.506	B	2
				1050.71	314217.3	– 409391.3	4–6	1.57+00	3.90−02	5.40−01	−0.807	B	*LS*
				1050.11	314217.3	– 409445.8	4–4	1.57+00	2.60−02	3.60−01	−0.983	B	*LS*
				1049.66	314217.3	– 409486.2	4–2	1.58+00	1.30−02	1.80−01	−1.284	B	*LS*
194.	$2s2p(^3P^\circ)3p$ – $2s2p(^3P^\circ)6d$	$^4S-^4P^\circ$		*929.32*	314217.3	– *421823*	4–12	1.50+00	5.82−02	7.12−01	−0.633	B	2
				929.568	314217.3	– 421794.2	4–6	1.50+00	2.91−02	3.56−01	−0.934	B	*LS*
				929.144	314217.3	– 421843.2	4–4	1.50+00	1.94−02	2.37−01	−1.110	B	*LS*
				928.937	314217.3	– 421867.2	4–2	1.50+00	9.70−03	1.19−01	−1.411	B	*LS*
195.	$2s2p(^3P^\circ)3p$ – $2s2p(^3P^\circ)7d$	$^4S-^4P^\circ$		867.536	314217.3	– 429486.3	4–12	4.22−01	1.43−02	1.63−01	−1.243	B	2
196.	$2s2p(^3P^\circ)3p$ – $2s2p(^3P^\circ)8d$	$^4S-^4P^\circ$		832.367	314217.3	– 434356.6	4–12	3.42−01	1.07−02	1.17−01	−1.370	B	2
197.	$2s2p(^3P^\circ)3p$ – $2s2p(^3P^\circ)3d$	$^4P-^4D^\circ$	*6462.3*	*6464.1*	*317359*	– *332829*	12–20	2.12−01	2.21−01	5.64+01	0.424	B	2
			6467.02	6468.81	317395.2	– 332854.0	6–8	2.11−01	1.77−01	2.26+01	0.025	B	*LS*
			6454.08	6455.86	317336.2	– 332826.0	4–6	1.49−01	1.39−01	1.18+01	−0.254	B	*LS*
			6445.34	6447.12	317293.6	– 332804.4	2–4	8.89−02	1.11−01	4.70+00	−0.654	B	*LS*
			6478.76	6480.55	317395.2	– 332826.0	6–6	6.31−02	3.97−02	5.08+00	−0.623	B	*LS*
			6463.09	6464.88	317336.2	– 332804.4	4–4	1.13−01	7.07−02	6.02+00	−0.548	B	*LS*
			6450.79	6452.57	317293.6	– 332791.3	2–2	1.77−01	1.11−01	4.70+00	−0.655	B	*LS*
			6487.84	6489.63	317395.2	– 332804.4	6–4	1.05−02	4.40−03	5.64−01	−1.578	B	*LS*
			6468.57	6470.36	317336.2	– 332791.3	4–2	3.52−02	1.10−02	9.40−01	−1.355	B	*LS*
198.		$^4P-^4P^\circ$	*5294.9*	*5296.3*	*317359*	– *336240*	12–12	1.64−01	6.91−02	1.45+01	−0.081	B	2
			5314.36	5315.84	317395.2	– 336206.9	6–6	1.14−01	4.82−02	5.06+00	−0.539	B	*LS*
			5282.43	5283.90	317336.2	– 336261.6	4–4	2.21−02	9.24−03	6.43−01	−1.432	B	*LS*
			5260.86	5262.33	317293.6	– 336296.6	2–2	2.80−02	1.16−02	4.02−01	−1.635	B	*LS*
			5298.95	5300.43	317395.2	– 336261.6	6–4	7.38−02	2.07−02	2.17+00	−0.905	B	*LS*
			5272.68	5274.15	317336.2	– 336296.6	4–2	1.39−01	2.89−02	2.01+00	−0.937	B	*LS*
			5297.75	5299.22	317336.2	– 336206.9	4–6	4.93−02	3.11−02	2.17+00	−0.905	B	*LS*
			5270.57	5272.04	317293.6	– 336261.6	2–4	6.95−02	5.79−02	2.01+00	−0.936	B	*LS*

N III: Allowed Transitions — Continued

No.	Transition Array	Mult.	λ_{air} (Å)	λ_{vac} (Å) or σ (cm^{-1})*	E_i (cm^{-1})	E_k (cm^{-1})	g_i-g_k	A_{ki} (10^8 s^{-1})	f_{ik}	S (at. u.)	log gf	Acc.	Source
199.	$2s2p(^3P^\circ)3p-2s2p(^3P^\circ)4s$	$^4P-^4P^\circ$		*1950.5*	*317359*	– *368627*	12–12	3.35+00	1.91–01	1.47+01	0.360	B	2
				1949.28	317395.2	– 368696.3	6–6	2.35+00	1.34–01	5.15+00	–0.096	B	*LS*
				1951.52	317336.2	– 368578.3	4–4	4.46–01	2.54–02	6.54–01	–0.992	B	*LS*
				1952.33	317293.6	– 368514.4	2–2	5.56–01	3.18–02	4.09–01	–1.197	B	*LS*
				1953.77	317395.2	– 368578.3	6–4	1.50+00	5.72–02	2.21+00	–0.465	B	*LS*
				1953.96	317336.2	– 368514.4	4–2	2.77+00	7.94–02	2.04+00	–0.498	B	*LS*
				1947.04	317336.2	– 368696.3	4–6	1.01+00	8.60–02	2.21+00	–0.463	B	*LS*
				1949.90	317293.6	– 368578.3	2–4	1.40+00	1.59–01	2.04+00	–0.497	B	*LS*
200.	$2s2p(^3P^\circ)3p-2s2p(^3P^\circ)4d$	$^4P-^4P^\circ$		*1451.1*	*317359*	– *386270*	12–12	1.87+00	5.91–02	3.39+00	–0.150	B	2
				1452.59	317395.2	– 386237.7	6–6	1.31+00	4.13–02	1.18+00	–0.606	B	*LS*
				1450.21	317336.2	– 386291.8	4–4	2.50–01	7.88–03	1.50–01	–1.501	B	*LS*
				1448.63	317293.6	– 386324.3	2–2	3.13–01	9.86–03	9.40–02	–1.705	B	*LS*
				1451.45	317395.2	– 386291.8	6–4	8.41–01	1.77–02	5.08–01	–0.974	B	*LS*
				1449.53	317336.2	– 386324.3	4–2	1.56+00	2.46–02	4.70–01	–1.006	B	*LS*
				1451.35	317336.2	– 386237.7	4–6	5.61–01	2.66–02	5.08–01	–0.974	B	*LS*
				1449.31	317293.6	– 386291.8	2–4	7.82–01	4.93–02	4.70–01	–1.006	B	*LS*
201.	$2s2p(^3P^\circ)3p-2s2p(^3P^\circ)5s$	$^4P-^4P^\circ$		*1195.8*	*317359*	– *400982*	12–12	1.39+00	2.99–02	1.41+00	–0.445	B	2
				1195.34	317395.2	– 401053.5	6–6	9.77–01	2.09–02	4.94–01	–0.901	B	*LS*
				1196.20	317336.2	– 400934.2	4–4	1.86–01	3.99–03	6.28–02	–1.797	B	*LS*
				1196.62	317293.6	– 400862.5	2–2	2.32–01	4.98–03	3.92–02	–2.002	B	*LS*
				1197.05	317395.2	– 400934.2	6–4	6.26–01	8.96–03	2.12–01	–1.270	B	*LS*
				1197.23	317336.2	– 400862.5	4–2	1.16+00	1.24–02	1.96–01	–1.303	B	*LS*
				1194.50	317336.2	– 401053.5	4–6	4.20–01	1.35–02	2.12–01	–1.269	B	*LS*
				1195.59	317293.6	– 400934.2	2–4	5.81–01	2.49–02	1.96–01	–1.302	B	*LS*
202.	$2s2p(^3P^\circ)3p-2s2p(^3P^\circ)5d$	$^4P-^4P^\circ$		*1086.2*	*317359*	– *409425*	12–12	1.27+00	2.25–02	9.64–01	–0.569	B	2
				1087.00	317395.2	– 409391.3	6–6	8.87–01	1.57–02	3.37–01	–1.026	B	*LS*
				1085.66	317336.2	– 409445.8	4–4	1.70–01	3.00–03	4.28–02	–1.921	B	*LS*
				1084.69	317293.6	– 409486.2	2–2	2.12–01	3.75–03	2.68–02	–2.125	B–	*LS*
				1086.36	317395.2	– 409445.8	6–4	5.71–01	6.74–03	1.45–01	–1.393	B	*LS*
				1085.19	317336.2	– 409486.2	4–2	1.06+00	9.37–03	1.34–01	–1.426	B	*LS*
				1086.31	317336.2	– 409391.3	4–6	3.81–01	1.01–02	1.45–01	–1.393	B	*LS*
				1085.16	317293.6	– 409445.8	2–4	5.30–01	1.87–02	1.34–01	–1.426	B	*LS*
203.	$2s2p(^3P^\circ)3p-2p^2(^3P)3p$	$^4P-^4D^\circ$		*976.11*	*317359*	– *419806*	12–20	4.97+00	1.18–01	4.56+00	0.152	B	2
				975.833	317395.2	– 419871.7	6–8	4.97+00	9.46–02	1.82+00	–0.246	B	*LS*
				975.948	317336.2	– 419800.7	4–6	3.48+00	7.45–02	9.57–01	–0.526	B	*LS*
				976.143	317293.6	– 419737.6	2–4	2.07+00	5.91–02	3.80–01	–0.927	B	*LS*
				976.510	317395.2	– 419800.7	6–6	1.49+00	2.13–02	4.10–01	–0.894	B	*LS*
				976.549	317336.2	– 419737.6	4–4	2.65+00	3.78–02	4.86–01	–0.820	B	*LS*
				976.499	317293.6	– 419700.3	2–2	4.13+00	5.91–02	3.80–01	–0.927	B	*LS*
				977.112	317395.2	– 419737.6	6–4	2.48–01	2.36–03	4.56–02	–1.849	B	*LS*
				976.905	317336.2	– 419700.3	4–2	8.26–01	5.91–03	7.60–02	–1.627	B	*LS*
204.	$2s2p(^3P^\circ)3p-2s2p(^3P^\circ)6d$	$^4P-^4P^\circ$		*957.27*	*317359*	– *421823*	12–12	1.08+00	1.49–02	5.62–01	–0.749	B	2
				957.864	317395.2	– 421794.2	6–6	7.56–01	1.04–02	1.97–01	–1.205	B	*LS*
				956.874	317336.2	– 421843.2	4–4	1.44–01	1.98–03	2.50–02	–2.101	B–	*LS*
				956.264	317293.6	– 421867.2	2–2	1.81–01	2.48–03	1.56–02	–2.304	B–	*LS*
				957.414	317395.2	– 421843.2	6–4	4.87–01	4.46–03	8.43–02	–1.573	B	*LS*
				956.654	317336.2	– 421867.2	4–2	9.04–01	6.20–03	7.81–02	–1.606	B	*LS*
				957.323	317336.2	– 421794.2	4–6	3.25–01	6.69–03	8.43–02	–1.572	B	*LS*
				956.484	317293.6	– 421843.2	2–4	4.52–01	1.24–02	7.81–02	–1.606	B	*LS*

N III: Allowed Transitions — Continued

No.	Transition Array	Mult.	λ_{air} (Å)	λ_{vac} (Å) or σ (cm^{-1})*	E_i (cm^{-1})	E_k (cm^{-1})	g_i-g_k	A_{ki} (10^8 s^{-1})	f_{ik}	S (at. u.)	log gf	Acc.	Source
205.	$2s2p(^3P^o)3p-2s2p(^3P^o)7d$	$^4P-^4D^o$		*892.43*	*317359*	– 429412.9	12–20	7.04–01	1.40–02	4.94–01	–0.774	B	2
				892.716	317395.2	– 429412.9	6–8	7.03–01	1.12–02	1.98–01	–1.173	B	*LS*
				892.246	317336.2	– 429412.9	4–6	4.93–01	8.83–03	1.04–01	–1.452	B	*LS*
				891.907	317293.6	– 429412.9	2–4	2.94–01	7.01–03	4.12–02	–1.853	B	*LS*
				892.716	317395.2	– 429412.9	6–6	2.11–01	2.52–03	4.44–02	–1.820	B	*LS*
				892.246	317336.2	– 429412.9	4–4	3.76–01	4.48–03	5.27–02	–1.746	B	*LS*
				891.907	317293.6	– 429412.9	2–2	5.88–01	7.01–03	4.12–02	–1.853	B	*LS*
				892.716	317395.2	– 429412.9	6–4	3.52–02	2.80–04	4.94–03	–2.775	C+	*LS*
				892.246	317336.2	– 429412.9	4–2	1.17–01	7.00–04	8.23–03	–2.553	C+	*LS*
206.		$^4P-^4P^o$		*891.85*	*317359*	– 429486.3	12–12	3.60–01	4.29–03	1.51–01	–1.288	B	2
				892.131	317395.2	– 429486.3	6–6	2.52–01	3.00–03	5.29–02	–1.744	B	*LS*
				891.662	317336.2	– 429486.3	4–4	4.80–02	5.72–04	6.72–03	–2.640	C+	*LS*
				891.324	317293.6	– 429486.3	2–2	6.01–02	7.16–04	4.20–03	–2.844	C	*LS*
				892.131	317395.2	– 429486.3	6–4	1.62–01	1.29–03	2.27–02	–2.112	B–	*LS*
				891.662	317336.2	– 429486.3	4–2	3.00–01	1.79–03	2.10–02	–2.145	B–	*LS*
				891.662	317336.2	– 429486.3	4–6	1.08–01	1.93–03	2.27–02	–2.112	B–	*LS*
				891.324	317293.6	– 429486.3	2–4	1.50–01	3.58–03	2.10–02	–2.145	B–	*LS*
207.	$2s2p(^3P^o)3p-2p^2(^3P)3p$	$^4P-^4S^o$		*887.57*	*317359*	– 430025.9	12–4	2.06+01	8.12–02	2.85+00	–0.011	B	2
				887.857	317395.2	– 430025.9	6–4	1.03+01	8.11–02	1.42+00	–0.313	B	*LS*
				887.393	317336.2	– 430025.9	4–4	6.88+00	8.12–02	9.49–01	–0.488	B	*LS*
				887.393	317336.2	– 430025.9	4–4	3.44+00	4.06–02	4.74–01	–0.790	B	*LS*
208.	$2s2p(^3P^o)3p-2s2p(^3P^o)8d$	$^4P-^4D^o$		*855.12*	*317359*	– 434301.2	12–20	5.18–01	9.47–03	3.20–01	–0.945	B	2
				855.388	317395.2	– 434301.2	6–8	5.18–01	7.57–03	1.28–01	–1.343	B	*LS*
				854.957	317336.2	– 434301.2	4–6	3.63–01	5.97–03	6.72–02	–1.622	B	*LS*
				854.645	317293.6	– 434301.2	2–4	2.16–01	4.74–03	2.67–02	–2.024	B–	*LS*
				855.388	317395.2	– 434301.2	6–6	1.55–01	1.70–03	2.88–02	–1.991	B–	*LS*
				854.957	317336.2	– 434301.2	4–4	2.76–01	3.03–03	3.41–02	–1.917	B	*LS*
				854.645	317293.6	– 434301.2	2–2	4.32–01	4.74–03	2.67–02	–2.024	B–	*LS*
				855.388	317395.2	– 434301.2	6–4	2.59–02	1.89–04	3.20–03	–2.945	C+	*LS*
				854.957	317336.2	– 434301.2	4–2	8.64–02	4.73–04	5.33–03	–2.723	C+	*LS*
209.		$^4P-^4P^o$		*854.72*	*317359*	– 434356.6	12–12	2.77–01	3.03–03	1.02–01	–1.439	B	2
				854.983	317395.2	– 434356.6	6–6	1.94–01	2.12–03	3.58–02	–1.895	B	*LS*
				854.552	317336.2	– 434356.6	4–4	3.69–02	4.04–04	4.55–03	–2.791	C+	*LS*
				854.241	317293.6	– 434356.6	2–2	4.62–02	5.06–04	2.84–03	–2.995	C	*LS*
				854.983	317395.2	– 434356.6	6–4	1.24–01	9.10–04	1.54–02	–2.263	B–	*LS*
				854.552	317336.2	– 434356.6	4–2	2.31–01	1.26–03	1.42–02	–2.296	B–	*LS*
				854.552	317336.2	– 434356.6	4–6	8.31–02	1.36–03	1.54–02	–2.263	B–	*LS*
				854.241	317293.6	– 434356.6	2–4	1.16–01	2.53–03	1.42–02	–2.296	B–	*LS*
210.	$2s^24d-2s^24f$	$^2D-^2F^o$	*39660*	*2521* cm^{-1}	*317767*	– *320288*	10–14	3.32–03	1.09–01	1.43+02	0.039	B	2
			39849	2508.8 cm^{-1}	317779.5	– 320288.3	6–8	3.27–03	1.04–01	8.17+01	–0.206	C+	*LS*
			39367	2539.5 cm^{-1}	317747.7	– 320287.2	4–6	3.16–03	1.10–01	5.72+01	–0.355	C+	*LS*
			39866	2507.7 cm^{-1}	317779.5	– 320287.2	6–6	2.18–04	5.19–03	4.08+00	–1.507	C+	*LS*
211.	$2s^24d-2s2p(^3P^o)3d$	$^2D-^2D^o$	*30630*	*3264* cm^{-1}	*317767*	– *321031*	10–10	1.05–04	1.47–03	1.48+00	–1.832	B	2
			30412	3287.3 cm^{-1}	317779.5	– 321066.8	6–6	9.97–05	1.38–03	8.31–01	–2.081	B	*LS*
			30947	3230.5 cm^{-1}	317747.7	– 320978.2	4–4	9.13–05	1.31–03	5.34–01	–2.281	B	*LS*
			31254	3198.7 cm^{-1}	317779.5	– 320978.2	6–4	9.84–06	9.61–05	5.93–02	–3.239	B	*LS*
			30120	3319.1 cm^{-1}	317747.7	– 321066.8	4–6	7.33–06	1.50–04	5.93–02	–3.223	B	*LS*

N III: Allowed Transitions — Continued

No.	Transition Array	Mult.	λ_{air} (Å)	λ_{vac} (Å) or σ (cm^{-1})*	E_i (cm^{-1})	E_k (cm^{-1})	g_i-g_k	A_{ki} ($10^8 s^{-1}$)	f_{ik}	S (at. u.)	log gf	Acc.	Source
212.	$2s^24d-2s^25p$	$^2D-^2P^o$	*4902.5*	*4903.9*	*317767*	*338159*	10−6	4.28−01	9.25−02	1.49+01	−0.034	B	2
			4904.78	4906.15	317779.5	338162.1	6−4	3.85−01	9.25−02	8.96+00	−0.256	B	*LS*
			4899.10	4900.47	317747.7	338153.9	4−2	4.29−01	7.71−02	4.98+00	−0.511	B	*LS*
			4897.14	4898.50	317747.7	338162.1	4−4	4.29−02	1.54−02	9.95−01	−1.209	B	*LS*
213.	$2s^24d-2s2p(^3P^o)3d$	$^2D-^2F^o$	*4535.7*	*4537.0*	*317767*	*339808*	10−14	7.00−02	3.02−02	4.51+00	−0.520	B	2
			4528.54	4529.81	317779.5	339855.5	6−8	7.03−02	2.88−02	2.58+00	−0.762	B	*LS*
			4544.82	4546.10	317747.7	339744.6	4−6	6.49−02	3.02−02	1.81+00	−0.918	B	*LS*
			4551.40	4552.68	317779.5	339744.6	6−6	4.62−03	1.43−03	1.29−01	−2.065	B	*LS*
214.		$^2D-^2P^o$	*4007.2*	*4008.3*	*317767*	*342715*	10−6	1.56−02	2.25−03	2.97−01	−1.648	B	2
			4013.00	4014.13	317779.5	342691.5	6−4	1.39−02	2.24−03	1.78−01	−1.871	B	*LS*
			3996.50	3997.63	317747.7	342762.5	4−2	1.57−02	1.88−03	9.88−02	−2.124	B	*LS*
			4007.88	4009.01	317747.7	342691.5	4−4	1.56−03	3.75−04	1.98−02	−2.824	B−	*LS*
215.	$2s^24d-2s^25f$	$^2D-^2F^o$	*4001.6*	*4002.7*	*317767*	*342750*	10−14	1.88+00	6.31−01	8.32+01	0.800	B	2
			4003.58	4004.71	317779.5	342750.1	6−8	1.88+00	6.01−01	4.75+01	0.557	C+	*LS*
			3998.63	3999.76	317747.7	342749.2	4−6	1.76+00	6.32−01	3.33+01	0.403	C+	*LS*
			4003.72	4004.85	317779.5	342749.2	6−6	1.25−01	3.00−02	2.38+00	−0.744	C+	*LS*
216.	$2s^24d-2s^26f$	$^2D-^2F^o$	*2688.3*	*2689.1*	*317767*	354954.1	10−14	9.91−01	1.50−01	1.33+01	0.177	B	2
			2689.21	2690.01	317779.5	354954.1	6−8	9.90−01	1.43−01	7.60+00	−0.066	C+	*LS*
			2686.91	2687.71	317747.7	354954.1	4−6	9.26−01	1.50−01	5.32+00	−0.221	C+	*LS*
			2689.21	2690.01	317779.5	354954.1	6−6	6.60−02	7.15−03	3.80−01	−1.367	C+	*LS*
217.	$2s^24d-2s2p(^1P^o)3s$	$^2D-^2P^o$	*2802.3*	*2803.1*	*317767*	*353442*	10−6	2.11−02	1.49−03	1.37−01	−1.827	B	2
			2803.20	2804.03	317779.5	353442.5	6−4	1.89−02	1.49−03	8.24−02	−2.049	B	*LS*
			2800.85	2801.68	317747.7	353440.6	4−2	2.11−02	1.24−03	4.58−02	−2.304	B	*LS*
			2800.70	2801.53	317747.7	353442.5	4−4	2.11−03	2.48−04	9.15−03	−3.003	C+	*LS*
218.	$2s^24d-2s^27f$	$^2D-^2F^o$	*2243.8*	*2244.5*	*317767*	362320.9	10−14	5.81−01	6.13−02	4.53+00	−0.212	B	2
			2244.41	2245.10	317779.5	362320.9	6−8	5.80−01	5.84−02	2.59+00	−0.455	C+	*LS*
			2242.80	2243.50	317747.7	362320.9	4−6	5.43−01	6.14−02	1.81+00	−0.610	C+	*LS*
			2244.41	2245.10	317779.5	362320.9	6−6	3.87−02	2.92−03	1.29−01	−1.756	C+	*LS*
219.	$2s^24d-2s2p(^3P^o)4d$	$^2D-^2D^o$		*1484.6*	*317767*	385126	10−10	2.64−01	8.72−03	4.26−01	−1.060	B	2
				1484.86	317779.5	385126	6−6	2.46−01	8.14−03	2.39−01	−1.311	B	*LS*
				1484.16	317747.7	385126	4−4	2.38−01	7.85−03	1.53−01	−1.503	B	*LS*
				1484.86	317779.5	385126	6−4	2.64−02	5.81−04	1.70−02	−2.458	B−	*LS*
				1484.16	317747.7	385126	4−6	1.76−02	8.72−04	1.70−02	−2.457	B−	*LS*
220.	$2s^24d-2s2p(^1P^o)3d$	$^2D-^2D^o$		*1268.8*	*317767*	*396581*	10−10	3.13−02	7.57−04	3.16−02	−2.121	B	2
				1268.95	317779.5	396584.8	6−6	2.92−02	7.06−04	1.77−02	−2.373	B−	*LS*
				1268.60	317747.7	396574.9	4−4	2.82−02	6.81−04	1.14−02	−2.565	B−	*LS*
				1269.11	317779.5	396574.9	6−4	3.13−03	5.04−05	1.26−03	−3.519	C	*LS*
				1268.44	317747.7	396584.8	4−6	2.09−03	7.57−05	1.26−03	−3.519	C	*LS*
221.	$2s^24d-2p^2(^3P)3p$	$^2D-^2D^o$		*1021.3*	*317767*	*415679*	10−10	1.66+00	2.59−02	8.72−01	−0.586	B	2
				1021.47	317779.5	415678.0	6−6	1.55+00	2.42−02	4.88−01	−0.838	B	*LS*
				1021.12	317747.7	415679.7	4−4	1.49+00	2.34−02	3.14−01	−1.030	B	*LS*
				1021.45	317779.5	415679.7	6−4	1.66−01	1.73−03	3.49−02	−1.984	B	*LS*
				1021.13	317747.7	415678.0	4−6	1.11−01	2.59−03	3.49−02	−1.984	B	*LS*

N III: Allowed Transitions — Continued

No.	Transition Array	Mult.	λ_{air} (Å)	λ_{vac} (Å) or σ (cm^{-1})*	E_i (cm^{-1})	E_k (cm^{-1})	g_i-g_k	A_{ki} (10^8 s^{-1})	f_{ik}	S (at. u.)	log gf	Acc.	Source
222.	$2s^24f-$ $2s2p(^3P^o)3p$	$^2F^o-^2D$		*743 cm^{-1}*	*320288*	*321031*	14–10	2.13–05	4.14–03	2.57+01	–1.237	B	2
				778.5 cm^{-1}	320288.3	321066.8	8–6	2.34–05	4.34–03	1.47+01	–1.460	C+	*LS*
				691.0 cm^{-1}	320287.2	320978.2	6–4	1.72–05	3.59–03	1.03+01	–1.667	C+	*LS*
				779.6 cm^{-1}	320287.2	321066.8	6–6	1.17–06	2.89–04	7.33–01	–2.760	C+	*LS*
223.	$2s^24f-$ $2s^25d$	$^2F^o-^2D$	*4615.5*	*4616.8*	*320288*	*341948*	14–10	7.12–02	1.62–02	3.45+00	–0.643	B	2
			4615.41	4616.70	320288.3	341948.8	8–6	6.78–02	1.62–02	1.97+00	–0.886	C+	*LS*
			4615.41	4616.70	320287.2	341947.7	6–4	7.12–02	1.52–02	1.38+00	–1.041	C+	*LS*
			4615.17	4616.46	320287.2	341948.8	6–6	3.39–03	1.08–03	9.87–02	–2.187	C+	*LS*
224.	$2s^24f-$ $2s^26d$	$^2F^o-^2D$	*2919.5*	*2920.3*	*320288*	354530.5	14–10	3.08–02	2.81–03	3.78–01	–1.405	B	2
			2919.52	2920.37	320288.3	354530.5	8–6	2.93–02	2.81–03	2.16–01	–1.648	C+	*LS*
			2919.42	2920.28	320287.2	354530.5	6–4	3.08–02	2.62–03	1.51–01	–1.803	C+	*LS*
			2919.42	2920.28	320287.2	354530.5	6–6	1.47–03	1.87–04	1.08–02	–2.949	C	*LS*
225.	$2s^24f-$ $2s^27d$	$^2F^o-^2D$	*2392.9*	*2393.6*	*320288*	362066	14–10	1.64–02	1.00–03	1.11–01	–1.853	B	2
			2392.89	2393.62	320288.3	362066	8–6	1.56–02	1.00–03	6.32–02	–2.096	C+	*LS*
			2392.83	2393.56	320287.2	362066	6–4	1.64–02	9.36–04	4.42–02	–2.251	C+	*LS*
			2392.83	2393.56	320287.2	362066	6–6	7.79–04	6.69–05	3.16–03	–3.397	C–	*LS*
226.	$2s^24f-$ $2s^28d$	$^2F^o-^2D$	*2142.9*	*2143.5*	*320288*	366940	14–10	9.90–03	4.87–04	4.81–02	–2.166	B	2
			2142.87	2143.54	320288.3	366940	8–6	9.43–03	4.87–04	2.75–02	–2.409	C	*LS*
			2142.82	2143.49	320287.2	366940	6–4	9.90–03	4.55–04	1.92–02	–2.564	C	*LS*
			2142.82	2143.49	320287.2	366940	6–6	4.72–04	3.25–05	1.37–03	–3.710	D	*LS*
227.	$2s2p(^3P^o)3p-$ $2s2p(^3P^o)3d$	$^2D-^2D^o$	*7391.1*	*7393.2*	*321031*	*334557*	10–10	3.89–02	3.19–02	7.76+00	–0.496	B	2
			7404.54	7406.58	321066.8	334568.3	6–6	3.61–02	2.97–02	4.35+00	–0.749	B	*LS*
			7371.51	7373.54	320978.2	334540.2	4–4	3.53–02	2.88–02	2.79+00	–0.939	B	*LS*
			7419.99	7422.03	321066.8	334540.2	6–4	3.85–03	2.12–03	3.10–01	–1.896	B	*LS*
			7356.27	7358.30	320978.2	334568.3	4–6	2.63–03	3.20–03	3.10–01	–1.892	B	*LS*
228.	$2s2p(^3P^o)3p-$ $2s^25p$	$^2D-^2P^o$	*5836.8*	*5838.4*	*321031*	*338159*	10–6	1.74–01	5.34–02	1.03+01	–0.273	B	2
			5847.94	5849.56	321066.8	338162.1	6–4	1.56–01	5.33–02	6.15+00	–0.495	B	*LS*
			5820.57	5822.18	320978.2	338153.9	4–2	1.76–01	4.46–02	3.42+00	–0.749	B	*LS*
			5817.79	5819.40	320978.2	338162.1	4–4	1.76–02	8.92–03	6.84–01	–1.447	B	*LS*
229.	$2s2p(^3P^o)3p-$ $2s2p(^3P^o)3d$	$^2D-^2F^o$	*5324.2*	*5325.7*	*321031*	*339808*	10–14	5.67–01	3.38–01	5.92+01	0.528	B	2
			5320.87	5322.35	321066.8	339855.5	6–8	5.68–01	3.22–01	3.38+01	0.286	B	*LS*
			5327.19	5328.67	320978.2	339744.6	4–6	5.29–01	3.37–01	2.37+01	0.130	B	*LS*
			5352.46	5353.95	321066.8	339744.6	6–6	3.72–02	1.60–02	1.69+00	–1.018	B	*LS*
230.	$2s2p(^3P^o)3p-$ $2s^25f$	$^2D-^2F^o$	*4603.0*	*4604.3*	*321031*	*342750*	10–14	2.91–01	1.30–01	1.96+01	0.112	B	2
			4610.55	4611.84	321066.8	342750.1	6–8	2.90–01	1.23–01	1.12+01	–0.131	C+	*LS*
			4591.98	4593.27	320978.2	342749.2	4–6	2.74–01	1.30–01	7.85+00	–0.285	C+	*LS*
			4610.74	4612.04	321066.8	342749.2	6–6	1.93–02	6.16–03	5.61–01	–1.432	C+	*LS*
231.	$2s2p(^3P^o)3p-$ $2s2p(^3P^o)3d$	$^2D-^2P^o$	*4610.4*	*4611.7*	*321031*	*342715*	10–6	6.32–02	1.21–02	1.83+00	–0.918	B	2
			4623.05	4624.34	321066.8	342691.5	6–4	5.64–02	1.21–02	1.10+00	–1.141	B	*LS*
			4589.18	4590.46	320978.2	342762.5	4–2	6.41–02	1.01–02	6.11–01	–1.393	B	*LS*
			4604.18	4605.47	320978.2	342691.5	4–4	6.35–03	2.02–03	1.22–01	–2.093	B	*LS*

N III: Allowed Transitions — Continued

No.	Transition Array	Mult.	λ_{air} (Å)	λ_{vac} (Å) or σ (cm^{-1})*	E_i (cm^{-1})	E_k (cm^{-1})	g_i–g_k	A_{ki} (10^8 s^{-1})	f_{ik}	S (at. u.)	log gf	Acc.	Source
232.	$2s2p(^3P^o)3p-2s2p(^1P^o)3s$	$^2D-^2P^o$	*3084.5*	*3085.4*	*321031*	*353442*	10–6	6.19–03	5.30–04	5.38–02	–2.276	B	2
			3087.84	3088.74	321066.8	353442.5	6–4	5.55–03	5.29–04	3.23–02	–2.498	B	*LS*
			3079.59	3080.49	320978.2	353440.6	4–2	6.22–03	4.42–04	1.79–02	–2.752	B–	*LS*
			3079.41	3080.31	320978.2	353442.5	4–4	6.22–04	8.84–05	3.59–03	–3.451	C+	*LS*
233.	$2s2p(^3P^o)3p-2s^26f$	$^2D-^2F^o$	*2947.0*	*2947.9*	*321031*	354954.1	10–14	1.45–01	2.65–02	2.57+00	–0.578	B	2
			2950.10	2950.96	321066.8	354954.1	6–8	1.45–01	2.52–02	1.47+00	–0.821	C+	*LS*
			2942.40	2943.26	320978.2	354954.1	4–6	1.36–01	2.65–02	1.03+00	–0.975	C+	*LS*
			2950.10	2950.96	321066.8	354954.1	6–6	9.64–03	1.26–03	7.33–02	–2.122	C+	*LS*
234.	$2s2p(^3P^o)3p-2s^27f$	$^2D-^2F^o$	*2421.2*	*2421.9*	*321031*	362320.9	10–14	7.32–02	9.00–03	7.18–01	–1.046	B	2
			2423.27	2424.00	321066.8	362320.9	6–8	7.30–02	8.57–03	4.10–01	–1.289	C+	*LS*
			2418.07	2418.81	320978.2	362320.9	4–6	6.86–02	9.01–03	2.87–01	–1.443	C+	*LS*
			2423.27	2424.00	321066.8	362320.9	6–6	4.87–03	4.28–04	2.05–02	–2.590	C	*LS*
235.	$2s2p(^3P^o)3p-2s2p(^3P^o)4d$	$^2D-^2D^o$		*1560.2*	*321031*	385126	10–10	1.18+00	4.32–02	2.22+00	–0.364	B	2
				1561.06	321066.8	385126	6–6	1.10+00	4.03–02	1.24+00	–0.616	B	*LS*
				1558.90	320978.2	385126	4–4	1.07+00	3.89–02	7.99–01	–0.808	B	*LS*
				1561.06	321066.8	385126	6–4	1.18–01	2.88–03	8.88–02	–1.763	B	*LS*
				1558.90	320978.2	385126	4–6	7.91–02	4.32–03	8.88–02	–1.762	B	*LS*
236.	$2s2p(^3P^o)3p-2s2p(^1P^o)3d$	$^2D-^2D^o$		*1323.6*	*321031*	*396581*	10–10	8.19–02	2.15–03	9.37–02	–1.667	B	2
				1324.19	321066.8	396584.8	6–6	7.63–02	2.01–03	5.25–02	–1.919	B	*LS*
				1322.81	320978.2	396574.9	4–4	7.38–02	1.94–03	3.37–02	–2.111	B	*LS*
				1324.36	321066.8	396574.9	6–4	8.17–03	1.43–04	3.75–03	–3.066	C+	*LS*
				1322.64	320978.2	396584.8	4–6	5.47–03	2.15–04	3.75–03	–3.065	C+	*LS*
237.	$2s2p(^3P^o)3p-2p^2(^3P)3p$	$^2D-^2D^o$		*1056.5*	*321031*	*415679*	10–10	6.52+00	1.09–01	3.80+00	0.038	B	2
				1056.96	321066.8	415678.0	6–6	6.08+00	1.02–01	2.13+00	–0.214	B	*LS*
				1055.95	320978.2	415679.7	4–4	5.88+00	9.83–02	1.37+00	–0.405	B	*LS*
				1056.94	321066.8	415679.7	6–4	6.51–01	7.27–03	1.52–01	–1.360	B	*LS*
				1055.97	320978.2	415678.0	4–6	4.35–01	1.09–02	1.52–01	–1.360	B	*LS*
238.	$2s2p(^3P^o)3p-2s^25p$	$^2S-^2P^o$	*9005.7*	*9008.2*	327058.0	*338159*	2–6	3.28–02	1.20–01	7.10+00	–0.621	B	2
			9003.21	9005.68	327058.0	338162.1	2–4	3.28–02	7.98–02	4.73+00	–0.797	B	*LS*
			9009.86	9012.34	327058.0	338153.9	2–2	3.28–02	3.99–02	2.37+00	–1.098	B	*LS*
239.	$2s2p(^3P^o)3p-2s2p(^3P^o)3d$	$^2S-^2P^o$	*6385.2*	*6386.9*	327058.0	*342715*	2–6	2.16–01	3.96–01	1.67+01	–0.101	B	2
			6394.75	6396.52	327058.0	342691.5	2–4	2.15–01	2.64–01	1.11+01	–0.278	B	*LS*
			6365.84	6367.60	327058.0	342762.5	2–2	2.18–01	1.32–01	5.55+00	–0.577	B	*LS*
240.	$2s2p(^3P^o)3p-2s2p(^1P^o)3s$	$^2S-^2P^o$	*3789.1*	*3790.2*	327058.0	*353442*	2–6	2.40–02	1.55–02	3.87–01	–1.509	B	2
			3789.03	3790.10	327058.0	353442.5	2–4	2.40–02	1.03–02	2.58–01	–1.685	B	*LS*
			3789.30	3790.38	327058.0	353440.6	2–2	2.40–02	5.17–03	1.29–01	–1.986	B	*LS*

N III: Allowed Transitions — Continued

No.	Transition Array	Mult.	λ_{air} (Å)	λ_{vac} (Å) or σ (cm^{-1})*	E_i (cm^{-1})	E_k (cm^{-1})	g_i-g_k	A_{ki} (10^8 s^{-1})	f_{ik}	S (at. u.)	log gf	Acc.	Source
241.	$2s2p(^3P^\circ)3d-$ $2s2p(^3P^\circ)4p$	$^4F^\circ-^4D$	*2147.6*	*2148.3*	*330321*	*– 376869*	28–20	1.10+00	5.44–02	1.08+01	0.183	B	2
			2147.31	2147.98	330389.9	– 376945.2	10–8	9.84–01	5.44–02	3.85+00	–0.264	B	*LS*
			2148.09	2148.77	330318.5	– 376856.8	8–6	8.99–01	4.66–02	2.64+00	–0.428	B	*LS*
			2148.51	2149.19	330266.5	– 376795.7	6–4	8.84–01	4.08–02	1.73+00	–0.611	B	*LS*
			2149.02	2149.69	330232.1	– 376750.4	4–2	1.10+00	3.81–02	1.08+00	–0.817	B	*LS*
			2144.02	2144.69	330318.5	– 376945.2	8–8	1.13–01	7.77–03	4.39–01	–1.207	B	*LS*
			2145.69	2146.37	330266.5	– 376856.8	6–6	1.92–01	1.33–02	5.62–01	–1.100	B	*LS*
			2146.92	2147.60	330232.1	– 376795.7	4–4	2.21–01	1.52–02	4.31–01	–1.215	B	*LS*
			2141.63	2142.30	330266.5	– 376945.2	6–8	5.75–03	5.27–04	2.23–02	–2.500	B–	*LS*
			2144.11	2144.79	330232.1	– 376856.8	4–6	1.05–02	1.09–03	3.08–02	–2.361	B	*LS*
242.	$2s2p(^3P^\circ)3d-$ $2s2p(^3P^\circ)4f$	$^4F^\circ-^4F$		*1764.4*	*330321*	*– 386997*	28–28	1.19+00	5.55–02	9.03+00	0.191	B	2
				1766.06	330389.9	– 387013.1	10–10	1.09+00	5.08–02	2.95+00	–0.294	C+	*LS*
				1764.22	330318.5	– 387000.9	8–8	9.07–01	4.23–02	1.97+00	–0.470	C+	*LS*
				1763.16	330266.5	– 386982.8	6–6	8.20–01	3.82–02	1.33+00	–0.640	C+	*LS*
				1762.44	330232.1	– 386971.6	4–4	9.55–01	4.45–02	1.03+00	–0.750	C+	*LS*
				1766.44	330389.9	– 387000.9	10–8	1.23–01	4.59–03	2.67–01	–1.338	C+	*LS*
				1764.78	330318.5	– 386982.8	8–6	2.12–01	7.43–03	3.45–01	–1.226	C+	*LS*
				1763.51	330266.5	– 386971.6	6–4	2.38–01	7.40–03	2.58–01	–1.352	C+	*LS*
				1763.84	330318.5	– 387013.1	8–10	9.86–02	5.75–03	2.67–01	–1.337	C+	*LS*
				1762.60	330266.5	– 387000.9	6–8	1.60–01	9.92–03	3.45–01	–1.225	C+	*LS*
				1762.09	330232.1	– 386982.8	4–6	1.59–01	1.11–02	2.58–01	–1.352	C+	*LS*
243.	$2s2p(^3P^\circ)3d-$ $^22s2p(^3P^\circ)4f$	$^4F^\circ-^4G$		*1730.1*	*330321*	*– 388122*	28–36	1.30+01	7.52–01	1.20+02	1.323	B	2
				1729.90	330389.9	– 388196.8	10–12	1.30+01	7.02–01	4.00+01	0.846	C+	*LS*
				1729.95	330318.5	– 388123.8	8–10	1.19+01	6.69–01	3.05+01	0.729	C+	*LS*
				1730.00	330266.5	– 388069.9	6–8	1.12+01	6.71–01	2.29+01	0.605	C+	*LS*
				1729.83	330232.1	– 388041.1	4–6	1.12+01	7.52–01	1.71+01	0.478	C+	*LS*
				1732.08	330389.9	– 388123.8	10–10	1.08+00	4.84–02	2.76+00	–0.315	C+	*LS*
				1731.56	330318.5	– 388069.9	8–8	1.76+00	7.93–02	3.62+00	–0.198	C+	*LS*
				1730.86	330266.5	– 388041.1	6–6	1.80+00	8.07–02	2.76+00	–0.315	C+	*LS*
				1733.70	330389.9	– 388069.9	10–8	3.93–02	1.42–03	8.09–02	–1.849	C+	*LS*
				1732.42	330318.5	– 388041.1	8–6	6.49–02	2.19–03	9.99–02	–1.756	C+	*LS*
244.		$^4F^\circ-^4D$		*1724.4*	*330321*	*– 388312*	28–20	1.28–01	4.07–03	6.47–01	–0.944	B	2
				1727.63	330389.9	– 388272.8	10–8	1.13–01	4.06–03	2.31–01	–1.392	C+	*LS*
				1724.41	330318.5	– 388309.3	8–6	1.04–01	3.49–03	1.58–01	–1.554	C+	*LS*
				1721.46	330266.5	– 388356.8	6–4	1.03–01	3.06–03	1.04–01	–1.737	C+	*LS*
				1719.58	330232.1	– 388385.8	4–2	1.29–01	2.86–03	6.47–02	–1.942	C+	*LS*
				1725.50	330318.5	– 388272.8	8–8	1.30–02	5.79–04	2.63–02	–2.334	C	*LS*
				1722.87	330266.5	– 388309.3	6–6	2.23–02	9.91–04	3.37–02	–2.226	C+	*LS*
				1720.44	330232.1	– 388356.8	4–4	2.57–02	1.14–03	2.59–02	–2.340	C	*LS*
				1723.95	330266.5	– 388272.8	6–8	6.62–04	3.93–05	1.34–03	–3.627	D	*LS*
				1721.85	330232.1	– 388309.3	4–6	1.22–03	8.15–05	1.85–03	–3.487	D	*LS*
245.	$2s2p(^3P^\circ)3d-$ $2s2p(^3P^\circ)4p$	$^4D^\circ-^4D$	*2270.0*	*2270.7*	*332829*	*– 376869*	20–20	2.23–01	1.72–02	2.57+00	–0.464	B	2
			2267.33	2268.03	332854.0	– 376945.2	8–8	1.91–01	1.48–02	8.81–01	–0.928	B	*LS*
			2270.44	2271.14	332826.0	– 376856.8	6–6	1.28–01	9.85–03	4.42–01	–1.228	B	*LS*
			2272.47	2273.18	332804.4	– 376795.7	4–4	8.87–02	6.87–03	2.06–01	–1.561	B	*LS*
			2274.14	2274.84	332791.3	– 376750.4	2–2	1.11–01	8.58–03	1.28–01	–1.766	B	*LS*
			2271.88	2272.58	332854.0	– 376856.8	8–6	4.22–02	2.45–03	1.46–01	–1.708	B	*LS*
			2273.59	2274.29	332826.0	– 376795.7	6–4	7.75–02	4.00–03	1.80–01	–1.619	B	*LS*
			2274.82	2275.52	332804.4	– 376750.4	4–2	1.11–01	4.29–03	1.28–01	–1.766	B	*LS*
			2265.89	2266.59	332826.0	– 376945.2	6–8	3.19–02	3.27–03	1.46–01	–1.707	B	*LS*
			2269.32	2270.02	332804.4	– 376856.8	4–6	5.20–02	6.02–03	1.80–01	–1.618	B	*LS*
			2271.80	2272.50	332791.3	– 376795.7	2–4	5.55–02	8.59–03	1.28–01	–1.765	B	*LS*

N III: Allowed Transitions — Continued

No.	Transition Array	Mult.	λ_{air} (Å)	λ_{vac} (Å) or σ (cm^{-1})*	E_i (cm^{-1})	E_k (cm^{-1})	g_i-g_k	A_{ki} (10^8 s^{-1})	f_{ik}	S (at. u.)	log gf	Acc.	Source
246.		^{4}D°–^{4}P	*2148.2*	*2148.9*	*332829*	*– 379364*	20–12	8.71–01	3.62–02	5.11+00	–0.141	B	2
			2147.85	2148.52	332854.0	– 379397.6	8–6	6.97–01	3.62–02	2.05+00	–0.539	B	*LS*
			2148.99	2149.67	332826.0	– 379344.8	6–4	5.48–01	2.53–02	1.07+00	–0.819	B	*LS*
			2150.06	2150.74	332804.4	– 379300.1	4–2	4.34–01	1.51–02	4.26–01	–1.220	B	*LS*
			2146.56	2147.23	332826.0	– 379397.6	6–6	1.57–01	1.09–02	4.60–01	–1.186	B	*LS*
			2147.99	2148.67	332804.4	– 379344.8	4–4	2.79–01	1.93–02	5.46–01	–1.113	B	*LS*
			2149.45	2150.13	332791.3	– 379300.1	2–2	4.35–01	3.01–02	4.26–01	–1.220	B	*LS*
			2145.56	2146.24	332804.4	– 379397.6	4–6	1.75–02	1.81–03	5.11–02	–2.140	B	*LS*
			2147.39	2148.07	332791.3	– 379344.8	2–4	4.36–02	6.03–03	8.52–02	–1.919	B	*LS*
247.	2s2p(^{3}P°)3d– 2s2p(^{3}P°)4f	^{4}D°–^{4}F		*1846.1*	*332829*	*– 386997*	20–28	1.10+01	7.89–01	9.59+01	1.198	B	2
				1846.41	332854.0	– 387013.1	8–10	1.10+01	7.05–01	3.43+01	0.751	C+	*LS*
				1845.87	332826.0	– 387000.9	6–8	9.47+00	6.45–01	2.35+01	0.588	C+	*LS*
				1845.75	332804.4	– 386982.8	4–6	8.28+00	6.34–01	1.54+01	0.404	C+	*LS*
				1845.69	332791.3	– 386971.6	2–4	7.73+00	7.90–01	9.59+00	0.198	C+	*LS*
				1846.83	332854.0	– 387000.9	8–8	1.57+00	8.03–02	3.91+00	–0.192	C+	*LS*
				1846.49	332826.0	– 386982.8	6–6	2.68+00	1.37–01	5.00+00	–0.085	C+	*LS*
				1846.14	332804.4	– 386971.6	4–4	3.09+00	1.58–01	3.84+00	–0.200	C+	*LS*
				1847.45	332854.0	– 386982.8	8–6	1.06–01	4.08–03	1.99–01	–1.486	C+	*LS*
				1846.87	332826.0	– 386971.6	6–4	2.20–01	7.51–03	2.74–01	–1.346	C+	*LS*
248.		^{4}D°–^{4}D		*1802.4*	*332829*	*– 388312*	20–20	2.01+00	9.80–02	1.16+01	0.292	B	2
				1804.44	332854.0	– 388272.8	8–8	1.72+00	8.39–02	3.99+00	–0.173	C+	*LS*
				1802.34	332826.0	– 388309.3	6–6	1.15+00	5.62–02	2.00+00	–0.472	C+	*LS*
				1800.10	332804.4	– 388356.8	4–4	8.08–01	3.92–02	9.30–01	–0.804	C+	*LS*
				1798.74	332791.3	– 388385.8	2–2	1.01+00	4.91–02	5.81–01	–1.008	C+	*LS*
				1803.25	332854.0	– 388309.3	8–6	3.82–01	1.40–02	6.63–01	–0.952	C+	*LS*
				1800.80	332826.0	– 388356.8	6–4	7.06–01	2.29–02	8.14–01	–0.862	C+	*LS*
				1799.16	332804.4	– 388385.8	4–2	1.01+00	2.45–02	5.81–01	–1.008	C+	*LS*
				1803.53	332826.0	– 388272.8	6–8	2.86–01	1.86–02	6.63–01	–0.952	C+	*LS*
				1801.64	332804.4	– 388309.3	4–6	4.70–01	3.43–02	8.14–01	–0.863	C+	*LS*
				1799.68	332791.3	– 388356.8	2–4	5.05–01	4.91–02	5.81–01	–1.008	C+	*LS*
249.	2s2p(^{3}P°)3d– 2p^2(^{3}P)3s	^{4}D°–^{4}P		*1474.0*	*332829*	*– 400671*	20–12	1.04–01	2.03–03	1.97–01	–1.392	B	2
				1473.23	332854.0	– 400732.2	8–6	8.32–02	2.03–03	7.88–02	–1.789	B	*LS*
				1474.78	332826.0	– 400632.9	6–4	6.53–02	1.42–03	4.14–02	–2.070	B	*LS*
				1475.76	332804.4	– 400566.2	4–2	5.17–02	8.45–04	1.64–02	–2.471	B–	*LS*
				1472.62	332826.0	– 400732.2	6–6	1.87–02	6.10–04	1.77–02	–2.437	B–	*LS*
				1474.31	332804.4	– 400632.9	4–4	3.32–02	1.08–03	2.10–02	–2.364	B–	*LS*
				1475.47	332791.3	– 400566.2	2–2	5.18–02	1.69–03	1.64–02	–2.471	B–	*LS*
				1472.15	332804.4	– 400732.2	4–6	2.09–03	1.02–04	1.97–03	–3.391	C	*LS*
				1474.02	332791.3	– 400632.9	2–4	5.19–03	3.38–04	3.28–03	–3.170	C+	*LS*
250.	2s2p(^{3}P°)3d– 2s2p(^{3}P°)4p	^{2}D°–^{2}P	*2485.1*	*2485.8*	*334557*	*– 374785*	10–6	1.03+00	5.75–02	4.70+00	–0.241	B	2
			2484.56	2485.31	334568.3	– 374804.8	6–4	9.32–01	5.75–02	2.82+00	–0.462	B	*LS*
			2486.43	2487.18	334540.2	– 374746.4	4–2	1.03+00	4.79–02	1.57+00	–0.718	B	*LS*
			2482.82	2483.57	334540.2	– 374804.8	4–4	1.04–01	9.59–03	3.13–01	–1.416	B	*LS*
251.	2s2p(^{3}P°)3d– 2s2p(^{3}P°)4f	^{2}D°–^{2}F		*1908.0*	*334557*	*– 386967*	10–14	1.04+01	7.92–01	4.97+01	0.899	B	2
				1908.08	334568.3	– 386977.0	6–8	1.04+01	7.54–01	2.84+01	0.656	C+	*LS*
				1907.92	334540.2	– 386953.4	4–6	9.68+00	7.92–01	1.99+01	0.501	C+	*LS*
				1908.94	334568.3	– 386953.4	6–6	6.90–01	3.77–02	1.42+00	–0.646	C+	*LS*
252.	2s2p(^{3}P°)3d– 2s2p(^{3}P°)5p	^{2}D°–^{2}P		*1468.0*	*334557*	*– 402679*	10–6	4.36–01	8.45–03	4.08–01	–1.073	B	2
				1467.62	334568.3	– 402705.6	6–4	3.93–01	8.45–03	2.45–01	–1.295	B	*LS*
				1468.77	334540.2	– 402624.6	4–2	4.35–01	7.04–03	1.36–01	–1.550	B	*LS*
				1467.02	334540.2	– 402705.6	4–4	4.37–02	1.41–03	2.72–02	–2.249	B–	*LS*

N III: Allowed Transitions — Continued

No.	Transition Array	Mult.	λ_{air} (Å)	λ_{vac} (Å) or σ (cm^{-1})*	E_i (cm^{-1})	E_k (cm^{-1})	g_i-g_k	A_{ki} (10^8 s^{-1})	f_{ik}	S (at. u.)	log gf	Acc.	Source
253.	$2s2p(^3P^o)3d-2s2p(^3P^o)6p$	$^2D^o-^2P$		*1168.3*	*334557*	*– 420154*	10–6	2.93–01	3.60–03	1.38–01	–1.444	B	2
				1168.09	334568.3	– 420178.4	6–4	2.64–01	3.60–03	8.30–02	–1.666	B	*LS*
				1168.69	334540.2	– 420106.1	4–2	2.93–01	3.00–03	4.61–02	–1.921	B	*LS*
				1167.70	334540.2	– 420178.4	4–4	2.94–02	6.00–04	9.23–03	–2.620	C+	*LS*
254.	$2s2p(^3P^o)3d-2s2p(^3P^o)7p$	$^2D^o-^2P$		*1069.4*	*334557*	*– 428069*	10–6	2.60–01	2.68–03	9.43–02	–1.572	B	2
				1069.38	334568.3	– 428080.8	6–4	2.34–01	2.68–03	5.66–02	–1.794	B	*LS*
				1069.46	334540.2	– 428045.5	4–2	2.60–01	2.23–03	3.14–02	–2.049	B	*LS*
				1069.05	334540.2	– 428080.8	4–4	2.61–02	4.47–04	6.29–03	–2.748	C+	*LS*
255.	$2s2p(^3P^o)3d-2s2p(^3P^o)8p$	$^2D^o-^2P$		*1012.5*	*334557*	*– 433322*	10–6	2.39–01	2.20–03	7.34–02	–1.657	B	2
				1012.48	334568.3	– 433335.6	6–4	2.15–01	2.20–03	4.40–02	–1.879	B	*LS*
				1012.60	334540.2	– 433296.2	4–2	2.39–01	1.83–03	2.45–02	–2.134	B–	*LS*
				1012.19	334540.2	– 433335.6	4–4	2.39–02	3.67–04	4.89–03	–2.833	C+	*LS*
256.	$2s2p(^3P^o)3d-2s2p(^3P^o)4p$	$^4P^o-^4D$	*2460.6*	*2461.3*	*336240*	*– 376869*	12–20	1.20–01	1.81–02	1.76+00	–0.662	B	2
			2453.95	2454.69	336206.9	– 376945.2	6–8	1.21–01	1.46–02	7.05–01	–1.059	B	*LS*
			2462.60	2463.35	336261.6	– 376856.8	4–6	8.37–02	1.14–02	3.70–01	–1.340	B	*LS*
			2468.44	2469.19	336296.6	– 376795.7	2–4	4.95–02	9.04–03	1.47–01	–1.743	B	*LS*
			2459.29	2460.03	336206.9	– 376856.8	6–6	3.60–02	3.27–03	1.59–01	–1.708	B	*LS*
			2466.31	2467.06	336261.6	– 376795.7	4–4	6.35–02	5.79–03	1.88–01	–1.635	B	*LS*
			2471.21	2471.96	336296.6	– 376750.4	2–2	9.87–02	9.03–03	1.47–01	–1.743	B	*LS*
			2462.99	2463.73	336206.9	– 376795.7	6–4	5.98–03	3.63–04	1.76–02	–2.663	B–	*LS*
			2469.07	2469.82	336261.6	– 376750.4	4–2	1.98–02	9.04–04	2.94–02	–2.442	B–	*LS*
257.		$^4P^o-^4S$	*2369.3*	*2370.1*	*336240*	– 378432.8	12–4	1.82+00	5.11–02	4.78+00	–0.213	B	2
			2367.49	2368.21	336206.9	– 378432.8	6–4	9.12–01	5.11–02	2.39+00	–0.514	B	*LS*
			2370.56	2371.29	336261.6	– 378432.8	4–4	6.06–01	5.10–02	1.59+00	–0.690	B	*LS*
			2372.53	2373.26	336296.6	– 378432.8	2–4	3.02–01	5.10–02	7.96–01	–0.992	B	*LS*
258.		$^4P^o-^4P$	*2318.2*	*2318.9*	*336240*	*– 379364*	12–12	1.00–01	8.06–03	7.38–01	–1.015	B	2
			2314.60	2315.31	336206.9	– 379397.6	6–6	7.03–02	5.65–03	2.58–01	–1.470	B	*LS*
			2320.38	2321.09	336261.6	– 379344.8	4–4	1.33–02	1.07–03	3.28–02	–2.367	B	*LS*
			2324.68	2325.39	336296.6	– 379300.1	2–2	1.65–02	1.34–03	2.05–02	–2.572	B–	*LS*
			2317.44	2318.15	336206.9	– 379344.8	6–4	4.50–02	2.42–03	1.11–01	–1.838	B	*LS*
			2322.79	2323.50	336261.6	– 379300.1	4–2	8.28–02	3.35–03	1.02–01	–1.873	B	*LS*
			2317.54	2318.25	336261.6	– 379397.6	4–6	3.00–02	3.63–03	1.11–01	–1.838	B	*LS*
			2322.26	2322.98	336296.6	– 379344.8	2–4	4.14–02	6.70–03	1.02–01	–1.873	B	*LS*
259.	$2s2p(^3P^o)3d-2s2p(^3P^o)4f$	$^4P^o-^4D$		*1920.4*	*336240*	*– 388312*	12–20	9.60+00	8.85–01	6.71+01	1.026	B	2
				1920.64	336206.9	– 388272.8	6–8	9.60+00	7.08–01	2.68+01	0.628	C+	*LS*
				1921.31	336261.6	– 388309.3	4–6	6.71+00	5.57–01	1.41+01	0.348	C+	*LS*
				1920.85	336296.6	– 388356.8	2–4	4.00+00	4.42–01	5.59+00	–0.053	C+	*LS*
				1919.30	336206.9	– 388309.3	6–6	2.89+00	1.59–01	6.04+00	–0.020	C+	*LS*
				1919.56	336261.6	– 388356.8	4–4	5.13+00	2.83–01	7.16+00	0.054	C+	*LS*
				1919.78	336296.6	– 388385.8	2–2	8.01+00	4.42–01	5.59+00	–0.053	C+	*LS*
				1917.55	336206.9	– 388356.8	6–4	4.82–01	1.77–02	6.71–01	–0.973	C+	*LS*
				1918.49	336261.6	– 388385.8	4–2	1.60+00	4.43–02	1.12+00	–0.752	C+	*LS*

N III: Allowed Transitions — Continued

No.	Transition Array	Mult.	λ_{air} (Å)	λ_{vac} (Å) or σ (cm^{-1})*	E_i (cm^{-1})	E_k (cm^{-1})	g_i-g_k	A_{ki} (10^8 s^{-1})	f_{ik}	S (at. u.)	log gf	Acc.	Source
260.	$2s2p(^3P^o)3d-2p^2(^3P)3s$	$^4P^o-^4P$		*1552.0*	*336240*	– *400671*	12–12	2.94–02	1.06–03	6.51–02	–1.895	B	2
				1549.78	336206.9	– 400732.2	6–6	2.07–02	7.44–04	2.28–02	–2.350	B–	*LS*
				1553.49	336261.6	– 400632.9	4–4	3.91–03	1.41–04	2.89–03	–3.247	C	*LS*
				1555.95	336296.6	– 400566.2	2–2	4.86–03	1.77–04	1.81–03	–3.452	C	*LS*
				1552.17	336206.9	– 400632.9	6–4	1.32–02	3.19–04	9.77–03	–2.719	C+	*LS*
				1555.10	336261.6	– 400566.2	4–2	2.44–02	4.42–04	9.04–03	–2.753	C+	*LS*
				1551.09	336261.6	– 400732.2	4–6	8.84–03	4.78–04	9.77–03	–2.718	C+	*LS*
				1554.33	336296.6	– 400632.9	2–4	1.22–02	8.83–04	9.04–03	–2.753	C+	*LS*
261.	$2s2p(^3P^o)3d-2s^25d$	$^2F^o-^2D$	*46700*	*2140* cm^{-1}	*339808*	– *341948*	14–10	5.11–05	1.19–03	2.57+00	–1.777	B	2
			47758	2093.3 cm^{-1}	339855.5	– 341948.8	8–6	4.55–05	1.17–03	1.47+00	–2.030	B	*LS*
			45378	2203.1 cm^{-1}	339744.6	– 341947.7	6–4	5.57–05	1.15–03	1.03+00	–2.162	B	*LS*
			45356	2204.2 cm^{-1}	339744.6	– 341948.8	6–6	2.66–06	8.19–05	7.34–02	–3.308	B	*LS*
262.	$2s2p(^3P^o)3d-2s^26d$	$^2F^o-^2D$	*6790.2*	*6792.1*	*339808*	– 354530.5	14–10	6.81–04	3.36–04	1.05–01	–2.327	B	2
			6812.43	6814.31	339855.5	– 354530.5	8–6	6.42–04	3.35–04	6.01–02	–2.572	B	*LS*
			6761.33	6763.20	339744.6	– 354530.5	6–4	6.90–04	3.15–04	4.21–02	–2.723	B	*LS*
			6761.33	6763.20	339744.6	– 354530.5	6–6	3.28–05	2.25–05	3.01–03	–3.870	C+	*LS*
263.	$2s2p(^3P^o)3d-2s^27d$	$^2F^o-^2D$	*4491.5*	*4492.8*	*339808*	– 362066	14–10	8.59–04	1.85–04	3.84–02	–2.586	B	2
			4501.11	4502.38	339855.5	– 362066	8–6	8.13–04	1.85–04	2.19–02	–2.829	B–	*LS*
			4478.75	4480.01	339744.6	– 362066	6–4	8.66–04	1.74–04	1.54–02	–2.982	B–	*LS*
			4478.75	4480.01	339744.6	– 362066	6–6	4.12–05	1.24–05	1.10–03	–4.128	C	*LS*
264.	$2s2p(^3P^o)3d-2s^28d$	$^2F^o-^2D$	*3684.6*	*3685.7*	*339808*	– 366940	14–10	9.72–04	1.41–04	2.40–02	–2.704	B–	2
			3691.10	3692.15	339855.5	– 366940	8–6	9.21–04	1.41–04	1.37–02	–2.947	B–	*LS*
			3676.05	3677.09	339744.6	– 366940	6–4	9.79–04	1.32–04	9.60–03	–3.101	C+	*LS*
			3676.05	3677.09	339744.6	– 366940	6–6	4.66–05	9.44–06	6.86–04	–4.247	D	*LS*
265.	$2s2p(^3P^o)3d-2s2p(^3P^o)4f$	$^2F^o-^2F$	*2119.8*	*2120.5*	*339808*	– *386967*	14–14	1.04+00	7.01–02	6.85+00	–0.008	B	2
			2121.50	2122.17	339855.5	– 386977.0	8–8	1.00+00	6.75–02	3.77+00	–0.267	C+	*LS*
			2117.58	2118.25	339744.6	– 386953.4	6–6	9.94–01	6.68–02	2.80+00	–0.397	C+	*LS*
			2122.57	2123.24	339855.5	– 386953.4	8–6	4.94–02	2.50–03	1.40–01	–1.699	C+	*LS*
			2116.52	2117.19	339744.6	– 386977.0	6–8	3.73–02	3.34–03	1.40–01	–1.698	C+	*LS*
266.	$2s2p(^3P^o)3d-2s^26s$	$^2P^o-^2S$	*14100*	*7110* cm^{-1}	*342715*	– 349824.8	6–2	1.54–04	1.52–04	4.23–02	–3.039	B	2
			14015	7133.3 cm^{-1}	342691.5	– 349824.8	4–2	1.04–04	1.53–04	2.82–02	–3.214	B–	*LS*
			14156	7062.3 cm^{-1}	342762.5	– 349824.8	2–2	5.04–05	1.51–04	1.41–02	–3.519	B–	*LS*
267.	$2s2p(^3P^o)3d-2s^26d$	$^2P^o-^2D$	*8460.8*	*8463.1*	*342715*	– 354530.5	6–10	5.40–04	9.66–04	1.61–01	–2.237	B	2
			8444.34	8446.66	342691.5	– 354530.5	4–6	5.43–04	8.71–04	9.68–02	–2.458	B	*LS*
			8495.29	8497.62	342762.5	– 354530.5	2–4	4.44–04	9.62–04	5.38–02	–2.716	B	*LS*
			8444.34	8446.66	342691.5	– 354530.5	4–4	9.05–05	9.68–05	1.08–02	–3.412	B–	*LS*
268.	$2s2p(^3P^o)3d-2s^27d$	$^2P^o-^2D$	*5166.3*	*5167.7*	*342715*	– 362066	6–10	4.31–04	2.87–04	2.93–02	–2.764	B–	2
			5159.99	5161.42	342691.5	– 362066	4–6	4.32–04	2.59–04	1.76–02	–2.985	B–	*LS*
			5178.97	5180.41	342762.5	– 362066	2–4	3.56–04	2.86–04	9.77–03	–3.242	C+	*LS*
			5159.99	5161.42	342691.5	– 362066	4–4	7.20–05	2.87–05	1.95–03	–3.939	C	*LS*
269.	$2s2p(^3P^o)3d-2s^28d$	$^2P^o-^2D$	*4126.8*	*4128.0*	*342715*	– 366940	6–10	4.06–04	1.73–04	1.41–02	–2.984	B–	2
			4122.80	4123.97	342691.5	– 366940	4–6	4.08–04	1.56–04	8.46–03	–3.205	C+	*LS*
			4134.91	4136.08	342762.5	– 366940	2–4	3.37–04	1.73–04	4.70–03	–3.462	C+	*LS*
			4122.80	4123.97	342691.5	– 366940	4–4	6.79–05	1.73–05	9.40–04	–4.160	D	*LS*

N III: Allowed Transitions — Continued

No.	Transition Array	Mult.	λ_{air} (Å)	λ_{vac} (Å) or σ (cm^{-1})*	E_i (cm^{-1})	E_k (cm^{-1})	$g_i - g_k$	A_{ki} ($10^8 s^{-1}$)	f_{ik}	S (at. u.)	log gf	Acc.	Source
270.	$2s2p(^3P^o)3d - 2s2p(^3P^o)4p$	$^2P^o - ^2P$	*3117.3*	*3118.2*	*342715*	*– 374785*	6–6	2.13–01	3.10–02	1.91+00	–0.730	B	2
			3113.07	3113.97	342691.5	– 374804.8	4–4	1.78–01	2.59–02	1.06+00	–0.985	B	*LS*
			3125.67	3126.57	342762.5	– 374746.4	2–2	1.41–01	2.06–02	4.24–01	–1.385	B	*LS*
			3118.74	3119.65	342691.5	– 374746.4	4–2	7.08–02	5.17–03	2.12–01	–1.685	B	*LS*
			3119.97	3120.87	342762.5	– 374804.8	2–4	3.54–02	1.03–02	2.12–01	–1.685	B	*LS*
271.	$2s2p(^3P^o)3d - 2s2p(^1P^o)3p$	$^2P^o - ^2P$	*2865.7*	*2866.6*	*342715*	*– 377600*	6–6	1.99–03	2.46–04	1.39–02	–2.832	B–	2
			2862.86	2863.70	342691.5	– 377611.3	4–4	1.67–03	2.05–04	7.72–03	–3.087	C+	*LS*
			2871.45	2872.29	342762.5	– 377577.9	2–2	1.32–03	1.63–04	3.09–03	–3.486	C+	*LS*
			2865.61	2866.45	342691.5	– 377577.9	4–2	6.65–04	4.09–05	1.54–03	–3.786	C	*LS*
			2868.70	2869.54	342762.5	– 377611.3	2–4	3.31–04	8.18–05	1.54–03	–3.786	C	*LS*
272.	$2s2p(^3P^o)3d - 2s2p(^3P^o)5p$	$^2P^o - ^2P$		*1667.7*	*342715*	*– 402679*	6–6	6.24–02	2.60–03	8.57–02	–1.807	B	2
				1666.28	342691.5	– 402705.6	4–4	5.21–02	2.17–03	4.76–02	–2.062	B	*LS*
				1670.51	342762.5	– 402624.6	2–2	4.14–02	1.73–03	1.90–02	–2.461	B–	*LS*
				1668.53	342691.5	– 402624.6	4–2	2.08–02	4.33–04	9.52–03	–2.761	C+	*LS*
				1668.25	342762.5	– 402705.6	2–4	1.04–02	8.67–04	9.52–03	–2.761	C+	*LS*
273.	$2s2p(^3P^o)3d - 2s2p(^3P^o)6p$	$^2P^o - ^2P$		*1291.3*	*342715*	*– 420154*	6–6	2.02–02	5.06–04	1.29–02	–2.518	B–	2
				1290.54	342691.5	– 420178.4	4–4	1.69–02	4.22–04	7.17–03	–2.773	C+	*LS*
				1292.93	342762.5	– 420106.1	2–2	1.34–02	3.37–04	2.87–03	–3.172	C	*LS*
				1291.75	342691.5	– 420106.1	4–2	6.74–03	8.43–05	1.43–03	–3.472	C	*LS*
				1291.72	342762.5	– 420178.4	2–4	3.37–03	1.69–04	1.43–03	–3.472	C	*LS*
274.	$2s2p(^1P^o)3s - 2s^27s$	$^2P^o - ^2S$	*17520*	*5706* cm^{-1}	*353442*	– 359148.1	6–2	9.68–02	1.49–01	5.14+01	–0.050	B	2
			17522	5705.6 cm^{-1}	353442.5	– 359148.1	4–2	6.45–02	1.49–01	3.43+01	–0.226	B	*LS*
			17516	5707.5 cm^{-1}	353440.6	– 359148.1	2–2	3.23–02	1.49–01	1.71+01	–0.527	B	*LS*
275.	$2s2p(^1P^o)3s - 2s^27d$	$^2P^o - ^2D$	*11590*	*8624* cm^{-1}	*353442*	– 362066	6–10	4.57–02	1.54–01	3.52+01	–0.036	B	2
			11590	8624 cm^{-1}	353442.5	– 362066	4–6	4.57–02	1.38–01	2.11+01	–0.257	B	*LS*
			11590	8625 cm^{-1}	353440.6	– 362066	2–4	3.81–02	1.54–01	1.17+01	–0.513	B	*LS*
			11590	8624 cm^{-1}	353442.5	– 362066	4–4	7.62–03	1.54–02	2.34+00	–1.212	B	*LS*
276.	$2s2p(^1P^o)3s - 2s^28s$	$^2P^o - ^2S$	*8654.0*	*8656.4*	*353442*	– 364994.2	6–2	2.53–02	9.49–03	1.62+00	–1.245	B	2
			8654.36	8656.73	353442.5	– 364994.2	4–2	1.69–02	9.49–03	1.08+00	–1.421	B	*LS*
			8652.93	8655.31	353440.6	– 364994.2	2–2	8.45–03	9.49–03	5.41–01	–1.722	B	*LS*
277.	$2s2p(^1P^o)3s - 2s^28d$	$^2P^o - ^2D$	*7406.5*	*7408.5*	*353442*	– 366940	6–10	2.33–02	3.19–02	4.67+00	–0.718	B	2
			7406.74	7408.78	353442.5	– 366940	4–6	2.33–02	2.87–02	2.80+00	–0.940	B	*LS*
			7405.70	7407.74	353440.6	– 366940	2–4	1.94–02	3.19–02	1.56+00	–1.195	B	*LS*
			7406.74	7408.78	353442.5	– 366940	4–4	3.88–03	3.19–03	3.11–01	–1.894	B	*LS*
278.	$2s2p(^1P^o)3s - 2s2p(^3P^o)4p$	$^2P^o - ^2P$	*4684.1*	*4685.4*	*353442*	*– 374785*	6–6	7.66–02	2.52–02	2.33+00	–0.820	B	2
			4679.83	4681.14	353442.5	– 374804.8	4–4	6.40–02	2.10–02	1.30+00	–1.075	B	*LS*
			4692.24	4693.56	353440.6	– 374746.4	2–2	5.08–02	1.68–02	5.18–01	–1.474	B	*LS*
			4692.66	4693.98	353442.5	– 374746.4	4–2	2.54–02	4.19–03	2.59–01	–1.776	B	*LS*
			4679.42	4680.73	353440.6	– 374804.8	2–4	1.28–02	8.41–03	2.59–01	–1.774	B	*LS*
279.	$2s2p(^1P^o)3s - 2s2p(^1P^o)3p$	$^2P^o - ^2P$	*4138.2*	*4139.4*	*353442*	*– 377600*	6–6	6.45–01	1.66–01	1.35+01	–0.003	B	2
			4136.40	4137.57	353442.5	– 377611.3	4–4	5.38–01	1.38–01	7.52+00	–0.258	B	*LS*
			4141.80	4142.97	353440.6	– 377577.9	2–2	4.29–01	1.10–01	3.01+00	–0.656	B	*LS*
			4142.12	4143.29	353442.5	– 377577.9	4–2	2.14–01	2.76–02	1.50+00	–0.957	B	*LS*
			4136.07	4137.24	353440.6	– 377611.3	2–4	1.08–01	5.52–02	1.50+00	–0.957	B	*LS*

N III: Allowed Transitions — Continued

No.	Transition Array	Mult.	λ_{air} (Å)	λ_{vac} (Å) or σ (cm^{-1})*	E_i (cm^{-1})	E_k (cm^{-1})	g_i-g_k	A_{ki} (10^8 s^{-1})	f_{ik}	S (at. u.)	log gf	Acc.	Source
280.	$2s2p(^1P^o)3s-2s2p(^3P^o)5p$	$^2P^o-^2P$	*2030.3*	*2031.0*	*353442*	– *402679*	6–6	2.37−02	1.47−03	5.88−02	−2.056	B	2
			2029.26	2029.92	353442.5	– 402705.6	4–4	1.98−02	1.22−03	3.27−02	−2.311	B	*LS*
			2032.53	2033.18	353440.6	– 402624.6	2–2	1.58−02	9.76−04	1.31−02	−2.709	B−	*LS*
			2032.61	2033.26	353442.5	– 402624.6	4–2	7.88−03	2.44−04	6.53−03	−3.010	C+	*LS*
			2029.19	2029.84	353440.6	– 402705.6	2–4	3.96−03	4.89−04	6.53−03	−3.010	C+	*LS*
281.	$2s2p(^1P^o)3s-2s2p(^3P^o)6p$	$^2P^o-^2P$		*1499.0*	*353442*	– *420154*	6–6	1.89−01	6.38−03	1.89−01	−1.417	B	2
				1498.44	353442.5	– 420178.4	4–4	1.58−01	5.32−03	1.05−01	−1.672	B	*LS*
				1500.03	353440.6	– 420106.1	2–2	1.26−01	4.25−03	4.20−02	−2.070	B	*LS*
				1500.07	353442.5	– 420106.1	4–2	6.30−02	1.06−03	2.10−02	−2.371	B−	*LS*
				1498.40	353440.6	– 420178.4	2–4	3.16−02	2.13−03	2.10−02	−2.371	B−	*LS*
282.	$2s2p(^1P^o)3s-2s2p(^3P^o)7p$	$^2P^o-^2P$		*1340.0*	*353442*	– *428069*	6–6	7.44−02	2.00−03	5.30−02	−1.920	B	2
				1339.79	353442.5	– 428080.8	4–4	6.20−02	1.67−03	2.94−02	−2.176	B−	*LS*
				1340.39	353440.6	– 428045.5	2–2	4.95−02	1.33−03	1.18−02	−2.574	B−	*LS*
				1340.43	353442.5	– 428045.5	4–2	2.48−02	3.34−04	5.89−03	−2.875	C+	*LS*
				1339.76	353440.6	– 428080.8	2–4	1.24−02	6.68−04	5.89−03	−2.874	C+	*LS*
283.	$2s2p(^1P^o)3s-2s2p(^3P^o)8p$	$^2P^o-^2P$		*1251.9*	*353442*	– *433322*	6–6	5.01−02	1.18−03	2.91−02	−2.151	B−	2
				1251.67	353442.5	– 433335.6	4–4	4.18−02	9.81−04	1.62−02	−2.406	B−	*LS*
				1252.26	353440.6	– 433296.2	2–2	3.34−02	7.84−04	6.47−03	−2.804	C+	*LS*
				1252.29	353442.5	– 433296.2	4–2	1.67−02	1.96−04	3.23−03	−3.106	C+	*LS*
				1251.64	353440.6	– 433335.6	2–4	8.35−03	3.92−04	3.23−03	−3.105	C+	*LS*
284.	$2s2p(^3P^o)4s-2s2p(^3P^o)4p$	$^4P^o-^4D$	*12130*	*8242* cm^{-1}	*368627*	– *376869*	12–20	1.40−01	5.16−01	2.47+02	0.792	B	2
			12120	8248.9 cm^{-1}	368696.3	– 376945.2	6–8	1.41−01	4.13−01	9.88+01	0.394	B	*LS*
			12076	8278.5 cm^{-1}	368578.3	– 376856.8	4–6	9.95−02	3.26−01	5.19+01	0.116	B	*LS*
			12072	8281.3 cm^{-1}	368514.4	– 376795.7	2–4	5.93−02	2.59−01	2.06+01	−0.286	B	*LS*
			12251	8160.5 cm^{-1}	368696.3	– 376856.8	6–6	4.08−02	9.19−02	2.22+01	−0.259	B	*LS*
			12166	8217.4 cm^{-1}	368578.3	– 376795.7	4–4	7.41−02	1.65−01	2.64+01	−0.182	B	*LS*
			12138	8236.0 cm^{-1}	368514.4	– 376750.4	2–2	1.17−01	2.58−01	2.06+01	−0.288	B	*LS*
			12343	8099.4 cm^{-1}	368696.3	– 376795.7	6–4	6.65−03	1.01−02	2.47+00	−1.216	B	*LS*
			12233	8172.1 cm^{-1}	368578.3	– 376750.4	4–2	2.28−02	2.56−02	4.12+00	−0.990	B	*LS*
285.		$^4P^o-^4S$	*10200*	*9806* cm^{-1}	*368627*	– 378432.8	12–4	1.91−01	9.91−02	3.99+01	0.075	B	2
			10268	9736.5 cm^{-1}	368696.3	– 378432.8	6–4	9.34−02	9.84−02	2.00+01	−0.229	B	*LS*
			10145	9854.5 cm^{-1}	368578.3	– 378432.8	4–4	6.45−02	9.96−02	1.33+01	−0.400	B	*LS*
			10080	9918.4 cm^{-1}	368514.4	– 378432.8	2–4	3.29−02	1.00−01	6.65+00	−0.698	B	*LS*
286.		$^4P^o-^4P$	*9311.0*	*9313.6*	*368627*	– *379364*	12–12	3.53−01	4.59−01	1.69+02	0.741	B	2
			9342.10	9344.66	368696.3	– 379397.6	6–6	2.45−01	3.20−01	5.91+01	0.284	B	*LS*
			9285.52	9288.07	368578.3	– 379344.8	4–4	4.75−02	6.14−02	7.51+00	−0.610	B	*LS*
			9268.99	9271.54	368514.4	– 379300.1	2–2	5.97−02	7.69−02	4.69+00	−0.813	B	*LS*
			9388.42	9390.99	368696.3	– 379344.8	6–4	1.55−01	1.37−01	2.53+01	−0.086	B	*LS*
			9324.23	9326.79	368578.3	– 379300.1	4–2	2.93−01	1.91−01	2.35+01	−0.117	B	*LS*
			9240.21	9242.74	368578.3	– 379397.6	4–6	1.08−01	2.08−01	2.53+01	−0.079	B	*LS*
			9230.74	9233.27	368514.4	– 379344.8	2–4	1.51−01	3.86−01	2.35+01	−0.112	B	*LS*

N III: Allowed Transitions — Continued

No.	Transition Array	Mult.	λ_{air} (Å)	λ_{vac} (Å) or σ (cm^{-1})*	E_i (cm^{-1})	E_k (cm^{-1})	g_i–g_k	A_{ki} (10^8 s^{-1})	f_{ik}	S (at. u.)	log gf	Acc.	Source
287.	$2s2p(^3P^\circ)4s-2s2p(^3P^\circ)4f$	$^4P^\circ-^4D$	*5078.6*	*5080.0*	*368627*	– *388312*	12–20	2.22–02	1.43–02	2.88+00	–0.764	B	2
			5106.74	5108.17	368696.3	– 388272.8	6–8	2.19–02	1.14–02	1.15+00	–1.165	C+	*LS*
			5066.75	5068.17	368578.3	– 388309.3	4–6	1.57–02	9.05–03	6.04–01	–1.441	C+	*LS*
			5038.31	5039.71	368514.4	– 388356.8	2–4	9.49–03	7.23–03	2.40–01	–1.840	C+	*LS*
			5097.24	5098.66	368696.3	– 388309.3	6–6	6.60–03	2.57–03	2.59–01	–1.812	C+	*LS*
			5054.59	5056.00	368578.3	– 388356.8	4–4	1.20–02	4.61–03	3.07–01	–1.734	C+	*LS*
			5030.96	5032.36	368514.4	– 388385.8	2–2	1.91–02	7.24–03	2.40–01	–1.839	C+	*LS*
			5084.92	5086.34	368696.3	– 388356.8	6–4	1.11–03	2.86–04	2.88–02	–2.765	C	*LS*
			5047.19	5048.59	368578.3	– 388385.8	4–2	3.78–03	7.21–04	4.79–02	–2.540	C+	*LS*
288.	$2s2p(^3P^\circ)4s-2p^2(^3P)3s$	$^4P^\circ-^4P$	*3119.8*	*3120.7*	*368627*	– *400671*	12–12	1.71–03	2.50–04	3.08–02	–2.523	B	2
			3120.59	3121.50	368696.3	– 400732.2	6–6	1.20–03	1.75–04	1.08–02	–2.979	B–	*LS*
			3118.77	3119.68	368578.3	– 400632.9	4–4	2.29–04	3.33–05	1.37–03	–3.875	C	*LS*
			3119.05	3119.95	368514.4	– 400566.2	2–2	2.86–04	4.17–05	8.56–04	–4.079	D	*LS*
			3130.30	3131.20	368696.3	– 400632.9	6–4	7.63–04	7.47–05	4.62–03	–3.348	C+	*LS*
			3125.28	3126.18	368578.3	– 400566.2	4–2	1.42–03	1.04–04	4.28–03	–3.381	C+	*LS*
			3109.14	3110.04	368578.3	– 400732.2	4–6	5.19–04	1.13–04	4.62–03	–3.345	C+	*LS*
			3112.57	3113.47	368514.4	– 400632.9	2–4	7.19–04	2.09–04	4.28–03	–3.379	C+	*LS*
289.	$2s2p(^3P^\circ)4p-2s2p(^3P^\circ)4d$	$^2P-^2D^\circ$	*9667.6*	*9670.2*	*374785*	– 385126	6–10	2.37–01	5.54–01	1.06+02	0.521	B	2
			9686.14	9688.80	374804.8	– 385126	4–6	2.36–01	4.97–01	6.34+01	0.299	B	*LS*
			9631.64	9634.28	374746.4	– 385126	2–4	2.00–01	5.56–01	3.52+01	0.046	B	*LS*
			9686.14	9688.80	374804.8	– 385126	4–4	3.93–02	5.53–02	7.05+00	–0.655	B	*LS*
290.	$2s2p(^3P^\circ)4p-2s2p(^1P^\circ)3d$	$^2P-^2D^\circ$	*4586.7*	*4588.0*	*374785*	– *396581*	6–10	1.21–02	6.35–03	5.75–01	–1.419	B	2
			4590.08	4591.37	374804.8	– 396584.8	4–6	1.21–02	5.71–03	3.45–01	–1.641	B	*LS*
			4579.88	4581.17	374746.4	– 396574.9	2–4	1.01–02	6.36–03	1.92–01	–1.896	B	*LS*
			4592.17	4593.46	374804.8	– 396574.9	4–4	2.01–03	6.34–04	3.84–02	–2.596	B	*LS*
291.	$2s2p(^3P^\circ)4p-2p^2(^3P)3p$	$^2P-^2S^\circ$		*1509.6*	*374785*	– 441026.6	6–2	4.45–02	5.06–04	1.51–02	–2.517	B–	2
				1510.08	374804.8	– 441026.6	4–2	2.96–02	5.06–04	1.01–02	–2.694	B–	*LS*
				1508.75	374746.4	– 441026.6	2–2	1.48–02	5.07–04	5.03–03	–2.994	C+	*LS*
292.	$2s2p(^3P^\circ)4p-2s2p(^3P^\circ)4d$	$^4D-^4P^\circ$	*10630*	*9401* cm^{-1}	*376869*	– *386270*	20–12	2.38–02	2.42–02	1.69+01	–0.315	B	2
			10758	9292.5 cm^{-1}	376945.2	– 386237.7	8–6	1.84–02	2.39–02	6.77+00	–0.718	B	*LS*
			10596	9435.0 cm^{-1}	376856.8	– 386291.8	6–4	1.51–02	1.70–02	3.56+00	–0.992	B	*LS*
			10492	9528.6 cm^{-1}	376795.7	– 386324.3	4–2	1.24–02	1.02–02	1.41+00	–1.389	B	*LS*
			10657	9380.9 cm^{-1}	376856.8	– 386237.7	6–6	4.25–03	7.24–03	1.52+00	–1.362	B	*LS*
			10528	9496.1 cm^{-1}	376795.7	– 386291.8	4–4	7.84–03	1.30–02	1.81+00	–1.283	B	*LS*
			10442	9573.9 cm^{-1}	376750.4	– 386324.3	2–2	1.26–02	2.05–02	1.41+00	–1.387	B	*LS*
			10588	9442.0 cm^{-1}	376795.7	– 386237.7	4–6	4.82–04	1.21–03	1.69–01	–2.314	B	*LS*
			10478	9541.4 cm^{-1}	376750.4	– 386291.8	2–4	1.24–03	4.09–03	2.82–01	–2.087	B	*LS*
293.	$2s2p(^3P^\circ)4p-2s2p(^3P^\circ)5s$	$^4D-^4P^\circ$	*4146.0*	*4147.1*	*376869*	– *400982*	20–12	1.03+00	1.60–01	4.36+01	0.505	B	2
			4146.78	4147.95	376945.2	– 401053.5	8–6	8.26–01	1.60–01	1.74+01	0.106	B	*LS*
			4152.10	4153.27	376856.8	– 400934.2	6–4	6.48–01	1.12–01	9.16+00	–0.174	B	*LS*
			4153.93	4155.10	376795.7	– 400862.5	4–2	5.14–01	6.64–02	3.63+00	–0.576	B	*LS*
			4131.63	4132.79	376856.8	– 401053.5	6–6	1.88–01	4.81–02	3.93+00	–0.540	B	*LS*
			4141.59	4142.76	376795.7	– 400934.2	4–4	3.32–01	8.53–02	4.65+00	–0.467	B	*LS*
			4146.13	4147.30	376750.4	– 400862.5	2–2	5.17–01	1.33–01	3.63+00	–0.575	B	*LS*
			4121.22	4122.39	376795.7	– 401053.5	4–6	2.10–02	8.04–03	4.36–01	–1.493	B	*LS*
			4133.83	4135.00	376750.4	– 400934.2	2–4	5.21–02	2.67–02	7.27–01	–1.272	B	*LS*

N III: Allowed Transitions — Continued

No.	Transition Array	Mult.	λ_{air} (Å)	λ_{vac} (Å) or σ (cm^{-1})*	E_i (cm^{-1})	E_k (cm^{-1})	g_i-g_k	A_{ki} (10^8 s^{-1})	f_{ik}	S (at. u.)	log gf	Acc.	Source
294.	$2s2p(^3P^o)4p-$ $2s2p(^3P^o)5d$	$^4D-^4P^o$	*3070.7*	*3071.6*	*376869*	*- 409425*	20-12	6.34-03	5.38-04	1.09-01	-1.968	B	2
			3081.14	3082.03	376945.2	- 409391.3	8-6	5.02-03	5.36-04	4.35-02	-2.367	B	*LS*
			3067.63	3068.52	376856.8	- 409445.8	6-4	4.01-03	3.77-04	2.28-02	-2.645	B-	*LS*
			3058.10	3058.99	376795.7	- 409486.2	4-2	3.21-03	2.25-04	9.07-03	-3.045	C+	*LS*
			3072.77	3073.66	376856.8	- 409391.3	6-6	1.14-03	1.61-04	9.79-03	-3.014	C+	*LS*
			3061.89	3062.78	376795.7	- 409445.8	4-4	2.05-03	2.88-04	1.16-02	-2.939	B-	*LS*
			3053.87	3054.76	376750.4	- 409486.2	2-2	3.22-03	4.51-04	9.07-03	-3.045	C+	*LS*
			3067.01	3067.90	376795.7	- 409391.3	4-6	1.27-04	2.69-05	1.09-03	-3.968	C	*LS*
			3057.65	3058.53	376750.4	- 409445.8	2-4	3.21-04	9.01-05	1.81-03	-3.744	C	*LS*
295.	$2s2p(^3P^o)4p-$ $2p^2(^3P)3p$	$^4D-^4D^o$	*2328.3*	*2329.0*	*376869*	*- 419806*	20-20	3.31-02	2.69-03	4.12-01	-1.269	B	2
			2328.85	2329.56	376945.2	- 419871.7	8-8	2.84-02	2.31-03	1.41-01	-1.734	B	*LS*
			2327.90	2328.62	376856.8	- 419800.7	6-6	1.90-02	1.54-03	7.09-02	-2.034	B	*LS*
			2328.01	2328.73	376795.7	- 419737.6	4-4	1.32-02	1.08-03	3.30-02	-2.366	B	*LS*
			2327.58	2328.29	376750.4	- 419700.3	2-2	1.66-02	1.35-03	2.06-02	-2.570	B-	*LS*
			2332.71	2333.42	376945.2	- 419800.7	8-6	6.25-03	3.83-04	2.35-02	-2.514	B-	*LS*
			2331.33	2332.05	376856.8	- 419737.6	6-4	1.15-02	6.27-04	2.89-02	-2.425	B-	*LS*
			2330.04	2330.75	376795.7	- 419700.3	4-2	1.65-02	6.72-04	2.06-02	-2.571	B-	*LS*
			2324.06	2324.78	376856.8	- 419871.7	6-8	4.74-03	5.12-04	2.35-02	-2.513	B-	*LS*
			2324.60	2325.31	376795.7	- 419800.7	4-6	7.76-03	9.43-04	2.89-02	-2.424	B-	*LS*
			2325.56	2326.27	376750.4	- 419737.6	2-4	8.30-03	1.35-03	2.06-02	-2.570	B-	*LS*
296.	$2s2p(^3P^o)4p-$ $2s2p(^3P^o)6d$	$^4D-^4D^o$	*2225.2*	*2225.9*	*376869*	- 421794.2	20-20	7.70-02	5.72-03	8.38-01	-0.942	B	2
			2229.01	2229.70	376945.2	- 421794.2	8-8	6.57-02	4.90-03	2.87-01	-1.407	B	*LS*
			2224.63	2225.32	376856.8	- 421794.2	6-6	4.42-02	3.28-03	1.44-01	-1.706	B	*LS*
			2221.60	2222.30	376795.7	- 421794.2	4-4	3.10-02	2.29-03	6.70-02	-2.038	B	*LS*
			2219.37	2220.06	376750.4	- 421794.2	2-2	3.88-02	2.87-03	4.19-02	-2.242	B	*LS*
			2229.01	2229.70	376945.2	- 421794.2	8-6	1.46-02	8.14-04	4.78-02	-2.187	B	*LS*
			2224.63	2225.32	376856.8	- 421794.2	6-4	2.70-02	1.33-03	5.87-02	-2.096	B	*LS*
			2221.60	2222.30	376795.7	- 421794.2	4-2	3.87-02	1.43-03	4.19-02	-2.242	B	*LS*
			2224.63	2225.32	376856.8	- 421794.2	6-8	1.10-02	1.09-03	4.78-02	-2.186	B	*LS*
			2221.60	2222.30	376795.7	- 421794.2	4-6	1.81-02	2.00-03	5.87-02	-2.096	B	*LS*
			2219.37	2220.06	376750.4	- 421794.2	2-4	1.94-02	2.87-03	4.19-02	-2.242	B	*LS*
297.		$^4D-^4P^o$	*2223.8*	*2224.5*	*376869*	*- 421823*	20-12	8.03-03	3.57-04	5.23-02	-2.146	B	2
			2229.01	2229.70	376945.2	- 421794.2	8-6	6.38-03	3.56-04	2.09-02	-2.545	B-	*LS*
			2222.20	2222.89	376856.8	- 421843.2	6-4	5.07-03	2.50-04	1.10-02	-2.824	B-	*LS*
			2218.01	2218.70	376795.7	- 421867.2	4-2	4.05-03	1.49-04	4.36-03	-3.224	C+	*LS*
			2224.63	2225.32	376856.8	- 421794.2	6-6	1.44-03	1.07-04	4.71-03	-3.192	C+	*LS*
			2219.19	2219.88	376795.7	- 421843.2	4-4	2.59-03	1.91-04	5.58-03	-3.117	C+	*LS*
			2215.78	2216.47	376750.4	- 421867.2	2-2	4.06-03	2.99-04	4.36-03	-3.224	C+	*LS*
			2221.60	2222.30	376795.7	- 421794.2	4-6	1.61-04	1.79-05	5.23-04	-4.146	D	*LS*
			2216.96	2217.65	376750.4	- 421843.2	2-4	4.05-04	5.97-05	8.72-04	-3.923	D	*LS*
298.	$2s2p(^3P^o)4p-$ $2s2p(^3P^o)7d$	$^4D-^4D^o$		*1903.2*	*376869*	- 429412.9	20-20	8.41-02	4.57-03	5.72-01	-1.039	B	2
				1905.93	376945.2	- 429412.9	8-8	7.18-02	3.91-03	1.96-01	-1.505	B	*LS*
				1902.73	376856.8	- 429412.9	6-6	4.83-02	2.62-03	9.84-02	-1.804	B	*LS*
				1900.52	376795.7	- 429412.9	4-4	3.38-02	1.83-03	4.58-02	-2.136	B	*LS*
				1898.88	376750.4	- 429412.9	2-2	4.23-02	2.29-03	2.86-02	-2.339	B-	*LS*
				1905.93	376945.2	- 429412.9	8-6	1.59-02	6.50-04	3.26-02	-2.284	B	*LS*
				1902.73	376856.8	- 429412.9	6-4	2.95-02	1.07-03	4.01-02	-2.194	B	*LS*
				1900.52	376795.7	- 429412.9	4-2	4.22-02	1.14-03	2.86-02	-2.340	B-	*LS*
				1902.73	376856.8	- 429412.9	6-8	1.20-02	8.68-04	3.26-02	-2.283	B	*LS*
				1900.52	376795.7	- 429412.9	4-6	1.97-02	1.60-03	4.01-02	-2.194	B	*LS*
				1898.88	376750.4	- 429412.9	2-4	2.12-02	2.29-03	2.86-02	-2.339	B-	*LS*

N III: Allowed Transitions — Continued

No.	Transition Array	Mult.	λ_{air} (Å)	λ_{vac} (Å) or σ (cm^{-1})*	E_i (cm^{-1})	E_k (cm^{-1})	g_i–g_k	A_{ki} (10^8 s^{-1})	f_{ik}	S (at. u.)	log gf	Acc.	Source
299.		^{4}D–^{4}P°		*1900.5*	*376869*	429486.3	20–12	3.89–03	1.26–04	1.58–02	–2.598	B–	2
				1903.27	376945.2	429486.3	8–6	3.10–03	1.26–04	6.32–03	–2.996	C+	*LS*
				1900.08	376856.8	429486.3	6–4	2.45–03	8.84–05	3.32–03	–3.275	C+	*LS*
				1897.87	376795.7	429486.3	4–2	1.95–03	5.27–05	1.32–03	–3.676	C	*LS*
				1900.08	376856.8	429486.3	6–6	7.00–04	3.79–05	1.42–03	–3.643	C	*LS*
				1897.87	376795.7	429486.3	4–4	1.25–03	6.74–05	1.69–03	–3.569	C	*LS*
				1896.24	376750.4	429486.3	2–2	1.96–03	1.05–04	1.32–03	–3.676	C	*LS*
				1897.87	376795.7	429486.3	4–6	7.80–05	6.32–06	1.58–04	–4.597	D	*LS*
				1896.24	376750.4	429486.3	2–4	1.96–04	2.11–05	2.63–04	–4.375	D	*LS*
300.	$2s2p(^3P^\circ)4p-$ $2s2p(^3P^\circ)8d$	^{4}D–^{4}D°		*1741.2*	*376869*	434301.2	20–20	6.07–02	2.76–03	3.16–01	–1.258	B	2
				1743.50	376945.2	434301.2	8–8	5.18–02	2.36–03	1.08–01	–1.723	B	*LS*
				1740.81	376856.8	434301.2	6–6	3.48–02	1.58–03	5.44–02	–2.023	B	*LS*
				1738.96	376795.7	434301.2	4–4	2.44–02	1.11–03	2.53–02	–2.355	B–	*LS*
				1737.60	376750.4	434301.2	2–2	3.05–02	1.38–03	1.58–02	–2.558	B–	*LS*
				1743.50	376945.2	434301.2	8–6	1.15–02	3.93–04	1.80–02	–2.503	B–	*LS*
				1740.81	376856.8	434301.2	6–4	2.13–02	6.44–04	2.21–02	–2.413	B–	*LS*
				1738.96	376795.7	434301.2	4–2	3.05–02	6.91–04	1.58–02	–2.559	B–	*LS*
				1740.81	376856.8	434301.2	6–8	8.66–03	5.24–04	1.80–02	–2.502	B–	*LS*
				1738.96	376795.7	434301.2	4–6	1.42–02	9.67–04	2.21–02	–2.413	B–	*LS*
				1737.60	376750.4	434301.2	2–4	1.53–02	1.38–03	1.58–02	–2.558	B–	*LS*
301.	$2s2p(^3P^\circ)4p-$ $2s2p(^3P^\circ)4d$	^{4}S–^{4}P°	*12760*	*7837* cm^{-1}	378432.8	*386270*	4–12	1.09–01	7.98–01	1.34+02	0.504	B	2
			12809	7804.9 cm^{-1}	378432.8	386237.7	4–6	1.08–01	3.97–01	6.70+01	0.201	B	*LS*
			12721	7859.0 cm^{-1}	378432.8	386291.8	4–4	1.10–01	2.67–01	4.47+01	0.028	B	*LS*
			12668	7891.5 cm^{-1}	378432.8	386324.3	4–2	1.11–01	1.34–01	2.23+01	–0.271	B	*LS*
302.	$2s2p(^3P^\circ)4p-$ $2s2p(^3P^\circ)5s$	^{4}S–^{4}P°	*4433.5*	*4434.7*	378432.8	*400982*	4–12	1.32–01	1.16–01	6.80+00	–0.332	B	2
			4419.49	4420.73	378432.8	401053.5	4–6	1.33–01	5.84–02	3.40+00	–0.632	B	*LS*
			4442.92	4444.17	378432.8	400934.2	4–4	1.31–01	3.87–02	2.27+00	–0.810	B	*LS*
			4457.12	4458.37	378432.8	400862.5	4–2	1.30–01	1.93–02	1.13+00	–1.112	B	*LS*
303.	$2s2p(^3P^\circ)4p-$ $2s2p(^3P^\circ)5d$	^{4}S–^{4}P°	*3197.8*	*3198.8*	378432.8	*409695*	4–12	3.22–01	1.48–01	6.23+00	–0.228	B	2
			3173.84	3174.75	378432.8	409931.3	4–6	3.29–01	7.45–02	3.11+00	–0.526	B	*LS*
			3223.52	3224.45	378432.8	409445.8	4–4	3.14–01	4.89–02	2.08+00	–0.709	B	*LS*
			3219.33	3220.26	378432.8	409486.2	4–2	3.15–01	2.45–02	1.04+00	–1.009	B	*LS*
304.	$2s2p(^3P^\circ)4p-$ $2s2p(^3P^\circ)6d$	^{4}S–^{4}P°	*2304.0*	*2304.7*	378432.8	*421823*	4–12	2.54–01	6.07–02	1.84+00	–0.615	B	2
			2305.49	2306.20	378432.8	421794.2	4–6	2.54–01	3.03–02	9.21–01	–0.916	B	*LS*
			2302.89	2303.60	378432.8	421843.2	4–4	2.55–01	2.02–02	6.14–01	–1.092	B	*LS*
			2301.61	2302.32	378432.8	421867.2	4–2	2.55–01	1.01–02	3.07–01	–1.393	B	*LS*
305.	$2s2p(^3P^\circ)4p-$ $2s2p(^3P^\circ)7d$	^{4}S–^{4}P°		1958.73	378432.8	429486.3	4–12	1.78–01	3.08–02	7.94–01	–0.910	B	2
306.	$2s2p(^3P^\circ)4p-$ $2s2p(^3P^\circ)8d$	^{4}S–^{4}P°		1788.15	378432.8	434356.6	4–12	1.27–01	1.83–02	4.31–01	–1.135	B	2
307.	$2s2p(^1P^\circ)3p-$ $2s2p(^3P^\circ)4d$	^{2}P–^{2}D°	*13280*	*7526* cm^{-1}	*377600*	385126	6–10	6.56–03	2.89–02	7.59+00	–0.760	B	2
			13300	7515 cm^{-1}	377611.3	385126	4–6	6.53–03	2.60–02	4.56+00	–0.983	B	*LS*
			13240	7548 cm^{-1}	377577.9	385126	2–4	5.52–03	2.90–02	2.53+00	–1.236	B	*LS*
			13300	7515 cm^{-1}	377611.3	385126	4–4	1.09–03	2.89–03	5.06–01	–1.937	B	*LS*

N III: Allowed Transitions — Continued

No.	Transition Array	Mult.	λ_{air} (Å)	λ_{vac} (Å) or σ (cm^{-1})*	E_i (cm^{-1})	E_k (cm^{-1})	g_i-g_k	A_{ki} (10^8 s^{-1})	f_{ik}	S (at. u.)	log gf	Acc.	Source
308.	$2s2p(^1P^\circ)3p-2s2p(^1P^\circ)3d$	$^2P-^2D^\circ$	*5267.0*	*5268.4*	*377600*	*396581*	6–10	4.05–01	2.80–01	2.92+01	0.226	B	2
			5269.04	5270.51	377611.3	396584.8	4–6	4.04–01	2.52–01	1.75+01	0.004	B	*LS*
			5262.52	5263.99	377577.9	396574.9	2–4	3.38–01	2.81–01	9.73+00	–0.251	B	*LS*
			5271.79	5273.26	377611.3	396574.9	4–4	6.72–02	2.80–02	1.95+00	–0.950	B	*LS*
309.	$2s2p(^1P^\circ)3p-2p^2(^3P)3p$	$^2P-^2S^\circ$		*1576.6*	*377600*	441026.6	6–2	1.55+00	1.93–02	6.00–01	–0.937	B	2
				1576.91	377611.3	441026.6	4–2	1.03+00	1.93–02	4.00–01	–1.113	B	*LS*
				1576.08	377577.9	441026.6	2–2	5.18–01	1.93–02	2.00–01	–1.414	B	*LS*
310.	$2s2p(^3P^\circ)4p-2s2p(^3P^\circ)4d$	$^4P-^4P^\circ$	*14480*	*6906* cm^{-1}	*379364*	*386270*	12–12	3.44–02	1.08–01	6.18+01	0.113	B	2
			14616	6840.1 cm^{-1}	379397.6	386237.7	6–6	2.34–02	7.49–02	2.16+01	–0.347	B	*LS*
			14391	6947.0 cm^{-1}	379344.8	386291.8	4–4	4.67–03	1.45–02	2.75+00	–1.237	B	*LS*
			14233	7024.2 cm^{-1}	379300.1	386324.3	2–2	6.03–03	1.83–02	1.72+00	–1.436	B	*LS*
			14501	6894.2 cm^{-1}	379397.6	386291.8	6–4	1.54–02	3.23–02	9.27+00	–0.712	B	*LS*
			14324	6979.5 cm^{-1}	379344.8	386324.3	4–2	2.96–02	4.55–02	8.58+00	–0.740	B	*LS*
			14504	6892.9 cm^{-1}	379344.8	386237.7	4–6	1.03–02	4.85–02	9.27+00	–0.712	B	*LS*
			14299	6991.7 cm^{-1}	379300.1	386291.8	2–4	1.49–02	9.11–02	8.58+00	–0.739	B	*LS*
311.	$2s2p(^3P^\circ)4p-2s2p(^3P^\circ)5s$	$^4P-^4P^\circ$	*4624.5*	*4625.8*	*379364*	*400982*	12–12	8.84–01	2.84–01	5.18+01	0.532	B	2
			4616.39	4617.68	379397.6	401053.5	6–6	6.22–01	1.99–01	1.81+01	0.077	B	*LS*
			4630.61	4631.90	379344.8	400934.2	4–4	1.17–01	3.78–02	2.30+00	–0.821	B	*LS*
			4636.40	4637.70	379300.1	400862.5	2–2	1.46–01	4.71–02	1.44+00	–1.026	B	*LS*
			4641.96	4643.26	379397.6	400934.2	6–4	3.93–01	8.47–02	7.77+00	–0.294	B	*LS*
			4646.04	4647.34	379344.8	400862.5	4–2	7.27–01	1.18–01	7.19+00	–0.328	B	*LS*
			4605.16	4606.45	379344.8	401053.5	4–6	2.69–01	1.28–01	7.77+00	–0.290	B	*LS*
			4621.04	4622.33	379300.1	400934.2	2–4	3.69–01	2.36–01	7.19+00	–0.325	B	*LS*
312.	$2s2p(^3P^\circ)4p-2s2p(^3P^\circ)5d$	$^4P-^4P^\circ$	*3325.6*	*3326.6*	*379364*	*409425*	12–12	2.69–01	4.46–02	5.86+00	–0.271	B	2
			3333.07	3334.03	379397.6	409391.3	6–6	1.87–01	3.12–02	2.05+00	–0.728	B	*LS*
			3321.19	3322.15	379344.8	409445.8	4–4	3.60–02	5.96–03	2.60–01	–1.623	B	*LS*
			3311.83	3312.78	379300.1	409486.2	2–2	4.54–02	7.47–03	1.63–01	–1.826	B	*LS*
			3327.03	3327.99	379397.6	409445.8	6–4	1.21–01	1.34–02	8.79–01	–1.095	B	*LS*
			3316.74	3317.70	379344.8	409486.2	4–2	2.26–01	1.86–02	8.14–01	–1.128	B	*LS*
			3327.22	3328.17	379344.8	409391.3	4–6	8.06–02	2.01–02	8.79–01	–1.095	B	*LS*
			3316.27	3317.22	379300.1	409445.8	2–4	1.13–01	3.73–02	8.14–01	–1.127	B	*LS*
313.	$2s2p(^3P^\circ)4p-2p^2(^3P)3p$	$^4P-^4D^\circ$	*2471.9*	*2472.7*	*379364*	*419806*	12–20	1.38–01	2.10–02	2.05+00	–0.598	B	2
			2469.97	2470.72	379397.6	419871.7	6–8	1.38–01	1.68–02	8.21–01	–0.996	B	*LS*
			2471.08	2471.83	379344.8	419800.7	4–6	9.65–02	1.33–02	4.31–01	–1.276	B	*LS*
			2472.20	2472.95	379300.1	419737.6	2–4	5.74–02	1.05–02	1.71–01	–1.677	B	*LS*
			2474.31	2475.06	379397.6	419800.7	6–6	4.12–02	3.78–03	1.85–01	–1.644	B	*LS*
			2474.94	2475.69	379344.8	419737.6	4–4	7.32–02	6.72–03	2.19–01	–1.571	B	*LS*
			2474.49	2475.24	379300.1	419700.3	2–2	1.14–01	1.05–02	1.71–01	–1.678	B	*LS*
			2478.18	2478.93	379397.6	419737.6	6–4	6.83–03	4.19–04	2.05–02	–2.599	B–	*LS*
			2477.23	2477.98	379344.8	419700.3	4–2	2.28–02	1.05–03	3.42–02	–2.377	B	*LS*
314.	$2s2p(^3P^\circ)4p-2s2p(^3P^\circ)6d$	$^4P-^4D^\circ$	*2356.1*	*2356.8*	*379364*	421794.2	12–20	2.82–01	3.91–02	3.64+00	–0.329	B	2
			2357.96	2358.68	379397.6	421794.2	6–8	2.81–01	3.12–02	1.45+00	–0.727	B	*LS*
			2355.03	2355.75	379344.8	421794.2	4–6	1.97–01	2.46–02	7.63–01	–1.007	B	*LS*
			2352.55	2353.27	379300.1	421794.2	2–4	1.18–01	1.96–02	3.03–01	–1.408	B	*LS*
			2357.96	2358.68	379397.6	421794.2	6–6	8.43–02	7.02–03	3.27–01	–1.375	B	*LS*
			2355.03	2355.75	379344.8	421794.2	4–4	1.50–01	1.25–02	3.88–01	–1.301	B	*LS*
			2352.55	2353.27	379300.1	421794.2	2–2	2.36–01	1.96–02	3.03–01	–1.408	B	*LS*
			2357.96	2358.68	379397.6	421794.2	6–4	1.40–02	7.80–04	3.64–02	–2.329	B	*LS*
			2355.03	2355.75	379344.8	421794.2	4–2	4.70–02	1.95–03	6.06–02	–2.107	B	*LS*

N III: Allowed Transitions — Continued

No.	Transition Array	Mult.	λ_{air} (Å)	λ_{vac} (Å) or σ (cm^{-1})*	E_i (cm^{-1})	E_k (cm^{-1})	g_i-g_k	A_{ki} (10^8 s^{-1})	f_{ik}	S (at. u.)	log gf	Acc.	Source
315.		$^4P-^4P^o$	*2354.5*	*2355.2*	*379364*	*421823*	12−12	1.97−01	1.64−02	1.53+00	−0.706	B	2
			2357.96	2358.68	379397.6	421794.2	6−6	1.37−01	1.15−02	5.34−01	−1.163	B	*LS*
			2352.31	2353.03	379344.8	421843.2	4−4	2.64−02	2.19−03	6.78−02	−2.058	B	*LS*
			2348.51	2349.23	379300.1	421867.2	2−2	3.31−02	2.74−03	4.24−02	−2.261	B	*LS*
			2355.24	2355.96	379397.6	421843.2	6−4	8.87−02	4.92−03	2.29−01	−1.530	B	*LS*
			2350.98	2351.70	379344.8	421867.2	4−2	1.65−01	6.84−03	2.12−01	−1.563	B	*LS*
			2355.03	2355.75	379344.8	421794.2	4−6	5.91−02	7.38−03	2.29−01	−1.530	B	*LS*
			2349.84	2350.56	379300.1	421843.2	2−4	8.27−02	1.37−02	2.12−01	−1.563	B	*LS*
316.	$2s2p(^3P^o)4p-$ $2s2p(^3P^o)7d$	$^4P-^4D^o$		*1998.0*	*379364*	429412.9	12−20	3.00−01	2.99−02	2.36+00	−0.445	B	2
				1999.39	379397.6	429412.9	6−8	2.99−01	2.39−02	9.45−01	−0.843	B	*LS*
				1997.28	379344.8	429412.9	4−6	2.10−01	1.89−02	4.96−01	−1.123	B	*LS*
				1995.50	379300.1	429412.9	2−4	1.25−01	1.50−02	1.97−01	−1.524	B	*LS*
				1999.39	379397.6	429412.9	6−6	8.98−02	5.38−03	2.13−01	−1.491	B	*LS*
				1997.28	379344.8	429412.9	4−4	1.60−01	9.58−03	2.52−01	−1.417	B	*LS*
				1995.50	379300.1	429412.9	2−2	2.51−01	1.50−02	1.97−01	−1.524	B	*LS*
				1999.39	379397.6	429412.9	6−4	1.50−02	5.98−04	2.36−02	−2.445	B−	*LS*
				1997.28	379344.8	429412.9	4−2	5.00−02	1.50−03	3.94−02	−2.223	B	*LS*
317.		$^4P-^4P^o$		*1995.1*	*379364*	429486.3	12−12	1.48−01	8.84−03	6.97−01	−0.974	B	2
				1996.46	379397.6	429486.3	6−6	1.04−01	6.19−03	2.44−01	−1.430	B	*LS*
				1994.36	379344.8	429486.3	4−4	1.98−02	1.18−03	3.10−02	−2.326	B	*LS*
				1992.58	379300.1	429486.3	2−2	2.48−02	1.48−03	1.94−02	−2.530	B−	*LS*
				1996.46	379397.6	429486.3	6−4	6.66−02	2.65−03	1.05−01	−1.798	B	*LS*
				1994.36	379344.8	429486.3	4−2	1.24−01	3.69−03	9.68−02	−1.831	B	*LS*
				1994.36	379344.8	429486.3	4−6	4.45−02	3.98−03	1.05−01	−1.798	B	*LS*
				1992.58	379300.1	429486.3	2−4	6.20−02	7.38−03	9.68−02	−1.831	B	*LS*
318.	$2s2p(^3P^o)4p-$ $2s2p(^3P^o)8d$	$^4P-^4D^o$		*1820.3*	*379364*	434301.2	12−20	2.09−01	1.73−02	1.25+00	−0.682	B	2
				1821.37	379397.6	434301.2	6−8	2.09−01	1.39−02	4.99−01	−1.080	B	*LS*
				1819.62	379344.8	434301.2	4−6	1.47−01	1.09−02	2.62−01	−1.359	B	*LS*
				1818.15	379300.1	434301.2	2−4	8.76−02	8.68−03	1.04−01	−1.761	B	*LS*
				1821.37	379397.6	434301.2	6−6	6.27−02	3.12−03	1.12−01	−1.728	B	*LS*
				1819.62	379344.8	434301.2	4−4	1.12−01	5.55−03	1.33−01	−1.654	B	*LS*
				1818.15	379300.1	434301.2	2−2	1.75−01	8.68−03	1.04−01	−1.761	B	*LS*
				1821.37	379397.6	434301.2	6−4	1.05−02	3.47−04	1.25−02	−2.682	B−	*LS*
				1819.62	379344.8	434301.2	4−2	3.49−02	8.67−04	2.08−02	−2.460	B−	*LS*
319.		$^4P-^4P^o$		*1818.4*	*379364*	434356.6	12−12	1.03−01	5.12−03	3.68−01	−1.212	B	2
				1819.54	379397.6	434356.6	6−6	7.22−02	3.58−03	1.29−01	−1.668	B	*LS*
				1817.79	379344.8	434356.6	4−4	1.38−02	6.83−04	1.63−02	−2.564	B−	*LS*
				1816.32	379300.1	434356.6	2−2	1.73−02	8.54−04	1.02−02	−2.767	B−	*LS*
				1819.54	379397.6	434356.6	6−4	4.64−02	1.54−03	5.52−02	−2.036	B	*LS*
				1817.79	379344.8	434356.6	4−2	8.62−02	2.13−03	5.11−02	−2.069	B	*LS*
				1817.79	379344.8	434356.6	4−6	3.10−02	2.30−03	5.52−02	−2.035	B	*LS*
				1816.32	379300.1	434356.6	2−4	4.32−02	4.27−03	5.11−02	−2.068	B	*LS*
320.	$2s2p(^3P^o)4p-$ $2p^2(^3P)4p$	$^4P-^4S^o$		*910.29*	*379364*	489219.0	12−4	1.87+01	7.76−02	2.79+00	−0.031	B	2
				910.569	379397.6	489219.0	6−4	9.36+00	7.75−02	1.39+00	−0.332	B	*LS*
				910.132	379344.8	489219.0	4−4	6.25+00	7.76−02	9.30−01	−0.508	B	*LS*
				909.762	379300.1	489219.0	2−4	3.13+00	7.76−02	4.65−01	−0.809	B	*LS*

*Wavelengths (Å) are always given unless cm^{-1} is indicated.

N III

Forbidden Transitions

We have selected the results of multi-configuration Hartree-Fock calculations with relativistic Breit-Pauli terms by Froese Fischer[1] for the M1 and E2 fine structure transitions in the $^2P^o$ term of the ground state configuration.

Reference

[1]C. Froese Fischer, J. Phys. B **16**, 157 (1983).

N III: Forbidden Transitions

No.	Transition Array	Mult.	λ_{air} (Å)	λ_{vac} (Å) or σ (cm^{-1})*	E_i (cm^{-1})		E_k (cm^{-1})	g_i-g_k	Type	A_{ki} (s^{-1})	S (at. u.)	Acc.	Source
1.	$2p-2p$	$^2P^o-^2P^o$											
				174.4 cm^{-1}	0.0	–	174.4	2–4	M1	4.79–05	1.34+00	A	1
				174.4 cm^{-1}	0.0	–	174.4	2–4	E2	7.38–12	1.63+00	B	1

*Wavelengths (Å) are always given unless cm^{-1} is indicated.

N IV

Beryllium Isoelectronic Sequence

Ground State: $1s^22s^2\ ^1S_0$

Ionization Energy: 77.472 eV = 624866 cm^{-1}

Allowed Transitions

List of tabulated lines

Wavelength (Å)	No.	Wavelength (Å)	No.	Wavelength (Å)	No.	Wavelength (Å)	No.
in vacuum		181.741	8	191.228	70	196.818	85
		183.146	72	191.649	33	196.866	85
161.202	45	185.235	40	191.651	33	196.920	29
161.219	45	185.257	40	191.674	33	196.944	29
161.256	45	185.306	40	191.702	33	197.000	29
162.319	44	185.552	39	191.727	33	197.228	6
162.336	44	185.573	39	191.748	33	198.739	28
162.374	44	185.623	39	191.875	32	198.764	28
163.916	43	186.218	71	191.898	32	198.821	28
163.933	43	186.687	38	191.951	32	199.081	84
163.972	43	186.709	38	192.823	7	199.110	84
164.794	15	186.759	38	192.854	31	199.159	84
165.945	14	187.121	37	192.859	31	200.340	69
166.320	42	187.144	37	192.864	31	202.514	83
166.337	42	187.194	37	192.887	31	202.544	83
166.377	42	188.582	36	192.908	31	202.595	83
167.074	13	188.605	36	192.941	31	205.939	27
167.709	12	188.656	36	193.137	30	205.966	27
170.074	11	188.744	35	193.160	30	206.026	27
170.189	41	188.767	35	193.214	30	206.027	27
170.207	41	188.818	35	195.183	86	208.113	82
170.249	41	189.363	34	195.210	86	208.131	82
172.171	10	189.385	34	195.258	86	208.144	82
174.602	9	189.437	34	196.789	85	208.150	82

List of tabulated lines — Continued

Wavelength (Å)	No.	Wavelength (Å)	No.	Wavelength (Å)	No.	Wavelength (Å)	No.
208.185	82	248.441	78	368.109	105	1086.27	146
208.199	82	248.461	78	387.356	49	1086.69	146
209.378	26	248.486	78	420.769	104	1104.54	127
209.406	26	248.540	78	433.930	73	1131.04	145
209.469	26	248.563	78	434.000	73	1131.49	145
209.842	68	248.654	101	434.067	73	1132.02	145
211.403	5	250.121	100	434.097	73	1132.22	145
211.679	67	250.566	99	434.235	73	1132.68	145
214.291	66	255.144	98	434.302	73	1132.94	145
214.843	65	258.320	97	460.217	117	1133.12	113
215.755	64	259.824	55	460.272	117	1135.25	113
217.218	63	260.448	54	460.394	117	1136.27	113
218.044	81	260.519	96	463.736	87	1168.48	151
218.067	81	270.994	53	476.909	124	1168.60	151
218.079	81	273.140	95	497.696	116	1169.01	151
218.088	81	274.451	52	498.106	116	1169.06	151
218.126	81	275.354	111	498.323	116	1169.18	151
218.138	81	276.740	94	518.600	141	1169.48	151
218.250	62	283.417	20	520.934	123	1169.53	151
220.124	61	283.465	20	521.845	140	1188.01	119
220.280	60	283.468	20	527.259	139	1195.40	150
224.629	59	283.574	20	536.082	138	1195.51	150
225.108	25	283.581	20	542.412	137	1195.57	150
225.137	25	283.584	20	554.137	136	1195.68	150
225.140	25	283.599	110	555.126	135	1195.73	150
225.209	25	285.559	51	583.601	134	1195.85	150
225.210	25	289.479	109	591.180	103	1224.95	144
225.213	25	296.418	108	594.895	18	1225.19	144
227.026	58	297.597	77	600.061	133	1225.72	144
232.107	24	297.634	77	619.663	122	1243.73	173
232.141	24	297.661	77	667.653	132	1244.91	173
232.218	24	297.707	77	703.191	131	1246.51	173
234.121	23	297.772	77	720.244	148	1270.27	143
234.172	23	297.817	77	720.327	148	1272.14	143
234.200	23	300.318	93	720.510	148	1272.72	143
234.203	23	303.013	76	744.566	115	1273.45	143
234.251	23	303.049	76	744.737	115	1273.71	143
236.068	57	303.080	76	745.020	115	1274.28	143
237.870	22	303.124	76	765.147	2	1284.22	154
237.905	22	303.163	76	770.678	121	1296.59	153
237.987	22	303.195	76	791.045	114	1309.56	126
238.656	80	303.280	92	791.095	114	1324.02	149
238.683	80	304.733	75	791.167	114	1325.68	149
238.698	80	304.793	75	822.299	17	1325.83	149
238.731	80	304.800	75	823.274	17	1326.96	149
238.769	80	304.811	75	846.215	120	1327.02	149
238.802	80	304.916	75	900.516	130	1327.17	149
239.146	79	304.927	75	908.057	129	1438.37	180
239.174	79	314.322	91	921.994	16	1446.12	152
239.188	79	315.058	90	922.519	16	1486.50	1
239.212	79	317.598	107	923.056	16	1687.57	185
239.245	79	322.502	19	923.220	16	1687.80	185
239.259	79	322.568	19	923.676	16	1688.12	185
239.619	21	322.718	19	924.284	16	1696.83	172
239.623	21	323.177	89	948.151	147	1699.03	172
239.642	21	335.047	50	948.240	147	1702.01	172
239.678	21	344.913	74	948.294	147	1718.55	47
239.705	21	345.023	74	948.540	147		
239.761	21	345.061	74	948.558	147	in air	
240.363	56	345.110	74	948.612	147		
244.107	102	345.204	74	955.334	48	2035.57	178
246.313	4	345.258	74	1050.60	128	2035.62	178
247.205	3	351.931	88	1078.71	155	2036.10	178
248.384	78	353.056	106	1086.08	146	2036.43	178

List of tabulated lines — Continued

Wavelength (Å)	No.	Wavelength (Å)	No.	Wavelength (Å)	No.	Wavelength (Å)	No.
2080.32	179	3127.43	169	4538.32	160	6212.39	170
2219.61	46	3128.23	169	4740.26	167	6215.45	170
2225.78	46	3141.19	169	4747.96	167	6219.89	170
2229.38	46	3142.30	169	4752.49	167	6380.75	118
2402.05	162	3143.11	169	4762.09	167	7103.24	142
2421.65	181	3443.61	157	4769.86	167	7109.35	142
2421.73	181	3445.22	157	4786.92	167	7111.28	142
2424.73	181	3454.65	157	4796.66	167	7122.98	142
2426.54	181	3461.36	157	5200.41	156	7127.25	142
2430.40	177	3463.36	157	5204.28	156	9165.07	166
2431.08	177	3474.53	157	5205.15	156	9182.16	166
2431.55	177	3478.72	112	5226.70	156	9222.99	166
2477.69	183	3483.00	112	5245.60	156	9247.04	166
2645.62	184	3484.93	112	5272.35	156	9311.55	166
2646.19	184	3689.94	168	5288.25	175	9378.29	166
2646.98	184	3694.14	168	5736.93	163	12727.4	159
2649.88	165	3701.13	168	5776.31	171	12870.5	176
2809.35	182	3707.39	168	5784.76	171	12873.8	176
3052.20	158	3714.43	168	5795.09	171	12883.7	176
3059.60	158	3735.43	168	5812.31	171	12892.9	176
3075.19	158	3747.54	161	5826.43	171	12902.9	176
3078.27	186	3823.93	164	5843.84	171	12916.0	176
3118.79	169	4057.76	125	6119.23	174		

For five singlets and one triplet, fully correlated superposition-of-configuration (SOC) wavefunctions have been calculated by Weiss,[1] in which several thousand interacting configurations are included for describing both upper and lower states. The dipole-length and dipole-velocity results agree typically within a few tenths of a percent—indicating a level of electron-correlation treatment which puts this work in a class by itself.

For most other transitions, we have utilized the results of the Opacity Project (OP), which is reviewed in the general introduction. For this Be-like ion, the OP calculations were carried out by Tully *et al.*[2] For a few transitions, data from similarly advanced configuration-interaction calculations by Hibbert,[3] carried out with the CIV 3 code, are available and are averaged with the OP data with which they are in very close agreement.

Allard *et al.*[4] have recently undertaken an extensive compilation of N IV transition probabilities, very similar in spirit to this one, and have also utilized the OP data as the main source of their tables. But they have averaged the OP results with less advanced earlier work, using the OP results with a weight factor of five and the earlier work with a weight of one, while our approach is to omit all less advanced work (weight of zero). Thus, our tabulated results differ, but usually only slightly, from those of Allard *et al.*[4] However, we utilize their tables for a few multiplets for which they had special OP data at their disposal and for one intercombination line. For this class of transitions, for which no OP results exist, we have otherwise utilized the new calculations by Weiss.[1]

For the neighboring Be-like ion O V, Hibbert[5] calculated the transition probabilities of individual lines in triplets involving s, p and d electrons in intermediate coupling. His line ratios within multiplets are very close to those obtained from LS-coupling, typically within 1 to 3%. Somewhat larger differences, up to 25%, are only encountered for some weak multiplet components. Therefore, we assume that LS-coupling is also a very good approximation for the spectrum of N IV, and we estimate that for the stronger lines in multiplets the same accuracy ratings as for the multiplet value apply. However, we expect some deterioration in accuracy for triplet lines involving f and g electrons and have accordingly lowered the ratings in those cases.

Engström *et al.*[6] measured the lifetime of the $2s2p\ ^1P^\circ_1$ level with the beam-foil method and applied the ANDC technique (see general introduction) for a reliable analysis and correction of the critical cascading effects. Their result, estimated by the authors to be accurate to $\pm 3.5\%$, agrees with our tabulated value based on Weiss' calculations[1] within 1.4%.

References

[1]A. W. Weiss, to be published.

[2]J. A. Tully, M. J. Seaton and K. A. Berrington, J. Phys. B **23**, 3811 (1990).

[3]A. Hibbert, J. Phys. B **9**, 2805 (1976).

[4]N. Allard, M.-C. Artru, T. Lanz, and M. Le Dourneuf, Astron. Astrophys., Suppl. Ser. **84**, 563 (1990) and **91**, 399 (1991).

[5]A. Hibbert, J. Phys. B **13**, 1721 (1980).

[6]L. Engström, B. Denne, J. O. Ekberg, K. W. Jones, C. Jupen, U. Litzen, Weng Tai Meng, A. Trigueiros and I. Martinson, Phys. Scr. **24**, 551 (1981).

N IV: Allowed Transitions

No.	Transition Array	Mult.	λ_{air} (Å)	λ_{vac} (Å) or σ (cm^{-1})*	E_i (cm^{-1})	E_k (cm^{-1})	g_i–g_k	A_{ki} (10^8 s^{-1})	f_{ik}	S (at. u.)	log gf	Acc.	Source
1.	$2s^2$–$2s2p$	$^1S-^3P^o$											
				1486.50	0	– 67272.3	1–3	5.95–06	5.91–07	2.89–06	–6.228	B	1
2.		$^1S-^1P^o$		765.147	0	– 130693.9	1–3	2.320+01	6.110–01	1.539+00	–0.2140	A+	1
3.	$2s^2$–$2s3p$	$^1S-^1P^o$		247.205	0	– 404522.4	1–3	1.19+02	3.27–01	2.66–01	–0.486	B	2
4.		$^1S-^3P^o$											
				246.313	0	– 405987.5	1–3	3.08–02	8.39–05	6.81–05	–4.076	C+	4
5.	$2s^2$–$2p3s$	$^1S-^1P^o$		211.403	0	– 473029.3	1–3	2.22+00	4.45–03	3.10–03	–2.351	C	2
6.	$2s^2$–$2s4p$	$^1S-^1P^o$		197.228	0	– 507027.9	1–3	6.01+01	1.05–01	6.83–02	–0.978	B	2
7.	$2s^2$–$2p3d$	$^1S-^1P^o$		192.823	0	– 518610.5	1–3	3.58+00	5.99–03	3.80–03	–2.223	C	2
8.	$2s^2$–$2s5p$	$^1S-^1P^o$		181.741	0	– 550232.6	1–3	2.80+01	4.16–02	2.49–02	–1.381	B–	2
9.	$2s^2$–$2s6p$	$^1S-^1P^o$		174.602	0	– 572731.1	1–3	1.70+01	2.33–02	1.34–02	–1.632	B–	2
10.	$2s^2$–$2p4s$	$^1S-^1P^o$		172.171	0	– 580817.5	1–3	5.85–02	7.80–05	4.42–05	–4.108	D	4
11.	$2s^2$–$2s7p$	$^1S-^1P^o$		170.074	0	– 587979.4	1–3	1.00+01	1.30–02	7.30–03	–1.885	C+	2
12.	$2s^2$–$2s8p$	$^1S-^1P^o$		167.709	0	– 596270.9	1–3	1.02+01	1.29–02	7.10–03	–1.891	C+	2
13.	$2s^2$–$2p4d$	$^1S-^1P^o$		167.074	0	– 598538.2	1–3	4.20+00	5.27–03	2.90–03	–2.278	C	2
14.	$2s^2$–$2s9p$	$^1S-^1P^o$		165.945	0	– 602609.3	1–3	3.29+00	4.07–03	2.23–03	–2.390	C	4
15.	$2s^2$–$2s10p$	$^1S-^1P^o$		164.794	0	– 606818.2	1–3	3.23+00	3.94–03	2.14–03	–2.404	C	4
16.	$2s2p$–$2p^2$	$^3P^o-^3P$		*923.16*	*67345*	– *175669*	9–9	1.759+01	2.247–01	6.147+00	0.3059	A+	1
				923.220	67416.3	– 175732.9	5–5	1.319+01	1.685–01	2.561+00	–0.0744	A+	1
				923.056	67272.3	– 175608.1	3–3	4.399+00	5.620–02	5.123–01	–0.7732	A+	1
				924.284	67416.3	– 175608.1	5–3	7.304+00	5.612–02	8.539–01	–0.5519	A+	1
				923.676	67272.3	– 175535.4	3–1	1.756+01	7.488–02	6.831–01	–0.6485	A+	1
				921.994	67272.3	– 175732.9	3–5	4.415+00	9.377–02	8.539–01	–0.5508	A+	1
				922.519	67209.2	– 175608.1	1–3	5.876+00	2.249–01	6.831–01	–0.6480	A+	1
17.		$^3P^o-^1D$											
				823.274	67416.3	– 188882.5	5–5	9.89–04	1.01–05	1.36–04	–4.220	B	1
				822.299	67272.3	– 188882.5	3–5	1.23–04	2.07–06	1.68–05	–6.207	B	1
18.		$^3P^o-^1S$											
				594.895	67272.3	– 235369.3	3–1	5.81–06	1.03–07	6.04–07	–6.511	C	1

N IV: Allowed Transitions — Continued

No.	Transition Array	Mult.	λ_{air} (Å)	λ_{vac} (Å) or σ (cm^{-1})*	E_i (cm^{-1})	E_k (cm^{-1})	g_i–g_k	A_{ki} (10^8 s^{-1})	f_{ik}	S (at. u.)	log gf	Acc.	Source
19.	$2s2p$– $2s3s$	^{3}P°–^{3}S		*322.64*	*67345*	– 377284.8	9–3	8.99+01	4.68–02	4.47–01	–0.376	B	2,3
				322.718	67416.3	– 377284.8	5–3	4.99+01	4.68–02	2.48–01	–0.631	B	*LS*
				322.568	67272.3	– 377284.8	3–3	3.00+01	4.68–02	1.49–01	–0.853	B	*LS*
				322.502	67209.2	– 377284.8	1–3	1.00+01	4.68–02	4.97–02	–1.330	B	*LS*
20.	$2s2p$– $2s3d$	^{3}P°–^{3}D		*283.52*	*67345*	– *420053*	9–15	3.05+02	6.12–01	5.14+00	0.741	B	2,3
				283.574	67416.3	– 420058.0	5–7	3.05+02	5.14–01	2.40+00	0.410	B	*LS*
				283.465	67272.3	– 420049.6	3–5	2.29+02	4.59–01	1.29+00	0.139	B	*LS*
				283.417	67209.2	– 420045.8	1–3	1.69+02	6.12–01	5.71–01	–0.213	B	*LS*
				283.581	67416.3	– 420049.6	5–5	7.61+01	9.18–02	4.28–01	–0.338	B	*LS*
				283.468	67272.3	– 420045.8	3–3	1.27+02	1.53–01	4.28–01	–0.338	B	*LS*
				283.584	67416.3	– 420045.8	5–3	8.46+00	6.12–03	2.86–02	–1.514	B–	*LS*
21.	$2s2p$– $2p3p$	^{3}P°–^{3}D		*239.64*	*67345*	– *484646*	9–15	3.19+01	4.58–02	3.25–01	–0.385	B	2
				239.619	67416.3	– 484746.2	5–7	3.19+01	3.85–02	1.52–01	–0.716	B	*LS*
				239.623	67272.3	– 484594.9	3–5	2.40+01	3.44–02	8.14–02	–0.987	B	*LS*
				239.642	67209.2	– 484498.2	1–3	1.77+01	4.58–02	3.62–02	–1.339	B	*LS*
				239.705	67416.3	– 484594.9	5–5	7.98+00	6.87–03	2.71–02	–1.464	B–	*LS*
				239.678	67272.3	– 484498.2	3–3	1.33+01	1.15–02	2.71–02	–1.464	B–	*LS*
				239.761	67416.3	– 484498.2	5–3	8.86–01	4.58–04	1.81–03	–2.640	C	*LS*
22.	$2s2p$– $2s4s$	^{3}P°–^{3}S		*237.95*	*67345*	– 487607.4	9–3	8.72+01	2.47–02	1.74–01	–0.654	B	4
				237.987	67416.3	– 487607.4	5–3	4.84+01	2.47–02	9.66–02	–0.909	B	4
				237.905	67272.3	– 487607.4	3–3	2.91+01	2.47–02	5.79–02	–1.131	B	4
				237.870	67209.2	– 487607.4	1–3	9.69+00	2.47–02	1.93–02	–1.608	B–	4
23.	$2s2p$– $2p3p$	^{3}P°–^{3}P		*234.19*	*67345*	– *494355*	9–9	4.76+01	3.91–02	2.71–01	–0.453	B	2
				234.200	67416.3	– 494402.0	5–5	3.57+01	2.93–02	1.13–01	–0.834	B	*LS*
				234.172	67272.3	– 494309.2	3–3	1.19+01	9.78–03	2.26–02	–1.533	B–	*LS*
				234.251	67416.3	– 494309.2	5–3	1.98+01	9.78–03	3.77–02	–1.311	B	*LS*
				234.203	67272.3	– 494253.1	3–1	4.76+01	1.30–02	3.02–02	–1.408	B	*LS*
				234.121	67272.3	– 494402.0	3–5	1.19+01	1.63–02	3.77–02	–1.311	B	*LS*
				234.172	67272.3	– 494309.2	1–3	1.59+01	3.91–02	3.02–02	–1.408	B	*LS*
24.		^{3}P°–^{3}S		*232.18*	*67345*	– 498045.5	9–3	9.23+01	2.49–02	1.71–01	–0.650	B	2
				232.218	67416.3	– 498045.5	5–3	5.12+01	2.49–02	9.50–02	–0.906	B	*LS*
				232.141	67272.3	– 498045.5	3–3	3.08+01	2.49–02	5.70–02	–1.127	B	*LS*
				232.107	67209.2	– 498045.5	1–3	1.03+01	2.49–02	1.90–02	–1.604	B–	*LS*
25.	$2s2p$– $2s4d$	^{3}P°–^{3}D		*225.17*	*67345*	– *511446*	9–15	1.00+02	1.27–01	8.47–01	0.058	B	2
				225.209	67416.3	– 511448	5–7	1.00+02	1.07–01	3.95–01	–0.273	B	*LS*
				225.137	67272.3	– 511446	3–5	7.52+01	9.53–02	2.12–01	–0.544	B	*LS*
				225.108	67209.2	– 511440	1–3	5.57+01	1.27–01	9.41–02	–0.896	B	*LS*
				225.210	67416.3	– 511446	5–5	2.50+01	1.90–02	7.06–02	–1.021	B	*LS*
				225.140	67272.3	– 511440	3–3	4.18+01	3.18–02	7.06–02	–1.021	B	*LS*
				225.213	67416.3	– 511440	5–3	2.78+00	1.27–03	4.71–03	–2.197	C+	*LS*
26.	$2s2p$– $2s5s$	^{3}P°–^{3}S		*209.44*	*67345*	– 544813.4	9–3	1.33+01	2.92–03	1.81–02	–1.581	B–	2
				209.469	67416.3	– 544813.4	5–3	7.39+00	2.92–03	1.01–02	–1.836	B–	*LS*
				209.406	67272.3	– 544813.4	3–3	4.44+00	2.92–03	6.03–03	–2.058	C+	*LS*
				209.378	67209.2	– 544813.4	1–3	1.48+00	2.92–03	2.01–03	–2.535	C	*LS*

N IV: Allowed Transitions — Continued

No.	Transition Array	Mult.	λ_{air} (Å)	λ_{vac} (Å) or σ (cm^{-1})*	E_i (cm^{-1})	E_k (cm^{-1})	g_i-g_k	A_{ki} (10^8 s^{-1})	f_{ik}	S (at. u.)	log gf	Acc.	Source
27.	$2s2p$–$2s5d$	^{3}P°–^{3}D		*206.00*	*67345*	– *552790*	9–15	4.98+01	5.28-02	3.22-01	-0.323	B	2
				206.026	67416.3	– 552790.9	5–7	4.98+01	4.43-02	1.50-01	-0.654	B	*LS*
				205.966	67272.3	– 552789.6	3–5	3.74+01	3.96-02	8.06-02	-0.925	B	*LS*
				205.939	67209.2	– 552789.6	1–3	2.77+01	5.28-02	3.58-02	-1.277	B	*LS*
				206.027	67416.3	– 552789.6	5–5	1.24+01	7.92-03	2.69-02	-1.402	B-	*LS*
				205.966	67272.3	– 552789.6	3–3	2.08+01	1.32-02	2.69-02	-1.402	B-	*LS*
				206.027	67416.3	– 552789.6	5–3	1.38+00	5.28-04	1.79-03	-2.579	C	*LS*
28.	$2s2p$–$2s6s$	^{3}P°–^{3}S		*198.79*	*67345*	– 570381.3	9–3	1.01+01	1.99-03	1.17-02	-1.748	B-	2
				198.821	67416.3	– 570381.3	5–3	5.59+00	1.99-03	6.50-03	-2.003	C+	*LS*
				198.764	67272.3	– 570381.3	3–3	3.35+00	1.99-03	3.90-03	-2.225	C+	*LS*
				198.739	67209.2	– 570381.3	1–3	1.12+00	1.99-03	1.30-03	-2.702	C	*LS*
29.	$2s2p$–$2s6d$	^{3}P°–^{3}D		*196.97*	*67345*	– 575030.5	9–15	2.86+01	2.77-02	1.62-01	-0.603	B	2
				197.000	67416.3	– 575030.5	5–7	2.86+01	2.33-02	7.56-02	-0.934	B	*LS*
				196.944	67272.3	– 575030.5	3–5	2.15+01	2.08-02	4.05-02	-1.205	B	*LS*
				196.920	67209.2	– 575030.5	1–3	1.59+01	2.77-02	1.80-02	-1.557	B-	*LS*
				197.000	67416.3	– 575030.5	5–5	7.15+00	4.16-03	1.35-02	-1.682	B-	*LS*
				196.944	67272.3	– 575030.5	3–3	1.19+01	6.94-03	1.35-02	-1.682	B-	*LS*
				197.000	67416.3	– 575030.5	5–3	7.95-01	2.77-04	8.99-04	-2.858	D	*LS*
30.	$2s2p$–$2s7s$	^{3}P°–^{3}S		*193.19*	*67345*	– 584977.1	9–3	1.43+01	2.67-03	1.53-02	-1.619	B-	2
				193.214	67416.3	– 584977.1	5–3	7.96+00	2.67-03	8.50-03	-1.874	C+	*LS*
				193.160	67272.3	– 584977.1	3–3	4.78+00	2.67-03	5.10-03	-2.096	C+	*LS*
				193.137	67209.2	– 584977.1	1–3	1.59+00	2.67-03	1.70-03	-2.573	C	*LS*
31.	$2s2p$–$2p4p$	^{3}P°–^{3}D		*192.87*	*67345*	– *585842*	9–15	1.96+01	1.83-02	1.04-01	-0.784	B	2
				192.859	67416.3	– 585929.8	5–7	1.96+01	1.53-02	4.87-02	-1.115	B	*LS*
				192.854	67272.3	– 585798.1	3–5	1.47+01	1.37-02	2.61-02	-1.386	B-	*LS*
				192.864	67209.2	– 585709.5	1–3	1.09+01	1.83-02	1.16-02	-1.739	B-	*LS*
				192.908	67416.3	– 585798.1	5–5	4.91+00	2.74-03	8.69-03	-1.864	C+	*LS*
				192.887	67272.3	– 585709.5	3–3	8.18+00	4.56-03	8.69-03	-1.864	C+	*LS*
				192.941	67416.3	– 585709.5	5–3	5.45-01	1.82-04	5.79-04	-3.040	D	*LS*
32.	$2s2p$–$2s7d$	^{3}P°–^{3}D		*191.92*	*67345*	– 588382.6	9–15	1.31+01	1.20-02	6.84-02	-0.966	B	2
				191.951	67416.3	– 588382.6	5–7	1.31+01	1.01-02	3.19-02	-1.297	B	*LS*
				191.898	67272.3	– 588382.6	3–5	9.81+00	9.02-03	1.71-02	-1.568	B-	*LS*
				191.875	67209.2	– 588382.6	1–3	7.27+00	1.20-02	7.60-03	-1.920	C+	*LS*
				191.951	67416.3	– 588382.6	5–5	3.27+00	1.80-03	5.70-03	-2.045	C+	*LS*
				191.898	67272.3	– 588382.6	3–3	5.45+00	3.01-03	5.70-03	-2.045	C+	*LS*
				191.951	67416.3	– 588382.6	5–3	3.63-01	1.20-04	3.80-04	-3.221	D	*LS*
33.	$2s2p$–$2p4p$	^{3}P°–^{3}P		*191.70*	*67345*	– *589007*	9–9	2.60+01	1.43-02	8.15-02	-0.889	B	2
				191.702	67416.3	– 589059.3	5–5	1.95+01	1.08-02	3.40-02	-1.269	B	*LS*
				191.674	67272.3	– 588991.2	3–3	6.51+00	3.59-03	6.79-03	-1.968	C+	*LS*
				191.727	67416.3	– 588991.2	5–3	1.08+01	3.59-03	1.13-02	-1.746	B-	*LS*
				191.748	67272.3	– 588790.1	3–1	2.60+01	4.78-03	9.06-03	-1.843	C+	*LS*
				191.649	67272.3	– 589059.3	3–5	6.52+00	5.98-03	1.13-02	-1.746	B-	*LS*
				191.651	67209.2	– 588991.2	1–3	8.69+00	1.44-02	9.06-03	-1.843	C+	*LS*
34.		^{3}P°–^{3}S		*189.41*	*67345*	– 595296.3	9–3	1.80+01	3.23-03	1.81-02	-1.537	B-	2
				189.437	67416.3	– 595296.3	5–3	9.99+00	3.22-03	1.01-02	-1.793	B-	*LS*
				189.385	67272.3	– 595296.3	3–3	6.00+00	3.23-03	6.03-03	-2.014	C+	*LS*
				189.363	67209.2	– 595296.3	1–3	2.00+00	3.23-03	2.01-03	-2.491	C	*LS*

N IV: Allowed Transitions — Continued

No.	Transition Array	Mult.	λ_{air} (Å)	λ_{vac} (Å) or σ (cm^{-1})*	E_i (cm^{-1})	E_k (cm^{-1})	g_i–g_k	A_{ki} (10^8 s^{-1})	f_{ik}	S (at. u.)	log gf	Acc.	Source
35.	$2s2p$– $2s8d$	$^3P^o$–^3D		*188.79*	*67345*	– 597026.8	9–15	8.91+00	7.94–03	4.44–02	–1.146	B	2
				188.818	67416.3	– 597026.8	5–7	8.91+00	6.67–03	2.07–02	–1.477	B–	*LS*
				188.767	67272.3	– 597026.8	3–5	6.69+00	5.95–03	1.11–02	–1.748	B–	*LS*
				188.744	67209.2	– 597026.8	1–3	4.96+00	7.94–03	4.93–03	–2.100	C+	*LS*
				188.818	67416.3	– 597026.8	5–5	2.23+00	1.19–03	3.70–03	–2.225	C+	*LS*
				188.767	67272.3	– 597026.8	3–3	3.72+00	1.98–03	3.70–03	–2.225	C+	*LS*
				188.818	67416.3	– 597026.8	5–3	2.47–01	7.94–05	2.47–04	–3.401	D	*LS*
36.	$2s2p$– $2s8s$	$^3P^o$–^3S		*188.63*	*67345*	– 597481.6	9–3	9.06–01	1.61–04	9.00–04	–2.839	D	2
				188.656	67416.3	– 597481.6	5–3	5.03–01	1.61–04	5.00–04	–3.094	D	*LS*
				188.605	67272.3	– 597481.6	3–3	3.02–01	1.61–04	3.00–04	–3.316	D	*LS*
				188.582	67209.2	– 597481.6	1–3	1.01–01	1.61–04	1.00–04	–3.793	D	*LS*
37.	$2s2p$– $2s9s$	$^3P^o$–^3S		*187.17*	*67345*	– 601621.5	9–3	1.40+00	2.44–04	1.36–03	–2.658	C	4
				187.194	67416.3	– 601621.5	5–3	7.81–01	2.46–04	7.58–04	–2.910	D	4
				187.144	67272.3	– 601621.5	3–3	4.70–01	2.47–04	4.56–04	–3.131	D	4
				187.121	67209.2	– 601621.5	1–3	1.57–01	2.47–04	1.52–04	–3.608	D	4
38.	$2s2p$– $2s9d$	$^3P^o$–^3D		*186.73*	*67345*	– 602865.7	9–15	7.52+00	6.56–03	3.63–02	–1.229	B	4
				186.759	67416.3	– 602865.7	5–7	7.56+00	5.53–03	1.70–02	–1.558	B–	4
				186.709	67272.3	– 602865.7	3–5	5.67+00	4.94–03	9.11–03	–1.829	C+	4
				186.687	67209.2	– 602865.7	1–3	4.21+00	6.59–03	4.05–03	–2.181	C	4
				186.759	67416.3	– 602865.7	5–5	1.88+00	9.83–04	3.02–03	–2.308	C	*LS*
				186.709	67272.3	– 602865.7	3–3	3.14+00	1.64–03	3.02–03	–2.308	C	*LS*
				186.759	67416.3	– 602865.7	5–3	2.09–01	6.55–05	2.02–04	–3.484	D	*LS*
39.	$2s2p$– $2s10s$	$^3P^o$–^3S		*185.60*	*67345*	– 606142.6	9–3	1.23+00	2.11–04	1.16–03	–2.721	C	4
				185.623	67416.3	– 606142.6	5–3	6.91–01	2.14–04	6.55–04	–2.970	D	4
				185.573	67272.3	– 606142.6	3–3	4.15–01	2.14–04	3.93–04	–3.192	D	4
				185.552	67209.2	– 606142.6	1–3	1.36–01	2.11–04	1.29–04	–3.675	D	*LS*
40.	$2s2p$– $2s10d$	$^3P^o$–^3D		*185.28*	*67345*	– 607064.2	9–15	5.57+00	4.78–03	2.62–02	–1.367	B–	4
				185.306	67416.3	– 607064.2	5–7	5.52+00	3.98–03	1.21–02	–1.701	B–	4
				185.257	67272.3	– 607064.2	3–5	4.15+00	3.56–03	6.51–03	–1.972	C+	4
				185.235	67209.2	– 607064.2	1–3	3.07+00	4.74–03	2.89–03	–2.324	C	4
				185.306	67416.3	– 607064.2	5–5	1.39+00	7.17–04	2.19–03	–2.446	C	*LS*
				185.257	67272.3	– 607064.2	3–3	2.32+00	1.19–03	2.19–03	–2.446	C	*LS*
				185.306	67416.3	– 607064.2	5–3	1.55–01	4.78–05	1.46–04	–3.622	D	*LS*
41.	$2s2p$– $2p6p$	$^3P^o$–^3P		*170.23*	*67345*	– 654791.3	9–9	8.53+00	3.71–03	1.87–02	–1.477	B–	2
				170.249	67416.3	– 654791.3	5–5	6.40+00	2.78–03	7.79–03	–1.857	C+	*LS*
				170.207	67272.3	– 654791.3	3–3	2.13+00	9.27–04	1.56–03	–2.556	C	*LS*
				170.249	67416.3	– 654791.3	5–3	3.55+00	9.27–04	2.60–03	–2.334	C	*LS*
				170.207	67272.3	– 654791.3	3–1	8.54+00	1.24–03	2.08–03	–2.431	C	*LS*
				170.207	67272.3	– 654791.3	3–5	2.13+00	1.55–03	2.60–03	–2.334	C	*LS*
				170.189	67209.2	– 654791.3	1–3	2.85+00	3.71–03	2.08–03	–2.431	C	*LS*
42.	$2s2p$– $2p7p$	$^3P^o$–^3P		*166.36*	*67345*	– 668460.9	9–9	5.43+00	2.25–03	1.11–02	–1.693	B–	2
				166.377	67416.3	– 668460.9	5–5	4.07+00	1.69–03	4.63–03	–2.073	C+	*LS*
				166.337	67272.3	– 668460.9	3–3	1.36+00	5.63–04	9.25–04	–2.772	D	*LS*
				166.377	67416.3	– 668460.9	5–3	2.26+00	5.63–04	1.54–03	–2.551	C	*LS*
				166.337	67272.3	– 668460.9	3–1	5.43+00	7.51–04	1.23–03	–2.647	C	*LS*
				166.337	67272.3	– 668460.9	3–5	1.36+00	9.38–04	1.54–03	–2.550	C	*LS*
				166.320	67209.2	– 668460.9	1–3	1.81+00	2.25–03	1.23–03	–2.647	C	*LS*

N IV: Allowed Transitions — Continued

No.	Transition Array	Mult.	λ_{air} (Å)	λ_{vac} (Å) or σ (cm^{-1})*	E_i (cm^{-1})	E_k (cm^{-1})	g_i–g_k	A_{ki} (10^8 s^{-1})	f_{ik}	S (at. u.)	log gf	Acc.	Source
43.	$2s2p$–$2p8p$	$^3P^o$–^3P		*163.95*	*67345*	– 677276.5	9–9	3.68+00	1.48–03	7.20–03	–1.875	C+	2
				163.972	67416.3	– 677276.5	5–5	2.76+00	1.11–03	3.00–03	–2.255	C+	*LS*
				163.933	67272.3	– 677276.5	3–3	9.20–01	3.71–04	6.00–04	–2.954	D	*LS*
				163.972	67416.3	– 677276.5	5–3	1.53+00	3.70–04	1.00–03	–2.732	C	*LS*
				163.933	67272.3	– 677276.5	3–1	3.68+00	4.94–04	8.00–04	–2.829	D	*LS*
				163.933	67272.3	– 677276.5	3–5	9.20–01	6.18–04	1.00–03	–2.732	C	*LS*
				163.916	67209.2	– 677276.5	1–3	1.23+00	1.48–03	8.00–04	–2.829	D	*LS*
44.	$2s2p$–$2p9p$	$^3P^o$–^3P		*162.36*	*67345*	– 683278.4	9–9	2.59+00	1.02–03	4.92–03	–2.036	C+	4
				162.374	67416.3	– 683278.4	5–5	1.94+00	7.67–04	2.05–03	–2.416	C	4
				162.336	67272.3	– 683278.4	3–3	6.47–01	2.56–04	4.10–04	–3.115	D	*LS*
				162.374	67416.3	– 683278.4	5–3	1.08+00	2.56–04	6.83–04	–2.894	D	*LS*
				162.336	67272.3	– 683278.4	3–1	2.59+00	3.41–04	5.46–04	–2.990	D	*LS*
				162.336	67272.3	– 683278.4	3–5	6.47–01	4.26–04	6.83–04	–2.893	D	*LS*
				162.319	67209.2	– 683278.4	1–3	8.63–01	1.02–03	5.46–04	–2.990	D	*LS*
45.	$2s2p$–$2p10p$	$^3P^o$–^3P		*161.24*	*67345*	– 687548.3	9–9	1.88+00	7.33–04	3.50–03	–2.180	C+	4
				161.256	67416.3	– 687548.3	5–5	1.42+00	5.53–04	1.47–03	–2.558	C	4
				161.219	67272.3	– 687548.3	3–3	4.71–01	1.83–04	2.92–04	–3.260	D	*LS*
				161.256	67416.3	– 687548.3	5–3	7.84–01	1.83–04	4.87–04	–3.038	D	*LS*
				161.219	67272.3	– 687548.3	3–1	1.88+00	2.44–04	3.89–04	–3.135	D	*LS*
				161.219	67272.3	– 687548.3	3–5	4.71–01	3.06–04	4.87–04	–3.038	D	*LS*
				161.202	67209.2	– 687548.3	1–3	6.28–01	7.34–04	3.89–04	–3.135	D	*LS*
46.	$2s2p$–$2p^2$	$^1P^o$–^3P											
			2219.61	2220.30	130693.9	– 175732.9	3–5	4.68–05	5.76–06	1.26–04	–4.762	B	1
			2225.78	2226.47	130693.9	– 175608.1	3–3	7.96–07	5.91–08	1.30–06	–6.751	B	1
			2229.38	2230.08	130693.9	– 175535.4	3–1	1.24–05	3.08–07	6.78–06	–6.034	B	1
47.		$^1P^o$–^1D		1718.55	130693.9	– 188882.5	3–5	2.321+00	1.713–01	2.907+00	–0.2892	A+	1
48.		$^1P^o$–^1S		955.334	130693.9	– 235369.3	3–1	2.919+01	1.331–01	1.256+00	–0.3986	A+	1
49.	$2s2p$–$2s3s$	$^1P^o$–^1S		387.356	130693.9	– 388854.6	3–1	2.55+01	1.91–02	7.32–02	–1.241	A	1
50.	$2s2p$–$2s3d$	$^1P^o$–^1D		335.047	130693.9	– 429159.6	3–5	1.845+02	5.176–01	1.713+00	0.1911	A+	1
51.	$2s2p$–$2p3p$	$^1P^o$–^1P		285.559	130693.9	– 480884.2	3–3	7.22+01	8.83–02	2.49–01	–0.577	B	2
52.	$2s2p$–$2s4s$	$^1P^o$–^1S		274.451	130693.9	– 495057.7	3–1	1.57+00	5.90–04	1.60–03	–2.752	C	2
53.	$2s2p$–$2p3p$	$^1P^o$–^1D		270.994	130693.9	– 499705.9	3–5	9.70+01	1.78–01	4.76–01	–0.272	B	2
54.	$2s2p$–$2s4d$	$^1P^o$–^1D		260.448	130693.9	– 514647.7	3–5	6.00+01	1.02–01	2.62–01	–0.515	B	2
55.	$2s2p$–$2p3p$	$^1P^o$–^1S		259.824	130693.9	– 515569.8	3–1	2.54+01	8.57–03	2.20–02	–1.590	B–	2

N IV: Allowed Transitions — Continued

No.	Transition Array	Mult.	λ_{air} (Å)	λ_{vac} (Å) or σ (cm^{-1})*	E_i (cm^{-1})	E_k (cm^{-1})	g_i-g_k	A_{ki} ($10^8 s^{-1}$)	f_{ik}	S (at. u.)	log gf	Acc.	Source
56.	$2s2p-$ $2s5s$	$^1P^o-^1S$		240.363	130693.9	– 546731.3	3–1	7.44+00	2.15–03	5.10–03	–2.191	C+	2
57.	$2s2p-$ $2s5d$	$^1P^o-^1D$		236.068	130693.9	– 554300.7	3–5	3.35+01	4.67–02	1.09–01	–0.854	B	2
58.	$2s2p-$ $2s6s$	$^1P^o-^1S$		227.026	130693.9	– 571172.1	3–1	2.25+00	5.80–04	1.30–03	–2.760	C	2
59.	$2s2p-$ $2s6d$	$^1P^o-^1D$		224.629	130693.9	– 575872.4	3–5	1.98+01	2.49–02	5.53–02	–1.126	B	2
60.	$2s2p-$ $2p4p$	$^1P^o-^1P$		220.280	130693.9	– 584661.6	3–3	3.03+01	2.20–02	4.79–02	–1.180	B	2
61.	$2s2p-$ $2s7s$	$^1P^o-^1S$		220.124	130693.9	– 584983.3	3–1	3.80–01	9.20–05	2.00–04	–3.559	D	2
62.	$2s2p-$ $2s7d$	$^1P^o-^1D$		218.250	130693.9	– 588884	3–5	1.54+01	1.83–02	3.94–02	–1.261	B	2
63.	$2s2p-$ $2p4p$	$^1P^o-^1D$		217.218	130693.9	– 591060.9	3–5	3.70+01	4.36–02	9.35–02	–0.884	B	2
64.		$^1P^o-^1S$		215.755	130693.9	– 594182.6	3–1	1.09+01	2.53–03	5.40–03	–2.119	C+	2
65.	$2s2p-$ $2s8s$	$^1P^o-^1S$		214.843	130693.9	– 596150.1	3–1	8.17–01	1.89–04	4.00–04	–3.248	D	2
66.	$2s2p-$ $2s8d$	$^1P^o-^1D$		214.291	130693.9	– 597349.1	3–5	2.88+00	3.31–03	7.00–03	–2.003	C+	2
67.	$2s2p-$ $2s9d$	$^1P^o-^1D$		211.679	130693.9	– 603107.3	3–5	5.01+00	5.61–03	1.17–02	–1.774	B–	4
68.	$2s2p-$ $2s10d$	$^1P^o-^1D$		209.842	130693.9	– 607242.9	3–5	3.70+00	4.07–03	8.44–03	–1.913	C+	4
69.	$2s2p-$ $2p5p$	$^1P^o-^1P$		200.340	130693.9	– 629845.3	3–3	1.54+01	9.25–03	1.83–02	–1.557	B–	2
70.	$2s2p-$ $2p6p$	$^1P^o-^1P$		191.228	130693.9	– 653629.9	3–3	8.79+00	4.82–03	9.10–03	–1.840	C+	2
71.	$2s2p-$ $2p7p$	$^1P^o-^1P$		186.218	130693.9	– 667698.9	3–3	5.44+00	2.83–03	5.20–03	–2.071	C+	2
72.	$2s2p-$ $2p8p$	$^1P^o-^1P$		183.146	130693.9	– 676706.4	3–3	3.63+00	1.82–03	3.30–03	–2.262	C+	2
73.	$2p^2-$ $2s3p$	$^3P-^3P^o$		*434.15*	*175669*	– *406005*	9–9	1.51–01	4.28–04	5.50–03	–2.415	C+	2
				434.235	175732.9	– 406022.8	5–5	1.13–01	3.21–04	2.29–03	–2.795	C	*LS*
				434.067	175608.1	– 405987.5	3–3	3.78–02	1.07–04	4.58–04	–3.494	D	*LS*
				434.302	175732.9	– 405987.5	5–3	6.30–02	1.07–04	7.64–04	–3.272	D	*LS*
				434.097	175608.1	– 405971.6	3–1	1.51–01	1.43–04	6.11–04	–3.369	D	*LS*
				434.000	175608.1	– 406022.8	3–5	3.79–02	1.78–04	7.64–04	–3.272	D	*LS*
				433.930	175535.4	– 405987.5	1–3	5.05–02	4.28–04	6.11–04	–3.369	D	*LS*

N IV: Allowed Transitions — Continued

No.	Transition Array	Mult.	λ_{air} (Å)	λ_{vac} (Å) or σ (cm^{-1})*	E_i (cm^{-1})	E_k (cm^{-1})	g_i–g_k	A_{ki} (10^8 s^{-1})	f_{ik}	S (at. u.)	log gf	Acc.	Source
74.	$2p^2$– $2p3s$	^3P–$^3P^o$		*345.08*	*175669*	– *465454*	9–9	6.58+01	1.17–01	1.20+00	0.024	B	2
				345.061	175732.9	– 465536.6	5–5	4.94+01	8.81–02	5.00–01	–0.356	B	*LS*
				345.110	175608.1	– 465371.0	3–3	1.64+01	2.94–02	1.00–01	–1.055	B	*LS*
				345.258	175732.9	– 465371.0	5–3	2.74+01	2.94–02	1.67–01	–0.833	B	*LS*
				345.204	175608.1	– 465291.8	3–1	6.57+01	3.91–02	1.33–01	–0.930	B	*LS*
				344.913	175608.1	– 465536.6	3–5	1.65+01	4.90–02	1.67–01	–0.833	B	*LS*
				345.023	175535.4	– 465371.0	1–3	2.19+01	1.17–01	1.33–01	–0.930	B	*LS*
75.	$2p^2$– $2s4p$	^3P–$^3P^o$		*304.86*	*175669*	– *503686*	9–9	4.63+01	6.45–02	5.82–01	–0.236	B	2
				304.927	175732.9	– 503680.4	5–5	3.47+01	4.83–02	2.43–01	–0.617	B	*LS*
				304.800	175608.1	– 503691.9	3–3	1.16+01	1.61–02	4.85–02	–1.316	B	*LS*
				304.916	175732.9	– 503691.9	5–3	1.93+01	1.61–02	8.09–02	–1.094	B	*LS*
				304.793	175608.1	– 503699.8	3–1	4.63+01	2.15–02	6.47–02	–1.191	B	*LS*
				304.811	175608.1	– 503680.4	3–5	1.16+01	2.69–02	8.09–02	–1.094	B	*LS*
				304.733	175535.4	– 503691.9	1–3	1.54+01	6.45–02	6.47–02	–1.190	B	*LS*
76.	$2p^2$– $2p3d$	^3P–$^3D^o$		*303.09*	*175669*	– *505601*	9–15	3.49+02	8.00–01	7.19+00	0.858	B	2
				303.124	175732.9	– 505630.6	5–7	3.49+02	6.72–01	3.35+00	0.526	B	*LS*
				303.049	175608.1	– 505588.2	3–5	2.62+02	6.00–01	1.80+00	0.256	B	*LS*
				303.013	175535.4	– 505554.0	1–3	1.94+02	8.01–01	7.99–01	–0.097	B	*LS*
				303.163	175732.9	– 505588.2	5–5	8.71+01	1.20–01	5.99–01	–0.222	B	*LS*
				303.080	175608.1	– 505554.0	3–3	1.45+02	2.00–01	5.99–01	–0.222	B	*LS*
				303.195	175732.9	– 505554.0	5–3	9.68+00	8.00–03	3.99–02	–1.398	B	*LS*
77.		^3P–$^3P^o$		*297.74*	*175669*	– *511535*	9–9	1.40+02	1.85–01	1.64+00	0.223	B	2
				297.817	175732.9	– 511509.3	5–5	1.05+02	1.39–01	6.82–01	–0.158	B	*LS*
				297.661	175608.1	– 511560.4	3–3	3.49+01	4.64–02	1.36–01	–0.857	B	*LS*
				297.772	175732.9	– 511560.4	5–3	5.81+01	4.64–02	2.27–01	–0.635	B	*LS*
				297.634	175608.1	– 511591.2	3–1	1.40+02	6.18–02	1.82–01	–0.732	B	*LS*
				297.707	175608.1	– 511509.3	3–5	3.49+01	7.73–02	2.27–01	–0.635	B	*LS*
				297.597	175535.4	– 511560.4	1–3	4.66+01	1.86–01	1.82–01	–0.732	B	*LS*
78.	$2p^2$– $2p4s$	^3P–$^3P^o$		*248.47*	*175669*	– *578127*	9–9	1.94+01	1.80–02	1.32–01	–0.791	B	2
				248.461	175732.9	– 578210.6	5–5	1.46+01	1.35–02	5.52–02	–1.171	B	*LS*
				248.486	175608.1	– 578045.4	3–3	4.86+00	4.50–03	1.10–02	–1.870	B–	*LS*
				248.563	175732.9	– 578045.4	5–3	8.09+00	4.49–03	1.84–02	–1.648	B–	*LS*
				248.540	175608.1	– 577957.8	3–1	1.94+01	5.99–03	1.47–02	–1.745	B–	*LS*
				248.384	175608.1	– 578210.6	3–5	4.86+00	7.50–03	1.84–02	–1.648	B–	*LS*
				248.441	175535.4	– 578045.4	1–3	6.48+00	1.80–02	1.47–02	–1.745	B–	*LS*
79.	$2p^2$– $2p4d$	^3P–$^3D^o$		*239.20*	*175669*	– *593736*	9–15	1.31+02	1.88–01	1.33+00	0.228	B	2
				239.212	175732.9	– 593772.1	5–7	1.31+02	1.58–01	6.21–01	–0.103	B	*LS*
				239.174	175608.1	– 593713.7	3–5	9.85+01	1.41–01	3.32–01	–0.374	B	*LS*
				239.146	175535.4	– 593690.0	1–3	7.30+01	1.88–01	1.48–01	–0.727	B	*LS*
				239.245	175732.9	– 593713.7	5–5	3.28+01	2.81–02	1.11–01	–0.852	B	*LS*
				239.188	175608.1	– 593690.0	3–3	5.47+01	4.69–02	1.11–01	–0.852	B	*LS*
				239.259	175732.9	– 593690.0	5–3	3.64+00	1.88–03	7.39–03	–2.028	C+	*LS*
80.		^3P–$^3P^o$		*238.75*	*175669*	– *594518*	9–9	5.23+01	4.47–02	3.16–01	–0.396	B	2
				238.802	175732.9	– 594489.9	5–5	3.92+01	3.35–02	1.32–01	–0.776	B	*LS*
				238.698	175608.1	– 594547.7	3–3	1.31+01	1.12–02	2.63–02	–1.475	B–	*LS*
				238.769	175732.9	– 594547.7	5–3	2.18+01	1.12–02	4.39–02	–1.253	B	*LS*
				238.683	175608.1	– 594573.8	3–1	5.23+01	1.49–02	3.51–02	–1.350	B	*LS*
				238.731	175608.1	– 594489.9	3–5	1.31+01	1.86–02	4.39–02	–1.253	B	*LS*
				238.656	175535.4	– 594547.7	1–3	1.74+01	4.47–02	3.51–02	–1.350	B	*LS*

N IV: Allowed Transitions — Continued

No.	Transition Array	Mult.	λ_{air} (Å)	λ_{vac} (Å) or σ (cm^{-1})*	E_i (cm^{-1})	E_k (cm^{-1})	g_i–g_k	A_{ki} (10^8 s^{-1})	f_{ik}	S (at. u.)	log gf	Acc.	Source
81.	$2p^2$– $2p5d$	^{3}P–^{3}D°		*218.08*	*175669*	– *634216*	9–15	6.37+01	7.57–02	4.89–01	–0.167	B	2
				218.088	175732.9	– 634263.4	5–7	6.37+01	6.36–02	2.28–01	–0.498	B	*LS*
				218.067	175608.1	– 634182.8	3–5	4.78+01	5.68–02	1.22–01	–0.769	B	*LS*
				218.044	175535.4	– 634158.4	1–3	3.54+01	7.57–02	5.43–02	–1.121	B	*LS*
				218.126	175732.9	– 634182.8	5–5	1.59+01	1.14–02	4.08–02	–1.246	B	*LS*
				218.079	175608.1	– 634158.4	3–3	2.65+01	1.89–02	4.08–02	–1.246	B	*LS*
				218.138	175732.9	– 634158.4	5–3	1.77+00	7.57–04	2.72–03	–2.422	C	*LS*
82.	$2p^2$– $2p6d$	^{3}P–^{3}D°		*208.14*	*175669*	– *656106*	9–15	3.57+01	3.86–02	2.38–01	–0.459	B	2
				208.150	175732.9	– 656155.7	5–7	3.57+01	3.24–02	1.11–01	–0.790	B	*LS*
				208.131	175608.1	– 656074.7	3–5	2.68+01	2.90–02	5.96–02	–1.061	B	*LS*
				208.113	175535.4	– 656043.6	1–3	1.98+01	3.86–02	2.65–02	–1.413	B–	*LS*
				208.185	175732.9	– 656074.7	5–5	8.91+00	5.79–03	1.99–02	–1.538	B–	*LS*
				208.144	175608.1	– 656043.6	3–3	1.49+01	9.66–03	1.99–02	–1.538	B–	*LS*
				208.199	175732.9	– 656043.6	5–3	9.92–01	3.86–04	1.32–03	–2.714	C	*LS*
83.	$2p^2$– $2p7d$	^{3}P–^{3}D°		*202.57*	*175669*	– 669328.5	9–15	2.20+01	2.25–02	1.35–01	–0.693	B	2
				202.595	175732.9	– 669328.5	5–7	2.20+01	1.89–02	6.31–02	–1.024	B	*LS*
				202.544	175608.1	– 669328.5	3–5	1.65+01	1.69–02	3.38–02	–1.295	B	*LS*
				202.514	175535.4	– 669328.5	1–3	1.22+01	2.25–02	1.50–02	–1.647	B–	*LS*
				202.595	175732.9	– 669328.5	5–5	5.49+00	3.38–03	1.13–02	–1.772	B–	*LS*
				202.544	175608.1	– 669328.5	3–3	9.16+00	5.64–03	1.13–02	–1.772	B–	*LS*
				202.595	175732.9	– 669328.5	5–3	6.10–01	2.25–04	7.52–04	–2.948	D	*LS*
84.	$2p^2$– $2p8d$	^{3}P–^{3}D°		*199.13*	*175669*	– 677844.3	9–15	1.45+01	1.44–02	8.50–02	–0.887	B	2
				199.159	175732.9	– 677844.3	5–7	1.45+01	1.21–02	3.97–02	–1.218	B	*LS*
				199.110	175608.1	– 677844.3	3–5	1.09+01	1.08–02	2.13–02	–1.489	B–	*LS*
				199.081	175535.4	– 677844.3	1–3	8.08+00	1.44–02	9.44–03	–1.841	C+	*LS*
				199.159	175732.9	– 677844.3	5–5	3.63+00	2.16–03	7.08–03	–1.966	C+	*LS*
				199.110	175608.1	– 677844.3	3–3	6.06+00	3.60–03	7.08–03	–1.966	C+	*LS*
				199.159	175732.9	– 677844.3	5–3	4.04–01	1.44–04	4.72–04	–3.143	D	*LS*
85.	$2p^2$– $2p9d$	^{3}P–^{3}D°		*196.84*	*175669*	– 683692.6	9–15	1.01+01	9.77–03	5.70–02	–1.056	B	4
				196.866	175732.9	– 683692.6	5–7	1.01+01	8.20–03	2.66–02	–1.387	B–	4
				196.818	175608.1	– 683692.6	3–5	7.57+00	7.33–03	1.42–02	–1.658	B–	*LS*
				196.789	175535.4	– 683692.6	1–3	5.61+00	9.77–03	6.33–03	–2.010	C+	*LS*
				196.866	175732.9	– 683692.6	5–5	2.52+00	1.47–03	4.75–03	–2.135	C+	*LS*
				196.818	175608.1	– 683692.6	3–3	4.21+00	2.44–03	4.75–03	–2.135	C+	*LS*
				196.866	175732.9	– 683692.6	5–3	2.80–01	9.77–05	3.16–04	–3.311	D	*LS*
86.	$2p^2$– $2p10d$	^{3}P–^{3}D°		*195.23*	*175669*	– 687875.8	9–15	7.31+00	6.96–03	4.03–02	–1.203	B	4
				195.258	175732.9	– 687875.8	5–7	7.31+00	5.85–03	1.88–02	–1.534	B–	4
				195.210	175608.1	– 687875.8	3–5	5.49+00	5.22–03	1.01–02	–1.805	B–	*LS*
				195.183	175535.4	– 687875.8	1–3	4.06+00	6.97–03	4.48–03	–2.157	C+	*LS*
				195.258	175732.9	– 687875.8	5–5	1.83+00	1.04–03	3.36–03	–2.282	C+	*LS*
				195.210	175608.1	– 687875.8	3–3	3.05+00	1.74–03	3.36–03	–2.282	C+	*LS*
				195.258	175732.9	– 687875.8	5–3	2.03–01	6.96–05	2.24–04	–3.458	D	*LS*
87.	$2p^2$– $2s3p$	^{1}D–^{1}P°		463.736	188882.5	– 404522.4	5–3	1.06+01	2.04–02	1.56–01	–0.991	B	2
88.	$2p^2$– $2p3s$	^{1}D–^{1}P°		351.931	188882.5	– 473029.3	5–3	5.33+01	5.94–02	3.44–01	–0.527	B	2

N IV: Allowed Transitions — Continued

No.	Transition Array	Mult.	λ_{air} (Å)	λ_{vac} (Å) or σ (cm^{-1})*	E_i (cm^{-1})	E_k (cm^{-1})	g_i-g_k	A_{ki} (10^8 s^{-1})	f_{ik}	S (at. u.)	log gf	Acc.	Source
89.	$2p^2-2p3d$	$^1D-^1D^o$		323.177	188882.5	498310.3	5–5	1.40+02	2.20–01	1.17+00	0.041	B	2
90.		$^1D-^1F^o$		315.058	188882.5	506284.8	5–7	1.36+02	2.84–01	1.47+00	0.153	B	2
91.	$2p^2-2s4p$	$^1D-^1P^o$		314.322	188882.5	507027.9	5–3	1.23–02	1.09–05	5.65–05	–4.263	D	4
92.	$2p^2-2p3d$	$^1D-^1P^o$		303.280	188882.5	518610.5	5–3	8.72+00	7.21–03	3.60–02	–1.443	B	2
93.	$2p^2-2s4f$	$^1D-^1F^o$		300.318	188882.5	521862.8	5–7	2.40+02	4.55–01	2.25+00	0.357	B	2
94.	$2p^2-2s5p$	$^1D-^1P^o$		276.740	188882.5	550232.6	5–3	3.82–01	2.63–04	1.20–03	–2.880	C	2
95.	$2p^2-2s5f$	$^1D-^1F^o$		273.140	188882.5	554995.7	5–7	1.35+00	2.11–03	9.50–03	–1.976	C+	2
96.	$2p^2-2s6p$	$^1D-^1P^o$		260.519	188882.5	572731.1	5–3	3.25+00	1.98–03	8.50–03	–2.004	C+	2
97.	$2p^2-2s6f$	$^1D-^1F^o$		258.320	188882.5	575999.3	5–7	5.54–01	7.76–04	3.30–03	–2.411	C+	2
98.	$2p^2-2p4s$	$^1D-^1P^o$		255.144	188882.5	580817.5	5–3	1.78+01	1.04–02	4.37–02	–1.284	B	2
99.	$2p^2-2s7p$	$^1D-^1P^o$		250.566	188882.5	587979.4	5–3	9.45–01	5.33–04	2.20–03	–2.574	C	2
100.	$2p^2-2p4d$	$^1D-^1D^o$		250.121	188882.5	588689	5–5	4.56+01	4.28–02	1.76–01	–0.670	B	2
101.		$^1D-^1F^o$		248.654	188882.5	591047.8	5–7	6.41+01	8.31–02	3.40–01	–0.381	B	2
102.		$^1D-^1P^o$		244.107	188882.5	598538.2	5–3	3.58+00	1.92–03	7.70–03	–2.019	C+	2
103.	$2p^2-2s3p$	$^1S-^1P^o$		591.180	235369.3	404522.4	1–3	6.67–01	1.05–02	2.04–02	–1.980	B–	2
104.	$2p^2-2p3s$	$^1S-^1P^o$		420.769	235369.3	473029.3	1–3	1.99+01	1.59–01	2.20–01	–0.799	B	2
105.	$2p^2-2s4p$	$^1S-^1P^o$		368.109	235369.3	507027.9	1–3	2.19+00	1.34–02	1.62–02	–1.874	B–	2
106.	$2p^2-2p3d$	$^1S-^1P^o$		353.056	235369.3	518610.5	1–3	1.80+02	1.01+00	1.17+00	0.003	B	2
107.	$2p^2-2s5p$	$^1S-^1P^o$		317.598	235369.3	550232.6	1–3	2.02+00	9.18–03	9.60–03	–2.037	C+	2
108.	$2p^2-2s6p$	$^1S-^1P^o$		296.418	235369.3	572731.1	1–3	7.78–02	3.07–04	3.00–04	–3.512	D	2

N IV: Allowed Transitions — Continued

No.	Transition Array	Mult.	λ_{air} (Å)	λ_{vac} (Å) or σ (cm^{-1})*	E_i (cm^{-1})	E_k (cm^{-1})	g_i–g_k	A_{ki} (10^8 s^{-1})	f_{ik}	S (at. u.)	log gf	Acc.	Source
109.	$2p^2$–$2p4s$	^{1}S–^{1}P°		289.479	235369.3	– 580817.5	1–3	7.16+00	2.70–02	2.57–02	–1.569	B–	2
110.	$2p^2$–$2s7p$	^{1}S–^{1}P°		283.599	235369.3	– 587979.4	1–3	7.11–01	2.57–03	2.40–03	–2.590	C	2
111.	$2p^2$–$2p4d$	^{1}S–^{1}P°		275.354	235369.3	– 598538.2	1–3	6.76+01	2.31–01	2.09–01	–0.637	B	2
112.	$2s3s$–$2s3p$	^{3}S–^{3}P°	*3480.8*	*3481.8*	377284.8	– *406005*	3–9	1.06+00	5.77–01	1.98+01	0.238	B	2,3
			3478.72	3479.71	377284.8	– 406022.8	3–5	1.06+00	3.21–01	1.10+01	–0.017	B	*LS*
			3483.00	3483.99	377284.8	– 405987.5	3–3	1.06+00	1.92–01	6.61+00	–0.239	B	*LS*
			3484.93	3485.92	377284.8	– 405971.6	3–1	1.06+00	6.41–02	2.20+00	–0.716	B	*LS*
113.	$2s3s$–$2p3s$	^{3}S–^{3}P°		*1134.2*	377284.8	– *465454*	3–9	4.06+00	2.35–01	2.63+00	–0.152	B	2
				1133.12	377284.8	– 465536.6	3–5	4.07+00	1.31–01	1.46+00	–0.407	B	*LS*
				1135.25	377284.8	– 465371.0	3–3	4.05+00	7.82–02	8.77–01	–0.630	B	*LS*
				1136.27	377284.8	– 465291.8	3–1	4.04+00	2.61–02	2.92–01	–1.107	B	*LS*
114.	$2s3s$–$2s4p$	^{3}S–^{3}P°		*791.13*	377284.8	– *503686*	3–9	4.28+00	1.21–01	9.42–01	–0.442	B	2
				791.167	377284.8	– 503680.4	3–5	4.29+00	6.70–02	5.24–01	–0.697	B	*LS*
				791.095	377284.8	– 503691.9	3–3	4.28+00	4.02–02	3.14–01	–0.919	B	*LS*
				791.045	377284.8	– 503699.8	3–1	4.28+00	1.34–02	1.05–01	–1.396	B	*LS*
115.	$2s3s$–$2p3d$	^{3}S–^{3}P°		*744.88*	377284.8	– *511535*	3–9	5.57–01	1.39–02	1.02–01	–1.380	B	2
				745.020	377284.8	– 511509.3	3–5	5.57–01	7.72–03	5.68–02	–1.635	B	*LS*
				744.737	377284.8	– 511560.4	3–3	5.58–01	4.64–03	3.41–02	–1.857	B	*LS*
				744.566	377284.8	– 511591.2	3–1	5.58–01	1.55–03	1.14–02	–2.334	B–	*LS*
116.	$2s3s$–$2p4s$	^{3}S–^{3}P°		*497.90*	377284.8	– *578127*	3–9	5.79–03	6.46–05	3.18–04	–3.713	D	4
				497.696	377284.8	– 578210.6	3–5	5.80–03	3.59–05	1.76–04	–3.968	D	4
				498.106	377284.8	– 578045.4	3–3	5.79–03	2.15–05	1.06–04	–4.190	D	4
				498.323	377284.8	– 577957.8	3–1	5.78–03	7.18–06	3.53–05	–4.667	D	4
117.	$2s3s$–$2p4d$	^{3}S–^{3}P°		*460.33*	377284.8	– *594518*	3–9	6.65–01	6.33–03	2.88–02	–1.721	B–	2
				460.394	377284.8	– 594489.9	3–5	6.64–01	3.52–03	1.60–02	–1.976	B–	*LS*
				460.272	377284.8	– 594547.7	3–3	6.65–01	2.11–03	9.60–03	–2.198	C+	*LS*
				460.217	377284.8	– 594573.8	3–1	6.65–01	7.04–04	3.20–03	–2.675	C+	*LS*
118.	$2s3s$–$2s3p$	^{1}S–^{1}P°	6380.75	6382.52	388854.6	– 404522.4	1–3	1.42–01	2.60–01	5.46+00	–0.585	B	2
119.	$2s3s$–$2p3s$	^{1}S–^{1}P°		1188.01	388854.6	– 473029.3	1–3	9.49+00	6.03–01	2.36+00	–0.220	B	2
120.	$2s3s$–$2s4p$	^{1}S–^{1}P°		846.215	388854.6	– 507027.9	1–3	1.63+00	5.25–02	1.46–01	–1.280	B	2
121.	$2s3s$–$2p3d$	^{1}S–^{1}P°		770.678	388854.6	– 518610.5	1–3	1.53+00	4.08–02	1.04–01	–1.389	B	2
122.	$2s3s$–$2s5p$	^{1}S–^{1}P°		619.663	388854.6	– 550232.6	1–3	3.05+00	5.26–02	1.07–01	–1.279	B	2

N IV: Allowed Transitions — Continued

No.	Transition Array	Mult.	λ_{air} (Å)	λ_{vac} (Å) or σ (cm^{-1})*	E_i (cm^{-1})	E_k (cm^{-1})	g_i-g_k	A_{ki} (10^8 s^{-1})	f_{ik}	S (at. u.)	log gf	Acc.	Source
123.	$2s3s-2p4s$	$^1S-^1P°$		520.934	388854.6	– 580817.5	1–3	1.67–01	2.04–03	3.50–03	–2.690	C+	2
124.	$2s3s-2p4d$	$^1S-^1P°$		476.909	388854.6	– 598538.2	1–3	6.85–02	7.01–04	1.10–03	–3.155	C	2
125.	$2s3p-2s3d$	$^1P°-^1D$	4057.76	4058.90	404522.4	– 429159.6	3–5	6.62–01	2.72–01	1.09+01	–0.088	B	2
126.	$2s3p-2p3p$	$^1P°-^1P$		1309.56	404522.4	– 480884.2	3–3	6.95+00	1.79–01	2.31+00	–0.271	B	2
127.	$2s3p-2s4s$	$^1P°-^1S$		1104.54	404522.4	– 495057.7	3–1	1.46+01	8.88–02	9.69–01	–0.574	B	2
128.	$2s3p-2p3p$	$^1P°-^1D$		1050.60	404522.4	– 499705.9	3–5	2.68–01	7.38–03	7.66–02	–1.655	B	2
129.	$2s3p-2s4d$	$^1P°-^1D$		908.057	404522.4	– 514647.7	3–5	1.41+01	2.90–01	2.60+00	–0.060	B	2
130.	$2s3p-2p3p$	$^1P°-^1S$		900.516	404522.4	– 515569.8	3–1	2.08–01	8.43–04	7.50–03	–2.597	C+	2
131.	$2s3p-2s5s$	$^1P°-^1S$		703.191	404522.4	– 546731.3	3–1	4.87+00	1.20–02	8.35–02	–1.443	B	2
132.	$2s3p-2s5d$	$^1P°-^1D$		667.653	404522.4	– 554300.7	3–5	8.58+00	9.55–02	6.30–01	–0.543	B	2
133.	$2s3p-2s6s$	$^1P°-^1S$		600.061	404522.4	– 571172.1	3–1	3.20+00	5.75–03	3.41–02	–1.763	B	2
134.	$2s3p-2s6d$	$^1P°-^1D$		583.601	404522.4	– 575872.4	3–5	5.22+00	4.44–02	2.56–01	–0.875	B	2
135.	$2s3p-2p4p$	$^1P°-^1P$		555.126	404522.4	– 584661.6	3–3	8.29–02	3.83–04	2.10–03	–2.940	C	2
136.	$2s3p-2s7s$	$^1P°-^1S$		554.137	404522.4	– 584983.3	3–1	2.11+00	3.23–03	1.77–02	–2.013	B–	2
137.	$2s3p-2s7d$	$^1P°-^1D$		542.412	404522.4	– 588884	3–5	3.29+00	2.42–02	1.30–01	–1.139	B	2
138.	$2s3p-2p4p$	$^1P°-^1D$		536.082	404522.4	– 591060.9	3–5	4.73–02	3.40–04	1.80–03	–2.991	D	2
139.		$^1P°-^1S$		527.259	404522.4	– 594182.6	3–1	2.98–02	4.14–05	2.16–04	–3.906	D	4
140.	$2s3p-2s8s$	$^1P°-^1S$		521.845	404522.4	– 596150.1	3–1	1.47+00	2.00–03	1.03–02	–2.222	C	2
141.	$2s3p-2s8d$	$^1P°-^1D$		518.600	404522.4	– 597349.1	3–5	9.99–01	6.72–03	3.44–02	–1.696	B	2

N IV: Allowed Transitions — Continued

No.	Transition Array	Mult.	λ_{air} (Å)	λ_{vac} (Å) or σ (cm^{-1})*	E_i (cm^{-1})	E_k (cm^{-1})	g_i-g_k	A_{ki} (10^8 s^{-1})	f_{ik}	S (at. u.)	log gf	Acc.	Source
142.	$2s3p-$ $2s3d$	$^3P^o-^3D$	*7116.8*	*7118.7*	*406005*	– *420053*	9–15	1.12–01	1.42–01	3.00+01	0.107	B	2,3
			7122.98	7124.94	406022.8	– 420058.0	5–7	1.12–01	1.19–01	1.40+01	–0.224	B	*LS*
			7109.35	7111.31	405987.5	– 420049.6	3–5	8.46–02	1.07–01	7.50+00	–0.494	B	*LS*
			7103.24	7105.20	405971.6	– 420045.8	1–3	6.28–02	1.43–01	3.33+00	–0.846	B	*LS*
			7127.25	7129.21	406022.8	– 420049.6	5–5	2.80–02	2.13–02	2.50+00	–0.972	B	*LS*
			7111.28	7113.24	405987.5	– 420045.8	3–3	4.70–02	3.56–02	2.50+00	–0.971	B	*LS*
			7127.25	7129.21	406022.8	– 420049.6	5–3	3.11–03	1.42–03	1.67–01	–2.149	B	*LS*
143.	$2s3p-$ $2p3p$	$^3P^o-^3D$		*1271.6*	*406005*	– *484646*	9–15	4.08+00	1.65–01	6.21+00	0.171	B	2
				1270.27	406022.8	– 484746.2	5–7	4.09+00	1.39–01	2.90+00	–0.160	B	*LS*
				1272.14	405987.5	– 484594.9	3–5	3.05+00	1.23–01	1.55+00	–0.431	B	*LS*
				1273.45	405971.6	– 484498.2	1–3	2.25+00	1.64–01	6.90–01	–0.784	B	*LS*
				1272.72	406022.8	– 484594.9	5–5	1.02+00	2.47–02	5.17–01	–0.909	B	*LS*
				1273.71	405987.5	– 484498.2	3–3	1.69+00	4.11–02	5.17–01	–0.909	B	*LS*
				1274.28	406022.8	– 484498.2	5–3	1.13–01	1.64–03	3.45–02	–2.085	B	*LS*
144.	$2s3p-$ $2s4s$	$^3P^o-^3S$		*1225.5*	*406005*	– 487607.4	9–3	2.70+00	2.03–02	7.37–01	–0.738	B	2
				1225.72	406022.8	– 487607.4	5–3	1.50+00	2.03–02	4.09–01	–0.994	B	*LS*
				1225.19	405987.5	– 487607.4	3–3	9.02–01	2.03–02	2.46–01	–1.216	B	*LS*
				1224.95	405971.6	– 487607.4	1–3	3.01–01	2.03–02	8.19–02	–1.693	B	*LS*
145.	$2s3p-$ $2p3p$	$^3P^o-^3P$		*1131.9*	*406005*	– *494355*	9–9	4.66+00	8.94–02	3.00+00	–0.094	B	2
				1131.49	406022.8	– 494402.0	5–5	3.50+00	6.71–02	1.25+00	–0.474	B	*LS*
				1132.22	405987.5	– 494309.2	3–3	1.16+00	2.23–02	2.50–01	–1.174	B	*LS*
				1132.68	406022.8	– 494309.2	5–3	1.94+00	2.23–02	4.16–01	–0.952	B	*LS*
				1132.94	405987.5	– 494253.1	3–1	4.64+00	2.98–02	3.33–01	–1.049	B	*LS*
				1131.04	405987.5	– 494402.0	3–5	1.17+00	3.73–02	4.16–01	–0.951	B	*LS*
				1132.02	405971.6	– 494309.2	1–3	1.55+00	8.94–02	3.33–01	–1.049	B	*LS*
146.		$^3P^o-^3S$		*1086.5*	*406005*	– 498045.5	9–3	2.06+01	1.21–01	3.91+00	0.038	B	2
				1086.69	406022.8	– 498045.5	5–3	1.14+01	1.21–01	2.17+00	–0.217	B	*LS*
				1086.27	405987.5	– 498045.5	3–3	6.86+00	1.21–01	1.30+00	–0.439	B	*LS*
				1086.08	405971.6	– 498045.5	1–3	2.29+00	1.21–01	4.34–01	–0.916	B	*LS*
147.	$2s3p-$ $2s4d$	$^3P^o-^3D$		*948.40*	*406005*	– *511446*	9–15	1.79+01	4.01–01	1.13+01	0.558	B	2
				948.540	406022.8	– 511448	5–7	1.78+01	3.37–01	5.26+00	0.227	B	*LS*
				948.240	405987.5	– 511446	3–5	1.34+01	3.01–01	2.82+00	–0.044	B	*LS*
				948.151	405971.6	– 511440	1–3	9.93+00	4.01–01	1.25+00	–0.396	B	*LS*
				948.558	406022.8	– 511446	5–5	4.46+00	6.02–02	9.40–01	–0.522	B	*LS*
				948.294	405987.5	– 511440	3–3	7.44+00	1.00–01	9.40–01	–0.521	B	*LS*
				948.612	406022.8	– 511440	5–3	4.96–01	4.01–03	6.26–02	–1.698	B	*LS*
148.	$2s3p-$ $2s5s$	$^3P^o-^3S$		*720.42*	*406005*	– 544813.4	9–3	5.88+00	1.53–02	3.26–01	–0.862	B	2
				720.510	406022.8	– 544813.4	5–3	3.27+00	1.53–02	1.81–01	–1.118	B	*LS*
				720.327	405987.5	– 544813.4	3–3	1.96+00	1.53–02	1.09–01	–1.339	B	*LS*
				720.244	405971.6	– 544813.4	1–3	6.54–01	1.53–02	3.62–02	–1.816	B	*LS*
149.	$2s3d-$ $2p3d$	$^3D-^3F^o$		*1325.3*	*420053*	– *495509*	15–21	3.96–01	1.46–02	9.56–01	–0.659	B	2
				1324.02	420058.0	– 495585.7	7–9	3.98–01	1.34–02	4.10–01	–1.027	B	*LS*
				1325.68	420049.6	– 495482.6	5–7	3.52–01	1.30–02	2.83–01	–1.188	B	*LS*
				1326.96	420045.8	– 495406.2	3–5	3.32–01	1.46–02	1.91–01	–1.359	B	*LS*
				1325.83	420058.0	– 495482.6	7–7	4.41–02	1.16–03	3.55–02	–2.089	B	*LS*
				1327.02	420049.6	– 495406.2	5–5	6.16–02	1.63–03	3.55–02	–2.090	B	*LS*
				1327.17	420058.0	– 495406.2	7–5	1.74–03	3.28–05	1.00–03	–3.640	C	*LS*

N IV: Allowed Transitions — Continued

No.	Transition Array	Mult.	λ_{air} (Å)	λ_{vac} (Å) or σ (cm^{-1})*	E_i (cm^{-1})	E_k (cm^{-1})	g_i–g_k	A_{ki} (10^8 s^{-1})	f_{ik}	S (at. u.)	log gf	Acc.	Source
150.	2s3d–2s4p	^{3}D–^{3}P°		*1195.7*	*420053*	– *503686*	15–9	5.43+00	6.98–02	4.12+00	0.020	B	2
				1195.85	420058.0	– 503680.4	7–5	4.56+00	6.98–02	1.92+00	–0.311	B	*LS*
				1195.57	420049.6	– 503691.9	5–3	4.07+00	5.24–02	1.03+00	–0.582	B	*LS*
				1195.40	420045.8	– 503699.8	3–1	5.43+00	3.88–02	4.58–01	–0.934	B	*LS*
				1195.73	420049.6	– 503680.4	5–5	8.14–01	1.75–02	3.44–01	–1.059	B	*LS*
				1195.51	420045.8	– 503691.9	3–3	1.36+00	2.91–02	3.44–01	–1.059	B	*LS*
				1195.68	420045.8	– 503680.4	3–5	5.43–02	1.94–03	2.29–02	–2.235	B–	*LS*
151.	2s3d–2p3d	^{3}D–^{3}D°		*1168.9*	*420053*	– *505601*	15–15	5.01+00	1.03–01	5.93+00	0.187	B	2
				1168.60	420058.0	– 505630.6	7–7	4.46+00	9.12–02	2.46+00	–0.195	B	*LS*
				1169.06	420049.6	– 505588.2	5–5	3.49+00	7.14–02	1.37+00	–0.447	B	*LS*
				1169.48	420045.8	– 505554.0	3–3	3.75+00	7.69–02	8.89–01	–0.637	B	*LS*
				1169.18	420058.0	– 505588.2	7–5	7.81–01	1.14–02	3.08–01	–1.097	B	*LS*
				1169.53	420049.6	– 505554.0	5–3	1.25+00	1.54–02	2.96–01	–1.114	B	*LS*
				1168.48	420049.6	– 505630.6	5–7	5.59–01	1.60–02	3.08–01	–1.096	B	*LS*
				1169.01	420045.8	– 505588.2	3–5	7.51–01	2.57–02	2.96–01	–1.114	B	*LS*
152.		^{1}D–^{1}D°		1446.12	429159.6	– 498310.3	5–5	2.00+00	6.28–02	1.49+00	–0.503	B	2
153.		^{1}D–^{1}F°		1296.59	429159.6	– 506284.8	5–7	1.87+01	6.61–01	1.41+01	0.519	B	2
154.	2s3d–2s4p	^{1}D–^{1}P°		1284.22	429159.6	– 507027.9	5–3	4.04+00	5.99–02	1.27+00	–0.524	B	2
155.	2s3d–2s4f	^{1}D–^{1}F°		1078.71	429159.6	– 521862.8	5–7	1.22+01	2.99–01	5.31+00	0.175	B	2
156.	2p3s–2p3p	^{3}P°–^{3}D	*5209.1*	*5210.5*	*465454*	– *484646*	9–15	3.54–01	2.40–01	3.70+01	0.334	B	2
			5204.28	5205.73	465536.6	– 484746.2	5–7	3.55–01	2.02–01	1.73+01	0.004	B	*LS*
			5200.41	5201.86	465371.0	– 484594.9	3–5	2.67–01	1.80–01	9.26+00	–0.267	B	*LS*
			5205.15	5206.60	465291.8	– 484498.2	1–3	1.97–01	2.40–01	4.11+00	–0.620	B	*LS*
			5245.60	5247.06	465536.6	– 484594.9	5–5	8.66–02	3.57–02	3.09+00	–0.748	B	*LS*
			5226.70	5228.16	465371.0	– 484498.2	3–3	1.46–01	5.98–02	3.09+00	–0.746	B	*LS*
			5272.35	5273.82	465536.6	– 484498.2	5–3	9.48–03	2.37–03	2.06–01	–1.926	B	*LS*
157.		^{3}P°–^{3}P	*3459.2*	*3460.2*	*465454*	– *494355*	9–9	1.36+00	2.45–01	2.51+01	0.343	B	2
			3463.36	3464.36	465536.6	– 494402.0	5–5	1.02+00	1.83–01	1.05+01	–0.038	B	*LS*
			3454.65	3455.64	465371.0	– 494309.2	3–3	3.42–01	6.13–02	2.09+00	–0.736	B	*LS*
			3474.53	3475.53	465536.6	– 494309.2	5–3	5.61–01	6.09–02	3.48+00	–0.516	B	*LS*
			3461.36	3462.35	465371.0	– 494253.1	3–1	1.36+00	8.15–02	2.79+00	–0.612	B	*LS*
			3443.61	3444.59	465371.0	– 494402.0	3–5	3.46–01	1.02–01	3.48+00	–0.512	B	*LS*
			3445.22	3446.21	465291.8	– 494309.2	1–3	4.60–01	2.46–01	2.79+00	–0.609	B	*LS*
158.		^{3}P°–^{3}S	*3067.4*	*3068.3*	*465454*	– 498045.5	9–3	1.18+00	5.53–02	5.02+00	–0.303	B	2
			3075.19	3076.08	465536.6	– 498045.5	5–3	6.48–01	5.51–02	2.79+00	–0.560	B	*LS*
			3059.60	3060.49	465371.0	– 498045.5	3–3	3.95–01	5.54–02	1.67+00	–0.779	B	*LS*
			3052.20	3053.09	465291.8	– 498045.5	1–3	1.33–01	5.55–02	5.58–01	–1.255	B	*LS*
159.		^{1}P°–^{1}P	12727.4	7854.9 cm^{-1}	473029.3	– 480884.2	3–3	1.81–02	4.38–02	5.51+00	–0.881	B	2
160.	2p3s–2s4s	^{1}P°–^{1}S	4538.32	4539.59	473029.3	– 495057.7	3–1	8.44–01	8.68–02	3.89+00	–0.584	B	2

N IV: Allowed Transitions — Continued

No.	Transition Array	Mult.	λ_{air} (Å)	λ_{vac} (Å) or σ (cm^{-1})*	E_i (cm^{-1})	E_k (cm^{-1})	g_i–g_k	A_{ki} (10^8 s^{-1})	f_{ik}	S (at. u.)	log gf	Acc.	Source
161.	$2p3s$–$2p3p$	^{1}P°–^{1}D	3747.54	3748.60	473029.3	– 499705.9	3–5	9.92–01	3.48–01	1.29+01	0.019	B	2
162.	$2p3s$–$2s4d$	^{1}P°–^{1}D	2402.05	2402.78	473029.3	– 514647.7	3–5	1.56+00	2.25–01	5.35+00	–0.170	B	2
163.	$2p3p$–$2p3d$	^{1}P–^{1}D°	5736.93	5738.52	480884.2	– 498310.3	3–5	1.84–01	1.51–01	8.56+00	–0.344	B	2
164.	$2p3p$–$2s4p$	^{1}P–^{1}P°	3823.93	3825.01	480884.2	– 507027.9	3–3	1.46–01	3.20–02	1.21+00	–1.017	B	2
165.	$2p3p$–$2p3d$	^{1}P–^{1}P°	2649.88	2650.67	480884.2	– 518610.5	3–3	1.07+00	1.13–01	2.95+00	–0.471	B	2
166.		^{3}D–^{3}F°	*9203.5*	*9206.0*	*484646*	*– 495509*	15–21	4.98–02	8.85–02	4.02+01	0.123	B	2
			9222.99	9225.52	484746.2	– 495585.7	7–9	4.95–02	8.11–02	1.72+01	–0.246	B	*LS*
			9182.16	9184.68	484594.9	– 495482.6	5–7	4.45–02	7.88–02	1.19+01	–0.404	B	*LS*
			9165.07	9167.58	484498.2	– 495406.2	3–5	4.23–02	8.89–02	8.04+00	–0.574	B	*LS*
			9311.55	9314.11	484746.2	– 495482.6	7–7	5.36–03	6.96–03	1.49+00	–1.312	B	*LS*
			9247.04	9249.58	484594.9	– 495406.2	5–5	7.66–03	9.82–03	1.49+00	–1.309	B	*LS*
			9378.29	9380.86	484746.2	– 495406.2	7–5	2.07–04	1.95–04	4.21–02	–2.865	B	*LS*
167.		^{3}D–^{3}D°	*4770.8*	*4772.1*	*484646*	*– 505601*	15–15	9.99–02	3.41–02	8.03+00	–0.291	B	2
			4786.92	4788.26	484746.2	– 505630.6	7–7	8.79–02	3.02–02	3.33+00	–0.675	B	*LS*
			4762.09	4763.42	484594.9	– 505588.2	5–5	6.99–02	2.38–02	1.86+00	–0.925	B	*LS*
			4747.96	4749.29	484498.2	– 505554.0	3–3	7.60–02	2.57–02	1.20+00	–1.113	B	*LS*
			4796.66	4798.00	484746.2	– 505588.2	7–5	1.53–02	3.78–03	4.18–01	–1.578	B	*LS*
			4769.86	4771.20	484594.9	– 505554.0	5–3	2.50–02	5.11–03	4.02–01	–1.592	B	*LS*
			4752.49	4753.82	484594.9	– 505630.6	5–7	1.13–02	5.34–03	4.18–01	–1.574	B	*LS*
			4740.26	4741.58	484498.2	– 505588.2	3–5	1.53–02	8.58–03	4.02–01	–1.590	B	*LS*
168.		^{3}D–^{3}P°	*3717.9*	*3719.0*	*484646*	*– 511535*	15–9	8.90–02	1.11–02	2.03+00	–0.780	B	2
			3735.43	3736.49	484746.2	– 511509.3	7–5	7.37–02	1.10–02	9.48–01	–1.113	B	*LS*
			3707.39	3708.44	484594.9	– 511560.4	5–3	6.73–02	8.32–03	5.08–01	–1.381	B	*LS*
			3689.94	3690.99	484498.2	– 511591.2	3–1	9.10–02	6.19–03	2.26–01	–1.731	B	*LS*
			3714.43	3715.48	484594.9	– 511509.3	5–5	1.34–02	2.77–03	1.69–01	–1.859	B	*LS*
			3694.14	3695.19	484498.2	– 511560.4	3–3	2.27–02	4.64–03	1.69–01	–1.856	B	*LS*
			3701.13	3702.18	484498.2	– 511509.3	3–5	9.02–04	3.09–04	1.13–02	–3.033	B–	*LS*
169.	$2p3p$–$2s4f$	^{3}D–^{3}F°	*3132.2*	*3133.1*	*484646*	*– 516564*	15–21	3.34–01	6.88–02	1.06+01	0.014	B	2
			3141.19	3142.10	484746.2	– 516572.0	7–9	3.31–01	6.30–02	4.56+00	–0.355	C+	*LS*
			3127.43	3128.33	484594.9	– 516560.8	5–7	2.98–01	6.12–02	3.15+00	–0.514	C+	*LS*
			3118.79	3119.70	484498.2	– 516552.6	3–5	2.84–01	6.91–02	2.13+00	–0.683	C+	*LS*
			3142.30	3143.21	484746.2	– 516560.8	7–7	3.69–02	5.46–03	3.95–01	–1.418	C+	*LS*
			3128.23	3129.14	484594.9	– 516552.6	5–5	5.23–02	7.68–03	3.95–01	–1.416	C+	*LS*
			3143.11	3144.02	484746.2	– 516552.6	7–5	1.46–03	1.54–04	1.12–02	–2.968	C	*LS*
170.	$2s4s$–$2s4p$	^{3}S–^{3}P°	*6217.6*	*6219.3*	487607.4	*– 503686*	3–9	2.62–01	4.56–01	2.80+01	0.136	B	2
			6219.89	6221.61	487607.4	– 503680.4	3–5	2.62–01	2.53–01	1.56+01	–0.119	B	*LS*
			6215.45	6217.17	487607.4	– 503691.9	3–3	2.63–01	1.52–01	9.34+00	–0.341	B	*LS*
			6212.39	6214.11	487607.4	– 503699.8	3–1	2.63–01	5.08–02	3.11+00	–0.817	B	*LS*

N IV: Allowed Transitions — Continued

No.	Transition Array	Mult.	λ_{air} (Å)	λ_{vac} (Å) or σ (cm^{-1})*	E_i (cm^{-1})	E_k (cm^{-1})	g_i–g_k	A_{ki} (10^8 s^{-1})	f_{ik}	S (at. u.)	log gf	Acc.	Source
171.	2p3p–2p3d	^{3}P–^{3}P°	*5818.8*	*5820.4*	*494355*	– *511535*	9–9	5.42–02	2.75–02	4.74+00	–0.606	B	2
			5843.84	5845.46	494402.0	– 511509.3	5–5	4.01–02	2.05–02	1.98+00	–0.989	B	*LS*
			5795.09	5796.70	494309.2	– 511560.4	3–3	1.37–02	6.90–03	3.95–01	–1.684	B	*LS*
			5826.43	5828.05	494402.0	– 511560.4	5–3	2.25–02	6.87–03	6.58–01	–1.464	B	*LS*
			5784.76	5786.37	494309.2	– 511591.2	3–1	5.51–02	9.22–03	5.27–01	–1.558	B	*LS*
			5812.31	5813.92	494309.2	– 511509.3	3–5	1.36–02	1.15–02	6.58–01	–1.463	B	*LS*
			5776.31	5777.91	494253.1	– 511560.4	1–3	1.85–02	2.77–02	5.27–01	–1.557	B	*LS*
172.	2p3d–2s5g	^{3}F°–^{3}G		*1699.8*	*495509*	– 554339.7	21–27	2.56+00	1.42–01	1.67+01	0.476	B	4
				1702.01	495585.7	– 554339.7	9–11	2.55+00	1.35–01	6.83+00	0.086	C+	4
				1699.03	495482.6	– 554339.7	7–9	2.40+00	1.33–01	5.22+00	–0.030	C+	4
				1696.83	495406.2	– 554339.7	5–7	2.35+00	1.42–01	3.97+00	–0.148	C+	4
				1702.01	495585.7	– 554339.7	9–9	1.59–01	6.92–03	3.49–01	–1.206	C+	*LS*
				1699.03	495482.6	– 554339.7	7–7	2.06–01	8.91–03	3.49–01	–1.205	C+	*LS*
				1702.01	495585.7	– 554339.7	9–7	3.12–03	1.05–04	5.31–03	–3.023	C–	*LS*
173.	2p3d–2s6g	^{3}F°–^{3}G		*1245.3*	*495509*	– 575809.4	21–27	2.12+00	6.34–02	5.46+00	0.125	B	4
				1246.51	495585.7	– 575809.4	9–11	2.12+00	6.04–02	2.23+00	–0.265	C+	4
				1244.91	495482.6	– 575809.4	7–9	1.99+00	5.94–02	1.70+00	–0.381	C+	4
				1243.73	495406.2	– 575809.4	5–7	1.95+00	6.34–02	1.30+00	–0.499	C+	4
				1246.51	495585.7	– 575809.4	9–9	1.32–01	3.08–03	1.14–01	–1.557	C+	*LS*
				1244.91	495482.6	– 575809.4	7–7	1.71–01	3.97–03	1.14–01	–1.556	C+	*LS*
				1246.51	495585.7	– 575809.4	9–7	2.59–03	4.69–05	1.73–03	–3.374	D	*LS*
174.	2p3d–2s4d	^{1}D°–^{1}D	6119.23	6120.92	498310.3	– 514647.7	5–5	6.99–04	3.92–04	3.95–02	–2.708	B	2
175.	2p3p–2p3d	^{1}D–^{1}P°	5288.25	5289.72	499705.9	– 518610.5	5–3	3.22–02	8.09–03	7.04–01	–1.393	B	2
176.	2s4p–2s4d	^{3}P°–^{3}D	*12884*	*7759* cm^{-1}	*503686*	– *511446*	9–15	7.11–02	2.95–01	1.13+02	0.424	B	2
			12870.5	7768 cm^{-1}	503680.4	– 511448	5–7	7.14–02	2.48–01	5.26+01	0.094	B	*LS*
			12892.9	7754 cm^{-1}	503691.9	– 511446	3–5	5.32–02	2.21–01	2.82+01	–0.178	B	*LS*
			12916.0	7740 cm^{-1}	503699.8	– 511440	1–3	3.92–02	2.94–01	1.25+01	–0.531	B	*LS*
			12873.8	7766 cm^{-1}	503680.4	– 511446	5–5	1.78–02	4.43–02	9.39+00	–0.655	B	*LS*
			12902.9	7748 cm^{-1}	503691.9	– 511440	3–3	2.95–02	7.37–02	9.39+00	–0.656	B	*LS*
			12883.7	7760 cm^{-1}	503680.4	– 511440	5–3	1.98–03	2.95–03	6.26–01	–1.831	B	*LS*
177.	2s4p–2s5s	^{3}P°–^{3}S	*2430.8*	*2431.5*	*503686*	– 544813.4	9–3	3.87+00	1.14–01	8.22+00	0.012	B	2
			2430.40	2431.14	503680.4	– 544813.4	5–3	2.15+00	1.14–01	4.57+00	–0.244	B	*LS*
			2431.08	2431.82	503691.9	– 544813.4	3–3	1.29+00	1.14–01	2.74+00	–0.466	B	*LS*
			2431.55	2432.29	503699.8	– 544813.4	1–3	4.29–01	1.14–01	9.13–01	–0.943	B	*LS*
178.	2s4p–2s5d	^{3}P°–^{3}D	*2035.8*	*2036.5*	*503686*	– *552790*	9–15	2.36+00	2.45–01	1.48+01	0.343	B	2
			2035.57	2036.22	503680.4	– 552790.9	5–7	2.37+00	2.06–01	6.89+00	0.012	B	*LS*
			2036.10	2036.76	503691.9	– 552789.6	3–5	1.77+00	1.84–01	3.69+00	–0.259	B	*LS*
			2036.43	2037.08	503699.8	– 552789.6	1–3	1.31+00	2.45–01	1.64+00	–0.611	B	*LS*
			2035.62	2036.28	503680.4	– 552789.6	5–5	5.91–01	3.67–02	1.23+00	–0.736	B	*LS*
			2036.10	2036.76	503691.9	– 552789.6	3–3	9.85–01	6.12–02	1.23+00	–0.736	B	*LS*
			2035.62	2036.28	503680.4	– 552789.6	5–3	6.57–02	2.45–03	8.21–02	–1.912	B	*LS*
179.	2p3d–2s5g	^{1}F°–^{1}G	2080.32	2080.98	506284.8	– 554339.1	7–9	7.36+00	6.14–01	2.94+01	0.633	B	4

N IV: Allowed Transitions — Continued

No.	Transition Array	Mult.	λ_{air} (Å)	λ_{vac} (Å) or σ (cm^{-1})*	E_i (cm^{-1})	E_k (cm^{-1})	g_i–g_k	A_{ki} (10^8 s^{-1})	f_{ik}	S (at. u.)	log gf	Acc.	Source
180.	$2p3d$–$2s6g$	$^1F^o$–^1G		1438.37	506284.8	– 575807.9	7–9	3.68+00	1.47–01	4.87+00	0.012	B	4
181.	$2p3d$–$2s5d$	$^3P^o$–^3D	*2423.2*	*2424.0*	*511535*	*– 552790*	9–15	1.13+00	1.65–01	1.19+01	0.173	B	2
			2421.65	2422.39	511509.3	– 552790.9	5–7	1.13+00	1.39–01	5.54+00	–0.158	B	*LS*
			2424.73	2425.47	511560.4	– 552789.6	3–5	8.44–01	1.24–01	2.97+00	–0.430	B	*LS*
			2426.54	2427.28	511591.2	– 552789.6	1–3	6.24–01	1.65–01	1.32+00	–0.782	B	*LS*
			2421.73	2422.46	511509.3	– 552789.6	5–5	2.82–01	2.48–02	9.90–01	–0.906	B	*LS*
			2424.73	2425.47	511560.4	– 552789.6	3–3	4.69–01	4.13–02	9.90–01	–0.907	B	*LS*
			2421.73	2422.46	511509.3	– 552789.6	5–3	3.14–02	1.65–03	6.60–02	–2.082	B	*LS*
182.	$2s4d$–$2s5p$	^1D–$^1P^o$	2809.35	2810.18	514647.7	– 550232.6	5–3	1.43+00	1.01–01	4.69+00	–0.295	B	2
183.	$2s4d$–$2s5f$	^1D–$^1F^o$	2477.69	2478.44	514647.7	– 554995.7	5–7	5.96+00	7.68–01	3.13+01	0.585	B	2
184.	$2s4f$–$2s5g$	$^3F^o$–^3G	*2646.4*	*2647.2*	*516564*	– 554339.7	21–27	8.63+00	1.17+00	2.13+02	1.389	B	4
			2646.98	2647.77	516572.0	– 554339.7	9–11	8.63+00	1.11+00	8.69+01	0.999	C+	4
			2646.19	2646.98	516560.8	– 554339.7	7–9	8.08+00	1.09+00	6.65+01	0.883	C+	4
			2645.62	2646.41	516552.6	– 554339.7	5–7	7.94+00	1.17+00	5.08+01	0.766	C+	4
			2646.98	2647.77	516572.0	– 554339.7	9–9	5.40–01	5.67–02	4.45+00	–0.292	C+	*LS*
			2646.19	2646.98	516560.8	– 554339.7	7–7	6.95–01	7.29–02	4.45+00	–0.292	C+	*LS*
			2646.98	2647.77	516572.0	– 554339.7	9–7	1.06–02	8.63–04	6.77–02	–2.110	C+	*LS*
185.	$2s4f$–$2s6g$	$^3F^o$–^3G		*1687.9*	*516564*	– 575809.4	21–27	1.77+00	9.71–02	1.13+01	0.310	B	4
				1688.12	516572.0	– 575809.4	9–11	1.77+00	9.24–02	4.62+00	–0.080	C+	4
				1687.80	516560.8	– 575809.4	7–9	1.66+00	9.10–02	3.54+00	–0.196	C+	4
				1687.57	516552.6	– 575809.4	5–7	1.62+00	9.71–02	2.70+00	–0.314	C+	4
				1688.12	516572.0	– 575809.4	9–9	1.11–01	4.73–03	2.36–01	–1.371	C+	*LS*
				1687.80	516560.8	– 575809.4	7–7	1.42–01	6.08–03	2.36–01	–1.371	C+	*LS*
				1688.12	516572.0	– 575809.4	9–7	2.17–03	7.20–05	3.60–03	–3.189	C–	*LS*
186.	$2s4f$–$2s5g$	$^1F^o$–^1G	3078.3	3079.2	521862.8	– 554339.1	7–9	3.91+00	7.14–01	5.07+01	0.699	B	4

*Wavelengths (Å) are always given unless cm^{-1} is indicated.

N IV

Forbidden Transitions

List of tabulated lines

Wavelength (Å)	No.	Wavelength (Å)	No.	Wavelength (Å)	No.	Wavenumber (cm^{-1})	No.
in vacuum		923.056	6	in air		63.1	4
		923.220	6			72.7	11
529.430	3	924.284	6	2150.47	14	124.8	11
569.045	2	924.906	6	2219.61	9	144.0	4
569.450	2	1483.32	1	2225.78	9		
595.405	8	1575.18	5	7490.20	12		
821.873	7	1576.75	5	7531.22	12		
822.299	7	1580.34	5	7602.70	12		
823.274	7	1673.33	13				
921.458	6	1676.83	13				
921.994	6	1718.55	10				

We have selected the data of Glass,[1-3] who calculated transition probabilities utilizing the CIV 3 code (see general introduction), i.e., by constructing advanced wavefunctions from configuration interaction expansions. He also included some relativistic corrections. The agreement with earlier, less comprehensive (and somewhat less advanced) calculations by Nussbaumer and Storey[4] is usually within a few percent, except for two weak M1 lines, $^1S_0$ – $^3P_1$ in the $2s^2$–$2p^2$ transition array and $^3P^o_1$ – $^1P^o_1$ in the $2s2p$ configuration, where the discrepancies are large.

For the M1 fine structure transitions in the ^3P terms of the $2s2p$ and $2p^2$ configurations, we have used the line strengths S obtained from general non-relativistic formulas for conditions of pure LS-coupling, given by Pasternak.[5] The A-values published by Glass, converted to S, are in close agreement, while the A-values of Nussbaumer and Storey[4] convert exactly to those of the formulas.

References

[1] R. Glass, Astrophys. Space Sci. **87**, 41 (1982).
[2] R. Glass, Astrophys. Space Sci. **91**, 417 (1983).
[3] R. Glass, Astrophys. Space Sci. **92**, 307 (1983).
[4] H. Nussbaumer and P. J. Storey, Astron. Astrophys. **74**, 244 (1979).
[5] S. Pasternak, Astrophys. J. **92**, 129 (1940).

N IV: Forbidden Transitions

No.	Transition Array	Mult.	λ_{air} (Å)	λ_{vac} (Å) or σ (cm^{-1})*	E_i (cm^{-1})	E_k (cm^{-1})	g_i–g_k	Type	A_{ki} (s^{-1})	S (at. u.)	Acc.	Source
1.	$2s^2$–$2s2p$	^1S–$^3P^o$										
				1483.32	0.0	67416.3	1–5	M2	1.15–02	2.77+01	B+	1,4
2.	$2s^2$–$2p^2$	^1S–^3P										
				569.450	0.0	175608.1	1–3	M1	6.63–03	1.36–07	D	2,4
				569.045	0.0	175732.9	1–5	E2	1.14–01	3.03–05	B	3,4
3.		^1S–^1D										
				529.430	0.0	188882.5	1–5	E2	2.68+03	4.97–01	B+	3,4
4.	$2s2p$–$2s2p$	$^3P^o$–$^3P^o$										
				63.1 cm^{-1}	67209.2	67272.3	1–3	M1	4.53–06	2.00+00	B+	4
				207.1 cm^{-1}	67209.2	67416.3	1–5	E2	6.53–12	7.66–01	D	3
				144.0 cm^{-1}	67272.3	67416.3	3–5	M1	4.03–05	2.50+00	B+	4
				144.0 cm^{-1}	67272.3	67416.3	3–5	E2	2.37–12	1.71+00	D	3
5.		$^3P^o$–$^1P^o$										
				1575.18	67209.2	130693.9	1–3	M1	1.24–02	5.40–06	B	2,4
				1576.75	67272.3	130693.9	3–3	M1	2.24–02	9.76–06	E	2,4
				1576.75	67272.3	130693.9	3–3	E2	3.26–04	8.52–06	D	3
				1580.34	67416.3	130693.9	5–3	M1	1.54–02	6.74–06	B	2,4
				1580.34	67416.3	130693.9	5–3	E2	2.25–04	5.95–06	D	3
6.	$2s2p$–$2p^2$	$^3P^o$–^3P										
				924.906	67416.3	175535.4	5–1	M2	7.65–02	3.47+00	C+	1,4
				923.056	67272.3	175608.1	3–3	M2	8.04–02	1.08+01	C+	1
				924.284	67416.3	175608.1	5–3	M2	1.08–04	1.46–02	C	1
				921.458	67209.2	175732.9	1–5	M2	1.73–02	3.85+00	C+	1
				921.994	67272.3	175732.9	3–5	M2	2.82–05	6.30–03	C	1
				923.220	67416.3	175732.9	5–5	M2	1.02–01	2.29+01	C+	1
7.		$^3P^o$–^1D										
				821.873	67209.2	188882.5	1–5	M2	7.21–02	9.07+00	C+	1
				822.299	67272.3	188882.5	3–5	M2	1.64–01	2.06+01	C+	1
				823.274	67416.3	188882.5	5–5	M2	1.29–01	1.63+01	C+	1
8.		$^3P^o$–^1S										
				595.405	67416.3	235369.3	5–1	M2	4.75–01	2.39+00	C+	1

N IV: Forbidden Transitions — Continued

No.	Transition Array	Mult.	λ_{air} (Å)	λ_{vac} (Å) or σ (cm^{-1})*	E_i (cm^{-1})	E_k (cm^{-1})	g_i-g_k	Type	A_{ki} (s^{-1})	S (at. u.)	Acc.	Source
9.		$^1P^o-^3P$										
			2225.78	2226.47	130693.9	175608.1	3−3	M2	6.12−04	6.73+00	C+	1
			2219.61	2220.30	130693.9	175732.9	3−5	M2	1.12−03	2.02+01	C+	1
10.		$^1P^o-^1D$										
				1718.55	130693.9	188882.5	3−5	M2	8.83−05	4.44−01	C	1
11.	$2p^2-2p^2$	$^3P-^3P$										
				72.7 cm^{-1}	175535.4	175608.1	1−3	M1	6.91−06	2.00+00	B+	5
				197.5 cm^{-1}	175535.4	175732.9	1−5	E2	4.69−12	6.96−01	D	3
				124.8 cm^{-1}	175608.1	175732.9	3−5	M1	2.62−05	2.50+00	B+	5
				124.8 cm^{-1}	175608.1	175732.9	3−5	E2	8.86−13	1.31+00	D	3
12.		$^3P-^1D$										
			7490.20	7492.26	175535.4	188882.5	1−5	E2	4.00−08	4.21−06	D	3
			7531.22	7533.30	175608.1	188882.5	3−5	M1	2.00−03	1.58−04	C	2
			7531.22	7533.30	175608.1	188882.5	3−5	E2	8.62−07	9.33−05	D	3
			7602.70	7604.79	175732.9	188882.5	5−5	M1	6.48−03	5.29−04	C	2
			7602.70	7604.79	175732.9	188882.5	5−5	E2	5.72−06	6.50−04	D+	3
13.		$^3P-^1S$										
				1673.33	175608.1	235369.3	3−1	M1	1.50−01	2.61−05	C	2
				1676.83	175732.9	235369.3	5−1	E2	7.04−03	8.33−05	C	3
14.		$^1D-^1S$										
			2150.47	2151.15	188882.5	235369.3	5−1	E2	5.88+01	2.42+00	B	3,4

*Wavelengths (Å) are always given unless cm^{-1} is indicated.

N V

Lithium Isoelectronic Sequence

Ground State: $1s^2 2s\ ^2S_{1/2}$

Ionization Energy: 97.888 eV = 789537.2 cm^{-1}

Allowed Transitions

List of tabulated lines

Wavelength (Å)	No.	Wavelength (Å)	No.	Wavelength (Å)	No.	Wavelength (Å)	No.
in vacuum		137.9	321	146.768	27	150.166	23
		138.6	332	146.773	27	150.167	23
129.1	324	139.2	320	146.860	26	150.423	22
129.2	323	140.356	5	146.916	26	150.481	22
129.812	11	140.358	5	147.424	4	151.5	316
130.431	10	140.9	319	147.427	4	153.127	21
131.248	9	141.4	318	148.116	25	153.187	21
131.7	322	141.6	317	148.168	25	153.188	21
132.378	8	144.96	29	148.173	25	153.619	20
133.993	7	145.4	331	148.321	24	153.680	20
136.424	6	145.72	28	148.378	24	155.3	315
136.425	6	146.717	27	150.108	23	158.024	19

List of tabulated lines — Continued

Wavelength (Å)	No.	Wavelength (Å)	No.	Wavelength (Å)	No.	Wavelength (Å)	No.
158.088	19	369.020	67	672.200	110	859.961	87
158.862	18	371.909	50	673.2	339	860.203	87
158.928	18	372.013	50	673.6	353	864.896	103
160.0	330	372.015	50	675.2	352	865.899	122
160.1	329	373.846	49	680.703	94	866.117	114
162.556	3	373.953	49	680.748	94	866.130	114
162.564	3	381.313	66	680.855	94	866.153	114
166.875	17	381.817	65	683.799	93	868.863	102
166.946	17	381.823	65	683.953	93	868.908	102
166.947	17	381.849	65	693.576	109	868.933	102
168.514	16	390.098	33	694.247	125	942.296	75
168.587	16	390.112	33	694.388	117	942.379	75
176.9	314	391.007	48	694.411	117	1018.97	86
178.1	313	391.121	48	694.435	108	1019.29	86
180.4	312	391.124	48	694.444	108	1019.31	86
186.063	15	394.229	47	694.479	108	1036.7	133
186.149	15	394.348	47	694.499	117	1048.2	101
186.152	15	401.405	64	711.911	92	1049.7	121
190.155	14	402.245	63	711.951	92	1050.18	113
190.249	14	402.254	63	712.077	92	1050.20	113
193.5	311	402.280	63	713.518	42	1050.23	113
202.6	328	424.604	46	713.860	42	1054.87	85
205.0	327	424.738	46	713.907	42	1055.24	85
209.274	2	424.742	46	714.484	77	1057.57	100
209.308	2	430.712	45	714.504	77	1057.67	100
211.1	310	430.854	45	716.661	91	1057.68	100
223.8	326	436.858	62	716.829	91	1077.6	132
244.2	325	438.465	61	723.7	345	1084.9	146
247.561	13	438.483	61	725.981	107	1102.6	165
247.706	13	438.507	61	726.0	350	1103.4	153
247.719	13	445.4	348	726.713	124	1129.2	145
266.196	12	445.7	336	726.883	116	1135.95	131
266.379	12	450.065	32	726.908	116	1135.98	131
318.25	39	450.098	32	727.015	116	1148.4	164
322.00	38	474.5	354	727.283	106	1149.8	152
327.020	37	475.2	340	727.298	106	1193.97	144
327.022	37	495.179	44	727.332	106	1194.06	144
334.127	36	495.355	44	740.0	349	1194.29	144
334.130	36	495.367	44	748.195	58	1203.63	143
341.686	56	499.2	351	748.291	58	1203.86	143
344.610	35	509.891	43	748.318	58	1213.8	158
344.614	35	510.089	43	760.413	90	1214.9	170
345.962	55	511.842	60	760.596	90	1215.1	175
349.737	72	515.697	59	760.603	90	1215.5	163
351.763	54	515.740	59	764.704	57	1216.46	151
351.829	54	515.756	59	764.833	57	1216.49	151
351.857	54	555.1	334	764.896	57	1216.53	151
352.588	53	595.0	344	768.553	89	1226.57	130
352.683	53	609.763	81	768.747	89	1226.61	130
354.271	71	619.1	347	776.715	105	1238.82	1
359.916	52	623.682	80	777.549	123	1242.80	1
359.983	52	626.2	335	777.663	115	1293.38	142
360.015	52	628.744	31	777.670	115	1293.53	142
360.121	70	628.874	31	777.692	115	1293.80	142
360.342	69	633.6	346	777.707	41	1309.30	141
360.345	69	642.800	79	778.169	41	1309.58	141
360.371	69	642.808	79	778.855	104	1316.6	157
361.126	51	643.832	96	778.879	104	1317.9	169
361.164	34	656.059	111	778.911	104	1318.1	174
361.172	34	656.078	119	789.513	76	1318.7	162
361.226	51	659.182	95	789.550	76	1320.98	150
368.665	68	670.846	78	844.777	88	1321.02	150
368.990	67	670.859	78	844.999	88	1321.06	150
368.994	67	672.025	118	845.011	88	1380.76	129

List of tabulated lines — Continued

Wavelength (Å)	No.	Wavelength (Å)	No.	Wavelength (Å)	No.	Wavelength (Å)	No.
1380.84	129	in air		3278.51	221	6547.92	265
1389.51	74			3278.72	221	6630.92	258
1389.82	74	2015.54	191	3439.96	230	6632.64	258
1463.10	140	2016.10	191	3441.98	230	6634.0	286
1463.42	140	2016.48	191	3442.68	230	6716.54	219
1463.44	140	2043.52	190	3497.3	240	6718.30	219
1475	338	2043.91	190	3500.4	248	6747.4	293
1492	341	2047.6	203	3501.2	251	6789.8	289
1493.54	139	2049.4	213	3505.4	243	7326.57	226
1493.6	156	2049.6	217	3519.76	235	7329.09	226
1493.90	139	2051.2	207	3519.98	235	7329.74	226
1495	357	2055.31	197	3520.01	235	7600.9	238
1495.2	168	2055.39	197	3522.68	229	7614.9	246
1495.5	173	2055.40	197	3523.41	229	7619.7	249
1495.7	161	2198.32	179	3984.3	256	7627.7	241
1501.56	149	2198.45	179	4135.4	263	7813.71	233
1501.65	149	2316.23	189	4167.42	220	7814.81	233
1501.68	149	2317.06	189	4167.91	220	7816.09	233
1548.67	84	2317.57	189	4198.1	273	8159.49	225
1549.34	84	2358.6	202	4218.6	267	8163.42	225
1549.45	84	2360.9	212	4333.42	177	8324.1	280
1554	333	2361.2	216	4334.52	177	8729.2	285
1580	343	2363.4	206	4418.88	228	8926.5	292
1587	359	2368.17	188	4422.62	228	9035.8	288
1611	337	2368.70	188	4423.77	228	9841.08	253
1616.3	99	2372.60	196	4513.4	239	9843.79	253
1619.7	120	2372.72	196	4518.5	247	10723.1	260
1621.84	112	2372.76	196	4520.0	250	10747.4	260
1621.97	112	2568.5	223	4527.7	242	10751.9	260
1648	358	2590.81	127	4565.16	234	11087	268
1655.4	182	2591.44	127	4565.54	234	11110	274
1655.62	98	2667.6	232	4565.74	234	11114	276
1655.88	98	2706.3	245	4603.74	30	11152	298
1656.07	98	2714.3	237	4612.59	227	11165	270
1691.39	128	2748.65	178	4613.85	227	11328.3	73
1691.56	128	2748.95	178	4619.97	30	11370.6	73
1702.27	83	2835.1	222	4664.7	255	11404.8	264
1703.22	83	2858.03	136	4748.77	185	11406.4	264
1711	356	2859.16	136	4750.48	185	11408.5	264
1716	355	2859.34	136	4750.89	185	11672	301
1723.7	193	2924.38	187	4863.1	262	11929	305
1749.5	209	2925.08	187	4933.9	200	11942.7	257
1752.3	199	2925.18	187	4943.2	210	11948.3	257
1762.2	181	2952.6	231	4945.7	214	12066	303
1811.10	138	2974.5	154	4949.9	272	13805.9	279
1811.57	138	2980.8	166	4951.1	204	13809.7	279
1811.63	138	2982.2	171	4989.2	266	15018.4	284
1838.4	192	2985.1	159	5067.04	194	15059.3	40
1857.8	155	2994.2	201	5067.60	194	15064.1	284
1860.2	167	2997.9	211	5068.55	194	15070.4	284
1860.8	172	2998.5	215	5272.18	184	15183.5	40
1861.3	160	2999.9	205	5274.80	184	15234.4	40
1867.7	208	3000.0	244	5999.77	254	15430	290
1872.4	198	3013.8	236	6000.49	254	15532	294
1876.30	148	3026.61	195	6342.98	261	15538	295
1876.47	148	3026.82	195	6350.68	261	15632	291
1876.50	148	3026.97	195	6352.25	261	15842.8	287
1882.35	137	3048.60	186	6379.9	281	15847.8	287
1882.92	137	3049.47	186	6469.5	269	15905.8	287
1923.85	180	3051.62	147	6477.5	275	16742.5	282
1923.92	180	3052.08	147	6478.8	277	16750.3	282
1964	342	3052.50	147	6494.8	271	18847	297
		3159.76	135	6547.07	265	20203	300
		3161.37	135	6547.58	265	20985	304

List of tabulated lines — Continued

Wavelength (Å)	No.	Wavenumber (cm^{-1})	No.	Wavenumber (cm^{-1})	No.	Wavenumber (cm^{-1})	No.
21615	302			219	283	806.7	183
22565.7	126	130	299	338.6	259	808.9	183
22648.0	126	132	299	339.8	259	816.1	183
26336	307	133.8	97	342.5	259	1050.4	252
28720	308	136	309	506.2	224	1054.3	252
36022.3	82	138.6	97	508.0	224	1396.2	134
36331.3	82	143.0	97	512.1	224	1401.1	134
36453.1	82	153	299	533.6	296	1412.3	134
39461.9	176	179	306	535.6	296	1577.9	218
39608.9	176	194	283	734.5	278	1583.8	218
		197	283	737.3	278		

Our tables are mainly based on the extensive results of the Opacity Project (OP), which is reviewed in the general introduction. For N v, the OP calculations were carried out by Peach *et al.*[1]

A few other comprehensive, but less sophisticated calculations exist. For N v—an ion with only three electrons, two of which are normally in the closed 1*s* shell—these are based on simpler calculational approaches, such as the single-configuration self-consistent field approximation (Biemont),[2] the semi-empirical Coulomb approximation (Lindgard and Nielsen),[3] and the Quantum Defect Orbital technique (Barrientos and Martin).[4] All of these calculations have produced quite accurate results, being in close agreement with each other and with the OP results, typically within ±10%. Therefore, they have been used for some highly excited transitions for which no OP data are available.

A compilation[5] performed in 1976 contains data for about 40 multiplets, which were critically selected from the then available literature. The OP data agree with these earlier results normally within a few percent. But for some higher *ms–np* transitions (m=2–4 and n=4–8), differences of up to ±20% are encountered.

Several calculations and measurements have also been carried out for transitions between the doubly excited quartet levels above the first ionization potential, which are metastable against autoionization. We have selected the results of Laughlin[6] which were calculated with the model-potential method. These data are estimated to be quite reliable, which is also indicated by good agreement (< 20%) with beam-foil lifetime data by Blanke *et al.*[7]

References

[1]G. Peach, H. E. Saraph, and M. J. Seaton, J. Phys. B **21**, 3669 (1988).
[2]E. Biémont, Astron. Astrophys., Suppl. Ser. **27**, 489 (1977).
[3]A. Lindgard and S. E. Nielsen, At. Data Nucl. Data Tables **19**, 533 (1977).
[4]C. Barrientos and I. Martin, J. Quant. Spectrosc. Radiat. Transfer **48**, 287 (1992).
[5]G. A. Martin and W. L. Wiese, J. Phys. Chem. Ref. Data **5**, 537 (1976).
[6]C. Laughlin, Atomic Spectra and Oscillator Strengths for Astrophysics and Fusion Research, 108–113 (Ed. J. E. Hansen, North-Holland, Amsterdam, 1990).
[7]J. H. Blanke, P. H. Heckmann, and E. Träbert, Phys. Scr. **32**, 509 (1985).

N v: Allowed Transitions

No.	Transition Array	Mult.	λ_{air} (Å)	λ_{vac} (Å) or σ (cm^{-1})*	E_i (cm^{-1})	E_k (cm^{-1})	g_i–g_k	A_{ki} (10^8 s^{-1})	f_{ik}	S (at. u.)	log gf	Acc.	Source
1.	$1s^22s$–$1s^22p$	^{2}S–^{2}P°		*1240.1*	0.0	*80636*	2–6	3.39+00	2.34−01	1.91+00	−0.329	A	1
				1238.82	0.0	80721.9	2–4	3.40+00	1.56−01	1.28+00	−0.505	A	*LS*
				1242.80	0.0	80463.2	2–2	3.37+00	7.79−02	6.38−01	−0.807	A	*LS*
2.	$1s^22s$–$1s^23p$	^{2}S–^{2}P°		*209.29*	0.0	*477817*	2–6	1.21+02	2.39−01	3.29−01	−0.321	A	1
				209.274	0.0	477842.0	2–4	1.21+02	1.59−01	2.19−01	−0.497	A	*LS*
				209.308	0.0	477765.7	2–2	1.21+02	7.96−02	1.10−01	−0.798	A	*LS*
3.	$1s^22s$–$1s^24p$	^{2}S–^{2}P°		*162.56*	0.0	*615163*	2–6	5.79+01	6.89−02	7.37−02	−0.861	A−	1
				162.556	0.0	615173.8	2–4	5.79+01	4.59−02	4.91−02	−1.037	A−	*LS*
				162.564	0.0	615141.0	2–2	5.79+01	2.30−02	2.46−02	−1.338	A−	*LS*

N v: Allowed Transitions — Continued

No.	Transition Array	Mult.	λ_{air} (Å)	λ_{vac} (Å) or σ (cm^{-1})*	E_i (cm^{-1})	E_k (cm^{-1})	g_i–g_k	A_{ki} (10^8 s^{-1})	f_{ik}	S (at. u.)	log gf	Acc.	Source
4.	$1s^22s-1s^25p$	$^2S-^2P^o$		*147.42*	0.0	– *678311*	2–6	3.07+01	3.00–02	2.91–02	–1.222	B+	1
				147.424	0.0	– 678316.5	2–4	3.07+01	2.00–02	1.94–02	–1.398	B+	*LS*
				147.427	0.0	– 678300.4	2–2	3.07+01	9.99–03	9.70–03	–1.699	B+	*LS*
5.	$1s^22s-1s^26p$	$^2S-^2P^o$		*140.36*	0.0	– *712470*	2–6	1.80+01	1.59–02	1.47–02	–1.497	B	1
				140.356	0.0	– 712472.6	2–4	1.80+01	1.06–02	9.80–03	–1.673	B	*LS*
				140.358	0.0	– 712463.2	2–2	1.80+01	5.30–03	4.90–03	–1.975	B	*LS*
6.	$1s^22s-1s^27p$	$^2S-^2P^o$		*136.42*	0.0	– *733007*	2–6	1.14+01	9.57–03	8.60–03	–1.718	B	1
				136.424	0.0	– 733009.2	2–4	1.14+01	6.38–03	5.73–03	–1.894	B	*LS*
				136.425	0.0	– 733003.3	2–2	1.14+01	3.19–03	2.87–03	–2.195	B	*LS*
7.	$1s^22s-1s^28p$	$^2S-^2P^o$		*133.99*	0.0	– *746310*	2–6	7.72+00	6.23–03	5.50–03	–1.904	B	1
				133.993	0.0	– 746309.9	2–4	7.72+00	4.16–03	3.67–03	–2.080	B	*LS*
				133.993	0.0	– 746306.0	2–2	7.72+00	2.08–03	1.83–03	–2.381	B	*LS*
8.	$1s^22s-1s^29p$	$^2S-^2P^o$		*132.38*	0.0	– *755413*	2–6	5.31+00	4.18–03	3.65–03	–2.078	B	3
				132.378	0.0	– 755414.3	2–4	5.31+00	2.79–03	2.43–03	–2.254	B	*LS*
				132.378	0.0	– 755411.5	2–2	5.31+00	1.39–03	1.22–03	–2.555	B	*LS*
9.	$1s^22s-1s^210p$	$^2S-^2P^o$		*131.25*	0.0	– *761918*	2–6	3.88+00	3.00–03	2.60–03	–2.221	B	3
				131.248	0.0	– 761918.3	2–4	3.88+00	2.00–03	1.73–03	–2.398	B	*LS*
				131.248	0.0	– 761916.3	2–2	3.88+00	1.00–03	8.65–04	–2.699	B	*LS*
10.	$1s^22s-1s^211p$	$^2S-^2P^o$		130.431	0.0	– 766687	2–6	3.01+00	2.30–03	1.98–03	–2.337	B	3
11.	$1s^22s-1s^212p$	$^2S-^2P^o$		129.812	0.0	– 770347	2–6	2.32+00	1.76–03	1.50–03	–2.455	B	3
12.	$1s^22p-1s^23s$	$^2P^o-^2S$		*266.32*	*80636*	– 456126.6	6–2	9.11+01	3.23–02	1.70–01	–0.713	A	1
				266.379	80721.9	– 456126.6	4–2	6.07+01	3.23–02	1.13–01	–0.889	A	*LS*
				266.196	80463.2	– 456126.6	2–2	3.04+01	3.23–02	5.66–02	–1.190	A	*LS*
13.	$1s^22p-1s^23d$	$^2P^o-^2D$		*247.66*	*80636*	– *484418*	6–10	4.26+02	6.53–01	3.19+00	0.593	A	1
				247.706	80721.9	– 484426.3	4–6	4.26+02	5.87–01	1.92+00	0.371	A	*LS*
				247.561	80463.2	– 484404.3	2–4	3.55+02	6.53–01	1.06+00	0.116	A	*LS*
				247.719	80721.9	– 484404.3	4–4	7.09+01	6.52–02	2.13–01	–0.583	A	*LS*
14.	$1s^22p-1s^24s$	$^2P^o-^2S$		*190.22*	*80636*	– 606348.8	6–2	3.46+01	6.25–03	2.35–02	–1.426	B+	1
				190.249	80721.9	– 606348.8	4–2	2.30+01	6.25–03	1.57–02	–1.602	B+	*LS*
				190.155	80463.2	– 606348.8	2–2	1.15+01	6.26–03	7.83–03	–1.903	B+	*LS*
15.	$1s^22p-1s^24d$	$^2P^o-^2D$		*186.12*	*80636*	– *617922*	6–10	1.42+02	1.23–01	4.51–01	–0.133	A	1
				186.149	80721.9	– 617925.5	4–6	1.42+02	1.10–01	2.70–01	–0.355	A	*LS*
				186.063	80463.2	– 617916.3	2–4	1.18+02	1.23–01	1.50–01	–0.610	A	*LS*
				186.152	80721.9	– 617916.3	4–4	2.36+01	1.23–02	3.01–02	–1.309	A	*LS*
16.	$1s^22p-1s^25s$	$^2P^o-^2S$		*168.56*	*80636*	– 673886.2	6–2	1.67+01	2.37–03	7.90–03	–1.847	B	1
				168.587	80721.9	– 673886.2	4–2	1.11+01	2.37–03	5.27–03	–2.023	B	*LS*
				168.514	80463.2	– 673886.2	2–2	5.57+00	2.37–03	2.63–03	–2.324	B	*LS*

N v: Allowed Transitions — Continued

No.	Transition Array	Mult.	λ_{air} (Å)	λ_{vac} (Å) or σ (cm^{-1})*	E_i (cm^{-1})	E_k (cm^{-1})	g_i-g_k	A_{ki} (10^8 s^{-1})	f_{ik}	S (at. u.)	log gf	Acc.	Source
17.	$1s^22p-1s^25d$	$^2P^o-^2D$		*166.92*	*80636*	*679716*	6−10	6.57+01	4.58−02	1.51−01	−0.561	A	1
				166.946	80721.9	679717.6	4−6	6.57+01	4.12−02	9.05−02	−0.783	A	*LS*
				166.875	80463.2	679712.7	2−4	5.48+01	4.58−02	5.03−02	−1.038	A	*LS*
				166.947	80721.9	679712.7	4−4	1.10+01	4.58−03	1.01−02	−1.737	A	*LS*
18.	$1s^22p-1s^26s$	$^2P^o-^2S$		*158.91*	*80636*	709939.2	6−2	9.34+00	1.18−03	3.70−03	−2.150	B	1
				158.928	80721.9	709939.2	4−2	6.23+00	1.18−03	2.47−03	−2.327	B	*LS*
				158.862	80463.2	709939.2	2−2	3.12+00	1.18−03	1.23−03	−2.627	B	*LS*
19.	$1s^22p-1s^26d$	$^2P^o-^2D$		*158.07*	*80636*	*713281*	6−10	3.62+01	2.26−02	7.05−02	−0.868	A−	1
				158.088	80721.9	713281.5	4−6	3.62+01	2.03−02	4.23−02	−1.090	A−	*LS*
				158.024	80463.2	713279.3	2−4	3.02+01	2.26−02	2.35−02	−1.345	A−	*LS*
				158.088	80721.9	713279.3	4−4	6.03+00	2.26−03	4.70−03	−2.044	A−	*LS*
20.	$1s^22p-1s^27s$	$^2P^o-^2S$		*153.66*	*80636*	731425.4	6−2	5.86+00	6.92−04	2.10−03	−2.382	B	1
				153.680	80721.9	731425.4	4−2	3.91+00	6.92−04	1.40−03	−2.558	B	*LS*
				153.619	80463.2	731425.4	2−2	1.96+00	6.92−04	7.00−04	−2.859	B	*LS*
21.	$1s^22p-1s^27d$	$^2P^o-^2D$		*153.17*	*80636*	*733516*	6−10	2.21+01	1.30−02	3.92−02	−1.109	B+	1
				153.187	80721.9	733517.2	4−6	2.21+01	1.17−02	2.35−02	−1.331	B+	*LS*
				153.127	80463.2	733515.4	2−4	1.84+01	1.30−02	1.31−02	−1.586	B+	*LS*
				153.188	80721.9	733515.4	4−4	3.68+00	1.30−03	2.61−03	−2.285	B+	*LS*
22.	$1s^22p-1s^28s$	$^2P^o-^2S$		*150.46*	*80636*	745255.6	6−2	3.87+00	4.37−04	1.30−03	−2.581	B	1
				150.481	80721.9	745255.6	4−2	2.58+00	4.37−04	8.67−04	−2.757	B	*LS*
				150.423	80463.2	745255.6	2−2	1.29+00	4.38−04	4.33−04	−3.058	B	*LS*
23.	$1s^22p-1s^28d$	$^2P^o-^2D$		*150.15*	*80636*	*746649*	6−10	1.45+01	8.19−03	2.43−02	−1.308	B+	1
				150.166	80721.9	746649.7	4−6	1.45+01	7.37−03	1.46−02	−1.530	B+	*LS*
				150.108	80463.2	746648.5	2−4	1.21+01	8.20−03	8.10−03	−1.785	B+	*LS*
				150.167	80721.9	746648.5	4−4	2.42+00	8.19−04	1.62−03	−2.485	B+	*LS*
24.	$1s^22p-1s^29s$	$^2P^o-^2S$		*148.36*	*80636*	754677.0	6−2	2.58+00	2.84−04	8.31−04	−2.769	B	3
				148.378	80721.9	754677.0	4−2	1.72+00	2.84−04	5.54−04	−2.945	B	*LS*
				148.321	80463.2	754677.0	2−2	8.60−01	2.84−04	2.77−04	−3.246	B	*LS*
25.	$1s^22p-1s^29d$	$^2P^o-^2D$		*148.15*	*80636*	*755623*	6−10	1.01+01	5.54−03	1.62−02	−1.478	B	3
				148.168	80721.9	755633	4−6	1.01+01	4.98−03	9.73−03	−1.700	B	*LS*
				148.116	80463.2	755608	2−4	8.42+00	5.54−03	5.40−03	−1.955	B	*LS*
				148.173	80721.9	755608	4−4	1.68+00	5.54−04	1.08−03	−2.655	B	*LS*
26.	$1s^22p-1s^210s$	$^2P^o-^2S$		*146.90*	*80636*	761382.7	6−2	1.87+00	2.01−04	5.84−04	−2.918	B	3
				146.916	80721.9	761382.7	4−2	1.24+00	2.01−04	3.89−04	−3.094	B	*LS*
				146.860	80463.2	761382.7	2−2	6.23−01	2.01−04	1.95−04	−3.395	B	*LS*
27.	$1s^22p-1s^210d$	$^2P^o-^2D$		*146.75*	*80636*	*762062*	6−10	7.28+00	3.92−03	1.14−02	−1.629	B	3
				146.768	80721.9	762071	4−6	7.28+00	3.53−03	6.82−03	−1.851	B	*LS*
				146.717	80463.2	762048	2−4	6.07+00	3.92−03	3.79−03	−2.106	B	*LS*
				146.773	80721.9	762048	4−4	1.21+00	3.92−04	7.57−04	−2.805	B	*LS*

N v: Allowed Transitions — Continued

No.	Transition Array	Mult.	λ_{air} (Å)	λ_{vac} (Å) or σ (cm^{-1})*	E_i (cm^{-1})	E_k (cm^{-1})	g_i-g_k	A_{ki} ($10^8 s^{-1}$)	f_{ik}	S (at. u.)	log gf	Acc.	Source
28.	$1s^22p-1s^211d$	$^2P^o-^2D$		145.72	80636	– 766866	6–10	5.52+00	2.93–03	8.43–03	–1.755	B	3
29.	$1s^22p-1s^212d$	$^2P^o-^2D$		144.96	80636	– 770483	6–10	4.21+00	2.21–03	6.33–03	–1.877	B	3
30.	$1s^23s-1s^23p$	$^2S-^2P^o$	*4609.1*	*4610.4*	456126.6	*– 477817*	2–6	4.13–01	3.95–01	1.20+01	–0.103	A	1
			4603.74	4605.03	456126.6	– 477842.0	2–4	4.14–01	2.63–01	7.98+00	–0.278	A	*LS*
			4619.97	4621.26	456126.6	– 477765.7	2–2	4.10–01	1.31–01	3.99+00	–0.581	A	*LS*
31.	$1s^23s-1s^24p$	$^2S-^2P^o$		*628.79*	456126.6	*– 615163*	2–6	1.38+01	2.46–01	1.02+00	–0.308	B	1
				628.744	456126.6	– 615173.8	2–4	1.38+01	1.64–01	6.78–01	–0.484	B	*LS*
				628.874	456126.6	– 615141.0	2–2	1.38+01	8.19–02	3.39–01	–0.786	B	*LS*
32.	$1s^23s-1s^25p$	$^2S-^2P^o$		*450.08*	456126.6	*– 678311*	2–6	8.27+00	7.54–02	2.23–01	–0.822	B	1
				450.065	456126.6	– 678316.5	2–4	8.27+00	5.02–02	1.49–01	–0.998	B	*LS*
				450.098	456126.6	– 678300.4	2–2	8.27+00	2.51–02	7.44–02	–1.299	B	*LS*
33.	$1s^23s-1s^26p$	$^2S-^2P^o$		*390.10*	456126.6	*– 712470*	2–6	5.02+00	3.44–02	8.83–02	–1.163	B	1
				390.098	456126.6	– 712472.6	2–4	5.02+00	2.29–02	5.89–02	–1.339	B	*LS*
				390.112	456126.6	– 712463.2	2–2	5.02+00	1.15–02	2.94–02	–1.640	B	*LS*
34.	$1s^23s-1s^27p$	$^2S-^2P^o$		*361.17*	456126.6	*– 733007*	2–6	3.23+00	1.90–02	4.51–02	–1.421	B	1
				361.164	456126.6	– 733009.2	2–4	3.23+00	1.26–02	3.01–02	–1.597	B	*LS*
				361.172	456126.6	– 733003.3	2–2	3.23+00	6.32–03	1.50–02	–1.898	B	*LS*
35.	$1s^23s-1s^28p$	$^2S-^2P^o$		*344.61*	456126.6	*– 746309*	2–6	2.19+00	1.17–02	2.66–02	–1.630	B	1
				344.610	456126.6	– 746309.9	2–4	2.19+00	7.82–03	1.77–02	–1.806	B	*LS*
				344.614	456126.6	– 746306.0	2–2	2.19+00	3.91–03	8.87–03	–2.107	B	*LS*
36.	$1s^23s-1s^29p$	$^2S-^2P^o$		*334.13*	456126.6	*– 755413*	2–6	1.55+00	7.76–03	1.71–02	–1.809	B	3
				334.127	456126.6	– 755414.3	2–4	1.55+00	5.17–03	1.14–02	–1.985	B	*LS*
				334.130	456126.6	– 755411.5	2–2	1.55+00	2.59–03	5.69–03	–2.286	B	*LS*
37.	$1s^23s-1s^210p$	$^2S-^2P^o$		*327.02*	456126.6	*– 761918*	2–6	1.13+00	5.45–03	1.17–02	–1.963	B	3
				327.020	456126.6	– 761918.3	2–4	1.13+00	3.63–03	7.82–03	–2.139	B	*LS*
				327.022	456126.6	– 761916.3	2–2	1.13+00	1.82–03	3.91–03	–2.440	B	*LS*
38.	$1s^23s-1s^211p$	$^2S-^2P^o$		322.00	456126.6	– 766687	2–6	8.87–01	4.14–03	8.77–03	–2.082	B	3
39.	$1s^23s-1s^212p$	$^2S-^2P^o$		318.25	456126.6	– 770347	2–6	6.85–01	3.12–03	6.54–03	–2.205	B	3
40.	$1s^23p-1s^23d$	$^2P^o-^2D$	*15145*	*6601* cm^{-1}	*477817*	*– 484418*	6–10	9.56–03	5.48–02	1.64+01	–0.483	A	1
			15183.5	6584.3 cm^{-1}	477842.0	– 484426.3	4–6	9.49–03	4.92–02	9.84+00	–0.706	A	*LS*
			15059.3	6638.6 cm^{-1}	477765.7	– 484404.3	2–4	8.11–03	5.51–02	5.46+00	–0.958	A	*LS*
			15234.4	6562.3 cm^{-1}	477842.0	– 484404.3	4–4	1.57–03	5.45–03	1.09+00	–1.662	A	*LS*
41.	$1s^23p-1s^24s$	$^2P^o-^2S$		*778.02*	*477817*	– 606348.8	6–2	2.34+01	7.08–02	1.09+00	–0.372	A	1
				778.169	477842.0	– 606348.8	4–2	1.56+01	7.08–02	7.26–01	–0.548	A	*LS*
				777.707	477765.7	– 606348.8	2–2	7.81+00	7.09–02	3.63–01	–0.849	A	*LS*

N v: Allowed Transitions — Continued

No.	Transition Array	Mult.	λ_{air} (Å) λ_{vac} (Å) or σ (cm^{-1})*	E_i (cm^{-1})	E_k (cm^{-1})	g_i-g_k	A_{ki} (10^8 s^{-1})	f_{ik}	S (at. u.)	log gf	Acc.	Source
42.	$1s^23p-1s^24d$	$^2P^o-^2D$	*713.75*	*477817*	*617922*	6–10	4.32+01	5.50–01	7.75+00	0.518	A	1
			713.860	477842.0	617925.5	4–6	4.32+01	4.95–01	4.65+00	0.296	A	*LS*
			713.518	477765.7	617916.3	2–4	3.60+01	5.50–01	2.58+00	0.041	A	*LS*
			713.907	477842.0	617916.3	4–4	7.19+00	5.50–02	5.17–01	–0.658	A	*LS*
43.	$1s^23p-1s^25s$	$^2P^o-^2S$	*510.02*	*477817*	673886.2	6–2	1.08+01	1.40–02	1.41–01	–1.075	A	1
			510.089	477842.0	673886.2	4–2	7.19+00	1.40–02	9.42–02	–1.251	A	*LS*
			509.891	477765.7	673886.2	2–2	3.60+00	1.40–02	4.71–02	–1.552	A	*LS*
44.	$1s^23p-1s^25d$	$^2P^o-^2D$	*495.30*	*477817*	*679716*	6–10	2.17+01	1.33–01	1.30+00	–0.098	A	1
			495.355	477842.0	679717.6	4–6	2.17+01	1.20–01	7.81–01	–0.320	A	*LS*
			495.179	477765.7	679712.7	2–4	1.81+01	1.33–01	4.34–01	–0.575	A	*LS*
			495.367	477842.0	679712.7	4–4	3.62+00	1.33–02	8.68–02	–1.274	A	*LS*
45.	$1s^23p-1s^26s$	$^2P^o-^2S$	*430.81*	*477817*	709939.2	6–2	5.90+00	5.48–03	4.66–02	–1.483	B+	1
			430.854	477842.0	709939.2	4–2	3.93+00	5.48–03	3.11–02	–1.660	B+	*LS*
			430.712	477765.7	709939.2	2–2	1.97+00	5.48–03	1.55–02	–1.960	B+	*LS*
46.	$1s^23p-1s^26d$	$^2P^o-^2D$	*424.69*	*477817*	*713281*	6–10	1.22+01	5.51–02	4.62–01	–0.481	A	1
			424.738	477842.0	713281.5	4–6	1.22+01	4.96–02	2.77–01	–0.703	A	*LS*
			424.604	477765.7	713279.3	2–4	1.02+01	5.51–02	1.54–01	–0.958	A	*LS*
			424.742	477842.0	713279.3	4–4	2.04+00	5.51–03	3.08–02	–1.657	A	*LS*
47.	$1s^23p-1s^27s$	$^2P^o-^2S$	*394.31*	*477817*	731425.4	6–2	3.59+00	2.79–03	2.17–02	–1.777	B+	1
			394.348	477842.0	731425.4	4–2	2.39+00	2.79–03	1.45–02	–1.953	B+	*LS*
			394.229	477765.7	731425.4	2–2	1.20+00	2.79–03	7.23–03	–2.254	B+	*LS*
48.	$1s^23p-1s^27d$	$^2P^o-^2D$	*391.08*	*477817*	*733516*	6–10	7.55+00	2.88–02	2.23–01	–0.762	A	1
			391.121	477842.0	733517.2	4–6	7.54+00	2.60–02	1.34–01	–0.984	A	*LS*
			391.007	477765.7	733515.4	2–4	6.29+00	2.88–02	7.43–02	–1.239	A	*LS*
			391.124	477842.0	733515.4	4–4	1.26+00	2.88–03	1.49–02	–1.938	A	*LS*
49.	$1s^23p-1s^28s$	$^2P^o-^2S$	*373.92*	*477817*	745255.6	6–2	2.34+00	1.64–03	1.21–02	–2.007	B+	1
			373.953	477842.0	745255.6	4–2	1.56+00	1.64–03	8.07–03	–2.184	B+	*LS*
			373.846	477765.7	745255.6	2–2	7.82–01	1.64–03	4.03–03	–2.484	B+	*LS*
50.	$1s^23p-1s^28d$	$^2P^o-^2D$	*371.98*	*477817*	*746649*	6–10	4.99+00	1.72–02	1.27–01	–0.985	A	1
			372.013	477842.0	746649.7	4–6	4.99+00	1.55–02	7.60–02	–1.207	A	*LS*
			371.909	477765.7	746648.5	2–4	4.16+00	1.72–02	4.22–02	–1.462	A	*LS*
			372.015	477842.0	746648.5	4–4	8.31–01	1.72–03	8.45–03	–2.161	A	*LS*
51.	$1s^23p-1s^29s$	$^2P^o-^2S$	*361.19*	*477817*	754677.0	6–2	1.62+00	1.06–03	7.54–03	–2.198	B	3
			361.226	477842.0	754677.0	4–2	1.08+00	1.06–03	5.03–03	–2.374	B	*LS*
			361.126	477765.7	754677.0	2–2	5.41–01	1.06–03	2.51–03	–2.675	B	*LS*
52.	$1s^23p-1s^29d$	$^2P^o-^2D$	*359.96*	*477817*	*755623*	6–10	3.47+00	1.12–02	7.98–02	–1.172	B	3
			359.983	477842.0	755633	4–6	3.46+00	1.01–02	4.79–02	–1.394	B	*LS*
			359.916	477765.7	755608	2–4	2.89+00	1.12–02	2.66–02	–1.649	B	*LS*
			360.015	477842.0	755608	4–4	5.77–01	1.12–03	5.32–03	–2.348	B	*LS*

N v: Allowed Transitions — Continued

No.	Transition Array	Mult.	λ_{air} (Å)	λ_{vac} (Å) or σ (cm^{-1})*	E_i (cm^{-1})	E_k (cm^{-1})	g_i-g_k	A_{ki} ($10^8\ s^{-1}$)	f_{ik}	S (at. u.)	log gf	Acc.	Source
53.	$1s^23p-1s^210s$	$^2P^o-^2S$		*352.65*	*477817*	– 761382.7	6–2	1.17+00	7.27–04	5.06–03	–2.360	B	3
				352.683	477842.0	– 761382.7	4–2	7.79–01	7.27–04	3.37–03	–2.537	B	*LS*
				352.588	477765.7	– 761382.7	2–2	3.90–01	7.27–04	1.69–03	–2.838	B	*LS*
54.	$1s^23p-1s^210d$	$^2P^o-^2D$		*351.81*	*477817*	– *762062*	6–10	2.51+00	7.76–03	5.39–02	–1.332	B	3
				351.829	477842.0	– 762071	4–6	2.51+00	6.98–03	3.24–02	–1.554	B	*LS*
				351.763	477765.7	– 762048	2–4	2.09+00	7.76–03	1.80–02	–1.809	B	*LS*
				351.857	477842.0	– 762048	4–4	4.18–01	7.76–04	3.59–03	–2.508	B	*LS*
55.	$1s^23p-1s^211d$	$^2P^o-^2D$		345.962	477817	– 766866	6–10	1.88+00	5.63–03	3.85–02	–1.471	B	3
56.	$1s^23p-1s^212d$	$^2P^o-^2D$		341.686	477817	– 770483	6–10	1.44+00	4.20–03	2.84–02	–1.598	B	3
57.	$1s^23d-1s^24p$	$^2D-^2P^o$		*764.85*	*484418*	– *615163*	10–6	3.01+00	1.58–02	3.99–01	–0.800	B+	1
				764.833	484426.3	– 615173.8	6–4	2.71+00	1.58–02	2.39–01	–1.022	B+	*LS*
				764.896	484404.3	– 615141.0	4–2	3.01+00	1.32–02	1.33–01	–1.277	B+	*LS*
				764.704	484404.3	– 615173.8	4–4	3.01–01	2.64–03	2.66–02	–1.976	B+	*LS*
58.	$1s^23d-1s^24f$	$^2D-^2F^o$		*748.25*	*484418*	– *618062*	10–14	8.65+01	1.02+00	2.50+01	1.007	B	2
				748.291	484426.3	– 618064.1	6–8	8.64+01	9.68–01	1.43+01	0.764	B	*LS*
				748.195	484404.3	– 618059.3	4–6	8.07+01	1.02+00	1.00+01	0.609	B	*LS*
				748.318	484426.3	– 618059.3	6–6	5.76+00	4.84–02	7.15–01	–0.537	B	*LS*
59.	$1s^23d-1s^25p$	$^2D-^2P^o$		*515.75*	*484418*	– *678311*	10–6	1.28+00	3.07–03	5.22–02	–1.512	B+	1
				515.756	484426.3	– 678316.5	6–4	1.16+00	3.07–03	3.13–02	–1.734	B+	*LS*
				515.740	484404.3	– 678300.4	4–2	1.28+00	2.56–03	1.74–02	–1.989	B+	*LS*
				515.697	484404.3	– 678316.5	4–4	1.29–01	5.12–04	3.48–03	–2.688	B+	*LS*
60.	$1s^23d-1s^25f$	$^2D-^2F^o$		511.842	484418	– 679790	10–14	2.85+01	1.57–01	2.64+00	0.195	B	2
61.	$1s^23d-1s^26p$	$^2D-^2P^o$		*438.50*	*484418*	– *712470*	10–6	6.73–01	1.16–03	1.68–02	–1.934	B	1
				438.507	484426.3	– 712472.6	6–4	6.06–01	1.16–03	1.01–02	–2.156	B	*LS*
				438.483	484404.3	– 712463.2	4–2	6.73–01	9.70–04	5.60–03	–2.411	B	*LS*
				438.465	484404.3	– 712472.6	4–4	6.73–02	1.94–04	1.12–03	–3.110	B	*LS*
62.	$1s^23d-1s^26f$	$^2D-^2F^o$		436.858	484418	– 713325	10–14	1.35+01	5.40–02	7.77–01	–0.268	B	2
63.	$1s^23d-1s^27p$	$^2D-^2P^o$		*402.27*	*484418*	– *733007*	10–6	3.99–01	5.81–04	7.70–03	–2.235	B–	1
				402.280	484426.3	– 733009.2	6–4	3.59–01	5.81–04	4.62–03	–2.457	B–	*LS*
				402.254	484404.3	– 733003.3	4–2	3.99–01	4.85–04	2.57–03	–2.713	B–	*LS*
				402.245	484404.3	– 733009.2	4–4	4.00–02	9.69–05	5.13–04	–3.412	B–	*LS*
64.	$1s^23d-1s^27f$	$^2D-^2F^o$		401.405	484418	– 733543	10–14	7.59+00	2.57–02	3.39–01	–0.591	B	2
65.	$1s^23d-1s^28p$	$^2D-^2P^o$		*381.84*	*484418*	– *746309*	10–6	2.55–01	3.34–04	4.20–03	–2.476	B–	1
				381.849	484426.3	– 746309.9	6–4	2.29–01	3.34–04	2.52–03	–2.698	B–	*LS*
				381.823	484404.3	– 746306.0	4–2	2.55–01	2.78–04	1.40–03	–2.953	B–	*LS*
				381.817	484404.3	– 746309.9	4–4	2.55–02	5.57–05	2.80–04	–3.652	B–	*LS*

N v: Allowed Transitions — Continued

No.	Transition Array	Mult.	λ_{air} (Å)	λ_{vac} (Å) or σ (cm^{-1})*	E_i (cm^{-1})	E_k (cm^{-1})	g_i–g_k	A_{ki} (10^8 s^{-1})	f_{ik}	S (at. u.)	log gf	Acc.	Source
66.	$1s^23d$–$1s^28f$	^{2}D–^{2}F°		381.313	484418	– 746669	10–14	4.74+00	1.45–02	1.82–01	–0.840	B	2
67.	$1s^23d$–$1s^29p$	^{2}D–^{2}P°		*369.01*	*484418*	– *755413*	10–6	1.77–01	2.17–04	2.64–03	–2.663	B	3
				369.020	484426.3	– 755414.3	6–4	1.60–01	2.17–04	1.58–03	–2.885	B	*LS*
				368.994	484404.3	– 755411.5	4–2	1.78–01	1.81–04	8.80–04	–3.140	B	*LS*
				368.990	484404.3	– 755414.3	4–4	1.78–02	3.62–05	1.76–04	–3.839	B	*LS*
68.	$1s^23d$–$1s^29f$	^{2}D–^{2}F°		368.665	484418	– 755667	10–14	3.18+00	9.07–03	1.10–01	–1.042	B	3
69.	$1s^23d$–$1s^210p$	^{2}D–^{2}P°		*360.36*	*484418*	– *761918*	10–6	1.27–01	1.48–04	1.76–03	–2.829	B	3
				360.371	484426.3	– 761918.3	6–4	1.14–01	1.48–04	1.06–03	–3.050	B	*LS*
				360.345	484404.3	– 761916.3	4–2	1.27–01	1.24–04	5.87–04	–3.306	B	*LS*
				360.342	484404.3	– 761918.3	4–4	1.27–02	2.47–05	1.17–04	–4.005	B	*LS*
70.	$1s^23d$–$1s^210f$	^{2}D–^{2}F°		360.121	484418	– 762102	10–14	2.25+00	6.11–03	7.25–02	–1.214	B	3
71.	$1s^23d$–$1s^211p$	^{2}D–^{2}P°		354.271	484418	– 766687	10–6	1.02–01	1.15–04	1.34–03	–2.938	B	3
72.	$1s^23d$–$1s^212p$	^{2}D–^{2}P°		349.737	484418	– 770347	10–6	7.80–02	8.59–05	9.89–04	–3.066	B	3
73.	$1s^24s$–$1s^24p$	^{2}S–^{2}P°	*11342*	*8814* cm^{-1}	606348.8	– *615163*	2–6	9.38–02	5.43–01	4.06+01	0.036	A	1
			11328.3	8825.0 cm^{-1}	606348.8	– 615173.8	2–4	9.42–02	3.62–01	2.70+01	–0.140	A	*LS*
			11370.6	8792.2 cm^{-1}	606348.8	– 615141.0	2–2	9.31–02	1.81–01	1.35+01	–0.442	A	*LS*
74.	$1s^24s$–$1s^25p$	^{2}S–^{2}P°		*1389.6*	606348.8	– *678311*	2–6	3.03+00	2.63–01	2.41+00	–0.279	B	1
				1389.51	606348.8	– 678316.5	2–4	3.03+00	1.75–01	1.61+00	–0.455	B	*LS*
				1389.82	606348.8	– 678300.4	2–2	3.03+00	8.77–02	8.03–01	–0.756	B	*LS*
75.	$1s^24s$–$1s^26p$	^{2}S–^{2}P°		*942.32*	606348.8	– *712470*	2–6	2.06+00	8.22–02	5.10–01	–0.784	B	1
				942.296	606348.8	– 712472.6	2–4	2.06+00	5.48–02	3.40–01	–0.960	B	*LS*
				942.379	606348.8	– 712463.2	2–2	2.06+00	2.74–02	1.70–01	–1.261	B	*LS*
76.	$1s^24s$–$1s^27p$	^{2}S–^{2}P°		*789.53*	606348.8	– *733007*	2–6	1.37+00	3.83–02	1.99–01	–1.116	B–	1
				789.513	606348.8	– 733009.2	2–4	1.37+00	2.55–02	1.33–01	–1.292	B–	*LS*
				789.550	606348.8	– 733003.3	2–2	1.37+00	1.28–02	6.63–02	–1.593	B–	*LS*
77.	$1s^24s$–$1s^28p$	^{2}S–^{2}P°		*714.49*	606348.8	– *746309*	2–6	9.37–01	2.15–02	1.01–01	–1.366	B–	1
				714.484	606348.8	– 746309.9	2–4	9.37–01	1.43–02	6.75–02	–1.542	B–	*LS*
				714.504	606348.8	– 746306.0	2–2	9.37–01	7.17–03	3.37–02	–1.843	B–	*LS*
78.	$1s^24s$–$1s^29p$	^{2}S–^{2}P°		*670.85*	606348.8	– *755413*	2–6	6.54–01	1.32–02	5.85–02	–1.577	B	3
				670.846	606348.8	– 755414.3	2–4	6.54–01	8.83–03	3.90–02	–1.753	B	*LS*
				670.859	606348.8	– 755411.5	2–2	6.54–01	4.41–03	1.95–02	–2.054	B	*LS*
79.	$1s^24s$–$1s^210p$	^{2}S–^{2}P°		*642.80*	606348.8	– *761918*	2–6	4.80–01	8.91–03	3.77–02	–1.749	B	3
				642.800	606348.8	– 761918.3	2–4	4.80–01	5.94–03	2.51–02	–1.925	B	*LS*
				642.808	606348.8	– 761916.3	2–2	4.80–01	2.97–03	1.26–02	–2.226	B	*LS*

N v: Allowed Transitions — Continued

No.	Transition Array	Mult.	λ_{air} (Å)	λ_{vac} (Å) or σ (cm^{-1})*	E_i (cm^{-1})	E_k (cm^{-1})	g_i-g_k	A_{ki} (10^8 s^{-1})	f_{ik}	S (at. u.)	log gf	Acc.	Source
80.	$1s^24s-1s^211p$	$^2S-^2P^\circ$		623.682	606348.8	– 766687	2–6	3.78–01	6.61–03	2.72–02	–1.879	B	3
81.	$1s^24s-1s^212p$	$^2S-^2P^\circ$		609.763	606348.8	– 770347	2–6	2.92–01	4.88–03	1.96–02	–2.011	B	3
82.	$1s^24p-1s^24d$	$^2P^\circ-^2D$	*36236*	*2759* cm^{-1}	*615163*	– *617922*	6–10	2.96–03	9.72–02	6.96+01	–0.234	A	1
			36331.3	2751.7 cm^{-1}	615173.8	– 617925.5	4–6	2.94–03	8.73–02	4.18+01	–0.457	A	*LS*
			36022.3	2775.3 cm^{-1}	615141.0	– 617916.3	2–4	2.51–03	9.78–02	2.32+01	–0.709	A	*LS*
			36453.1	2742.5 cm^{-1}	615173.8	– 617916.3	4–4	4.85–04	9.67–03	4.64+00	–1.413	A	*LS*
83.	$1s^24p-1s^25s$	$^2P^\circ-^2S$		*1702.9*	*615163*	– 673886.2	6–2	7.71+00	1.12–01	3.76+00	–0.174	A	1
				1703.22	615173.8	– 673886.2	4–2	5.14+00	1.12–01	2.50+00	–0.350	A	*LS*
				1702.27	615141.0	– 673886.2	2–2	2.57+00	1.12–01	1.25+00	–0.651	A	*LS*
84.	$1s^24p-1s^25d$	$^2P^\circ-^2D$		*1549.1*	*615163*	– *679716*	6–10	8.76+00	5.25–01	1.61+01	0.499	A	1
				1549.34	615173.8	– 679717.6	4–6	8.76+00	4.73–01	9.65+00	0.277	A	*LS*
				1548.67	615141.0	– 679712.7	2–4	7.31+00	5.26–01	5.36+00	0.022	A	*LS*
				1549.45	615173.8	– 679712.7	4–4	1.46+00	5.25–02	1.07+00	–0.678	A	*LS*
85.	$1s^24p-1s^26s$	$^2P^\circ-^2S$		*1055.1*	*615163*	– 709939.2	6–2	4.01+00	2.23–02	4.65–01	–0.874	A–	1
				1055.24	615173.8	– 709939.2	4–2	2.67+00	2.23–02	3.10–01	–1.050	A–	*LS*
				1054.87	615141.0	– 709939.2	2–2	1.34+00	2.23–02	1.55–01	–1.351	A–	*LS*
86.	$1s^24p-1s^26d$	$^2P^\circ-^2D$		*1019.2*	*615163*	– *713281*	6–10	5.30+00	1.38–01	2.77+00	–0.084	A	1
				1019.29	615173.8	– 713281.5	4–6	5.30+00	1.24–01	1.66+00	–0.305	A	*LS*
				1018.97	615141.0	– 713279.3	2–4	4.42+00	1.38–01	9.23–01	–0.561	A	*LS*
				1019.31	615173.8	– 713279.3	4–4	8.83–01	1.37–02	1.85–01	–1.260	A	*LS*
87.	$1s^24p-1s^27s$	$^2P^\circ-^2S$		*860.12*	*615163*	– 731425.4	6–2	2.38+00	8.79–03	1.49–01	–1.278	A–	1
				860.203	615173.8	– 731425.4	4–2	1.58+00	8.79–03	9.95–02	–1.454	A–	*LS*
				859.961	615141.0	– 731425.4	2–2	7.93–01	8.79–03	4.98–02	–1.755	A–	*LS*
88.	$1s^24p-1s^27d$	$^2P^\circ-^2D$		*844.93*	*615163*	– *733516*	6–10	3.34+00	5.95–02	9.93–01	–0.447	A–	1
				844.999	615173.8	– 733517.2	4–6	3.33+00	5.35–02	5.96–01	–0.669	A–	*LS*
				844.777	615141.0	– 733515.4	2–4	2.78+00	5.95–02	3.31–01	–0.924	A–	*LS*
				845.011	615173.8	– 733515.4	4–4	5.56–01	5.95–03	6.62–02	–1.623	A–	*LS*
89.	$1s^24p-1s^28s$	$^2P^\circ-^2S$		*768.68*	*615163*	– 745255.6	6–2	1.53+00	4.52–03	6.87–02	–1.566	B+	1
				768.747	615173.8	– 745255.6	4–2	1.02+00	4.52–03	4.58–02	–1.742	B+	*LS*
				768.553	615141.0	– 745255.6	2–2	5.11–01	4.53–03	2.29–02	–2.043	B+	*LS*
90.	$1s^24p-1s^28d$	$^2P^\circ-^2D$		*760.54*	*615163*	– *746649*	6–10	2.22+00	3.21–02	4.83–01	–0.715	A–	1
				760.596	615173.8	– 746649.7	4–6	2.22+00	2.89–02	2.90–01	–0.937	A–	*LS*
				760.413	615141.0	– 746648.5	2–4	1.85+00	3.21–02	1.61–01	–1.192	A–	*LS*
				760.603	615173.8	– 746648.5	4–4	3.71–01	3.21–03	3.22–02	–1.891	A–	*LS*
91.	$1s^24p-1s^29s$	$^2P^\circ-^2S$		*716.77*	*615163*	– 754677.0	6–2	1.03+00	2.63–03	3.73–02	–1.802	B	3
				716.829	615173.8	– 754677.0	4–2	6.83–01	2.63–03	2.48–02	–1.978	B	*LS*
				716.661	615141.0	– 754677.0	2–2	3.42–01	2.63–03	1.24–02	–2.279	B	*LS*

N v: Allowed Transitions — Continued

No.	Transition Array	Mult.	λ_{air} (Å)	λ_{vac} (Å) or σ (cm^{-1})*	E_i (cm^{-1})	E_k (cm^{-1})	g_i–g_k	A_{ki} (10^8 s^{-1})	f_{ik}	S (at. u.)	log gf	Acc.	Source
92.	$1s^24p$– $1s^29d$	$^2P^o$–^2D		*711.95*	*615163*	– *755623*	6–10	1.53+00	1.94–02	2.73–01	–0.933	B	3
				711.951	615173.8	– 755633	4–6	1.53+00	1.75–02	1.64–01	–1.155	B	*LS*
				711.911	615141.0	– 755608	2–4	1.28+00	1.94–02	9.11–02	–1.410	B	*LS*
				712.077	615173.8	– 755608	4–4	2.56–01	1.94–03	1.82–02	–2.110	B	*LS*
93.	$1s^24p$– $1s^210s$	$^2P^o$–^2S		*683.90*	*615163*	– 761382.7	6–2	7.31–01	1.71–03	2.31–02	–1.989	B	3
				683.953	615173.8	– 761382.7	4–2	4.87–01	1.71–03	1.54–02	–2.165	B	*LS*
				683.799	615141.0	– 761382.7	2–2	2.44–01	1.71–03	7.69–03	–2.466	B	*LS*
94.	$1s^24p$– $1s^210d$	$^2P^o$–^2D		*680.74*	*615163*	– *762062*	6–10	1.11+00	1.29–02	1.73–01	–1.112	B	3
				680.748	615173.8	– 762071	4–6	1.11+00	1.16–02	1.04–01	–1.334	B	*LS*
				680.703	615141.0	– 762048	2–4	9.27–01	1.29–02	5.77–02	–1.589	B	*LS*
				680.855	615173.8	– 762048	4–4	1.85–01	1.29–03	1.15–02	–2.288	B	*LS*
95.	$1s^24p$– $1s^211d$	$^2P^o$–^2D		659.182	615163	– 766866	6–10	8.30–01	9.01–03	1.17–01	–1.267	B	3
96.	$1s^24p$– $1s^212d$	$^2P^o$–^2D		643.832	615163	– 770483	6–10	6.36–01	6.59–03	8.37–02	–1.403	B	3
97.	$1s^24d$– $1s^24f$	^2D–$^2F^o$		*140* cm^{-1}	*617922*	– *618062*	10–14	2.39–07	2.55–03	5.98+01	–1.594	B	3
				138.6 cm^{-1}	617925.5	– 618064.1	6–8	2.30–07	2.40–03	3.42+01	–1.842	B	*LS*
				143.0 cm^{-1}	617916.3	– 618059.3	4–6	2.36–07	2.60–03	2.39+01	–1.983	B	*LS*
				133.8 cm^{-1}	617925.5	– 618059.3	6–6	1.38–08	1.16–04	1.71+00	–3.158	B	*LS*
98.	$1s^24d$– $1s^25p$	^2D–$^2P^o$		*1655.9*	*617922*	– *678311*	10–6	1.57+00	3.88–02	2.11+00	–0.412	A–	1
				1655.88	617925.5	– 678316.5	6–4	1.41+00	3.88–02	1.27+00	–0.634	A–	*LS*
				1656.07	617916.3	– 678300.4	4–2	1.57+00	3.23–02	7.04–01	–0.889	A–	*LS*
				1655.62	617916.3	– 678316.5	4–4	1.57–01	6.46–03	1.41–01	–1.588	A–	*LS*
99.	$1s^24d$– $1s^25f$	^2D–$^2F^o$		1616.3	617922	– 679790	10–14	1.62+01	8.88–01	4.72+01	0.948	B	2
100.	$1s^24d$– $1s^26p$	^2D–$^2P^o$		*1057.7*	*617922*	– *712470*	10–6	7.77–01	7.82–03	2.72–01	–1.107	B+	1
				1057.67	617925.5	– 712472.6	6–4	6.99–01	7.82–03	1.63–01	–1.329	B+	*LS*
				1057.68	617916.3	– 712463.2	4–2	7.77–01	6.51–03	9.07–02	–1.584	B+	*LS*
				1057.57	617916.3	– 712472.6	4–4	7.77–02	1.30–03	1.81–02	–2.283	B+	*LS*
101.	$1s^24d$– $1s^26f$	^2D–$^2P^o$		1048.2	617922	– 713325	10–6	1.88+01	1.86–01	6.42+00	0.270	B	2
102.	$1s^24d$– $1s^27p$	^2D–$^2P^o$		*868.92*	*617922*	– *733007*	10–6	4.44–01	3.02–03	8.63–02	–1.520	B	1
				868.933	617925.5	– 733009.2	6–4	4.00–01	3.02–03	5.18–02	–1.742	B	*LS*
				868.908	617916.3	– 733003.3	4–2	4.44–01	2.51–03	2.88–02	–1.998	B	*LS*
				868.863	617916.3	– 733009.2	4–4	4.44–02	5.03–04	5.75–03	–2.697	B	*LS*
103.	$1s^24d$– $1s^27f$	^2D–$^2F^o$		864.896	617922	– 733543	10–14	4.61+00	7.23–02	2.06+00	–0.141	B	2
104.	$1s^24d$– $1s^28p$	^2D–$^2P^o$		*778.90*	*617922*	– *746309*	10–6	2.79–01	1.52–03	3.91–02	–1.817	B–	1
				778.911	617925.5	– 746309.9	6–4	2.51–01	1.52–03	2.35–02	–2.039	B–	*LS*
				778.879	617916.3	– 746306.0	4–2	2.79–01	1.27–03	1.30–02	–2.294	B–	*LS*
				778.855	617916.3	– 746309.9	4–4	2.79–02	2.54–04	2.61–03	–2.993	B–	*LS*

N V: Allowed Transitions — Continued

No.	Transition Array	Mult.	λ_{air} (Å)	λ_{vac} (Å) or σ (cm^{-1})*	E_i (cm^{-1})	E_k (cm^{-1})	g_i-g_k	A_{ki} (10^8 s^{-1})	f_{ik}	S (at. u.)	log gf	Acc.	Source
105.	$1s^24d-$ $1s^28f$	$^2D-^2F^\circ$		776.715	617922	– 746669	10–14	2.89+00	3.67–02	9.37–01	–0.436	B	2
106.	$1s^24d-$ $1s^29p$	$^2D-^2P^\circ$		*727.32*	*617922*	*– 755413*	10–6	1.83–01	8.71–04	2.08–02	–2.060	B	3
				727.332	617925.5	– 755414.3	6–4	1.65–01	8.71–04	1.25–02	–2.282	B	*LS*
				727.298	617916.3	– 755411.5	4–2	1.83–01	7.26–04	6.95–03	–2.537	B	*LS*
				727.283	617916.3	– 755414.3	4–4	1.83–02	1.45–04	1.39–03	–3.236	B	*LS*
107.	$1s^24d-$ $1s^29f$	$^2D-^2F^\circ$		725.981	617922	– 755667	10–14	1.93+00	2.14–02	5.10–01	–0.671	B	3
108.	$1s^24d-$ $1s^210p$	$^2D-^2P^\circ$		*694.46*	*617922*	*– 761918*	10–6	1.29–01	5.58–04	1.28–02	–2.253	B	3
				694.479	617925.5	– 761918.3	6–4	1.16–01	5.58–04	7.66–03	–2.475	B	*LS*
				694.444	617916.3	– 761916.3	4–2	1.29–01	4.65–04	4.25–03	–2.730	B	*LS*
				694.435	617916.3	– 761918.3	4–4	1.29–02	9.30–05	8.51–04	–3.429	B	*LS*
109.	$1s^24d-$ $1s^210f$	$^2D-^2F^\circ$		693.576	617922	– 762102	10–14	1.37+00	1.38–02	3.15–01	–0.860	B	3
110.	$1s^24d-$ $1s^211p$	$^2D-^2P^\circ$		672.200	617922	– 766687	10–6	1.01–01	4.12–04	9.11–03	–2.385	B	3
111.	$1s^24d-$ $1s^212p$	$^2D-^2P^\circ$		656.059	617922	– 770347	10–6	7.67–02	2.97–04	6.41–03	–2.528	B	3
112.	$1s^24f-$ $1s^25d$	$^2F^\circ-^2D$		*1622.0*	*618062*	*– 679716*	14–10	3.20–01	9.01–03	6.73–01	–0.899	B	2
				1621.97	618064.1	– 679717.6	8–6	3.04–01	9.01–03	3.85–01	–1.142	B	*LS*
				1621.97	618059.3	– 679712.7	6–4	3.20–01	8.41–03	2.69–01	–1.297	B	*LS*
				1621.84	618059.3	– 679717.6	6–6	1.52–02	6.01–04	1.92–02	–2.443	B	*LS*
113.	$1s^24f-$ $1s^26d$	$^2F^\circ-^2D$		*1050.2*	*618062*	*– 713281*	14–10	1.36–01	1.61–03	7.78–02	–1.648	B	2
				1050.23	618064.1	– 713281.5	8–6	1.30–01	1.61–03	4.45–02	–1.891	B	*LS*
				1050.20	618059.3	– 713279.3	6–4	1.36–01	1.50–03	3.11–02	–2.046	B	*LS*
				1050.18	618059.3	– 713281.5	6–6	6.48–03	1.07–04	2.22–03	–3.192	B	*LS*
114.	$1s^24f-$ $1s^27d$	$^2F^\circ-^2D$		*866.14*	*618062*	*– 733516*	14–10	7.14–02	5.73–04	2.29–02	–2.096	B	2
				866.153	618064.1	– 733517.2	8–6	6.80–02	5.73–04	1.31–02	–2.339	B	*LS*
				866.130	618059.3	– 733515.4	6–4	7.14–02	5.35–04	9.15–03	–2.493	B	*LS*
				866.117	618059.3	– 733517.2	6–6	3.40–03	3.82–05	6.54–04	–3.640	B	*LS*
115.	$1s^24f-$ $1s^28d$	$^2F^\circ-^2D$		*777.68*	*618062*	*– 746649*	14–10	4.26–02	2.76–04	9.89–03	–2.413	B	2
				777.692	618064.1	– 746649.7	8–6	4.06–02	2.76–04	5.65–03	–2.656	B	*LS*
				777.670	618059.3	– 746648.5	6–4	4.26–02	2.58–04	3.96–03	–2.811	B	*LS*
				777.663	618059.3	– 746649.7	6–6	2.03–03	1.84–05	2.83–04	–3.957	B	*LS*
116.	$1s^24f-$ $1s^29d$	$^2F^\circ-^2D$		*726.95*	*618062*	*– 755623*	14–10	2.73–02	1.54–04	5.17–03	–2.666	B	3
				726.908	618064.1	– 755633	8–6	2.60–02	1.54–04	2.95–03	–2.909	B	*LS*
				727.015	618059.3	– 755608	6–4	2.73–02	1.44–04	2.07–03	–3.063	B	*LS*
				726.883	618059.3	– 755633	6–6	1.30–03	1.03–05	1.48–04	–4.210	B	*LS*
117.	$1s^24f-$ $1s^210d$	$^2F^\circ-^2D$		*694.45*	*618062*	*– 762062*	14–10	1.90–02	9.80–05	3.14–03	–2.863	B	3
				694.411	618064.1	– 762071	8–6	1.81–02	9.80–05	1.79–03	–3.106	B	*LS*
				694.499	618059.3	– 762048	6–4	1.90–02	9.15–05	1.25–03	–3.261	B	*LS*
				694.388	618059.3	– 762071	6–6	9.04–04	6.53–06	8.96–05	–4.407	B	*LS*

N V: Allowed Transitions — Continued

No.	Transition Array	Mult.	λ_{air} (Å)	λ_{vac} (Å) or σ (cm^{-1})*	E_i (cm^{-1})	E_k (cm^{-1})	g_i-g_k	A_{ki} ($10^8 s^{-1}$)	f_{ik}	S (at. u.)	log gf	Acc.	Source
118.	$1s^24f$– $1s^211d$	$^2F^o$–^2D		672.025	618062	– 766866	14–10	1.26−02	6.08−05	1.88−03	−3.070	B	3
119.	$1s^24f$– $1s^212d$	$^2F^o$–^2D		656.078	618062	– 770483	14–10	9.49−03	4.37−05	1.32−03	−3.213	B	3
120.	$1s^24f$– $1s^25g$	$^2F^o$–^2G		1619.7	618062	– 679802	14–18	2.66+01	1.35+00	1.00+02	1.275	B	3
121.	$1s^24f$– $1s^26g$	$^2F^o$–^2G		1049.7	618062	– 713329	14–18	8.58+00	1.82−01	8.82+00	0.407	B	3
122.	$1s^24f$– $1s^27g$	$^2F^o$–^2G		865.899	618062	– 733549	14–18	4.03+00	5.83−02	2.33+00	−0.088	B	3
123.	$1s^24f$– $1s^28g$	$^2F^o$–^2G		777.549	618062	– 746671	14–18	2.28+00	2.66−02	9.53−01	−0.429	B	3
124.	$1s^24f$– $1s^29g$	$^2F^o$–^2G		726.713	618062	– 755668	14–18	1.43+00	1.46−02	4.88−01	−0.690	B	3
125.	$1s^24f$– $1s^210g$	$^2F^o$–^2G		694.247	618062	– 762103	14–18	9.68−01	9.00−03	2.88−01	−0.900	B	3
126.	$1s^25s$– $1s^25p$	^2S–$^2P^o$	*22593*	*4425* cm^{-1}	673886.2	– *678311*	2–6	3.00−02	6.88−01	1.02+02	0.138	A	1
			22565.7	4430.3 cm^{-1}	673886.2	– 678316.5	2–4	3.01−02	4.59−01	6.82+01	−0.037	A	*LS*
			22648.0	4414.2 cm^{-1}	673886.2	– 678300.4	2–2	2.97−02	2.29−01	3.41+01	−0.340	A	*LS*
127.	$1s^25s$– $1s^26p$	^2S–$^2P^o$	*2591.0*	*2591.8*	673886.2	– *712470*	2–6	9.44−01	2.85−01	4.86+00	−0.244	A−	1
			2590.81	2591.59	673886.2	– 712472.6	2–4	9.44−01	1.90−01	3.24+00	−0.420	A−	*LS*
			2591.44	2592.22	673886.2	– 712463.2	2–2	9.43−01	9.50−02	1.62+00	−0.721	A−	*LS*
128.	$1s^25s$– $1s^27p$	^2S–$^2P^o$		*1691.4*	673886.2	– *733007*	2–6	6.95−01	8.94−02	9.96−01	−0.748	B+	1
				1691.39	673886.2	– 733009.2	2–4	6.95−01	5.96−02	6.64−01	−0.924	B+	*LS*
				1691.56	673886.2	– 733003.3	2–2	6.95−01	2.98−02	3.32−01	−1.225	B+	*LS*
129.	$1s^25s$– $1s^28p$	^2S–$^2P^o$		*1380.8*	673886.2	– *746309*	2–6	4.89−01	4.20−02	3.82−01	−1.076	B	1
				1380.76	673886.2	– 746309.9	2–4	4.90−01	2.80−02	2.54−01	−1.252	B	*LS*
				1380.84	673886.2	– 746306.0	2–2	4.89−01	1.40−02	1.27−01	−1.553	B	*LS*
130.	$1s^25s$– $1s^29p$	^2S–$^2P^o$		*1226.6*	673886.2	– *755413*	2–6	3.49−01	2.36−02	1.91−01	−1.326	B	3
				1226.57	673886.2	– 755414.3	2–4	3.49−01	1.57−02	1.27−01	−1.502	B	*LS*
				1226.61	673886.2	– 755411.5	2–2	3.49−01	7.87−03	6.36−02	−1.803	B	*LS*
131.	$1s^25s$– $1s^210p$	^2S–$^2P^o$		*1136.0*	673886.2	– *761918*	2–6	2.60−01	1.51−02	1.13−01	−1.521	B	3
				1135.95	673886.2	– 761918.3	2–4	2.60−01	1.00−02	7.51−02	−1.697	B	*LS*
				1135.98	673886.2	– 761916.3	2–2	2.60−01	5.02−03	3.76−02	−1.998	B	*LS*
132.	$1s^25s$– $1s^211p$	^2S–$^2P^o$		1077.6	673886.2	– 766687	2–6	2.08−01	1.09−02	7.70−02	−1.664	B	3
133.	$1s^25s$– $1s^212p$	^2S–$^2P^o$		1036.7	673886.2	– 770347	2–6	1.62−01	7.81−03	5.33−02	−1.806	B	3

N v: Allowed Transitions — Continued

No.	Transition Array	Mult.	λ_{air} (Å)	λ_{vac} (Å) or σ (cm^{-1})*	E_i (cm^{-1})	E_k (cm^{-1})	g_i–g_k	A_{ki} (10^8 s^{-1})	f_{ik}	S (at. u.)	log gf	Acc.	Source
134.	$1s^25p$–$1s^25d$	$^2P^o$–^2D		*1405* cm^{-1}	*678311*	– *679716*	6–10	1.07–03	1.35–01	1.90+02	–0.091	A	1
				1401.1 cm^{-1}	678316.5	– 679717.6	4–6	1.06–03	1.21–01	1.14+02	–0.314	A	*LS*
				1412.3 cm^{-1}	678300.4	– 679712.7	2–4	9.04–04	1.36–01	6.33+01	–0.566	A	*LS*
				1396.2 cm^{-1}	678316.5	– 679712.7	4–4	1.75–04	1.34–02	1.27+01	–1.270	A	*LS*
135.	$1s^25p$–$1s^26s$	$^2P^o$–^2S	*3160.8*	*3161.7*	*678311*	– 709939.2	6–2	3.08+00	1.54–01	9.60+00	–0.035	A	1
			3161.37	3162.29	678316.5	– 709939.2	4–2	2.05+00	1.54–01	6.40+00	–0.211	A	*LS*
			3159.76	3160.68	678300.4	– 709939.2	2–2	1.03+00	1.54–01	3.20+00	–0.512	A	*LS*
136.	$1s^25p$–$1s^26d$	$^2P^o$–^2D	*2858.8*	*2859.6*	*678311*	– *713281*	6–10	2.59+00	5.29–01	2.98+01	0.501	A	1
			2859.16	2860.00	678316.5	– 713281.5	4–6	2.59+00	4.76–01	1.79+01	0.279	A	*LS*
			2858.03	2858.87	678300.4	– 713279.3	2–4	2.16+00	5.29–01	9.95+00	0.024	A	*LS*
			2859.34	2860.18	678316.5	– 713279.3	4–4	4.31–01	5.28–02	1.99+00	–0.675	A	*LS*
137.	$1s^25p$–$1s^27s$	$^2P^o$–^2S		*1882.7*	*678311*	– 731425.4	6–2	1.74+00	3.08–02	1.14+00	–0.734	A–	1
				1882.92	678316.5	– 731425.4	4–2	1.16+00	3.08–02	7.63–01	–0.910	A–	*LS*
				1882.35	678300.4	– 731425.4	2–2	5.80–01	3.08–02	3.82–01	–1.211	A–	*LS*
138.	$1s^25p$–$1s^27d$	$^2P^o$–^2D		*1811.4*	*678311*	– *733516*	6–10	1.74+00	1.42–01	5.10+00	–0.068	A	1
				1811.57	678316.5	– 733517.2	4–6	1.74+00	1.28–01	3.06+00	–0.290	A	*LS*
				1811.10	678300.4	– 733515.4	2–4	1.45+00	1.43–01	1.70+00	–0.545	A	*LS*
				1811.63	678316.5	– 733515.4	4–4	2.90–01	1.42–02	3.40–01	–1.244	A	*LS*
139.	$1s^25p$–$1s^28s$	$^2P^o$–^2S		*1493.8*	*678311*	– 745255.6	6–2	1.09+00	1.22–02	3.60–01	–1.136	B+	1
				1493.90	678316.5	– 745255.6	4–2	7.28–01	1.22–02	2.40–01	–1.312	B+	*LS*
				1493.54	678300.4	– 745255.6	2–2	3.64–01	1.22–02	1.20–01	–1.613	B+	*LS*
140.	$1s^25p$–$1s^28d$	$^2P^o$–^2D		*1463.3*	*678311*	– *746649*	6–10	1.18+00	6.30–02	1.82+00	–0.422	A–	1
				1463.42	678316.5	– 746649.7	4–6	1.18+00	5.67–02	1.09+00	–0.644	A–	*LS*
				1463.10	678300.4	– 746648.5	2–4	9.82–01	6.30–02	6.07–01	–0.900	A–	*LS*
				1463.44	678316.5	– 746648.5	4–4	1.96–01	6.30–03	1.21–01	–1.599	A–	*LS*
141.	$1s^25p$–$1s^29s$	$^2P^o$–^2S		*1309.5*	*678311*	– 754677.0	6–2	7.33–01	6.29–03	1.63–01	–1.424	B	3
				1309.58	678316.5	– 754677.0	4–2	4.89–01	6.28–03	1.08–01	–1.600	B	*LS*
				1309.30	678300.4	– 754677.0	2–2	2.45–01	6.29–03	5.42–02	–1.901	B	*LS*
142.	$1s^25p$–$1s^29d$	$^2P^o$–^2D		*1293.5*	*678311*	– *755623*	6–10	8.31–01	3.47–02	8.87–01	–0.681	B	3
				1293.38	678316.5	– 755633	4–6	8.31–01	3.13–02	5.32–01	–0.903	B	*LS*
				1293.53	678300.4	– 755608	2–4	6.92–01	3.47–02	2.96–01	–1.158	B	*LS*
				1293.80	678316.5	– 755608	4–4	1.38–01	3.47–03	5.92–02	–1.857	B	*LS*
143.	$1s^25p$–$1s^210s$	$^2P^o$–^2S		*1203.8*	*678311*	– 761382.7	6–2	5.24–01	3.80–03	9.03–02	–1.643	B	3
				1203.86	678316.5	– 761382.7	4–2	3.49–01	3.80–03	6.02–02	–1.819	B	*LS*
				1203.63	678300.4	– 761382.7	2–2	1.75–01	3.80–03	3.01–02	–2.120	B	*LS*
144.	$1s^25p$–$1s^210d$	$^2P^o$–^2D		*1194.0*	*678311*	– *762062*	6–10	6.09–01	2.17–02	5.12–01	–0.885	B	3
				1193.97	678316.5	– 762071	4–6	6.10–01	1.95–02	3.07–01	–1.107	B	*LS*
				1194.06	678300.4	– 762048	2–4	5.08–01	2.17–02	1.71–01	–1.362	B	*LS*
				1194.29	678316.5	– 762048	4–4	1.02–01	2.17–03	3.41–02	–2.061	B	*LS*

N v: Allowed Transitions — Continued

No.	Transition Array	Mult.	λ_{air} (Å)	λ_{vac} (Å) or σ (cm^{-1})*	E_i (cm^{-1})	E_k (cm^{-1})	g_i–g_k	A_{ki} (10^8 s^{-1})	f_{ik}	S (at. u.)	log gf	Acc.	Source
145.	$1s^25p-1s^211d$	$^2P^o-^2D$		1129.2	678311	– 766866	6–10	4.55–01	1.45–02	3.24–01	–1.060	B	3
146.	$1s^25p-1s^212d$	$^2P^o-^2D$		1084.9	678311	– 770483	6–10	3.51–01	1.03–02	2.21–01	–1.208	B	3
147.	$1s^25d-1s^26p$	$^2D-^2P^o$	*3052.2*	*3053.1*	*679716*	*– 712470*	10–6	7.82–01	6.55–02	6.58+00	–0.184	A	1
			3052.08	3052.97	679717.6	– 712472.6	6–4	7.04–01	6.55–02	3.95+00	–0.405	A	*LS*
			3052.50	3053.39	679712.7	– 712463.2	4–2	7.82–01	5.46–02	2.19+00	–0.661	A	*LS*
			3051.62	3052.51	679712.7	– 712472.6	4–4	7.82–02	1.09–02	4.39–01	–1.360	A	*LS*
148.	$1s^25d-1s^27p$	$^2D-^2P^o$		*1876.5*	*679716*	*– 733007*	10–6	4.27–01	1.35–02	8.35–01	–0.869	B+	1
				1876.47	679717.6	– 733009.2	6–4	3.84–01	1.35–02	5.01–01	–1.091	B+	*LS*
				1876.50	679712.7	– 733003.3	4–2	4.27–01	1.13–02	2.78–01	–1.346	B+	*LS*
				1876.30	679712.7	– 733009.2	4–4	4.27–02	2.25–03	5.57–02	–2.045	B+	*LS*
149.	$1s^25d-1s^28p$	$^2D-^2P^o$		*1501.7*	*679716*	*– 746309*	10–6	2.60–01	5.28–03	2.61–01	–1.277	B+	1
				1501.68	679717.6	– 746309.9	6–4	2.34–01	5.28–03	1.57–01	–1.499	B+	*LS*
				1501.65	679712.7	– 746306.0	4–2	2.60–01	4.40–03	8.71–02	–1.754	B+	*LS*
				1501.56	679712.7	– 746309.9	4–4	2.61–02	8.81–04	1.74–02	–2.453	B+	*LS*
150.	$1s^25d-1s^29p$	$^2D-^2P^o$		*1321.0*	*679716*	*– 755413*	10–6	1.72–01	2.69–03	1.17–01	–1.570	B	3
				1321.06	679717.6	– 755414.3	6–4	1.54–01	2.69–03	7.03–02	–1.792	B	*LS*
				1321.02	679712.7	– 755411.5	4–2	1.72–01	2.24–03	3.90–02	–2.047	B	*LS*
				1320.98	679712.7	– 755414.3	4–4	1.72–02	4.49–04	7.81–03	–2.746	B	*LS*
151.	$1s^25d-1s^210p$	$^2D-^2P^o$		*1216.5*	*679716*	*– 761918*	10–6	1.21–01	1.61–03	6.46–02	–1.792	B	3
				1216.53	679717.6	– 761918.3	6–4	1.09–01	1.61–03	3.88–02	–2.014	B	*LS*
				1216.49	679712.7	– 761916.3	4–2	1.21–01	1.35–03	2.15–02	–2.269	B	*LS*
				1216.46	679712.7	– 761918.3	4–4	1.21–02	2.69–04	4.31–03	–2.968	B	*LS*
152.	$1s^25d-1s^211p$	$^2D-^2P^o$		1149.8	679716	– 766687	10–6	9.55–02	1.14–03	4.30–02	–1.945	B	3
153.	$1s^25d-1s^212p$	$^2D-^2P^o$		1103.4	679716	– 770347	10–6	7.26–02	7.95–04	2.89–02	–2.100	B	3
154.	$1s^25d-1s^26f$	$^2D-^2F^o$	2974.5	2975.4	679716	– 713325	10–14	4.52+00	8.41–01	8.23+01	0.925	B	2
155.	$1s^25d-1s^27f$	$^2D-^2F^o$		1857.8	679716	– 733543	10–14	2.71+00	1.97–01	1.20+01	0.293	B	2
156.	$1s^25d-1s^28f$	$^2D-^2F^o$		1493.6	679716	– 746669	10–14	1.73+00	8.08–02	3.97+00	–0.093	B	2
157.	$1s^25d-1s^29f$	$^2D-^2F^o$		1316.6	679716	– 755667	10–14	1.18+00	4.28–02	1.86+00	–0.368	B	3
158.	$1s^25d-1s^210f$	$^2D-^2F^o$		1213.8	679716	– 762102	10–14	8.39–01	2.60–02	1.04+00	–0.586	B	3
159.	$1s^25f-1s^26d$	$^2F^o-^2D$	2985.1	2985.9	679790	– 713281	14–10	2.47–01	2.36–02	3.25+00	–0.481	B	2

N v: Allowed Transitions — Continued

No.	Transition Array	Mult.	λ_{air} (Å)	λ_{vac} (Å) or σ (cm^{-1})*	E_i (cm^{-1})	E_k (cm^{-1})	g_i–g_k	A_{ki} (10^8 s^{-1})	f_{ik}	S (at. u.)	log gf	Acc.	Source
160.	$1s^25f$–$1s^27d$	$^2F^o$–^2D		1861.3	679790	733516	14–10	1.21–01	4.49–03	3.85–01	–1.201	B	2
161.	$1s^25f$–$1s^28d$	$^2F^o$–^2D		1495.7	679790	746649	14–10	6.90–02	1.65–03	1.14–01	–1.636	B	2
162.	$1s^25f$–$1s^29d$	$^2F^o$–^2D		1318.7	679790	755623	14–10	4.65–02	8.65–04	5.26–02	–1.917	B	3
163.	$1s^25f$–$1s^210d$	$^2F^o$–^2D		1215.5	679790	762062	14–10	3.26–02	5.16–04	2.89–02	–2.141	B	3
164.	$1s^25f$–$1s^211d$	$^2F^o$–^2D		1148.4	679790	766866	14–10	2.22–02	3.13–04	1.66–02	–2.358	B	3
165.	$1s^25f$–$1s^212d$	$^2F^o$–^2D		1102.6	679790	770483	14–10	1.68–02	2.19–04	1.11–02	–2.513	B	3
166.	$1s^25f$–$1s^26g$	$^2F^o$–^2G	2980.8	2981.6	679790	713329	14–18	6.92+00	1.19+00	1.63+02	1.220	B	3
167.	$1s^25f$–$1s^27g$	$^2F^o$–^2G		1860.2	679790	733549	14–18	3.44+00	2.29–01	1.97+01	0.507	B	3
168.	$1s^25f$–$1s^28g$	$^2F^o$–^2G		1495.2	679790	746671	14–18	1.98+00	8.53–02	5.88+00	0.077	B	3
169.	$1s^25f$–$1s^29g$	$^2F^o$–^2G		1317.9	679790	755668	14–18	1.25+00	4.19–02	2.55+00	–0.231	B	3
170.	$1s^25f$–$1s^210g$	$^2F^o$–^2G		1214.9	679790	762103	14–18	8.50–01	2.42–02	1.35+00	–0.470	B	3
171.	$1s^25g$–$1s^26f$	^2G–$^2F^o$	2982.2	2983.1	679802	713325	18–14	7.14–02	7.41–03	1.31+00	–0.875	B	3
172.	$1s^25g$–$1s^27f$	^2G–$^2F^o$		1860.8	679802	733543	18–14	2.99–02	1.21–03	1.33–01	–1.663	B	3
173.	$1s^25g$–$1s^28f$	^2G–$^2F^o$		1495.5	679802	746669	18–14	1.60–02	4.16–04	3.69–02	–2.125	B	3
174.	$1s^25g$–$1s^29f$	^2G–$^2F^o$		1318.1	679802	755667	18–14	9.78–03	1.98–04	1.55–02	–2.448	B	3
175.	$1s^25g$–$1s^210f$	^2G–$^2F^o$		1215.1	679802	762102	18–14	6.53–03	1.12–04	8.09–03	–2.694	B	3
176.	$1s^26s$–$1s^26p$	^2S–$^2P^o$	*39510*	*2530* cm^{-1}	709939.2	*712470*	2–6	1.18–02	8.31–01	2.16+02	0.220	A	1
			39461.9	2533.4 cm^{-1}	709939.2	712472.6	2–4	1.19–02	5.54–01	1.44+02	0.045	A	*LS*
			39608.9	2524.0 cm^{-1}	709939.2	712463.2	2–2	1.17–02	2.76–01	7.20+01	–0.258	A	*LS*
177.	$1s^26s$–$1s^27p$	^2S–$^2P^o$	*4333.8*	*4335.0*	709939.2	*733007*	2–6	3.66–01	3.09–01	8.81+00	–0.209	B	1
			4333.42	4334.63	709939.2	733009.2	2–4	3.66–01	2.06–01	5.87+00	–0.385	B	*LS*
			4334.52	4335.74	709939.2	733003.3	2–2	3.65–01	1.03–01	2.94+00	–0.687	B	*LS*
178.	$1s^26s$–$1s^28p$	^2S–$^2P^o$	*2748.8*	*2749.6*	709939.2	*746309*	2–6	2.85–01	9.68–02	1.75+00	–0.713	B	1
			2748.65	2749.47	709939.2	746309.9	2–4	2.85–01	6.45–02	1.17+00	–0.889	B	*LS*
			2748.95	2749.76	709939.2	746306.0	2–2	2.85–01	3.23–02	5.84–01	–1.190	B	*LS*

N v: Allowed Transitions — Continued

No.	Transition Array	Mult.	λ_{air} (Å)	λ_{vac} (Å) or σ (cm^{-1})*	E_i (cm^{-1})	E_k (cm^{-1})	g_i–g_k	A_{ki} (10^8 s^{-1})	f_{ik}	S (at. u.)	log gf	Acc.	Source
179.	$1s^26s$–$1s^29p$	^2S–$^2P^o$	*2198.4*	*2199.1*	709939.2	*755413*	2–6	2.11–01	4.60–02	6.65–01	–1.037	B	3
			2198.32	2199.01	709939.2	755414.3	2–4	2.11–01	3.06–02	4.44–01	–1.213	B	*LS*
			2198.45	2199.14	709939.2	755411.5	2–2	2.11–01	1.53–02	2.22–01	–1.514	B	*LS*
180.	$1s^26s$–$1s^210p$	^2S–$^2P^o$		*1923.9*	709939.2	*761918*	2–6	1.56–01	2.60–02	3.29–01	–1.284	B	3
				1923.85	709939.2	761918.3	2–4	1.56–01	1.73–02	2.20–01	–1.460	B	*LS*
				1923.92	709939.2	761916.3	2–2	1.56–01	8.67–03	1.10–01	–1.761	B	*LS*
181.	$1s^26s$–$1s^211p$	^2S–$^2P^o$		1762.2	709939.2	766687	2–6	1.24–01	1.73–02	2.01–01	–1.460	B	3
182.	$1s^26s$–$1s^212p$	^2S–$^2P^o$		1655.4	709939.2	770347	2–6	9.57–02	1.18–02	1.29–01	–1.627	B	3
183.	$1s^26p$–$1s^26d$	$^2P^o$–^2D		*811* cm^{-1}	*712470*	*713281*	6–10	4.50–04	1.71–01	4.16+02	0.011	A	1
				808.9 cm^{-1}	712472.6	713281.5	4–6	4.46–04	1.53–01	2.50+02	–0.212	A	*LS*
				816.1 cm^{-1}	712463.2	713279.3	2–4	3.82–04	1.72–01	1.39+02	–0.464	A	*LS*
				806.7 cm^{-1}	712472.6	713279.3	4–4	7.38–05	1.70–02	2.77+01	–1.168	A	*LS*
184.	$1s^26p$–$1s^27s$	$^2P^o$–^2S	*5273.9*	*5275.4*	*712470*	731425.4	6–2	1.41+00	1.96–01	2.04+01	0.071	A	1
			5274.80	5276.27	712472.6	731425.4	4–2	9.41–01	1.96–01	1.36+01	–0.105	A	*LS*
			5272.18	5273.65	712463.2	731425.4	2–2	4.71–01	1.96–01	6.82+00	–0.406	A	*LS*
185.	$1s^26p$–$1s^27d$	$^2P^o$–^2D	*4749.9*	*4751.3*	*712470*	*733516*	6–10	9.65–01	5.44–01	5.10+01	0.514	A	1
			4750.48	4751.81	712472.6	733517.2	4–6	9.64–01	4.89–01	3.06+01	0.292	A	*LS*
			4748.77	4750.10	712463.2	733515.4	2–4	8.05–01	5.44–01	1.70+01	0.037	A	*LS*
			4750.89	4752.22	712472.6	733515.4	4–4	1.61–01	5.44–02	3.40+00	–0.663	A	*LS*
186.	$1s^26p$–$1s^28s$	$^2P^o$–^2S	*3049.2*	*3050.1*	*712470*	745255.6	6–2	8.47–01	3.94–02	2.37+00	–0.627	A	1
			3049.47	3050.36	712472.6	745255.6	4–2	5.65–01	3.94–02	1.58+00	–0.803	A	*LS*
			3048.60	3049.49	712463.2	745255.6	2–2	2.83–01	3.94–02	7.90–01	–1.104	A	*LS*
187.	$1s^26p$–$1s^28d$	$^2P^o$–^2D	*2924.9*	*2925.7*	*712470*	*746649*	6–10	6.94–01	1.48–01	8.57+00	–0.050	A–	1
			2925.08	2925.94	712472.6	746649.7	4–6	6.94–01	1.34–01	5.14+00	–0.272	A–	*LS*
			2924.38	2925.23	712463.2	746648.5	2–4	5.79–01	1.48–01	2.86+00	–0.527	A–	*LS*
			2925.18	2926.04	712472.6	746648.5	4–4	1.16–01	1.48–02	5.72–01	–1.226	A–	*LS*
188.	$1s^26p$–$1s^29s$	$^2P^o$–^2S	*2368.5*	*2369.2*	*712470*	754677.0	6–2	5.61–01	1.57–02	7.36–01	–1.025	B	3
			2368.70	2369.42	712472.6	754677.0	4–2	3.74–01	1.57–02	4.91–01	–1.201	B	*LS*
			2368.17	2368.89	712463.2	754677.0	2–2	1.87–01	1.57–02	2.45–01	–1.502	B	*LS*
189.	$1s^26p$–$1s^29d$	$^2P^o$–^2D	*2316.6*	*2317.3*	*712470*	*755623*	6–10	4.99–01	6.69–02	3.06+00	–0.396	B	3
			2316.23	2316.94	712472.6	755633	4–6	4.99–01	6.02–02	1.84+00	–0.618	B	*LS*
			2317.06	2317.78	712463.2	755608	2–4	4.15–01	6.69–02	1.02+00	–0.874	B	*LS*
			2317.57	2318.28	712472.6	755608	4–4	8.30–02	6.69–03	2.04–01	–1.573	B	*LS*
190.	$1s^26p$–$1s^210s$	$^2P^o$–^2S	*2043.8*	*2044.4*	*712470*	761382.7	6–2	3.87–01	8.09–03	3.27–01	–1.314	B	3
			2043.91	2044.57	712472.6	761382.7	4–2	2.58–01	8.09–03	2.18–01	–1.490	B	*LS*
			2043.52	2044.17	712463.2	761382.7	2–2	1.29–01	8.09–03	1.09–01	–1.791	B	*LS*

N v: Allowed Transitions — Continued

No.	Transition Array	Mult.	λ_{air} (Å)	λ_{vac} (Å) or σ (cm^{-1})*	E_i (cm^{-1})	E_k (cm^{-1})	g_i-g_k	A_{ki} (10^8 s^{-1})	f_{ik}	S (at. u.)	log gf	Acc.	Source
191.	$1s^26p-$ $1s^210d$	^{2}P°–^{2}D	*2015.8*	*2016.4*	*712470*	– *762062*	6–10	3.63–01	3.69–02	1.47+00	–0.655	B	3
			2015.54	2016.19	712472.6	– 762071	4–6	3.63–01	3.32–02	8.82–01	–0.877	B	*LS*
			2016.10	2016.75	712463.2	– 762048	2–4	3.03–01	3.69–02	4.90–01	–1.132	B	*LS*
			2016.48	2017.13	712472.6	– 762048	4–4	6.05–02	3.69–03	9.80–02	–1.831	B	*LS*
192.	$1s^26p-$ $1s^211d$	^{2}P°–^{2}D		1838.4	712470	– 766866	6–10	2.68–01	2.27–02	8.23–01	–0.867	B	3
193.	$1s^26p-$ $1s^212d$	^{2}P°–^{2}D		1723.7	712470	– 770483	6–10	2.06–01	1.53–02	5.20–01	–1.038	B	3
194.	$1s^26d-$ $1s^27p$	^{2}D–^{2}P°	*5067.9*	*5069.3*	*713281*	– *733007*	10–6	4.09–01	9.46–02	1.58+01	–0.024	A	1
			5067.60	5069.01	713281.5	– 733009.2	6–4	3.69–01	9.46–02	9.47+00	–0.246	A	*LS*
			5068.55	5069.97	713279.3	– 733003.3	4–2	4.09–01	7.88–02	5.26+00	–0.501	A	*LS*
			5067.04	5068.45	713279.3	– 733009.2	4–4	4.10–02	1.58–02	1.05+00	–1.200	A	*LS*
195.	$1s^26d-$ $1s^28p$	^{2}D–^{2}P°	*3026.9*	*3027.7*	*713281*	– *746309*	10–6	2.41–01	1.98–02	1.98+00	–0.703	B+	1
			3026.82	3027.70	713281.5	– 746309.9	6–4	2.16–01	1.98–02	1.19+00	–0.925	B+	*LS*
			3026.97	3027.85	713279.3	– 746306.0	4–2	2.41–01	1.65–02	6.58–01	–1.180	B+	*LS*
			3026.61	3027.50	713279.3	– 746309.9	4–4	2.41–02	3.30–03	1.32–01	–1.879	B+	*LS*
196.	$1s^26d-$ $1s^29p$	^{2}D–^{2}P°	*2372.7*	*2373.4*	*713281*	– *755413*	10–6	1.57–01	7.97–03	6.23–01	–1.098	B	3
			2372.72	2373.45	713281.5	– 755414.3	6–4	1.42–01	7.97–03	3.74–01	–1.320	B	*LS*
			2372.76	2373.48	713279.3	– 755411.5	4–2	1.57–01	6.64–03	2.08–01	–1.575	B	*LS*
			2372.60	2373.32	713279.3	– 755414.3	4–4	1.57–02	1.33–03	4.15–02	–2.274	B	*LS*
197.	$1s^26d-$ $1s^210p$	^{2}D–^{2}P°	*2055.4*	*2056.0*	*713281*	– *761918*	10–6	1.06–01	4.04–03	2.74–01	–1.393	B	3
			2055.40	2056.06	713281.5	– 761918.3	6–4	9.57–02	4.04–03	1.64–01	–1.615	B	*LS*
			2055.39	2056.05	713279.3	– 761916.3	4–2	1.06–01	3.37–03	9.12–02	–1.870	B	*LS*
			2055.31	2055.96	713279.3	– 761918.3	4–4	1.06–02	6.74–04	1.82–02	–2.569	B	*LS*
198.	$1s^26d-$ $1s^211p$	^{2}D–^{2}P°		1872.4	713281	– 766687	10–6	8.02–02	2.53–03	1.56–01	–1.597	B	3
199.	$1s^26d-$ $1s^212p$	^{2}D–^{2}P°		1752.3	713281	– 770347	10–6	5.92–02	1.64–03	9.44–02	–1.786	B	3
200.	$1s^26d-$ $1s^27f$	^{2}D–^{2}F°	4933.9	4935.3	713281	– 733543	10–14	1.62+00	8.29–01	1.35+02	0.918	B	2
201.	$1s^26d-$ $1s^28f$	^{2}D–^{2}F°	2994.2	2995.0	713281	– 746669	10–14	1.08+00	2.03–01	2.00+01	0.308	B	2
202.	$1s^26d-$ $1s^29f$	^{2}D–^{2}F°	2358.6	2359.3	713281	– 755667	10–14	7.35–01	8.59–02	6.67+00	–0.066	B	3
203.	$1s^26d-$ $1s^210f$	^{2}D–^{2}F°	2047.6	2048.3	713281	– 762102	10–14	5.19–01	4.57–02	3.08+00	–0.340	B	3
204.	$1s^26f-$ $1s^27d$	^{2}F°–^{2}D	4951.1	4952.5	713325	– 733516	14–10	1.60–01	4.19–02	9.57+00	–0.231	B	2
205.	$1s^26f-$ $1s^28d$	^{2}F°–^{2}D	2999.9	3000.8	713325	– 746649	14–10	8.62–02	8.31–03	1.15+00	–0.934	B	2
206.	$1s^26f-$ $1s^29d$	^{2}F°–^{2}D	2363.4	2364.2	713325	– 755623	14–10	5.46–02	3.27–03	3.56–01	–1.339	B	3

N v: Allowed Transitions — Continued

No.	Transition Array	Mult.	λ_{air} (Å)	λ_{vac} (Å) or σ (cm^{-1})*	E_i (cm^{-1})	E_k (cm^{-1})	g_i-g_k	A_{ki} (10^8 s^{-1})	f_{ik}	S (at. u.)	log gf	Acc.	Source
207.	$1s^26f-1s^210d$	$^2F^o-^2D$	2051.2	2051.8	713325	– 762062	14–10	3.56–02	1.60–03	1.52–01	–1.649	B	3
208.	$1s^26f-1s^211d$	$^2F^o-^2D$		1867.7	713325	– 766866	14–10	2.29–02	8.55–04	7.36–02	–1.922	B	3
209.	$1s^26f-1s^212d$	$^2F^o-^2D$		1749.5	713325	– 770483	14–10	1.66–02	5.43–04	4.38–02	–2.119	B	3
210.	$1s^26f-1s^27g$	$^2F^o-^2G$	4943.2	4944.5	713325	– 733549	14–18	2.36+00	1.11+00	2.53+02	1.192	B	3
211.	$1s^26f-1s^28g$	$^2F^o-^2G$	2997.9	2998.8	713325	– 746671	14–18	1.41+00	2.45–01	3.39+01	0.536	B	3
212.	$1s^26f-1s^29g$	$^2F^o-^2G$	2360.9	2361.7	713325	– 755668	14–18	8.96–01	9.63–02	1.05+01	0.130	B	3
213.	$1s^26f-1s^210g$	$^2F^o-^2G$	2049.4	2050.1	713325	– 762103	14–18	6.06–01	4.91–02	4.64+00	–0.163	B	3
214.	$1s^26g-1s^27f$	$^2G-^2F^o$	4945.7	4947.1	713329	– 733543	18–14	7.07–02	2.02–02	5.92+00	–0.440	B	3
215.	$1s^26g-1s^28f$	$^2G-^2F^o$	2998.5	2999.4	713329	– 746669	18–14	3.31–02	3.47–03	6.16–01	–1.205	B	3
216.	$1s^26g-1s^29f$	$^2G-^2F^o$	2361.2	2362.0	713329	– 755667	18–14	1.76–02	1.15–03	1.60–01	–1.686	B	3
217.	$1s^26g-1s^210f$	$^2G-^2F^o$	2049.6	2050.3	713329	– 762102	18–14	1.04–02	5.12–04	6.22–02	–2.035	B	3
218.	$1s^27s-1s^27p$	$^2S-^2P^o$		*1582* cm^{-1}	731425.4	– *733007*	2–6	5.44–03	9.77–01	4.07+02	0.291	B	2
				1583.8 cm^{-1}	731425.4	– 733009.2	2–4	5.46–03	6.52–01	2.71+02	0.115	B	*LS*
				1577.9 cm^{-1}	731425.4	– 733003.3	2–2	5.40–03	3.25–01	1.36+02	–0.187	B	*LS*
219.	$1s^27s-1s^28p$	$^2S-^2P^o$	*6717.1*	*6719.0*	731425.4	– *746309*	2–6	1.63–01	3.31–01	1.47+01	–0.179	B	2
			6716.54	6718.40	731425.4	– 746309.9	2–4	1.63–01	2.21–01	9.77+00	–0.355	B	*LS*
			6718.30	6720.16	731425.4	– 746306.0	2–2	1.63–01	1.10–01	4.89+00	–0.656	B	*LS*
220.	$1s^27s-1s^29p$	$^2S-^2P^o$	*4167.6*	*4168.8*	731425.4	– *755413*	2–6	1.32–01	1.03–01	2.83+00	–0.686	B	3
			4167.42	4168.59	731425.4	– 755414.3	2–4	1.32–01	6.87–02	1.88+00	–0.862	B	*LS*
			4167.91	4169.08	731425.4	– 755411.5	2–2	1.32–01	3.43–02	9.42–01	–1.163	B	*LS*
221.	$1s^27s-1s^210p$	$^2S-^2P^o$	*3278.6*	*3279.5*	731425.4	– *761918*	2–6	1.03–01	4.96–02	1.07+00	–1.003	B	3
			3278.51	3279.45	731425.4	– 761918.3	2–4	1.03–01	3.31–02	7.14–01	–1.180	B	*LS*
			3278.72	3279.67	731425.4	– 761916.3	2–2	1.03–01	1.65–02	3.57–01	–1.481	B	*LS*
222.	$1s^27s-1s^211p$	$^2S-^2P^o$	2835.1	2835.9	731425.4	– 766687	2–6	8.47–02	3.06–02	5.72–01	–1.213	B	3
223.	$1s^27s-1s^212p$	$^2S-^2P^o$	2568.5	2569.3	731425.4	– 770347	2–6	6.67–02	1.98–02	3.35–01	–1.403	B	3

N v: Allowed Transitions — Continued

No.	Transition Array	Mult.	λ_{air} (Å)	λ_{vac} (Å) or σ (cm^{-1})*	E_i (cm^{-1})	E_k (cm^{-1})	g_i-g_k	A_{ki} (10^8 s^{-1})	f_{ik}	S (at. u.)	log gf	Acc.	Source
224.	$1s^27p-1s^27d$	$^2P^o-^2D$		*509* cm^{-1}	*733007*	*733516*	6–10	2.03–04	1.95–01	7.58+02	0.069	B	2
				508.0 cm^{-1}	733009.2	733517.2	4–6	2.01–04	1.75–01	4.55+02	–0.154	B	*LS*
				512.1 cm^{-1}	733003.3	733515.4	2–4	1.72–04	1.96–01	2.53+02	–0.406	B	*LS*
				506.2 cm^{-1}	733009.2	733515.4	4–4	3.32–05	1.94–02	5.05+01	–1.110	B	*LS*
225.	$1s^27p-1s^28s$	$^2P^o-^2S$	*8162.1*	*8164.4*	*733007*	745255.6	6–2	7.21–01	2.40–01	3.87+01	0.159	B	2
			8163.42	8165.67	733009.2	745255.6	4–2	4.80–01	2.40–01	2.58+01	–0.017	B	*LS*
			8159.49	8161.73	733003.3	745255.6	2–2	2.41–01	2.40–01	1.29+01	–0.318	B	*LS*
226.	$1s^27p-1s^28d$	$^2P^o-^2D$	*7328.3*	*7330.3*	*733007*	*746649*	6–10	4.23–01	5.68–01	8.23+01	0.533	B	2
			7329.09	7331.11	733009.2	746649.7	4–6	4.23–01	5.11–01	4.94+01	0.311	B	*LS*
			7326.57	7328.58	733003.3	746648.5	2–4	3.53–01	5.68–01	2.74+01	0.056	B	*LS*
			7329.74	7331.75	733009.2	746648.5	4–4	7.05–02	5.68–02	5.48+00	–0.644	B	*LS*
227.	$1s^27p-1s^29s$	$^2P^o-^2S$	*4613.4*	*4614.7*	*733007*	754677.0	6–2	4.48–01	4.77–02	4.35+00	–0.543	B	3
			4613.85	4615.14	733009.2	754677.0	4–2	2.99–01	4.77–02	2.90+00	–0.719	B	*LS*
			4612.59	4613.89	733003.3	754677.0	2–2	1.49–01	4.77–02	1.45+00	–1.020	B	*LS*
228.	$1s^27p-1s^29d$	$^2P^o-^2D$	*4420.5*	*4421.7*	*733007*	*755623*	6–10	3.18–01	1.55–01	1.36+01	–0.031	B	3
			4418.88	4420.12	733009.2	755633	4–6	3.18–01	1.40–01	8.14+00	–0.252	B	*LS*
			4422.62	4423.86	733003.3	755608	2–4	2.65–01	1.55–01	4.52+00	–0.508	B	*LS*
			4423.77	4425.01	733009.2	755608	4–4	5.29–02	1.55–02	9.04–01	–1.207	B	*LS*
229.	$1s^27p-1s^210s$	$^2P^o-^2S$	*3523.2*	*3524.2*	*733007*	761382.7	6–2	3.12–01	1.93–02	1.35+00	–0.935	B	3
			3523.41	3524.42	733009.2	761382.7	4–2	2.08–01	1.93–02	8.98–01	–1.112	B	*LS*
			3522.68	3523.68	733003.3	761382.7	2–2	1.04–01	1.93–02	4.49–01	–1.412	B	*LS*
230.	$1s^27p-1s^210d$	$^2P^o-^2D$	*3440.8*	*3441.8*	*733007*	*762062*	6–10	2.40–01	7.10–02	4.83+00	–0.370	B	3
			3439.96	3440.94	733009.2	762071	4–6	2.40–01	6.40–02	2.90+00	–0.592	B	*LS*
			3441.98	3442.97	733003.3	762048	2–4	2.00–01	7.10–02	1.61+00	–0.848	B	*LS*
			3442.68	3443.67	733009.2	762048	4–4	3.99–02	7.10–03	3.22–01	–1.547	B	*LS*
231.	$1s^27p-1s^211d$	$^2P^o-^2D$	2952.6	2953.4	733007	766866	6–10	1.81–01	3.94–02	2.30+00	–0.626	B	3
232.	$1s^27p-1s^212d$	$^2P^o-^2D$	2667.6	2668.4	733007	770483	6–10	1.41–01	2.50–02	1.32+00	–0.823	B	3
233.	$1s^27d-1s^28p$	$^2D-^2P^o$	*7815.2*	*7817.3*	*733516*	*746309*	10–6	2.26–01	1.24–01	3.20+01	0.094	B	2
			7814.81	7816.96	733517.2	746309.9	6–4	2.04–01	1.24–01	1.92+01	–0.127	B	*LS*
			7816.09	7818.24	733515.4	746306.0	4–2	2.26–01	1.04–01	1.07+01	–0.383	B	*LS*
			7813.71	7815.86	733515.4	746309.9	4–4	2.26–02	2.07–02	2.13+00	–1.082	B	*LS*
234.	$1s^27d-1s^29p$	$^2D-^2P^o$	*4565.6*	*4566.9*	*733516*	*755413*	10–6	1.40–01	2.63–02	3.95+00	–0.580	B	3
			4565.54	4566.81	733517.2	755414.3	6–4	1.26–01	2.63–02	2.37+00	–0.802	B	*LS*
			4565.74	4567.02	733515.4	755411.5	4–2	1.40–01	2.19–02	1.32+00	–1.057	B	*LS*
			4565.16	4566.44	733515.4	755414.3	4–4	1.40–02	4.38–03	2.64–01	–1.756	B	*LS*
235.	$1s^27d-1s^210p$	$^2D-^2P^o$	*3520.0*	*3521.0*	*733516*	*761918*	10–6	9.60–02	1.07–02	1.24+00	–0.970	B	3
			3519.98	3520.99	733517.2	761918.3	6–4	8.64–02	1.07–02	7.45–01	–1.192	B	*LS*
			3520.01	3521.02	733515.4	761916.3	4–2	9.60–02	8.92–03	4.14–01	–1.447	B	*LS*
			3519.76	3520.77	733515.4	761918.3	4–4	9.61–03	1.79–03	8.28–02	–2.146	B	*LS*

N V: Allowed Transitions — Continued

No.	Transition Array	Mult.	λ_{air} (Å)	λ_{vac} (Å) or σ (cm^{-1})*	E_i (cm^{-1})	E_k (cm^{-1})	g_i-g_k	A_{ki} (10^8 s^{-1})	f_{ik}	S (at. u.)	log gf	Acc.	Source
236.	$1s^27d-$ $1s^211p$	$^2D-^2P^o$	3013.8	3014.7	733516	766687	10–6	7.38–02	6.03–03	5.99–01	–1.219	B	3
237.	$1s^27d-$ $1s^212p$	$^2D-^2P^o$	2714.3	2715.1	733516	770347	10–6	5.54–02	3.67–03	3.28–01	–1.435	B	3
238.	$1s^27d-$ $1s^28f$	$^2D-^2F^o$	7600.9	7603.0	733516	746669	10–14	6.88–01	8.34–01	2.09+02	0.921	B	2
239.	$1s^27d-$ $1s^29f$	$^2D-^2F^o$	4513.4	4514.7	733516	755667	10–14	4.89–01	2.09–01	3.11+01	0.320	B	3
240.	$1s^27d-$ $1s^210f$	$^2D-^2F^o$	3497.3	3498.3	733516	762102	10–14	3.56–01	9.13–02	1.05+01	–0.039	B	3
241.	$1s^27f-$ $1s^28d$	$^2F^o-^2D$	7627.7	7629.8	733543	746649	14–10	1.01–01	6.28–02	2.21+01	–0.056	B	2
242.	$1s^27f-$ $1s^29d$	$^2F^o-^2D$	4527.7	4528.9	733543	755623	14–10	5.93–02	1.30–02	2.72+00	–0.739	B	3
243.	$1s^27f-$ $1s^210d$	$^2F^o-^2D$	3505.4	3506.4	733543	762062	14–10	3.95–02	5.20–03	8.41–01	–1.138	B	3
244.	$1s^27f-$ $1s^211d$	$^2F^o-^2D$	3000.0	3000.9	733543	766866	14–10	2.65–02	2.55–03	3.53–01	–1.447	B	3
245.	$1s^27f-$ $1s^212d$	$^2F^o-^2D$	2706.3	2707.1	733543	770483	14–10	1.96–02	1.54–03	1.92–01	–1.666	B	3
246.	$1s^27f-$ $1s^28g$	$^2F^o-^2G$	7614.9	7617.0	733543	746671	14–18	9.57–01	1.07+00	3.76+02	1.176	B	3
247.	$1s^27f-$ $1s^29g$	$^2F^o-^2G$	4518.5	4519.7	733543	755668	14–18	6.45–01	2.54–01	5.29+01	0.551	B	3
248.	$1s^27f-$ $1s^210g$	$^2F^o-^2G$	3500.4	3501.4	733543	762103	14–18	4.49–01	1.06–01	1.71+01	0.172	B	3
249.	$1s^27g-$ $1s^28f$	$^2G-^2F^o$	7619.7	7621.8	733549	746669	18–14	5.21–02	3.53–02	1.60+01	–0.197	B	3
250.	$1s^27g-$ $1s^29f$	$^2G-^2F^o$	4520.0	4521.3	733549	755667	18–14	2.75–02	6.56–03	1.76+00	–0.928	B	3
251.	$1s^27g-$ $1s^210f$	$^2G-^2F^o$	3501.2	3502.2	733549	762102	18–14	1.72–02	2.46–03	5.10–01	–1.354	B	3
252.	$1s^28s-$ $1s^28p$	$^2S-^2P^o$		*1053* cm^{-1}	745255.6	*746309*	2–6	2.76–03	1.12+00	7.00+02	0.350	B	2
				1054.3 cm^{-1}	745255.6	746309.9	2–4	2.77–03	7.47–01	4.66+02	0.174	B	*LS*
				1050.4 cm^{-1}	745255.6	746306.0	2–2	2.74–03	3.72–01	2.33+02	–0.128	B	*LS*
253.	$1s^28s-$ $1s^29p$	$^2S-^2P^o$	*9842.0*	*9844.7*	745255.6	*755413*	2–6	8.31–02	3.62–01	2.35+01	–0.140	B	3
			9841.08	9843.78	745255.6	755414.3	2–4	8.31–02	2.42–01	1.57+01	–0.316	B	*LS*
			9843.79	9846.49	745255.6	755411.5	2–2	8.31–02	1.21–01	7.83+00	–0.617	B	*LS*
254.	$1s^28s-$ $1s^210p$	$^2S-^2P^o$	*6000.0*	*6001.7*	745255.6	*761918*	2–6	6.86–02	1.11–01	4.39+00	–0.653	B	3
			5999.77	6001.43	745255.6	761918.3	2–4	6.86–02	7.41–02	2.93+00	–0.829	B	*LS*
			6000.49	6002.15	745255.6	761916.3	2–2	6.86–02	3.71–02	1.46+00	–1.130	B	*LS*

N v: Allowed Transitions — Continued

No.	Transition Array	Mult.	λ_{air} (Å)	λ_{vac} (Å) or σ (cm^{-1})*	E_i (cm^{-1})	E_k (cm^{-1})	g_i–g_k	A_{ki} ($10^8\ s^{-1}$)	f_{ik}	S (at. u.)	log gf	Acc.	Source
255.	$1s^28s$–$1s^211p$	^2S–$^2P^o$	4664.7	4666.1	745255.6	766687	2–6	5.57–02	5.46–02	1.68+00	–0.962	B	3
256.	$1s^28s$–$1s^212p$	^2S–$^2P^o$	3984.3	3985.4	745255.6	770347	2–6	4.29–02	3.06–02	8.04–01	–1.213	B	3
257.	$1s^28p$–$1s^29s$	$^2P^o$–^2S	*11946*	*8368* cm^{-1}	*746309*	754677.0	6–2	3.99–01	2.84–01	6.71+01	0.232	B	3
			11948.3	8367.1 cm^{-1}	746309.9	754677.0	4–2	2.66–01	2.84–01	4.48+01	0.056	B	*LS*
			11942.7	8371.0 cm^{-1}	746306.0	754677.0	2–2	1.33–01	2.84–01	2.24+01	–0.245	B	*LS*
258.	$1s^28p$–$1s^210s$	$^2P^o$–^2S	*6632.1*	*6633.9*	*746309*	761382.7	6–2	2.56–01	5.63–02	7.38+00	–0.471	B	3
			6632.64	6634.47	746309.9	761382.7	4–2	1.71–01	5.63–02	4.92+00	–0.647	B	*LS*
			6630.92	6632.75	746306.0	761382.7	2–2	8.54–02	5.63–02	2.46+00	–0.948	B	*LS*
259.	$1s^28p$–$1s^28d$	$^2P^o$–^2D		*341* cm^{-1}	*746309*	*746649*	6–10	1.06–04	2.28–01	1.32+03	0.135	B	2
				339.8 cm^{-1}	746309.9	746649.7	4–6	1.05–04	2.04–01	7.92+02	–0.088	B	*LS*
				342.5 cm^{-1}	746306.0	746648.5	2–4	8.95–05	2.29–01	4.40+02	–0.340	B	*LS*
				338.6 cm^{-1}	746309.9	746648.5	4–4	1.73–05	2.26–02	8.80+01	–1.044	B	*LS*
260.	$1s^28p$–$1s^29d$	$^2P^o$–^2D	*10733*	*9314* cm^{-1}	*746309*	*755623*	6–10	2.09–01	6.03–01	1.28+02	0.558	B	3
			10723.1	9323 cm^{-1}	746309.9	755633	4–6	2.10–01	5.43–01	7.67+01	0.337	B	*LS*
			10747.4	9302 cm^{-1}	746306.0	755608	2–4	1.74–01	6.02–01	4.26+01	0.081	B	*LS*
			10751.9	9298 cm^{-1}	746309.9	755608	4–4	3.47–02	6.02–02	8.52+00	–0.619	B	*LS*
261.	$1s^28p$–$1s^210d$	$^2P^o$–^2D	*6346.2*	*6347.9*	*746309*	*762062*	6–10	1.63–01	1.64–01	2.05+01	–0.008	B	3
			6342.98	6344.73	746309.9	762071	4–6	1.63–01	1.47–01	1.23+01	–0.229	B	*LS*
			6350.68	6352.43	746306.0	762048	2–4	1.35–01	1.64–01	6.84+00	–0.485	B	*LS*
			6352.25	6354.01	746309.9	762048	4–4	2.70–02	1.64–02	1.37+00	–1.184	B	*LS*
262.	$1s^28p$–$1s^211d$	$^2P^o$–^2D	4863.1	4864.4	746309	766866	6–10	1.20–01	7.07–02	6.79+00	–0.372	B	3
263.	$1s^28p$–$1s^212d$	$^2P^o$–^2D	4135.4	4136.6	746309	770483	6–10	9.16–02	3.92–02	3.20+00	–0.629	B	3
264.	$1s^28d$–$1s^29p$	^2D–$^2P^o$	*11407*	*8764* cm^{-1}	*746649*	*755413*	10–6	1.36–01	1.59–01	5.98+01	0.202	B	3
			11406.4	8764.6 cm^{-1}	746649.7	755414.3	6–4	1.22–01	1.59–01	3.59+01	–0.020	B	*LS*
			11408.5	8763.0 cm^{-1}	746648.5	755411.5	4–2	1.36–01	1.33–01	1.99+01	–0.276	B	*LS*
			11404.8	8765.8 cm^{-1}	746648.5	755414.3	4–4	1.36–02	2.65–02	3.98+00	–0.974	B	*LS*
265.	$1s^28d$–$1s^210p$	^2D–$^2P^o$	*6547.7*	*6549.5*	*746649*	*761918*	10–6	8.67–02	3.34–02	7.21+00	–0.476	B	3
			6547.58	6549.39	746649.7	761918.3	6–4	7.80–02	3.34–02	4.33+00	–0.698	B	*LS*
			6547.92	6549.73	746648.5	761916.3	4–2	8.67–02	2.79–02	2.40+00	–0.953	B	*LS*
			6547.07	6548.87	746648.5	761918.3	4–4	8.67–03	5.57–03	4.81–01	–1.652	B	*LS*
266.	$1s^28d$–$1s^211p$	^2D–$^2P^o$	4989.2	4990.6	746649	766687	10–6	6.12–02	1.37–02	2.25+00	–0.863	B	3
267.	$1s^28d$–$1s^212p$	^2D–$^2P^o$	4218.6	4219.8	746649	770347	10–6	4.33–02	6.93–03	9.63–01	–1.159	B	3
268.	$1s^28d$–$1s^29f$	^2D–$^2F^o$	11087	9017 cm^{-1}	746649	755667	10–14	3.30–01	8.51–01	3.11+02	0.930	B	3

N V: Allowed Transitions — Continued

No.	Transition Array	Mult.	λ_{air} (Å)	λ_{vac} (Å) or σ (cm^{-1})*	E_i (cm^{-1})	E_k (cm^{-1})	g_i-g_k	A_{ki} (10^8 s^{-1})	f_{ik}	S (at. u.)	log gf	Acc.	Source
269.	$1s^28d-$ $1s^210f$	$^2D-^2F^\circ$	6469.5	6471.3	746649	– 762102	10–14	2.44–01	2.14–01	4.57+01	0.331	B	3
270.	$1s^28f-$ $1s^29d$	$^2F^\circ-^2D$	11165	8954 cm^{-1}	746669	– 755623	14–10	6.70–02	8.95–02	4.61+01	0.098	B	3
271.	$1s^28f-$ $1s^210d$	$^2F^\circ-^2D$	6494.8	6496.6	746669	– 762062	14–10	4.05–02	1.83–02	5.48+00	–0.591	B	3
272.	$1s^28f-$ $1s^211d$	$^2F^\circ-^2D$	4949.9	4951.3	746669	– 766866	14–10	2.43–02	6.38–03	1.45+00	–1.049	B	3
273.	$1s^28f-$ $1s^212d$	$^2F^\circ-^2D$	4198.1	4199.2	746669	– 770483	14–10	1.66–02	3.13–03	6.06–01	–1.358	B	3
274.	$1s^28f-$ $1s^29g$	$^2F^\circ-^2G$	11110	8999 cm^{-1}	746669	– 755668	14–18	4.47–01	1.06+00	5.44+02	1.173	B	3
275.	$1s^28f-$ $1s^210g$	$^2F^\circ-^2G$	6477.5	6479.2	746669	– 762103	14–18	3.18–01	2.57–01	7.68+01	0.557	B	3
276.	$1s^28g-$ $1s^29f$	$^2G-^2F^\circ$	11114	8995 cm^{-1}	746671	– 755667	18–14	3.83–02	5.52–02	3.63+01	–0.003	B	3
277.	$1s^28g-$ $1s^210f$	$^2G-^2F^\circ$	6478.8	6480.5	746671	– 762102	18–14	2.06–02	1.01–02	3.87+00	–0.741	B	3
278.	$1s^29s-$ $1s^29p$	$^2S-^2P^\circ$		*736* cm^{-1}	754677.0	– *755413*	2–6	1.51–03	1.26+00	1.12+03	0.400	B	3
				737.3 cm^{-1}	754677.0	– 755414.3	2–4	1.52–03	8.38–01	7.48+02	0.224	B	*LS*
				734.5 cm^{-1}	754677.0	– 755411.5	2–2	1.50–03	4.17–01	3.74+02	–0.079	B	*LS*
279.	$1s^29s-$ $1s^210p$	$^2S-^2P^\circ$	*13807*	*7241* cm^{-1}	754677.0	– *761918*	2–6	4.47–02	3.84–01	3.49+01	–0.115	B	3
			13805.9	7241.3 cm^{-1}	754677.0	– 761918.3	2–4	4.47–02	2.56–01	2.32+01	–0.291	B	*LS*
			13809.7	7239.3 cm^{-1}	754677.0	– 761916.3	2–2	4.47–02	1.28–01	1.16+01	–0.592	B	*LS*
280.	$1s^29s-$ $1s^211p$	$^2S-^2P^\circ$	8324.1	8326.4	754677.0	– 766687	2–6	4.16–02	1.30–01	7.11+00	–0.586	B	3
281.	$1s^29s-$ $1s^212p$	$^2S-^2P^\circ$	6379.9	6381.6	754677.0	– 770347	2–6	3.28–02	6.00–02	2.52+00	–0.921	B	3
282.	$1s^29p-$ $1s^210s$	$^2P^\circ-^2S$	*16748*	*5969* cm^{-1}	*755413*	– 761382.7	6–2	2.32–01	3.26–01	1.08+02	0.291	B	3
			16750.3	5968.4 cm^{-1}	755414.3	– 761382.7	4–2	1.55–01	3.26–01	7.18+01	0.115	B	*LS*
			16742.5	5971.2 cm^{-1}	755411.5	– 761382.7	2–2	7.75–02	3.26–01	3.59+01	–0.186	B	*LS*
283.	$1s^29p-$ $1s^29d$	$^2P^\circ-^2D$		*210* cm^{-1}	*755413*	– *755623*	6–10	4.37–05	2.49–01	2.34+03	0.174	B	3
				219 cm^{-1}	755414.3	– 755633	4–6	4.97–05	2.34–01	1.41+03	–0.030	B	*LS*
				197 cm^{-1}	755411.5	– 755608	2–4	3.00–05	2.33–01	7.81+02	–0.331	B	*LS*
				194 cm^{-1}	755414.3	– 755608	4–4	5.75–06	2.30–02	1.56+02	–1.037	B	*LS*
284.	$1s^29p-$ $1s^210d$	$^2P^\circ-^2D$	*15037*	*6648* cm^{-1}	*755413*	– *762062*	6–10	1.12–01	6.32–01	1.88+02	0.579	B	3
			15018.4	6657 cm^{-1}	755414.3	– 762071	4–6	1.12–01	5.70–01	1.13+02	0.358	B	*LS*
			15064.1	6637 cm^{-1}	755411.5	– 762048	2–4	9.27–02	6.31–01	6.26+01	0.101	B	*LS*
			15070.4	6634 cm^{-1}	755414.3	– 762048	4–4	1.85–02	6.31–02	1.25+01	–0.598	B	*LS*

N V: Allowed Transitions — Continued

No.	Transition Array	Mult.	λ_{air} (Å)	λ_{vac} (Å) or σ (cm^{-1})*	E_i (cm^{-1})	E_k (cm^{-1})	g_i–g_k	A_{ki} (10^8 s^{-1})	f_{ik}	S (at. u.)	log gf	Acc.	Source
285.	$1s^29p$–$1s^211d$	$^2P^o$–^2D	8729.2	8731.6	755413	– 766866	6–10	8.83–02	1.68–01	2.90+01	0.004	B	3
286.	$1s^29p$–$1s^212d$	$^2P^o$–^2D	6634.0	6635.9	755413	– 770483	6–10	6.86–02	7.55–02	9.90+00	−0.344	B	3
287.	$1s^29d$–$1s^210p$	^2D–$^2P^o$	*15882*	*6295* cm^{-1}	*755623*	– *761918*	10–6	8.17–02	1.85–01	9.70+01	0.268	B	3
			15905.8	6285 cm^{-1}	755633	– 761918.3	6–4	7.32–02	1.85–01	5.82+01	0.046	B	*LS*
			15847.8	6308 cm^{-1}	755608	– 761916.3	4–2	8.22–02	1.55–01	3.23+01	−0.208	B	*LS*
			15842.8	6310 cm^{-1}	755608	– 761918.3	4–4	8.23–03	3.10–02	6.46+00	−0.907	B	*LS*
288.	$1s^29d$–$1s^211p$	^2D–$^2P^o$	9035.8	9038.3	755623	– 766687	10–6	5.85–02	4.30–02	1.28+01	−0.366	B	3
289.	$1s^29d$–$1s^212p$	^2D–$^2P^o$	6789.8	6791.6	755623	– 770347	10–6	4.09–02	1.70–02	3.79+00	−0.771	B	3
290.	$1s^29d$–$1s^210f$	^2D–$^2F^o$	15430	6479 cm^{-1}	755623	– 762102	10–14	1.71–01	8.57–01	4.35+02	0.933	B	3
291.	$1s^29f$–$1s^210d$	$^2F^o$–^2D	15632	6395 cm^{-1}	755667	– 762062	14–10	4.37–02	1.15–01	8.25+01	0.205	B	3
292.	$1s^29f$–$1s^211d$	$^2F^o$–^2D	8926.5	8929.0	755667	– 766866	14–10	2.71–02	2.32–02	9.53+00	−0.489	B	3
293.	$1s^29f$–$1s^212d$	$^2F^o$–^2D	6747.4	6749.2	755667	– 770483	14–10	1.84–02	8.98–03	2.79+00	−0.900	B	3
294.	$1s^29f$–$1s^210g$	$^2F^o$–^2G	15532	6437 cm^{-1}	755667	– 762103	14–18	2.29–01	1.07+00	7.63+02	1.173	B	3
295.	$1s^29g$–$1s^210f$	^2G–$^2F^o$	15538	6434 cm^{-1}	755668	– 762102	18–14	2.69–02	7.57–02	6.97+01	0.134	B	3
296.	$1s^210s$–$1s^210p$	^2S–$^2P^o$		*535* cm^{-1}	761382.7	– *761918*	2–6	8.88–04	1.40+00	1.72+03	0.446	B	3
				535.6 cm^{-1}	761382.7	– 761918.3	2–4	8.91–04	9.31–01	1.14+03	0.270	B	*LS*
				533.6 cm^{-1}	761382.7	– 761916.3	2–2	8.81–04	4.64–01	5.72+02	−0.033	B	*LS*
297.	$1s^210s$–$1s^211p$	^2S–$^2P^o$	18847	5304 cm^{-1}	761382.7	– 766687	2–6	2.78–02	4.45–01	5.53+01	−0.051	B	3
298.	$1s^210s$–$1s^212p$	^2S–$^2P^o$	11152	8964 cm^{-1}	761382.7	– 770347	2–6	2.45–02	1.37–01	1.01+01	−0.562	B	3
299.	$1s^210p$–$1s^210d$	$^2P^o$–^2D		*144* cm^{-1}	*761918*	– *762062*	6–10	2.22–05	2.67–01	3.66+03	0.204	B	3
				153 cm^{-1}	761918.3	– 762071	4–6	2.64–05	2.54–01	2.19+03	0.008	B	*LS*
				132 cm^{-1}	761916.3	– 762048	2–4	1.41–05	2.44–01	1.22+03	−0.312	B	*LS*
				130 cm^{-1}	761918.3	– 762048	4–4	2.69–06	2.40–02	2.44+02	−1.018	B	*LS*
300.	$1s^210p$–$1s^211d$	$^2P^o$–^2D	20203	4948 cm^{-1}	761918	– 766866	6–10	6.16–02	6.28–01	2.51+02	0.576	B	3
301.	$1s^210p$–$1s^212d$	$^2P^o$–^2D	11672	8565 cm^{-1}	761918	– 770483	6–10	5.15–02	1.75–01	4.04+01	0.022	B	3
302.	$1s^210d$–$1s^211p$	^2D–$^2P^o$	21615	4625 cm^{-1}	762062	– 766687	10–6	5.46–02	2.30–01	1.63+02	0.361	B	3

N v: Allowed Transitions — Continued

No.	Transition Array	Mult.	λ_{air} (Å)	λ_{vac} (Å) or σ (cm^{-1})*	E_i (cm^{-1})	E_k (cm^{-1})	g_i-g_k	A_{ki} ($10^8 s^{-1}$)	f_{ik}	S (at. u.)	log gf	Acc.	Source
303.	$1s^210d-1s^212p$	$^2D-^2P^o$	12066	8285 cm^{-1}	762062	– 770347	10–6	3.80–02	4.98–02	1.98+01	–0.303	B	3
304.	$1s^210f-1s^211d$	$^2F^o-^2D$	20985	4764 cm^{-1}	762102	– 766866	14–10	2.76–02	1.30–01	1.26+02	0.260	B	3
305.	$1s^210f-1s^212d$	$^2F^o-^2D$	11929	8381 cm^{-1}	762102	– 770483	14–10	1.88–02	2.87–02	1.58+01	–0.396	B	3
306.	$1s^211p-1s^211d$	$^2P^o-^2D$		179 cm^{-1}	766687	– 766866	6–10	5.93–05	4.62–01	5.10+03	0.443	B	3
307.	$1s^211p-1s^212d$	$^2P^o-^2D$	26336	3796 cm^{-1}	766687	– 770483	6–10	3.64–02	6.31–01	3.28+02	0.578	B	3
308.	$1s^211d-1s^212p$	$^2D-^2P^o$	28720	3481 cm^{-1}	766866	– 770347	10–6	3.84–02	2.85–01	2.70+02	0.455	B	3
309.	$1s^212p-1s^212d$	$^2P^o-^2D$		136 cm^{-1}	770347	– 770483	6–10	3.70–05	5.00–01	7.26+03	0.477	B	3
310.	$1s2s2p-1s2s3s$	$^4P^o-^4S$		211.1			12–4	1.78+02	3.96–02	3.31–01	–0.323	B	6
311.	$1s2s2p-1s2s3d$	$^4P^o-^4D$		193.5			12–20	7.60+02	7.10–01	5.43+00	0.931	B	6
312.	$1s2s2p-1s2p3p$	$^4P^o-^4D$		180.4			12–20	3.19+01	2.59–02	1.85–01	–0.507	B	6
313.		$^4P^o-^4S$		178.1			12–4	1.42+02	2.25–02	1.58–01	–0.569	B	6
314.		$^4P^o-^4P$		176.9			12–12	1.48+02	6.97–02	4.87–01	–0.078	B	6
315.	$1s2s2p-1s2s4s$	$^4P^o-^4S$		155.3			12–4	4.45+01	5.36–03	3.29–02	–1.191	B	6
316.	$1s2s2p-1s2s4d$	$^4P^o-^4D$		151.5			12–20	2.38+02	1.36–01	8.15–01	0.214	B	6
317.	$1s2s2p-1s2p4p$	$^4P^o-^4D$		141.6			12–20	5.21+01	2.61–02	1.46–01	–0.504	B	6
318.		$^4P^o-^4S$		141.4			12–4	9.36+01	9.35–03	5.22–02	–0.950	B	6
319.		$^4P^o-^4P$		140.9			12–12	7.36+01	2.19–02	1.22–01	–0.581	B	6
320.	$1s2s2p-1s2s5s$	$^4P^o-^4S$		139.2			12–4	1.20+00	1.16–04	6.39–04	–2.856	B	6
321.	$1s2s2p-1s2s5d$	$^4P^o-^4D$		137.9			12–20	9.91+01	4.71–02	2.57–01	–0.248	B	6
322.	$1s2s2p-1s2s6d$	$^4P^o-^4D$		131.7			12–20	6.28+01	2.72–02	1.42–01	–0.486	B	6
323.	$1s2s2p-1s2p5p$	$^4P^o-^4D$		129.2			12–20	2.84+01	1.18–02	6.05–02	–0.847	B	6

N v: Allowed Transitions — Continued

No.	Transition Array	Mult.	λ_{air} (Å) λ_{vac} (Å) or σ (cm^{-1})*	E_i (cm^{-1})	E_k (cm^{-1})	g_i–g_k	A_{ki} (10^8 s^{-1})	f_{ik}	S (at. u.)	log gf	Acc.	Source
324.		^{4}P°–^{4}S	129.1			12–4	4.51+01	3.76–03	1.92–02	–1.346	B	6
325.	$1s2p^2$– $1s2s3p$	^{4}P–^{4}P°	244.2			12–12	2.40+00	2.15–03	2.07–02	–1.589	B	6
326.	$1s2p^2$– $1s2p3s$	^{4}P–^{4}P°	223.8			12–12	1.30+02	9.75–02	8.62–01	0.068	B	6
327.	$1s2p^2$– $1s2p3d$	^{4}P–^{4}D°	205.0			12–20	7.99+02	8.39–01	6.80+00	1.003	B	6
328.		^{4}P–^{4}P°	202.6			12–12	4.11+02	2.53–01	2.03+00	0.483	B	6
329.	$1s2p^2$– $1s2p4d$	^{4}P–^{4}D°	160.1			12–20	2.94+02	1.88–01	1.19+00	0.354	B	6
330.		^{4}P–^{4}P°	160.0			12–12	1.24+02	4.77–02	3.01–01	–0.242	B	6
331.	$1s2p^2$– $1s2p5d$	^{4}P–^{4}D°	145.4			12–20	1.41+02	7.44–02	4.28–01	–0.049	B	6
332.	$1s2p^2$– $1s2p6d$	^{4}P–^{4}D°	138.6			12–20	7.90+01	3.79–02	2.07–01	–0.342	B	6
333.	$1s2s3s$– $1s2p3s$	^{4}S–^{4}P°	1554			4–12	1.00+00	1.09–01	2.22+00	–0.362	B	6
334.	$1s2s3s$– $1s2s4p$	^{4}S–^{4}P°	555.1			4–12	1.71+01	2.37–01	1.73+00	–0.023	B	6
335.	$1s2s3p$– $1s2s4d$	^{4}P°–^{4}D	626.2			12–20	4.52+01	4.43–01	1.10+01	0.725	B	6
336.	$1s2s3p$– $1s2s5d$	^{4}P°–^{4}D	445.7			12–20	2.18+01	1.08–01	1.90+00	0.113	B	6
337.	$1s2s3d$– $1s2p3d$	^{4}D–^{4}D°	1611			20–20	1.80+00	7.00–02	7.43+00	0.146	B	6
338.		^{4}D–^{4}P°	1475			20–12	1.60+00	3.13–02	3.04+00	–0.203	B	6
339.	$1s2s3d$– $1s2s4f$	^{4}D–^{4}F°	673.2			20–28	9.28+01	8.83–01	3.91+01	1.247	B	6
340.	$1s2s3d$– $1s2s5f$	^{4}D–^{4}F°	475.2			20–28	3.51+01	1.66–01	5.21+00	0.522	B	6
341.	$1s2s4d$– $1s2s5f$	^{4}D–^{4}F°	1492			20–28	1.64+01	7.67–01	7.53+01	1.186	B	6
342.	$1s2s4f$– $1s2p4f$	^{4}F°–^{4}G	1964			28–36	4.00–01	2.97–02	5.39+00	–0.079	B	6
343.	$1s2s4f$– $1s2s5g$	^{4}F°–^{4}G	1580			28–36	2.80+01	1.35+00	1.96+02	1.577	B	6
344.	$1s2p3s$– $1s2p4p$	^{4}P°–^{4}D	595.0			12–20	1.84+01	1.63–01	3.83+00	0.291	B	6

N V: Allowed Transitions — Continued

No.	Transition Array	Mult.	λ_{air} (Å) λ_{vac} (Å) or σ (cm^{-1})*	E_i (cm^{-1})	E_k (cm^{-1})	g_i–g_k	A_{ki} (10^8 s^{-1})	f_{ik}	S (at. u.)	log gf	Acc.	Source
345.	$1s2p3p$–$1s2p4s$	^4S–$^4P^o$	723.7			4–12	1.30+00	3.06–02	2.92–01	–0.912	B	6
346.	$1s2p3p$–$1s2p4d$	^4D–$^4F^o$	633.6			20–28	4.64+01	3.91–01	1.63+01	0.893	B	6
347.		^4D–$^4D^o$	619.1			20–20	9.60+00	5.52–02	2.25+00	0.043	B	6
348.	$1s2p3p$–$1s2p5d$	^4D–$^4D^o$	445.4			20–20	5.60+00	1.67–02	4.89–01	–0.477	B	6
349.	$1s2p3d$–$1s2p4f$	$^4P^o$–^4D	740.0			12–20	6.78+01	9.28–01	2.71+01	1.047	B	6
350.		$^4D^o$–^4F	726.0			20–28	7.91+01	8.75–01	4.18+01	1.243	B	6
351.	$1s2p3d$–$1s2p5f$	$^4D^o$–^4F	499.2			20–28	2.55+01	1.33–01	4.38+00	0.426	B	6
352.	$1s2p3d$–$1s2p4f$	$^4F^o$–^4G	675.2			28–36	8.48+01	7.45–01	4.64+01	1.319	B	6
353.		$^4F^o$–^4F	673.6			28–28	1.34+01	9.11–02	5.66+00	0.407	B	6
354.	$1s2p3d$–$1s2p5f$	$^4F^o$–^4G	474.5			28–36	2.64+01	1.15–01	5.01+00	0.506	B	6
355.	$1s2p4d$–$1s2p5p$	$^4D^o$–^4D	1716			20–20	4.00–01	1.77–02	2.00+00	–0.452	B	6
356.		$^4P^o$–^4S	1711			12–4	4.50+00	6.58–02	4.45+00	–0.103	B	6
357.	$1s2p4d$–$1s2p5f$	$^4F^o$–^4G	1495			28–36	1.25+01	5.38–01	7.42+01	1.178	B	6
358.	$1s2p4f$–$1s2p5g$	^4D–$^4F^o$	1648			20–28	2.33+01	1.33+00	1.44+02	1.424	B	6
359.		^4F–$^4G^o$	1587			28–36	2.57+01	1.25+00	1.83+02	1.543	B	6

*Wavelengths (Å) are always given unless cm^{-1} is indicated.

N VI

Helium Isoelectronic Sequence

Ground State: $1s^2\ ^1S_0$

Ionization Energy: 552.057 eV = 4452758 cm^{-1}

Allowed Transitions

List of tabulated lines

Wavelength (Å)	No.	Wavelength (Å)	No.	Wavelength (Å)	No.	Wavenumber (cm^{-1})	No.
in vacuum		173.871	11	710.379	59	980	66
		173.958	11	727.484	68	1420	79
23.0240	5	180.650	10	1003.1	77	1490	78
23.2770	4	180.655	10	1020.0	82		
23.7710	3	180.750	10	1038.6	56		
24.8980	2	185.192	24	1068.7	58		
28.7870	1	187.896	23	1087.5	64		
104.884	22	258.592	37	1107.9	67		
107.151	21	263.894	50	1138.2	72		
110.230	9	272.822	36	1142.5	63		
110.618	20	278.730	49	1146.0	65		
116.711	19	296.481	35	1160.0	71		
116.786	15	303.472	48	1257.5	76		
116.788	15	331.983	31	1284.2	81		
116.828	15	344.721	34	1896.8	6		
117.392	14	346.729	42	1907.3	6		
117.394	14	352.125	41	1907.9	6		
117.434	14	354.208	47	1989.3	75		
121.820	28	357.628	44				
122.066	27	359.350	54	in air			
122.440	8	361.494	53				
130.287	18	474.473	30	2056.1	80		
130.290	13	497.983	33	2896.4	16		
130.293	13	500.851	40	6991.1	29		
130.342	13	518.027	46	9622.0	32		
131.840	12	524.549	39	13695	38		
131.843	12	525.790	61	17177	55		
131.893	12	527.872	52	23304	57		
136.605	26	528.653	43	31241	62		
137.225	25	535.103	70	32564	74		
161.220	7	537.259	51	34120	73		
173.275	17	588.166	60	38158	45		
173.866	11	599.844	69				

We have mainly utilized the results of Cann and Thakkar,[1] who calculated the oscillator strengths of about 100 $n\mathrm{S} - m\mathrm{P}$ and $n\mathrm{P} - m\mathrm{D}$ transitions ($n, m < 7$) with variationally determined, highly correlated wavefunctions. This work is an expansion of, and is similar in spirit to, the classic variational calculations on He-like ions by Schiff and Pekeris.[2] Cann and Thakkar improved this variational approach by the use of exponential correlation factors which yield more accurate wavefunctions with a similar number of term expansions. Comparisons between their results, the earlier data by Schiff and Pekeris[2] and similar work by Kono and Hattori[3] in the 1980s show that the differences do not exceed 0.1%.

The Cann and Thakkar[1] data tabulated here are the arithmetic mean values of the "dipole length" and "dipole velocity" results. Their numerical values are given to at least three, but most often five or six significant figures, and the differences between the length and velocity data do not exceed seven units in the last digit shown. These differences, which they consider to be their main indicator of accuracy, are smaller than 0.1% for all but a few transitions.

For some transitions involving higher excited states, which their material does not cover, we have used the results of the Opacity Project (OP), which is reviewed in the general introduction. The calculations for He-like ions were carried out by Fernley *et al.*[4] and also are estimated to be very accurate, as comparisons with the above work of Cann and Thakkar show when there is overlap.

It should be noted that all variational calculations[1-3] have been done in a non-relativistic framework. This is an excellent approximation for neutral helium, but for highly charged

He-like ions, relativistic effects are expected to become significant. For C v, N vi and O vii, some test calculations by Kim and Weiss[5] show that such effects are still very small. They become noticeable only for the line strengths of P-D transitions, where deviations from non-relativistic values may sometimes reach 0.5%.

References

[1]N. M. Cann and A. J. Thakkar, Phys. Rev. A **46**, 5397 (1992).
[2]B. Schiff and C. L. Pekeris, Phys. Rev. **134**, A638 (1964).
[3]A. Kono and S. Hattori, Phys. Rev. A **30**, 2093 (1984).
[4]J. A. Fernley, K. T. Taylor and M. J. Seaton, J. Phys. B **20**, 6457 (1987).
[5]Y.-K. Kim and A. W. Weiss (private communication), 1993.

N vi: Allowed Transitions

No.	Transition Array	Mult.	λ_{air} (Å)	λ_{vac} (Å) or σ (cm^{-1})*	E_i (cm^{-1})	E_k (cm^{-1})	g_i-g_k	A_{ki} (10^8 s^{-1})	f_{ik}	S (at. u.)	log gf	Acc.	Source
1.	$1s^2-1s2p$	$^1S-^1P^o$		28.7870	0.0	3473790	1–3	1.809+04	6.742−01	6.389−02	−0.1712	AA	1
2.	$1s^2-1s3p$	$^1S-^1P^o$		24.8980	0.0	4016390	1–3	5.158+03	1.438−01	1.179−02	−0.8422	AA	1
3.	$1s^2-1s4p$	$^1S-^1P^o$		23.7710	0.0	4206810	1–3	2.145+03	5.451−02	4.265−03	−1.2636	AA	1
4.	$1s^2-1s5p$	$^1S-^1P^o$		23.2770	0.0	4296090	1–3	1.091+03	2.659−02	2.038−03	−1.5753	AA	1
5.	$1s^2-1s6p$	$^1S-^1P^o$		23.0240	0.0	4343290	1–3	6.290+02	1.500−02	1.137−03	−1.8240	AA	1
6.	$1s2s-1s2p$	$^3S-^3P^o$		*1901*	3385890	*3438500*	3–9	6.780−01	1.102−01	2.068+00	−0.4808	AA	1
				1896.8	3385890	3438610	3–5	6.823−01	6.134−02	1.149+00	−0.7352	AA	*LS*
				1907.3	3385890	3438320	3–3	6.711−01	3.660−02	6.894−01	−0.9594	AA	*LS*
				1907.9	3385890	3438304	3–1	6.705−01	1.220−02	2.298−01	−1.4367	AA	*LS*
7.	$1s2s-1s3p$	$^3S-^3P^o$		161.220	3385890	4006160	3–9	2.859+02	3.342−01	5.322−01	0.0012	AA	1
8.	$1s2s-1s4p$	$^3S-^3P^o$		122.440	3385890	4202620	3–9	1.287+02	8.680−02	1.050−01	−0.5843	AA	1
9.	$1s2s-1s5p$	$^3S-^3P^o$		110.230	3385890	4293080	3–9	6.684+01	3.653−02	3.976−02	−0.9603	AA	1
10.	$1s2p-1s3s$	$^3P^o-^3S$		*180.71*	*3438500*	3991860	9–3	1.286+02	2.098−02	1.124−01	−0.7239	AA	1
				180.750	3438610	3991860	5–3	7.139+01	2.098−02	6.242−02	−0.9792	AA	*LS*
				180.655	3438320	3991860	3–3	4.290+01	2.099−02	3.745−02	−1.2009	AA	*LS*
				180.650	3438304	3991860	1–3	1.430+01	2.099−02	1.248−02	−1.6780	AA	*LS*
11.	$1s2p-1s3d$	$^3P^o-^3D$		*173.93*	*3438500*	4013460	9–15	8.756+02	6.618−01	3.410+00	0.7750	AA	1
				173.958	3438610	4013460	5–7	8.751+02	5.558−01	1.592+00	0.4439	AA	*LS*
				173.871	3438320	4013460	3–5	6.573+02	4.965−01	8.526−01	0.1730	AA	*LS*
				173.866	3438304	4013460	1–3	4.869+02	6.620−01	3.789−01	−0.1791	AA	*LS*
				173.958	3438610	4013460	5–5	2.188+02	9.925−02	2.842−01	−0.3043	AA	*LS*
				173.871	3438320	4013460	3–3	3.652+02	1.655−01	2.842−01	−0.3041	AA	*LS*
				173.958	3438610	4013460	5–3	2.431+01	6.617−03	1.895−02	−1.4804	AA	*LS*
12.	$1s2p-1s4s$	$^3P^o-^3S$		*131.87*	*3438500*	4196800	9–3	5.036+01	4.377−03	1.710−02	−1.4046	AA	1
				131.893	3438610	4196800	5–3	2.797+01	4.376−03	9.501−03	−1.6599	AA	*LS*
				131.843	3438320	4196800	3–3	1.680+01	4.378−03	5.701−03	−1.8816	AA	*LS*
				131.840	3438304	4196800	1–3	5.600+00	4.378−03	1.900−03	−2.3587	AA	*LS*

N VI: Allowed Transitions — Continued

No.	Transition Array	Mult.	λ_{air} (Å)	λ_{vac} (Å) or σ (cm^{-1})*	E_i (cm^{-1})	E_k (cm^{-1})	g_i–g_k	A_{ki} (10^8 s^{-1})	f_{ik}	S (at. u.)	log gf	Acc.	Source
13.	1s2p– 1s4d	^{3}P°–^{3}D		*130.32*	*3438500*	4205820	9–15	2.898+02	1.230–01	4.749–01	0.0441	AA	1
				130.342	3438610	4205820	5–7	2.897+02	1.033–01	2.216–01	–0.2869	AA	*LS*
				130.293	3438320	4205820	3–5	2.175+02	9.227–02	1.187–01	–0.5578	AA	*LS*
				130.290	3438304	4205820	1–3	1.611+02	1.230–01	5.277–02	–0.9100	AA	*LS*
				130.342	3438610	4205820	5–5	7.243+01	1.845–02	3.958–02	–1.0351	AA	*LS*
				130.293	3438320	4205820	3–3	1.209+02	3.076–02	3.958–02	–1.0349	AA	*LS*
				130.342	3438610	4205820	5–3	8.048+00	1.230–03	2.639–03	–2.2112	AA	*LS*
14.	1s2p– 1s5s	^{3}P°–^{3}S		*117.42*	*3438500*	4290150	9–3	2.477+01	1.707–03	5.938–03	–1.8136	AA	1
				117.434	3438610	4290150	5–3	1.376+01	1.706–03	3.299–03	–2.0689	AA	*LS*
				117.394	3438320	4290150	3–3	8.262+00	1.707–03	1.979–03	–2.2906	AA	*LS*
				117.392	3438304	4290150	1–3	2.754+00	1.707–03	6.597–04	–2.7678	AA	*LS*
15.	1s2p– 1s5d	^{3}P°–^{3}D		*116.81*	*3438500*	4294570	9–15	1.342+02	4.575–02	1.583–01	–0.3854	AA	1
				116.828	3438610	4294570	5–7	1.341+02	3.842–02	7.389–02	–0.7164	AA	*LS*
				116.788	3438320	4294570	3–5	1.007+02	3.432–02	3.958–02	–0.9873	AA	*LS*
				116.786	3438304	4294570	1–3	7.460+01	4.576–02	1.759–02	–1.3395	AA	*LS*
				116.828	3438610	4294570	5–5	3.353+01	6.861–03	1.319–02	–1.4646	AA	*LS*
				116.788	3438320	4294570	3–3	5.594+01	1.144–02	1.319–02	–1.4645	AA	*LS*
				116.828	3438610	4294570	5–3	3.726+00	4.574–04	8.797–04	–2.6407	AA	*LS*
16.	1s2s– 1s2p	^{1}S–^{1}P°	2896.4	2897.2	3439274	3473790	1–3	2.079–01	7.848–02	7.485–01	–1.1052	AA	1
17.	1s2s– 1s3p	^{1}S–^{1}P°		173.275	3439274	4016390	1–3	2.697+02	3.642–01	2.078–01	–0.4386	AA	1
18.	1s2s– 1s4p	^{1}S–^{1}P°		130.287	3439274	4206810	1–3	1.207+02	9.212–02	3.951–02	–1.0356	AA	1
19.	1s2s– 1s5p	^{1}S–^{1}P°		116.711	3439274	4296090	1–3	6.275+01	3.844–02	1.477–02	–1.4152	AA	1
20.	1s2s– 1s6p	^{1}S–^{1}P°		110.618	3439274	4343290	1–3	3.642+01	2.004–02	7.299–03	–1.6980	AA	1
21.	1s2s– 1s7p	^{1}S–^{1}P°		107.151	3439274	4372540	1–3	2.31+01	1.19–02	4.20–03	–1.924	A	4
22.	1s2s– 1s8p	^{1}S–^{1}P°		104.884	3439274	4392710	1–3	1.58+01	7.82–03	2.70–03	–2.107	A	4
23.	1s2p– 1s3s	^{1}P°–^{1}S		187.896	3473790	4006000	3–1	1.125+02	1.984–02	3.682–02	–1.2253	AA	1
24.	1s2p– 1s3d	^{1}P°–^{1}D		185.192	3473790	4013770	3–5	8.205+02	7.031–01	1.286+00	0.3242	AA	1
25.	1s2p– 1s4s	^{1}P°–^{1}S		137.225	3473790	4202520	3–1	4.534+01	4.267–03	5.782–03	–1.8928	AA	1
26.	1s2p– 1s4d	^{1}P°–^{1}D		136.605	3473790	4205830	3–5	2.567+02	1.197–01	1.615–01	–0.4449	AA	1
27.	1s2p– 1s5s	^{1}P°–^{1}S		122.066	3473790	4293020	3–1	2.259+01	1.682–03	2.028–03	–2.2970	AA	1
28.	1s2p– 1s5d	^{1}P°–^{1}D		121.820	3473790	4294670	3–5	1.163+02	4.313–02	5.189–02	–0.8881	AA	1

N VI: Allowed Transitions — Continued

No.	Transition Array	Mult.	λ_{air} (Å)	λ_{vac} (Å) or σ (cm^{-1})*	E_i (cm^{-1})		E_k (cm^{-1})	g_i–g_k	A_{ki} (10^8 s^{-1})	f_{ik}	S (at. u.)	log gf	Acc.	Source
29.	1s3s–1s3p	^{3}S–^{3}P°	6991.1	6993.0	3991860	–	4006160	3–9	8.384–02	1.844–01	1.274+01	–0.2571	AA	1
30.	1s3s–1s4p	^{3}S–^{3}P°		474.473	3991860	–	4202620	3–9	3.576+01	3.621–01	1.697+00	0.0360	AA	1
31.	1s3s–1s5p	^{3}S–^{3}P°		331.983	3991860	–	4293080	3–9	2.005+01	9.937–02	3.258–01	–0.5256	AA	1
32.	1s3s–1s3p	^{1}S–^{1}P°	9622.0	9624.6	4006000	–	4016390	1–3	3.276–02	1.365–01	4.325+00	–0.8649	AA	1
33.	1s3s–1s4p	^{1}S–^{1}P°		497.983	4006000	–	4206810	1–3	3.540+01	3.949–01	6.474–01	–0.4035	AA	1
34.	1s3s–1s5p	^{1}S–^{1}P°		344.721	4006000	–	4296090	1–3	1.978+01	1.057–01	1.200–01	–0.9758	AA	1
35.	1s3s–1s6p	^{1}S–^{1}P°		296.481	4006000	–	4343290	1–3	1.165+01	4.605–02	4.495–02	–1.3368	AA	1
36.	1s3s–1s7p	^{1}S–^{1}P°		272.822	4006000	–	4372540	1–3	7.45+00	2.49–02	2.24–02	–1.603	A	4
37.	1s3s–1s8p	^{1}S–^{1}P°		258.592	4006000	–	4392710	1–3	5.04+00	1.52–02	1.29–02	–1.819	A	4
38.	1s3p–1s3d	^{3}P°–^{3}D	13695	7300 cm^{-1}	4006160	–	4013460	9–15	8.90–03	4.17–02	1.69+01	–0.426	A	1
39.	1s3p–1s4s	^{3}P°–^{3}S		524.549	4006160	–	4196800	9–3	3.494+01	4.804–02	7.467–01	–0.3641	AA	1
40.	1s3p–1s4d	^{3}P°–^{3}D		500.851	4006160	–	4205820	9–15	8.998+01	5.640–01	8.370+00	0.7055	AA	1
41.	1s3p–1s5s	^{3}P°–^{3}S		352.125	4006160	–	4290150	9–3	1.663+01	1.031–02	1.075–01	–1.0326	AA	1
42.	1s3p–1s5d	^{3}P°–^{3}D		346.729	4006160	–	4294570	9–15	4.477+01	1.345–01	1.382+00	0.0829	AA	1
43.	1s3d–1s4p	^{3}D–^{3}P°		528.653	4013460	–	4202620	15–9	5.838+00	1.468–02	3.832–01	–0.6573	AA	1
44.	1s3d–1s5p	^{3}D–^{3}P°		357.628	4013460	–	4293080	15–9	2.492+00	2.867–03	5.064–02	–1.3664	AA	1
45.	1s3d–1s3p	^{1}D–^{1}P°	38158	2620 cm^{-1}	4013770	–	4016390	5–3	6.08–04	7.96–03	5.00+00	–1.400	A	1
46.	1s3d–1s4p	^{1}D–^{1}P°		518.027	4013770	–	4206810	5–3	4.119+00	9.942–03	8.478–02	–1.3036	AA	1
47.	1s3d–1s5p	^{1}D–^{1}P°		354.208	4013770	–	4296090	5–3	1.790+00	2.020–03	1.178–02	–1.9958	AA	1
48.	1s3d–1s6p	^{1}D–^{1}P°		303.472	4013770	–	4343290	5–3	9.314–01	7.716–04	3.854–03	–2.4136	AA	1
49.	1s3d–1s7p	^{1}D–^{1}P°		278.730	4013770	–	4372540	5–3	5.61–01	3.92–04	1.80–03	–2.707	A	4

N VI: Allowed Transitions — Continued

No.	Transition Array	Mult.	λ_{air} (Å)	λ_{vac} (Å) or σ (cm^{-1})*	E_i (cm^{-1})	E_k (cm^{-1})	g_i–g_k	A_{ki} (10^8 s^{-1})	f_{ik}	S (at. u.)	log gf	Acc.	Source
50.	1s3d–1s8p	^{1}D–^{1}P°		263.894	4013770	4392710	5–3	3.67–01	2.30–04	1.00–03	–2.939	A	4
51.	1s3p–1s4s	^{1}P°–^{1}S		537.259	4016390	4202520	3–1	3.091+01	4.459–02	2.366–01	–0.8736	AA	1
52.	1s3p–1s4d	^{1}P°–^{1}D		527.872	4016390	4205830	3–5	9.130+01	6.357–01	3.314+00	0.2804	AA	1
53.	1s3p–1s5s	^{1}P°–^{1}S		361.494	4016390	4293020	3–1	1.497+01	9.776–03	3.490–02	–1.5327	AA	1
54.	1s3p–1s5d	^{1}P°–^{1}D		359.350	4016390	4294670	3–5	4.349+01	1.403–01	4.980–01	–0.3758	AA	1
55.	1s4s–1s4p	^{3}S–^{3}P°	17177	5820 cm^{-1}	4196800	4202620	3–9	1.91–02	2.54–01	4.31+01	–0.118	A	1
56.	1s4s–1s5p	^{3}S–^{3}P°		1038.6	4196800	4293080	3–9	8.237+00	3.997–01	4.100+00	0.0788	AA	1
57.	1s4s–1s4p	^{1}S–^{1}P°	23304	4290 cm^{-1}	4202520	4206810	1–3	7.77–03	1.90–01	1.46+01	–0.721	A	1
58.	1s4s–1s5p	^{1}S–^{1}P°		1068.7	4202520	4296090	1–3	8.495+00	4.364–01	1.535+00	–0.3601	AA	1
59.	1s4s–1s6p	^{1}S–^{1}P°		710.379	4202520	4343290	1–3	5.237+00	1.189–01	2.780–01	–0.9250	AA	1
60.	1s4s–1s7p	^{1}S–^{1}P°		588.166	4202520	4372540	1–3	3.40+00	5.29–02	1.02–01	–1.277	A	4
61.	1s4s–1s8p	^{1}S–^{1}P°		525.790	4202520	4392710	1–3	2.35+00	2.92–02	5.05–02	–1.535	A	4
62.	1s4p–1s4d	^{3}P°–^{3}D	31241	3200 cm^{-1}	4202620	4205820	9–15	2.99–03	7.31–02	6.76+01	–0.182	A	1
63.	1s4p–1s5s	^{3}P°–^{3}S		1142.5	4202620	4290150	9–3	1.186+01	7.733–02	2.618+00	–0.1574	AA	1
64.	1s4p–1s5d	^{3}P°–^{3}D		1087.5	4202620	4294570	9–15	1.840+01	5.438–01	1.752+01	0.6897	AA	1
65.	1s4d–1s5p	^{3}D–^{3}P°		1146.0	4205820	4293080	15–9	3.052+00	3.605–02	2.040+00	–0.2670	AA	1
66.	1s4d–1s4p	^{1}D–^{1}P°		980 cm^{-1}	4205830	4206810	5–3	1.45–04	1.35–02	2.27+01	–1.170	A	1
67.	1s4d–1s5p	^{1}D–^{1}P°		1107.9	4205830	4296090	5–3	2.310+00	2.550–02	4.651–01	–0.8944	AA	1
68.	1s4d–1s6p	^{1}D–^{1}P°		727.484	4205830	4343290	5–3	1.136+00	5.407–03	6.475–02	–1.5681	AA	1
69.	1s4d–1s7p	^{1}D–^{1}P°		599.844	4205830	4372540	5–3	6.51–01	2.11–03	2.08–02	–1.977	A	4
70.	1s4d–1s8p	^{1}D–^{1}P°		535.103	4205830	4392710	5–3	4.19–01	1.08–03	9.50–03	–2.268	A	4

N VI: Allowed Transitions — Continued

No.	Transition Array	Mult.	λ_{air} (Å)	λ_{vac} (Å) or σ (cm^{-1})*	E_i (cm^{-1})	E_k (cm^{-1})	g_i-g_k	A_{ki} (10^8 s^{-1})	f_{ik}	S (at. u.)	log gf	Acc.	Source
71.	1s4p–1s5s	$^1P^o-^1S$		1160.0	4206810	4293020	3–1	1.065+01	7.158–02	8.200–01	–0.6681	AA	1
72.	1s4p–1s5d	$^1P^o-^1D$		1138.2	4206810	4294670	3–5	1.947+01	6.303–01	7.085+00	0.2766	AA	1
73.	1s5s–1s5p	$^3S-^3P^o$	34120	2930 cm^{-1}	4290150	4293080	3–9	6.14–03	3.22–01	1.09+02	–0.015	A	1
74.		$^1S-^1P^o$	32564	3070 cm^{-1}	4293020	4296090	1–3	5.07–03	2.42–01	2.59+01	–0.616	A	1
75.	1s5s–1s6p	$^1S-^1P^o$		1989.3	4293020	4343290	1–3	2.711+00	4.825–01	3.160+00	–0.3165	AA	1
76.	1s5s–1s7p	$^1S-^1P^o$		1257.5	4293020	4372540	1–3	1.87+00	1.33–01	5.50–01	–0.877	A	4
77.	1s5s–1s8p	$^1S-^1P^o$		1003.1	4293020	4392710	1–3	1.33+00	6.00–02	1.98–01	–1.222	A	4
78.	1s5p–1s5d	$^3P^o-^3D$		1490 cm^{-1}	4293080	4294570	9–15	8.97–04	1.01–01	2.01+02	–0.042	A	1
79.	1s5d–1s5p	$^1D-^1P^o$		1420 cm^{-1}	4294670	4296090	5–3	4.13–04	1.84–02	2.14+01	–1.035	A	1
80.	1s5d–1s6p	$^1D-^1P^o$	2056.1	2056.8	4294670	4343290	5–3	1.164+00	4.431–02	1.500+00	–0.6545	AA	1
81.	1s5d–1s7p	$^1D-^1P^o$		1284.2	4294670	4372540	5–3	6.51–01	9.65–03	2.04–01	–1.316	A	4
82.	1s5d–1s8p	$^1D-^1P^o$		1020.0	4294670	4392710	5–3	4.14–01	3.87–03	6.50–02	–1.713	A	4

*Wavelengths (Å) are always given unless cm^{-1} is indicated.

N VI

Forbidden Transitions

Most tabulated data are from calculations which are based on correlated, but nonrelativistic, wavefunctions.[1,3–5] Drake has used either a nuclear charge expansion calculation[1] with the perturbation coefficient evaluated through ninth order or he has used 50-term variational wavefunctions.[3] For the helium-like carbon and nitrogen spectra, where high precision experimental data for the M1 lines are available[2] (see also the C V introduction on forbidden lines), the agreement between the experiments and Drake's results is almost perfect. Godefroid and Verhaegen[5] use an extensive multiconfiguration approach and have calculated their data both in the length and velocity forms. These two forms usually agree within four significant figures.

For the magnetic dipole transition from $1s2s$ $^3S_1$ to the ground state, a high-precision measurement of the decay rate has been made by Schmidt *et al.*[2] on an ion storage ring. Their result, estimated to be accurate within 1.2% (one standard deviation), agrees perfectly with Drake's work.

References

[1]G. W. F. Drake, Phys. Rev. A **3**, 908 (1971).
[2]H. T. Schmidt, P. Forck, M. Grieser, D. Habs, J. Kenntner, G. Miersch, R. Repnow, U. Schramm, T. Schüssler, D. Schwalm, and A. Wolf, Phys. Rev. Lett. **72**, 1616 (1994).
[3]G. W. F. Drake, Astrophys. J. **158**, 1199 (1969).
[4]B. Kundu and P. K. Mukherjee, Astrophys. J. **298**, 844 (1985).
[5]M. Godefroid and G. Verhaegen, J. Phys. B **13**, 3081 (1980).

N VI: Forbidden Transitions

No.	Transition Array	Mult.	λ_{air} (Å)	λ_{vac} (Å) or σ (cm^{-1})*	E_i (cm^{-1})	E_k (cm^{-1})	g_i–g_k	Type	A_{ki} (s^{-1})	S (at. u.)	Acc.	Source
1.	$1s^2$–$1s2s$	^{1}S–^{3}S		29.5343	0	3385890	1–3	M1	2.55+02	7.30–07	A+	1,2
2.	$1s^2$–$1s2p$	^{1}S–^{3}P°		29.0815	0	3438610	1–5	M2	1.03+05	7.18–01	A	3
3.	$1s^2$–$1s3s$	^{1}S–^{3}S		25.0510	0	3991860	1–3	M1	8.36+01	1.46–07	C	4
4.	$1s^2$–$1s3d$	^{1}S–^{1}D		24.9142	0	4013770	1–5	E2	3.847+07	1.649–03	AA	5
5.	$1s^2$–$1s4s$	^{1}S–^{3}S		23.8277	0	4196800	1–3	M1	4.59+01	6.91–08	C	4
6.	$1s2s$–$1s3d$	^{1}S–^{1}D		174.066	3439274	4013770	1–5	E2	1.801+06	1.285+00	AA	5
7.	$1s2p$–$1s4f$	^{1}P–^{1}F°		136.606	3473790	4205820	3–7	E2	1.919+06	5.708–01	AA	5

*Wavelengths (Å) are always given unless cm^{-1} is indicated.

N VII

Ground State: $1s\ ^2S_{1/2}$

Ionization Energy: 667.029 eV = 5380089 cm^{-1}

Allowed Transitions

For this hydrogen-like spectrum, we have obtained all data from a computer program of Storey and Hummer,[1] developed for the radiative properties of non-relativistic hydrogenic ions. (According to the work by Younger and Weiss,[2] relativistic effects are negligibly small for this ion.)

A special tabular arrangement has been used for this spectrum. In Table A, we present "average" transition probabilities for transitions between different principal quantum numbers $n_i \rightarrow n_k$, i.e., for lines of the well-known Lyman, Balmer, Paschen, etc. spectral series. Because of the near-degeneracy of levels of the same n, but different orbital and total quantum numbers ℓ and j, normally only a single spectral line is observed for all possible combinations between two atomic states of different principal quantum numbers. These "average" values are applicable for the spectra of laboratory and astrophysical plasmas of moderate and high densities, in which the observed lines are significantly Stark-broadened and where the atomic substates are occupied according to their statistical weights.

Table B contains data for the fine-structure transitions of a smaller group of lines, limited to principal quantum numbers n = 6 or lower. For given higher n, the individual levels, characterized by $n\,\ell\,j$, are so closely spaced that fine-structure lines are difficult to resolve. Furthermore, the knowledge of the structure data is incomplete. The energy levels and wavelengths have been taken from the calculations of Erikson[3] and Johnson and Soff.[4]

References

[1]P. J. Storey and D. G. Hummer, Comput. Phys. Commun. **66**, 129 (1991).

[2]S. M. Younger and A. W. Weiss, J. Res. Natl. Bur. Stand. Sect. A **79**, 629 (1975).

[3]G. W. Erickson, J. Phys. Chem. Ref. Data **6**, 831 (1977).

[4]W. R. Johnson and G. Soff, At. Data Nucl. Data Tables **33**, 405 (1985).

List of tabulated lines (Table A)

Wavelength (Å)	No.	Wavelength (Å)	No.	Wavelength (Å)	No.	Wavenumber (cm^{-1})	No.
in vacuum		354.182	60	2478.73	145	1460 cm^{-1}	190
		370.744	59	2522.52	86	1700 cm^{-1}	188
18.6336	19	382.437	38	2525.87	115		
18.6387	18	396.678	58	2566.34	129		
18.6446	17	441.750	57	2571.84	144		
18.6516	16	495.883	85	2689.47	143		
18.6599	15	499.500	84	2843.22	142		
18.6700	14	503.777	83	2892.31	128		
18.6823	13	508.931	82	3050.59	141		
18.6976	12	515.197	81	3224.88	154		
18.7170	11	523.011	80	3305.12	114		
18.7419	10	532.849	79	3346.77	140		
18.7747	9	535.500	56	3384.27	153		
18.8193	8	545.612	78	3442.03	127		
18.8819	7	562.555	77	3590.93	152		
18.9741	6	585.959	76	3795.29	139		
19.1178	5	619.850	75	3870.37	151		
19.3611	4	672.422	74	3886.73	100		
19.8256	3	735.669	99	4182.92	162		
20.9096	2	743.656	98	4265.01	150		
24.7818	1	753.178	97	4455.08	161		
75.1068	37	762.881	73	4554.32	138		
75.1893	36	764.756	96	4555.71	126		
75.2855	35	778.994	95	4820.25	160		
75.3996	34	796.999	94	4867.19	149		
75.5357	33	820.071	93	5337.54	159		
75.7016	32	826.367	55	5439.18	169		
75.9044	31	850.697	92	5669.26	113		
76.1582	30	892.612	91	5877.27	148		
76.4797	29	949.120	72	5908.53	168		
76.8972	28	953.010	90	6084.37	137		
77.4530	27	1038.42	112	6118.26	158		
78.2171	26	1046.03	89	6568.49	167		
79.3111	25	1054.41	111	7145.99	175		
80.9627	24	1073.66	110	7438.43	157		
83.6466	23	1097.34	109	7567.94	166		
88.5131	22	1126.89	108	7921.75	147		
99.1311	21	1164.96	107	7925.96	125		
133.818	20	1205.02	88	7978.65	174		
171.188	54	1214.92	106	9231.08	173		
171.617	53	1283.37	105	9239.61	165		
172.119	52	1381.22	104	9557.61	180		
172.717	51	1416.86	124	10088.1	156		
173.433	50	1446.79	123	10714.0	136		
174.309	49	1483.27	122	11108.1	179		
175.388	48	1521.38	71	11334.8	172		
176.749	47	1528.86	121	12622.8	164		
178.491	46	1530.17	87	13223.9	184		
180.782	45	1531.40	103	13694.9	178		
183.884	44	1586.84	120	14080.7	146		
188.250	43	1663.38	119	15547.9	171		
194.714	42	1767.15	118	16389.0	183		
204.980	41	1786.71	102	18111.0	155		
223.104	40	1888.78	135	18898.4	177		
261.443	39	1915.77	117	19337.1	187		
309.912	70	1942.34	134	22721.1	182		
311.321	69			22772.8	163		
312.977	68	in air		26946.8	186		
314.959	67			28320.9	170		
317.347	66	2008.02	133	31636.9	189		
320.295	65	2092.51	132	34473.4	176		
323.958	64	2141.64	116	41829.6	181		
328.632	63	2202.72	131	49737.7	185		
334.703	62	2305.81	101				
342.851	61	2353.09	130				

N VII: Table A. $(n)_i - (n)_k$ Transitions (Average Values)

No.	Transition	λ_{air} (Å)	λ_{vac} (Å) or σ (cm^{-1})*	E_i (cm^{-1})	E_k (cm^{-1})	g_i–g_k	A_{ki} (10^8 s^{-1})	f_{ik}	S (at. u.)	log gf	Acc.	Source
1.	1–2(L_α)		24.7818	0.0	– 4035213.3	2–8	1.129+04	4.157–01	6.783–02	–0.0802	AA	1
2.	1–3(L_β)		20.9096	0.0	– 4782497.1	2–18	1.339+03	7.901–02	1.088–02	–0.8013	AA	1
3.	1–4(L_γ)		19.8256	0.0	– 5043978.0	2–32	3.071+02	2.896–02	3.780–03	–1.2372	AA	1
4.	1–5(L_δ)		19.3611	0.0	– 5164989.6	2–50	9.909+01	1.392–02	1.775–03	–1.5553	AA	1
5.	1–6(L_ϵ)		19.1178	0.0	– 5230719.3	2–72	3.949+01	7.790–03	9.806–04	–1.8074	AA	1
6.	1–7		18.9741	0.0	– 5270350.3	2–98	1.818+01	4.808–03	6.007–04	–2.0170	AA	1
7.	1–8		18.8819	0.0	– 5296071.6	2–128	9.295+00	3.180–03	3.953–04	–2.1966	AA	1
8.	1–9		18.8193	0.0	– 5313705.7	2–162	5.147+00	2.214–03	2.743–04	–2.3539	AA	1
9.	1–10		18.7747	0.0	– 5326319.0	2–200	3.034+00	1.604–03	1.982–04	–2.4939	AA	1
10.	1–11		18.7419	0.0	– 5335650	2–242	1.882+00	1.199–03	1.480–04	–2.6201	AA	1
11.	1–12		18.7170	0.0	– 5342750	2–288	1.217+00	9.204–04	1.134–04	–2.7350	AA	1
12.	1–13		18.6976	0.0	– 5348270	2–338	8.150–01	7.219–04	8.887–05	–2.8405	AA	1
13.	1–14		18.6823	0.0	– 5352660	2–392	5.623–01	5.767–04	7.094–05	–2.9380	AA	1
14.	1–15		18.6700	0.0	– 5356190	2–450	3.981–01	4.681–04	5.754–05	–3.0286	AA	1
15.	1–16		18.6599	0.0	– 5359090	2–512	2.882–01	3.851–04	4.732–05	–3.1134	AA	1
16.	1–17		18.6516	0.0	– 5361480	2–578	2.128–01	3.207–04	3.939–05	–3.1929	AA	1
17.	1–18		18.6446	0.0	– 5363490	2–648	1.598–01	2.699–04	3.313–05	–3.2678	AA	1
18.	1–19		18.6387	0.0	– 5365190	2–722	1.220–01	2.293–04	2.814–05	–3.3386	AA	1
19.	1–20		18.6336	0.0	– 5366650	2–800	9.435–02	1.964–04	2.410–05	–3.4057	AA	1
20.	2–3(H_α)		133.818	4035213.3	– 4782497.1	8–18	1.059+03	6.399–01	2.255+00	0.7092	AA	1
21.	2–4(H_β)		99.1311	4035213.3	– 5043978.0	8–32	2.022+02	1.192–01	3.112–01	–0.0207	AA	1
22.	2–5(H_γ)		88.5131	4035213.3	– 5164989.6	8–50	6.079+01	4.462–02	1.040–01	–0.4474	AA	1
23.	2–6(H_δ)		83.6466	4035213.3	– 5230719.3	8–72	2.338+01	2.207–02	4.862–02	–0.7531	AA	1
24.	2–7(H_ϵ)		80.9627	4035213.3	– 5270350.3	8–98	1.054+01	1.269–02	2.706–02	–0.9934	AA	1
25.	2–8		79.3111	4035213.3	– 5296071.6	8–128	5.320+00	8.028–03	1.677–02	–1.1923	AA	1
26.	2–9		78.2171	4035213.3	– 5313705.7	8–162	2.920+00	5.424–03	1.117–02	–1.3626	AA	1
27.	2–10		77.4530	4035213.3	– 5326319.0	8–200	1.711+00	3.847–03	7.847–03	–1.5118	AA	1
28.	2–11		76.8972	4035213.3	– 5335650	8–242	1.056+00	2.833–03	5.737–03	–1.6447	AA	1
29.	2–12		76.4797	4035213.3	– 5342750	8–288	6.807–01	2.149–03	4.328–03	–1.7647	AA	1
30.	2–13		76.1582	4035213.3	– 5348270	8–338	4.547–01	1.670–03	3.350–03	–1.8741	AA	1
31.	2–14		75.9044	4035213.3	– 5352660	8–392	3.131–01	1.325–03	2.649–03	–1.9747	AA	1
32.	2–15		75.7016	4035213.3	– 5356190	8–450	2.212–01	1.069–03	2.132–03	–2.0678	AA	1

N VII: Table A. $(n)_i - (n)_k$ Transitions (Average Values) — Continued

No.	Transition	λ_{air} (Å)	λ_{vac} (Å) or σ (cm^{-1})*	E_i (cm^{-1})	E_k (cm^{-1})	g_i–g_k	A_{ki} (10^8 s^{-1})	f_{ik}	S (at. u.)	log gf	Acc.	Source
33.	2–16		75.5357	4035213.3	5359090	8–512	1.599–01	8.756–04	1.742–03	–2.1546	AA	1
34.	2–17		75.3996	4035213.3	5361480	8–578	1.180–01	7.263–04	1.442–03	–2.2358	AA	1
35.	2–18		75.2855	4035213.3	5363490	8–648	8.852–02	6.093–04	1.208–03	–2.3121	AA	1
36.	2–19		75.1893	4035213.3	5365190	8–722	6.749–02	5.162–04	1.022–03	–2.3841	AA	1
37.	2–20		75.1068	4035213.3	5366650	8–800	5.217–02	4.412–04	8.728–04	–2.4522	AA	1
38.	3–4(P_α)		382.437	4782497.1	5043978.0	18–32	2.159+02	8.415–01	1.907+01	1.1803	AA	1
39.	3–5(P_β)		261.443	4782497.1	5164989.6	18–50	5.287+01	1.505–01	2.331+00	0.4328	AA	1
40.	3–6(P_γ)		223.104	4782497.1	5230719.3	18–72	1.870+01	5.581–02	7.378–01	0.0020	AA	1
41.	3–7(P_δ)		204.980	4782497.1	5270350.3	18–98	8.068+00	2.767–02	3.361–01	–0.3027	AA	1
42.	3–8(P_ϵ)		194.714	4782497.1	5296071.6	18–128	3.965+00	1.603–02	1.849–01	–0.5399	AA	1
43.	3–9		188.250	4782497.1	5313705.7	18–162	2.139+00	1.023–02	1.141–01	–0.7349	AA	1
44.	3–10		183.884	4782497.1	5326319.0	18–200	1.239+00	6.976–03	7.601–02	–0.9011	AA	1
45.	3–11		180.782	4782497.1	5335650	18–242	7.581–01	4.994–03	5.350–02	–1.0463	AA	1
46.	3–12		178.491	4782497.1	5342750	18–288	4.854–01	3.710–03	3.924–02	–1.1754	AA	1
47.	3–13		176.749	4782497.1	5348270	18–338	3.226–01	2.837–03	2.972–02	–1.2918	AA	1
48.	3–14		175.388	4782497.1	5352660	18–392	2.213–01	2.222–03	2.310–02	–1.3979	AA	1
49.	3–15		174.309	4782497.1	5356190	18–450	1.559–01	1.775–03	1.834–02	–1.4954	AA	1
50.	3–16		173.433	4782497.1	5359090	18–512	1.124–01	1.442–03	1.482–02	–1.5857	AA	1
51.	3–17		172.717	4782497.1	5361480	18–578	8.274–02	1.188–03	1.216–02	–1.6699	AA	1
52.	3–18		172.119	4782497.1	5363490	18–648	6.199–02	9.911–04	1.011–02	–1.7486	AA	1
53.	3–19		171.617	4782497.1	5365190	18–722	4.719–02	8.357–04	8.499–03	–1.8227	AA	1
54.	3–20		171.188	4782497.1	5366650	18–800	3.644–02	7.114–04	7.217–03	–1.8926	AA	1
55.	4–5		826.367	5043978.0	5164989.6	32–50	6.484+01	1.037+00	9.030+01	1.5210	AA	1
56.	4–6		535.500	5043978.0	5230719.3	32–72	1.852+01	1.792–01	1.011+01	0.7584	AA	1
57.	4–7		441.750	5043978.0	5270350.3	32–98	7.306+00	6.546–02	3.046+00	0.3211	AA	1
58.	4–8		396.678	5043978.0	5296071.6	32–128	3.421+00	3.228–02	1.349+00	0.0141	AA	1
59.	4–9		370.744	5043978.0	5313705.7	32–162	1.792+00	1.869–02	7.301–01	–0.2232	AA	1
60.	4–10		354.182	5043978.0	5326319.0	32–200	1.017+00	1.196–02	4.461–01	–0.4172	AA	1
61.	4–11		342.851	5043978.0	5335650	32–242	6.141–01	8.185–03	2.956–01	–0.5819	AA	1
62.	4–12		334.703	5043978.0	5342750	32–288	3.893–01	5.884–03	2.075–01	–0.7252	AA	1
63.	4–13		328.632	5043978.0	5348270	32–338	2.568–01	4.391–03	1.520–01	–0.8523	AA	1
64.	4–14		323.958	5043978.0	5352660	32–392	1.751–01	3.374–03	1.152–01	–0.9667	AA	1

N VII: Table A. $(n)_i - (n)_k$ Transitions (Average Values) — Continued

No.	Transition	λ_{air} (Å)	λ_{vac} (Å) or σ (cm^{-1})*	E_i (cm^{-1})	E_k (cm^{-1})	$g_i - g_k$	A_{ki} (10^8 s^{-1})	f_{ik}	S (at. u.)	log gf	Acc.	Source
65.	4−15		320.295	5043978.0	5356190	32−450	1.228−01	2.655−03	8.959−02	−1.0707	AA	1
66.	4−16		317.347	5043978.0	5359090	32−512	8.820−02	2.131−03	7.123−02	−1.1664	AA	1
67.	4−17		314.959	5043978.0	5361480	32−578	6.470−02	1.738−03	5.767−02	−1.2548	AA	1
68.	4−18		312.977	5043978.0	5363490	32−648	4.835−02	1.438−03	4.741−02	−1.3371	AA	1
69.	4−19		311.321	5043978.0	5365190	32−722	3.673−02	1.204−03	3.949−02	−1.4142	AA	1
70.	4−20		309.912	5043978.0	5366650	32−800	2.831−02	1.019−03	3.327−02	−1.4867	AA	1
71.	5−6		1521.38	5164989.6	5230719.3	50−72	2.463+01	1.231+00	3.082+02	1.7892	AA	1
72.	5−7		949.120	5164989.6	5270350.3	50−98	7.814+00	2.068−01	3.231+01	1.0146	AA	1
73.	5−8		762.881	5164989.6	5296071.6	50−128	3.334+00	7.446−02	9.350+00	0.5709	AA	1
74.	5−9		672.422	5164989.6	5313705.7	50−162	1.659+00	3.644−02	4.034+00	0.2606	AA	1
75.	5−10		619.850	5164989.6	5326319.0	50−200	9.128−01	2.103−02	2.146+00	0.0218	AA	1
76.	5−11		585.959	5164989.6	5335650	50−242	5.395−01	1.344−02	1.296+00	−0.1726	AA	1
77.	5−12		562.555	5164989.6	5342750	50−288	3.369−01	9.207−03	8.525−01	−0.3369	AA	1
78.	5−13		545.612	5164989.6	5348270	50−338	2.198−01	6.630−03	5.954−01	−0.4795	AA	1
79.	5−14		532.849	5164989.6	5352660	50−392	1.486−01	4.958−03	4.349−01	−0.6057	AA	1
80.	5−15		523.011	5164989.6	5356190	50−450	1.035−01	3.820−03	3.289−01	−0.7190	AA	1
81.	5−16		515.197	5164989.6	5359090	50−512	7.396−02	3.014−03	2.556−01	−0.8219	AA	1
82.	5−17		508.931	5164989.6	5361480	50−578	5.403−02	2.425−03	2.032−01	−0.9163	AA	1
83.	5−18		503.777	5164989.6	5363490	50−648	4.023−02	1.984−03	1.645−01	−1.0035	AA	1
84.	5−19		499.500	5164989.6	5365190	50−722	3.047−02	1.646−03	1.353−01	−1.0847	AA	1
85.	5−20		495.883	5164989.6	5366650	50−800	2.342−02	1.382−03	1.128−01	−1.1606	AA	1
86.	6−7	2522.52	2523.28	5230719.3	5270350.3	72−98	1.096+01	1.423+00	8.513+02	2.0107	AA	1
87.	6−8		1530.17	5230719.3	5296071.6	72−128	3.750+00	2.340−01	8.487+01	1.2265	AA	1
88.	6−9		1205.02	5230719.3	5313705.7	72−162	1.697+00	8.313−02	2.374+01	0.7771	AA	1
89.	6−10		1046.03	5230719.3	5326319.0	72−200	8.860−01	4.037−02	1.001+01	0.4634	AA	1
90.	6−11		953.010	5230719.3	5335650	72−242	5.068−01	2.319−02	5.239+00	0.2227	AA	1
91.	6−12		892.612	5230719.3	5342750	72−288	3.095−01	1.479−02	3.129+00	0.0272	AA	1
92.	6−13		850.697	5230719.3	5348270	72−338	1.987−01	1.012−02	2.041+00	−0.1375	AA	1
93.	6−14		820.071	5230719.3	5352660	72−392	1.328−01	7.287−03	1.417+00	−0.2801	AA	1
94.	6−15		796.999	5230719.3	5356190	72−450	9.165−02	5.455−03	1.030+00	−0.4059	AA	1
95.	6−16		778.994	5230719.3	5359090	72−512	6.502−02	4.207−03	7.767−01	−0.5187	AA	1

N VII: Table A. $(n)_i - (n)_k$ Transitions (Average Values) — Continued

No.	Transition	λ_{air} (Å)	λ_{vac} (Å) or σ (cm⁻¹)*	E_i (cm^{-1})		E_k (cm^{-1})	g_i-g_k	A_{ki} (10^8 s^{-1})	f_{ik}	S (at. u.)	log gf	Acc.	Source
96.	6−17		764.756	5230719.3	−	5361480	72−578	4.723−02	3.324−03	6.026−01	−0.6210	AA	1
97.	6−18		753.178	5230719.3	−	5363490	72−648	3.500−02	2.679−03	4.783−01	−0.7147	AA	1
98.	6−19		743.656	5230719.3	−	5365190	72−722	2.641−02	2.195−03	3.870−01	−0.8011	AA	1
99.	6−20		735.669	5230719.3	−	5366650	72−800	2.024−02	1.824−03	3.181−01	−0.8816	AA	1
100.	7−8	3886.73	3887.83	5270350.3	−	5296071.6	98−128	5.458+00	1.615+00	2.026+03	2.1995	AA	1
101.	7−9	2305.81	2306.52	5270350.3	−	5313705.7	98−162	1.979+00	2.609−01	1.941+02	1.4077	AA	1
102.	7−10		1786.71	5270350.3	−	5326319.0	98−200	9.380−01	9.162−02	5.281+01	0.9532	AA	1
103.	7−11		1531.40	5270350.3	−	5335650	98−242	5.086−01	4.416−02	2.182+01	0.6363	AA	1
104.	7−12		1381.22	5270350.3	−	5342750	98−288	3.004−01	2.525−02	1.125+01	0.3934	AA	1
105.	7−13		1283.37	5270350.3	−	5348270	98−338	1.885−01	1.605−02	6.646+00	0.1967	AA	1
106.	7−14		1214.92	5270350.3	−	5352660	98−392	1.239−01	1.096−02	4.297+00	0.0312	AA	1
107.	7−15		1164.96	5270350.3	−	5356190	98−450	8.446−02	7.890−03	2.966+00	−0.1117	AA	1
108.	7−16		1126.89	5270350.3	−	5359090	98−512	5.936−02	5.904−03	2.146+00	−0.2377	AA	1
109.	7−17		1097.34	5270350.3	−	5361480	98−578	4.279−02	4.556−03	1.613+00	−0.3502	AA	1
110.	7−18		1073.66	5270350.3	−	5363490	98−648	3.152−02	3.602−03	1.248+00	−0.4522	AA	1
111.	7−19		1054.41	5270350.3	−	5365190	98−722	2.366−02	2.906−03	9.884−01	−0.5455	AA	1
112.	7−20		1038.42	5270350.3	−	5366650	98−800	1.806−02	2.383−03	7.984−01	−0.6317	AA	1
113.	8−9	5669.26	5670.83	5296071.6	−	5313705.7	128−162	2.962+00	1.807+00	4.318+03	2.3642	AA	1
114.	8−10	3305.12	3306.07	5296071.6	−	5326319.0	128−200	1.123+00	2.876−01	4.007+02	1.5660	AA	1
115.	8−11	2525.87	2526.63	5296071.6	−	5335650	128−242	5.527−01	1.000−01	1.065+02	1.1072	AA	1
116.	8−12	2141.64	2142.32	5296071.6	−	5342750	128−288	3.092−01	4.786−02	4.321+01	0.7872	AA	1
117.	8−13		1915.77	5296071.6	−	5348270	128−338	1.875−01	2.724−02	2.199+01	0.5424	AA	1
118.	8−14		1767.15	5296071.6	−	5352660	128−392	1.203−01	1.725−02	1.285+01	0.3441	AA	1
119.	8−15		1663.38	5296071.6	−	5356190	128−450	8.068−02	1.177−02	8.247+00	0.1778	AA	1
120.	8−16		1586.84	5296071.6	−	5359090	128−512	5.599−02	8.454−03	5.653+00	0.0343	AA	1
121.	8−17		1528.86	5296071.6	−	5361480	128−578	3.996−02	6.323−03	4.074+00	−0.0918	AA	1
122.	8−18		1483.27	5296071.6	−	5363490	128−648	2.921−02	4.877−03	3.049+00	−0.2046	AA	1
123.	8−19		1446.79	5296071.6	−	5365190	128−722	2.179−02	3.857−03	2.351+00	−0.3066	AA	1
124.	8−20		1416.86	5296071.6	−	5366650	128−800	1.654−02	3.111−03	1.858+00	−0.3998	AA	1
125.	9−10	7925.96	7928.14	5313705.7	−	5326319.0	162−200	1.718+00	1.999+00	8.450+03	2.5102	AA	1
126.	9−11	4555.71	4556.99	5313705.7	−	5335650	162−242	6.758−01	3.143−01	7.638+02	1.7068	AA	1
127.	9−12	3442.03	3443.02	5313705.7	−	5342750	162−288	3.428−01	1.083−01	1.989+02	1.2441	AA	1

N VII: Table A. $(n)_i - (n)_k$ Transitions (Average Values) — Continued

No.	Transition	λ_{air} (Å)	λ_{vac} (Å) or σ (cm^{-1})*	E_i (cm^{-1})		E_k (cm^{-1})	g_i-g_k	A_{ki} (10^8 s^{-1})	f_{ik}	S (at. u.)	log gf	Acc.	Source
128.	9–13	2892.31	2893.16	5313705.7	–	5348270	162–338	1.968–01	5.152–02	7.950+01	0.9215	AA	1
129.	9–14	2566.34	2567.11	5313705.7	–	5352660	162–392	1.220–01	2.917–02	3.994+01	0.6745	AA	1
130.	9–15	2353.09	2353.81	5313705.7	–	5356190	162–450	7.988–02	1.843–02	2.314+01	0.4751	AA	1
131.	9–16	2202.72	2203.41	5313705.7	–	5359090	162–512	5.448–02	1.253–02	1.473+01	0.3076	AA	1
132.	9–17	2092.51	2093.18	5313705.7	–	5361480	162–578	3.838–02	8.996–03	1.004+01	0.1635	AA	1
133.	9–18	2008.02	2008.67	5313705.7	–	5363490	162–648	2.777–02	6.720–03	7.199+00	0.0369	AA	1
134.	9–19		1942.34	5313705.7	–	5365190	162–722	2.055–02	5.180–03	5.366+00	–0.0761	AA	1
135.	9–20		1888.78	5313705.7	–	5366650	162–800	1.550–02	4.094–03	4.124+00	–0.1784	AA	1
136.	10–11	10714.0	9331 cm^{-1}	5326319.0	–	5335650	200–242	1.051+00	2.190+00	1.546+04	2.6416	AA	1
137.	10–12	6084.37	6086.06	5326319.0	–	5342750	200–288	4.262–01	3.408–01	1.366+03	1.8335	AA	1
138.	10–13	4554.32	4555.60	5326319.0	–	5348270	200–338	2.217–01	1.166–01	3.497+02	1.3677	AA	1
139.	10–14	3795.29	3796.36	5326319.0	–	5352660	200–392	1.301–01	5.511–02	1.378+02	1.0423	AA	1
140.	10–15	3346.77	3347.73	5326319.0	–	5356190	200–450	8.225–02	3.110–02	6.854+01	0.7937	AA	1
141.	10–16	3050.59	3051.48	5326319.0	–	5359090	200–512	5.476–02	1.957–02	3.932+01	0.5926	AA	1
142.	10–17	2843.22	2844.06	5326319.0	–	5361480	200–578	3.791–02	1.329–02	2.488+01	0.4244	AA	1
143.	10–18	2689.47	2690.27	5326319.0	–	5363490	200–648	2.707–02	9.517–03	1.686+01	0.2795	AA	1
144.	10–19	2571.84	2572.61	5326319.0	–	5365190	200–722	1.982–02	7.101–03	1.203+01	0.1523	AA	1
145.	10–20	2478.73	2479.48	5326319.0	–	5366650	200–800	1.483–02	5.467–03	8.925+00	0.0388	AA	1
146.	11–12	14080.7	7100 cm^{-1}	5335650	–	5342750	242–288	6.724–01	2.380+00	2.670+04	2.7603	AA	1
147.	11–13	7921.75	7923.93	5335650	–	5348270	242–338	2.795–01	3.674–01	2.319+03	1.9490	AA	1
148.	11–14	5877.27	5878.89	5335650	–	5352660	242–392	1.486–01	1.247–01	5.841+02	1.4797	AA	1
149.	11–15	4867.19	4868.55	5335650	–	5356190	242–450	8.886–02	5.872–02	2.278+02	1.1526	AA	1
150.	11–16	4265.01	4266.21	5335650	–	5359090	242–512	5.711–02	3.297–02	1.121+02	0.9019	AA	1
151.	11–17	3870.37	3871.47	5335650	–	5361480	242–578	3.858–02	2.071–02	6.387+01	0.6999	AA	1
152.	11–18	3590.93	3591.95	5335650	–	5363490	242–648	2.707–02	1.402–02	4.012+01	0.5305	AA	1
153.	11–19	3384.27	3385.24	5335650	–	5365190	242–722	1.956–02	1.002–02	2.704+01	0.3849	AA	1
154.	11–20	3224.88	3225.81	5335650	–	5366650	242–800	1.448–02	7.466–03	1.919+01	0.2569	AA	1
155.	12–13	18111.0	5520 cm^{-1}	5342750	–	5348270	288–338	4.461–01	2.576+00	4.424+04	2.8703	AA	1
156.	12–14	10088.1	9910 cm^{-1}	5342750	–	5352660	288–392	1.894–01	3.935–01	3.765+03	2.0544	AA	1
157.	12–15	7438.43	7440.48	5342750	–	5356190	288–450	1.026–01	1.330–01	9.386+02	1.5834	AA	1
158.	12–16	6118.26	6119.95	5342750	–	5359090	288–512	6.237–02	6.226–02	3.612+02	1.2536	AA	1
159.	12–17	5337.54	5339.03	5342750	–	5361480	288–578	4.067–02	3.488–02	1.766+02	1.0019	AA	1

N VII: Table A. $(n)_i - (n)_k$ Transitions (Average Values) — Continued

No.	Transition	λ_{air} (Å)	λ_{vac} (Å) or σ (cm^{-1})*	E_i (cm^{-1})		E_k (cm^{-1})	g_i–g_k	A_{ki} (10^8 s^{-1})	f_{ik}	S (at. u.)	log gf	Acc.	Source
160.	12–18	4820.25	4821.60	5342750	–	5363490	288–648	2.783–02	2.183–02	9.979+01	0.7984	AA	1
161.	12–19	4455.08	4456.33	5342750	–	5365190	288–722	1.975–02	1.474–02	6.230+01	0.6280	AA	1
162.	12–20	4182.92	4184.10	5342750	–	5366650	288–800	1.443–02	1.052–02	4.172+01	0.4813	AA	1
163.	13–14	22772.8	4390 cm^{-1}	5348270	–	5352660	338–392	3.053–01	2.754+00	6.982+04	2.9689	AA	1
164.	13–15	12622.8	7920 cm^{-1}	5348270	–	5356190	338–450	1.320–01	4.201–01	5.902+03	2.1523	AA	1
165.	13–16	9239.61	9242.14	5348270	–	5359090	338–512	7.268–02	1.410–01	1.450+03	1.6781	AA	1
166.	13–17	7567.94	7570.02	5348270	–	5361480	338–578	4.482–02	6.585–02	5.547+02	1.3475	AA	1
167.	13–18	6568.49	6570.30	5348270	–	5363490	338–648	2.960–02	3.673–02	2.685+02	1.0939	AA	1
168.	13–19	5908.53	5910.17	5348270	–	5365190	338–722	2.050–02	2.293–02	1.508+02	0.8893	AA	1
169.	13–20	5439.18	5440.70	5348270	–	5366650	338–800	1.470–02	1.544–02	9.347+01	0.7175	AA	1
170.	14–15	28320.9	3530 cm^{-1}	5352660	–	5356190	392–450	2.146–01	2.964+00	1.084+05	3.0652	AA	1
171.	14–16	15547.9	6430 cm^{-1}	5352660	–	5359090	392–512	9.431–02	4.466–01	8.964+03	2.2432	AA	1
172.	14–17	11334.8	8820 cm^{-1}	5352660	–	5361480	392–578	5.266–02	1.496–01	2.189+03	1.7683	AA	1
173.	14–18	9231.08	9233.61	5352660	–	5363490	392–648	3.288–02	6.948–02	8.280+02	1.4352	AA	1
174.	14–19	7978.65	7980.85	5352660	–	5365190	392–722	2.197–02	3.864–02	3.979+02	1.1803	AA	1
175.	14–20	7145.99	7147.96	5352660	–	5366650	392–800	1.537–02	2.402–02	2.216+02	0.9739	AA	1
176.	15–16	34473.4	2900 cm^{-1}	5356190	–	5359090	450–512	1.544–01	3.132+00	1.600+05	3.1490	AA	1
177.	15–17	18898.4	5290 cm^{-1}	5356190	–	5361480	450–578	6.881–02	4.735–01	1.326+04	2.3285	AA	1
178.	15–18	13694.9	7300 cm^{-1}	5356190	–	5363490	450–648	3.890–02	1.576–01	3.198+03	1.8508	AA	1
179.	15–19	11108.1	9000 cm^{-1}	5356190	–	5365190	450–722	2.457–02	7.296–02	1.201+03	1.5163	AA	1
180.	15–20	9557.61	9560.23	5356190	–	5366650	450–800	1.658–02	4.039–02	5.721+02	1.2595	AA	1
181.	16–17	41829.6	2390 cm^{-1}	5359090	–	5361480	512–578	1.134–01	3.360+00	2.369+05	3.2356	AA	1
182.	16–18	22721.1	4400 cm^{-1}	5359090	–	5363490	512–648	5.116–02	5.014–01	1.921+04	2.4094	AA	1
183.	16–19	16389.0	6100 cm^{-1}	5359090	–	5365190	512–722	2.925–02	1.662–01	4.591+03	1.9298	AA	1
184.	16–20	13223.9	7560 cm^{-1}	5359090	–	5366650	512–800	1.866–02	7.647–02	1.705+03	1.5928	AA	1
185.	17–18	49737.7	2010 cm^{-1}	5361480	–	5363490	578–648	8.477–02	3.527+00	3.339+05	3.3093	AA	1
186.	17–19	26946.8	3710 cm^{-1}	5361480	–	5365190	578–722	3.867–02	5.261–01	2.698+04	2.4830	AA	1
187.	17–20	19337.1	5170 cm^{-1}	5361480	–	5366650	578–800	2.233–02	1.733–01	6.380+03	2.0008	AA	1
188.	18–19		1700 cm^{-1}	5363490	–	5365190	648–722	6.440–02	3.722+00	4.671+05	3.3824	AA	1
189.	18–20	31636.9	3160 cm^{-1}	5363490	–	5366650	648–800	2.967–02	5.499–01	3.712+04	2.5518	AA	1
190.	19–20		1460 cm^{-1}	5365190	–	5366650	722–800	4.963–02	3.868+00	6.296+05	3.4460	AA	1

*Wavelengths (Å) are always given unless cm^{-1} is indicated.

List of tabulated lines (Table B)

Wavelength (Å)	No.	Wavelength (Å)	No.	Wavelength (Å)	No.	Wavelength (Å)	No.
in vacuum		133.776	10	382.444	30	826.500	42
		133.787	6	382.445	24	826.501	41
19.1179	5	133.874	14	382.497	27	826.523	44
19.1180	5	133.890	14	382.498	24	826.562	44
19.3612	4	133.934	10	382.545	30	826.625	42
19.3614	4	222.994	26	382.571	30	826.628	39
19.8257	3	223.001	20	382.625	27	826.628	42
19.8261	3	223.009	23	382.651	21	826.744	35
20.9095	2	223.017	20	382.659	27	826.763	39
20.9106	2	223.118	32	535.181	38	1520.01	48
24.7792	1	223.118	26	535.198	34	1520.08	46
24.7846	1	223.124	29	535.270	36	1520.72	47
83.6159	17	223.124	26	535.291	34	1520.83	46
83.6181	13	223.139	23	535.465	38	1521.06	50
83.6191	9	223.140	29	535.496	40	1521.06	48
83.6214	9	223.159	32	535.497	38	1521.31	49
83.6767	17	223.161	32	535.554	45	1521.31	48
83.6775	17	223.167	29	535.570	43	1521.37	52
83.6797	13	261.295	25	535.585	36	1521.37	50
88.4790	16	261.304	19	535.589	40	1521.49	51
88.4825	8	261.331	22	535.598	45	1521.49	50
88.4831	12	261.342	19	535.601	40	1521.51	54
88.4869	8	261.460	31	535.601	43	1521.51	52
88.5465	16	261.460	25	535.607	45	1521.58	53
88.5479	16	261.472	28	535.622	43	1521.58	52
88.5521	12	261.473	25	825.629	37	1521.59	54
99.0882	15	261.509	22	825.668	33	1521.64	54
99.0926	7	261.511	28	825.993	35	1521.71	51
99.0984	11	261.513	31	826.052	33	1521.71	53
99.1034	7	261.519	31	826.251	41	1521.74	49
99.1711	15	261.532	28	826.252	37	1521.74	51
99.1747	15	382.117	24	826.378	39	1522.03	47
99.1849	11	382.137	18	826.380	37	1522.06	49
133.732	14	382.269	21	826.437	44		
133.740	6	382.298	18	826.437	41		

N VII: Table B. $(nlj)_i - (nlj)_k$ Transitions (Fine Structure Lines)

No.	Transition Array	Mult.	λ_{air} (Å)	λ_{vac} (Å) or σ (cm^{-1})*	E_i (cm^{-1})	E_k (cm^{-1})	g_i–g_k	A_{ki} (10^8 s^{-1})	f_{ik}	S (at. u.)	log gf	Acc.	Source
1.	1*s*–2*p*	^{2}S–^{2}P°		*24.781*	0.00	– *4035348.7*	2–6	1.505+04	4.157–01	6.782–02	–0.0802	AA	1
				24.7792	0.00	– 4035642.12	2–4	1.505+04	2.771–01	4.521–02	–0.2563	AA	*LS*
				24.7846	0.00	– 4034761.80	2–2	1.504+04	1.385–01	2.261–02	–0.5574	AA	*LS*
2.	1*s*–3*p*	^{2}S–^{2}P°		*20.910*	0.00	– *4782437.9*	2–6	4.018+03	7.901–02	1.088–02	–0.8013	AA	1
				20.9095	0.00	– 4782524.86	2–4	4.018+03	5.267–02	7.252–03	–0.9774	AA	*LS*
				20.9106	0.00	– 4782264.01	2–2	4.017+03	2.633–02	3.626–03	–1.2784	AA	*LS*
3.	1*s*–4*p*	^{2}S–^{2}P°		*19.826*	0.00	– *5043927.4*	2–6	1.638+03	2.896–02	3.780–03	–1.2372	AA	1
				19.8257	0.00	– 5043964.05	2–4	1.638+03	1.930–02	2.520–03	–1.4133	AA	*LS*
				19.8261	0.00	– 5043854.01	2–2	1.638+03	9.652–03	1.260–03	–1.7143	AA	*LS*
4.	1*s*–5*p*	^{2}S–^{2}P°		*19.361*	0.00	– *5164955.1*	2–6	8.258+02	1.392–02	1.775–03	–1.5553	AA	1
				19.3612	0.00	– 5164973.86	2–4	8.258+02	9.281–03	1.183–03	–1.7314	AA	*LS*
				19.3614	0.00	– 5164917.53	2–2	8.258+02	4.641–03	5.916–04	–2.0324	AA	*LS*

N VII: Table B. $(nlj)_i - (nlj)_k$ Transitions (Fine Structure Lines) — Continued

No.	Transition Array	Mult.	λ_{air} (Å)	λ_{vac} (Å) or σ (cm^{-1})*	E_i (cm^{-1})	E_k (cm^{-1})	$g_i - g_k$	A_{ki} (10^8 s^{-1})	f_{ik}	S (at. u.)	log gf	Acc.	Source
5.	1s–6p	^{2}S–^{2}P°		*19.118*	0.00	*5230695.8*	2–6	4.739+02	7.790–03	9.806–04	–1.8074	AA	1
				19.1179	0.00	5230706.65	2–4	4.739+02	5.194–03	6.538–04	–1.9835	AA	*LS*
				19.1180	0.00	5230674.05	2–2	4.739+02	2.597–03	3.269–04	–2.2845	AA	*LS*
6.	2s–3p	^{2}S–^{2}P°		*133.76*	4034807.21	*4782437.9*	2–6	5.393+02	4.339–01	3.821–01	–0.0616	AA	1
				133.740	4034807.21	4782524.86	2–4	5.394+02	2.893–01	2.548–01	–0.2376	AA	*LS*
				133.787	4034807.21	4782264.01	2–2	5.389+02	1.446–01	1.274–01	–0.5388	AA	*LS*
7.	2s–4p	^{2}S–^{2}P°		*99.096*	4034807.21	*5043927.4*	2–6	2.322+02	1.026–01	6.693–02	–0.6879	AA	1
				99.0926	4034807.21	5043964.05	2–4	2.323+02	6.839–02	4.462–02	–0.8640	AA	*LS*
				99.1034	4034807.21	5043854.01	2–2	2.322+02	3.419–02	2.231–02	–1.1651	AA	*LS*
8.	2s–5p	^{2}S–^{2}P°		*88.484*	4034807.21	*5164955.1*	2–6	1.189+02	4.186–02	2.439–02	–1.0772	AA	1
				88.4825	4034807.21	5164973.86	2–4	1.189+02	2.791–02	1.626–02	–1.2533	AA	*LS*
				88.4869	4034807.21	5164917.53	2–2	1.189+02	1.395–02	8.129–03	–1.5543	AA	*LS*
9.	2s–6p	^{2}S–^{2}P°		*83.620*	4034807.21	*5230695.8*	2–6	6.866+01	2.159–02	1.189–02	–1.3647	AA	1
				83.6191	4034807.21	5230706.65	2–4	6.866+01	1.440–02	7.926–03	–1.5407	AA	*LS*
				83.6214	4034807.21	5230674.05	2–2	6.866+01	7.198–03	3.963–03	–1.8418	AA	*LS*
10.	2p–3s	^{2}P°–^{2}S		*133.88*	*4035348.7*	4782277.62	6–2	1.517+02	1.359–02	3.593–02	–1.0888	AA	1
				133.934	4035642.12	4782277.62	4–2	1.010+02	1.358–02	2.395–02	–1.2650	AA	*LS*
				133.776	4034761.80	4782277.62	2–2	5.067+01	1.360–02	1.198–02	–1.5656	AA	*LS*
11.	2p–4s	^{2}P°–^{2}S		*99.156*	*4035348.7*	5043859.77	6–2	6.193+01	3.043–03	5.960–03	–1.7386	AA	1
				99.1849	4035642.12	5043859.77	4–2	4.125+01	3.042–03	3.973–03	–1.9148	AA	*LS*
				99.0984	4034761.80	5043859.77	2–2	2.068+01	3.045–03	1.987–03	–2.2154	AA	*LS*
12.	2p–5s	^{2}P°–^{2}S		*88.529*	*4035348.7*	5164920.48	6–2	3.096+01	1.212–03	2.120–03	–2.1382	AA	1
				88.5521	4035642.12	5164920.48	4–2	2.062+01	1.212–03	1.413–03	–2.3144	AA	*LS*
				88.4831	4034761.80	5164920.48	2–2	1.033+01	1.213–03	7.067–04	–2.6151	AA	*LS*
13.	2p–6s	^{2}P°–^{2}S		*83.659*	*4035348.7*	5230675.76	6–2	1.766+01	6.175–04	1.020–03	–2.4312	AA	1
				83.6797	4035642.12	5230675.76	4–2	1.176+01	6.174–04	6.803–04	–2.6074	AA	*LS*
				83.6181	4034761.80	5230675.76	2–2	5.894+00	6.178–04	3.402–04	–2.9081	AA	*LS*
14.	2p–3d	^{2}P°–^{2}D		*133.83*	*4035348.7*	*4782576.5*	6–10	1.553+03	6.950–01	1.837+00	0.6201	AA	1
				133.874	4035642.12	4782611.27	4–6	1.551+03	6.253–01	1.102+00	0.3981	AA	*LS*
				133.732	4034761.80	4782524.42	2–4	1.297+03	6.955–01	6.124–01	0.1433	AA	*LS*
				133.890	4035642.12	4782524.42	4–4	2.585+02	6.947–02	1.225–01	–0.5562	AA	*LS*
15.	2p–4d	^{2}P°–^{2}D		*99.144*	*4035348.7*	*5043985.8*	6–10	4.955+02	1.217–01	2.383–01	–0.1366	AA	1
				99.1711	4035642.12	5044000.50	4–6	4.951+02	1.095–01	1.430–01	–0.3586	AA	*LS*
				99.0882	4034761.80	5043963.86	2–4	4.136+02	1.218–01	7.944–02	–0.6135	AA	*LS*
				99.1747	4035642.12	5043963.86	4–4	8.250+01	1.216–02	1.589–02	–1.3128	AA	*LS*
16.	2p–5d	^{2}P°–^{2}D		*88.524*	*4035348.7*	*5164985.0*	6–10	2.264+02	4.433–02	7.752–02	–0.5751	AA	1
				88.5465	4035642.12	5164992.53	4–6	2.262+02	3.989–02	4.651–02	–0.7971	AA	*LS*
				88.4790	4034761.80	5164973.77	2–4	1.890+02	4.436–02	2.584–02	–1.0520	AA	*LS*
				88.5479	4035642.12	5164973.77	4–4	3.771+01	4.432–03	5.168–03	–1.7513	AA	*LS*

N VII: Table B. $(nlj)_i - (nlj)_k$ Transitions (Fine Structure Lines) — Continued

No.	Transition Array	Mult.	λ_{air} (Å)	λ_{vac} (Å) or σ (cm^{-1})*	E_i (cm^{-1})	E_k (cm^{-1})	g_i-g_k	A_{ki} (10^8 s^{-1})	f_{ik}	S (at. u.)	log gf	Acc.	Source
17.	$2p-6d$	$^2P^o-^2D$		*83.656*	*4035348.7*	*5230713.1*	6–10	1.236+02	2.161–02	3.571–02	–0.8871	AA	1
				83.6767	4035642.12	5230717.45	4–6	1.235+02	1.945–02	2.143–02	–1.1091	AA	*LS*
				83.6159	4034761.80	5230706.59	2–4	1.031+02	2.162–02	1.190–02	–1.3641	AA	*LS*
				83.6775	4035642.12	5230706.59	4–4	2.058+01	2.161–03	2.381–03	–2.0633	AA	*LS*
18.	$3s-4p$	$^2S-^2P^o$		*382.19*	4782277.62	*5043927.4*	2–6	7.363+01	4.837–01	1.217+00	–0.0144	AA	1
				382.137	4782277.62	5043964.05	2–4	7.366+01	3.225–01	8.115–01	–0.1904	AA	*LS*
				382.298	4782277.62	5043854.01	2–2	7.357+01	1.612–01	4.058–01	–0.4916	AA	*LS*
19.	$3s-5p$	$^2S-^2P^o$		*261.32*	4782277.62	*5164955.1*	2–6	3.934+01	1.208–01	2.079–01	–0.6168	AA	1
				261.304	4782277.62	5164973.86	2–4	3.935+01	8.056–02	1.386–01	–0.7929	AA	*LS*
				261.342	4782277.62	5164917.53	2–2	3.933+01	4.027–02	6.930–02	–1.0940	AA	*LS*
20.	$3s-6p$	$^2S-^2P^o$		*223.01*	4782277.62	*5230695.8*	2–6	2.294+01	5.132–02	7.535–02	–0.9887	AA	1
				223.001	4782277.62	5230706.65	2–4	2.294+01	3.421–02	5.023–02	–1.1648	AA	*LS*
				223.017	4782277.62	5230674.05	2–2	2.294+01	1.711–02	2.512–02	–1.4658	AA	*LS*
21.	$3p-4s$	$^2P^o-^2S$		*382.52*	*4782437.9*	5043859.77	6–2	4.409+01	3.224–02	2.436–01	–0.7134	AA	1
				382.651	4782524.86	5043859.77	4–2	2.936+01	3.223–02	1.624–01	–0.8897	AA	*LS*
				382.269	4782264.01	5043859.77	2–2	1.473+01	3.226–02	8.120–02	–1.1903	AA	*LS*
22.	$3p-5s$	$^2P^o-^2S$		*261.45*	*4782437.9*	5164920.48	6–2	2.173+01	7.424–03	3.834–02	–1.3512	AA	1
				261.509	4782524.86	5164920.48	4–2	1.448+01	7.422–03	2.556–02	–1.5274	AA	*LS*
				261.331	4782264.01	5164920.48	2–2	7.254+00	7.427–03	1.278–02	–1.8281	AA	*LS*
23.	$3p-6s$	$^2P^o-^2S$		*223.10*	*4782437.9*	5230675.76	6–2	1.218+01	3.030–03	1.335–02	–1.7404	AA	1
				223.139	4782524.86	5230675.76	4–2	8.117+00	3.030–03	8.902–03	–1.9165	AA	*LS*
				223.009	4782264.01	5230675.76	2–2	4.066+00	3.031–03	4.451–03	–2.2173	AA	*LS*
24.	$3p-4d$	$^2P^o-^2D$		*382.34*	*4782437.9*	*5043985.8*	6–10	1.691+02	6.175–01	4.664+00	0.5688	AA	1
				382.445	4782524.86	5044000.50	4–6	1.689+02	5.556–01	2.798+00	0.3468	AA	*LS*
				382.117	4782264.01	5043963.86	2–4	1.411+02	6.179–01	1.555+00	0.0919	AA	*LS*
				382.498	4782524.86	5043963.86	4–4	2.814+01	6.172–02	3.109–01	–0.6075	AA	*LS*
25.	$3p-5d$	$^2P^o-^2D$		*261.41*	*4782437.9*	*5164985.0*	6–10	8.147+01	1.391–01	7.183–01	–0.0785	AA	1
				261.460	4782524.86	5164992.53	4–6	8.142+01	1.252–01	4.310–01	–0.3005	AA	*LS*
				261.295	4782264.01	5164973.77	2–4	6.798+01	1.392–01	2.394–01	–0.5554	AA	*LS*
				261.473	4782524.86	5164973.77	4–4	1.357+01	1.391–02	4.788–02	–1.2547	AA	*LS*
26.	$3p-6d$	$^2P^o-^2D$		*223.08*	*4782437.9*	*5230713.1*	6–10	4.511+01	5.609–02	2.471–01	–0.4730	AA	1
				223.118	4782524.86	5230717.45	4–6	4.508+01	5.047–02	1.483–01	–0.6949	AA	*LS*
				222.994	4782264.01	5230706.59	2–4	3.763+01	5.611–02	8.238–02	–0.9499	AA	*LS*
				223.124	4782524.86	5230706.59	4–4	7.513+00	5.608–03	1.648–02	–1.6492	AA	*LS*
27.	$3d-4p$	$^2D-^2P^o$		*382.63*	*4782576.5*	*5043927.4*	10–6	8.349+00	1.099–02	1.385–01	–0.9588	AA	1
				382.625	4782611.27	5043964.05	6–4	7.514+00	1.099–02	8.310–02	–1.1807	AA	*LS*
				382.659	4782524.42	5043854.01	4–2	8.347+00	9.161–03	4.616–02	–1.4360	AA	*LS*
				382.497	4782524.42	5043964.05	4–4	8.357–01	1.833–03	9.233–03	–2.1348	AA	*LS*

N VII: Table B. $(nlj)_i - (nlj)_k$ Transitions (Fine Structure Lines) — Continued

No.	Transition Array	Mult.	λ_{air} (Å)	λ_{vac} (Å) or σ (cm^{-1})*	E_i (cm^{-1})	E_k (cm^{-1})	g_i-g_k	A_{ki} (10^8 s^{-1})	f_{ik}	S (at. u.)	log gf	Acc.	Source
28.	$3d-5p$	^{2}D–^{2}P°		*261.52*	*4782576.5*	*5164955.1*	10–6	3.592+00	2.210–03	1.903–02	–1.6556	AA	1
				261.532	4782611.27	5164973.86	6–4	3.233+00	2.210–03	1.142–02	–1.8775	AA	*LS*
				261.511	4782524.42	5164917.53	4–2	3.593+00	1.842–03	6.342–03	–2.1327	AA	*LS*
				261.472	4782524.42	5164973.86	4–4	3.594–01	3.684–04	1.268–03	–2.8316	AA	*LS*
29.	$3d-6p$	^{2}D–^{2}P°		*223.15*	*4782576.5*	*5230695.8*	10–6	1.880+00	8.419–04	6.185–03	–2.0747	AA	1
				223.167	4782611.27	5230706.65	6–4	1.691+00	8.419–04	3.711–03	–2.2966	AA	*LS*
				223.140	4782524.42	5230674.05	4–2	1.880+00	7.016–04	2.062–03	–2.5518	AA	*LS*
				223.124	4782524.42	5230706.65	4–4	1.880–01	1.403–04	4.123–04	–3.2508	AA	*LS*
30.	$3d-4f$	^{2}D–^{2}F°		*382.51*	*4782576.5*	*5044010.9*	10–14	3.312+02	1.017+00	1.281+01	1.0074	AA	1
				382.545	4782611.27	5044018.75	6–8	3.311+02	9.686–01	7.319+00	0.7643	AA	*LS*
				382.444	4782524.42	5044000.44	4–6	3.093+02	1.017+00	5.123+00	0.6095	AA	*LS*
				382.571	4782611.27	5044000.44	6–6	2.207+01	4.843–02	3.659–01	–0.5368	AA	*LS*
31.	$3d-5f$	^{2}D–^{2}F°		*261.49*	*4782576.5*	*5164997.9*	10–14	1.091+02	1.566–01	1.348+00	0.1948	AA	1
				261.513	4782611.27	5165001.87	6–8	1.091+02	1.491–01	7.703–01	–0.0483	AA	*LS*
				261.460	4782524.42	5164992.49	4–6	1.019+02	1.566–01	5.392–01	–0.2031	AA	*LS*
				261.519	4782611.27	5164992.49	6–6	7.272+00	7.456–03	3.852–02	–1.3493	AA	*LS*
32.	$3d-6f$	^{2}D–^{2}F°		*223.14*	*4782576.5*	*5230720.5*	10–14	5.155+01	5.388–02	3.958–01	–0.2686	AA	1
				223.159	4782611.27	5230722.86	6–8	5.154+01	5.131–02	2.262–01	–0.5117	AA	*LS*
				223.118	4782524.42	5230717.43	4–6	4.813+01	5.388–02	1.583–01	–0.6665	AA	*LS*
				223.161	4782611.27	5230717.43	6–6	3.436+00	2.565–03	1.131–02	–1.8127	AA	*LS*
33.	$4s-5p$	^{2}S–^{2}P°		*825.80*	5043859.77	*5164955.1*	2–6	1.771+01	5.431–01	2.953+00	0.0359	AA	1
				825.668	5043859.77	5164973.86	2–4	1.772+01	3.622–01	1.969+00	–0.1401	AA	*LS*
				826.052	5043859.77	5164917.53	2–2	1.769+01	1.810–01	9.844–01	–0.4413	AA	*LS*
34.	$4s-6p$	^{2}S–^{2}P°		*535.23*	5043859.77	*5230695.8*	2–6	1.070+01	1.379–01	4.860–01	–0.5593	AA	1
				535.198	5043859.77	5230706.65	2–4	1.071+01	9.195–02	3.240–01	–0.7354	AA	*LS*
				535.291	5043859.77	5230674.05	2–2	1.070+01	4.597–02	1.620–01	–1.0365	AA	*LS*
35.	$4p-5s$	^{2}P°–^{2}S		*826.49*	*5043927.4*	5164920.48	6–2	1.550+01	5.290–02	8.636–01	–0.4984	AA	1
				826.744	5043964.05	5164920.48	4–2	1.032+01	5.288–02	5.757–01	–0.6746	AA	*LS*
				825.993	5043854.01	5164920.48	2–2	5.175+00	5.293–02	2.879–01	–0.9753	AA	*LS*
36.	$4p-6s$	^{2}P°–^{2}S		*535.48*	*5043927.4*	5230675.76	6–2	8.606+00	1.233–02	1.304–01	–1.1308	AA	1
				535.585	5043964.05	5230675.76	4–2	5.734+00	1.233–02	8.695–02	–1.3070	AA	*LS*
				535.270	5043854.01	5230675.76	2–2	2.872+00	1.234–02	4.348–02	–1.6078	AA	*LS*
37.	$4p-5d$	^{2}P°–^{2}D		*826.05*	*5043927.4*	*5164985.0*	6–10	3.569+01	6.085–01	9.929+00	0.5624	AA	1
				826.252	5043964.05	5164992.53	4–6	3.567+01	5.476–01	5.958+00	0.3405	AA	*LS*
				825.629	5043854.01	5164973.77	2–4	2.979+01	6.089–01	3.310+00	0.0855	AA	*LS*
				826.380	5043964.05	5164973.77	4–4	5.942+00	6.083–02	6.620–01	–0.6138	AA	*LS*
38.	$4p-6d$	^{2}P°–^{2}D		*535.37*	*5043927.4*	*5230713.1*	6–10	2.071+01	1.483–01	1.569+00	–0.0506	AA	1
				535.465	5043964.05	5230717.45	4–6	2.070+01	1.335–01	9.412–01	–0.2725	AA	*LS*
				535.181	5043854.01	5230706.59	2–4	1.728+01	1.484–01	5.229–01	–0.5276	AA	*LS*
				535.497	5043964.05	5230706.59	4–4	3.450+00	1.483–02	1.046–01	–1.2268	AA	*LS*

N VII: Table B. $(nlj)_i - (nlj)_k$ Transitions (Fine Structure Lines) — Continued

No.	Transition Array	Mult.	λ_{air} (Å)	λ_{vac} (Å) or σ (cm^{-1})*	E_i (cm^{-1})	E_k (cm^{-1})	g_i-g_k	A_{ki} (10^8 s^{-1})	f_{ik}	S (at. u.)	log gf	Acc.	Source
39.	4*d*–5*p*	^{2}D–^{2}P°		*826.66*	*5043985.8*	– *5164955.1*	10–6	4.527+00	2.783–02	7.573–01	–0.5555	AA	1
				826.628	5044000.50	– 5164973.86	6–4	4.075+00	2.783–02	4.544–01	–0.7773	AA	*LS*
				826.763	5043963.86	– 5164917.53	4–2	4.525+00	2.319–02	2.524–01	–1.0327	AA	*LS*
				826.378	5043963.86	– 5164973.86	4–4	4.532–01	4.640–03	5.049–02	–1.7315	AA	*LS*
40.	4*d*–6*p*	^{2}D–^{2}P°		*535.59*	*5043985.8*	– *5230695.8*	10–6	2.262+00	5.837–03	1.029–01	–1.2338	AA	1
				535.601	5044000.50	– 5230706.65	6–4	2.036+00	5.837–03	6.175–02	–1.4557	AA	*LS*
				535.589	5043963.86	– 5230674.05	4–2	2.262+00	4.864–03	3.431–02	–1.7109	AA	*LS*
				535.496	5043963.86	– 5230706.65	4–4	2.263–01	9.730–04	6.861–03	–2.4098	AA	*LS*
41.	4*d*–5*f*	^{2}D–^{2}F°		*826.36*	*5043985.8*	– *5164997.9*	10–14	6.208+01	8.898–01	2.421+01	0.9493	AA	1
				826.437	5044000.50	– 5165001.87	6–8	6.207+01	8.474–01	1.383+01	0.7062	AA	*LS*
				826.251	5043963.86	– 5164992.49	4–6	5.797+01	8.899–01	9.683+00	0.5514	AA	*LS*
				826.501	5044000.50	– 5164992.49	6–6	4.137+00	4.237–02	6.916–01	–0.5948	AA	*LS*
42.	4*f*–5*d*	^{2}F°–^{2}D		*826.62*	*5044010.9*	– *5164985.03*	14–10	1.213+00	8.873–03	3.380–01	–0.9058	AA	1
				826.625	5044018.75	– 5164992.53	8–6	1.155+00	8.873–03	1.932–01	–1.1489	AA	*LS*
				826.628	5044000.44	– 5164973.77	6–4	1.213+00	8.281–03	1.352–01	–1.3038	AA	*LS*
				826.500	5044000.44	– 5164992.53	6–6	5.777–02	5.916–04	9.658–03	–2.4498	AA	*LS*
43.	4*f*–6*d*	^{2}F°–^{2}D		*535.61*	*5044010.9*	– *5230713.1*	14–10	5.153–01	1.583–03	3.908–02	–1.6544	AA	1
				535.622	5044018.75	– 5230717.45	8–6	4.907–01	1.583–03	2.233–02	–1.8975	AA	*LS*
				535.601	5044000.44	– 5230706.59	6–4	5.153–01	1.477–03	1.563–02	–2.0523	AA	*LS*
				535.570	5044000.44	– 5230717.45	6–6	2.454–02	1.055–04	1.116–03	–3.1984	AA	*LS*
44.	4*f*–5*g*	^{2}F°–^{2}G		*826.49*	*5044010.9*	– *5165005.0*	14–18	1.022+02	1.346+00	5.126+01	1.2750	AA	1
				826.523	5044018.75	– 5165007.48	8–10	1.022+02	1.308+00	2.848+01	1.0197	AA	*LS*
				826.437	5044000.44	– 5165001.85	6–8	9.856+01	1.346+00	2.197+01	0.9071	AA	*LS*
				826.562	5044018.75	– 5165001.85	8–8	3.649+00	3.737–02	8.136–01	–0.5244	AA	*LS*
45.	4*f*–6*g*	^{2}F°–^{2}G		*535.58*	*5044010.9*	– *5230724.7*	14–18	3.298+01	1.823–01	4.501+00	0.4070	AA	1
				535.598	5044018.75	– 5230726.10	8–10	3.297+01	1.773–01	2.500+00	0.1517	AA	*LS*
				535.554	5044000.44	– 5230722.85	6–8	3.180+01	1.823–01	1.929+00	0.0390	AA	*LS*
				535.607	5044018.75	– 5230722.85	8–8	1.178+00	5.065–03	7.144–02	–1.3924	AA	*LS*
46.	5*s*–6*p*	^{2}S–^{2}P°		*1520.3*	5164920.48	– *5230695.8*	2–6	5.836+00	6.067–01	6.073+00	0.0840	AA	1
				1520.08	5164920.48	– 5230706.65	2–4	5.839+00	4.046–01	4.049+00	–0.0920	AA	*LS*
				1520.83	5164920.48	– 5230674.05	2–2	5.831+00	2.022–01	2.024+00	–0.3932	AA	*LS*
47.	5*p*–6*s*	^{2}P°–^{2}S		*1521.6*	*5164955.1*	– 5230675.76	6–2	6.442+00	7.453–02	2.240+00	–0.3495	AA	1
				1522.03	5164973.86	– 5230675.76	4–2	4.291+00	7.451–02	1.493+00	–0.5257	AA	*LS*
				1520.72	5164917.53	– 5230675.76	2–2	2.151+00	7.458–02	7.467–01	–0.8264	AA	*LS*
48.	5*p*–6*d*	^{2}P°–^{2}D		*1520.7*	*5164955.1*	– *5230713.1*	6–10	1.080+01	6.239–01	1.874+01	0.5733	AA	1
				1521.06	5164973.86	– 5230717.45	4–6	1.079+01	5.614–01	1.125+01	0.3513	AA	*LS*
				1520.01	5164917.53	– 5230706.59	2–4	9.011+00	6.242–01	6.247+00	0.0964	AA	*LS*
				1521.31	5164973.86	– 5230706.59	4–4	1.798+00	6.237–02	1.249+00	–0.6030	AA	*LS*
49.	5*d*–6*p*	^{2}D–^{2}P°		*1521.8*	*5164985.0*	– *5230695.8*	10–6	2.305+00	4.801–02	2.405+00	–0.3187	AA	1
				1521.74	5164992.53	– 5230706.65	6–4	2.074+00	4.801–02	1.443+00	–0.5405	AA	*LS*
				1522.06	5164973.77	– 5230674.05	4–2	2.303+00	4.000–02	8.017–01	–0.7959	AA	*LS*
				1521.31	5164973.77	– 5230706.65	4–4	2.307–01	8.004–03	1.603–01	–1.4946	AA	*LS*

N VII: Table B. $(nlj)_i - (nlj)_k$ Transitions (Fine Structure Lines) — Continued

No.	Transition Array	Mult.	λ_{air} (Å)	λ_{vac} (Å) or σ (cm⁻¹)*	E_i (cm⁻¹)	E_k (cm⁻¹)	g_i-g_k	A_{ki} (10^8 s⁻¹)	f_{ik}	S (at. u.)	log gf	Acc.	Source
50.	5d−6f	^{2}D−^{2}F°		*1521.2*	*5164985.03*	*5230720.5*	10−14	1.737+01	8.439−01	4.226+01	0.9263	AA	1
				1521.37	5164992.53	5230722.86	6−8	1.737+01	8.036−01	2.415+01	0.6832	AA	*LS*
				1521.06	5164973.77	5230717.43	4−6	1.622+01	8.440−01	1.691+01	0.5284	AA	*LS*
				1521.49	5164992.53	5230717.43	6−6	1.158+00	4.018−02	1.208+00	−0.6178	AA	*LS*
51.	5f−6d	^{2}F°−^{2}D		*1521.7*	*5164997.9*	*5230713.1*	14−10	9.388−01	2.328−02	1.633+00	−0.4869	AA	1
				1521.71	5165001.87	5230717.45	8−6	8.941−01	2.328−02	9.329−01	−0.7299	AA	*LS*
				1521.74	5164992.49	5230706.59	6−4	9.387−01	2.173−02	6.531−01	−0.8849	AA	*LS*
				1521.49	5164992.49	5230717.45	6−6	4.472−02	1.552−03	4.665−02	−2.0309	AA	*LS*
52.	5f−6g	^{2}F°−^{2}G		*1521.4*	*5164997.9*	*5230724.7*	14−18	2.656+01	1.185+00	8.310+01	1.2199	AA	1
				1521.51	5165001.87	5230726.10	8−10	2.656+01	1.152+00	4.617+01	0.9646	AA	*LS*
				1521.37	5164992.49	5230722.85	6−8	2.562+01	1.185+00	3.562+01	0.8519	AA	*LS*
				1521.58	5165001.87	5230722.85	8−8	9.484−01	3.292−02	1.319+00	−0.5795	AA	*LS*
53.	5g−6f	^{2}G−^{2}F°		*1521.7*	*5165005.0*	*5230720.5*	18−14	2.732−01	7.377−03	6.652−01	−0.8769	AA	1
				1521.71	5165007.48	5230722.86	10−8	2.656−01	7.377−03	3.695−01	−1.1321	AA	*LS*
				1521.71	5165001.85	5230717.43	8−6	2.732−01	7.113−03	2.851−01	−1.2448	AA	*LS*
				1521.58	5165001.85	5230722.86	8−8	7.591−03	2.635−04	1.056−02	−2.6762	AA	*LS*
54.	5g−6h	^{2}G−^{2}H°		*1521.6*	*5165005.0*	*5230727.3*	18−22	3.951+01	1.676+00	1.511+02	1.4796	AA	1
				1521.59	5165007.48	5230728.27	10−12	3.951+01	1.646+00	8.243+01	1.2163	AA	*LS*
				1521.51	5165001.85	5230726.10	8−10	3.864+01	1.676+00	6.717+01	1.1274	AA	*LS*
				1521.64	5165007.48	5230726.10	10−10	8.779−01	3.047−02	1.527+00	−0.5161	AA	*LS*

*Wavelengths (Å) are always given unless cm⁻¹ is indicated.

Oxygen

O I

Ground State: $1s^2 2s^2 2p^4\ ^3P_2$

Ionization Energy: 13.618 eV = 109837.02 cm^{-1}

Allowed Transitions

List of tabulated lines

Wavelength (Å)	No.	Wavelength (Å)	No.	Wavelength (Å)	No.	Wavelength (Å)	No.
in vacuum		843.589	52	978.617	11	2611.47	148
697.528	32	844.330	50	988.578	8	2799.85	92
698.299	32	849.451	49	988.655	8	2799.89	92
698.634	32	849.480	48	988.773	8	2799.90	92
724.835	31	850.679	47	990.127	8	2805.53	91
725.667	31	859.312	46	990.204	8	2805.57	91
726.029	31	859.349	45	990.801	8	2805.58	91
769.353	30	859.586	44	999.497	34	2806.02	91
769.408	30	861.563	43	1025.76	7	2806.07	91
770.260	30	877.798	22	1027.43	7	2806.90	91
770.291	30	877.879	22	1028.16	7	2876.29	90
770.346	30	878.201	42	1039.23	6	2878.93	90
770.699	30	878.247	41	1040.94	6	2878.98	90
770.793	29	878.618	40	1041.69	6	2878.99	90
770.992	28	878.972	22	1152.15	13	2883.81	90
771.025	28	879.019	22	1152.15	33	2883.86	90
771.056	28	879.100	22	1172.50	9	3298.99	147
771.734	29	879.551	22	1172.61	9	3299.55	147
771.934	28	882.889	39	1172.78	9	3299.56	147
771.967	28	922.008	38	1217.65	55	3299.59	147
772.143	29	922.073	37	1302.17	2	3300.57	147
772.344	28	922.457	36	1304.86	2	3300.58	147
791.514	27	924.950	21	1306.03	2	3326.21	146
791.973	27	926.306	21	1355.60	1	3326.22	146
792.233	27	926.896	21	1358.51	1	3326.24	146
792.506	27	929.517	20	1484.50	10	3329.93	145
792.938	27	930.257	19	1641.31	4	3329.97	145
792.967	27	930.886	20	1727.11	3	3330.54	145
804.270	26	931.482	20	1827.06	97	3330.55	145
804.738	25	931.628	19	1827.08	97	3331.13	145
804.779	25	932.225	19	in air		3692.36	62
804.848	25	935.193	35	2026.43	96	3692.39	62
805.295	26	936.629	18	2026.46	96	3822.58	156
805.741	26	937.841	17	2288.54	149	3823.41	156
805.765	25	938.020	18	2288.55	149	3823.87	156
805.805	25	938.625	18	2288.56	149	3824.35	156
806.211	25	939.235	17	2324.74	5	3825.02	156
810.665	24	939.841	17	2417.41	95	3825.19	156
811.051	24	948.686	16	2417.72	95	3825.50	156
811.497	24	950.112	16	2417.75	95	3855.01	161
811.706	24	950.733	16	2417.76	95	3947.29	58
812.094	24	950.885	15	2418.26	95	3947.48	58
812.159	24	952.318	15	2418.30	95	3947.59	58
816.720	23	952.941	15	2432.00	94	3951.93	89
816.772	23	971.737	14	2432.03	94	3952.89	89
816.862	23	971.738	14	2432.04	94	3952.98	89
817.777	23	973.234	14	2433.99	93	3953.00	89
817.829	23	973.885	14	2434.03	93	3954.52	89
818.237	23	974.070	12	2434.31	93	3954.61	89
839.827	53	975.574	12	2434.35	93	3997.95	160
839.827	54	976.448	11	2434.65	93	4054.77	144
843.589	51	977.959	11	2611.45	148	4054.79	144

List of tabulated lines — Continued

Wavelength (Å)	No.	Wavelength (Å)	No.	Wavelength (Å)	No.	Wavelength (Å)	No.
4054.83	144	6155.96	67	9265.83	64	11947.8	109
4066.70	143	6155.97	67	9265.93	64	11968.4	128
4066.71	143	6155.99	67	9266.01	64	11969.3	128
4066.75	143	6156.74	67	9482.89	157	11971.1	128
4067.74	143	6156.76	67	9622.11	174	12266.8	99
4067.75	143	6156.78	67	9622.16	174	12267.5	99
4069.51	143	6158.15	67	9625.26	174	12268.0	99
4217.09	142	6158.17	67	9625.30	174	12510.6	108
4222.76	142	6158.19	67	9625.32	174	12510.8	108
4222.77	142	6323.39	120	9625.36	174	12511.0	108
4222.82	142	6323.41	120	9625.37	174	12511.1	108
4233.26	142	6324.77	120	9694.66	100	12511.2	108
4233.27	142	6324.80	120	9694.91	100	12569.9	116
4368.19	61	6324.82	120	9695.06	100	12570.0	116
4368.24	61	6324.84	120	9825.79	111	12570.1	116
4368.26	61	6324.85	120	9825.87	111	12651.9	139
4654.12	75	6453.60	66	9825.99	111	12652.0	139
4654.56	75	6454.44	66	9826.09	111	12652.4	139
4655.35	75	6455.98	66	9826.14	111	12790.4	138
4655.36	75	6726.28	57	9891.65	119	12790.5	138
4772.45	74	6726.54	57	9891.72	119	12790.9	138
4772.91	74	7001.90	81	9891.80	119	13053.8	127
4773.75	74	7001.92	81	10167.3	59	13054.9	127
4801.67	73	7002.17	81	10169.3	59	13055.0	127
4802.13	73	7002.20	81	10320.3	104	13057.0	127
4802.98	73	7002.23	81	10675.7	110	13076.5	103
4967.38	72	7002.25	81	10675.8	110	13076.9	103
4967.88	72	7017.30	141	10675.9	110	13077.0	103
4968.79	72	7020.33	141	10676.0	110	13163.9	76
5018.78	71	7020.37	141	10676.1	110	13164.9	76
5019.29	71	7020.50	141	10753.4	118	13165.1	76
5020.22	71	7025.47	141	10753.5	118	13414.9	126
5131.14	88	7025.51	141	10753.6	118	13416.1	126
5131.29	88	7156.70	159	11091.9	130	13418.2	126
5131.33	88	7254.15	80	11092.7	130	14110.8	137
5274.97	87	7254.45	80	11094.2	130	14111.0	137
5275.12	87	7254.53	80	11286.3	78	14111.5	137
5275.17	87	7771.94	56	11286.4	78	14390.8	136
5298.89	86	7774.17	56	11286.9	78	14391.0	136
5299.04	86	7775.39	56	11287.0	78	14391.5	136
5299.09	86	7981.94	79	11287.1	78	15476.3	154
5329.10	70	7982.30	79	11287.3	78	15505.4	154
5329.11	70	7982.40	79	11295.1	63	15524.4	154
5329.67	70	7986.98	79	11297.7	63	15665.1	115
5329.68	70	7987.33	79	11302.4	63	15665.6	115
5329.69	70	7995.07	79	11373.9	77	15665.8	115
5330.73	70	8221.80	155	11374.0	77	15666.0	115
5330.74	70	8221.82	155	11788.5	129	15666.2	115
5435.18	69	8227.65	155	11789.4	129	15887.6	125
5435.77	69	8230.00	155	11791.1	129	15887.7	125
5436.86	69	8230.03	155	11854.9	140	15889.4	125
5512.60	85	8233.00	155	11855.0	140	15889.5	125
5512.77	85	8235.35	155	11855.4	140	15892.5	125
5512.82	85	8446.25	60	11868.1	117	15892.6	125
5554.83	84	8446.36	60	11868.2	117	16109.1	107
5555.00	84	8446.76	60	11868.3	117	16109.3	107
5555.05	84	8820.42	158	11946.7	109	16109.6	107
5958.39	83	9204.93	65	11946.8	109	16110.6	107
5958.58	83	9260.81	64	11947.0	109	16110.9	107
5958.64	83	9260.85	64	11947.2	109	16111.2	107
6046.23	82	9260.94	64	11947.4	109	16111.9	107
6046.44	82	9262.58	64	11947.5	109	16112.1	107
6046.49	82	9262.67	64	11947.6	109	16112.2	107
6140.58	68	9262.78	64	11947.7	109	16869.2	124

List of tabulated lines — Continued

Wavelength (Å)	No.	Wavelength (Å)	No.	Wavelength (Å)	No.	Wavenumber (cm^{-1})	No.
16871.1	124	25506.5	166	33283.7	171	59.854	168
16874.5	124	25506.6	166	33284.4	171	59.900	168
17453.2	135	25787.3	172	33284.9	171	59.969	168
17453.5	135	25787.7	172	34198.0	165	102.570	162
17454.3	135	25788.0	172	34198.5	165	102.594	162
17694.9	153	26201.9	123	34199.2	165	102.643	162
17733.0	153	26207.0	123	34201.5	165	102.702	162
17757.8	153	26326.2	180	34202.2	165	102.743	162
18020.8	106	26425.1	176	34202.8	165	760.075	163
18021.1	106	26425.2	176	34204.1	165	760.099	163
18021.5	106	26425.3	176	34204.7	165	760.148	163
18021.8	106	26425.4	176	34205.0	165	760.456	163
18022.0	106	26425.5	176	34850.8	101	760.505	163
18229.0	134	26498.9	122	34859.0	101	760.564	163
18229.3	134	26499.1	122	36559.0	151	761.005	163
18230.1	134	26499.4	122	36606.3	131	761.064	163
18241.4	129	26503.8	122	36607.4	131	761.105	163
18243.3	114	26504.1	122	36610.9	131	961.525	169
18243.5	114	26504.6	122	36720.9	151	961.539	169
18243.8	114	26512.6	122	36721.7	151	961.585	169
18454.8	175	26513.0	122	36824.7	151	961.594	169
19428.6	182	26513.3	122	36827.7	151	961.763	169
19458.1	178	27631.2	98	38807.2	164	1605.103	112
19458.2	178	27640.3	98	38807.8	164	1671.977	105
21137.5	167	27645.5	98	38808.7	164	1672.026	105
21137.7	167	28925.1	102	38809.4	164	1672.129	105
21138.0	167	28927.3	102	38809.8	164	1672.699	105
21138.2	167	28928.0	102	39462.7	170	1672.802	105
21138.3	167	30975.1	133	39463.7	170	1672.925	105
21330.8	173	30975.5	133	39464.5	170	1673.998	105
21331.1	173	30975.9	133	40961.3	179	1674.121	105
21331.3	173	30976.3	133	41358.9	175	1674.207	105
21665.2	181	30977.0	133	41359.1	175	1812.921	150
21708.4	177	30978.4	133	41360.1	175	1820.817	150
21708.5	177	31392.2	132	41360.5	175	1832.936	150
23215.4	152	31393.0	132	45588.6	113		
23280.9	152	31393.6	132	45594.0	113		
23323.8	152	31395.1	132	45595.5	113		
25505.5	166	33062.7	121	45595.7	113		
25505.8	166	33070.1	121	45597.2	113		
25506.1	166	33083.2	121	45597.4	113		

Several advanced calculations have recently been carried out for this spectrum.[1-9] The most extensive data have been produced by the Opacity project[1] (OP) which is reviewed in the general introduction. K. Butler and C. J. Zeippen[1] did the OP calculations for O I.

Four other recent calculations by Hibbert *et al.*,[2] Bell and Hibbert,[3] Pradhan and Saraph,[4] and Biemont and Zeippen[5] have also provided numerous data, mainly for transitions between states of quantum numbers $n = 2$, 3 and 4. Hibbert *et al.*[2] calculated the wavefunctions with the CIV 3 code and included about 450 configuration state functions for both upper and lower states in their configuration-interaction treatment. Relativistic corrections of the Breit-Pauli type were also applied, and results were reported in intermediate coupling for individual lines, including intercombination lines. The calculations by Bell and Hibbert[3] were also undertaken with the CIV 3 code—utilized on a less extensive scale than by Hibbert *et al.*[2]—as well as with the R-Matrix code, which also has been applied in the OP calculations.[1] The calculations by Pradhan and Saraph[4] were performed with a close-coupling method in the frozen-core approximation—an approach similar to but less elaborate than the R-matrix method used in the Opacity Project. Biemont and Zeippen[5] used the atomic structure program SUPERSTRUCTURE, in which configuration interaction effects are also treated extensively.

For O I, the differences between *LS*-coupling and intermediate coupling have been found to be very small (i.e., typically within 1–3%) by both Hibbert *et al.*[2] and Biemont and Zeippen[5]. The available experimental data also indicate close adherence to *LS*-coupling. Therefore, for the OP, Bell and Hibbert, and Pradhan and Saraph calculations, where only multiplet values have been calculated, we applied *LS*-coupling fractions to obtain individual line data.

The five above-cited calculations[1-5] often yield very similar results, within about ±10%. We have selected the results of three of these calculations for our tables: the OP multiplet data combined with LS-coupling, the CIV 3 data by Hibbert *et al.*, and the SUPERSTRUCTURE data by Biemont and Zeippen, These three calculations represent very recent applications of three sophisticated but rather different atomic structure theory approaches, and feature the most comprehensive sets of results. The agreement is usually excellent, as seen in the graphical OP/CIV 3 comparison of Fig. 1. We have therefore almost exclusively relied on theoretical results and utilized experimental data only in two special cases discussed below. But for three multiplets, disagreements occur among the calculations which are in the 20–50% range. In these cases we have chosen the theoretical data in closest agreement with experiment. In addition, for the far infrared $4p\ ^3\mathrm{P} - 3s'\ ^3\mathrm{D}^\circ$ multiplet, the calculations by OP and Hibbert *et al.* disagree by about a factor of six; therefore, this multiplet is not included in our compilation.

Other advanced calculations for this spectrum[6–9]—applied to relatively few lines—partly agree well with the comprehensive approaches, but also show some disagreements with them as well as with recent experimental results. For the $2s^22p^4 - 2s2p^5$ transition array, special multiconfiguration calculations were carried out by Fischer and Saha,[8] and Baluja and Zeippen.[9] While they are in close agreement with each other, they disagree with other calculations[1,4] as well as with the electron-scattering experiment of Doering *et al.*[10] and therefore have not been utilized.

The experimental transition probability data for this spectrum are unfortunately not extensive. Five emission experiments with wall-stabilized arcs by Solarski and Wiese,[11] Ott,[12] Wiese and Shumaker,[13] Goldbach and Nollez[14], and Veres *et al.*[15] are usually in good agreement with theory, as is the electron scattering experiment by Doering *et al.*[10] and an absorption experiment by Jenkins[16] for the principal resonance transition. Fig. 2 shows that the experimental data are generally in good agreement with the selected theoretical results.

Among the lifetime experiments, selective excitation techniques with tunable lasers, which produce cascade-free lifetimes, have been used in the work by Bischel *et al.*[17] and Kröll *et al.*[18] who applied the laser-induced fluorescence method. Fig. 3 shows a comparison of their results with the data of the Opacity project. The experimental lifetimes are usually somewhat longer than the calculated ones, and the disagreements become large for the levels with the shortest lifetimes—the $2p^35s\ ^3\mathrm{S}^\circ$ and $2p^34d\ ^3\mathrm{D}^\circ$ levels—measured by Kröll *et al.*[18] We have no plausible explanation for these apparently systematic differences, but have decided to use the theoretical data, because they have generally given excellent results for this spectrum and are strongly supported by other experimental results, as shown in Fig. 2. We should note that in their lifetime experiments, Bischel *et al.* as well as Kröll *et al.* worked with fairly high pressures in their interaction chambers and noticed radiation trapping and collisional effects. Therefore, they performed extended extrapolations of their data with Stern-Vollmer plots to zero pressure, which may have introduced significant uncertainties. Also, neither group has discussed possible distortions of the lifetime results due to possible quantum beats and alignment effects on account of the non-isotropic laser excitation. But we should also point out that the lifetime result by Kröll *et al.*[18] for the $3p\ ^3\mathrm{P}$ term agrees within 10% with the emission data of Solarski and Wiese[11] and the f-value of Biemont and Zeippen[5] obtained with the SUPERSTRUCTURE code. These results differ, however, by about 20% from the OP and CIV 3 data, in the same direction as the other disagreements in Fig. 3.

In an emission experiment, Veres *et al.*[15] studied the weak $3s\ ^3\mathrm{S}^\circ - 4p\ ^3\mathrm{P}$ multiplet at 4368 Å, for which calculations have encountered considerable problems because of appreciable cancellation effects in the transition integral. By measuring the strength of this multiplet relative to four other O I multiplets, for which excellent theoretical data of "A" or "B" quality are available (multiplets No. 56, 66, 67, and 159), Veres *et al.* obtained a result of about equal quality for this transition, i.e. approximately ±10%, which we have tabulated instead of the more uncertain theoretical data.

Experimentally, much attention has been given to the two intersystem lines at 1355.60 Å and 1358.52 Å, which connect the quasi-metastable $3s\ ^5\mathrm{S}^\circ_2$ level to the ground state levels $2p^4\ ^3\mathrm{P}_2$ and $^3\mathrm{P}_1$. To obtain their absolute scale we have selected the arithmetic average of four measurements of the lifetime of the $3s\ ^5\mathrm{S}^\circ_2$ level with time-of-flight techniques.[19–22] These results fall within a narrow band of only ±5%, and thus define the average value quite accurately. Individual data for the two lines were obtained from the calculation by Biemont and Zeippen,[5] which differs by only 8% from another earlier result of Zeippen *et al.*[23]

We should draw attention to the fact that for some $^3\mathrm{P} - {}^3\mathrm{D}^\circ$ multiplets (e.g., Nos. 93, 145, 148, 149), no multiplet values are given since Hibbert *et al.*[2] did not calculate the f-value for the normally weak J= 2 to J= 1 line.

References

[1] K. Butler and C. J. Zeippen, J. Phys. IV, Colloq. C1, 1, C1–141 (1991) and to be published.

[2] A. Hibbert, E. Biémont, M. Godefroid, and N. Vaeck, J. Phys. B **24**, 3943 (1991).

[3] K. L. Bell and A. Hibbert, J. Phys. B **23**, 2673 (1990).

[4] A. K. Pradhan and H. E. Saraph, J. Phys. B **10**, 3365 (1977).

[5] E. Biémont and C. J. Zeippen, Astron. Astrophys. **265**, 850 (1992).

[6] S. S. Tayal and R. J. W. Henry, Phys. Rev. A **39**, 4531 (1989).

[7] C. Froese Fischer, J. Phys. B. **20**, 1193 (1987).

[8] C. Froese Fischer and H. P. Saha, J. Phys. B **17**, 943 (1984).

[9] K. L. Baluja and C. J. Zeippen, J. Phys. B **21**, 15 (1988).

[10] J. P. Doering, E. E. Gulcicek, and S. O. Vaughan, J. Geophys. Res., Space Phys. **90**, 5279 (1985).

[11] J. E. Solarski and W. L. Wiese, Phys. Rev. **135**, A1236 (1964).

[12] W. R. Ott, Phys. Rev. A **4**, 245 (1971).

[13] W. L. Wiese and J. B. Shumaker, Jr., J. Opt. Soc. Am. **51**, 937 (1961).

[14] C. Goldbach and G. Nollez, Astron. Astrophys. **284**, 307 (1994).

[15] G. Veres, J. Musielok and W. L. Wiese (to be published).

[16] D. B. Jenkins, J. Quant. Spectrosc. Radiat. Transfer **34**, 55 (1985).

[17] W. K. Bischel, B. E. Perry, and D. R. Crosley, Chem. Phys. Lett. **82**, 85 (1981).

[18] S. Kröll, H. Lundberg, A. Persson, and S. Svanberg, Phys. Rev. Lett. **55**, 284 (1985).

[19] C. E. Johnson, Phys. Rev. A **5**, 2688 (1972).

[20] W. C. Wells and E. C. Zipf, Phys. Rev. A **9**, 568 (1974).

[21] G. Nowak, W. L. Borst, and J. Fricke, Phys. Rev. A **17**, 1921 (1978).

[22] N. J. Mason, Meas. Sci. Technol. **1**, 596 (1990).

[23] C. J. Zeippen, M. J. Seaton, and D. C. Morton, Mon. Not. R. Astron. Soc. **181**, 527 (1977).

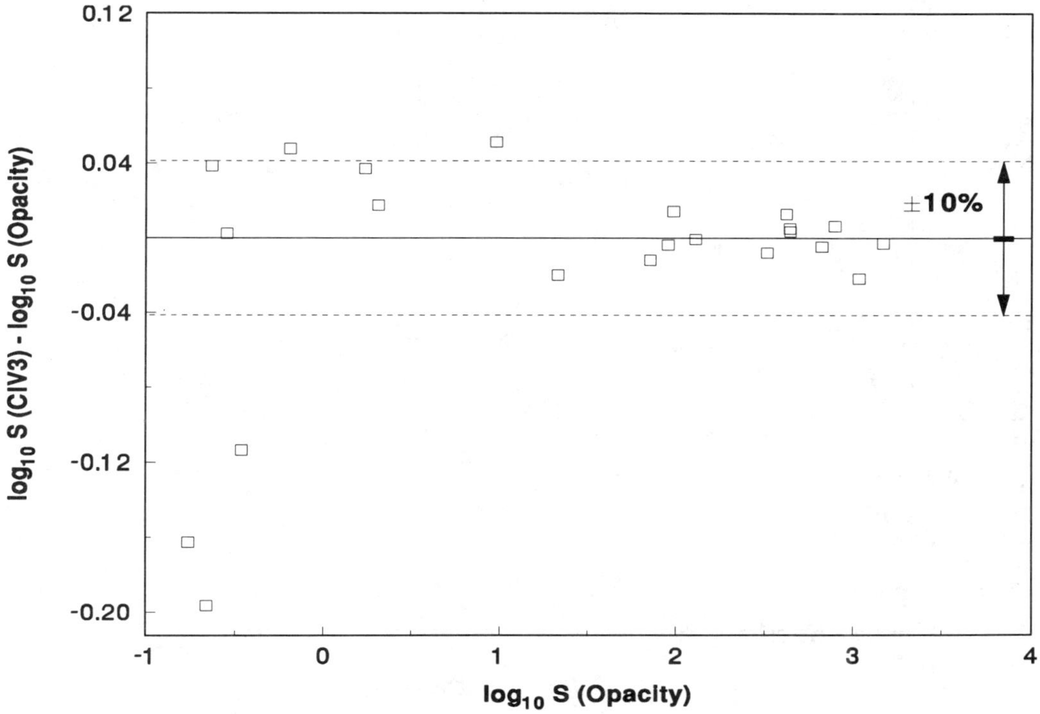

FIG. 1. Comparison of the OP multiplet data by Butler and Zeippen[1] with the CIV 3 results of Hibbert *et al.*[2] The ratios of the multiplet line strengths, S(CIV 3)/S(OP), are plotted on a logarithmic scale versus the log S(OP). The figure indicates that for all but three weak multiplets the range of deviations for the ratios stays within ±10%, as indicated by the broken lines.

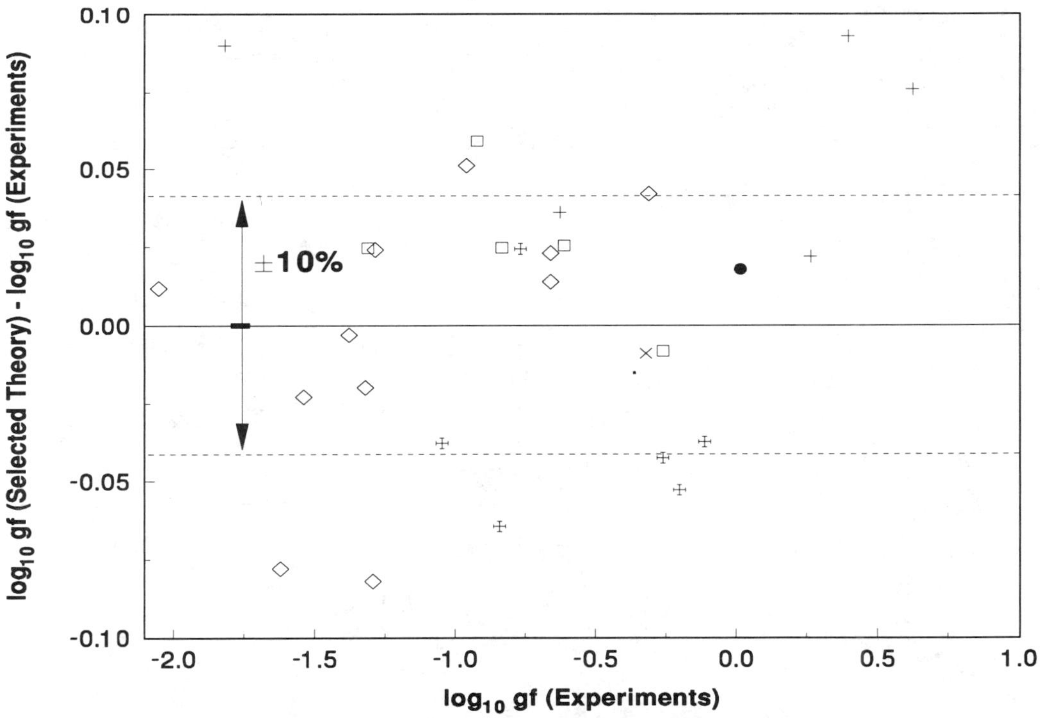

FIG. 2. Comparison of the experimental log *gf* data with the selected theoretical results. The theory/experiment ratios of the *gf*-values are plotted on a logarithmic scale versus the log *gf* of the various experiments. The data from the *emission* experiments by Solarski and Wiese[11] are denoted by plus signs (+), those of Ott[12] by squares; Wiese and Shumaker[13] by a filled circle; and Goldbach and Nollez[14] by diamonds. Also shown is a data point (×) from an *absorption* experiment by Jenkins,[16] and the results of an *electron scattering* experiment by Doering *et al.*[10] (crosses). It is seen that about two-thirds of the ratios do not deviate by more than ±10% (shown by the broken lines) from a perfect ratio of 1.00.

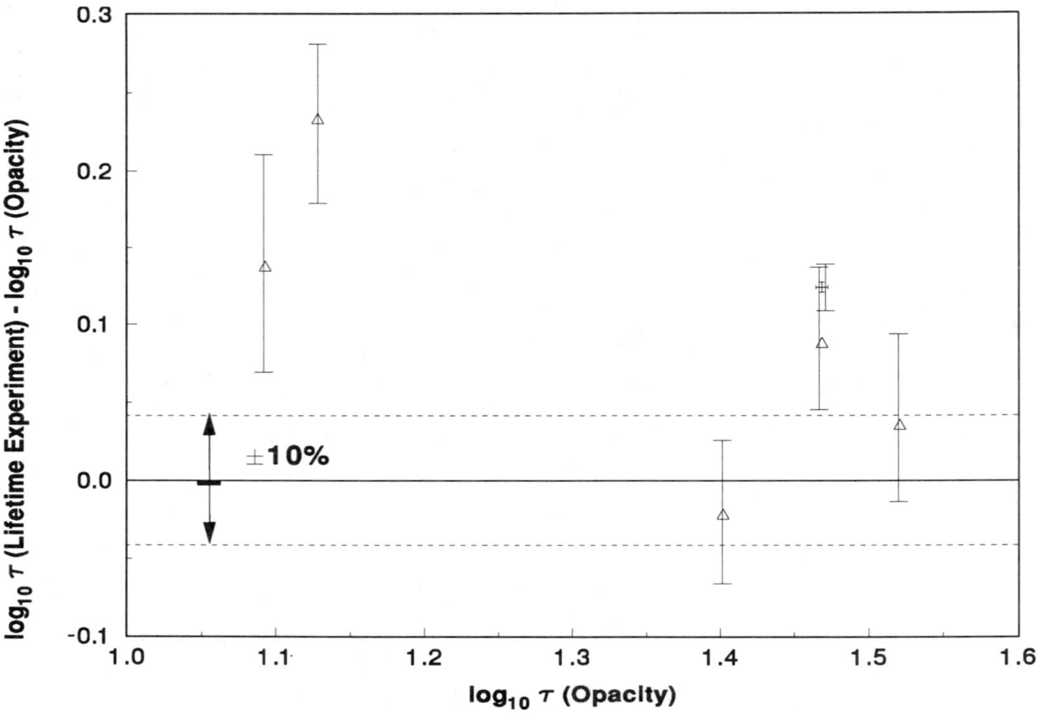

FIG. 3. Comparison of the lifetime results by Bischel *et al.*[17] (cross) and Kröll *et al.*[18] (triangles) with the OP data by Butler and Zeippen.[1] The ratios of the experimental to the theoretical lifetimes are plotted on a logarithmic scale versus the log τ (OP) (τ in *ns*). The experimental uncertainties are indicated by error flags.

O I: Allowed Transitions

No.	Transition Array	Mult.	λ_{air} (Å)	λ_{vac} (Å) or σ (cm^{-1})*	E_i (cm^{-1})	E_k (cm^{-1})	g_i–g_k	A_{ki} (10^8 s^{-1})	f_{ik}	S (at. u.)	log gf	Acc.	Source
1.	$2p^4$–$2p^3(^4S^o)3s$	^3P–$^5S^o$											
				1355.60	0 –	73768.200	5–5	4.20−05	1.16−06	2.58−05	−5.238	B	5,19-22
				1358.51	158.265 –	73768.200	3–5	1.36−05	6.27−07	8.41−06	−5.726	B	5,19-22
2.		^3P–$^3S^o$		*1303.5*	*77.97* –	76794.978	9–3	6.11+00	5.19−02	2.00+00	−0.330	A	1,2,5
				1302.17	0 –	76794.978	5–3	3.41+00	5.20−02	1.11+00	−0.585	A	2,5,*LS*
				1304.86	158.265 –	76794.978	3–3	2.03+00	5.19−02	6.68−01	−0.808	A	2,5,*LS*
				1306.03	226.977 –	76794.978	1–3	6.76−01	5.19−02	2.23−01	−1.285	A	2,5,*LS*
3.		^1D–$^5S^o$											
				1727.11	15867.862 –	73768.200	5–5	5.32−11	2.38−12	6.77−11	−10.924	D	5
4.		^1D–$^3S^o$											
				1641.31	15867.862 –	76794.978	5–3	1.83−05	4.44−07	1.20−05	−5.654	C	5
5.		^1S–$^3S^o$											
			2324.74	2325.45	33792.583 –	76794.978	1–3	4.61−08	1.12−08	8.57−08	−7.951	C	5
6.	$2p^4$–$2p^3(^4S^o)4s$	^3P–$^3S^o$		*1040.1*	*77.97* –	96225.049	9–3	1.70+00	9.16−03	2.82−01	−1.084	A	1,2
				1039.23	0 –	96225.049	5–3	9.43−01	9.16−03	1.57−01	−1.339	A	2,*LS*
				1040.94	158.265 –	96225.049	3–3	5.64−01	9.16−03	9.42−02	−1.561	A	2,*LS*
				1041.69	226.977 –	96225.049	1–3	1.88−01	9.16−03	3.14−02	−2.038	A	2,*LS*

O I: Allowed Transitions — Continued

No.	Transition Array	Mult.	λ_{air} (Å)	λ_{vac} (Å) or σ (cm^{-1})*	E_i (cm^{-1})	E_k (cm^{-1})	g_i–g_k	A_{ki} (10^8 s^{-1})	f_{ik}	S (at. u.)	log gf	Acc.	Source
7.	$2p^4$– $2p^3(^4S^o)3d$	^{3}P–^{3}D°		*1026.6*	*77.97*	*97488.48*	9–15	7.63–01	2.01–02	6.11–01	–0.743	B	1,2
				1025.76	0	97488.538	5–7	7.66–01	1.69–02	2.85–01	–1.073	B	2,*LS*
				1027.43	158.265	97488.448	3–5	5.71–01	1.51–02	1.53–01	–1.345	B	2,*LS*
				1028.16	226.977	97488.378	1–3	4.22–01	2.01–02	6.79–02	–1.698	B	2,*LS*
				1025.76	0	97488.448	5–5	1.91–01	3.01–03	5.08–02	–1.822	B	2,*LS*
				1027.43	158.265	97488.378	3–3	3.17–01	5.01–03	5.09–02	–1.823	B	2,*LS*
				1025.76	0	97488.378	5–3	2.11–02	2.00–04	3.38–03	–3.000	B	2,*LS*
8.	$2p^4$– $2p^3(^2D^o)3s$	^{3}P–^{3}D°		*989.46*	*77.97*	*101143.45*	9–15	2.26+00	5.53–02	1.62+00	–0.303	B+	1,2,5
				988.773	0	101135.407	5–7	2.26+00	4.64–02	7.56–01	–0.634	B	2,5,*LS*
				990.204	158.265	101147.526	3–5	1.68+00	4.12–02	4.03–01	–0.908	B	2,5,*LS*
				990.801	226.977	101155.422	1–3	1.25+00	5.50–02	1.79–01	–1.260	B	2,5,*LS*
				988.655	0	101147.526	5–5	5.77–01	8.45–03	1.38–01	–1.374	B	2,5,*LS*
				990.127	158.265	101155.422	3–3	9.47–01	1.39–02	1.36–01	–1.379	B	2,5,*LS*
				988.578	0	101155.422	5–3	6.47–02	5.69–04	9.26–03	–2.546	B	2,5,*LS*
9.		^{1}D–^{3}D°											
				1172.78	15867.862	101135.407	5–7	7.97–05	2.30–06	4.44–05	–4.939	C	5
				1172.61	15867.862	101147.526	5–5	3.66–06	7.54–08	1.46–06	–6.424	C	5
				1172.50	15867.862	101155.422	5–3	4.98–04	6.16–06	1.19–04	–4.511	C	5
10.		^{1}S–^{3}D°											
				1484.50	33792.583	101155.422	1–3	9.05–05	8.97–06	4.38–05	–5.047	C	5
11.	$2p^4$– $2p^3(^4S^o)5s$	^{3}P–^{3}S°		*977.19*	*77.97*	102411.995	9–3	6.93–01	3.31–03	9.57–02	–1.527	B	1
				976.448	0	102411.995	5–3	3.86–01	3.31–03	5.32–02	–1.781	C+	*LS*
				977.959	158.265	102411.995	3–3	2.30–01	3.30–03	3.19–02	–2.004	C+	*LS*
				978.617	226.977	102411.995	1–3	7.66–02	3.30–03	1.06–02	–2.481	C	*LS*
12.	$2p^4$– $2p^3(^2D^o)3s$	^{3}P–^{1}D°											
				974.070	0	102662.026	5–5	1.10–03	1.56–05	2.50–04	–4.108	C	5
				975.574	158.265	102662.026	3–5	2.22–04	5.27–06	5.08–05	–4.801	C	5
13.		^{1}D–^{1}D°		1152.15	15867.862	102662.026	5–5	5.28+00	1.05–01	1.99+00	–0.280	B	5
14.	$2p^4$– $2p^3(^4S^o)4d$	^{3}P–^{3}D°		*972.47*	*77.97*	*102908.42*	9–15	5.84–01	1.38–02	3.97–01	–0.906	C+	1,2
				971.738	0	102908.374	5–7	5.85–01	1.16–02	1.85–01	–1.237	C+	2,*LS*
				973.234	158.265	102908.443	3–5	4.37–01	1.03–02	9.93–02	–1.509	C+	2,*LS*
				973.885	226.977	102908.489	1–3	3.23–01	1.38–02	4.42–02	–1.861	C+	2,*LS*
				971.738	0	102908.443	5–5	1.46–01	2.07–03	3.31–02	–1.985	C+	2,*LS*
				973.234	158.265	102908.489	3–3	2.42–01	3.44–03	3.31–02	–1.986	C+	2,*LS*
				971.737	0	102908.489	5–3	1.62–02	1.38–04	2.21–03	–3.161	C+	2,*LS*
15.	$2p^4$– $2p^3(^4S^o)6s$	^{3}P–^{3}S°		*951.59*	*77.97*	105165.232	9–3	3.48–01	1.57–03	4.44–02	–1.849	B	1
				950.885	0	105165.232	5–3	1.94–01	1.58–03	2.47–02	–2.103	C	*LS*
				952.318	158.265	105165.232	3–3	1.16–01	1.57–03	1.48–02	–2.326	C	*LS*
				952.941	226.977	105165.232	1–3	3.85–02	1.57–03	4.93–03	–2.803	C–	*LS*

O I: Allowed Transitions — Continued

No.	Transition Array	Mult.	λ_{air} (Å)	λ_{vac} (Å) or σ (cm^{-1})*	E_i (cm^{-1})	E_k (cm^{-1})	g_i-g_k	A_{ki} (10^8 s^{-1})	f_{ik}	S (at. u.)	log gf	Acc.	Source
16.	$2p^4-$ $2p^3(^4S^o)5d$	$^3P-^3D^o$		*949.39*	*77.97*	– 105409.008	9–15	2.80–01	6.31–03	1.78–01	–1.246	B	1
				948.686	0	– 105409.008	5–7	2.81–01	5.30–03	8.28–02	–1.576	C+	*LS*
				950.112	158.265	– 105409.008	3–5	2.10–01	4.73–03	4.44–02	–1.848	C+	*LS*
				950.733	226.977	– 105409.008	1–3	1.55–01	6.30–03	1.97–02	–2.201	C	*LS*
				948.686	0	– 105409.008	5–5	7.02–02	9.47–04	1.48–02	–2.325	C	*LS*
				950.112	158.265	– 105409.008	3–3	1.16–01	1.58–03	1.48–02	–2.325	C	*LS*
				948.686	0	– 105409.008	5–3	7.80–03	6.31–05	9.86–04	–3.501	D–	*LS*
17.	$2p^4-$ $2p^3(^4S^o)7s$	$^3P-^3S^o$		*938.53*	*77.97*	– 106627.934	9–3	1.99–01	8.77–04	2.44–02	–2.103	B–	1
				937.841	0	– 106627.934	5–3	1.11–01	8.78–04	1.36–02	–2.357	C	*LS*
				939.235	158.265	– 106627.934	3–3	6.63–02	8.77–04	8.13–03	–2.580	C–	*LS*
				939.841	226.977	– 106627.934	1–3	2.21–02	8.76–04	2.71–03	–3.057	D	*LS*
18.	$2p^4-$ $2p^3(^4S^o)6d$	$^3P-^3D^o$		*937.31*	*77.97*	– 106765.803	9–15	1.66–01	3.63–03	1.01–01	–1.485	B	1
				936.629	0	– 106765.803	5–7	1.66–01	3.05–03	4.71–02	–1.816	C+	*LS*
				938.020	158.265	– 106765.803	3–5	1.24–01	2.72–03	2.52–02	–2.088	C	*LS*
				938.625	226.977	– 106765.803	1–3	9.16–02	3.63–03	1.12–02	–2.440	C	*LS*
				936.629	0	– 106765.803	5–5	4.15–02	5.45–04	8.41–03	–2.564	C–	*LS*
				938.020	158.265	– 106765.803	3–3	6.88–02	9.08–04	8.41–03	–2.565	C–	*LS*
				936.629	0	– 106765.803	5–3	4.61–03	3.64–05	5.61–04	–3.740	D–	*LS*
19.	$2p^4-$ $2p^3(^4S^o)8s$	$^3P-^3S^o$		*930.93*	*77.97*	– 107497.224	9–3	1.24–01	5.37–04	1.48–02	–2.316	B–	1
				930.257	0	– 107497.224	5–3	6.90–02	5.37–04	8.22–03	–2.571	C–	*LS*
				931.628	158.265	– 107497.224	3–3	4.12–02	5.36–04	4.93–03	–2.794	C–	*LS*
				932.225	226.977	– 107497.224	1–3	1.37–02	5.36–04	1.64–03	–3.271	D	*LS*
20.	$2p^4-$ $2p^3(^4S^o)7d$	$^3P-^3D^o$		*930.19*	*77.97*	– 107582.777	9–15	1.06–01	2.30–03	6.33–02	–1.685	B	1
				929.517	0	– 107582.777	5–7	1.06–01	1.93–03	2.95–02	–2.015	C	*LS*
				930.886	158.265	– 107582.777	3–5	7.95–02	1.72–03	1.58–02	–2.287	C	*LS*
				931.482	226.977	– 107582.777	1–3	5.88–02	2.29–03	7.03–03	–2.639	C–	*LS*
				929.517	0	– 107582.777	5–5	2.66–02	3.45–04	5.28–03	–2.764	C–	*LS*
				930.886	158.265	– 107582.777	3–3	4.42–02	5.74–04	5.28–03	–2.764	C–	*LS*
				929.517	0	– 107582.777	5–3	2.96–03	2.30–05	3.52–04	–3.940	D–	*LS*
21.	$2p^4-$ $2p^3(^4S^o)8d$	$^3P-^3D^o$		*925.62*	*77.97*	– 108114.0	9–15	7.20–02	1.54–03	4.23–02	–1.858	B	1
				924.950	0	– 108114.0	5–7	7.22–02	1.30–03	1.97–02	–2.188	C	*LS*
				926.306	158.265	– 108114.0	3–5	5.39–02	1.16–03	1.06–02	–2.460	C	*LS*
				926.896	226.977	– 108114.0	1–3	3.99–02	1.54–03	4.70–03	–2.812	C–	*LS*
				924.950	0	– 108114.0	5–5	1.81–02	2.32–04	3.53–03	–2.936	C–	*LS*
				926.306	158.265	– 108114.0	3–3	3.00–02	3.85–04	3.53–03	–2.937	C–	*LS*
				924.950	0	– 108114.0	5–3	2.01–03	1.54–05	2.35–04	–4.113	D–	*LS*
22.	$2p^4-$ $2p^3(^2P^o)3s$	$^3P-^3P^o$		*878.44*	*77.97*	*– 113916.28*	9–9	6.82+00	7.89–02	2.05+00	–0.149	B	1
				877.879	0	– 113910.957	5–5	5.12+00	5.92–02	8.55–01	–0.529	C+	*LS*
				879.019	158.265	– 113921.391	3–3	1.70+00	1.97–02	1.71–01	–1.228	C+	*LS*
				877.798	0	– 113921.391	5–3	2.85+00	1.97–02	2.85–01	–1.006	C+	*LS*
				878.972	158.265	– 113927.534	3–1	6.81+00	2.63–02	2.28–01	–1.103	C+	*LS*
				879.100	158.265	– 113910.957	3–5	1.70+00	3.28–02	2.85–01	–1.006	C+	*LS*
				879.551	226.977	– 113921.391	1–3	2.26+00	7.88–02	2.28–01	–1.104	C+	*LS*

O I: Allowed Transitions — Continued

No.	Transition Array	Mult.	λ_{air} (Å)	λ_{vac} (Å) or σ (cm^{-1})*	E_i (cm^{-1})	E_k (cm^{-1})	g_i–g_k	A_{ki} (10^8 s^{-1})	f_{ik}	S (at. u.)	log gf	Acc.	Source
23.	$2p^4$– $2p^3(^2D^o)4s$	^3P–$^3D^o$		*817.32*	*77.97*	*122428.5*	9–15	5.88–01	9.81–03	2.37–01	–1.054	B–	2
				816.862	0	122419.7	5–7	5.89–01	8.24–03	1.11–01	–1.385	B–	2
				817.829	158.265	122433.2	3–5	4.40–01	7.35–03	5.94–02	–1.656	B–	2
				818.237	226.977	122441.0	1–3	3.25–01	9.80–03	2.64–02	–2.009	C+	2
				816.772	0	122433.2	5–5	1.47–01	1.47–03	1.98–02	–2.133	C+	2
				817.777	158.265	122441.0	3–3	2.44–01	2.45–03	1.98–02	–2.134	C+	2
				816.720	0	122441.0	5–3	1.63–02	9.80–05	1.32–03	–3.310	C	2
24.	$2p^4$– $2p^3(^2D^o)3d$	^3P–$^3P^o$		*811.37*	*77.97*	*123326.42*	9–9	3.27–01	3.23–03	7.76–02	–1.537	C+	1
				811.051	0	123296.777	5–5	2.46–01	2.42–03	3.23–02	–1.917	C–	*LS*
				811.706	158.265	123355.512	3–3	8.17–02	8.07–04	6.47–03	–2.616	D–	*LS*
				810.665	0	123355.512	5–3	1.37–01	8.08–04	1.08–02	–2.394	D	*LS*
				811.497	158.265	123387.339	3–1	3.27–01	1.08–03	8.62–03	–2.491	D–	*LS*
				812.094	158.265	123296.777	3–5	8.15–02	1.34–03	1.08–02	–2.395	D	*LS*
				812.159	226.977	123355.512	1–3	1.09–01	3.22–03	8.62–03	–2.491	D–	*LS*
25.		^3P–$^3D^o$		*805.31*	*77.97*	*124254.0*	9–15	4.41–01	7.15–03	1.71–01	–1.192	B–	2
				804.848	0	124247.1	5–7	4.47–01	6.07–03	8.04–02	–1.518	B–	2
				805.805	158.265	124257.7	3–5	3.29–01	5.33–03	4.24–02	–1.796	B–	2
				806.211	226.977	124264.0	1–3	2.40–01	7.03–03	1.87–02	–2.153	C+	2
				804.779	0	124257.7	5–5	1.10–01	1.07–03	1.41–02	–2.274	C+	2
				805.765	158.265	124264.0	3–3	1.80–01	1.76–03	1.40–02	–2.278	C+	2
				804.738	0	124264.0	5–3	1.21–02	7.02–05	9.30–04	–3.455	D	2
26.		^3P–$^3S^o$		*804.78*	*77.97*	124336.3	9–3	1.26+00	4.09–03	9.75–02	–1.434	C	2
				804.270	0	124336.3	5–3	7.02–01	4.09–03	5.41–02	–1.690	C	2
				805.295	158.265	124336.3	3–3	4.21–01	4.09–03	3.25–02	–1.911	C	2
				805.741	226.977	124336.3	1–3	1.40–01	4.09–03	1.09–02	–2.388	C	2
27.	$2s^22p^4$– $2s2p^5$	^3P–$^3P^o$		*792.23*	*77.97*	*126304.32*	9–9	6.59+00	6.20–02	1.45+00	–0.254	B	1
				791.973	0	126266.896	5–5	4.94+00	4.65–02	6.06–01	–0.634	B	*LS*
				792.506	158.265	126340.225	3–3	1.64+00	1.55–02	1.21–01	–1.333	B	*LS*
				791.514	0	126340.225	5–3	2.75+00	1.55–02	2.02–01	–1.110	B	*LS*
				792.233	158.265	126383.751	3–1	6.59+00	2.07–02	1.62–01	–1.208	B	*LS*
				792.967	158.265	126266.896	3–5	1.64+00	2.58–02	2.02–01	–1.111	B	*LS*
				792.938	226.977	126340.225	1–3	2.19+00	6.19–02	1.62–01	–1.208	B	*LS*
28.	$2p^4$– $2p^3(^2D^o)4d$	^3P–$^3D^o$		*771.50*	*77.97*	*129696.2*	9–15	1.76–01	2.62–03	6.00–02	–1.627	B	2
				771.056	0	129692.3	5–7	1.79–01	2.23–03	2.83–02	–1.952	C+	2
				771.967	158.265	129697.5	3–5	1.31–01	1.95–03	1.49–02	–2.232	C+	2
				772.344	226.977	129703.0	1–3	9.58–02	2.57–03	6.53–03	–2.590	C	2
				771.025	0	129697.5	5–5	4.40–02	3.92–04	4.97–03	–2.708	C	2
				771.934	158.265	129703.0	3–3	7.20–02	6.43–04	4.91–03	–2.714	C	2
				770.992	0	129703.0	5–3	4.83–03	2.58–05	3.28–04	–3.889	C–	2
29.		^3P–$^3S^o$		*771.26*	*77.97*	129736.6	9–3	4.64–01	1.38–03	3.15–02	–1.906	C	2
				770.793	0	129736.6	5–3	2.58–01	1.38–03	1.75–02	–2.161	C	2
				771.734	158.265	129736.6	3–3	1.54–01	1.38–03	1.05–02	–2.384	C	2
				772.143	226.977	129736.6	1–3	5.12–02	1.37–03	3.49–03	–2.862	C	2

O I: Allowed Transitions — Continued

No.	Transition Array	Mult.	λ_{air} (Å)	λ_{vac} (Å) or σ (cm^{-1})*	E_i (cm^{-1})	E_k (cm^{-1})	g_i–g_k	A_{ki} (10^8 s^{-1})	f_{ik}	S (at. u.)	log gf	Acc.	Source
30.		^{3}P–^{3}P°		*769.84*	*77.97*	*129974.75*	9–9	2.12+00	1.88–02	4.29–01	–0.771	B	1
				769.408	0	129970.000	5–5	1.59+00	1.41–02	1.79–01	–1.151	C+	*LS*
				770.291	158.265	129979.384	3–3	5.29–01	4.70–03	3.58–02	–1.851	C+	*LS*
				769.353	0	129979.384	5–3	8.84–01	4.71–03	5.96–02	–1.628	C+	*LS*
				770.260	158.265	129984.553	3–1	2.11+00	6.27–03	4.77–02	–1.726	C+	*LS*
				770.346	158.265	129970.000	3–5	5.29–01	7.84–03	5.96–02	–1.629	C+	*LS*
				770.699	226.977	129979.384	1–3	7.04–01	1.88–02	4.77–02	–1.726	C+	*LS*
31.	$2p^4$–$2p^3(^2$P°$)3d$	^{3}P–^{3}D°		*725.24*	*77.97*	137962.5	9–15	4.68–01	6.15–03	1.32–01	–1.257	B–	2
				724.835	0	137962.5	5–7	4.60–01	5.07–03	6.05–02	–1.596	C+	2
				725.667	158.265	137962.5	3–5	3.53–01	4.65–03	3.33–02	–1.856	C+	2
				726.029	226.977	137962.5	1–3	2.66–01	6.31–03	1.51–02	–2.200	C	2
				724.835	0	137962.5	5–5	1.19–01	9.37–04	1.12–02	–2.330	C	2
				725.667	158.265	137962.5	3–3	2.01–01	1.58–03	1.14–02	–2.323	C	2
				724.835	0	137962.5	5–3	1.35–02	6.38–05	7.62–04	–3.496	D	2
32.	$2p^4$–$2p^3(^2$P°$)4d$	^{3}P–^{3}D°		*697.91*	*77.97*	143363.4	9–15	2.09–01	2.55–03	5.26–02	–1.640	B–	2
				697.528	0	143363.4	5–7	2.05–01	2.10–03	2.41–02	–1.980	C	2
				698.299	158.265	143363.4	3–5	1.58–01	1.93–03	1.33–02	–2.238	C	2
				698.634	226.977	143363.4	1–3	1.19–01	2.62–03	6.02–03	–2.582	C	2
				697.528	0	143363.4	5–5	5.30–02	3.87–04	4.44–03	–2.714	D	2
				698.299	158.265	143363.4	3–3	8.97–02	6.56–04	4.52–03	–2.706	C	2
				697.528	0	143363.4	5–3	6.02–03	2.63–05	3.02–04	–3.880	D	2
33.	$2p^4$–$2p^3(^2$D°$)3s$	^{1}D–^{1}D°		1152.15	15867.862	102662.026	5–5	5.42+00	1.08–01	2.05+00	–0.268	B	1
34.	$2p^4$–$2p^3(^2$P°$)3s$	^{1}D–^{1}P°		999.497	15867.862	115918.143	5–3	5.06+00	4.55–02	7.48–01	–0.643	B	1
35.	$2p^4$–$2p^3(^2$D°$)4s$	^{1}D–^{1}D°		935.193	15867.862	122797.661	5–5	1.33+00	1.74–02	2.68–01	–1.061	B	1
36.	$2p^4$–$2p^3(^2$D°$)3d$	^{1}D–^{1}P°		922.457	15867.862	124274	5–3	2.04–01	1.56–03	2.37–02	–2.108	B–	1
37.	$2p^4$–$2p^3(^2$D°$)3d$	^{1}D–^{1}D°		922.073	15867.862	124319.175	5–5	1.17+00	1.49–02	2.27–01	–1.127	B	1
38.		^{1}D–^{1}F°		922.008	15867.862	124326.779	5–7	1.23+00	2.19–02	3.33–01	–0.960	B	1
39.	$2p^4$–$2p^3(^2$D°$)5s$	^{1}D–^{1}D°		882.889	15867.862	129132.323	5–5	5.22–01	6.10–03	8.86–02	–1.516	B	1
40.	$2p^4$–$2p^3(^2$D°$)4d$	^{1}D–^{1}P°		878.618	15867.862	129683	5–3	1.70–01	1.18–03	1.71–02	–2.228	B–	1
41.	$2p^4$–$2p^3(^2$D°$)4d$	^{1}D–^{1}D°		878.247	15867.862	129731	5–5	6.68–01	7.72–03	1.12–01	–1.413	B	1

O I: Allowed Transitions — Continued

No.	Transition Array	Mult.	λ_{air} (Å)	λ_{vac} (Å) or σ (cm^{-1})*	E_i (cm^{-1})	E_k (cm^{-1})	g_i-g_k	A_{ki} ($10^8 s^{-1}$)	f_{ik}	S (at. u.)	log gf	Acc.	Source
42.		$^1D-^1F^o$		878.201	15867.862	129737.052	5–7	7.05–01	1.14–02	1.65–01	–1.244	B	1
43.	$2p^4-$ $2p^3(^2D^o)6s$	$^1D-^1D^o$		861.563	15867.862	131936	5–5	2.58–01	2.87–03	4.07–02	–1.843	B	1
44.	$2p^4-$ $2p^3(^2D^o)5d$	$^1D-^1P^o$		859.586	15867.862	132203	5–3	1.10–01	7.28–04	1.03–02	–2.439	B–	1
45.	$2p^4-$ $2p^3(^2D^o)5d$	$^1D-^1D^o$		859.349	15867.862	132235	5–5	3.82–01	4.23–03	5.98–02	–1.675	B	1
46.		$^1D-^1F^o$		859.312	15867.862	132240	5–7	4.06–01	6.29–03	8.90–02	–1.502	B	1
47.	$2p^4-$ $2p^3(^2D^o)7s$	$^1D-^1D^o$		850.679	15867.862	133421	5–5	1.46–01	1.59–03	2.22–02	–2.101	B–	1
48.	$2p^4-$ $2p^3(^2D^o)6d$	$^1D-^1D^o$		849.480	15867.862	133587	5–5	2.33–01	2.52–03	3.53–02	–1.899	B	1
49.		$^1D-^1F^o$		849.451	15867.862	133591	5–7	2.49–01	3.78–03	5.28–02	–1.724	B	1
50.	$2p^4-$ $2p^3(^2D^o)8s$	$^1D-^1D^o$		844.330	15867.862	134305	5–5	9.09–02	9.71–04	1.35–02	–2.314	B–	1
51.	$2p^4-$ $2p^3(^2D^o)7d$	$^1D-^1D^o$		843.589	15867.862	134409	5–5	1.51–01	1.61–03	2.24–02	–2.093	B	1
52.		$^1D-^1F^o$		843.589	15867.862	134409	5–7	1.63–01	2.43–03	3.38–02	–1.915	B–	1
53.	$2p^4-$ $2p^3(^2D^o)8d$	$^1D-^1D^o$		839.827	15867.862	134940	5–5	1.03–01	1.09–03	1.51–02	–2.263	B–	1
54.		$^1D-^1F^o$		839.827	15867.862	134940	5–7	1.12–01	1.66–03	2.29–02	–2.082	B–	1
55.	$2p^4-$ $2p^3(^2P^o)3s$	$^1S-^1P^o$		1217.65	33792.583	115918.143	1–3	2.06+00	1.37–01	5.51–01	–0.862	B	1
56.	$2p^3(^4S^o)3s-$ $2p^3(^4S^o)3p$	$^5S^o-^5P$	*7773.4*	*7775.5*	73768.200	*86629.09*	5–15	3.69–01	1.00+00	1.28+02	0.700	A	1,2,5
			7771.94	7774.08	73768.200	86631.454	5–7	3.69–01	4.68–01	5.99+01	0.369	A	2,5,*LS*
			7774.17	7776.30	73768.200	86627.778	5–5	3.69–01	3.34–01	4.28+01	0.223	A	2,5,*LS*
			7775.39	7777.53	73768.200	86625.757	5–3	3.69–01	2.00–01	2.57+01	0.001	A	2,5,*LS*
57.		$^5S^o-^3P$											
			6726.28	6728.14	73768.200	88631.146	5–5	1.18–05	8.00–06	8.86–04	–4.398	C	5
			6726.54	6728.39	73768.200	88630.587	5–3	6.44–06	2.62–06	2.90–04	–4.883	C	5

O I: Allowed Transitions — Continued

No.	Transition Array	Mult.	λ_{air} (Å)	λ_{vac} (Å) or σ (cm^{-1})*	E_i (cm^{-1})	E_k (cm^{-1})	g_i–g_k	A_{ki} (10^8 s^{-1})	f_{ik}	S (at. u.)	log gf	Acc.	Source
58.	$2p^3(^4S^\circ)3s$–$2p^3(^4S^\circ)4p$	$^5S^\circ$–^5P	*3947.4*	*3948.5*	73768.200 –	*99094.06*	5–15	4.89–03	3.43–03	2.23–01	–1.766	B	1,2
			3947.29	3948.41	73768.200 –	99094.837	5–7	4.91–03	1.60–03	1.04–01	–2.096	B	2,*LS*
			3947.48	3948.60	73768.200 –	99093.641	5–5	4.88–03	1.14–03	7.41–02	–2.244	B	2,*LS*
			3947.59	3948.70	73768.200 –	99092.968	5–3	4.87–03	6.83–04	4.44–02	–2.467	B	2,*LS*
59.	$2p^3(^4S^\circ)3s$–$2p^3(^4S^\circ)3p$	$^3S^\circ$–^5P											
			10167.3	9832.800 cm^{-1}	76794.978 –	86627.778	3–5	3.61–06	9.33–06	9.37–04	–4.553	C	5
			10169.3	9830.779 cm^{-1}	76794.978 –	86625.757	3–3	1.97–06	3.06–06	3.07–04	–5.037	C	5
60.		$^3S^\circ$–^3P	*8446.5*	*8448.8*	76794.978 –	*88630.98*	3–9	3.22–01	1.03+00	8.63+01	0.492	B	1,2,5
			8446.36	8448.68	76794.978 –	88631.146	3–5	3.22–01	5.75–01	4.79+01	0.236	B	2,5,*LS*
			8446.76	8449.08	76794.978 –	88630.587	3–3	3.22–01	3.45–01	2.87+01	0.014	B	2,5,*LS*
			8446.25	8448.57	76794.978 –	88631.303	3–1	3.22–01	1.15–01	9.59+00	–0.463	B	2,5,*LS*
61.	$2p^3(^4S^\circ)3s$–$2p^3(^4S^\circ)4p$	$^3S^\circ$–^3P	*4368.2*	*4369.5*	76794.978 –	*99681.05*	3–9	7.58–03	6.51–03	2.81–01	–1.709	B	15
			4368.24	4369.47	76794.978 –	99681.049	3–5	7.59–03	3.62–03	1.56–01	–1.964	B–	2*n*,*LS*
			4368.26	4369.49	76794.978 –	99680.968	3–3	7.58–03	2.17–03	9.37–02	–2.186	B–	2*n*,*LS*
			4368.19	4369.42	76794.978 –	99681.309	3–1	7.56–03	7.21–04	3.11–02	–2.665	B–	2*n*,*LS*
62.	$2p^3(^4S^\circ)3s$–$2p^3(^4S^\circ)5p$	$^3S^\circ$–^3P	*3692.4*	*3693.4*	76794.978 –	*103870.02*	3–9	9.30–04	5.70–04	2.08–02	–2.767	B–	1
			3692.39	3693.45	76794.978 –	103869.968	3–5	9.30–04	3.17–04	1.16–02	–3.022	C	*LS*
			3692.39	3693.44	76794.978 –	103870.028	3–3	9.30–04	1.90–04	6.93–03	–3.244	C–	*LS*
			3692.36	3693.41	76794.978 –	103870.252	3–1	9.30–04	6.34–05	2.31–03	–3.721	D	*LS*
63.	$2p^3(^4S^\circ)3p$–$2p^3(^4S^\circ)4s$	^5P–$^5S^\circ$	*11299*	*8847.64* cm^{-1}	*86629.09* –	95476.728	15–5	2.67–01	1.70–01	9.50+01	0.407	A	1,2
			11302.4	8845.274 cm^{-1}	86631.454 –	95476.728	7–5	1.25–01	1.70–01	4.44+01	0.076	A	2,*LS*
			11297.7	8848.950 cm^{-1}	86627.778 –	95476.728	5–5	8.90–02	1.70–01	3.17+01	–0.070	A	2,*LS*
			11295.1	8850.971 cm^{-1}	86625.757 –	95476.728	3–5	5.34–02	1.70–01	1.90+01	–0.292	A	2,*LS*
64.	$2p^3(^4S^\circ)3p$–$2p^3(^4S^\circ)3d$	^5P–$^5D^\circ$	*9263.9*	*9266.4*	*86629.09* –	*97420.75*	15–25	4.46–01	9.55–01	4.37+02	1.156	A	1,2
			9266.01	9268.55	86631.454 –	97420.630	7–9	4.45–01	7.37–01	1.57+02	0.712	A	2,*LS*
			9262.78	9265.32	86627.778 –	97420.716	5–7	2.97–01	5.35–01	8.16+01	0.427	A	2,*LS*
			9260.94	9263.48	86625.757 –	97420.839	3–5	1.56–01	3.34–01	3.06+01	0.001	A	2,*LS*
			9265.93	9268.47	86631.454 –	97420.716	7–7	1.48–01	1.91–01	4.08+01	0.126	A	2,*LS*
			9262.67	9265.21	86627.778 –	97420.839	5–5	2.60–01	3.34–01	5.10+01	0.223	A	2,*LS*
			9260.85	9263.39	86625.757 –	97420.942	3–3	3.34–01	4.30–01	3.93+01	0.110	A	2,*LS*
			9265.83	9268.37	86631.454 –	97420.839	7–5	2.97–02	2.73–02	5.83+00	–0.719	A	2,*LS*
			9262.58	9265.12	86627.778 –	97420.942	5–3	1.11–01	8.60–02	1.31+01	–0.367	A	2,*LS*
			9260.81	9263.35	86625.757 –	97420.991	3–1	4.46–01	1.91–01	1.75+01	–0.242	A	2,*LS*
65.		^5P–$^3D^\circ$											
			9204.93	9207.46	86627.778 –	97488.538	5–7	1.66–05	2.95–05	4.47–03	–3.831	C	2
66.	$2p^3(^4S^\circ)3p$–$2p^3(^4S^\circ)5s$	^5P–$^5S^\circ$	*6455.0*	*6456.8*	*86629.09* –	102116.698	15–5	8.26–02	1.72–02	5.48+00	–0.589	B	1
			6455.98	6457.76	86631.454 –	102116.698	7–5	3.85–02	1.72–02	2.56+00	–0.920	C+	*LS*
			6454.44	6456.23	86627.778 –	102116.698	5–5	2.75–02	1.72–02	1.83+00	–1.066	C+	*LS*
			6453.60	6455.39	86625.757 –	102116.698	3–5	1.65–02	1.72–02	1.10+00	–1.288	C+	*LS*

O I: Allowed Transitions — Continued

No.	Transition Array	Mult.	λ_{air} (Å)	λ_{vac} (Å) or σ (cm^{-1})*	E_i (cm^{-1})	E_k (cm^{-1})	g_i-g_k	A_{ki} (10^8 s^{-1})	f_{ik}	S (at. u.)	log gf	Acc.	Source
67.	$2p^3(^4S^o)3p-2p^3(^4S^o)4d$	$^5P-^5D^o$	*6157.3*	*6159.0*	*86629.09*	*102865.56*	15–25	7.62–02	7.22–02	2.19+01	0.034	B+	1,2
			6158.19	6159.89	86631.454	102865.506	7–9	7.62–02	5.57–02	7.90+00	–0.409	B+	2,*LS*
			6156.78	6158.48	86627.778	102865.547	5–7	5.08–02	4.04–02	4.10+00	–0.694	B+	2,*LS*
			6155.99	6157.69	86625.757	102865.606	3–5	2.67–02	2.53–02	1.54+00	–1.120	B+	2,*LS*
			6158.17	6159.88	86631.454	102865.547	7–7	2.54–02	1.44–02	2.05+00	–0.996	B+	2,*LS*
			6156.76	6158.46	86627.778	102865.606	5–5	4.45–02	2.53–02	2.56+00	–0.899	B+	2,*LS*
			6155.97	6157.67	86625.757	102865.655	3–3	5.72–02	3.25–02	1.98+00	–1.011	B+	2,*LS*
			6158.15	6159.85	86631.454	102865.606	7–5	5.07–03	2.06–03	2.92–01	–1.841	B+	2,*LS*
			6156.74	6158.44	86627.778	102865.655	5–3	1.91–02	6.50–03	6.58–01	–1.488	B+	2,*LS*
			6155.96	6157.67	86625.757	102865.679	3–1	7.62–02	1.44–02	8.78–01	–1.363	B+	2,*LS*
68.		$^5P-^3D^o$											
			6140.58	6142.28	86627.778	102908.374	5–7	9.62–06	7.61–06	7.69–04	–4.420	C	2
69.	$2p^3(^4S^o)3p-2p^3(^4S^o)6s$	$^5P-^5S^o$	*5436.2*	*5437.7*	*86629.09*	105019.307	15–5	3.87–02	5.71–03	1.53+00	–1.067	B	1
			5436.86	5438.37	86631.454	105019.307	7–5	1.80–02	5.71–03	7.16–01	–1.398	C+	*LS*
			5435.77	5437.29	86627.778	105019.307	5–5	1.29–02	5.71–03	5.11–01	–1.544	C+	*LS*
			5435.18	5436.69	86625.757	105019.307	3–5	7.74–03	5.71–03	3.07–01	–1.766	C+	*LS*
70.	$2p^3(^4S^o)3p-2p^3(^4S^o)5d$	$^5P-^5D^o$	*5330.1*	*5331.5*	*86629.09*	*105385.39*	15–25	2.71–02	1.92–02	5.06+00	–0.540	B	1
			5330.74	5332.22	86631.454	105385.354	7–9	2.71–02	1.48–02	1.82+00	–0.984	C+	*LS*
			5329.69	5331.17	86627.778	105385.377	5–7	1.81–02	1.08–02	9.44–01	–1.269	C+	*LS*
			5329.11	5330.59	86625.757	105385.409	3–5	9.48–03	6.73–03	3.54–01	–1.695	C+	*LS*
			5330.73	5332.22	86631.454	105385.377	7–7	9.02–03	3.84–03	4.72–01	–1.570	C+	*LS*
			5329.68	5331.16	86627.778	105385.409	5–5	1.58–02	6.73–03	5.90–01	–1.473	C+	*LS*
			5329.10	5330.58	86625.757	105385.436	3–3	2.03–02	8.65–03	4.55–01	–1.586	C+	*LS*
			5330.73	5332.21	86631.454	105385.409	7–5	1.80–03	5.49–04	6.74–02	–2.415	C+	*LS*
			5329.67	5331.16	86627.778	105385.436	5–3	6.77–03	1.73–03	1.52–01	–2.063	C+	*LS*
			5329.10	5330.58	86625.757	105385.449	3–1	2.71–02	3.84–03	2.02–01	–1.938	C+	*LS*
71.	$2p^3(^4S^o)3p-2p^3(^4S^o)7s$	$^5P-^5S^o$	*5019.6*	*5021.0*	*86629.09*	106545.354	15–5	2.14–02	2.69–03	6.68–01	–1.394	B	1
			5020.22	5021.62	86631.454	106545.354	7–5	9.98–03	2.69–03	3.12–01	–1.725	C+	*LS*
			5019.29	5020.69	86627.778	106545.354	5–5	7.13–03	2.69–03	2.23–01	–1.871	C+	*LS*
			5018.78	5020.18	86625.757	106545.354	3–5	4.28–03	2.69–03	1.34–01	–2.092	C+	*LS*
72.	$2p^3(^4S^o)3p-2p^3(^4S^o)6d$	$^5P-^5D^o$	*4968.2*	*4969.6*	*86629.09*	*106751.46*	15–25	1.27–02	7.81–03	1.92+00	–0.931	B	1
			4968.79	4970.18	86631.454	106751.447	7–9	1.27–02	6.02–03	6.89–01	–1.375	C+	*LS*
			4967.88	4969.27	86627.778	106751.458	5–7	8.44–03	4.37–03	3.57–01	–1.660	C+	*LS*
			4967.38	4968.77	86625.757	106751.474	3–5	4.43–03	2.73–03	1.34–01	–2.086	C+	*LS*
			4968.79	4970.18	86631.454	106751.458	7–7	4.22–03	1.56–03	1.79–01	–1.961	C+	*LS*
			4967.88	4969.27	86627.778	106751.474	5–5	7.38–03	2.73–03	2.23–01	–1.865	C+	*LS*
			4967.38	4968.76	86625.757	106751.487	3–3	9.50–03	3.51–03	1.72–01	–1.977	C+	*LS*
			4968.79	4970.17	86631.454	106751.474	7–5	8.43–04	2.23–04	2.55–02	–2.807	C	*LS*
			4967.88	4969.26	86627.778	106751.487	5–3	3.16–03	7.03–04	5.75–02	–2.454	C+	*LS*
			4967.38	4968.76	86625.757	106751.494	3–1	1.27–02	1.56–03	7.66–02	–2.329	C+	*LS*
73.	$2p^3(^4S^o)3p-2p^3(^4S^o)8s$	$^5P-^5S^o$	*4802.4*	*4803.8*	*86629.09*	107446.036	15–5	1.31–02	1.51–03	3.58–01	–1.645	B	1
			4802.98	4804.32	86631.454	107446.036	7–5	6.11–03	1.51–03	1.67–01	–1.976	C+	*LS*
			4802.13	4803.48	86627.778	107446.036	5–5	4.37–03	1.51–03	1.19–01	–2.122	C+	*LS*
			4801.67	4803.01	86625.757	107446.036	3–5	2.62–03	1.51–03	7.16–02	–2.344	C+	*LS*

O I: Allowed Transitions — Continued

No.	Transition Array	Mult.	λ_{air} (Å)	λ_{vac} (Å) or σ (cm^{-1})*	E_i (cm^{-1})	E_k (cm^{-1})	g_i-g_k	A_{ki} (10^8 s^{-1})	f_{ik}	S (at. u.)	log gf	Acc.	Source
74.	$2p^3(^4S^o)3p-$ $2p^3(^4S^o)7d$	$^5P-^5D^o$	*4773.2*	*4774.5*	*86629.09*	*107573.49*	15−25	7.00−03	3.98−03	9.39−01	−1.224	B	1
			4773.75	4775.09	86631.454	107573.476	7−9	7.00−03	3.07−03	3.38−01	−1.667	C+	*LS*
			4772.91	4774.25	86627.778	107573.484	5−7	4.67−03	2.23−03	1.75−01	−1.953	C+	*LS*
			4772.45	4773.79	86625.757	107573.495	3−5	2.45−03	1.39−03	6.57−02	−2.378	C+	*LS*
			4773.75	4775.09	86631.454	107573.484	7−7	2.33−03	7.97−04	8.76−02	−2.254	C+	*LS*
			4772.91	4774.25	86627.778	107573.495	5−5	4.08−03	1.39−03	1.10−01	−2.157	C+	*LS*
			4772.45	4773.78	86625.757	107573.504	3−3	5.25−03	1.79−03	8.45−02	−2.269	C+	*LS*
			4773.75	4775.08	86631.454	107573.495	7−5	4.66−04	1.14−04	1.25−02	−3.099	C	*LS*
			4772.91	4774.24	86627.778	107573.504	5−3	1.75−03	3.59−04	2.82−02	−2.746	C+	*LS*
			4772.45	4773.78	86625.757	107573.508	3−1	7.00−03	7.97−04	3.76−02	−2.621	C+	*LS*
75.	$2p^3(^4S^o)3p-$ $2p^3(^4S^o)8d$	$^5P-^5D^o$	*4654.8*	*4656.1*	*86629.09*	*108106.079*	15−25	4.30−03	2.33−03	5.36−01	−1.457	B	1
			4655.36	4656.66	86631.454	108106.072	7−9	4.30−03	1.80−03	1.93−01	−1.900	C+	*LS*
			4654.56	4655.86	86627.778	108106.077	5−7	2.87−03	1.30−03	1.00−01	−2.186	C+	*LS*
			4654.12	4655.42	86625.757	108106.085	3−5	1.51−03	8.16−04	3.75−02	−2.611	C+	*LS*
			4655.36	4656.66	86631.454	108106.077	7−7	1.43−03	4.66−04	5.00−02	−2.487	C+	*LS*
			4654.56	4655.86	86627.778	108106.085	5−5	2.51−03	8.15−04	6.25−02	−2.390	C+	*LS*
			4654.12	4655.42	86625.757	108106.091	3−3	3.23−03	1.05−03	4.82−02	−2.502	C+	*LS*
			4655.35	4656.66	86631.454	108106.085	7−5	2.87−04	6.66−05	7.14−03	−3.332	C−	*LS*
			4654.56	4655.86	86627.778	108106.091	5−3	1.08−03	2.10−04	1.61−02	−2.979	C	*LS*
			4654.12	4655.42	86625.757	108106.094	3−1	4.30−03	4.66−04	2.14−02	−2.854	C	*LS*
76.	$2p^3(^4S^o)3p-$ $2p^3(^4S^o)4s$	$^3P-^3S^o$	*13165*	*7594.07* cm^{-1}	*88630.98*	96225.049	9−3	2.14−01	1.85−01	7.23+01	0.222	B+	1,2
			13164.9	7593.903 cm^{-1}	88631.146	96225.049	5−3	1.19−01	1.85−01	4.02+01	−0.033	B+	2,*LS*
			13163.9	7594.462 cm^{-1}	88630.587	96225.049	3−3	7.14−02	1.85−01	2.41+01	−0.255	B+	2,*LS*
			13165.1	7593.746 cm^{-1}	88631.303	96225.049	1−3	2.38−02	1.85−01	8.03+00	−0.732	B+	2,*LS*
77.	$2p^3(^4S^o)3p-$ $2p^3(^4S^o)3d$	$^3P-^5D^o$											
			11374.0	8789.570 cm^{-1}	88631.146	97420.716	5−7	1.01−05	2.75−05	5.15−03	−3.861	C	2
			11373.9	8789.639 cm^{-1}	88631.303	97420.942	1−3	3.98−06	2.31−05	8.66−04	−4.636	C	2
78.		$^3P-^3D^o$	*11287*	*8857.50* cm^{-1}	*88630.98*	*97488.48*	9−15	3.09−01	9.85−01	3.29+02	0.948	A	1,2
			11286.9	8857.392 cm^{-1}	88631.146	97488.538	5−7	3.09−01	8.27−01	1.54+02	0.617	A	2,*LS*
			11286.3	8857.861 cm^{-1}	88630.587	97488.448	3−5	2.32−01	7.39−01	8.24+01	0.346	A	2,*LS*
			11287.3	8857.075 cm^{-1}	88631.303	97488.378	1−3	1.72−01	9.85−01	3.66+01	−0.007	A	2,*LS*
			11287.0	8857.302 cm^{-1}	88631.146	97488.448	5−5	7.74−02	1.48−01	2.75+01	−0.131	A	2,*LS*
			11286.4	8857.791 cm^{-1}	88630.587	97488.378	3−3	1.29−01	2.46−01	2.75+01	−0.131	A	2,*LS*
			11287.1	8857.232 cm^{-1}	88631.146	97488.378	5−3	8.60−03	9.85−03	1.83+00	−1.308	A	2,*LS*
79.	$2p^3(^4S^o)3p-$ $2p^3(^2D^o)3s$	$^3P-^3D^o$	*7989.8*	*7992.0*	*88630.98*	*101143.45*	9−15	5.62−04	8.96−04	2.12−01	−2.094	C	1,2
			7995.07	7997.27	88631.146	101135.407	5−7	5.63−04	7.56−04	9.94−02	−2.423	C	2,*LS*
			7986.98	7989.17	88630.587	101147.526	3−5	4.19−04	6.68−04	5.27−02	−2.698	C	2,*LS*
			7982.40	7984.59	88631.303	101155.422	1−3	3.09−04	8.86−04	2.33−02	−3.052	C	2,*LS*
			7987.33	7989.53	88631.146	101147.526	5−5	1.41−04	1.35−04	1.77−02	−3.172	C	2,*LS*
			7981.94	7984.14	88630.587	101155.422	3−3	2.33−04	2.23−04	1.76−02	−3.175	C	2,*LS*
			7982.30	7984.49	88631.146	101155.422	5−3	1.80−05	1.03−05	1.36−03	−4.288	C	2,*LS*
80.	$2p^3(^4S^o)3p-$ $2p^3(^4S^o)5s$	$^3P-^3S^o$	*7254.4*	*7256.4*	*88630.98*	102411.995	9−3	6.71−02	1.76−02	3.79+00	−0.799	B	1
			7254.45	7256.45	88631.146	102411.995	5−3	3.73−02	1.76−02	2.11+00	−1.054	C+	*LS*
			7254.15	7256.15	88630.587	102411.995	3−3	2.24−02	1.76−02	1.26+00	−1.276	C+	*LS*
			7254.53	7256.53	88631.303	102411.995	1−3	7.45−03	1.76−02	4.21−01	−1.753	C+	*LS*

O I: Allowed Transitions — Continued

No.	Transition Array	Mult.	λ_{air} (Å)	λ_{vac} (Å) or σ (cm^{-1})*	E_i (cm^{-1})	E_k (cm^{-1})	g_i–g_k	A_{ki} (10^8 s^{-1})	f_{ik}	S (at. u.)	log *gf*	Acc.	Source
81.	$2p^3(^4S^o)3p-$ $2p^3(^4S^o)4d$	$^3P-^3D^o$	*7002.1*	*7004.1*	*88630.98*	*102908.42*	9–15	3.53–02	4.32–02	8.97+00	–0.410	B	1,2
			7002.23	7004.16	88631.146	102908.374	5–7	3.53–02	3.63–02	4.19+00	–0.741	B	2,*LS*
			7001.92	7003.85	88630.587	102908.443	3–5	2.65–02	3.24–02	2.24+00	–1.012	B	2,*LS*
			7002.25	7004.18	88631.303	102908.489	1–3	1.96–02	4.33–02	9.97–01	–1.364	B	2,*LS*
			7002.20	7004.13	88631.146	102908.443	5–5	8.83–03	6.49–03	7.48–01	–1.489	B	2,*LS*
			7001.90	7003.83	88630.587	102908.489	3–3	1.47–02	1.08–02	7.48–01	–1.489	B	2,*LS*
			7002.17	7004.10	88631.146	102908.489	5–3	9.83–04	4.33–04	5.00–02	–2.664	B	2,*LS*
82.	$2p^3(^4S^o)3p-$ $2p^3(^4S^o)6s$	$^3P-^3S^o$	*6046.4*	*6048.1*	*88630.98*	105165.232	9–3	3.15–02	5.76–03	1.03+00	–1.286	B	1
			6046.44	6048.11	88631.146	105165.232	5–3	1.75–02	5.76–03	5.73–01	–1.541	C+	*LS*
			6046.23	6047.91	88630.587	105165.232	3–3	1.05–02	5.76–03	3.44–01	–1.763	C+	*LS*
			6046.49	6048.17	88631.303	105165.232	1–3	3.50–03	5.76–03	1.15–01	–2.240	C+	*LS*
83.	$2p^3(^4S^o)3p-$ $2p^3(^4S^o)5d$	$^3P-^3D^o$	*5958.5*	*5960.2*	*88630.98*	105409.008	9–15	9.06–03	8.04–03	1.42+00	–1.140	B	1
			5958.58	5960.23	88631.146	105409.008	5–7	9.06–03	6.75–03	6.62–01	–1.471	C+	*LS*
			5958.39	5960.04	88630.587	105409.008	3–5	6.80–03	6.03–03	3.55–01	–1.742	C+	*LS*
			5958.64	5960.29	88631.303	105409.008	1–3	5.04–03	8.04–03	1.58–01	–2.095	C+	*LS*
			5958.58	5960.23	88631.146	105409.008	5–5	2.27–03	1.21–03	1.18–01	–2.220	C+	*LS*
			5958.39	5960.04	88630.587	105409.008	3–3	3.78–03	2.01–03	1.18–01	–2.220	C+	*LS*
			5958.58	5960.23	88631.146	105409.008	5–3	2.52–04	8.04–05	7.89–03	–3.396	C–	*LS*
84.	$2p^3(^4S^o)3p-$ $2p^3(^4S^o)7s$	$^3P-^3S^o$	*5555.0*	*5556.5*	*88630.98*	106627.934	9–3	1.75–02	2.70–03	4.44–01	–1.615	B	1
			5555.00	5556.55	88631.146	106627.934	5–3	9.71–03	2.70–03	2.47–01	–1.870	C+	*LS*
			5554.83	5556.37	88630.587	106627.934	3–3	5.83–03	2.70–03	1.48–01	–2.092	C+	*LS*
			5555.05	5556.60	88631.303	106627.934	1–3	1.94–03	2.70–03	4.93–02	–2.569	C+	*LS*
85.	$2p^3(^4S^o)3p-$ $2p^3(^4S^o)6d$	$^3P-^3D^o$	*5512.7*	*5514.3*	*88630.98*	106765.803	9–15	3.58–03	2.72–03	4.44–01	–1.611	B	1
			5512.77	5514.30	88631.146	106765.803	5–7	3.58–03	2.28–03	2.07–01	–1.942	C+	*LS*
			5512.60	5514.13	88630.587	106765.803	3–5	2.69–03	2.04–03	1.11–01	–2.213	C+	*LS*
			5512.82	5514.35	88631.303	106765.803	1–3	1.99–03	2.72–03	4.93–02	–2.566	C+	*LS*
			5512.77	5514.30	88631.146	106765.803	5–5	8.95–04	4.08–04	3.70–02	–2.691	C+	*LS*
			5512.60	5514.13	88630.587	106765.803	3–3	1.49–03	6.80–04	3.70–02	–2.691	C+	*LS*
			5512.77	5514.30	88631.146	106765.803	5–3	9.95–05	2.72–05	2.47–03	–3.867	D	*LS*
86.	$2p^3(^4S^o)3p-$ $2p^3(^4S^o)8s$	$^3P-^3S^o$	*5299.0*	*5300.5*	*88630.98*	107497.224	9–3	1.07–02	1.51–03	2.37–01	–1.868	B	1
			5299.04	5300.52	88631.146	107497.224	5–3	5.96–03	1.51–03	1.31–01	–2.123	C+	*LS*
			5298.89	5300.36	88630.587	107497.224	3–3	3.58–03	1.51–03	7.88–02	–2.345	C+	*LS*
			5299.09	5300.56	88631.303	107497.224	1–3	1.19–03	1.51–03	2.63–02	–2.822	C	*LS*
87.	$2p^3(^4S^o)3p-$ $2p^3(^4S^o)7d$	$^3P-^3D^o$	*5275.1*	*5276.5*	*88630.98*	107582.777	9–15	1.76–03	1.22–03	1.91–01	–1.958	B	1
			5275.12	5276.59	88631.146	107582.777	5–7	1.76–03	1.03–03	8.93–02	–2.289	C+	*LS*
			5274.97	5276.44	88630.587	107582.777	3–5	1.32–03	9.18–04	4.79–02	–2.560	C+	*LS*
			5275.17	5276.63	88631.303	107582.777	1–3	9.78–04	1.22–03	2.13–02	–2.912	C	*LS*
			5275.12	5276.59	88631.146	107582.777	5–5	4.40–04	1.84–04	1.60–02	–3.037	C	*LS*
			5274.97	5276.44	88630.587	107582.777	3–3	7.34–04	3.06–04	1.60–02	–3.037	C	*LS*
			5275.12	5276.59	88631.146	107582.777	5–3	4.89–05	1.22–05	1.06–03	–4.213	D	*LS*
88.	$2p^3(^4S^o)3p-$ $2p^3(^4S^o)8d$	$^3P-^3D^o$	*5131.2*	*5132.7*	*88630.98*	108114.0	9–15	9.98–04	6.56–04	9.98–02	–2.229	B	1
			5131.29	5132.72	88631.146	108114.0	5–7	9.98–04	5.51–04	4.66–02	–2.560	C+	*LS*
			5131.14	5132.57	88630.587	108114.0	3–5	7.48–04	4.92–04	2.50–02	–2.831	C	*LS*
			5131.33	5132.76	88631.303	108114.0	1–3	5.54–04	6.56–04	1.11–02	–3.183	C	*LS*
			5131.29	5132.72	88631.146	108114.0	5–5	2.49–04	9.85–05	8.32–03	–3.308	C–	*LS*
			5131.14	5132.57	88630.587	108114.0	3–3	4.16–04	1.64–04	8.32–03	–3.308	C–	*LS*
			5131.29	5132.72	88631.146	108114.0	5–3	2.77–05	6.56–06	5.54–04	–4.484	D–	*LS*

O I: Allowed Transitions — Continued

No.	Transition Array	Mult.	λ_{air} (Å)	λ_{vac} (Å) or σ (cm^{-1})*	E_i (cm^{-1})	E_k (cm^{-1})	g_i-g_k	A_{ki} ($10^8\ s^{-1}$)	f_{ik}	S (at. u.)	log gf	Acc.	Source
89.	$2p^3(^4S^o)3p-$ $2p^3(^2P^o)3s$	$^3P-^3P^o$	*3953.7*	*3954.9*	*88630.98*	*– 113916.28*	9–9	3.09–03	7.25–04	8.49–02	–2.186	B	1
			3954.61	3955.73	88631.146	– 113910.957	5–5	2.32–03	5.43–04	3.54–02	–2.566	C+	*LS*
			3952.89	3954.01	88630.587	– 113921.391	3–3	7.74–04	1.81–04	7.08–03	–3.265	C–	*LS*
			3952.98	3954.09	88631.146	– 113921.391	5–3	1.29–03	1.81–04	1.18–02	–3.043	C	*LS*
			3951.93	3953.05	88630.587	– 113927.534	3–1	3.10–03	2.42–04	9.43–03	–3.140	C–	*LS*
			3954.52	3955.64	88630.587	– 113910.957	3–5	7.73–04	3.02–04	1.18–02	–3.043	C	*LS*
			3953.00	3954.12	88631.303	– 113921.391	1–3	1.03–03	7.25–04	9.43–03	–3.140	C–	*LS*
90.	$2p^3(^4S^o)3p-$ $2p^3(^2D^o)3d$	$^3P-^3P^o$	*2881.4*	*2882.2*	*88630.98*	*– 123326.42*	9–9	1.47–02	1.83–03	1.56–01	–1.784	B	1
			2883.86	2884.70	88631.146	– 123296.777	5–5	1.10–02	1.37–03	6.50–02	–2.165	C+	*LS*
			2878.93	2879.78	88630.587	– 123355.512	3–3	3.68–03	4.57–04	1.30–02	–2.863	C	*LS*
			2878.98	2879.82	88631.146	– 123355.512	5–3	6.13–03	4.57–04	2.17–02	–2.641	C	*LS*
			2876.29	2877.14	88630.587	– 123387.339	3–1	1.47–02	6.10–04	1.73–02	–2.738	C	*LS*
			2883.81	2884.66	88630.587	– 123296.777	3–5	3.66–03	7.60–04	2.17–02	–2.642	C	*LS*
			2878.99	2879.84	88631.303	– 123355.512	1–3	4.90–03	1.83–03	1.73–02	–2.738	C	*LS*
91.		$^3P-^3D^o$	*2806.3*	*2807.2*	*88630.98*	*– 124254.0*	9–15	4.31–03	8.48–04	7.05–02	–2.118	B–	2
			2806.90	2807.73	88631.146	– 124247.1	5–7	4.35–03	7.19–04	3.32–02	–2.444	C	2
			2806.02	2806.85	88630.587	– 124257.7	3–5	3.22–03	6.33–04	1.75–02	–2.722	C	2
			2805.58	2806.41	88631.303	– 124264.0	1–3	2.36–03	8.36–04	7.72–03	–3.078	C	2
			2806.07	2806.90	88631.146	– 124257.7	5–5	1.07–03	1.26–04	5.83–03	–3.200	C	2
			2805.53	2806.35	88630.587	– 124264.0	3–3	1.77–03	2.09–04	5.79–03	–3.203	C	2
			2805.57	2806.40	88631.146	– 124264.0	5–3	1.18–04	8.35–06	3.86–04	–4.379	D	2
92.		$^3P-^3S^o$	*2799.9*	*2800.7*	*88630.98*	– 124336.3	9–3	1.45–02	5.70–04	4.73–02	–2.290	B–	2
			2799.89	2800.72	88631.146	– 124336.3	5–3	8.08–03	5.70–04	2.63–02	–2.545	C+	2
			2799.85	2800.67	88630.587	– 124336.3	3–3	4.85–03	5.70–04	1.57–02	–2.767	C+	2
			2799.90	2800.73	88631.303	– 124336.3	1–3	1.61–03	5.69–04	5.25–03	–3.245	C	2
93.	$2p^3(^4S^o)3p-$ $2p^3(^2D^o)4d$	$^3P-^3D^o$											
			2434.65	2435.39	88631.146	– 129692.3	5–7	3.13–03	3.89–04	1.56–02	–2.711	C	2
			2434.31	2435.05	88630.587	– 129697.5	3–5	2.32–03	3.43–04	8.26–03	–2.987	C	2
			2434.03	2434.77	88631.303	– 129703.0	1–3	1.70–03	4.54–04	3.63–03	–3.343	C	2
			2434.35	2435.08	88631.146	– 129697.5	5–5	7.70–04	6.84–05	2.74–03	–3.466	D	2
			2433.99	2434.72	88630.587	– 129703.0	3–3	1.28–03	1.13–04	2.72–03	–3.469	D	2
94.		$^3P-^3S^o$	*2432.0*	*2432.8*	*88630.98*	– 129736.6	9–3	1.11–02	3.29–04	2.37–02	–2.529	C+	2
			2432.03	2432.77	88631.146	– 129736.6	5–3	6.19–03	3.29–04	1.32–02	–2.783	C	2
			2432.00	2432.73	88630.587	– 129736.6	3–3	3.71–03	3.29–04	7.90–03	–3.006	C	2
			2432.04	2432.78	88631.303	– 129736.6	1–3	1.23–03	3.28–04	2.63–03	–3.484	D	2
95.		$^3P-^3P^o$	*2418.0*	*2418.7*	*88630.98*	*– 129974.75*	9–9	2.98–03	2.61–04	1.87–02	–2.629	B–	1
			2418.30	2419.03	88631.146	– 129970.000	5–5	2.23–03	1.96–04	7.79–03	–3.009	C–	*LS*
			2417.72	2418.45	88630.587	– 129979.384	3–3	7.45–04	6.53–05	1.56–03	–3.708	D	*LS*
			2417.75	2418.48	88631.146	– 129979.384	5–3	1.24–03	6.53–05	2.60–03	–3.486	D	*LS*
			2417.41	2418.15	88630.587	– 129984.553	3–1	2.98–03	8.70–05	2.08–03	–3.583	D	*LS*
			2418.26	2419.00	88630.587	– 129970.000	3–5	7.44–04	1.09–04	2.60–03	–3.486	D	*LS*
			2417.76	2418.49	88631.303	– 129979.384	1–3	9.93–04	2.61–04	2.08–03	–3.583	D	*LS*
96.	$2p^3(^4S^o)3p-$ $2p^3(^2P^o)3d$	$^3P-^3D^o$											
			2026.46	2027.11	88631.146	– 137962.5	5–7	3.70–03	3.19–04	1.06–02	–2.797	C	2
			2026.43	2027.09	88630.587	– 137962.5	3–5	2.82–03	2.90–04	5.80–03	–3.061	C	2
			2026.46	2027.11	88631.303	– 137962.5	1–3	1.92–03	3.55–04	2.37–03	–3.450	D	2
			2026.46	2027.11	88631.146	– 137962.5	5–5	4.85–04	2.98–05	9.95–04	–3.826	D	2
			2026.43	2027.09	88630.587	– 137962.5	3–3	1.05–03	6.45–05	1.29–03	–3.714	D	2

O I: Allowed Transitions — Continued

No.	Transition Array	Mult.	λ_{air} (Å)	λ_{vac} (Å) or σ (cm^{-1})*	E_i (cm^{-1})	E_k (cm^{-1})	g_i–g_k	A_{ki} (10^8 s^{-1})	f_{ik}	S (at. u.)	log gf	Acc.	Source
97.	$2p^3(^4S°)3p$ – $2p^3(^2P°)4d$	$^3P–^3D°$											
				1827.08	88631.146	143363.4	5–7	1.96–03	1.37–04	4.13–03	–3.163	D	2
				1827.06	88630.587	143363.4	3–5	1.52–03	1.27–04	2.29–03	–3.420	D	2
				1827.08	88631.303	143363.4	1–3	1.12–03	1.69–04	1.02–03	–3.773	D	2
				1827.08	88631.146	143363.4	5–5	4.32–04	2.16–05	6.50–04	–3.966	D	2
				1827.06	88630.587	143363.4	3–3	7.79–04	3.90–05	7.03–04	–3.932	D	2
98.	$2p^3(^4S°)4s$ – $2p^3(^4S°)4p$	$^5S°–^5P$	*27637*	*3617.33* cm^{-1}	95476.728	*99094.06*	5–15	4.30–02	1.48+00	6.71+02	0.868	A	1,2
			27631.2	3618.109 cm^{-1}	95476.728	99094.837	5–7	4.30–02	6.89–01	3.13+02	0.537	A	2,*LS*
			27640.3	3616.913 cm^{-1}	95476.728	99093.641	5–5	4.29–02	4.92–01	2.24+02	0.391	A	2,*LS*
			27645.5	3616.240 cm^{-1}	95476.728	99092.968	5–3	4.29–02	2.95–01	1.34+02	0.169	A	2,*LS*
99.	$2p^3(^4S°)4s$ – $2p^3(^4S°)5p$	$^5S°–^5P$	*12267*	*8149.54* cm^{-1}	95476.728	*103626.27*	5–15	2.35–03	1.59–02	3.21+00	–1.100	B	1
			12266.8	8149.883 cm^{-1}	95476.728	103626.611	5–7	2.35–03	7.41–03	1.50+00	–1.431	C+	*LS*
			12267.5	8149.383 cm^{-1}	95476.728	103626.111	5–5	2.35–03	5.29–03	1.07+00	–1.577	C+	*LS*
			12268.0	8149.026 cm^{-1}	95476.728	103625.754	5–3	2.35–03	3.18–03	6.41–01	–1.799	C+	*LS*
100.	$2p^3(^4S°)4s$ – $2p^3(^4S°)6p$	$^5S°–^5P$	*9694.8*	*9697.5*	95476.728	*105788.68*	5–15	4.54–04	1.92–03	3.06–01	–2.018	B	1
			9694.66	9697.32	95476.728	105788.856	5–7	4.54–04	8.95–04	1.43–01	–2.349	C+	*LS*
			9694.91	9697.56	95476.728	105788.595	5–5	4.54–04	6.39–04	1.02–01	–2.495	C+	*LS*
			9695.06	9697.72	95476.728	105788.431	5–3	4.54–04	3.84–04	6.12–02	–2.717	C+	*LS*
101.	$2p^3(^4S°)4s$ – $2p^3(^4S°)4p$	$^3S°–^5P$											
			34850.8	2868.592 cm^{-1}	96225.049	99093.641	3–5	6.20–07	1.88–05	6.48–03	–4.248	D	2
			34859.0	2867.919 cm^{-1}	96225.049	99092.968	3–3	4.72–07	8.59–06	2.96–03	–4.589	D	2
102.		$^3S°–^3P$	*28927*	*3456.00* cm^{-1}	96225.049	*99681.05*	3–9	4.10–02	1.54+00	4.41+02	0.666	A	1,2
			28927.3	3456.000 cm^{-1}	96225.049	99681.049	3–5	4.10–02	8.57–01	2.45+02	0.410	A	2,*LS*
			28928.0	3455.919 cm^{-1}	96225.049	99680.968	3–3	4.10–02	5.14–01	1.47+02	0.188	A	2,*LS*
			28925.1	3456.260 cm^{-1}	96225.049	99681.309	3–1	4.10–02	1.71–01	4.90+01	–0.289	A	2,*LS*
103.	$2p^3(^4S°)4s$ – $2p^3(^4S°)5p$	$^3S°–^3P$	*13077*	*7644.97* cm^{-1}	96225.049	*103870.02*	3–9	2.86–03	2.20–02	2.84+00	–1.180	B	1
			13077.0	7644.919 cm^{-1}	96225.049	103869.968	3–5	2.86–03	1.22–02	1.58+00	–1.435	C+	*LS*
			13076.9	7644.979 cm^{-1}	96225.049	103870.028	3–3	2.86–03	7.34–03	9.48–01	–1.657	C+	*LS*
			13076.5	7645.203 cm^{-1}	96225.049	103870.252	3–1	2.86–03	2.45–03	3.16–01	–2.134	C+	*LS*
104.	$2p^3(^4S°)4s$ – $2p^3(^4S°)6p$	$^3S°–^3P$	10320.3	9686.982 cm^{-1}	96225.049	105912.031	3–9	6.13–04	2.94–03	2.99–01	–2.055	B	1
105.	$2p^3(^4S°)3d$ – $2p^3(^4S°)4p$	$^5D°–^5P$		*1673.31* cm^{-1}	*97420.75*	*99094.06*	25–15	4.97–03	1.60–01	7.85+02	0.601	A	1,2
				1674.207 cm^{-1}	97420.630	99094.837	9–7	3.84–03	1.60–01	2.82+02	0.157	A	2,*LS*
				1672.925 cm^{-1}	97420.716	99093.641	7–5	2.78–03	1.06–01	1.46+02	–0.128	A	2,*LS*
				1672.129 cm^{-1}	97420.839	99092.968	5–3	1.74–03	5.58–02	5.49+01	–0.554	A	2,*LS*
				1674.121 cm^{-1}	97420.716	99094.837	7–7	9.95–04	5.32–02	7.32+01	–0.429	A	2,*LS*
				1672.802 cm^{-1}	97420.839	99093.641	5–5	1.74–03	9.30–02	9.15+01	–0.332	A	2,*LS*
				1672.026 cm^{-1}	97420.942	99092.968	3–3	2.23–03	1.20–01	7.06+01	–0.445	A	2,*LS*
				1673.998 cm^{-1}	97420.839	99094.837	5–7	1.42–04	1.06–02	1.05+01	–1.274	A	2,*LS*
				1672.699 cm^{-1}	97420.942	99093.641	3–5	4.47–04	3.99–02	2.35+01	–0.922	A	2,*LS*
				1671.977 cm^{-1}	97420.991	99092.968	1–3	9.92–04	1.59–01	3.14+01	–0.797	A	2,*LS*

O I: Allowed Transitions — Continued

No.	Transition Array	Mult.	λ_{air} (Å)	λ_{vac} (Å) or σ (cm^{-1})*	E_i (cm^{-1})	E_k (cm^{-1})	g_i-g_k	A_{ki} (10^8 s^{-1})	f_{ik}	S (at. u.)	log gf	Acc.	Source
106.	$2p^3(^4S^\circ)3d-$ $2p^3(^4S^\circ)4f$	$^5D^\circ-^5F$	*18021*	*5547.50* cm^{-1}	*97420.75*	102968.249	25−35	1.48−01	1.01+00	1.49+03	1.401	B	1
			18020.8	5547.619 cm^{-1}	97420.630	102968.249	9−11	1.48−01	8.79−01	4.70+02	0.898	C+	*LS*
			18021.1	5547.533 cm^{-1}	97420.716	102968.249	7−9	1.23−01	7.71−01	3.20+02	0.732	C+	*LS*
			18021.5	5547.410 cm^{-1}	97420.839	102968.249	5−7	1.01−01	6.91−01	2.05+02	0.538	C+	*LS*
			18021.8	5547.307 cm^{-1}	97420.942	102968.249	3−5	8.27−02	6.72−01	1.20+02	0.304	C+	*LS*
			18022.0	5547.258 cm^{-1}	97420.991	102968.249	1−3	6.90−02	1.01+00	5.98+01	0.003	C+	*LS*
			18020.8	5547.619 cm^{-1}	97420.630	102968.249	9−9	2.46−02	1.20−01	6.40+01	0.033	C+	*LS*
			18021.1	5547.533 cm^{-1}	97420.716	102968.249	7−7	4.43−02	2.16−01	8.96+01	0.179	C+	*LS*
			18021.5	5547.410 cm^{-1}	97420.839	102968.249	5−5	5.91−02	2.88−01	8.54+01	0.158	C+	*LS*
			18021.8	5547.307 cm^{-1}	97420.942	102968.249	3−3	6.90−02	3.36−01	5.98+01	0.003	C+	*LS*
			18020.8	5547.619 cm^{-1}	97420.630	102968.249	9−7	2.11−03	8.00−03	4.27+00	−1.143	C+	*LS*
			18021.1	5547.533 cm^{-1}	97420.716	102968.249	7−5	5.91−03	2.06−02	8.54+00	−0.842	C+	*LS*
			18021.5	5547.410 cm^{-1}	97420.839	102968.249	5−3	9.85−03	2.88−02	8.54+00	−0.842	C+	*LS*
107.	$2p^3(^4S^\circ)3d-$ $2p^3(^4S^\circ)5p$	$^5D^\circ-^5P$	*16110*	*6205.52* cm^{-1}	*97420.75*	*103626.27*	25−15	4.04−04	9.43−04	1.25+00	−1.628	B	1
			16109.1	6205.981 cm^{-1}	97420.630	103626.611	9−7	3.12−04	9.43−04	4.50−01	−2.071	C+	*LS*
			16110.6	6205.395 cm^{-1}	97420.716	103626.111	7−5	2.26−04	6.28−04	2.33−01	−2.357	C+	*LS*
			16111.9	6204.915 cm^{-1}	97420.839	103625.754	5−3	1.41−04	3.30−04	8.75−02	−2.783	C+	*LS*
			16109.3	6205.895 cm^{-1}	97420.716	103626.611	7−7	8.08−05	3.14−04	1.17−01	−2.658	C+	*LS*
			16110.9	6205.272 cm^{-1}	97420.839	103626.111	5−5	1.41−04	5.50−04	1.46−01	−2.561	C+	*LS*
			16112.1	6204.812 cm^{-1}	97420.942	103625.754	3−3	1.82−04	7.07−04	1.12−01	−2.674	C+	*LS*
			16109.6	6205.772 cm^{-1}	97420.839	103626.611	5−7	1.15−05	6.28−05	1.67−02	−3.503	C	*LS*
			16111.2	6205.169 cm^{-1}	97420.942	103626.111	3−5	3.63−05	2.36−04	3.75−02	−3.151	C+	*LS*
			16112.2	6204.763 cm^{-1}	97420.991	103625.754	1−3	8.07−05	9.43−04	5.00−02	−3.026	C+	*LS*
108.	$2p^3(^4S^\circ)3d-$ $2p^3(^4S^\circ)5f$	$^5D^\circ-^5F$	*12511*	*7990.90* cm^{-1}	*97420.75*	105411.645	25−35	4.91−02	1.61−01	1.66+02	0.606	B	1
			12510.6	7991.015 cm^{-1}	97420.630	105411.645	9−11	4.91−02	1.41−01	5.22+01	0.103	C+	*LS*
			12510.8	7990.929 cm^{-1}	97420.716	105411.645	7−9	4.09−02	1.24−01	3.56+01	−0.063	C+	*LS*
			12511.0	7990.806 cm^{-1}	97420.839	105411.645	5−7	3.37−02	1.11−01	2.28+01	−0.257	C+	*LS*
			12511.1	7990.703 cm^{-1}	97420.942	105411.645	3−5	2.75−02	1.08−01	1.33+01	−0.491	C+	*LS*
			12511.2	7990.654 cm^{-1}	97420.991	105411.645	1−3	2.29−02	1.61−01	6.65+00	−0.792	C+	*LS*
			12510.6	7991.015 cm^{-1}	97420.630	105411.645	9−9	8.19−03	1.92−02	7.12+00	−0.762	C+	*LS*
			12510.8	7990.929 cm^{-1}	97420.716	105411.645	7−7	1.47−02	3.46−02	9.97+00	−0.616	C+	*LS*
			12511.0	7990.806 cm^{-1}	97420.839	105411.645	5−5	1.96−02	4.61−02	9.49+00	−0.637	C+	*LS*
			12511.1	7990.703 cm^{-1}	97420.942	105411.645	3−3	2.29−02	5.38−02	6.65+00	−0.792	C+	*LS*
			12510.6	7991.015 cm^{-1}	97420.630	105411.645	9−7	7.02−04	1.28−03	4.75−01	−1.938	C+	*LS*
			12510.8	7990.929 cm^{-1}	97420.716	105411.645	7−5	1.96−03	3.29−03	9.49−01	−1.637	C+	*LS*
			12511.0	7990.806 cm^{-1}	97420.839	105411.645	5−3	3.27−03	4.61−03	9.49−01	−1.637	C+	*LS*
109.	$2p^3(^4S^\circ)3d-$ $2p^3(^4S^\circ)6p$	$^5D^\circ-^5P$	*11947*	*8367.93* cm^{-1}	*97420.75*	*105788.68*	25−15	2.08−04	2.68−04	2.63−01	−2.175	B	1
			11946.7	8368.226 cm^{-1}	97420.630	105788.856	9−7	1.61−04	2.68−04	9.47−02	−2.618	C+	*LS*
			11947.2	8367.879 cm^{-1}	97420.716	105788.595	7−5	1.17−04	1.78−04	4.91−02	−2.904	C+	*LS*
			11947.6	8367.592 cm^{-1}	97420.839	105788.431	5−3	7.29−05	9.36−05	1.84−02	−3.330	C	*LS*
			11946.8	8368.140 cm^{-1}	97420.716	105788.856	7−7	4.17−05	8.92−05	2.46−02	−3.205	C	*LS*
			11947.4	8367.756 cm^{-1}	97420.839	105788.595	5−5	7.29−05	1.56−04	3.07−02	−3.108	C+	*LS*
			11947.7	8367.489 cm^{-1}	97420.942	105788.431	3−3	9.38−05	2.01−04	2.37−02	−3.220	C	*LS*
			11947.0	8368.017 cm^{-1}	97420.839	105788.856	5−7	5.95−06	1.78−05	3.51−03	−4.050	C−	*LS*
			11947.5	8367.653 cm^{-1}	97420.942	105788.595	3−5	1.88−05	6.69−05	7.89−03	−3.698	C−	*LS*
			11947.8	8367.440 cm^{-1}	97420.991	105788.431	1−3	4.17−05	2.68−04	1.05−02	−3.573	C	*LS*

O I: Allowed Transitions — Continued

No.	Transition Array	Mult.	λ_{air} (Å)	λ_{vac} (Å) or σ (cm^{-1})*	E_i (cm^{-1})	E_k (cm^{-1})	g_i-g_k	A_{ki} (10^8 s^{-1})	f_{ik}	S (at. u.)	log gf	Acc.	Source
110.	$2p^3(^4S^o)3d-2p^3(^4S^o)6f$	$^5D^o-^5F$	*10676*	*9364.41* cm^{-1}	*97420.75*	− 106785.160	25−35	2.37−02	5.68−02	4.99+01	0.152	B	1
			10675.7	9364.530 cm^{-1}	97420.630	− 106785.160	9−11	2.37−02	4.96−02	1.57+01	−0.351	C+	*LS*
			10675.8	9364.444 cm^{-1}	97420.716	− 106785.160	7−9	1.98−02	4.34−02	1.07+01	−0.517	C+	*LS*
			10675.9	9364.321 cm^{-1}	97420.839	− 106785.160	5−7	1.63−02	3.89−02	6.84+00	−0.711	C+	*LS*
			10676.0	9364.218 cm^{-1}	97420.942	− 106785.160	3−5	1.33−02	3.78−02	3.99+00	−0.945	C+	*LS*
			10676.1	9364.169 cm^{-1}	97420.991	− 106785.160	1−3	1.11−02	5.68−02	1.99+00	−1.246	C+	*LS*
			10675.7	9364.530 cm^{-1}	97420.630	− 106785.160	9−9	3.95−03	6.76−03	2.14+00	−1.216	C+	*LS*
			10675.8	9364.444 cm^{-1}	97420.716	− 106785.160	7−7	7.12−03	1.22−02	2.99+00	−1.070	C+	*LS*
			10675.9	9364.321 cm^{-1}	97420.839	− 106785.160	5−5	9.49−03	1.62−02	2.85+00	−1.091	C+	*LS*
			10676.0	9364.218 cm^{-1}	97420.942	− 106785.160	3−3	1.11−02	1.89−02	1.99+00	−1.246	C+	*LS*
			10675.7	9364.530 cm^{-1}	97420.630	− 106785.160	9−7	3.39−04	4.50−04	1.42−01	−2.392	C+	*LS*
			10675.8	9364.444 cm^{-1}	97420.716	− 106785.160	7−5	9.49−04	1.16−03	2.85−01	−2.091	C+	*LS*
			10675.9	9364.321 cm^{-1}	97420.839	− 106785.160	5−3	1.58−03	1.62−03	2.85−01	−2.091	C+	*LS*
111.	$2p^3(^4S^o)3d-2p^3(^4S^o)7f$	$^5D^o-^5F$	*9825.9*	*9828.6*	*97420.75*	− 107595.140	25−35	1.34−02	2.72−02	2.20+01	−0.167	B	1
			9825.79	9828.48	97420.630	− 107595.140	9−11	1.34−02	2.38−02	6.92+00	−0.670	C+	*LS*
			9825.87	9828.57	97420.716	− 107595.140	7−9	1.12−02	2.08−02	4.72+00	−0.836	C+	*LS*
			9825.99	9828.69	97420.839	− 107595.140	5−7	9.21−03	1.87−02	3.02+00	−1.030	C+	*LS*
			9826.09	9828.78	97420.942	− 107595.140	3−5	7.52−03	1.81−02	1.76+00	−1.264	C+	*LS*
			9826.14	9828.83	97420.991	− 107595.140	1−3	6.27−03	2.72−02	8.80−01	−1.565	C+	*LS*
			9825.79	9828.48	97420.630	− 107595.140	9−9	2.24−03	3.24−03	9.43−01	−1.535	C+	*LS*
			9825.87	9828.57	97420.716	− 107595.140	7−7	4.03−03	5.83−03	1.32+00	−1.389	C+	*LS*
			9825.99	9828.69	97420.839	− 107595.140	5−5	5.37−03	7.78−03	1.26+00	−1.410	C+	*LS*
			9826.09	9828.78	97420.942	− 107595.140	3−3	6.27−03	9.07−03	8.80−01	−1.565	C+	*LS*
			9825.79	9828.48	97420.630	− 107595.140	9−7	1.92−04	2.16−04	6.29−02	−2.711	C+	*LS*
			9825.87	9828.57	97420.716	− 107595.140	7−5	5.37−04	5.55−04	1.26−01	−2.410	C+	*LS*
			9825.99	9828.69	97420.839	− 107595.140	5−3	8.95−04	7.78−04	1.26−01	−2.410	C+	*LS*
112.	$2p^3(^4S^o)3d-2p^3(^4S^o)4p$	$^3D^o-^5P$											
				1605.103 cm^{-1}	97488.538	− 99093.641	7−5	1.71−07	7.10−06	1.02−02	−4.304	C	2
113.		$^3D^o-^3P$	*45596*	*2192.57* cm^{-1}	*97488.48*	− *99681.05*	15−9	9.85−03	1.84−01	4.15+02	0.442	A	1,2
			45597.4	2192.511 cm^{-1}	97488.538	− 99681.049	7−5	8.28−03	1.84−01	1.94+02	0.111	A	2,*LS*
			45597.2	2192.520 cm^{-1}	97488.448	− 99680.968	5−3	7.39−03	1.38−01	1.04+02	−0.160	A	2,*LS*
			45588.6	2192.931 cm^{-1}	97488.378	− 99681.309	3−1	9.86−03	1.02−01	4.61+01	−0.512	A	2,*LS*
			45595.5	2192.601 cm^{-1}	97488.448	− 99681.049	5−5	1.48−03	4.61−02	3.46+01	−0.637	A	2,*LS*
			45595.7	2192.590 cm^{-1}	97488.378	− 99680.968	3−3	·2.46−03	7.68−02	3.46+01	−0.637	A	2,*LS*
			45594.0	2192.671 cm^{-1}	97488.378	− 99681.049	3−5	9.86−05	5.12−03	2.31+00	−1.813	A	2,*LS*
114.	$2p^3(^4S^o)3d-2p^3(^4S^o)4f$	$^3D^o-^3F$	*18244*	*5479.86* cm^{-1}	*97488.48*	− 102968.343	15−21	1.47−01	1.03+00	9.27+02	1.188	B	1
			18243.8	5479.805 cm^{-1}	97488.538	− 102968.343	7−9	1.47−01	9.45−01	3.97+02	0.820	C+	*LS*
			18243.5	5479.895 cm^{-1}	97488.448	− 102968.343	5−7	1.31−01	9.14−01	2.75+02	0.660	C+	*LS*
			18243.3	5479.965 cm^{-1}	97488.378	− 102968.343	3−5	1.24−01	1.03+00	1.85+02	0.489	C+	*LS*
			18243.8	5479.805 cm^{-1}	97488.538	− 102968.343	7−7	1.64−02	8.19−02	3.44+01	−0.242	C+	*LS*
			18243.5	5479.895 cm^{-1}	97488.448	− 102968.343	5−5	2.30−02	1.15−01	3.44+01	−0.242	C+	*LS*
			18243.8	5479.805 cm^{-1}	97488.538	− 102968.343	7−5	6.48−04	2.31−03	9.71−01	−1.791	C+	*LS*
115.	$2p^3(^4S^o)3d-2p^3(^4S^o)5p$	$^3D^o-^3P$	*15666*	*6381.54* cm^{-1}	*97488.48*	− *103870.02*	15−9	1.83−03	4.04−03	3.12+00	−1.218	B	1
			15666.2	6381.430 cm^{-1}	97488.538	− 103869.968	7−5	1.54−03	4.04−03	1.46+00	−1.549	C+	*LS*
			15665.8	6381.580 cm^{-1}	97488.448	− 103870.028	5−3	1.37−03	3.03−03	7.81−01	−1.820	C+	*LS*
			15665.1	6381.874 cm^{-1}	97488.378	− 103870.252	3−1	1.83−03	2.24−03	3.47−01	−2.172	C+	*LS*
			15666.0	6381.520 cm^{-1}	97488.448	− 103869.968	5−5	2.74−04	1.01−03	2.60−01	−2.297	C+	*LS*
			15665.6	6381.650 cm^{-1}	97488.378	− 103870.028	3−3	4.57−04	1.68−03	2.60−01	−2.297	C+	*LS*
			15665.8	6381.590 cm^{-1}	97488.378	− 103869.968	3−5	1.83−05	1.12−04	1.74−02	−3.473	C	*LS*

O I: Allowed Transitions — Continued

No.	Transition Array	Mult.	λ_{air} (Å)	λ_{vac} (Å) or σ (cm^{-1})*	E_i (cm^{-1})	E_k (cm^{-1})	g_i-g_k	A_{ki} ($10^8\ s^{-1}$)	f_{ik}	S (at. u.)	log gf	Acc.	Source
116.	$2p^3(^4S^\circ)3d-2p^3(^4S^\circ)5f$	$^3D^\circ-^3F$	*12570*	*7953.24* cm^{-1}	*97488.48*	105441.724	15–21	4.82–02	1.60–01	9.93+01	0.380	B	1
			12570.1	7953.186 cm^{-1}	97488.538	105441.724	7–9	4.82–02	1.47–01	4.25+01	0.012	C+	*LS*
			12570.0	7953.276 cm^{-1}	97488.448	105441.724	5–7	4.28–02	1.42–01	2.94+01	–0.148	C+	*LS*
			12569.9	7953.346 cm^{-1}	97488.378	105441.724	3–5	4.05–02	1.60–01	1.99+01	–0.319	C+	*LS*
			12570.1	7953.186 cm^{-1}	97488.538	105441.724	7–7	5.37–03	1.27–02	3.69+00	–1.050	C+	*LS*
			12570.0	7953.276 cm^{-1}	97488.448	105441.724	5–5	7.52–03	1.78–02	3.69+00	–1.050	C+	*LS*
			12570.1	7953.186 cm^{-1}	97488.538	105441.724	7–5	2.12–04	3.59–04	1.04–01	–2.600	C+	*LS*
117.	$2p^3(^4S^\circ)3d-2p^3(^4S^\circ)6p$	$^3D^\circ-^3P$	*11868*	*8423.55* cm^{-1}	*97488.48*	105912.031	15–9	8.52–04	1.08–03	6.33–01	–1.791	B	1
			11868.3	8423.493 cm^{-1}	97488.538	105912.031	7–5	7.15–04	1.08–03	2.95–01	–2.122	C+	*LS*
			11868.2	8423.583 cm^{-1}	97488.448	105912.031	5–3	6.39–04	8.09–04	1.58–01	–2.393	C+	*LS*
			11868.1	8423.653 cm^{-1}	97488.378	105912.031	3–1	8.52–04	6.00–04	7.03–02	–2.745	C+	*LS*
			11868.2	8423.583 cm^{-1}	97488.448	105912.031	5–5	1.28–04	2.70–04	5.27–02	–2.870	C+	*LS*
			11868.1	8423.653 cm^{-1}	97488.378	105912.031	3–3	2.13–04	4.50–04	5.27–02	–2.870	C+	*LS*
			11868.1	8423.653 cm^{-1}	97488.378	105912.031	3–5	8.52–06	3.00–05	3.51–03	–4.046	C–	*LS*
118.	$2p^3(^4S^\circ)3d-2p^3(^4S^\circ)6f$	$^3D^\circ-^3F$	*10754*	*9296.72* cm^{-1}	*97488.48*	106785.201	15–21	2.28–02	5.53–02	2.93+01	–0.082	B	1
			10753.6	9296.663 cm^{-1}	97488.538	106785.201	7–9	2.28–02	5.07–02	1.26+01	–0.450	C+	*LS*
			10753.5	9296.753 cm^{-1}	97488.448	106785.201	5–7	2.02–02	4.91–02	8.69+00	–0.610	C+	*LS*
			10753.4	9296.823 cm^{-1}	97488.378	106785.201	3–5	1.91–02	5.53–02	5.87+00	–0.781	C+	*LS*
			10753.6	9296.663 cm^{-1}	97488.538	106785.201	7–7	2.54–03	4.40–03	1.09+00	–1.512	C+	*LS*
			10753.5	9296.753 cm^{-1}	97488.448	106785.201	5–5	3.55–03	6.16–03	1.09+00	–1.512	C+	*LS*
			10753.6	9296.663 cm^{-1}	97488.538	106785.201	7–5	1.00–04	1.24–04	3.07–02	–3.061	C+	*LS*
119.	$2p^3(^4S^\circ)3d-2p^3(^4S^\circ)7f$	$^3D^\circ-^3F$	*9891.7*	*9894.5*	*97488.48*	107595.147	15–21	1.28–02	2.63–02	1.28+01	–0.404	B	1
			9891.80	9894.52	97488.538	107595.147	7–9	1.28–02	2.41–02	5.50+00	–0.772	C+	*LS*
			9891.72	9894.43	97488.448	107595.147	5–7	1.14–02	2.34–02	3.80+00	–0.932	C+	*LS*
			9891.65	9894.36	97488.378	107595.147	3–5	1.08–02	2.63–02	2.57+00	–1.103	C+	*LS*
			9891.80	9894.52	97488.538	107595.147	7–7	1.43–03	2.09–03	4.77–01	–1.834	C+	*LS*
			9891.72	9894.43	97488.448	107595.147	5–5	2.00–03	2.93–03	4.77–01	–1.834	C+	*LS*
			9891.80	9894.52	97488.538	107595.147	7–5	5.63–05	5.90–05	1.35–02	–3.384	C	*LS*
120.	$2p^3(^4S^\circ)3d-2p^3(^2D^\circ)3p$	$^3D^\circ-^3D$	*6324.5*	*6326.3*	*97488.48*	*113295.53*	15–15	2.41–04	1.44–04	4.51–02	–2.664	B–	1
			6324.85	6326.60	97488.538	113294.816	7–7	2.14–04	1.28–04	1.87–02	–3.047	C	*LS*
			6324.80	6326.55	97488.448	113294.854	5–5	1.68–04	1.01–04	1.05–02	–3.299	C	*LS*
			6323.39	6325.13	97488.378	113298.320	3–3	1.81–04	1.08–04	6.77–03	–3.488	C–	*LS*
			6324.84	6326.58	97488.538	113294.854	7–5	3.76–05	1.61–05	2.35–03	–3.948	D	*LS*
			6323.41	6325.16	97488.448	113298.320	5–3	6.02–05	2.17–05	2.26–03	–3.965	D	*LS*
			6324.82	6326.56	97488.448	113294.816	5–7	2.68–05	2.25–05	2.35–03	–3.948	D	*LS*
			6324.77	6326.52	97488.378	113294.854	3–5	3.61–05	3.61–05	2.26–03	–3.965	D	*LS*
121.	$2p^3(^4S^\circ)4p-2p^3(^4S^\circ)5s$	$^5P-^5S^\circ$	*33075*	*3022.64* cm^{-1}	*99094.06*	102116.698	15–5	5.51–02	3.01–01	4.92+02	0.655	B	1
			33083.2	3021.861 cm^{-1}	99094.837	102116.698	7–5	2.57–02	3.01–01	2.30+02	0.324	C+	*LS*
			33070.1	3023.057 cm^{-1}	99093.641	102116.698	5–5	1.84–02	3.01–01	1.64+02	0.178	C+	*LS*
			33062.7	3023.730 cm^{-1}	99092.968	102116.698	3–5	1.10–02	3.01–01	9.84+01	–0.044	C+	*LS*
122.	$2p^3(^4S^\circ)4p-2p^3(^4S^\circ)4d$	$^5P-^5D^\circ$	*26507*	*3771.50* cm^{-1}	*99094.06*	*102865.56*	15–25	6.46–02	1.13+00	1.48+03	1.231	A	1,2
			26513.3	3770.669 cm^{-1}	99094.837	102865.506	7–9	6.44–02	8.73–01	5.33+02	0.786	A	2,*LS*
			26504.6	3771.906 cm^{-1}	99093.641	102865.547	5–7	4.30–02	6.33–01	2.76+02	0.501	A	2,*LS*
			26499.4	3772.638 cm^{-1}	99092.968	102865.606	3–5	2.26–02	3.96–01	1.04+02	0.075	A	2,*LS*
			26513.0	3770.710 cm^{-1}	99094.837	102865.547	7–7	2.15–02	2.26–01	1.38+02	0.200	A	2,*LS*
			26504.1	3771.965 cm^{-1}	99093.641	102865.606	5–5	3.76–02	3.96–01	1.73+02	0.297	A	2,*LS*
			26499.1	3772.687 cm^{-1}	99092.968	102865.655	3–3	4.84–02	5.09–01	1.33+02	0.184	A	2,*LS*
			26512.6	3770.769 cm^{-1}	99094.837	102865.606	7–5	4.29–03	3.23–02	1.97+01	–0.646	A	2,*LS*
			26503.8	3772.014 cm^{-1}	99093.641	102865.655	5–3	1.61–02	1.02–01	4.44+01	–0.293	A	2,*LS*
			26498.9	3772.711 cm^{-1}	99092.968	102865.679	3–1	6.45–02	2.26–01	5.92+01	–0.168	A	2,*LS*

O I: Allowed Transitions — Continued

No.	Transition Array	Mult.	λ_{air} (Å)	λ_{vac} (Å) or σ (cm^{-1})*	E_i (cm^{-1})	E_k (cm^{-1})	g_i-g_k	A_{ki} (10^8 s^{-1})	f_{ik}	S (at. u.)	log gf	Acc.	Source
123.		$^5P-^3D^\circ$											
			26207.0	3814.733 cm^{-1}	99093.641	102908.374	5–7	8.47–06	1.22–04	5.26–02	–3.215	C	2
			26201.9	3815.475 cm^{-1}	99092.968	102908.443	3–5	9.23–07	1.58–05	4.10–03	–4.323	C	2
124.	$2p^3(^4S^\circ)4p-$ $2p^3(^4S^\circ)6s$	$^5P-^5S^\circ$	*16872*	*5925.25* cm^{-1}	*99094.06*	105019.307	15–5	1.99–02	2.84–02	2.36+01	–0.371	B	1
			16874.5	5924.470 cm^{-1}	99094.837	105019.307	7–5	9.30–03	2.83–02	1.10+01	–0.702	C+	*LS*
			16871.1	5925.666 cm^{-1}	99093.641	105019.307	5–5	6.64–03	2.84–02	7.87+00	–0.848	C+	*LS*
			16869.2	5926.339 cm^{-1}	99092.968	105019.307	3–5	3.99–03	2.84–02	4.72+00	–1.070	C+	*LS*
125.	$2p^3(^4S^\circ)4p-$ $2p^3(^4S^\circ)5d$	$^5P-^5D^\circ$	*15891*	*6291.33* cm^{-1}	*99094.06*	*105385.39*	15–25	1.93–02	1.22–01	9.55+01	0.261	B	1
			15892.6	6290.517 cm^{-1}	99094.837	105385.354	7–9	1.93–02	9.39–02	3.44+01	–0.182	C+	*LS*
			15889.5	6291.736 cm^{-1}	99093.641	105385.377	5–7	1.29–02	6.81–02	1.78+01	–0.468	C+	*LS*
			15887.7	6292.441 cm^{-1}	99092.968	105385.409	3–5	6.75–03	4.26–02	6.68+00	–0.893	C+	*LS*
			15892.5	6290.540 cm^{-1}	99094.837	105385.377	7–7	6.43–03	2.43–02	8.91+00	–0.769	C+	*LS*
			15889.4	6291.768 cm^{-1}	99093.641	105385.409	5–5	1.13–02	4.26–02	1.11+01	–0.672	C+	*LS*
			15887.7	6292.468 cm^{-1}	99092.968	105385.436	3–3	1.45–02	5.48–02	8.59+00	–0.784	C+	*LS*
			15892.5	6290.572 cm^{-1}	99094.837	105385.409	7–5	1.29–03	3.48–03	1.27+00	–1.614	C+	*LS*
			15889.4	6291.795 cm^{-1}	99093.641	105385.436	5–3	4.82–03	1.10–02	2.86+00	–1.262	C+	*LS*
			15887.6	6292.481 cm^{-1}	99092.968	105385.449	3–1	1.93–02	2.43–02	3.82+00	–1.137	C+	*LS*
126.	$2p^3(^4S^\circ)4p-$ $2p^3(^4S^\circ)7s$	$^5P-^5S^\circ$	*13417*	*7451.29* cm^{-1}	*99094.06*	106545.354	15–5	1.04–02	9.33–03	6.18+00	–0.854	B	1
			13418.2	7450.517 cm^{-1}	99094.837	106545.354	7–5	4.84–03	9.33–03	2.88+00	–1.185	C+	*LS*
			13416.1	7451.713 cm^{-1}	99093.641	106545.354	5–5	3.46–03	9.33–03	2.06+00	–1.331	C+	*LS*
			13414.9	7452.386 cm^{-1}	99092.968	106545.354	3–5	2.08–03	9.33–03	1.24+00	–1.553	C+	*LS*
127.	$2p^3(^4S^\circ)4p-$ $2p^3(^4S^\circ)6d$	$^5P-^5D^\circ$	*13056*	*7657.40* cm^{-1}	*99094.06*	*106751.46*	15–25	8.65–03	3.68–02	2.38+01	–0.258	B	1
			13057.0	7656.610 cm^{-1}	99094.837	106751.447	7–9	8.65–03	2.84–02	8.55+00	–0.701	C+	*LS*
			13055.0	7657.817 cm^{-1}	99093.641	106751.458	5–7	5.77–03	2.06–02	4.43+00	–0.986	C+	*LS*
			13053.8	7658.506 cm^{-1}	99092.968	106751.474	3–5	3.03–03	1.29–02	1.66+00	–1.412	C+	*LS*
			13057.0	7656.621 cm^{-1}	99094.837	106751.458	7–7	2.88–03	7.37–03	2.22+00	–1.288	C+	*LS*
			13055.0	7657.833 cm^{-1}	99093.641	106751.474	5–5	5.05–03	1.29–02	2.77+00	–1.191	C+	*LS*
			13053.8	7658.519 cm^{-1}	99092.968	106751.487	3–3	6.49–03	1.66–02	2.14+00	–1.303	C+	*LS*
			13057.0	7656.637 cm^{-1}	99094.837	106751.474	7–5	5.77–04	1.05–03	3.17–01	–2.133	C+	*LS*
			13054.9	7657.846 cm^{-1}	99093.641	106751.487	5–3	2.16–03	3.32–03	7.13–01	–1.780	C+	*LS*
			13053.8	7658.526 cm^{-1}	99092.968	106751.494	3–1	8.65–03	7.37–03	9.50–01	–1.655	C+	*LS*
128.	$2p^3(^4S^\circ)4p-$ $2p^3(^4S^\circ)8s$	$^5P-^5S^\circ$	*11970*	*8351.98* cm^{-1}	*99094.06*	107446.036	15–5	6.18–03	4.43–03	2.62+00	–1.178	B	1
			11971.1	8351.199 cm^{-1}	99094.837	107446.036	7–5	2.89–03	4.43–03	1.22+00	–1.509	C+	*LS*
			11969.3	8352.395 cm^{-1}	99093.641	107446.036	5–5	2.06–03	4.43–03	8.73–01	–1.655	C+	*LS*
			11968.4	8353.068 cm^{-1}	99092.968	107446.036	3–5	1.24–03	4.43–03	5.24–01	–1.877	C+	*LS*
129.	$2p^3(^4S^\circ)4p-$ $2p^3(^4S^\circ)7d$	$^5P-^5D^\circ$	*11790*	*8479.43* cm^{-1}	*99094.06*	*107573.49*	15–25	4.70–03	1.63–02	9.51+00	–0.611	B	1
			11791.1	8478.639 cm^{-1}	99094.837	107573.476	7–9	4.70–03	1.26–02	3.42+00	–1.055	C+	*LS*
			11789.4	8479.843 cm^{-1}	99093.641	107573.484	5–7	3.13–03	9.14–03	1.77+00	–1.340	C+	*LS*
			11788.5	8480.527 cm^{-1}	99092.968	107573.495	3–5	1.65–03	5.72–03	6.65–01	–1.766	C+	*LS*
			11791.1	8478.647 cm^{-1}	99094.837	107573.484	7–7	1.57–03	3.27–03	8.87–01	–1.641	C+	*LS*
			11789.4	8479.854 cm^{-1}	99093.641	107573.495	5–5	2.74–03	5.72–03	1.11+00	–1.544	C+	*LS*
			11788.5	8480.536 cm^{-1}	99092.968	107573.504	3–3	3.53–03	7.35–03	8.56–01	–1.657	C+	*LS*
			11791.1	8478.658 cm^{-1}	99094.837	107573.495	7–5	3.13–04	4.66–04	1.27–01	–2.486	C+	*LS*
			11789.4	8479.863 cm^{-1}	99093.641	107573.504	5–3	1.18–03	1.47–03	2.85–01	–2.134	C+	*LS*
			18241.4	5480.540 cm^{-1}	99092.968	104573.508	3–1	1.27–03	2.11–03	3.80–01	–2.198	C+	*LS*

O I: Allowed Transitions — Continued

No.	Transition Array	Mult.	λ_{air} (Å)	λ_{vac} (Å) or σ (cm^{-1})*	E_i (cm^{-1})	E_k (cm^{-1})	g_i-g_k	A_{ki} (10^8 s^{-1})	f_{ik}	S (at. u.)	log gf	Acc.	Source
130.	$2p^3(^4S^o)4p-2p^3(^4S^o)8d$	$^5P-^5D^o$	*11093*	*9012.02* cm^{-1}	*99094.06*	*108106.08*	15–25	2.87–03	8.81–03	4.83+00	–0.879	B	1
			11094.2	9011.235 cm^{-1}	99094.837	108106.072	7–9	2.87–03	6.80–03	1.74+00	–1.323	C+	*LS*
			11092.7	9012.436 cm^{-1}	99093.641	108106.077	5–7	1.91–03	4.93–03	9.01–01	–1.608	C+	*LS*
			11091.9	9013.117 cm^{-1}	99092.968	108106.085	3–5	1.00–03	3.08–03	3.38–01	–2.034	C+	*LS*
			11094.2	9011.240 cm^{-1}	99094.837	108106.077	7–7	9.55–04	1.76–03	4.51–01	–1.909	C+	*LS*
			11092.7	9012.444 cm^{-1}	99093.641	108106.085	5–5	1.67–03	3.08–03	5.63–01	–1.812	C+	*LS*
			11091.9	9013.123 cm^{-1}	99092.968	108106.091	3–3	2.15–03	3.97–03	4.34–01	–1.925	C+	*LS*
			11094.2	9011.248 cm^{-1}	99094.837	108106.085	7–5	1.91–04	2.52–04	6.44–02	–2.754	C+	*LS*
			11092.7	9012.450 cm^{-1}	99093.641	108106.091	5–3	7.17–04	7.93–04	1.45–01	–2.402	C+	*LS*
			11091.9	9013.126 cm^{-1}	99092.968	108106.094	3–1	2.87–03	1.76–03	1.93–01	–2.277	C+	*LS*
131.	$2p^3(^4S^o)4p-2p^3(^4S^o)5s$	$^3P-^3S^o$	*36607*	*2730.95* cm^{-1}	*99681.05*	102411.995	9–3	4.82–02	3.23–01	3.50+02	0.463	B	1
			36607.4	2730.946 cm^{-1}	99681.049	102411.995	5–3	2.68–02	3.23–01	1.95+02	0.208	C+	*LS*
			36606.3	2731.027 cm^{-1}	99680.968	102411.995	3–3	1.61–02	3.23–01	1.17+02	–0.014	C+	*LS*
			36610.9	2730.686 cm^{-1}	99681.309	102411.995	1–3	5.35–03	3.23–01	3.89+01	–0.491	C+	*LS*
132.	$2p^3(^4S^o)4p-2p^3(^4S^o)4d$	$^3P-^5D^o$											
			31393.6	3184.498 cm^{-1}	99681.049	102865.547	5–7	6.97–06	1.44–04	7.45–02	–3.142	C	2
			31392.2	3184.638 cm^{-1}	99680.968	102865.606	3–5	3.25–07	8.01–06	2.48–03	–4.619	C	2
			31395.1	3184.346 cm^{-1}	99681.309	102865.655	1–3	6.27–07	2.78–05	2.87–03	–4.556	C	2
			31393.0	3184.557 cm^{-1}	99681.049	102865.606	5–5	7.24–07	1.07–05	5.53–03	–4.272	C	2
133.		$^3P-^3D^o$	*30977*	*3227.37* cm^{-1}	*99681.05*	*102908.42*	9–15	5.05–02	1.21+00	1.11+03	1.037	B+	1,2
			30977.0	3227.325 cm^{-1}	99681.049	102908.374	5–7	5.04–02	1.01+00	5.18+02	0.705	B+	2,*LS*
			30975.5	3227.475 cm^{-1}	99680.968	102908.443	3–5	3.78–02	9.06–01	2.77+02	0.434	B+	2,*LS*
			30978.4	3227.180 cm^{-1}	99681.309	102908.489	1–3	2.80–02	1.21+00	1.23+02	0.082	B+	2,*LS*
			30976.3	3227.394 cm^{-1}	99681.049	102908.443	5–5	1.26–02	1.81–01	9.24+01	–0.043	B+	2,*LS*
			30975.1	3227.521 cm^{-1}	99680.968	102908.489	3–3	2.10–02	3.02–01	9.24+01	–0.043	B+	2,*LS*
			30975.9	3227.440 cm^{-1}	99681.049	102908.489	5–3	1.40–03	1.21–02	6.16+00	–1.219	B+	2,*LS*
134.	$2p^3(^4S^o)4p-2p^3(^4S^o)6s$	$^3P-^3S^o$	*18229*	*5484.18* cm^{-1}	*99681.05*	105165.232	9–3	1.71–02	2.83–02	1.53+01	–0.594	B	1
			18229.3	5484.183 cm^{-1}	99681.049	105165.232	5–3	9.47–03	2.83–02	8.50+00	–0.849	C+	*LS*
			18229.0	5484.264 cm^{-1}	99680.968	105165.232	3–3	5.68–03	2.83–02	5.10+00	–1.071	C+	*LS*
			18230.1	5483.923 cm^{-1}	99681.309	105165.232	1–3	1.89–03	2.83–02	1.70+00	–1.548	C+	*LS*
135.	$2p^3(^4S^o)4p-2p^3(^4S^o)5d$	$^3P-^3D^o$	*17453*	*5727.96* cm^{-1}	*99681.05*	105409.008	9–15	1.18–02	8.98–02	4.65+01	–0.092	B	1
			17453.5	5727.959 cm^{-1}	99681.049	105409.008	5–7	1.18–02	7.55–02	2.17+01	–0.423	C+	*LS*
			17453.2	5728.040 cm^{-1}	99680.968	105409.008	3–5	8.85–03	6.74–02	1.16+01	–0.694	C+	*LS*
			17454.3	5727.699 cm^{-1}	99681.309	105409.008	1–3	6.56–03	8.98–02	5.16+00	–1.047	C+	*LS*
			17453.5	5727.959 cm^{-1}	99681.049	105409.008	5–5	2.95–03	1.35–02	3.87+00	–1.171	C+	*LS*
			17453.2	5728.040 cm^{-1}	99680.968	105409.008	3–3	4.92–03	2.25–02	3.87+00	–1.171	C+	*LS*
			17453.5	5727.959 cm^{-1}	99681.049	105409.008	5–3	3.28–04	8.98–04	2.58–01	–2.348	C+	*LS*
136.	$2p^3(^4S^o)4p-2p^3(^4S^o)7s$	$^3P-^3S^o$	*14391*	*6946.88* cm^{-1}	*99681.05*	106627.934	9–3	8.88–03	9.19–03	3.92+00	–1.083	B	1
			14391.0	6946.885 cm^{-1}	99681.049	106627.934	5–3	4.93–03	9.19–03	2.18+00	–1.338	C+	*LS*
			14390.8	6946.966 cm^{-1}	99680.968	106627.934	3–3	2.96–03	9.19–03	1.31+00	–1.560	C+	*LS*
			14391.5	6946.625 cm^{-1}	99681.309	106627.934	1–3	9.86–04	9.19–03	4.35–01	–2.037	C+	*LS*
137.	$2p^3(^4S^o)4p-2p^3(^4S^o)6d$	$^3P-^3D^o$	*14111*	*7084.75* cm^{-1}	*99681.05*	106765.803	9–15	4.65–03	2.32–02	9.68+00	–0.681	B	1
			14111.0	7084.754 cm^{-1}	99681.049	106765.803	5–7	4.65–03	1.94–02	4.52+00	–1.012	C+	*LS*
			14110.8	7084.835 cm^{-1}	99680.968	106765.803	3–5	3.49–03	1.74–02	2.42+00	–1.283	C+	*LS*
			14111.5	7084.494 cm^{-1}	99681.309	106765.803	1–3	2.58–03	2.31–02	1.08+00	–1.635	C+	*LS*
			14111.0	7084.754 cm^{-1}	99681.049	106765.803	5–5	1.16–03	3.47–03	8.07–01	–1.760	C+	*LS*
			14110.8	7084.835 cm^{-1}	99680.968	106765.803	3–3	1.94–03	5.79–03	8.07–01	–1.760	C+	*LS*
			14111.0	7084.754 cm^{-1}	99681.049	106765.803	5–3	1.29–04	2.32–04	5.38–02	–2.936	C+	*LS*

O I: Allowed Transitions — Continued

No.	Transition Array	Mult.	λ_{air} (Å)	λ_{vac} (Å) or σ (cm^{-1})*	E_i (cm^{-1})	E_k (cm^{-1})	g_i-g_k	A_{ki} (10^8 s^{-1})	f_{ik}	S (at. u.)	log gf	Acc.	Source
138.	$2p^3(^4S^o)4p-$ $2p^3(^4S^o)8s$	$^3P-^3S^o$	*12790*	*7816.17* cm^{-1}	*99681.05*	107497.224	9−3	5.30−03	4.34−03	1.64+00	−1.409	B	1
			12790.5	7816.175 cm^{-1}	99681.049	107497.224	5−3	2.95−03	4.34−03	9.13−01	−1.664	C+	LS
			12790.4	7816.256 cm^{-1}	99680.968	107497.224	3−3	1.77−03	4.34−03	5.48−01	−1.886	C+	LS
			12790.9	7815.915 cm^{-1}	99681.309	107497.224	1−3	5.89−04	4.34−03	1.83−01	−2.363	C+	LS
139.	$2p^3(^4S^o)4p-$ $2p^3(^4S^o)7d$	$^3P-^3D^o$	*12652*	*7901.73* cm^{-1}	*99681.05*	107582.777	9−15	2.34−03	9.36−03	3.51+00	−1.074	B	1
			12652.0	7901.728 cm^{-1}	99681.049	107582.777	5−7	2.34−03	7.86−03	1.64+00	−1.405	C+	LS
			12651.9	7901.809 cm^{-1}	99680.968	107582.777	3−5	1.76−03	7.02−03	8.77−01	−1.677	C+	LS
			12652.4	7901.468 cm^{-1}	99681.309	107582.777	1−3	1.30−03	9.36−03	3.90−01	−2.029	C+	LS
			12652.0	7901.728 cm^{-1}	99681.049	107582.777	5−5	5.85−04	1.40−03	2.92−01	−2.154	C+	LS
			12651.9	7901.809 cm^{-1}	99680.968	107582.777	3−3	9.75−04	2.34−03	2.92−01	−2.154	C+	LS
			12652.0	7901.728 cm^{-1}	99681.049	107582.777	5−3	6.50−05	9.36−05	1.95−02	−3.330	C	LS
140.	$2p^3(^4S^o)4p-$ $2p^3(^4S^o)8d$	$^3P-^3D^o$	*11855*	*8432.95* cm^{-1}	*99681.05*	108114.0	9−15	1.36−03	4.76−03	1.67+00	−1.368	B	1
			11855.0	8432.951 cm^{-1}	99681.049	108114.0	5−7	1.36−03	4.00−03	7.81−01	−1.699	C+	LS
			11854.9	8433.032 cm^{-1}	99680.968	108114.0	3−5	1.02−03	3.57−03	4.18−01	−1.970	C+	LS
			11855.4	8432.691 cm^{-1}	99681.309	108114.0	1−3	7.54−04	4.76−03	1.86−01	−2.322	C+	LS
			11855.0	8432.951 cm^{-1}	99681.049	108114.0	5−5	3.39−04	7.15−04	1.39−01	−2.447	C+	LS
			11854.9	8433.032 cm^{-1}	99680.968	108114.0	3−3	5.65−04	1.19−03	1.39−01	−2.447	C+	LS
			11855.0	8432.951 cm^{-1}	99681.049	108114.0	5−3	3.77−05	4.76−05	9.30−03	−3.623	C−	LS
141.	$2p^3(^4S^o)4p-$ $2p^3(^2P^o)3s$	$^3P-^3P^o$	*7022.9*	*7024.8*	*99681.05*	*113916.28*	9−9	3.22−04	2.38−04	4.95−02	−2.669	B−	1
			7025.51	7027.45	99681.049	113910.957	5−5	2.41−04	1.78−04	2.06−02	−3.050	C	LS
			7020.33	7022.26	99680.968	113921.391	3−3	8.05−05	5.95−05	4.13−03	−3.748	C−	LS
			7020.37	7022.30	99681.049	113921.391	5−3	1.34−04	5.95−05	6.88−03	−3.527	C−	LS
			7017.30	7019.24	99680.968	113927.534	3−1	3.22−04	7.94−05	5.50−03	−3.623	C−	LS
			7025.47	7027.41	99680.968	113910.957	3−5	8.03−05	9.91−05	6.88−03	−3.527	C−	LS
			7020.50	7022.43	99681.309	113921.391	1−3	1.07−04	2.38−04	5.50−03	−3.623	C−	LS
142.	$2p^3(^4S^o)4p-$ $2p^3(^2D^o)3d$	$^3P-^3P^o$	*4228.0*	*4229.2*	*99681.05*	*123326.42*	9−9	5.40−03	1.45−03	1.81−01	−1.885	B	1
			4233.27	4234.47	99681.049	123296.777	5−5	4.04−03	1.08−03	7.55−02	−2.266	C+	LS
			4222.76	4223.95	99680.968	123355.512	3−3	1.36−03	3.62−04	1.51−02	−2.964	C	LS
			4222.77	4223.96	99681.049	123355.512	5−3	2.26−03	3.62−04	2.52−02	−2.742	C	LS
			4217.09	4218.28	99680.968	123387.339	3−1	5.44−03	4.84−04	2.01−02	−2.838	C	LS
			4233.26	4234.45	99680.968	123296.777	3−5	1.35−03	6.02−04	2.52−02	−2.743	C	LS
			4222.82	4224.01	99681.309	123355.512	1−3	1.81−03	1.45−03	2.01−02	−2.839	C	LS
143.		$^3P-^3D^o$	*4068.4*	*4069.5*	*99681.05*	*124254.0*	9−15	2.33−03	9.63−04	1.16−01	−2.062	B−	2
			4069.51	4070.66	99681.049	124247.1	5−7	2.35−03	8.15−04	5.46−02	−2.390	C+	2
			4067.74	4068.89	99680.968	124257.7	3−5	1.74−03	7.19−04	2.89−02	−2.666	C+	2
			4066.75	4067.90	99681.309	124264.0	1−3	1.28−03	9.50−04	1.27−02	−3.022	C	2
			4067.75	4068.90	99681.049	124257.7	5−5	5.84−04	1.45−04	9.70−03	−3.140	C	2
			4066.70	4067.85	99680.968	124264.0	3−3	9.60−04	2.38−04	9.56−03	−3.146	C	2
			4066.71	4067.86	99681.049	124264.0	5−3	6.43−05	9.57−06	6.40−04	−4.320	D	2
144.		$^3P-^3S^o$	*4054.8*	*4055.9*	*99681.05*	124336.3	9−3	8.19−03	6.73−04	8.08−02	−2.218	B−	2
			4054.79	4055.93	99681.049	124336.3	5−3	4.56−03	6.74−04	4.50−02	−2.472	B−	2
			4054.77	4055.92	99680.968	124336.3	3−3	2.72−03	6.71−04	2.69−02	−2.696	C+	2
			4054.83	4055.97	99681.309	124336.3	1−3	9.05−04	6.69−04	8.93−03	−3.174	C	2
145.	$2p^3(^4S^o)4p-$ $2p^3(^2D^o)4d$	$^3P-^3D^o$											
			3331.13	3332.08	99681.049	129692.3	5−7	2.81−04	6.54−05	3.58−03	−3.486	D	2
			3330.54	3331.50	99680.968	129697.5	3−5	2.01−04	5.57−05	1.83−03	−3.777	D	2
			3329.97	3330.92	99681.309	129703.0	1−3	1.45−04	7.24−05	7.94−04	−4.140	D	2
			3330.55	3331.51	99681.049	129697.5	5−5	6.72−05	1.12−05	6.13−04	−4.253	D	2
			3329.93	3330.89	99680.968	129703.0	3−3	1.10−04	1.84−05	6.04−04	−4.259	D	2

O I: Allowed Transitions — Continued

No.	Transition Array	Mult.	λ_{air} (Å)	λ_{vac} (Å) or σ (cm^{-1})*	E_i (cm^{-1})	E_k (cm^{-1})	g_i-g_k	A_{ki} (10^8 s^{-1})	f_{ik}	S (at. u.)	log gf	Acc.	Source
146.		$^3P-^3S^o$	*3326.2*	*3327.2*	*99681.05*	129736.6	9–3	6.53−03	3.61−04	3.56−02	−2.488	B−	2
			3326.22	3327.17	99681.049	129736.6	5–3	3.64−03	3.62−04	1.98−02	−2.743	C+	2
			3326.21	3327.16	99680.968	129736.6	3–3	2.17−03	3.60−04	1.18−02	−2.966	C	2
			3326.24	3327.20	99681.309	129736.6	1–3	7.22−04	3.59−04	3.93−03	−3.445	C	2
147.		$^3P-^3P^o$	*3300.1*	*3301.0*	*99681.05*	*129974.75*	9–9	1.43−03	2.33−04	2.28−02	−2.678	B−	1
			3300.58	3301.53	99681.049	129970.000	5–5	1.07−03	1.75−04	9.50−03	−3.058	C−	*LS*
			3299.55	3300.50	99680.968	129979.384	3–3	3.57−04	5.83−05	1.90−03	−3.757	D	*LS*
			3299.56	3300.51	99681.049	129979.384	5–3	5.95−04	5.83−05	3.17−03	−3.535	C−	*LS*
			3298.99	3299.94	99680.968	129984.553	3–1	1.43−03	7.78−05	2.53−03	−3.632	D	*LS*
			3300.57	3301.53	99680.968	129970.000	3–5	3.57−04	9.71−05	3.17−03	−3.535	C−	*LS*
			3299.59	3300.54	99681.309	129979.384	1–3	4.76−04	2.33−04	2.53−03	−3.632	D	*LS*
148.	$2p^3(^4S^o)4p-$ $2p^3(^2P^o)3d$	$^3P-^3D^o$											
			2611.45	2612.23	99681.049	137962.5	5–7	1.48−03	2.11−04	9.08−03	−2.976	C	2
			2611.45	2612.23	99680.968	137962.5	3–5	1.08−03	1.84−04	4.76−03	−3.257	D	2
			2611.47	2612.25	99681.309	137962.5	1–3	9.00−04	2.76−04	2.37−03	−3.559	D	2
			2611.45	2612.23	99681.049	137962.5	5–5	7.10−04	7.26−05	3.12−03	−3.440	D	2
			2611.45	2612.23	99680.968	137962.5	3–3	9.57−04	9.78−05	2.52−03	−3.532	D	2
149.	$2p^3(^4S^o)4p-$ $2p^3(^2P^o)4d$	$^3P-^3D^o$											
			2288.55	2289.25	99681.049	143363.4	5–7	1.67−03	1.83−04	6.91−03	−3.037	D	2
			2288.54	2289.25	99680.968	143363.4	3–5	1.35−03	1.76−04	3.98−03	−3.277	D	2
			2288.56	2289.27	99681.309	143363.4	1–3	8.29−04	1.95−04	1.47−03	−3.709	D	2
			2288.54	2289.25	99680.968	143363.4	3–3	2.48−04	1.95−05	4.40−04	−4.233	D	2
150.	$2p^3(^2D^o)3s-$ $2p^3(^4S^o)4f$	$^3D^o-^3F$		*1824.89* cm^{-1}	*101143.45*	102968.343	15–21	1.56−04	9.84−03	2.66+01	−0.831	B	1
				1832.936 cm^{-1}	101135.407	102968.343	7–9	1.58−04	9.08−03	1.14+01	−1.197	C+	*LS*
				1820.817 cm^{-1}	101147.526	102968.343	5–7	1.38−04	8.73−03	7.89+00	−1.360	C+	*LS*
				1812.921 cm^{-1}	101155.422	102968.343	3–5	1.29−04	9.78−03	5.33+00	−1.533	C+	*LS*
				1832.936 cm^{-1}	101135.407	102968.343	7–7	1.76−05	7.87−04	9.89−01	−2.259	C+	*LS*
				1820.817 cm^{-1}	101147.526	102968.343	5–5	2.42−05	1.09−03	9.89−01	−2.262	C+	*LS*
				1832.936 cm^{-1}	101135.407	102968.343	7–5	6.97−07	2.22−05	2.79−02	−3.809	C	*LS*
151.	$2p^3(^2D^o)3s-$ $2p^3(^4S^o)5p$	$^3D^o-^3P$	*36666*	*2726.57* cm^{-1}	*101143.45*	*103870.02*	15–9	2.59−04	3.13−03	5.67+00	−1.328	B	1
			36559.0	2734.561 cm^{-1}	101135.407	103869.968	7–5	2.20−04	3.14−03	2.65+00	−1.658	C+	*LS*
			36720.9	2722.502 cm^{-1}	101147.526	103870.028	5–3	1.93−04	2.35−03	1.42+00	−1.931	C+	*LS*
			36824.7	2714.830 cm^{-1}	101155.422	103870.252	3–1	2.56−04	1.73−03	6.30−01	−2.284	C+	*LS*
			36721.7	2722.442 cm^{-1}	101147.526	103869.968	5–5	3.87−05	7.82−04	4.73−01	−2.408	C+	*LS*
			36827.7	2714.606 cm^{-1}	101155.422	103870.028	3–3	6.39−05	1.30−03	4.73−01	−2.409	C+	*LS*
			36720.9	2722.502 cm^{-1}	101147.526	103870.028	5–3	4.30−06	5.22−05	3.15−02	−3.584	C+	*LS*
152.	$2p^3(^2D^o)3s-$ $2p^3(^4S^o)5f$	$^3D^o-^3F$	*23259*	*4298.27* cm^{-1}	*101143.45*	105441.724	15–21	4.34−04	4.93−03	5.66+00	−1.131	B	1
			23215.4	4306.317 cm^{-1}	101135.407	105441.724	7–9	4.37−04	4.54−03	2.43+00	−1.498	C+	*LS*
			23280.9	4294.198 cm^{-1}	101147.526	105441.724	5–7	3.85−04	4.38−03	1.68+00	−1.660	C+	*LS*
			23323.8	4286.302 cm^{-1}	101155.422	105441.724	3–5	3.62−04	4.92−03	1.13+00	−1.831	C+	*LS*
			23215.4	4306.317 cm^{-1}	101135.407	105441.724	7–7	4.87−05	3.93−04	2.10−01	−2.560	C+	*LS*
			23280.9	4294.198 cm^{-1}	101147.526	105441.724	5–5	6.76−05	5.49−04	2.10−01	−2.561	C+	*LS*
			23215.4	4306.317 cm^{-1}	101135.407	105441.724	7–5	1.92−06	1.11−05	5.93−03	−4.110	C−	*LS*

O I: Allowed Transitions — Continued

No.	Transition Array	Mult.	λ_{air} (Å)	λ_{vac} (Å) or σ (cm^{-1})*	E_i (cm^{-1})	E_k (cm^{-1})	g_i-g_k	A_{ki} (10^8 s^{-1})	f_{ik}	S (at. u.)	log gf	Acc.	Source
153.	$2p^3(^2D^o)3s-2p^3(^4S^o)6f$	$^3D^o-^3F$	*17720*	*5641.75* cm^{-1}	*101143.45*	106785.201	15–21	2.51–04	1.65–03	1.45+00	–1.606	B	1
			17694.9	5649.794 cm^{-1}	101135.407	106785.201	7–9	2.52–04	1.52–03	6.20–01	–1.973	C+	*LS*
			17733.0	5637.675 cm^{-1}	101147.526	106785.201	5–7	2.22–04	1.47–03	4.28–01	–2.134	C+	*LS*
			17757.8	5629.779 cm^{-1}	101155.422	106785.201	3–5	2.09–04	1.65–03	2.89–01	–2.306	C+	*LS*
			17694.9	5649.794 cm^{-1}	101135.407	106785.201	7–7	2.81–05	1.32–04	5.37–02	–3.035	C+	*LS*
			17733.0	5637.675 cm^{-1}	101147.526	106785.201	5–5	3.90–05	1.84–04	5.37–02	–3.036	C+	*LS*
			17694.9	5649.794 cm^{-1}	101135.407	106785.201	7–5	1.11–06	3.72–06	1.52–03	–4.585	D	*LS*
154.	$2p^3(^2D^o)3s-2p^3(^4S^o)7f$	$^3D^o-^3F$	*15496*	*6451.70* cm^{-1}	*101143.45*	107595.147	15–21	1.54–04	7.78–04	5.96–01	–1.933	B	1
			15476.3	6459.740 cm^{-1}	101135.407	107595.147	7–9	1.55–04	7.16–04	2.55–01	–2.300	C+	*LS*
			15505.4	6447.621 cm^{-1}	101147.526	107595.147	5–7	1.37–04	6.91–04	1.76–01	–2.461	C+	*LS*
			15524.4	6439.725 cm^{-1}	101155.422	107595.147	3–5	1.29–04	7.77–04	1.19–01	–2.633	C+	*LS*
			15476.3	6459.740 cm^{-1}	101135.407	107595.147	7–7	1.73–05	6.20–05	2.21–02	–3.362	C	*LS*
			15505.4	6447.621 cm^{-1}	101147.526	107595.147	5–5	2.40–05	8.67–05	2.21–02	–3.363	C	*LS*
			15476.3	6459.740 cm^{-1}	101135.407	107595.147	7–5	6.82–07	1.75–06	6.24–04	–4.912	D–	*LS*
155.	$2p^3(^2D^o)3s-2p^3(^2D^o)3p$	$^3D^o-^3D$	*8226.8*	*8229.0*	*101143.45*	*113295.53*	15–15	3.25–01	3.30–01	1.34+02	0.695	B	1
			8221.82	8224.08	101135.407	113294.816	7–7	2.89–01	2.93–01	5.56+01	0.313	C+	*LS*
			8230.00	8232.26	101147.526	113294.854	5–5	2.26–01	2.30–01	3.11+01	0.060	C+	*LS*
			8233.00	8235.27	101155.422	113298.320	3–3	2.43–01	2.47–01	2.01+01	–0.130	C+	*LS*
			8221.80	8224.06	101135.407	113294.854	7–5	5.08–02	3.68–02	6.97+00	–0.589	C+	*LS*
			8227.65	8229.91	101147.526	113298.320	5–3	8.13–02	4.95–02	6.70+00	–0.607	C+	*LS*
			8230.03	8232.29	101147.526	113294.816	5–7	3.62–02	5.15–02	6.97+00	–0.590	C+	*LS*
			8235.35	8237.62	101155.422	113294.854	3–5	4.86–02	8.24–02	6.70+00	–0.607	C+	*LS*
156.	$2p^3(^2D^o)3s-2p^3(^2P^o)3p$	$^3D^o-^3D$	*3824.0*	*3825.1*	*101143.45*	*127286.34*	15–15	7.46–03	1.64–03	3.09–01	–1.610	B	1
			3823.41	3824.50	101135.407	127282.626	7–7	6.63–03	1.45–03	1.28–01	–1.992	C+	*LS*
			3824.35	3825.43	101147.526	127288.349	5–5	5.19–03	1.14–03	7.17–02	–2.245	C+	*LS*
			3825.02	3826.11	101155.422	127291.638	3–3	5.59–03	1.23–03	4.63–02	–2.434	C+	*LS*
			3822.58	3823.66	101135.407	127288.349	7–5	1.17–03	1.82–04	1.61–02	–2.894	C	*LS*
			3823.87	3824.95	101147.526	127291.638	5–3	1.87–03	2.45–04	1.54–02	–2.911	C	*LS*
			3825.19	3826.27	101147.526	127282.626	5–7	8.31–04	2.55–04	1.61–02	–2.894	C	*LS*
			3825.50	3826.59	101155.422	127288.349	3–5	1.12–03	4.09–04	1.54–02	–2.911	C	*LS*
157.	$2p^3(^2D^o)3s-2p^3(^2D^o)3p$	$^1D^o-^1P$	9482.89	9485.49	102662.026	113204.445	5–3	2.34–01	1.89–01	2.96+01	–0.023	B	1
158.		$^1D^o-^1F$	8820.42	8822.84	102662.026	113996.239	5–7	2.93–01	4.79–01	6.95+01	0.379	B	1
159.		$^1D^o-^1D$	7156.70	7158.67	102662.026	116631.094	5–5	5.05–01	3.88–01	4.57+01	0.288	B	1
160.	$2p^3(^2D^o)3s-2p^3(^2P^o)3p$	$^1D^o-^1P$	3997.95	3999.08	102662.026	127667.754	5–3	2.41–02	3.46–03	2.28–01	–1.762	B	1
161.		$^1D^o-^1D$	3855.01	3856.11	102662.026	128594.916	5–5	1.63–02	3.63–03	2.30–01	–1.741	B	1

O I: Allowed Transitions — Continued

No.	Transition Array	Mult.	λ_{air} (Å)	λ_{vac} (Å) or σ (cm^{-1})*	E_i (cm^{-1})	E_k (cm^{-1})	g_i-g_k	A_{ki} ($10^8 s^{-1}$)	f_{ik}	S (at. u.)	log gf	Acc.	Source
162.	$2p^3(^4S^\circ)4d-$ $2p^3(^4S^\circ)4f$	$^5D^\circ-^5F$		*102.69* cm^{-1}	*102865.56*	102968.249	25–35	2.40–06	4.77–02	3.82+03	0.077	B	1
				102.743 cm^{-1}	102865.506	102968.249	9–11	2.40–06	4.17–02	1.20+03	–0.426	C+	*LS*
				102.702 cm^{-1}	102865.547	102968.249	7–9	2.00–06	3.65–02	8.19+02	–0.592	C+	*LS*
				102.643 cm^{-1}	102865.606	102968.249	5–7	1.64–06	3.27–02	5.24+02	–0.786	C+	*LS*
				102.594 cm^{-1}	102865.655	102968.249	3–5	1.34–06	3.18–02	3.06+02	–1.021	C+	*LS*
				102.570 cm^{-1}	102865.679	102968.249	1–3	1.12–06	4.77–02	1.53+02	–1.322	C+	*LS*
				102.743 cm^{-1}	102865.506	102968.249	9–9	4.00–07	5.68–03	1.64+02	–1.291	C+	*LS*
				102.702 cm^{-1}	102865.547	102968.249	7–7	7.20–07	1.02–02	2.29+02	–1.145	C+	*LS*
				102.643 cm^{-1}	102865.606	102968.249	5–5	9.58–07	1.36–02	2.18+02	–1.167	C+	*LS*
				102.594 cm^{-1}	102865.655	102968.249	3–3	1.12–06	1.59–02	1.53+02	–1.322	C+	*LS*
				102.743 cm^{-1}	102865.506	102968.249	9–7	3.43–08	3.79–04	1.09+01	–2.467	C+	*LS*
				102.702 cm^{-1}	102865.547	102968.249	7–5	9.60–08	9.74–04	2.18+01	–2.166	C+	*LS*
				102.643 cm^{-1}	102865.606	102968.249	5–3	1.60–07	1.36–03	2.18+01	–2.167	C+	*LS*
163.	$2p^3(^4S^\circ)4d-$ $2p^3(^4S^\circ)5p$	$^5D^\circ-^5P$		*760.71* cm^{-1}	*102865.56*	*103626.27*	25–15	2.02–03	3.13–01	3.39+03	0.894	B	1
				761.105 cm^{-1}	102865.506	103626.611	9–7	1.56–03	3.14–01	1.22+03	0.451	C+	*LS*
				760.564 cm^{-1}	102865.547	103626.111	7–5	1.13–03	2.09–01	6.33+02	0.165	C+	*LS*
				760.148 cm^{-1}	102865.606	103625.754	5–3	7.04–04	1.10–01	2.37+02	–0.261	C+	*LS*
				761.064 cm^{-1}	102865.547	103626.611	7–7	4.04–04	1.05–01	3.16+02	–0.136	C+	*LS*
				760.505 cm^{-1}	102865.606	103626.111	5–5	7.05–04	1.83–01	3.95+02	–0.039	C+	*LS*
				760.099 cm^{-1}	102865.655	103625.754	3–3	9.06–04	2.35–01	3.05+02	–0.152	C+	*LS*
				761.005 cm^{-1}	102865.606	103626.611	5–7	5.77–05	2.09–02	4.52+01	–0.981	C+	*LS*
				760.456 cm^{-1}	102865.655	103626.111	3–5	1.81–04	7.83–02	1.02+02	–0.629	C+	*LS*
				760.075 cm^{-1}	102865.679	103625.754	1–3	4.02–04	3.13–01	1.36+02	–0.504	C+	*LS*
164.	$2p^3(^4S^\circ)4d-$ $2p^3(^4S^\circ)5f$	$^5D^\circ-^5F$	*38808*	*2576.09* cm^{-1}	*102865.56*	105441.645	25–35	2.68–02	8.48–01	2.71+03	1.326	B	1
			38807.2	2576.139 cm^{-1}	102865.506	105441.645	9–11	2.68–02	7.40–01	8.51+02	0.824	C+	*LS*
			38807.8	2576.098 cm^{-1}	102865.547	105441.645	7–9	2.24–02	6.49–01	5.80+02	0.657	C+	*LS*
			38808.7	2576.039 cm^{-1}	102865.606	105441.645	5–7	1.84–02	5.81–01	3.71+02	0.464	C+	*LS*
			38809.4	2575.990 cm^{-1}	102865.655	105441.645	3–5	1.50–02	5.65–01	2.17+02	0.229	C+	*LS*
			38809.8	2575.966 cm^{-1}	102865.679	105441.645	1–3	1.25–02	8.48–01	1.08+02	–0.072	C+	*LS*
			38807.2	2576.139 cm^{-1}	102865.506	105441.645	9–9	4.47–03	1.01–01	1.16+02	–0.042	C+	*LS*
			38807.8	2576.098 cm^{-1}	102865.547	105441.645	7–7	8.05–03	1.82–01	1.63+02	0.105	C+	*LS*
			38808.7	2576.039 cm^{-1}	102865.606	105441.645	5–5	1.07–02	2.42–01	1.55+02	0.083	C+	*LS*
			38809.4	2575.990 cm^{-1}	102865.655	105441.645	3–3	1.25–02	2.83–01	1.08+02	–0.072	C+	*LS*
			38807.2	2576.139 cm^{-1}	102865.506	105441.645	9–7	3.83–04	6.73–03	7.74+00	–1.218	C+	*LS*
			38807.8	2576.098 cm^{-1}	102865.547	105441.645	7–5	1.07–03	1.73–02	1.55+01	–0.917	C+	*LS*
			38808.7	2576.039 cm^{-1}	102865.606	105441.645	5–3	1.79–03	2.42–02	1.55+01	–0.917	C+	*LS*
165.	$2p^3(^4S^\circ)4d-$ $2p^3(^4S^\circ)6p$	$^5D^\circ-^5P$	*34201*	*2923.12* cm^{-1}	*102865.56*	*105788.68*	25–15	1.35–04	1.42–03	4.01+00	–1.449	B	1
			34198.0	2923.350 cm^{-1}	102865.506	105788.856	9–7	1.04–04	1.42–03	1.44+00	–1.893	C+	*LS*
			34201.5	2923.048 cm^{-1}	102865.547	105788.595	7–5	7.57–05	9.49–04	7.48–01	–2.178	C+	*LS*
			34204.1	2922.825 cm^{-1}	102865.606	105788.431	5–3	4.73–05	4.98–04	2.80–01	–2.604	C+	*LS*
			34198.5	2923.309 cm^{-1}	102865.547	105788.856	7–7	2.71–05	4.74–04	3.74–01	–2.479	C+	*LS*
			34202.2	2922.989 cm^{-1}	102865.606	105788.595	5–5	4.73–05	8.30–04	4.67–01	–2.382	C+	*LS*
			34204.7	2922.776 cm^{-1}	102865.655	105788.431	3–3	6.08–05	1.07–03	3.60–01	–2.495	C+	*LS*
			34199.2	2923.250 cm^{-1}	102865.606	105788.856	5–7	3.86–06	9.49–05	5.34–02	–3.324	C+	*LS*
			34202.8	2922.940 cm^{-1}	102865.655	105788.595	3–5	1.22–05	3.56–04	1.20–01	–2.972	C+	*LS*
			34205.0	2922.752 cm^{-1}	102865.679	105788.431	1–3	2.70–05	1.42–03	1.60–01	–2.847	C+	*LS*

O I: Allowed Transitions — Continued

No.	Transition Array	Mult.	λ_{air} (Å)	λ_{vac} (Å) or σ (cm^{-1})*	E_i (cm^{-1})	E_k (cm^{-1})	g_i-g_k	A_{ki} (10^8 s^{-1})	f_{ik}	S (at. u.)	log gf	Acc.	Source
166.	$2p^3(^4S^\circ)4d-2p^3(^4S^\circ)6f$	$^5D^\circ-^5F$	*25506*	*3919.60* cm^{-1}	*102865.56*	106785.160	25−35	1.37−02	1.87−01	3.92+02	0.669	B	1
			25505.5	3919.654 cm^{-1}	102865.506	106785.160	9−11	1.37−02	1.63−01	1.23+02	0.166	C+	*LS*
			25505.8	3919.613 cm^{-1}	102865.547	106785.160	7−9	1.14−02	1.43−01	8.39+01	−0.000	C+	*LS*
			25506.1	3919.554 cm^{-1}	102865.606	106785.160	5−7	9.37−03	1.28−01	5.37+01	−0.194	C+	*LS*
			25506.5	3919.505 cm^{-1}	102865.655	106785.160	3−5	7.65−03	1.24−01	3.13+01	−0.428	C+	*LS*
			25506.6	3919.481 cm^{-1}	102865.679	106785.160	1−3	6.37−03	1.87−01	1.57+01	−0.729	C+	*LS*
			25505.5	3919.654 cm^{-1}	102865.506	106785.160	9−9	2.28−03	2.22−02	1.68+01	−0.699	C+	*LS*
			25505.8	3919.613 cm^{-1}	102865.547	106785.160	7−7	4.10−03	4.00−02	2.35+01	−0.553	C+	*LS*
			25506.1	3919.554 cm^{-1}	102865.606	106785.160	5−5	5.46−03	5.33−02	2.24+01	−0.574	C+	*LS*
			25506.5	3919.505 cm^{-1}	102865.655	106785.160	3−3	6.37−03	6.22−02	1.57+01	−0.729	C+	*LS*
			25505.5	3919.654 cm^{-1}	102865.506	106785.160	9−7	1.95−04	1.48−03	1.12+00	−1.875	C+	*LS*
			25505.8	3919.613 cm^{-1}	102865.547	106785.160	7−5	5.46−04	3.81−03	2.24+00	−1.574	C+	*LS*
			25506.1	3919.554 cm^{-1}	102865.606	106785.160	5−3	9.11−04	5.33−03	2.24+00	−1.574	C+	*LS*
167.	$2p^3(^4S^\circ)4d-2p^3(^4S^\circ)7f$	$^5D^\circ-^5F$	*21138*	*4729.58* cm^{-1}	*102865.56*	107595.140	25−35	7.88−03	7.39−02	1.29+02	0.267	B	1
			21137.5	4729.634 cm^{-1}	102865.506	107595.140	9−11	7.88−03	6.45−02	4.04+01	−0.236	C+	*LS*
			21137.7	4729.593 cm^{-1}	102865.547	107595.140	7−9	6.57−03	5.65−02	2.75+01	−0.402	C+	*LS*
			21138.0	4729.534 cm^{-1}	102865.606	107595.140	5−7	5.40−03	5.07−02	1.76+01	−0.596	C+	*LS*
			21138.2	4729.485 cm^{-1}	102865.655	107595.140	3−5	4.41−03	4.93−02	1.03+01	−0.830	C+	*LS*
			21138.3	4729.461 cm^{-1}	102865.679	107595.140	1−3	3.68−03	7.39−02	5.14+00	−1.131	C+	*LS*
			21137.5	4729.634 cm^{-1}	102865.506	107595.140	9−9	1.31−03	8.80−03	5.51+00	−1.101	C+	*LS*
			21137.7	4729.593 cm^{-1}	102865.547	107595.140	7−7	2.36−03	1.58−02	7.71+00	−0.955	C+	*LS*
			21138.0	4729.534 cm^{-1}	102865.606	107595.140	5−5	3.15−03	2.11−02	7.35+00	−0.977	C+	*LS*
			21138.2	4729.485 cm^{-1}	102865.655	107595.140	3−3	3.68−03	2.46−02	5.14+00	−1.131	C+	*LS*
			21137.5	4729.634 cm^{-1}	102865.506	107595.140	9−7	1.13−04	5.86−04	3.67−01	−2.278	C+	*LS*
			21137.7	4729.593 cm^{-1}	102865.547	107595.140	7−5	3.15−04	1.51−03	7.35−01	−1.977	C+	*LS*
			21138.0	4729.534 cm^{-1}	102865.606	107595.140	5−3	5.25−04	2.11−03	7.35−01	−1.977	C+	*LS*
168.	$2p^3(^4S^\circ)4d-2p^3(^4S^\circ)4f$	$^3D^\circ-^3F$		*59.92* cm^{-1}	*102908.42*	102968.343	15−21	4.68−07	2.73−02	2.25+03	−0.387	B	1
				59.969 cm^{-1}	102908.374	102968.343	7−9	4.69−07	2.51−02	9.65+02	−0.755	C+	*LS*
				59.900 cm^{-1}	102908.443	102968.343	5−7	4.15−07	2.43−02	6.67+02	−0.916	C+	*LS*
				59.854 cm^{-1}	102908.489	102968.343	3−5	3.92−07	2.73−02	4.50+02	−1.087	C+	*LS*
				59.969 cm^{-1}	102908.374	102968.343	7−7	5.23−08	2.18−03	8.37+01	−1.817	C+	*LS*
				59.900 cm^{-1}	102908.443	102968.343	5−5	7.29−08	3.05−03	8.37+01	−1.817	C+	*LS*
				59.969 cm^{-1}	102908.374	102968.343	7−5	2.06−09	6.14−05	2.36+00	−3.367	C+	*LS*
169.	$2p^3(^4S^\circ)4d-2p^3(^4S^\circ)5p$	$^3D^\circ-^3P$		*961.60* cm^{-1}	*102908.42*	*103870.02*	15−9	3.64−03	3.54−01	1.82+03	0.725	B	1
				961.594 cm^{-1}	102908.374	103869.968	7−5	3.06−03	3.54−01	8.48+02	0.394	C+	*LS*
				961.585 cm^{-1}	102908.443	103870.028	5−3	2.73−03	2.65−01	4.54+02	0.123	C+	*LS*
				961.763 cm^{-1}	102908.489	103870.252	3−1	3.64−03	1.97−01	2.02+02	−0.229	C+	*LS*
				961.525 cm^{-1}	102908.443	103869.968	5−5	5.46−04	8.84−02	1.51+02	−0.354	C+	*LS*
				961.539 cm^{-1}	102908.489	103870.028	3−3	9.09−04	1.47−01	1.51+02	−0.354	C+	*LS*
				961.585 cm^{-1}	102908.443	103870.028	5−3	6.06−05	5.90−03	1.01+01	−1.530	C+	*LS*
170.	$2p^3(^4S^\circ)4d-2p^3(^4S^\circ)5f$	$^3D^\circ-^3F$	*39463*	*2533.30* cm^{-1}	*102908.42*	105441.724	15−21	2.72−02	8.90−01	1.73+03	1.126	B	1
			39462.7	2533.350 cm^{-1}	102908.374	105441.724	7−9	2.72−02	8.18−01	7.43+02	0.758	C+	*LS*
			39463.7	2533.281 cm^{-1}	102908.443	105441.724	5−7	2.42−02	7.91−01	5.14+02	0.597	C+	*LS*
			39464.5	2533.235 cm^{-1}	102908.489	105441.724	3−5	2.29−02	8.90−01	3.47+02	0.427	C+	*LS*
			39462.7	2533.350 cm^{-1}	102908.374	105441.724	7−7	3.03−03	7.09−02	6.44+01	−0.305	C+	*LS*
			39463.7	2533.281 cm^{-1}	102908.443	105441.724	5−5	4.25−03	9.92−02	6.44+01	−0.305	C+	*LS*
			39462.7	2533.350 cm^{-1}	102908.374	105441.724	7−5	1.20−04	2.00−03	1.82+00	−1.854	C+	*LS*

O I: Allowed Transitions — Continued

No.	Transition Array	Mult.	λ_{air} (Å)	λ_{vac} (Å) or σ (cm^{-1})*	E_i (cm^{-1})	E_k (cm^{-1})	g_i-g_k	A_{ki} ($10^8\ s^{-1}$)	f_{ik}	S (at. u.)	log gf	Acc.	Source
171.	$2p^3(^4S^\circ)4d-2p^3(^4S^\circ)6p$	$^3D^\circ-^3P$	*33284*	*3003.61* cm^{-1}	*102908.42*	105912.031	15–9	6.79–04	6.76–03	1.11+01	–0.994	B	1
			33283.7	3003.657 cm^{-1}	102908.374	105912.031	7–5	5.70–04	6.76–03	5.19+00	–1.325	C+	*LS*
			33284.4	3003.588 cm^{-1}	102908.443	105912.031	5–3	5.09–04	5.07–03	2.78+00	–1.596	C+	*LS*
			33284.9	3003.542 cm^{-1}	102908.489	105912.031	3–1	6.78–04	3.76–03	1.23+00	–1.948	C+	*LS*
			33284.4	3003.588 cm^{-1}	102908.443	105912.031	5–5	1.02–04	1.69–03	9.26–01	–2.073	C+	*LS*
			33284.9	3003.542 cm^{-1}	102908.489	105912.031	3–3	1.70–04	2.82–03	9.26–01	–2.073	C+	*LS*
			33284.4	3003.588 cm^{-1}	102908.443	105912.031	5–3	1.13–05	1.13–04	6.17–02	–3.249	C+	*LS*
172.	$2p^3(^4S^\circ)4d-2p^3(^4S^\circ)6f$	$^3D^\circ-^3F$	*25788*	*3876.78* cm^{-1}	*102908.42*	106785.201	15–21	1.34–02	1.87–01	2.38+02	0.448	B	1
			25787.3	3876.827 cm^{-1}	102908.374	106785.201	7–9	1.34–02	1.72–01	1.02+02	0.080	C+	*LS*
			25787.7	3876.758 cm^{-1}	102908.443	106785.201	5–7	1.19–02	1.66–01	7.05+01	–0.081	C+	*LS*
			25788.0	3876.712 cm^{-1}	102908.489	106785.201	3–5	1.12–02	1.87–01	4.76+01	–0.251	C+	*LS*
			25787.3	3876.827 cm^{-1}	102908.374	106785.201	7–7	1.49–03	1.49–02	8.84+00	–0.983	C+	*LS*
			25787.7	3876.758 cm^{-1}	102908.443	106785.201	5–5	2.09–03	2.08–02	8.84+00	–0.983	C+	*LS*
			25787.3	3876.827 cm^{-1}	102908.374	106785.201	7–5	5.89–05	4.19–04	2.49–01	–2.532	C+	*LS*
173.	$2p^3(^4S^\circ)4d-2p^3(^4S^\circ)7f$	$^3D^\circ-^3F$	*21331*	*4686.73* cm^{-1}	*102908.42*	107595.147	15–21	7.60–03	7.26–02	7.65+01	0.037	B	1
			21330.8	4686.773 cm^{-1}	102908.374	107595.147	7–9	7.60–03	6.67–02	3.28+01	–0.331	C+	*LS*
			21331.1	4686.704 cm^{-1}	102908.443	107595.147	5–7	6.75–03	6.45–02	2.26+01	–0.491	C+	*LS*
			21331.3	4686.658 cm^{-1}	102908.489	107595.147	3–5	6.38–03	7.26–02	1.53+01	–0.662	C+	*LS*
			21330.8	4686.773 cm^{-1}	102908.374	107595.147	7–7	8.47–04	5.78–03	2.84+00	–1.393	C+	*LS*
			21331.1	4686.704 cm^{-1}	102908.443	107595.147	5–5	1.19–03	8.09–03	2.84+00	–1.393	C+	*LS*
			21330.8	4686.773 cm^{-1}	102908.374	107595.147	7–5	3.34–05	1.63–04	8.01–02	–2.943	C+	*LS*
174.	$2p^3(^4S^\circ)4d-2p^3(^2D^\circ)3p$	$^3D^\circ-^3D$	*9624.7*	*9627.3*	*102908.42*	*113295.53*	15–15	2.09–03	2.90–03	1.38+00	–1.362	B	1
			9625.30	9627.94	102908.374	113294.816	7–7	1.85–03	2.57–03	5.71–01	–1.744	C+	*LS*
			9625.32	9627.96	102908.443	113294.854	5–5	1.45–03	2.02–03	3.19–01	–1.996	C+	*LS*
			9622.16	9624.80	102908.489	113298.320	3–3	1.57–03	2.17–03	2.07–01	–2.186	C+	*LS*
			9625.26	9627.90	102908.374	113294.854	7–5	3.25–04	3.23–04	7.16–02	–2.646	C+	*LS*
			9622.11	9624.75	102908.443	113298.320	5–3	5.22–04	4.35–04	6.89–02	–2.663	C+	*LS*
			9625.36	9628.00	102908.443	113294.816	5–7	2.32–04	4.52–04	7.16–02	–2.646	C+	*LS*
			9625.37	9628.01	102908.489	113294.854	3–5	3.13–04	7.24–04	6.89–02	–2.663	C+	*LS*
175.	$2p^3(^4S^\circ)4f-2p^3(^4S^\circ)5d$	$^5F-^5D^\circ$	*33135*	*3017.14* cm^{-1}	102968.249	*105985.39*	35–25	1.25–03	1.47–02	5.62+01	–0.288	B	1
			41360.5	2417.105 cm^{-1}	102968.249	105385.354	11–9	5.62–04	1.18–02	1.77+01	–0.887	C+	*LS*
			41360.1	2417.128 cm^{-1}	102968.249	105385.377	9–7	4.93–04	9.84–03	1.21+01	–1.053	C+	*LS*
			18454.8	5417.160 cm^{-1}	102968.249	108385.409	7–5	4.97–03	1.81–02	7.71+00	–0.896	C+	*LS*
			41359.1	2417.187 cm^{-1}	102968.249	105385.436	5–3	4.30–04	6.61–03	4.50+00	–1.481	C+	*LS*
			41358.9	2417.200 cm^{-1}	102968.249	105385.449	3–1	6.44–04	5.51–03	2.25+00	–1.782	C+	*LS*
			41360.5	2417.105 cm^{-1}	102968.249	105385.354	9–9	7.67–05	1.97–03	2.41+00	–1.752	C+	*LS*
			41360.1	2417.128 cm^{-1}	102968.249	105385.377	7–7	1.38–04	3.54–03	3.37+00	–1.606	C+	*LS*
			18454.8	5417.160 cm^{-1}	102968.249	108385.409	5–5	2.07–03	1.06–02	3.21+00	–1.277	C+	*LS*
			41359.1	2417.187 cm^{-1}	102968.249	105385.436	3–3	2.15–04	5.51–03	2.25+00	–1.782	C+	*LS*
			41360.5	2417.105 cm^{-1}	102968.249	105385.354	7–9	5.11–06	1.69–04	1.61–01	–2.928	C+	*LS*
			41360.1	2417.128 cm^{-1}	102968.249	105385.377	5–7	1.31–05	4.72–04	3.21–01	–2.627	C+	*LS*
			18454.8	5417.160 cm^{-1}	102968.249	108385.409	3–5	2.07–04	1.76–03	3.21–01	–2.277	C+	*LS*

O I: Allowed Transitions — Continued

No.	Transition Array	Mult.	λ_{air} (Å)	λ_{vac} (Å) or σ (cm^{-1})*	E_i (cm^{-1})	E_k (cm^{-1})	g_i-g_k	A_{ki} (10^8 s^{-1})	f_{ik}	S (at. u.)	log gf	Acc.	Source
176.	$2p^3(^4S^o)4f-$ $2p^3(^4S^o)6d$	$^5F-^5D^o$	*26425*	*3783.21* cm^{-1}	102968.249	*106751.46*	35–25	2.81–04	2.10–03	6.40+00	–1.133	B	1
			26425.5	3783.198 cm^{-1}	102968.249	106751.447	11–9	2.45–04	2.10–03	2.01+00	–1.636	C+	*LS*
			26425.4	3783.209 cm^{-1}	102968.249	106751.458	9–7	2.15–04	1.75–03	1.37+00	–1.802	C+	*LS*
			26425.3	3783.225 cm^{-1}	102968.249	106751.474	7–5	1.93–04	1.44–03	8.78–01	–1.996	C+	*LS*
			26425.2	3783.238 cm^{-1}	102968.249	106751.487	5–3	1.87–04	1.18–03	5.12–01	–2.230	C+	*LS*
			26425.1	3783.245 cm^{-1}	102968.249	106751.494	3–1	2.81–04	9.81–04	2.56–01	–2.531	C+	*LS*
			26425.5	3783.198 cm^{-1}	102968.249	106751.447	9–9	3.35–05	3.50–04	2.74–01	–2.501	C+	*LS*
			26425.4	3783.209 cm^{-1}	102968.249	106751.458	7–7	6.03–05	6.31–04	3.84–01	–2.355	C+	*LS*
			26425.3	3783.225 cm^{-1}	102968.249	106751.474	5–5	8.03–05	8.41–04	3.66–01	–2.376	C+	*LS*
			26425.2	3783.238 cm^{-1}	102968.249	106751.487	3–3	9.37–05	9.81–04	2.56–01	–2.531	C+	*LS*
			26425.5	3783.198 cm^{-1}	102968.249	106751.447	7–9	2.23–06	3.00–05	1.83–02	–3.677	C	*LS*
			26425.4	3783.209 cm^{-1}	102968.249	106751.458	5–7	5.74–06	8.41–05	3.66–02	–3.376	C+	*LS*
			26425.3	3783.225 cm^{-1}	102968.249	106751.474	3–5	8.03–06	1.40–04	3.66–02	–3.376	C+	*LS*
177.	$2p^3(^4S^o)4f-$ $2p^3(^4S^o)7d$	$^5F-^5D^o$	*21708*	*4605.24* cm^{-1}	102968.249	*107573.49*	35–25	1.50–04	7.55–04	1.89+00	–1.578	B	1
			21708.5	4605.227 cm^{-1}	102968.249	107573.476	11–9	1.31–04	7.55–04	5.94–01	–2.080	C+	*LS*
			21708.5	4605.235 cm^{-1}	102968.249	107573.484	9–7	1.15–04	6.29–04	4.05–01	–2.247	C+	*LS*
			21708.4	4605.246 cm^{-1}	102968.249	107573.495	7–5	1.03–04	5.18–04	2.59–01	–2.441	C+	*LS*
			21708.4	4605.255 cm^{-1}	102968.249	107573.504	5–3	9.98–05	4.23–04	1.51–01	–2.675	C+	*LS*
			21708.4	4605.259 cm^{-1}	102968.249	107573.508	3–1	1.50–04	3.52–04	7.56–02	–2.976	C+	*LS*
			21708.5	4605.227 cm^{-1}	102968.249	107573.476	9–9	1.78–05	1.26–04	8.10–02	–2.946	C+	*LS*
			21708.5	4605.235 cm^{-1}	102968.249	107573.484	7–7	3.21–05	2.27–04	1.13–01	–2.800	C+	*LS*
			21708.4	4605.246 cm^{-1}	102968.249	107573.495	5–5	4.28–05	3.02–04	1.08–01	–2.821	C+	*LS*
			21708.4	4605.255 cm^{-1}	102968.249	107573.504	3–3	4.99–05	3.52–04	7.56–02	–2.976	C+	*LS*
			21708.5	4605.227 cm^{-1}	102968.249	107573.476	7–9	1.19–06	1.08–05	5.40–03	–4.122	C–	*LS*
			21708.5	4605.235 cm^{-1}	102968.249	107573.484	5–7	3.05–06	3.02–05	1.08–02	–3.821	C	*LS*
			21708.4	4605.246 cm^{-1}	102968.249	107573.495	3–5	4.28–06	5.04–05	1.08–02	–3.821	C	*LS*
178.	$2p^3(^4S^o)4f-$ $2p^3(^4S^o)8d$	$^5F-^5D^o$	*19458*	*5137.83* cm^{-1}	102968.249	*108106.08*	35–25	9.02–05	3.66–04	8.20–01	–1.893	B	1
			19458.2	5137.823 cm^{-1}	102968.249	108106.072	11–9	7.88–05	3.66–04	2.58–01	–2.395	C+	*LS*
			19458.2	5137.828 cm^{-1}	102968.249	108106.077	9–7	6.91–05	3.05–04	1.76–01	–2.562	C+	*LS*
			19458.1	5137.836 cm^{-1}	102968.249	108106.085	7–5	6.19–05	2.51–04	1.12–01	–2.755	C+	*LS*
			19458.1	5137.842 cm^{-1}	102968.249	108106.091	5–3	6.02–05	2.05–04	6.56–02	–2.990	C+	*LS*
			19458.1	5137.845 cm^{-1}	102968.249	108106.094	3–1	9.02–05	1.71–04	3.28–02	–3.291	C+	*LS*
			19458.2	5137.823 cm^{-1}	102968.249	108106.072	9–9	1.07–05	6.10–05	3.52–02	–3.261	C+	*LS*
			19458.2	5137.828 cm^{-1}	102968.249	108106.077	7–7	1.93–05	1.10–04	4.92–02	–3.115	C+	*LS*
			19458.1	5137.836 cm^{-1}	102968.249	108106.085	5–5	2.58–05	1.46–04	4.69–02	–3.136	C+	*LS*
			19458.1	5137.842 cm^{-1}	102968.249	108106.091	3–3	3.01–05	1.71–04	3.28–02	–3.291	C+	*LS*
			19458.2	5137.823 cm^{-1}	102968.249	108106.072	7–9	7.16–07	5.23–06	2.34–03	–4.437	D	*LS*
			19458.2	5137.828 cm^{-1}	102968.249	108106.077	5–7	1.84–06	1.46–05	4.69–03	–4.136	C–	*LS*
			19458.1	5137.836 cm^{-1}	102968.249	108106.085	3–5	2.58–06	2.44–05	4.69–03	–4.136	C–	*LS*
179.	$2p^3(^4S^o)4f-$ $2p^3(^4S^o)5d$	$^3F-^3D^o$	40961.3	2440.665 cm^{-1}	102968.343	105409.008	21–15	5.32–04	9.55–03	2.70+01	–0.698	B	1
180.	$2p^3(^4S^o)4f-$ $2p^3(^4S^o)6d$	$^3F-^3D^o$	26326.2	3797.460 cm^{-1}	102968.343	106765.803	21–15	2.31–04	1.72–03	3.12+00	–1.443	B	1
181.	$2p^3(^4S^o)4f-$ $2p^3(^4S^o)7d$	$^3F-^3D^o$	21665.2	4614.434 cm^{-1}	102968.343	107582.777	21–15	1.22–04	6.15–04	9.21–01	–1.889	B	1
182.	$2p^3(^4S^o)4f-$ $2p^3(^4S^o)8d$	$^3F-^3D^o$	19428.6	5145.7 cm^{-1}	102968.343	108114.0	21–15	7.34–05	2.97–04	3.99–01	–2.205	B	1

*Wavelengths (Å) are always given unless cm^{-1} is indicated.

O I

Forbidden Transitions

We have selected the results of two sophisticated calculations by Froese Fischer and Saha[1], and Baluja and Zeippen[2], which were performed with the multi-configuration Hartree-Fock method and with the CIV 3 code (see general introduction), respectively, both with Breit-Pauli corrections. In these two approaches, configuration interaction is addressed through extensive expansions of configuration state wave-functions containing contributions from numerous configuration states in the n= 2, 3 and 4 electron shells. Our tabulated data are the averages of the line strengths, and we use – as always – experimental transition energies to convert to the A-values.

For the M1 transitions between two levels of the ground state term, $2p^4\ ^3P$, the line strengths S of Refs. 1 and 2 are the same as obtained from simple non-relativistic formulas for conditions of pure LS-coupling, given by Pasternak.[3]

O I is one of the few cases with available experimental data for forbidden lines. The results of an emission[4] and a lifetime[5] experiment are in excellent agreement with the calculations. Table 1 shows the comparisons.

TABLE 1. Comparisons of recent experimental and theoretical determinations of the transition probability sums from the $^1D_2$ and $^1S_0$ levels.

Author	$2p^4\ ^1D_2$ (in $10^{-3}s^{-1}$)	$2p^4\ ^1S_0$ (in s^{-1})
Froese Fischer and Saha[1]	7.50	1.38
Baluja and Zeippen[2]	7.44	1.29
Kernahan and Pang[4]	6.81 ± 1.32	1.11 ± 0.34
Corney and Williams[5]	—	1.31 ± 0.05

References

[1]C. Froese Fischer and H. P. Saha, Phys. Rev. A. **28**, 3169 (1983).
[2]K. L. Baluja and C. J. Zeippen, J. Phys. B. **21**, 1455 (1988).
[3]S. Pasternak, Astrophys J. **92**, 129 (1940).
[4]J. A. Kernahan and P. H.-L. Pang, Can. J. Phys. **53**, 455 (1975).
[5]A. Corney and O. M. Williams, J. Phys. B. **5**, 686 (1972).

O I: Forbidden Transitions

No.	Transition Array	Mult.	λ_{air} (Å)	λ_{vac} (Å) or σ (cm^{-1})*	E_i (cm^{-1})	E_k (cm^{-1})	g_i–g_k	Type	A_{ki} (s^{-1})	S (at. u.)	Acc.	Source
1.	$2p^4$– $2p^4$	^3P–^3P										
				158.265 cm^{-1}	0.000 –	158.265	5–3	M1	8.91–05	2.50+00	A	1
				158.265 cm^{-1}	0.000 –	158.265	5–3	E2	1.66–11	4.48+00	C+	1,2
				226.977 cm^{-1}	0.000 –	226.977	5–1	E2	1.34–10	1.98+00	C+	1,2
				68.712 cm^{-1}	158.265 –	226.977	3–1	M1	1.75–05	2.00+00	A	1
2.		^3P–^1D										
			6300.30	6302.05	0.000 –	15867.862	5–5	M1	5.63–03	2.61–04	B+	1,2
			6300.30	6302.05	0.000 –	15867.862	5–5	E2	2.11–05	9.38–04	C+	1,2
			6363.78	6365.54	158.265 –	15867.862	3–5	M1	1.82–03	8.70–05	B+	1,2
			6363.78	6365.54	158.265 –	15867.862	3–5	E2	3.39–06	1.58–04	C+	1,2
			6391.73	6393.50	226.977 –	15867.862	1–5	E2	8.60–07	4.10–05	B+	1,2
3.		^3P–^1S										
			2958.36	2959.23	0.000 –	33792.583	5–1	E2	2.42–04	4.90–05	C+	1,2
			2972.29	2973.15	158.265 –	33792.583	3–1	M1	7.54–02	7.34–05	B+	1,2
4.		^1D–^1S										
			5577.34	5578.89	15867.862 –	33792.583	5–1	E2	1.26+00	6.08+00	B+	1,2

*Wavelengths (Å) are always given unless cm^{-1} is indicated.

O II

Nitrogen Isoelectronic Sequence

Ground State: $1s^2 2s^2 2p^3\ {}^4S^\circ_{3/2}$

Ionization Energy: 35.1211 eV = 283270.9 cm^{-1}

Allowed Transitions

List of tabulated lines

Wavelength (Å)	No.	Wavelength (Å)	No.	Wavelength (Å)	No.	Wavelength (Å)	No.
in vacuum		440.603	24	499.877	44	796.645	33
		440.607	24	499.882	44	796.683	33
387.649	10	442.012	23	500.345	44	832.758	1
387.898	10	442.016	23	500.350	44	833.330	1
388.055	10	442.051	23	515.499	43	834.465	1
391.906	9	442.055	23	515.637	43	1083.13	56
391.938	9	445.605	22	515.643	43	1085.05	56
391.995	9	445.610	22	517.929	42	1086.02	56
401.180	32	445.644	22	517.935	42	1128.07	55
401.212	32	456.996	51	518.235	42	1129.24	55
401.327	32	456.997	51	518.241	42	1130.15	55
401.359	32	457.001	51	519.944	41	1131.33	55
403.035	31	458.425	50	520.080	40	1131.92	55
403.054	31	458.429	50	520.086	40	1132.38	55
403.067	31	458.835	50	520.173	40	1132.97	55
403.087	31	458.839	50	520.179	40	1147.25	54
403.273	30	464.189	49	520.474	39	1148.89	54
403.339	30	464.307	49	520.649	39	1149.36	62
403.372	30	464.311	49	520.848	39	1149.47	62
418.596	8	464.782	48	525.924	38	1149.55	62
418.879	8	464.786	48	525.930	38	1150.10	54
419.063	8	465.519	47	537.832	14	1151.05	54
424.435	53	465.523	47	538.262	14	1152.26	54
424.600	53	465.754	47	538.320	14	1153.00	54
424.603	53	465.758	47	539.086	2	1153.36	54
426.512	52	467.931	21	539.547	2	1154.10	54
426.522	7	467.975	21	539.854	2	1373.43	78
426.534	52	468.385	21	555.055	13	1373.60	78
426.537	52	468.770	46	555.059	13	1376.84	78
428.393	6	468.774	46	555.117	13	1433.77	69
429.560	5	468.775	46	555.120	13	1435.94	69
429.650	4	468.779	46	580.402	37	1439.21	69
429.653	4	470.409	45	580.409	37	1502.89	61
429.716	4	470.414	45	580.970	37	1503.07	61
429.918	3	470.418	45	580.977	37	1504.12	61
430.041	3	481.593	20	600.584	36	1559.94	77
430.149	29	481.640	20	600.587	36	1560.09	77
430.150	29	481.714	20	600.591	36	1564.34	77
430.176	3	481.760	20	616.303	12	1564.49	77
430.186	29	483.760	19	616.379	12	1577.14	129
430.187	29	483.980	19	617.063	12	1577.44	129
431.414	28	484.027	19	644.154	35	1581.81	128
431.451	28	485.087	18	644.162	35	1581.90	129
431.815	28	485.470	18	672.945	34	1582.20	129
436.515	27	485.518	18	672.954	34	1585.57	128
436.553	27	485.571	17	673.761	34	1586.60	128
436.620	27	485.585	17	673.770	34	1593.40	60
436.658	27	485.589	17	718.463	11	1594.00	60
437.339	26	485.633	17	718.505	11	1594.20	60
437.653	26	485.717	17	718.567	11	1616.93	158
437.691	26	485.975	16	718.608	11	1617.76	158
437.730	25	486.085	16	739.950	57	1619.32	158
437.899	25	486.305	16	740.844	57	1646.20	157
437.938	25	488.717	15	741.297	57	1647.49	157
440.564	24	488.951	15	796.632	33	1647.82	157

List of tabulated lines — Continued

Wavelength (Å)	No.	Wavelength (Å)	No.	Wavelength (Å)	No.	Wavelength (Å)	No.
1650.67	68	2182.57	137	2526.87	151	3639.68	205
1651.18	68	2186.34	58	2527.03	151	3646.56	205
1651.99	68	2190.47	137	2530.11	122	3662.50	205
1652.07	68	2195.48	137	2530.28	122	3669.47	205
1653.55	68	2195.88	58	2571.46	150	3678.97	164
1654.87	68	2215.71	156	2575.28	150	3682.17	164
1655.52	68	2218.65	156	2575.42	150	3686.39	164
1657.89	68	2218.67	156	2715.36	79	3712.74	67
1691.43	76	2222.47	181	2718.81	79	3713.96	224
1691.53	76	2223.07	181	2733.29	87	3715.62	224
1696.59	76	2223.53	181	2747.37	87	3727.32	67
1745.72	81	2224.12	181	2783.03	141	3728.36	94
1745.91	81	2229.74	180	2789.93	141	3729.22	192
1821.55	105	2231.77	180	2796.14	141	3735.72	192
1821.59	105	2232.83	180	2803.11	141	3735.79	192
1827.25	105	2249.71	155	2808.73	140	3736.10	94
1827.28	105	2252.74	155	2822.72	140	3738.34	163
1829.36	89	2259.63	155	2836.21	140	3738.46	224
1835.91	89	2262.69	155	3113.62	95	3739.76	135
1893.77	80	2283.44	125	3122.52	95	3739.99	223
1896.96	80	2284.83	125	3123.91	95	3740.14	224
1947.81	88	2290.85	75	3129.34	95	3741.64	163
1951.93	88	2293.30	75	3134.21	95	3744.29	94
1953.93	96	2293.43	125	3134.73	95	3749.48	67
1957.44	96	2295.60	194	3138.34	95	3752.09	94
1958.13	96	2297.96	194	3139.68	95	3759.06	223
1960.26	96	2300.33	75	3270.86	165	3761.40	163
1962.13	96	2300.42	194	3270.90	165	3762.47	135
1962.22	96	2302.81	75	3273.43	165	3764.75	163
1963.79	96	2324.80	136	3277.56	112	3764.83	223
1964.27	96	2325.92	136	3281.99	86	3765.03	94
		2327.95	136	3287.47	112	3774.00	233
in air		2355.79	193	3294.99	112	3777.42	135
		2357.72	193	3301.41	112	3784.98	233
2016.58	113	2358.39	193	3305.00	112	3786.70	233
2020.33	113	2365.01	124	3306.45	112	3794.36	149
2021.44	113	2365.14	124	3377.15	85	3802.98	149
2023.32	113	2375.72	124	3390.21	85	3813.73	66
2025.70	113	2398.91	153	3407.22	179	3821.53	149
2027.10	113	2399.20	153	3407.28	179	3830.28	149
2027.60	113	2402.35	153	3409.71	179	3833.07	93
2028.73	142	2406.38	103	3409.76	179	3835.86	111
2035.80	142	2406.44	103	3438.56	197	3836.70	111
2042.81	142	2407.36	123	3457.93	197	3842.81	92
2072.26	59	2407.48	123	3458.92	197	3843.58	93
2072.60	59	2411.60	103	3470.28	121	3847.89	92
2075.17	59	2411.64	154	3470.67	121	3850.80	92
2092.88	127	2415.13	154	3470.78	84	3851.03	92
2092.90	127	2418.46	123	3474.93	84	3851.47	93
2101.26	127	2418.58	123	3488.16	83	3852.39	177
2101.28	127	2425.57	102	3488.27	178	3855.57	177
2111.20	167	2425.63	102	3490.88	178	3856.13	92
2112.25	167	2431.64	152	3496.28	83	3857.16	93
2115.28	167	2433.54	74	3514.82	178	3860.53	177
2123.18	126	2435.18	152	3515.24	201	3863.50	92
2131.81	126	2435.99	102	3536.53	201	3863.72	177
2132.02	126	2436.06	102	3550.86	201	3864.13	91
2170.51	166	2438.07	152	3562.13	225	3864.43	92
2171.62	166	2441.63	152	3573.76	225	3864.67	92
2172.21	104	2444.25	74	3584.66	225	3872.44	91
2172.26	104	2445.53	74	3596.44	225	3874.09	91
2172.44	166	2517.96	122	3627.41	234	3875.80	93
2172.55	104	2523.06	151	3627.74	234	3882.19	92
2173.55	166	2523.21	151	3639.48	234	3882.45	91

List of tabulated lines — Continued

Wavelength (Å)	No.	Wavelength (Å)	No.	Wavelength (Å)	No.	Wavelength (Å)	No.
3883.14	92	4327.46	173	4705.35	118	6666.66	199
3893.52	91	4327.85	173	4710.01	117	6677.87	199
3896.30	91	4328.59	191	4719.80	213	6717.75	199
3907.45	91	4331.47	173	4741.70	118	6721.39	70
3911.96	101	4331.86	173	4751.28	117	6774.98	198
3912.12	101	4336.86	65	4752.69	117	6810.48	183
3917.53	176	4345.56	65	4774.08	116	6844.10	183
3919.27	101	4347.22	100	4844.92	132	6846.80	183
3926.58	91	4347.41	100	4856.39	131	6856.96	198
3942.87	176	4349.43	65	4856.76	131	6869.48	183
3945.04	73	4351.26	100	4860.97	187	6884.88	183
3946.20	176	4351.46	100	4864.88	131	6895.10	183
3949.35	175	4354.59	106	4871.52	187	6906.44	183
3954.36	73	4357.26	106	4872.02	187	6907.87	183
3962.96	175	4359.40	120	4890.86	130	6910.56	183
3966.32	175	4366.89	65	4906.83	130	6911.03	198
3967.37	110	4369.27	120	4924.53	130	7083.96	203
3973.26	73	4384.45	189	4941.07	148	7110.08	203
3982.71	73	4393.44	189	4943.01	148	7128.86	203
3985.41	110	4395.93	120	4955.71	148	7252.87	202
3992.76	110	4405.38	189	5015.34	139	7280.25	202
4069.62	90	4405.98	120	5041.97	115	7346.90	202
4069.88	90	4414.46	189	5070.05	139	7375.00	202
4072.15	90	4414.90	72	5070.81	115	7527.07	195
4075.86	90	4416.97	72	5110.30	200	7530.61	195
4078.84	90	4439.48	184	5111.91	200	7598.28	195
4084.65	109	4443.01	160	5137.96	71	7599.20	195
4085.11	90	4443.52	160	5140.27	200	7602.81	195
4092.93	90	4447.68	160	5141.90	200	7628.54	195
4094.14	90	4448.19	160	5149.04	71	7648.58	195
4096.53	109	4452.38	72	5159.94	147	7649.51	195
4097.22	108	4453.97	184	5175.90	147	7656.78	195
4103.00	108	4466.24	207	5185.93	71	7676.97	195
4104.72	108	4467.21	184	5190.50	147	7894.61	212
4104.99	108	4467.46	207	5206.65	147	7898.46	212
4106.02	90	4484.50	134	5344.10	63	8013.16	212
4109.84	162	4539.54	119	5374.36	63	8017.12	212
4110.19	162	4557.73	243	5377.57	204	8229.27	215
4110.79	108	4558.94	243	5379.35	204	8288.30	215
4112.02	109	4560.22	243	5381.78	145	8309.19	215
4113.83	162	4563.18	119	5384.51	146	8342.70	215
4119.22	108	4590.97	99	5391.74	145	8375.84	215
4120.28	108	4595.96	99	5392.61	204	8381.77	215
4120.55	108	4596.18	99	5398.68	145	8403.38	215
4121.46	107	4603.23	119	5399.14	145	8415.88	215
4129.32	107	4638.86	64	5409.17	145	8864.51	222
4132.80	107	4641.81	64	5423.67	144	8864.87	222
4140.70	107	4649.13	64	5461.09	144	9005.38	222
4153.30	107	4650.84	64	5483.02	144	9005.75	222
4156.53	107	4661.63	64	5583.22	206	9204.17	232
4169.22	107	4673.73	64	5611.07	206	9280.12	232
4185.44	161	4676.23	64	5753.93	138	9280.51	232
4189.58	161	4678.42	213	5761.01	138	9389.08	235
4189.79	161	4681.96	213	5810.25	138	9391.37	235
4192.51	174	4690.89	188	6081.22	143	9434.83	221
4196.27	174	4691.42	188	6103.40	143	9457.89	171
4196.70	174	4696.35	64	6112.99	209	9477.05	171
4277.69	190	4698.44	172	6152.56	209	9544.87	235
4279.17	190	4699.01	172	6214.25	209	9594.56	221
4287.73	190	4699.22	118	6495.79	196	9611.80	221
4317.14	65	4700.44	133	6565.28	196	9648.40	171
4319.63	65	4701.18	188	6571.11	196	9820.55	231
4319.87	191	4701.71	188	6627.37	199	9907.06	231
4325.76	65	4703.16	172	6641.03	70	10012.4	231

List of tabulated lines — Continued

Wavelength (Å)	No.	Wavelength (Å)	No.	Wavelength (Å)	No.	Wavenumber (cm^{-1})	No.
10102.4	231	14000.0	208	21086.6	97	881.09	114
10641.4	214	14010.9	208	21091.2	97	903.44	185
10664.8	214	14050.0	219	21637.6	226	950.09	185
10696.4	214	14065.8	230	21969.9	97	1017.50	185
10720.1	214	14112.4	208	21974.9	97	1064.15	185
10763.3	214	14209.3	208	22051.9	226	1071.70	114
10820.5	214	14262.9	208	22062.0	226	1240.13	114
10864.5	214	14441.2	208	22178.9	170	1870.99	182
11015.4	214	14582.8	228	22284.5	170	1875.71	186
11330.7	236	14774.4	228	22436.9	170	1917.64	182
11333.3	236	14817.0	228	22545.0	170	1922.36	186
11334.8	82	14929.4	245	23736.6	159	1927.54	186
11554.8	236	15014.8	228	24548.5	159		
11555.5	82	15221.6	246	31100.0	216		
11557.6	236	15321.1	245	32905.8	216		
12213.7	211	15579.8	245	32934.0	216		
12347.8	211	15611.4	246	33189.9	169		
12376.6	237	15628.9	246	33507.9	244		
12499.8	211	15866.6	242	34423.5	239		
12640.3	211	15882.1	242	35008.4	239		
12640.7	237	15883.3	242	35084.1	239		
12644.6	237	16587.4	247	35087.7	169		
12675.0	220	16739.3	247	35352.8	169		
12763.5	220	17072.3	247	35471.7	244		
12965.0	220	17233.2	247	35618.7	168		
13016.9	210	17289.8	218	35698.3	168		
13035.5	210	17297.0	218	35972.8	168		
13057.6	220	17791.5	241	36833.4	244		
13342.4	210	17811.0	241	37932.1	240		
13479.8	229	17833.9	218	38021.0	240		
13542.2	229	18431.4	241	38714.2	248		
13643.3	229	18452.4	241	38721.1	248		
13710.1	219	18630.8	227	41462.9	248		
13733.5	219	18639.2	227	41470.8	248		
13774.2	98	18944.6	227	43566.8	238		
13776.1	98	18953.3	227	43684.1	238		
13890.5	98	19841.4	217	45725.9	238		
13892.1	230	19849.6	217	45855.1	238		
13905.9	208	20561.3	217				
13942.1	208	20570.1	217				

We have utilized the results of two sophisticated large-scale calculations for this nitrogen-like spectrum: (a) Opacity Project (OP) calculations, carried out by Bell and Berrington[1] (OP is reviewed in the general introduction), and (b) configuration interaction calculations by Bell, Hibbert, Stafford, and McLaughlin[2] with the CIV 3 code. The latter work is limited to transitions involving principal quantum numbers $n = 2$ and 3.

We did not include any results of similar, smaller calculations by Becker and Butler,[3] based on the OP approach, or by Bell *et al.*[4] and Ho and Henry[5] based on the CIV 3 code, because these earlier calculations contain less extensive treatments of configuration interaction, but use the same approaches.

For the $2s$-$2p$, $2p$-$3s$, $2p$-$3d$, and $3s$-$3p$ multiplet data, the agreement between Refs. 1 and 2 as well as with earlier calculations (Refs. 3–5) is very good—typically of the order of ±10%. Only a few larger disagreements occur. However, for $3p$-$3d$ transitions, the disagreements between the multiplet data of Refs. 1 and 2 are substantial. For some of these, as well as for several $3s$-$3p$ transitions, emission measurements by Veres *et al.*[6] are available and have been applied. These experimental data, whose absolute scale is based on a recent cascade-free lifetime measurement of Coetzer *et al.*,[7] are in better agreement with the CIV 3 than with the OP multiplet data. Also, for individual lines, the emission data are in much better agreement with the intermediate coupling data of Bell *et al.*[2] than with *LS*-coupling ratios. We found that they usually agree within a range from ±5% to ±15%. When OP multiplet values were the only data source, we used *LS*-coupling fractions for the individual lines, but applied very conservative error estimates. We were also guided by the observation that the emission data for strong lines show better agreement with the *LS*-coupling predictions, while the weaker lines deviate appreciably.

Some lifetime data,[7–9] obtained with state-selective excitation techniques, are also available for comparisons. Principally, transitions originating from $3p$ states are involved, for which our two main data sources[1,2] are in very good agreement. For OP, comparisons for 10 levels are possible and give a mean value of 1.08 ± 3% for the ratio τ_{expt}/τ_{op}; for the CIV 3 calculations (Ref. 2), the analogous ratio is 1.07 ± 2.1% for 7 levels.

References

[1] K. L. Bell and K. A. Berrington, J. Phys. B **24**, 933 (1991).

[2] K. L. Bell, A. Hibbert, R. P. Stafford, and B. M. McLaughlin, Phys. Scr. **50**, 343 (1994).

[3] S. R. Becker and K. Butler, Astron. Astrophys. **201**, 232 (1988).

[4] K. L. Bell, A. Hibbert, B. M. McLaughlin, and K. Higgins, J. Phys. B **24**, 2665 (1991).

[5] Y. K. Ho and R. J. W. Henry, J. Quant. Spectrosc. Radiat. Transfer **31**, 57 (1984).

[6] G. Veres and W. L. Wiese, to be published.

[7] F. J. Coetzer, T. C. Kotze, F. J. Mostert, and P. van der Westhuizen, Phys. Scr. **34**, 328 (1986).

[8] T. C. Kotze, F. J. Coetzer, and P. van der Westhuizen, J. Quant. Spectrosc. Radiat. Transfer **36**, 249 (1986).

[9] F. J. Coetzer, T. C. Kotze and P. van der Westhuizen, Z. Phys. A **322**, 357 (1985).

O II: Allowed Transitions

No.	Transition Array	Mult.	λ_{air} (Å)	λ_{vac} (Å) or σ (cm^{-1})*	E_i (cm^{-1})	E_k (cm^{-1})	g_i–g_k	A_{ki} (10^8 s^{-1})	f_{ik}	S (at. u.)	log gf	Acc.	Source
1.	$2s^22p^3$– $2s2p^4$	$^4S^o$–^4P		*833.80*	0.00	*119932.6*	4–12	8.63+00	2.70–01	2.96+00	0.033	B+	1,2
				834.465	0.00	119837.21	4–6	8.61+00	1.35–01	1.48+00	–0.268	B+	2*n*
				833.330	0.00	120000.43	4–4	8.65+00	9.01–02	9.88–01	–0.443	B+	2*n*
				832.758	0.00	120082.86	4–2	8.67+00	4.51–02	4.94–01	–0.744	B+	2*n*
2.	$2s^22p^3$– $2s^22p^2(^3P)3s$	$^4S^o$–^4P		*539.37*	0.00	*185402.30*	4–12	9.82+00	1.29–01	9.13–01	–0.289	B+	1,2
				539.086	0.00	185499.124	4–6	9.83+00	6.43–02	4.56–01	–0.590	B+	2*n*
				539.547	0.00	185340.577	4–4	9.81+00	4.28–02	3.04–01	–0.766	B+	2*n*
				539.854	0.00	185235.281	4–2	9.81+00	2.14–02	1.52–01	–1.067	B+	2*n*
3.	$2s^22p^3$– $2s^22p^2(^3P)3d$	$^4S^o$–^4P		*430.09*	0.00	*232510.43*	4–12	4.26+01	3.55–01	2.01+00	0.152	B+	1,2
				430.176	0.00	232462.724	4–6	4.36+01	1.81–01	1.03+00	–0.139	B+	2*n*
				430.041	0.00	232535.949	4–4	4.13+01	1.15–01	6.48–01	–0.339	B+	2*n*
				429.918	0.00	232602.492	4–2	4.25+01	5.89–02	3.34–01	–0.627	B+	2*n*
4.		$^4S^o$–^4D											
				429.716	0.00	232711.642	4–2	5.21+00	7.22–03	4.08–02	–1.540	C	2
				429.653	0.00	232745.981	4–4	6.35+00	1.76–02	9.95–02	–1.153	C	2
				429.650	0.00	232747.562	4–6	3.49+00	1.45–02	8.20–02	–1.237	C	2
5.		$^4S^o$–^2F											
				429.560	0.00	232796.298	4–6	6.52–01	2.71–03	1.53–02	–1.966	D	2
6.		$^4S^o$–^2P											
				428.393	0.00	233430.53	4–4	3.06–02	8.42–05	4.75–04	–3.472	C	2
7.		$^4S^o$–^2D											
				426.522	0.00	234454.634	4–6	2.06–02	8.42–05	4.73–04	–3.472	C	2
8.	$2s^22p^3$– $2s^22p^2(^3P)4s$	$^4S^o$–^4P		*418.77*	0.00	*238795.8*	4–12	1.86+00	1.47–02	8.09–02	–1.231	B	1
				418.596	0.00	238893.96	4–6	1.86+00	7.34–03	4.05–02	–1.532	C+	*LS*
				418.879	0.00	238732.65	4–4	1.86+00	4.89–03	2.70–02	–1.709	C+	*LS*
				419.063	0.00	238627.46	4–2	1.86+00	2.44–03	1.35–02	–2.010	C+	*LS*

O II: Allowed Transitions — Continued

No.	Transition Array	Mult.	λ_{air} (Å)	λ_{vac} (Å) or σ (cm^{-1})*	E_i (cm^{-1})	E_k (cm^{-1})	g_i-g_k	A_{ki} (10^8 s^{-1})	f_{ik}	S (at. u.)	log gf	Acc.	Source
9.	$2s^22p^3-$ $2s^22p^2(^3P)4d$	$^4S^o-^4P$		*391.96*	0.00	– *255127.2*	4–12	2.38+01	1.64−01	8.48−01	−0.182	B	1
				391.995	0.00	– 255105.01	4–6	2.38+01	8.21−02	4.24−01	−0.484	C+	*LS*
				391.938	0.00	– 255142.41	4–4	2.38+01	5.47−02	2.83−01	−0.660	C+	*LS*
				391.906	0.00	– 255163.08	4–2	2.38+01	2.74−02	1.41−01	−0.961	C+	*LS*
10.	$2s^22p^3-$ $2s^22p^2(^3P)5s$	$^4S^o-^4P$		*387.80*	0.00	– *257865.2*	4–12	5.59−01	3.78−03	1.93−02	−1.821	B	1
				387.649	0.00	– 257965.11	4–6	5.59−01	1.89−03	9.65−03	−2.121	C+	*LS*
				387.898	0.00	– 257799.93	4–4	5.58−01	1.26−03	6.43−03	−2.298	C+	*LS*
				388.055	0.00	– 257695.74	4–2	5.58−01	6.29−04	3.22−03	−2.599	C+	*LS*
11.	$2s^22p^3-$ $2s2p^4$	$^2D^o-^2D$		*718.53*	*26818.6*	– *165991.7*	10–10	2.04+01	1.58−01	3.73+00	0.198	B	1,2
				718.505	26810.55	– 165988.46	6–6	1.90+01	1.47−01	2.09+00	−0.054	B	2*n*
				718.567	26830.57	– 165996.50	4–4	1.85+01	1.43−01	1.35+00	−0.242	B	2*n*
				718.463	26810.55	– 165996.50	6–4	2.01+00	1.04−02	1.47−01	−1.207	B	2*n*
				718.608	26830.57	– 165988.46	4–6	1.30+00	1.51−02	1.43−01	−1.218	B	2*n*
12.	$2s^22p^3-$ $2s^22p^2(^3P)3s$	$^2D^o-^2P$		*616.56*	*26818.6*	– *189008.52*	10–6	3.43+01	1.17−01	2.38+00	0.069	B	1,2
				616.303	26810.55	– 189068.514	6–4	3.11+01	1.18−01	1.44+00	−0.150	B	2*n*
				617.063	26830.57	– 188888.543	4–2	3.43+01	9.80−02	7.96−01	−0.407	B	2*n*
				616.379	26830.57	– 189068.514	4–4	3.24+00	1.85−02	1.50−01	−1.132	B	2*n*
13.	$2s^22p^3-$ $2s^22p^2(^1D)3s$	$^2D^o-^2D$		*555.08*	*26818.6*	– *206972.1*	10–10	1.11+01	5.15−02	9.40−01	−0.288	B	1,2
				555.059	26810.55	– 206971.68	6–6	1.04+01	4.80−02	5.27−01	−0.540	B	2*n*
				555.117	26830.57	– 206972.72	4–4	1.01+01	4.66−02	3.41−01	−0.730	B	2*n*
				555.055	26810.55	– 206972.72	6–4	9.67−01	2.98−03	3.26−02	−1.748	B−	2*n*
				555.120	26830.57	– 206971.68	4–6	8.01−01	5.55−03	4.06−02	−1.654	B	2*n*
14.	$2s^22p^3-$ $2s2p^4$	$^2D^o-^2P$		*538.12*	*26818.6*	– *212650.0*	10–6	5.75+01	1.50−01	2.65+00	0.176	B+	1,2
				538.262	26810.55	– 212593.82	6–4	5.18+01	1.50−01	1.59+00	−0.046	B+	2*n*
				537.832	26830.57	– 212762.25	4–2	5.73+01	1.24−01	8.79−01	−0.304	B+	2*n*
				538.320	26830.57	– 212593.82	4–4	5.87+00	2.55−02	1.81−01	−0.992	B+	2*n*
15.	$2s^22p^3-$ $2s^22p^2(^3P)3d$	$^2D^o-^4F$											
				488.717	26810.55	– 231427.970	6–8	3.28−02	1.57−04	1.51−03	−3.027	C	2
				488.951	26830.57	– 231350.087	4–6	2.43−02	1.31−04	8.41−04	−3.282	C	2
16.		$^2D^o-^4P$											
				486.085	26810.55	– 232535.949	6–4	4.50−02	1.06−04	1.02−03	−3.196	C	2
				485.975	26830.57	– 232602.492	4–2	3.10−02	5.49−05	3.51−04	−3.658	C	2
				486.305	26830.57	– 232462.724	4–6	3.53−02	1.88−04	1.20−03	−3.125	C	2
17.		$^2D^o-^4D$											
				485.571	26810.55	– 232753.816	6–8	1.52+00	7.16−03	6.87−02	−1.367	C	2
				485.633	26830.57	– 232747.562	4–6	6.15+00	3.26−02	2.09−01	−0.884	C	2
				485.585	26810.55	– 232747.562	6–6	3.30−01	1.17−03	1.12−02	−2.154	C	2
				485.589	26810.55	– 232745.981	6–4	4.09−02	9.64−05	9.24−04	−3.238	C	2
				485.717	26830.57	– 232711.642	4–2	8.13−02	1.44−04	9.20−04	−3.240	C	2
18.		$^2D^o-^2F$		*485.27*	*26818.6*	– *232889.39*	10–14	2.36+01	1.17−01	1.87+00	0.068	B	1,2
				485.087	26810.55	– 232959.210	6–8	2.60+01	1.22−01	1.17+00	−0.135	B	2*n*
				485.518	26830.57	– 232796.298	4–6	1.93+01	1.03−01	6.55−01	−0.387	C+	2*n*
				485.470	26810.55	– 232796.298	6–6	1.20+00	4.24−03	4.07−02	−1.594	C	2*n*

O II: Allowed Transitions — Continued

No.	Transition Array	Mult.	λ_{air} (Å)	λ_{vac} (Å) or σ (cm^{-1})*	E_i (cm^{-1})	E_k (cm^{-1})	g_i-g_k	A_{ki} (10^8 s^{-1})	f_{ik}	S (at. u.)	log gf	Acc.	Source
19.		$^2D^o-^2P$		*483.91*	*26818.6*	*– 233468.6*	10–6	2.10+01	4.42−02	7.04−01	−0.354	B+	1,2
				483.980	26810.55	– 233430.53	6–4	1.80+01	4.21−02	4.03−01	−0.597	B	2*n*
				483.760	26830.57	– 233544.59	4–2	2.05+01	3.60−02	2.29−01	−0.841	B	2*n*
				484.027	26830.57	– 233430.53	4–4	3.22+00	1.13−02	7.22−02	−1.344	C	2*n*
20.		$^2D^o-^2D$		*481.66*	*26818.6*	*– 234433.90*	10–10	7.85+00	2.73−02	4.33−01	−0.564	B	1,2
				481.593	26810.55	– 234454.634	6–6	7.60+00	2.64−02	2.52−01	−0.800	B	2*n*
				481.760	26830.57	– 234402.797	4–4	6.00+00	2.09−02	1.32−01	−1.078	B	2*n*
				481.714	26810.55	– 234402.797	6–4	1.65+00	3.82−03	3.63−02	−1.640	C	2*n*
				481.640	26830.57	– 234454.634	4–6	3.75−01	1.96−03	1.24−02	−2.106	C	2*n*
21.	$2s^22p^3-$ $2s^22p^2(^3P)4s$	$^2D^o-^2P$		*468.09*	*26818.6*	*– 240454.9*	10–6	8.49−01	1.67−03	2.58−02	−1.776	B−	1
				467.931	26810.55	– 240517.35	6–4	7.65−01	1.67−03	1.55−02	−1.998	C	*LS*
				468.385	26830.57	– 240330.01	4–2	8.48−01	1.39−03	8.60−03	−2.254	C−	*LS*
				467.975	26830.57	– 240517.35	4–4	8.50−02	2.79−04	1.72−03	−2.952	D	*LS*
22.	$2s^22p^3-$ $2s^22p^2(^1D)3d$	$^2D^o-^2F$		*445.62*	*26818.6*	*– 251223.3*	10–14	1.44+01	5.99−02	8.78−01	−0.223	C+	1
				445.610	26810.55	– 251222.19	6–8	1.44+01	5.70−02	5.02−01	−0.466	C−	*LS*
				445.644	26830.57	– 251224.79	4–6	1.34+01	5.99−02	3.51−01	−0.621	C−	*LS*
				445.605	26810.55	– 251224.79	6–6	9.58−01	2.85−03	2.51−02	−1.767	D	*LS*
23.		$^2D^o-^2D$		*442.03*	*26818.6*	*– 253048.0*	10–10	2.79+01	8.18−02	1.19+00	−0.087	C+	1
				442.012	26810.55	– 253048.82	6–6	2.61+01	7.63−02	6.66−01	−0.339	D	*LS*
				442.055	26830.57	– 253046.74	4–4	2.51+01	7.36−02	4.28−01	−0.531	D	*LS*
				442.016	26810.55	– 253046.74	6–4	2.79+00	5.45−03	4.76−02	−1.485	D	*LS*
				442.051	26830.57	– 253048.82	4–6	1.86+00	8.18−03	4.76−02	−1.485	D	*LS*
24.		$^2D^o-^2P$		*440.58*	*26818.6*	*– 253791.6*	10–6	1.53+01	2.67−02	3.87−01	−0.574	C	1
				440.564	26810.55	– 253792.40	6–4	1.37+01	2.67−02	2.32−01	−0.796	D	*LS*
				440.607	26830.57	– 253789.99	4–2	1.53+01	2.22−02	1.29−01	−1.051	D	*LS*
				440.603	26830.57	– 253792.40	4–4	1.53+00	4.44−03	2.58−02	−1.750	D−	*LS*
25.	$2s^22p^3-$ $2s^22p^2(^3P)4d$	$^2D^o-^2P$		*437.85*	*26818.6*	*– 255209.7*	10–6	9.49−01	1.64−03	2.36−02	−1.786	C	1
				437.899	26810.55	– 255173.58	6–4	8.54−01	1.64−03	1.42−02	−2.008	D	*LS*
				437.730	26830.57	– 255281.93	4–2	9.50−01	1.36−03	7.87−03	−2.263	D−	*LS*
				437.938	26830.57	– 255173.58	4–4	9.49−02	2.73−04	1.57−03	−2.962	D−	*LS*
26.		$^2D^o-^2F$		*437.49*	*26818.6*	*– 255395.8*	10–14	2.04+01	8.18−02	1.18+00	−0.087	C+	1
				437.339	26810.55	– 255466.10	6–8	2.04+01	7.79−02	6.73−01	−0.330	C−	*LS*
				437.691	26830.57	– 255302.11	4–6	1.90+01	8.18−02	4.71−01	−0.485	C−	*LS*
				437.653	26810.55	– 255302.11	6–6	1.36+00	3.89−03	3.37−02	−1.632	D	*LS*
27.		$^2D^o-^2D$		*436.57*	*26818.6*	*– 255875.7*	10–10	3.10+00	8.86−03	1.27−01	−1.053	C	1
				436.515	26810.55	– 255897.59	6–6	2.89+00	8.27−03	7.13−02	−1.304	D	*LS*
				436.658	26830.57	– 255842.91	4–4	2.79+00	7.97−03	4.58−02	−1.496	D	*LS*
				436.620	26810.55	– 255842.91	6–4	3.10−01	5.90−04	5.09−03	−2.451	D−	*LS*
				436.553	26830.57	– 255897.59	4–6	2.07−01	8.86−04	5.09−03	−2.451	D−	*LS*
28.	$2s^22p^3-$ $2s^22p^2(^3P)5s$	$^2D^o-^2P$		*431.55*	*26818.6*	*– 258541.3*	10–6	9.33−01	1.56−03	2.22−02	−1.806	C	1
				431.414	26810.55	– 258606.35	6–4	8.40−01	1.56−03	1.33−02	−2.028	D	*LS*
				431.815	26830.57	– 258411.26	4–2	9.31−01	1.30−03	7.40−03	−2.284	D−	*LS*
				431.451	26830.57	– 258606.35	4–4	9.33−02	2.60−04	1.48−03	−2.982	D−	*LS*

O II: Allowed Transitions — Continued

No.	Transition Array	Mult.	λ_{air} (Å)	λ_{vac} (Å) or σ (cm^{-1})*	E_i (cm^{-1})	E_k (cm^{-1})	g_i-g_k	A_{ki} (10^8 s^{-1})	f_{ik}	S (at. u.)	log gf	Acc.	Source
29.	$2s^22p^3-$ $2s^22p^2(^1D)4s$	$^2D°-^2D$		*430.16*	*26818.6*	*– 259287.8*	10–10	2.74+00	7.61–03	1.08–01	–1.119	C	1
				430.150	26810.55	– 259287.61	6–6	2.56+00	7.10–03	6.03–02	–1.371	D	*LS*
				430.186	26830.57	– 259288.07	4–4	2.47+00	6.84–03	3.88–02	–1.563	D	*LS*
				430.149	26810.55	– 259288.07	6–4	2.74–01	5.07–04	4.31–03	–2.517	D–	*LS*
				430.187	26830.57	– 259287.61	4–6	1.83–01	7.60–04	4.31–03	–2.517	D–	*LS*
30.	$2s^22p^3-$ $2s^22p^2(^1D)4d$	$^2D°-^2F$		*403.31*	*26818.6*	*– 274764*	10–14	1.24+01	4.22–02	5.60–01	–0.375	C	1
				403.273	26810.55	– 274781.5	6–8	1.24+01	4.02–02	3.20–01	–0.618	D	*LS*
				403.372	26830.57	– 274740.7	4–6	1.15+01	4.22–02	2.24–01	–0.773	D	*LS*
				403.339	26810.55	– 274740.7	6–6	8.24–01	2.01–03	1.60–02	–1.919	D–	*LS*
31.		$^2D°-^2D$		*403.06*	*26818.6*	*– 274923*	10–10	1.20+01	2.93–02	3.89–01	–0.533	C	1
				403.035	26810.55	– 274928.0	6–6	1.12+01	2.74–02	2.18–01	–0.784	D	*LS*
				403.087	26830.57	– 274916.0	4–4	1.08+01	2.64–02	1.40–01	–0.976	D	*LS*
				403.054	26810.55	– 274916.0	6–4	1.20+00	1.96–03	1.56–02	–1.931	D–	*LS*
				403.067	26830.57	– 274928.0	4–6	8.03–01	2.93–03	1.56–02	–1.931	D–	*LS*
32.	$2s^22p^3-$ $2s^22p^2(^1S)3d$	$^2D°-^2D$		*401.25*	*26818.6*	*– 276038.8*	10–10	2.01+00	4.85–03	6.41–02	–1.314	D	1
				401.180	26810.55	– 276075.32	6–6	1.88+00	4.53–03	3.59–02	–1.566	D–	*LS*
				401.359	26830.57	– 275983.95	4–4	1.81+00	4.37–03	2.31–02	–1.758	D–	*LS*
				401.327	26810.55	– 275983.95	6–4	2.01–01	3.23–04	2.56–03	–2.712	E	*LS*
				401.212	26830.57	– 276075.32	4–6	1.34–01	4.85–04	2.56–03	–2.712	E	*LS*
33.	$2s^22p^3-$ $2s2p^4$	$^2P°-^2D$		*796.67*	*40468.7*	*– 165991.7*	6–10	1.98+00	3.15–02	4.95–01	–0.724	B+	1,2
				796.683	40468.01	– 165988.46	4–6	2.02+00	2.89–02	3.03–01	–0.937	B+	2*n*
				796.645	40470.00	– 165996.50	2–4	1.65+00	3.13–02	1.64–01	–1.203	B+	2*n*
				796.632	40468.01	– 165996.50	4–4	2.79–01	2.66–03	2.79–02	–1.974	B	2*n*
34.	$2s^22p^3-$ $2s^22p^2(^3P)3s$	$^2P°-^2P$		*673.22*	*40468.7*	*– 189008.52*	6–6	5.67+00	3.85–02	5.12–01	–0.636	B+	1,2
				672.945	40468.01	– 189068.514	4–4	4.78+00	3.25–02	2.88–01	–0.886	B+	2*n*
				673.770	40470.00	– 188888.543	2–2	3.63+00	2.47–02	1.10–01	–1.306	B+	2*n*
				673.761	40468.01	– 188888.543	4–2	1.86+00	6.33–03	5.61–02	–1.597	B+	2*n*
				672.954	40470.00	– 189068.514	2–4	9.68–01	1.31–02	5.83–02	–1.580	B+	2*n*
35.	$2s^22p^3-$ $2s2p^4$	$^2P°-^2S$		*644.16*	*40468.7*	– 195710.47	6–2	5.27+01	1.09–01	1.39+00	–0.184	B+	1,2
				644.154	40468.01	– 195710.47	4–2	3.48+01	1.08–01	9.17–01	–0.364	B+	2*n*
				644.162	40470.00	– 195710.47	2–2	1.79+01	1.11–01	4.72–01	–0.652	B+	2*n*
36.	$2s^22p^3-$ $2s^22p^2(^1D)3s$	$^2P°-^2D$		*600.59*	*40468.7*	*– 206972.1*	6–10	4.76+00	4.29–02	5.09–01	–0.590	B	1,2
				600.587	40468.01	– 206971.68	4–6	4.71+00	3.82–02	3.02–01	–0.816	B	2*n*
				600.591	40470.00	– 206972.72	2–4	3.96+00	4.28–02	1.69–01	–1.067	B	2*n*
				600.584	40468.01	– 206972.72	4–4	8.72–01	4.71–03	3.73–02	–1.725	C+	2*n*
37.	$2s^22p^3-$ $2s2p^4$	$^2P°-^2P$		*580.78*	*40468.7*	*– 212650.0*	6–6	1.59+01	8.02–02	9.21–01	–0.317	B	1,2
				580.970	40468.01	– 212593.82	4–4	1.30+01	6.60–02	5.05–01	–0.578	B	2*n*
				580.409	40470.00	– 212762.25	2–2	1.03+01	5.19–02	1.98–01	–0.984	B	2*n*
				580.402	40468.01	– 212762.25	4–2	5.97+00	1.51–02	1.15–01	–1.220	C+	2*n*
				580.977	40470.00	– 212593.82	2–4	2.63+00	2.66–02	1.02–01	–1.273	B	2*n*
38.	$2s^22p^3-$ $2s^22p^2(^1S)3s$	$^2P°-^2S$		*525.93*	*40468.7*	– 230609.45	6–2	9.45+00	1.31–02	1.36–01	–1.106	D	1,2
				525.924	40468.01	– 230609.45	4–2	6.21+00	1.29–02	8.92–02	–1.288	D	2*n*
				525.930	40470.00	– 230609.45	2–2	3.24+00	1.34–02	4.65–02	–1.571	D	2*n*

O II: Allowed Transitions — Continued

No.	Transition Array	Mult.	λ_{air} (Å)	λ_{vac} (Å) or σ (cm^{-1})*	E_i (cm^{-1})	E_k (cm^{-1})	g_i-g_k	A_{ki} (10^8 s^{-1})	f_{ik}	S (at. u.)	log gf	Acc.	Source
39.	$2s^22p^3-$ $2s^22p^2(^3P)3d$	$^2P^o-^4P$											
				520.848	40468.01	– 232462.724	4–6	1.42–02	8.66–05	5.94–04	−3.461	C	2
				520.649	40468.01	– 232535.949	4–4	6.62–02	2.69–04	1.84–03	−2.968	C	2
				520.474	40470.00	– 232602.492	2–2	2.99–02	1.21–04	4.16–04	−3.615	C	2
40.		$^2P^o-^4D$											
				520.086	40470.00	– 232745.981	2–4	8.03–03	6.51–05	2.23–04	−3.885	C	2
				520.080	40468.01	– 232745.981	4–4	4.15–02	1.68–04	1.15–03	−3.172	C	2
				520.179	40470.00	– 232711.642	2–2	6.82–02	2.77–04	9.48–04	−3.257	C	2
				520.173	40468.01	– 232711.642	4–2	3.49–02	7.07–05	4.85–04	−3.548	C	2
41.		$^2P^o-^2F$											
				519.944	40468.01	– 232796.298	4–6	1.18–02	7.18–05	4.92–04	−3.542	C	2
42.		$^2P^o-^2P$		*518.14*	*40468.7*	*– 233468.6*	6–6	2.70+01	1.09–01	1.11+00	−0.186	C+	1,2
				518.235	40468.01	– 233430.53	4–4	2.07+01	8.32–02	5.68–01	−0.478	C+	2*n*
				517.935	40470.00	– 233544.59	2–2	1.82+01	7.33–02	2.50–01	−0.834	C+	2*n*
				517.929	40468.01	– 233544.59	4–2	9.35+00	1.88–02	1.28–01	−1.124	C+	2*n*
				518.241	40470.00	– 233430.53	2–4	6.05+00	4.87–02	1.66–01	−1.011	D	2*n*
43.		$^2P^o-^2D$		*515.56*	*40468.7*	*– 234433.90*	6–10	1.84+01	1.22–01	1.25+00	−0.134	B	1,2
				515.499	40468.01	– 234454.634	4–6	1.83+01	1.09–01	7.43–01	−0.359	B	2*n*
				515.643	40470.00	– 234402.797	2–4	1.36+01	1.08–01	3.68–01	−0.664	B	2*n*
				515.637	40468.01	– 234402.797	4–4	5.03+00	2.01–02	1.36–01	−1.096	D	2*n*
44.	$2s^22p^3-$ $2s^22p^2(^3P)4s$	$^2P^o-^2P$		*500.03*	*40468.7*	*– 240454.9*	6–6	8.52+00	3.19–02	3.16–01	−0.717	C	1
				499.877	40468.01	– 240517.35	4–4	7.11+00	2.66–02	1.75–01	−0.973	D	*LS*
				500.350	40470.00	– 240330.01	2–2	5.67+00	2.13–02	7.01–02	−1.371	D	*LS*
				500.345	40468.01	– 240330.01	4–2	2.84+00	5.32–03	3.51–02	−1.672	D	*LS*
				499.882	40470.00	– 240517.35	2–4	1.42+00	1.07–02	3.51–02	−1.672	D	*LS*
45.	$2s^22p^3-$ $2s^22p^2(^1D)3d$	$^2P^o-^2D$		*470.41*	*40468.7*	*– 253048.0*	6–10	8.06+00	4.46–02	4.14–01	−0.573	C	1
				470.409	40468.01	– 253048.82	4–6	8.06+00	4.01–02	2.48–01	−0.795	D	*LS*
				470.418	40470.00	– 253046.74	2–4	6.71+00	4.46–02	1.38–01	−1.050	D	*LS*
				470.414	40468.01	– 253046.74	4–4	1.34+00	4.46–03	2.76–02	−1.749	D−	*LS*
46.		$^2P^o-^2P$		*468.77*	*40468.7*	*– 253791.6*	6–6	2.74+00	9.04–03	8.37–02	−1.266	D	1
				468.770	40468.01	– 253792.40	4–4	2.29+00	7.53–03	4.65–02	−1.521	D−	*LS*
				468.779	40470.00	– 253789.99	2–2	1.83+00	6.03–03	1.86–02	−1.919	D−	*LS*
				468.775	40468.01	– 253789.99	4–2	9.15–01	1.51–03	9.30–03	−2.220	E	*LS*
				468.774	40470.00	– 253792.40	2–4	4.57–01	3.01–03	9.30–03	−2.220	E	*LS*
47.	$2s^22p^3-$ $2s^22p^2(^3P)4d$	$^2P^o-^2P$		*465.68*	*40468.7*	*– 255209.7*	6–6	1.67+01	5.42–02	4.99–01	−0.488	C−	1
				465.754	40468.01	– 255173.58	4–4	1.39+01	4.52–02	2.77–01	−0.743	D	*LS*
				465.523	40470.00	– 255281.93	2–2	1.11+01	3.61–02	1.11–01	−1.141	D	*LS*
				465.519	40468.01	– 255281.93	4–2	5.56+00	9.04–03	5.54–02	−1.442	E	*LS*
				465.758	40470.00	– 255173.58	2–4	2.78+00	1.81–02	5.54–02	−1.442	E	*LS*
48.	$2s^22p^3-$ $2s^22p^2(^1D)3d$	$^2P^o-^2S$		*464.78*	*40468.7*	– 255622.80	6–2	1.77+01	1.91–02	1.75–01	−0.942	C−	1
				464.782	40468.01	– 255622.80	4–2	1.18+01	1.91–02	1.17–01	−1.118	D−	*LS*
				464.786	40470.00	– 255622.80	2–2	5.89+00	1.91–02	5.83–02	−1.419	D−	*LS*

O II: Allowed Transitions — Continued

No.	Transition Array	Mult.	λ_{air} (Å)	λ_{vac} (Å) or σ (cm^{-1})*	E_i (cm^{-1})	E_k (cm^{-1})	g_i–g_k	A_{ki} (10^8 s^{-1})	f_{ik}	S (at. u.)	log gf	Acc.	Source
49.	$2s^22p^3$– $2s^22p^2(^3P)4d$	$^2P^o$–^2D		*464.24*	*40468.7*	– *255875.7*	6–10	8.97+00	4.83–02	4.43–01	–0.538	C–	1
				464.189	40468.01	– 255897.59	4–6	8.98+00	4.35–02	2.66–01	–0.760	D–	*LS*
				464.311	40470.00	– 255842.91	2–4	7.47+00	4.83–02	1.48–01	–1.015	D–	*LS*
				464.307	40468.01	– 255842.91	4–4	1.49+00	4.83–03	2.95–02	–1.714	E	*LS*
50.	$2s^22p^3$– $2s^22p^2(^3P)5s$	$^2P^o$–^2P		*458.56*	*40468.7*	– *258541.3*	6–6	3.32+00	1.05–02	9.47–02	–1.203	D	1
				458.425	40468.01	– 258606.35	4–4	2.77+00	8.72–03	5.26–02	–1.458	D–	*LS*
				458.839	40470.00	– 258411.26	2–2	2.21+00	6.97–03	2.10–02	–1.856	D–	*LS*
				458.835	40468.01	– 258411.26	4–2	1.10+00	1.74–03	1.05–02	–2.157	E	*LS*
				458.429	40470.00	– 258606.35	2–4	5.53–01	3.49–03	1.05–02	–2.157	E	*LS*
51.	$2s^22p^3$– $2s^22p^2(^1D)4s$	$^2P^o$–^2D		*457.00*	*40468.7*	– *259287.8*	6–10	2.50+00	1.31–02	1.18–01	–1.106	C–	1
				456.997	40468.01	– 259287.61	4–6	2.50+00	1.18–02	7.07–02	–1.328	D–	*LS*
				457.001	40470.00	– 259288.07	2–4	2.09+00	1.31–02	3.93–02	–1.583	D–	*LS*
				456.996	40468.01	– 259288.07	4–4	4.17–01	1.31–03	7.86–03	–2.282	E	*LS*
52.	$2s^22p^3$– $2s^22p^2(^1D)4d$	$^2P^o$–^2D		*426.52*	*40468.7*	– *274923*	6–10	1.23+01	5.57–02	4.70–01	–0.476	C–	1
				426.512	40468.01	– 274928.0	4–6	1.23+01	5.02–02	2.82–01	–0.698	D	*LS*
				426.537	40470.00	– 274916.0	2–4	1.02+01	5.57–02	1.57–01	–0.953	D	*LS*
				426.534	40468.01	– 274916.0	4–4	2.04+00	5.57–03	3.13–02	–1.652	E	*LS*
53.	$2s^22p^3$– $2s^22p^2(^1S)3d$	$^2P^o$–^2D		*424.50*	*40468.7*	– *276038.8*	6–10	7.99+00	3.60–02	3.02–01	–0.666	C–	1
				424.435	40468.01	– 276075.32	4–6	7.99+00	3.24–02	1.81–01	–0.888	D	*LS*
				424.603	40470.00	– 275983.95	2–4	6.65+00	3.60–02	1.01–01	–1.143	D	*LS*
				424.600	40468.01	– 275983.95	4–4	1.33+00	3.60–03	2.01–02	–1.842	E	*LS*
54.	$2s2p^4$– $2s^22p^2(^3P)3p$	^4P–$^4D^o$		*1149.9*	*119932.6*	– *206894.69*	12–20	5.80–02	1.92–03	8.71–02	–1.638	C+	1,2
				1147.25	119837.21	– 207002.482	6–8	5.78–02	1.52–03	3.45–02	–2.040	C+	2*n*
				1151.05	120000.43	– 206877.865	4–6	4.61–02	1.37–03	2.08–02	–2.260	C	2*n*
				1153.36	120082.86	– 206786.286	2–4	2.89–02	1.15–03	8.76–03	–2.637	C	2*n*
				1148.89	119837.21	– 206877.865	6–6	1.22–02	2.41–04	5.46–03	–2.840	C+	2*n*
				1152.26	120000.43	– 206786.286	4–4	2.78–02	5.54–04	8.40–03	–2.654	C+	2*n*
				1154.10	120082.86	– 206730.762	2–2	5.05–02	1.01–03	7.66–03	–2.696	C	2*n*
				1150.10	119837.21	– 206786.286	6–4	1.52–03	2.01–05	4.56–04	–3.919	D	2*n*
				1153.00	120000.43	– 206730.762	4–2	7.27–03	7.24–05	1.10–03	–3.538	C+	2*n*
55.		^4P–$^4P^o$		*1130.0*	*119932.6*	– *208430.54*	12–12	4.64–01	8.89–03	3.97–01	–0.972	B	1,2
				1128.07	119837.21	– 208484.202	6–6	3.32–01	6.33–03	1.41–01	–1.421	B	2*n*
				1131.33	120000.43	– 208392.258	4–4	6.05–02	1.16–03	1.73–02	–2.333	B	2*n*
				1132.97	120082.86	– 208346.104	2–2	7.45–02	1.43–03	1.07–02	–2.542	B	2*n*
				1129.24	119837.21	– 208392.258	6–4	2.21–01	2.82–03	6.29–02	–1.771	B	2*n*
				1131.92	120000.43	– 208346.104	4–2	3.89–01	3.74–03	5.57–02	–1.826	B	2*n*
				1130.15	120000.43	– 208484.202	4–6	1.33–01	3.83–03	5.70–02	–1.815	B	2*n*
				1132.38	120082.86	– 208392.258	2–4	1.82–01	7.00–03	5.22–02	–1.854	B	2*n*
56.		^4P–$^4S^o$		*1084.3*	*119932.6*	– 212161.881	12–4	5.31–01	3.12–03	1.34–01	–1.427	B	1,2
				1083.13	119837.21	– 212161.881	6–4	2.55–01	2.99–03	6.40–02	–1.746	B	2*n*
				1085.05	120000.43	– 212161.881	4–4	1.82–01	3.21–03	4.58–02	–1.892	B	2*n*
				1086.02	120082.86	– 212161.881	2–4	9.44–02	3.34–03	2.39–02	–2.175	C+	2*n*
57.	$2s2p^4$– $2s2p^3(^5S^o)3s$	^4P–$^4S^o$		*740.47*	*119932.6*	– 254981.55	12–4	1.40+01	3.85–02	1.13+00	–0.336	C	1
				739.950	119837.21	– 254981.55	6–4	7.03+00	3.85–02	5.63–01	–0.636	D	*LS*
				740.844	120000.43	– 254981.55	4–4	4.67+00	3.84–02	3.75–01	–0.813	D	*LS*
				741.297	120082.86	– 254981.55	2–4	2.33+00	3.84–02	1.88–01	–1.114	D–	*LS*

O II: Allowed Transitions — Continued

No.	Transition Array	Mult.	λ_{air} (Å)	λ_{vac} (Å) or σ (cm^{-1})*	E_i (cm^{-1})	E_k (cm^{-1})	g_i-g_k	A_{ki} (10^8 s^{-1})	f_{ik}	S (at. u.)	log gf	Acc.	Source
58.	$2s2p^4-$ $2s^22p^2(^3P)3p$	$^2D-^2D^\circ$											
			2186.34	2187.02	165988.46	211712.732	6−6	9.17−04	6.58−05	2.84−03	−3.404	C	2
			2195.88	2196.57	165996.50	211522.117	4−4	8.36−04	6.05−05	1.75−03	−3.616	C	2
59.		$^2D-^2P^\circ$	*2073.2*	*2073.9*	*165991.7*	*214209.75*	10−6	3.63−02	1.40−03	9.58−02	−1.853	C+	1,2
			2072.26	2072.92	165988.46	214229.671	6−4	3.30−02	1.42−03	5.80−02	−2.070	C+	2*n*
			2075.17	2075.83	165996.50	214169.920	4−2	3.55−02	1.15−03	3.14−02	−2.338	C+	2*n*
			2072.60	2073.26	165996.50	214229.671	4−4	3.61−03	2.33−04	6.35−03	−3.031	C+	2*n*
60.	$2s2p^4-$ $2s^22p^2(^1D)3p$	$^2D-^2F^\circ$		*1593.7*	*165991.7*	*228737.3*	10−14	2.26−01	1.21−02	6.33−01	−0.918	C+	1,2
				1593.40	165988.46	228747.45	6−8	2.27−01	1.15−02	3.62−01	−1.161	C+	2*n*
				1594.20	165996.50	228723.84	4−6	2.11−01	1.21−02	2.53−01	−1.317	C+	2*n*
				1594.00	165988.46	228723.84	6−6	1.51−02	5.75−04	1.81−02	−2.462	C+	2*n*
61.		$^2D-^2P^\circ$		*1503.3*	*165991.7*	*232511.5*	10−6	5.55−01	1.13−02	5.58−01	−0.948	C+	1,2
				1502.89	165988.46	232527.09	6−4	4.99−01	1.13−02	3.34−01	−1.170	C+	2*n*
				1504.12	165996.50	232480.44	4−2	5.56−01	9.42−03	1.87−01	−1.424	C+	2*n*
				1503.07	165996.50	232527.09	4−4	5.53−02	1.87−03	3.70−02	−2.126	C+	2*n*
62.	$2s2p^4-$ $2s^22p^2(^1S)3p$	$^2D-^2P^\circ$		*1149.4*	*165991.7*	*252991.3*	10−6	7.78−02	9.25−04	3.50−02	−2.034	D	1
				1149.36	165988.46	252993.39	6−4	7.01−02	9.25−04	2.10−02	−2.256	D−	*LS*
				1149.55	165996.50	252987.23	4−2	7.78−02	7.71−04	1.17−02	−2.511	D−	*LS*
				1149.47	165996.50	252993.39	4−4	7.78−03	1.54−04	2.33−03	−3.210	E	*LS*
63.	$2s^22p^2(^3P)3s-$ $2s^22p^2(^3P)3p$	$^4P-^2S^\circ$											
			5374.36	5375.85	185340.577	203942.288	4−2	4.15−04	8.99−05	6.36−03	−3.444	C	2
			5344.10	5345.59	185235.281	203942.288	2−2	1.46−04	6.25−05	2.20−03	−3.903	C	2
64.		$^4P-^4D^\circ$	*4651.5*	*4652.8*	*185402.30*	*206894.69*	12−20	7.89−01	4.27−01	7.84+01	0.709	B	2,6
			4649.13	4650.44	185499.124	207002.482	6−8	7.84−01	3.39−01	3.11+01	0.308	B	2,6
			4641.81	4643.11	185340.577	206877.865	4−6	5.85−01	2.83−01	1.73+01	0.054	B	2,6
			4638.86	4640.15	185235.281	206786.286	2−4	3.61−01	2.33−01	7.12+00	−0.332	B	2,6
			4676.23	4677.54	185499.124	206877.865	6−6	2.05−01	6.71−02	6.20+00	−0.395	B	2,6
			4661.63	4662.94	185340.577	206786.286	4−4	4.04−01	1.32−01	8.09+00	−0.278	B	2,6
			4650.84	4652.14	185235.281	206730.762	2−2	6.70−01	2.18−01	6.66+00	−0.361	B	2,6
			4696.35	4697.67	185499.124	206786.286	6−4	3.15−02	6.95−03	6.45−01	−1.380	B	2,6
			4673.73	4675.04	185340.577	206730.762	4−2	1.24−01	2.04−02	1.25+00	−1.089	B	2,6
65.		$^4P-^4P^\circ$	*4341.3*	*4342.5*	*185402.30*	*208430.54*	12−12	9.22−01	2.61−01	4.47+01	0.495	B	2,6
			4349.43	4350.65	185499.124	208484.202	6−6	6.75−01	1.91−01	1.65+01	0.060	B	2,6
			4336.86	4338.08	185340.577	208392.258	4−4	1.53−01	4.32−02	2.47+00	−0.762	B	2,6
			4325.76	4326.98	185235.281	208346.104	2−2	1.42−01	3.98−02	1.13+00	−1.099	B	2,6
			4366.89	4368.12	185499.124	208392.258	6−4	3.92−01	7.48−02	6.46+00	−0.348	B	2,6
			4345.56	4346.78	185340.577	208346.104	4−2	7.95−01	1.13−01	6.45+00	−0.346	B	2,6
			4319.63	4320.84	185340.577	208484.202	4−6	2.48−01	1.04−01	5.92+00	−0.380	B	2,6
			4317.14	4318.35	185235.281	208392.258	2−4	3.68−01	2.06−01	5.84+00	−0.386	B	2,6
66.		$^4P-^2D^\circ$											
			3813.73	3814.81	185499.124	211712.732	6−6	3.74−04	8.15−05	6.14−03	−3.311	C	2
67.		$^4P-^4S^\circ$	*3735.9*	*3737.0*	*185402.30*	212161.881	12−4	1.74+00	1.21−01	1.79+01	0.163	B	2,6
			3749.48	3750.55	185499.124	212161.881	6−4	8.97−01	1.26−01	9.34+00	−0.121	B	2,6
			3727.32	3728.38	185340.577	212161.881	4−4	5.64−01	1.18−01	5.78+00	−0.327	B	2,6
			3712.74	3713.80	185235.281	212161.881	2−4	2.75−01	1.14−01	2.78+00	−0.644	B	2,6

O II: Allowed Transitions — Continued

No.	Transition Array	Mult.	λ_{air} (Å)	λ_{vac} (Å) or σ (cm^{-1})*	E_i (cm^{-1})	E_k (cm^{-1})	g_i–g_k	A_{ki} (10^8 s^{-1})	f_{ik}	S (at. u.)	log gf	Acc.	Source
68.	$2s^22p^2(^3P)3s$–$2s^22p^2(^3P)4p$	^{4}P–^{4}D°		*1652.3*	*185402.30*	– *245922.9*	12–20	2.16–01	1.47–02	9.61–01	–0.753	B	1
				1652.07	185499.124	– 246029.295	6–8	2.16–01	1.18–02	3.84–01	–1.151	C+	*LS*
				1651.18	185340.577	– 245903.224	4–6	1.51–01	9.28–03	2.02–01	–1.430	C+	*LS*
				1650.67	185235.281	– 245816.70	2–4	9.02–02	7.37–03	8.01–02	–1.832	C+	*LS*
				1655.52	185499.124	– 245903.224	6–6	6.43–02	2.64–03	8.65–02	–1.800	C+	*LS*
				1653.55	185340.577	– 245816.70	4–4	1.15–01	4.71–03	1.02–01	–1.725	C+	*LS*
				1651.99	185235.281	– 245768.37	2–2	1.80–01	7.36–03	8.01–02	–1.832	C+	*LS*
				1657.89	185499.124	– 245816.70	6–4	1.07–02	2.93–04	9.61–03	–2.754	C–	*LS*
				1654.87	185340.577	– 245768.37	4–2	3.58–02	7.35–04	1.60–02	–2.532	C	*LS*
69.	$2s^22p^2(^3P)3s$–$2s2p^3(^5S°)3s$	^{4}P–^{4}S°		*1437.2*	*185402.30*	– 254981.55	12–4	3.04–01	3.13–03	1.78–01	–1.425	C+	1
				1439.21	185499.124	– 254981.55	6–4	1.51–01	3.13–03	8.90–02	–1.726	C–	*LS*
				1435.94	185340.577	– 254981.55	4–4	1.01–01	3.14–03	5.93–02	–1.902	C–	*LS*
				1433.77	185235.281	– 254981.55	2–4	5.10–02	3.14–03	2.97–02	–2.202	D	*LS*
70.	$2s^22p^2(^3P)3s$–$2s^22p^2(^3P)3p$	^{2}P–^{2}S°	*6694.4*	*6696.2*	*189008.52*	– 203942.288	6–2	2.80–01	6.28–02	8.30+00	–0.424	B+	1,2
			6721.39	6723.24	189068.514	– 203942.288	4–2	1.81–01	6.15–02	5.44+00	–0.609	B+	2*n*
			6641.03	6642.87	188888.543	– 203942.288	2–2	9.88–02	6.54–02	2.86+00	–0.884	B+	2*n*
71.		^{2}P–^{4}P°											
			5149.04	5150.47	189068.514	– 208484.202	4–6	1.13–04	6.73–05	4.56–03	–3.570	C	2
			5137.96	5139.39	188888.543	– 208346.104	2–2	1.89–04	7.47–05	2.53–03	–3.826	C	2
			5185.93	5187.37	189068.514	– 208346.104	4–2	4.23–04	8.52–05	5.82–03	–3.467	C	2
72.		^{2}P–^{2}D°	*4418.1*	*4419.3*	*189008.52*	– *211636.49*	6–10	8.49–01	4.14–01	3.62+01	0.395	B	2,6
			4414.90	4416.14	189068.514	– 211712.732	4–6	8.47–01	3.71–01	2.16+01	0.172	B	2,6
			4416.97	4418.22	188888.543	– 211522.117	2–4	7.16–01	4.19–01	1.22+01	–0.077	B	2,6
			4452.38	4453.63	189068.514	– 211522.117	4–4	1.37–01	4.06–02	2.38+00	–0.789	B	2,6
73.		^{2}P–^{2}P°	*3966.9*	*3968.1*	*189008.52*	– *214209.75*	6–6	1.24+00	2.92–01	2.29+01	0.243	B	2,6
			3973.26	3974.38	189068.514	– 214229.671	4–4	1.02+00	2.40–01	1.26+01	–0.017	B	2,6
			3954.36	3955.48	188888.543	– 214169.920	2–2	8.57–01	2.01–01	5.23+00	–0.396	B	2
			3982.71	3983.84	189068.514	– 214169.920	4–2	4.16–01	4.94–02	2.59+00	–0.704	B	2,6
			3945.04	3946.15	188888.543	– 214229.671	2–4	2.01–01	9.39–02	2.44+00	–0.726	B	2,6
74.	$2s^22p^2(^3P)3s$–$2s^22p^2(^1D)3p$	^{2}P–^{2}D°	*2441.4*	*2442.2*	*189008.52*	– *229955.6*	6–10	4.98–01	7.42–02	3.58+00	–0.352	B	1,2
			2445.53	2446.27	189068.514	– 229947.07	4–6	4.98–01	6.71–02	2.16+00	–0.571	B–	2*n*
			2433.54	2434.28	188888.543	– 229968.44	2–4	4.21–01	7.48–02	1.20+00	–0.825	B–	2*n*
			2444.25	2444.99	189068.514	– 229968.44	4–4	7.56–02	6.78–03	2.18–01	–1.567	B	2*n*
75.		^{2}P–^{2}P°	*2298.0*	*2298.7*	*189008.52*	– *232511.5*	6–6	4.92–01	3.89–02	1.77+00	–0.631	C	1,2
			2300.33	2301.04	189068.514	– 232527.09	4–4	4.17–01	3.31–02	1.00+00	–0.878	C	2*n*
			2293.30	2294.00	188888.543	– 232480.44	2–2	3.25–01	2.56–02	3.87–01	–1.290	C	2*n*
			2302.81	2303.51	189068.514	– 232480.44	4–2	1.67–01	6.64–03	2.02–01	–1.575	C	2*n*
			2290.85	2291.55	188888.543	– 232527.09	2–4	7.41–02	1.17–02	1.76–01	–1.632	C	2*n*
76.	$2s^22p^2(^3P)3s$–$2s^22p^2(^3P)4p$	^{2}P–^{2}D°		*1691.8*	*189008.52*	– *248116.1*	6–10	7.45–03	5.33–04	1.78–02	–2.495	C	1
				1691.53	189068.514	– 248186.64	4–6	7.45–03	4.79–04	1.07–02	–2.717	D	*LS*
				1691.43	188888.543	– 248010.23	2–4	6.21–03	5.33–04	5.93–03	–2.972	D–	*LS*
				1696.59	189068.514	– 248010.23	4–4	1.23–03	5.31–05	1.19–03	–3.673	D–	*LS*

O II: Allowed Transitions — Continued

No.	Transition Array	Mult.	λ_{air} (Å)	λ_{vac} (Å) or σ (cm^{-1})*	E_i (cm^{-1})	E_k (cm^{-1})	g_i–g_k	A_{ki} (10^8 s^{-1})	f_{ik}	S (at. u.)	log gf	Acc.	Source
77.	$2s^22p^2(^3P)3s-$ $2s^22p^2(^1S)3p$	$^2P-^2P^o$		*1562.9*	*189008.52*	*252991.3*	6–6	1.37–01	5.03–03	1.55–01	–1.520	C+	1
				1564.34	189068.514	252993.39	4–4	1.14–01	4.19–03	8.63–02	–1.776	C–	*LS*
				1560.09	188888.543	252987.23	2–2	9.21–02	3.36–03	3.45–02	–2.172	C–	*LS*
				1564.49	189068.514	252987.23	4–2	4.57–02	8.38–04	1.73–02	–2.475	D	*LS*
				1559.94	188888.543	252993.39	2–4	2.30–02	1.68–03	1.73–02	–2.473	D	*LS*
78.	$2s^22p^2(^3P)3s-$ $2s^22p^2(^3P)5p$	$^2P-^2D^o$		*1373.8*	*189008.52*	*261801.5*	6–10	6.48–02	3.06–03	8.29–02	–1.737	C+	1
				1373.60	189068.514	261869.94	4–6	6.48–02	2.75–03	4.97–02	–1.959	C–	*LS*
				1373.43	188888.543	261698.75	2–4	5.40–02	3.06–03	2.76–02	–2.214	D	*LS*
				1376.84	189068.514	261698.75	4–4	1.07–02	3.05–04	5.53–03	–2.914	D–	*LS*
79.	$2s2p^4-$ $2s^22p^2(^1D)3p$	$^2S-^2P^o$	*2716.5*	*2717.3*	195710.47	*232511.5*	2–6	1.19–02	3.95–03	7.07–02	–2.102	C+	1,2
			2715.36	2716.16	195710.47	232527.09	2–4	1.24–02	2.73–03	4.89–02	–2.262	C	2*n*
			2718.81	2719.61	195710.47	232480.44	2–2	1.10–02	1.22–03	2.18–02	–2.614	C+	2*n*
80.	$2s2p^4-$ $2s^22p^2(^3P)4p$	$^2S-^2P^o$		*1894.8*	195710.47	*248485.7*	2–6	2.99–02	4.83–03	6.02–02	–2.015	C	1
				1893.77	195710.47	248515.30	2–4	2.99–02	3.22–03	4.01–02	–2.191	D	*LS*
				1896.96	195710.47	248426.41	2–2	2.98–02	1.61–03	2.01–02	–2.493	D	*LS*
81.	$2s2p^4-$ $2s^22p^2(^1S)3p$	$^2S-^2P^o$		*1745.8*	195710.47	*252991.3*	2–6	4.15–01	5.69–02	6.54–01	–0.944	C+	1
				1745.72	195710.47	252993.39	2–4	4.15–01	3.79–02	4.36–01	–1.120	C–	*LS*
				1745.91	195710.47	252987.23	2–2	4.15–01	1.90–02	2.18–01	–1.421	D	*LS*
82.	$2s^22p^2(^3P)3p-$ $2s2p^4$	$^2S^o-^2P$	*11481*	*8707.7* cm^{-1}	203942.288	*212650.0*	2–6	1.02–02	6.05–02	4.57+00	–0.918	B+	1,2
			11555.5	8651.53 cm^{-1}	203942.288	212593.82	2–4	1.00–02	4.01–02	3.06+00	–1.095	B+	2*n*
			11334.8	8819.96 cm^{-1}	203942.288	212762.25	2–2	1.05–02	2.03–02	1.52+00	–1.391	B+	2*n*
83.	$2s^22p^2(^3P)3p-$ $2s^22p^2(^3P)3d$	$^2S^o-^4P$											
			3496.28	3497.28	203942.288	232535.949	2–4	5.69–03	2.09–03	4.80–02	–2.380	C	2
			3488.16	3489.16	203942.288	232602.492	2–2	3.56–03	6.50–04	1.49–02	–2.886	C	2
84.		$^2S^o-^4D$											
			3470.78	3471.78	203942.288	232745.981	2–4	3.43–03	1.24–03	2.83–02	–2.606	C	2
			3474.93	3475.92	203942.288	232711.642	2–2	7.19–03	1.30–03	2.98–02	–2.584	C	2
85.		$^2S^o-^2P$	*3385.8*	*3386.8*	203942.288	*233468.6*	2–6	1.32+00	6.79–01	1.51+01	0.133	C+	2,6
			3390.21	3391.18	203942.288	233430.53	2–4	1.31+00	4.52–01	1.01+01	–0.044	C+	2,6
			3377.15	3378.12	203942.288	233544.59	2–2	1.33+00	2.27–01	5.05+00	–0.343	C+	2,6
86.		$^2S^o-^2D$											
			3281.99	3282.94	203942.288	234402.797	2–4	1.16–02	3.75–03	8.10–02	–2.125	C	2
87.	$2s^22p^2(^3P)3p-$ $2s^22p^2(^3P)4s$	$^2S^o-^2P$	*2738.0*	*2738.8*	203942.288	*240454.9*	2–6	5.90–01	1.99–01	3.59+00	–0.400	C+	1
			2733.29	2734.10	203942.288	240517.35	2–4	5.93–01	1.33–01	2.39+00	–0.575	C–	*LS*
			2747.37	2748.18	203942.288	240330.01	2–2	5.84–01	6.61–02	1.20+00	–0.879	C–	*LS*
88.	$2s^22p^2(^3P)3p-$ $2s^22p^2(^3P)4d$	$^2S^o-^2P$		*1950.6*	203942.288	*255209.7*	2–6	1.23–02	2.10–03	2.70–02	–2.376	D	1
				1951.93	203942.288	255173.58	2–4	1.23–02	1.40–03	1.80–02	–2.553	D–	*LS*
				1947.81	203942.288	255281.93	2–2	1.23–02	7.02–04	9.00–03	–2.853	E	*LS*

O II: Allowed Transitions — Continued

No.	Transition Array	Mult.	λ_{air} (Å)	λ_{vac} (Å) or σ (cm^{-1})*	E_i (cm^{-1})	E_k (cm^{-1})	g_i-g_k	A_{ki} (10^8 s^{-1})	f_{ik}	S (at. u.)	log gf	Acc.	Source
89.	$2s^22p^2(^3P)3p-$ $2s^22p^2(^3P)5s$	$^2S^o-^2P$		*1831.5*	203942.288	*258541.3*	2−6	1.84−01	2.78−02	3.36−01	−1.255	C	1
				1829.36	203942.288	258606.35	2−4	1.85−01	1.86−02	2.24−01	−1.430	D	*LS*
				1835.91	203942.288	258411.26	2−2	1.83−01	9.25−03	1.12−01	−1.733	D	*LS*
90.	$2s^22p^2(^3P)3p-$ $2s^22p^2(^3P)3d$	$^4D^o-^4F$	*4074.8*	*4075.9*	*206894.69*	*231428.97*	20−28	1.98+00	6.92−01	1.86+02	1.141	B	2,6
			4075.86	4077.01	207002.482	231530.246	8−10	1.98+00	6.15−01	6.61+01	0.692	B+	2,6
			4072.15	4073.30	206877.865	231427.970	6−8	1.79+00	5.94−01	4.78+01	0.552	B+	2,6
			4069.88	4071.03	206786.286	231350.087	4−6	1.48+00	5.52−01	2.96+01	0.344	B+	2,6
			4069.62	4070.77	206730.762	231296.126	2−4	1.42+00	7.07−01	1.90+01	0.150	B+	2,6
			4092.93	4094.08	207002.482	231427.970	8−8	2.45−01	6.15−02	6.63+00	−0.308	B+	2,6
			4085.11	4086.27	206877.865	231350.087	6−6	4.31−01	1.08−01	8.72+00	−0.188	B+	2,6
			4078.84	4079.99	206786.286	231296.126	4−4	5.21−01	1.30−01	6.99+00	−0.284	B+	2,6
			4106.02	4107.18	207002.482	231350.087	8−6	1.51−02	2.87−03	3.10−01	−1.639	B+	2,6
			4094.14	4095.30	206877.865	231296.126	6−4	3.85−02	6.45−03	5.21−01	−1.413	C+	2,6
91.		$^4D^o-^4P$	3902.7	3903.8	206894.69	232510.43	20−12	1.11−01	1.53−02	3.92+00	−0.515	C	2
			3896.30	3897.41	206877.865	232535.949	6−4	1.62−03	2.46−04	1.89−02	−2.831	C	2
			3926.58	3927.69	207002.482	232462.724	8−6	5.05−03	8.76−04	9.06−02	−2.154	C	2
			3872.44	3873.54	206786.286	232602.492	4−2	4.84−03	5.44−04	2.78−02	−2.662	C	2
			3907.45	3908.56	206877.865	232462.724	6−6	8.64−02	1.98−02	1.53+00	−0.925	C	2
			3882.45	3883.55	206786.286	232535.949	4−4	8.94−02	2.02−02	1.03+00	−1.092	C	2
			3864.13	3865.22	206730.762	232602.492	2−2	9.12−02	2.04−02	5.20−01	−1.389	C	2
			3893.52	3894.62	206786.286	232462.724	4−6	1.89−02	6.45−03	3.31−01	−1.588	C	2
			3874.09	3875.19	206730.762	232535.949	2−4	3.26−02	1.47−02	3.74−01	−1.533	C	2
92.		$^4D^o-^4D$	*3867.2*	*3868.3*	*206894.69*	*232746.16*	20−20	4.61−01	1.03−01	2.63+01	0.316	B	2,6
			3882.19	3883.29	207002.482	232753.816	8−8	5.10−01	1.15−01	1.18+01	−0.035	B+	2,6
			3864.43	3865.53	206877.865	232747.562	6−6	2.06−01	4.61−02	3.52+00	−0.558	B+	2,6
			3851.03	3852.13	206786.286	232745.981	4−4	1.52−01	3.38−02	1.71+00	−0.870	B+	2,6
			3847.89	3848.98	206730.762	232711.642	2−2	1.87−01	4.15−02	1.05+00	−1.081	B+	2,6
			3883.14	3884.24	207002.482	232747.562	8−6	1.07−01	1.81−02	1.85+00	−0.839	B+	2,6
			3864.67	3865.76	206877.865	232745.981	6−4	1.91−01	2.86−02	2.18+00	−0.766	C+	2,6
			3856.13	3857.23	206786.286	232711.642	4−2	2.45−01	2.73−02	1.39+00	−0.962	C+	2,6
			3863.50	3864.59	206877.865	232753.816	6−8	5.99−02	1.79−02	1.37+00	−0.969	B+	2,6
			3850.80	3851.89	206786.286	232747.562	4−6	3.40−02	1.14−02	5.76−01	−1.343	C	2,6
			3842.81	3843.90	206730.762	232745.981	2−4	8.13−02	3.60−02	9.12−01	−1.142	C	2,6
93.		$^4D^o-^2F$											
			3851.47	3852.57	207002.482	232959.210	8−8	2.72−02	6.05−03	6.14−01	−1.315	C	2
			3857.16	3858.26	206877.865	232796.298	6−6	6.59−02	1.47−02	1.12+00	−1.055	C	2
			3875.80	3876.90	207002.482	232796.298	8−6	3.38−02	5.71−03	5.83−01	−1.340	C	2
			3833.07	3834.16	206877.865	232959.210	6−8	1.02−02	3.00−03	2.27−01	−1.745	C	2
			3843.58	3844.67	206786.286	232796.298	4−6	3.55−02	1.18−02	5.98−01	−1.326	C	2
94.		$^4D^o-^2P$											
			3752.09	3753.16	206786.286	233430.53	4−4	4.64−04	9.80−05	4.84−03	−3.407	C	2
			3728.36	3729.42	206730.762	233544.59	2−2	7.22−04	1.51−04	3.70−03	−3.521	C	2
			3765.03	3766.10	206877.865	233430.53	6−4	3.88−04	5.50−05	4.09−03	−3.482	C	2
			3736.10	3737.16	206786.286	233544.59	4−2	8.99−04	9.41−05	4.63−03	−3.424	C	2
			3744.29	3745.35	206730.762	233430.53	2−4	6.76−04	2.84−04	7.01−03	−3.245	C	2

O II: Allowed Transitions — Continued

No.	Transition Array	Mult.	λ_{air} (Å)	λ_{vac} (Å) or σ (cm^{-1})*	E_i (cm^{-1})	E_k (cm^{-1})	g_i–g_k	A_{ki} (10^8 s^{-1})	f_{ik}	S (at. u.)	log gf	Acc.	Source
95.	$2s^22p^2(^3P)3p$– $2s^22p^2(^3P)4s$	$^4D^o$–^4P	*3133.8*	*3134.7*	*206894.69*	– *238795.8*	20–12	1.54+00	1.36–01	2.80+01	0.434	C+	1
			3134.73	3135.63	207002.482	– 238893.96	8–6	1.23+00	1.36–01	1.12+01	0.036	C–	*LS*
			3138.34	3139.25	206877.865	– 238732.65	6–4	9.64–01	9.49–02	5.89+00	–0.244	C–	*LS*
			3139.68	3140.59	206786.286	– 238627.46	4–2	7.64–01	5.65–02	2.34+00	–0.646	C–	*LS*
			3122.52	3123.43	206877.865	– 238893.96	6–6	2.80–01	4.09–02	2.52+00	–0.610	C–	*LS*
			3129.34	3130.25	206786.286	– 238732.65	4–4	4.94–01	7.25–02	2.99+00	–0.537	C–	*LS*
			3134.21	3135.12	206730.762	– 238627.46	2–2	7.68–01	1.13–01	2.34+00	–0.645	C–	*LS*
			3113.62	3114.52	206786.286	– 238893.96	4–6	3.14–02	6.84–03	2.80–01	–1.563	C–	*LS*
			3123.91	3124.82	206730.762	– 238732.65	2–4	7.76–02	2.27–02	4.67–01	–1.343	C–	*LS*
96.	$2s^22p^2(^3P)3p$– $2s^22p^2(^3P)5s$	$^4D^o$–^4P		*1961.9*	*206894.69*	– *257865.2*	20–12	5.90–01	2.04–02	2.64+00	–0.389	C+	1
				1962.22	207002.482	– 257965.11	8–6	4.72–01	2.04–02	1.06+00	–0.787	C–	*LS*
				1963.79	206877.865	– 257799.93	6–4	3.71–01	1.43–02	5.54–01	–1.067	C–	*LS*
				1964.27	206786.286	– 257695.74	4–2	2.94–01	8.50–03	2.20–01	–1.468	C–	*LS*
				1957.44	206877.865	– 257965.11	6–6	1.07–01	6.14–03	2.37–01	–1.434	C–	*LS*
				1960.26	206786.286	– 257799.93	4–4	1.89–01	1.09–02	2.81–01	–1.360	C–	*LS*
				1962.13	206730.762	– 257695.74	2–2	2.95–01	1.70–02	2.20–01	–1.468	C–	*LS*
				1953.93	206786.286	– 257965.11	4–6	1.19–02	1.03–03	2.64–02	–2.387	D	*LS*
				1958.13	206730.762	– 257799.93	2–4	2.97–02	3.41–03	4.40–02	–2.166	C–	*LS*
97.	$2s^22p^2(^1D)3s$– $2s^22p^2(^3P)3p$	^2D–$^2D^o$	*21433*	*4664.4* cm^{-1}	*206972.1*	– *211636.49*	10–10	6.95–04	4.79–03	3.38+00	–1.320	C+	1,2
			21086.6	4741.05 cm^{-1}	206971.68	– 211712.732	6–6	6.86–04	4.58–03	1.91+00	–1.561	C+	2*n*
			21974.9	4549.40 cm^{-1}	206972.72	– 211522.117	4–4	5.75–04	4.16–03	1.20+00	–1.779	C+	2*n*
			21969.9	4550.44 cm^{-1}	206971.68	– 211522.117	6–4	6.58–05	3.18–04	1.38–01	–2.720	C+	2*n*
			21091.2	4740.01 cm^{-1}	206972.72	– 211712.732	4–6	4.70–05	4.71–04	1.31–01	–2.725	C+	2*n*
98.		^2D–$^2P^o$	*13813*	*7237.6* cm^{-1}	*206972.1*	– *214209.75*	10–6	1.88–03	3.23–03	1.47+00	–1.491	B	1,2
			13774.2	7257.99 cm^{-1}	206971.68	– 214229.671	6–4	1.69–03	3.20–03	8.72–01	–1.716	B	2*n*
			13890.5	7197.20 cm^{-1}	206972.72	– 214169.920	4–2	1.85–03	2.68–03	4.90–01	–1.970	B	2*n*
			13776.1	7256.95 cm^{-1}	206972.72	– 214229.671	4–4	2.07–04	5.90–04	1.07–01	–2.627	C+	2*n*
99.	$2s^22p^2(^1D)3s$– $2s^22p^2(^1D)3p$	^2D–$^2F^o$	*4593.2*	*4594.5*	*206972.1*	– *228737.3*	10–14	8.84–01	3.92–01	5.93+01	0.593	B+	1,2
			4590.97	4592.26	206971.68	– 228747.45	6–8	8.85–01	3.73–01	3.39+01	0.350	B+	2*n*
			4596.18	4597.46	206972.72	– 228723.84	4–6	8.34–01	3.96–01	2.40+01	0.200	B+	2*n*
			4595.96	4597.24	206971.68	– 228723.84	6–6	4.87–02	1.54–02	1.40+00	–1.033	B	2*n*
100.		^2D–$^2D^o$	*4349.7*	*4350.9*	*206972.1*	– *229955.6*	10–10	1.05+00	2.98–01	4.26+01	0.474	B+	1,2
			4351.26	4352.48	206971.68	– 229947.07	6–6	9.89–01	2.81–01	2.41+01	0.226	B+	2*n*
			4347.41	4348.64	206972.72	– 229968.44	4–4	9.32–01	2.64–01	1.51+01	0.024	B+	2*n*
			4347.22	4348.44	206971.68	– 229968.44	6–4	1.19–01	2.24–02	1.93+00	–0.871	B+	2*n*
			4351.46	4352.68	206972.72	– 229947.07	4–6	5.82–02	2.48–02	1.42+00	–1.003	B	2*n*
101.		^2D–$^2P^o$	*3914.4*	*3915.5*	*206972.1*	– *232511.5*	10–6	1.23+00	1.70–01	2.19+01	0.229	B+	1,2
			3911.96	3913.07	206971.68	– 232527.09	6–4	1.09+00	1.67–01	1.29+01	0.000	B+	2*n*
			3919.27	3920.38	206972.72	– 232480.44	4–2	1.22+00	1.41–01	7.29+00	–0.248	B+	2*n*
			3912.12	3913.23	206972.72	– 232527.09	4–4	1.41–01	3.25–02	1.67+00	–0.886	B	2*n*
102.	$2s^22p^2(^1D)3s$– $2s^22p^2(^3P)4p$	^2D–$^2D^o$	*2429.8*	*2430.5*	*206972.1*	– *248116.1*	10–10	1.89–01	1.67–02	1.34+00	–0.777	B	1
			2425.57	2426.30	206971.68	– 248186.64	6–6	1.77–01	1.56–02	7.49–01	–1.028	C+	*LS*
			2436.06	2436.80	206972.72	– 248010.23	4–4	1.69–01	1.50–02	4.82–01	–1.221	C+	*LS*
			2435.99	2436.73	206971.68	– 248010.23	6–4	1.88–02	1.11–03	5.35–02	–2.176	C+	*LS*
			2425.63	2426.36	206972.72	– 248186.64	4–6	1.27–02	1.68–03	5.35–02	–2.174	C+	*LS*

O II: Allowed Transitions — Continued

No.	Transition Array	Mult.	λ_{air} (Å)	λ_{vac} (Å) or σ (cm^{-1})*	E_i (cm^{-1})	E_k (cm^{-1})	g_i-g_k	A_{ki} (10^8 s^{-1})	f_{ik}	S (at. u.)	log gf	Acc.	Source
103.		$^2D-^2P^o$	*2408.1*	*2408.9*	*206972.1*	*– 248485.7*	10–6	2.05−01	1.07−02	8.50−01	−0.970	B	1
			2406.38	2407.11	206971.68	– 248515.30	6–4	1.85−01	1.07−02	5.10−01	−1.191	C+	*LS*
			2411.60	2412.33	206972.72	– 248426.41	4–2	2.05−01	8.92−03	2.83−01	−1.448	C+	*LS*
			2406.44	2407.17	206972.72	– 248515.30	4–4	2.06−02	1.79−03	5.67−02	−2.146	C+	*LS*
104.	$2s^22p^2(^1D)3s-2s^22p^2(^1S)3p$	$^2D-^2P^o$	*2172.3*	*2173.0*	*206972.1*	*– 252991.3*	10–6	8.43−03	3.58−04	2.56−02	−2.446	C+	1
			2172.21	2172.89	206971.68	– 252993.39	6–4	7.59−03	3.58−04	1.54−02	−2.668	C−	*LS*
			2172.55	2173.23	206972.72	– 252987.23	4–2	8.43−03	2.98−04	8.53−03	−2.923	D	*LS*
			2172.26	2172.94	206972.72	– 252993.39	4–4	8.43−04	5.97−05	1.71−03	−3.622	D−	*LS*
105.	$2s^22p^2(^1D)3s-2s^22p^2(^3P)5p$	$^2D-^2D^o$		*1823.8*	*206972.1*	*– 261801.5*	10–10	1.09−01	5.42−03	3.25−01	−1.266	B−	1
				1821.55	206971.68	– 261869.94	6–6	1.02−01	5.06−03	1.82−01	−1.518	C	*LS*
				1827.28	206972.72	– 261698.75	4–4	9.72−02	4.87−03	1.17−01	−1.711	C	*LS*
				1827.25	206971.68	– 261698.75	6–4	1.08−02	3.60−04	1.30−02	−2.665	D	*LS*
				1821.59	206972.72	– 261869.94	4–6	7.27−03	5.42−04	1.30−02	−2.664	D	*LS*
106.	$2s^22p^2(^3P)3p-2s^22p^2(^3P)3d$	$^4P^o-^4F$											
			4357.26	4358.48	208484.202	– 231427.970	6–8	2.72−04	1.03−04	8.91−03	−3.207	C	2
			4354.59	4355.81	208392.258	– 231350.087	4–6	2.19−04	9.34−05	5.36−03	−3.427	C	2
107.		$^4P^o-^4P$	*4151.7*	*4152.8*	*208430.54*	*– 232510.43*	12–12	9.62−01	2.49−01	4.08+01	0.475	B	2,6
			4169.22	4170.40	208484.202	– 232462.724	6–6	2.49−01	6.50−02	5.35+00	−0.409	B	6*n*
			4140.70	4141.87	208392.258	– 232535.949	4–4	3.76−02	9.68−03	5.28−01	−1.412	B	6*n*
			4121.46	4122.63	208346.104	– 232602.492	2–2	5.60−01	1.43−01	3.87+00	−0.544	C	2*n*
			4156.53	4157.70	208484.202	– 232535.949	6–4	1.94−01	3.36−02	2.76+00	−0.696	B	6*n*
			4129.32	4130.48	208392.258	– 232602.492	4–2	1.65−01	2.11−02	1.15+00	−1.073	B	6*n*
			4153.30	4154.47	208392.258	– 232462.724	4–6	7.28−01	2.82−01	1.55+01	0.053	B	6*n*
			4132.80	4133.97	208346.104	– 232535.949	2–4	8.40−01	4.30−01	1.17+01	−0.065	B	6*n*
108.		$^4P^o-^4D$	*4111.4*	*4112.6*	*208430.54*	*– 232746.16*	12–20	1.18+00	5.00−01	8.12+01	0.778	C+	2,6
			4119.22	4120.38	208484.202	– 232753.816	6–8	1.39+00	4.71−01	3.83+01	0.451	C+	6*n*
			4104.72	4105.88	208392.258	– 232747.562	4–6	3.29−01	1.25−01	6.74+00	−0.302	C+	6*n*
			4097.22	4098.38	208346.104	– 232745.981	2–4	3.79−01	1.91−01	5.15+00	−0.418	C+	6*n*
			4120.28	4121.44	208484.202	– 232747.562	6–6	2.25−01	5.74−02	4.67+00	−0.463	C+	6*n*
			4104.99	4106.15	208392.258	– 232745.981	4–4	9.55−01	2.42−01	1.31+01	−0.015	C+	6*n*
			4103.00	4104.16	208346.104	– 232711.642	2–2	5.32−01	1.34−01	3.63+00	−0.570	C+	6*n*
			4120.55	4121.71	208484.202	– 232745.981	6–4	2.95−01	5.02−02	4.08+00	−0.522	C	2*n*
			4110.79	4111.95	208392.258	– 232711.642	4–2	8.05−01	1.02−01	5.53+00	−0.389	C+	6*n*
109.		$^4P^o-^2F$											
			4084.65	4085.80	208484.202	– 232959.21	6–8	7.28−02	2.43−02	1.96+00	−0.836	C	2
			4096.53	4097.68	208392.258	– 232796.298	4–6	1.73−01	6.54−02	3.53+00	−0.582	C	2
			4112.02	4113.18	208484.202	– 232796.298	6–6	1.81−01	4.58−02	3.72+00	−0.561	C	2
110.		$^4P^o-^2P$											
			3992.76	3993.89	208392.258	– 233430.53	4–4	1.27−03	3.03−04	1.59−02	−2.917	C	2
			3967.37	3968.49	208346.104	– 233544.59	2–2	1.00−02	2.36−03	6.18−02	−2.325	C	2
			3985.41	3986.54	208346.104	– 233430.53	2–4	6.68−03	3.18−03	8.35−02	−2.196	C	2
111.		$^4P^o-^2D$											
			3835.86	3836.95	208392.258	– 234454.634	4–6	2.45−04	8.12−05	4.10−03	−3.488	C	2
			3836.70	3837.79	208346.104	– 234402.797	2–4	1.64−04	7.22−05	1.82−03	−3.840	C	2

O II: Allowed Transitions — Continued

No.	Transition Array	Mult.	λ_{air} (Å)	λ_{vac} (Å) or σ (cm^{-1})*	E_i (cm^{-1})	E_k (cm^{-1})	g_i-g_k	A_{ki} (10^8 s^{-1})	f_{ik}	S (at. u.)	log *gf*	Acc.	Source
112.	$2s^22p^2(^3P)3p-$ $2s^22p^2(^3P)4s$	$^4P^o-^4P$											
			3287.47	3288.42	208484.202	238893.96	6−6	4.92−01	7.98−02	5.18+00	−0.320	B	6
			3294.99	3295.94	208392.258	238732.65	4−4	1.00−01	1.63−02	7.07−01	−1.186	B	6
			3301.41	3302.36	208346.104	238627.46	2−2	1.06−01	1.73−02	3.77−01	−1.460	B	6
			3305.00	3305.95	208484.202	238732.65	6−4	2.89−01	3.16−02	2.06+00	−0.723	B	6
			3306.45	3307.40	208392.258	238627.46	4−2	5.55−01	4.55−02	1.98+00	−0.740	B	6
			3277.56	3278.51	208392.258	238893.96	4−6	2.02−01	4.88−02	2.11+00	−0.709	B	6
113.	$2s^22p^2(^3P)3p-$ $2s^22p^2(^3P)5s$	$^4P^o-^4P$	*2022.2*	*2022.9*	*208430.54*	*257865.2*	12−12	2.83−01	1.74−02	1.39+00	−0.681	C+	1
			2020.33	2020.98	208484.202	257965.11	6−6	1.99−01	1.22−02	4.86−01	−1.136	C−	*LS*
			2023.32	2023.98	208392.258	257799.93	4−4	3.77−02	2.32−03	6.17−02	−2.033	C−	*LS*
			2025.70	2026.36	208346.104	257695.74	2−2	4.70−02	2.89−03	3.86−02	−2.238	C−	*LS*
			2027.10	2027.75	208484.202	257799.93	6−4	1.27−01	5.20−03	2.08−01	−1.506	C−	*LS*
			2027.60	2028.25	208392.258	257695.74	4−2	2.34−01	7.22−03	1.93−01	−1.539	C−	*LS*
			2016.58	2017.23	208392.258	257965.11	4−6	8.57−02	7.84−03	2.08−01	−1.504	C−	*LS*
			2021.44	2022.09	208346.104	257799.93	2−4	1.18−01	1.45−02	1.93−01	−1.538	C−	*LS*
114.	$2s^22p^2(^3P)3p-$ $2s2p^4$	$^2D^o-^2P$		*1013.5* cm^{-1}	*211636.49*	*212650.0*	10−6	1.18−05	1.04−03	3.36+00	−1.985	C+	1,2
				881.09 cm^{-1}	211712.732	212593.82	6−4	7.05−06	9.07−04	2.03+00	−2.264	C+	2*n*
				1240.13 cm^{-1}	211522.117	212762.25	4−2	2.14−05	1.04−03	1.11+00	−2.380	C+	2*n*
				1071.70 cm^{-1}	211522.117	212593.82	4−4	1.40−06	1.83−04	2.25−01	−3.136	C+	2*n*
115.	$2s^22p^2(^3P)3p-$ $2s^22p^2(^3P)3d$	$^2D^o-^4F$											
			5070.81	5072.22	211712.732	231427.970	6−8	7.38−04	3.80−04	3.80−02	−2.642	C	2
			5041.97	5043.38	211522.117	231350.087	4−6	5.98−04	3.42−04	2.27−02	−2.864	C	2
116.		$^2D^o-^4P$											
			4774.08	4775.41	211522.117	232462.724	4−6	1.36−03	6.97−04	4.38−02	−2.555	D	2
117.		$^2D^o-^4D$											
			4751.28	4752.61	211712.732	232753.816	6−8	6.39−02	2.89−02	2.71+00	−0.761	C	2
			4710.01	4711.33	211522.117	232747.562	4−6	2.98−01	1.48−01	9.21+00	−0.226	C	2
			4752.69	4754.02	211712.732	232747.562	6−6	1.45−02	4.91−03	4.61−01	−1.531	C	2
118.		$^2D^o-^2F$	*4703.9*	*4705.2*	*211636.49*	*232889.39*	10−14	1.07+00	4.95−01	7.67+01	0.695	B	2,6
			4705.35	4706.66	211712.732	232959.210	6−8	1.13+00	4.98−01	4.63+01	0.476	B	2,6
			4699.22	4700.53	211522.117	232796.298	4−6	9.36−01	4.65−01	2.88+01	0.270	B	2
			4741.70	4743.03	211712.732	232796.298	6−6	5.07−02	1.71−02	1.60+00	−0.989	C+	2,6
119.		$^2D^o-^2P$	4579.1	4580.4	211636.49	233468.6	10−6	9.37−03	1.77−03	2.67−01	−1.753	C	2
			4603.23	4604.52	211712.732	233430.53	6−4	3.63−03	7.68−04	6.99−02	−2.336	C	2
			4539.54	4540.82	211522.117	233544.59	4−2	6.69−03	1.03−03	6.19−02	−2.383	C	2
			4563.18	4564.46	211522.117	233430.53	4−4	7.18−03	2.24−03	1.35−01	−2.047	C	2
120.		$^2D^o-^2D$	*4385.2*	*4386.5*	*211636.49*	*234433.90*	10−10	4.03−01	1.16−01	1.68+01	0.066	B	1,2
			4395.93	4397.17	211712.732	234454.634	6−6	3.91−01	1.13−01	9.83+00	−0.168	B	2*n*
			4369.27	4370.50	211522.117	234402.797	4−4	3.57−01	1.02−01	5.88+00	−0.388	B	2*n*
			4405.98	4407.22	211712.732	234402.797	6−4	4.30−02	8.34−03	7.26−01	−1.301	B	2*n*
			4359.40	4360.62	211522.117	234454.634	4−6	1.44−02	6.16−03	3.54−01	−1.608	C+	2*n*
121.	$2s^22p^2(^3P)3p-$ $2s^22p^2(^3P)4s$	$^2D^o-^2P$											
			3470.67	3471.67	211712.732	240517.35	6−4	8.63−01	1.04−01	7.13+00	−0.205	B	6
			3470.28	3471.27	211522.117	240330.01	4−2	1.09+00	9.85−02	4.50+00	−0.405	B	6

O II: Allowed Transitions — Continued

No.	Transition Array	Mult.	λ_{air} (Å)	λ_{vac} (Å) or σ (cm^{-1})*	E_i (cm^{-1})	E_k (cm^{-1})	g_i-g_k	A_{ki} ($10^8 s^{-1}$)	f_{ik}	S (at. u.)	log gf	Acc.	Source
122.	$2s^22p^2(^3P)3p-2s^22p^2(^1D)3d$	$^2D^o-^2F$	*2525.3*	*2526.1*	*211636.49*	*251223.3*	10–14	8.20–02	1.10–02	9.13–01	−0.959	C	1
			2530.28	2531.04	211712.732	251222.19	6–8	8.16–02	1.04–02	5.22–01	−1.203	D	*LS*
			2517.96	2518.72	211522.117	251224.79	4–6	7.72–02	1.10–02	3.65–01	−1.356	D	*LS*
			2530.11	2530.87	211712.732	251224.79	6–6	5.44–03	5.22–04	2.61–02	−2.504	D−	*LS*
123.		$^2D^o-^2D$	*2414.1*	*2414.8*	*211636.49*	*253048.0*	10–10	2.48–01	2.17–02	1.72+00	−0.664	C	1
			2418.46	2419.19	211712.732	253048.82	6–6	2.30–01	2.02–02	9.64–01	−0.917	D	*LS*
			2407.48	2408.21	211522.117	253046.74	4–4	2.25–01	1.96–02	6.20–01	−1.107	D	*LS*
			2418.58	2419.32	211712.732	253046.74	6–4	2.47–02	1.44–03	6.89–02	−2.063	D−	*LS*
			2407.36	2408.09	211522.117	253048.82	4–6	1.67–02	2.17–03	6.89–02	−2.061	D−	*LS*
124.		$^2D^o-^2P$	*2371.5*	*2372.2*	*211636.49*	*253791.6*	10–6	1.51–01	7.65–03	5.97–01	−1.116	C	1
			2375.72	2376.44	211712.732	253792.40	6–4	1.35–01	7.63–03	3.58–01	−1.339	D	*LS*
			2365.14	2365.86	211522.117	253789.99	4–2	1.52–01	6.39–03	1.99–01	−1.592	D	*LS*
			2365.01	2365.73	211522.117	253792.40	4–4	1.52–02	1.28–03	3.98–02	−2.291	D−	*LS*
125.	$2s^22p^2(^3P)3p-2s^22p^2(^3P)4d$	$^2D^o-^2F$	*2284.5*	*2285.2*	*211636.49*	*255395.8*	10–14	7.69–02	8.42–03	6.33–01	−1.075	C	1
			2284.83	2285.54	211712.732	255466.10	6–8	7.69–02	8.02–03	3.62–01	−1.318	D	*LS*
			2283.44	2284.15	211522.117	255302.11	4–6	7.19–02	8.43–03	2.53–01	−1.472	D	*LS*
			2293.43	2294.14	211712.732	255302.11	6–6	5.07–03	3.99–04	1.81–02	−2.620	D−	*LS*
126.	$2s^22p^2(^3P)3p-2s^22p^2(^3P)5s$	$^2D^o-^2P$	*2131.3*	*2132.0*	*211636.49*	*258541.3*	10–6	4.94–01	2.02–02	1.42+00	−0.695	C+	1
			2131.81	2132.49	211712.732	258606.35	6–4	4.45–01	2.02–02	8.50–01	−0.917	C−	*LS*
			2132.02	2132.69	211522.117	258411.26	4–2	4.94–01	1.68–02	4.72–01	−1.172	C−	*LS*
			2123.18	2123.85	211522.117	258606.35	4–4	5.00–02	3.38–03	9.45–02	−1.869	C−	*LS*
127.	$2s^22p^2(^3P)3p-2s^22p^2(^1D)4s$	$^2D^o-^2D$	*2097.9*	*2098.6*	*211636.49*	*259287.8*	10–10	7.38–02	4.87–03	3.37–01	−1.312	C+	1
			2101.28	2101.95	211712.732	259287.61	6–6	6.86–02	4.54–03	1.88–01	−1.565	C−	*LS*
			2092.88	2093.54	211522.117	259288.07	4–4	6.69–02	4.40–03	1.21–01	−1.755	C−	*LS*
			2101.26	2101.93	211712.732	259288.07	6–4	7.35–03	3.24–04	1.35–02	−2.711	D	*LS*
			2092.90	2093.56	211522.117	259287.61	4–6	4.96–03	4.88–04	1.35–02	−2.709	D	*LS*
128.	$2s^22p^2(^3P)3p-2s^22p^2(^1D)4d$	$^2D^o-^2F$		*1584.1*	*211636.49*	*274764*	10–14	1.08–02	5.68–04	2.96–02	−2.246	D	1
				1585.57	211712.732	274781.5	6–8	1.07–02	5.40–04	1.69–02	−2.489	D−	*LS*
				1581.81	211522.117	274740.7	4–6	1.01–02	5.68–04	1.18–02	−2.643	D−	*LS*
				1586.60	211712.732	274740.7	6–6	7.15–04	2.70–05	8.46–04	−3.791	E	*LS*
129.		$^2D^o-^2D$		*1580.1*	*211636.49*	*274923*	10–10	2.35–02	8.80–04	4.58–02	−2.055	D	1
				1581.90	211712.732	274928.0	6–6	2.19–02	8.21–04	2.56–02	−2.308	D−	*LS*
				1577.44	211522.117	274916.0	4–4	2.13–02	7.94–04	1.65–02	−2.498	D−	*LS*
				1582.20	211712.732	274916.0	6–4	2.34–03	5.86–05	1.83–03	−3.454	E	*LS*
				1577.14	211522.117	274928.0	4–6	1.58–03	8.82–05	1.83–03	−3.452	E	*LS*
130.	$2s^22p^2(^3P)3p-2s^22p^2(^3P)3d$	$^4S^o-^4P$	*4913.0*	*4914.4*	212161.881	*232510.43*	4–12	5.16–01	5.61–01	3.63+01	0.351	C+	2,6
			4924.53	4925.90	212161.881	232462.724	4–6	5.43–01	2.96–01	1.92+01	0.074	C+	2
			4906.83	4908.20	212161.881	232535.949	4–4	4.78–01	1.73–01	1.12+01	−0.160	C+	2,6
			4890.86	4892.22	212161.881	232602.492	4–2	5.10–01	9.15–02	5.90+00	−0.436	C+	2,6
131.		$^4S^o-^4D$											
			4856.39	4857.75	212161.881	232747.562	4–6	5.58–02	2.96–02	1.89+00	−0.926	C	2
			4856.76	4858.12	212161.881	232745.981	4–4	1.00–01	3.55–02	2.27+00	−0.848	C	2
			4864.88	4866.24	212161.881	232711.642	4–2	8.07–02	1.43–02	9.18–01	−1.242	C	2

O II: Allowed Transitions — Continued

No.	Transition Array	Mult.	λ_{air} (Å)	λ_{vac} (Å) or σ (cm^{-1})*	E_i (cm^{-1})	E_k (cm^{-1})	g_i–g_k	A_{ki} (10^8 s^{-1})	f_{ik}	S (at. u.)	log gf	Acc.	Source
132.		$^4S^o-^2F$											
			4844.92	4846.27	212161.881	232796.298	4–6	1.02–02	5.40–03	3.45–01	–1.665	D	2
133.		$^4S^o-^2P$											
			4700.44	4701.76	212161.881	233430.53	4–4	3.80–04	1.26–04	7.80–03	–3.298	C	2
134.		$^4S^o-^2D$											
			4484.50	4485.76	212161.881	234454.634	4–6	1.93–04	8.72–05	5.15–03	–3.457	C	2
135.	$2s^22p^2(^3P)3p-$ $2s^22p^2(^3P)4s$	$^4S^o-^4P$	*3753.5*	*3754.6*	212161.881	*238795.8*	4–12	2.89–01	1.83–01	9.06+00	–0.135	B	6
			3739.76	3740.82	212161.881	238893.96	4–6	2.97–01	9.35–02	4.60+00	–0.427	B	6
			3762.47	3763.53	212161.881	238732.65	4–4	3.10–01	6.58–02	3.26+00	–0.580	B	6
			3777.42	3778.49	212161.881	238627.46	4–2	2.25–01	2.41–02	1.20+00	–1.016	B	6
136.	$2s^22p^2(^3P)3p-$ $2s^22p^2(^3P)4d$	$^4S^o-^4P$	*2326.7*	*2327.5*	212161.881	*255127.2*	4–12	6.69–02	1.63–02	4.99–01	–1.186	C	1
			2327.95	2328.66	212161.881	255105.01	4–6	6.68–02	8.14–03	2.49–01	–1.488	D	*LS*
			2325.92	2326.63	212161.881	255142.41	4–4	6.69–02	5.43–03	1.66–01	–1.663	D	*LS*
			2324.80	2325.52	212161.881	255163.08	4–2	6.70–02	2.72–03	8.31–02	–1.964	D–	*LS*
137.	$2s^22p^2(^3P)3p-$ $2s^22p^2(^3P)5s$	$^4S^o-^4P$	*2187.3*	*2188.0*	212161.881	*257865.2*	4–12	1.14–01	2.45–02	7.06–01	–1.009	C+	1
			2182.57	2183.25	212161.881	257965.11	4–6	1.15–01	1.23–02	3.53–01	–1.309	C–	*LS*
			2190.47	2191.15	212161.881	257799.93	4–4	1.13–01	8.16–03	2.35–01	–1.487	C–	*LS*
			2195.48	2196.17	212161.881	257695.74	4–2	1.13–01	4.07–03	1.18–01	–1.789	C–	*LS*
138.	$2s2p^4-$ $2s^22p^2(^1D)3p$	$^2P-^2D^o$	*5776.9*	*5778.5*	*212650.0*	*229955.6*	6–10	6.44–03	5.37–03	6.13–01	–1.492	C+	1,2
			5761.01	5762.61	212593.82	229947.07	4–6	6.39–03	4.77–03	3.62–01	–1.719	C+	2*n*
			5810.25	5811.86	212762.25	229968.44	2–4	5.24–03	5.31–03	2.03–01	–1.974	C+	2*n*
			5753.93	5755.52	212593.82	229968.44	4–4	1.27–03	6.31–04	4.78–02	–2.598	C+	2*n*
139.		$^2P-^2P^o$											
			5015.34	5016.74	212593.82	232527.09	4–4	6.18–04	2.33–04	1.54–02	–3.030	C	2
			5070.05	5071.46	212762.25	232480.44	2–2	3.80–04	1.46–04	4.89–03	–3.534	C	2
140.	$2s2p^4-$ $2s^22p^2(^3P)4p$	$^2P-^2D^o$	*2818.8*	*2819.6*	*212650.0*	*248116.1*	6–10	4.43–02	8.80–03	4.90–01	–1.278	C	1
			2808.73	2809.56	212593.82	248186.64	4–6	4.48–02	7.94–03	2.94–01	–1.498	D	*LS*
			2836.21	2837.04	212762.25	248010.23	2–4	3.62–02	8.74–03	1.63–01	–1.757	D	*LS*
			2822.72	2823.55	212593.82	248010.23	4–4	7.35–03	8.78–04	3.26–02	–2.454	D–	*LS*
141.		$^2P-^2P^o$	*2789.7*	*2790.5*	*212650.0*	*248485.7*	6–6	3.04–02	3.55–03	1.95–01	–1.672	C	1
			2783.03	2783.85	212593.82	248515.30	4–4	2.55–02	2.96–03	1.09–01	–1.926	D	*LS*
			2803.11	2803.94	212762.25	248426.41	2–2	2.00–02	2.35–03	4.34–02	–2.327	D	*LS*
			2789.93	2790.76	212593.82	248426.41	4–2	1.01–02	5.91–04	2.17–02	–2.626	D–	*LS*
			2796.14	2796.96	212762.25	248515.30	2–4	5.03–03	1.18–03	2.17–02	–2.627	D–	*LS*
142.	$2s2p^4-$ $2s^22p^2(^3P)5p$	$^2P-^2D^o$	*2033.9*	*2034.5*	*212650.0*	*261801.5*	6–10	2.25–02	2.32–03	9.33–02	–1.856	C–	1
			2028.73	2029.38	212593.82	261869.94	4–6	2.26–02	2.10–03	5.60–02	–2.077	D–	*LS*
			2042.81	2043.46	212762.25	261698.75	2–4	1.85–02	2.31–03	3.11–02	–2.335	D–	*LS*
			2035.80	2036.46	212593.82	261698.75	4–4	3.73–03	2.32–04	6.22–03	–3.032	E	*LS*

O II: Allowed Transitions — Continued

No.	Transition Array	Mult.	λ_{air} (Å)	λ_{vac} (Å) or σ (cm^{-1})*	E_i (cm^{-1})	E_k (cm^{-1})	$g_i - g_k$	A_{ki} (10^8 s^{-1})	f_{ik}	S (at. u.)	log gf	Acc.	Source
143.	$2s^22p^2(^3P)3p-$ $2s^22p^2(^1S)3s$	$^2P^o-^2S$	*6096.0*	*6097.7*	*214209.75*	230609.45	6–2	1.26–02	2.34–03	2.82–01	–1.852	D	1,2
			6103.40	6105.09	214229.671	230609.45	4–2	8.36–03	2.34–03	1.88–01	–2.029	D	2*n*
			6081.22	6082.90	214169.920	230609.45	2–2	4.25–03	2.36–03	9.45–02	–2.326	D	2*n*
144.	$2s^22p^2(^3P)3p-$ $2s^22p^2(^3P)3d$	$^2P^o-^4P$											
			5483.02	5484.55	214229.671	232462.724	4–6	2.12–04	1.43–04	1.04–02	–3.241	C	2
			5461.09	5462.61	214229.671	232535.949	4–4	6.53–04	2.92–04	2.10–02	–2.933	C	2
			5423.67	5425.18	214169.920	232602.492	2–2	2.81–04	1.24–04	4.43–03	–3.605	C	2
145.		$^2P^o-^4D$											
			5398.68	5400.18	214229.671	232747.562	4–6	2.49–04	1.63–04	1.16–02	–3.184	C	2
			5381.78	5383.27	214169.920	232745.981	2–4	7.55–05	6.56–05	2.32–03	–3.882	C	2
			5399.14	5400.64	214229.671	232745.981	4–4	4.06–04	1.77–04	1.26–02	–3.149	C	2
			5391.74	5393.24	214169.920	232711.642	2–2	6.59–04	2.87–04	1.02–02	–3.240	C	2
			5409.17	5410.68	214229.671	232711.642	4–2	3.10–04	6.79–05	4.84–03	–3.566	C	2
146.		$^2P^o-^2F$											
			5384.51	5386.01	214229.671	232796.298	4–6	5.93–04	3.87–04	2.74–02	–2.811	C	2
147.		$^2P^o-^2P$	*5191.0*	*5192.4*	*214209.75*	*233468.6*	6–6	4.53–01	1.83–01	1.88+01	0.041	B	2,6
			5206.65	5208.10	214229.671	233430.53	4–4	3.33–01	1.36–01	9.30+00	–0.266	B	2,6
			5159.94	5161.38	214169.920	233544.59	2–2	3.12–01	1.25–01	4.23+00	–0.603	B	2,6
			5175.90	5177.35	214229.671	233544.59	4–2	1.46–01	2.94–02	2.01+00	–0.929	B	2,6
			5190.50	5191.94	214169.920	233430.53	2–4	1.18–01	9.53–02	3.26+00	–0.720	B	2,6
148.		$^2P^o-^2D$	*4943.2*	*4944.6*	*214209.75*	*234433.90*	6–10	7.87–01	4.81–01	4.69+01	0.460	B	2,6
			4943.01	4944.38	214229.671	234454.634	4–6	7.88–01	4.33–01	2.82+01	0.239	B	2,6
			4941.07	4942.45	214169.920	234402.797	2–4	6.04–01	4.42–01	1.44+01	–0.054	B	2,6
			4955.71	4957.09	214229.671	234402.797	4–4	1.81–01	6.68–02	4.36+00	–0.573	B	2,6
149.	$2s^22p^2(^3P)3p-$ $2s^22p^2(^3P)4s$	$^2P^o-^2P$	*3809.1*	*3810.2*	*214209.75*	*240454.9*	6–6	4.07–01	8.85–02	6.66+00	–0.275	C+	1
			3802.98	3804.06	214229.671	240517.35	4–4	3.41–01	7.39–02	3.70+00	–0.529	C–	*LS*
			3821.53	3822.62	214169.920	240330.01	2–2	2.69–01	5.88–02	1.48+00	–0.929	C–	*LS*
			3830.28	3831.37	214229.671	240330.01	4–2	1.33–01	1.47–02	7.40–01	–1.232	C–	*LS*
			3794.36	3795.44	214169.920	240517.35	2–4	6.86–02	2.96–02	7.40–01	–1.227	C–	*LS*
150.	$2s^22p^2(^3P)3p-$ $2s^22p^2(^1D)3d$	$^2P^o-^2D$	*2574.0*	*2574.8*	*214209.75*	*253048.0*	6–10	1.37–01	2.27–02	1.16+00	–0.865	C	1
			2575.28	2576.05	214229.671	253048.82	4–6	1.37–01	2.05–02	6.94–01	–1.087	D	*LS*
			2571.46	2572.23	214169.920	253046.74	2–4	1.15–01	2.28–02	3.85–01	–1.342	D	*LS*
			2575.42	2576.19	214229.671	253046.74	4–4	2.29–02	2.27–03	7.71–02	–2.041	D–	*LS*
151.		$^2P^o-^2P$	*2525.7*	*2526.4*	*214209.75*	*253791.6*	6–6	1.44–01	1.38–02	6.87–01	–1.083	C	1
			2526.87	2527.63	214229.671	253792.40	4–4	1.20–01	1.15–02	3.82–01	–1.338	D	*LS*
			2523.21	2523.97	214169.920	253789.99	2–2	9.63–02	9.19–03	1.53–01	–1.736	D	*LS*
			2527.03	2527.79	214229.671	253789.99	4–2	4.79–02	2.29–03	7.63–02	–2.037	D	*LS*
			2523.06	2523.82	214169.920	253792.40	2–4	2.41–02	4.59–03	7.63–02	–2.037	D	*LS*
152.	$2s^22p^2(^3P)3p-$ $2s^22p^2(^3P)4d$	$^2P^o-^2P$	*2438.3*	*2439.0*	*214209.75*	*255209.7*	6–6	1.21–01	1.07–02	5.18–01	–1.191	C	1
			2441.63	2442.37	214229.671	255173.58	4–4	1.00–01	8.94–03	2.88–01	–1.447	D	*LS*
			2431.64	2432.38	214169.920	255281.93	2–2	8.10–02	7.18–03	1.15–01	–1.843	D	*LS*
			2435.18	2435.92	214229.671	255281.93	4–2	4.03–02	1.79–03	5.75–02	–2.144	D	*LS*
			2438.07	2438.81	214169.920	255173.58	2–4	2.01–02	3.58–03	5.75–02	–2.145	D	*LS*

O II: Allowed Transitions — Continued

No.	Transition Array	Mult.	λ_{air} (Å)	λ_{vac} (Å) or σ (cm^{-1})*	E_i (cm^{-1})	E_k (cm^{-1})	g_i-g_k	A_{ki} (10^8 s^{-1})	f_{ik}	S (at. u.)	log gf	Acc.	Source
153.		$^2P^o-^2D$	*2399.3*	*2400.0*	*214209.75*	*255875.7*	6−10	2.37−02	3.41−03	1.61−01	−1.690	C−	1
			2399.20	2399.93	214229.671	255897.59	4−6	2.37−02	3.07−03	9.68−02	−1.911	D−	*LS*
			2398.91	2399.64	214169.920	255842.91	2−4	1.97−02	3.41−03	5.38−02	−2.167	D−	*LS*
			2402.35	2403.08	214229.671	255842.91	4−4	3.93−03	3.40−04	1.08−02	−2.866	E	*LS*
154.	$2s^22p^2(^3P)3p-$ $2s^22p^2(^1D)3d$	$^2P^o-^2S$	*2414.0*	*2414.7*	*214209.75*	255622.80	6−2	3.30−01	9.61−03	4.58−01	−1.239	C	1
			2415.13	2415.86	214229.671	255622.80	4−2	2.20−01	9.61−03	3.06−01	−1.415	D	*LS*
			2411.64	2412.38	214169.920	255622.80	2−2	1.10−01	9.62−03	1.53−01	−1.716	D	*LS*
155.	$2s^22p^2(^3P)3p-$ $2s^22p^2(^3P)5s$	$^2P^o-^2P$	*2255.0*	*2255.7*	*214209.75*	*258541.3*	6−6	2.14−01	1.63−02	7.26−01	−1.010	C	1
			2252.74	2253.44	214229.671	258606.35	4−4	1.79−01	1.36−02	4.03−01	−1.264	D	*LS*
			2259.63	2260.33	214169.920	258411.26	2−2	1.42−01	1.08−02	1.61−01	−1.664	D	*LS*
			2262.69	2263.39	214229.671	258411.26	4−2	7.06−02	2.71−03	8.07−02	−1.965	D	*LS*
			2249.71	2250.41	214169.920	258606.35	2−4	3.59−02	5.45−03	8.07−02	−1.963	D	*LS*
156.	$2s^22p^2(^3P)3p-$ $2s^22p^2(^1D)4s$	$^2P^o-^2D$	*2217.7*	*2218.4*	*214209.75*	*259287.8*	6−10	1.38−02	1.70−03	7.43−02	−1.992	C	1
			2218.67	2219.36	214229.671	259287.61	4−6	1.38−02	1.53−03	4.46−02	−2.214	D	*LS*
			2215.71	2216.40	214169.920	259288.07	2−4	1.15−02	1.70−03	2.48−02	−2.469	D−	*LS*
			2218.65	2219.34	214229.671	259288.07	4−4	2.30−03	1.70−04	4.95−03	−3.169	E	*LS*
157.	$2s^22p^2(^3P)3p-$ $2s^22p^2(^1D)4d$	$^2P^o-^2D$		*1647.1*	*214209.75*	*274923*	6−10	4.34−02	2.94−03	9.58−02	−1.753	D	1
				1647.49	214229.671	274928.0	4−6	4.34−02	2.65−03	5.75−02	−1.975	D−	*LS*
				1646.20	214169.920	274916.0	2−4	3.63−02	2.95−03	3.19−02	−2.230	D−	*LS*
				1647.82	214229.671	274916.0	4−4	7.23−03	2.94−04	6.39−03	−2.929	E	*LS*
158.	$2s^22p^2(^3P)3p-$ $2s^22p^2(^1S)3d$	$^2P^o-^2D$		*1617.4*	*214209.75*	*276038.8*	6−10	3.84−02	2.51−03	8.01−02	−1.823	D	1
				1616.93	214229.671	276075.32	4−6	3.84−02	2.26−03	4.81−02	−2.044	D−	*LS*
				1617.76	214169.920	275983.95	2−4	3.19−02	2.51−03	2.67−02	−2.300	D−	*LS*
				1619.32	214229.671	275983.95	4−4	6.37−03	2.50−04	5.34−03	−2.999	E	*LS*
159.	$2s^22p^2(^1D)3p-$ $2s^22p^2(^3P)3d$	$^2F^o-^2F$											
			23736.6	4211.76 cm^{-1}	228747.45	232959.210	8−8	8.87−06	7.49−05	4.69−02	−3.222	C	2
			24548.5	4072.46 cm^{-1}	228723.84	232796.298	6−6	7.27−06	6.57−05	3.19−02	−3.404	C	2
160.	$2s^22p^2(^1D)3p-$ $2s^22p^2(^1D)3d$	$^2F^o-^2F$	*4446.0*	*4447.2*	*228737.3*	*251223.3*	14−14	5.29−01	1.57−01	3.21+01	0.342	C	1
			4448.19	4449.44	228747.45	251222.19	8−8	5.10−01	1.51−01	1.77+01	0.083	D	*LS*
			4443.01	4444.26	228723.84	251224.79	6−6	5.05−01	1.49−01	1.31+01	−0.047	D	*LS*
			4447.68	4448.93	228747.45	251224.79	8−6	2.52−02	5.60−03	6.56−01	−1.349	D	*LS*
			4443.52	4444.77	228723.84	251222.19	6−8	1.89−02	7.47−03	6.56−01	−1.348	D	*LS*
161.		$^2F^o-^2G$	*4187.9*	*4189.1*	*228737.3*	*252608.8*	14−18	1.98+00	6.69−01	1.29+02	0.972	C	1
			4189.79	4190.97	228747.45	252608.28	8−10	1.98+00	6.50−01	7.18+01	0.716	D	*LS*
			4185.44	4186.62	228723.84	252609.46	6−8	1.91+00	6.70−01	5.54+01	0.604	D	*LS*
			4189.58	4190.76	228747.45	252609.46	8−8	7.06−02	1.86−02	2.05+00	−0.828	D	*LS*
162.		$^2F^o-^2D$	*4112.3*	*4113.4*	*228737.3*	*253048.0*	14−10	2.54−01	4.59−02	8.70+00	−0.192	C	1
			4113.83	4114.99	228747.45	253048.82	8−6	2.41−01	4.59−02	4.97+00	−0.435	D	*LS*
			4110.19	4111.35	228723.84	253046.74	6−4	2.54−01	4.29−02	3.48+00	−0.590	D	*LS*
			4109.84	4111.00	228723.84	253048.82	6−6	1.21−02	3.06−03	2.49−01	−1.736	D	*LS*

O II: Allowed Transitions — Continued

No.	Transition Array	Mult.	λ_{air} (Å)	λ_{vac} (Å) or σ (cm^{-1})*	E_i (cm^{-1})	E_k (cm^{-1})	g_i-g_k	A_{ki} (10^8 s^{-1})	f_{ik}	S (at. u.)	log gf	Acc.	Source
163.	$2s^22p^2(^1D)3p-$ $2s^22p^2(^3P)4d$	$^2F^o-^2F$	*3750.1*	*3751.2*	*228737.3*	*– 255395.8*	14–14	4.63–02	9.77–03	1.69+00	–0.864	C	1
			3741.64	3742.70	228747.45	– 255466.10	8–8	4.50–02	9.44–03	9.31–01	–1.122	D	*LS*
			3761.40	3762.47	228723.84	– 255302.11	6–6	4.37–02	9.28–03	6.89–01	–1.254	D	*LS*
			3764.75	3765.82	228747.45	– 255302.11	8–6	2.18–03	3.48–04	3.45–02	–2.556	D–	*LS*
			3738.34	3739.40	228723.84	– 255466.10	6–8	1.67–03	4.67–04	3.45–02	–2.553	D–	*LS*
164.		$^2F^o-^2D$	*3683.8*	*3684.8*	*228737.3*	*– 255875.7*	14–10	2.39–03	3.48–04	5.90–02	–2.313	D	1
			3682.17	3683.22	228747.45	– 255897.59	8–6	2.28–03	3.48–04	3.37–02	–2.556	D–	*LS*
			3686.39	3687.44	228723.84	– 255842.91	6–4	2.39–03	3.24–04	2.36–02	–2.711	D–	*LS*
			3678.97	3680.02	228723.84	– 255897.59	6–6	1.14–04	2.32–05	1.69–03	–3.856	E	*LS*
165.	$2s^22p^2(^1D)3p-$ $2s^22p^2(^1D)4s$	$^2F^o-^2D$	*3272.3*	*3273.3*	*228737.3*	*– 259287.8*	14–10	1.05+00	1.20–01	1.82+01	0.227	C+	1
			3273.43	3274.38	228747.45	– 259287.61	8–6	9.99–01	1.20–01	1.04+01	–0.017	C–	*LS*
			3270.86	3271.80	228723.84	– 259288.07	6–4	1.05+00	1.12–01	7.26+00	–0.171	C–	*LS*
			3270.90	3271.85	228723.84	– 259287.61	6–6	5.00–02	8.03–03	5.19–01	–1.317	C–	*LS*
166.	$2s^22p^2(^1D)3p-$ $2s^22p^2(^1D)4d$	$^2F^o-^2F$	*2172.0*	*2172.7*	*228737.3*	*– 274764*	14–14	1.03–02	7.26–04	7.27–02	–1.993	D	1
			2171.62	2172.31	228747.45	– 274781.5	8–8	9.91–03	7.00–04	4.01–02	–2.252	D–	*LS*
			2172.44	2173.12	228723.84	– 274740.7	6–6	9.77–03	6.92–04	2.97–02	–2.382	D–	*LS*
			2173.55	2174.23	228747.45	– 274740.7	8–6	4.88–04	2.59–05	1.48–03	–3.683	E	*LS*
			2170.51	2171.19	228723.84	– 274781.5	6–8	3.67–04	3.46–05	1.48–03	–3.683	E	*LS*
167.	$2s^22p^2(^1D)3p-$ $2s^22p^2(^1S)3d$	$^2F^o-^2D$	*2113.4*	*2114.1*	*228737.3*	*– 276038.8*	14–10	2.26–02	1.08–03	1.05–01	–1.820	D	1
			2112.25	2112.92	228747.45	– 276075.32	8–6	2.15–02	1.08–03	6.01–02	–2.063	D–	*LS*
			2115.28	2115.95	228723.84	– 275983.95	6–4	2.25–02	1.01–03	4.21–02	–2.219	D–	*LS*
			2111.20	2111.87	228723.84	– 276075.32	6–6	1.08–03	7.21–05	3.01–03	–3.364	E	*LS*
168.	$2s^22p^2(^1D)3p-$ $2s^22p^2(^3P)3d$	$^2D^o-^4D$											
			35618.7	2806.75 cm^{-1}	229947.07	– 232753.816	6–8	1.44–05	3.65–04	2.57–01	–2.660	C	2
			35972.8	2779.12 cm^{-1}	229968.44	– 232747.562	4–6	5.49–05	1.60–03	7.58–01	–2.194	C	2
			35698.3	2800.49 cm^{-1}	229947.07	– 232747.562	6–6	3.21–06	6.13–05	4.32–02	–3.434	C	2
169.		$^2D^o-^2F$	*34077*	*2933.8* cm^{-1}	*229955.6*	*– 232889.39*	10–14	3.34–04	8.14–03	9.13+00	–1.089	D	1,2
			33189.9	3012.14 cm^{-1}	229947.07	– 232959.210	6–8	3.98–04	8.76–03	5.75+00	–1.279	D	2*n*
			35352.8	2827.86 cm^{-1}	229968.44	– 232796.298	4–6	2.44–04	6.87–03	3.20+00	–1.561	D	2*n*
			35087.7	2849.23 cm^{-1}	229947.07	– 232796.298	6–6	1.46–05	2.69–04	1.87–01	–2.792	D–	2*n*
170.		$^2D^o-^2D$	*22324*	*4478.3* cm^{-1}	*229955.6*	*– 234433.90*	10–10	4.68–04	3.50–03	2.57+00	–1.457	C+	1,2
			22178.9	4507.56 cm^{-1}	229947.07	– 234454.634	6–6	4.64–04	3.42–03	1.50+00	–1.688	C+	2*n*
			22545.0	4434.36 cm^{-1}	229968.44	– 234402.797	4–4	3.92–04	2.99–03	8.87–01	–1.923	C+	2*n*
			22436.9	4455.73 cm^{-1}	229947.07	– 234402.797	6–4	4.96–05	2.49–04	1.11–01	–2.825	C+	2*n*
			22284.5	4486.19 cm^{-1}	229968.44	– 234454.634	4–6	2.23–05	2.49–04	7.30–02	–3.002	C–	2*n*
171.	$2s^22p^2(^1D)3p-$ $2s^22p^2(^3P)4s$	$^2D^o-^2P$	*9521.9*	*9524.5*	*229955.6*	*– 240454.9*	10–6	4.10–02	3.34–02	1.05+01	–0.476	B	1
			9457.89	9460.49	229947.07	– 240517.35	6–4	3.77–02	3.37–02	6.29+00	–0.695	C+	*LS*
			9648.40	9651.05	229968.44	– 240330.01	4–2	3.94–02	2.75–02	3.49+00	–0.959	C+	*LS*
			9477.05	9479.65	229968.44	– 240517.35	4–4	4.16–03	5.60–03	6.99–01	–1.650	C+	*LS*
172.	$2s^22p^2(^1D)3p-$ $2s^22p^2(^1D)3d$	$^2D^o-^2F$	*4700.7*	*4702.0*	*229955.6*	*– 251223.3*	10–14	9.87–01	4.58–01	7.08+01	0.661	C	1
			4699.01	4700.33	229947.07	– 251222.19	6–8	9.88–01	4.36–01	4.05+01	0.418	D	*LS*
			4703.16	4704.48	229968.44	– 251224.79	4–6	9.20–01	4.58–01	2.83+01	0.262	D	*LS*
			4698.44	4699.75	229947.07	– 251224.79	6–6	6.59–02	2.18–02	2.02+00	–0.883	D	*LS*

O II: Allowed Transitions — Continued

No.	Transition Array	Mult.	λ_{air} (Å)	λ_{vac} (Å) or σ (cm^{-1})*	E_i (cm^{-1})	E_k (cm^{-1})	g_i-g_k	A_{ki} (10^8 s^{-1})	f_{ik}	S (at. u.)	log gf	Acc.	Source
173.		$^2D^o-^2D$	*4329.2*	*4330.4*	*229955.6*	*− 253048.0*	10−10	7.23−01	2.03−01	2.90+01	0.308	C	1
			4327.46	4328.68	229947.07	− 253048.82	6−6	6.76−01	1.90−01	1.62+01	0.056	D	*LS*
			4331.86	4333.07	229968.44	− 253046.74	4−4	6.50−01	1.83−01	1.04+01	−0.136	D	*LS*
			4327.85	4329.07	229947.07	− 253046.74	6−4	7.24−02	1.36−02	1.16+00	−1.090	D	*LS*
			4331.47	4332.68	229968.44	− 253048.82	4−6	4.82−02	2.03−02	1.16+00	−1.090	D	*LS*
174.		$^2D^o-^2P$	*4194.2*	*4195.3*	*229955.6*	*− 253791.6*	10−6	3.57−01	5.64−02	7.79+00	−0.248	C	1
			4192.51	4193.69	229947.07	− 253792.40	6−4	3.21−01	5.65−02	4.67+00	−0.470	D	*LS*
			4196.70	4197.88	229968.44	− 253789.99	4−2	3.56−01	4.70−02	2.60+00	−0.726	D	*LS*
			4196.27	4197.45	229968.44	− 253792.40	4−4	3.56−02	9.40−03	5.19−01	−1.425	D	*LS*
175.	$2s^22p^2(^1D)3p-$ $2s^22p^2(^3P)4d$	$^2D^o-^2P$	*3958.6*	*3959.8*	*229955.6*	*− 255209.7*	10−6	9.73−02	1.37−02	1.79+00	−0.863	C	1
			3962.96	3964.08	229947.07	− 255173.58	6−4	8.73−02	1.37−02	1.07+00	−1.085	D	*LS*
			3949.35	3950.46	229968.44	− 255281.93	4−2	9.80−02	1.15−02	5.96−01	−1.339	D	*LS*
			3966.32	3967.44	229968.44	− 255173.58	4−4	9.67−03	2.28−03	1.19−01	−2.040	D	*LS*
176.		$^2D^o-^2F$	*3929.7*	*3930.8*	*229955.6*	*− 255395.8*	10−14	7.57−03	2.46−03	3.18−01	−1.610	D	1
			3917.53	3918.64	229947.07	− 255466.10	6−8	7.64−03	2.35−03	1.81−01	−1.852	D−	*LS*
			3946.20	3947.32	229968.44	− 255302.11	4−6	6.98−03	2.44−03	1.27−01	−2.010	D−	*LS*
			3942.87	3943.99	229947.07	− 255302.11	6−6	5.00−04	1.17−04	9.07−03	−3.155	E	*LS*
177.		$^2D^o-^2D$	*3856.9*	*3858.0*	*229955.6*	*− 255875.7*	10−10	3.18−02	7.09−03	9.00−01	−1.149	D+	1
			3852.39	3853.49	229947.07	− 255897.59	6−6	2.98−02	6.62−03	5.04−01	−1.401	D	*LS*
			3863.72	3864.81	229968.44	− 255842.91	4−4	2.85−02	6.37−03	3.24−01	−1.594	D	*LS*
			3860.53	3861.62	229947.07	− 255842.91	6−4	3.17−03	4.72−04	3.60−02	−2.548	D−	*LS*
			3855.57	3856.66	229968.44	− 255897.59	4−6	2.12−03	7.09−04	3.60−02	−2.547	D−	*LS*
178.	$2s^22p^2(^1D)3p-$ $2s^22p^2(^3P)5s$	$^2D^o-^2P$	*3497.3*	*3498.3*	*229955.6*	*− 258541.3*	10−6	2.25−03	2.48−04	2.85−02	−2.606	C	1
			3488.27	3489.27	229947.07	− 258606.35	6−4	2.04−03	2.48−04	1.71−02	−2.827	D	*LS*
			3514.82	3515.83	229968.44	− 258411.26	4−2	2.22−03	2.05−04	9.50−03	−3.086	D−	*LS*
			3490.88	3491.87	229968.44	− 258606.35	4−4	2.26−04	4.13−05	1.90−03	−3.782	E	*LS*
179.	$2s^22p^2(^1D)3p-$ $2s^22p^2(^1D)4s$	$^2D^o-^2D$	*3408.2*	*3409.2*	*229955.6*	*− 259287.8*	10−10	1.09+00	1.89−01	2.12+01	0.277	C+	1
			3407.28	3408.25	229947.07	− 259287.61	6−6	1.02+00	1.77−01	1.19+01	0.025	C−	*LS*
			3409.71	3410.68	229968.44	− 259288.07	4−4	9.77−01	1.70−01	7.65+00	−0.167	C−	*LS*
			3407.22	3408.20	229947.07	− 259288.07	6−4	1.09−01	1.26−02	8.49−01	−1.121	C−	*LS*
			3409.76	3410.74	229968.44	− 259287.61	4−6	7.24−02	1.89−02	8.49−01	−1.121	C−	*LS*
180.	$2s^22p^2(^1D)3p-$ $2s^22p^2(^1D)4d$	$^2D^o-^2F$	*2231.0*	*2231.7*	*229955.6*	*− 274764*	10−14	6.34−02	6.63−03	4.87−01	−1.179	D	1
			2229.74	2230.43	229947.07	− 274781.5	6−8	6.36−02	6.32−03	2.78−01	−1.421	D−	*LS*
			2232.83	2233.53	229968.44	− 274740.7	4−6	5.91−02	6.62−03	1.95−01	−1.577	D−	*LS*
			2231.77	2232.46	229947.07	− 274740.7	6−6	4.23−03	3.16−04	1.39−02	−2.723	E	*LS*
181.		$^2D^o-^2D$	*2223.1*	*2223.8*	*229955.6*	*− 274923*	10−10	3.27−02	2.42−03	1.77−01	−1.615	D	1
			2222.47	2223.16	229947.07	− 274928.0	6−6	3.06−02	2.26−03	9.93−02	−1.867	D−	*LS*
			2224.12	2224.81	229968.44	− 274916.0	4−4	2.94−02	2.18−03	6.39−02	−2.059	D−	*LS*
			2223.07	2223.76	229947.07	− 274916.0	6−4	3.27−03	1.62−04	7.10−03	−3.013	E	*LS*
			2223.53	2224.22	229968.44	− 274928.0	4−6	2.18−03	2.42−04	7.10−03	−3.013	E	*LS*
182.	$2s^22p^2(^1S)3s-$ $2s^22p^2(^1D)3p$	$^2S-^2P^o$		*1902.1* cm^{-1}	230609.45	*− 232511.5*	2−6	1.03−05	1.29−03	4.45−01	−2.590	B	2
				1917.64 cm^{-1}	230609.45	− 232527.09	2−4	1.06−05	8.63−04	2.96−01	−2.763	B	2
				1870.99 cm^{-1}	230609.45	− 232480.44	2−2	9.86−06	4.22−04	1.49−01	−3.074	B	2

O II: Allowed Transitions — Continued

No.	Transition Array	Mult.	λ_{air} (Å)	λ_{vac} (Å) or σ (cm^{-1})*	E_i (cm^{-1})	E_k (cm^{-1})	g_i–g_k	A_{ki} (10^8 s^{-1})	f_{ik}	S (at. u.)	log gf	Acc.	Source
183.	$2s^22p^2(^3P)3d$–$2s^22p^2(^3P)4p$	^4F–$^4D^o$	*6897.6*	*6899.5*	*231428.97*	*– 245922.9*	28–20	3.04–01	1.55–01	9.86+01	0.637	C+	1
			6895.10	6897.00	231530.246	– 246029.295	10–8	2.72–01	1.55–01	3.52+01	0.190	C–	*LS*
			6906.44	6908.34	231427.970	– 245903.224	8–6	2.48–01	1.33–01	2.41+01	0.026	C–	*LS*
			6910.56	6912.47	231350.087	– 245816.70	6–4	2.43–01	1.16–01	1.58+01	–0.157	C–	*LS*
			6907.87	6909.78	231296.126	– 245768.37	4–2	3.03–01	1.08–01	9.86+00	–0.363	C–	*LS*
			6846.80	6848.69	231427.970	– 246029.295	8–8	3.17–02	2.23–02	4.01+00	–0.750	C–	*LS*
			6869.48	6871.37	231350.087	– 245903.224	6–6	5.35–02	3.79–02	5.14+00	–0.644	C–	*LS*
			6884.88	6886.78	231296.126	– 245816.70	4–4	6.12–02	4.35–02	3.94+00	–0.760	C–	*LS*
			6810.48	6812.36	231350.087	– 246029.295	6–8	1.64–03	1.52–03	2.04–01	–2.041	D	*LS*
			6844.10	6845.99	231296.126	– 245903.224	4–6	2.97–03	3.12–03	2.82–01	–1.903	D	*LS*
184.	$2s^22p^2(^3P)3d$–$2s2p^3(^5S^o)3s$	^4P–$^4S^o$	*4448.9*	*4450.2*	*232510.43*	– 254981.55	12–4	2.87–03	2.84–04	4.99–02	–2.468	D	1
			4439.48	4440.73	232462.724	– 254981.55	6–4	1.44–03	2.85–04	2.50–02	–2.768	D–	*LS*
			4453.97	4455.22	232535.949	– 254981.55	4–4	9.54–04	2.84–04	1.66–02	–2.945	D–	*LS*
			4467.21	4468.46	232602.492	– 254981.55	2–4	4.73–04	2.83–04	8.32–03	–3.248	E	*LS*
185.	$2s^22p^2(^1D)3p$–$2s^22p^2(^3P)3d$	$^2P^o$–^2P		957.0 cm^{-1}	232511.5	– 233468.6	6–6	5.04–06	8.24–04	1.70+00	–2.306	D	2
				903.44 cm^{-1}	232527.09	– 233430.53	4–4	3.13–06	5.74–04	8.37–01	–2.639	D	2
				1064.15 cm^{-1}	232480.44	– 233544.59	2–2	4.69–06	6.21–04	3.84–01	–2.906	D	2
				1017.50 cm^{-1}	232527.09	– 233544.59	4–2	2.02–06	1.46–04	1.89–01	–3.234	D	2
				950.09 cm^{-1}	232480.44	– 233430.53	2–4	1.27–06	4.21–04	2.92–01	–3.074	D	2
186.		$^2P^o$–^2D		*1922.4* cm^{-1}	*232511.5*	*– 234433.90*	6–10	6.96–05	4.70–03	4.83+00	–1.549	D	2
				1927.54 cm^{-1}	232527.09	– 234454.634	4–6	6.96–05	4.21–03	2.88+00	–1.773	D	2
				1922.36 cm^{-1}	232480.44	– 234402.797	2–4	5.37–05	4.36–03	1.49+00	–2.060	D	2
				1875.71 cm^{-1}	232527.09	– 234402.797	4–4	1.54–05	6.57–04	4.61–01	–2.580	D	2
187.	$2s^22p^2(^1D)3p$–$2s^22p^2(^1D)3d$	$^2P^o$–^2D	*4868.0*	*4869.4*	*232511.5*	*– 253048.0*	6–10	5.62–01	3.33–01	3.20+01	0.300	C+	1
			4871.52	4872.88	232527.09	– 253048.82	4–6	5.60–01	2.99–01	1.92+01	0.078	C–	*LS*
			4860.97	4862.32	232480.44	– 253046.74	2–4	4.70–01	3.33–01	1.07+01	–0.176	C–	*LS*
			4872.02	4873.38	232527.09	– 253046.74	4–4	9.34–02	3.32–02	2.13+00	–0.876	D	*LS*
188.		$^2P^o$–^2P	*4697.9*	*4699.2*	*232511.5*	*– 253791.6*	6–6	1.11+00	3.67–01	3.41+01	0.343	C+	1
			4701.18	4702.49	232527.09	– 253792.40	4–4	9.23–01	3.06–01	1.89+01	0.088	C–	*LS*
			4691.42	4692.73	232480.44	– 253789.99	2–2	7.43–01	2.45–01	7.57+00	–0.309	D	*LS*
			4701.71	4703.03	232527.09	– 253789.99	4–2	3.69–01	6.12–02	3.79+00	–0.611	D	*LS*
			4690.89	4692.20	232480.44	– 253792.40	2–4	1.86–01	1.23–01	3.79+00	–0.610	D	*LS*
189.	$2s^22p^2(^1D)3p$–$2s^22p^2(^3P)4d$	$^2P^o$–^2P	*4404.4*	*4405.6*	*232511.5*	*– 255209.7*	6–6	3.40–02	9.87–03	8.59–01	–1.227	D	1
			4414.46	4415.70	232527.09	– 255173.58	4–4	2.81–02	8.21–03	4.77–01	–1.484	D–	*LS*
			4384.45	4385.68	232480.44	– 255281.93	2–2	2.29–02	6.61–03	1.91–01	–1.879	D–	*LS*
			4393.44	4394.67	232527.09	– 255281.93	4–2	1.14–02	1.65–03	9.54–02	–2.181	E	*LS*
			4405.38	4406.62	232480.44	– 255173.58	2–4	5.65–03	3.29–03	9.54–02	–2.182	E	*LS*
190.		$^2P^o$–^2D	*4278.9*	*4280.1*	*232511.5*	*– 255875.7*	6–10	6.92–02	3.17–02	2.68+00	–0.721	C	1
			4277.69	4278.90	232527.09	– 255897.59	4–6	6.93–02	2.85–02	1.61+00	–0.943	D	*LS*
			4279.17	4280.37	232480.44	– 255842.91	2–4	5.77–02	3.17–02	8.92–01	–1.199	D	*LS*
			4287.73	4288.93	232527.09	– 255842.91	4–4	1.15–02	3.16–03	1.78–01	–1.898	D	*LS*
191.	$2s^22p^2(^1D)3p$–$2s^22p^2(^1D)3d$	$^2P^o$–^2S	*4325.7*	*4326.9*	*232511.5*	– 255622.8	6–2	1.68+00	1.57–01	1.34+01	–0.026	C	1
			4328.59	4329.81	232527.09	– 255622.8	4–2	1.12+00	1.57–01	8.94+00	–0.202	D	*LS*
			4319.87	4321.08	232480.44	– 255622.8	2–2	5.62–01	1.57–01	4.47+00	–0.502	D	*LS*

O II: Allowed Transitions — Continued

No.	Transition Array	Mult.	λ_{air} (Å)	λ_{vac} (Å) or σ (cm^{-1})*	E_i (cm^{-1})	E_k (cm^{-1})	g_i-g_k	A_{ki} (10^8 s^{-1})	f_{ik}	S (at. u.)	log gf	Acc.	Source
192.	$2s^22p^2(^1D)3p-2s^22p^2(^1D)4s$	$^2P^o-^2D$	*3733.6*	*3734.7*	*232511.5*	*259287.8*	6−10	2.92−01	1.02−01	7.51+00	−0.214	C+	1
			3735.79	3736.85	232527.09	259287.61	4−6	2.92−01	9.16−02	4.51+00	−0.436	C−	*LS*
			3729.22	3730.28	232480.44	259288.07	2−4	2.44−01	1.02−01	2.50+00	−0.691	C−	*LS*
			3735.72	3736.78	232527.09	259288.07	4−4	4.86−02	1.02−02	5.01−01	−1.390	D	*LS*
193.	$2s^22p^2(^1D)3p-2s^22p^2(^1D)4d$	$^2P^o-^2D$	*2357.1*	*2357.9*	*232511.5*	*274923*	6−10	1.26−02	1.75−03	8.13−02	−1.980	D	1
			2357.72	2358.44	232527.09	274928.0	4−6	1.26−02	1.57−03	4.88−02	−2.202	D−	*LS*
			2355.79	2356.51	232480.44	274916.0	2−4	1.05−02	1.75−03	2.71−02	−2.457	D−	*LS*
			2358.39	2359.11	232527.09	274916.0	4−4	2.09−03	1.75−04	5.42−03	−3.156	E	*LS*
194.	$2s^22p^2(^1D)3p-2s^22p^2(^1S)3d$	$^2P^o-^2D$	*2296.7*	*2297.4*	*232511.5*	*276038.8*	6−10	5.28−03	6.97−04	3.16−02	−2.379	D	1
			2295.60	2296.30	232527.09	276075.32	4−6	5.29−03	6.27−04	1.90−02	−2.601	D−	*LS*
			2297.96	2298.67	232480.44	275983.95	2−4	4.40−03	6.96−04	1.05−02	−2.856	D−	*LS*
			2300.42	2301.13	232527.09	275983.95	4−4	8.77−04	6.95−05	2.11−03	−3.556	E	*LS*
195.	$2s^22p^2(^3P)3d-2s^22p^2(^3P)4p$	$^4D-^4D^o$	*7587.1*	*7589.2*	*232746.16*	*245922.9*	20−20	4.74−02	4.09−02	2.04+01	−0.088	C	1
			7530.61	7532.68	232753.816	246029.295	8−8	4.15−02	3.53−02	7.00+00	−0.549	D	*LS*
			7599.20	7601.29	232747.562	245903.224	6−6	2.70−02	2.34−02	3.51+00	−0.853	D	*LS*
			7648.58	7650.69	232745.981	245816.70	4−4	1.85−02	1.62−02	1.63+00	−1.188	D	*LS*
			7656.78	7658.89	232711.642	245768.37	2−2	2.30−02	2.03−02	1.02+00	−1.393	D	*LS*
			7602.81	7604.91	232753.816	245903.224	8−6	8.94−03	5.81−03	1.16+00	−1.333	D	*LS*
			7649.51	7651.61	232747.562	245816.70	6−4	1.62−02	9.46−03	1.43+00	−1.246	D	*LS*
			7676.97	7679.08	232745.981	245768.37	4−2	2.29−02	1.01−02	1.02+00	−1.394	D	*LS*
			7527.07	7529.14	232747.562	246029.295	6−8	6.91−03	7.83−03	1.16+00	−1.328	D	*LS*
			7598.28	7600.38	232745.981	245903.224	4−6	1.10−02	1.43−02	1.43+00	−1.243	D	*LS*
			7628.54	7630.64	232711.642	245816.70	2−4	1.16−02	2.03−02	1.02+00	−1.391	D	*LS*
196.		$^2F-^2D^o$	*6565.6*	*6567.4*	*232889.39*	*248116.1*	14−10	3.00−01	1.39−01	4.19+01	0.288	C	1
			6565.28	6567.10	232959.210	248186.64	8−6	2.86−01	1.39−01	2.40+01	0.045	D	*LS*
			6571.11	6572.92	232796.298	248010.23	6−4	2.99−01	1.29−01	1.68+01	−0.111	D	*LS*
			6495.79	6497.58	232796.298	248186.64	6−6	1.48−02	9.34−03	1.20+00	−1.252	D	*LS*
197.	$2s^22p^2(^3P)3d-2s^22p^2(^3P)5p$	$^2F-^2D^o$	*3457.8*	*3458.8*	*232889.39*	*261801.5*	14−10	1.02−01	1.30−02	2.08+00	−0.739	C	1
			3457.93	3458.92	232959.210	261869.94	8−6	9.69−02	1.30−02	1.19+00	−0.982	D	*LS*
			3458.92	3459.91	232796.298	261698.75	6−4	1.02−01	1.22−02	8.31−01	−1.137	D	*LS*
			3438.56	3439.54	232796.298	261869.94	6−6	4.93−03	8.74−04	5.93−02	−2.281	D−	*LS*
198.	$2s^22p^2(^3P)3d-2s^22p^2(^3P)4p$	$^2P-^2D^o$	*6825.2*	*6827.1*	*233468.6*	*248116.1*	6−10	1.71−02	1.99−02	2.69+00	−0.922	C	1
			6774.98	6776.85	233430.53	248186.64	4−6	1.75−02	1.81−02	1.61+00	−1.141	D	*LS*
			6911.03	6912.93	233544.59	248010.23	2−4	1.37−02	1.97−02	8.96−01	−1.405	D	*LS*
			6856.96	6858.85	233430.53	248010.23	4−4	2.82−03	1.98−03	1.79−01	−2.100	D	*LS*
199.		$^2P-^2P^o$	*6657.2*	*6659.1*	*233468.6*	*248485.7*	6−6	2.04−01	1.36−01	1.78+01	−0.089	C	1
			6627.37	6629.20	233430.53	248515.30	4−4	1.73−01	1.14−01	9.92+00	−0.342	D	*LS*
			6717.75	6719.61	233544.59	248426.41	2−2	1.33−01	8.97−02	3.97+00	−0.746	D	*LS*
			6666.66	6668.50	233430.53	248426.41	4−2	6.78−02	2.26−02	1.98+00	−1.044	D	*LS*
			6677.87	6679.71	233544.59	248515.30	2−4	3.37−02	4.51−02	1.98+00	−1.045	D	*LS*
200.	$2s^22p^2(^3P)3d-2s^22p^2(^1S)3p$	$^2P-^2P^o$	*5120.8*	*5122.2*	*233468.6*	*252991.3*	6−6	8.62−03	3.39−03	3.43−01	−1.692	D	1
			5110.30	5111.73	233430.53	252993.39	4−4	7.22−03	2.83−03	1.90−01	−1.946	D−	*LS*
			5141.90	5143.33	233544.59	252987.23	2−2	5.67−03	2.25−03	7.61−02	−2.347	D−	*LS*
			5111.91	5113.34	233430.53	252987.23	4−2	2.89−03	5.65−04	3.81−02	−2.646	E	*LS*
			5140.27	5141.71	233544.59	252993.39	2−4	1.42−03	1.12−03	3.81−02	−2.648	E	*LS*

O II: Allowed Transitions — Continued

No.	Transition Array	Mult.	λ_{air} (Å)	λ_{vac} (Å) or σ (cm^{-1})*	E_i (cm^{-1})	E_k (cm^{-1})	$g_i - g_k$	A_{ki} (10^8 s^{-1})	f_{ik}	S (at. u.)	log gf	Acc.	Source
201.	$2s^22p^2(^3P)3d-2s^22p^2(^3P)5p$	$^2P-^2D^\circ$	*3528.5*	*3529.5*	*233468.6*	*261801.5*	6–10	1.24–03	3.86–04	2.69–02	–2.635	D	1
			3515.24	3516.25	233430.53	261869.94	4–6	1.25–03	3.49–04	1.61–02	–2.856	D–	*LS*
			3550.86	3551.87	233544.59	261698.75	2–4	1.01–03	3.84–04	8.97–03	–3.115	E	*LS*
			3536.53	3537.54	233430.53	261698.75	4–4	2.05–04	3.85–05	1.79–03	–3.812	E	*LS*
202.	$2s^22p^2(^3P)3d-2s^22p^2(^3P)4p$	$^2D-^2D^\circ$	*7306.8*	*7308.8*	*234433.90*	*248116.1*	10–10	4.86–02	3.89–02	9.36+00	–0.410	C	1
			7280.25	7282.26	234454.634	248186.64	6–6	4.59–02	3.64–02	5.24+00	–0.660	D	*LS*
			7346.90	7348.92	234402.797	248010.23	4–4	4.30–02	3.48–02	3.37+00	–0.856	D	*LS*
			7375.00	7377.03	234454.634	248010.23	6–4	4.73–03	2.57–03	3.74–01	–1.812	D–	*LS*
			7252.87	7254.87	234402.797	248186.64	4–6	3.31–03	3.92–03	3.74–01	–1.805	D–	*LS*
203.		$^2D-^2P^\circ$	*7114.6*	*7116.5*	*234433.90*	*248485.7*	10–6	2.25–01	1.02–01	2.40+01	0.011	C	1
			7110.08	7112.04	234454.634	248515.30	6–4	2.03–01	1.03–01	1.44+01	–0.211	D	*LS*
			7128.86	7130.83	234402.797	248426.41	4–2	2.24–01	8.52–02	8.00+00	–0.467	D	*LS*
			7083.96	7085.92	234402.797	248515.30	4–4	2.28–02	1.72–02	1.60+00	–1.164	D	*LS*
204.	$2s^22p^2(^3P)3d-2s^22p^2(^1S)3p$	$^2D-^2P^\circ$	*5387.2*	*5388.7*	*234433.90*	*252991.3*	10–6	9.63–03	2.51–03	4.46–01	–1.600	D	1
			5392.61	5394.11	234454.634	252993.39	6–4	8.64–03	2.51–03	2.68–01	–1.822	D–	*LS*
			5379.35	5380.85	234402.797	252987.23	4–2	9.67–03	2.10–03	1.49–01	–2.076	D–	*LS*
			5377.57	5379.06	234402.797	252993.39	4–4	9.68–04	4.20–04	2.97–02	–2.775	E	*LS*
205.	$2s^22p^2(^3P)3d-2s^22p^2(^3P)5p$	$^2D-^2D^\circ$	*3652.9*	*3654.0*	*234433.90*	*261801.5*	10–10	1.34–02	2.68–03	3.22–01	–1.572	D	1
			3646.56	3647.60	234454.634	261869.94	6–6	1.26–02	2.50–03	1.80–01	–1.823	D–	*LS*
			3662.50	3663.55	234402.797	261698.75	4–4	1.20–02	2.40–03	1.16–01	–2.017	D–	*LS*
			3669.47	3670.52	234454.634	261698.75	6–4	1.32–03	1.78–04	1.29–02	–2.972	E	*LS*
			3639.68	3640.71	234402.797	261869.94	4–6	9.02–04	2.69–04	1.29–02	–2.969	E	*LS*
206.	$2s^22p^2(^1S)3s-2s^22p^2(^3P)4p$	$^2S-^2P^\circ$	*5592.5*	*5594.0*	230609.45	*248485.7*	2–6	2.16–02	3.04–02	1.12+00	–1.217	C	1
			5583.22	5584.77	230609.45	248515.30	2–4	2.17–02	2.03–02	7.45–01	–1.392	D	*LS*
			5611.07	5612.63	230609.45	248426.41	2–2	2.14–02	1.01–02	3.73–01	–1.695	D	*LS*
207.	$2s^22p^2(^1S)3s-2s^22p^2(^1S)3p$	$^2S-^2P^\circ$	*4466.6*	*4467.9*	230609.45	*252991.3*	2–6	9.00–01	8.08–01	2.38+01	0.208	B	1
			4466.24	4467.49	230609.45	252993.39	2–4	9.00–01	5.39–01	1.58+01	0.032	C+	*LS*
			4467.46	4468.72	230609.45	252987.23	2–2	9.00–01	2.69–01	7.92+00	–0.269	C+	*LS*
208.	$2s^22p^2(^3P)4s-2s^22p^2(^3P)4p$	$^4P-^4D^\circ$	*14027*	*7127.1* cm^{-1}	*238795.8*	*245922.9*	12–20	1.44–01	7.08–01	3.92+02	0.929	B	1
			14010.9	7135.34 cm^{-1}	238893.96	246029.295	6–8	1.44–01	5.67–01	1.57+02	0.531	C+	*LS*
			13942.1	7170.57 cm^{-1}	238732.65	245903.224	4–6	1.03–01	4.48–01	8.23+01	0.254	C+	*LS*
			13905.9	7189.24 cm^{-1}	238627.46	245816.70	2–4	6.15–02	3.57–01	3.27+01	–0.146	C+	*LS*
			14262.9	7009.26 cm^{-1}	238893.96	245903.224	6–6	4.11–02	1.25–01	3.53+01	–0.124	C+	*LS*
			14112.4	7084.05 cm^{-1}	238732.65	245816.70	4–4	7.54–02	2.25–01	4.18+01	–0.046	C+	*LS*
			14000.0	7140.91 cm^{-1}	238627.46	245768.37	2–2	1.21–01	3.54–01	3.27+01	–0.149	C+	*LS*
			14441.2	6922.74 cm^{-1}	238893.96	245816.70	6–4	6.59–03	1.37–02	3.92+00	–1.084	C	*LS*
			14209.3	7035.72 cm^{-1}	238732.65	245768.37	4–2	2.31–02	3.49–02	6.53+00	–0.855	C	*LS*
209.	$2s^22p^2(^3P)4s-2s2p^3(^5S^\circ)3s$	$^4P-^4S^\circ$	*6176.6*	*6178.3*	*238795.8*	254981.55	12–4	2.12–01	4.03–02	9.84+00	–0.315	C+	1
			6214.25	6215.97	238893.96	254981.55	6–4	1.04–01	4.01–02	4.92+00	–0.619	C–	*LS*
			6152.56	6154.26	238732.65	254981.55	4–4	7.14–02	4.05–02	3.28+00	–0.790	C–	*LS*
			6112.99	6114.68	238627.46	254981.55	2–4	3.64–02	4.08–02	1.64+00	–1.089	C–	*LS*
210.	$2s^22p^2(^3P)4s-2s^22p^2(^3P)4p$	$^2P-^2D^\circ$	*13049*	*7661.2* cm^{-1}	*240454.9*	*248116.1*	6–10	1.79–01	7.60–01	1.96+02	0.659	B	1
			13035.5	7669.29 cm^{-1}	240517.35	248186.64	4–6	1.79–01	6.85–01	1.18+02	0.438	C+	*LS*
			13016.9	7680.22 cm^{-1}	240330.01	248010.23	2–4	1.50–01	7.62–01	6.53+01	0.183	C+	*LS*
			13342.4	7492.88 cm^{-1}	240517.35	248010.23	4–4	2.79–02	7.44–02	1.31+01	–0.527	C+	*LS*

O II: Allowed Transitions — Continued

No.	Transition Array	Mult.	λ_{air} (Å)	λ_{vac} (Å) or σ (cm^{-1})*	E_i (cm^{-1})	E_k (cm^{-1})	g_i–g_k	A_{ki} (10^8 s^{-1})	f_{ik}	S (at. u.)	log gf	Acc.	Source
211.		^{2}P–^{2}P°	*12449*	*8030.8* cm^{-1}	*240454.9*	– *248485.7*	6–6	1.51–01	3.51–01	8.62+01	0.323	B	1
			12499.8	7997.95 cm^{-1}	240517.35	– 248515.30	4–4	1.24–01	2.91–01	4.79+01	0.066	C+	*LS*
			12347.8	8096.40 cm^{-1}	240330.01	– 248426.41	2–2	1.03–01	2.36–01	1.92+01	–0.327	C+	*LS*
			12640.3	7909.06 cm^{-1}	240517.35	– 248426.41	4–2	4.81–02	5.76–02	9.58+00	–0.638	C+	*LS*
			12213.7	8185.29 cm^{-1}	240330.01	– 248515.30	2–4	2.66–02	1.19–01	9.58+00	–0.623	C+	*LS*
212.	$2s^22p^2(^3P)4s$– $2s^22p^2(^1S)3p$	^{2}P–^{2}P°	*7974.6*	*7976.7*	*240454.9*	– *252991.3*	6–6	3.42–02	3.27–02	5.14+00	–0.708	C+	1
			8013.16	8015.36	240517.35	– 252993.39	4–4	2.81–02	2.71–02	2.86+00	–0.965	C–	*LS*
			7898.46	7900.63	240330.01	– 252987.23	2–2	2.35–02	2.20–02	1.14+00	–1.357	C–	*LS*
			8017.12	8019.32	240517.35	– 252987.23	4–2	1.12–02	5.41–03	5.72–01	–1.664	C–	*LS*
			7894.61	7896.79	240330.01	– 252993.39	2–4	5.88–03	1.10–02	5.72–01	–1.658	C–	*LS*
213.	$2s^22p^2(^3P)4s$– $2s^22p^2(^3P)5p$	^{2}P–^{2}D°	*4683.3*	*4684.6*	*240454.9*	– *261801.5*	6–10	5.26–03	2.89–03	2.67–01	–1.762	C+	1
			4681.96	4683.27	240517.35	– 261869.94	4–6	5.27–03	2.60–03	1.60–01	–1.983	C–	*LS*
			4678.42	4679.73	240330.01	– 261698.75	2–4	4.40–03	2.89–03	8.90–02	–2.238	C–	*LS*
			4719.80	4721.12	240517.35	– 261698.75	4–4	8.57–04	2.86–04	1.78–02	–2.941	D	*LS*
214.	$2s^22p^2(^3P)4p$– $2s^22p^2(^3P)4d$	^{4}D°–^{4}P	*10862*	*9204.3* cm^{-1}	*245922.9*	– *255127.2*	20–12	3.59–03	3.81–03	2.72+00	–1.118	C	1
			11015.4	9075.72 cm^{-1}	246029.295	– 255105.01	8–6	2.75–03	3.75–03	1.09+00	–1.523	D	*LS*
			10820.5	9239.19 cm^{-1}	245903.224	– 255142.41	6–4	2.29–03	2.67–03	5.72–01	–1.795	D	*LS*
			10696.4	9346.38 cm^{-1}	245816.70	– 255163.08	4–2	1.88–03	1.61–03	2.27–01	–2.191	D	*LS*
			10864.5	9201.79 cm^{-1}	245903.224	– 255105.01	6–6	6.45–04	1.14–03	2.45–01	–2.164	D	*LS*
			10720.1	9325.71 cm^{-1}	245816.70	– 255142.41	4–4	1.19–03	2.06–03	2.90–01	–2.085	D	*LS*
			10641.4	9394.71 cm^{-1}	245768.37	– 255163.08	2–2	1.91–03	3.24–03	2.27–01	–2.189	D	*LS*
			10763.3	9288.31 cm^{-1}	245816.70	– 255105.01	4–6	7.37–05	1.92–04	2.72–02	–3.115	D–	*LS*
			10664.8	9374.04 cm^{-1}	245768.37	– 255142.41	2–4	1.89–04	6.46–04	4.54–02	–2.889	D–	*LS*
215.	$2s^22p^2(^3P)4p$– $2s^22p^2(^3P)5s$	^{4}D°–^{4}P	*8371.3*	*8373.6*	*245922.9*	– *257865.2*	20–12	3.70–01	2.33–01	1.29+02	0.669	C+	1
			8375.84	8378.15	246029.295	– 257965.11	8–6	2.96–01	2.33–01	5.14+01	0.271	C–	*LS*
			8403.38	8405.69	245903.224	– 257799.93	6–4	2.30–01	1.63–01	2.70+01	–0.011	C–	*LS*
			8415.88	8418.19	245816.70	– 257695.74	4–2	1.82–01	9.67–02	1.07+01	–0.413	C–	*LS*
			8288.30	8290.58	245903.224	– 257965.11	6–6	6.86–02	7.07–02	1.16+01	–0.373	C–	*LS*
			8342.70	8345.00	245816.70	– 257799.93	4–4	1.20–01	1.25–01	1.37+01	–0.302	C–	*LS*
			8381.77	8384.08	245768.37	– 257695.74	2–2	1.84–01	1.94–01	1.07+01	–0.411	C–	*LS*
			8229.27	8231.53	245816.70	– 257965.11	4–6	7.79–03	1.19–02	1.29+00	–1.324	C–	*LS*
			8309.19	8311.47	245768.37	– 257799.93	2–4	1.89–02	3.92–02	2.14+00	–1.106	C–	*LS*
216.	$2s^22p^2(^3P)4p$– $2s^22p^2(^1D)3d$	^{2}D°–^{2}F	*32174*	*3107.2* cm^{-1}	*248116.1*	– *251223.3*	10–14	4.16–04	9.03–03	9.56+00	–1.044	C	1
			32934.0	3035.55 cm^{-1}	248186.64	– 251222.19	6–8	3.87–04	8.40–03	5.46+00	–1.298	D	*LS*
			31100.0	3214.56 cm^{-1}	248010.23	– 251224.79	4–6	4.29–04	9.34–03	3.83+00	–1.428	D	*LS*
			32905.8	3038.15 cm^{-1}	248186.64	– 251224.79	6–6	2.59–05	4.20–04	2.73–01	–2.598	D	*LS*
217.		^{2}D°–^{2}D	*20271*	*4931.9* cm^{-1}	*248116.1*	– *253048.0*	10–10	5.42–04	3.34–03	2.23+00	–1.476	C	1
			20561.3	4862.18 cm^{-1}	248186.64	– 253048.82	6–6	4.85–04	3.07–03	1.25+00	–1.734	D	*LS*
			19849.6	5036.51 cm^{-1}	248010.23	– 253046.74	4–4	5.20–04	3.07–03	8.03–01	–1.911	D	*LS*
			20570.1	4860.10 cm^{-1}	248186.64	– 253046.74	6–4	5.19–05	2.19–04	8.92–02	–2.881	D–	*LS*
			19841.4	5038.59 cm^{-1}	248010.23	– 253048.82	4–6	3.85–05	3.41–04	8.92–02	–2.865	D–	*LS*
218.		^{2}D°–^{2}P	*17615*	*5675.5* cm^{-1}	*248116.1*	– *253791.6*	10–6	3.85–04	1.08–03	6.23–01	–1.969	C	1
			17833.9	5605.76 cm^{-1}	248186.64	– 253792.40	6–4	3.34–04	1.06–03	3.74–01	–2.196	D	*LS*
			17297.0	5779.76 cm^{-1}	248010.23	– 253789.99	4–2	4.07–04	9.12–04	2.08–01	–2.438	D	*LS*
			17289.8	5782.17 cm^{-1}	248010.23	– 253792.40	4–4	4.07–05	1.83–04	4.16–02	–3.137	E	*LS*

O II: Allowed Transitions — Continued

No.	Transition Array	Mult.	λ_{air} (Å)	λ_{vac} (Å) or σ (cm^{-1})*	E_i (cm^{-1})	E_k (cm^{-1})	g_i-g_k	A_{ki} (10^8 s^{-1})	f_{ik}	S (at. u.)	log gf	Acc.	Source
219.	$2s^22p^2(^3P)4p-$ $2s^22p^2(^3P)4d$	$^2D^o-^2F$	*13733*	*7279.7 cm⁻¹*	*248116.1*	*– 255395.8*	10–14	2.43–01	9.64–01	4.36+02	0.984	C	1
			13733.5	7279.46 cm⁻¹	248186.64	– 255466.10	6–8	2.43–01	9.18–01	2.49+02	0.741	D	*LS*
			13710.1	7291.88 cm⁻¹	248010.23	– 255302.11	4–6	2.28–01	9.65–01	1.74+02	0.587	D	*LS*
			14050.0	7115.47 cm⁻¹	248186.64	– 255302.11	6–6	1.52–02	4.49–02	1.24+01	–0.570	D	*LS*
220.		$^2D^o-^2D$	*12884*	*7759.6 cm⁻¹*	*248116.1*	*– 255875.7*	10–10	7.17–02	1.78–01	7.56+01	0.251	C	1
			12965.0	7710.95 cm⁻¹	248186.64	– 255897.59	6–6	6.56–02	1.65–01	4.24+01	–0.003	D	*LS*
			12763.5	7832.68 cm⁻¹	248010.23	– 255842.91	4–4	6.63–02	1.62–01	2.72+01	–0.188	D	*LS*
			13057.6	7656.27 cm⁻¹	248186.64	– 255842.91	6–4	6.88–03	1.17–02	3.03+00	–1.153	D	*LS*
			12675.0	7887.36 cm⁻¹	248010.23	– 255897.59	4–6	5.02–03	1.81–02	3.03+00	–1.140	D	*LS*
221.	$2s^22p^2(^3P)4p-$ $2s^22p^2(^3P)5s$	$^2D^o-^2P$	*9589.5*	*9592.1*	*248116.1*	*– 258541.3*	10–6	3.35–01	2.77–01	8.76+01	0.443	C+	1
			9594.56	9597.20	248186.64	– 258606.35	6–4	3.01–01	2.77–01	5.25+01	0.221	C–	*LS*
			9611.80	9614.43	248010.23	– 258411.26	4–2	3.33–01	2.31–01	2.92+01	–0.035	C–	*LS*
			9434.83	9437.42	248010.23	– 258606.35	4–4	3.52–02	4.70–02	5.84+00	–0.726	C–	*LS*
222.	$2s^22p^2(^3P)4p-$ $2s^22p^2(^1D)4s$	$^2D^o-^2D$	*8948.7*	*8951.2*	*248116.1*	*– 259287.8*	10–10	8.19–03	9.83–03	2.90+00	–1.007	C+	1
			9005.75	9008.22	248186.64	– 259287.61	6–6	7.50–03	9.12–03	1.62+00	–1.262	C–	*LS*
			8864.51	8866.95	248010.23	– 259288.07	4–4	7.58–03	8.93–03	1.04+00	–1.447	C–	*LS*
			9005.38	9007.85	248186.64	– 259288.07	6–4	8.03–04	6.51–04	1.16–01	–2.408	C–	*LS*
			8864.87	8867.31	248010.23	– 259287.61	4–6	5.62–04	9.92–04	1.16–01	–2.401	C–	*LS*
223.	$2s^22p^2(^3P)4p-$ $2s^22p^2(^1D)4d$	$^2D^o-^2F$	*3751.6*	*3752.6*	*248116.1*	*– 274764*	10–14	8.56–03	2.53–03	3.12–01	–1.597	D	1
			3759.06	3760.13	248186.64	– 274781.5	6–8	8.51–03	2.40–03	1.78–01	–1.841	D–	*LS*
			3739.99	3741.05	248010.23	– 274740.7	4–6	8.06–03	2.54–03	1.25–01	–1.994	D–	*LS*
			3764.83	3765.90	248186.64	– 274740.7	6–6	5.64–04	1.20–04	8.92–03	–3.143	E	*LS*
224.		$^2D^o-^2D$	*3729.3*	*3730.4*	*248116.1*	*– 274923*	10–10	2.32–02	4.83–03	5.93–01	–1.316	D	1
			3738.46	3739.53	248186.64	– 274928.0	6–6	2.15–02	4.50–03	3.32–01	–1.569	D–	*LS*
			3715.62	3716.67	248010.23	– 274916.0	4–4	2.11–02	4.36–03	2.13–01	–1.758	D–	*LS*
			3740.14	3741.20	248186.64	– 274916.0	6–4	2.30–03	3.21–04	2.37–02	–2.715	E	*LS*
			3713.96	3715.02	248010.23	– 274928.0	4–6	1.56–03	4.85–04	2.37–02	–2.712	E	*LS*
225.	$2s^22p^2(^3P)4p-$ $2s^22p^2(^1S)3d$	$^2D^o-^2D$	*3580.3*	*3581.3*	*248116.1*	*– 276038.8*	10–10	5.56–04	1.07–04	1.26–02	–2.971	D	1
			3584.66	3585.68	248186.64	– 276075.32	6–6	5.17–04	9.97–05	7.06–03	–3.223	D–	*LS*
			3573.76	3574.78	248010.23	– 275983.95	4–4	5.03–04	9.64–05	4.54–03	–3.414	D–	*LS*
			3596.44	3597.47	248186.64	– 275983.95	6–4	5.49–05	7.09–06	5.04–04	–4.371	E	*LS*
			3562.13	3563.15	248010.23	– 276075.32	4–6	3.77–05	1.07–05	5.04–04	–4.367	E	*LS*
226.	$2s^22p^2(^3P)4p-$ $2s^22p^2(^1D)3d$	$^2P^o-^2D$	*21913*	*4562.3 cm⁻¹*	*248485.7*	*– 253048.0*	6–10	1.36–04	1.64–03	7.08–01	–2.008	D	1
			22051.9	4533.52 cm⁻¹	248515.30	– 253048.82	4–6	1.34–04	1.46–03	4.25–01	–2.232	D–	*LS*
			21637.6	4620.33 cm⁻¹	248426.41	– 253046.74	2–4	1.18–04	1.66–03	2.36–01	–2.480	D–	*LS*
			22062.0	4531.44 cm⁻¹	248515.30	– 253046.74	4–4	2.23–05	1.63–04	4.72–02	–3.187	E	*LS*
227.		$^2P^o-^2P$	*18842*	*5305.9 cm⁻¹*	*248485.7*	*– 253791.6*	6–6	2.80–03	1.49–02	5.55+00	–1.048	C	1
			18944.6	5277.10 cm⁻¹	248515.30	– 253792.40	4–4	2.30–03	1.24–02	3.08+00	–1.306	D	*LS*
			18639.2	5363.58 cm⁻¹	248426.41	– 253789.99	2–2	1.93–03	1.01–02	1.23+00	–1.697	D	*LS*
			18953.3	5274.69 cm⁻¹	248515.30	– 253789.99	4–2	9.18–04	2.47–03	6.17–01	–2.005	D	*LS*
			18630.8	5365.99 cm⁻¹	248426.41	– 253792.40	2–4	4.83–04	5.03–03	6.17–01	–1.998	D	*LS*

O II: Allowed Transitions — Continued

No.	Transition Array	Mult.	λ_{air} (Å)	λ_{vac} (Å) or σ (cm^{-1})*	E_i (cm^{-1})	E_k (cm^{-1})	g_i-g_k	A_{ki} ($10^8 s^{-1}$)	f_{ik}	S (at. u.)	log gf	Acc.	Source
228.	$2s^22p^2(^3P)4p-$ $2s^22p^2(^3P)4d$	$^2P°-^2P$	14868	6724.0 cm^{-1}	248485.7	– 255209.7	6–6	1.07–01	3.56–01	1.04+02	0.329	C	1
			15014.8	6658.28 cm^{-1}	248515.30	– 255173.58	4–4	8.68–02	2.94–01	5.80+01	0.070	D	LS
			14582.8	6855.52 cm^{-1}	248426.41	– 255281.93	2–2	7.58–02	2.42–01	2.32+01	–0.316	D	LS
			14774.4	6766.63 cm^{-1}	248515.30	– 255281.93	4–2	3.65–02	5.97–02	1.16+01	–0.622	D	LS
			14817.0	6747.17 cm^{-1}	248426.41	– 255173.58	2–4	1.81–02	1.19–01	1.16+01	–0.624	D	LS
229.		$^2P°-^2D$	13528	7390.0 cm^{-1}	248485.7	– 255875.7	6–10	1.89–01	8.62–01	2.30+02	0.714	C	1
			13542.2	7382.29 cm^{-1}	248515.30	– 255897.59	4–6	1.88–01	7.75–01	1.38+02	0.491	D	LS
			13479.8	7416.50 cm^{-1}	248426.41	– 255842.91	2–4	1.59–01	8.65–01	7.68+01	0.238	D	LS
			13643.3	7327.61 cm^{-1}	248515.30	– 255842.91	4–4	3.06–02	8.55–02	1.54+01	–0.466	D	LS
230.	$2s^22p^2(^3P)4p-$ $2s^22p^2(^1D)3d$	$^2P°-^2S$	14007	7137.1 cm^{-1}	248485.7	– 255622.80	6–2	5.27–03	5.16–03	1.43+00	–1.509	C	1
			14065.8	7107.50 cm^{-1}	248515.30	– 255622.80	4–2	3.47–03	5.14–03	9.52–01	–1.687	D	LS
			13892.1	7196.39 cm^{-1}	248426.41	– 255622.80	2–2	1.80–03	5.21–03	4.76–01	–1.982	D	LS
231.	$2s^22p^2(^3P)4p-$ $2s^22p^2(^3P)5s$	$^2P°-^2P$	9941.9	9944.7	248485.7	– 258541.3	6–6	1.25–01	1.86–01	3.65+01	0.048	C+	1
			9907.06	9909.77	248515.30	– 258606.35	4–4	1.06–01	1.55–01	2.03+01	–0.206	C–	LS
			10012.4	9984.85 cm^{-1}	248426.41	– 258411.26	2–2	8.19–02	1.23–01	8.11+00	–0.609	C–	LS
			10102.4	9895.96 cm^{-1}	248515.30	– 258411.26	4–2	3.99–02	3.05–02	4.06+00	–0.914	C–	LS
			9820.55	9823.24	248426.41	– 258606.35	2–4	2.17–02	6.27–02	4.06+00	–0.901	C–	LS
232.	$2s^22p^2(^3P)4p-$ $2s^22p^2(^1D)4s$	$^2P°-^2D$	9254.9	9257.4	248485.7	– 259287.8	6–10	2.09–03	4.48–03	8.18–01	–1.571	C	1
			9280.51	9283.06	248515.30	– 259287.61	4–6	2.07–03	4.02–03	4.91–01	–1.794	D	LS
			9204.17	9206.70	248426.41	– 259288.07	2–4	1.77–03	4.50–03	2.73–01	–2.046	D	LS
			9280.12	9282.66	248515.30	– 259288.07	4–4	3.46–04	4.46–04	5.45–02	–2.748	D–	LS
233.	$2s^22p^2(^3P)4p-$ $2s^22p^2(^1D)4d$	$^2P°-^2D$	3781.5	3782.5	248485.7	– 274923	6–10	5.85–02	2.09–02	1.56+00	–0.902	C	1
			3784.98	3786.06	248515.30	– 274928.0	4–6	5.83–02	1.88–02	9.36–01	–1.124	D	LS
			3774.00	3775.07	248426.41	– 274916.0	2–4	4.90–02	2.09–02	5.20–01	–1.378	D	LS
			3786.70	3787.78	248515.30	– 274916.0	4–4	9.70–03	2.09–03	1.04–01	–2.079	D	LS
234.	$2s^22p^2(^3P)4p-$ $2s^22p^2(^1S)3d$	$^2P°-^2D$	3628.3	3629.4	248485.7	– 276038.8	6–10	6.30–02	2.07–02	1.49+00	–0.905	C	1
			3627.41	3628.44	248515.30	– 276075.32	4–6	6.31–02	1.87–02	8.91–01	–1.127	D	LS
			3627.74	3628.77	248426.41	– 275983.95	2–4	5.25–02	2.07–02	4.95–01	–1.382	D	LS
			3639.48	3640.51	248515.30	– 275983.95	4–4	1.04–02	2.07–03	9.91–02	–2.083	D–	LS
235.	$2s^22p^2(^1D)3d-$ $2s^22p^2(^3P)5p$	$^2F-^2D°$	9450.8	9453.4	251223.3	– 261801.5	14–10	2.52–02	2.42–02	1.05+01	–0.471	C	1
			9389.08	9391.66	251222.19	– 261869.94	8–6	2.58–02	2.56–02	6.32+00	–0.689	D	LS
			9544.87	9547.49	251224.79	– 261698.75	6–4	2.05–02	1.86–02	3.51+00	–0.952	D	LS
			9391.37	9393.95	251224.79	– 261869.94	6–6	2.86–03	3.79–03	7.02–01	–1.644	D–	LS
236.		$^2D-^2D°$	11421	8753.5 cm^{-1}	253048.0	– 261801.5	10–10	1.16–03	2.28–03	8.56–01	–1.643	C–	1
			11333.3	8821.12 cm^{-1}	253048.82	– 261869.94	6–6	1.11–03	2.14–03	4.79–01	–1.891	D–	LS
			11554.8	8652.01 cm^{-1}	253046.74	– 261698.75	4–4	1.01–03	2.03–03	3.08–01	–2.091	D–	LS
			11557.6	8649.93 cm^{-1}	253048.82	– 261698.75	6–4	1.12–04	1.50–04	3.42–02	–3.046	E	LS
			11330.7	8823.20 cm^{-1}	253046.74	– 261869.94	4–6	7.95–05	2.29–04	3.42–02	–3.037	E	LS
237.		$^2P-^2D°$	12481	8009.9 cm^{-1}	253791.6	– 261801.5	6–10	1.15–03	4.48–03	1.11+00	–1.570	C	1
			12376.6	8077.54 cm^{-1}	253792.40	– 261869.94	4–6	1.18–03	4.07–03	6.63–01	–1.788	D	LS
			12640.7	7908.76 cm^{-1}	253789.99	– 261698.75	2–4	9.24–04	4.43–03	3.68–01	–2.053	D	LS
			12644.6	7906.35 cm^{-1}	253792.40	– 261698.75	4–4	1.85–04	4.43–04	7.37–02	–2.752	D–	LS

O II: Allowed Transitions — Continued

No.	Transition Array	Mult.	λ_{air} (Å)	λ_{vac} (Å) or σ (cm^{-1})*	E_i (cm^{-1})	E_k (cm^{-1})	g_i-g_k	A_{ki} ($10^8 s^{-1}$)	f_{ik}	S (at. u.)	log gf	Acc.	Source
238.	$2s^22p^2(^1S)3p-$ $2s^22p^2(^3P)4d$	$^2P^o-^2P$	*45066*	*2218.4* cm^{-1}	*252991.3*	*– 255209.7*	6–6	1.05–04	7.83–03	1.09+01	–1.328	C	1
			45855.1	2180.19 cm^{-1}	252993.39	– 255173.58	4–4	3.18–04	1.00–02	6.06+00	–1.397	D	*LS*
			43566.8	2294.70 cm^{-1}	252987.23	– 255281.93	2–2	2.97–04	8.44–03	2.42+00	–1.772	D	*LS*
			43684.1	2288.54 cm^{-1}	252993.39	– 255281.93	4–2	1.47–04	2.11–03	1.21+00	–2.075	D	*LS*
			45725.9	2186.35 cm^{-1}	252987.23	– 255173.58	2–4	6.42–05	4.02–03	1.21+00	–2.094	D	*LS*
239.		$^2P^o-^2D$	*34660*	*2884.4* cm^{-1}	*252991.3*	*– 255875.7*	6–10	5.18–04	2.98–02	2.82+01	–0.748	C	1
			34423.5	2904.20 cm^{-1}	252993.39	– 255897.59	4–6	1.40–03	3.74–02	1.69+01	–0.825	D	*LS*
			35008.4	2855.68 cm^{-1}	252987.23	– 255842.91	2–4	1.11–03	4.08–02	9.41+00	–1.088	D	*LS*
			35084.1	2849.52 cm^{-1}	252993.39	– 255842.91	4–4	2.21–04	4.07–03	1.88+00	–1.788	D	*LS*
240.	$2s^22p^2(^1S)3p-$ $2s^22p^2(^1D)3d$	$^2P^o-^2S$	*37991*	*2631.5* cm^{-1}	*252991.3*	– 255622.80	6–2	7.68–05	1.14–03	1.23+00	–2.164	C	1
			38021.0	2629.41 cm^{-1}	252993.39	– 255622.80	4–2	1.51–04	1.64–03	8.22–01	–2.183	D	*LS*
			37932.1	2635.57 cm^{-1}	252987.23	– 255622.80	2–2	7.63–05	1.65–03	4.11–01	–2.483	D	*LS*
241.	$2s^22p^2(^1S)3p-$ $2s^22p^2(^3P)5s$	$^2P^o-^2P$	*18013*	*5550.0* cm^{-1}	*252991.3*	*– 258541.3*	6–6	5.06–03	3.36–02	1.40+01	–0.695	C+	1
			17811.0	5612.96 cm^{-1}	252993.39	– 258606.35	4–4	6.96–03	3.31–02	7.76+00	–0.878	C–	*LS*
			18431.4	5424.03 cm^{-1}	252987.23	– 258411.26	2–2	5.02–03	2.56–02	3.11+00	–1.291	C–	*LS*
			18452.4	5417.87 cm^{-1}	252993.39	– 258411.26	4–2	2.50–03	6.39–03	1.55+00	–1.592	C–	*LS*
			17791.5	5619.12 cm^{-1}	252987.23	– 258606.35	2–4	1.40–03	1.33–02	1.55+00	–1.577	C–	*LS*
242.	$2s^22p^2(^1S)3p-$ $2s^22p^2(^1D)4s$	$^2P^o-^2D$	*15878*	*6296.5* cm^{-1}	*252991.3*	*– 259287.8*	6–10	1.34–04	1.11–03	3.99–01	–2.176	C	1
			15883.3	6294.22 cm^{-1}	252993.39	– 259287.61	4–6	2.02–04	1.15–03	2.40–01	–2.339	D	*LS*
			15866.6	6300.84 cm^{-1}	252987.23	– 259288.07	2–4	1.69–04	1.27–03	1.33–01	–2.594	D	*LS*
			15882.1	6294.68 cm^{-1}	252993.39	– 259288.07	4–4	3.37–05	1.27–04	2.66–02	–3.293	D–	*LS*
243.	$2s^22p^2(^1S)3p-$ $2s^22p^2(^1D)4d$	$^2P^o-^2D$	*4558.3*	*4559.6*	*252991.3*	*– 274923*	6–10	2.27–01	1.27–01	1.19+01	–0.118	C	1
			4557.73	4559.01	252993.39	– 274928.0	4–6	2.54–01	1.19–01	7.13+00	–0.323	D	*LS*
			4558.94	4560.22	252987.23	– 274916.0	2–4	2.12–01	1.32–01	3.96+00	–0.579	D	*LS*
			4560.22	4561.50	252993.39	– 274916.0	4–4	4.23–02	1.32–02	7.92–01	–1.278	D–	*LS*
244.	$2s2p^3(^5S^o)3s-$ $2s^22p^2(^3P)5s$	$^4S^o-^4P$	*34669*	*2883.6* cm^{-1}	254981.55	*– 257865.2*	4–12	2.87–03	1.55–01	7.09+01	–0.207	C+	1
			33507.9	2983.56 cm^{-1}	254981.55	– 257965.11	4–6	3.18–03	8.03–02	3.54+01	–0.493	C–	*LS*
			35471.7	2818.38 cm^{-1}	254981.55	– 257799.93	4–4	2.68–03	5.06–02	2.36+01	–0.694	C–	*LS*
			36833.4	2714.19 cm^{-1}	254981.55	– 257695.74	4–2	2.39–03	2.44–02	1.18+01	–1.011	C–	*LS*
245.	$2s^22p^2(^3P)4d-$ $2s^22p^2(^3P)5p$	$^2P-^2D^o$	*15166*	*6591.8* cm^{-1}	*255209.7*	*– 261801.5*	6–10	8.16–04	4.69–03	1.40+00	–1.551	C	1
			14929.4	6696.36 cm^{-1}	255173.58	– 261869.94	4–6	8.55–04	4.29–03	8.43–01	–1.766	D	*LS*
			15579.8	6416.82 cm^{-1}	255281.93	– 261698.75	2–4	6.27–04	4.56–03	4.68–01	–2.040	D	*LS*
			15321.1	6525.17 cm^{-1}	255173.58	– 261698.75	4–4	1.32–04	4.64–04	9.36–02	–2.731	D–	*LS*
246.		$^2F-^2D^o$	*15607*	*6405.6* cm^{-1}	*255395.8*	*– 261801.5*	14–10	1.03–01	2.70–01	1.94+02	0.577	C	1
			15611.4	6403.84 cm^{-1}	255466.10	– 261869.94	8–6	1.03–01	2.83–01	1.16+02	0.355	D	*LS*
			15628.9	6396.64 cm^{-1}	255302.11	– 261698.75	6–4	8.58–02	2.10–01	6.47+01	0.099	D	*LS*
			15221.6	6567.83 cm^{-1}	255302.11	– 261869.94	6–6	1.24–02	4.30–02	1.29+01	–0.588	D	*LS*
247.		$^2D-^2D^o$	*16871*	*5925.7* cm^{-1}	*255875.7*	*– 261801.5*	10–10	1.95–02	8.34–02	4.63+01	–0.079	C	1
			16739.3	5972.35 cm^{-1}	255897.59	– 261869.94	6–6	1.87–02	7.84–02	2.59+01	–0.327	D	*LS*
			17072.3	5855.84 cm^{-1}	255842.91	– 261698.75	4–4	1.70–02	7.41–02	1.67+01	–0.528	D	*LS*
			17233.2	5801.16 cm^{-1}	255897.59	– 261698.75	6–4	1.83–03	5.44–03	1.85+00	–1.486	D	*LS*
			16587.4	6027.03 cm^{-1}	255842.91	– 261869.94	4–6	1.37–03	8.48–03	1.85+00	–1.470	D	*LS*

O II: Allowed Transitions — Continued

No.	Transition Array	Mult.	λ_{air} (Å)	λ_{vac} (Å) or σ (cm^{-1})*	E_i (cm^{-1})	E_k (cm^{-1})	g_i-g_k	A_{ki} (10^8 s^{-1})	f_{ik}	S (at. u.)	log gf	Acc.	Source
248.	$2s^22p^2(^1D)4s-$ $2s^22p^2(^3P)5p$	$^2D-^2D^o$	*39772*	*2513.7* cm^{-1}	*259287.8*	*– 261801.5*	10–10	4.22–04	1.00–02	1.31+01	–1.000	C+	1
			38714.2	2582.33 cm^{-1}	259287.61	– 261869.94	6–6	4.27–04	9.60–03	7.34+00	–1.240	C–	*LS*
			41470.8	2410.68 cm^{-1}	259288.07	– 261698.75	4–4	3.35–04	8.64–03	4.72+00	–1.461	C–	*LS*
			41462.9	2411.14 cm^{-1}	259287.61	– 261698.75	6–4	3.73–05	6.40–04	5.24–01	–2.416	D	*LS*
			38721.1	2581.87 cm^{-1}	259288.07	– 261869.94	4–6	3.05–05	1.03–03	5.24–01	–2.386	D	*LS*

*Wavelengths (Å) are always given unless cm^{-1} is indicated.

O II

Forbidden Transitions

We have selected the data of two recent calculations and present the arithmetic averages. Zeippen[1] applied the well-known SUPERSTRUCTURE code, and Godefroid and Froese-Fischer[2] used a multi-configuration Hartree-Fock program to obtain their atomic wavefunctions. Both calculations contain extensive treatments of configuration interaction, but Ref. 2 does not include relativistic terms in the M1 transition operator which affect especially the $^4S-^2D$ transitions in the p^3 configuration. We have replaced the wavelength data used in Ref. 2 with slightly more accurate values based on the latest experimental term energies. The two calculations agree typically within 10–20%, except for the very weak E2 lines at 1.99 cm^{-1} and 20.02 cm^{-1}, where they disagree by factors of 160 and 2.3, respectively.

For the M1 $^2D_{5/2}-^2D_{3/2}$ and $^2P_{3/2}-^2P_{1/2}$ transitions, we have applied the line strengths obtained from general nonrelativistic expansion formulas first developed by Shortley *et al.*[3] For near *LS*-coupling conditions, one obtains S=2.40 and S=1.33, respectively. We used these values rather than re-converting the published *A*-values of Ref. 1 and 2 to *S* with inaccurate wavenumber data. For some neighboring lower ions of the nitrogen-sequence for which—in contrast to O II—more accurate wavelength data are available, such line strengths are indeed also derived from the published A_{ki} values.

References

[1]C. J. Zeippen, Astron. Astrophys. **173**, 410 (1987).
[2]M. Godefroid and C. Froese Fischer, J. Phys. B **17**, 681 (1984).
[3]G. H. Shortley, L. H. Aller, J. G. Baker, and D. H. Menzel, Astrophys. J. **93**, 178 (1941).

O II: Forbidden Transitions

No.	Transition Array	Mult.	λ_{air} (Å)	λ_{vac} (Å) or σ (cm^{-1})*	E_i (cm^{-1})	E_k (cm^{-1})	g_i-g_k	Type	A_{ki} (s^{-1})	S (at. u.)	Acc.	Source
1.	$2p^3-$ $2p^3$	$^4S^o-^2D^o$										
			3728.81	3729.87	0.00	– 26810.55	4–6	M1	1.98–06	2.29–08	C	1
			3728.81	3729.87	0.00	– 26810.55	4–6	E2	2.86–05	1.11–04	C	1,2
			3726.03	3727.09	0.00	– 26830.57	4–4	M1	1.59–04	1.22–06	C	1
			3726.03	3727.09	0.00	– 26830.57	4–4	E2	1.86–05	4.77–05	C	1,2
2.		$^4S^o-^2P^o$										
			2470.34	2471.09	0.00	– 40468.01	4–4	M1	5.22–02	1.17–04	C+	1,2
			2470.34	2471.09	0.00	– 40468.01	4–4	E2	2.43–07	8.01–08	C+	1,2
			2470.34	2471.09	0.00	– 40468.01	4–2	M1	2.12–02	2.37–05	C+	1,2
			2470.34	2471.09	0.00	– 40468.01	4–2	E2	3.72–07	6.13–08	D	1,2
3.		$^2D^o-^2D^o$										
				20.02 cm^{-1}	26810.55	– 26830.57	6–4	M1	1.30–07	2.40+00	B+	3
				20.02 cm^{-1}	26810.55	– 26830.57	6–4	E2	1.49–20	1.66–04	E	1,2

O II: Forbidden Transitions — Continued

No.	Transition Array	Mult.	λ_{air} (Å)	λ_{vac} (Å) or σ (cm^{-1})*	E_i (cm^{-1})	E_k (cm^{-1})	g_i-g_k	Type	A_{ki} (s^{-1})	S (at. u.)	Acc.	Source
4.		$^2D^o-^2P^o$										
			7319.99	7322.01	26810.55	– 40468.01	6–4	M1	8.37–03	4.87–04	B	1,2
			7319.99	7322.01	26810.55	– 40468.01	6–4	E2	9.07–02	6.82+00	B	1,2
			7318.92	7320.94	26810.55	– 40470.00	6–2	E2	5.19–02	1.95+00	B	1,2
			7330.73	7332.75	26830.57	– 40468.01	4–4	M1	1.49–02	8.72–04	B	1,2
			7330.73	7332.75	26830.57	– 40468.01	4–4	E2	3.85–02	2.92+00	B	1,2
			7329.67	7331.68	26830.57	– 40470.00	4–2	M1	9.32–03	2.72–04	B	1,2
			7329.67	7331.68	26830.57	– 40470.00	4–2	E2	7.74–02	2.93+00	B	1,2
5.		$^2P^o-^2P^o$										
				1.99 cm^{-1}	40468.01	– 40470.00	4–2	M1	1.41–10	1.33+00	B	3
				1.99 cm^{-1}	40468.01	– 40470.00	4–2	E2	4.24–24	2.42–03	E	1,2

*Wavelengths (Å) are always given unless cm^{-1} is indicated.

O III

Carbon Isoelectronic Sequence

Ground State: $1s^2 2s^2 2p^2\ ^3P_0$

Ionization Energy: 54.934 eV = 443084.7 cm^{-1}

Allowed Transitions

List of tabulated lines

Wavelength (Å)	No.	Wavelength (Å)	No.	Wavelength (Å)	No.	Wavelength (Å)	No.
in vacuum		264.472	13	303.517	10	355.337	50
		264.481	13	303.622	10	355.429	50
240.859	15	266.972	12	303.695	10	355.465	50
240.925	15	266.987	12	303.800	10	355.471	50
241.037	15	267.031	12	305.596	9	356.561	48
246.265	31	267.052	12	305.656	9	356.624	48
248.320	30	267.124	12	305.702	9	356.630	48
256.423	43	267.190	12	305.767	9	356.727	48
256.459	43	271.395	42	305.836	9	356.763	48
256.507	43	271.516	42	305.882	9	356.769	48
261.024	29	271.609	42	320.978	25	359.017	40
262.111	28	275.282	11	328.448	24	359.225	40
263.694	14	275.368	11	345.312	38	359.386	40
263.727	14	275.515	11	349.828	51	364.734	49
263.773	14	277.386	27	349.863	51	364.771	49
263.817	14	279.788	26	349.889	51	364.828	49
263.861	14	295.621	41	349.923	51	364.865	49
263.907	14	295.655	41	349.929	51	364.871	49
264.258	13	295.716	41	349.959	51	364.929	49
264.325	13	295.942	39	349.964	51	364.935	49
264.337	13	303.413	10	355.138	50	366.191	8
264.346	13	303.461	10	355.302	50	366.343	8

List of tabulated lines — Continued

Wavelength (Å)	No.	Wavelength (Å)	No.	Wavelength (Å)	No.	Wavelength (Å)	No.
366.602	8	574.880	45	859.159	139	1033.12	137
373.803	7	574.895	45	865.294	110	1033.49	92
374.004	7	575.331	45	866.317	110	1033.69	92
374.073	7	575.347	45	867.610	110	1033.81	92
374.162	7	597.814	35	867.972	110	1034.70	92
374.328	7	599.590	19	868.160	110	1040.32	66
374.432	7	609.705	44	868.637	110	1050.36	136
379.508	59	610.039	44	870.019	128	1056.43	91
379.580	59	610.057	44	870.564	128	1056.51	116
379.581	59	610.745	44	871.555	128	1056.57	91
379.621	59	610.850	44	898.957	60	1058.67	91
379.622	59	610.868	44	902.255	98	1058.87	91
379.639	59	632.665	55	916.404	127	1059.00	91
387.397	57	632.668	55	917.502	127	1059.78	116
387.477	57	633.188	55	918.394	127	1059.93	91
387.478	57	633.191	55	919.002	127	1060.76	116
387.496	57	633.238	55	919.619	127	1079.38	87
387.641	57	633.480	55	920.106	127	1088.17	126
387.642	57	644.423	54	928.309	109	1089.72	126
395.557	23	644.426	54	928.763	109	1090.78	126
397.114	58	644.475	54	928.908	109	1091.63	126
397.225	58	649.764	34	930.086	109	1092.67	126
397.227	58	658.579	53	930.211	109	1093.19	126
397.301	58	659.536	53	931.995	109	1098.47	90
397.302	58	659.539	53	967.540	62	1100.06	90
397.321	58	660.129	53	967.670	67	1101.18	90
404.268	22	660.133	53	968.765	62	1102.70	90
404.688	22	660.184	53	969.176	138	1103.82	90
434.267	47	702.337	3	969.372	79	1104.98	90
434.320	47	702.838	3	969.441	62	1114.44	144
434.329	47	702.896	3	969.468	79	1138.53	65
434.648	47	702.900	3	970.485	79	1149.58	171
434.657	47	703.850	3	971.326	79	1149.63	61
434.846	47	703.854	3	971.888	79	1150.88	61
434.980	37	705.766	52	972.911	79	1150.91	171
446.047	36	706.233	52	982.094	97	1153.02	96
475.146	6	706.237	52	992.408	162	1153.21	171
475.401	6	706.295	52	992.666	162	1153.78	61
475.838	6	707.320	52	992.917	162	1160.15	135
480.960	56	707.324	52	993.096	78	1170.60	107
480.962	56	719.069	101	993.638	162	1171.50	107
481.363	56	738.652	100	994.148	162	1171.59	107
481.365	56	752.760	68	994.255	162	1172.47	107
481.392	56	759.111	118	994.265	78	1173.37	107
481.597	56	788.515	80	995.230	162	1174.62	107
491.576	63	788.953	80	996.811	78	1175.50	107
491.707	63	789.229	117	1002.16	17	1179.55	185
491.971	63	789.252	80	1002.20	17	1180.04	185
507.388	5	790.115	80	1002.49	17	1180.79	185
507.680	5	790.556	80	1007.76	77	1180.99	185
508.178	5	790.855	80	1007.88	77	1181.75	185
525.794	21	808.291	99	1008.10	33	1182.77	185
534.929	4	818.944	18	1008.38	77	1183.15	185
535.481	4	818.949	18	1008.97	77	1184.17	185
554.270	46	824.310	111	1010.49	77	1185.96	185
554.357	46	825.238	111	1011.59	77	1196.75	106
554.372	46	826.741	111	1016.85	108	1197.24	106
554.759	46	832.929	2	1018.26	108	1197.33	106
554.773	46	833.715	2	1019.87	108	1199.29	106
554.995	46	833.749	2	1020.55	108	1199.91	106
565.565	20	835.059	2	1020.78	108	1202.24	115
574.060	45	835.092	2	1021.29	108	1202.46	106
574.153	45	835.289	2	1031.37	92	1202.51	115
574.786	45	844.495	140	1031.49	92	1202.68	115

List of tabulated lines — Continued

Wavelength (Å)	No.	Wavelength (Å)	No.	Wavelength (Å)	No.	Wavelength (Å)	No.
1205.96	76	1588.53	122	2292.03	175	2534.06	150
1206.77	76	1588.84	184	2293.27	88	2539.47	150
1208.50	76	1588.97	184	2297.16	88	2542.64	150
1209.71	76	1589.91	184	2297.80	175	2546.41	150
1209.96	76	1589.98	105	2299.75	88	2547.42	150
1212.26	76	1590.57	184	2306.42	175	2549.59	150
1223.00	205	1590.59	105	2308.71	88	2549.67	74
1223.81	205	1591.29	105	2308.91	151	2558.04	156
1223.86	205	1591.83	122	2311.33	88	2562.30	215
1224.67	205	1649.80	192	2311.52	151	2566.49	74
1225.76	205	1660.81	1	2312.25	175	2569.42	215
1226.18	205	1661.24	204	2314.69	151	2572.29	224
1227.28	205	1664.03	204	2315.48	151	2576.78	215
1236.45	114	1664.21	204	2316.39	88	2582.71	215
1236.72	114	1665.80	204	2317.32	151	2587.60	215
1236.90	114	1666.15	1	2319.48	151	2589.95	215
1242.29	64	1667.72	204	2327.81	200	2597.70	154
1266.03	125	1668.61	204	2333.64	200	2605.42	154
1266.13	125	1669.31	204	2344.65	200	2609.59	154
1267.18	125	1679.03	113	2344.97	200	2622.29	158
1267.20	125	1686.73	113	2348.68	200	2665.68	161
1268.22	125	1689.83	113	2350.57	200	2674.58	161
1269.28	125	1760.41	121	2361.24	145	2677.81	153
1279.13	124	1764.46	121	2372.22	145	2683.66	180
1280.03	124	1766.63	121	2372.83	145	2686.15	161
1280.31	124	1768.88	121	2375.41	169	2687.55	180
1280.51	124	1772.28	121	2378.89	145	2692.73	153
1282.46	124	1772.97	121	2382.29	145	2695.48	180
1282.66	124	1803.82	227	2383.92	145	2700.54	153
1300.86	32	1848.50	228	2385.27	169	2701.03	153
1344.94	123	1897.10	234	2386.29	169	2708.89	153
1344.96	123	1903.05	234	2390.43	94	2713.39	153
1345.32	123	1907.63	234	2392.00	169	2726.89	248
1346.63	123	1920.02	132	2396.99	169	2735.06	248
1347.25	235	1920.24	226	2401.99	169	2739.09	248
1347.33	123			2410.06	209	2771.96	189
1347.78	134	in air		2420.52	244	2794.14	104
1348.24	235			2421.14	244	2798.93	104
1349.00	123	2037.45	176	2425.90	244	2807.90	104
1349.57	235	2040.77	176	2426.31	244	2809.66	104
1406.45	133	2041.97	176	2426.57	244	2818.70	104
1434.04	146	2042.28	191	2426.93	244	2836.31	104
1435.49	146	2050.08	176	2428.11	209	2845.75	199
1436.48	146	2052.17	176	2429.36	244	2848.93	199
1438.08	146	2053.09	201	2429.61	244	2853.75	199
1439.54	146	2053.38	176	2429.69	244	2862.52	199
1441.78	146	2057.62	201	2432.43	209	2862.60	199
1476.87	95	2066.42	201	2434.92	155	2879.57	199
1486.18	143	2153.30	75	2438.81	147	2959.69	93
1499.63	217	2165.39	148	2441.04	245	2983.78	85
1501.63	217	2181.57	89	2441.35	155	2992.08	103
1504.08	217	2187.02	89	2441.70	155	2996.48	103
1534.79	170	2189.05	89	2445.23	245	2997.69	103
1537.15	170	2191.99	89	2446.93	245	3004.34	103
1541.27	170	2196.54	89	2448.17	155	3008.78	103
1581.01	105	2197.48	89	2451.13	245	3017.62	103
1583.24	184	2228.16	142	2451.85	155	3023.14	183
1584.43	105	2228.92	216	2453.45	245	3023.43	73
1585.66	184	2233.36	216	2453.53	155	3024.36	103
1585.76	184	2238.77	216	2454.24	245	3024.54	73
1586.71	122	2269.97	190	2454.97	86	3026.33	183
1587.38	184	2283.77	225	2456.56	245	3029.86	183
1587.84	105	2286.33	175	2496.03	16	3032.61	183
1587.96	184	2289.20	175	2529.33	150	3034.22	157

List of tabulated lines — Continued

Wavelength (Å)	No.	Wavelength (Å)	No.	Wavelength (Å)	No.	Wavelength (Å)	No.
3035.41	73	3377.26	202	3761.17	231	4747.34	239
3036.16	183	3382.61	202	3774.03	71	4748.66	251
3042.07	73	3383.31	202	3791.28	71	4761.75	239
3042.64	183	3383.81	202	3810.98	71	4764.63	251
3045.43	183	3384.90	202	3816.75	131	4778.43	82
3047.10	73	3394.22	202	3853.47	229	4799.75	152
3051.95	183	3395.43	202	3861.50	229	4809.78	82
3059.28	73	3406.88	120	3875.25	229	4826.19	152
3063.85	183	3408.13	120	3877.89	229	4840.17	152
3064.98	182	3415.26	120	3886.76	198	4867.06	152
3068.13	182	3423.36	167	3891.76	229	4881.61	152
3068.26	182	3427.36	167	3903.04	198	4890.86	152
3068.67	182	3428.63	120	3911.80	229	5049.87	242
3074.14	182	3430.57	120	3931.35	198	5060.73	242
3074.72	182	3435.12	167	3934.82	198	5063.66	242
3075.13	182	3440.36	167	3944.85	198	5068.56	242
3075.95	182	3444.05	120	3948.01	198	5088.92	242
3078.13	223	3446.68	181	3961.57	130	5091.88	242
3083.65	182	3447.15	181	4072.64	165	5112.18	230
3084.64	182	3447.92	167	4073.98	165	5117.75	230
3088.04	182	3447.97	181	4081.02	165	5124.74	230
3095.79	182	3450.91	181	4084.26	83	5136.32	230
3115.67	112	3451.30	181	4089.30	165	5148.99	230
3121.63	112	3454.84	181	4103.07	165	5152.47	230
3132.79	112	3454.99	181	4118.60	165	5171.29	230
3198.18	232	3455.78	167	4121.59	208	5268.30	141
3201.14	232	3459.48	181	4171.77	208	5508.24	129
3202.51	232	3459.94	181	4186.98	208	5592.25	81
3207.61	232	3466.13	181	4218.30	70	5946.37	236
3210.58	232	3466.85	181	4239.48	70	5971.45	236
3216.07	232	3475.24	181	4286.18	70	5979.67	69
3221.21	232	3520.94	166	4333.58	149	5988.23	69
3236.41	222	3531.22	166	4342.81	149	5988.58	236
3257.09	168	3533.38	166	4356.79	149	6009.54	236
3260.86	102	3534.90	166	4366.50	149	6022.32	69
3265.33	102	3555.24	166	4375.88	149	6035.16	236
3265.69	168	3556.78	166	4385.65	149	6077.59	236
3267.20	102	3653.77	84	4440.09	233	6102.35	69
3267.73	168	3671.29	84	4447.69	233	6116.98	69
3268.95	168	3681.03	84	4461.61	233	6199.57	69
3281.83	102	3695.16	188	4493.37	241	6330.48	238
3284.45	102	3695.38	159	4506.64	241	6358.91	238
3286.41	168	3698.72	159	4513.83	241	6363.83	238
3287.65	168	3703.36	159	4519.62	241	6378.34	238
3299.39	72	3704.12	119	4524.22	237	6383.30	238
3305.73	102	3704.75	159	4527.29	241	6390.36	238
3312.33	72	3707.27	119	4532.78	237	6507.55	187
3326.06	203	3709.54	159	4535.29	237	6510.86	250
3330.30	160	3712.49	159	4538.85	237	6531.24	250
3330.32	203	3714.03	119	4540.40	241	6561.50	250
3332.41	203	3715.09	119	4555.39	237	6644.32	250
3332.93	160	3720.89	159	4557.91	237	6665.54	250
3336.67	160	3721.95	159	4569.26	221	6676.76	250
3336.69	203	3725.31	119	4587.72	214	6707.36	240
3340.76	72	3728.51	231	4610.61	214	6709.49	240
3344.20	160	3728.84	231	4630.77	214	6731.13	240
3344.51	203	3729.80	231	4649.97	214	6736.16	240
3347.98	203	3732.13	119	4668.88	214	6739.29	240
3350.62	160	3734.83	159	4673.49	214	6741.94	247
3350.92	160	3742.63	231	4696.23	239	6761.12	240
3355.86	203	3746.90	231	4708.15	239	6768.36	240
3362.31	160	3754.70	71	4722.80	239	6792.06	247
3376.61	202	3757.23	71	4732.54	239	6817.02	247
3376.76	202	3759.88	71	4737.88	251	7016.32	197

List of tabulated lines — Continued

Wavelength (Å)	No.	Wavelength (Å)	No.	Wavelength (Å)	No.	Wavelength (Å)	No.
7025.73	197	8368.21	195	9623.17	213	12497.3	211
7031.37	197	8380.36	195	9688.50	213	12508.3	193
7079.12	197	8388.91	195	9706.45	213	12516.4	211
7084.84	197	8456.42	195	9717.22	213	12534.4	193
7190.25	197	8465.13	195	9809.48	213	12668.6	211
7306.85	174	8616.05	195	9813.69	220	12688.2	211
7307.12	174	8912.12	173	9820.47	213	12717.7	193
7365.35	174	8999.30	173	10118.7	219	12991.4	249
7410.09	243	9076.59	253	10251.3	172	13157.2	249
7414.04	174	9116.14	194	10252.8	172	13280.6	249
7449.07	243	9119.40	253	10321.8	172	13436.5	249
7455.36	174	9121.38	253	10366.9	172	13523.5	249
7455.41	243	9133.29	194	10547.0	172	13653.9	249
7475.76	243	9164.61	253	10667.8	172	13764.4	179
7482.14	243	9174.21	194	11007.1	246	13972.5	179
7492.84	243	9221.34	252	11141.4	246	14148.8	179
7515.99	174	9225.21	173	11235.4	212	14164.7	177
7551.42	164	9226.47	253	11264.9	212	14297.0	163
7608.88	164	9245.08	252	11321.6	212	14657.7	163
7710.95	164	9248.12	253	11349.1	212	14661.7	163
7807.56	196	9265.45	194	11394.0	246	14735.2	218
7820.68	207	9265.53	252	11403.9	212	14875.7	163
7832.37	207	9291.40	194	11490.2	212	15120.2	163
7834.23	196	9311.11	253	11537.9	246	15270.9	163
7839.38	207	9330.06	252	11610.1	246	23420.2	210
7847.94	196	9375.30	252	11636.1	246	23549.6	210
7873.54	196	9388.06	252	12192.7	193	23564.8	210
7914.61	196	9446.56	194	12229.0	193	24070.6	210
7963.32	196	9475.71	252	12359.6	211	24165.3	210
8003.94	196	9538.22	206	12368.0	193	24714.2	210
8172.15	178	9554.01	206	12391.6	193		
8358.86	186	9565.12	206	12470.3	211		

As usual, by far the largest amount of the selected data for O III originates from the Opacity Project (OP) which is reviewed in the general introduction. For this carbon-like ion, the numerical calculations have been performed by Luo *et al.*[1] Another fairly large configuration interaction calculation comprising 59 multiplets was recently carried out by Aggarwal and Hibbert[2] with the CIV 3 computer code,–an approach fundamentally different from the OP calculation. As Fig. 1 shows, the results of the two calculations agree within ±15% for the large majority of multiplets but for weak multiplets some larger disagreements appear. Accordingly, we have reduced the accuracy ratings for weak multiplets (and lines) both for the averaged data from Refs. 1 and 2, and for those where only Ref. 1 is available.

For a few strong multiplets, data are also available from four other sophisticated calculations and have been utilized: (a) multi-configuration calculations by Nussbaumer and Storey[3] with the SUPERSTRUCTURE code, partly in intermediate coupling, (b) multi-configuration Hartree-Fock calculations by Froese Fischer and Saha[4] with Breit-Pauli relativistic corrections for individual lines of the $2s^2 2p^2 - 2s2p^3$ transition array, (c) superposition-of-configurations (SOC) calculations in intermediate coupling by Weiss,[5] and (d) calculations by Aggarwal and Hibbert,[6] similar in spirit to their above-cited work (Ref. 2), but with a considerably extended configuration interaction treatment. This latter work was specifically done for the downward transitions from the $2p3s$ $^1P^0$ and $2s2p$ $^1P^0$ levels, in order to address a significant difference between their earlier result and a recent accurate lifetime measurement (Ref. 8, see discussion below).

Since all the above-cited calculations[1–6] are of similar levels of sophistication, we have selected arithmetic averages of the results for these tables whenever appropriate and available. A good indication of their mutual consistency may be gained from Fig. 1, where the ratios of the data from Refs. 2–6 are plotted versus the OP data[1] for all multiplets for which two or more data sources are available.

For the two principal intercombination lines out of the ground state, we have selected, with equal weights, two theoretical values[3,4] and a measurement obtained from an ion-trap experiment by Johnson *et al.*[7] The three results agree within ±12%, provided experimental energies are used. For numerous other intersystem lines, data from Refs. 3, 4 and 5 are available. The data of Refs. 3 and 4 overlap for most lines, and the differences in the results vary from a few percent to about 60%, as indicated by our varying error estimates.

The selected theoretical data may be tested against some experimental lifetime and emission data. Among the available lifetime measurements, especially the beam-foil-laser experiment by Baudinet-Robinet *et al.*,[8] and four beam-foil experiments by Kotze *et al.*[9–11] and Coetzer *et al.*[12] need to be considered. In these experiments, the critical cascading problem either has been eliminated by selective excitation[8] or taken care of with the well-proven ANDC method[9–12] (see general introduction). We compared lifetime data for 18 different levels with the selected theoretical data and found that the mean value of the ratios "experiment/selected data" is 0.985, with a two-standard deviation of ±0.076. The rather large deviation indicates that some appreciable differences between experiments and calculations exist, which are as high as 36% in one instance — the $2s^2 2p3p$ ^3S level. Also, a difference of 27% occurs between the single result of the beam-foil-laser experiment of Baudinet-Robinet *et al.*[8] and the closely agreeing calculational results of Refs. 1 and 6. In Fig. 2, the above cited lifetime data are compared with the selected theoretical results.

A test of the validity of the *LS* coupling scheme was performed by Glenzer *et al.*[13] with a gas liner pinch for some $3s$-$3p$ and $3p$-$3d$ lines. Relative line intensities in five multiplets were measured and compared with *LS*-intensity fractions and, for the $3s$-$3p$ multiplets, with intermediate coupling data calculated by Weiss.[5] The emission data tend to be slightly closer to the intermediate coupling data. Also, most measured deviations from *LS* coupling are not large, only in the 3% to 25% range, and thus are usually within the estimated experimental uncertainties.

References

[1] D. Luo, A. K. Pradhan, H. E. Saraph, P. J. Storey, and Y. Yu, J. Phys. B **22**, 389 (1989).

[2] K. M. Aggarwal and A. Hibbert, J. Phys. B **24**, 3445 (1991).

[3] H. Nussbaumer and P. J. Storey, Astron. Astrophys. **99**, 177 (1981).

[4] C. Froese Fischer and H. P. Saha, Phys. Scr. **32**, 181 (1985).

[5] A. W. Weiss, to be published.

[6] K. M. Aggarwal and A. Hibbert, J. Phys. B **24**, 4685 (1991).

[7] B. C. Johnson, P. L. Smith, and R. D. Knight, Astrophys. J. **281**, 477 (1984).

[8] Y. Baudinet-Robinet, P. D. Dumont, and H. P. Garnir, Phys. Rev. A **43**, 4022 (1991).

[9] T. C. Kotze, F. J. Coetzer, and P. van der Westhuizen, J. Quant. Spectrosc. Radiat. Transfer **36**, 249 (1986).

[10] T. C. Kotze, P. van der Westhuizen, and F. J. Coetzer, Phys. Rev. A **39**, 2956 (1989).

[11] T. C. Kotze and P. van der Westhuizen, Spectrochim. Acta, Part B **45**, 421 (1990).

[12] F. J. Coetzer, T. C. Kotze, F. J. Mostert, and P. van der Westhuizen, Spectrochim. Acta, Part B **41**, 847 (1986).

[13] S. Glenzer, J. Musielok, H.-J. Kunze, and W. L. Wiese (to be published).

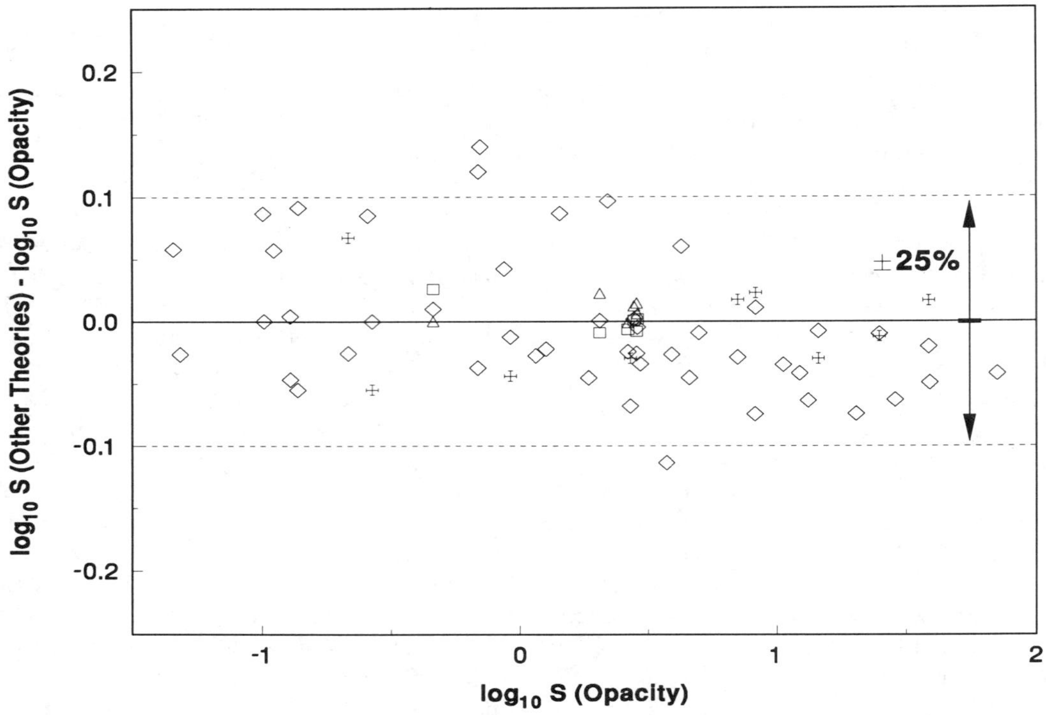

FIG. 1 Comparison of the multiplet line strengths S from the OP calculations of Luo *et al.*[1] with other advanced theoretical results by Aggarwal and Hibbert[2,6] (diamonds), Nussbaumer and Storey[3] (squares), Froese Fischer and Saha[4] (triangles), and Weiss[5] (crosses). The ratios S(other theory)/S(OP) are plotted on a logarithmic scale versus log S(OP). The figure illustrates that these ratios are essentially within a band which deviates no more than ±25% from a perfect ratio of 1.00 (broken lines).

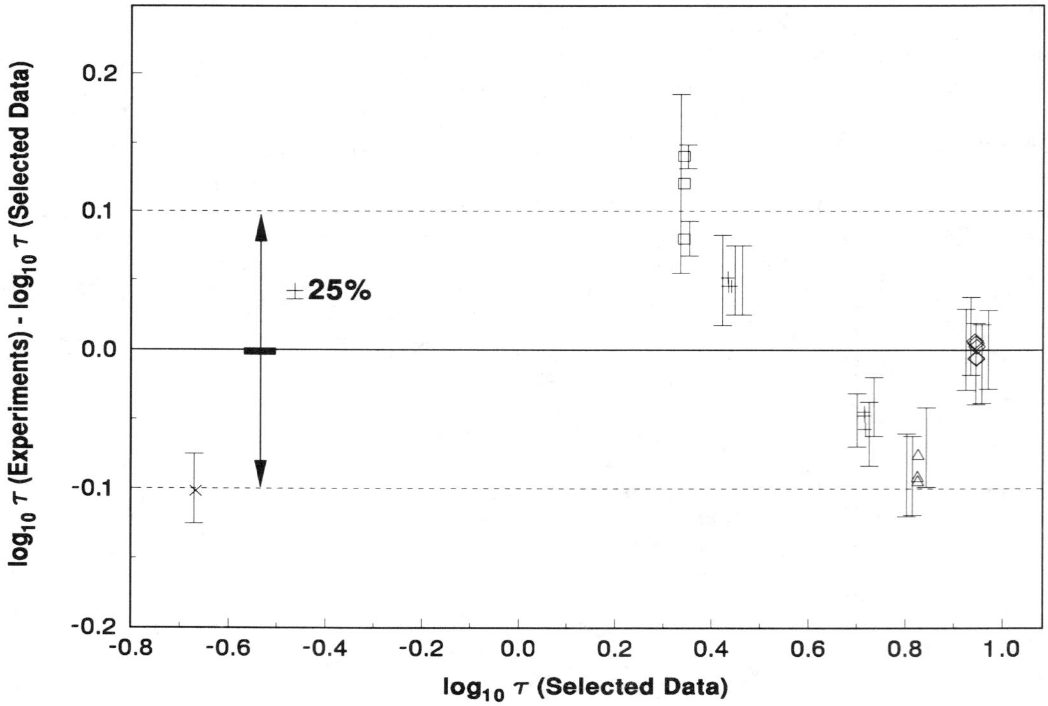

FIG. 2. Comparison of measured lifetimes τ (in ns) versus lifetimes derived from the selected theoretical data. The comparison is carried out in a manner analogous to Fig. 1. The symbols for the experimental data, presented with their estimated errors, are as follows: ×, Baudinet-Robinet *et al.*;[8] squares, Kotze *et al.*;[9] diamonds, Kotze *et al.*;[10] triangles, Kotze *et al.*;[11] and plus signs, Coetzer *et al.*[12]

O III: Allowed Transitions

No.	Transition Array	Mult.	λ_{air} (Å)	λ_{vac} (Å) or σ (cm^{-1})*	E_i (cm^{-1})	E_k (cm^{-1})	g_i-g_k	A_{ki} (10^8 s^{-1})	f_{ik}	S (at. u.)	log gf	Acc.	Source
1.	$2s^22p^2-2s2p^3$	$^3P-^5S^o$											
				1660.81	113.178	60324.79	3–5	2.20–06	1.52–07	2.49–06	–6.342	B	3,4,7
				1666.15	306.174	60324.79	5–5	5.48–06	2.28–07	6.25–06	–5.943	B	3,4,7
2.		$^3P-^3D^o$		*834.49*	*207.82*	*120041*	9–15	6.01+00	1.05–01	2.59+00	–0.026	A	1,2,3,4
				835.289	306.174	120025.2	5–7	5.99+00	8.77–02	1.21+00	–0.358	A	3*n*, 4*n*
				833.749	113.178	120053.4	3–5	4.58+00	7.95–02	6.55–01	–0.622	A	3*n*, 4*n*
				832.929	0.000	120058.2	1–3	3.41+00	1.06–01	2.92–01	–0.973	A	3*n*, 4*n*
				835.092	306.174	120053.4	5–5	1.44+00	1.51–02	2.08–01	–1.122	A	3*n*, 4*n*
				833.715	113.178	120058.2	3–3	2.48+00	2.59–02	2.13–01	–1.110	A	3*n*, 4*n*
				835.059	306.174	120058.2	5–3	1.56–01	9.81–04	1.35–02	–2.309	B–	3*n*, 4*n*
3.		$^3P-^3P^o$		*703.36*	*207.82*	*142383*	9–9	1.82+01	1.35–01	2.82+00	0.085	A	1,2,3,4
				703.854	306.174	142381.0	5–5	1.37+01	1.02–01	1.18+00	–0.293	A	3*n*, 4*n*
				702.896	113.178	142381.8	3–3	4.66+00	3.45–02	2.40–01	–0.984	A	3*n*, 4*n*
				703.850	306.174	142381.8	5–3	7.54+00	3.36–02	3.89–01	–0.775	A	3*n*, 4*n*
				702.838	113.178	142393.5	3–1	1.83+01	4.52–02	3.14–01	–0.868	A	3*n*, 4*n*
				702.900	113.178	142381.0	3–5	4.47+00	5.52–02	3.83–01	–0.781	A	3*n*, 4*n*
				702.337	0.000	142381.8	1–3	6.06+00	1.35–01	3.11–01	–0.871	A	3*n*, 4*n*
4.		$^3P-^1D^o$											
				535.481	306.174	187054.0	5–5	2.58–03	1.11–05	9.77–05	–4.256	B	3,4
				534.929	113.178	187054.0	3–5	1.21–04	8.62–07	4.55–06	–5.587	B	3,4
5.		$^3P-^3S^o$		*507.92*	*207.82*	197087.7	9–3	1.45+02	1.87–01	2.81+00	0.225	A	1,2,3,4
				508.178	306.174	197087.7	5–3	8.04+01	1.87–01	1.56+00	–0.030	A	3*n*, 4*n*
				507.680	113.178	197087.7	3–3	4.82+01	1.86–01	9.34–01	–0.253	A	3*n*, 4*n*
				507.388	0.000	197087.7	1–3	1.61+01	1.86–01	3.11–01	–0.730	A	3*n*, 4*n*
6.		$^3P-^1P^o$											
				475.838	306.174	210461.8	5–3	1.45–03	2.96–06	2.32–05	–4.830	D	3,4
				475.401	113.178	210461.8	3–3	7.85–03	2.66–05	1.25–04	–4.098	C	3,4
				475.146	0.000	210461.8	1–3	4.86–04	4.94–06	7.73–06	–5.306	C	3,4
7.	$2s^22p^2-2s^22p3s$	$^3P-^3P^o$		*374.11*	*207.82*	*267506.7*	9–9	3.79+01	7.96–02	8.83–01	–0.145	B	1,2,5
				374.073	306.174	267634.00	5–5	2.85+01	5.97–02	3.68–01	–0.525	C+	5*n*
				374.162	113.178	267377.11	3–3	9.46+00	1.99–02	7.34–02	–1.225	C+	5*n*
				374.432	306.174	267377.11	5–3	1.58+01	1.99–02	1.23–01	–1.001	C+	5*n*
				374.328	113.178	267258.71	3–1	37.919021	2.66–02	9.82–02	–1.099	C+	5*n*
				373.803	113.178	267634.00	3–5	9.50+00	3.32–02	1.22–01	–1.002	C+	5*n*
				374.004	0.000	267377.11	1–3	1.26+01	7.96–02	9.80–02	–1.099	C+	5*n*
8.		$^3P-^1P^o$											
				366.602	306.174	273081.33	5–3	5.03–03	6.08–06	3.67–05	–4.517	C	5
				366.343	113.178	273081.33	3–3	6.43–03	1.29–05	4.68–05	–4.411	C	5
				366.191	0.000	273081.33	1–3	4.41–03	2.66–05	3.21–05	–4.575	C	5
9.	$2s^22p^2-2s^22p3d$	$^3P-^3D^o$		*305.72*	*207.82*	*327303.0*	9–15	2.16+02	5.05–01	4.57+00	0.657	B	1,2
				305.767	306.174	327352.17	5–7	2.16+02	4.24–01	2.13+00	0.326	C+	*LS*
				305.656	113.178	327278.30	3–5	1.62+02	3.79–01	1.14+00	0.055	C+	*LS*
				305.596	0.000	327229.25	1–3	1.20+02	5.05–01	5.08–01	–0.297	C+	*LS*
				305.836	306.174	327278.30	5–5	5.40+01	7.57–02	3.81–01	–0.422	C+	*LS*
				305.702	113.178	327229.25	3–3	9.01+01	1.26–01	3.81–01	–0.422	C+	*LS*
				305.882	306.174	327229.25	5–3	6.00+00	5.05–03	2.54–02	–1.598	C	*LS*

O III: Allowed Transitions — Continued

No.	Transition Array	Mult.	λ_{air} (Å)	λ_{vac} (Å) or σ (cm^{-1})*	E_i (cm^{-1})	E_k (cm^{-1})	g_i-g_k	A_{ki} (10^8 s^{-1})	f_{ik}	S (at. u.)	log gf	Acc.	Source
10.		^{3}P–^{3}P°		*303.66*	*207.82*	– *329527.3*	9–9	1.28+02	1.77–01	1.60+00	0.203	C+	1,2
				303.800	306.174	– 329469.80	5–5	9.61+01	1.33–01	6.65–01	–0.177	C–	*LS*
				303.517	113.178	– 329583.89	3–3	3.21+01	4.44–02	1.33–01	–0.876	C–	*LS*
				303.695	306.174	– 329583.89	5–3	5.34+01	4.43–02	2.22–01	–0.654	C–	*LS*
				303.461	113.178	– 329645.14	3–1	1.29+02	5.92–02	1.77–01	–0.751	C–	*LS*
				303.622	113.178	– 329469.80	3–5	3.21+01	7.39–02	2.22–01	–0.654	C–	*LS*
				303.413	0.000	– 329583.89	1–3	4.29+01	1.77–01	1.77–01	–0.751	C–	*LS*
11.	$2s^22p^2$–$2s2p^23p$	^{3}P–^{3}S°		*275.44*	*207.82*	– 363263.38	9–3	3.65+01	1.38–02	1.13–01	–0.905	C+	1
				275.515	306.174	– 363263.38	5–3	2.02+01	1.38–02	6.27–02	–1.161	C–	*LS*
				275.368	113.178	– 363263.38	3–3	1.22+01	1.38–02	3.76–02	–1.382	C–	*LS*
				275.282	0.000	– 363263.38	1–3	4.06+00	1.38–02	1.25–02	–1.859	D	*LS*
12.		^{3}P–^{3}D°		*267.02*	*207.82*	– *374706.6*	9–15	6.63+01	1.18–01	9.35–01	0.027	B	1
				267.031	306.174	– 374795.14	5–7	6.63+01	9.93–02	4.36–01	–0.304	C+	*LS*
				266.987	113.178	– 374663.52	3–5	4.98+01	8.86–02	2.34–01	–0.575	C+	*LS*
				266.972	0.000	– 374571.64	1–3	3.69+01	1.18–01	1.04–01	–0.927	C+	*LS*
				267.124	306.174	– 374663.52	5–5	1.66+01	1.77–02	7.79–02	–1.053	C+	*LS*
				267.052	113.178	– 374571.64	3–3	2.76+01	2.95–02	7.79–02	–1.052	C+	*LS*
				267.190	306.174	– 374571.64	5–3	1.84+00	1.18–03	5.19–03	–2.229	C–	*LS*
13.		^{3}P–^{3}P°		*264.41*	*207.82*	– *378413.0*	9–9	7.02+01	7.36–02	5.76–01	–0.179	B	1
				264.481	306.174	– 378405.68	5–5	5.26+01	5.52–02	2.40–01	–0.559	C+	*LS*
				264.337	113.178	– 378417.84	3–3	1.76+01	1.84–02	4.80–02	–1.258	C+	*LS*
				264.472	306.174	– 378417.84	5–3	2.92+01	1.84–02	8.00–02	–1.037	C+	*LS*
				264.325	113.178	– 378435.16	3–1	7.03+01	2.45–02	6.40–02	–1.133	C+	*LS*
				264.346	113.178	– 378405.68	3–5	1.76+01	3.07–02	8.00–02	–1.036	C+	*LS*
				264.258	0.000	– 378417.84	1–3	2.34+01	7.36–02	6.40–02	–1.133	C+	*LS*
14.	$2s^22p^2$–$2s^22p4d$	^{3}P–^{3}D°		*263.78*	*207.82*	– *379309.6*	9–15	5.97+01	1.04–01	8.11–01	–0.030	B	1
				263.817	306.174	– 379356.75	5–7	5.97+01	8.72–02	3.79–01	–0.361	C+	*LS*
				263.727	113.178	– 379293.03	3–5	4.48+01	7.79–02	2.03–01	–0.632	C+	*LS*
				263.694	0.000	– 379227.15	1–3	3.32+01	1.04–01	9.01–02	–0.984	C+	*LS*
				263.861	306.174	– 379293.03	5–5	1.49+01	1.56–02	6.76–02	–1.109	C+	*LS*
				263.773	113.178	– 379227.15	3–3	2.49+01	2.59–02	6.76–02	–1.109	C+	*LS*
				263.907	306.174	– 379227.15	5–3	1.66+00	1.04–03	4.51–03	–2.285	C–	*LS*
15.	$2s^22p^2$–$2s^22p6d$	^{3}P–^{3}D°		*240.98*	*207.82*	– 415180	9–15	2.57+01	3.72–02	2.66–01	–0.475	C+	1
				241.037	306.174	– 415180	5–7	2.56+01	3.13–02	1.24–01	–0.806	C–	*LS*
				240.925	113.178	– 415180	3–5	1.93+01	2.79–02	6.65–02	–1.077	C–	*LS*
				240.859	0.000	– 415180	1–3	1.43+01	3.73–02	2.95–02	–1.429	C–	*LS*
				241.037	306.174	– 415180	5–5	6.41+00	5.58–03	2.22–02	–1.554	C–	*LS*
				240.925	113.178	– 415180	3–3	1.07+01	9.31–03	2.22–02	–1.554	C–	*LS*
				241.037	306.174	– 415180	5–3	7.12–01	3.72–04	1.48–03	–2.730	D	*LS*
16.	$2s^22p^2$–$2s2p^3$	^{1}D–^{5}S°											
			2496.03	2496.78	20273.27	– 60324.79	5–5	4.74–05	4.43–06	1.82–04	–4.655	C	4
17.		^{1}D–^{3}D°											
				1002.49	20273.27	– 120025.2	5–7	1.80–04	3.79–06	6.25–05	–4.723	B	3,4
				1002.20	20273.27	– 120053.4	5–5	3.66–05	5.51–07	9.10–06	–5.560	B	3,4
				1002.16	20273.27	– 120058.2	5–3	2.29–05	2.07–07	3.41–06	–5.986	B	3,4

O III: Allowed Transitions — Continued

No.	Transition Array	Mult.	λ_{air} (Å)	λ_{vac} (Å) or σ (cm^{-1})*	E_i (cm^{-1})	E_k (cm^{-1})	g_i-g_k	A_{ki} (10^8 s^{-1})	f_{ik}	S (at. u.)	log gf	Acc.	Source
18.		$^1D-^3P^o$											
				818.949	20273.27	– 142381.0	5–5	6.69−05	6.72−07	9.06−06	−5.473	B	3,4
				818.944	20273.27	– 142381.8	5–3	2.65−04	1.60−06	2.16−05	−5.097	B	3,4
19.		$^1D-^1D^o$		599.590	20273.27	– 187054.0	5–5	5.41+01	2.91−01	2.88+00	0.164	A	1,2,3,4
20.		$^1D-^3S^o$											
				565.565	20273.27	– 197087.7	5–3	2.58−03	7.42−06	6.91−05	−4.431	C+	3,4
21.		$^1D-^1P^o$		525.794	20273.27	– 210461.8	5–3	9.60+01	2.39−01	2.07+00	0.077	A	1,3,4,6
22.	$2s^22p^2-$ $2s^22p3s$	$^1D-^3P^o$											
				404.268	20273.27	– 267634.00	5–5	8.53−04	2.09−06	1.39−05	−4.981	C	5
				404.688	20273.27	– 267377.11	5–3	1.27−02	1.88−05	1.25−04	−4.028	C	5
23.		$^1D-^1P^o$		395.557	20273.27	– 273081.33	5–3	2.80+01	3.94−02	2.56−01	−0.706	B+	1,5,6
24.	$2s^22p^2-$ $2s^22p3d$	$^1D-^1D^o$		328.448	20273.27	– 324735.65	5–5	1.04+02	1.69−01	9.13−01	−0.073	B	1,2
25.		$^1D-^1F^o$		320.978	20273.27	– 331821.44	5–7	2.17+02	4.69−01	2.48+00	0.370	C+	1,2
26.	$2s^22p^2-$ $2s^22p4d$	$^1D-^1D^o$		279.788	20273.27	– 377686.83	5–5	4.25+01	4.98−02	2.30−01	−0.604	C+	1
27.		$^1D-^1F^o$		277.386	20273.27	– 380782.17	5–7	9.43+01	1.52−01	6.96−01	−0.118	B	1
28.	$2s^22p^2-$ $2s^22p5d$	$^1D-^1D^o$		262.111	20273.27	– 401791.7	5–5	2.26+01	2.33−02	1.01−01	−0.933	C+	1
29.		$^1D-^1F^o$		261.024	20273.27	– 403380.0	5–7	5.17+01	7.39−02	3.17−01	−0.433	C+	1
30.	$2s^22p^2-$ $2s^22p7d$	$^1D-^1F^o$		248.320	20273.27	– 422979	5–7	2.52+01	3.26−02	1.33−01	−0.788	C+	1
31.	$2s^22p^2-$ $2s2p^2(^2D)3p$	$^1D-^1D^o$		246.265	20273.27	– 426340	5–5	3.71+01	3.37−02	1.37−01	−0.773	C+	1
32.	$2s^22p^2-$ $2s2p^3$	$^1S-^3D^o$											
				1300.86	43185.74	– 120058.2	1–3	1.67−05	1.27−06	5.43−06	−5.897	B	3,4
33.		$^1S-^3P^o$											
				1008.10	43185.74	– 142381.8	1–3	9.22−05	4.22−06	1.40−05	−5.375	B	3,4
34.		$^1S-^3S^o$											
				649.764	43185.74	– 197087.7	1–3	5.46−04	1.04−05	2.22−05	−4.984	D	3,4
35.		$^1S-^1P^o$		597.814	43185.74	– 210461.8	1–3	1.49+01	2.40−01	4.72−01	−0.621	A−	1,3,4,6

O III: Allowed Transitions — Continued

No.	Transition Array	Mult.	λ_{air} (Å)	λ_{vac} (Å) or σ (cm^{-1})*	E_i (cm^{-1})	E_k (cm^{-1})	g_i-g_k	A_{ki} (10^8 s^{-1})	f_{ik}	S (at. u.)	log gf	Acc.	Source
36.	$2s^22p^2-$ $2s^22p3s$	$^1S-^3P^o$											
				446.047	43185.74	– 267377.11	1–3	1.02–02	9.13–05	1.34–04	–4.040	C	5
37.		$^1S-^1P^o$		434.980	43185.74	– 273081.33	1–3	1.84+01	1.56–01	2.24–01	–0.806	B	1,5,6
38.	$2s^22p^2-$ $2s^22p3d$	$^1S-^1P^o$		345.312	43185.74	– 332778.94	1–3	1.35+02	7.26–01	8.26–01	–0.139	C	1,2
39.	$2s^22p^2-$ $2s^22p4d$	$^1S-^1P^o$		295.942	43185.74	– 381089.27	1–3	5.56+01	2.19–01	2.13–01	–0.660	C+	1
40.	$2s2p^3-$ $2s2p^23s$	$^5S^o-^5P$		*359.16*	60324.79	– *338752.2*	5–15	3.29+01	1.91–01	1.13+00	–0.020	B	1
				359.017	60324.79	– 338863.03	5–7	3.30+01	8.92–02	5.27–01	–0.351	C+	*LS*
				359.225	60324.79	– 338701.98	5–5	3.29+01	6.37–02	3.76–01	–0.497	C+	*LS*
				359.386	60324.79	– 338577.25	5–3	3.29+01	3.82–02	2.26–01	–0.719	C+	*LS*
41.	$2s2p^3-$ $2s2p^23d$	$^5S^o-^5P$		*295.68*	60324.79	– *398532.2*	5–15	2.16+02	8.49–01	4.13+00	0.628	B	1
				295.716	60324.79	– 398487.08	5–7	2.16+02	3.96–01	1.93+00	0.297	C+	*LS*
				295.655	60324.79	– 398557.17	5–5	2.16+02	2.83–01	1.38+00	0.151	C+	*LS*
				295.621	60324.79	– 398595.65	5–3	2.16+02	1.70–01	8.26–01	–0.071	C+	*LS*
42.	$2s2p^3-$ $2s2p^24s$	$^5S^o-^5P$		*271.48*	60324.79	– *428678.8*	5–15	7.90+00	2.62–02	1.17–01	–0.883	C+	1
				271.395	60324.79	– 428791.89	5–7	7.91+00	1.22–02	5.46–02	–1.214	C–	*LS*
				271.516	60324.79	– 428627.20	5–5	7.90+00	8.73–03	3.90–02	–1.360	C–	*LS*
				271.609	60324.79	– 428500.87	5–3	7.89+00	5.23–03	2.34–02	–1.582	C–	*LS*
43.	$2s2p^3-$ $2s2p^24d$	$^5S^o-^5P$		*256.47*	60324.79	– *450227.5*	5–15	9.94+01	2.94–01	1.24+00	0.168	B	1
				256.507	60324.79	– 450177.71	5–7	9.94+01	1.37–01	5.80–01	–0.163	C+	*LS*
				256.459	60324.79	– 450250.84	5–5	9.94+01	9.81–02	4.14–01	–0.310	C+	*LS*
				256.423	60324.79	– 450304.93	5–3	9.95+01	5.88–02	2.48–01	–0.531	C+	*LS*
44.	$2s2p^3-$ $2p^4$	$^3D^o-^3P$		*610.40*	*120041*	– *283867*	15–9	4.88+01	1.64–01	4.93+00	0.390	B	1,2
				610.745	120025.2	– 283759.7	7–5	4.09+01	1.63–01	2.30+00	0.059	C+	*LS*
				610.039	120053.4	– 283977.4	5–3	3.67+01	1.23–01	1.23+00	–0.212	C+	*LS*
				609.705	120058.2	– 284071.9	3–1	4.90+01	9.10–02	5.48–01	–0.564	C+	*LS*
				610.850	120053.4	– 283759.7	5–5	7.30+00	4.09–02	4.11–01	–0.690	C+	*LS*
				610.057	120058.2	– 283977.4	3–3	1.22+01	6.82–02	4.11–01	–0.689	C+	*LS*
				610.868	120058.2	– 283759.7	3–5	4.87–01	4.54–03	2.74–02	–1.866	C	*LS*
45.	$2s2p^3-$ $2s^22p3p$	$^3D^o-^3D$		*574.59*	*120041*	– *294078.4*	15–15	3.33–01	1.65–03	4.68–02	–1.606	B	1,2
				574.060	120025.2	– 294223.07	7–7	2.97–01	1.47–03	1.94–02	–1.988	C	*LS*
				574.880	120053.4	– 294002.86	5–5	2.32–01	1.15–03	1.09–02	–2.241	C	*LS*
				575.347	120058.2	– 293866.49	3–3	2.49–01	1.24–03	7.02–03	–2.431	C–	*LS*
				574.786	120025.2	– 294002.86	7–5	5.20–02	1.84–04	2.43–03	–2.891	D	*LS*
				575.331	120053.4	– 293866.49	5–3	8.30–02	2.47–04	2.34–03	–2.908	D	*LS*
				574.153	120053.4	– 294223.07	5–7	3.72–02	2.58–04	2.43–03	–2.890	D	*LS*
				574.895	120058.2	– 294002.86	3–5	4.99–02	4.12–04	2.34–03	–2.908	D	*LS*
46.		$^3D^o-^3P$		*554.52*	*120041*	– *300376.5*	15–9	1.48+00	4.09–03	1.12–01	–1.212	C+	1,2
				554.270	120025.2	– 300442.55	7–5	1.24+00	4.09–03	5.23–02	–1.543	C–	*LS*
				554.759	120053.4	– 300311.96	5–3	1.11+00	3.07–03	2.80–02	–1.814	D	*LS*
				554.995	120058.2	– 300239.93	3–1	1.48+00	2.27–03	1.25–02	–2.167	D	*LS*
				554.357	120053.4	– 300442.55	5–5	2.22–01	1.02–03	9.34–03	–2.291	D–	*LS*
				554.773	120058.2	– 300311.96	3–3	3.69–01	1.70–03	9.34–03	–2.291	D–	*LS*
				554.372	120058.2	– 300442.55	3–5	1.48–02	1.14–04	6.23–04	–3.467	D–	*LS*

O III: Allowed Transitions — Continued

No.	Transition Array	Mult.	λ_{air} (Å)	λ_{vac} (Å) or σ (cm^{-1})*	E_i (cm^{-1})	E_k (cm^{-1})	g_i-g_k	A_{ki} (10^8 s^{-1})	f_{ik}	S (at. u.)	log gf	Acc.	Source
47.	$2s2p^3-$ $2s2p^23s$	$^3D^o-^3P$		*434.46*	*120041*	*– 350210.0*	15–9	2.27+01	3.85–02	8.26–01	–0.238	B	1
				434.267	120025.2	– 350298.38	7–5	1.91+01	3.85–02	3.85–01	–0.569	C+	*LS*
				434.648	120053.4	– 350124.45	5–3	1.70+01	2.89–02	2.07–01	–0.841	C+	*LS*
				434.846	120058.2	– 350024.49	3–1	2.26+01	2.14–02	9.18–02	–1.193	C+	*LS*
				434.320	120053.4	– 350298.38	5–5	3.40+00	9.63–03	6.88–02	–1.317	C+	*LS*
				434.657	120058.2	– 350124.45	3–3	5.66+00	1.60–02	6.88–02	–1.318	C+	*LS*
				434.329	120058.2	– 350298.38	3–5	2.27–01	1.07–03	4.59–03	–2.494	C–	*LS*
48.	$2s2p^3-$ $2s2p^23d$	$^3D^o-^3P$		*356.68*	*120041*	*– 400406.2*	15–9	1.46+01	1.67–02	2.95–01	–0.600	C+	1
				356.727	120025.2	– 400351.56	7–5	1.23+01	1.67–02	1.38–01	–0.931	C–	*LS*
				356.624	120053.4	– 400460.99	5–3	1.10+01	1.26–02	7.38–02	–1.202	C–	*LS*
				356.561	120058.2	– 400514.89	3–1	1.47+01	9.31–03	3.28–02	–1.554	D	*LS*
				356.763	120053.4	– 400351.56	5–5	2.19+00	4.19–03	2.46–02	–1.679	D	*LS*
				356.630	120058.2	– 400460.99	3–3	3.66+00	6.98–03	2.46–02	–1.679	D	*LS*
				356.769	120058.2	– 400351.56	3–5	1.46–01	4.65–04	1.64–03	–2.855	D–	*LS*
49.	$2s2p^3-$ $2s2p^2(^2D)3s$	$^3D^o-^3D$		*364.82*	*120041*	*– 394151*	15–15	2.40+01	4.80–02	8.64–01	–0.143	B	1
				364.734	120025.2	– 394197.9	7–7	2.14+01	4.26–02	3.58–01	–0.525	C+	*LS*
				364.865	120053.4	– 394127.3	5–5	1.67+01	3.34–02	2.00–01	–0.778	C+	*LS*
				364.935	120058.2	– 394079.4	3–3	1.80+01	3.60–02	1.30–01	–0.967	C+	*LS*
				364.828	120025.2	– 394127.3	7–5	3.75+00	5.34–03	4.49–02	–1.427	C+	*LS*
				364.929	120053.4	– 394079.4	5–3	6.00+00	7.19–03	4.32–02	–1.444	C+	*LS*
				364.771	120053.4	– 394197.9	5–7	2.68+00	7.48–03	4.49–02	–1.427	C+	*LS*
				364.871	120058.2	– 394127.3	3–5	3.60+00	1.20–02	4.32–02	–1.444	C+	*LS*
50.	$2s2p^3-$ $2s2p^23d$	$^3D^o-^3F$		*355.28*	*120041*	*– 401507.6*	15–21	8.97+01	2.38–01	4.17+00	0.552	B	1
				355.138	120025.2	– 401605.52	7–9	8.99+01	2.18–01	1.79+00	0.184	C+	*LS*
				355.337	120053.4	– 401476.29	5–7	7.97+01	2.11–01	1.24+00	0.024	C+	*LS*
				355.471	120058.2	– 401375.09	3–5	7.53+01	2.38–01	8.34–01	–0.147	C+	*LS*
				355.302	120025.2	– 401476.29	7–7	1.00+01	1.89–02	1.55–01	–0.878	C+	*LS*
				355.465	120053.4	– 401375.09	5–5	1.40+01	2.65–02	1.55–01	–0.878	C+	*LS*
				355.429	120025.2	– 401375.09	7–5	3.94–01	5.34–04	4.37–03	–2.428	C–	*LS*
51.		$^3D^o-^3D$		*349.89*	*120041*	*– 405847.7*	15–15	2.54+01	4.66–02	8.06–01	–0.155	B	1
				349.828	120025.2	– 405879.97	7–7	2.26+01	4.15–02	3.34–01	–0.537	C+	*LS*
				349.923	120053.4	– 405830.38	5–5	1.77+01	3.25–02	1.87–01	–0.790	C+	*LS*
				349.964	120058.2	– 405801.50	3–3	1.90+01	3.50–02	1.21–01	–0.979	C+	*LS*
				349.889	120025.2	– 405830.38	7–5	3.96+00	5.20–03	4.19–02	–1.439	C+	*LS*
				349.959	120053.4	– 405801.50	5–3	6.35+00	7.00–03	4.03–02	–1.456	C+	*LS*
				349.863	120053.4	– 405879.97	5–7	2.83+00	7.28–03	4.19–02	–1.439	C+	*LS*
				349.929	120058.2	– 405830.38	3–5	3.81+00	1.17–02	4.03–02	–1.456	C+	*LS*
52.	$2s2p^3-$ $2p^4$	$^3P^o-^3P$		*706.79*	*142381*	*– 283867*	9–9	7.16+00	5.36–02	1.12+00	–0.317	B	1,2
				707.320	142381.0	– 283759.7	5–5	5.36+00	4.02–02	4.68–01	–0.697	C+	*LS*
				706.237	142381.8	– 283977.4	3–3	1.79+00	1.34–02	9.35–02	–1.395	C+	*LS*
				706.233	142381.0	– 283977.4	5–3	2.99+00	1.34–02	1.56–01	–1.174	C+	*LS*
				705.766	142381.8	– 284071.9	3–1	7.19+00	1.79–02	1.25–01	–1.270	C+	*LS*
				707.324	142381.8	– 283759.7	3–5	1.79+00	2.23–02	1.56–01	–1.174	C+	*LS*
				706.295	142393.5	– 283977.4	1–3	2.39+00	5.36–02	1.25–01	–1.271	C+	*LS*
53.	$2s2p^3-$ $2s^22p3p$	$^3P^o-^3D$		*659.21*	*142383*	*– 294078.4*	9–15	6.09–01	6.62–03	1.29–01	–1.225	B–	1,2
				658.579	142381.0	– 294223.07	5–7	6.11–01	5.56–03	6.03–02	–1.556	C	*LS*
				659.539	142381.8	– 294002.86	3–5	4.56–01	4.96–03	3.23–02	–1.827	C	*LS*
				660.184	142393.5	– 293866.49	1–3	3.37–01	6.61–03	1.44–02	–2.180	C	*LS*
				659.536	142381.0	– 294002.86	5–5	1.52–01	9.92–04	1.08–02	–2.304	C	*LS*
				660.133	142381.8	– 293866.49	3–3	2.53–01	1.65–03	1.08–02	–2.305	C	*LS*
				660.129	142381.0	– 293866.49	5–3	1.69–02	6.61–05	7.18–04	–3.481	D–	*LS*

O III: Allowed Transitions — Continued

No.	Transition Array	Mult.	λ_{air} (Å)	λ_{vac} (Å) or σ (cm^{-1})*	E_i (cm^{-1})	E_k (cm^{-1})	g_i–g_k	A_{ki} (10^8 s^{-1})	f_{ik}	S (at. u.)	log gf	Acc.	Source
54.		$^3P^o-^3S$		*644.43*	*142383*	– 297558.66	9–3	3.26+00	6.77–03	1.29–01	–1.215	B	1,2
				644.423	142381.0	– 297558.66	5–3	1.81+00	6.77–03	7.18–02	–1.471	C+	*LS*
				644.426	142381.8	– 297558.66	3–3	1.09+00	6.77–03	4.31–02	–1.692	C+	*LS*
				644.475	142393.5	– 297558.66	1–3	3.62–01	6.77–03	1.44–02	–2.170	C	*LS*
55.		$^3P^o-^3P$		*632.94*	*142383*	*– 300376.5*	9–9	2.35–01	1.41–03	2.65–02	–1.895	D	1,2
				632.665	142381.0	– 300442.55	5–5	1.77–01	1.06–03	1.11–02	–2.275	D–	*LS*
				633.191	142381.8	– 300311.96	3–3	5.88–02	3.53–04	2.21–03	–2.975	D–	*LS*
				633.188	142381.0	– 300311.96	5–3	9.80–02	3.53–04	3.68–03	–2.753	D–	*LS*
				633.480	142381.8	– 300239.93	3–1	2.35–01	4.71–04	2.95–03	–2.850	D–	*LS*
				632.668	142381.8	– 300442.55	3–5	5.89–02	5.90–04	3.68–03	–2.752	D–	*LS*
				633.238	142393.5	– 300311.96	1–3	7.84–02	1.41–03	2.95–03	–2.850	D–	*LS*
56.	$2s2p^3$– $2s2p^23s$	$^3P^o-^3P$		*481.17*	*142381*	*– 350210.0*	9–9	1.86+01	6.47–02	9.23–01	–0.235	B	1
				480.960	142381.0	– 350298.38	5–5	1.40+01	4.86–02	3.84–01	–0.615	C+	*LS*
				481.365	142381.8	– 350124.45	3–3	4.65+00	1.62–02	7.69–02	–1.314	C+	*LS*
				481.363	142381.0	– 350124.45	5–3	7.76+00	1.62–02	1.28–01	–1.092	C+	*LS*
				481.597	142381.8	– 350024.49	3–1	1.86+01	2.16–02	1.03–01	–1.189	C+	*LS*
				480.962	142381.8	– 350298.38	3–5	4.67+00	2.70–02	1.28–01	–1.092	C+	*LS*
				481.392	142393.5	– 350124.45	1–3	6.21+00	6.47–02	1.03–01	–1.189	C+	*LS*
57.	$2s2p^3$– $2s2p^23d$	$^3P^o-^3P$		*387.56*	*142381*	*– 400406.2*	9–9	5.68+01	1.28–01	1.47+00	0.061	B	1
				387.641	142381.0	– 400351.56	5–5	4.26+01	9.59–02	6.12–01	–0.319	C+	*LS*
				387.478	142381.8	– 400460.99	3–3	1.42+01	3.20–02	1.22–01	–1.018	C+	*LS*
				387.477	142381.0	– 400460.99	5–3	2.37+01	3.20–02	2.04–01	–0.796	C+	*LS*
				387.397	142381.8	– 400514.89	3–1	5.69+01	4.26–02	1.63–01	–0.893	C+	*LS*
				387.642	142381.8	– 400351.56	3–5	1.42+01	5.33–02	2.04–01	–0.796	C+	*LS*
				387.496	142393.5	– 400460.99	1–3	1.89+01	1.28–01	1.63–01	–0.893	C+	*LS*
58.	$2s2p^3$– $2s2p^2(^2D)3s$	$^3P^o-^3D$		*397.19*	*142383*	*– 394151*	9–15	9.38+00	3.70–02	4.35–01	–0.478	C+	1
				397.114	142381.0	– 394197.9	5–7	9.38+00	3.11–02	2.03–01	–0.809	C–	*LS*
				397.227	142381.8	– 394127.3	3–5	7.03+00	2.77–02	1.09–01	–1.080	C–	*LS*
				397.321	142393.5	– 394079.4	1–3	5.20+00	3.70–02	4.83–02	–1.432	C–	*LS*
				397.225	142381.0	– 394127.3	5–5	2.34+00	5.54–03	3.63–02	–1.557	C–	*LS*
				397.302	142381.8	– 394079.4	3–3	3.90+00	9.24–03	3.63–02	–1.557	C–	*LS*
				397.301	142381.0	– 394079.4	5–3	2.60–01	3.70–04	2.42–03	–2.733	D–	*LS*
59.	$2s2p^3$– $2s2p^23d$	$^3P^o-^3D$		*379.56*	*142383*	*– 405847.7*	9–15	5.88+01	2.12–01	2.38+00	0.280	B	1
				379.508	142381.0	– 405879.97	5–7	5.88+01	1.78–01	1.11+00	–0.051	C+	*LS*
				379.581	142381.8	– 405830.38	3–5	4.41+01	1.59–01	5.95–01	–0.322	C+	*LS*
				379.639	142393.5	– 405801.50	1–3	3.26+01	2.12–01	2.64–01	–0.675	C+	*LS*
				379.580	142381.0	– 405830.38	5–5	1.47+01	3.17–02	1.98–01	–0.799	C+	*LS*
				379.622	142381.8	– 405801.50	3–3	2.45+01	5.29–02	1.98–01	–0.799	C+	*LS*
				379.621	142381.0	– 405801.50	5–3	1.63+00	2.12–03	1.32–02	–1.976	C	*LS*
60.	$2s2p^3$– $2p^4$	$^1D^o-^1D$		898.957	187054.0	– 298294.0	5–5	2.11+01	2.56–01	3.79+00	0.107	B	1,2
61.		$^3S^o-^3P$		*1152.3*	197087.7	*– 283867*	3–9	4.17+00	2.49–01	2.84+00	–0.126	B	1,2
				1153.78	197087.7	– 283759.7	3–5	4.16+00	1.38–01	1.58+00	–0.382	C+	*LS*
				1150.88	197087.7	– 283977.4	3–3	4.19+00	8.32–02	9.46–01	–0.603	C+	*LS*
				1149.63	197087.7	– 284071.9	3–1	4.20+00	2.78–02	3.15–01	–1.079	C+	*LS*

O III: Allowed Transitions — Continued

No.	Transition Array	Mult.	λ_{air} (Å)	λ_{vac} (Å) or σ (cm^{-1})*	E_i (cm^{-1})	E_k (cm^{-1})	g_i–g_k	A_{ki} (10^8 s^{-1})	f_{ik}	S (at. u.)	log gf	Acc.	Source
62.	$2s2p^3$– $2s^22p3p$	$^3S^o$–^3P		*968.16*	197087.7	– *300376.5*	3–9	1.19–02	5.00–04	4.78–03	–2.824	C+	1
				967.540	197087.7	– 300442.55	3–5	1.19–02	2.78–04	2.66–03	–3.079	D	*LS*
				968.765	197087.7	– 300311.96	3–3	1.18–02	1.67–04	1.59–03	–3.301	D	*LS*
				969.441	197087.7	– 300239.93	3–1	1.18–02	5.55–05	5.31–04	–3.779	D–	*LS*
63.	$2s2p^3$– $2s2p^23d$	$^3S^o$–^3P		*491.84*	197087.7	– *400406.2*	3–9	1.16+01	1.26–01	6.13–01	–0.422	B	1
				491.971	197087.7	– 400351.56	3–5	1.16+01	7.01–02	3.41–01	–0.677	C+	*LS*
				491.707	197087.7	– 400460.99	3–3	1.16+01	4.21–02	2.04–01	–0.899	C+	*LS*
				491.576	197087.7	– 400514.89	3–1	1.16+01	1.40–02	6.81–02	–1.376	C+	*LS*
64.	$2s2p^3$– $2s^22p3p$	$^1P^o$–^1P		1242.29	210461.8	– 290958.25	3–3	3.59–01	8.30–03	1.02–01	–1.604	B–	1,2
65.	$2s2p^3$– $2p^4$	$^1P^o$–^1D		1138.53	210461.8	– 298294.0	3–5	1.81+00	5.86–02	6.59–01	–0.755	B	1,2
66.	$2s2p^3$– $2s^22p3p$	$^1P^o$–^1D		1040.32	210461.8	– 306586.08	3–5	1.02+00	2.77–02	2.85–01	–1.080	C+	1,2
67.		$^1P^o$–^1S		967.670	210461.8	– 313802.77	3–1	2.73+00	1.28–02	1.22–01	–1.416	B–	1,2
68.	$2s2p^3$– $2p^4$	$^1P^o$–^1S		752.760	210461.8	– 343306.3	3–1	5.87+01	1.66–01	1.24+00	–0.302	B	1,2
69.	$2s^22p3s$– $2p^4$	$^3P^o$–^3P	*6110.7*	*6112.4*	*267506.7*	– *283867*	9–9	1.17–03	6.57–04	1.19–01	–2.228	B–	1,2
			6199.57	6201.28	267634.00	– 283759.7	5–5	8.43–04	4.86–04	4.95–02	–2.615	C	*LS*
			6022.32	6023.99	267377.11	– 283977.4	3–3	3.06–04	1.67–04	9.91–03	–3.301	D	*LS*
			6116.98	6118.68	267634.00	– 283977.4	5–3	4.87–04	1.64–04	1.65–02	–3.086	C–	*LS*
			5988.23	5989.89	267377.11	– 284071.9	3–1	1.25–03	2.23–04	1.32–02	–3.174	C–	*LS*
			6102.35	6104.04	267377.11	– 283759.7	3–5	2.95–04	2.74–04	1.65–02	–3.085	C–	*LS*
			5979.67	5981.33	267258.71	– 283977.4	1–3	4.17–04	6.71–04	1.32–02	–3.173	C–	*LS*
70.	$2s^22p3s$– $2s^22p3p$	$^3P^o$–^1P											
			4286.18	4287.38	267634.00	– 290958.25	5–3	3.20–07	5.29–08	3.73–06	–6.578	C	5
			4239.48	4240.68	267377.11	– 290958.25	3–3	3.62–03	9.77–04	4.09–02	–2.533	C	5
			4218.30	4219.49	267258.71	– 290958.25	1–3	4.36–04	3.49–04	4.85–03	–3.457	C	5
71.		$^3P^o$–^3D	*3762.3*	*3763.4*	*267506.7*	– *294078.4*	9–15	9.76–01	3.45–01	3.85+01	0.493	B	1,2,5
			3759.88	3760.94	267634.00	– 294223.07	5–7	9.79–01	2.91–01	1.80+01	0.162	C+	5*n*
			3754.70	3755.76	267377.11	– 294002.86	3–5	7.53–01	2.65–01	9.84+00	–0.099	C+	5*n*
			3757.23	3758.30	267258.71	– 293866.49	1–3	5.56–01	3.53–01	4.37+00	–0.452	C+	5*n*
			3791.28	3792.35	267634.00	– 294002.86	5–5	2.24–01	4.82–02	3.01+00	–0.618	C+	5*n*
			3774.03	3775.10	267377.11	– 293866.49	3–3	3.91–01	8.35–02	3.11+00	–0.601	C+	5*n*
			3810.98	3812.07	267634.00	– 293866.49	5–3	2.37–02	3.10–03	1.95–01	–1.810	C+	5*n*
72.		$^3P^o$–^3S	*3326.6*	*3327.6*	*267506.7*	– 297558.66	9–3	1.28+00	7.08–02	6.98+00	–0.196	B	1,2,5
			3340.76	3341.73	267634.00	– 297558.66	5–3	6.57–01	6.60–02	3.63+00	–0.481	C+	5*n*
			3312.33	3313.28	267377.11	– 297558.66	3–3	4.60–01	7.57–02	2.48+00	–0.644	C+	5*n*
			3299.39	3300.34	267258.71	– 297558.66	1–3	1.64–01	8.03–02	8.72–01	–1.095	C+	5*n*

O III: Allowed Transitions — Continued

No.	Transition Array	Mult.	λ_{air} (Å)	λ_{vac} (Å) or σ (cm^{-1})*	E_i (cm^{-1})	E_k (cm^{-1})	g_i-g_k	A_{ki} ($10^8\ s^{-1}$)	f_{ik}	S (at. u.)	log gf	Acc.	Source
73.		$^3P°-^3P$	*3041.4*	*3042.3*	*267506.7*	– *300376.5*	9–9	1.96+00	2.72−01	2.45+01	0.389	B	1,2,5
			3047.10	3047.99	267634.00	– 300442.55	5–5	1.49+00	2.08−01	1.04+01	0.016	C+	5*n*
			3035.41	3036.30	267377.11	– 300311.96	3–3	4.59−01	6.34−02	1.90+00	−0.720	C+	5*n*
			3059.28	3060.17	267634.00	– 300311.96	5–3	8.72−01	7.35−02	3.70+00	−0.435	C+	5*n*
			3042.07	3042.95	267377.11	– 300239.93	3–1	1.94+00	8.97−02	2.70+00	−0.570	C+	5*n*
			3023.43	3024.31	267377.11	– 300442.55	3–5	4.79−01	1.10−01	3.27+00	−0.483	C+	5*n*
			3024.54	3025.42	267258.71	– 300311.96	1–3	6.16−01	2.54−01	2.53+00	−0.596	C+	5*n*
74.		$^3P°-^1D$											
			2566.49	2567.26	267634.00	– 306586.08	5–5	2.39−04	2.36−05	9.99−04	−3.927	C	5
			2549.67	2550.44	267377.11	– 306586.08	3–5	3.69−04	5.99−05	1.51−03	−3.745	C	5
75.		$^3P°-^1S$											
			2153.30	2153.98	267377.11	– 313802.77	3–1	9.49−04	2.20−05	4.68−04	−4.180	C	5
76.	$2s^22p\,3s-$ $2s\,2p^23s$	$^3P°-^3P$		*1209.1*	*267506.7*	– *350210.0*	9–9	1.04+00	2.28−02	8.17−01	−0.688	B	1
				1209.71	267634.00	– 350298.38	5–5	7.80−01	1.71−02	3.41−01	−1.068	C+	*LS*
				1208.50	267377.11	– 350124.45	3–3	2.61−01	5.71−03	6.81−02	−1.766	C+	*LS*
				1212.26	267634.00	– 350124.45	5–3	4.30−01	5.69−03	1.14−01	−1.546	C+	*LS*
				1209.96	267377.11	– 350024.49	3–1	1.04+00	7.60−03	9.08−02	−1.642	C+	*LS*
				1205.96	267377.11	– 350298.38	3–5	2.62−01	9.53−03	1.14−01	−1.544	C+	*LS*
				1206.77	267258.71	– 350124.45	1–3	3.49−01	2.29−02	9.08−02	−1.641	C+	*LS*
77.	$2s^22p\,3s-$ $2s^22p\,4p$	$^3P°-^3D$		*1008.4*	*267506.7*	– *366670.8*	9–15	2.37+00	6.03−02	1.80+00	−0.265	B	1
				1008.38	267634.00	– 366802.62	5–7	2.37+00	5.07−02	8.41−01	−0.596	C+	*LS*
				1007.88	267377.11	– 366595.76	3–5	1.78+00	4.53−02	4.51−01	−0.867	C+	*LS*
				1007.76	267258.71	– 366488.45	1–3	1.32+00	6.04−02	2.00−01	−1.219	C+	*LS*
				1010.49	267634.00	– 366595.76	5–5	5.90−01	9.03−03	1.50−01	−1.345	C+	*LS*
				1008.97	267377.11	– 366488.45	3–3	9.87−01	1.51−02	1.50−01	−1.345	C+	*LS*
				1011.59	267634.00	– 366488.45	5–3	6.53−02	6.01−04	1.00−02	−2.522	C	*LS*
78.		$^3P°-^3S$		*995.55*	*267506.7*	– 367953.90	9–3	1.61+00	7.98−03	2.35−01	−1.144	C+	1
				996.811	267634.00	– 367953.90	5–3	8.91−01	7.97−03	1.31−01	−1.400	C−	*LS*
				994.265	267377.11	– 367953.90	3–3	5.39−01	7.99−03	7.84−02	−1.620	C−	*LS*
				993.096	267258.71	– 367953.90	1–3	1.80−01	8.00−03	2.61−02	−2.097	C−	*LS*
79.		$^3P°-^3P$		*971.23*	*267506.7*	– *370468.5*	9–9	4.93−01	6.97−03	2.01−01	−1.202	C+	1
				971.888	267634.00	– 370526.49	5–5	3.69−01	5.22−03	8.36−02	−1.583	C−	*LS*
				970.485	267377.11	– 370418.32	3–3	1.24−01	1.74−03	1.67−02	−2.281	D	*LS*
				972.911	267634.00	– 370418.32	5–3	2.04−01	1.74−03	2.79−02	−2.061	D	*LS*
				971.326	267377.11	– 370329.18	3–1	4.93−01	2.32−03	2.23−02	−2.157	D	*LS*
				969.468	267377.11	– 370526.49	3–5	1.24−01	2.91−03	2.79−02	−2.059	D	*LS*
				969.372	267258.71	– 370418.32	1–3	1.65−01	6.98−03	2.23−02	−2.156	D	*LS*
80.	$2s^22p\,3s-$ $2s\,2p^2(^2D)3s$	$^3P°-^3D$		*789.62*	*267506.7*	– *394151*	9–15	2.45+00	3.81−02	8.92−01	−0.465	B	1
				790.115	267634.00	– 394197.9	5–7	2.44+00	3.20−02	4.16−01	−0.796	C+	*LS*
				788.953	267377.11	– 394127.3	3–5	1.84+00	2.86−02	2.23−01	−1.066	C+	*LS*
				788.515	267258.71	– 394079.4	1–3	1.37+00	3.82−02	9.91−02	−1.418	C+	*LS*
				790.556	267634.00	– 394127.3	5–5	6.10−01	5.71−03	7.43−02	−1.544	C+	*LS*
				789.252	267377.11	– 394079.4	3–3	1.02+00	9.54−03	7.43−02	−1.544	C+	*LS*
				790.855	267634.00	– 394079.4	5–3	6.77−02	3.81−04	4.96−03	−2.721	C−	*LS*

O III: Allowed Transitions — Continued

No.	Transition Array	Mult.	λ_{air} (Å)	λ_{vac} (Å) or σ (cm^{-1})*	E_i (cm^{-1})	E_k (cm^{-1})	g_i–g_k	A_{ki} (10^8 s^{-1})	f_{ik}	S (at. u.)	log gf	Acc.	Source
81.	$2s^22p3s$–$2s^22p3p$	^{1}P°–^{1}P	5592.25	5593.80	273081.33	– 290958.25	3–3	3.27–01	1.53–01	8.47+00	−0.337	B	1,2,5
82.		^{1}P°–^{3}D											
			4778.43	4779.77	273081.33	– 294002.86	3–5	8.57–05	4.89–05	2.31–03	−3.833	C	5
			4809.78	4811.12	273081.33	– 293866.49	3–3	4.52–03	1.57–03	7.46–02	−2.327	C	5
83.		^{1}P°–^{3}S											
			4084.26	4085.41	273081.33	– 297558.66	3–3	9.02–04	2.26–04	9.11–03	−3.169	C	5
84.		^{1}P°–^{3}P											
			3653.77	3654.81	273081.33	– 300442.55	3–5	3.97–06	1.32–06	4.78–05	−5.401	C	5
			3671.29	3672.34	273081.33	– 300311.96	3–3	4.20–04	8.49–05	3.08–03	−3.594	C	5
			3681.03	3682.07	273081.33	– 300239.93	3–1	2.52–04	1.71–05	6.22–04	−4.290	C	5
85.		^{1}P°–^{1}D	2983.78	2984.65	273081.33	– 306586.08	3–5	2.15+00	4.78–01	1.41+01	0.156	B	1,2,5
86.		^{1}P°–^{1}S	2454.97	2455.71	273081.33	– 313802.77	3–1	3.43+00	1.03–01	2.50+00	−0.509	B–	1,2,5
87.	$2s^22p3s$–$2s^22p4p$	^{1}P°–^{1}P		1079.38	273081.33	– 365726.76	3–3	3.72+00	6.50–02	6.93–01	−0.710	B	1
88.	$2p^4$–$2s^22p3d$	^{3}P–^{3}D°	*2301.5*	*2302.2*	*283867*	*– 327303.0*	9–15	1.71–02	2.26–03	1.54–01	−1.691	C+	1,2
			2293.27	2293.97	283759.7	– 327352.17	5–7	1.73–02	1.91–03	7.20–02	−2.021	C–	*LS*
			2308.71	2309.42	283977.4	– 327278.30	3–5	1.27–02	1.69–03	3.86–02	−2.295	C–	*LS*
			2316.39	2317.10	284071.9	– 327229.25	1–3	9.32–03	2.25–03	1.71–02	−2.648	D	*LS*
			2297.16	2297.87	283759.7	– 327278.30	5–5	4.30–03	3.40–04	1.29–02	−2.770	D	*LS*
			2311.33	2312.04	283977.4	– 327229.25	3–3	7.03–03	5.63–04	1.29–02	−2.772	D	*LS*
			2299.75	2300.46	283759.7	– 327229.25	5–3	4.76–04	2.26–05	8.57–04	−3.946	D–	*LS*
89.		^{3}P–^{3}P°	*2189.4*	*2190.1*	*283867*	*– 329527.3*	9–9	1.04–02	7.50–04	4.87–02	−2.171	B–	1,2
			2187.02	2187.70	283759.7	– 329469.80	5–5	7.85–03	5.63–04	2.03–02	−2.550	C	*LS*
			2191.99	2192.67	283977.4	– 329583.89	3–3	2.60–03	1.87–04	4.05–03	−3.250	C–	*LS*
			2181.57	2182.25	283759.7	– 329583.89	5–3	4.40–03	1.88–04	6.76–03	−3.026	C–	*LS*
			2189.05	2189.73	283977.4	– 329645.14	3–1	1.04–02	2.50–04	5.41–03	−3.125	C–	*LS*
			2197.48	2198.17	283977.4	– 329469.80	3–5	2.58–03	3.11–04	6.76–03	−3.030	C–	*LS*
			2196.54	2197.22	284071.9	– 329583.89	1–3	3.45–03	7.48–04	5.41–03	−3.126	C–	*LS*
90.	$2p^4$–$2s2p^23p$	^{3}P–^{3}D°		*1100.8*	*283867*	*– 374706.6*	9–15	2.92–01	8.84–03	2.88–01	−1.099	C+	1
				1098.47	283759.7	– 374795.14	5–7	2.94–01	7.44–03	1.35–01	−1.429	C–	*LS*
				1102.70	283977.4	– 374663.52	3–5	2.18–01	6.62–03	7.21–02	−1.702	C–	*LS*
				1104.98	284071.9	– 374571.64	1–3	1.60–01	8.81–03	3.20–02	−2.055	C–	*LS*
				1100.06	283759.7	– 374663.52	5–5	7.32–02	1.33–03	2.40–02	−2.178	D	*LS*
				1103.82	283977.4	– 374571.64	3–3	1.21–01	2.20–03	2.40–02	−2.180	D	*LS*
				1101.18	283759.7	– 374571.64	5–3	8.10–03	8.84–05	1.60–03	−3.355	D–	*LS*
91.		^{3}P–^{3}P°		*1057.7*	*283867*	*– 378413.0*	9–9	1.04+00	1.75–02	5.48–01	−0.803	B	1
				1056.57	283759.7	– 378405.68	5–5	7.84–01	1.31–02	2.28–01	−1.183	C+	*LS*
				1058.87	283977.4	– 378417.84	3–3	2.60–01	4.37–03	4.57–02	−1.883	C+	*LS*
				1056.43	283759.7	– 378417.84	5–3	4.36–01	4.38–03	7.61–02	−1.660	C+	*LS*
				1058.67	283977.4	– 378435.16	3–1	1.04+00	5.82–03	6.09–02	−1.758	C+	*LS*
				1059.00	283977.4	– 378405.68	3–5	2.60–01	7.27–03	7.61–02	−1.661	C+	*LS*
				1059.93	284071.9	– 378417.84	1–3	3.45–01	1.74–02	6.09–02	−1.758	C+	*LS*

O III: Allowed Transitions — Continued

No.	Transition Array	Mult.	λ_air (Å)	λ_vac (Å) or σ (cm⁻¹)*	E_i (cm⁻¹)	E_k (cm⁻¹)	g_i–g_k	A_ki (10⁸ s⁻¹)	f_ik	S (at. u.)	log gf	Acc.	Source
92.	2p4– 2s22p4d	3P–3P°		*1032.6*	*283867*	*– 380713.7*	9–9	4.23–01	6.77–03	2.07–01	–1.215	C+	1
				1031.49	283759.7	– 380706.51	5–5	3.18–01	5.08–03	8.63–02	–1.595	C–	*LS*
				1033.69	283977.4	– 380717.92	3–3	1.05–01	1.69–03	1.73–02	–2.295	D	*LS*
				1031.37	283759.7	– 380717.92	5–3	1.77–01	1.69–03	2.88–02	–2.072	D	*LS*
				1033.49	283977.4	– 380737.00	3–1	4.22–01	2.25–03	2.30–02	–2.170	D	*LS*
				1033.81	283977.4	– 380706.51	3–5	1.05–01	2.82–03	2.88–02	–2.073	D	*LS*
				1034.70	284071.9	– 380717.92	1–3	1.40–01	6.75–03	2.30–02	–2.171	D	*LS*
93.	2s22p3p– 2s22p3d	1P–1D°	2959.69	2960.56	290958.25	– 324735.65	3–5	1.83+00	3.99–01	1.17+01	0.079	B	1,2
94.		1P–1P°	2390.43	2391.16	290958.25	– 332778.94	3–3	1.62+00	1.38–01	3.27+00	–0.381	C	1,2
95.	2s22p3p– 2s22p4s	1P–1P°		1476.87	290958.25	– 358668.90	3–3	3.05+00	9.98–02	1.46+00	–0.524	B	1
96.	2s22p3p– 2s22p4d	1P–1D°		1153.02	290958.25	– 377686.83	3–5	2.25+00	7.46–02	8.49–01	–0.650	B	1
97.	2s22p3p– 2s22p5s	1P–1P°		982.094	290958.25	– 392781.47	3–3	1.34+00	1.94–02	1.88–01	–1.235	C+	1
98.	2s22p3p– 2s22p5d	1P–1D°		902.255	290958.25	– 401791.7	3–5	1.99+00	4.04–02	3.60–01	–0.917	C+	1
99.	2s22p3p– 2s22p6d	1P–1D°		808.291	290958.25	– 414676	3–5	1.58+00	2.59–02	2.07–01	–1.110	C+	1
100.	2s22p3p– 2s2p2(2D)3p	1P–1D°		738.652	290958.25	– 426340	3–5	7.63+00	1.04–01	7.59–01	–0.506	B	1
101.		1P–1P°		719.069	290958.25	– 430027	3–3	4.36+00	3.38–02	2.40–01	–0.994	C+	1
102.	2s22p3p– 2s22p3d	3D–3F°	*3265.7*	*3266.7*	*294078.4*	*– 324690.5*	15–21	1.88+00	4.21–01	6.79+01	0.800	B	1,2
			3265.33	3266.27	294223.07	– 324839.03	7–9	1.88+00	3.87–01	2.91+01	0.432	C+	*LS*
			3260.86	3261.80	294002.86	– 324660.80	5–7	1.68+00	3.75–01	2.01+01	0.272	C+	*LS*
			3267.20	3268.15	293866.49	– 324464.88	3–5	1.58+00	4.21–01	1.36+01	0.101	C+	*LS*
			3284.45	3285.40	294223.07	– 324660.80	7–7	2.06–01	3.33–02	2.52+00	–0.632	C+	*LS*
			3281.83	3282.78	294002.86	– 324464.88	5–5	2.89–01	4.67–02	2.52+00	–0.632	C+	*LS*
			3305.73	3306.68	294223.07	– 324464.88	7–5	7.98–03	9.33–04	7.11–02	–2.185	C+	*LS*
103.		3D–3D°	*3008.9*	*3009.8*	*294078.4*	*– 327303.0*	15–15	6.11–01	8.29–02	1.23+01	0.095	B–	1,2
			3017.62	3018.49	294223.07	– 327352.17	7–7	5.38–01	7.35–02	5.11+00	–0.289	C	*LS*
			3004.34	3005.22	294002.86	– 327278.30	5–5	4.27–01	5.78–02	2.86+00	–0.539	C	*LS*
			2996.48	2997.35	293866.49	– 327229.25	3–3	4.64–01	6.25–02	1.85+00	–0.727	C	*LS*
			3024.36	3025.24	294223.07	– 327278.30	7–5	9.39–02	9.19–03	6.41–01	–1.191	C	*LS*
			3008.78	3009.66	294002.86	– 327229.25	5–3	1.53–01	1.24–02	6.16–01	–1.206	C	*LS*
			2997.69	2998.56	294002.86	– 327352.17	5–7	6.88–02	1.30–02	6.41–01	–1.188	C	*LS*
			2992.08	2992.95	293866.49	– 327278.30	3–5	9.32–02	2.08–02	6.16–01	–1.204	C	*LS*
104.		3D–3P°	*2820.1*	*2821.0*	*294078.4*	*– 329527.3*	15–9	1.77–01	1.27–02	1.76+00	–0.722	B	1,2
			2836.31	2837.14	294223.07	– 329469.80	7–5	1.46–01	1.26–02	8.23–01	–1.055	C+	*LS*
			2809.66	2810.49	294002.86	– 329583.89	5–3	1.34–01	9.53–03	4.41–01	–1.322	C+	*LS*
			2794.14	2794.96	293866.49	– 329645.14	3–1	1.82–01	7.10–03	1.96–01	–1.672	C+	*LS*
			2818.70	2819.53	294002.86	– 329469.80	5–5	2.66–02	3.17–03	1.47–01	–1.801	C+	*LS*
			2798.93	2799.76	293866.49	– 329583.89	3–3	4.52–02	5.31–03	1.47–01	–1.797	C+	*LS*
			2807.90	2808.73	293866.49	– 329469.80	3–5	1.79–03	3.53–04	9.79–03	–2.975	C–	*LS*

O III: Allowed Transitions — Continued

No.	Transition Array	Mult.	λ_{air} (Å)	λ_{vac} (Å) or σ (cm⁻¹)*	E_i (cm⁻¹)	E_k (cm⁻¹)	g_i–g_k	A_{ki} (10^8 s⁻¹)	f_{ik}	S (at. u.)	log gf	Acc.	Source
105.	$2s^22p3p-2s^22p4s$	$^3D-^3P°$		*1589.7*	*294078.4*	*– 356984.0*	15–9	4.72+00	1.07–01	8.42+00	0.207	B	1
				1589.98	294223.07	– 357117.01	7–5	3.96+00	1.07–01	3.93+00	–0.125	C+	*LS*
				1591.29	294002.86	– 356844.98	5–3	3.53+00	8.04–02	2.11+00	–0.396	C+	*LS*
				1590.59	293866.49	– 356736.30	3–1	4.71+00	5.96–02	9.36–01	–0.748	C+	*LS*
				1584.43	294002.86	– 357117.01	5–5	7.15–01	2.69–02	7.02–01	–0.871	C+	*LS*
				1587.84	293866.49	– 356844.98	3–3	1.18+00	4.47–02	7.02–01	–0.872	C+	*LS*
				1581.01	293866.49	– 357117.01	3–5	4.80–02	3.00–03	4.68–02	–2.046	C+	*LS*
106.	$2s^22p3p-2s^22p4d$	$^3D-^3F°$		*1197.3*	*294078.4*	*– 377600.1*	15–21	3.31+00	9.96–02	5.89+00	0.175	B	1
				1197.24	294223.07	– 377748.57	7–9	3.31+00	9.15–02	2.52+00	–0.193	C+	*LS*
				1196.75	294002.86	– 377562.31	5–7	2.95+00	8.86–02	1.75+00	–0.354	C+	*LS*
				1197.33	293866.49	– 377385.58	3–5	2.78+00	9.96–02	1.18+00	–0.524	C+	*LS*
				1199.91	294223.07	– 377562.31	7–7	3.67–01	7.91–03	2.19–01	–1.257	C+	*LS*
				1199.29	294002.86	– 377385.58	5–5	5.14–01	1.11–02	2.19–01	–1.256	C+	*LS*
				1202.46	294223.07	– 377385.58	7–5	1.44–02	2.23–04	6.17–03	–2.807	C–	*LS*
107.		$^3D-^3D°$		*1173.3*	*294078.4*	*– 379309.6*	15–15	5.50–01	1.13–02	6.58–01	–0.769	B	1
				1174.62	294223.07	– 379356.75	7–7	4.87–01	1.01–02	2.73–01	–1.152	C+	*LS*
				1172.47	294002.86	– 379293.03	5–5	3.84–01	7.90–03	1.53–01	–1.403	C+	*LS*
				1171.50	293866.49	– 379227.15	3–3	4.14–01	8.52–03	9.86–02	–1.592	C+	*LS*
				1175.50	294223.07	– 379293.03	7–5	8.53–02	1.26–03	3.42–02	–2.054	C+	*LS*
				1173.37	294002.86	– 379227.15	5–3	1.37–01	1.70–03	3.29–02	–2.070	C+	*LS*
				1171.59	294002.86	– 379356.75	5–7	6.15–02	1.77–03	3.42–02	–2.052	C+	*LS*
				1170.60	293866.49	– 379293.03	3–5	8.30–02	2.84–03	3.29–02	–2.069	C+	*LS*
108.	$2s^22p3p-2s^22p5s$	$^3D-^3P°$		*1020.5*	*294078.4*	*– 392070.2*	15–9	2.04+00	1.91–02	9.65–01	–0.542	B	1
				1020.55	294223.07	– 392209.53	7–5	1.72+00	1.91–02	4.50–01	–0.873	C+	*LS*
				1021.29	294002.86	– 391917.80	5–3	1.53+00	1.43–02	2.41–01	–1.144	C+	*LS*
				1020.78	293866.49	– 391830.76	3–1	2.04+00	1.06–02	1.07–01	–1.496	C+	*LS*
				1018.26	294002.86	– 392209.53	5–5	3.09–01	4.80–03	8.04–02	–1.620	C+	*LS*
				1019.87	293866.49	– 391917.80	3–3	5.12–01	7.98–03	8.04–02	–1.621	C+	*LS*
				1016.85	293866.49	– 392209.53	3–5	2.07–02	5.34–04	5.36–03	–2.796	C–	*LS*
109.	$2s^22p3p-2s^22p5d$	$^3D-^3F°$		*928.76*	*294078.4*	*– 401748*	15–21	2.52+00	4.57–02	2.10+00	–0.164	B	1
				928.763	294223.07	– 401893.2	7–9	2.52+00	4.20–02	8.98–01	–0.532	C+	*LS*
				928.309	294002.86	– 401725.6	5–7	2.25+00	4.06–02	6.21–01	–0.692	C+	*LS*
				928.908	293866.49	– 401519.8	3–5	2.12+00	4.57–02	4.19–01	–0.863	C+	*LS*
				930.211	294223.07	– 401725.6	7–7	2.80–01	3.63–03	7.79–02	–1.595	C+	*LS*
				930.086	294002.86	– 401519.8	5–5	3.92–01	5.09–03	7.79–02	–1.595	C+	*LS*
				931.995	294223.07	– 401519.8	7–5	1.10–02	1.02–04	2.20–03	–3.145	D–	*LS*
110.	$2s^22p3p-2s^22p6s$	$^3D-^3P°$		*867.98*	*294078.4*	*– 409289*	15–9	1.14+00	7.72–03	3.31–01	–0.937	C+	1
				867.972	294223.07	– 409434.1	7–5	9.56–01	7.72–03	1.54–01	–1.268	C–	*LS*
				868.637	294002.86	– 409125.7	5–3	8.52–01	5.78–03	8.27–02	–1.539	C–	*LS*
				868.160	293866.49	– 409052.6	3–1	1.14+00	4.29–03	3.67–02	–1.891	C–	*LS*
				866.317	294002.86	– 409434.1	5–5	1.72–01	1.93–03	2.76–02	–2.015	D	*LS*
				867.610	293866.49	– 409125.7	3–3	2.85–01	3.22–03	2.76–02	–2.016	D	*LS*
				865.294	293866.49	– 409434.1	3–5	1.15–02	2.15–04	1.84–03	–3.190	D–	*LS*
111.	$2s^22p3p-2s^22p6d$	$^3D-^3D°$		*825.75*	*294078.4*	*– 415180*	15–15	3.87–01	3.96–03	1.61–01	–1.227	C+	1
				826.741	294223.07	– 415180	7–7	3.43–01	3.51–03	6.69–02	–1.610	C–	*LS*
				825.238	294002.86	– 415180	5–5	2.70–01	2.75–03	3.74–02	–1.861	C–	*LS*
				824.310	293866.49	– 415180	3–3	2.92–01	2.97–03	2.42–02	–2.050	D	*LS*
				826.741	294223.07	– 415180	7–5	6.01–02	4.40–04	8.39–03	–2.511	D–	*LS*
				825.238	294002.86	– 415180	5–3	9.69–02	5.94–04	8.07–03	–2.527	D–	*LS*
				825.238	294002.86	– 415180	5–7	4.32–02	6.17–04	8.39–03	–2.510	D–	*LS*
				824.310	293866.49	– 415180	3–5	5.83–02	9.91–04	8.07–03	–2.527	D–	*LS*

O III: Allowed Transitions — Continued

No.	Transition Array	Mult.	λ_{air} (Å)	λ_{vac} (Å) or σ (cm^{-1})*	E_i (cm^{-1})	E_k (cm^{-1})	g_i-g_k	A_{ki} (10^8 s^{-1})	f_{ik}	S (at. u.)	log gf	Acc.	Source
112.	$2s^22p3p-$ $2s^22p3d$	^{3}S–^{3}P°	*3127.2*	*3128.1*	297558.66	– *329527.3*	3–9	1.38+00	6.06–01	1.87+01	0.259	B–	1,2
			3132.79	3133.70	297558.66	– 329469.80	3–5	1.37+00	3.36–01	1.04+01	0.003	C	*LS*
			3121.63	3122.54	297558.66	– 329583.89	3–3	1.38+00	2.02–01	6.24+00	–0.217	C	*LS*
			3115.67	3116.58	297558.66	– 329645.14	3–1	1.39+00	6.75–02	2.08+00	–0.693	C	*LS*
113.	$2s^22p3p-$ $2s^22p4s$	^{3}S–^{3}P°		*1682.8*	297558.66	– *356984.0*	3–9	6.53–01	8.31–02	1.38+00	–0.603	B	1
				1679.03	297558.66	– 357117.01	3–5	6.57–01	4.63–02	7.67–01	–0.858	C+	*LS*
				1686.73	297558.66	– 356844.98	3–3	6.48–01	2.76–02	4.60–01	–1.081	C+	*LS*
				1689.83	297558.66	– 356736.30	3–1	6.44–01	9.20–03	1.53–01	–1.559	C+	*LS*
114.	$2s^22p3p-$ $2s2p^23p$	^{3}S–^{3}P°		*1236.8*	297558.66	– *378413.0*	3–9	5.46–01	3.76–02	4.59–01	–0.948	C+	1
				1236.90	297558.66	– 378405.68	3–5	5.46–01	2.09–02	2.55–01	–1.203	C–	*LS*
				1236.72	297558.66	– 378417.84	3–3	5.46–01	1.25–02	1.53–01	–1.425	C–	*LS*
				1236.45	297558.66	– 378435.16	3–1	5.47–01	4.18–03	5.10–02	–1.902	C–	*LS*
115.	$2s^22p3p-$ $2s^22p4d$	^{3}S–^{3}P°		*1202.6*	297558.66	– *380713.7*	3–9	6.53–01	4.25–02	5.05–01	–0.895	B	1
				1202.68	297558.66	– 380706.51	3–5	6.53–01	2.36–02	2.80–01	–1.150	C+	*LS*
				1202.51	297558.66	– 380717.92	3–3	6.53–01	1.42–02	1.68–01	–1.372	C+	*LS*
				1202.24	297558.66	– 380737.00	3–1	6.54–01	4.72–03	5.61–02	–1.849	C+	*LS*
116.	$2s^22p3p-$ $2s^22p5s$	^{3}S–^{3}P°		*1058.1*	297558.66	– *392070.2*	3–9	3.17–01	1.60–02	1.67–01	–1.320	C+	1
				1056.51	297558.66	– 392209.53	3–5	3.18–01	8.88–03	9.27–02	–1.574	C–	*LS*
				1059.78	297558.66	– 391917.80	3–3	3.15–01	5.31–03	5.56–02	–1.798	C–	*LS*
				1060.76	297558.66	– 391830.76	3–1	3.15–01	1.77–03	1.85–02	–2.275	D	*LS*
117.	$2p^4-$ $2s2p^2(^2D)3p$	^{1}D–^{1}F°		789.229	298294.0	– 425000	5–7	7.19–01	9.41–03	1.22–01	–1.328	C+	1
118.		^{1}D–^{1}P°		759.111	298294.0	– 430027	5–3	2.12+00	1.10–02	1.38–01	–1.259	C+	1
119.	$2s^22p3p-$ $2s^22p3d$	^{3}P–^{3}D°	*3712.8*	*3713.8*	*300376.5*	– *327303.0*	9–15	9.75–01	3.36–01	3.69+01	0.480	B	1,2
			3715.09	3716.14	300442.55	– 327352.17	5–7	9.73–01	2.82–01	1.72+01	0.149	C+	*LS*
			3707.27	3708.33	300311.96	– 327278.30	3–5	7.34–01	2.52–01	9.23+00	–0.121	C+	*LS*
			3704.12	3705.17	300239.93	– 327229.25	1–3	5.45–01	3.36–01	4.10+00	–0.473	C+	*LS*
			3725.31	3726.37	300442.55	– 327278.30	5–5	2.41–01	5.02–02	3.08+00	–0.600	C+	*LS*
			3714.03	3715.08	300311.96	– 327229.25	3–3	4.06–01	8.39–02	3.08+00	–0.599	C+	*LS*
			3732.13	3733.20	300442.55	– 327229.25	5–3	2.67–02	3.34–03	2.05–01	–1.777	C+	*LS*
120.		^{3}P–^{3}P°	*3429.5*	*3430.4*	*300376.5*	– *329527.3*	9–9	5.68–01	1.00–01	1.02+01	–0.045	B	1,2
			3444.05	3445.04	300442.55	– 329469.80	5–5	4.21–01	7.48–02	4.24+00	–0.427	C+	*LS*
			3415.26	3416.24	300311.96	– 329583.89	3–3	1.44–01	2.52–02	8.48–01	–1.122	C+	*LS*
			3430.57	3431.55	300442.55	– 329583.89	5–3	2.37–01	2.50–02	1.41+00	–0.902	C+	*LS*
			3408.13	3409.11	300311.96	– 329645.14	3–1	5.79–01	3.36–02	1.13+00	–0.996	C+	*LS*
			3428.63	3429.61	300311.96	– 329469.80	3–5	1.42–01	4.18–02	1.41+00	–0.902	C+	*LS*
			3406.88	3407.86	300239.93	– 329583.89	1–3	1.93–01	1.01–01	1.13+00	–0.996	C+	*LS*
121.	$2s^22p3p-$ $2s^22p4s$	^{3}P–^{3}P°		*1766.5*	*300376.5*	– *356984.0*	9–9	3.32+00	1.55–01	8.12+00	0.145	B	1
				1764.46	300442.55	– 357117.01	5–5	2.50+00	1.17–01	3.38+00	–0.235	C+	*LS*
				1768.88	300311.96	– 356844.98	3–3	8.26–01	3.87–02	6.77–01	–0.935	C+	*LS*
				1772.97	300442.55	– 356844.98	5–3	1.37+00	3.87–02	1.13+00	–0.714	C+	*LS*
				1772.28	300311.96	– 356736.30	3–1	3.29+00	5.16–02	9.03–01	–0.811	C+	*LS*
				1760.41	300311.96	– 357117.01	3–5	8.38–01	6.49–02	1.13+00	–0.711	C+	*LS*
				1766.63	300239.93	– 356844.98	1–3	1.11+00	1.55–01	9.03–01	–0.809	C+	*LS*

O III: Allowed Transitions — Continued

No.	Transition Array	Mult.	λ_{air} (Å)	λ_{vac} (Å) or σ (cm^{-1})*	E_i (cm^{-1})	E_k (cm^{-1})	g_i-g_k	A_{ki} ($10^8 s^{-1}$)	f_{ik}	S (at. u.)	log *gf*	Acc.	Source
122.	$2s^22p3p-$ $2s2p^23p$	$^3P-^3S^o$		*1590.2*	*300376.5*	– 363263.38	9–3	5.34–01	6.75–03	3.18–01	–1.216	C+	1
				1591.83	300442.55	– 363263.38	5–3	2.96–01	6.74–03	1.77–01	–1.472	C–	*LS*
				1588.53	300311.96	– 363263.38	3–3	1.79–01	6.76–03	1.06–01	–1.693	C–	*LS*
				1586.71	300239.93	– 363263.38	1–3	5.98–02	6.77–03	3.53–02	–2.170	C–	*LS*
123.		$^3P-^3D^o$		*1345.4*	*300376.5*	*– 374706.6*	9–15	2.13+00	9.61–02	3.83+00	–0.063	B	1
				1344.94	300442.55	– 374795.14	5–7	2.13+00	8.08–02	1.79+00	–0.394	C+	*LS*
				1344.96	300311.96	– 374663.52	3–5	1.60+00	7.21–02	9.58–01	–0.665	C+	*LS*
				1345.32	300239.93	– 374571.64	1–3	1.18+00	9.61–02	4.26–01	–1.017	C+	*LS*
				1347.33	300442.55	– 374663.52	5–5	5.29–01	1.44–02	3.19–01	–1.143	C+	*LS*
				1346.63	300311.96	– 374571.64	3–3	8.83–01	2.40–02	3.19–01	–1.143	C+	*LS*
				1349.00	300442.55	– 374571.64	5–3	5.86–02	9.59–04	2.13–02	–2.319	C	*LS*
124.		$^3P-^3P^o$		*1281.5*	*300376.5*	*– 378413.0*	9–9	2.24+00	5.52–02	2.10+00	–0.304	B	1
				1282.66	300442.55	– 378405.68	5–5	1.68+00	4.14–02	8.73–01	–0.684	C+	*LS*
				1280.31	300311.96	– 378417.84	3–3	5.62–01	1.38–02	1.75–01	–1.383	C+	*LS*
				1282.46	300442.55	– 378417.84	5–3	9.32–01	1.38–02	2.91–01	–1.161	C+	*LS*
				1280.03	300311.96	– 378435.16	3–1	2.25+00	1.84–02	2.33–01	–1.257	C+	*LS*
				1280.51	300311.96	– 378405.68	3–5	5.62–01	2.30–02	2.91–01	–1.161	C+	*LS*
				1279.13	300239.93	– 378417.84	1–3	7.52–01	5.53–02	2.33–01	–1.257	C+	*LS*
125.	$2s^22p3p-$ $2s^22p4d$	$^3P-^3D^o$		*1266.9*	*300376.5*	*– 379309.6*	9–15	1.48+00	5.95–02	2.23+00	–0.271	B	1
				1267.20	300442.55	– 379356.75	5–7	1.48+00	5.00–02	1.04+00	–0.602	C+	*LS*
				1266.13	300311.96	– 379293.03	3–5	1.11+00	4.46–02	5.58–01	–0.873	C+	*LS*
				1266.03	300239.93	– 379227.15	1–3	8.26–01	5.95–02	2.48–01	–1.225	C+	*LS*
				1268.22	300442.55	– 379293.03	5–5	3.70–01	8.91–03	1.86–01	–1.351	C+	*LS*
				1267.18	300311.96	– 379227.15	3–3	6.18–01	1.49–02	1.86–01	–1.351	C+	*LS*
				1269.28	300442.55	– 379227.15	5–3	4.10–02	5.94–04	1.24–02	–2.527	C	*LS*
126.	$2s^22p3p-$ $2s^22p5s$	$^3P-^3P^o$		*1090.6*	*300376.5*	*– 392070.2*	9–9	1.62+00	2.88–02	9.31–01	–0.586	B	1
				1089.72	300442.55	– 392209.53	5–5	1.22+00	2.16–02	3.88–01	–0.966	C+	*LS*
				1091.63	300311.96	– 391917.80	3–3	4.03–01	7.20–03	7.76–02	–1.666	C+	*LS*
				1093.19	300442.55	– 391917.80	5–3	6.69–01	7.19–03	1.29–01	–1.444	C+	*LS*
				1092.67	300311.96	– 391830.76	3–1	1.61+00	9.59–03	1.03–01	–1.541	C+	*LS*
				1088.17	300311.96	– 392209.53	3–5	4.07–01	1.20–02	1.29–01	–1.442	C+	*LS*
				1090.78	300239.93	– 391917.80	1–3	5.39–01	2.88–02	1.03–01	–1.540	C+	*LS*
127.	$2s^22p3p-$ $2s^22p6s$	$^3P-^3P^o$		*918.17*	*300376.5*	*– 409288.9*	9–9	8.22–01	1.04–02	2.83–01	–1.029	C+	1
				917.502	300442.55	– 409434.1	5–5	6.18–01	7.79–03	1.18–01	–1.409	C–	*LS*
				919.002	300311.96	– 409125.7	3–3	2.05–01	2.59–03	2.35–02	–2.109	D	*LS*
				920.106	300442.55	– 409125.7	5–3	3.40–01	2.59–03	3.92–02	–1.888	C–	*LS*
				919.619	300311.96	– 409052.6	3–1	8.18–01	3.46–03	3.14–02	–1.984	C–	*LS*
				916.404	300311.96	– 409434.1	3–5	2.07–01	4.34–03	3.92–02	–1.886	C–	*LS*
				918.394	300239.93	– 409125.7	1–3	2.74–01	1.04–02	3.14–02	–1.984	C–	*LS*
128.	$2s^22p3p-$ $2s^22p6d$	$^3P-^3D^o$		*871.05*	*300376.5*	– 415180	9–15	1.04+00	1.97–02	5.08–01	–0.752	B	1
				871.555	300442.55	– 415180	5–7	1.04+00	1.65–02	2.37–01	–1.083	C+	*LS*
				870.564	300311.96	– 415180	3–5	7.80–01	1.48–02	1.27–01	–1.353	C+	*LS*
				870.019	300239.93	– 415180	1–3	5.79–01	1.97–02	5.64–02	–1.705	C+	*LS*
				871.555	300442.55	– 415180	5–5	2.59–01	2.95–03	4.23–02	–1.831	C+	*LS*
				870.564	300311.96	– 415180	3–3	4.33–01	4.92–03	4.23–02	–1.831	C+	*LS*
				871.555	300442.55	– 415180	5–3	2.88–02	1.97–04	2.82–03	–3.007	D	*LS*

O III: Allowed Transitions — Continued

No.	Transition Array	Mult.	λ_{air} (Å)	λ_{vac} (Å) or σ (cm^{-1})*	E_i (cm^{-1})	E_k (cm^{-1})	g_i-g_k	A_{ki} (10^8 s^{-1})	f_{ik}	S (at. u.)	log gf	Acc.	Source
129.	$2s^22p3p-$ $2s^22p3d$	$^1D-^1D^o$	5508.24	5509.77	306586.08	– 324735.65	5–5	1.06–01	4.80–02	4.35+00	–0.620	B	1,2
130.		$^1D-^1F^o$	3961.57	3962.69	306586.08	– 331821.44	5–7	1.25+00	4.10–01	2.67+01	0.312	B–	1,2
131.		$^1D-^1P^o$	3816.75	3817.83	306586.08	– 332778.94	5–3	9.63–02	1.26–02	7.93–01	–1.200	C	1,2
132.	$2s^22p3p-$ $2s^22p4s$	$^1D-^1P^o$		1920.02	306586.08	– 358668.90	5–3	4.65+00	1.54–01	4.88+00	–0.113	B	1
133.	$2s^22p3p-$ $2s^22p4d$	$^1D-^1D^o$		1406.45	306586.08	– 377686.83	5–5	1.35+00	4.00–02	9.25–01	–0.699	B	1
134.		$^1D-^1F^o$		1347.78	306586.08	– 380782.17	5–7	2.86+00	1.09–01	2.42+00	–0.263	B	1
135.	$2s^22p3p-$ $2s^22p5s$	$^1D-^1P^o$		1160.15	306586.08	– 392781.47	5–3	1.94+00	2.35–02	4.49–01	–0.930	C+	1
136.	$2s^22p3p-$ $2s^22p5d$	$^1D-^1D^o$		1050.36	306586.08	– 401791.7	5–5	8.08–01	1.34–02	2.31–01	–1.175	C+	1
137.		$^1D-^1F^o$		1033.12	306586.08	– 403380.0	5–7	2.30+00	5.15–02	8.75–01	–0.589	B	1
138.	$2s^22p3p-$ $2s^22p6s$	$^1D-^1P^o$		969.176	306586.08	– 409766.5	5–3	1.04+00	8.82–03	1.41–01	–1.356	C+	1
139.	$2s^22p3p-$ $2s^22p7d$	$^1D-^1F^o$		859.159	306586.08	– 422979	5–7	1.66+00	2.58–02	3.64–01	–0.890	C+	1
140.	$2s^22p3p-$ $2s2p^2(^2D)3p$	$^1D-^1F^o$		844.495	306586.08	– 425000	5–7	4.39+00	6.58–02	9.15–01	–0.483	B	1
141.	$2s^22p3p-$ $2s^22p3d$	$^1S-^1P^o$	5268.30	5269.77	313802.77	– 332778.94	1–3	3.50–01	4.37–01	7.57+00	–0.360	B	1,2
142.	$2s^22p3p-$ $2s^22p4s$	$^1S-^1P^o$	2228.16	2228.85	313802.77	– 358668.90	1–3	6.47–01	1.45–01	1.06+00	–0.840	B	1
143.	$2s^22p3p-$ $2s^22p4d$	$^1S-^1P^o$		1486.18	313802.77	– 381089.27	1–3	2.02+00	2.01–01	9.84–01	–0.696	B	1
144.	$2s^22p3p-$ $2s^22p5d$	$^1S-^1P^o$		1114.44	313802.77	– 403533.7	1–3	1.36+00	7.61–02	2.79–01	–1.118	C+	1
145.	$2s^22p3d-$ $2s^22p4p$	$^3F^o-^3D$	*2381.3*	*2382.1*	*324690.5*	*– 366670.8*	21–15	1.44+00	8.76–02	1.44+01	0.265	B	1
			2382.29	2383.02	324839.03	– 366802.62	9–7	1.32+00	8.76–02	6.18+00	–0.103	C+	*LS*
			2383.92	2384.65	324660.80	– 366595.76	7–5	1.28+00	7.78–02	4.27+00	–0.264	C+	*LS*
			2378.89	2379.62	324464.88	– 366488.45	5–3	1.45+00	7.37–02	2.88+00	–0.434	C+	*LS*
			2372.22	2372.94	324660.80	– 366802.62	7–7	1.16–01	9.80–03	5.36–01	–1.164	C+	*LS*
			2372.83	2373.56	324464.88	– 366595.76	5–5	1.63–01	1.37–02	5.36–01	–1.164	C+	*LS*
			2361.24	2361.96	324464.88	– 366802.62	5–7	3.32–03	3.89–04	1.51–02	–2.711	C	*LS*

O III: Allowed Transitions — Continued

No.	Transition Array	Mult.	λ_{air} (Å)	λ_{vac} (Å) or σ (cm^{-1})*	E_i (cm^{-1})	E_k (cm^{-1})	g_i-g_k	A_{ki} (10^8 s^{-1})	f_{ik}	S (at. u.)	log gf	Acc.	Source
146.	$2s^22p\,3d-$ $2s2p^2(^2D)3s$	$^3F^\circ-^3D$		*1439.7*	*324690.5*	*– 394151*	21–15	2.12–01	4.71–03	4.69–01	–1.005	C+	1
				1441.78	324839.03	– 394197.9	9–7	1.94–01	4.70–03	2.01–01	–1.374	C–	*LS*
				1439.54	324660.80	– 394127.3	7–5	1.89–01	4.18–03	1.39–01	–1.533	C–	*LS*
				1436.48	324464.88	– 394079.4	5–3	2.14–01	3.96–03	9.37–02	–1.703	C–	*LS*
				1438.08	324660.80	– 394197.9	7–7	1.69–02	5.25–04	1.74–02	–2.435	D	*LS*
				1435.49	324464.88	– 394127.3	5–5	2.38–02	7.37–04	1.74–02	–2.434	D	*LS*
				1434.04	324464.88	– 394197.9	5–7	4.82–04	2.08–05	4.91–04	–3.983	D–	*LS*
147.	$2s^22p\,3d-$ $2s^22p\,4p$	$^1D^\circ-^1P$	2438.81	2439.55	324735.65	– 365726.76	5–3	1.40+00	7.51–02	3.02+00	–0.425	B	1
148.		$^1D^\circ-^1D$	2165.39	2166.07	324735.65	– 370902.22	5–5	1.68–01	1.18–02	4.21–01	–1.229	C+	1
149.	$2s^22p\,3d-$ $2s2p^23s$	$^3D^\circ-^3P$	*4364.2*	*4365.5*	*327303.0*	*– 350210.0*	15–9	1.19–01	2.03–02	4.38+00	–0.516	B	1
			4356.79	4358.02	327352.17	– 350298.38	7–5	1.00–01	2.04–02	2.05+00	–0.846	C+	*LS*
			4375.88	4377.11	327278.30	– 350124.45	5–3	8.83–02	1.52–02	1.10+00	–1.119	C+	*LS*
			4385.65	4386.88	327229.25	– 350024.49	3–1	1.17–01	1.12–02	4.87–01	–1.472	C+	*LS*
			4342.81	4344.03	327278.30	– 350298.38	5–5	1.81–02	5.11–03	3.65–01	–1.593	C+	*LS*
			4366.50	4367.73	327229.25	– 350124.45	3–3	2.96–02	8.47–03	3.65–01	–1.595	C+	*LS*
			4333.58	4334.80	327229.25	– 350298.38	3–5	1.21–03	5.69–04	2.44–02	–2.768	C	*LS*
150.	$2s^22p\,3d-$ $2s^22p\,4p$	$^3D^\circ-^3D$	*2539.4*	*2540.1*	*327303.0*	*– 366670.8*	15–15	2.67–01	2.58–02	3.24+00	–0.412	B	1
			2534.06	2534.83	327352.17	– 366802.62	7–7	2.39–01	2.30–02	1.34+00	–0.793	C+	*LS*
			2542.64	2543.40	327278.30	– 366595.76	5–5	1.85–01	1.80–02	7.52–01	–1.047	C+	*LS*
			2546.41	2547.17	327229.25	– 366488.45	3–3	1.99–01	1.93–02	4.86–01	–1.237	C+	*LS*
			2547.42	2548.19	327352.17	– 366595.76	7–5	4.13–02	2.87–03	1.69–01	–1.697	C+	*LS*
			2549.59	2550.36	327278.30	– 366488.45	5–3	6.60–02	3.86–03	1.62–01	–1.714	C+	*LS*
			2529.33	2530.09	327278.30	– 366802.62	5–7	3.01–02	4.05–03	1.69–01	–1.694	C+	*LS*
			2539.47	2540.23	327229.25	– 366595.76	3–5	4.01–02	6.46–03	1.62–01	–1.713	C+	*LS*
151.		$^3D^\circ-^3P$	*2316.0*	*2316.7*	*327303.0*	*– 370468.5*	15–9	1.01+00	4.89–02	5.59+00	–0.135	B	1
			2315.48	2316.19	327352.17	– 370526.49	7–5	8.51–01	4.89–02	2.61+00	–0.466	C+	*LS*
			2317.32	2318.03	327278.30	– 370418.32	5–3	7.58–01	3.66–02	1.40+00	–0.737	C+	*LS*
			2319.48	2320.19	327229.25	– 370329.18	3–1	1.01+00	2.71–02	6.21–01	–1.090	C+	*LS*
			2311.52	2312.24	327278.30	– 370526.49	5–5	1.53–01	1.22–02	4.66–01	–1.213	C+	*LS*
			2314.69	2315.40	327229.25	– 370418.32	3–3	2.54–01	2.04–02	4.66–01	–1.214	C+	*LS*
			2308.91	2309.62	327229.25	– 370526.49	3–5	1.02–02	1.36–03	3.10–02	–2.389	C+	*LS*
152.	$2s^22p\,3d-$ $2s2p^23s$	$^3P^\circ-^3P$	*4833.6*	*4835.0*	*329527.3*	*– 350210.0*	9–9	2.42–02	8.48–03	1.21+00	–1.117	B	1
			4799.75	4801.10	329469.80	– 350298.38	5–5	1.85–02	6.41–03	5.06–01	–1.494	C+	*LS*
			4867.06	4868.42	329583.89	– 350124.45	3–3	5.93–03	2.11–03	1.01–01	–2.199	C+	*LS*
			4840.17	4841.52	329469.80	– 350124.45	5–3	1.00–02	2.12–03	1.69–01	–1.975	C+	*LS*
			4890.86	4892.22	329583.89	– 350024.49	3–1	2.34–02	2.79–03	1.35–01	–2.077	C+	*LS*
			4826.19	4827.54	329583.89	– 350298.38	3–5	6.08–03	3.54–03	1.69–01	–1.974	C+	*LS*
			4881.61	4882.98	329645.14	– 350124.45	1–3	7.84–03	8.40–03	1.35–01	–2.076	C+	*LS*
153.	$2s^22p\,3d-$ $2s^22p\,4p$	$^3P^\circ-^3D$	*2691.5*	*2692.3*	*329527.3*	*– 366670.8*	9–15	9.52–02	1.72–02	1.37+00	–0.810	B	1
			2677.81	2678.61	329469.80	– 366802.62	5–7	9.66–02	1.45–02	6.41–01	–1.138	C+	*LS*
			2701.03	2701.84	329583.89	– 366595.76	3–5	7.06–02	1.29–02	3.43–01	–1.413	C+	*LS*
			2713.39	2714.20	329645.14	– 366488.45	1–3	5.16–02	1.71–02	1.53–01	–1.767	C+	*LS*
			2692.73	2693.53	329469.80	– 366595.76	5–5	2.38–02	2.58–03	1.14–01	–1.889	C+	*LS*
			2708.89	2709.69	329583.89	– 366488.45	3–3	3.89–02	4.28–03	1.14–01	–1.892	C+	*LS*
			2700.54	2701.34	329469.80	– 366488.45	5–3	2.62–03	1.72–04	7.63–03	–3.066	C–	*LS*

O III: Allowed Transitions — Continued

No.	Transition Array	Mult.	λ_{air} (Å)	λ_{vac} (Å) or σ (cm^{-1})*	E_i (cm^{-1})	E_k (cm^{-1})	g_i–g_k	A_{ki} (10^8 s^{-1})	f_{ik}	S (at. u.)	log gf	Acc.	Source
154.		$^3P^o$–^3S	*2601.6*	*2602.4*	*329527.3*	– 367953.90	9–3	2.15+00	7.28–02	5.61+00	–0.184	B	1
			2597.70	2598.48	329469.80	– 367953.90	5–3	1.20+00	7.29–02	3.12+00	–0.438	C+	*LS*
			2605.42	2606.20	329583.89	– 367953.90	3–3	7.14–01	7.27–02	1.87+00	–0.662	C+	*LS*
			2609.59	2610.37	329645.14	– 367953.90	1–3	2.37–01	7.25–02	6.23–01	–1.139	C+	*LS*
155.		$^3P^o$–^3P	*2441.8*	*2442.5*	*329527.3*	*– 370468.5*	9–9	1.80–01	1.61–02	1.16+00	–0.839	B	1
			2434.92	2435.66	329469.80	– 370526.49	5–5	1.36–01	1.21–02	4.85–01	–1.218	C+	*LS*
			2448.17	2448.91	329583.89	– 370418.32	3–3	4.46–02	4.01–03	9.70–02	–1.920	C+	*LS*
			2441.35	2442.09	329469.80	– 370418.32	5–3	7.50–02	4.02–03	1.62–01	–1.697	C+	*LS*
			2453.53	2454.27	329583.89	– 370329.18	3–1	1.77–01	5.33–03	1.29–01	–1.796	C+	*LS*
			2441.70	2442.44	329583.89	– 370526.49	3–5	4.50–02	6.70–03	1.62–01	–1.697	C+	*LS*
			2451.85	2452.59	329645.14	– 370418.32	1–3	5.92–02	1.60–02	1.29–01	–1.795	C+	*LS*
156.		$^1F^o$–^1D	2558.04	2558.80	331821.44	– 370902.22	7–5	1.38+00	9.68–02	5.71+00	–0.169	B	1
157.		$^1P^o$–^1P	3034.22	3035.10	332778.94	– 365726.76	3–3	2.93–01	4.05–02	1.21+00	–0.916	B	1
158.		$^1P^o$–^1D	2622.29	2623.07	332778.94	– 370902.22	3–5	1.21–01	2.08–02	5.38–01	–1.206	B	1
159.	$2s2p^23s$– $2s2p^23p$	^5P–$^5D^o$	*3706.0*	*3707.0*	*338752.2*	*– 365728.0*	15–25	1.14+00	3.90–01	7.13+01	0.767	B	1
			3703.36	3704.41	338863.03	– 365857.89	7–9	1.14+00	3.01–01	2.57+01	0.323	C+	*LS*
			3698.72	3699.77	338701.98	– 365730.68	5–7	7.62–01	2.19–01	1.33+01	0.039	C+	*LS*
			3695.38	3696.43	338577.25	– 365630.40	3–5	4.01–01	1.37–01	4.99+00	–0.387	C+	*LS*
			3720.89	3721.95	338863.03	– 365730.68	7–7	3.74–01	7.76–02	6.66+00	–0.265	C+	*LS*
			3712.49	3713.55	338701.98	– 365630.40	5–5	6.59–01	1.36–01	8.32+00	–0.167	C+	*LS*
			3704.75	3705.80	338577.25	– 365561.95	3–3	8.53–01	1.75–01	6.42+00	–0.279	C+	*LS*
			3734.83	3735.89	338863.03	– 365630.40	7–5	7.40–02	1.10–02	9.51–01	–1.112	C+	*LS*
			3721.95	3723.01	338701.98	– 365561.95	5–3	2.80–01	3.49–02	2.14+00	–0.758	C+	*LS*
			3709.54	3710.60	338577.25	– 365527.08	3–1	1.13+00	7.79–02	2.85+00	–0.631	C+	*LS*
160.		^5P–$^5P^o$	*3345.8*	*3346.8*	*338752.2*	*– 368631.6*	15–15	1.49+00	2.51–01	4.14+01	0.575	B	1
			3350.92	3351.88	338863.03	– 368697.00	7–7	9.91–01	1.67–01	1.29+01	0.067	C+	*LS*
			3344.20	3345.16	338701.98	– 368595.93	5–5	1.25–01	2.09–02	1.15+00	–0.981	C+	*LS*
			3336.67	3337.63	338577.25	– 368538.65	3–3	3.76–01	6.28–02	2.07+00	–0.725	C+	*LS*
			3362.31	3363.28	338863.03	– 368595.93	7–5	6.87–01	8.31–02	6.44+00	–0.235	C+	*LS*
			3350.62	3351.58	338701.98	– 368538.65	5–3	1.12+00	1.13–01	6.21+00	–0.249	C+	*LS*
			3332.93	3333.89	338701.98	– 368697.00	5–7	5.04–01	1.17–01	6.44+00	–0.231	C+	*LS*
			3330.30	3331.26	338577.25	– 368595.93	3–5	6.81–01	1.89–01	6.21+00	–0.247	C+	*LS*
161.		^5P–$^5S^o$	*2678.2*	*2679.0*	*338752.2*	– 376079.92	15–5	3.33+00	1.19–01	1.58+01	0.253	B	1
			2686.15	2686.95	338863.03	– 376079.92	7–5	1.54+00	1.19–01	7.36+00	–0.080	C+	*LS*
			2674.58	2675.37	338701.98	– 376079.92	5–5	1.11+00	1.19–01	5.26+00	–0.224	C+	*LS*
			2665.68	2666.48	338577.25	– 376079.92	3–5	6.75–01	1.20–01	3.15+00	–0.444	C+	*LS*
162.	$2s2p^23s$– $2s2p^24p$	^5P–$^5P^o$		*993.78*	*338752.2*	*– 439378*	15–15	1.08+00	1.60–02	7.85–01	–0.620	B	1
				994.255	338863.03	– 439440.8	7–7	7.19–01	1.07–02	2.44–01	–1.127	C+	*LS*
				993.638	338701.98	– 439342.3	5–5	9.00–02	1.33–03	2.18–02	–2.176	C	*LS*
				992.917	338577.25	– 439290.6	3–3	2.71–01	4.00–03	3.92–02	–1.921	C+	*LS*
				995.230	338863.03	– 439342.3	7–5	5.02–01	5.32–03	1.22–01	–1.429	C+	*LS*
				994.148	338701.98	– 439290.6	5–3	8.09–01	7.19–03	1.18–01	–1.444	C+	*LS*
				992.666	338701.98	– 439440.8	5–7	3.61–01	7.47–03	1.22–01	–1.428	C+	*LS*
				992.408	338577.25	– 439342.3	3–5	4.88–01	1.20–02	1.18–01	–1.443	C+	*LS*

O III: Allowed Transitions — Continued

No.	Transition Array	Mult.	λ_{air} (Å)	λ_{vac} (Å) or σ (cm^{-1})*	E_i (cm^{-1})	E_k (cm^{-1})	g_i-g_k	A_{ki} ($10^8 s^{-1}$)	f_{ik}	S (at. u.)	log *gf*	Acc.	Source
163.	$2s2p^23s-2s^22p4s$	$^3P-^3P^o$	*14758*	*6774.1* cm^{-1}	*350210.0*	*– 356984.0*	9–9	7.36–03	2.40–02	1.05+01	–0.665	B	1
			14661.7	6818.63 cm^{-1}	350298.38	– 357117.01	5–5	5.63–03	1.81–02	4.38+00	–1.042	C+	*LS*
			14875.7	6720.53 cm^{-1}	350124.45	– 356844.98	3–3	1.80–03	5.96–03	8.76–01	–1.747	C+	*LS*
			15270.9	6546.60 cm^{-1}	350298.38	– 356844.98	5–3	2.77–03	5.81–03	1.46+00	–1.537	C+	*LS*
			15120.2	6611.85 cm^{-1}	350124.45	– 356736.30	3–1	6.85–03	7.82–03	1.17+00	–1.630	C+	*LS*
			14297.0	6992.56 cm^{-1}	350124.45	– 357117.01	3–5	2.02–03	1.03–02	1.46+00	–1.508	C+	*LS*
			14657.7	6820.49 cm^{-1}	350024.49	– 356844.98	1–3	2.50–03	2.42–02	1.17+00	–1.616	C+	*LS*
164.	$2s2p^23s-2s2p^23p$	$^3P-^3S^o$	*7658.7*	*7660.8*	*350210.0*	– 363263.38	9–3	1.28–01	3.75–02	8.51+00	–0.472	B	1
			7710.95	7713.07	350298.38	– 363263.38	5–3	6.96–02	3.72–02	4.73+00	–0.730	C+	*LS*
			7608.88	7610.97	350124.45	– 363263.38	3–3	4.35–02	3.77–02	2.84+00	–0.946	C+	*LS*
			7551.42	7553.50	350024.49	– 363263.38	1–3	1.48–02	3.80–02	9.45–01	–1.420	C+	*LS*
165.		$^3P-^3D^o$	*4081.0*	*4082.2*	*350210.0*	*– 374706.6*	9–15	6.02–01	2.51–01	3.03+01	0.353	B	1
			4081.02	4082.17	350298.38	– 374795.14	5–7	6.02–01	2.11–01	1.41+01	0.022	C+	*LS*
			4073.98	4075.13	350124.45	– 374663.52	3–5	4.54–01	1.88–01	7.58+00	–0.248	C+	*LS*
			4072.64	4073.79	350024.49	– 374571.64	1–3	3.37–01	2.51–01	3.37+00	–0.600	C+	*LS*
			4103.07	4104.22	350298.38	– 374663.52	5–5	1.48–01	3.74–02	2.53+00	–0.728	C+	*LS*
			4089.30	4090.45	350124.45	– 374571.64	3–3	2.49–01	6.25–02	2.53+00	–0.727	C+	*LS*
			4118.60	4119.76	350298.38	– 374571.64	5–3	1.63–02	2.48–03	1.68–01	–1.906	C+	*LS*
166.		$^3P-^3P^o$	*3544.7*	*3545.7*	*350210.0*	*– 378413.0*	9–9	4.40–01	8.28–02	8.70+00	–0.128	B	1
			3556.78	3557.79	350298.38	– 378405.68	5–5	3.26–01	6.19–02	3.62+00	–0.509	C+	*LS*
			3533.38	3534.39	350124.45	– 378417.84	3–3	1.11–01	2.08–02	7.25–01	–1.205	C+	*LS*
			3555.24	3556.26	350298.38	– 378417.84	5–3	1.82–01	2.06–02	1.21+00	–0.986	C+	*LS*
			3531.22	3532.23	350124.45	– 378435.16	3–1	4.45–01	2.77–02	9.66–01	–1.080	C+	*LS*
			3534.90	3535.91	350124.45	– 378405.68	3–5	1.11–01	3.46–02	1.21+00	–0.984	C+	*LS*
			3520.94	3521.95	350024.49	– 378417.84	1–3	1.50–01	8.34–02	9.66–01	–1.079	C+	*LS*
167.	$2s2p^23s-2s^22p4d$	$^3P-^3D^o$	*3435.5*	*3436.5*	*350210.0*	*– 379309.6*	9–15	6.45–01	1.90–01	1.94+01	0.234	B	1
			3440.36	3441.35	350298.38	– 379356.75	5–7	6.42–01	1.60–01	9.04+00	–0.098	C+	*LS*
			3427.36	3428.35	350124.45	– 379293.03	3–5	4.87–01	1.43–01	4.84+00	–0.367	C+	*LS*
			3423.36	3424.35	350024.49	– 379227.15	1–3	3.62–01	1.91–01	2.15+00	–0.719	C+	*LS*
			3447.92	3448.91	350298.38	– 379293.03	5–5	1.60–01	2.84–02	1.61+00	–0.847	C+	*LS*
			3435.12	3436.11	350124.45	– 379227.15	3–3	2.69–01	4.76–02	1.61+00	–0.846	C+	*LS*
			3455.78	3456.77	350298.38	– 379227.15	5–3	1.76–02	1.89–03	1.08–01	–2.024	C+	*LS*
168.		$^3P-^3P^o$	*3277.3*	*3278.3*	*350210.0*	*– 380713.7*	9–9	9.58–01	1.54–01	1.50+01	0.142	B	1
			3287.65	3288.59	350298.38	– 380706.51	5–5	7.12–01	1.15–01	6.24+00	–0.239	C+	*LS*
			3267.73	3268.67	350124.45	– 380717.92	3–3	2.42–01	3.87–02	1.25+00	–0.936	C+	*LS*
			3286.41	3287.36	350298.38	– 380717.92	5–3	3.96–01	3.84–02	2.08+00	–0.716	C+	*LS*
			3265.69	3266.63	350124.45	– 380737.00	3–1	9.68–01	5.16–02	1.66+00	–0.810	C+	*LS*
			3268.95	3269.89	350124.45	– 380706.51	3–5	2.41–01	6.44–02	2.08+00	–0.714	C+	*LS*
			3257.09	3258.03	350024.49	– 380717.92	1–3	3.25–01	1.55–01	1.66+00	–0.809	C+	*LS*
169.	$2s2p^23s-2s^22p5s$	$^3P-^3P^o$	*2388.2*	*2388.9*	*350210.0*	*– 392070.2*	9–9	4.86–02	4.15–03	2.94–01	–1.427	C+	1
			2385.27	2386.00	350298.38	– 392209.53	5–5	3.66–02	3.12–03	1.23–01	–1.807	C–	*LS*
			2392.00	2392.73	350124.45	– 391917.80	3–3	1.21–02	1.04–03	2.45–02	–2.507	D	*LS*
			2401.99	2402.72	350298.38	– 391917.80	5–3	1.99–02	1.03–03	4.08–02	–2.287	C–	*LS*
			2396.99	2397.72	350124.45	– 391830.76	3–1	4.81–02	1.38–03	3.27–02	–2.383	C–	*LS*
			2375.41	2376.14	350124.45	– 392209.53	3–5	1.23–02	1.74–03	4.08–02	–2.282	C–	*LS*
			2386.29	2387.02	350024.49	– 391917.80	1–3	1.62–02	4.16–03	3.27–02	–2.381	C–	*LS*

O III: Allowed Transitions — Continued

No.	Transition Array	Mult.	λ_{air} (Å)	λ_{vac} (Å) or σ (cm^{-1})*	E_i (cm^{-1})	E_k (cm^{-1})	g_i-g_k	A_{ki} (10^8 s^{-1})	f_{ik}	S (at. u.)	log gf	Acc.	Source
170.	$2s2p^23s-$ $2s^22p6d$	$^3P-^3D^o$		*1539.2*	*350210.0*	– 415180	9–15	5.02–02	2.97–03	1.36–01	–1.572	C+	1
				1541.27	350298.38	– 415180	5–7	5.00–02	2.49–03	6.33–02	–1.904	C–	*LS*
				1537.15	350124.45	– 415180	3–5	3.78–02	2.23–03	3.39–02	–2.174	C–	*LS*
				1534.79	350024.49	– 415180	1–3	2.81–02	2.98–03	1.51–02	–2.526	D	*LS*
				1541.27	350298.38	– 415180	5–5	1.25–02	4.45–04	1.13–02	–2.652	D	*LS*
				1537.15	350124.45	– 415180	3–3	2.10–02	7.44–04	1.13–02	–2.651	D	*LS*
				1541.27	350298.38	– 415180	5–3	1.39–03	2.97–05	7.53–04	–3.828	D–	*LS*
171.	$2s2p^23s-$ $2s2p^24p$	$^3P-^3S^o$		*1152.0*	*350210.0*	– 437012.5	9–3	4.65+00	3.09–02	1.05+00	–0.556	B	1
				1153.21	350298.38	– 437012.5	5–3	2.58+00	3.08–02	5.85–01	–0.812	C+	*LS*
				1150.91	350124.45	– 437012.5	3–3	1.56+00	3.09–02	3.51–01	–1.033	C+	*LS*
				1149.58	350024.49	– 437012.5	1–3	5.20–01	3.09–02	1.17–01	–1.510	C+	*LS*
172.	$2s^22p4s-$ $2s^22p4p$	$^3P^o-^3D$	*10320*	*9686.8* cm^{-1}	*356984.0*	– *366670.8*	9–15	2.03–01	5.40–01	1.65+02	0.687	B	1
			10321.8	9685.61 cm^{-1}	357117.01	– 366802.62	5–7	2.03–01	4.53–01	7.70+01	0.355	C+	*LS*
			10252.8	9750.78 cm^{-1}	356844.98	– 366595.76	3–5	1.55–01	4.08–01	4.13+01	0.087	C+	*LS*
			10251.3	9752.15 cm^{-1}	356736.30	– 366488.45	1–3	1.15–01	5.44–01	1.83+01	–0.265	C+	*LS*
			10547.0	9478.75 cm^{-1}	357117.01	– 366595.76	5–5	4.75–02	7.92–02	1.38+01	–0.402	C+	*LS*
			10366.9	9643.47 cm^{-1}	356844.98	– 366488.45	3–3	8.34–02	1.34–01	1.38+01	–0.395	C+	*LS*
			10667.8	9371.44 cm^{-1}	357117.01	– 366488.45	5–3	5.10–03	5.22–03	9.17–01	–1.583	C+	*LS*
173.		$^3P^o-^3S$	*9113.4*	*9115.9*	*356984.0*	– 367953.90	9–3	2.66–01	1.10–01	2.98+01	–0.003	B	1
			9225.21	9227.74	357117.01	– 367953.90	5–3	1.42–01	1.09–01	1.65+01	–0.264	C+	*LS*
			8999.30	9001.78	356844.98	– 367953.90	3–3	9.19–02	1.12–01	9.92+00	–0.475	C+	*LS*
			8912.12	8914.56	356736.30	– 367953.90	1–3	3.16–02	1.13–01	3.31+00	–0.948	C+	*LS*
174.		$^3P^o-^3P$	*7413.9*	*7415.9*	*356984.0*	– *370468.5*	9–9	5.38–01	4.44–01	9.74+01	0.601	B	1
			7455.36	7457.41	357117.01	– 370526.49	5–5	3.97–01	3.31–01	4.06+01	0.219	C+	*LS*
			7365.35	7367.38	356844.98	– 370418.32	3–3	1.37–01	1.12–01	8.12+00	–0.475	C+	*LS*
			7515.99	7518.06	357117.01	– 370418.32	5–3	2.15–01	1.09–01	1.35+01	–0.262	C+	*LS*
			7414.04	7416.09	356844.98	– 370329.18	3–1	5.38–01	1.48–01	1.08+01	–0.353	C+	*LS*
			7307.12	7309.13	356844.98	– 370526.49	3–5	1.41–01	1.88–01	1.35+01	–0.250	C+	*LS*
			7306.85	7308.86	356736.30	– 370418.32	1–3	1.87–01	4.50–01	1.08+01	–0.347	C+	*LS*
175.	$2s^22p4s-$ $2s2p^23d$	$^3P^o-^3P$	*2302.3*	*2303.0*	*356984.0*	– *400406.2*	9–9	1.89–02	1.50–03	1.02–01	–1.870	C+	1
			2312.25	2312.96	357117.01	– 400351.56	5–5	1.40–02	1.12–03	4.26–02	–2.252	C–	*LS*
			2292.03	2292.74	356844.98	– 400460.99	3–3	4.78–03	3.77–04	8.53–03	–2.947	D–	*LS*
			2306.42	2307.13	357117.01	– 400460.99	5–3	7.82–03	3.74–04	1.42–02	–2.728	D	*LS*
			2289.20	2289.91	356844.98	– 400514.89	3–1	1.92–02	5.03–04	1.14–02	–2.822	D	*LS*
			2297.80	2298.50	356844.98	– 400351.56	3–5	4.75–03	6.26–04	1.42–02	–2.726	D	*LS*
			2286.33	2287.04	356736.30	– 400460.99	1–3	6.42–03	1.51–03	1.14–02	–2.821	D	*LS*
176.		$^3P^o-^3D$	*2045.9*	*2046.5*	*356984.0*	– *405847.7*	9–15	4.03–02	4.21–03	2.55–01	–1.421	C+	1
			2050.08	2050.74	357117.01	– 405879.97	5–7	4.00–02	3.53–03	1.19–01	–1.753	C–	*LS*
			2040.77	2041.42	356844.98	– 405830.38	3–5	3.04–02	3.17–03	6.38–02	–2.022	C–	*LS*
			2037.45	2038.10	356736.30	– 405801.50	1–3	2.26–02	4.23–03	2.84–02	–2.374	D	*LS*
			2052.17	2052.82	357117.01	– 405830.38	5–5	9.97–03	6.30–04	2.13–02	–2.502	D	*LS*
			2041.97	2042.63	356844.98	– 405801.50	3–3	1.69–02	1.05–03	2.13–02	–2.500	D	*LS*
			2053.38	2054.04	357117.01	– 405801.50	5–3	1.11–03	4.19–05	1.42–03	–3.678	D–	*LS*
177.	$2s^22p4s-$ $2s^22p4p$	$^1P^o-^1P$	14164.7	7057.86 cm^{-1}	358668.90	– 365726.76	3–3	8.83–02	2.65–01	3.71+01	–0.099	B	1

O III: Allowed Transitions — Continued

No.	Transition Array	Mult.	λ_{air} (Å)	λ_{vac} (Å) or σ (cm^{-1})*	E_i (cm^{-1})	E_k (cm^{-1})	g_i–g_k	A_{ki} (10^8 s^{-1})	f_{ik}	S (at. u.)	log gf	Acc.	Source
178.		$^{1}P^{\circ}-{}^{1}D$	8172.15	8174.40	358668.90	– 370902.22	3–5	4.25–01	7.09–01	5.72+01	0.328	B	1
179.	$2s2p^23p-2s^22p4p$	$^{3}S^{\circ}-{}^{3}P$	*13875*	*7205.1* cm^{-1}	363263.38	– *370468.5*	3–9	4.84–04	4.19–03	5.74–01	–1.901	B	1
			13764.4	7263.11 cm^{-1}	363263.38	– 370526.49	3–5	4.96–04	2.35–03	3.19–01	–2.153	C+	*LS*
			13972.5	7154.94 cm^{-1}	363263.38	– 370418.32	3–3	4.74–04	1.39–03	1.91–01	–2.381	C+	*LS*
			14148.8	7065.80 cm^{-1}	363263.38	– 370329.18	3–1	4.56–04	4.56–04	6.38–02	–2.864	C+	*LS*
180.	$2s2p^23p-2s2p^23d$	$^{3}S^{\circ}-{}^{3}P$	*2691.5*	*2692.3*	363263.38	– *400406.2*	3–9	1.83+00	5.96–01	1.59+01	0.253	B	1
			2695.48	2696.28	363263.38	– 400351.56	3–5	1.82+00	3.31–01	8.81+00	–0.003	C+	*LS*
			2687.55	2688.34	363263.38	– 400460.99	3–3	1.84+00	1.99–01	5.28+00	–0.224	C+	*LS*
			2683.66	2684.45	363263.38	– 400514.89	3–1	1.85+00	6.64–02	1.76+00	–0.700	C+	*LS*
181.		$^{5}D^{\circ}-{}^{5}F$	*3452.9*	*3453.9*	*365728.0*	– *394680.6*	25–35	1.72+00	4.32–01	1.23+02	1.033	B	1
			3454.99	3455.98	365857.89	– 394793.28	9–11	1.72+00	3.77–01	3.85+01	0.530	C+	*LS*
			3450.91	3451.90	365730.68	– 394700.27	7–9	1.44+00	3.30–01	2.63+01	0.364	C+	*LS*
			3447.97	3448.96	365630.40	– 394624.68	5–7	1.19+00	2.96–01	1.68+01	0.171	C+	*LS*
			3446.68	3447.67	365561.95	– 394567.05	3–5	9.71–01	2.88–01	9.81+00	–0.063	C+	*LS*
			3447.15	3448.14	365527.08	– 394528.20	1–3	8.09–01	4.32–01	4.91+00	–0.364	C+	*LS*
			3466.13	3467.12	365857.89	– 394700.27	9–9	2.84–01	5.12–02	5.26+00	–0.337	C+	*LS*
			3459.94	3460.93	365730.68	– 394624.68	7–7	5.14–01	9.23–02	7.36+00	–0.190	C+	*LS*
			3454.84	3455.83	365630.40	– 394567.05	5–5	6.89–01	1.23–01	7.01+00	–0.210	C+	*LS*
			3451.30	3452.29	365561.95	– 394528.20	3–3	8.06–01	1.44–01	4.91+00	–0.365	C+	*LS*
			3475.24	3476.23	365857.89	– 394624.68	9–7	2.42–02	3.40–03	3.50–01	–1.514	C+	*LS*
			3466.85	3467.84	365730.68	– 394567.05	7–5	6.82–02	8.77–03	7.01–01	–1.212	C+	*LS*
			3459.48	3460.47	365630.40	– 394528.20	5–3	1.14–01	1.23–02	7.01–01	–1.211	C+	*LS*
182.		$^{5}D^{\circ}-{}^{5}D$	*3080.9*	*3081.8*	*365728.0*	– *398176.6*	25–25	6.41–01	9.12–02	2.31+01	0.358	B	1
			3088.04	3088.94	365857.89	– 398231.48	9–9	5.30–01	7.58–02	6.94+00	–0.166	C+	*LS*
			3083.65	3084.54	365730.68	– 398150.40	7–7	3.20–01	4.55–02	3.24+00	–0.496	C+	*LS*
			3075.13	3076.02	365630.40	– 398139.92	5–5	1.61–01	2.28–02	1.16+00	–0.942	C+	*LS*
			3068.26	3069.15	365561.95	– 398144.29	3–3	5.41–02	7.63–03	2.31–01	–1.640	C+	*LS*
			3095.79	3096.69	365857.89	– 398150.40	9–7	1.35–01	1.51–02	1.39+00	–0.866	C+	*LS*
			3084.64	3085.54	365730.68	– 398139.92	7–5	2.55–01	2.60–02	1.85+00	–0.740	C+	*LS*
			3074.72	3075.61	365630.40	– 398144.29	5–3	3.76–01	3.20–02	1.62+00	–0.796	C+	*LS*
			3068.13	3069.02	365561.95	– 398145.63	3–1	6.49–01	3.05–02	9.25–01	–1.038	C+	*LS*
			3075.95	3076.85	365730.68	– 398231.48	7–9	1.07–01	1.96–02	1.39+00	–0.863	C+	*LS*
			3074.14	3075.03	365630.40	– 398150.40	5–7	1.84–01	3.66–02	1.85+00	–0.738	C+	*LS*
			3068.67	3069.56	365561.95	– 398139.92	3–5	2.27–01	5.34–02	1.62+00	–0.795	C+	*LS*
			3064.98	3065.87	365527.08	– 398144.29	1–3	2.17–01	9.17–02	9.25–01	–1.038	C+	*LS*
183.		$^{5}D^{\circ}-{}^{5}P$	*3047.5*	*3048.4*	*365728.0*	– *398532.2*	25–15	1.20–02	1.00–03	2.51–01	–1.601	C+	1
			3063.85	3064.74	365857.89	– 398487.08	9–7	9.10–03	9.97–04	9.05–02	–2.047	C–	*LS*
			3045.43	3046.32	365730.68	– 398557.17	7–5	6.73–03	6.68–04	4.69–02	–2.330	C–	*LS*
			3032.61	3033.50	365630.40	– 398595.65	5–3	4.26–03	3.52–04	1.76–02	–2.754	D	*LS*
			3051.95	3052.84	365730.68	– 398487.08	7–7	2.39–03	3.33–04	2.35–02	–2.632	D	*LS*
			3036.16	3037.04	365630.40	– 398557.17	5–5	4.24–03	5.87–04	2.93–02	–2.533	D	*LS*
			3026.33	3027.21	365561.95	– 398595.65	3–3	5.51–03	7.57–04	2.26–02	–2.644	D	*LS*
			3042.64	3043.52	365630.40	– 398487.08	5–7	3.44–04	6.69–05	3.35–03	–3.476	D–	*LS*
			3029.86	3030.74	365561.95	– 398557.17	3–5	1.10–03	2.52–04	7.54–03	–3.122	D–	*LS*
			3023.14	3024.02	365527.08	– 398595.65	1–3	2.46–03	1.01–03	1.01–02	–2.996	D	*LS*

O III: Allowed Transitions — Continued

No.	Transition Array	Mult.	λ_{air} (Å)	λ_{vac} (Å) or σ (cm^{-1})*	E_i (cm^{-1})	E_k (cm^{-1})	g_i-g_k	A_{ki} ($10^8 s^{-1}$)	f_{ik}	S (at. u.)	log gf	Acc.	Source
184.	$2s2p^23p-2s2p^24s$	$^5D°-^5P$		*1588.5*	*365728.0*	*− 428678.8*	25−15	4.96+00	1.13−01	1.47+01	0.449	B	1
				1588.97	365857.89	− 428791.89	9−7	3.82+00	1.13−01	5.30+00	0.005	C+	*LS*
				1589.91	365730.68	− 428627.20	7−5	2.77+00	7.50−02	2.75+00	−0.280	C+	*LS*
				1590.57	365630.40	− 428500.87	5−3	1.73+00	3.93−02	1.03+00	−0.706	C+	*LS*
				1585.76	365730.68	− 428791.89	7−7	9.97−01	3.76−02	1.37+00	−0.580	C+	*LS*
				1587.38	365630.40	− 428627.20	5−5	1.74+00	6.57−02	1.72+00	−0.483	C+	*LS*
				1588.84	365561.95	− 428500.87	3−3	2.23+00	8.44−02	1.32+00	−0.597	C+	*LS*
				1583.24	365630.40	− 428791.89	5−7	1.43−01	7.53−03	1.96−01	−1.424	C+	*LS*
				1585.66	365561.95	− 428627.20	3−5	4.49−01	2.82−02	4.41−01	−1.073	C+	*LS*
				1587.96	365527.08	− 428500.87	1−3	9.93−01	1.13−01	5.89−01	−0.949	C+	*LS*
185.	$2s2p^23p-2s2p^24d$	$^5D°-^5P$		*1183.4*	*365728.0*	*− 450227.5*	25−15	1.64−01	2.07−03	2.01−01	−1.287	C+	1
				1185.96	365857.89	− 450177.71	9−7	1.26−01	2.06−03	7.25−02	−1.731	C−	*LS*
				1183.15	365730.68	− 450250.84	7−5	9.20−02	1.38−03	3.76−02	−2.015	C−	*LS*
				1180.99	365630.40	− 450304.93	5−3	5.78−02	7.25−04	1.41−02	−2.441	D	*LS*
				1184.17	365730.68	− 450177.71	7−7	3.28−02	6.89−04	1.88−02	−2.317	D	*LS*
				1181.75	365630.40	− 450250.84	5−5	5.77−02	1.21−03	2.35−02	−2.219	D	*LS*
				1180.04	365561.95	− 450304.93	3−3	7.45−02	1.56−03	1.81−02	−2.331	D	*LS*
				1182.77	365630.40	− 450177.71	5−7	4.70−03	1.38−04	2.69−03	−3.161	D−	*LS*
				1180.79	365561.95	− 450250.84	3−5	1.49−02	5.18−04	6.04−03	−2.808	D−	*LS*
				1179.55	365527.08	− 450304.93	1−3	3.32−02	2.07−03	8.06−03	−2.683	D−	*LS*
186.	$2s^22p4p-2s^22p4d$	$^1P-^1D°$	8358.86	8361.16	365726.76	− 377686.83	3−5	3.93−01	6.87−01	5.67+01	0.314	B	1
187.		$^1P-^1P°$	6507.55	6509.35	365726.76	− 381089.27	3−3	3.98−01	2.53−01	1.63+01	−0.120	B	1
188.	$2s^22p4p-2s^22p5s$	$^1P-^1P°$	3695.16	3696.21	365726.76	− 392781.47	3−3	7.76−01	1.59−01	5.80+00	−0.322	B	1
189.	$2s^22p4p-2s^22p5d$	$^1P-^1D°$	2771.96	2772.78	365726.76	− 401791.7	3−5	1.91−01	3.67−02	1.01+00	−0.958	B	1
190.	$2s^22p4p-2s^22p6s$	$^1P-^1P°$	2269.97	2270.68	365726.76	− 409766.5	3−3	3.78−01	2.92−02	6.54−01	−1.058	B	1
191.	$2s^22p4p-2s^22p6d$	$^1P-^1D°$	2042.28	2042.93	365726.76	− 414676	3−5	2.18−01	2.27−02	4.58−01	−1.167	C+	1
192.	$2s^22p4p-2s2p^2(^2D)3p$	$^1P-^1D°$		1649.80	365726.76	− 426340	3−5	9.05−02	6.16−03	1.00−01	−1.734	C+	1
193.	$2s^22p4p-2s2p^23p$	$^3D-^3D°$	*12441*	*8035.7* cm^{-1}	*366670.8*	*− 374706.6*	15−15	2.15−03	4.99−03	3.06+00	−1.126	B	1
			12508.3	7992.52 cm^{-1}	366802.62	− 374795.14	7−7	1.88−03	4.41−03	1.27+00	−1.511	C+	*LS*
			12391.6	8067.76 cm^{-1}	366595.76	− 374663.52	5−5	1.51−03	3.48−03	7.11−01	−1.759	C+	*LS*
			12368.0	8083.19 cm^{-1}	366488.45	− 374571.64	3−3	1.64−03	3.76−03	4.59−01	−1.947	C+	*LS*
			12717.7	7860.90 cm^{-1}	366802.62	− 374663.52	7−5	3.14−04	5.44−04	1.59−01	−2.420	C+	*LS*
			12534.4	7975.88 cm^{-1}	366595.76	− 374571.64	5−3	5.25−04	7.42−04	1.53−01	−2.430	C+	*LS*
			12192.7	8199.38 cm^{-1}	366595.76	− 374795.14	5−7	2.54−04	7.94−04	1.59−01	−2.401	C+	*LS*
			12229.0	8175.07 cm^{-1}	366488.45	− 374663.52	3−5	3.39−04	1.27−03	1.53−01	−2.420	C+	*LS*
194.	$2s^22p4p-2s^22p4d$	$^3D-^3F°$	*9147.3*	*9149.8*	*366670.8*	*− 377600.1*	15−21	4.06−01	7.13−01	3.22+02	1.029	B	1
			9133.29	9135.80	366802.62	− 377748.57	7−9	4.08−01	6.56−01	1.38+02	0.662	C+	*LS*
			9116.14	9118.64	366595.76	− 377562.31	5−7	3.64−01	6.36−01	9.54+01	0.502	C+	*LS*
			9174.21	9176.73	366488.45	− 377385.58	3−5	3.38−01	7.11−01	6.44+01	0.329	C+	*LS*
			9291.40	9293.95	366802.62	− 377562.31	7−7	4.32−02	5.59−02	1.20+01	−0.408	C+	*LS*
			9265.45	9268.00	366595.76	− 377385.58	5−5	6.09−02	7.84−02	1.20+01	−0.407	C+	*LS*
			9446.56	9449.15	366802.62	− 377385.58	7−5	1.62−03	1.55−03	3.37−01	−1.965	C+	*LS*

O III: Allowed Transitions — Continued

No.	Transition Array	Mult.	λ_{air} (Å)	λ_{vac} (Å) or σ (cm^{-1})*	E_i (cm^{-1})	E_k (cm^{-1})	g_i-g_k	A_{ki} (10^8 s^{-1})	f_{ik}	S (at. u.)	log gf	Acc.	Source
195.	$2s^22p4p-$ $2s2p^23p$	$^3D-^3P^o$	*8514.0*	*8516.3*	*366670.8*	*– 378413.0*	15–9	4.94–04	3.22–04	1.36–01	–2.316	C+	1
			8616.05	8618.42	366802.62	– 378405.68	7–5	4.01–04	3.18–04	6.32–02	–2.652	C–	*LS*
			8456.42	8458.75	366595.76	– 378417.84	5–3	3.78–04	2.43–04	3.39–02	–2.915	C–	*LS*
			8368.21	8370.51	366488.45	– 378435.16	3–1	5.21–04	1.82–04	1.51–02	–3.262	D	*LS*
			8465.13	8467.46	366595.76	– 378405.68	5–5	7.54–05	8.10–05	1.13–02	–3.392	D	*LS*
			8380.36	8382.66	366488.45	– 378417.84	3–3	1.30–04	1.36–04	1.13–02	–3.388	D	*LS*
			8388.91	8391.21	366488.45	– 378405.68	3–5	5.17–06	9.09–06	7.53–04	–4.565	D–	*LS*
196.	$2s^22p4p-$ $2s^22p4d$	$^3D-^3D^o$	*7910.0*	*7912.2*	*366670.8*	*– 379309.6*	15–15	1.47–01	1.38–01	5.40+01	0.317	B	1
			7963.32	7965.51	366802.62	– 379356.75	7–7	1.28–01	1.22–01	2.24+01	–0.068	C+	*LS*
			7873.54	7875.71	366595.76	– 379293.03	5–5	1.04–01	9.67–02	1.25+01	–0.316	C+	*LS*
			7847.94	7850.09	366488.45	– 379227.15	3–3	1.13–01	1.05–01	8.10+00	–0.504	C+	*LS*
			8003.94	8006.14	366802.62	– 379293.03	7–5	2.22–02	1.52–02	2.81+00	–0.972	C+	*LS*
			7914.61	7916.79	366595.76	– 379227.15	5–3	3.68–02	2.07–02	2.70+00	–0.984	C+	*LS*
			7834.23	7836.38	366595.76	– 379356.75	5–7	1.69–02	2.18–02	2.81+00	–0.963	C+	*LS*
			7807.56	7809.71	366488.45	– 379293.03	3–5	2.30–02	3.50–02	2.70+00	–0.978	C+	*LS*
197.		$^3D-^3P^o$	*7119.1*	*7121.1*	*366670.8*	*– 380713.7*	15–9	6.27–02	2.86–02	1.00+01	–0.368	B	1
			7190.25	7192.23	366802.62	– 380706.51	7–5	5.11–02	2.83–02	4.69+00	–0.703	C+	*LS*
			7079.12	7081.07	366595.76	– 380717.92	5–3	4.78–02	2.16–02	2.51+00	–0.967	C+	*LS*
			7016.32	7018.26	366488.45	– 380737.00	3–1	6.55–02	1.61–02	1.12+00	–1.316	C+	*LS*
			7084.84	7086.80	366595.76	– 380706.51	5–5	9.54–03	7.18–03	8.37–01	–1.445	C+	*LS*
			7025.73	7027.67	366488.45	– 380717.92	3–3	1.63–02	1.21–02	8.37–01	–1.441	C+	*LS*
			7031.37	7033.31	366488.45	– 380706.51	3–5	6.51–04	8.04–04	5.58–02	–2.618	C+	*LS*
198.	$2s^22p4p-$ $2s^22p5s$	$^3D-^3P^o$	*3936.0*	*3937.1*	*366670.8*	*– 392070.2*	15–9	1.18+00	1.65–01	3.20+01	0.392	B	1
			3934.82	3935.94	366802.62	– 392209.53	7–5	9.93–01	1.65–01	1.49+01	0.062	C+	*LS*
			3948.01	3949.13	366595.76	– 391917.80	5–3	8.78–01	1.23–01	8.00+00	–0.211	C+	*LS*
			3944.85	3945.97	366488.45	– 391830.76	3–1	1.17+00	9.12–02	3.55+00	–0.563	C+	*LS*
			3903.04	3904.15	366595.76	– 392209.53	5–5	1.82–01	4.15–02	2.67+00	–0.683	C+	*LS*
			3931.35	3932.46	366488.45	– 391917.80	3–3	2.96–01	6.87–02	2.67+00	–0.686	C+	*LS*
			3886.76	3887.86	366488.45	– 392209.53	3–5	1.23–02	4.63–03	1.78–01	–1.857	C+	*LS*
199.	$2s^22p4p-$ $2s^22p5d$	$^3D-^3F^o$	*2850.0*	*2850.8*	*366670.8*	*– 401748*	15–21	3.25–01	5.55–02	7.80+00	–0.080	B	1
			2848.93	2849.77	366802.62	– 401893.2	7–9	3.26–01	5.09–02	3.34+00	–0.448	C+	*LS*
			2845.75	2846.58	366595.76	– 401725.6	5–7	2.90–01	4.93–02	2.31+00	–0.608	C+	*LS*
			2853.75	2854.59	366488.45	– 401519.8	3–5	2.72–01	5.54–02	1.56+00	–0.780	C+	*LS*
			2862.60	2863.44	366802.62	– 401725.6	7–7	3.58–02	4.39–03	2.90–01	–1.512	C+	*LS*
			2862.52	2863.36	366595.76	– 401519.8	5–5	5.01–02	6.15–03	2.90–01	–1.512	C+	*LS*
			2879.57	2880.42	366802.62	– 401519.8	7–5	1.39–03	1.23–04	8.18–03	–3.064	C–	*LS*
200.	$2s^22p4p-$ $2s^22p6s$	$^3D-^3P^o$	*2345.7*	*2346.4*	*366670.8*	*– 409289*	15–9	5.67–01	2.81–02	3.25+00	–0.376	B	1
			2344.97	2345.68	366802.62	– 409434.1	7–5	4.77–01	2.81–02	1.52+00	–0.707	C+	*LS*
			2350.57	2351.28	366595.76	– 409125.7	5–3	4.23–01	2.10–02	8.12–01	–0.979	C+	*LS*
			2348.68	2349.39	366488.45	– 409052.6	3–1	5.65–01	1.56–02	3.61–01	–1.331	C+	*LS*
			2333.64	2334.36	366595.76	– 409434.1	5–5	8.64–02	7.05–03	2.71–01	–1.453	C+	*LS*
			2344.65	2345.37	366488.45	– 409125.7	3–3	1.42–01	1.17–02	2.71–01	–1.455	C+	*LS*
			2327.81	2328.52	366488.45	– 409434.1	3–5	5.80–03	7.85–04	1.81–02	–2.628	C	*LS*
201.	$2s^22p4p-$ $2s^22p6d$	$^3D-^3D^o$	*2060.8*	*2061.5*	*366670.8*	– 415180	15–15	5.23–02	3.33–03	3.39–01	–1.302	C+	1
			2066.42	2067.08	366802.62	– 415180	7–7	4.61–02	2.95–03	1.40–01	–1.685	C–	*LS*
			2057.62	2058.28	366595.76	– 415180	5–5	3.66–02	2.32–03	7.86–02	–1.935	C–	*LS*
			2053.09	2053.74	366488.45	– 415180	3–3	3.97–02	2.51–03	5.08–02	–2.124	C–	*LS*
			2066.42	2067.08	366802.62	– 415180	7–5	8.09–03	3.70–04	1.76–02	–2.587	D	*LS*
			2057.62	2058.28	366595.76	– 415180	5–3	1.31–02	5.00–04	1.69–02	–2.602	D	*LS*
			2057.62	2058.28	366595.76	– 415180	5–7	5.85–03	5.20–04	1.76–02	–2.585	D	*LS*
			2053.09	2053.74	366488.45	– 415180	3–5	7.93–03	8.35–04	1.69–02	–2.601	D	*LS*

O III: Allowed Transitions — Continued

No.	Transition Array	Mult.	λ_{air} (Å)	λ_{vac} (Å) or σ (cm^{-1})*	E_i (cm^{-1})	E_k (cm^{-1})	g_i-g_k	A_{ki} (10^8 s^{-1})	f_{ik}	S (at. u.)	log gf	Acc.	Source
202.	$2s2p^23p-$ $2s2p^23d$	$^5P^o-^5D$	*3383.7*	*3384.7*	*368631.6*	*– 398176.6*	15–25	1.48+00	4.23–01	7.06+01	0.802	B	1
			3384.90	3385.87	368697.00	– 398231.48	7–9	1.48+00	3.26–01	2.54+01	0.358	C+	*LS*
			3382.61	3383.58	368595.93	– 398150.40	5–7	9.86–01	2.37–01	1.32+01	0.073	C+	*LS*
			3377.26	3378.23	368538.65	– 398139.92	3–5	5.20–01	1.48–01	4.95+00	–0.352	C+	*LS*
			3394.22	3395.19	368697.00	– 398150.40	7–7	4.88–01	8.43–02	6.59+00	–0.229	C+	*LS*
			3383.81	3384.78	368595.93	– 398139.92	5–5	8.62–01	1.48–01	8.24+00	–0.131	C+	*LS*
			3376.76	3377.73	368538.65	– 398144.29	3–3	1.12+00	1.91–01	6.36+00	–0.243	C+	*LS*
			3395.43	3396.40	368697.00	– 398139.92	7–5	9.75–02	1.20–02	9.42–01	–1.074	C+	*LS*
			3383.31	3384.28	368595.93	– 398144.29	5–3	3.70–01	3.81–02	2.12+00	–0.721	C+	*LS*
			3376.61	3377.58	368538.65	– 398145.63	3–1	1.49+00	8.47–02	2.83+00	–0.595	C+	*LS*
203.		$^5P^o-^5P$	*3343.5*	*3344.4*	*368631.6*	*– 398532.2*	15–15	1.05+00	1.75–01	2.89+01	0.420	B	1
			3355.86	3356.82	368697.00	– 398487.08	7–7	6.89–01	1.16–01	9.00+00	–0.089	C+	*LS*
			3336.69	3337.65	368595.93	– 398557.17	5–5	8.77–02	1.46–02	8.04–01	–1.136	C+	*LS*
			3326.06	3327.01	368538.65	– 398595.65	3–3	2.65–01	4.40–02	1.45+00	–0.879	C+	*LS*
			3347.98	3348.94	368697.00	– 398557.17	7–5	4.86–01	5.83–02	4.50+00	–0.389	C+	*LS*
			3332.41	3333.36	368595.93	– 398595.65	5–3	7.92–01	7.91–02	4.34+00	–0.403	C+	*LS*
			3344.51	3345.47	368595.93	– 398487.08	5–7	3.48–01	8.17–02	4.50+00	–0.389	C+	*LS*
			3330.32	3331.28	368538.65	– 398557.17	3–5	4.76–01	1.32–01	4.34+00	–0.403	C+	*LS*
204.	$2s2p^23p-$ $2s2p^24s$	$^5P^o-^5P$		*1665.4*	*368631.6*	*– 428678.8*	15–15	2.15+00	8.94–02	7.35+00	0.127	B	1
				1664.03	368697.00	– 428791.89	7–7	1.44+00	5.96–02	2.29+00	–0.380	C+	*LS*
				1665.80	368595.93	– 428627.20	5–5	1.79–01	7.44–03	2.04–01	–1.429	C+	*LS*
				1667.72	368538.65	– 428500.87	3–3	5.35–01	2.23–02	3.67–01	–1.174	C+	*LS*
				1668.61	368697.00	– 428627.20	7–5	9.97–01	2.97–02	1.14+00	–0.682	C+	*LS*
				1669.31	368595.93	– 428500.87	5–3	1.60+00	4.01–02	1.10+00	–0.698	C+	*LS*
				1661.24	368595.93	– 428791.89	5–7	7.22–01	4.18–02	1.14+00	–0.680	C+	*LS*
				1664.21	368538.65	– 428627.20	3–5	9.69–01	6.71–02	1.10+00	–0.696	C+	*LS*
205.	$2s2p^23p-$ $2s2p^24d$	$^5P^o-^5P$		*1225.6*	*368631.6*	*– 450227.5*	15–15	9.96–01	2.24–02	1.36+00	–0.473	B	1
				1227.28	368697.00	– 450177.71	7–7	6.61–01	1.49–02	4.22–01	–0.981	C+	*LS*
				1224.67	368595.93	– 450250.84	5–5	8.32–02	1.87–03	3.77–02	–2.029	C	*LS*
				1223.00	368538.65	– 450304.93	3–3	2.51–01	5.62–03	6.79–02	–1.773	C+	*LS*
				1226.18	368697.00	– 450250.84	7–5	4.64–01	7.47–03	2.11–01	–1.281	C+	*LS*
				1223.86	368595.93	– 450304.93	5–3	7.50–01	1.01–02	2.04–01	–1.296	C+	*LS*
				1225.76	368595.93	– 450177.71	5–7	3.32–01	1.05–02	2.11–01	–1.281	C+	*LS*
				1223.81	368538.65	– 450250.84	3–5	4.50–01	1.68–02	2.04–01	–1.296	C+	*LS*
206.	$2s^22p4p-$ $2s2p^23p$	$^3S-^3P^o$	*9558.4*	*9561.0*	367953.90	*– 378413.0*	3–9	5.37–02	2.21–01	2.08+01	–0.179	B	1
			9565.12	9567.75	367953.90	– 378405.68	3–5	5.36–02	1.23–01	1.16+01	–0.435	C+	*LS*
			9554.01	9556.63	367953.90	– 378417.84	3–3	5.38–02	7.36–02	6.95+00	–0.656	C+	*LS*
			9538.22	9540.84	367953.90	– 378435.16	3–1	5.41–02	2.46–02	2.32+00	–1.132	C+	*LS*
207.	$2s^22p4p-$ $2s^22p4d$	$^3S-^3P^o$	*7835.0*	*7837.1*	367953.90	*– 380713.7*	3–9	3.08–01	8.52–01	6.59+01	0.407	B	1
			7839.38	7841.53	367953.90	– 380706.51	3–5	3.08–01	4.73–01	3.66+01	0.152	C+	*LS*
			7832.37	7834.52	367953.90	– 380717.92	3–3	3.09–01	2.84–01	2.20+01	–0.070	C+	*LS*
			7820.68	7822.83	367953.90	– 380737.00	3–1	3.10–01	9.48–02	7.32+00	–0.546	C+	*LS*
208.	$2s^22p4p-$ $2s^22p5s$	$^3S-^3P^o$	*4145.4*	*4146.6*	367953.90	*– 392070.2*	3–9	1.70–01	1.32–01	5.38+00	–0.404	B	1
			4121.59	4122.75	367953.90	– 392209.53	3–5	1.73–01	7.35–02	2.99+00	–0.657	C+	*LS*
			4171.77	4172.94	367953.90	– 391917.80	3–3	1.67–01	4.36–02	1.79+00	–0.884	C+	*LS*
			4186.98	4188.16	367953.90	– 391830.76	3–1	1.65–01	1.45–02	5.98–01	–1.362	C+	*LS*

O III: Allowed Transitions — Continued

No.	Transition Array	Mult.	λ_{air} (Å)	λ_{vac} (Å) or σ (cm^{-1})*	E_i (cm^{-1})	E_k (cm^{-1})	g_i-g_k	A_{ki} ($10^8 s^{-1}$)	f_{ik}	S (at. u.)	log gf	Acc.	Source
209.	$2s^22p4p-2s^22p6s$	$^3S-^3P^o$	2418.5	2419.3	367953.90	– 409289	3–9	8.97–02	2.36–02	5.64–01	–1.150	B	1
			2410.06	2410.79	367953.90	– 409434.1	3–5	9.07–02	1.32–02	3.13–01	–1.404	C+	LS
			2428.11	2428.85	367953.90	– 409125.7	3–3	8.87–02	7.84–03	1.88–01	–1.629	C+	LS
			2432.43	2433.17	367953.90	– 409052.6	3–1	8.82–02	2.61–03	6.26–02	–2.107	C+	LS
210.	$2s^22p4p-2s2p^23p$	$^3P-^3D^o$	23589	4238.1 cm^{-1}	370468.5	– 374706.6	9–15	2.15–03	2.99–02	2.09+01	–0.570	B	1
			23420.2	4268.65 cm^{-1}	370526.49	– 374795.14	5–7	2.20–03	2.53–02	9.75+00	–0.898	C+	LS
			23549.6	4245.20 cm^{-1}	370418.32	– 374663.52	3–5	1.62–03	2.24–02	5.22+00	–1.172	C+	LS
			23564.8	4242.46 cm^{-1}	370329.18	– 374571.64	1–3	1.20–03	2.99–02	2.32+00	–1.524	C+	LS
			24165.3	4137.03 cm^{-1}	370526.49	– 374663.52	5–5	5.00–04	4.38–03	1.74+00	–1.660	C+	LS
			24070.6	4153.32 cm^{-1}	370418.32	– 374571.64	3–3	8.43–04	7.32–03	1.74+00	–1.658	C+	LS
			24714.2	4045.15 cm^{-1}	370526.49	– 374571.64	5–3	5.19–05	2.85–04	1.16–01	–2.846	C+	LS
211.		$^3P-^3P^o$	12584	7944.5 cm^{-1}	370468.5	– 378413.0	9–9	3.98–02	9.45–02	3.52+01	–0.071	B	1
			12688.2	7879.19 cm^{-1}	370526.49	– 378405.68	5–5	2.91–02	7.03–02	1.47+01	–0.454	C+	LS
			12497.3	7999.52 cm^{-1}	370418.32	– 378417.84	3–3	1.02–02	2.38–02	2.93+00	–1.147	C+	LS
			12668.6	7891.35 cm^{-1}	370526.49	– 378417.84	5–3	1.62–02	2.35–02	4.89+00	–0.931	C+	LS
			12470.3	8016.84 cm^{-1}	370418.32	– 378435.16	3–1	4.09–02	3.18–02	3.91+00	–1.021	C+	LS
			12516.4	7987.36 cm^{-1}	370418.32	– 378405.68	3–5	1.01–02	3.96–02	4.89+00	–0.926	C+	LS
			12359.6	8088.66 cm^{-1}	370329.18	– 378417.84	1–3	1.40–02	9.62–02	3.91+00	–1.017	C+	LS
212.	$2s^22p4p-2s^22p4d$	$^3P-^3D^o$	11308	8841.1 cm^{-1}	370468.5	– 379309.6	9–15	1.50–01	4.80–01	1.61+02	0.636	B	1
			11321.6	8830.26 cm^{-1}	370526.49	– 379356.75	5–7	1.50–01	4.03–01	7.51+01	0.304	C+	LS
			11264.9	8874.71 cm^{-1}	370418.32	– 379293.03	3–5	1.14–01	3.62–01	4.02+01	0.035	C+	LS
			11235.4	8897.97 cm^{-1}	370329.18	– 379227.15	1–3	8.51–02	4.83–01	1.79+01	–0.316	C+	LS
			11403.9	8766.54 cm^{-1}	370526.49	– 379293.03	5–5	3.66–02	7.14–02	1.34+01	–0.447	C+	LS
			11349.1	8808.83 cm^{-1}	370418.32	– 379227.15	3–3	6.20–02	1.20–01	1.34+01	–0.445	C+	LS
			11490.2	8700.66 cm^{-1}	370526.49	– 379227.15	5–3	3.98–03	4.73–03	8.94–01	–1.626	C+	LS
213.		$^3P-^3P^o$	9758.0	9760.7	370468.5	– 380713.7	9–9	5.02–02	7.16–02	2.07+01	–0.191	B	1
			9820.47	9823.16	370526.49	– 380706.51	5–5	3.69–02	5.34–02	8.63+00	–0.574	C+	LS
			9706.45	9709.11	370418.32	– 380717.92	3–3	1.27–02	1.80–02	1.73+00	–1.267	C+	LS
			9809.48	9812.17	370526.49	– 380717.92	5–3	2.06–02	1.78–02	2.88+00	–1.050	C+	LS
			9688.50	9691.16	370418.32	– 380737.00	3–1	5.13–02	2.41–02	2.30+00	–1.142	C+	LS
			9717.22	9719.88	370418.32	– 380706.51	3–5	1.27–02	3.00–02	2.88+00	–1.046	C+	LS
			9623.17	9625.81	370329.18	– 380717.92	1–3	1.74–02	7.26–02	2.30+00	–1.139	C+	LS
214.	$2s^22p4p-2s^22p5s$	$^3P-^3P^o$	4628.0	4629.3	370468.5	– 392070.2	9–9	8.71–01	2.80–01	3.83+01	0.401	B	1
			4610.61	4611.90	370526.49	– 392209.53	5–5	6.60–01	2.10–01	1.60+01	0.022	C+	LS
			4649.97	4651.28	370418.32	– 391917.80	3–3	2.15–01	6.96–02	3.19+00	–0.681	C+	LS
			4673.49	4674.80	370526.49	– 391917.80	5–3	3.52–01	6.92–02	5.32+00	–0.461	C+	LS
			4668.88	4670.18	370418.32	– 391830.76	3–1	8.48–01	9.24–02	4.26+00	–0.557	C+	LS
			4587.72	4589.01	370418.32	– 392209.53	3–5	2.23–01	1.17–01	5.32+00	–0.453	C+	LS
			4630.77	4632.07	370329.18	– 391917.80	1–3	2.90–01	2.79–01	4.26+00	–0.554	C+	LS
215.	$2s^22p4p-2s^22p6s$	$^3P-^3P^o$	2575.2	2576.0	370468.5	– 409289	9–9	3.69–01	3.66–02	2.80+00	–0.482	B	1
			2569.42	2570.19	370526.49	– 409434.1	5–5	2.78–01	2.75–02	1.16+00	–0.861	C+	LS
			2582.71	2583.49	370418.32	– 409125.7	3–3	9.13–02	9.13–03	2.33–01	–1.562	C+	LS
			2589.95	2590.73	370526.49	– 409125.7	5–3	1.51–01	9.11–03	3.88–01	–1.342	C+	LS
			2587.60	2588.37	370418.32	– 409052.6	3–1	3.63–01	1.22–02	3.11–01	–1.438	C+	LS
			2562.30	2563.07	370418.32	– 409434.1	3–5	9.35–02	1.53–02	3.88–01	–1.337	C+	LS
			2576.78	2577.55	370329.18	– 409125.7	1–3	1.23–01	3.66–02	3.11–01	–1.436	C+	LS

O III: Allowed Transitions — Continued

No.	Transition Array	Mult.	λ_{air} (Å)	λ_{vac} (Å) or σ (cm^{-1})*	E_i (cm^{-1})	E_k (cm^{-1})	g_i–g_k	A_{ki} (10^8 s^{-1})	f_{ik}	S (at. u.)	log gf	Acc.	Source
216.	$2s^22p4p$–$2s^22p6d$	$^3P-^3D^{\circ}$	*2235.9*	*2236.6*	*370468.5*	– 415180	9–15	3.13–01	3.91–02	2.59+00	–0.454	B	1
			2238.77	2239.47	370526.49	– 415180	5–7	3.12–01	3.28–02	1.21+00	–0.785	C+	*LS*
			2233.36	2234.05	370418.32	– 415180	3–5	2.35–01	2.93–02	6.47–01	–1.055	C+	*LS*
			2228.92	2229.61	370329.18	– 415180	1–3	1.75–01	3.92–02	2.88–01	–1.407	C+	*LS*
			2238.77	2239.47	370526.49	– 415180	5–5	7.79–02	5.85–03	2.16–01	–1.534	C+	*LS*
			2233.36	2234.05	370418.32	– 415180	3–3	1.31–01	9.78–03	2.16–01	–1.533	C+	*LS*
			2238.77	2239.47	370526.49	– 415180	5–3	8.66–03	3.90–04	1.44–02	–2.710	C	*LS*
217.	$2s^22p4p$–$2s2p^24p$	$^3P-^3S^{\circ}$		*1502.8*	*370468.5*	– 437012.5	9–3	5.07–01	5.72–03	2.55–01	–1.289	C+	1
				1504.08	370526.49	– 437012.5	5–3	2.81–01	5.71–03	1.41–01	–1.544	C–	*LS*
				1501.63	370418.32	– 437012.5	3–3	1.69–01	5.72–03	8.49–02	–1.765	C–	*LS*
				1499.63	370329.18	– 437012.5	1–3	5.67–02	5.73–03	2.83–02	–2.242	D	*LS*
218.	$2s^22p4p$–$2s^22p4d$	$^1D-^1D^{\circ}$	14735.2	6784.61 cm^{-1}	370902.22	– 377686.83	5–5	2.50–02	8.14–02	1.97+01	–0.391	B	1
219.		$^1D-^1F^{\circ}$	10118.7	9879.95 cm^{-1}	370902.22	– 380782.17	5–7	3.37–01	7.25–01	1.21+02	0.559	B	1
220.		$^1D-^1P^{\circ}$	9813.69	9816.38	370902.22	– 381089.27	5–3	2.30–02	2.00–02	3.22+00	–1.001	B	1
221.	$2s^22p4p$–$2s^22p5s$	$^1D-^1P^{\circ}$	4569.26	4570.54	370902.22	– 392781.47	5–3	1.27+00	2.39–01	1.80+01	0.077	B	1
222.	$2s^22p4p$–$2s^22p5d$	$^1D-^1D^{\circ}$	3236.41	3237.35	370902.22	– 401791.7	5–5	1.99–01	3.13–02	1.67+00	–0.806	B	1
223.		$^1D-^1F^{\circ}$	3078.13	3079.03	370902.22	– 403380.0	5–7	2.89–01	5.75–02	2.91+00	–0.542	B	1
224.	$2s^22p4p$–$2s^22p6s$	$^1D-^1P^{\circ}$	2572.29	2573.06	370902.22	– 409766.5	5–3	5.77–01	3.43–02	1.45+00	–0.766	B	1
225.	$2s^22p4p$–$2s^22p6d$	$^1D-^1D^{\circ}$	2283.77	2284.47	370902.22	– 414676	5–5	1.62–01	1.26–02	4.75–01	–1.200	C+	1
226.	$2s^22p4p$–$2s^22p7d$	$^1D-^1F^{\circ}$		1920.24	370902.22	– 422979	5–7	2.67–01	2.07–02	6.54–01	–0.985	B	1
227.	$2s^22p4p$–$2s2p^2(^2D)3p$	$^1D-^1D^{\circ}$		1803.82	370902.22	– 426340	5–5	1.83–01	8.94–03	2.66–01	–1.350	C+	1
228.		$^1D-^1F^{\circ}$		1848.50	370902.22	– 425000	5–7	4.95–02	3.55–03	1.08–01	–1.751	C+	1
229.	$2s2p^23p$–$2s2p^23d$	$^3D^{\circ}-^3P$	*3890.0*	*3891.1*	*374706.6*	*– 400406.2*	15–9	5.05–03	6.88–04	1.32–01	–1.987	C+	1
			3911.80	3912.91	374795.14	– 400351.56	7–5	4.17–03	6.84–04	6.16–02	–2.320	C–	*LS*
			3875.25	3876.35	374663.52	– 400460.99	5–3	3.83–03	5.18–04	3.30–02	–2.587	C–	*LS*
			3853.47	3854.57	374571.64	– 400514.89	3–1	5.20–03	3.86–04	1.47–02	–2.937	D	*LS*
			3891.76	3892.86	374663.52	– 400351.56	5–5	7.57–04	1.72–04	1.10–02	–3.066	D	*LS*
			3861.50	3862.59	374571.64	– 400460.99	3–3	1.29–03	2.89–04	1.10–02	–3.063	D	*LS*
			3877.89	3878.99	374571.64	– 400351.56	3–5	5.10–05	1.92–05	7.34–04	–4.240	D–	*LS*

O III: Allowed Transitions — Continued

No.	Transition Array	Mult.	λ_{air} (Å)	λ_{vac} (Å) or σ (cm^{-1})*	E_i (cm^{-1})	E_k (cm^{-1})	g_i–g_k	A_{ki} (10^8 s^{-1})	f_{ik}	S (at. u.)	log gf	Acc.	Source
230.	$2s2p^23p-2s2p^2(^2D)3s$	$^3D°-^3D$	*5141.5*	*5142.9*	*374706.6*	*– 394151*	15–15	5.78–03	2.29–03	5.81–01	–1.464	B	1
			5152.47	5153.91	374795.14	– 394197.9	7–7	5.10–03	2.03–03	2.41–01	–1.847	C+	*LS*
			5136.32	5137.75	374663.52	– 394127.3	5–5	4.03–03	1.59–03	1.35–01	–2.098	C+	*LS*
			5124.74	5126.17	374571.64	– 394079.4	3–3	4.37–03	1.72–03	8.72–02	–2.287	C+	*LS*
			5171.29	5172.73	374795.14	– 394127.3	7–5	8.86–04	2.54–04	3.02–02	–2.751	C+	*LS*
			5148.99	5150.42	374663.52	– 394079.4	5–3	1.44–03	3.43–04	2.91–02	–2.766	C	*LS*
			5117.75	5119.18	374663.52	– 394197.9	5–7	6.53–04	3.59–04	3.02–02	–2.746	C+	*LS*
			5112.18	5113.61	374571.64	– 394127.3	3–5	8.81–04	5.76–04	2.91–02	–2.763	C	*LS*
231.	$2s2p^23p-2s2p^23d$	$^3D°-^3F$	*3730.1*	*3731.2*	*374706.6*	*– 401507.6*	15–21	1.45+00	4.23–01	7.80+01	0.803	B	1
			3728.84	3729.90	374795.14	– 401605.52	7–9	1.45+00	3.89–01	3.34+01	0.435	C+	*LS*
			3728.51	3729.57	374663.52	– 401476.29	5–7	1.29+00	3.76–01	2.31+01	0.274	C+	*LS*
			3729.80	3730.86	374571.64	– 401375.09	3–5	1.22+00	4.23–01	1.56+01	0.104	C+	*LS*
			3746.90	3747.96	374795.14	– 401476.29	7–7	1.59–01	3.35–02	2.90+00	–0.629	C+	*LS*
			3742.63	3743.70	374663.52	– 401375.09	5–5	2.24–01	4.70–02	2.90+00	–0.629	C+	*LS*
			3761.17	3762.23	374795.14	– 401375.09	7–5	6.22–03	9.42–04	8.17–02	–2.181	C+	*LS*
232.		$^3D°-^3D$	*3210.3*	*3211.2*	*374706.6*	*– 405847.7*	15–15	6.31–01	9.75–02	1.55+01	0.165	B	1
			3216.07	3217.00	374795.14	– 405879.97	7–7	5.58–01	8.65–02	6.41+00	–0.218	C+	*LS*
			3207.61	3208.54	374663.52	– 405830.38	5–5	4.40–01	6.79–02	3.59+00	–0.469	C+	*LS*
			3201.14	3202.06	374571.64	– 405801.50	3–3	4.77–01	7.33–02	2.32+00	–0.658	C+	*LS*
			3221.21	3222.14	374795.14	– 405830.38	7–5	9.75–02	1.08–02	8.04–01	–1.120	C+	*LS*
			3210.58	3211.51	374663.52	– 405801.50	5–3	1.58–01	1.46–02	7.73–01	–1.136	C+	*LS*
			3202.51	3203.44	374663.52	– 405879.97	5–7	7.08–02	1.52–02	8.04–01	–1.118	C+	*LS*
			3198.18	3199.11	374571.64	– 405830.38	3–5	9.57–02	2.45–02	7.73–01	–1.134	C+	*LS*
233.		$^5S°-^5P$	*4452.6*	*4453.9*	376079.92	*– 398532.2*	5–15	4.39–01	3.91–01	2.87+01	0.291	B	1
			4461.61	4462.86	376079.92	– 398487.08	5–7	4.36–01	1.82–01	1.34+01	–0.040	C+	*LS*
			4447.69	4448.94	376079.92	– 398557.17	5–5	4.40–01	1.31–01	9.56+00	–0.185	C+	*LS*
			4440.09	4441.34	376079.92	– 398595.65	5–3	4.42–01	7.85–02	5.74+00	–0.406	C+	*LS*
234.	$2s2p^23p-2s2p^24s$	$^5S°-^5P$		*1901.2*	376079.92	*– 428678.8*	5–15	1.48+00	2.40–01	7.51+00	0.079	B	1
				1897.10	376079.92	– 428791.89	5–7	1.49+00	1.12–01	3.51+00	–0.251	C+	*LS*
				1903.05	376079.92	– 428627.20	5–5	1.47+00	7.99–02	2.50+00	–0.398	C+	*LS*
				1907.63	376079.92	– 428500.87	5–3	1.46+00	4.78–02	1.50+00	–0.621	C+	*LS*
235.	$2s2p^23p-2s2p^24d$	$^5S°-^5P$		*1348.7*	376079.92	*– 450227.5*	5–15	2.23+00	1.82–01	4.05+00	–0.040	B	1
				1349.57	376079.92	– 450177.71	5–7	2.22+00	8.50–02	1.89+00	–0.371	C+	*LS*
				1348.24	376079.92	– 450250.84	5–5	2.23+00	6.08–02	1.35+00	–0.517	C+	*LS*
				1347.25	376079.92	– 450304.93	5–3	2.24+00	3.65–02	8.10–01	–0.739	C+	*LS*
236.	$2s^22p4d-2s2p^2(^2D)3s$	$^3F°-^3D$	*6040.4*	*6042.1*	*377600.1*	*– 394151*	21–15	7.65–02	2.99–02	1.25+01	–0.202	B	1
			6077.59	6079.27	377748.57	– 394197.9	9–7	6.90–02	2.97–02	5.35+00	–0.573	C+	*LS*
			6035.16	6036.83	377562.31	– 394127.3	7–5	6.82–02	2.66–02	3.70+00	–0.730	C+	*LS*
			5988.58	5990.24	377385.58	– 394079.4	5–3	7.85–02	2.53–02	2.50+00	–0.897	C+	*LS*
			6009.54	6011.21	377562.31	– 394197.9	7–7	6.18–03	3.35–03	4.64–01	–1.630	C+	*LS*
			5971.45	5973.10	377385.58	– 394127.3	5–5	8.82–03	4.72–03	4.64–01	–1.627	C+	*LS*
			5946.37	5948.02	377385.58	– 394197.9	5–7	1.80–04	1.34–04	1.31–02	–3.175	C	*LS*
237.	$2s2p^23p-2s2p^23d$	$^3P°-^3P$	*4545.6*	*4546.9*	*378413.0*	*– 400406.2*	9–9	3.34–01	1.03–01	1.39+01	–0.031	B	1
			4555.39	4556.66	378405.68	– 400351.56	5–5	2.49–01	7.74–02	5.80+00	–0.413	C+	*LS*
			4535.29	4536.56	378417.84	– 400460.99	3–3	8.40–02	2.59–02	1.16+00	–1.110	C+	*LS*
			4532.78	4534.06	378405.68	– 400460.99	5–3	1.40–01	2.59–02	1.93+00	–0.887	C+	*LS*
			4524.22	4525.49	378417.84	– 400514.89	3–1	3.38–01	3.46–02	1.55+00	–0.984	C+	*LS*
			4557.91	4559.19	378417.84	– 400351.56	3–5	8.27–02	4.30–02	1.93+00	–0.890	C+	*LS*
			4538.85	4540.12	378435.16	– 400460.99	1–3	1.12–01	1.04–01	1.55+00	–0.985	C+	*LS*

O III: Allowed Transitions — Continued

No.	Transition Array	Mult.	λ_{air} (Å)	λ_{vac} (Å) or σ (cm^{-1})*	E_i (cm^{-1})	E_k (cm^{-1})	g_i-g_k	A_{ki} (10^8 s^{-1})	f_{ik}	S (at. u.)	log gf	Acc.	Source
238.	$2s2p^23p-$ $2s2p^2(^2D)3s$	$^3P^o-^3D$	*6352.4*	*6354.2*	*378413.0*	*- 394151*	9–15	9.34–04	9.41–04	1.77–01	–2.072	C+	1
			6330.48	6332.23	378405.68	- 394197.9	5–7	9.43–04	7.94–04	8.27–02	–2.401	C–	*LS*
			6363.83	6365.59	378417.84	- 394127.3	3–5	6.97–04	7.05–04	4.43–02	–2.675	C–	*LS*
			6390.36	6392.13	378435.16	- 394079.4	1–3	5.10–04	9.36–04	1.97–02	–3.029	D	*LS*
			6358.91	6360.67	378405.68	- 394127.3	5–5	2.33–04	1.41–04	1.48–02	–3.152	D	*LS*
			6383.30	6385.06	378417.84	- 394079.4	3–3	3.83–04	2.34–04	1.48–02	–3.153	D	*LS*
			6378.34	6380.11	378405.68	- 394079.4	5–3	2.56–05	9.38–06	9.84–04	–4.329	D–	*LS*
239.	$2s^22p4d-$ $2s2p^23d$	$^3D^o-^3P$	*4738.8*	*4740.1*	*379309.6*	*- 400406.2*	15–9	5.61–03	1.13–03	2.65–01	–1.770	C+	1
			4761.75	4763.08	379356.75	- 400351.56	7–5	4.64–03	1.13–03	1.24–01	–2.103	C–	*LS*
			4722.80	4724.12	379293.03	- 400460.99	5–3	4.25–03	8.52–04	6.63–02	–2.370	C–	*LS*
			4696.23	4697.54	379227.15	- 400514.89	3–1	5.76–03	6.35–04	2.94–02	–2.720	D	*LS*
			4747.34	4748.67	379293.03	- 400351.56	5–5	8.36–04	2.83–04	2.21–02	–2.850	D	*LS*
			4708.15	4709.46	379227.15	- 400460.99	3–3	1.43–03	4.75–04	2.21–02	–2.846	D	*LS*
			4732.54	4733.86	379227.15	- 400351.56	3–5	5.63–05	3.15–05	1.47–03	–4.025	D–	*LS*
240.	$2s^22p4d-$ $2s2p^2(^2D)3s$	$^3D^o-^3D$	*6736.2*	*6738.1*	*379309.6*	*- 394151*	15–15	1.22–02	8.28–03	2.75+00	–0.906	B	1
			6736.16	6738.02	379356.75	- 394197.9	7–7	1.08–02	7.36–03	1.14+00	–1.288	C+	*LS*
			6739.29	6741.15	379293.03	- 394127.3	5–5	8.46–03	5.76–03	6.39–01	–1.541	C+	*LS*
			6731.13	6732.99	379227.15	- 394079.4	3–3	9.15–03	6.22–03	4.13–01	–1.729	C+	*LS*
			6768.36	6770.23	379356.75	- 394127.3	7–5	1.87–03	9.18–04	1.43–01	–2.192	C+	*LS*
			6761.12	6762.99	379293.03	- 394079.4	5–3	3.01–03	1.24–03	1.38–01	–2.208	C+	*LS*
			6707.36	6709.22	379293.03	- 394197.9	5–7	1.37–03	1.30–03	1.43–01	–2.188	C+	*LS*
			6709.49	6711.34	379227.15	- 394127.3	3–5	1.85–03	2.08–03	1.38–01	–2.205	C+	*LS*
241.	$2s^22p4d-$ $2s2p^23d$	$^3D^o-^3F$	*4503.6*	*4504.9*	*379309.6*	*- 401507.6*	15–21	9.10–02	3.88–02	8.62+00	–0.236	B	1
			4493.37	4494.63	379356.75	- 401605.52	7–9	9.17–02	3.57–02	3.69+00	–0.603	C+	*LS*
			4506.64	4507.90	379293.03	- 401476.29	5–7	8.07–02	3.44–02	2.55+00	–0.764	C+	*LS*
			4513.83	4515.09	379227.15	- 401375.09	3–5	7.60–02	3.87–02	1.72+00	–0.935	C+	*LS*
			4519.62	4520.89	379356.75	- 401476.29	7–7	1.00–02	3.07–03	3.20–01	–1.667	C+	*LS*
			4527.29	4528.56	379293.03	- 401375.09	5–5	1.40–02	4.30–03	3.20–01	–1.668	C+	*LS*
			4540.40	4541.67	379356.75	- 401375.09	7–5	3.91–04	8.63–05	9.03–03	–3.219	C–	*LS*
242.		$^3P^o-^3P$	*5076.7*	*5078.1*	*380713.7*	*- 400406.2*	9–9	8.27–02	3.19–02	4.80+00	–0.541	B	1
			5088.92	5090.34	380706.51	- 400351.56	5–5	6.15–02	2.39–02	2.00+00	–0.923	C+	*LS*
			5063.66	5065.07	380717.92	- 400460.99	3–3	2.08–02	8.01–03	4.00–01	–1.620	C+	*LS*
			5060.73	5062.14	380706.51	- 400460.99	5–3	3.48–02	8.01–03	6.67–01	–1.397	C+	*LS*
			5049.87	5051.28	380717.92	- 400514.89	3–1	8.40–02	1.07–02	5.34–01	–1.493	C+	*LS*
			5091.88	5093.30	380717.92	- 400351.56	3–5	2.05–02	1.33–02	6.67–01	–1.400	C+	*LS*
			5068.56	5069.97	380737.00	- 400460.99	1–3	2.77–02	3.20–02	5.34–01	–1.495	C+	*LS*
243.	$2s^22p4d-$ $2s2p^2(^2D)3s$	$^3P^o-^3D$	*7440.1*	*7442.2*	*380713.7*	*- 394151*	9–15	6.61–03	9.14–03	2.01+00	–1.085	B	1
			7410.09	7412.13	380706.51	- 394197.9	5–7	6.69–03	7.71–03	9.40–01	–1.414	C+	*LS*
			7455.41	7457.47	380717.92	- 394127.3	3–5	4.92–03	6.84–03	5.04–01	–1.688	C+	*LS*
			7492.84	7494.90	380737.00	- 394079.4	1–3	3.59–03	9.07–03	2.24–01	–2.042	C+	*LS*
			7449.07	7451.13	380706.51	- 394127.3	5–5	1.65–03	1.37–03	1.68–01	–2.165	C+	*LS*
			7482.14	7484.20	380717.92	- 394079.4	3–3	2.71–03	2.27–03	1.68–01	–2.166	C+	*LS*
			7475.76	7477.82	380706.51	- 394079.4	5–3	1.81–04	9.10–05	1.12–02	–3.342	C	*LS*

O III: Allowed Transitions — Continued

No.	Transition Array	Mult.	λ_{air} (Å)	λ_{vac} (Å) or σ (cm^{-1})*	E_i (cm^{-1})	E_k (cm^{-1})	g_i-g_k	A_{ki} (10^8 s^{-1})	f_{ik}	S (at. u.)	log gf	Acc.	Source
244.	$2s2p^23d-$ $2s2p^24p$	$^5D-^5P^\circ$	*2426.4*	*2427.1*	*398176.6*	*– 439378*	25–15	1.17+00	6.21–02	1.24+01	0.191	B	1
			2425.90	2426.64	398231.48	– 439440.8	9–7	9.05–01	6.21–02	4.46+00	–0.253	C+	*LS*
			2426.93	2427.66	398150.40	– 439342.3	7–5	6.56–01	4.14–02	2.31+00	–0.538	C+	*LS*
			2429.36	2430.09	398139.92	– 439290.6	5–3	4.09–01	2.17–02	8.68–01	–0.965	C+	*LS*
			2421.14	2421.87	398150.40	– 439440.8	7–7	2.36–01	2.07–02	1.16+00	–0.838	C+	*LS*
			2426.31	2427.04	398139.92	– 439342.3	5–5	4.10–01	3.62–02	1.45+00	–0.742	C+	*LS*
			2429.61	2430.35	398144.29	– 439290.6	3–3	5.25–01	4.65–02	1.12+00	–0.856	C+	*LS*
			2420.52	2421.26	398139.92	– 439440.8	5–7	3.37–02	4.15–03	1.65–01	–1.683	C+	*LS*
			2426.57	2427.30	398144.29	– 439342.3	3–5	1.05–01	1.55–02	3.72–01	–1.332	C+	*LS*
			2429.69	2430.43	398145.63	– 439290.6	1–3	2.33–01	6.20–02	4.96–01	–1.208	C+	*LS*
245.		$^5P-^5P^\circ$	*2447.5*	*2448.2*	*398532.2*	*– 439378*	15–15	7.20–01	6.46–02	7.81+00	–0.013	B	1
			2441.04	2441.78	398487.08	– 439440.8	7–7	4.84–01	4.32–02	2.43+00	–0.519	C+	*LS*
			2451.13	2451.87	398557.17	– 439342.3	5–5	5.97–02	5.38–03	2.17–01	–1.570	C+	*LS*
			2456.56	2457.31	398595.65	– 439290.6	3–3	1.78–01	1.61–02	3.91–01	–1.316	C+	*LS*
			2446.93	2447.67	398487.08	– 439342.3	7–5	3.36–01	2.15–02	1.22+00	–0.821	C+	*LS*
			2454.24	2454.99	398557.17	– 439290.6	5–3	5.35–01	2.90–02	1.17+00	–0.839	C+	*LS*
			2445.23	2445.97	398557.17	– 439440.8	5–7	2.41–01	3.02–02	1.22+00	–0.821	C+	*LS*
			2453.45	2454.19	398595.65	– 439342.3	3–5	3.21–01	4.84–02	1.17+00	–0.838	C+	*LS*
246.	$2s2p^23d-$ $2s^22p6s$	$^3P-^3P^\circ$	*11255*	*8883* cm^{-1}	*400406.2*	*– 409289*	9–9	2.09–03	3.96–03	1.32+00	–1.448	B	1
			11007.1	9082.5 cm^{-1}	400351.56	– 409434.1	5–5	1.67–03	3.04–03	5.50–01	–1.819	C+	*LS*
			11537.9	8664.7 cm^{-1}	400460.99	– 409125.7	3–3	4.84–04	9.66–04	1.10–01	–2.538	C+	*LS*
			11394.0	8774.1 cm^{-1}	400351.56	– 409125.7	5–3	8.37–04	9.78–04	1.83–01	–2.311	C+	*LS*
			11636.1	8591.6 cm^{-1}	400460.99	– 409052.6	3–1	1.89–03	1.28–03	1.47–01	–2.417	C+	*LS*
			11141.4	8973.1 cm^{-1}	400460.99	– 409434.1	3–5	5.37–04	1.67–03	1.83–01	–2.301	C+	*LS*
			11610.1	8610.8 cm^{-1}	400514.89	– 409125.7	1–3	6.33–04	3.84–03	1.47–01	–2.416	C+	*LS*
247.	$2s2p^23d-$ $2s^22p6d$	$^3P-^3D^\circ$	*6766.9*	*6768.7*	*400406.2*	– 415180	9–15	7.13–04	8.15–04	1.64–01	–2.134	C+	1
			6741.94	6743.80	400351.56	– 415180	5–7	7.21–04	6.88–04	7.63–02	–2.464	C–	*LS*
			6792.06	6793.94	400460.99	– 415180	3–5	5.29–04	6.09–04	4.09–02	–2.738	C–	*LS*
			6817.02	6818.91	400514.89	– 415180	1–3	3.87–04	8.09–04	1.82–02	–3.092	D	*LS*
			6741.94	6743.80	400351.56	– 415180	5–5	1.80–04	1.23–04	1.36–02	–3.212	D	*LS*
			6792.06	6793.94	400460.99	– 415180	3–3	2.94–04	2.03–04	1.36–02	–3.215	D	*LS*
			6741.94	6743.80	400351.56	– 415180	5–3	2.00–05	8.19–06	9.08–04	–4.388	D–	*LS*
248.	$2s2p^23d-$ $2s2p^24p$	$^3P-^3S^\circ$	*2731.0*	*2731.8*	*400406.2*	– 437012.5	9–3	1.44+00	5.35–02	4.33+00	–0.317	B	1
			2726.89	2727.70	400351.56	– 437012.5	5–3	8.01–01	5.36–02	2.41+00	–0.572	C+	*LS*
			2735.06	2735.87	400460.99	– 437012.5	3–3	4.76–01	5.34–02	1.44+00	–0.795	C+	*LS*
			2739.09	2739.91	400514.89	– 437012.5	1–3	1.58–01	5.34–02	4.81–01	–1.273	C+	*LS*
249.	$2s2p^2(^2D)3s-$ $2s^22p5d$	$^3D-^3F^\circ$	*13158*	*7598* cm^{-1}	*394151*	*– 401748*	15–21	1.73–01	6.30–01	4.09+02	0.975	B	1
			12991.4	7695.3 cm^{-1}	394197.9	– 401893.2	7–9	1.80–01	5.86–01	1.75+02	0.613	C+	*LS*
			13157.2	7598.3 cm^{-1}	394127.3	– 401725.6	5–7	1.54–01	5.60–01	1.21+02	0.447	C+	*LS*
			13436.5	7440.4 cm^{-1}	394079.4	– 401519.8	3–5	1.37–01	6.17–01	8.19+01	0.267	C+	*LS*
			13280.6	7527.7 cm^{-1}	394197.9	– 401725.6	7–7	1.88–02	4.97–02	1.52+01	–0.459	C+	*LS*
			13523.5	7392.5 cm^{-1}	394127.3	– 401519.8	5–5	2.49–02	6.83–02	1.52+01	–0.467	C+	*LS*
			13653.9	7321.9 cm^{-1}	394197.9	– 401519.8	7–5	6.83–04	1.36–03	4.29–01	–2.021	C+	*LS*
250.	$2s2p^2(^2D)3s-$ $2s^22p6s$	$^3D-^3P^\circ$	*6604.0*	*6605.8*	*394151*	*– 409289*	15–9	6.90–01	2.71–01	8.83+01	0.609	B	1
			6561.50	6563.32	394197.9	– 409434.1	7–5	5.91–01	2.73–01	4.12+01	0.281	C+	*LS*
			6665.54	6667.38	394127.3	– 409125.7	5–3	5.04–01	2.01–01	2.21+01	0.003	C+	*LS*
			6676.76	6678.60	394079.4	– 409052.6	3–1	6.68–01	1.49–01	9.81+00	–0.350	C+	*LS*
			6531.24	6533.04	394127.3	– 409434.1	5–5	1.07–01	6.85–02	7.36+00	–0.466	C+	*LS*
			6644.32	6646.15	394079.4	– 409125.7	3–3	1.69–01	1.12–01	7.36+00	–0.473	C+	*LS*
			6510.86	6512.66	394079.4	– 409434.1	3–5	7.20–03	7.63–03	4.91–01	–1.640	C+	*LS*

O III: Allowed Transitions — Continued

No.	Transition Array	Mult.	λ_{air} (Å)	λ_{vac} (Å) or σ (cm^{-1})*	E_i (cm^{-1})	E_k (cm^{-1})	g_i-g_k	A_{ki} (10^8 s^{-1})	f_{ik}	S (at. u.)	log gf	Acc.	Source
251.	$2s2p^2(^2D)3s-2s^22p6d$	$^3D-^3D^o$	*4753.9*	*4755.3*	*394151*	– 415180	15–15	3.61–02	1.22–02	2.87+00	–0.737	B	1
			4764.63	4765.97	394197.9	– 415180	7–7	3.19–02	1.08–02	1.19+00	–1.120	C+	*LS*
			4748.66	4749.98	394127.3	– 415180	5–5	2.52–02	8.52–03	6.66–01	–1.371	C+	*LS*
			4737.88	4739.20	394079.4	– 415180	3–3	2.73–02	9.20–03	4.31–01	–1.559	C+	*LS*
			4764.63	4765.97	394197.9	– 415180	7–5	5.59–03	1.36–03	1.49–01	–2.021	C+	*LS*
			4748.66	4749.98	394127.3	– 415180	5–3	9.05–03	1.84–03	1.44–01	–2.037	C+	*LS*
			4748.66	4749.98	394127.3	– 415180	5–7	4.04–03	1.91–03	1.49–01	–2.020	C+	*LS*
			4737.88	4739.20	394079.4	– 415180	3–5	5.47–03	3.07–03	1.44–01	–2.036	C+	*LS*
252.	$2s2p^24s-2s2p^24p$	$^5P-^5P^o$	*9344.0*	*9346.5*	*428678.8*	*– 439378*	15–15	2.66–01	3.48–01	1.61+02	0.718	B	1
			9388.06	9390.63	428791.89	– 439440.8	7–7	1.75–01	2.31–01	5.00+01	0.209	C+	*LS*
			9330.06	9332.62	428627.20	– 439342.3	5–5	2.23–02	2.91–02	4.47+00	–0.838	C+	*LS*
			9265.53	9268.07	428500.87	– 439290.6	3–3	6.82–02	8.78–02	8.04+00	–0.579	C+	*LS*
			9475.71	9478.30	428791.89	– 439342.3	7–5	1.19–01	1.15–01	2.50+01	–0.096	C+	*LS*
			9375.30	9377.87	428627.20	– 439290.6	5–3	1.98–01	1.56–01	2.41+01	–0.107	C+	*LS*
			9245.08	9247.61	428627.20	– 439440.8	5–7	9.16–02	1.64–01	2.50+01	–0.085	C+	*LS*
			9221.34	9223.88	428500.87	– 439342.3	3–5	1.25–01	2.65–01	2.41+01	–0.100	C+	*LS*
253.	$2s2p^24p-2s2p^24d$	$^5P^o-^5P$	*9214.4*	*9216.9*	*439378*	*– 450227.5*	15–15	2.09–01	2.66–01	1.21+02	0.601	B	1
			9311.11	9313.67	439440.8	– 450177.71	7–7	1.35–01	1.75–01	3.77+01	0.089	C+	*LS*
			9164.61	9167.13	439342.3	– 450250.84	5–5	1.77–02	2.23–02	3.36+00	–0.953	C+	*LS*
			9076.59	9079.08	439290.6	– 450304.93	3–3	5.47–02	6.75–02	6.05+00	–0.693	C+	*LS*
			9248.12	9250.66	439440.8	– 450250.84	7–5	9.65–02	8.83–02	1.88+01	–0.209	C+	*LS*
			9119.40	9121.90	439342.3	– 450304.93	5–3	1.62–01	1.21–01	1.82+01	–0.218	C+	*LS*
			9226.47	9229.00	439342.3	– 450177.71	5–7	6.94–02	1.24–01	1.88+01	–0.208	C+	*LS*
			9121.38	9123.89	439290.6	– 450250.84	3–5	9.69–02	2.02–01	1.82+01	–0.219	C+	*LS*

*Wavelengths (Å) are always given unless cm^{-1} is indicated.

O III

Forbidden Transitions

The tabulated data are selected from three advanced calculations, which include extensive treatments of electron correlation: the multi-configuration calculations by Baluja and Doyle[1] with the CIV 3 code, the multi-configuration Hartree-Fock calculations by Froese Fischer and Saha,[2] and the multi-configuration calculations by Nussbaumer and Storey[3] with the SUPERSTRUCTURE code. The wavefunction expansions developed for these calculations include numerous configurations that are expected to mix with the studied levels, and the results of the three papers typically agree within ±10–20%. For the M1 fine structure transitions within the ^3P term of the ground state configuration, the authors obtain the same line strengths as derived from general expansion formulas first developed by Shortley *et al.*[4] for the case of *LS*-coupling conditions.

References

[1]K. L. Baluja and J. G. Doyle, J. Phys. B **14**, L11 (1981).

[2]C. Froese Fischer and H. P. Saha, Phys. Scr. **32**, 181 (1985).

[3]H. Nussbaumer and P. J. Storey, Astron. Astrophys. **99**, 177 (1981).

[4]G. H. Shortley, L. H. Aller, J. G. Baker, and D. H. Menzel, Astrophys. J. **93**, 178 (1961).

O III: Forbidden Transitions

No.	Transition Array	Mult.	λ_{air} (Å)	λ_{vac} (Å) or σ (cm^{-1})*	E_i (cm^{-1})	E_k (cm^{-1})	$g_i - g_k$	Type	A_{ki} (s^{-1})	S (at. u.)	Acc.	Source
1.	$2p^2-$ $2p^2$	^{3}P–^{3}P										
				113.400 cm^{-1}	0.000	113.400	1–3	M1	2.62–05	2.00+00	A	1,2,3
				306.174 cm^{-1}	0.000	306.174	1–5	E2	3.17–11	5.26–01	B+	1,2,3
				192.996 cm^{-1}	113.178	306.174	3–5	M1	9.76–05	2.50+00	A	1,2,3
				192.996 cm^{-1}	113.178	306.174	3–5	E2	7.34–12	1.22+00	B	2
2.		^{3}P–^{1}D										
			4931.23	4932.60	0.000	20273.27	1–5	E2	2.41–06	3.14–05	C+	1,2,3
			4958.91	4960.29	113.178	20273.27	3–5	M1	6.21–03	1.40–04	B	1,2,3
			4958.91	4960.29	113.178	20273.27	3–5	E2	4.57–06	6.13–05	C+	2
			5006.84	5008.24	306.174	20273.27	5–5	M1	1.81–02	4.21–04	B	1,2,3
			5006.84	5008.24	306.174	20273.27	5–5	E2	3.52–05	4.95–04	C+	2
3.		^{3}P–^{1}S										
			2320.95	2321.66	113.178	43185.74	3–1	M1	2.15–01	9.97–05	B+	1,2,3
			2331.40	2332.11	306.174	43185.74	5–1	E2	6.34–04	3.90–05	C	1,2,3
4.		^{1}D–^{1}S										
			4363.21	4364.44	20273.27	43185.74	5–1	E2	1.71+00	2.41+00	B	1,2,3

*Wavelengths (Å) are always given unless cm^{-1} is indicated.

O IV

Boron Isoelectronic Sequence

Ground State: $1s^2 2s^2 2p\ ^2P^\circ_{1/2}$

Ionization Energy: 77.416 eV = 624382.0 cm^{-1}

Allowed Transitions

List of tabulated lines

Wavelength (Å)	No.	Wavelength (Å)	No.	Wavelength (Å)	No.	Wavelength (Å)	No.
in vacuum		177.662	28	184.636	47	192.138	24
		177.693	28	184.637	47	192.163	24
169.472	18	180.354	14	185.402	10	192.186	24
169.583	18	180.480	14	185.535	10	192.200	24
171.074	17	181.150	13	186.868	26	192.231	24
171.123	17	181.276	13	186.873	26	192.255	24
171.187	17	181.866	12	186.884	26	195.860	9
173.800	16	181.887	12	186.913	26	196.006	9
173.852	16	181.994	12	186.930	26	196.008	9
173.916	16	182.015	12	186.937	26	198.029	46
173.969	16	182.706	11	186.978	26	198.034	46
174.104	15	182.826	11	191.559	25	200.919	23
174.221	15	182.835	11	191.583	25	200.966	23
177.553	28	183.338	27	191.607	25	200.994	23
177.562	28	183.382	27	191.631	25	201.019	23
177.594	28	183.444	27	191.680	25	201.021	23
177.604	28	184.631	47	191.699	25	201.073	23
177.635	28	184.632	47	191.748	25	201.094	23

List of tabulated lines — Continued

Wavelength (Å)	No.	Wavelength (Å)	No.	Wavelength (Å)	No.	Wavelength (Å)	No.
202.885	8	239.671	65	289.470	70	349.211	90
203.044	8	239.725	65	289.590	70	349.221	90
204.698	45	239.811	65	289.897	81	349.422	90
204.704	45	239.939	66	289.909	81	349.432	90
204.711	45	240.079	66	289.921	81	366.850	57
204.716	45	242.045	64	289.933	81	367.178	57
204.886	69	242.140	64	291.052	96	367.207	57
204.988	69	242.188	64	291.059	96	367.536	57
204.990	69	244.180	52	291.197	96	376.721	89
207.183	7	244.302	52	291.203	96	376.732	89
207.239	7	246.465	71	292.704	80	379.778	31
207.348	7	246.502	71	292.729	80	379.798	31
212.975	44	246.563	71	292.824	80	379.923	31
213.057	44	247.010	85	292.849	80	383.377	88
213.063	44	247.028	85	294.653	79	383.389	88
213.975	6	249.214	63	294.678	79	383.470	88
214.028	6	249.227	63	294.850	79	383.482	88
214.152	6	249.365	63	295.057	78	439.048	73
214.205	6	249.379	63	295.104	78	439.104	73
214.237	43	252.550	84	295.130	78	439.541	73
214.244	43	252.562	84	295.658	60	439.597	73
214.275	43	252.581	84	295.667	60	442.710	49
214.281	43	252.948	62	295.871	60	442.880	49
216.960	42	253.082	62	295.880	60	470.856	134
216.967	42	253.104	62	299.499	59	470.893	134
221.517	68	255.250	37	299.635	59	471.300	72
221.637	68	255.259	37	299.718	59	471.365	72
222.758	41	255.266	37	299.853	59	471.599	72
222.765	41	258.117	36	303.393	77	476.347	119
222.775	41	258.208	36	303.420	77	476.513	119
222.782	41	258.218	36	303.596	77	476.544	119
223.278	40	258.892	35	303.623	77	489.292	118
223.285	40	258.899	35	306.621	32	489.500	118
223.731	39	258.908	35	306.634	32	489.767	118
223.738	39	260.389	34	306.884	32	489.975	118
223.841	39	260.547	34	311.499	58	490.495	87
229.896	67	260.556	34	311.682	58	490.515	87
230.025	67	264.962	83	311.735	58	496.703	167
230.040	67	265.062	83	312.286	76	496.988	167
230.679	54	265.082	83	312.314	76	496.998	167
230.758	54	266.710	82	312.483	76	500.231	117
231.031	22	266.731	82	321.458	95	500.448	117
231.070	22	266.931	33	321.464	95	514.104	116
231.100	22	266.941	33	321.473	95	514.457	116
231.140	22	266.971	33	324.904	94	514.687	116
231.200	22	266.981	33	324.913	94	515.735	86
231.239	22	269.554	97	325.061	94	516.394	86
231.299	22	269.560	97	327.308	93	516.415	86
233.451	21	271.990	20	327.317	93	528.529	103
233.466	21	272.076	20	327.520	93	528.572	103
233.496	21	272.127	20	327.529	93	542.866	115
233.522	21	272.173	20	332.167	92	543.122	115
233.537	21	272.176	20	332.176	92	553.329	3
233.562	21	272.273	20	338.138	91	554.076	3
233.597	21	272.310	20	338.380	91	554.513	3
233.623	21	272.972	61	338.389	91	555.263	3
234.985	38	272.983	61	339.329	75	556.203	154
234.993	38	273.154	61	339.363	75	556.733	154
238.360	5	273.165	61	339.435	75	557.024	154
238.570	5	279.631	4	342.232	74	557.555	154
238.579	5	279.933	4	342.266	74	565.213	133
239.585	65	285.710	51	342.300	74	565.735	133
239.592	53	285.834	51	346.374	50	565.789	133
239.604	53	289.292	70	346.692	50	574.138	132

List of tabulated lines — Continued

Wavelength (Å)	No.	Wavelength (Å)	No.	Wavelength (Å)	No.	Wavelength (Å)	No.
574.193	132	639.925	129	728.831	224	809.821	172
574.405	132	641.830	188	751.217	249	810.298	172
574.460	132	641.958	188	751.506	249	810.712	172
594.120	131	642.219	188	751.776	249	811.122	172
594.179	131	642.347	188	751.857	249	811.675	172
601.267	175	642.756	188	751.921	249	815.871	150
601.379	175	642.880	188	752.099	126	817.024	150
601.547	175	643.290	188	752.194	126	817.638	150
601.659	175	644.024	128	752.272	249	818.796	150
602.018	175	644.093	128	752.417	249	822.625	149
602.151	175	647.806	127	753.047	204	824.421	149
602.510	175	647.877	127	753.211	204	824.872	318
603.271	175	648.668	127	754.498	204	824.899	318
607.225	166	670.211	110	754.663	204	825.069	318
607.651	166	670.601	110	756.339	179	825.096	318
608.397	2	674.624	165	757.003	179	830.506	109
608.632	114	675.151	165	757.278	179	831.075	109
608.954	114	675.283	165	759.103	151	831.105	109
609.829	2	685.976	293	759.362	151	837.871	270
616.488	130	686.687	293	760.633	151	839.505	270
616.552	130	694.485	153	762.847	245	840.372	178
616.572	130	695.765	153	763.844	245	840.846	178
616.952	30	696.443	153	764.166	245	841.787	178
617.005	30	702.131	100	765.464	164	841.999	327
617.036	30	702.343	100	766.142	164	842.446	327
617.786	113	705.560	152	766.338	164	842.474	327
618.107	113	706.547	152	767.018	164	844.762	292
618.117	113	706.881	152	769.085	163	845.840	292
620.753	102	707.719	306	769.769	163	845.984	292
620.791	102	707.769	306	777.691	244	847.065	292
624.318	180	707.872	152	778.025	244	855.375	279
624.438	180	712.392	271	779.736	29	855.456	279
624.619	19	712.489	271	779.820	29	868.034	203
624.825	180	713.573	271	779.912	29	869.158	203
624.988	236	713.670	271	779.997	29	871.092	203
625.127	19	723.175	174	782.772	186	872.415	185
625.156	236	723.376	174	782.905	186	872.925	185
625.853	19	723.458	138	783.199	186	873.133	185
626.081	112	723.628	174	783.484	186	873.644	185
626.421	112	723.782	174	783.564	186	874.630	185
626.446	112	723.861	138	783.778	186	874.660	185
626.786	112	723.885	174	784.277	186	875.649	185
630.959	246	724.034	174	785.640	125	877.979	305
631.179	246	724.468	187	785.734	125	878.056	305
632.962	262	724.494	174	785.743	125	879.299	305
632.978	262	724.569	138	787.710	1	879.376	305
632.980	262	724.617	138	790.112	1	880.058	202
632.996	262	724.753	138	790.199	1	880.643	202
634.046	139	724.963	187	791.154	162	882.041	202
634.263	139	724.984	174	791.370	162	884.561	184
634.589	139	725.006	280	791.878	162	884.806	184
634.806	139	725.065	280	799.133	173	885.159	184
635.082	139	725.106	280	799.561	173	885.300	184
635.803	139	725.165	280	799.628	173	885.818	184
636.079	139	725.326	138	800.057	173	886.313	184
636.275	111	725.642	187	800.909	173	892.480	211
636.509	111	725.868	138	800.927	173	893.669	211
636.537	139	727.237	224	801.781	173	900.561	201
636.626	111	727.388	224	802.201	48	901.218	201
637.930	101	727.565	224	802.253	48	902.638	201
638.545	101	727.716	224	803.129	173	903.298	201
639.717	129	727.964	224	809.313	172	920.898	161
639.785	129	728.240	224	809.408	172	921.296	56
639.856	129	728.489	224	809.747	172	921.365	56

List of tabulated lines — Continued

Wavelength (Å)	No.	Wavelength (Å)	No.	Wavelength (Å)	No.	Wavelength (Å)	No.
921.590	161	1015.44	218	1088.30	198	1237.14	276
921.879	161	1033.02	234	1093.77	122	1237.31	276
923.367	56	1033.07	223	1096.07	197	1237.55	215
923.436	56	1033.37	223	1096.36	122	1242.18	121
937.456	160	1033.48	234	1096.56	122	1242.43	121
938.473	160	1033.64	223	1098.54	197	1242.84	121
939.261	160	1033.95	223	1099.15	197	1244.93	337
940.282	160	1034.45	223	1106.93	183	1245.22	337
950.716	200	1034.68	223	1108.51	183	1256.45	159
952.580	217	1034.98	234	1108.80	183	1256.69	159
953.031	200	1035.18	223	1109.96	183	1258.28	159
953.934	217	1035.87	223	1110.20	233	1258.52	159
973.428	108	1036.39	223	1110.46	183	1277.51	370
974.252	108	1037.09	223	1110.73	233	1277.87	370
983.696	148	1042.52	210	1111.56	183	1280.42	303
986.221	220	1042.76	210	1111.63	183	1280.59	303
986.266	148	1045.36	107	1125.47	216	1282.89	303
986.744	220	1046.31	107	1125.74	216	1291.77	221
987.411	220	1047.59	99	1135.34	209	1292.55	221
987.495	220	1049.44	242	1137.99	209	1293.63	221
988.163	220	1050.05	242	1155.25	145	1293.99	260
988.523	219	1050.51	99	1158.80	145	1294.06	260
988.573	219	1050.98	242	1162.07	269	1294.84	221
988.627	219	1051.59	242	1163.96	269	1295.31	221
988.708	219	1054.80	177	1164.13	261	1295.61	105
989.267	220	1057.55	177	1164.32	268	1296.10	221
989.272	220	1057.98	146	1164.34	291	1297.07	105
989.277	219	1059.07	177	1164.53	291	1297.26	221
989.684	219	1060.63	146	1164.55	268	1297.55	105
990.127	219	1060.96	146	1166.52	261	1297.74	221
990.379	220	1061.78	222	1166.58	261	1299.02	105
990.389	219	1061.95	222	1166.58	291	1318.42	289
991.187	219	1062.10	222	1167.12	269	1321.04	289
991.394	171	1062.13	222	1167.48	268	1326.61	231
991.885	220	1062.27	222	1180.31	232	1326.70	302
992.729	171	1062.48	222	1182.21	317	1326.88	302
993.062	171	1062.63	222	1182.62	317	1327.37	231
993.827	171	1062.86	222	1182.79	232	1327.67	302
994.396	171	1063.21	222	1183.39	232	1327.84	302
994.797	171	1067.77	123	1195.47	277	1328.39	361
995.163	171	1067.83	123	1196.01	278	1328.66	361
995.169	171	1067.96	123	1198.02	278	1328.99	361
999.030	235	1073.11	199	1198.18	278	1329.66	231
999.318	235	1073.50	199	1198.40	277	1330.42	231
999.458	235	1074.94	241	1198.56	277	1338.61	55
1007.83	147	1075.57	241	1200.06	267	1342.99	55
1008.32	124	1076.06	199	1200.85	267	1343.51	55
1008.42	124	1076.46	199	1201.18	239	1344.56	230
1008.49	124	1077.23	241	1201.97	239	1345.34	230
1010.50	147	1079.06	106	1203.42	267	1347.40	395
1010.53	147	1080.97	240	1204.21	267	1351.00	395
1012.22	218	1080.97	240	1204.36	239	1351.06	266
1012.35	218	1081.02	106	1205.16	239	1351.08	395
1012.69	218	1081.61	240	1210.98	290	1353.81	238
1013.01	218	1082.04	106	1213.03	290	1354.82	238
1013.14	218	1082.06	248	1213.20	290	1355.09	266
1013.32	218	1082.75	248	1217.72	326	1355.31	266
1013.86	218	1082.86	248	1218.65	326	1356.98	238
1014.18	218	1083.38	248	1228.55	304	1367.74	247
1014.23	243	1083.61	248	1228.61	304	1369.86	247
1014.38	243	1084.18	248	1228.76	304	1371.72	247
1014.80	243	1084.19	248	1234.41	215	1373.85	247
1014.90	218	1084.24	248	1236.30	276	1375.24	247
1014.94	243	1085.29	198	1236.47	276	1376.57	247

List of tabulated lines — Continued

Wavelength (Å)	No.	Wavelength (Å)	No.	Wavelength (Å)	No.	Wavelength (Å)	No.
1377.97	247	1730.10	366	2180.54	237	2674.79	394
1397.16	275	1731.03	366	2181.92	237	2730.68	263
1397.38	275	1731.21	144	2183.17	237	2732.84	263
1401.15	301	1734.42	366	2184.55	237	2741.32	386
1401.35	301	1735.36	366	2243.12	417	2747.42	386
1401.47	275	1738.34	366	2244.58	417	2748.12	263
1405.02	259	1739.05	144	2275.18	381	2757.88	254
1405.11	259	1739.18	144	2278.86	381	2758.22	254
1405.46	259	1740.01	388	2279.64	322	2759.05	254
1417.62	288	1741.77	366	2280.32	381	2773.65	194
1422.06	288	1741.89	388	2280.58	322	2774.42	194
1425.12	288	1742.46	388	2282.09	381	2781.22	170
1432.07	196	1744.77	366	2282.35	381	2786.91	170
1437.33	196	1763.33	355	2283.86	322	2787.22	170
1460.60	412	1768.72	355	2285.58	381	2792.94	170
1460.69	412	1796.31	299	2287.26	381	2793.46	194
1460.90	412	1796.63	299	2303.17	330	2795.21	298
1460.98	412	1824.09	402	2311.11	330	2795.99	298
1462.57	340	1824.72	402	2384.46	255	2801.39	416
1463.89	340	1830.09	402	2384.71	255	2801.80	170
1465.46	340	1830.73	402	2421.31	264	2802.89	416
1466.79	340	1842.43	360	2421.47	264	2803.57	170
1467.70	316	1843.59	360	2435.01	264	2805.17	416
1468.32	316	1854.33	214	2466.93	312	2805.43	283
1468.92	340	1854.86	214	2468.69	312	2805.87	136
1468.95	316	1921.79	229	2483.23	354	2810.40	283
1472.13	340	1923.28	229	2484.12	285	2812.50	170
1474.27	340	1923.38	229	2493.39	137	2816.53	136
1513.96	300	1942.51	158	2493.47	285	2817.37	283
1518.79	300	1942.89	158	2493.75	137	2822.38	283
1519.02	300	1946.89	158	2493.95	354	2828.58	405
1522.81	325	1947.26	158	2493.99	143	2829.17	170
1524.27	325	1978.51	315	2495.75	354	2835.67	404
1524.95	325	1984.36	315	2497.50	365	2836.27	136
1556.71	258	1985.51	315	2499.27	137	2842.34	405
1556.82	258			2501.81	137	2846.97	404
1558.78	195	in air		2506.52	365	2866.38	408
1559.16	195			2507.73	137	2878.68	408
1565.02	195	2004.96	228	2509.22	137	2880.50	408
1571.74	265	2006.69	228	2510.58	143	2892.92	408
1577.50	265	2007.50	228	2517.37	137	2904.69	311
1579.00	257	2009.23	228	2519.91	365	2906.55	311
1579.11	257	2035.18	314	2539.04	226	2907.14	311
1584.13	257	2041.45	314	2541.81	226	2916.31	156
1593.78	406	2042.66	314	2565.31	387	2921.46	156
1605.70	409	2079.33	323	2573.56	273	2925.88	272
1609.55	409	2082.06	323	2573.74	273	2926.18	156
1634.48	274	2085.80	323	2574.29	273	2926.83	272
1634.78	274	2088.54	323	2576.61	387	2928.36	272
1635.20	287	2106.28	324	2582.01	387	2957.72	401
1639.24	287	2109.08	324	2617.57	397	2973.55	401
1639.42	208	2110.28	376	2618.53	397	2980.86	120
1639.84	208	2112.11	227	2621.10	397	2982.34	120
1639.84	287	2112.73	227	2623.23	397	2999.37	297
1643.91	287	2114.02	227	2624.80	397	3000.03	297
1660.54	256	2120.13	376	2625.34	397	3000.27	297
1660.66	256	2120.58	157	2626.94	397	3004.59	120
1662.56	286	2125.79	157	2630.06	397	3017.11	282
1666.74	286	2127.39	157	2638.15	391	3019.76	282
1686.23	336	2132.64	157	2641.99	391	3028.04	142
1697.76	348	2155.33	313	2643.88	284	3033.59	282
1702.71	348	2156.68	313	2644.07	284	3039.80	207
1703.69	348	2159.10	313	2654.67	284	3040.73	207
1708.67	348	2160.46	313	2660.69	394	3047.10	155

List of tabulated lines — Continued

Wavelength (Å)	No.	Wavelength (Å)	No.	Wavelength (Å)	No.	Wavelength (Å)	No.
3051.09	155	3410.24	335	4146.66	319	6015.18	363
3052.53	142	3411.30	104	4157.52	319	6039.53	363
3057.88	155	3411.69	104	4182.22	319	6067.75	363
3061.90	155	3411.87	335	4193.27	319	6092.53	363
3063.43	98	3425.55	135	4260.47	310	6105.93	344
3071.60	98	3489.89	295	4265.74	310	6142.32	363
3128.89	321	3492.21	295	4293.77	332	6172.27	363
3131.04	321	3493.43	295	4303.94	332	6183.35	344
3135.07	321	3504.45	309	4327.38	346	6223.38	363
3137.23	321	3508.01	309	4366.12	346	6350.39	206
3141.67	411	3508.63	309	4392.96	346	6372.74	206
3143.05	411	3512.21	309	4410.46	353	6505.23	410
3153.14	193	3560.39	191	4439.65	353	6511.16	410
3168.22	193	3563.33	191	4444.38	353	6517.53	410
3177.89	169	3593.08	191	4474.02	353	6523.48	410
3180.77	169	3614.93	339	4533.46	352	6644.45	362
3180.99	169	3620.66	339	4569.31	352	6680.41	362
3185.74	169	3632.66	339	4654.10	212	6708.66	362
3188.22	169	3637.75	308	4662.66	390	6738.05	362
3188.64	169	3638.45	339	4668.97	390	6776.41	362
3194.09	193	3641.59	308	4681.00	390	6805.47	362
3194.78	169	3647.53	339	4687.03	212	6817.54	414
3199.58	169	3664.78	329	4733.52	393	6831.05	414
3208.73	296	3667.74	329	4765.29	358	6876.49	140
3209.65	169	3671.46	339	4772.60	181	6931.60	140
3209.76	296	3680.70	339	4773.03	358	7004.11	140
3215.23	296	3696.08	339	4778.32	393	7032.34	205
3216.26	296	3725.89	168	4779.10	181	7053.62	205
3216.31	169	3725.94	168	4783.42	181	7061.30	140
3233.32	375	3729.03	168	4784.95	393	7156.22	331
3236.36	375	3736.68	168	4794.18	181	7270.72	331
3251.73	413	3736.85	168	4796.66	345	7500.94	374
3256.50	375	3744.89	168	4797.28	358	7585.74	374
3256.81	413	3755.73	168	4798.27	181	7714.52	374
3258.30	413	3758.39	168	4800.74	181	7799.99	369
3259.58	375	3761.87	307	4805.12	358	7813.40	369
3263.41	413	3765.98	307	4813.15	181	7922.38	369
3273.99	413	3774.49	168	4824.05	181	7936.22	369
3275.48	192	3787.38	347	4836.81	294	8505.41	373
3275.81	413	3791.11	307	4839.15	294	8534.45	373
3286.47	413	3814.98	347	4842.67	345	8667.64	373
3301.92	192	3817.02	347	4844.31	345	9205.58	343
3303.13	192	3836.71	333	4891.24	345	9218.31	343
3348.06	141	3841.07	320	4902.27	418	9395.92	343
3349.11	141	3844.00	334	4917.22	418	9453.85	225
3354.27	176	3845.36	320	4941.29	418	9488.39	225
3362.55	176	3850.39	320	5143.39	328	9492.36	225
3363.21	364	3864.79	334	5198.22	328	9647.07	396
3366.38	364	3868.74	333	5290.01	190	9753.42	396
3368.65	364	3872.15	213	5305.51	190	9818.95	396
3375.40	176	3873.65	213	5362.51	190	9865.07	396
3378.02	141	3930.68	182	5378.44	190	9932.10	396
3379.58	364	3942.06	182	5425.90	351	9979.69	396
3381.21	135	3945.31	182	5453.13	351	9985.48	396
3381.30	135	3956.77	182	5477.33	351	10033.6	396
3385.07	364	3974.58	182	5486.65	357	10905.7	372
3385.52	135	3975.40	281	5492.98	357	11002.9	372
3390.19	135	3977.09	182	5503.26	357	11173.9	372
3390.58	364	3995.08	182	5754.89	189	11180.1	356
3396.80	135	3999.40	281	5769.14	189	11222.8	356
3401.23	104	4116.27	359	5833.71	415	11276.0	372
3403.97	364	4118.64	359	5840.80	189	11462.1	356
3405.77	135	4122.04	359	5882.45	415	11507.0	356
3409.70	135	4124.42	359	5892.51	415	11975.6	368

List of tabulated lines — Continued

Wavelength (Å)	No.	Wavelength (Å)	No.	Wavenumber (cm^{-1})	No.
12007.3	368	29577.7	392	156.900	398
12232.1	338	30012.8	392	170.800	250
12282.7	253	30212.4	371	175.200	250
12289.4	253	30861.5	380	183.300	399
12297.8	253	31418.2	392	311.000	382
12348.4	338	31525.2	384	336.900	398
12557.8	338	31527.2	380	363.300	399
12572.0	338	31735.4	380	366.000	382
12613.3	338	32182.8	380	372.600	398
12789.1	338	32351.5	384	392.000	382
12817.0	377	32364.1	371	403.100	250
12959.9	338	32425.0	380	444.000	349
13066.6	377	32692.2	380	467.700	341
15284.0	350	32892.3	380	478.000	349
15335.6	350	32907.4	380	503.200	399
15430.3	385	32980.2	384	617.000	349
15463.7	385	33125.5	367	672.700	341
15625.6	385	33885.6	384	713.000	383
15659.9	385	35451.3	367	830.000	383
15699.2	350	35730.1	367	911.000	383
15753.7	350	38096.5	251	921.800	403
21466.3	379	38194.0	251	1061.70	403
21670.1	379	38258.3	251	1070.80	403
21970.1	379	40705.5	389	1301.00	342
21996.2	379	41638.0	389	1474.00	342
22305.4	379	41899.7	407	1506.00	342
22731.4	379	42147.0	407	1679.00	342
22882.3	379	43924.8	389	1915.90	378
23330.9	379	44972.0	407	1946.80	378
26232.7	400	45012.5	389	1986.70	378
26297.5	252	48462.2	378	1998.90	378
26329.4	400	48774.3	378	1999.90	378
27384.5	252	49228.3	378		
27417.6	252				
27533.1	400				
28328.9	392				

For all multiplet data of this boron-like ion, we have used the results of extensive calculations by Yu Yan *et al.*[1] These are part of the Opacity Project (OP), which is reviewed in the general introduction. Comparisons with earlier theoretical data by Saraph,[2] which were based on similar but less extensive calculations than OP and utilized the SUPERSTRUCTURE code for the description of the O v target ion, show that about 80% of the 174 multiplets common to both calculations agree within ±25%.

An experimental test of the line ratios for the $3s$-$3p$ and $3p$-$3d$ doublets, performed by Glenzer *et al.*[3] with emission measurements in a gas liner pinch, shows very close adherence to LS-coupling ratios (deviations ≤2%). We have therefore used LS-coupling ratios throughout this spectrum and have applied the same uncertainty estimates for stronger individual lines as for the multiplet data—as we have also done for other boron-like ions. But we have made an exception to this rule for the comparatively few multiplets which contain an f electron in an upper or lower state. In these cases, significant deviations from LS-coupling are expected, so that our error estimates are increased accordingly.

Other experimental comparison data are available from six beam-foil lifetime measurements,[4–9] in which the effects of electron cascading from higher levels were approximately taken into account by applying multi-exponential fits to the decay curves. This procedure has been shown to have a systematic tendency to yield lifetimes that are too long.[10] Indeed, we have found that of the 16 ratios of beam-foil lifetimes versus OP results, all but two are above unity, usually by about 20–30%.

References

[1]Yu Yan, K. T. Taylor, and M. J. Seaton, J. Phys. B **20**, 6399 (1987), and to be published.
[2]H. E. Saraph, J. Phys. B **13**, 3129 (1980).
[3]S. Glenzer, H.-J. Kunze, J. Musielok, Y.-K. Kim, and W. L. Wiese, Phys. Rev. A. **49**, 221 (1994).
[4]J. A. Kernahan and F. R. Simpson, Nucl. Instrum. Methods **202**, 49 (1982).
[5]B. Denne, L. Engstroem, S. Huldt, J. O. Ekberg, L. J. Curtis, K. Ishii, E. Veje, and I. Martinson, Phys. Scr. **21**, 151 (1980).
[6]K. Ishii, M. Suzuki, and J. Takahashi, J. Phys. Soc. Jpn. **54**, 3742 (1985).
[7]R. Mythili, P. Singh, P. K. Bhattacharya, A. B. Parui, M. G. Betigeri, P. M. Raja Rao, G. Krishnamurthy, and V. S. Raju, Z. Phys. D **7**, 311 (1988).
[8]F. J. Coetzer, T. C. Kotze and P. van der Westhuizen, Spectrochim. Acta B **41**, 243 (1986).
[9]L. J. Ryan, L. A. Rayburn, and A. J. Cunningham, J. Quant. Spectrosc. Radiat. Transfer **42**, 295 (1989).
[10]See e.g., S. M. Younger and W. L. Wiese, Phys. Rev. A **17**, 1944 (1978).

O IV: Allowed Transitions

No.	Transition Array	Mult.	λ_{air} (Å)	λ_{vac} (Å) or σ (cm^{-1})*	E_i (cm^{-1})	E_k (cm^{-1})	g_i–g_k	A_{ki} (10^8 s^{-1})	f_{ik}	S (at. u.)	log gf	Acc.	Source
1.	$2s^22p$–$2s2p^2$	$^2P^o$–^2D		*789.36*	*257*	– *126942*	6–10	7.10+00	1.11–01	1.72+00	–0.178	B	1
				790.199	385.9	– 126936.3	4–6	7.08+00	9.94–02	1.03+00	–0.401	B	*LS*
				787.710	0	– 126950.2	2–4	5.95+00	1.11–01	5.74–01	–0.655	B	*LS*
				790.112	385.9	– 126950.2	4–4	1.18+00	1.10–02	1.15–01	–1.355	B	*LS*
2.		$^2P^o$–^2S		*609.35*	*257*	– 164366.4	6–2	3.61+01	6.69–02	8.06–01	–0.396	B	1
				609.829	385.9	– 164366.4	4–2	2.40+01	6.69–02	5.37–01	–0.573	B	*LS*
				608.397	0	– 164366.4	2–2	1.21+01	6.70–02	2.69–01	–0.873	B	*LS*
3.		$^2P^o$–^2P		*554.37*	*257*	– *180643*	6–6	7.28+01	3.35–01	3.67+00	0.303	B	1
				554.513	385.9	– 180724.2	4–4	6.06+01	2.79–01	2.04+00	0.048	B	*LS*
				554.076	0	– 180480.8	2–2	4.86+01	2.24–01	8.16–01	–0.350	B	*LS*
				555.263	385.9	– 180480.8	4–2	2.41+01	5.58–02	4.08–01	–0.651	B	*LS*
				553.329	0	– 180724.2	2–4	1.22+01	1.12–01	4.08–01	–0.650	B	*LS*
4.	$2s^22p$–$2s^23s$	$^2P^o$–^2S		*279.83*	*257*	– 357614.3	6–2	8.02+01	3.14–02	1.74–01	–0.7251	B	1
				279.933	385.9	– 357614.3	4–2	5.34+01	3.14–02	1.16–01	–0.9013	B	*LS*
				279.631	0	– 357614.3	2–2	2.68+01	3.14–02	5.78–02	–1.2019	B	*LS*
5.	$2s^22p$–$2s^23d$	$^2P^o$–^2D		*238.50*	*257*	– *419544*	6–10	3.54+02	5.04–01	2.37+00	0.480	B	1
				238.570	385.9	– 419550.6	4–6	3.54+02	4.53–01	1.42+00	0.258	B	*LS*
				238.360	0	– 419533.9	2–4	2.96+02	5.04–01	7.91–01	0.003	B	*LS*
				238.579	385.9	– 419533.9	4–4	5.90+01	5.04–02	1.58–01	–0.696	B	*LS*
6.	$2s^22p$–$2s2p(^3P^o)3p$	$^2P^o$–^2P		*214.11*	*257*	– *467306*	6–6	1.09+02	7.51–02	3.18–01	–0.346	B	1
				214.152	385.9	– 467344.9	4–4	9.10+01	6.26–02	1.77–01	–0.601	B	*LS*
				214.028	0	– 467229.3	2–2	7.30+01	5.01–02	7.06–02	–0.999	B	*LS*
				214.205	385.9	– 467229.3	4–2	3.64+01	1.25–02	3.53–02	–1.301	B	*LS*
				213.975	0	– 467344.9	2–4	1.83+01	2.51–02	3.53–02	–1.300	B	*LS*
7.		$^2P^o$–^2D		*207.23*	*257*	– *482819*	6–10	1.18+02	1.26–01	5.16–01	–0.121	B	1
				207.239	385.9	– 482921.6	4–6	1.18+02	1.13–01	3.10–01	–0.343	B	*LS*
				207.183	0	– 482666.1	2–4	9.80+01	1.26–01	1.72–01	–0.598	B	*LS*
				207.348	385.9	– 482666.1	4–4	1.96+01	1.26–02	3.44–02	–1.297	B	*LS*
8.		$^2P^o$–^2S		*202.99*	*257*	– 492890.9	6–2	1.06+02	2.18–02	8.74–02	–0.883	B	1
				203.044	385.9	– 492890.9	4–2	7.05+01	2.18–02	5.83–02	–1.060	B	*LS*
				202.885	0	– 492890.9	2–2	3.53+01	2.18–02	2.91–02	–1.360	B–	*LS*
9.	$2s^22p$–$2s^24d$	$^2P^o$–^2D		*195.96*	*257*	– *510572*	6–10	1.20+02	1.15–01	4.45–01	–0.162	B	1
				196.006	385.9	– 510574.1	4–6	1.20+02	1.03–01	2.67–01	–0.384	B	*LS*
				195.860	0	– 510569.7	2–4	9.99+01	1.15–01	1.48–01	–0.639	B	*LS*
				196.008	385.9	– 510569.7	4–4	1.99+01	1.15–02	2.96–02	–1.338	B–	*LS*
10.	$2s^22p$–$2s^25s$	$^2P^o$–^2S		*185.49*	*257*	– 539368	6–2	3.17+01	5.46–03	2.00–02	–1.485	B–	1
				185.535	385.9	– 539368	4–2	2.11+01	5.46–03	1.33–02	–1.661	B–	*LS*
				185.402	0	– 539368	2–2	1.06+01	5.46–03	6.67–03	–1.962	C+	*LS*
11.	$2s^22p$–$2s2p(^1P^o)3p$	$^2P^o$–^2D		*182.79*	*257*	– *547343*	6–10	2.09+01	1.74–02	6.29–02	–0.981	B	1
				182.826	385.9	– 547355	4–6	2.09+01	1.57–02	3.77–02	–1.203	B	*LS*
				182.706	0	– 547326	2–4	1.74+01	1.74–02	2.10–02	–1.458	B–	*LS*
				182.835	385.9	– 547326	4–4	3.48+00	1.74–03	4.19–03	–2.157	C+	*LS*

O IV: Allowed Transitions — Continued

No.	Transition Array	Mult.	λ_{air} (Å)	λ_{vac} (Å) or σ (cm^{-1})*	E_i (cm^{-1})	E_k (cm^{-1})	g_i-g_k	A_{ki} (10^8 s^{-1})	f_{ik}	S (at. u.)	log gf	Acc.	Source
12.		$^2P^o-^2P$		*181.96*	*257*	– *549834*	6–6	3.14+01	1.56–02	5.61–02	–1.028	B	1
				181.994	385.9	– 549855	4–4	2.62+01	1.30–02	3.12–02	–1.284	B	*LS*
				181.887	0	– 549792	2–2	2.10+01	1.04–02	1.25–02	–1.682	B–	*LS*
				182.015	385.9	– 549792	4–2	1.05+01	2.60–03	6.23–03	–1.983	C+	*LS*
				181.866	0	– 549855	2–4	5.25+00	5.21–03	6.23–03	–1.983	C+	*LS*
13.	$2s^22p-$ $2s^25d$	$^2P^o-^2D$		*181.23*	*257*	– *552031*	6–10	5.99+01	4.92–02	1.76–01	–0.530	B	1
				181.276	385.9	– 552032.3	4–6	5.99+01	4.43–02	1.06–01	–0.752	B	*LS*
				181.150	0	– 552029.6	2–4	5.00+01	4.92–02	5.87–02	–1.007	B	*LS*
				181.276	385.9	– 552029.6	4–4	9.98+00	4.92–03	1.17–02	–1.706	B–	*LS*
14.	$2s^22p-$ $2s2p(^1P^o)3p$	$^2P^o-^2S$		*180.44*	*257*	– 554464	6–2	4.24+01	6.90–03	2.46–02	–1.383	B–	1
				180.480	385.9	– 554464	4–2	2.83+01	6.90–03	1.64–02	–1.559	B–	*LS*
				180.354	0	– 554464	2–2	1.42+01	6.91–03	8.20–03	–1.860	C+	*LS*
15.	$2s^22p-$ $2s^26d$	$^2P^o-^2D$		*174.18*	*257*	– 574368.7	6–10	3.80+01	2.88–02	9.90–02	–0.763	B	1
				174.221	385.9	– 574368.7	4–6	3.79+01	2.59–02	5.94–02	–0.985	B	*LS*
				174.104	0	– 574368.7	2–4	3.17+01	2.88–02	3.30–02	–1.240	B	*LS*
				174.221	385.9	– 574368.7	4–4	6.32+00	2.88–03	6.60–03	–1.939	C+	*LS*
16.	$2s^22p-$ $2s2p(^3P^o)4p$	$^2P^o-^2P$		*173.89*	*257*	– *575317*	6–6	4.57+01	2.07–02	7.12–02	–0.905	B	1
				173.916	385.9	– 575375	4–4	3.81+01	1.73–02	3.96–02	–1.161	B	*LS*
				173.852	0	– 575202	2–2	3.05+01	1.38–02	1.58–02	–1.558	B–	*LS*
				173.969	385.9	– 575202	4–2	1.52+01	3.45–03	7.91–03	–1.860	C+	*LS*
				173.800	0	– 575375	2–4	7.63+00	6.91–03	7.91–03	–1.859	C+	*LS*
17.		$^2P^o-^2D$		*171.11*	*257*	– *584673*	6–10	4.93+01	3.61–02	1.22–01	–0.665	B	1
				171.123	385.9	– 584761	4–6	4.93+01	3.25–02	7.31–02	–0.887	B	*LS*
				171.074	0	– 584541	2–4	4.11+01	3.61–02	4.06–02	–1.142	B	*LS*
				171.187	385.9	– 584541	4–4	8.21+00	3.61–03	8.13–03	–1.841	C+	*LS*
18.		$^2P^o-^2S$		*169.55*	*257*	– 590069	6–2	2.04+01	2.93–03	9.80–03	–1.756	C+	1
				169.583	385.9	– 590069	4–2	1.36+01	2.93–03	6.53–03	–1.932	C+	*LS*
				169.472	0	– 590069	2–2	6.80+00	2.93–03	3.27–03	–2.232	C+	*LS*
19.	$2s2p^2-$ $2p^3$	$^4P-^4S^o$		*625.40*	*71641*	– 231537.5	12–4	6.39+01	1.25–01	3.08+00	0.176	B	1
				625.853	71755.5	– 231537.5	6–4	3.19+01	1.25–01	1.54+00	–0.126	B	*LS*
				625.127	71570.1	– 231537.5	4–4	2.13+01	1.25–01	1.03+00	–0.301	B	*LS*
				624.619	71439.8	– 231537.5	2–4	1.07+01	1.25–01	5.14–01	–0.602	B	*LS*
20.	$2s2p^2-$ $2s2p(^3P^o)3s$	$^4P-^4P^o$		*272.15*	*71641*	– *439085*	12–12	1.03+02	1.15–01	1.23+00	0.139	B	1
				272.127	71755.5	– 439230.9	6–6	7.23+01	8.03–02	4.32–01	–0.317	B	*LS*
				272.173	71570.1	– 438983.9	4–4	1.38+01	1.53–02	5.48–02	–1.213	B	*LS*
				272.176	71439.8	– 438849.0	2–2	1.72+01	1.91–02	3.43–02	–1.418	B	*LS*
				272.310	71755.5	– 438983.9	6–4	4.64+01	3.44–02	1.85–01	–0.685	B	*LS*
				272.273	71570.1	– 438849.0	4–2	8.60+01	4.78–02	1.71–01	–0.719	B	*LS*
				271.990	71570.1	– 439230.9	4–6	3.10+01	5.16–02	1.85–01	–0.685	B	*LS*
				272.076	71439.8	– 438983.9	2–4	4.31+01	9.56–02	1.71–01	–0.718	B	*LS*

O IV: Allowed Transitions — Continued

No.	Transition Array	Mult.	λ_{air} (Å)	λ_{vac} (Å) or σ (cm^{-1})*	E_i (cm^{-1})	E_k (cm^{-1})	g_i-g_k	A_{ki} (10^8 s^{-1})	f_{ik}	S (at. u.)	log gf	Acc.	Source
21.	$2s2p^2-$ $2s2p(^3P^o)3d$	$^4P-^4D^o$		*233.53*	*71641*	*499851*	12–20	5.27+02	7.18–01	6.63+00	0.935	B	1
				233.562	71755.5	499907.0	6–8	5.27+02	5.75–01	2.65+00	0.537	B	*LS*
				233.496	71570.1	499842.6	4–6	3.69+02	4.53–01	1.39+00	0.258	B	*LS*
				233.451	71439.8	499795.7	2–4	2.20+02	3.59–01	5.52–01	–0.144	B	*LS*
				233.597	71755.5	499842.6	6–6	1.58+02	1.29–01	5.96–01	–0.110	B	*LS*
				233.522	71570.1	499795.7	4–4	2.81+02	2.30–01	7.07–01	–0.036	B	*LS*
				233.466	71439.8	499767.2	2–2	4.40+02	3.59–01	5.52–01	–0.144	B	*LS*
				233.623	71755.5	499795.7	6–4	2.63+01	1.44–02	6.63–02	–1.065	B	*LS*
				233.537	71570.1	499767.2	4–2	8.78+01	3.59–02	1.10–01	–0.843	B	*LS*
22.		$^4P-^4P^o$		*231.20*	*71641*	*504165*	12–12	2.86+02	2.29–01	2.09+00	0.439	B	1
				231.299	71755.5	504095.7	6–6	2.00+02	1.60–01	7.32–01	–0.017	B	*LS*
				231.140	71570.1	504208.9	4–4	3.81+01	3.05–02	9.30–02	–0.913	B	*LS*
				231.031	71439.8	504282.3	2–2	4.77+01	3.82–02	5.81–02	–1.117	B	*LS*
				231.239	71755.5	504208.9	6–4	1.29+02	6.87–02	3.14–01	–0.385	B	*LS*
				231.100	71570.1	504282.3	4–2	2.38+02	9.55–02	2.91–01	–0.418	B	*LS*
				231.200	71570.1	504095.7	4–6	8.57+01	1.03–01	3.14–01	–0.385	B	*LS*
				231.070	71439.8	504208.9	2–4	1.19+02	1.91–01	2.91–01	–0.418	B	*LS*
23.	$2s2p^2-$ $2s2p(^3P^o)4s$	$^4P-^4P^o$		*201.01*	*71641*	*569137*	12–12	3.67+01	2.22–02	1.77–01	–0.574	B	1
				200.994	71755.5	569283	6–6	2.57+01	1.56–02	6.18–02	–1.030	B	*LS*
				201.019	71570.1	569036	4–4	4.89+00	2.97–03	7.85–03	–1.926	C+	*LS*
				201.021	71439.8	568901	2–2	6.12+00	3.71–03	4.91–03	–2.130	C+	*LS*
				201.094	71755.5	569036	6–4	1.65+01	6.67–03	2.65–02	–1.398	B–	*LS*
				201.073	71570.1	568901	4–2	3.06+01	9.26–03	2.45–02	–1.431	B–	*LS*
				200.919	71570.1	569283	4–6	1.10+01	1.00–02	2.65–02	–1.397	B–	*LS*
				200.966	71439.8	569036	2–4	1.53+01	1.85–02	2.45–02	–1.431	B–	*LS*
24.	$2s2p^2-$ $2s2p(^3P^o)4d$	$^4P-^4D^o$		*192.18*	*71641*	*591977*	12–20	2.20+02	2.03–01	1.54+00	0.386	B	1
				192.200	71755.5	592046	6–8	2.20+02	1.62–01	6.16–01	–0.012	B	*LS*
				192.163	71570.1	591962	4–6	1.54+02	1.28–01	3.23–01	–0.291	B	*LS*
				192.138	71439.8	591899	2–4	9.17+01	1.01–01	1.28–01	–0.693	B	*LS*
				192.231	71755.5	591962	6–6	6.59+01	3.65–02	1.39–01	–0.659	B	*LS*
				192.186	71570.1	591899	4–4	1.17+02	6.49–02	1.64–01	–0.586	B	*LS*
				192.138	71439.8	591899	2–2	1.83+02	1.01–01	1.28–01	–0.693	B	*LS*
				192.255	71755.5	591899	6–4	1.10+01	4.06–03	1.54–02	–1.614	B–	*LS*
				192.186	71570.1	591899	4–2	3.66+01	1.01–02	2.57–02	–1.392	B–	*LS*
25.		$^4P-^4P^o$		*191.68*	*71641*	*593351*	12–12	1.19+02	6.56–02	4.97–01	–0.104	B	1
				191.748	71755.5	593273	6–6	8.33+01	4.59–02	1.74–01	–0.560	B	*LS*
				191.631	71570.1	593406	4–4	1.59+01	8.75–03	2.21–02	–1.456	B–	*LS*
				191.559	71439.8	593473	2–2	1.99+01	1.09–02	1.38–02	–1.660	B–	*LS*
				191.699	71755.5	593406	6–4	5.36+01	1.97–02	7.45–02	–0.928	B	*LS*
				191.607	71570.1	593473	4–2	9.93+01	2.73–02	6.90–02	–0.961	B	*LS*
				191.680	71570.1	593273	4–6	3.57+01	2.95–02	7.45–02	–0.928	B	*LS*
				191.583	71439.8	593406	2–4	4.97+01	5.47–02	6.90–02	–0.961	B	*LS*
26.	$2s2p^2-$ $2p^2(^3P)3p$	$^4P-^4P^o$		*186.92*	*71641*	*606628*	12–12	3.66+01	1.91–02	1.41–01	–0.639	B	1
				186.937	71755.5	606694	6–6	2.56+01	1.34–02	4.95–02	–1.095	B	*LS*
				186.913	71570.1	606578	4–4	4.87+00	2.55–03	6.28–03	–1.991	C+	*LS*
				186.884	71439.8	606530	2–2	6.10+00	3.19–03	3.93–03	–2.195	C+	*LS*
				186.978	71755.5	606578	6–4	1.64+01	5.74–03	2.12–02	–1.463	B–	*LS*
				186.930	71570.1	606530	4–2	3.05+01	7.98–03	1.96–02	–1.496	B–	*LS*
				186.873	71570.1	606694	4–6	1.10+01	8.62–03	2.12–02	–1.462	B–	*LS*
				186.868	71439.8	606578	2–4	1.52+01	1.60–02	1.96–02	–1.496	B–	*LS*

O IV: Allowed Transitions — Continued

No.	Transition Array	Mult.	λ_{air} (Å)	λ_{vac} (Å) or σ (cm^{-1})*	E_i (cm^{-1})	E_k (cm^{-1})	g_i–g_k	A_{ki} (10^8 s^{-1})	f_{ik}	S (at. u.)	log gf	Acc.	Source
27.		$^4P-^4S^\circ$		*183.41*	*71641*	– 616881	12–4	4.77+01	8.02–03	5.81–02	–1.017	B	1
				183.444	71755.5	– 616881	6–4	2.38+01	8.02–03	2.91–02	–1.318	B–	*LS*
				183.382	71570.1	– 616881	4–4	1.59+01	8.02–03	1.94–02	–1.494	B–	*LS*
				183.338	71439.8	– 616881	2–4	7.96+00	8.02–03	9.68–03	–1.795	C+	*LS*
28.	$2s2p^2-$ $2s2p(^3P^\circ)5d$	$^4P-^4P^\circ$		*177.64*	*71641*	– *634578*	12–12	5.33+01	2.52–02	1.77–01	–0.519	B	1
				177.693	71755.5	– 634523	6–6	3.73+01	1.76–02	6.19–02	–0.975	B	*LS*
				177.604	71570.1	– 634622	4–4	7.11+00	3.36–03	7.86–03	–1.871	C+	*LS*
				177.553	71439.8	– 634653	2–2	8.89+00	4.20–03	4.91–03	–2.075	C+	*LS*
				177.662	71755.5	– 634622	6–4	2.40+01	7.56–03	2.65–02	–1.343	B–	*LS*
				177.594	71570.1	– 634653	4–2	4.44+01	1.05–02	2.46–02	–1.377	B–	*LS*
				177.635	71570.1	– 634523	4–6	1.60+01	1.13–02	2.65–02	–1.343	B–	*LS*
				177.562	71439.8	– 634622	2–4	2.22+01	2.10–02	2.46–02	–1.376	B–	*LS*
29.	$2s2p^2-$ $2p^3$	$^2D-^2D^\circ$		*779.88*	*126942*	– *255168*	10–10	1.46+01	1.33–01	3.41+00	0.123	B	1
				779.912	126936.3	– 255155.9	6–6	1.36+01	1.24–01	1.91+00	–0.129	B	*LS*
				779.820	126950.2	– 255184.9	4–4	1.31+01	1.19–01	1.23+00	–0.321	B	*LS*
				779.736	126936.3	– 255184.9	6–4	1.46+00	8.85–03	1.36–01	–1.275	B	*LS*
				779.997	126950.2	– 255155.9	4–6	9.70–01	1.33–02	1.36–01	–1.275	B	*LS*
30.		$^2D-^2P^\circ$		*616.98*	*126942*	– *289021*	10–6	2.89+01	9.91–02	2.01+00	–0.004	B	1
				616.952	126936.3	– 289023.5	6–4	2.60+01	9.91–02	1.21+00	–0.226	B	*LS*
				617.036	126950.2	– 289015.4	4–2	2.89+01	8.25–02	6.71–01	–0.481	B	*LS*
				617.005	126950.2	– 289023.5	4–4	2.89+00	1.65–02	1.34–01	–1.180	B	*LS*
31.	$2s2p^2-$ $2s^23p$	$^2D-^2P^\circ$		*379.83*	*126942*	– *390219*	10–6	4.91+00	6.37–03	7.97–02	–1.196	B	1
				379.778	126936.3	– 390248.0	6–4	4.42+00	6.37–03	4.78–02	–1.417	B	*LS*
				379.923	126950.2	– 390161.2	4–2	4.91+00	5.31–03	2.66–02	–1.673	B–	*LS*
				379.798	126950.2	– 390248.0	4–4	4.91–01	1.06–03	5.31–03	–2.372	C+	*LS*
32.	$2s2p^2-$ $2s2p(^3P^\circ)3s$	$^2D-^2P^\circ$		*306.71*	*126942*	– *452983*	10–6	5.02+01	4.25–02	4.29–01	–0.372	B	1
				306.621	126936.3	– 453071.5	6–4	4.52+01	4.25–02	2.57–01	–0.593	B	*LS*
				306.884	126950.2	– 452806.6	4–2	5.01+01	3.54–02	1.43–01	–0.849	B	*LS*
				306.634	126950.2	– 453071.5	4–4	5.03+00	7.08–03	2.86–02	–1.548	B–	*LS*
33.	$2s2p^2-$ $2s2p(^3P^\circ)3d$	$^2D-^2D^\circ$		*266.95*	*126942*	– *501542*	10–10	1.50+02	1.60–01	1.41+00	0.204	B	1
				266.931	126936.3	– 501564.4	6–6	1.40+02	1.49–01	7.87–01	–0.048	B	*LS*
				266.981	126950.2	– 501509.2	4–4	1.35+02	1.44–01	5.06–01	–0.240	B	*LS*
				266.971	126936.3	– 501509.2	6–4	1.50+01	1.07–02	5.62–02	–1.194	B	*LS*
				266.941	126950.2	– 501564.4	4–6	9.98+00	1.60–02	5.62–02	–1.194	B	*LS*
34.		$^2D-^2F^\circ$		*260.46*	*126942*	– *510878*	10–14	3.13+02	4.46–01	3.83+00	0.650	B	1
				260.389	126936.3	– 510977.2	6–8	3.14+02	4.25–01	2.19+00	0.407	B	*LS*
				260.556	126950.2	– 510744.9	4–6	2.92+02	4.46–01	1.53+00	0.252	B	*LS*
				260.547	126936.3	– 510744.9	6–6	2.09+01	2.12–02	1.09–01	–0.894	B	*LS*
35.	$2s2p^2-$ $2s^24f$	$^2D-^2F^\circ$		*258.90*	*126942*	– *513194*	10–14	4.44+00	6.25–03	5.33–02	–1.204	B	1
				258.892	126936.3	– 513198.3	6–8	4.45+00	5.96–03	3.05–02	–1.447	C+	*LS*
				258.908	126950.2	– 513187.2	4–6	4.15+00	6.25–03	2.13–02	–1.602	C	*LS*
				258.899	126936.3	– 513187.2	6–6	2.96–01	2.98–04	1.52–03	–2.748	D	*LS*

O IV: Allowed Transitions — Continued

No.	Transition Array	Mult.	λ_{air} (Å)	λ_{vac} (Å) or σ (cm^{-1})*	E_i (cm^{-1})	E_k (cm^{-1})	g_i-g_k	A_{ki} (10^8 s^{-1})	f_{ik}	S (at. u.)	log gf	Acc.	Source
36.	$2s2p^2-$ $2s2p(^3P^o)3d$	$^2D-^2P^o$		*258.18*	*126942*	– *514271*	10–6	1.63+01	9.75–03	8.29–02	–1.011	B	1
				258.208	126936.3	– 514220.4	6–4	1.46+01	9.75–03	4.97–02	–1.233	B	*LS*
				258.117	126950.2	– 514371.3	4–2	1.63+01	8.13–03	2.76–02	–1.488	B–	*LS*
				258.218	126950.2	– 514220.4	4–4	1.63+00	1.63–03	5.53–03	–2.187	C+	*LS*
37.	$2s2p^2-$ $2s2p(^1P^o)3s$	$^2D-^2P^o$		*255.26*	*126942*	– *518706*	10–6	3.13+01	1.83–02	1.54–01	–0.737	B	1
				255.250	126936.3	– 518709	6–4	2.81+01	1.83–02	9.24–02	–0.959	B	*LS*
				255.266	126950.2	– 518699	4–2	3.13+01	1.53–02	5.13–02	–1.214	B	*LS*
				255.259	126950.2	– 518709	4–4	3.13+00	3.05–03	1.03–02	–1.913	B–	*LS*
38.	$2s2p^2-$ $2s^25f$	$^2D-^2F^o$		*234.99*	*126942*	– 552495.1	10–14	1.45+01	1.68–02	1.30–01	–0.775	B	1
				234.985	126936.3	– 552495.1	6–8	1.45+01	1.60–02	7.42–02	–1.018	C+	*LS*
				234.993	126950.2	– 552495.1	4–6	1.35+01	1.68–02	5.19–02	–1.173	C+	*LS*
				234.985	126936.3	– 552495.1	6–6	9.65–01	7.99–04	3.71–03	–2.319	C–	*LS*
39.	$2s2p^2-$ $2s2p(^3P^o)4s$	$^2D-^2P^o$		*223.77*	*126942*	– *573833*	10–6	1.39+01	6.24–03	4.60–02	–1.205	B	1
				223.731	126936.3	– 573901	6–4	1.25+01	6.25–03	2.76–02	–1.426	B–	*LS*
				223.841	126950.2	– 573696	4–2	1.38+01	5.20–03	1.53–02	–1.682	B–	*LS*
				223.738	126950.2	– 573901	4–4	1.39+00	1.04–03	3.07–03	–2.381	C+	*LS*
40.	$2s2p^2-$ $2s^26f$	$^2D-^2F^o$		*223.28*	*126942*	– 574807.7	10–14	9.30+01	9.74–02	7.16–01	–0.012	B	1
				223.278	126936.3	– 574807.7	6–8	9.31+01	9.27–02	4.09–01	–0.255	C+	*LS*
				223.285	126950.2	– 574807.7	4–6	8.68+01	9.74–02	2.86–01	–0.410	C+	*LS*
				223.278	126936.3	– 574807.7	6–6	6.20+00	4.64–03	2.04–02	–1.556	C	*LS*
41.	$2s2p^2-$ $2s2p(^1P^o)3d$	$^2D-^2D^o$		*222.77*	*126942*	– *575839*	10–10	8.21+01	6.11–02	4.48–01	–0.214	B	1
				222.758	126936.3	– 575853	6–6	7.66+01	5.70–02	2.51–01	–0.466	B	*LS*
				222.782	126950.2	– 575819	4–4	7.38+01	5.50–02	1.61–01	–0.658	B	*LS*
				222.775	126936.3	– 575819	6–4	8.21+00	4.07–03	1.79–02	–1.612	B–	*LS*
				222.765	126950.2	– 575853	4–6	5.47+00	6.11–03	1.79–02	–1.612	B–	*LS*
42.	$2s2p^2-$ $2s^27f$	$^2D-^2F^o$		*216.96*	*126942*	– 587850	10–14	2.59+00	2.56–03	1.83–02	–1.591	B–	1
				216.960	126936.3	– 587850	6–8	2.59+00	2.44–03	1.05–02	–1.834	C	*LS*
				216.967	126950.2	– 587850	4–6	2.42+00	2.56–03	7.32–03	–1.989	C–	*LS*
				216.960	126936.3	– 587850	6–6	1.73–01	1.22–04	5.23–04	–3.135	D–	*LS*
43.	$2s2p^2-$ $2s2p(^3P^o)4d$	$^2D-^2D^o$		*214.25*	*126942*	– *593676*	10–10	2.67+01	1.84–02	1.30–01	–0.735	B	1
				214.237	126936.3	– 593708	6–6	2.50+01	1.72–02	7.27–02	–0.987	B	*LS*
				214.281	126950.2	– 593627	4–4	2.41+01	1.66–02	4.67–02	–1.179	B	*LS*
				214.275	126936.3	– 593627	6–4	2.67+00	1.23–03	5.19–03	–2.133	C+	*LS*
				214.244	126950.2	– 593708	4–6	1.78+00	1.84–03	5.19–03	–2.133	C+	*LS*
44.		$^2D-^2F^o$		*213.01*	*126942*	– *596398*	10–14	1.05+02	1.00–01	7.02–01	0.000	B	1
				212.975	126936.3	– 596475	6–8	1.05+02	9.53–02	4.01–01	–0.243	B	*LS*
				213.063	126950.2	– 596295	4–6	9.80+01	1.00–01	2.81–01	–0.398	B	*LS*
				213.057	126936.3	– 596295	6–6	7.00+00	4.76–03	2.00–02	–1.544	B–	*LS*
45.	$2s2p^2-$ $2p^2(^3P)3p$	$^2D-^2D^o$		*204.71*	*126942*	– *615443*	10–10	7.01+00	4.41–03	2.97–02	–1.356	B–	1
				204.711	126936.3	– 615431	6–6	6.55+00	4.11–03	1.66–02	–1.608	B–	*LS*
				204.704	126950.2	– 615460	4–4	6.31+00	3.97–03	1.07–02	–1.800	B–	*LS*
				204.698	126936.3	– 615460	6–4	7.02–01	2.94–04	1.19–03	–2.754	C	*LS*
				204.716	126950.2	– 615431	4–6	4.68–01	4.41–04	1.19–03	–2.754	C	*LS*

O IV: Allowed Transitions — Continued

No.	Transition Array	Mult.	λ_{air} (Å)	λ_{vac} (Å) or σ (cm^{-1})*	E_i (cm^{-1})	E_k (cm^{-1})	g_i–g_k	A_{ki} (10^8 s^{-1})	f_{ik}	S (at. u.)	log gf	Acc.	Source
46.	$2s2p^2$– $2p^2(^1D)3p$	^2D–$^2D^o$		*198.03*	*126942*	– 631913	10–10	6.56+01	3.86–02	2.51–01	–0.414	B	1
				198.029	126936.3	– 631913	6–6	6.12+01	3.60–02	1.41–01	–0.666	B	*LS*
				198.034	126950.2	– 631913	4–4	5.90+01	3.47–02	9.05–02	–0.858	B	*LS*
				198.029	126936.3	– 631913	6–4	6.56+00	2.57–03	1.01–02	–1.812	B–	*LS*
				198.034	126950.2	– 631913	4–6	4.37+00	3.86–03	1.01–02	–1.812	B–	*LS*
47.	$2s2p^2$– $2s2p(^1P^o)4d$	^2D–$^2D^o$		*184.63*	*126942*	– *668555*	10–10	3.09+00	1.58–03	9.60–03	–1.802	C+	1
				184.632	126936.3	– 668553	6–6	2.88+00	1.47–03	5.38–03	–2.053	C+	*LS*
				184.636	126950.2	– 668557	4–4	2.78+00	1.42–03	3.46–03	–2.245	C+	*LS*
				184.631	126936.3	– 668557	6–4	3.09–01	1.05–04	3.84–04	–3.199	D	*LS*
				184.637	126950.2	– 668553	4–6	2.06–01	1.58–04	3.84–04	–3.199	D	*LS*
48.	$2s2p^2$– $2p^3$	^2S–$^2P^o$		*802.22*	164366.4	– *289021*	2–6	4.05+00	1.17–01	6.19–01	–0.630	B	1
				802.201	164366.4	– 289023.5	2–4	4.05+00	7.82–02	4.13–01	–0.806	B	*LS*
				802.253	164366.4	– 289015.4	2–2	4.05+00	3.91–02	2.06–01	–1.107	B	*LS*
49.	$2s2p^2$– $2s^23p$	^2S–$^2P^o$		*442.77*	164366.4	– *390219*	2–6	1.07+00	9.47–03	2.76–02	–1.723	B–	1
				442.710	164366.4	– 390248.0	2–4	1.07+00	6.31–03	1.84–02	–1.899	B–	*LS*
				442.880	164366.4	– 390161.2	2–2	1.07+00	3.16–03	9.20–03	–2.200	C+	*LS*
50.	$2s2p^2$– $2s2p(^3P^o)3s$	^2S–$^2P^o$		*346.48*	164366.4	– *452983*	2–6	1.43+01	7.71–02	1.76–01	–0.812	B	1
				346.374	164366.4	– 453071.5	2–4	1.43+01	5.14–02	1.17–01	–0.988	B	*LS*
				346.692	164366.4	– 452806.6	2–2	1.43+01	2.57–02	5.86–02	–1.289	B	*LS*
51.	$2s2p^2$– $2s2p(^3P^o)3d$	^2S–$^2P^o$		*285.79*	164366.4	– *514271*	2–6	1.97+02	7.25–01	1.36+00	0.161	B	1
				285.834	164366.4	– 514220.4	2–4	1.97+02	4.83–01	9.09–01	–0.015	B	*LS*
				285.710	164366.4	– 514371.3	2–2	1.97+02	2.42–01	4.55–01	–0.316	B	*LS*
52.	$2s2p^2$– $2s2p(^3P^o)4s$	^2S–$^2P^o$		*244.22*	164366.4	– *573833*	2–6	4.15+00	1.11–02	1.79–02	–1.652	B–	1
				244.180	164366.4	– 573901	2–4	4.15+00	7.42–03	1.19–02	–1.828	B–	*LS*
				244.302	164366.4	– 573696	2–2	4.15+00	3.71–03	5.97–03	–2.130	C+	*LS*
53.	$2s2p^2$– $2s2p(^1P^o)3d$	^2S–$^2P^o$		*239.60*	164366.4	– *581736*	2–6	6.36+01	1.64–01	2.59–01	–0.484	B	1
				239.592	164366.4	– 581743	2–4	6.36+01	1.09–01	1.73–01	–0.660	B	*LS*
				239.604	164366.4	– 581721	2–2	6.36+01	5.47–02	8.63–02	–0.961	B	*LS*
54.	$2s2p^2$– $2s2p(^3P^o)4d$	^2S–$^2P^o$		*230.73*	164366.4	– *597770*	2–6	8.34+01	2.00–01	3.03–01	–0.399	B	1
				230.758	164366.4	– 597720	2–4	8.34+01	1.33–01	2.02–01	–0.575	B	*LS*
				230.679	164366.4	– 597869	2–2	8.34+01	6.66–02	1.01–01	–0.876	B	*LS*
55.	$2s2p^2$– $2p^3$	^2P–$^2D^o$		*1341.8*	*180643*	– *255168*	6–10	2.58+00	1.16–01	3.08+00	–0.157	B	1
				1343.51	180724.2	– 255155.9	4–6	2.57+00	1.04–01	1.85+00	–0.379	B	*LS*
				1338.61	180480.8	– 255184.9	2–4	2.17+00	1.17–01	1.03+00	–0.633	B	*LS*
				1342.99	180724.2	– 255184.9	4–4	4.29–01	1.16–02	2.05–01	–1.333	B	*LS*
56.		^2P–$^2P^o$		*922.70*	*180643*	– *289021*	6–6	1.32+01	1.68–01	3.07+00	0.005	B	1
				923.367	180724.2	– 289023.5	4–4	1.10+01	1.40–01	1.71+00	–0.251	B	*LS*
				921.365	180480.8	– 289015.4	2–2	8.83+00	1.12–01	6.82–01	–0.648	B	*LS*
				923.436	180724.2	– 289015.4	4–2	4.39+00	2.80–02	3.41–01	–0.950	B	*LS*
				921.296	180480.8	– 289023.5	2–4	2.21+00	5.62–02	3.41–01	–0.949	B	*LS*

O IV: Allowed Transitions — Continued

No.	Transition Array	Mult.	λ_{air} (Å)	λ_{vac} (Å) or σ (cm^{-1})*	E_i (cm^{-1})	E_k (cm^{-1})	g_i-g_k	A_{ki} (10^8 s^{-1})	f_{ik}	S (at. u.)	log gf	Acc.	Source
57.	$2s2p^2-$ $2s2p(^3P^\circ)3s$	$^2P-^2P^\circ$		*367.19*	*180643*	– *452983*	6–6	4.68+00	9.46–03	6.86–02	–1.246	B	1
				367.178	180724.2	– 453071.5	4–4	3.90+00	7.88–03	3.81–02	–1.501	B	*LS*
				367.207	180480.8	– 452806.6	2–2	3.12+00	6.31–03	1.52–02	–1.899	B–	*LS*
				367.536	180724.2	– 452806.6	4–2	1.56+00	1.57–03	7.62–03	–2.201	C+	*LS*
				366.850	180480.8	– 453071.5	2–4	7.82–01	3.16–03	7.62–03	–2.200	C+	*LS*
58.	$2s2p^2-$ $2s2p(^3P^\circ)3d$	$^2P-^2D^\circ$		*311.62*	*180643*	– *501542*	6–10	4.88+01	1.19–01	7.30–01	–0.148	B	1
				311.682	180724.2	– 501564.4	4–6	4.88+01	1.07–01	4.38–01	–0.370	B	*LS*
				311.499	180480.8	– 501509.2	2–4	4.08+01	1.19–01	2.43–01	–0.625	B	*LS*
				311.735	180724.2	– 501509.2	4–4	8.13+00	1.18–02	4.86–02	–1.324	B	*LS*
59.		$^2P-^2P^\circ$		*299.74*	*180643*	– *514271*	6–6	6.33+01	8.52–02	5.05–01	–0.291	B	1
				299.853	180724.2	– 514220.4	4–4	5.27+01	7.10–02	2.80–01	–0.547	B	*LS*
				299.499	180480.8	– 514371.3	2–2	4.23+01	5.69–02	1.12–01	–0.944	B	*LS*
				299.718	180724.2	– 514371.3	4–2	2.11+01	1.42–02	5.61–02	–1.245	B	*LS*
				299.635	180480.8	– 514220.4	2–4	1.06+01	2.84–02	5.61–02	–1.245	B	*LS*
60.	$2s2p^2-$ $2s2p(^1P^\circ)3s$	$^2P-^2P^\circ$		*295.80*	*180643*	– *518706*	6–6	2.77+01	3.63–02	2.12–01	–0.662	B	1
				295.871	180724.2	– 518709	4–4	2.30+01	3.02–02	1.18–01	–0.918	B	*LS*
				295.667	180480.8	– 518699	2–2	1.85+01	2.42–02	4.71–02	–1.315	B	*LS*
				295.880	180724.2	– 518699	4–2	9.21+00	6.05–03	2.36–02	–1.616	B–	*LS*
				295.658	180480.8	– 518709	2–4	4.62+00	1.21–02	2.36–02	–1.616	B–	*LS*
61.	$2s2p^2-$ $2s^25p$	$^2P-^2P^\circ$		*273.10*	*180643*	– *546814*	6–6	1.87+00	2.09–03	1.13–02	–1.901	B–	1
				273.154	180724.2	– 546818.7	4–4	1.56+00	1.75–03	6.28–03	–2.156	C+	*LS*
				272.983	180480.8	– 546803.4	2–2	1.25+00	1.40–03	2.51–03	–2.554	C	*LS*
				273.165	180724.2	– 546803.4	4–2	6.24–01	3.49–04	1.26–03	–2.855	C	*LS*
				272.972	180480.8	– 546818.7	2–4	3.13–01	6.99–04	1.26–03	–2.855	C	*LS*
62.	$2s2p^2-$ $2s2p(^1P^\circ)3d$	$^2P-^2D^\circ$		*253.04*	*180643*	– *575839*	6–10	2.85+02	4.56–01	2.28+00	0.437	B	1
				253.082	180724.2	– 575853	4–6	2.85+02	4.10–01	1.37+00	0.215	B	*LS*
				252.948	180480.8	– 575819	2–4	2.38+02	4.56–01	7.59–01	–0.040	B	*LS*
				253.104	180724.2	– 575819	4–4	4.74+01	4.56–02	1.52–01	–0.739	B	*LS*
63.		$^2P-^2P^\circ$		*249.32*	*180643*	– *581736*	6–6	1.39+02	1.29–01	6.37–01	–0.110	B	1
				249.365	180724.2	– 581743	4–4	1.16+02	1.08–01	3.54–01	–0.366	B	*LS*
				249.227	180480.8	– 581721	2–2	9.26+01	8.62–02	1.41–01	–0.763	B	*LS*
				249.379	180724.2	– 581721	4–2	4.62+01	2.15–02	7.07–02	–1.065	B	*LS*
				249.214	180480.8	– 581743	2–4	2.32+01	4.31–02	7.07–02	–1.064	B	*LS*
64.	$2s2p^2-$ $2s2p(^3P^\circ)4d$	$^2P-^2D^\circ$		*242.11*	*180643*	– *593676*	6–10	9.82+01	1.44–01	6.88–01	–0.064	B	1
				242.140	180724.2	– 593708	4–6	9.81+01	1.29–01	4.13–01	–0.286	B	*LS*
				242.045	180480.8	– 593627	2–4	8.19+01	1.44–01	2.29–01	–0.541	B	*LS*
				242.188	180724.2	– 593627	4–4	1.63+01	1.44–02	4.58–02	–1.240	B	*LS*
65.		$^2P-^2P^\circ$		*239.74*	*180643*	– *597770*	6–6	1.48+01	1.27–02	6.03–02	–1.117	B	1
				239.811	180724.2	– 597720	4–4	1.23+01	1.06–02	3.35–02	–1.372	B	*LS*
				239.585	180480.8	– 597869	2–2	9.87+00	8.49–03	1.34–02	–1.770	B–	*LS*
				239.725	180724.2	– 597869	4–2	4.93+00	2.12–03	6.70–03	–2.071	C+	*LS*
				239.671	180480.8	– 597720	2–4	2.47+00	4.25–03	6.70–03	–2.071	C+	*LS*

O IV: Allowed Transitions — Continued

No.	Transition Array	Mult.	λ_{air} (Å)	λ_{vac} (Å) or σ (cm^{-1})*	E_i (cm^{-1})	E_k (cm^{-1})	g_i–g_k	A_{ki} (10^8 s^{-1})	f_{ik}	S (at. u.)	log gf	Acc.	Source
66.	$2s2p^2$– $2p^2(^3P)3p$	$^2P-^2S^o$		*240.03*	*180643*	– 597254	6–2	8.72+01	2.51–02	1.19–01	–0.822	B	1
				240.079	180724.2	– 597254	4–2	5.81+01	2.51–02	7.94–02	–0.998	B	*LS*
				239.939	180480.8	– 597254	2–2	2.91+01	2.51–02	3.97–02	–1.299	B	*LS*
67.		$^2P-^2D^o$		*229.99*	*180643*	– *615443*	6–10	6.11+01	8.08–02	3.67–01	–0.315	B	1
				230.040	180724.2	– 615431	4–6	6.11+01	7.27–02	2.20–01	–0.537	B	*LS*
				229.896	180480.8	– 615460	2–4	5.10+01	8.08–02	1.22–01	–0.792	B	*LS*
				230.025	180724.2	– 615460	4–4	1.02+01	8.08–03	2.45–02	–1.491	B–	*LS*
68.	$2s2p^2$– $2p^2(^1D)3p$	$^2P-^2D^o$		*221.60*	*180643*	– 631913	6–10	4.33+01	5.31–02	2.32–01	–0.497	B	1
				221.637	180724.2	– 631913	4–6	4.32+01	4.78–02	1.39–01	–0.719	B	*LS*
				221.517	180480.8	– 631913	2–4	3.61+01	5.31–02	7.74–02	–0.974	B	*LS*
				221.637	180724.2	– 631913	4–4	7.20+00	5.31–03	1.55–02	–1.673	B–	*LS*
69.	$2s2p^2$– $2s2p(^1P^o)4d$	$^2P-^2D^o$		*204.96*	*180643*	– *668555*	6–10	1.03+02	1.08–01	4.37–01	–0.189	B	1
				204.990	180724.2	– 668553	4–6	1.03+02	9.70–02	2.62–01	–0.411	B	*LS*
				204.886	180480.8	– 668557	2–4	8.57+01	1.08–01	1.46–01	–0.666	B	*LS*
				204.988	180724.2	– 668557	4–4	1.71+01	1.08–02	2.91–02	–1.365	B–	*LS*
70.	$2p^3$– $2p^2(^3P)3s$	$^4S^o-^4P$		*289.40*	231537.5	– *577079*	4–12	4.90+01	1.84–01	7.03–01	–0.132	B	1
				289.292	231537.5	– 577209	4–6	4.90+01	9.23–02	3.51–01	–0.433	B	*LS*
				289.470	231537.5	– 576997	4–4	4.89+01	6.15–02	2.34–01	–0.609	B	*LS*
				289.590	231537.5	– 576853	4–2	4.89+01	3.07–02	1.17–01	–0.911	B	*LS*
71.	$2p^3$– $2p^2(^3P)3d$	$^4S^o-^4P$		*246.53*	231537.5	– *637174*	4–12	4.40+02	1.20+00	3.91+00	0.683	B	1
				246.563	231537.5	– 637114	4–6	4.40+02	6.02–01	1.95+00	0.381	B	*LS*
				246.502	231537.5	– 637213	4–4	4.40+02	4.01–01	1.30+00	0.206	B	*LS*
				246.465	231537.5	– 637275	4–2	4.41+02	2.01–01	6.51–01	–0.095	B	*LS*
72.	$2p^3$– $2s2p(^3P^o)3p$	$^2D^o-^2P$		*471.40*	*255168*	– *467300*	10–6	4.25+00	8.50–03	1.32–01	–1.071	B	1
				471.300	255155.9	– 467334.9	6–4	3.83+00	8.50–03	7.91–02	–1.292	B	*LS*
				471.599	255184.9	– 467229.3	4–2	4.25+00	7.08–03	4.40–02	–1.548	B	*LS*
				471.365	255184.9	– 467334.9	4–4	4.25–01	1.42–03	8.79–03	–2.247	C+	*LS*
73.		$^2D^o-^2D$		*439.27*	*255168*	– *482819*	10–10	1.50+00	4.33–03	6.26–02	–1.364	B	1
				439.048	255155.9	– 482921.6	6–6	1.40+00	4.04–03	3.51–02	–1.615	B	*LS*
				439.597	255184.9	– 482666.1	4–4	1.34+00	3.89–03	2.25–02	–1.808	B–	*LS*
				439.541	255155.9	– 482666.1	6–4	1.49–01	2.88–04	2.50–03	–2.762	C	*LS*
				439.104	255184.9	– 482921.6	4–6	9.99–02	4.33–04	2.50–03	–2.761	C	*LS*
74.	$2p^3$– $2s2p(^1P^o)3p$	$^2D^o-^2D$		*342.26*	*255168*	– *547343*	10–10	1.05+00	1.84–03	2.07–02	–1.736	B–	1
				342.232	255155.9	– 547355	6–6	9.77–01	1.71–03	1.16–02	–1.988	B–	*LS*
				342.300	255184.9	– 547326	4–4	9.41–01	1.65–03	7.45–03	–2.180	C+	*LS*
				342.266	255155.9	– 547326	6–4	1.05–01	1.22–04	8.28–04	–3.134	D	*LS*
				342.266	255184.9	– 547355	4–6	6.97–02	1.84–04	8.28–04	–3.134	D	*LS*
75.		$^2D^o-^2P$		*339.37*	*255168*	– *549834*	10–6	1.17+01	1.21–02	1.35–01	–0.917	B	1
				339.329	255155.9	– 549855	6–4	1.05+01	1.21–02	8.12–02	–1.138	B	*LS*
				339.435	255184.9	– 549792	4–2	1.17+01	1.01–02	4.51–02	–1.394	B	*LS*
				339.363	255184.9	– 549855	4–4	1.17+00	2.02–03	9.03–03	–2.093	C+	*LS*

O IV: Allowed Transitions — Continued

No.	Transition Array	Mult.	λ_{air} (Å)	λ_{vac} (Å) or σ (cm^{-1})*	E_i (cm^{-1})	E_k (cm^{-1})	g_i–g_k	A_{ki} (10^8 s^{-1})	f_{ik}	S (at. u.)	log gf	Acc.	Source
76.	$2p^3-$ $2s2p(^3P^o)4p$	$^2D^o-^2P$		*312.35*	*255168*	– *575317*	10–6	1.90+01	1.66–02	1.71–01	–0.779	B	1
				312.286	255155.9	– 575375	6–4	1.71+01	1.66–02	1.03–01	–1.001	B	*LS*
				312.483	255184.9	– 575202	4–2	1.89+01	1.39–02	5.70–02	–1.256	B	*LS*
				312.314	255184.9	– 575375	4–4	1.90+00	2.77–03	1.14–02	–1.955	B–	*LS*
77.		$^2D^o-^2D$		*303.49*	*255168*	– *584673*	10–10	8.57+00	1.18–02	1.18–01	–0.927	B	1
				303.393	255155.9	– 584761	6–6	8.01+00	1.11–02	6.62–02	–1.178	B	*LS*
				303.623	255184.9	– 584541	4–4	7.71+00	1.07–02	4.26–02	–1.371	B	*LS*
				303.596	255155.9	– 584541	6–4	8.57–01	7.89–04	4.73–03	–2.325	C+	*LS*
				303.420	255184.9	– 584761	4–6	5.72–01	1.18–03	4.73–03	–2.324	C+	*LS*
78.	$2p^3-$ $2s2p(^3P^o)4f$	$^2D^o-^2F$		*295.09*	*255168*	– *594050*	10–14	1.55+01	2.83–02	2.75–01	–0.549	B	1
				295.057	255155.9	– 594074	6–8	1.55+01	2.69–02	1.57–01	–0.792	C+	*LS*
				295.130	255184.9	– 594019	4–6	1.44+01	2.83–02	1.10–01	–0.947	C+	*LS*
				295.104	255155.9	– 594019	6–6	1.03+00	1.35–03	7.85–03	–2.093	C–	*LS*
79.	$2p^3-$ $2p^2(^3P)3s$	$^2D^o-^2P$		*294.72*	*255168*	– *594472*	10–6	3.94+01	3.08–02	2.99–01	–0.511	B	1
				294.653	255155.9	– 594538	6–4	3.55+01	3.08–02	1.79–01	–0.733	B	*LS*
				294.850	255184.9	– 594340	4–2	3.94+01	2.57–02	9.96–02	–0.989	B	*LS*
				294.678	255184.9	– 594538	4–4	3.94+00	5.13–03	1.99–02	–1.688	B–	*LS*
80.	$2p^3-$ $2s2p(^3P^o)4f$	$^2D^o-^2D$		*292.79*	*255168*	– *596714*	10–10	2.42+00	3.11–03	3.00–02	–1.507	B	1
				292.824	255155.9	– 596658.3	6–6	2.26+00	2.90–03	1.68–02	–1.759	C	*LS*
				292.729	255184.9	– 596798.2	4–4	2.18+00	2.80–03	1.08–02	–1.951	C	*LS*
				292.704	255155.9	– 596798.2	6–4	2.42–01	2.08–04	1.20–03	–2.905	D	*LS*
				292.849	255184.9	– 596658.3	4–6	1.61–01	3.11–04	1.20–03	–2.905	D	*LS*
81.	$2p^3-$ $2p^2(^1D)3s$	$^2D^o-^2D$		*289.91*	*255168*	– *600100*	10–10	5.16+01	6.50–02	6.21–01	–0.187	B	1
				289.897	255155.9	– 600106	6–6	4.82+01	6.07–02	3.47–01	–0.439	B	*LS*
				289.933	255184.9	– 600092	4–4	4.64+01	5.85–02	2.23–01	–0.631	B	*LS*
				289.909	255155.9	– 600092	6–4	5.16+00	4.33–03	2.48–02	–1.585	B–	*LS*
				289.921	255184.9	– 600106	4–6	3.44+00	6.50–03	2.48–02	–1.585	B–	*LS*
82.	$2p^3-$ $2p^2(^3P)3d$	$^2D^o-^2F$		*266.72*	*255168*	– 630095	10–14	6.03+01	9.01–02	7.91–01	–0.045	B	1
				266.710	255155.9	– 630095	6–8	6.03+01	8.58–02	4.52–01	–0.288	B	*LS*
				266.731	255184.9	– 630095	4–6	5.63+01	9.01–02	3.16–01	–0.443	B	*LS*
				266.710	255155.9	– 630095	6–6	4.02+00	4.29–03	2.26–02	–1.589	B–	*LS*
83.		$^2D^o-^2P$		*265.03*	*255168*	– *632483*	10–6	3.60+01	2.27–02	1.98–01	–0.644	B	1
				265.062	255155.9	– 632426	6–4	3.23+01	2.27–02	1.19–01	–0.866	B	*LS*
				264.962	255184.9	– 632597	4–2	3.60+01	1.89–02	6.61–02	–1.121	B	*LS*
				265.082	255184.9	– 632426	4–4	3.59+00	3.79–03	1.32–02	–1.820	B–	*LS*
84.	$2p^3-$ $2p^2(^1D)3d$	$^2D^o-^2F$		*252.56*	*255168*	– *651109*	10–14	4.00+02	5.35–01	4.45+00	0.728	B	1
				252.550	255155.9	– 651117	6–8	4.00+02	5.10–01	2.54+00	0.485	B	*LS*
				252.581	255184.9	– 651098	4–6	3.73+02	5.35–01	1.78+00	0.330	B	*LS*
				252.562	255155.9	– 651098	6–6	2.66+01	2.55–02	1.27–01	–0.816	B	*LS*
85.		$^2D^o-^2P$		*247.02*	*255168*	– 659998	10–6	4.47+01	2.45–02	2.00–01	–0.610	B	1
				247.010	255155.9	– 659998	6–4	4.02+01	2.45–02	1.20–01	–0.832	B	*LS*
				247.028	255184.9	– 659998	4–2	4.47+01	2.04–02	6.65–02	–1.087	B	*LS*
				247.028	255184.9	– 659998	4–4	4.47+00	4.09–03	1.33–02	–1.786	B–	*LS*

O IV: Allowed Transitions — Continued

No.	Transition Array	Mult.	λ_{air} (Å)	λ_{vac} (Å) or σ (cm^{-1})*	E_i (cm^{-1})	E_k (cm^{-1})	g_i-g_k	A_{ki} (10^8 s^{-1})	f_{ik}	S (at. u.)	log gf	Acc.	Source
86.	$2p^3-$ $2s2p(^3P^o)3p$	$^2P^o-^2D$		*516.00*	*289021*	*482819*	6−10	1.46+00	9.73−03	9.92−02	−1.234	B	1
				515.735	289023.5	482921.6	4−6	1.47+00	8.76−03	5.95−02	−1.455	B	*LS*
				516.394	289015.4	482666.1	2−4	1.22+00	9.73−03	3.31−02	−1.711	B	*LS*
				516.415	289023.5	482666.1	4−4	2.43−01	9.73−04	6.61−03	−2.410	C+	*LS*
87.		$^2P^o-^2S$		*490.51*	*289021*	492890.9	6−2	7.97+00	9.59−03	9.29−02	−1.240	B	1
				490.515	289023.5	492890.9	4−2	5.32+00	9.59−03	6.19−02	−1.416	B	*LS*
				490.495	289015.4	492890.9	2−2	2.66+00	9.59−03	3.10−02	−1.717	B	*LS*
88.	$2p^3-$ $2s2p(^1P^o)3p$	$^2P^o-^2P$		*383.42*	*289021*	*549834*	6−6	3.08+00	6.79−03	5.14−02	−1.390	B	1
				383.389	289023.5	549855	4−4	2.57+00	5.66−03	2.86−02	−1.645	B−	*LS*
				383.470	289015.4	549792	2−2	2.05+00	4.52−03	1.14−02	−2.043	B−	*LS*
				383.482	289023.5	549792	4−2	1.03+00	1.13−03	5.71−03	−2.344	C+	*LS*
				383.377	289015.4	549855	2−4	5.13−01	2.26−03	5.71−03	−2.344	C+	*LS*
89.		$^2P^o-^2S$		*376.73*	*289021*	554464	6−2	2.37+00	1.68−03	1.25−02	−1.997	B−	1
				376.732	289023.5	554464	4−2	1.58+00	1.68−03	8.33−03	−2.173	C+	*LS*
				376.721	289015.4	554464	2−2	7.90−01	1.68−03	4.17−03	−2.474	C+	*LS*
90.	$2p^3-$ $2s2p(^3P^o)4p$	$^2P^o-^2P$		*349.29*	*289021*	*575317*	6−6	1.19+01	2.19−02	1.51−01	−0.882	B	1
				349.221	289023.5	575375	4−4	9.96+00	1.82−02	8.38−02	−1.137	B	*LS*
				349.422	289015.4	575202	2−2	7.96+00	1.46−02	3.35−02	−1.536	B	*LS*
				349.432	289023.5	575202	4−2	3.98+00	3.64−03	1.68−02	−1.837	B−	*LS*
				349.211	289015.4	575375	2−4	1.99+00	7.29−03	1.68−02	−1.836	B−	*LS*
91.		$^2P^o-^2D$		*338.24*	*289021*	*584673*	6−10	1.16+00	3.32−03	2.22−02	−1.700	B−	1
				338.138	289023.5	584761	4−6	1.16+00	2.99−03	1.33−02	−1.922	B−	*LS*
				338.380	289015.4	584541	2−4	9.67−01	3.32−03	7.40−03	−2.178	C+	*LS*
				338.389	289023.5	584541	4−4	1.93−01	3.32−04	1.48−03	−2.877	C	*LS*
92.		$^2P^o-^2S$		*332.17*	*289021*	590069	6−2	2.63+00	1.45−03	9.50−03	−2.061	C+	1
				332.176	289023.5	590069	4−2	1.75+00	1.45−03	6.33−03	−2.237	C+	*LS*
				332.167	289015.4	590069	2−2	8.75−01	1.45−03	3.17−03	−2.538	C+	*LS*
93.	$2p^3-$ $2p^2(^3P)3s$	$^2P^o-^2P$		*327.38*	*289021*	*594472*	6−6	3.23+01	5.19−02	3.36−01	−0.507	B	1
				327.317	289023.5	594538	4−4	2.69+01	4.32−02	1.86−01	−0.762	B	*LS*
				327.520	289015.4	594340	2−2	2.15+01	3.46−02	7.46−02	−1.160	B	*LS*
				327.529	289023.5	594340	4−2	1.07+01	8.64−03	3.73−02	−1.461	B	*LS*
				327.308	289015.4	594538	2−4	5.38+00	1.73−02	3.73−02	−1.461	B	*LS*
94.	$2p^3-$ $2s2p(^3P^o)4f$	$^2P^o-^2D$		*325.00*	*289021*	*596714*	6−10	4.76+00	1.26−02	8.06−02	−1.123	B	1
				325.061	289023.5	596658.3	4−6	4.75+00	1.13−02	4.84−02	−1.345	C+	*LS*
				324.904	289015.4	596798.2	2−4	3.97+00	1.26−02	2.69−02	−1.600	C	*LS*
				324.913	289023.5	596798.2	4−4	7.93−01	1.26−03	5.37−03	−2.299	C−	*LS*
95.	$2p^3-$ $2p^2(^1D)3s$	$^2P^o-^2D$		*321.46*	*289021*	*600100*	6−10	1.47+01	3.80−02	2.42−01	−0.642	B	1
				321.458	289023.5	600106	4−6	1.47+01	3.42−02	1.45−01	−0.863	B	*LS*
				321.464	289015.4	600092	2−4	1.23+01	3.80−02	8.05−02	−1.119	B	*LS*
				321.473	289023.5	600092	4−4	2.46+00	3.80−03	1.61−02	−1.818	B−	*LS*

O IV: Allowed Transitions — Continued

No.	Transition Array	Mult.	λ_{air} (Å)	λ_{vac} (Å) or σ (cm^{-1})*	E_i (cm^{-1})	E_k (cm^{-1})	g_i–g_k	A_{ki} (10^8 s^{-1})	f_{ik}	S (at. u.)	log gf	Acc.	Source
96.	$2p^3$– $2p^2(^3P)3d$	$^2P^o$–^2P		*291.15*	*289021*	*632483*	6–6	9.69+01	1.23–01	7.08–01	–0.131	B	1
				291.203	289023.5	632426	4–4	8.07+01	1.03–01	3.94–01	–0.387	B	*LS*
				291.052	289015.4	632597	2–2	6.47+01	8.21–02	1.57–01	–0.784	B	*LS*
				291.059	289023.5	632597	4–2	3.23+01	2.05–02	7.87–02	–1.085	B	*LS*
				291.197	289015.4	632426	2–4	1.61+01	4.11–02	7.87–02	–1.086	B	*LS*
97.	$2p^3$– $2p^2(^1D)3d$	$^2P^o$–^2P		*269.56*	*289021*	659998	6–6	9.53+01	1.04–01	5.53–01	–0.205	B	1
				269.560	289023.5	659998	4–4	7.94+01	8.65–02	3.07–01	–0.461	B	*LS*
				269.554	289015.4	659998	2–2	6.36+01	6.92–02	1.23–01	–0.859	B	*LS*
				269.560	289023.5	659998	4–2	3.18+01	1.73–02	6.14–02	–1.160	B	*LS*
				269.554	289015.4	659998	2–4	1.59+01	3.46–02	6.14–02	–1.160	B	*LS*
98.	$2s^23s$– $2s^23p$	^2S–$^2P^o$	*3066.1*	*3067.0*	357614.3	*390219*	2–6	1.30+00	5.49–01	1.11+01	0.040	B	1
			3063.43	3064.32	357614.3	390248.0	2–4	1.30+00	3.66–01	7.38+00	–0.135	B	*LS*
			3071.60	3072.49	357614.3	390161.2	2–2	1.29+00	1.83–01	3.69+00	–0.438	B	*LS*
99.	$2s^23s$– $2s2p(^3P^o)3s$	^2S–$^2P^o$		*1048.6*	357614.3	*452983*	2–6	8.76–01	4.33–02	2.99–01	–1.062	B	1
				1047.59	357614.3	453071.5	2–4	8.79–01	2.89–02	1.99–01	–1.238	B	*LS*
				1050.51	357614.3	452806.6	2–2	8.71–01	1.44–02	9.97–02	–1.540	B	*LS*
100.	$2s^23s$– $2s^24p$	^2S–$^2P^o$		*702.20*	357614.3	*500024*	2–6	7.74+00	1.72–01	7.93–01	–0.464	B	1
				702.131	357614.3	500037.8	2–4	7.74+00	1.14–01	5.29–01	–0.641	B	*LS*
				702.343	357614.3	499994.9	2–2	7.73+00	5.72–02	2.64–01	–0.942	B	*LS*
101.	$2s^23s$– $2s2p(^3P^o)3d$	^2S–$^2P^o$		*638.34*	357614.3	*514271*	2–6	2.70–01	4.95–03	2.08–02	–2.004	B–	1
				638.545	357614.3	514220.4	2–4	2.70–01	3.30–03	1.39–02	–2.181	B–	*LS*
				637.930	357614.3	514371.3	2–2	2.71–01	1.65–03	6.93–03	–2.481	C+	*LS*
102.	$2s^23s$– $2s2p(^1P^o)3s$	^2S–$^2P^o$		*620.76*	357614.3	*518706*	2–6	2.86+01	4.95–01	2.02+00	–0.004	B	1
				620.753	357614.3	518709	2–4	2.86+01	3.30–01	1.35+00	–0.181	B	*LS*
				620.791	357614.3	518699	2–2	2.85+01	1.65–01	6.74–01	–0.482	B	*LS*
103.	$2s^23s$– $2s^25p$	^2S–$^2P^o$		*528.54*	357614.3	*546814*	2–6	1.95+00	2.45–02	8.51–02	–1.311	B	1
				528.529	357614.3	546818.7	2–4	1.95+00	1.63–02	5.67–02	–1.487	B	*LS*
				528.572	357614.3	546803.4	2–2	1.95+00	8.15–03	2.84–02	–1.788	B–	*LS*
104.	$2s^23p$– $2s^23d$	$^2P^o$–^2D	*3408.2*	*3409.2*	*390219*	*419552*	6–10	1.02+00	2.96–01	1.99+01	0.249	B	1
			3411.69	3412.67	390248.0	419550.6	4–6	1.02+00	2.66–01	1.20+01	0.027	B	*LS*
			3401.23	3402.21	390161.2	419553.9	2–4	8.55–01	2.96–01	6.64+00	–0.227	B	*LS*
			3411.30	3412.28	390248.0	419553.9	4–4	1.69–01	2.96–02	1.33+00	–0.927	B	*LS*
105.	$2s^23p$– $2s2p(^3P^o)3p$	$^2P^o$–^2P		*1297.2*	*390219*	*467306*	6–6	4.79–01	1.21–02	3.10–01	–1.139	B	1
				1297.07	390248.0	467344.9	4–4	4.00–01	1.01–02	1.72–01	–1.394	B	*LS*
				1297.55	390161.2	467229.3	2–2	3.19–01	8.06–03	6.89–02	–1.793	B	*LS*
				1299.02	390248.0	467229.3	4–2	1.59–01	2.01–03	3.44–02	–2.094	B	*LS*
				1295.61	390161.2	467344.9	2–4	8.02–02	4.04–03	3.44–02	–2.093	B	*LS*
106.		$^2P^o$–^2D		*1079.9*	*390219*	*482819*	6–10	1.97+00	5.75–02	1.23+00	–0.462	B	1
				1079.06	390248.0	482921.6	4–6	1.98+00	5.18–02	7.36–01	–0.684	B	*LS*
				1081.02	390161.2	482666.1	2–4	1.64+00	5.74–02	4.09–01	–0.940	B	*LS*
				1082.04	390248.0	482666.1	4–4	3.27–01	5.74–03	8.17–02	–1.639	B	*LS*

O IV: Allowed Transitions — Continued

No.	Transition Array	Mult.	λ_{air} (Å)	λ_{vac} (Å) or σ (cm⁻¹)*	E_i (cm⁻¹)	E_k (cm⁻¹)	g_i–g_k	A_{ki} (10^8 s⁻¹)	f_{ik}	S (at. u.)	log gf	Acc.	Source
107.	$2s^23p-$ $2s^24s$	$^2P^o-^2S$		*1046.0*	*390219*	– 485821.6	6–2	1.68+01	9.20–02	1.90+00	–0.258	B	1
				1046.31	390248.0	– 485821.6	4–2	1.12+01	9.20–02	1.27+00	–0.434	B	*LS*
				1045.36	390161.2	– 485821.6	2–2	5.62+00	9.20–02	6.34–01	–0.735	B	*LS*
108.	$2s^23p-$ $2s2p(^3P^o)3p$	$^2P^o-^2S$		*973.98*	*390219*	– 492890.9	6–2	4.75+00	2.25–02	4.33–01	–0.870	B	1
				974.252	390248.0	– 492890.9	4–2	3.16+00	2.25–02	2.89–01	–1.046	B	*LS*
				973.428	390161.2	– 492890.9	2–2	1.59+00	2.25–02	1.44–01	–1.346	B	*LS*
109.	$2s^23p-$ $2s^24d$	$^2P^o-^2D$		*830.89*	*390219*	– *510572*	6–10	1.48+01	2.55–01	4.18+00	0.184	B	1
				831.075	390248.0	– 510574.1	4–6	1.48+01	2.29–01	2.51+00	–0.038	B	*LS*
				830.506	390161.2	– 510569.7	2–4	1.23+01	2.55–01	1.39+00	–0.293	B	*LS*
				831.105	390248.0	– 510569.7	4–4	2.46+00	2.55–02	2.79–01	–0.992	B	*LS*
110.	$2s^23p-$ $2s^25s$	$^2P^o-^2S$		*670.47*	*390219*	– 539368	6–2	1.53+01	3.43–02	4.54–01	–0.687	B	1
				670.601	390248.0	– 539368	4–2	1.02+01	3.43–02	3.03–01	–0.863	B	*LS*
				670.211	390161.2	– 539368	2–2	5.09+00	3.43–02	1.51–01	–1.164	B	*LS*
111.	$2s^23p-$ $2s2p(^1P^o)3p$	$^2P^o-^2D$		*636.44*	*390219*	– *547343*	6–10	2.21+01	2.24–01	2.81+00	0.128	B	1
				636.509	390248.0	– 547355	4–6	2.21+01	2.01–01	1.69+00	–0.094	B	*LS*
				636.275	390161.2	– 547326	2–4	1.84+01	2.24–01	9.38–01	–0.349	B	*LS*
				636.626	390248.0	– 547326	4–4	3.68+00	2.24–02	1.88–01	–1.048	B	*LS*
112.		$^2P^o-^2P$		*626.43*	*390219*	– *549854*	6–6	3.13+01	1.84–01	2.28+00	0.043	B	1
				626.421	390248.0	– 549885	4–4	2.61+01	1.53–01	1.27+00	–0.212	B	*LS*
				626.446	390161.2	– 549792	2–2	2.09+01	1.23–01	5.06–01	–0.610	B	*LS*
				626.786	390248.0	– 549792	4–2	1.04+01	3.07–02	2.53–01	–0.911	B	*LS*
				626.081	390161.2	– 549885	2–4	5.23+00	6.14–02	2.53–01	–0.911	B	*LS*
113.	$2s^23p-$ $2s^25d$	$^2P^o-^2D$		*618.00*	*390219*	– *552031*	6–10	7.78+00	7.43–02	9.07–01	–0.351	B	1
				618.107	390248.0	– 552032.3	4–6	7.78+00	6.68–02	5.44–01	–0.573	B	*LS*
				617.786	390161.2	– 552029.6	2–4	6.49+00	7.43–02	3.02–01	–0.828	B	*LS*
				618.117	390248.0	– 552029.6	4–4	1.30+00	7.43–03	6.05–02	–1.527	B	*LS*
114.	$2s^23p-$ $2s2p(^1P^o)3p$	$^2P^o-^2S$		*608.85*	*390219*	– 554464	6–2	1.83+01	3.39–02	4.08–01	–0.692	B	1
				608.954	390248.0	– 554464	4–2	1.22+01	3.39–02	2.72–01	–0.868	B	*LS*
				608.632	390161.2	– 554464	2–2	6.10+00	3.39–02	1.36–01	–1.169	B	*LS*
115.	$2s^23p-$ $2s^26d$	$^2P^o-^2D$		*543.04*	*390219*	– 574368.7	6–10	5.47+00	4.03–02	4.32–01	–0.617	B	1
				543.122	390248.0	– 574368.7	4–6	5.47+00	3.63–02	2.59–01	–0.839	B	*LS*
				542.866	390161.2	– 574368.7	2–4	4.56+00	4.03–02	1.44–01	–1.094	B	*LS*
				543.122	390248.0	– 574368.7	4–4	9.11–01	4.03–03	2.88–02	–1.793	B–	*LS*
116.	$2s^23p-$ $2s2p(^3P^o)4p$	$^2P^o-^2D$		*514.26*	*390219*	– *584673*	6–10	4.60–01	3.04–03	3.09–02	–1.739	B–	1
				514.104	390248.0	– 584761	4–6	4.61–01	2.74–03	1.85–02	–1.960	B–	*LS*
				514.457	390161.2	– 584541	2–4	3.83–01	3.04–03	1.03–02	–2.216	B–	*LS*
				514.687	390248.0	– 584541	4–4	7.65–02	3.04–04	2.06–03	–2.915	C	*LS*
117.		$^2P^o-^2S$		*500.38*	*390219*	– 590069	6–2	1.13+00	1.42–03	1.40–02	–2.071	B–	1
				500.448	390248.0	– 590069	4–2	7.54–01	1.42–03	9.33–03	–2.247	C+	*LS*
				500.231	390161.2	– 590069	2–2	3.78–01	1.42–03	4.67–03	–2.548	C+	*LS*

O IV: Allowed Transitions — Continued

No.	Transition Array	Mult.	λ_{air} (Å)	λ_{vac} (Å) or σ (cm^{-1})*	E_i (cm^{-1})	E_k (cm^{-1})	g_i-g_k	A_{ki} (10^8 s^{-1})	f_{ik}	S (at. u.)	log gf	Acc.	Source
118.	$2s^23p-$ $2p^2(^3P)3s$	$^2P^o-^2P$		*489.59*	*390219*	*594472*	6–6	5.52–01	1.99–03	1.92–02	–1.924	B–	1
				489.500	390248.0	594538	4–4	4.61–01	1.65–03	1.07–02	–2.179	B–	*LS*
				489.767	390161.2	594340	2–2	3.68–01	1.32–03	4.27–03	–2.577	C+	*LS*
				489.975	390248.0	594340	4–2	1.84–01	3.31–04	2.13–03	–2.879	C	*LS*
				489.292	390161.2	594538	2–4	9.22–02	6.62–04	2.13–03	–2.878	C	*LS*
119.	$2s^23p-$ $2p^2(^1D)3s$	$^2P^o-^2D$		*476.46*	*390219*	*600100*	6–10	1.72+00	9.75–03	9.18–02	–1.233	B	1
				476.513	390248.0	600106	4–6	1.72+00	8.78–03	5.51–02	–1.455	B	*LS*
				476.347	390161.2	600092	2–4	1.43+00	9.76–03	3.06–02	–1.710	B	*LS*
				476.544	390248.0	600092	4–4	2.86–01	9.75–04	6.12–03	–2.409	C+	*LS*
120.	$2s^23d-$ $2s2p(^3P^o)3s$	$^2D-^2P^o$	*2989.6*	*2990.5*	*419544*	*452983*	10–6	1.30–01	1.04–02	1.03+00	–0.981	B	1
			2982.34	2983.21	419550.6	453071.5	6–4	1.18–01	1.05–02	6.16–01	–1.202	B	*LS*
			3004.59	3005.47	419533.9	452806.6	4–2	1.28–01	8.65–03	3.42–01	–1.461	B	*LS*
			2980.86	2981.73	419533.9	453071.5	4–4	1.31–02	1.74–03	6.85–02	–2.156	B	*LS*
121.	$2s^23d-$ $2s^24p$	$^2D-^2P^o$		*1242.6*	*419544*	*500024*	10–6	3.56+00	4.95–02	2.02+00	–0.305	B	1
				1242.43	419550.6	500037.8	6–4	3.21+00	4.95–02	1.21+00	–0.527	B	*LS*
				1242.84	419533.9	499994.9	4–2	3.56+00	4.12–02	6.75–01	–0.783	B	*LS*
				1242.18	419533.9	500037.8	4–4	3.57–01	8.25–03	1.35–01	–1.481	B	*LS*
122.	$2s^23d-$ $2s2p(^3P^o)3d$	$^2D-^2F^o$		*1094.9*	*419544*	*510878*	10–14	1.58+00	3.96–02	1.43+00	–0.402	B	1
				1093.77	419550.6	510977.2	6–8	1.58+00	3.78–02	8.16–01	–0.645	B	*LS*
				1096.36	419533.9	510744.9	4–6	1.46+00	3.96–02	5.71–01	–0.800	B	*LS*
				1096.56	419550.6	510744.9	6–6	1.05–01	1.88–03	4.08–02	–1.947	B	*LS*
123.	$2s^23d-$ $2s^24f$	$^2D-^2F^o$		*1067.8*	*419544*	*513194*	10–14	3.33+01	7.96–01	2.80+01	0.901	B	1
				1067.83	419550.6	513198.3	6–8	3.33+01	7.58–01	1.60+01	0.658	C+	*LS*
				1067.77	419533.9	513187.2	4–6	3.11+01	7.96–01	1.12+01	0.503	C+	*LS*
				1067.96	419550.6	513187.2	6–6	2.22+00	3.79–02	8.00–01	–0.643	C+	*LS*
124.	$2s^23d-$ $2s2p(^1P^o)3s$	$^2D-^2P^o$		*1008.5*	*419544*	*518706*	10–6	9.19–02	8.40–04	2.79–02	–2.076	B–	1
				1008.49	419550.6	518709	6–4	8.27–02	8.40–04	1.67–02	–2.297	B–	*LS*
				1008.42	419533.9	518699	4–2	9.19–02	7.00–04	9.30–03	–2.553	C+	*LS*
				1008.32	419533.9	518709	4–4	9.19–03	1.40–04	1.86–03	–3.252	C	*LS*
125.	$2s^23d-$ $2s^25p$	$^2D-^2P^o$		*785.73*	*419544*	*546814*	10–6	2.20+00	1.22–02	3.16–01	–0.913	B	1
				785.743	419550.6	546818.7	6–4	1.98+00	1.22–02	1.90–01	–1.135	B	*LS*
				785.734	419533.9	546803.4	4–2	2.20+00	1.02–02	1.05–01	–1.390	B	*LS*
				785.640	419533.9	546818.7	4–4	2.20–01	2.04–03	2.11–02	–2.089	B–	*LS*
126.	$2s^23d-$ $2s^25f$	$^2D-^2F^o$		*752.16*	*419544*	552495.1	10–14	6.48+00	7.70–02	1.91+00	–0.114	B	1
				752.194	419550.6	552495.1	6–8	6.48+00	7.33–02	1.09+00	–0.357	C+	*LS*
				752.099	419533.9	552495.1	4–6	6.05+00	7.70–02	7.63–01	–0.511	C+	*LS*
				752.194	419550.6	552495.1	6–6	4.32–01	3.67–03	5.45–02	–1.658	C+	*LS*
127.	$2s^23d-$ $2s2p(^3P^o)4s$	$^2D-^2P^o$		*648.13*	*419544*	*573833*	10–6	3.20–01	1.21–03	2.58–02	–1.918	B–	1
				647.877	419550.6	573901	6–4	2.88–01	1.21–03	1.55–02	–2.139	B–	*LS*
				648.668	419533.9	573696	4–2	3.19–01	1.01–03	8.60–03	–2.395	C+	*LS*
				647.806	419533.9	573901	4–4	3.20–02	2.02–04	1.72–03	–3.093	C	*LS*

O IV: Allowed Transitions — Continued

No.	Transition Array	Mult.	λ_{air} (Å)	λ_{vac} (Å) or σ (cm^{-1})*	E_i (cm^{-1})	E_k (cm^{-1})	g_i-g_k	A_{ki} (10^8 s^{-1})	f_{ik}	S (at. u.)	log gf	Acc.	Source
128.	$2s^23d-$ $2s^26f$	^{2}D$-^2$F°		*644.07*	*419544*	– 574807.7	10–14	5.37+00	4.68–02	9.92–01	–0.330	B	1
				644.093	419550.6	– 574807.7	6–8	5.37+00	4.45–02	5.67–01	–0.573	C+	*LS*
				644.024	419533.9	– 574807.7	4–6	5.01+00	4.68–02	3.97–01	–0.728	C+	*LS*
				644.093	419550.6	– 574807.7	6–6	3.58–01	2.23–03	2.83–02	–1.874	C	*LS*
129.	$2s^23d-$ $2s2p(^1$P°$)3d$	^{2}D$-^2$D°		*639.82*	*419544*	– *575839*	10–10	2.66+01	1.63–01	3.44+00	0.212	B	1
				639.785	419550.6	– 575853	6–6	2.48+01	1.52–01	1.92+00	–0.039	B	*LS*
				639.856	419533.9	– 575819	4–4	2.39+01	1.47–01	1.24+00	–0.231	B	*LS*
				639.925	419550.6	– 575819	6–4	2.66+00	1.09–02	1.37–01	–1.186	B	*LS*
				639.717	419533.9	– 575853	4–6	1.77+00	1.63–02	1.37–01	–1.185	B	*LS*
130.		^{2}D$-^2$P°		*616.55*	*419544*	– *581736*	10–6	2.42+01	8.26–02	1.68+00	–0.083	B	1
				616.552	419550.6	– 581743	6–4	2.17+01	8.26–02	1.01+00	–0.305	B	*LS*
				616.572	419533.9	– 581721	4–2	2.42+01	6.88–02	5.59–01	–0.560	B	*LS*
				616.488	419533.9	– 581743	4–4	2.42+00	1.38–02	1.12–01	–1.259	B	*LS*
131.	$2s^23d-$ $2s^27f$	^{2}D$-^2$F°		*594.16*	*419544*	– 587850	10–14	1.06+01	7.83–02	1.53+00	–0.106	B	1
				594.179	419550.6	– 587850	6–8	1.06+01	7.46–02	8.75–01	–0.349	C+	*LS*
				594.120	419533.9	– 587850	4–6	9.86+00	7.83–02	6.13–01	–0.504	C+	*LS*
				594.179	419550.6	– 587850	6–6	7.04–01	3.73–03	4.38–02	–1.650	C+	*LS*
132.	$2s^23d-$ $2s2p(^3$P°$)4d$	^{2}D$-^2$D°		*574.28*	*419544*	– *593676*	10–10	2.87+00	1.42–02	2.69–01	–0.847	B	1
				574.193	419550.6	– 593708	6–6	2.68+00	1.33–02	1.50–01	–1.099	B	*LS*
				574.405	419533.9	– 593627	4–4	2.59+00	1.28–02	9.67–02	–1.291	B	*LS*
				574.460	419550.6	– 593627	6–4	2.87–01	9.47–04	1.07–02	–2.245	B–	*LS*
				574.138	419533.9	– 593708	4–6	1.92–01	1.42–03	1.07–02	–2.245	B–	*LS*
133.		^{2}D$-^2$F°		*565.44*	*419544*	– *596398*	10–14	4.52–01	3.04–03	5.65–02	–1.518	B	1
				565.213	419550.6	– 596475	6–8	4.53–01	2.89–03	3.23–02	–1.761	B	*LS*
				565.735	419533.9	– 596295	4–6	4.21–01	3.03–03	2.26–02	–1.916	B–	*LS*
				565.789	419550.6	– 596295	6–6	3.01–02	1.44–04	1.61–03	–3.062	C	*LS*
134.	$2s^23d-$ $2p^2(^1$D$)3p$	^{2}D$-^2$D°		*470.88*	*419544*	– 631913	10–10	4.44–01	1.48–03	2.29–02	–1.831	B–	1
				470.893	419550.6	– 631913	6–6	4.15–01	1.38–03	1.28–02	–2.082	B–	*LS*
				470.856	419533.9	– 631913	4–4	4.00–01	1.33–03	8.24–03	–2.274	C+	*LS*
				470.893	419550.6	– 631913	6–4	4.44–02	9.85–05	9.16–04	–3.228	D	*LS*
				470.856	419533.9	– 631913	4–6	2.96–02	1.48–04	9.16–04	–3.228	D	*LS*
135.	$2s2p(^3$P°$)3s-$ $2s2p(^3$P°$)3p$	^{4}P°$-^4$D	*3388.7*	*3389.7*	*439085*	– *468586*	12–20	1.02+00	2.93–01	3.92+01	0.546	B	1
			3385.52	3386.49	439230.9	– 468760.0	6–8	1.02+00	2.34–01	1.57+01	0.148	B	*LS*
			3381.21	3382.18	438983.9	– 468550.6	4–6	7.19–01	1.85–01	8.23+00	–0.131	B	*LS*
			3381.30	3382.27	438849.0	– 468414.9	2–4	4.28–01	1.47–01	3.26+00	–0.533	B	*LS*
			3409.70	3410.68	439230.9	– 468550.6	6–6	3.00–01	5.24–02	3.53+00	–0.503	B	*LS*
			3396.80	3397.78	438983.9	– 468414.9	4–4	5.40–01	9.34–02	4.18+00	–0.427	B	*LS*
			3390.19	3391.16	438849.0	– 468337.4	2–2	8.49–01	1.46–01	3.26+00	–0.534	B	*LS*
			3425.55	3426.54	439230.9	– 468414.9	6–4	4.94–02	5.79–03	3.92–01	–1.459	B	*LS*
			3405.77	3406.75	438983.9	– 468337.4	4–2	1.67–01	1.46–02	6.53–01	–1.235	B	*LS*
136.		^{4}P°$-^4$S	*2824.6*	*2825.4*	*439085*	– 474478.1	12–4	1.71+00	6.81–02	7.59+00	–0.088	B	1
			2836.27	2837.10	439230.9	– 474478.1	6–4	8.43–01	6.78–02	3.80+00	–0.391	B	*LS*
			2816.53	2817.36	438983.9	– 474478.1	4–4	5.74–01	6.83–02	2.53+00	–0.564	B	*LS*
			2805.87	2806.69	438849.0	– 474478.1	2–4	2.90–01	6.85–02	1.27+00	–0.863	B	*LS*

O IV: Allowed Transitions — Continued

No.	Transition Array	Mult.	λ_{air} (Å)	λ_{vac} (Å) or σ (cm^{-1})*	E_i (cm^{-1})	E_k (cm^{-1})	g_i–g_k	A_{ki} (10^8 s^{-1})	f_{ik}	S (at. u.)	log gf	Acc.	Source
137.		^{4}P°–^{4}P	*2505.1*	*2505.8*	*439085*	*– 478992*	12–12	2.79+00	2.62–01	2.60+01	0.498	B	1
			2509.22	2509.97	439230.9	– 479072.0	6–6	1.94+00	1.83–01	9.09+00	0.041	B	*LS*
			2501.81	2502.57	438983.9	– 478942.9	4–4	3.73–01	3.50–02	1.15+00	–0.854	B	*LS*
			2499.27	2500.03	438849.0	– 478848.6	2–2	4.68–01	4.38–02	7.21–01	–1.057	B	*LS*
			2517.37	2518.13	439230.9	– 478942.9	6–4	1.24+00	7.83–02	3.89+00	–0.328	B	*LS*
			2507.73	2508.48	438983.9	– 478848.6	4–2	2.32+00	1.09–01	3.61+00	–0.360	B	*LS*
			2493.75	2494.51	438983.9	– 479072.0	4–6	8.48–01	1.19–01	3.89+00	–0.324	B	*LS*
			2493.39	2494.14	438849.0	– 478942.9	2–4	1.18+00	2.20–01	3.61+00	–0.357	B	*LS*
138.	$2s2p(^3$P°$)3s$– $2p^2(^3$P$)3s$	^{4}P°–^{4}P		*724.67*	*439085*	*– 577079*	12–12	2.70+01	2.13–01	6.09+00	0.407	B	1
				724.753	439230.9	– 577209	6–6	1.89+01	1.49–01	2.13+00	–0.049	B	*LS*
				724.569	438983.9	– 576997	4–4	3.60+00	2.83–02	2.70–01	–0.945	B	*LS*
				724.617	438849.0	– 576853	2–2	4.50+00	3.54–02	1.69–01	–1.150	B	*LS*
				725.868	439230.9	– 576997	6–4	1.21+01	6.37–02	9.13–01	–0.418	B	*LS*
				725.326	438983.9	– 576853	4–2	2.24+01	8.85–02	8.45–01	–0.451	B	*LS*
				723.458	438983.9	– 577209	4–6	8.14+00	9.58–02	9.13–01	–0.416	B	*LS*
				723.861	438849.0	– 576997	2–4	1.13+01	1.77–01	8.45–01	–0.450	B	*LS*
139.	$2s2p(^3$P°$)3s$– $2s2p(^3$P°$)4f$	^{4}P°–^{4}D		*635.57*	*439085*	*– 596425*	12–20	3.73–01	3.76–03	9.45–02	–1.345	B	1
				636.537	439230.9	– 596331	6–8	3.71–01	3.01–03	3.78–02	–1.744	C+	*LS*
				635.082	438983.9	– 596444	4–6	2.62–01	2.37–03	1.98–02	–2.023	C	*LS*
				634.263	438849.0	– 596512.4	2–4	1.56–01	1.89–03	7.88–03	–2.423	C–	*LS*
				636.079	439230.9	– 596444	6–6	1.12–01	6.77–04	8.51–03	–2.391	C–	*LS*
				634.806	438983.9	– 596512.4	4–4	2.00–01	1.21–03	1.01–02	–2.317	C	*LS*
				634.046	438849.0	– 596566.2	2–2	3.13–01	1.89–03	7.88–03	–2.423	C–	*LS*
				635.803	439230.9	– 596512.4	6–4	1.86–02	7.52–05	9.45–04	–3.345	D–	*LS*
				634.589	438983.9	– 596566.2	4–2	6.24–02	1.88–04	1.58–03	–3.123	D	*LS*
140.	$2s2p(^3$P°$)3s$– $2s2p(^3$P°$)3p$	^{2}P°–^{2}P	*6979.8*	*6981.7*	*452983*	*– 467306*	6–6	1.08–01	7.89–02	1.09+01	–0.325	B	1
			7004.11	7006.04	453071.5	– 467344.9	4–4	8.90–02	6.55–02	6.04+00	–0.582	B	*LS*
			6931.60	6933.51	452806.6	– 467229.3	2–2	7.35–02	5.29–02	2.42+00	–0.975	B	*LS*
			7061.30	7063.24	453071.5	– 467229.3	4–2	3.48–02	1.30–02	1.21+00	–1.284	B	*LS*
			6876.49	6878.38	452806.6	– 467344.9	2–4	1.88–02	2.67–02	1.21+00	–1.273	B	*LS*
141.		^{2}P°–^{2}D	*3350.7*	*3351.6*	*452983*	*– 482819*	6–10	1.02+00	2.86–01	1.89+01	0.234	B	1
			3349.11	3350.07	453071.5	– 482921.6	4–6	1.02+00	2.57–01	1.14+01	0.013	B	*LS*
			3348.06	3349.02	452806.6	– 482666.1	2–4	8.51–01	2.86–01	6.31+00	–0.242	B	*LS*
			3378.02	3378.99	453071.5	– 482666.1	4–4	1.66–01	2.84–02	1.26+00	–0.945	B	*LS*
142.	$2s2p(^3$P°$)3s$– $2s^24s$	^{2}P°–^{2}S	*3044.3*	*3045.2*	*452983*	– 485821.7	6–2	7.79–01	3.61–02	2.17+00	–0.665	B	1
			3052.53	3053.42	453071.5	– 485821.7	4–2	5.15–01	3.60–02	1.45+00	–0.842	B	*LS*
			3028.04	3028.92	452806.6	– 485821.7	2–2	2.64–01	3.63–02	7.23–01	–1.139	B	*LS*
143.	$2s2p(^3$P°$)3s$– $2s2p(^3$P°$)3p$	^{2}P°–^{2}S	*2505.0*	*2505.8*	*452983*	– 492890.9	6–2	1.80+00	5.66–02	2.80+00	–0.469	B	1
			2510.58	2511.34	453071.5	– 492890.9	4–2	1.19+00	5.65–02	1.87+00	–0.646	B	*LS*
			2493.99	2494.74	452806.6	– 492890.9	2–2	6.09–01	5.68–02	9.33–01	–0.944	B	*LS*
144.	$2s2p(^3$P°$)3s$– $2s^24d$	^{2}P°–^{2}D		*1736.4*	*452983*	*– 510572*	6–10	9.18–01	6.91–02	2.37+00	–0.382	B	1
				1739.05	453071.5	– 510574.1	4–6	9.14–01	6.21–02	1.42+00	–0.605	B	*LS*
				1731.21	452806.6	– 510569.7	2–4	7.72–01	6.93–02	7.90–01	–0.858	B	*LS*
				1739.18	453071.5	– 510569.7	4–4	1.52–01	6.90–03	1.58–01	–1.559	B	*LS*

O IV: Allowed Transitions — Continued

No.	Transition Array	Mult.	λ_{air} (Å)	λ_{vac} (Å) or σ (cm^{-1})*	E_i (cm^{-1})	E_k (cm^{-1})	g_i–g_k	A_{ki} (10^8 s^{-1})	f_{ik}	S (at. u.)	log gf	Acc.	Source
145.	$2s2p(^3P°)3s-2s^25s$	$^2P°-^2S$		*1157.6*	*452983*	– 539368	6–2	1.33–01	8.88–04	2.03–02	–2.274	B–	1
				1158.80	453071.5	– 539368	4–2	8.81–02	8.87–04	1.35–02	–2.450	B–	*LS*
				1155.25	452806.6	– 539368	2–2	4.45–02	8.90–04	6.77–03	–2.750	C+	*LS*
146.	$2s2p(^3P°)3s-2s2p(^1P°)3p$	$^2P°-^2D$		*1059.8*	*452983*	– *547343*	6–10	1.75+00	4.90–02	1.03+00	–0.532	B	1
				1060.63	453071.5	– 547355	4–6	1.74+00	4.40–02	6.15–01	–0.754	B	*LS*
				1057.98	452806.6	– 547326	2–4	1.46+00	4.91–02	3.42–01	–1.008	B	*LS*
				1060.96	453071.5	– 547326	4–4	2.90–01	4.89–03	6.83–02	–1.708	B	*LS*
147.	$2s2p(^3P°)3s-2s^25d$	$^2P°-^2D$		*1009.6*	*452983*	– *552031*	6–10	1.74–01	4.44–03	8.85–02	–1.575	B	1
				1010.50	453071.5	– 552032.3	4–6	1.74–01	3.99–03	5.31–02	–1.797	B	*LS*
				1007.83	452806.6	– 552029.6	2–4	1.46–01	4.45–03	2.95–02	–2.051	B–	*LS*
				1010.53	453071.5	– 552029.6	4–4	2.90–02	4.43–04	5.90–03	–2.751	C+	*LS*
148.	$2s2p(^3P°)3s-2s2p(^1P°)3p$	$^2P°-^2S$		*985.41*	*452983*	– 554464	6–2	2.40+00	1.17–02	2.27–01	–1.155	B	1
				986.266	453071.5	– 554464	4–2	1.60+00	1.16–02	1.51–01	–1.332	B	*LS*
				983.696	452806.6	– 554464	2–2	8.05–01	1.17–02	7.56–02	–1.632	B	*LS*
149.	$2s2p(^3P°)3s-2s^26d$	$^2P°-^2D$		*823.82*	*452983*	– 574368.7	6–10	1.26–01	2.14–03	3.49–02	–1.890	B	1
				824.421	453071.5	– 574368.7	4–6	1.26–01	1.93–03	2.09–02	–2.113	B–	*LS*
				822.625	452806.6	– 574368.7	2–4	1.06–01	2.15–03	1.16–02	–2.367	B–	*LS*
				824.421	453071.5	– 574368.7	4–4	2.10–02	2.14–04	2.33–03	–3.067	C	*LS*
150.	$2s2p(^3P°)3s-2s2p(^3P°)4p$	$^2P°-^2P$		*817.44*	*452983*	– *575317*	6–6	2.46+01	2.47–01	3.99+00	0.171	B	1
				817.638	453071.5	– 575375	4–4	2.05+01	2.06–01	2.21+00	–0.085	B	*LS*
				817.024	452806.6	– 575202	2–2	1.65+01	1.65–01	8.86–01	–0.482	B	*LS*
				818.796	453071.5	– 575202	4–2	8.17+00	4.11–02	4.43–01	–0.784	B	*LS*
				815.871	452806.6	– 575375	2–4	4.13+00	8.24–02	4.43–01	–0.783	B	*LS*
151.		$^2P°-^2D$		*759.36*	*452983*	– *584673*	6–10	3.58+00	5.16–02	7.74–01	–0.509	B	1
				759.362	453071.5	– 584761	4–6	3.58+00	4.64–02	4.64–01	–0.731	B	*LS*
				759.103	452806.6	– 584541	2–4	2.99+00	5.16–02	2.58–01	–0.986	B	*LS*
				760.633	453071.5	– 584541	4–4	5.94–01	5.15–03	5.16–02	–1.686	B	*LS*
152.	$2s2p(^3P°)3s-2p^2(^3P)3s$	$^2P°-^2P$		*706.77*	*452983*	– *594472*	6–6	1.41+01	1.05–01	1.47+00	–0.199	B	1
				706.881	453071.5	– 594538	4–4	1.17+01	8.79–02	8.18–01	–0.454	B	*LS*
				706.547	452806.6	– 594340	2–2	9.40+00	7.03–02	3.27–01	–0.852	B	*LS*
				707.872	453071.5	– 594340	4–2	4.67+00	1.75–02	1.64–01	–1.154	B	*LS*
				705.560	452806.6	– 594538	2–4	2.36+00	3.52–02	1.64–01	–1.152	B	*LS*
153.	$2s2p(^3P°)3s-2s2p(^3P°)4f$	$^2P°-^2D$		*695.74*	*452983*	– *596714*	6–10	3.32–01	4.02–03	5.52–02	–1.618	B	1
				696.443	453071.5	– 596658.3	4–6	3.31–01	3.61–03	3.31–02	–1.840	B	*LS*
				694.485	452806.6	– 596798.2	2–4	2.78–01	4.02–03	1.84–02	–2.094	B–	*LS*
				695.765	453071.5	– 596798.2	4–4	5.53–02	4.02–04	3.68–03	–2.794	C+	*LS*
154.	$2s2p(^3P°)3s-2p^2(^3P)3d$	$^2P°-^2P$		*557.10*	*452983*	– *632483*	6–6	5.49–01	2.55–03	2.81–02	–1.815	B–	1
				557.555	453071.5	– 632426	4–4	4.56–01	2.13–03	1.56–02	–2.070	B–	*LS*
				556.203	452806.6	– 632597	2–2	3.68–01	1.71–03	6.24–03	–2.467	C+	*LS*
				557.024	453071.5	– 632597	4–2	1.83–01	4.26–04	3.12–03	–2.769	C+	*LS*
				556.733	452806.6	– 632426	2–4	9.16–02	8.52–04	3.12–03	–2.769	C+	*LS*

O IV: Allowed Transitions — Continued

No.	Transition Array	Mult.	λ_{air} (Å)	λ_{vac} (Å) or σ (cm^{-1})*	E_i (cm^{-1})	E_k (cm^{-1})	g_i-g_k	A_{ki} (10^8 s^{-1})	f_{ik}	S (at. u.)	log gf	Acc.	Source
155.	$2s2p(^3P^\circ)3p-2s^24p$	$^2P-^2P^\circ$	*3055.6*	*3056.5*	*467306*	*500024*	6–6	2.38–02	3.34–03	2.02–01	–1.698	B	1
			3057.88	3058.77	467344.9	500037.8	4–4	1.98–02	2.78–03	1.12–01	–1.954	B	LS
			3051.09	3051.98	467229.3	499994.9	2–2	1.60–02	2.23–03	4.48–02	–2.351	B	LS
			3061.90	3062.79	467344.9	499994.9	4–2	7.90–03	5.55–04	2.24–02	–2.653	B–	LS
			3047.10	3047.99	467229.3	500037.8	2–4	4.01–03	1.12–03	2.24–02	–2.651	B–	LS
156.	$2s2p(^3P^\circ)3p-2s2p(^3P^\circ)3d$	$^2P-^2D^\circ$	*2920.1*	*2920.9*	*467306*	*501542*	6–10	1.27+00	2.71–01	1.56+01	0.211	B	1
			2921.46	2922.31	467344.9	501564.4	4–6	1.27+00	2.44–01	9.38+00	–0.011	B	LS
			2916.31	2917.16	467229.3	501509.2	2–4	1.06+00	2.71–01	5.21+00	–0.266	B	LS
			2926.18	2927.03	467344.9	501509.2	4–4	2.11–01	2.70–02	1.04+00	–0.966	B	LS
157.		$^2P-^2P^\circ$	*2128.6*	*2129.3*	*467306*	*514271*	6–6	1.56+00	1.06–01	4.46+00	–0.196	B	1
			2132.64	2133.31	467344.9	514220.4	4–4	1.29+00	8.82–02	2.48+00	–0.452	B	LS
			2120.58	2121.25	467229.3	514371.3	2–2	1.05+00	7.10–02	9.91–01	–0.848	B	LS
			2125.79	2126.47	467344.9	514371.3	4–2	5.23–01	1.77–02	4.96–01	–1.150	B	LS
			2127.39	2128.06	467229.3	514220.4	2–4	2.61–01	3.54–02	4.96–01	–1.150	B	LS
158.	$2s2p(^3P^\circ)3p-2s2p(^1P^\circ)3s$	$^2P-^2P^\circ$		*1945.5*	*467306*	*518706*	6–6	2.29–01	1.30–02	5.00–01	–1.108	B	1
				1946.89	467344.9	518709	4–4	1.91–01	1.08–02	2.78–01	–1.363	B	LS
				1942.89	467229.3	518699	2–2	1.53–01	8.69–03	1.11–01	–1.760	B	LS
				1947.26	467344.9	518699	4–2	7.62–02	2.17–03	5.56–02	–2.062	B	LS
				1942.51	467229.3	518709	2–4	3.84–02	4.34–03	5.56–02	–2.061	B	LS
159.	$2s2p(^3P^\circ)3p-2s^25p$	$^2P-^2P^\circ$		*1257.7*	*467306*	*546814*	6–6	1.06–01	2.50–03	6.22–02	–1.823	B	1
				1258.28	467344.9	546818.7	4–4	8.79–02	2.09–03	3.46–02	–2.079	B	LS
				1256.69	467229.3	546803.4	2–2	7.06–02	1.67–03	1.38–02	–2.476	B–	LS
				1258.52	467344.9	546803.4	4–2	3.51–02	4.17–04	6.91–03	–2.778	C+	LS
				1256.45	467229.3	546818.7	2–4	1.76–02	8.35–04	6.91–03	–2.777	C+	LS
160.	$2s2p(^3P^\circ)3p-2s2p(^3P^\circ)4s$	$^2P-^2P^\circ$		*938.73*	*467306*	*573833*	6–6	4.28+00	5.65–02	1.05+00	–0.470	B	1
				938.473	467344.9	573901	4–4	3.57+00	4.71–02	5.82–01	–0.725	B	LS
				939.261	467229.3	573696	2–2	2.85+00	3.77–02	2.33–01	–1.123	B	LS
				940.282	467344.9	573696	4–2	1.42+00	9.41–03	1.16–01	–1.424	B	LS
				937.456	467229.3	573901	2–4	7.16–01	1.89–02	1.16–01	–1.423	B	LS
161.	$2s2p(^3P^\circ)3p-2s2p(^1P^\circ)3d$	$^2P-^2D^\circ$		*921.38*	*467306*	*575839*	6–10	9.07–01	1.92–02	3.50–01	–0.938	B	1
				921.590	467344.9	575853	4–6	9.06–01	1.73–02	2.10–01	–1.160	B	LS
				920.898	467229.3	575819	2–4	7.57–01	1.92–02	1.17–01	–1.415	B	LS
				921.879	467344.9	575819	4–4	1.51–01	1.92–03	2.33–02	–2.114	B–	LS
162.	$2s2p(^3P^\circ)3p-2s2p(^3P^\circ)4d$	$^2P-^2D^\circ$		*791.33*	*467306*	*593676*	6–10	9.27+00	1.45–01	2.27+00	–0.060	B	1
				791.370	467344.9	593708	4–6	9.27+00	1.30–01	1.36+00	–0.282	B	LS
				791.154	467229.3	593627	2–4	7.73+00	1.45–01	7.56–01	–0.537	B	LS
				791.878	467344.9	593627	4–4	1.54+00	1.45–02	1.51–01	–1.237	B	LS
163.	$2s2p(^3P^\circ)3p-2p^2(^3P)3p$	$^2P-^2S^\circ$		*769.54*	*467306*	597254	6–2	3.21+01	9.51–02	1.45+00	–0.243	B	1
				769.769	467344.9	597254	4–2	2.14+01	9.51–02	9.64–01	–0.420	B	LS
				769.085	467229.3	597254	2–2	1.07+01	9.52–02	4.82–01	–0.720	B	LS
164.	$2s2p(^3P^\circ)3p-2s2p(^3P^\circ)4d$	$^2P-^2P^\circ$		*766.50*	*467306*	*597770*	6–6	2.81+00	2.47–02	3.74–01	–0.829	B	1
				767.018	467344.9	597720	4–4	2.33+00	2.06–02	2.08–01	–1.085	B	LS
				765.464	467229.3	597869	2–2	1.88+00	1.65–02	8.31–02	–1.482	B	LS
				766.142	467344.9	597869	4–2	9.36–01	4.12–03	4.16–02	–1.783	B	LS
				766.338	467229.3	597720	2–4	4.68–01	8.24–03	4.16–02	–1.783	B	LS

O IV: Allowed Transitions — Continued

No.	Transition Array	Mult.	λ_{air} (Å)	λ_{vac} (Å) or σ (cm^{-1})*	E_i (cm^{-1})	E_k (cm^{-1})	g_i-g_k	A_{ki} ($10^8 s^{-1}$)	f_{ik}	S (at. u.)	log gf	Acc.	Source
165.	$2s2p(^3P^o)3p-$ $2p^2(^3P)3p$	$^2P-^2D^o$		*675.05*	*467306*	– *615443*	6–10	1.99+00	2.26–02	3.02–01	–0.867	B	1
				675.283	467344.9	– 615431	4–6	1.99+00	2.04–02	1.81–01	–1.089	B	*LS*
				674.624	467229.3	– 615460	2–4	1.66+00	2.27–02	1.01–01	–1.344	B	*LS*
				675.151	467344.9	– 615460	4–4	3.31–01	2.26–03	2.01–02	–2.043	B–	*LS*
166.	$2s2p(^3P^o)3p-$ $2p^2(^1D)3p$	$^2P-^2D^o$		*607.51*	*467306*	– 631913	6–10	2.78+00	2.57–02	3.08–01	–0.812	B	1
				607.651	467344.9	– 631913	4–6	2.78+00	2.31–02	1.85–01	–1.034	B	*LS*
				607.225	467229.3	– 631913	2–4	2.32+00	2.57–02	1.03–01	–1.289	B	*LS*
				607.651	467344.9	– 631913	4–4	4.64–01	2.57–03	2.05–02	–1.989	B–	*LS*
167.	$2s2p(^3P^o)3p-$ $2s2p(^1P^o)4d$	$^2P-^2D^o$		*496.90*	*467306*	– *668555*	6–10	6.90–01	4.26–03	4.18–02	–1.593	B	1
				496.998	467344.9	– 668553	4–6	6.90–01	3.83–03	2.51–02	–1.814	B–	*LS*
				496.703	467229.3	– 668557	2–4	5.76–01	4.26–03	1.39–02	–2.070	B–	*LS*
				496.988	467344.9	– 668557	4–4	1.15–01	4.26–04	2.79–03	–2.769	C	*LS*
168.	$2s2p(^3P^o)3p-$ $2s2p(^3P^o)3d$	$^4D-^4F^o$	*3733.5*	*3734.6*	*468586*	– *495363*	20–28	7.97–01	2.33–01	5.73+01	0.669	B	1
			3736.85	3737.91	468760.0	– 495512.9	8–10	7.95–01	2.08–01	2.05+01	0.221	B	*LS*
			3729.03	3730.09	468550.6	– 495359.6	6–8	6.86–01	1.91–01	1.40+01	0.058	B	*LS*
			3725.94	3727.00	468414.9	– 495246.1	4–6	6.01–01	1.88–01	9.21+00	–0.124	B	*LS*
			3725.89	3726.95	468337.4	– 495169	2–4	5.61–01	2.34–01	5.73+00	–0.331	B	*LS*
			3758.39	3759.46	468760.0	– 495359.6	8–8	1.11–01	2.36–02	2.33+00	–0.725	B	*LS*
			3744.89	3745.95	468550.6	– 495246.1	6–6	1.92–01	4.04–02	2.99+00	–0.616	B	*LS*
			3736.68	3737.74	468414.9	– 495169	4–4	2.23–01	4.66–02	2.29+00	–0.730	B	*LS*
			3774.49	3775.57	468760.0	– 495246.1	8–6	7.45–03	1.19–03	1.19–01	–2.020	B	*LS*
			3755.73	3756.80	468550.6	– 495169	6–4	1.57–02	2.21–03	1.64–01	–1.878	B	*LS*
169.		$^4D-^4D^o$	*3197.5*	*3198.4*	*468586*	– *499851*	20–20	2.98–01	4.57–02	9.61+00	–0.039	B	1
			3209.65	3210.58	468760.0	– 499907.0	8–8	2.53–01	3.90–02	3.30+00	–0.506	B	*LS*
			3194.78	3195.70	468550.6	– 499842.6	6–6	1.71–01	2.62–02	1.65+00	–0.803	B	*LS*
			3185.74	3186.66	468414.9	– 499795.7	4–4	1.21–01	1.83–02	7.69–01	–1.135	B	*LS*
			3180.77	3181.69	468337.4	– 499767.2	2–2	1.51–01	2.30–02	4.81–01	–1.338	B	*LS*
			3216.31	3217.23	468760.0	– 499842.6	8–6	5.56–02	6.47–03	5.48–01	–1.286	B	*LS*
			3199.58	3200.50	468550.6	– 499795.7	6–4	1.04–01	1.06–02	6.73–01	–1.195	B	*LS*
			3188.64	3189.56	468414.9	– 499767.2	4–2	1.50–01	1.14–02	4.81–01	–1.339	B	*LS*
			3188.22	3189.14	468550.6	– 499907.0	6–8	4.28–02	8.70–03	5.48–01	–1.282	B	*LS*
			3180.99	3181.91	468414.9	– 499842.6	4–6	7.06–02	1.61–02	6.73–01	–1.192	B	*LS*
			3177.89	3178.81	468337.4	– 499795.7	2–4	7.59–02	2.30–02	4.81–01	–1.338	B	*LS*
170.		$^4D-^4P^o$	*2809.9*	*2810.7*	*468586*	– *504165*	20–12	1.99–01	1.42–02	2.62+00	–0.548	B	1
			2829.17	2830.00	468760.0	– 504095.7	8–6	1.56–01	1.41–02	1.05+00	–0.949	B	*LS*
			2803.57	2804.40	468550.6	– 504208.9	6–4	1.26–01	9.93–03	5.50–01	–1.225	B	*LS*
			2787.22	2788.05	468414.9	– 504282.3	4–2	1.02–01	5.95–03	2.18–01	–1.624	B	*LS*
			2812.50	2813.33	468550.6	– 504095.7	6–6	3.58–02	4.24–03	2.36–01	–1.594	B	*LS*
			2792.94	2793.76	468414.9	– 504208.9	4–4	6.50–02	7.60–03	2.79–01	–1.517	B	*LS*
			2781.22	2782.04	468337.4	– 504282.3	2–2	1.03–01	1.19–02	2.18–01	–1.623	B	*LS*
			2801.80	2802.63	468414.9	– 504095.7	4–6	4.02–03	7.10–04	2.62–02	–2.547	B–	*LS*
			2786.91	2787.73	468337.4	– 504208.9	2–4	1.02–02	2.38–03	4.37–02	–2.323	B	*LS*
171.	$2s2p(^3P^o)3p-$ $2s2p(^3P^o)4s$	$^4D-^4P^o$		*994.52*	*468586*	– *569137*	20–12	1.03+01	9.12–02	5.97+00	0.261	B	1
				994.797	468760.0	– 569283	8–6	8.20+00	9.12–02	2.39+00	–0.137	B	*LS*
				995.169	468550.6	– 569036	6–4	6.45+00	6.38–02	1.25+00	–0.417	B	*LS*
				995.163	468414.9	– 568901	4–2	5.12+00	3.80–02	4.98–01	–0.818	B	*LS*
				992.729	468550.6	– 569283	6–6	1.86+00	2.74–02	5.38–01	–0.784	B	*LS*
				993.827	468414.9	– 569036	4–4	3.29+00	4.87–02	6.37–01	–0.711	B	*LS*
				994.396	468337.4	– 568901	2–2	5.13+00	7.60–02	4.98–01	–0.818	B	*LS*
				991.394	468414.9	– 569283	4–6	2.07–01	4.58–03	5.97–02	–1.738	B	*LS*
				993.062	468337.4	– 569036	2–4	5.15–01	1.52–02	9.96–02	–1.516	B	*LS*

O IV: Allowed Transitions — Continued

No.	Transition Array	Mult.	λ_{air} (Å)	λ_{vac} (Å) or σ (cm^{-1})*	E_i (cm^{-1})	E_k (cm^{-1})	g_i–g_k	A_{ki} (10^8 s^{-1})	f_{ik}	S (at. u.)	log gf	Acc.	Source
172.	$2s2p(^3P°)3p$ – $2s2p(^3P°)4d$	^4D–$^4D°$		*810.43*	*468586*	– *591977*	20–20	6.35+00	6.25–02	3.34+00	0.097	B	1
				811.122	468760.0	– 592046	8–8	5.43+00	5.36–02	1.14+00	–0.368	B	*LS*
				810.298	468550.6	– 591962	6–6	3.64+00	3.59–02	5.74–01	–0.667	B	*LS*
				809.821	468414.9	– 591899	4–4	2.55+00	2.50–02	2.67–01	–1.000	B	*LS*
				809.313	468337.4	– 591899	2–2	3.19+00	3.13–02	1.67–01	–1.203	B	*LS*
				811.675	468760.0	– 591962	8–6	1.20+00	8.90–03	1.90–01	–1.148	B	*LS*
				810.712	468550.6	– 591899	6–4	2.22+00	1.46–02	2.34–01	–1.058	B	*LS*
				809.821	468414.9	– 591899	4–2	3.18+00	1.56–02	1.67–01	–1.204	B	*LS*
				809.747	468550.6	– 592046	6–8	9.07–01	1.19–02	1.90–01	–1.147	B	*LS*
				809.408	468414.9	– 591962	4–6	1.49+00	2.19–02	2.34–01	–1.057	B	*LS*
				809.313	468337.4	– 591899	2–4	1.59+00	3.13–02	1.67–01	–1.203	B	*LS*
173.		^4D–$^4P°$		*801.51*	*468586*	– *593351*	20–12	1.51–01	8.72–04	4.60–02	–1.759	B	1
				803.129	468760.0	– 593273	8–6	1.20–01	8.70–04	1.84–02	–2.157	B–	*LS*
				800.927	468550.6	– 593406	6–4	9.52–02	6.11–04	9.66–03	–2.436	C+	*LS*
				799.628	468414.9	– 593473	4–2	7.60–02	3.64–04	3.83–03	–2.837	C+	*LS*
				801.781	468550.6	– 593273	6–6	2.71–02	2.61–04	4.14–03	–2.805	C	*LS*
				800.057	468414.9	– 593406	4–4	4.85–02	4.66–04	4.91–03	–2.730	C	*LS*
				799.133	468337.4	– 593473	2–2	7.61–02	7.29–04	3.83–03	–2.837	C	*LS*
				800.909	468414.9	– 593273	4–6	3.02–03	4.36–05	4.60–04	–3.758	D	*LS*
				799.561	468337.4	– 593406	2–4	7.60–03	1.46–04	7.67–04	–3.536	D	*LS*
174.	$2s2p(^3P°)3p$ – $2p^2(^3P)3p$	^4D–$^4P°$		*724.42*	*468586*	– *606628*	20–12	1.06+01	4.98–02	2.38+00	–0.002	B	1
				724.984	468760.0	– 606694	8–6	8.42+00	4.98–02	9.50–01	–0.400	B	*LS*
				724.494	468550.6	– 606578	6–4	6.65+00	3.49–02	4.99–01	–0.679	B	*LS*
				724.034	468414.9	– 606530	4–2	5.28+00	2.08–02	1.98–01	–1.081	B	*LS*
				723.885	468550.6	– 606694	6–6	1.90+00	1.50–02	2.14–01	–1.047	B	*LS*
				723.782	468414.9	– 606578	4–4	3.39+00	2.66–02	2.53–01	–0.973	B	*LS*
				723.628	468337.4	– 606530	2–2	5.29+00	4.16–02	1.98–01	–1.080	B	*LS*
				723.175	468414.9	– 606694	4–6	2.12–01	2.49–03	2.38–02	–2.001	B–	*LS*
				723.376	468337.4	– 606578	2–4	5.30–01	8.31–03	3.96–02	–1.779	B	*LS*
175.	$2s2p(^3P°)3p$ – $2s2p(^3P°)5d$	^4D–$^4P°$		*602.44*	*468586*	– *634578*	20–12	2.63–01	8.60–04	3.41–02	–1.765	B	1
				603.271	468760.0	– 634523	8–6	2.10–01	8.59–04	1.36–02	–2.163	B–	*LS*
				602.151	468550.6	– 634622	6–4	1.66–01	6.02–04	7.16–03	–2.442	C+	*LS*
				601.547	468414.9	– 634653	4–2	1.32–01	3.59–04	2.84–03	–2.843	C	*LS*
				602.510	468550.6	– 634523	6–6	4.74–02	2.58–04	3.07–03	–2.810	C+	*LS*
				601.659	468414.9	– 634622	4–4	8.46–02	4.59–04	3.64–03	–2.736	C+	*LS*
				601.267	468337.4	– 634653	2–2	1.32–01	7.18–04	2.84–03	–2.843	C	*LS*
				602.018	468414.9	– 634523	4–6	5.28–03	4.30–05	3.41–04	–3.764	D	*LS*
				601.379	468337.4	– 634622	2–4	1.32–02	1.44–04	5.68–04	–3.542	D	*LS*
176.	$2s2p(^3P°)3p$ – $2s2p(^3P°)3d$	^4S–$^4P°$	*3367.6*	*3368.5*	474478.1	– *504165*	4–12	7.62–01	3.89–01	1.72+01	0.192	B	1
			3375.40	3376.37	474478.1	– 504095.7	4–6	7.56–01	1.94–01	8.62+00	–0.111	B	*LS*
			3362.55	3363.52	474478.1	– 504208.9	4–4	7.65–01	1.30–01	5.74+00	–0.285	B	*LS*
			3354.27	3355.23	474478.1	– 504282.3	4–2	7.71–01	6.50–02	2.87+00	–0.585	B	*LS*
177.	$2s2p(^3P°)3p$ – $2s2p(^3P°)4s$	^4S–$^4P°$		*1056.4*	474478.1	– *569137*	4–12	7.25–01	3.64–02	5.06–01	–0.837	B	1
				1054.80	474478.1	– 569283	4–6	7.28–01	1.82–02	2.53–01	–1.138	B	*LS*
				1057.55	474478.1	– 569036	4–4	7.22–01	1.21–02	1.69–01	–1.315	B	*LS*
				1059.07	474478.1	– 568901	4–2	7.19–01	6.05–03	8.43–02	–1.616	B	*LS*
178.	$2s2p(^3P°)3p$ – $2s2p(^3P°)4d$	^4S–$^4P°$		*841.23*	474478.1	– *593351*	4–12	9.13+00	2.91–01	3.22+00	0.066	B	1
				841.787	474478.1	– 593273	4–6	9.12+00	1.45–01	1.61+00	–0.236	B	*LS*
				840.846	474478.1	– 593406	4–4	9.15+00	9.70–02	1.07+00	–0.411	B	*LS*
				840.372	474478.1	– 593473	4–2	9.16+00	4.85–02	5.37–01	–0.712	B	*LS*

O IV: Allowed Transitions — Continued

No.	Transition Array	Mult.	λ_{air} (Å)	λ_{vac} (Å) or σ (cm^{-1})*	E_i (cm^{-1})	E_k (cm^{-1})	g_i-g_k	A_{ki} ($10^8 s^{-1}$)	f_{ik}	S (at. u.)	log gf	Acc.	Source
179.	$2s2p(^3P^o)3p-$ $2p^2(^3P)3p$	$^4S-^4P^o$		*756.72*	474478.1	*606628*	4–12	6.73+00	1.73−01	1.73+00	−0.159	B	1
				756.339	474478.1	606694	4–6	6.74+00	8.67−02	8.63−01	−0.460	B	*LS*
				757.003	474478.1	606578	4–4	6.72+00	5.77−02	5.76−01	−0.636	B	*LS*
				757.278	474478.1	606530	4–2	6.71+00	2.89−02	2.88−01	−0.938	B	*LS*
180.	$2s2p(^3P^o)3p-$ $2s2p(^3P^o)5d$	$^4S-^4P^o$		*624.61*	474478.1	*634578*	4–12	4.82+00	8.45−02	6.95−01	−0.471	B	1
				624.825	474478.1	634523	4–6	4.81+00	4.22−02	3.48−01	−0.772	B	*LS*
				624.438	474478.1	634622	4–4	4.82+00	2.82−02	2.32−01	−0.948	B	*LS*
				624.318	474478.1	634653	4–2	4.82+00	1.41−02	1.16−01	−1.249	B	*LS*
181.	$2s2p(^3P^o)3p-$ $2s2p(^3P^o)3d$	$^4P-^4D^o$	*4792.6*	*4793.9*	*478992*	*499851*	12–20	2.92−01	1.68−01	3.17+01	0.303	B	1
			4798.27	4799.62	479072.0	499907.0	6–8	2.91−01	1.34−01	1.27+01	−0.095	B	*LS*
			4783.42	4784.76	478942.9	499842.6	4–6	2.06−01	1.06−01	6.66+00	−0.373	B	*LS*
			4772.60	4773.93	478848.6	499795.7	2–4	1.23−01	8.42−02	2.64+00	−0.774	B	*LS*
			4813.15	4814.50	479072.0	499842.6	6–6	8.65−02	3.00−02	2.86+00	−0.744	B	*LS*
			4794.18	4795.52	478942.9	499795.7	4–4	1.56−01	5.36−02	3.38+00	−0.669	B	*LS*
			4779.10	4780.43	478848.6	499767.2	2–2	2.45−01	8.40−02	2.64+00	−0.775	B	*LS*
			4824.05	4825.39	479072.0	499795.7	6–4	1.43−02	3.33−03	3.17−01	−1.699	B	*LS*
			4800.74	4802.08	478942.9	499767.2	4–2	4.84−02	8.37−03	5.29−01	−1.475	B	*LS*
182.		$^4P-^4P^o$	*3971.4*	*3972.5*	*478992*	*504165*	12–12	2.21−01	5.23−02	8.20+00	−0.203	B	1
			3995.08	3996.21	479072.0	504095.7	6–6	1.52−01	3.64−02	2.87+00	−0.661	B	*LS*
			3956.77	3957.89	478942.9	504208.9	4–4	2.98−02	7.00−03	3.64−01	−1.553	B	*LS*
			3930.68	3931.79	478848.6	504282.3	2–2	3.80−02	8.80−03	2.28−01	−1.754	B	*LS*
			3977.09	3978.22	479072.0	504208.9	6–4	9.91−02	1.57−02	1.23+00	−1.027	B	*LS*
			3945.31	3946.42	478942.9	504282.3	4–2	1.88−01	2.19−02	1.14+00	−1.057	B	*LS*
			3974.58	3975.70	478942.9	504095.7	4–6	6.62−02	2.35−02	1.23+00	−1.027	B	*LS*
			3942.06	3943.17	478848.6	504208.9	2–4	9.42−02	4.39−02	1.14+00	−1.057	B	*LS*
183.	$2s2p(^3P^o)3p-$ $2s2p(^3P^o)4s$	$^4P-^4P^o$		*1109.3*	*478992*	*569137*	12–12	9.13+00	1.69−01	7.38+00	0.306	B	1
				1108.51	479072.0	569283	6–6	6.41+00	1.18−01	2.58+00	−0.150	B	*LS*
				1109.96	478942.9	569036	4–4	1.22+00	2.25−02	3.28−01	−1.047	B	*LS*
				1110.46	478848.6	568901	2–2	1.52+00	2.81−02	2.05−01	−1.251	B	*LS*
				1111.56	479072.0	569036	6–4	4.09+00	5.04−02	1.11+00	−0.519	B	*LS*
				1111.63	478942.9	568901	4–2	7.56+00	7.01−02	1.03+00	−0.552	B	*LS*
				1106.93	478942.9	569283	4–6	2.76+00	7.60−02	1.11+00	−0.517	B	*LS*
				1108.80	478848.6	569036	2–4	3.81+00	1.40−01	1.03+00	−0.551	B	*LS*
184.	$2s2p(^3P^o)3p-$ $2s2p(^3P^o)4d$	$^4P-^4D^o$		*885.07*	*478992*	*591977*	12–20	1.53+01	3.00−01	1.05+01	0.556	B	1
				885.159	479072.0	592046	6–8	1.53+01	2.40−01	4.19+00	0.158	B	*LS*
				884.806	478942.9	591962	4–6	1.07+01	1.89−01	2.20+00	−0.122	B	*LS*
				884.561	478848.6	591899	2–4	6.39+00	1.50−01	8.74−01	−0.523	B	*LS*
				885.818	479072.0	591962	6–6	4.58+00	5.39−02	9.44−01	−0.490	B	*LS*
				885.300	478942.9	591899	4–4	8.16+00	9.59−02	1.12+00	−0.416	B	*LS*
				884.561	478848.6	591899	2–2	1.28+01	1.50−01	8.74−01	−0.523	B	*LS*
				886.313	479072.0	591899	6–4	7.63−01	5.99−03	1.05−01	−1.444	B	*LS*
				885.300	478942.9	591899	4–2	2.55+00	1.50−02	1.75−01	−1.222	B	*LS*
185.		$^4P-^4P^o$		*874.44*	*478992*	*593351*	12–12	7.01+00	8.03−02	2.77+00	−0.016	B	1
				875.649	479072.0	593273	6–6	4.88+00	5.61−02	9.71−01	−0.473	B	*LS*
				873.644	478942.9	593406	4–4	9.37−01	1.07−02	1.23−01	−1.368	B	*LS*
				872.415	478848.6	593473	2–2	1.18+00	1.34−02	7.71−02	−1.571	B	*LS*
				874.630	479072.0	593406	6–4	3.15+00	2.41−02	4.16−01	−0.840	B	*LS*
				873.133	478942.9	593473	4–2	5.86+00	3.35−02	3.85−01	−0.873	B	*LS*
				874.660	478942.9	593273	4–6	2.10+00	3.61−02	4.16−01	−0.840	B	*LS*
				872.925	478848.6	593406	2–4	2.93+00	6.70−02	3.85−01	−0.873	B	*LS*

O IV: Allowed Transitions — Continued

No.	Transition Array	Mult.	λ_{air} (Å)	λ_{vac} (Å) or σ (cm^{-1})*	E_i (cm^{-1})	E_k (cm^{-1})	g_i–g_k	A_{ki} (10^8 s^{-1})	f_{ik}	S (at. u.)	log gf	Acc.	Source
186.	$2s2p(^3P^o)3p$ – $2p^2(^3P)3p$	^4P–$^4P^o$		*783.48*	*478992*	*606628*	12–12	4.35+00	4.01–02	1.24+00	–0.318	B	1
				783.564	479072.0	606694	6–6	3.05+00	2.80–02	4.34–01	–0.774	B	*LS*
				783.484	478942.9	606578	4–4	5.80–01	5.34–03	5.51–02	–1.670	B	*LS*
				783.199	478848.6	606530	2–2	7.26–01	6.68–03	3.44–02	–1.874	B	*LS*
				784.277	479072.0	606578	6–4	1.95+00	1.20–02	1.86–01	–1.142	B	*LS*
				783.778	478942.9	606530	4–2	3.62+00	1.67–02	1.72–01	–1.176	B	*LS*
				782.772	478942.9	606694	4–6	1.31+00	1.80–02	1.86–01	–1.142	B	*LS*
				782.905	478848.6	606578	2–4	1.82+00	3.34–02	1.72–01	–1.175	B	*LS*
187.		^4P–$^4S^o$		*725.22*	*478992*	616881	12–4	2.63+01	6.92–02	1.98+00	–0.080	B	1
				725.642	479072.0	616881	6–4	1.31+01	6.92–02	9.92–01	–0.382	B	*LS*
				724.963	478942.9	616881	4–4	8.79+00	6.93–02	6.61–01	–0.557	B	*LS*
				724.468	478848.6	616881	2–4	4.40+00	6.93–02	3.31–01	–0.858	B	*LS*
188.	$2s2p(^3P^o)3p$ – $2s2p(^3P^o)5d$	^4P–$^4P^o$		*642.73*	*478992*	*634578*	12–12	3.71+00	2.29–02	5.83–01	–0.560	B	1
				643.290	479072.0	634523	6–6	2.59+00	1.61–02	2.04–01	–1.016	B	*LS*
				642.347	478942.9	634622	4–4	4.95–01	3.06–03	2.59–02	–1.912	B–	*LS*
				641.830	478848.6	634653	2–2	6.20–01	3.83–03	1.62–02	–2.116	B–	*LS*
				642.880	479072.0	634622	6–4	1.67+00	6.88–03	8.74–02	–1.384	B	*LS*
				642.219	478942.9	634653	4–2	3.10+00	9.57–03	8.09–02	–1.417	B	*LS*
				642.756	478942.9	634523	4–6	1.11+00	1.03–02	8.74–02	–1.384	B	*LS*
				641.958	478848.6	634622	2–4	1.55+00	1.91–02	8.09–02	–1.417	B	*LS*
189.	$2s2p(^3P^o)3p$ – $2s^24p$	^2D–$^2P^o$	*5811.0*	*5812.6*	*482819*	*500024*	10–6	2.23–02	6.76–03	1.29+00	–1.170	B	1
			5840.80	5842.42	482921.6	500037.8	6–4	1.97–02	6.73–03	7.76–01	–1.394	B	*LS*
			5769.14	5770.74	482666.1	499994.9	4–2	2.28–02	5.68–03	4.31–01	–1.644	B	*LS*
			5754.89	5756.49	482666.1	500037.8	4–4	2.29–03	1.14–03	8.63–02	–2.342	B	*LS*
190.	$2s2p(^3P^o)3p$ – $2s2p(^3P^o)3d$	^2D–$^2D^o$	*5339.6*	*5341.1*	*482819*	*501542*	10–10	6.64–02	2.84–02	4.99+00	–0.547	B	1
			5362.51	5364.00	482921.6	501564.4	6–6	6.12–02	2.64–02	2.80+00	–0.800	B	*LS*
			5305.51	5306.98	482666.1	501509.2	4–4	6.10–02	2.57–02	1.80+00	–0.988	B	*LS*
			5378.44	5379.93	482921.6	501509.2	6–4	6.50–03	1.88–03	2.00–01	–1.948	B	*LS*
			5290.01	5291.48	482666.1	501564.4	4–6	4.55–03	2.87–03	2.00–01	–1.941	B	*LS*
191.		^2D–$^2F^o$	*3563.0*	*3564.0*	*482819*	*510878*	10–14	1.10+00	2.93–01	3.44+01	0.467	B	1
			3563.33	3564.35	482921.6	510977.2	6–8	1.10+00	2.79–01	1.97+01	0.224	B	*LS*
			3560.39	3561.41	482666.1	510744.9	4–6	1.03+00	2.93–01	1.38+01	0.070	B	*LS*
			3593.08	3594.11	482921.6	510744.9	6–6	7.15–02	1.38–02	9.83–01	–1.081	B	*LS*
192.	$2s2p(^3P^o)3p$ – $2s^24f$	^2D–$^2F^o$	*3291.3*	*3292.3*	*482819*	*513194*	10–14	2.48–02	5.65–03	6.12–01	–1.248	B	1
			3301.92	3302.87	482921.6	513198.3	6–8	2.46–02	5.36–03	3.50–01	–1.493	C+	*LS*
			3275.48	3276.42	482666.1	513187.2	4–6	2.35–02	5.67–03	2.45–01	–1.644	C+	*LS*
			3303.13	3304.08	482921.6	513187.2	6–6	1.64–03	2.68–04	1.75–02	–2.794	C	*LS*
193.	$2s2p(^3P^o)3p$ – $2s2p(^3P^o)3d$	^2D–$^2P^o$	*3178.6*	*3179.5*	*482819*	*514271*	10–6	9.63–02	8.75–03	9.16–01	–1.058	B	1
			3194.09	3195.01	482921.6	514220.4	6–4	8.54–02	8.71–03	5.50–01	–1.282	B	*LS*
			3153.14	3154.06	482666.1	514371.3	4–2	9.87–02	7.35–03	3.05–01	–1.531	B	*LS*
			3168.22	3169.14	482666.1	514220.4	4–4	9.73–03	1.46–03	6.11–02	–2.233	B	*LS*
194.	$2s2p(^3P^o)3p$ – $2s2p(^1P^o)3s$	^2D–$^2P^o$	*2785.7*	*2786.6*	*482819*	*518706*	10–6	5.02–02	3.50–03	3.21–01	–1.456	B	1
			2793.46	2794.28	482921.6	518709	6–4	4.48–02	3.49–03	1.93–01	–1.679	B	*LS*
			2774.42	2775.24	482666.1	518699	4–2	5.08–02	2.93–03	1.07–01	–1.931	B	*LS*
			2773.65	2774.47	482666.1	518709	4–4	5.08–03	5.86–04	2.14–02	–2.630	B–	*LS*

O IV: Allowed Transitions — Continued

No.	Transition Array	Mult.	λ_{air} (Å)	λ_{vac} (Å) or σ (cm^{-1})*	E_i (cm^{-1})	E_k (cm^{-1})	g_i-g_k	A_{ki} ($10^8 s^{-1}$)	f_{ik}	S (at. u.)	log gf	Acc.	Source
195.	$2s2p(^3P^o)3p-2s^25p$	$^2D-^2P^o$		*1562.6*	*482819*	– *546814*	10–6	9.87–02	2.17–03	1.12–01	–1.664	B	1
				1565.02	482921.6	– 546818.7	6–4	8.84–02	2.16–03	6.69–02	–1.887	B	LS
				1559.16	482666.1	– 546803.4	4–2	9.93–02	1.81–03	3.72–02	–2.140	B	LS
				1558.78	482666.1	– 546818.7	4–4	9.94–03	3.62–04	7.43–03	–2.839	C+	LS
196.	$2s2p(^3P^o)3p-2s^25f$	$^2D-^2F^o$		*1435.2*	*482819*	– 552495.1	10–14	3.27–02	1.41–03	6.68–02	–1.850	B	1
				1437.33	482921.6	– 552495.1	6–8	3.26–02	1.34–03	3.82–02	–2.093	C+	LS
				1432.07	482666.1	– 552495.1	4–6	3.07–02	1.42–03	2.67–02	–2.247	C	LS
				1437.33	482921.6	– 552495.1	6–6	2.17–03	6.72–05	1.91–03	–3.394	D	LS
197.	$2s2p(^3P^o)3p-2s2p(^3P^o)4s$	$^2D-^2P^o$		*1098.7*	*482819*	– *573833*	10–6	7.07+00	7.67–02	2.78+00	–0.115	B	1
				1099.15	482921.6	– 573901	6–4	6.35+00	7.67–02	1.67+00	–0.337	B	LS
				1098.54	482666.1	– 573696	4–2	7.07+00	6.40–02	9.25–01	–0.592	B	LS
				1096.07	482666.1	– 573901	4–4	7.12–01	1.28–02	1.85–01	–1.290	B	LS
198.	$2s2p(^3P^o)3p-2s^26f$	$^2D-^2F^o$		*1087.1*	*482819*	– 574807.7	10–14	4.75–02	1.18–03	4.22–02	–1.928	B–	1
				1088.30	482921.6	– 574807.7	6–8	4.74–02	1.12–03	2.41–02	–2.172	C	LS
				1085.29	482666.1	– 574807.7	4–6	4.46–02	1.18–03	1.69–02	–2.326	C	LS
				1088.30	482921.6	– 574807.7	6–6	3.16–03	5.61–05	1.21–03	–3.473	D	LS
199.	$2s2p(^3P^o)3p-2s2p(^1P^o)3d$	$^2D-^2D^o$		*1075.0*	*482819*	– *575839*	10–10	1.18+00	2.05–02	7.26–01	–0.688	B	1
				1076.06	482921.6	– 575853	6–6	1.10+00	1.91–02	4.07–01	–0.940	B	LS
				1073.50	482666.1	– 575819	4–4	1.07+00	1.85–02	2.62–01	–1.131	B	LS
				1076.46	482921.6	– 575819	6–4	1.18–01	1.37–03	2.91–02	–2.086	B–	LS
				1073.11	482666.1	– 575853	4–6	7.94–02	2.06–03	2.91–02	–2.085	B–	LS
200.	$2s2p(^3P^o)3p-2s^27f$	$^2D-^2F^o$		*952.10*	*482819*	– 587850	10–14	9.24–02	1.76–03	5.51–02	–1.755	B	1
				953.031	482921.6	– 587850	6–8	9.21–02	1.67–03	3.15–02	–1.998	C+	LS
				950.716	482666.1	– 587850	4–6	8.66–02	1.76–03	2.20–02	–2.152	C	LS
				953.031	482921.6	– 587850	6–6	6.14–03	8.36–05	1.57–03	–3.299	D	LS
201.	$2s2p(^3P^o)3p-2s2p(^3P^o)4d$	$^2D-^2D^o$		*902.07*	*482819*	– *593676*	10–10	3.27+00	3.99–02	1.19+00	–0.399	B	1
				902.638	482921.6	– 593708	6–6	3.05+00	3.72–02	6.64–01	–0.651	B	LS
				901.218	482666.1	– 593627	4–4	2.95+00	3.60–02	4.27–01	–0.842	B	LS
				903.298	482921.6	– 593627	6–4	3.26–01	2.66–03	4.74–02	–1.797	B	LS
				900.561	482666.1	– 593708	4–6	2.19–01	4.00–03	4.74–02	–1.796	B	LS
202.		$^2D-^2F^o$		*880.45*	*482819*	– *596398*	10–14	8.66+00	1.41–01	4.08+00	0.149	B	1
				880.643	482921.6	– 596475	6–8	8.66+00	1.34–01	2.33+00	–0.094	B	LS
				880.058	482666.1	– 596295	4–6	8.10+00	1.41–01	1.63+00	–0.249	B	LS
				882.041	482921.6	– 596295	6–6	5.74–01	6.70–03	1.17–01	–1.396	B	LS
203.		$^2D-^2P^o$		*869.94*	*482819*	– *597770*	10–6	1.77–01	1.20–03	3.45–02	–1.919	B	1
				871.092	482921.6	– 597720	6–4	1.59–01	1.20–03	2.07–02	–2.142	B–	LS
				868.034	482666.1	– 597869	4–2	1.78–01	1.01–03	1.15–02	–2.395	B–	LS
				869.158	482666.1	– 597720	4–4	1.77–02	2.01–04	2.30–03	–3.095	C	LS
204.	$2s2p(^3P^o)3p-2p^2(^3P)3p$	$^2D-^2D^o$		*754.01*	*482819*	– *615443*	10–10	1.57+01	1.33–01	3.31+00	0.125	B	1
				754.663	482921.6	– 615431	6–6	1.46+01	1.24–01	1.85+00	–0.127	B	LS
				753.047	482666.1	– 615460	4–4	1.41+01	1.20–01	1.19+00	–0.318	B	LS
				754.498	482921.6	– 615460	6–4	1.56+00	8.89–03	1.32–01	–1.273	B	LS
				753.211	482666.1	– 615431	4–6	1.05+00	1.34–02	1.32–01	–1.272	B	LS

O IV: Allowed Transitions — Continued

No.	Transition Array	Mult.	λ_{air} (Å)	λ_{vac} (Å) or σ (cm^{-1})*	E_i (cm^{-1})	E_k (cm^{-1})	g_i-g_k	A_{ki} (10^8 s^{-1})	f_{ik}	S (at. u.)	log gf	Acc.	Source
205.	$2s^24s-$ $2s^24p$	$^2S-^2P^o$	*7039.4*	*7041.4*	485821.7	*500024*	2–6	3.88–01	8.66–01	4.01+01	0.238	B	1
			7032.34	7034.28	485821.7	500037.8	2–4	3.90–01	5.78–01	2.68+01	0.063	B	*LS*
			7053.62	7055.57	485821.7	499994.9	2–2	3.86–01	2.88–01	1.34+01	–0.240	B	*LS*
206.	$2s^24s-$ $2s2p(^3P^o)3d$	$^2S-^2P^o$	*6357.8*	*6359.6*	485821.7	*501546*	2–6	4.34–03	7.89–03	3.30–01	–1.802	B	1
			6350.39	6352.15	485821.7	501564.4	2–4	4.36–03	5.27–03	2.20–01	–1.977	B	*LS*
			6372.74	6374.50	485821.7	501509.2	2–2	4.31–03	2.62–03	1.10–01	–2.280	B	*LS*
207.	$2s^24s-$ $2s2p(^1P^o)3s$	$^2S-^2P^o$	*3040.1*	*3041.0*	485821.7	*518706*	2–6	1.09–01	4.53–02	9.06–01	–1.043	B	1
			3039.80	3040.69	485821.7	518709	2–4	1.09–01	3.02–02	6.04–01	–1.219	B	*LS*
			3040.73	3041.61	485821.7	518699	2–2	1.09–01	1.51–02	3.02–01	–1.520	B	*LS*
208.	$2s^24s-$ $2s^25p$	$^2S-^2P^o$		*1639.6*	485821.7	*546814*	2–6	7.36–01	8.90–02	9.60–01	–0.750	B	1
				1639.42	485821.7	546818.7	2–4	7.36–01	5.93–02	6.40–01	–0.926	B	*LS*
				1639.84	485821.7	546803.4	2–2	7.35–01	2.97–02	3.20–01	–1.227	B	*LS*
209.	$2s^24s-$ $2s2p(^3P^o)4s$	$^2S-^2P^o$		*1136.2*	485821.7	*573833*	2–6	2.51–02	1.46–03	1.09–02	–2.536	B–	1
				1135.34	485821.7	573901	2–4	2.52–02	9.72–04	7.27–03	–2.711	C+	*LS*
				1137.99	485821.7	573696	2–2	2.50–02	4.85–04	3.63–03	–3.013	C+	*LS*
210.	$2s^24s-$ $2s2p(^1P^o)3d$	$^2S-^2P^o$		*1042.6*	485821.7	*581736*	2–6	1.29–01	6.32–03	4.34–02	–1.898	B	1
				1042.52	485821.7	581743	2–4	1.29–01	4.22–03	2.89–02	–2.074	B–	*LS*
				1042.76	485821.7	581721	2–2	1.29–01	2.11–03	1.45–02	–2.375	B–	*LS*
211.	$2s^24s-$ $2s2p(^3P^o)4d$	$^2S-^2P^o$		*893.27*	485821.7	*597770*	2–6	5.63–01	2.02–02	1.19–01	–1.393	B	1
				893.669	485821.7	597720	2–4	5.63–01	1.35–02	7.93–02	–1.570	B	*LS*
				892.480	485821.7	597869	2–2	5.65–01	6.74–03	3.96–02	–1.870	B	*LS*
212.	$2s2p(^3P^o)3p-$ $2s2p(^3P^o)3d$	$^2S-^2P^o$	*4676.0*	*4677.3*	492890.9	*514271*	2–6	2.81–01	2.76–01	8.51+00	–0.257	B	1
			4687.03	4688.34	492890.9	514220.4	2–4	2.79–01	1.84–01	5.67+00	–0.435	B	*LS*
			4654.10	4655.41	492890.9	514371.3	2–2	2.85–01	9.26–02	2.84+00	–0.733	B	*LS*
213.	$2s2p(^3P^o)3p-$ $2s2p(^1P^o)3s$	$^2S-^2P^o$	*3872.6*	*3873.7*	492890.9	*518706*	2–6	2.99–02	2.02–02	5.14–01	–1.394	B	1
			3872.15	3873.25	492890.9	518709	2–4	2.99–02	1.34–02	3.43–01	–1.570	B	*LS*
			3873.65	3874.75	492890.9	518699	2–2	2.99–02	6.72–03	1.71–01	–1.871	B	*LS*
214.	$2s2p(^3P^o)3p-$ $2s^25p$	$^2S-^2P^o$		*1854.5*	492890.9	*546814*	2–6	8.19–02	1.27–02	1.55–01	–1.596	B	1
				1854.33	492890.9	546818.7	2–4	8.19–02	8.45–03	1.03–01	–1.772	B	*LS*
				1854.86	492890.9	546803.4	2–2	8.19–02	4.22–03	5.16–02	–2.073	B	*LS*
215.	$2s2p(^3P^o)3p-$ $2s2p(^3P^o)4s$	$^2S-^2P^o$		*1235.5*	492890.9	*573833*	2–6	7.84–01	5.38–02	4.38–01	–0.968	B	1
				1234.41	492890.9	573901	2–4	7.86–01	3.59–02	2.92–01	–1.144	B	*LS*
				1237.55	492890.9	573696	2–2	7.80–01	1.79–02	1.46–01	–1.446	B	*LS*
216.	$2s2p(^3P^o)3p-$ $2s2p(^1P^o)3d$	$^2S-^2P^o$		*1125.6*	492890.9	*581736*	2–6	1.93–01	1.10–02	8.17–02	–1.657	B	1
				1125.47	492890.9	581743	2–4	1.94–01	7.35–03	5.45–02	–1.833	B	*LS*
				1125.74	492890.9	581721	2–2	1.93–01	3.67–03	2.72–02	–2.134	B–	*LS*

O IV: Allowed Transitions — Continued

No.	Transition Array	Mult.	λ_{air} (Å)	λ_{vac} (Å) or σ (cm^{-1})*	E_i (cm^{-1})	E_k (cm^{-1})	g_i–g_k	A_{ki} (10^8 s^{-1})	f_{ik}	S (at. u.)	log gf	Acc.	Source
217.	$2s2p(^3P^o)3p$–$2s2p(^3P^o)4d$	^2S–$^2P^o$		*953.48*	492890.9	*597770*	2–6	6.98+00	2.85–01	1.79+00	–0.243	B	1
				953.934	492890.9	597720	2–4	6.97+00	1.90–01	1.19+00	–0.420	B	*LS*
				952.580	492890.9	597869	2–2	7.00+00	9.52–02	5.97–01	–0.720	B	*LS*
218.	$2s2p(^3P^o)3d$–$2s2p(^3P^o)4f$	$^4F^o$–^4F		*1013.8*	*495363*	*593999*	28–28	2.55+00	3.93–02	3.68+00	0.042	B	1
				1014.90	495512.9	594044.9	10–10	2.33+00	3.60–02	1.20+00	–0.444	C+	*LS*
				1013.86	495359.6	593992.8	8–8	1.95+00	3.00–02	8.01–01	–0.620	C+	*LS*
				1013.01	495246.1	593961.9	6–6	1.76+00	2.71–02	5.42–01	–0.789	C+	*LS*
				1012.35	495169	593948.7	4–4	2.05+00	3.15–02	4.20–01	–0.899	C+	*LS*
				1015.44	495512.9	593992.8	10–8	2.63–01	3.25–03	1.09–01	–1.488	C+	*LS*
				1014.18	495359.6	593961.9	8–6	4.55–01	5.27–03	1.41–01	–1.375	C+	*LS*
				1013.14	495246.1	593948.7	6–4	5.11–01	5.25–03	1.05–01	–1.502	C+	*LS*
				1013.32	495359.6	594044.9	8–10	2.12–01	4.08–03	1.09–01	–1.487	C+	*LS*
				1012.69	495246.1	593992.8	6–8	3.43–01	7.03–03	1.41–01	–1.375	C+	*LS*
				1012.22	495169	593961.9	4–6	3.42–01	7.88–03	1.05–01	–1.501	C+	*LS*
219.		$^4F^o$–^4G		*988.72*	*495363*	*596504*	28–36	4.10+01	7.73–01	7.04+01	1.335	B	1
				988.708	495512.9	596655	10–12	4.10+01	7.21–01	2.35+01	0.858	C+	*LS*
				988.627	495359.6	596510	8–10	3.76+01	6.88–01	1.79+01	0.741	C+	*LS*
				988.573	495246.1	596402	6–8	3.53+01	6.90–01	1.35+01	0.617	C+	*LS*
				988.523	495169	596330	4–6	3.52+01	7.73–01	1.01+01	0.490	C+	*LS*
				990.127	495512.9	596510	10–10	3.38+00	4.97–02	1.62+00	–0.303	C+	*LS*
				989.684	495359.6	596402	8–8	5.55+00	8.15–02	2.12+00	–0.186	C+	*LS*
				989.277	495246.1	596330	6–6	5.65+00	8.30–02	1.62+00	–0.303	C+	*LS*
				991.187	495512.9	596402	10–8	1.24–01	1.46–03	4.75–02	–1.837	C+	*LS*
				990.389	495359.6	596330	8–6	2.04–01	2.25–03	5.87–02	–1.745	C+	*LS*
220.		$^4F^o$–^4D		*989.49*	*495363*	*596425*	28–20	6.54–01	6.86–03	6.26–01	–0.716	B	1
				991.885	495512.9	596331	10–8	5.80–01	6.84–03	2.24–01	–1.165	C+	*LS*
				989.272	495359.6	596444	8–6	5.35–01	5.88–03	1.53–01	–1.327	C+	*LS*
				987.495	495246.1	596512.4	6–4	5.29–01	5.16–03	1.01–01	–1.510	C+	*LS*
				986.221	495169	596566.2	4–2	6.61–01	4.82–03	6.26–02	–1.715	C+	*LS*
				990.379	495359.6	596331	8–8	6.64–02	9.77–04	2.55–02	–2.107	C	*LS*
				988.163	495246.1	596444	6–6	1.14–01	1.67–03	3.26–02	–1.999	C+	*LS*
				986.744	495169	596512.4	4–4	1.32–01	1.93–03	2.50–02	–2.113	C	*LS*
				989.267	495246.1	596331	6–8	3.39–03	6.63–05	1.30–03	–3.400	D	*LS*
				987.411	495169	596444	4–6	6.27–03	1.38–04	1.79–03	–3.260	D	*LS*
221.	$2s2p(^3P^o)3d$–$2p^2(^3P)3s$	$^4D^o$–^4P		*1294.9*	*499851*	*577079*	20–12	1.72–02	2.59–04	2.21–02	–2.285	B–	1
				1293.63	499907.0	577209	8–6	1.38–02	2.59–04	8.84–03	–2.683	C+	*LS*
				1296.10	499842.6	576997	6–4	1.08–02	1.81–04	4.64–03	–2.963	C+	*LS*
				1297.74	499795.7	576853	4–2	8.54–03	1.08–04	1.84–03	–3.365	C	*LS*
				1292.55	499842.6	577209	6–6	3.11–03	7.79–05	1.99–03	–3.330	C	*LS*
				1295.31	499795.7	576997	4–4	5.49–03	1.38–04	2.36–03	–3.257	C	*LS*
				1297.26	499767.2	576853	2–2	8.55–03	2.16–04	1.84–03	–3.365	C	*LS*
				1291.77	499795.7	577209	4–6	3.46–04	1.30–05	2.21–04	–4.284	D	*LS*
				1294.84	499767.2	576997	2–4	8.59–04	4.32–05	3.68–04	–4.063	D	*LS*
222.	$2s2p(^3P^o)3d$–$2s2p(^3P^o)4f$	$^4D^o$–^4F		*1062.2*	*499851*	*593999*	20–28	3.12+01	7.38–01	5.16+01	1.169	B	1
				1062.27	499907.0	594044.9	8–10	3.12+01	6.59–01	1.84+01	0.722	C+	*LS*
				1062.13	499842.6	593992.8	6–8	2.67+01	6.03–01	1.27+01	0.559	C+	*LS*
				1061.95	499795.7	593961.9	4–6	2.34+01	5.94–01	8.30+00	0.375	C+	*LS*
				1061.78	499767.2	593948.7	2–4	2.19+01	7.39–01	5.16+00	0.170	C+	*LS*
				1062.86	499907.0	593992.8	8–8	4.44+00	7.51–02	2.10+00	–0.221	C+	*LS*
				1062.48	499842.6	593961.9	6–6	7.58+00	1.28–01	2.69+00	–0.114	C+	*LS*
				1062.10	499795.7	593948.7	4–4	8.73+00	1.48–01	2.07+00	–0.229	C+	*LS*
				1063.21	499907.0	593961.9	8–6	3.01–01	3.82–03	1.07–01	–1.515	C+	*LS*
				1062.63	499842.6	593948.7	6–4	6.23–01	7.03–03	1.48–01	–1.375	C+	*LS*

O IV: Allowed Transitions — Continued

No.	Transition Array	Mult.	λ_{air} (Å)	λ_{vac} (Å) or σ (cm^{-1})*	E_i (cm^{-1})	E_k (cm^{-1})	g_i–g_k	A_{ki} (10^8 s^{-1})	f_{ik}	S (at. u.)	log gf	Acc.	Source
223.		$^4D^o-^4D$		*1035.5*	*499851*	– *596425*	20–20	5.97+00	9.60–02	6.55+00	0.284	B	1
				1037.09	499907.0	– 596331	8–8	5.10+00	8.22–02	2.25+00	–0.182	C+	*LS*
				1035.18	499842.6	– 596444	6–6	3.43+00	5.51–02	1.13+00	–0.481	C+	*LS*
				1033.95	499795.7	– 596512.4	4–4	2.40+00	3.85–02	5.24–01	–0.813	C+	*LS*
				1033.07	499767.2	– 596566.2	2–2	3.01+00	4.81–02	3.27–01	–1.017	C+	*LS*
				1035.87	499907.0	– 596444	8–6	1.13+00	1.37–02	3.73–01	–0.961	C+	*LS*
				1034.45	499842.6	– 596512.4	6–4	2.10+00	2.24–02	4.58–01	–0.871	C+	*LS*
				1033.37	499795.7	– 596566.2	4–2	3.01+00	2.41–02	3.27–01	–1.017	C+	*LS*
				1036.39	499842.6	– 596331	6–8	8.49–01	1.82–02	3.73–01	–0.961	C+	*LS*
				1034.68	499795.7	– 596444	4–6	1.40+00	3.36–02	4.58–01	–0.871	C+	*LS*
				1033.64	499767.2	– 596512.4	2–4	1.50+00	4.81–02	3.27–01	–1.017	C+	*LS*
224.	$2s2p(^3P^o)3d-$ $2p^2(^3P)3d$	$^4D^o-^4P$		*728.22*	*499851*	– *637173*	20–12	1.77+01	8.45–02	4.05+00	0.228	B	1
				728.831	499907.0	– 637113	8–6	1.41+01	8.44–02	1.62+00	–0.170	B	*LS*
				727.964	499842.6	– 637212	6–4	1.12+01	5.92–02	8.51–01	–0.450	B	*LS*
				727.388	499795.7	– 637274	4–2	8.89+00	3.52–02	3.38–01	–0.851	B	*LS*
				728.489	499842.6	– 637113	6–6	3.18+00	2.53–02	3.65–01	–0.818	B	*LS*
				727.716	499795.7	– 637212	4–4	5.68+00	4.51–02	4.32–01	–0.744	B	*LS*
				727.237	499767.2	– 637274	2–2	8.89+00	7.05–02	3.38–01	–0.851	B	*LS*
				728.240	499795.7	– 637113	4–6	3.54–01	4.22–03	4.05–02	–1.772	B	*LS*
				727.565	499767.2	– 637212	2–4	8.88–01	1.41–02	6.75–02	–1.550	B	*LS*
225.	$2s^24p-$ $2s^24d$	$^2P^o-^2D$	*9477.2*	*9479.8*	*500024*	– *510572*	6–10	2.23–01	5.01–01	9.37+01	0.478	B	1
			9488.39	9491.00	500037.8	– 510574.1	4–6	2.22–01	4.50–01	5.62+01	0.255	B	*LS*
			9453.85	9456.44	499994.9	– 510569.7	2–4	1.87–01	5.02–01	3.12+01	0.002	B	*LS*
			9492.36	9494.96	500037.8	– 510569.7	4–4	3.70–02	5.00–02	6.25+00	–0.699	B	*LS*
226.	$2s^24p-$ $2s^25s$	$^2P^o-^2S$	*2540.9*	*2541.7*	*500024*	– 539368	6–2	5.29+00	1.71–01	8.57+00	0.011	B	1
			2541.81	2542.58	500037.8	– 539368	4–2	3.53+00	1.71–01	5.71+00	–0.166	B	*LS*
			2539.04	2539.81	499994.9	– 539368	2–2	1.77+00	1.71–01	2.86+00	–0.466	B	*LS*
227.	$2s^24p-$ $2s2p(^1P^o)3p$	$^2P^o-^2D$	*2112.6*	*2113.3*	*500024*	– *547343*	6–10	2.36–03	2.64–04	1.10–02	–2.801	B–	1
			2112.73	2113.40	500037.8	– 547355	4–6	2.36–03	2.37–04	6.60–03	–3.023	C+	*LS*
			2112.11	2112.78	499994.9	– 547326	2–4	1.97–03	2.64–04	3.67–03	–3.278	C+	*LS*
			2114.02	2114.69	500037.8	– 547326	4–4	3.93–04	2.63–05	7.33–04	–3.977	D	*LS*
228.		$^2P^o-^2P$	*2007.0*	*2007.6*	*500024*	– *549834*	6–6	5.11–02	3.09–03	1.22–01	–1.733	B	1
			2006.69	2007.34	500037.8	– 549855	4–4	4.26–02	2.57–03	6.79–02	–1.988	B	*LS*
			2007.50	2008.15	499994.9	– 549792	2–2	3.40–02	2.06–03	2.72–02	–2.386	B–	*LS*
			2009.23	2009.88	500037.8	– 549792	4–2	1.70–02	5.14–04	1.36–02	–2.687	B–	*LS*
			2004.96	2005.61	499994.9	– 549855	2–4	8.54–03	1.03–03	1.36–02	–2.686	B–	*LS*
229.	$2s^24p-$ $2s^25d$	$^2P^o-^2D$		*1922.8*	*500024*	– *552031*	6–10	2.43+00	2.24–01	8.51+00	0.129	B	1
				1923.28	500037.8	– 552032.3	4–6	2.43+00	2.02–01	5.11+00	–0.093	B	*LS*
				1921.79	499994.9	– 552029.6	2–4	2.03+00	2.24–01	2.84+00	–0.348	B	*LS*
				1923.38	500037.8	– 552029.6	4–4	4.04–01	2.24–02	5.68–01	–1.047	B	*LS*
230.	$2s^24p-$ $2s^26d$	$^2P^o-^2D$		*1345.1*	*500024*	– 574368.7	6–10	1.83+00	8.26–02	2.19+00	–0.305	B	1
				1345.34	500037.8	– 574368.7	4–6	1.83+00	7.43–02	1.32+00	–0.527	B	*LS*
				1344.56	499994.9	– 574368.7	2–4	1.52+00	8.26–02	7.31–01	–0.782	B	*LS*
				1345.34	500037.8	– 574368.7	4–4	3.04–01	8.26–03	1.46–01	–1.481	B	*LS*

O IV: Allowed Transitions — Continued

No.	Transition Array	Mult.	λ_{air} (Å)	λ_{vac} (Å) or σ (cm^{-1})*	E_i (cm^{-1})	E_k (cm^{-1})	g_i–g_k	A_{ki} (10^8 s^{-1})	f_{ik}	S (at. u.)	log gf	Acc.	Source
231.	$2s^24p$– $2s2p(^3P^o)4p$	$^2P^o$–^2P		*1328.1*	*500024*	– *575317*	6–6	3.37–01	8.90–03	2.34–01	–1.272	B	1
				1327.37	500037.8	– 575375	4–4	2.81–01	7.42–03	1.30–01	–1.527	B	*LS*
				1329.66	499994.9	– 575202	2–2	2.24–01	5.93–03	5.19–02	–1.926	B	*LS*
				1330.42	500037.8	– 575202	4–2	1.12–01	1.48–03	2.59–02	–2.227	B–	*LS*
				1326.61	499994.9	– 575375	2–4	5.63–02	2.97–03	2.59–02	–2.226	B–	*LS*
232.		$^2P^o$–^2D		*1181.3*	*500024*	– *584673*	6–10	6.27–01	2.19–02	5.10–01	–0.882	B	1
				1180.31	500037.8	– 584761	4–6	6.28–01	1.97–02	3.06–01	–1.104	B	*LS*
				1182.79	499994.9	– 584541	2–4	5.20–01	2.18–02	1.70–01	–1.360	B	*LS*
				1183.39	500037.8	– 584541	4–4	1.04–01	2.18–03	3.40–02	–2.059	B	*LS*
233.		$^2P^o$–^2S		*1110.5*	*500024*	– 590069	6–2	1.31–01	8.07–04	1.77–02	–2.315	B–	1
				1110.73	500037.8	– 590069	4–2	8.72–02	8.07–04	1.18–02	–2.491	B–	*LS*
				1110.20	499994.9	– 590069	2–2	4.37–02	8.07–04	5.90–03	–2.792	C+	*LS*
234.	$2s^24p$– $2s2p(^3P^o)4f$	$^2P^o$–^2D		*1034.2*	*500024*	– *596714*	6–10	1.65–01	4.42–03	9.03–02	–1.576	B	1
				1034.98	500037.8	– 596658.3	4–6	1.65–01	3.98–03	5.42–02	–1.799	C+	*LS*
				1033.02	499994.9	– 596798.2	2–4	1.38–01	4.43–03	3.01–02	–2.053	C+	*LS*
				1033.48	500037.8	– 596798.2	4–4	2.76–02	4.42–04	6.02–03	–2.752	C–	*LS*
235.	$2s^24p$– $2p^2(^1D)3s$	$^2P^o$–^2D		*999.24*	*500024*	– *600100*	6–10	2.66–02	6.64–04	1.31–02	–2.400	B–	1
				999.318	500037.8	– 600106	4–6	2.66–02	5.97–04	7.86–03	–2.622	C+	*LS*
				999.030	499994.9	– 600092	2–4	2.22–02	6.64–04	4.37–03	–2.877	C+	*LS*
				999.458	500037.8	– 600092	4–4	4.43–03	6.64–05	8.73–04	–3.576	D	*LS*
236.	$2s^24p$– $2p^2(^1D)3d$	$^2P^o$–^2P		*625.10*	*500024*	– 659998	6–6	8.37+00	4.90–02	6.05–01	–0.531	B	1
				625.156	500037.8	– 659998	4–4	6.97+00	4.08–02	3.36–01	–0.787	B	*LS*
				624.988	499994.9	– 659998	2–2	5.58+00	3.27–02	1.35–01	–1.185	B	*LS*
				625.156	500037.8	– 659998	4–2	2.79+00	8.17–03	6.73–02	–1.486	B	*LS*
				624.988	499994.9	– 659998	2–4	1.40+00	1.63–02	6.73–02	–1.486	B	*LS*
237.	$2s2p(^3P^o)3d$– $2s2p(^1P^o)3p$	$^2D^o$–^2D	*2182.7*	*2183.4*	*501542*	– *547343*	10–10	5.38–02	3.84–03	2.76–01	–1.415	B	1
			2183.17	2183.85	501564.4	– 547355	6–6	5.02–02	3.59–03	1.55–01	–1.667	B	*LS*
			2181.92	2182.61	501509.2	– 547326	4–4	4.85–02	3.46–03	9.94–02	–1.859	B	*LS*
			2184.55	2185.24	501564.4	– 547326	6–4	5.37–03	2.56–04	1.10–02	–2.814	B–	*LS*
			2180.54	2181.22	501509.2	– 547355	4–6	3.60–03	3.85–04	1.10–02	–2.813	B–	*LS*
238.	$2s2p(^3P^o)3d$– $2s2p(^3P^o)4p$	$^2D^o$–^2P		*1355.5*	*501542*	– *575317*	10–6	2.86+00	4.73–02	2.11+00	–0.326	B	1
				1354.82	501564.4	– 575375	6–4	2.58+00	4.73–02	1.27+00	–0.547	B	*LS*
				1356.98	501509.2	– 575202	4–2	2.85+00	3.93–02	7.03–01	–0.803	B	*LS*
				1353.81	501509.2	– 575375	4–4	2.87–01	7.89–03	1.41–01	–1.501	B	*LS*
239.		$^2D^o$–^2D		*1202.9*	*501542*	– *584673*	10–10	2.86–01	6.21–03	2.46–01	–1.207	B	1
				1201.97	501564.4	– 584761	6–6	2.68–01	5.80–03	1.38–01	–1.459	B	*LS*
				1204.36	501509.2	– 584541	4–4	2.57–01	5.58–03	8.85–02	–1.651	B	*LS*
				1205.16	501564.4	– 584541	6–4	2.85–02	4.13–04	9.83–03	–2.606	C+	*LS*
				1201.18	501509.2	– 584761	4–6	1.92–02	6.22–04	9.83–03	–2.604	C+	*LS*
240.	$2s2p(^3P^o)3d$– $2s2p(^3P^o)4f$	$^2D^o$–^2F		*1081.0*	*501542*	– *594050*	10–14	3.00+01	7.37–01	2.62+01	0.867	B	1
				1080.97	501564.4	– 594074	6–8	3.00+01	7.02–01	1.50+01	0.624	C+	*LS*
				1080.97	501509.2	– 594019	4–6	2.80+01	7.37–01	1.05+01	0.470	C+	*LS*
				1081.61	501564.4	– 594019	6–6	2.00+00	3.51–02	7.49–01	–0.677	C+	*LS*

O IV: Allowed Transitions — Continued

No.	Transition Array	Mult.	λ_{air} (Å)	λ_{vac} (Å) or σ (cm^{-1})*	E_i (cm^{-1})	E_k (cm^{-1})	g_i-g_k	A_{ki} (10^8 s^{-1})	f_{ik}	S (at. u.)	log gf	Acc.	Source
241.	$2s2p(^3P^\circ)3d-2p^2(^3P)3s$	$^2D^\circ-^2P$		*1076.1*	*501542*	– *594472*	10–6	3.24–01	3.38–03	1.20–01	–1.472	B	1
				1075.57	501564.4	– 594538	6–4	2.92–01	3.38–03	7.18–02	–1.693	B	*LS*
				1077.23	501509.2	– 594340	4–2	3.23–01	2.81–03	3.99–02	–1.949	B	*LS*
				1074.94	501509.2	– 594538	4–4	3.25–02	5.63–04	7.97–03	–2.647	C+	*LS*
242.	$2s2p(^3P^\circ)3d-2s2p(^3P^\circ)4f$	$^2D^\circ-^2D$		*1050.7*	*501542*	– *596714*	10–10	5.44+00	9.01–02	3.12+00	–0.045	B	1
				1051.59	501564.4	– 596658.3	6–6	5.07+00	8.40–02	1.74+00	–0.298	C+	*LS*
				1049.44	501509.2	– 596798.2	4–4	4.92+00	8.12–02	1.12+00	–0.489	C+	*LS*
				1050.05	501564.4	– 596798.2	6–4	5.45–01	6.01–03	1.25–01	–1.443	C+	*LS*
				1050.98	501509.2	– 596658.3	4–6	3.63–01	9.01–03	1.25–01	–1.443	C+	*LS*
243.	$2s2p(^3P^\circ)3d-2p^2(^1D)3s$	$^2D^\circ-^2D$		*1014.6*	*501542*	– *600100*	10–10	4.66–02	7.19–04	2.40–02	–2.144	B–	1
				1014.80	501564.4	– 600106	6–6	4.34–02	6.70–04	1.34–02	–2.395	B–	*LS*
				1014.38	501509.2	– 600092	4–4	4.19–02	6.47–04	8.64–03	–2.587	C+	*LS*
				1014.94	501564.4	– 600092	6–4	4.65–03	4.79–05	9.60–04	–3.542	D	*LS*
				1014.23	501509.2	– 600106	4–6	3.11–03	7.19–05	9.60–04	–3.541	D	*LS*
244.	$2s2p(^3P^\circ)3d-2p^2(^3P)3d$	$^2D^\circ-^2F$		*777.89*	*501542*	– 630095	10–14	3.11–02	3.94–04	1.01–02	–2.404	B–	1
				778.025	501564.4	– 630095	6–8	3.10–02	3.76–04	5.77–03	–2.647	C+	*LS*
				777.691	501509.2	– 630095	4–6	2.90–02	3.94–04	4.04–03	–2.802	C+	*LS*
				778.025	501564.4	– 630095	6–6	2.07–03	1.88–05	2.89–04	–3.948	D	*LS*
245.		$^2D^\circ-^2P$		*763.70*	*501542*	– *632483*	10–6	1.07+01	5.60–02	1.41+00	–0.252	B	1
				764.166	501564.4	– 632426	6–4	9.59+00	5.60–02	8.45–01	–0.474	B	*LS*
				762.847	501509.2	– 632597	4–2	1.07+01	4.67–02	4.69–01	–0.729	B	*LS*
				763.844	501509.2	– 632426	4–4	1.07+00	9.33–03	9.38–02	–1.428	B	*LS*
246.	$2s2p(^3P^\circ)3d-2p^2(^1D)3d$	$^2D^\circ-^2P$		*631.09*	*501542*	– 659998	10–6	7.15–01	2.56–03	5.32–02	–1.592	B	1
				631.179	501564.4	– 659998	6–4	6.43–01	2.56–03	3.19–02	–1.814	B	*LS*
				630.959	501509.2	– 659998	4–2	7.15–01	2.13–03	1.77–02	–2.069	B–	*LS*
				630.959	501509.2	– 659998	4–4	7.15–02	4.27–04	3.55–03	–2.768	C	*LS*
247.	$2s2p(^3P^\circ)3d-2p^2(^3P)3s$	$^4P^\circ-^4P$		*1371.5*	*504165*	– *577079*	12–12	1.30–02	3.65–04	1.98–02	–2.358	B–	1
				1367.74	504095.7	– 577209	6–6	9.15–03	2.57–04	6.93–03	–2.813	C+	*LS*
				1373.85	504208.9	– 576997	4–4	1.72–03	4.86–05	8.80–04	–3.711	D	*LS*
				1377.97	504282.3	– 576853	2–2	2.13–03	6.06–05	5.50–04	–3.916	D	*LS*
				1371.72	504095.7	– 576997	6–4	5.83–03	1.10–04	2.97–03	–3.182	C	*LS*
				1376.57	504208.9	– 576853	4–2	1.07–02	1.52–04	2.75–03	–3.217	C	*LS*
				1369.86	504208.9	– 577209	4–6	3.90–03	1.65–04	2.97–03	–3.181	C	*LS*
				1375.24	504282.3	– 576997	2–4	5.36–03	3.04–04	2.75–03	–3.217	C	*LS*
248.	$2s2p(^3P^\circ)3d-2s2p(^3P^\circ)4f$	$^4P^\circ-^4D$		*1083.9*	*504165*	– *596425*	12–20	2.84+01	8.35–01	3.58+01	1.001	B	1
				1084.18	504095.7	– 596331	6–8	2.84+01	6.68–01	1.43+01	0.603	C+	*LS*
				1084.19	504208.9	– 596444	4–6	1.99+01	5.26–01	7.51+00	0.323	C+	*LS*
				1084.24	504282.3	– 596512.4	2–4	1.18+01	4.17–01	2.98+00	–0.079	C+	*LS*
				1082.86	504095.7	– 596444	6–6	8.56+00	1.50–01	3.22+00	–0.045	C+	*LS*
				1083.38	504208.9	– 596512.4	4–4	1.52+01	2.67–01	3.81+00	0.029	C+	*LS*
				1083.61	504282.3	– 596566.2	2–2	2.37+01	4.18–01	2.98+00	–0.078	C+	*LS*
				1082.06	504095.7	– 596512.4	6–4	1.43+00	1.67–02	3.58–01	–0.998	C+	*LS*
				1082.75	504208.9	– 596566.2	4–2	4.76+00	4.18–02	5.96–01	–0.777	C+	*LS*

O IV: Allowed Transitions — Continued

No.	Transition Array	Mult.	λ_{air} (Å)	λ_{vac} (Å) or σ (cm^{-1})*	E_i (cm^{-1})	E_k (cm^{-1})	g_i-g_k	A_{ki} ($10^8 s^{-1}$)	f_{ik}	S (at. u.)	log *gf*	Acc.	Source
249.	$2s2p(^3P^o)3d-$ $2p^2(^3P)3d$	$^4P^o-^4P$		*751.83*	*504165*	*637174*	12–12	7.12+00	6.03–02	1.79+00	–0.140	B	1
				751.776	504095.7	637114	6–6	4.98+00	4.22–02	6.27–01	–0.596	B	*LS*
				751.857	504208.9	637213	4–4	9.49–01	8.04–03	7.96–02	–1.492	B	*LS*
				751.921	504282.3	637275	2–2	1.19+00	1.01–02	4.98–02	–1.697	B	*LS*
				751.217	504095.7	637213	6–4	3.21+00	1.81–02	2.69–01	–0.964	B	*LS*
				751.506	504208.9	637275	4–2	5.94+00	2.51–02	2.49–01	–0.997	B	*LS*
				752.417	504208.9	637114	4–6	2.13+00	2.71–02	2.69–01	–0.965	B	*LS*
				752.272	504282.3	637213	2–4	2.96+00	5.02–02	2.49–01	–0.998	B	*LS*
250.	$2s^24d-$ $2s2p(^3P^o)3d$	$^2D-^2F^o$		*305.30* cm^{-1}	*510572*	*510878*	10–14	1.54–07	3.47–04	3.75+00	–2.459	B	1
				403.100 cm^{-1}	510574.1	510977.2	6–8	3.55–07	4.37–04	2.14+00	–2.582	B	*LS*
				175.200 cm^{-1}	510569.7	510744.9	4–6	2.72–08	1.99–04	1.50+00	–3.098	B	*LS*
				170.800 cm^{-1}	510574.1	510744.9	6–6	1.80–09	9.26–06	1.07–01	–4.255	B	*LS*
251.	$2s^24d-$ $2s^24f$	$^2D-^2F^o$	*38140*	*2621.2* cm^{-1}	*510572*	*513194*	10–14	2.31–03	7.05–02	8.85+01	–0.152	B	1
			38096.5	2624.20 cm^{-1}	510574.1	513198.3	6–8	2.32–03	6.72–02	5.06+01	–0.394	C+	*LS*
			38194.0	2617.50 cm^{-1}	510569.7	513187.2	4–6	2.15–03	7.04–02	3.54+01	–0.550	C+	*LS*
			38258.3	2613.10 cm^{-1}	510574.1	513187.2	6–6	1.53–04	3.35–03	2.53+00	–1.697	C+	*LS*
252.	$2s^24d-$ $2s2p(^3P^o)3d$	$^2D-^2P^o$	*27031*	*3698.4* cm^{-1}	*510572*	*514271*	10–6	3.22–05	2.12–04	1.88–01	–2.674	B	1
			27417.6	3646.30 cm^{-1}	510574.1	514220.4	6–4	2.78–05	2.09–04	1.13–01	–2.902	B	*LS*
			26297.5	3801.60 cm^{-1}	510569.7	514371.3	4–2	3.50–05	1.81–04	6.28–02	–3.139	B	*LS*
			27384.5	3650.70 cm^{-1}	510569.7	514220.4	4–4	3.10–06	3.48–05	1.26–02	–3.856	B–	*LS*
253.	$2s^24d-$ $2s2p(^1P^o)3s$	$^2D-^2P^o$	*12291*	*8133.7* cm^{-1}	*510572*	*518706*	10–6	3.20–03	4.35–03	1.76+00	–1.362	B	1
			12289.4	8134.90 cm^{-1}	510574.1	518709	6–4	2.88–03	4.35–03	1.06+00	–1.584	B	*LS*
			12297.8	8129.30 cm^{-1}	510569.7	518699	4–2	3.19–03	3.62–03	5.86–01	–1.839	B	*LS*
			12282.7	8139.30 cm^{-1}	510569.7	518709	4–4	3.20–04	7.25–04	1.17–01	–2.538	B	*LS*
254.	$2s^24d-$ $2s^25p$	$^2D-^2P^o$	*2758.5*	*2759.3*	*510572*	*546814*	10–6	1.61+00	1.10–01	1.00+01	0.042	B	1
			2758.22	2759.03	510574.1	546818.7	6–4	1.45+00	1.10–01	6.00+00	–0.180	B	*LS*
			2759.05	2759.86	510569.7	546803.4	4–2	1.61+00	9.18–02	3.34+00	–0.435	B	*LS*
			2757.88	2758.70	510569.7	546818.7	4–4	1.61–01	1.84–02	6.67–01	–1.134	B	*LS*
255.	$2s^24d-$ $2s^25f$	$^2D-^2F^o$	*2384.6*	*2385.3*	*510572*	552495.1	10–14	6.65+00	7.93–01	6.23+01	0.900	B	1
			2384.71	2385.44	510574.1	552495.1	6–8	6.65+00	7.56–01	3.56+01	0.656	C+	*LS*
			2384.46	2385.19	510569.7	552495.1	4–6	6.21+00	7.94–01	2.49+01	0.502	C+	*LS*
			2384.71	2385.44	510574.1	552495.1	6–6	4.43–01	3.78–02	1.78+00	–0.645	C+	*LS*
256.	$2s^24d-$ $2s2p(^1P^o)3d$	$^2D-^2F^o$		*1660.6*	*510572*	570791	10–14	7.98–01	4.62–02	2.53+00	–0.335	B	1
				1660.66	510574.1	570791	6–8	7.98–01	4.40–02	1.44+00	–0.578	B	*LS*
				1660.54	510569.7	570791	4–6	7.45–01	4.62–02	1.01+00	–0.733	B	*LS*
				1660.66	510574.1	570791	6–6	5.32–02	2.20–03	7.22–02	–1.879	B	*LS*
257.	$2s^24d-$ $2s2p(^3P^o)4s$	$^2D-^2P^o$		*1580.8*	*510572*	*573833*	10–6	1.82–02	4.09–04	2.13–02	–2.388	B–	1
				1579.11	510574.1	573901	6–4	1.64–02	4.10–04	1.28–02	–2.609	B–	*LS*
				1584.13	510569.7	573696	4–2	1.81–02	3.40–04	7.10–03	–2.866	C+	*LS*
				1579.00	510569.7	573901	4–4	1.83–03	6.83–05	1.42–03	–3.564	C	*LS*
258.	$2s^24d-$ $2s^26f$	$^2D-^2F^o$		*1556.8*	*510572*	574807.7	10–14	1.62+00	8.23–02	4.22+00	–0.085	B	1
				1556.82	510574.1	574807.7	6–8	1.62+00	7.84–02	2.41+00	–0.328	C+	*LS*
				1556.71	510569.7	574807.7	4–6	1.51+00	8.23–02	1.69+00	–0.483	C+	*LS*
				1556.82	510574.1	574807.7	6–6	1.08–01	3.92–03	1.21–01	–1.629	C+	*LS*

O IV: Allowed Transitions — Continued

No.	Transition Array	Mult.	λ_{air} (Å)	λ_{vac} (Å) or σ (cm^{-1})*	E_i (cm^{-1})	E_k (cm^{-1})	g_i–g_k	A_{ki} (10^8 s^{-1})	f_{ik}	S (at. u.)	log gf	Acc.	Source
259.	$2s^2 4d$– $2s2p(^1P^\circ)3d$	^2D–$^2P^\circ$		*1405.2*	*510572*	– *581736*	10–6	1.11–01	1.97–03	9.12–02	–1.705	B	1
				1405.11	510574.1	– 581743	6–4	9.99–02	1.97–03	5.47–02	–1.927	B	*LS*
				1405.46	510569.7	– 581721	4–2	1.11–01	1.64–03	3.04–02	–2.182	B	*LS*
				1405.02	510569.7	– 581743	4–4	1.11–02	3.29–04	6.08–03	–2.881	C+	*LS*
260.	$2s^2 4d$– $2s^2 7f$	^2D–$^2F^\circ$		*1294.0*	*510572*	– 587850	10–14	1.80+00	6.33–02	2.70+00	–0.199	B	1
				1294.06	510574.1	– 587850	6–8	1.80+00	6.03–02	1.54+00	–0.442	C+	*LS*
				1293.99	510569.7	– 587850	4–6	1.68+00	6.33–02	1.08+00	–0.597	C+	*LS*
				1294.06	510574.1	– 587850	6–6	1.20–01	3.01–03	7.70–02	–1.743	C+	*LS*
261.	$2s^2 4d$– $2s2p(^3P^\circ)4d$	^2D–$^2F^\circ$		*1165.2*	*510572*	– *596398*	10–14	2.02–01	5.76–03	2.21–01	–1.240	B	1
				1164.13	510574.1	– 596475	6–8	2.03–01	5.49–03	1.26–01	–1.482	B	*LS*
				1166.52	510569.7	– 596295	4–6	1.88–01	5.75–03	8.84–02	–1.638	B	*LS*
				1166.58	510574.1	– 596295	6–6	1.34–02	2.74–04	6.31–03	–2.784	C+	*LS*
262.	$2s^2 4d$– $2s2p(^1P^\circ)4d$	^2D–$^2D^\circ$		*632.98*	*510572*	– *668555*	10–10	1.72+01	1.03–01	2.16+00	0.015	B	1
				632.996	510574.1	– 668553	6–6	1.61+01	9.66–02	1.21+00	–0.237	B	*LS*
				632.962	510569.7	– 668557	4–4	1.55+01	9.31–02	7.76–01	–0.429	B	*LS*
				632.980	510574.1	– 668557	6–4	1.72+00	6.90–03	8.63–02	–1.383	B	*LS*
				632.978	510569.7	– 668553	4–6	1.15+00	1.03–02	8.63–02	–1.383	B	*LS*
263.	$2s2p(^3P^\circ)3d$– $2s2p(^1P^\circ)3p$	$^2F^\circ$–^2D	*2741.5*	*2742.3*	*510878*	– *547343*	14–10	5.80–02	4.67–03	5.90–01	–1.185	B	1
			2748.12	2748.93	510977.2	– 547355	8–6	5.48–02	4.66–03	3.37–01	–1.429	B	*LS*
			2732.84	2733.65	510744.9	– 547326	6–4	5.85–02	4.37–03	2.36–01	–1.581	B	*LS*
			2730.68	2731.49	510744.9	– 547355	6–6	2.79–03	3.12–04	1.68–02	–2.727	B–	*LS*
264.	$2s2p(^3P^\circ)3d$– $2s^2 5d$	$^2F^\circ$–^2D	*2429.2*	*2429.9*	*510878*	– *552031*	14–10	1.64–02	1.04–03	1.16–01	–1.837	B	1
			2435.01	2435.75	510977.2	– 552032.3	8–6	1.55–02	1.04–03	6.65–02	–2.081	B	*LS*
			2421.47	2422.20	510744.9	– 552029.6	6–4	1.66–02	9.73–04	4.65–02	–2.234	B	*LS*
			2421.31	2422.05	510744.9	– 552032.3	6–6	7.90–04	6.95–05	3.32–03	–3.380	C+	*LS*
265.	$2s2p(^3P^\circ)3d$– $2s^2 6d$	$^2F^\circ$–^2D		*1575.0*	*510878*	– 574368.7	14–10	2.10–02	5.58–04	4.05–02	–2.107	B	1
				1577.50	510977.2	– 574368.7	8–6	1.99–02	5.57–04	2.31–02	–2.351	B–	*LS*
				1571.74	510744.9	– 574368.7	6–4	2.11–02	5.22–04	1.62–02	–2.504	B–	*LS*
				1571.74	510744.9	– 574368.7	6–6	1.01–03	3.73–05	1.16–03	–3.650	C	*LS*
266.	$2s2p(^3P^\circ)3d$– $2s2p(^3P^\circ)4p$	$^2F^\circ$–^2D		*1355.1*	*510878*	– *584673*	14–10	3.01+00	5.93–02	3.70+00	–0.081	B	1
				1355.31	510977.2	– 584761	8–6	2.87+00	5.93–02	2.12+00	–0.324	B	*LS*
				1355.09	510744.9	– 584541	6–4	3.01+00	5.53–02	1.48+00	–0.479	B	*LS*
				1351.06	510744.9	– 584761	6–6	1.45–01	3.96–03	1.06–01	–1.624	B	*LS*
267.	$2s2p(^3P^\circ)3d$– $2s2p(^3P^\circ)4f$	$^2F^\circ$–^2F		*1202.3*	*510878*	– *594050*	14–14	2.36+00	5.11–02	2.83+00	–0.145	B	1
				1203.42	510977.2	– 594074	8–8	2.27+00	4.93–02	1.56+00	–0.404	C+	*LS*
				1200.85	510744.9	– 594019	6–6	2.26+00	4.88–02	1.16+00	–0.534	C+	*LS*
				1204.21	510977.2	– 594019	8–6	1.12–01	1.82–03	5.78–02	–1.836	C+	*LS*
				1200.06	510744.9	– 594074	6–8	8.47–02	2.44–03	5.78–02	–1.835	C+	*LS*
268.		$^2F^\circ$–^2G		*1164.5*	*510878*	– *596752*	14–18	3.30+01	8.61–01	4.62+01	1.081	B	1
				1164.55	510977.2	– 596847.6	8–10	3.30+01	8.38–01	2.57+01	0.826	C+	*LS*
				1164.32	510744.9	– 596631.9	6–8	3.18+01	8.62–01	1.98+01	0.713	C+	*LS*
				1167.48	510977.2	– 596631.9	8–8	1.17+00	2.39–02	7.34–01	–0.719	C+	*LS*

O IV: Allowed Transitions — Continued

No.	Transition Array	Mult.	λ_{air} (Å)	λ_{vac} (Å) or σ (cm^{-1})*	E_i (cm^{-1})	E_k (cm^{-1})	g_i-g_k	A_{ki} (10^8 s^{-1})	f_{ik}	S (at. u.)	log gf	Acc.	Source
269.		$^2F^o-^2D$		*1165.0*	*510878*	– *596714*	14–10	2.92–01	4.24–03	2.28–01	–1.227	B	1
				1167.12	510977.2	– 596658.3	8–6	2.76–01	4.23–03	1.30–01	–1.471	C+	*LS*
				1162.07	510744.9	– 596798.2	6–4	2.94–01	3.96–03	9.10–02	–1.624	C+	*LS*
				1163.96	510744.9	– 596658.3	6–6	1.39–02	2.83–04	6.50–03	–2.770	C–	*LS*
270.	$2s2p(^3P^o)3d-2p^2(^3P)3d$	$^2F^o-^2F$		*838.80*	*510878*	– 630095	14–14	2.69+00	2.84–02	1.10+00	–0.401	B	1
				839.505	510977.2	– 630095	8–8	2.59+00	2.74–02	6.05–01	–0.660	B	*LS*
				837.871	510744.9	– 630095	6–6	2.57+00	2.71–02	4.48–01	–0.789	B	*LS*
				839.505	510977.2	– 630095	8–6	1.28–01	1.01–03	2.24–02	–2.091	B–	*LS*
				837.871	510744.9	– 630095	6–8	9.64–02	1.35–03	2.24–02	–2.090	B–	*LS*
271.	$2s2p(^3P^o)3d-2p^2(^1D)3d$	$^2F^o-^2F$		*713.11*	*510878*	– *651109*	14–14	4.20+00	3.20–02	1.05+00	–0.349	B	1
				713.573	510977.2	– 651117	8–8	4.04+00	3.08–02	5.80–01	–0.608	B	*LS*
				712.489	510744.9	– 651098	6–6	4.01+00	3.05–02	4.29–01	–0.737	B	*LS*
				713.670	510977.2	– 651098	8–6	1.99–01	1.14–03	2.15–02	–2.039	B–	*LS*
				712.392	510744.9	– 651117	6–8	1.50–01	1.53–03	2.15–02	–2.038	B–	*LS*
272.	$2s^24f-2s2p(^1P^o)3p$	$^2F^o-^2D$	*2741.5*	*2742.3*	*510878*	– *547343*	14–10	1.60–02	1.29–03	1.63–01	–1.743	B	1
			2926.83	2927.68	513198.3	– 547355	8–6	1.26–02	1.21–03	9.32–02	–2.014	C+	*LS*
			2928.36	2929.22	513187.2	– 547326	6–4	1.32–02	1.13–03	6.52–02	–2.170	C+	*LS*
			2925.88	2926.73	513187.2	– 547355	6–6	6.28–04	8.06–05	4.66–03	–3.315	C–	*LS*
273.	$2s^24f-2s^25d$	$^2F^o-^2D$	*2429.2*	*2429.9*	*510878*	– *552031*	14–10	2.22–01	1.40–02	1.57+00	–0.707	B	1
			2574.29	2575.06	513198.3	– 552032.3	8–6	1.78–01	1.32–02	8.98–01	–0.975	C+	*LS*
			2573.74	2574.51	513187.2	– 552029.6	6–4	1.87–01	1.24–02	6.29–01	–1.130	C+	*LS*
			2573.56	2574.33	513187.2	– 552032.3	6–6	8.90–03	8.83–04	4.49–02	–2.276	C+	*LS*
274.	$2s^24f-2s^26d$	$^2F^o-^2D$		*1575.0*	*510878*	– 574368.7	14–10	8.15–02	2.17–03	1.57–01	–1.518	B	1
				1634.78	513198.3	– 574368.7	8–6	6.94–02	2.09–03	8.98–02	–1.778	C+	*LS*
				1634.48	513187.2	– 574368.7	6–4	7.29–02	1.95–03	6.29–02	–1.932	C+	*LS*
				1634.48	513187.2	– 574368.7	6–6	3.47–03	1.39–04	4.49–03	–3.078	C–	*LS*
275.	$2s^24f-2s2p(^3P^o)4p$	$^2F^o-^2D$		*1355.1*	*510878*	– *584673*	14–10	2.14–01	4.20–03	2.62–01	–1.231	B	1
				1397.38	513198.3	– 584761	8–6	1.85–01	4.07–03	1.50–01	–1.487	C+	*LS*
				1401.47	513187.2	– 584541	6–4	1.93–01	3.79–03	1.05–01	–1.643	C+	*LS*
				1397.16	513187.2	– 584761	6–6	9.28–03	2.72–04	7.49–03	–2.788	C–	*LS*
276.	$2s^24f-2s2p(^3P^o)4f$	$^2F^o-^2F$		*1202.3*	*510878*	– *594050*	14–14	1.49–01	3.23–03	1.79–01	–1.345	B	1
				1236.47	513198.3	– 594074	8–8	1.32–01	3.03–03	9.86–02	–1.616	B	*LS*
				1237.14	513187.2	– 594019	6–6	1.30–01	2.99–03	7.30–02	–1.746	B	*LS*
				1237.31	513198.3	– 594019	8–6	6.51–03	1.12–04	3.65–03	–3.048	C	*LS*
				1236.30	513187.2	– 594074	6–8	4.89–03	1.50–04	3.65–03	–3.047	C	*LS*
277.		$^2F^o-^2G$		*1164.5*	*510878*	– *596752*	14–18	2.02+00	5.28–02	2.83+00	–0.132	B	1
				1195.47	513198.3	– 596847.6	8–10	1.87+00	5.00–02	1.57+00	–0.398	B	*LS*
				1198.40	513187.2	– 596631.9	6–8	1.79+00	5.13–02	1.21+00	–0.512	B	*LS*
				1198.56	513198.3	– 596631.9	8–8	6.61–02	1.42–03	4.49–02	–1.943	B	*LS*
278.		$^2F^o-^2D$		*1165.0*	*510878*	– *596714*	14–10	2.51–02	3.65–04	1.96–02	–2.292	B–	1
				1198.18	513198.3	– 596658.3	8–6	2.20–02	3.55–04	1.12–02	–2.547	B–	*LS*
				1196.01	513187.2	– 596798.2	6–4	2.32–02	3.32–04	7.84–03	–2.701	C+	*LS*
				1198.02	513187.2	– 596658.3	6–6	1.10–03	2.37–05	5.60–04	–3.848	D	*LS*

O IV: Allowed Transitions — Continued

No.	Transition Array	Mult.	λ_{air} (Å)	λ_{vac} (Å) or σ (cm^{-1})*	E_i (cm^{-1})	E_k (cm^{-1})	g_i–g_k	A_{ki} (10^8 s^{-1})	f_{ik}	S (at. u.)	log gf	Acc.	Source
279.	$2s^24f$– $2p^2(^3P)3d$	$^2F^o$–^2F		*838.80*	*510878*	– 630095	14–14	3.62–01	3.82–03	1.48–01	–1.272	B	1
				855.456	513198.3	– 630095	8–8	3.29–01	3.61–03	8.13–02	–1.539	C+	*LS*
				855.375	513187.2	– 630095	6–6	3.25–01	3.57–03	6.02–02	–1.670	C+	*LS*
				855.456	513198.3	– 630095	8–6	1.62–02	1.34–04	3.01–03	–2.971	C–	*LS*
				855.375	513187.2	– 630095	6–8	1.22–02	1.78–04	3.01–03	–2.971	C–	*LS*
280.	$2s^24f$– $2p^2(^1D)3d$	$^2F^o$–^2F		*713.11*	*510878*	– *651109*	14–14	5.34–01	4.07–03	1.34–01	–1.244	B	1
				725.065	513198.3	– 651117	8–8	4.89–01	3.86–03	7.37–02	–1.511	C+	*LS*
				725.106	513187.2	– 651098	6–6	4.83–01	3.81–03	5.46–02	–1.641	C+	*LS*
				725.165	513198.3	– 651098	8–6	2.42–02	1.43–04	2.73–03	–2.942	D	*LS*
				725.006	513187.2	– 651117	6–8	1.81–02	1.91–04	2.73–03	–2.942	D	*LS*
281.	$2s2p(^3P^o)3d$– $2s^25s$	$^2P^o$–^2S	*3983.4*	*3984.5*	*514271*	– 539368	6–2	3.55–02	2.81–03	2.21–01	–1.773	B	1
			3975.40	3976.52	514220.4	– 539368	4–2	2.38–02	2.82–03	1.48–01	–1.948	B	*LS*
			3999.40	4000.53	514371.3	– 539368	2–2	1.17–02	2.80–03	7.38–02	–2.252	B	*LS*
282.	$2s2p(^3P^o)3d$– $2s2p(^1P^o)3p$	$^2P^o$–^2D	*3022.8*	*3023.7*	*514271*	– *547343*	6–10	8.55–02	1.95–02	1.17+00	–0.931	B	1
			3017.11	3017.99	514220.4	– 547355	4–6	8.60–02	1.76–02	7.00–01	–1.152	B	*LS*
			3033.59	3034.47	514371.3	– 547326	2–4	7.05–02	1.95–02	3.89–01	–1.410	B	*LS*
			3019.76	3020.64	514220.4	– 547326	4–4	1.43–02	1.96–03	7.77–02	–2.107	B	*LS*
283.		$^2P^o$–^2P	*2811.1*	*2811.9*	*514271*	– *549834*	6–6	1.39–01	1.65–02	9.16–01	–1.005	B	1
			2805.43	2806.26	514220.4	– 549855	4–4	1.17–01	1.38–02	5.09–01	–1.259	B	*LS*
			2822.38	2823.21	514371.3	– 549792	2–2	9.17–02	1.10–02	2.04–01	–1.659	B	*LS*
			2810.40	2811.23	514220.4	– 549792	4–2	4.64–02	2.75–03	1.02–01	–1.959	B	*LS*
			2817.37	2818.20	514371.3	– 549855	2–4	2.30–02	5.49–03	1.02–01	–1.960	B	*LS*
284.	$2s2p(^3P^o)3d$– $2s^25d$	$^2P^o$–^2D	*2647.5*	*2648.3*	*514271*	– *552031*	6–10	1.64–02	2.88–03	1.50–01	–1.763	B	1
			2643.88	2644.67	514220.4	– 552032.3	4–6	1.65–02	2.59–03	9.02–02	–1.984	B	*LS*
			2654.67	2655.46	514371.3	– 552029.6	2–4	1.36–02	2.87–03	5.01–02	–2.241	B	*LS*
			2644.07	2644.86	514220.4	– 552029.6	4–4	2.75–03	2.88–04	1.00–02	–2.939	B–	*LS*
285.	$2s2p(^3P^o)3d$– $2s2p(^1P^o)3p$	$^2P^o$–^2S	*2487.2*	*2488.0*	*514271*	– 554464	6–2	5.77–01	1.78–02	8.76–01	–0.971	B	1
			2484.12	2484.87	514220.4	– 554464	4–2	3.86–01	1.79–02	5.84–01	–1.146	B	*LS*
			2493.47	2494.22	514371.3	– 554464	2–2	1.91–01	1.78–02	2.92–01	–1.449	B	*LS*
286.	$2s2p(^3P^o)3d$– $2s^26d$	$^2P^o$–^2D		*1663.9*	*514271*	– 574368.7	6–10	1.74–02	1.20–03	3.96–02	–2.141	B	1
				1662.56	514220.4	– 574368.7	4–6	1.75–02	1.09–03	2.38–02	–2.362	B–	*LS*
				1666.74	514371.3	– 574368.7	2–4	1.44–02	1.20–03	1.32–02	–2.619	B–	*LS*
				1662.56	514220.4	– 574368.7	4–4	2.91–03	1.21–04	2.64–03	–3.317	C	*LS*
287.	$2s2p(^3P^o)3d$– $2s2p(^3P^o)4p$	$^2P^o$–^2P		*1638.1*	*514271*	– *575317*	6–6	6.12–01	2.46–02	7.96–01	–0.831	B	1
				1635.20	514220.4	– 575375	4–4	5.13–01	2.05–02	4.42–01	–1.085	B	*LS*
				1643.91	514371.3	– 575202	2–2	4.04–01	1.64–02	1.77–01	–1.485	B	*LS*
				1639.84	514220.4	– 575202	4–2	2.03–01	4.10–03	8.85–02	–1.785	B	*LS*
				1639.24	514371.3	– 575375	2–4	1.02–01	8.20–03	8.85–02	–1.785	B	*LS*
288.		$^2P^o$–^2D		*1420.4*	*514271*	– *584673*	6–10	1.32–01	6.65–03	1.87–01	–1.399	B	1
				1417.62	514220.4	– 584761	4–6	1.33–01	5.99–03	1.12–01	–1.620	B	*LS*
				1425.12	514371.3	– 584541	2–4	1.09–01	6.63–03	6.22–02	–1.878	B	*LS*
				1422.06	514220.4	– 584541	4–4	2.19–02	6.64–04	1.24–02	–2.576	B–	*LS*

O IV: Allowed Transitions — Continued

No.	Transition Array	Mult.	λ_{air} (Å)	λ_{vac} (Å) or σ (cm^{-1})*	E_i (cm^{-1})	E_k (cm^{-1})	g_i-g_k	A_{ki} (10^8 s^{-1})	f_{ik}	S (at. u.)	log gf	Acc.	Source
289.		$^2P^o-^2S$		*1319.3*	*514271*	– 590069	6–2	2.12+00	1.84–02	4.81–01	–0.956	B	1
				1318.42	514220.4	– 590069	4–2	1.42+00	1.85–02	3.20–01	–1.132	B	*LS*
				1321.04	514371.3	– 590069	2–2	7.04–01	1.84–02	1.60–01	–1.434	B	*LS*
290.	$2s2p(^3P^o)3d-$ $2s2p(^3P^o)4f$	$^2P^o-^2D$		*1213.0*	*514271*	– *596714*	6–10	2.30+01	8.46–01	2.03+01	0.706	B	1
				1213.03	514220.4	– 596658.3	4–6	2.30+01	7.62–01	1.22+01	0.484	C+	*LS*
				1213.20	514371.3	– 596798.2	2–4	1.92+01	8.46–01	6.76+00	0.229	C+	*LS*
				1210.98	514220.4	– 596798.2	4–4	3.86+00	8.48–02	1.35+00	–0.470	C+	*LS*
291.	$2s2p(^3P^o)3d-$ $2p^2(^1D)3s$	$^2P^o-^2D$		*1165.1*	*514271*	– *600100*	6–10	5.97–01	2.03–02	4.66–01	–0.915	B	1
				1164.34	514220.4	– 600106	4–6	5.98–01	1.82–02	2.80–01	–1.137	B	*LS*
				1166.58	514371.3	– 600092	2–4	4.96–01	2.02–02	1.55–01	–1.393	B	*LS*
				1164.53	514220.4	– 600092	4–4	9.97–02	2.03–03	3.11–02	–2.091	B	*LS*
292.	$2s2p(^3P^o)3d-$ $2p^2(^3P)3d$	$^2P^o-^2P$		*845.94*	*514271*	– *632483*	6–6	3.83+00	4.11–02	6.87–01	–0.608	B	1
				845.984	514220.4	– 632426	4–4	3.19+00	3.42–02	3.81–01	–0.863	B	*LS*
				845.840	514371.3	– 632597	2–2	2.55+00	2.74–02	1.53–01	–1.261	B	*LS*
				844.762	514220.4	– 632597	4–2	1.28+00	6.86–03	7.63–02	–1.562	B	*LS*
				847.065	514371.3	– 632426	2–4	6.36–01	1.37–02	7.63–02	–1.563	B	*LS*
293.	$2s2p(^3P^o)3d-$ $2p^2(^1D)3d$	$^2P^o-^2P$		*686.21*	*514271*	– 659998	6–6	2.33–01	1.65–03	2.23–02	–2.006	B–	1
				685.976	514220.4	– 659998	4–4	1.94–01	1.37–03	1.24–02	–2.261	B–	*LS*
				686.687	514371.3	– 659998	2–2	1.55–01	1.10–03	4.96–03	–2.659	C+	*LS*
				685.976	514220.4	– 659998	4–2	7.78–02	2.74–04	2.48–03	–2.960	C	*LS*
				686.687	514371.3	– 659998	2–4	3.88–02	5.48–04	2.48–03	–2.960	C	*LS*
294.	$2s2p(^1P^o)3s-$ $2s^25s$	$^2P^o-^2S$	*4838.5*	*4839.8*	*518706*	– 539368	6–2	1.80–01	2.10–02	2.01+00	–0.899	B	1
			4839.15	4840.51	518709	– 539368	4–2	1.20–01	2.10–02	1.34+00	–1.076	B	*LS*
			4836.81	4838.16	518699	– 539368	2–2	5.99–02	2.10–02	6.69–01	–1.376	B	*LS*
295.	$2s2p(^1P^o)3s-$ $2s2p(^1P^o)3p$	$^2P^o-^2D$	*3491.0*	*3492.0*	*518706*	– *547343*	6–10	7.28–01	2.22–01	1.53+01	0.124	B	1
			3489.89	3490.89	518709	– 547355	4–6	7.29–01	2.00–01	9.17+00	–0.098	B	*LS*
			3492.21	3493.21	518699	– 547326	2–4	6.06–01	2.22–01	5.10+00	–0.353	B	*LS*
			3493.43	3494.43	518709	– 547326	4–4	1.21–01	2.22–02	1.02+00	–1.052	B	*LS*
296.		$^2P^o-^2P$	*3211.6*	*3212.5*	*518706*	– *549834*	6–6	1.05+00	1.62–01	1.03+01	–0.012	B	1
			3209.76	3210.69	518709	– 549855	4–4	8.75–01	1.35–01	5.71+00	–0.267	B	*LS*
			3215.23	3216.16	518699	– 549792	2–2	6.96–01	1.08–01	2.28+00	–0.666	B	*LS*
			3216.26	3217.19	518709	– 549792	4–2	3.48–01	2.70–02	1.14+00	–0.967	B	*LS*
			3208.73	3209.65	518699	– 549855	2–4	1.75–01	5.41–02	1.14+00	–0.966	B	*LS*
297.	$2s2p(^1P^o)3s-$ $2s^25d$	$^2P^o-^2D$	*2999.9*	*3000.7*	*518706*	– *552031*	6–10	8.25–02	1.86–02	1.10+00	–0.953	B	1
			3000.03	3000.90	518709	– 552032.3	4–6	8.25–02	1.67–02	6.60–01	–1.175	B	*LS*
			2999.37	3000.25	518699	– 552029.6	2–4	6.88–02	1.86–02	3.67–01	–1.430	B	*LS*
			3000.27	3001.15	518709	– 552029.6	4–4	1.38–02	1.86–03	7.33–02	–2.129	B	*LS*
298.	$2s2p(^1P^o)3s-$ $2s2p(^1P^o)3p$	$^2P^o-^2S$	*2795.8*	*2796.6*	*518706*	– 554464	6–2	6.79–01	2.65–02	1.46+00	–0.798	B	1
			2795.99	2796.81	518709	– 554464	4–2	4.53–01	2.65–02	9.77–01	–0.974	B	*LS*
			2795.21	2796.03	518699	– 554464	2–2	2.26–01	2.65–02	4.88–01	–1.275	B	*LS*

O IV: Allowed Transitions — Continued

No.	Transition Array	Mult.	λ_{air} (Å)	λ_{vac} (Å) or σ (cm^{-1})*	E_i (cm^{-1})	E_k (cm^{-1})	g_i–g_k	A_{ki} (10^8 s^{-1})	f_{ik}	S (at. u.)	log gf	Acc.	Source
299.	$2s2p(^1P^\circ)3s-2s^26d$	$^2P^\circ-^2D$		*1796.5*	*518706*	– 574368.7	6–10	3.43-02	2.77-03	9.82-02	-1.780	B	1
				1796.63	518709	– 574368.7	4–6	3.43-02	2.49-03	5.89-02	-2.002	B	*LS*
				1796.31	518699	– 574368.7	2–4	2.86-02	2.77-03	3.27-02	-2.257	B	*LS*
				1796.63	518709	– 574368.7	4–4	5.72-03	2.77-04	6.55-03	-2.956	C+	*LS*
300.	$2s2p(^1P^\circ)3s-2s2p(^3P^\circ)4p$	$^2P^\circ-^2D$		*1515.9*	*518706*	– *584673*	6–10	1.17+00	6.70-02	2.01+00	-0.396	B	1
				1513.96	518709	– 584761	4–6	1.17+00	6.04-02	1.20+00	-0.617	B	*LS*
				1518.79	518699	– 584541	2–4	9.67-01	6.69-02	6.69-01	-0.873	B	*LS*
				1519.02	518709	– 584541	4–4	1.93-01	6.69-03	1.34-01	-1.573	B	*LS*
301.		$^2P^\circ-^2S$		*1401.3*	*518706*	– 590069	6–2	1.52+00	1.49-02	4.12-01	-1.049	B	1
				1401.35	518709	– 590069	4–2	1.01+00	1.49-02	2.75-01	-1.225	B	*LS*
				1401.15	518699	– 590069	2–2	5.06-01	1.49-02	1.37-01	-1.526	B	*LS*
302.	$2s2p(^1P^\circ)3s-2p^2(^3P)3s$	$^2P^\circ-^2P$		*1327.1*	*518706*	– *594056*	6–6	1.09+00	2.89-02	7.57-01	-0.762	B	1
				1326.88	518709	– 594074	4–4	9.11-01	2.41-02	4.20-01	-1.017	B	*LS*
				1327.67	518699	– 594019	2–2	7.28-01	1.92-02	1.68-01	-1.415	B	*LS*
				1327.84	518709	– 594019	4–2	3.64-01	4.81-03	8.41-02	-1.716	B	*LS*
				1326.70	518699	– 594074	2–4	1.82-01	9.62-03	8.41-02	-1.716	B	*LS*
303.	$2s2p(^1P^\circ)3s-2s2p(^3P^\circ)4f$	$^2P^\circ-^2D$		*1281.9*	*518706*	– *596714*	6–10	1.27+00	5.20-02	1.32+00	-0.506	B	1
				1282.89	518709	– 596658.3	4–6	1.26+00	4.68-02	7.90-01	-0.728	C+	*LS*
				1280.42	518699	– 596798.2	2–4	1.06+00	5.21-02	4.39-01	-0.982	C+	*LS*
				1280.59	518709	– 596798.2	4–4	2.12-01	5.21-03	8.78-02	-1.681	C+	*LS*
304.	$2s2p(^1P^\circ)3s-2p^2(^1D)3s$	$^2P^\circ-^2D$		*1228.6*	*518706*	– *600100*	6–10	4.56+00	1.72-01	4.17+00	0.013	B	1
				1228.55	518709	– 600106	4–6	4.56+00	1.55-01	2.50+00	-0.208	B	*LS*
				1228.61	518699	– 600092	2–4	3.80+00	1.72-01	1.39+00	-0.464	B	*LS*
				1228.76	518709	– 600092	4–4	7.59-01	1.72-02	2.78-01	-1.163	B	*LS*
305.	$2s2p(^1P^\circ)3s-2p^2(^3P)3d$	$^2P^\circ-^2P$		*878.91*	*518706*	– *632483*	6–6	1.39-01	1.61-03	2.79-02	-2.016	B–	1
				879.376	518709	– 632426	4–4	1.15-01	1.34-03	1.55-02	-2.271	B–	*LS*
				877.979	518699	– 632597	2–2	9.28-02	1.07-03	6.20-03	-2.669	C+	*LS*
				878.056	518709	– 632597	4–2	4.64-02	2.68-04	3.10-03	-2.970	C+	*LS*
				879.299	518699	– 632426	2–4	2.31-02	5.35-04	3.10-03	-2.970	C+	*LS*
306.	$2s2p(^1P^\circ)3s-2p^2(^1D)3d$	$^2P^\circ-^2P$		*707.75*	*518706*	– 659998	6–6	2.70+00	2.03-02	2.84-01	-0.914	B	1
				707.769	518709	– 659998	4–4	2.25+00	1.69-02	1.58-01	-1.170	B	*LS*
				707.719	518699	– 659998	2–2	1.80+00	1.35-02	6.31-02	-1.568	B	*LS*
				707.769	518709	– 659998	4–2	9.01-01	3.38-03	3.15-02	-1.869	B	*LS*
				707.719	518699	– 659998	2–4	4.51-01	6.77-03	3.15-02	-1.869	B	*LS*
307.	$2s2p(^1P^\circ)3p-2s2p(^3P^\circ)4s$	$^2D-^2P^\circ$	*3773.9*	*3775.0*	*547343*	– *573833*	10–6	5.35-02	6.85-03	8.51-01	-1.164	B	1
			3765.98	3767.05	547355	– 573901	6–4	4.84-02	6.86-03	5.11-01	-1.385	B	*LS*
			3791.11	3792.19	547326	– 573696	4–2	5.27-02	5.68-03	2.84-01	-1.643	B	*LS*
			3761.87	3762.94	547326	– 573901	4–4	5.40-03	1.15-03	5.67-02	-2.339	B	*LS*
308.	$2s2p(^1P^\circ)3p-2s^26f$	$^2D-^2F^\circ$	*3640.0*	*3641.0*	*547343*	– 574807.7	10–14	4.29-01	1.19-01	1.43+01	0.076	B	1
			3641.59	3642.63	547355	– 574807.7	6–8	4.28-01	1.13-01	8.16+00	-0.167	C+	*LS*
			3637.75	3638.79	547326	– 574807.7	4–6	4.01-01	1.19-01	5.71+00	-0.321	C+	*LS*
			3641.59	3642.63	547355	– 574807.7	6–6	2.85-02	5.67-03	4.08-01	-1.468	C+	*LS*

O IV: Allowed Transitions — Continued

No.	Transition Array	Mult.	λ_{air} (Å)	λ_{vac} (Å) or σ (cm^{-1})*	E_i (cm^{-1})	E_k (cm^{-1})	g_i-g_k	A_{ki} (10^8 s^{-1})	f_{ik}	S (at. u.)	log gf	Acc.	Source
309.	$2s2p(^1P^o)3p-2s2p(^1P^o)3d$	$^2D-^2D^o$	*3508.3*	*3509.3*	*547343*	*575839*	10−10	1.91−01	3.52−02	4.06+00	−0.454	B	1
			3508.01	3509.02	547355	575853	6−6	1.78−01	3.28−02	2.28+00	−0.705	B	*LS*
			3508.63	3509.63	547326	575819	4−4	1.72−01	3.17−02	1.46+00	−0.897	B	*LS*
			3512.21	3513.21	547355	575819	6−4	1.90−02	2.34−03	1.63−01	−1.852	B	*LS*
			3504.45	3505.45	547326	575853	4−6	1.28−02	3.52−03	1.63−01	−1.851	B	*LS*
310.		$^2D-^2F^o$	*4263.6*	*4264.8*	*547343*	570791	10−14	2.06−01	7.84−02	1.10+01	−0.105	B	1
			4265.74	4266.94	547355	570791	6−8	2.05−01	7.47−02	6.29+00	−0.349	B	*LS*
			4260.47	4261.67	547326	570791	4−6	1.92−01	7.85−02	4.40+00	−0.503	B	*LS*
			4265.74	4266.94	547355	570791	6−6	1.37−02	3.73−03	3.15−01	−1.650	B	*LS*
311.		$^2D-^2P^o$	*2906.7*	*2907.6*	*547343*	*581736*	10−6	3.66−02	2.78−03	2.66−01	−1.556	B	1
			2907.14	2907.99	547355	581743	6−4	3.29−02	2.78−03	1.60−01	−1.778	B	*LS*
			2906.55	2907.40	547326	581721	4−2	3.66−02	2.32−03	8.87−02	−2.033	B	*LS*
			2904.69	2905.54	547326	581743	4−4	3.67−03	4.64−04	1.77−02	−2.732	B−	*LS*
312.	$2s2p(^1P^o)3p-2s^27f$	$^2D-^2F^o$	*2468.0*	*2468.7*	*547343*	587850	10−14	1.17−01	1.50−02	1.22+00	−0.825	B	1
			2468.69	2469.44	547355	587850	6−8	1.17−01	1.42−02	6.95−01	−1.068	C+	*LS*
			2466.93	2467.67	547326	587850	4−6	1.09−01	1.50−02	4.86−01	−1.223	C+	*LS*
			2468.69	2469.44	547355	587850	6−6	7.79−03	7.12−04	3.47−02	−2.369	C+	*LS*
313.	$2s2p(^1P^o)3p-2s2p(^3P^o)4d$	$^2D-^2D^o$	*2157.6*	*2158.3*	*547343*	*593676*	10−10	3.16−01	2.20−02	1.56+00	−0.657	B	1
			2156.68	2157.36	547355	593708	6−6	2.95−01	2.06−02	8.76−01	−0.909	B	*LS*
			2159.10	2159.78	547326	593627	4−4	2.83−01	1.98−02	5.63−01	−1.101	B	*LS*
			2160.46	2161.13	547355	593627	6−4	3.14−02	1.47−03	6.26−02	−2.055	B	*LS*
			2155.33	2156.01	547326	593708	4−6	2.11−02	2.21−03	6.26−02	−2.054	B	*LS*
314.		$^2D-^2F^o$	*2037.9*	*2038.5*	*547343*	*596398*	10−14	2.06−01	1.80−02	1.20+00	−0.746	B	1
			2035.18	2035.83	547355	596475	6−8	2.07−01	1.71−02	6.88−01	−0.988	B	*LS*
			2041.45	2042.11	547326	596295	4−6	1.91−01	1.79−02	4.82−01	−1.144	B	*LS*
			2042.66	2043.32	547355	596295	6−6	1.36−02	8.53−04	3.44−02	−2.291	B	*LS*
315.		$^2D-^2P^o$		*1983.1*	*547343*	*597770*	10−6	3.11−02	1.10−03	7.18−02	−1.959	B	1
				1985.51	547355	597720	6−4	2.79−02	1.10−03	4.31−02	−2.181	B	*LS*
				1978.51	547326	597869	4−2	3.13−02	9.19−04	2.39−02	−2.435	B−	*LS*
				1984.36	547326	597720	4−4	3.10−03	1.83−04	4.79−03	−3.135	C+	*LS*
316.	$2s2p(^1P^o)3p-2p^2(^3P)3p$	$^2D-^2D^o$		*1468.4*	*547343*	*615443*	10−10	1.35−01	4.35−03	2.11−01	−1.361	B	1
				1468.95	547355	615431	6−6	1.26−01	4.06−03	1.18−01	−1.613	B	*LS*
				1467.70	547326	615460	4−4	1.21−01	3.92−03	7.58−02	−1.805	B	*LS*
				1468.32	547355	615460	6−4	1.35−02	2.90−04	8.42−03	−2.759	C+	*LS*
				1468.32	547326	615431	4−6	8.98−03	4.35−04	8.42−03	−2.759	C+	*LS*
317.	$2s2p(^1P^o)3p-2p^2(^1D)3p$	$^2D-^2D^o$		*1182.5*	*547343*	631913	10−10	8.21−01	1.72−02	6.70−01	−0.764	B	1
				1182.62	547355	631913	6−6	7.66−01	1.61−02	3.75−01	−1.016	B	*LS*
				1182.21	547326	631913	4−4	7.40−01	1.55−02	2.41−01	−1.208	B	*LS*
				1182.62	547355	631913	6−4	8.21−02	1.15−03	2.68−02	−2.162	B−	*LS*
				1182.21	547326	631913	4−6	5.48−02	1.72−03	2.68−02	−2.162	B−	*LS*
318.	$2s2p(^1P^o)3p-2s2p(^1P^o)4d$	$^2D-^2D^o$		*825.00*	*547343*	*668555*	10−10	1.96+00	2.00−02	5.45−01	−0.698	B	1
				825.096	547355	668553	6−6	1.83+00	1.87−02	3.05−01	−0.950	B	*LS*
				824.872	547326	668557	4−4	1.77+00	1.80−02	1.96−01	−1.142	B	*LS*
				825.069	547355	668557	6−4	1.96−01	1.34−03	2.18−02	−2.096	B−	*LS*
				824.899	547326	668553	4−6	1.31−01	2.01−03	2.18−02	−2.096	B−	*LS*

O IV: Allowed Transitions — Continued

No.	Transition Array	Mult.	λ_{air} (Å)	λ_{vac} (Å) or σ (cm^{-1})*	E_i (cm^{-1})	E_k (cm^{-1})	g_i–g_k	A_{ki} (10^8 s^{-1})	f_{ik}	S (at. u.)	log gf	Acc.	Source
319.	$2s2p(^1P^\circ)3p$ – $2s2p(^3P^\circ)4s$	$^2P-^2P^\circ$	*4165.7*	*4166.8*	*549834*	*573833*	6–6	5.69–03	1.48–03	1.22–01	–2.052	B	1
			4157.52	4158.70	549855	573901	4–4	4.77–03	1.23–03	6.76–02	–2.306	B	*LS*
			4182.22	4183.40	549792	573696	2–2	3.75–03	9.82–04	2.70–02	–2.707	B–	*LS*
			4193.27	4194.45	549855	573696	4–2	1.86–03	2.45–04	1.35–02	–3.009	B–	*LS*
			4146.66	4147.83	549792	573901	2–4	9.61–04	4.95–04	1.35–02	–3.004	B–	*LS*
320.	$2s2p(^1P^\circ)3p$ – $2s2p(^1P^\circ)3d$	$^2P-^2D^\circ$	*3844.3*	*3845.4*	*549834*	*575839*	6–10	4.68–01	1.73–01	1.31+01	0.016	B	1
			3845.36	3846.45	549855	575853	4–6	4.68–01	1.56–01	7.88+00	–0.206	B	*LS*
			3841.07	3842.16	549792	575819	2–4	3.91–01	1.73–01	4.38+00	–0.461	B	*LS*
			3850.39	3851.49	549855	575819	4–4	7.77–02	1.73–02	8.75–01	–1.161	B	*LS*
321.		$^2P-^2P^\circ$	*3133.7*	*3134.6*	*549834*	*581736*	6–6	3.19–01	4.70–02	2.91+00	–0.549	B	1
			3135.07	3135.98	549855	581743	4–4	2.66–01	3.92–02	1.62+00	–0.805	B	*LS*
			3131.04	3131.95	549792	581721	2–2	2.13–01	3.14–02	6.47–01	–1.202	B	*LS*
			3137.23	3138.14	549855	581721	4–2	1.06–01	7.83–03	3.23–01	–1.504	B	*LS*
			3128.89	3129.79	549792	581743	2–4	5.35–02	1.57–02	3.23–01	–1.503	B	*LS*
322.	$2s2p(^1P^\circ)3p$ – $2s2p(^3P^\circ)4d$	$^2P-^2D^\circ$	*2280.2*	*2280.9*	*549834*	*593676*	6–10	7.38–02	9.58–03	4.32–01	–1.240	B	1
			2279.64	2280.35	549855	593708	4–6	7.38–02	8.63–03	2.59–01	–1.462	B	*LS*
			2280.58	2281.28	549792	593627	2–4	6.15–02	9.58–03	1.44–01	–1.717	B	*LS*
			2283.86	2284.57	549855	593627	4–4	1.22–02	9.57–04	2.88–02	–2.417	B–	*LS*
323.		$^2P-^2P^\circ$	*2085.5*	*2086.1*	*549834*	*597770*	6–6	2.54–02	1.66–03	6.83–02	–2.002	B	1
			2088.54	2089.21	549855	597720	4–4	2.11–02	1.38–03	3.79–02	–2.258	B	*LS*
			2079.33	2080.00	549792	597869	2–2	1.71–02	1.11–03	1.52–02	–2.654	B–	*LS*
			2082.06	2082.73	549855	597869	4–2	8.52–03	2.77–04	7.59–03	–2.956	C+	*LS*
			2085.80	2086.46	549792	597720	2–4	4.24–03	5.53–04	7.59–03	–2.957	C+	*LS*
324.	$2s2p(^1P^\circ)3p$ – $2p^2(^3P)3p$	$^2P-^2S^\circ$	*2108.1*	*2108.8*	*549834*	597254	6–2	3.76–01	8.36–03	3.48–01	–1.300	B	1
			2109.08	2109.75	549855	597254	4–2	2.51–01	8.35–03	2.32–01	–1.476	B	*LS*
			2106.28	2106.95	549792	597254	2–2	1.26–01	8.36–03	1.16–01	–1.777	B	*LS*
325.		$^2P-^2D^\circ$		*1524.2*	*549834*	*615443*	6–10	2.36+00	1.37–01	4.13+00	–0.085	B	1
				1524.95	549855	615431	4–6	2.36+00	1.23–01	2.48+00	–0.307	B	*LS*
				1522.81	549792	615460	2–4	1.97+00	1.37–01	1.38+00	–0.562	B	*LS*
				1524.27	549855	615460	4–4	3.94–01	1.37–02	2.75–01	–1.261	B	*LS*
326.	$2s2p(^1P^\circ)3p$ – $2p^2(^1D)3p$	$^2P-^2D^\circ$		*1218.3*	*549834*	631913	6–10	9.79–01	3.63–02	8.74–01	–0.662	B	1
				1218.65	549855	631913	4–6	9.78–01	3.27–02	5.24–01	–0.884	B	*LS*
				1217.72	549792	631913	2–4	8.17–01	3.63–02	2.91–01	–1.139	B	*LS*
				1218.65	549855	631913	4–4	1.63–01	3.63–03	5.83–02	–1.838	B	*LS*
327.	$2s2p(^1P^\circ)3p$ – $2s2p(^1P^\circ)4d$	$^2P-^2D^\circ$		*842.31*	*549834*	*668555*	6–10	6.21+00	1.10–01	1.83+00	–0.180	B	1
				842.474	549855	668553	4–6	6.20+00	9.90–02	1.10+00	–0.402	B	*LS*
				841.999	549792	668557	2–4	5.18+00	1.10–01	6.10–01	–0.657	B	*LS*
				842.446	549855	668557	4–4	1.03+00	1.10–02	1.22–01	–1.357	B	*LS*
328.	$2s2p(^1P^\circ)3p$ – $2s2p(^3P^\circ)4s$	$^2S-^2P^\circ$	*5161.5*	*5162.9*	554464	*573833*	2–6	1.22–01	1.46–01	4.96+00	–0.535	B	1
			5143.39	5144.83	554464	573901	2–4	1.23–01	9.77–02	3.31+00	–0.709	B	*LS*
			5198.22	5199.67	554464	573696	2–2	1.19–01	4.83–02	1.65+00	–1.015	B	*LS*

O IV: Allowed Transitions — Continued

No.	Transition Array	Mult.	λ_{air} (Å)	λ_{vac} (Å) or σ (cm^{-1})*	E_i (cm^{-1})	E_k (cm^{-1})	$g_i - g_k$	A_{ki} ($10^8 s^{-1}$)	f_{ik}	S (at. u.)	log gf	Acc.	Source
329.	$2s2p(^1P^o)3p-2s2p(^1P^o)3d$	$^2S-^2P^o$	*3665.7*	*3666.8*	554464	– *581736*	2–6	3.91–01	2.37–01	5.71+00	–0.325	B	1
			3664.78	3665.82	554464	– 581743	2–4	3.92–01	1.58–01	3.81+00	–0.501	B	*LS*
			3667.74	3668.78	554464	– 581721	2–2	3.91–01	7.88–02	1.90+00	–0.802	B	*LS*
330.	$2s2p(^1P^o)3p-2s2p(^3P^o)4d$	$^2S-^2P^o$	*2308.4*	*2309.1*	554464	– *597770*	2–6	3.13–01	7.50–02	1.14+00	–0.824	B	1
			2311.11	2311.82	554464	– 597720	2–4	3.12–01	4.99–02	7.60–01	–1.001	B	*LS*
			2303.17	2303.88	554464	– 597869	2–2	3.15–01	2.50–02	3.80–01	–1.300	B	*LS*
331.	$2s2p(^1P^o)3d-2s2p(^3P^o)4p$	$^2F^o-^2D$	*7201.6*	*7203.6*	570791	– *584673*	14–10	7.37–03	4.10–03	1.36+00	–1.242	B	1
			7156.22	7158.20	570791	– 584761	8–6	7.16–03	4.12–03	7.77–01	–1.482	B	*LS*
			7270.72	7272.73	570791	– 584541	6–4	7.17–03	3.79–03	5.44–01	–1.644	B	*LS*
			7156.22	7158.20	570791	– 584761	6–6	3.58–04	2.75–04	3.88–02	–2.783	B	*LS*
332.	$2s2p(^1P^o)3d-2s2p(^3P^o)4f$	$^2F^o-^2F$	*4298.2*	*4299.4*	570791	– *594050*	14–14	2.34–03	6.49–04	1.29–01	–2.042	B	1
			4293.77	4294.98	570791	– 594074	8–8	2.27–03	6.26–04	7.08–02	–2.300	C+	*LS*
			4303.94	4305.15	570791	– 594019	6–6	2.22–03	6.17–04	5.24–02	–2.432	C+	*LS*
			4303.94	4305.15	570791	– 594019	8–6	1.11–04	2.31–05	2.62–03	–3.733	D	*LS*
			4293.77	4294.98	570791	– 594074	6–8	8.39–05	3.09–05	2.62–03	–3.732	D	*LS*
333.		$^2F^o-^2G$	*3850.9*	*3852.0*	570791	– *596752*	14–18	2.13–02	6.10–03	1.08+00	–1.069	B	1
			3836.71	3837.80	570791	– 596847.6	8–10	2.16–02	5.95–03	6.01–01	–1.322	C+	*LS*
			3868.74	3869.83	570791	– 596631.9	6–8	2.03–02	6.07–03	4.64–01	–1.439	C+	*LS*
			3868.74	3869.83	570791	– 596631.9	8–8	7.51–04	1.69–04	1.72–02	–2.870	C	*LS*
334.		$^2F^o-^2D$	*3856.4*	*3857.5*	570791	– *596714*	14–10	1.35–02	2.15–03	3.83–01	–1.520	B	1
			3864.79	3865.88	570791	– 596658.3	8–6	1.28–02	2.15–03	2.19–01	–1.764	C+	*LS*
			3844.00	3845.09	570791	– 596798.2	6–4	1.37–02	2.02–03	1.53–01	–1.917	C+	*LS*
			3864.79	3865.88	570791	– 596658.3	6–6	6.40–04	1.43–04	1.09–02	–3.065	C	*LS*
335.	$2s2p(^1P^o)3d-2p^2(^1D)3s$	$^2F^o-^2D$	*3410.9*	*3411.9*	570791	– *600100*	14–10	5.52–03	6.88–04	1.08–01	–2.017	B	1
			3410.24	3411.22	570791	– 600106	8–6	5.26–03	6.88–04	6.18–02	–2.259	B	*LS*
			3411.87	3412.85	570791	– 600092	6–4	5.51–03	6.42–04	4.32–02	–2.415	B	*LS*
			3410.24	3411.22	570791	– 600106	6–6	2.63–04	4.59–05	3.09–03	–3.560	C+	*LS*
336.	$2s2p(^1P^o)3d-2p^2(^3P)3d$	$^2F^o-^2F$		1686.23	570791	– 630095	14–14	1.16–02	4.94–04	3.84–02	–2.160	B–	1
337.	$2s2p(^1P^o)3d-2p^2(^1D)3d$	$^2F^o-^2F$		*1245.1*	570791	– *651109*	14–14	1.19+00	2.77–02	1.59+00	–0.411	B	1
				1244.93	570791	– 651117	8–8	1.15+00	2.67–02	8.76–01	–0.670	B	*LS*
				1245.22	570791	– 651098	6–6	1.13+00	2.64–02	6.49–01	–0.801	B	*LS*
				1245.22	570791	– 651098	8–6	5.67–02	9.89–04	3.24–02	–2.102	B	*LS*
				1244.93	570791	– 651117	6–8	4.26–02	1.32–03	3.24–02	–2.102	B	*LS*
338.	$2s2p(^3P^o)4s-2p^2(^3P)3s$	$^4P^o-^4P$	*12588*	*7942.0* cm^{-1}	*569137*	– *577079*	12–12	4.51–03	1.07–02	5.33+00	–0.891	B	1
			12613.3	7926.00 cm^{-1}	569283	– 577209	6–6	3.14–03	7.49–03	1.87+00	–1.347	B	*LS*
			12557.8	7961.00 cm^{-1}	569036	– 576997	4–4	6.06–04	1.43–03	2.37–01	–2.242	B	*LS*
			12572.0	7952.00 cm^{-1}	568901	– 576853	2–2	7.55–04	1.79–03	1.48–01	–2.446	B	*LS*
			12959.9	7714.00 cm^{-1}	569283	– 576997	6–4	1.86–03	3.12–03	8.00–01	–1.727	B	*LS*
			12789.1	7817.00 cm^{-1}	569036	– 576853	4–2	3.59–03	4.40–03	7.41–01	–1.755	B	*LS*
			12232.1	8173.00 cm^{-1}	569036	– 577209	4–6	1.48–03	4.97–03	8.00–01	–1.702	B	*LS*
			12348.4	8096.00 cm^{-1}	568901	– 576997	2–4	1.99–03	9.11–03	7.41–01	–1.740	B	*LS*

O IV: Allowed Transitions — Continued

No.	Transition Array	Mult.	λ_{air} (Å)	λ_{vac} (Å) or σ (cm^{-1})*	E_i (cm^{-1})	E_k (cm^{-1})	g_i-g_k	A_{ki} (10^8 s^{-1})	f_{ik}	S (at. u.)	log gf	Acc.	Source
339.	$2s2p(^3P^\circ)4s-2s2p(^3P^\circ)4f$	$^4P^\circ-^4D$	*3663.7*	*3664.7*	*569137*	*596424*	12–20	4.28–02	1.44–02	2.08+00	–0.764	B	1
			3696.08	3697.13	569283	596331	6–8	4.17–02	1.14–02	8.31–01	–1.165	C+	*LS*
			3647.53	3648.57	569036	596444	4–6	3.04–02	9.09–03	4.36–01	–1.440	C+	*LS*
			3620.66	3621.69	568901	596512.4	2–4	1.85–02	7.26–03	1.73–01	–1.838	C+	*LS*
			3680.70	3681.75	569283	596444	6–6	1.27–02	2.57–03	1.87–01	–1.811	C+	*LS*
			3638.45	3639.49	569036	596512.4	4–4	2.33–02	4.63–03	2.22–01	–1.733	C+	*LS*
			3614.93	3615.96	568901	596556.2	2–2	3.71–02	7.28–03	1.73–01	–1.837	C+	*LS*
			3671.46	3672.50	569283	596512.4	6–4	2.13–03	2.87–04	2.08–02	–2.765	C	*LS*
			3632.66	3633.69	569036	596556.2	4–2	7.32–03	7.24–04	3.46–02	–2.538	C+	*LS*
340.	$2s2p(^3P^\circ)4s-2p^2(^3P)3d$	$^4P^\circ-^4P$		*1469.8*	*569137*	*637173*	12–12	8.77–03	2.84–04	1.65–02	–2.467	B–	1
				1474.27	569283	637113	6–6	6.09–03	1.98–04	5.78–03	–2.924	C+	*LS*
				1466.79	569036	637212	4–4	1.18–03	3.80–05	7.33–04	–3.819	D	*LS*
				1462.57	568901	637274	2–2	1.48–03	4.76–05	4.58–04	–4.021	D	*LS*
				1472.13	569283	637212	6–4	3.93–03	8.51–05	2.48–03	–3.292	C	*LS*
				1465.46	569036	637274	4–2	7.38–03	1.19–04	2.29–03	–3.323	C	*LS*
				1468.92	569036	637113	4–6	2.64–03	1.28–04	2.48–03	–3.291	C	*LS*
				1463.89	568901	637212	2–4	3.70–03	2.38–04	2.29–03	–3.323	C	*LS*
341.	$2s2p(^3P^\circ)4s-2s^26d$	$^2P^\circ-^2D$		*535.70* cm^{-1}	*573833*	574368.7	6–10	5.24–05	4.56–02	1.68+02	–0.562	B	1
				467.700 cm^{-1}	573901	574368.7	4–6	3.49–05	3.59–02	1.01+02	–0.843	B	*LS*
				672.700 cm^{-1}	573696	574368.7	2–4	8.65–05	5.73–02	5.61+01	–0.941	B	*LS*
				467.700 cm^{-1}	573901	574368.7	4–4	5.82–06	3.98–03	1.12+01	–1.798	B	*LS*
342.	$2s2p(^3P^\circ)4s-2s2p(^3P^\circ)4p$	$^2P^\circ-^2P$		*1484.0* cm^{-1}	*573833*	*575317*	6–6	3.31–04	2.26–02	3.00+01	–0.869	B	1
				1474.00 cm^{-1}	573901	575375	4–4	2.71–04	1.87–02	1.67+01	–1.127	B	*LS*
				1506.00 cm^{-1}	573696	575202	2–2	2.31–04	1.53–02	6.67+00	–1.515	B	*LS*
				1301.00 cm^{-1}	573901	575202	4–2	7.44–05	3.30–03	3.33+00	–1.880	B	*LS*
				1679.00 cm^{-1}	573696	575375	2–4	8.00–05	8.51–03	3.33+00	–1.769	B	*LS*
343.		$^2P^\circ-^2D$	*9222.6*	*9225.1*	*573833*	*584673*	6–10	1.05–01	2.23–01	4.06+01	0.126	B	1
			9205.58	9208.10	573901	584761	4–6	1.05–01	2.01–01	2.43+01	–0.095	B	*LS*
			9218.31	9220.84	573696	584541	2–4	8.75–02	2.23–01	1.35+01	–0.351	B	*LS*
			9395.92	9398.50	573901	584541	4–4	1.65–02	2.19–02	2.71+00	–1.058	B	*LS*
344.		$^2P^\circ-^2S$	*6157.4*	*6159.2*	*573833*	590069	6–2	4.37–01	8.29–02	1.01+01	–0.304	B	1
			6183.35	6185.06	573901	590069	4–2	2.88–01	8.25–02	6.72+00	–0.481	B	*LS*
			6105.93	6107.62	573696	590069	2–2	1.49–01	8.36–02	3.36+00	–0.777	B	*LS*
345.	$2s2p(^3P^\circ)4s-2p^2(^3P)3s$	$^2P^\circ-^2P$	*4843.8*	*4845.2*	*573833*	*594472*	6–6	2.27–01	7.99–02	7.65+00	–0.319	B	1
			4844.31	4845.67	573901	594538	4–4	1.89–01	6.66–02	4.25+00	–0.575	B	*LS*
			4842.67	4844.02	573696	594340	2–2	1.52–01	5.33–02	1.70+00	–0.972	B	*LS*
			4891.24	4892.61	573901	594340	4–2	7.36–02	1.32–02	8.50–01	–1.278	B	*LS*
			4796.66	4798.00	573696	594538	2–4	3.90–02	2.69–02	8.50–01	–1.269	B	*LS*
346.	$2s2p(^3P^\circ)4s-2s2p(^3P^\circ)4f$	$^2P^\circ-^2D$	*4369.2*	*4370.4*	*573833*	*596714*	6–10	4.38–03	2.09–03	1.80–01	–1.902	B	1
			4392.96	4394.19	573901	596658.3	4–6	4.30–03	1.87–03	1.08–01	–2.127	C+	*LS*
			4327.38	4328.59	573696	596798.2	2–4	3.75–03	2.11–03	6.00–02	–2.375	C+	*LS*
			4366.12	4367.35	573901	596798.2	4–4	7.31–04	2.09–04	1.20–02	–3.078	C	*LS*
347.	$2s2p(^3P^\circ)4s-2p^2(^1D)3s$	$^2P^\circ-^2D$	*3806.0*	*3807.1*	*573833*	*600100*	6–10	9.75–02	3.53–02	2.65+00	–0.674	B	1
			3814.98	3816.07	573901	600106	4–6	9.68–02	3.17–02	1.59+00	–0.897	B	*LS*
			3787.38	3788.45	573696	600092	2–4	8.25–02	3.55–02	8.84–01	–1.149	B	*LS*
			3817.02	3818.11	573901	600092	4–4	1.61–02	3.52–03	1.77–01	–1.852	B	*LS*

O IV: Allowed Transitions — Continued

No.	Transition Array	Mult.	λ_{air} (Å)	λ_{vac} (Å) or σ (cm^{-1})*	E_i (cm^{-1})	E_k (cm^{-1})	g_i–g_k	A_{ki} (10^8 s^{-1})	f_{ik}	S (at. u.)	log gf	Acc.	Source
348.	$2s2p(^3P^\circ)4s-$ $2p^2(^3P)3d$	$^2P^\circ-^2P$		*1705.0*	*573833*	*632483*	6–6	1.27−01	5.54−03	1.87−01	−1.478	B	1
				1708.67	573901	632426	4–4	1.05−01	4.61−03	1.04−01	−1.734	B	LS
				1697.76	573696	632597	2–2	8.59−02	3.71−03	4.15−02	−2.129	B	LS
				1703.69	573901	632597	4–2	4.25−02	9.25−04	2.07−02	−2.432	B−	LS
				1702.71	573696	632426	2–4	2.13−02	1.85−03	2.07−02	−2.432	B−	LS
349.	$2s2p(^3P^\circ)4p-$ $2s2p(^1P^\circ)3d$	$^2P-^2D^\circ$		*522.00* cm^{-1}	*575317*	*575839*	6–10	1.52−06	1.39−03	5.26+00	−2.079	B	1
				478.000 cm^{-1}	575375	575853	4–6	1.16−06	1.14−03	3.15+00	−2.339	B	LS
				617.000 cm^{-1}	575202	575819	2–4	2.09−06	1.64−03	1.75+00	−2.484	B	LS
				444.000 cm^{-1}	575375	575819	4–4	1.55−07	1.18−04	3.50−01	−3.325	B	LS
350.		$^2P-^2P^\circ$	*15574*	*6419.0* cm^{-1}	*575317*	*581736*	6–6	4.24−04	1.54−03	4.74−01	−2.034	B	1
			15699.2	6368.00 cm^{-1}	575375	581743	4–4	3.45−04	1.27−03	2.63−01	−2.293	B	LS
			15335.6	6519.00 cm^{-1}	575202	581721	2–2	2.96−04	1.04−03	1.05−01	−2.681	B	LS
			15753.7	6346.00 cm^{-1}	575375	581721	4–2	1.36−04	2.54−04	5.26−02	−2.993	B	LS
			15284.0	6541.00 cm^{-1}	575202	581743	2–4	7.47−05	5.23−04	5.26−02	−2.980	B	LS
351.	$2s2p(^3P^\circ)4p-$ $2s2p(^3P^\circ)4d$	$^2P-^2D^\circ$	*5445.4*	*5446.9*	*575317*	*593676*	6–10	6.15−01	4.55−01	4.90+01	0.437	B	1
			5453.13	5454.64	575375	593708	4–6	6.12−01	4.09−01	2.94+01	0.214	B	LS
			5425.90	5427.41	575202	593627	2–4	5.18−01	4.57−01	1.63+01	−0.039	B	LS
			5477.33	5478.85	575375	593627	4–4	1.01−01	4.53−02	3.27+00	−0.742	B	LS
352.	$2s2p(^3P^\circ)4p-$ $2p^2(^3P)3p$	$^2P-^2S^\circ$	*4557.2*	*4558.5*	*575317*	597254	6–2	7.89−02	8.19−03	7.38−01	−1.308	B	1
			4569.31	4570.59	575375	597254	4–2	5.22−02	8.17−03	4.92−01	−1.486	B	LS
			4533.46	4534.74	575202	597254	2–2	2.67−02	8.24−03	2.46−01	−1.783	B	LS
353.	$2s2p(^3P^\circ)4p-$ $2s2p(^3P^\circ)4d$	$^2P-^2P^\circ$	*4452.5*	*4453.7*	*575317*	*597770*	6–6	6.21−01	1.84−01	1.62+01	0.044	B	1
			4474.02	4475.27	575375	597720	4–4	5.10−01	1.53−01	9.01+00	−0.213	B	LS
			4410.46	4411.70	575202	597869	2–2	4.26−01	1.24−01	3.61+00	−0.605	B	LS
			4444.38	4445.63	575375	597869	4–2	2.08−01	3.08−02	1.80+00	−0.909	B	LS
			4439.65	4440.89	575202	597720	2–4	1.04−01	6.17−02	1.80+00	−0.909	B	LS
354.	$2s2p(^3P^\circ)4p-$ $2p^2(^3P)3p$	$^2P-^2D^\circ$	*2491.4*	*2492.1*	*575317*	*615443*	6–10	5.27−01	8.17−02	4.02+00	−0.310	B	1
			2495.75	2496.50	575375	615431	4–6	5.24−01	7.34−02	2.41+00	−0.532	B	LS
			2483.23	2483.98	575202	615460	2–4	4.43−01	8.19−02	1.34+00	−0.785	B	LS
			2493.95	2494.70	575375	615460	4–4	8.75−02	8.16−03	2.68−01	−1.486	B	LS
355.	$2s2p(^3P^\circ)4p-$ $2p^2(^1D)3p$	$^2P-^2D^\circ$		*1766.9*	*575317*	631913	6–10	1.71+00	1.33−01	4.65+00	−0.097	B	1
				1768.72	575375	631913	4–6	1.70+00	1.20−01	2.79+00	−0.319	B	LS
				1763.33	575202	631913	2–4	1.43+00	1.34−01	1.55+00	−0.573	B	LS
				1768.72	575375	631913	4–4	2.84−01	1.33−02	3.10−01	−1.273	B	LS
356.	$2s2p(^1P^\circ)3d-$ $2s2p(^3P^\circ)4p$	$^2D^\circ-^2D$	*11317*	*8834.0* cm^{-1}	*575839*	*584673*	10–10	2.61−03	5.01−03	1.87+00	−1.300	B	1
			11222.8	8908.00 cm^{-1}	575853	584761	6–6	2.50−03	4.71−03	1.05+00	−1.548	B	LS
			11462.1	8722.00 cm^{-1}	575819	584541	4–4	2.26−03	4.45−03	6.72−01	−1.749	B	LS
			11507.0	8688.00 cm^{-1}	575853	584541	6–4	2.48−04	3.28−04	7.46−02	−2.705	B	LS
			11180.1	8942.00 cm^{-1}	575819	584761	4–6	1.80−04	5.07−04	7.46−02	−2.693	B	LS
357.	$2s2p(^1P^\circ)3d-$ $2s2p(^3P^\circ)4f$	$^2D^\circ-^2F$	*5489.7*	*5491.2*	*575839*	*594050*	10–14	6.29−02	3.98−02	7.19+00	−0.400	B	1
			5486.65	5488.17	575853	594074	6–8	6.30−02	3.79−02	4.11+00	−0.643	C+	LS
			5492.98	5494.51	575819	594019	4–6	5.86−02	3.98−02	2.88+00	−0.798	C+	LS
			5503.26	5504.79	575853	594019	6–6	4.16−03	1.89−03	2.06−01	−1.945	C+	LS

O IV: Allowed Transitions — Continued

No.	Transition Array	Mult.	λ_{air} (Å)	λ_{vac} (Å) or σ (cm^{-1})*	E_i (cm^{-1})	E_k (cm^{-1})	g_i-g_k	A_{ki} (10^8 s^{-1})	f_{ik}	S (at. u.)	log gf	Acc.	Source
358.		^{2}D°–^{2}D	*4789.0*	*4790.4*	*575839*	– *596714*	10–10	2.09–02	7.20–03	1.13+00	–1.143	B	1
			4805.12	4806.47	575853	– 596658.3	6–6	1.93–02	6.69–03	6.35–01	–1.396	C+	*LS*
			4765.29	4766.63	575819	– 596798.2	4–4	1.91–02	6.51–03	4.08–01	–1.584	C+	*LS*
			4773.03	4774.36	575853	– 596798.2	6–4	2.11–03	4.81–04	4.54–02	–2.539	C+	*LS*
			4797.28	4798.63	575819	– 596658.3	4–6	1.39–03	7.18–04	4.54–02	–2.542	C+	*LS*
359.	$2s2p(^1P^o)3d-2p^2(^1D)3s$	^{2}D°–^{2}D	*4120.7*	*4121.8*	*575839*	– *600100*	10–10	4.30–03	1.09–03	1.48–01	–1.961	B	1
			4122.04	4123.20	575853	– 600106	6–6	4.01–03	1.02–03	8.31–02	–2.213	B	*LS*
			4118.64	4119.80	575819	– 600092	4–4	3.87–03	9.85–04	5.34–02	–2.404	B	*LS*
			4124.42	4125.58	575853	– 600092	6–4	4.29–04	7.29–05	5.94–03	–3.359	C+	*LS*
			4116.27	4117.43	575819	– 600106	4–6	2.87–04	1.10–04	5.94–03	–3.358	C+	*LS*
360.	$2s2p(^1P^o)3d-2p^2(^3P)3d$	^{2}D°–^{2}F		*1843.1*	*575839*	– 630095	10–14	3.23–01	2.30–02	1.40+00	–0.638	B	1
				1843.59	575853	– 630095	6–8	3.23–01	2.19–02	7.99–01	–0.881	B	*LS*
				1842.43	575819	– 630095	4–6	3.02–01	2.30–02	5.59–01	–1.035	B	*LS*
				1843.59	575853	– 630095	6–6	2.15–02	1.10–03	3.99–02	–2.182	B	*LS*
361.	$2s2p(^1P^o)3d-2p^2(^1D)3d$	^{2}D°–^{2}F		*1328.6*	*575839*	– *651109*	10–14	1.30–01	4.81–03	2.10–01	–1.318	B	1
				1328.66	575853	– 651117	6–8	1.30–01	4.58–03	1.20–01	–1.561	B	*LS*
				1328.39	575819	– 651098	4–6	1.21–01	4.81–03	8.41–02	–1.716	B	*LS*
				1328.99	575853	– 651098	6–6	8.64–03	2.29–04	6.01–03	–2.862	C+	*LS*
362.	$2p^2(^3P)3s-2s2p(^3P^o)4d$	^{4}P–^{4}D°	*6710.5*	*6712.3*	*577079*	– *591977*	12–20	3.29–02	3.70–02	9.81+00	–0.352	B	1
			6738.05	6739.91	577209	– 592046	6–8	3.25–02	2.95–02	3.93+00	–0.752	B	*LS*
			6680.41	6682.26	576997	– 591962	4–6	2.33–02	2.34–02	2.06+00	–1.028	B	*LS*
			6644.45	6646.28	576853	– 591899	2–4	1.41–02	1.87–02	8.18–01	–1.427	B	*LS*
			6776.41	6778.28	577209	– 591962	6–6	9.58–03	6.60–03	8.83–01	–1.402	B	*LS*
			6708.66	6710.51	576997	– 591899	4–4	1.76–02	1.18–02	1.05+00	–1.324	B	*LS*
			6644.45	6646.28	576853	– 591899	2–2	2.82–02	1.87–02	8.18–01	–1.427	B	*LS*
			6805.47	6807.35	577209	– 591899	6–4	1.58–03	7.30–04	9.81–02	–2.359	B	*LS*
			6708.66	6710.51	576997	– 591899	4–2	5.49–03	1.85–03	1.64–01	–2.130	B	*LS*
363.		^{4}P–^{4}P°	*6143.8*	*6145.5*	*577079*	– *593351*	12–12	1.48–02	8.39–03	2.04+00	–0.997	B	1
			6223.38	6225.10	577209	– 593273	6–6	9.98–03	5.80–03	7.13–01	–1.459	B	*LS*
			6092.53	6094.22	576997	– 593406	4–4	2.03–03	1.13–03	9.05–02	–2.346	B	*LS*
			6015.18	6016.85	576853	– 593473	2–2	2.63–03	1.43–03	5.66–02	–2.544	B	*LS*
			6172.27	6173.98	577209	– 593406	6–4	6.58–03	2.50–03	3.05–01	–1.823	B	*LS*
			6067.75	6069.43	576997	– 593473	4–2	1.28–02	3.54–03	2.83–01	–1.849	B	*LS*
			6142.32	6144.02	576997	– 593273	4–6	4.45–03	3.78–03	3.05–01	–1.821	B	*LS*
			6039.53	6041.20	576853	– 593406	2–4	6.50–03	7.11–03	2.83–01	–1.847	B	*LS*
364.	$2p^2(^3P)3s-2p^2(^3P)3p$	^{4}P–^{4}P°	*3383.2*	*3384.2*	*577079*	– *606628*	12–12	9.40–01	1.61–01	2.16+01	0.287	B	1
			3390.58	3391.56	577209	– 606694	6–6	6.54–01	1.13–01	7.54+00	–0.170	B	*LS*
			3379.58	3380.55	576997	– 606578	4–4	1.26–01	2.15–02	9.58–01	–1.065	B	*LS*
			3368.65	3369.61	576853	– 606530	2–2	1.59–01	2.70–02	5.99–01	–1.268	B	*LS*
			3403.97	3404.95	577209	– 606578	6–4	4.15–01	4.81–02	3.23+00	–0.540	B	*LS*
			3385.07	3386.04	576997	– 606530	4–2	7.82–01	6.72–02	2.99+00	–0.571	B	*LS*
			3366.38	3367.34	576997	– 606694	4–6	2.86–01	7.29–02	3.23+00	–0.535	B	*LS*
			3363.21	3364.17	576853	– 606578	2–4	3.99–01	1.35–01	2.99+00	–0.568	B	*LS*
365.		^{4}P–^{4}S°	*2511.7*	*2512.4*	*577079*	– 616881	12–4	3.14+00	9.91–02	9.83+00	0.075	B	1
			2519.91	2520.67	577209	– 616881	6–4	1.56+00	9.88–02	4.92+00	–0.227	B	*LS*
			2506.52	2507.27	576997	– 616881	4–4	1.05+00	9.93–02	3.28+00	–0.401	B	*LS*
			2497.50	2498.25	576853	– 616881	2–4	5.33–01	9.97–02	1.64+00	–0.700	B	*LS*

O IV: Allowed Transitions — Continued

No.	Transition Array	Mult.	λ_{air} (Å)	λ_{vac} (Å) or σ (cm^{-1})*	E_i (cm^{-1})	E_k (cm^{-1})	g_i-g_k	A_{ki} (10^8 s^{-1})	f_{ik}	S (at. u.)	log gf	Acc.	Source
366.	$2p^2(^3P)3s-2s2p(^3P^o)5d$	$^4P-^4P^o$		*1739.2*	*577079*	*634578*	12–12	3.82–02	1.73–03	1.19–01	–1.682	B	1
				1744.77	577209	634523	6–6	2.65–02	1.21–03	4.17–02	–2.139	B	*LS*
				1735.36	576997	634622	4–4	5.13–03	2.32–04	5.29–03	–3.033	C+	*LS*
				1730.10	576853	634653	2–2	6.47–03	2.90–04	3.31–03	–3.236	C+	*LS*
				1741.77	577209	634622	6–4	1.71–02	5.19–04	1.79–02	–2.506	B–	*LS*
				1734.42	576997	634653	4–2	3.21–02	7.24–04	1.65–02	–2.538	B–	*LS*
				1738.34	576997	634523	4–6	1.15–02	7.80–04	1.79–02	–2.506	B–	*LS*
				1731.03	576853	634622	2–4	1.62–02	1.45–03	1.65–02	–2.537	B–	*LS*
367.	$2s2p(^1P^o)3d-2s2p(^3P^o)4p$	$^2P^o-^2D$	*34039*	*2937.0* cm^{-1}	*581736*	*584673*	6–10	3.33–04	9.63–03	6.47+00	–1.238	B	1
			33125.5	3018.00 cm^{-1}	581743	584761	4–6	3.61–04	8.90–03	3.88+00	–1.448	B	*LS*
			35451.3	2820.00 cm^{-1}	581721	584541	2–4	2.45–04	9.25–03	2.16+00	–1.733	B	*LS*
			35730.1	2798.00 cm^{-1}	581743	584541	4–4	4.79–05	9.17–04	4.32–01	–2.435	B	*LS*
368.		$^2P^o-^2S$	*11997*	*8333.0* cm^{-1}	*581736*	590069	6–2	3.20–02	2.30–02	5.46+00	–0.859	B	1
			12007.3	8326.00 cm^{-1}	581743	590069	4–2	2.13–02	2.30–02	3.64+00	–1.036	B	*LS*
			11975.6	8348.00 cm^{-1}	581721	590069	2–2	1.07–02	2.31–02	1.82+00	–1.336	B	*LS*
369.	$2s2p(^1P^o)3d-2p^2(^3P)3s$	$^2P^o-^2P$	*7849.6*	*7851.8*	*581736*	*594472*	6–6	8.81–04	8.14–04	1.26–01	–2.311	B	1
			7813.40	7815.55	581743	594538	4–4	7.45–04	6.81–04	7.01–02	–2.565	B	*LS*
			7922.38	7924.56	581721	594340	2–2	5.71–04	5.38–04	2.80–02	–2.968	B–	*LS*
			7936.22	7938.40	581743	594340	4–2	2.84–04	1.34–04	1.40–02	–3.270	B–	*LS*
			7799.99	7802.14	581721	594538	2–4	1.50–04	2.73–04	1.40–02	–3.263	B–	*LS*
370.	$2s2p(^1P^o)3d-2p^2(^1D)3d$	$^2P^o-^2P$		*1277.8*	*581736*	659998	6–6	7.16–01	1.75–02	4.43–01	–0.978	B	1
				1277.87	581743	659998	4–4	5.97–01	1.46–02	2.46–01	–1.233	B	*LS*
				1277.51	581721	659998	2–2	4.78–01	1.17–02	9.83–02	–1.631	B	*LS*
				1277.87	581743	659998	4–2	2.39–01	2.92–03	4.92–02	–1.932	B	*LS*
				1277.51	581721	659998	2–4	1.19–01	5.85–03	4.92–02	–1.932	B	*LS*
371.	$2s2p(^3P^o)4p-2s^27f$	$^2D-^2F^o$	*31468*	*3177.0* cm^{-1}	*584673*	587850	10–14	1.14–03	2.38–02	2.46+01	–0.624	B	1
			32364.1	3089.00 cm^{-1}	584761	587850	6–8	1.05–03	2.20–02	1.41+01	–0.879	C+	*LS*
			30212.4	3309.00 cm^{-1}	584541	587850	4–6	1.21–03	2.47–02	9.85+00	–1.004	C+	*LS*
			32364.1	3089.00 cm^{-1}	584761	587850	6–6	7.01–05	1.10–03	7.03–01	–2.180	C+	*LS*
372.	$2s2p(^3P^o)4p-2s2p(^3P^o)4d$	$^2D-^2D^o$	*11104*	*9003.0* cm^{-1}	*584673*	*593676*	10–10	2.73–02	5.05–02	1.85+01	–0.297	B	1
			11173.9	8947.00 cm^{-1}	584761	593708	6–6	2.50–02	4.68–02	1.03+01	–0.551	B	*LS*
			11002.9	9086.00 cm^{-1}	584541	593627	4–4	2.53–02	4.59–02	6.65+00	–0.736	B	*LS*
			11276.0	8866.00 cm^{-1}	584761	593627	6–4	2.61–03	3.32–03	7.38–01	–1.701	B	*LS*
			10905.7	9167.00 cm^{-1}	584541	593708	4–6	1.92–03	5.14–03	7.38–01	–1.687	B	*LS*
373.		$^2D-^2F^o$	*8526.4*	*8528.8*	*584673*	*596398*	10–14	1.89–01	2.89–01	8.10+01	0.460	B	1
			8534.45	8536.79	584761	596475	6–8	1.89–01	2.75–01	4.63+01	0.217	B	*LS*
			8505.41	8507.74	584541	596295	4–6	1.78–01	2.89–01	3.24+01	0.063	B	*LS*
			8667.64	8670.02	584761	596295	6–6	1.20–02	1.35–02	2.31+00	–1.091	B	*LS*
374.		$^2D-^2P^o$	*7633.2*	*7635.3*	*584673*	*597770*	10–6	3.54–02	1.85–02	4.66+00	–0.732	B	1
			7714.52	7716.64	584761	597720	6–4	3.08–02	1.83–02	2.80+00	–0.958	B	*LS*
			7500.94	7503.00	584541	597869	4–2	3.73–02	1.57–02	1.55+00	–1.201	B	*LS*
			7585.74	7587.83	584541	597720	4–4	3.61–03	3.11–03	3.11–01	–1.905	B	*LS*

O IV: Allowed Transitions — Continued

No.	Transition Array	Mult.	λ_{air} (Å)	λ_{vac} (Å) or σ (cm^{-1})*	E_i (cm^{-1})	E_k (cm^{-1})	g_i-g_k	A_{ki} (10^8 s^{-1})	f_{ik}	S (at. u.)	log gf	Acc.	Source
375.	$2s2p(^3P^o)4p-2p^2(^3P)3p$	$^2D-^2D^o$	*3249.0*	*3249.9*	*584673*	– *615443*	10–10	9.50–03	1.50–03	1.61–01	–1.823	B	1
			3259.58	3260.52	584761	– 615431	6–6	8.78–03	1.40–03	9.00–02	–2.076	B	*LS*
			3233.32	3234.26	584541	– 615460	4–4	8.67–03	1.36–03	5.79–02	–2.265	B	*LS*
			3256.50	3257.44	584761	– 615460	6–4	9.43–04	1.00–04	6.43–03	–3.222	C+	*LS*
			3236.36	3237.29	584541	– 615431	4–6	6.41–04	1.51–04	6.43–03	–3.219	C+	*LS*
376.	$2s2p(^3P^o)4p-2p^2(^1D)3p$	$^2D-^2D^o$	*2116.2*	*2116.9*	*584673*	– 631913	10–10	9.68–01	6.50–02	4.53+00	–0.187	B	1
			2120.13	2120.80	584761	– 631913	6–6	8.99–01	6.06–02	2.54+00	–0.440	B	*LS*
			2110.28	2110.95	584541	– 631913	4–4	8.79–01	5.87–02	1.63+00	–0.629	B	*LS*
			2120.13	2120.80	584761	– 631913	6–4	9.63–02	4.33–03	1.81–01	–1.586	B	*LS*
			2110.28	2110.95	584541	– 631913	4–6	6.51–02	6.52–03	1.81–01	–1.584	B	*LS*
377.	$2s2p(^3P^o)4p-2s2p(^3P^o)4d$	$^2S-^2P^o$	*12982*	*7701.0* cm^{-1}	590069	– *597770*	2–6	5.21–02	3.95–01	3.37+01	–0.103	B	1
			13066.6	7651.00 cm^{-1}	590069	– 597720	2–4	5.11–02	2.61–01	2.25+01	–0.282	B	*LS*
			12817.0	7800.00 cm^{-1}	590069	– 597869	2–2	5.41–02	1.33–01	1.12+01	–0.574	B	*LS*
378.	$2s2p(^3P^o)4d-2s2p(^3P^o)4f$	$^4D^o-^4F$	*49455*	*2021.5* cm^{-1}	*591977*	– *593999*	20–28	8.90–04	4.57–02	1.49+02	–0.039	B	1
				1998.90 cm^{-1}	592046	– 594044.9	8–10	8.60–04	4.03–02	5.31+01	–0.491	C+	*LS*
			49228.3	2030.80 cm^{-1}	591962	– 593992.8	6–8	7.74–04	3.75–02	3.64+01	–0.648	C+	*LS*
			48462.2	2062.90 cm^{-1}	591899	– 593961.9	4–6	7.09–04	3.75–02	2.39+01	–0.824	C+	*LS*
			48774.3	2049.70 cm^{-1}	591899	– 593948.7	2–4	6.49–04	4.63–02	1.49+01	–1.033	C+	*LS*
				1946.80 cm^{-1}	592046	– 593992.8	8–8	1.13–04	4.48–03	6.06+00	–1.446	C+	*LS*
				1999.90 cm^{-1}	591962	– 593961.9	6–6	2.10–04	7.86–03	7.76+00	–1.327	C+	*LS*
			48774.3	2049.70 cm^{-1}	591899	– 593948.7	4–4	2.60–04	9.26–03	5.95+00	–1.431	C+	*LS*
				1915.90 cm^{-1}	592046	– 593961.9	8–6	7.32–06	2.24–04	3.08–01	–2.746	C+	*LS*
				1986.70 cm^{-1}	591962	– 593948.7	6–4	1.69–05	4.28–04	4.25–01	–2.591	C+	*LS*
379.		$^4D^o-^4D$	*22481*	*4447.0* cm^{-1}	*591977*	– *596424*	20–20	2.04–03	1.54–02	2.29+01	–0.510	B	1
			23330.9	4285.00 cm^{-1}	592046	– 596331	8–8	1.56–03	1.28–02	7.84+00	–0.991	C+	*LS*
			22305.4	4482.00 cm^{-1}	591962	– 596444	6–6	1.20–03	8.92–03	3.93+00	–1.271	C+	*LS*
			21670.1	4613.40 cm^{-1}	591899	– 596512.4	4–4	9.10–04	6.41–03	1.83+00	–1.591	C+	*LS*
			21466.3	4657.20 cm^{-1}	591899	– 596556.2	2–2	1.17–03	8.09–03	1.14+00	–1.791	C+	*LS*
			22731.4	4398.00 cm^{-1}	592046	– 596444	8–6	3.75–04	2.18–03	1.30+00	–1.759	C+	*LS*
			21970.1	4550.40 cm^{-1}	591962	– 596512.4	6–4	7.64–04	3.69–03	1.60+00	–1.655	C+	*LS*
			21466.3	4657.20 cm^{-1}	591899	– 596556.2	4–2	1.17–03	4.04–03	1.14+00	–1.791	C+	*LS*
			22882.3	4369.00 cm^{-1}	591962	– 596331	6–8	2.75–04	2.88–03	1.30+00	–1.762	C+	*LS*
			21996.2	4545.00 cm^{-1}	591899	– 596444	4–6	5.08–04	5.52–03	1.60+00	–1.656	C+	*LS*
			21670.1	4613.40 cm^{-1}	591899	– 596512.4	2–4	5.69–04	8.01–03	1.14+00	–1.795	C+	*LS*
380.		$^4P^o-^4D$	*32533*	*3073.0* cm^{-1}	*593351*	– *596424*	12–20	3.24–03	8.57–02	1.10+02	0.012	B	1
			32692.2	3058.00 cm^{-1}	593273	– 596331	6–8	3.20–03	6.83–02	4.41+01	–0.388	C+	*LS*
			32907.4	3038.00 cm^{-1}	593406	– 596444	4–6	2.19–03	5.34–02	2.31+01	–0.670	C+	*LS*
			32892.3	3039.40 cm^{-1}	593473	– 596512.4	2–4	1.31–03	4.24–02	9.18+00	–1.072	C+	*LS*
			31527.2	3171.00 cm^{-1}	593273	– 596444	6–6	1.07–03	1.59–02	9.92+00	–1.020	C+	*LS*
			32182.8	3106.40 cm^{-1}	593406	– 596512.4	4–4	1.79–03	2.77–02	1.18+01	–0.955	C+	*LS*
			32425.0	3083.20 cm^{-1}	593473	– 596556.2	2–2	2.73–03	4.30–02	9.18+00	–1.065	C+	*LS*
			30861.5	3239.40 cm^{-1}	593273	– 596512.4	6–4	1.90–04	1.81–03	1.10+00	–1.965	C+	*LS*
			31735.4	3150.20 cm^{-1}	593406	– 596556.2	4–2	5.82–04	4.40–03	1.84+00	–1.755	C+	*LS*
381.	$2s2p(^3P^o)4d-2p^2(^3P)3d$	$^4P^o-^4P$	*2281.3*	*2282.0*	*593351*	– *637173*	12–12	7.61–03	5.94–04	5.35–02	–2.147	B	1
			2280.32	2281.02	593273	– 637113	6–6	5.33–03	4.16–04	1.87–02	–2.603	B–	*LS*
			2282.09	2282.79	593406	– 637212	4–4	1.01–03	7.91–05	2.38–03	–3.500	C	*LS*
			2282.35	2283.05	593473	– 637274	2–2	1.27–03	9.89–05	1.49–03	–3.704	C	*LS*
			2275.18	2275.88	593273	– 637212	6–4	3.45–03	1.79–04	8.03–03	–2.970	C+	*LS*
			2278.86	2279.57	593406	– 637274	4–2	6.36–03	2.48–04	7.43–03	–3.004	C+	*LS*
			2287.26	2287.96	593406	– 637113	4–6	2.26–03	2.66–04	8.03–03	–2.972	C+	*LS*
			2285.58	2286.29	593473	– 637212	2–4	3.15–03	4.94–04	7.43–03	–3.005	C+	*LS*

O IV: Allowed Transitions — Continued

No.	Transition Array	Mult.	λ_{air} (Å)	λ_{vac} (Å) or σ (cm^{-1})*	E_i (cm^{-1})	E_k (cm^{-1})	g_i–g_k	A_{ki} (10^8 s^{-1})	f_{ik}	S (at. u.)	log gf	Acc.	Source
382.	$2s2p(^3P^o)4d$– $2s2p(^3P^o)4f$	$^2D^o$–^2F		*374.00* cm^{-1}	*593676*	– *594050*	10–14	5.31–06	7.97–03	7.01+01	–1.099	B	1
				366.000 cm^{-1}	593708	– 594074	6–8	4.98–06	7.42–03	4.01+01	–1.351	C+	LS
				392.000 cm^{-1}	593627	– 594019	4–6	5.71–06	8.35–03	2.80+01	–1.476	C+	LS
				311.000 cm^{-1}	593708	– 594019	6–6	2.04–07	3.15–04	2.00+00	–2.723	C+	LS
383.	$2s2p(^3P^o)4d$– $2p^2(^3P)3s$	$^2D^o$–^2P		*796.00* cm^{-1}	*593676*	– *594472*	10–6	4.06–05	5.76–03	2.38+01	–1.240	B	1
				830.000 cm^{-1}	593708	– 594538	6–4	4.14–05	6.00–03	1.43+01	–1.444	B	LS
				713.000 cm^{-1}	593627	– 594340	4–2	2.92–05	4.30–03	7.93+00	–1.765	B	LS
				911.000 cm^{-1}	593627	– 594538	4–4	6.08–06	1.10–03	1.59+00	–2.357	B	LS
384.	$2s2p(^3P^o)4d$– $2s2p(^3P^o)4f$	$^2D^o$–^2D	*32904*	*3038.3* cm^{-1}	*593676*	– *596714*	10–10	6.27–04	1.02–02	1.10+01	–0.992	B	1
			33885.6	2950.30 cm^{-1}	593708	– 596658.3	6–6	5.36–04	9.23–03	6.18+00	–1.257	C+	LS
			31525.2	3171.20 cm^{-1}	593627	– 596798.2	4–4	6.42–04	9.56–03	3.97+00	–1.417	C+	LS
			32351.5	3090.20 cm^{-1}	593708	– 596798.2	6–4	6.60–05	6.90–04	4.41–01	–2.383	C+	LS
			32980.2	3031.30 cm^{-1}	593627	– 596658.3	4–6	4.15–05	1.02–03	4.41–01	–2.391	C+	LS
385.	$2s2p(^3P^o)4d$– $2p^2(^1D)3s$	$^2D^o$–^2D	*15562*	*6424.0* cm^{-1}	*593676*	– *600100*	10–10	1.10–03	3.98–03	2.04+00	–1.400	B	1
			15625.6	6398.00 cm^{-1}	593708	– 600106	6–6	1.01–03	3.70–03	1.14+00	–1.654	B	LS
			15463.7	6465.00 cm^{-1}	593627	– 600092	4–4	1.01–03	3.60–03	7.34–01	–1.841	B	LS
			15659.9	6384.00 cm^{-1}	593708	– 600092	6–4	1.08–04	2.64–04	8.16–02	–2.801	B	LS
			15430.3	6479.00 cm^{-1}	593627	– 600106	4–6	7.50–05	4.01–04	8.16–02	–2.794	B	LS
386.	$2s2p(^3P^o)4d$– $2p^2(^3P)3d$	$^2D^o$–^2F	*2745.0*	*2745.8*	*593676*	– 630095	10–14	3.12+00	4.94–01	4.46+01	0.693	B	1
			2747.42	2748.23	593708	– 630095	6–8	3.11+00	4.70–01	2.55+01	0.450	B	LS
			2741.32	2742.13	593627	– 630095	4–6	2.93+00	4.94–01	1.78+01	0.296	B	LS
			2747.42	2748.23	593708	– 630095	6–6	2.08–01	2.35–02	1.27+00	–0.851	B	LS
387.		$^2D^o$–^2P	*2576.1*	*2576.9*	*593676*	– *632483*	10–6	7.76–02	4.63–03	3.93–01	–1.334	B	1
			2582.01	2582.78	593708	– 632426	6–4	6.94–02	4.62–03	2.36–01	–1.557	B	LS
			2565.31	2566.08	593627	– 632597	4–2	7.86–02	3.88–03	1.31–01	–1.810	B	LS
			2576.61	2577.39	593627	– 632426	4–4	7.75–03	7.72–04	2.62–02	–2.510	B–	LS
388.	$2s2p(^3P^o)4d$– $2p^2(^1D)3d$	$^2D^o$–^2F		*1741.2*	*593676*	– *651109*	10–14	1.33–01	8.46–03	4.85–01	–1.073	B	1
				1741.89	593708	– 651117	6–8	1.33–01	8.05–03	2.77–01	–1.316	B	LS
				1740.01	593627	– 651098	4–6	1.24–01	8.46–03	1.94–01	–1.470	B	LS
				1742.46	593708	– 651098	6–6	8.84–03	4.02–04	1.38–02	–2.617	B–	LS
389.	$2s2p(^3P^o)4f$– $2s2p(^3P^o)4d$	^2F–$^2F^o$	*42578*	*2348.0* cm^{-1}	*594050*	– *596398*	14–14	1.27–04	3.46–03	6.78+00	–1.315	B	1
			41638.0	2401.00 cm^{-1}	594074	– 596475	8–8	1.31–04	3.41–03	3.74+00	–1.564	C+	LS
			43924.8	2276.00 cm^{-1}	594019	– 596295	6–6	1.10–04	3.19–03	2.77+00	–1.718	C+	LS
			45012.5	2221.00 cm^{-1}	594074	– 596295	8–6	5.12–06	1.17–04	1.38–01	–3.030	C+	LS
			40705.5	2456.00 cm^{-1}	594019	– 596475	6–8	5.20–06	1.72–04	1.38–01	–2.986	C+	LS
390.	$2s2p(^3P^o)4f$– $2p^2(^3P)3p$	^2F–$^2D^o$	*4673.1*	*4674.4*	*594050*	– *615443*	14–10	9.97–03	2.33–03	5.02–01	–1.486	B	1
			4681.00	4682.31	594074	– 615431	8–6	9.44–03	2.33–03	2.87–01	–1.730	C+	LS
			4662.66	4663.96	594019	– 615460	6–4	1.00–02	2.18–03	2.01–01	–1.883	C+	LS
			4668.97	4670.28	594019	– 615431	6–6	4.76–04	1.56–04	1.43–02	–3.030	C	LS
391.	$2s2p(^3P^o)4f$– $2p^2(^1D)3p$	^2F–$^2D^o$	*2640.3*	*2641.1*	*594050*	– 631913	14–10	5.02–02	3.75–03	4.56–01	–1.280	B	1
			2641.99	2642.78	594074	– 631913	8–6	4.77–02	3.75–03	2.61–01	–1.523	C+	LS
			2638.15	2638.94	594019	– 631913	6–4	5.03–02	3.50–03	1.82–01	–1.678	C+	LS
			2638.15	2638.94	594019	– 631913	6–6	2.40–03	2.50–04	1.30–02	–2.824	C	LS

O IV: Allowed Transitions — Continued

No.	Transition Array	Mult.	λ_{air} (Å)	λ_{vac} (Å) or σ (cm^{-1})*	E_i (cm^{-1})	E_k (cm^{-1})	g_i-g_k	A_{ki} (10^8 s^{-1})	f_{ik}	S (at. u.)	log gf	Acc.	Source
392.	$2p^2(^3P)3s-2s2p(^3P^o)4d$	$^2P-^2P^o$	*30313*	*3298.0 cm^{-1}*	*594472*	*597770*	6–6	7.02–04	9.68–03	5.79+00	−1.236	B	1
			31418.2	3182.00 cm^{-1}	594538	597720	4–4	5.26–04	7.78–03	3.22+00	−1.507	B	*LS*
			28328.9	3529.00 cm^{-1}	594340	597869	2–2	5.74–04	6.90–03	1.29+00	−1.860	B	*LS*
			30012.8	3331.00 cm^{-1}	594538	597869	4–2	2.41–04	1.63–03	6.44–01	−2.186	B	*LS*
			29577.7	3380.00 cm^{-1}	594340	597720	2–4	1.26–04	3.31–03	6.44–01	−2.180	B	*LS*
393.	$2p^2(^3P)3s-2p^2(^3P)3p$	$^2P-^2D^o$	*4767.2*	*4768.5*	*594472*	*615443*	6–10	1.92–01	1.09–01	1.03+01	−0.185	B	1
			4784.95	4786.29	594538	615431	4–6	1.90–01	9.76–02	6.15+00	−0.408	B	*LS*
			4733.52	4734.85	594340	615460	2–4	1.63–01	1.10–01	3.42+00	−0.659	B	*LS*
			4778.32	4779.66	594538	615460	4–4	3.17–02	1.09–02	6.84–01	−1.362	B	*LS*
394.	$2p^2(^3P)3s-2p^2(^1D)3p$	$^2P-^2D^o$	*2670.1*	*2670.9*	*594472*	*631913*	6–10	1.40–02	2.49–03	1.31–01	−1.826	B	1
			2674.79	2675.59	594538	631913	4–6	1.39–02	2.23–03	7.87–02	−2.049	B	*LS*
			2660.69	2661.49	594340	631913	2–4	1.18–02	2.49–03	4.37–02	−2.302	B	*LS*
			2674.79	2675.59	594538	631913	4–4	2.31–03	2.48–04	8.74–03	−3.003	C+	*LS*
395.	$2p^2(^3P)3s-2s2p(^1P^o)4d$	$^2P-^2D^o$		*1349.8*	*594472*	*668555*	6–10	1.22+00	5.58–02	1.49+00	−0.476	B	1
				1351.08	594538	668553	4–6	1.22+00	5.01–02	8.92–01	−0.698	B	*LS*
				1347.40	594340	668557	2–4	1.03+00	5.59–02	4.96–01	−0.952	B	*LS*
				1351.00	594538	668557	4–4	2.04–01	5.57–03	9.91–02	−1.652	B	*LS*
396.	$2s2p(^3P^o)4f-2p^2(^3P)3p$	$^4D-^4P^o$	*9798.4*	*9801.0*	*596425*	*606628*	20–12	4.31–03	3.72–03	2.40+00	−1.129	B	1
			9647.07	9649.72	596331	606694	8–6	3.61–03	3.78–03	9.60–01	−1.520	C+	*LS*
			9865.07	9867.77	596444	606578	6–4	2.66–03	2.59–03	5.04–01	−1.809	C+	*LS*
			9979.69	9982.43	596512.4	606530	4–2	2.04–03	1.52–03	2.00–01	−2.216	C+	*LS*
			9753.42	9756.10	596444	606694	6–6	7.86–04	1.12–03	2.16–01	−2.172	C+	*LS*
			9932.10	9934.83	596512.4	606578	4–4	1.32–03	1.96–03	2.56–01	−2.106	C+	*LS*
			10033.6	9963.80 cm^{-1}	596566.2	606530	2–2	2.00–03	3.03–03	2.00–01	−2.218	C+	*LS*
			9818.95	9821.64	596512.4	606694	4–6	8.56–05	1.86–04	2.40–02	−3.130	C	*LS*
			9985.48	9988.21	596566.2	606578	2–4	2.03–04	6.08–04	4.00–02	−2.915	C+	*LS*
397.	$2s2p(^3P^o)4f-2s2p(^3P^o)5d$	$^4D-^4P^o$	*2620.2*	*2621.0*	*596425*	*634578*	20–12	2.05–01	1.26–02	2.18+00	−0.597	B	1
			2617.57	2618.35	596331	634523	8–6	1.64–01	1.27–02	8.72–01	−0.995	C+	*LS*
			2618.53	2619.31	596444	634622	6–4	1.29–01	8.85–03	4.58–01	−1.275	C+	*LS*
			2621.10	2621.88	596512.4	634653	4–2	1.02–01	5.27–03	1.82–01	−1.676	C+	*LS*
			2625.34	2626.12	596444	634523	6–6	3.66–02	3.79–03	1.96–01	−1.644	C+	*LS*
			2623.23	2624.01	596512.4	634622	4–4	6.53–02	6.73–03	2.33–01	−1.570	C+	*LS*
			2624.80	2625.58	596566.2	634653	2–2	1.02–01	1.05–02	1.82–01	−1.677	C+	*LS*
			2630.06	2630.85	596512.4	634523	4–6	4.05–03	6.30–04	2.18–02	−2.599	C	*LS*
			2626.94	2627.72	596566.2	634622	2–4	1.02–02	2.10–03	3.63–02	−2.376	C+	*LS*
398.	$2s2p(^3P^o)4d-2s2p(^3P^o)4f$	$^2F^o-^2G$		*353.70 cm^{-1}*	*596398*	*596752*	14–18	5.08–06	7.82–03	1.02+02	−0.961	B	1
				372.600 cm^{-1}	596475	596847.6	8–10	5.93–06	8.01–03	5.66+01	−1.194	C+	*LS*
				336.900 cm^{-1}	596295	596631.9	6–8	4.23–06	7.45–03	4.36+01	−1.350	C+	*LS*
				156.900 cm^{-1}	596475	596631.9	8–8	1.58–08	9.63–05	1.62+00	−3.113	C+	*LS*
399.		$^2F^o-^2D$		*316.30 cm^{-1}*	*596398*	*596714*	14–10	3.95–06	4.22–03	6.15+01	−1.228	B	1
				183.300 cm^{-1}	596475	596658.3	8–6	7.32–07	2.45–03	3.52+01	−1.708	C+	*LS*
				503.200 cm^{-1}	596295	596798.2	6–4	1.59–05	6.27–03	2.46+01	−1.424	C+	*LS*
				363.300 cm^{-1}	596295	596658.3	6–6	2.85–07	3.24–04	1.76+00	−2.712	C+	*LS*
400.	$2s2p(^3P^o)4d-2p^2(^1D)3s$	$^2F^o-^2D$	*27005*	*3702.0 cm^{-1}*	*596398*	*600100*	14–10	7.92–05	6.19–04	7.70–01	−2.062	B	1
			27533.1	3631.00 cm^{-1}	596475	600106	8–6	7.12–05	6.07–04	4.40–01	−2.314	B	*LS*
			26329.4	3797.00 cm^{-1}	596295	600092	6–4	8.55–05	5.92–04	3.08–01	−2.449	B	*LS*
			26232.7	3811.00 cm^{-1}	596295	600106	6–6	4.12–06	4.25–05	2.20–02	−3.594	B−	*LS*

O IV: Allowed Transitions — Continued

No.	Transition Array	Mult.	λ_{air} (Å)	λ_{vac} (Å) or σ (cm^{-1})*	E_i (cm^{-1})	E_k (cm^{-1})	g_i-g_k	A_{ki} (10^8 s^{-1})	f_{ik}	S (at. u.)	log gf	Acc.	Source
401.	$2s2p(^3P^\circ)4d-2p^2(^3P)3d$	$^2F^\circ-^2F$	*2966.8*	*2967.6*	*596398*	630095	14–14	2.48–01	3.27–02	4.48+00	–0.339	B	1
			2973.55	2974.42	596475	630095	8–8	2.38–01	3.15–02	2.47+00	–0.599	B	LS
			2957.72	2958.58	596295	630095	6–6	2.38–01	3.13–02	1.83+00	–0.727	B	LS
			2973.55	2974.42	596475	630095	8–6	1.17–02	1.17–03	9.14–02	–2.030	B	LS
			2957.72	2958.58	596295	630095	6–8	8.94–03	1.56–03	9.14–02	–2.028	B	LS
402.	$2s2p(^3P^\circ)4d-2p^2(^1D)3d$	$^2F^\circ-^2F$		*1827.8*	*596398*	*651109*	14–14	6.33–03	3.17–04	2.67–02	–2.353	B–	1
				1830.09	596475	651117	8–8	6.08–03	3.05–04	1.47–02	–2.612	B–	LS
				1824.72	596295	651098	6–6	6.06–03	3.02–04	1.09–02	–2.741	B–	LS
				1830.73	596475	651098	8–6	3.00–04	1.13–05	5.45–04	–4.044	D	LS
				1824.09	596295	651117	6–8	2.27–04	1.51–05	5.45–04	–4.042	D	LS
403.	$2s2p(^3P^\circ)4f-2s2p(^3P^\circ)4d$	$^2D-^2P^\circ$		*1055.7* cm^{-1}	*596714*	*597770*	10–6	1.79–04	1.44–02	4.50+01	–0.841	B	1
				1061.70 cm^{-1}	596658.3	597720	6–4	1.64–04	1.45–02	2.70+01	–1.060	C+	LS
				1070.80 cm^{-1}	596798.2	597869	4–2	1.87–04	1.22–02	1.50+01	–1.312	C+	LS
				921.800 cm^{-1}	596798.2	597720	4–4	1.19–05	2.10–03	3.00+00	–2.076	C+	LS
404.	$2s2p(^3P^\circ)4f-2p^2(^1D)3p$	$^2D-^2D^\circ$	*2840.2*	*2841.0*	*596714*	631913	10–10	2.36–02	2.86–03	2.67–01	–1.544	B	1
			2835.67	2836.50	596658.3	631913	6–6	2.22–02	2.67–03	1.50–01	–1.795	C+	LS
			2846.97	2847.80	596798.2	631913	4–4	2.11–02	2.57–03	9.63–02	–1.988	C+	LS
			2835.67	2836.50	596658.3	631913	6–4	2.38–03	1.91–04	1.07–02	–2.941	C	LS
			2846.97	2847.80	596798.2	631913	4–6	1.57–03	2.85–04	1.07–02	–2.943	C	LS
405.	$2p^2(^3P)3p-2p^2(^3P)3d$	$^2S^\circ-^2P$	*2837.7*	*2838.6*	597254	*632483*	2–6	7.85–01	2.84–01	5.31+00	–0.245	B	1
			2842.34	2843.17	597254	632426	2–4	7.81–01	1.89–01	3.54+00	–0.422	B	LS
			2828.58	2829.41	597254	632597	2–2	7.93–01	9.51–02	1.77+00	–0.721	B	LS
406.	$2p^2(^3P)3p-2p^2(^1D)3d$	$^2S^\circ-^2P$		1593.78	597254	659998	2–6	1.20–01	1.37–02	1.44–01	–1.562	B	1
407.	$2s2p(^3P^\circ)4d-2p^2(^1D)3s$	$^2P^\circ-^2D$	*42907*	*2330.0* cm^{-1}	*597770*	*600100*	6–10	2.17–06	9.98–05	8.46–02	–3.223	B	1
			41899.7	2386.00 cm^{-1}	597720	600106	4–6	2.33–06	9.20–05	5.08–02	–3.434	B	LS
			44972.0	2223.00 cm^{-1}	597869	600092	2–4	1.57–06	9.52–05	2.82–02	–3.720	B–	LS
			42147.0	2372.00 cm^{-1}	597720	600092	4–4	3.82–07	1.02–05	5.64–03	–4.391	C+	LS
408.	$2s2p(^3P^\circ)4d-2p^2(^3P)3d$	$^2P^\circ-^2P$	*2879.9*	*2880.8*	*597770*	*632483*	6–6	3.86–02	4.80–03	2.73–01	–1.540	B	1
			2880.50	2881.35	597720	632426	4–4	3.22–02	4.00–03	1.52–01	–1.796	B	LS
			2878.68	2879.52	597869	632597	2–2	2.58–02	3.20–03	6.07–02	–2.193	B	LS
			2866.38	2867.22	597720	632597	4–2	1.31–02	8.05–04	3.04–02	–2.492	B	LS
			2892.92	2893.77	597869	632426	2–4	6.35–03	1.59–03	3.04–02	–2.496	B	LS
409.	$2s2p(^3P^\circ)4d-2p^2(^1D)3d$	$^2P^\circ-^2P$		*1607.0*	*597770*	659998	6–6	4.21–02	1.63–03	5.17–02	–2.010	B	1
				1605.70	597720	659998	4–4	3.51–02	1.36–03	2.87–02	–2.265	B–	LS
				1609.55	597869	659998	2–2	2.79–02	1.08–03	1.15–02	–2.664	B–	LS
				1605.70	597720	659998	4–2	1.41–02	2.72–04	5.74–03	–2.964	C+	LS
				1609.55	597869	659998	2–4	6.98–03	5.42–04	5.74–03	–2.965	C+	LS
410.	$2p^2(^1D)3s-2p^2(^3P)3p$	$^2D-^2D^\circ$	*6515.8*	*6517.6*	*600100*	*615443*	10–10	3.00–02	1.91–02	4.10+00	–0.719	B	1
			6523.48	6525.29	600106	615431	6–6	2.79–02	1.78–02	2.29+00	–0.971	B	LS
			6505.23	6507.03	600092	615460	4–4	2.71–02	1.72–02	1.47+00	–1.162	B	LS
			6511.16	6512.96	600106	615460	6–4	3.01–03	1.27–03	1.64–01	–2.117	B	LS
			6517.53	6519.33	600092	615431	4–6	2.00–03	1.91–03	1.64–01	–2.117	B	LS

O IV: Allowed Transitions — Continued

No.	Transition Array	Mult.	λ_{air} (Å)	λ_{vac} (Å) or σ (cm^{-1})*	E_i (cm^{-1})	E_k (cm^{-1})	g_i-g_k	A_{ki} (10^8 s^{-1})	f_{ik}	S (at. u.)	log gf	Acc.	Source
411.	$2p^2(^1D)3s-2p^2(^1D)3p$	$^2D-^2D^o$	*3142.5*	*3143.4*	*600100*	631913	10–10	2.15–01	3.18–02	3.29+00	–0.497	B	1
			3143.05	3143.96	600106	631913	6–6	2.00–01	2.97–02	1.84+00	–0.749	B	*LS*
			3141.67	3142.58	600092	631913	4–4	1.94–01	2.86–02	1.18+00	–0.941	B	*LS*
			3143.05	3143.96	600106	631913	6–4	2.15–02	2.12–03	1.32–01	–1.895	B	*LS*
			3141.67	3142.58	600092	631913	4–6	1.43–02	3.18–03	1.32–01	–1.895	B	*LS*
412.	$2p^2(^1D)3s-2s2p(^1P^o)4d$	$^2D-^2D^o$		*1460.8*	*600100*	*668555*	10–10	9.84–02	3.15–03	1.51–01	–1.502	B	1
				1460.98	600106	668553	6–6	9.18–02	2.94–03	8.48–02	–1.754	B	*LS*
				1460.60	600092	668557	4–4	8.86–02	2.83–03	5.45–02	–1.946	B	*LS*
				1460.90	600106	668557	6–4	9.84–03	2.10–04	6.06–03	–2.900	C+	*LS*
				1460.69	600092	668553	4–6	6.56–03	3.15–04	6.06–03	–2.900	C+	*LS*
413.	$2p^2(^3P)3p-2p^2(^3P)3d$	$^4P^o-^4P$	*3272.9*	*3273.9*	*606628*	*637173*	12–12	5.95–01	9.56–02	1.24+01	0.059	B	1
			3286.47	3287.42	606694	637113	6–6	4.11–01	6.66–02	4.32+00	–0.398	B	*LS*
			3263.41	3264.35	606578	637212	4–4	8.00–02	1.28–02	5.49–01	–1.291	B	*LS*
			3251.73	3252.67	606530	637274	2–2	1.01–01	1.60–02	3.43–01	–1.494	B	*LS*
			3275.81	3276.75	606694	637212	6–4	2.67–01	2.86–02	1.85+00	–0.765	B	*LS*
			3256.81	3257.75	606578	637274	4–2	5.03–01	4.00–02	1.72+00	–0.796	B	*LS*
			3273.99	3274.93	606578	637113	4–6	1.78–01	4.30–02	1.85+00	–0.765	B	*LS*
			3258.30	3259.24	606530	637212	2–4	2.51–01	8.00–02	1.72+00	–0.796	B	*LS*
414.		$^2D^o-^2F$	*6823.1*	*6825.0*	*615443*	630095	10–14	7.33–02	7.16–02	1.61+01	–0.145	B	1
			6817.54	6819.42	615431	630095	6–8	7.34–02	6.82–02	9.19+00	–0.388	B	*LS*
			6831.05	6832.93	615460	630095	4–6	6.81–02	7.15–02	6.43+00	–0.544	B	*LS*
			6817.54	6819.42	615431	630095	6–6	4.90–03	3.41–03	4.59–01	–1.689	B	*LS*
415.		$^2D^o-^2P$	*5866.9*	*5868.5*	*615443*	*632483*	10–6	1.30–03	4.01–04	7.75–02	–2.397	B	1
			5882.45	5884.08	615431	632426	6–4	1.16–03	4.00–04	4.65–02	–2.620	B	*LS*
			5833.71	5835.33	615460	632597	4–2	1.32–03	3.36–04	2.58–02	–2.871	B–	*LS*
			5892.51	5894.14	615460	632426	4–4	1.28–04	6.66–05	5.17–03	–3.575	C+	*LS*
416.	$2p^2(^3P)3p-2p^2(^1D)3d$	$^2D^o-^2F$	*2803.0*	*2803.8*	*615443*	*651109*	10–14	2.27–02	3.75–03	3.46–01	–1.426	B	1
			2801.39	2802.22	615431	651117	6–8	2.28–02	3.57–03	1.98–01	–1.669	B	*LS*
			2805.17	2805.99	615460	651098	4–6	2.12–02	3.75–03	1.38–01	–1.824	B	*LS*
			2802.89	2803.71	615431	651098	6–6	1.52–03	1.79–04	9.88–03	–2.970	C+	*LS*
417.		$^2D^o-^2P$	*2243.7*	*2244.4*	*615443*	659998	10–6	1.65–01	7.49–03	5.53–01	–1.126	B	1
			2243.12	2243.81	615431	659998	6–4	1.49–01	7.49–03	3.32–01	–1.347	B	*LS*
			2244.58	2245.27	615460	659998	4–2	1.65–01	6.24–03	1.84–01	–1.603	B	*LS*
			2244.58	2245.27	615460	659998	4–4	1.65–02	1.25–03	3.69–02	–2.302	B	*LS*
418.	$2p^2(^3P)3p-2p^2(^3P)3d$	$^4S^o-^4P$	*4926.7*	*4928.1*	616881	*637173*	4–12	1.93–01	2.11–01	1.37+01	–0.074	B	1
			4941.29	4942.67	616881	637113	4–6	1.92–01	1.05–01	6.84+00	–0.376	B	*LS*
			4917.22	4918.60	616881	637212	4–4	1.94–01	7.05–02	4.56+00	–0.550	B	*LS*
			4902.27	4903.64	616881	637274	4–2	1.96–01	3.53–02	2.28+00	–0.850	B	*LS*

*Wavelengths (Å) are always given unless cm^{-1} is indicated.

O IV

Forbidden Transitions

We have selected the results of multi-configuration Hartree-Fock calculations with relativistic Breit-Pauli terms by Froese Fischer[1] for the M1 and E2 fine structure transitions in the $^2P^o$ term of the ground state configuration.

Reference

[1]C. Froese Fischer, J. Phys. B **16**, 157 (1983).

O IV: Forbidden Transitions

No.	Transition Array	Mult.	λ_{air} (Å)	λ_{vac} (Å) or σ (cm^{-1})*	E_i (cm^{-1})		E_k (cm^{-1})	g_i-g_k	Type	A_{ki} (s^{-1})	S (at. u.)	Acc.	Source
1.	$2p-2p$	$^2P^o-^2P^o$											
				385.9 cm^{-1}	0.0	–	385.9	2–4	M1	5.19–04	1.34+00	A	1
				385.9 cm^{-1}	0.0	–	385.9	2–4	E2	1.70–10	7.10–01	B	1

*Wavelengths (Å) are always given unless cm^{-1} is indicated.

O V

Beryllium Isoelectronic Sequence

Ground State: $1s^22s^2\ ^1S_0$

Ionization Energy: 113.896 eV = 918657 cm^{-1}

Allowed Transitions

List of tabulated lines

Wavelength (Å)	No.	Wavelength (Å)	No.	Wavelength (Å)	No.	Wavelength (Å)	No.
in vacuum		126.139	28	133.335	24	138.110	22
		128.232	27	133.389	24	139.029	4
114.358	10	128.255	27	133.520	23	140.031	21
116.161	9	128.302	27	133.520	23	140.058	21
118.000	8	128.305	27	133.525	23	140.118	21
119.102	7	128.817	46	133.549	23	142.122	44
122.087	29	131.736	26	133.575	23	144.809	55
122.088	29	131.760	26	133.604	23	144.837	55
122.108	29	131.813	26	135.175	45	144.842	55
122.133	29	132.791	25	135.523	5	144.865	55
122.154	29	132.797	25	138.025	22	144.898	55
124.616	6	132.821	25	138.051	22	146.351	43
126.069	28	132.845	25	138.051	22	147.263	42
126.090	28	132.875	25	138.109	22	149.038	67
126.135	28	133.310	24	138.110	22	149.076	41

List of tabulated lines — Continued

Wavelength (Å)	No.	Wavelength (Å)	No.	Wavelength (Å)	No.	Wavelength (Å)	No.
151.447	20	193.006	61	562.471	99	in air	
151.477	20	194.592	35	563.045	99		
151.478	20	202.158	50	563.559	99	2695.30	123
151.547	20	202.189	50	566.237	79	2700.02	125
151.548	20	202.222	50	604.412	84	2706.82	123
151.549	20	202.281	50	627.223	90	2706.98	123
153.952	40	202.332	50	627.349	90	2711.53	123
156.119	19	202.391	50	627.367	90	2711.60	123
156.152	19	203.782	49	627.635	90	2711.76	123
156.227	19	203.822	49	627.653	90	2729.31	98
158.792	54	203.847	49	627.671	90	2731.45	98
158.818	54	203.890	49	629.732	2	2743.61	98
158.831	54	203.934	49	654.212	109	2752.23	98
158.858	54	203.959	49	654.719	109	2754.70	120
158.899	54	205.106	60	654.998	109	2755.13	98
158.926	54	207.796	59	655.042	109	2755.91	120
159.327	53	215.040	14	655.730	109	2756.14	120
159.343	53	215.103	14	656.010	109	2756.98	120
159.367	53	215.245	14	662.930	83	2757.13	120
159.380	53	216.019	58	668.258	12	2757.36	120
159.411	53	220.353	34	669.630	12	2769.69	98
159.435	53	222.237	71	677.963	108	2781.01	74
160.022	52	227.373	48	678.610	108	2786.99	74
160.062	52	227.470	48	678.807	108	2789.85	74
160.131	52	227.512	48	679.132	108	3058.68	100
162.492	66	227.551	48	679.696	108	3078.95	121
164.177	65	227.635	48	679.893	108	3144.66	81
164.574	18	227.690	48	715.955	89	3156.11	122
164.589	18	231.073	57	716.143	89	3168.10	122
164.626	18	231.822	70	716.539	89	3168.32	122
164.657	18	248.460	33	728.731	93	3172.31	122
164.657	18	265.551	69	751.998	113	3172.40	122
164.709	18	270.723	47	758.677	11	3172.62	122
164.986	64	270.781	47	759.442	11	3179.60	124
166.113	17	270.838	47	760.227	11	3219.24	104
166.150	17	270.865	47	760.446	11	3222.29	97
166.235	17	270.978	47	761.128	11	3227.54	104
167.988	16	271.035	47	762.004	11	3239.21	97
167.991	16	286.448	56	774.518	32	3242.61	104
168.008	16	341.396	68	784.796	112	3248.28	104
168.047	16	343.168	77	788.578	82	3263.54	104
168.078	16	382.757	86	1055.47	88	3275.64	97
168.134	16	390.774	76	1058.14	88	3297.62	104
168.759	63	390.799	76	1059.01	88	3690.17	103
170.219	39	414.612	80	1059.95	88	3698.36	103
172.169	3	438.201	95	1060.36	88	3702.65	114
173.252	62	439.508	85	1061.23	88	3702.72	103
174.559	38	447.147	92	1218.34	1	3717.31	103
178.715	73	447.217	92	1371.30	31	3725.63	103
182.203	37	447.221	92	1417.74	116	3746.64	103
185.745	36	447.365	92	1417.91	116	3761.58	103
186.438	72	447.371	92	1417.95	116	4119.37	96
191.402	51	447.375	92	1418.39	116	4120.49	105
191.450	51	468.901	91	1418.45	116	4123.96	96
191.459	51	468.982	91	1418.49	116	4125.49	96
191.463	51	469.152	91	1419.01	117	4134.11	105
191.548	51	485.513	13	1596.38	118	4153.27	96
191.558	51	509.421	94	1708.00	119	4158.86	105
192.750	15	529.167	75	1815.23	30	4178.46	96
192.797	15	529.243	75	1824.14	30	4213.35	96
192.801	15	529.272	75	1829.33	30	4522.66	111
192.904	15	561.959	99			4548.48	127
192.911	15	561.959	99			4554.53	101
192.915	15	562.080	99			4628.87	126

List of tabulated lines — Continued

Wavelength (Å)	No.	Wavelength (Å)	No.	Wavelength (Å)	No.	Wavelength (Å)	No.
5114.06	78	5580.12	87	6500.24	102	6878.76	106
5339.94	107	5583.23	87	6543.77	102	6907.27	106
5349.74	107	5597.89	87	6601.28	102	7422.36	115
5372.71	107	5604.27	87	6682.20	102	7437.32	115
5414.59	107	5607.41	87	6764.72	106	7443.08	115
5428.38	107	6330.05	110	6789.62	106		
5471.12	107	6460.12	102	6817.40	106		
5571.81	87	6466.14	102	6828.95	106		

For five singlets and one triplet, fully correlated superposition-of-configuration (SOC) wavefunctions have been calculated by Weiss,[1] in which several thousand interacting configurations are included for describing both upper and lower states. The dipole-length and dipole-velocity results agree typically within a few tenths of a percent—indicating a level of electron-correlation treatment which puts this work in a class by itself.

For most other transitions, we have utilized the results of the Opacity Project (OP), which is reviewed in the general introduction. For this Be-like ion, the OP calculations were carried out by Tully *et al.*[2] For a considerable number of lines, Hibbert[3] also carried out similarly advanced configuration-interaction calculations with the CIV 3 code in intermediate coupling. We have averaged the multiplet line strengths of these two data sets which are always in very close agreement.

For the individual lines, or multiplet components, we have used Hibbert's[3] intermediate-coupling line strengths when available, and otherwise *LS*-coupling data. We have compared Hibbert's intermediate coupling data with *LS*-coupling line strengths for triplets involving s, p and d electrons and have found them to be in excellent agreement, i.e., typically within a range of ±3%, except for some weak multiplet components where differences in the 10–25% range occur. However, for the $2s3p\ ^3P^o_2 - 2p3p\ ^3D_1$ and $2p3p\ ^3P_1$ -$2p3d\ ^3P^o_1$ lines, Hibbert's *f*-values are smaller than the *LS*-values by about three and two orders of magnitude respectively, which is probably due to a numerical error. A test calculation by Kim[4] closely reproduces the *LS*-coupling data, so that we have used the latter in these two cases. For lines in triplets involving *f* and *g* electrons, we expect somewhat larger deviations from *LS*-coupling, and we have therefore lowered the accuracy ratings in those few cases.

Allard *et al.*[5] have recently undertaken an extensive compilation of O V transition probabilities, very similar in spirit to this one, and have also utilized the OP data as the main source of their tables. But they have averaged the OP results with less advanced earlier work, using the OP results with a weight factor of five and the earlier work with a weight of one, while our approach is to omit all less advanced work. Thus, our tables often differ somewhat, but usually only slightly, from those of Allard *et al.*[5] However, we utilize their tables for a few multiplets for which they had special OP data at their disposal. All intercombination line data (no OP results exist for these) are from the recent calculations of Weiss.[1]

Engström *et al.*[6] measured the lifetime of the $2s2p\ ^1P^o_1$ level with the beam-foil method and applied the ANDC technique (see general introduction) for a reliable analysis and correction of the critical cascading effects. Their result, estimated by the authors to be uncertain by ±3.8%, differs from our tabulated value (from Weiss' calculations[1]) by only 3.9%.

References

[1]A. W. Weiss, to be published.

[2]J. A. Tully, M. J. Seaton and K. A. Berrington, J. Phys. B **23**, 3811 (1990).

[3]A. Hibbert, J. Phys. B **13**, 1721 (1980).

[4]Y.-K. Kim, private communication (1993).

[5]N. Allard, M.-C. Artru, T. Lanz, and M. LeDourneuf, Astron. Astrophys., Suppl. Ser. **84**, 563 (1990).

[6]L. Engström, B. Denne, J. O. Ekberg, K. W. Jones, C. Jupen, U. Litzen, Weng Tai Meng, A. Trigueiros and I. Martinson, Phys. Scr. **24**, 551 (1981).

O v: Allowed Transitions

No.	Transition Array	Mult.	λ_{air} (Å)	λ_{vac} (Å) or σ (cm^{-1})*	E_i (cm^{-1})	E_k (cm^{-1})	g_i–g_k	A_{ki} (10^8 s^{-1})	f_{ik}	S (at. u.)	log gf	Acc.	Source
1.	$2s^2$–$2s2p$	^{1}S–^{3}P°											
				1218.34	0	82078.6	1–3	2.34–05	1.56–06	6.27–06	–5.806	B	1
2.		^{1}S–^{1}P°		629.732	0	158797.7	1–3	2.872+01	5.123–01	1.062+00	–0.2905	A+	1
3.	$2s^2$–$2s3p$	^{1}S–^{1}P°		172.169	0	580824.9	1–3	2.94+02	3.92–01	2.22–01	–0.407	B	2,3
4.	$2s^2$–$2p3d$	^{1}S–^{1}P°		139.029	0	719274.9	1–3	5.13+01	4.46–02	2.04–02	–1.351	B–	2,3
5.	$2s^2$–$2s4p$	^{1}S–^{1}P°		135.523	0	737880.8	1–3	9.42+01	7.78–02	3.47–02	–1.109	B	2
6.	$2s^2$–$2s5p$	^{1}S–^{1}P°		124.616	0	802466	1–3	7.15+01	5.00–02	2.05–02	–1.301	B–	2
7.	$2s^2$–$2s6p$	^{1}S–^{1}P°		119.102	0	839616	1–3	4.64+01	2.96–02	1.16–02	–1.529	B–	2
8.	$2s^2$–$2p4d$	^{1}S–^{1}P°		118.000	0	847460	1–3	9.04+00	5.66–03	2.20–03	–2.247	C	2
9.	$2s^2$–$2s7p$	^{1}S–^{1}P°		116.161	0	860874	1–3	2.28+01	1.39–02	5.30–03	–1.858	C+	2
10.	$2s^2$–$2s8p$	^{1}S–^{1}P°		114.358	0	874447	1–3	1.72+01	1.01–02	3.80–03	–1.996	C+	2
11.	$2s2p$–$2p^2$	^{3}P°–^{3}P		*760.36*	*82234*	*213750*	9–9	2.204+01	1.910–01	4.304+00	0.2354	A+	1
				760.446	82385.3	213887.0	5–5	1.652+01	1.432–01	1.793+00	–0.1450	A+	1
				760.227	82078.6	213618.2	3–3	5.514+00	4.777–02	3.587–01	–0.8437	A+	1
				762.004	82385.3	213618.2	5–3	9.125+00	4.766–02	5.978–01	–0.6229	A+	1
				761.128	82078.6	213462.5	3–1	2.197+01	6.361–02	4.782–01	–0.7193	A+	1
				758.677	82078.6	213887.0	3–5	5.547+00	7.978–02	5.978–01	–0.6210	A+	1
				759.442	81942.5	213618.2	1–3	7.373+00	1.913–01	4.782–01	–0.7184	A+	1
12.		^{3}P°–^{1}D											
				669.630	82385.3	231721.4	5–5	3.22–03	2.16–05	2.39–04	–3.966	B	1
				668.258	82078.6	231721.4	3–5	3.31–04	3.69–06	2.44–05	–4.956	B	1
13.		^{3}P°–^{1}S											
				485.513	81942.5	287910.3	1–3	6.97–05	7.39–07	1.18–06	–6.131	B	1
14.	$2s2p$–$2s3s$	^{3}P°–^{3}S		*215.17*	*82234*	546972.7	9–3	1.83+02	4.23–02	2.70–01	–0.420	B	2,3
				215.245	82385.3	546972.7	5–3	1.02+02	4.23–02	1.50–01	–0.674	B	3*n*
				215.103	82078.6	546972.7	3–3	6.09+01	4.22–02	8.97–02	–0.897	B	3*n*
				215.040	81942.5	546972.7	1–3	2.03+01	4.23–02	2.99–02	–1.374	B–	3*n*
15.	$2s2p$–$2s3d$	^{3}P°–^{3}D		*192.85*	*82234*	*600766*	9–15	6.90+02	6.41–01	3.66+00	0.761	B	2,3
				192.904	82385.3	600779.2	5–7	6.89+02	5.38–01	1.71+00	0.430	B	3*n*
				192.797	82078.6	600758.9	3–5	5.17+02	4.80–01	9.15–01	0.159	B	3*n*
				192.750	81942.5	600748.9	1–3	3.83+02	6.41–01	4.07–01	–0.193	B	3*n*
				192.911	82385.3	600758.9	5–5	1.72+02	9.60–02	3.05–01	–0.319	B	3*n*
				192.801	82078.6	600748.9	3–3	2.88+02	1.60–01	3.05–01	–0.318	B	3*n*
				192.915	82385.3	600748.9	5–3	1.91+01	6.40–03	2.03–02	–1.495	B–	3*n*

O v: Allowed Transitions — Continued

No.	Transition Array	Mult.	λ_{air} (Å)	λ_{vac} (Å) or σ (cm^{-1})*	E_i (cm^{-1})	E_k (cm^{-1})	g_i-g_k	A_{ki} (10^8 s^{-1})	f_{ik}	S (at. u.)	log gf	Acc.	Source
16.	$2s2p-$ $2p3p$	$^3P^o-^3D$		*168.00*	*82234*	*677457*	9−15	7.45+01	5.25−02	2.61−01	−0.325	B	2,3
				167.988	82385.3	677665.0	5−7	7.46+01	4.42−02	1.22−01	−0.656	B	3*n*
				167.991	82078.6	677348.5	3−5	5.66+01	3.99−02	6.62−02	−0.922	B	3*n*
				168.008	81942.5	677150.7	1−3	4.19+01	5.31−02	2.94−02	−1.275	B−	3*n*
				168.078	82385.3	677348.5	5−5	1.79+01	7.58−03	2.10−02	−1.421	B−	3*n*
				168.047	82078.6	677150.7	3−3	3.04+01	1.29−02	2.14−02	−1.413	B−	3*n*
				168.134	82385.3	677150.7	5−3	1.95+00	4.97−04	1.38−03	−2.605	C	3*n*
17.		$^3P^o-^3S$		*166.19*	*82234*	683942.9	9−3	1.50+02	2.07−02	1.02−01	−0.730	B	2,3
				166.235	82385.3	683942.9	5−3	7.97+01	1.98−02	5.42−02	−1.004	B	3*n*
				166.150	82078.6	683942.9	3−3	5.22+01	2.16−02	3.54−02	−1.189	B	3*n*
				166.113	81942.5	683942.9	1−3	1.82+01	2.25−02	1.23−02	−1.647	B−	3*n*
18.		$^3P^o-^3P$		*164.64*	*82234*	*689611*	9−9	1.29+02	5.25−02	2.56−01	−0.326	B	2,3
				164.657	82385.3	689708.5	5−5	9.77+01	3.97−02	1.08−01	−0.702	B	3*n*
				164.626	82078.6	689517.7	3−3	3.06+01	1.24−02	2.02−02	−1.428	B−	3*n*
				164.709	82385.3	689517.7	5−3	5.76+01	1.40−02	3.81−02	−1.153	B	3*n*
				164.657	82078.6	689403.5	3−1	1.29+02	1.75−02	2.85−02	−1.279	B−	3*n*
				164.574	82078.6	689708.5	3−5	3.14+01	2.12−02	3.45−02	−1.196	B	3*n*
				164.589	81942.5	689517.7	1−3	4.10+01	4.99−02	2.71−02	−1.302	B−	3*n*
19.	$2s2p-$ $2s4s$	$^3P^o-^3S$		*156.19*	*82234*	722480.0	9−3	3.28+01	4.00−03	1.85−02	−1.444	B−	2
				156.227	82385.3	722480.0	5−3	1.82+01	4.00−03	1.03−02	−1.699	B−	*LS*
				156.152	82078.6	722480.0	3−3	1.09+01	4.00−03	6.17−03	−1.921	C+	*LS*
				156.119	81942.5	722480.0	1−3	3.65+00	4.00−03	2.06−03	−2.398	C	*LS*
20.	$2s2p-$ $2s4d$	$^3P^o-^3D$		*151.51*	*82234*	*742245*	9−15	2.27+02	1.30−01	5.85−01	0.069	B	2
				151.547	82385.3	742248.6	5−7	2.27+02	1.09−01	2.73−01	−0.262	B	*LS*
				151.477	82078.6	742244.1	3−5	1.70+02	9.77−02	1.46−01	−0.533	B	*LS*
				151.447	81942.5	742239.4	1−3	1.26+02	1.30−01	6.50−02	−0.885	B	*LS*
				151.548	82385.3	742244.1	5−5	5.67+01	1.95−02	4.87−02	−1.010	B	*LS*
				151.478	82078.6	742239.4	3−3	9.47+01	3.26−02	4.87−02	−1.010	B	*LS*
				151.549	82385.3	742239.4	5−3	6.30+00	1.30−03	3.25−03	−2.186	C+	*LS*
21.	$2s2p-$ $2s5s$	$^3P^o-^3S$		*140.09*	*82234*	796071	9−3	3.29+01	3.23−03	1.34−02	−1.537	B−	2
				140.118	82385.3	796071	5−3	1.83+01	3.23−03	7.44−03	−1.792	C+	*LS*
				140.058	82078.6	796071	3−3	1.10+01	3.23−03	4.47−03	−2.014	C+	*LS*
				140.031	81942.5	796071	1−3	3.66+00	3.23−03	1.49−03	−2.491	C	*LS*
22.	$2s2p-$ $2s5d$	$^3P^o-^3D$		*138.08*	*82234*	*806449*	9−15	1.12+02	5.36−02	2.19−01	−0.317	B	2
				138.109	82385.3	806451.4	5−7	1.12+02	4.50−02	1.02−01	−0.648	B	*LS*
				138.051	82078.6	806448.4	3−5	8.44+01	4.02−02	5.48−02	−0.919	B	*LS*
				138.025	81942.5	806446.4	1−3	6.25+01	5.36−02	2.43−02	−1.271	B−	*LS*
				138.110	82385.3	806448.4	5−5	2.81+01	8.03−03	1.83−02	−1.396	B−	*LS*
				138.051	82078.6	806446.4	3−3	4.69+01	1.34−02	1.83−02	−1.396	B−	*LS*
				138.110	82385.3	806446.4	5−3	3.12+00	5.35−04	1.22−03	−2.572	C	*LS*
23.	$2s2p-$ $2p4p$	$^3P^o-^3D$		*133.53*	*82234*	*831139*	9−15	4.74+01	2.11−02	8.36−02	−0.721	B	2
				133.520	82385.3	831334	5−7	4.74+01	1.78−02	3.90−02	−1.052	B	*LS*
				133.520	82078.6	831029	3−5	3.56+01	1.58−02	2.09−02	−1.323	B−	*LS*
				133.525	81942.5	830867	1−3	2.64+01	2.11−02	9.29−03	−1.675	C+	*LS*
				133.575	82385.3	831029	5−5	1.18+01	3.17−03	6.97−03	−1.800	C+	*LS*
				133.549	82078.6	830867	3−3	1.98+01	5.28−03	6.97−03	−1.800	C+	*LS*
				133.604	82385.3	830867	5−3	1.32+00	2.11−04	4.64−04	−2.976	C	*LS*

O v: Allowed Transitions — Continued

No.	Transition Array	Mult.	λ_{air} (Å)	λ_{vac} (Å) or σ (cm^{-1})*	E_i (cm^{-1})	E_k (cm^{-1})	g_i-g_k	A_{ki} (10^8 s^{-1})	f_{ik}	S (at. u.)	log *gf*	Acc.	Source
24.		$^3P^o-^3S$		*133.36*	*82234*	832072	9–3	7.49+01	6.66–03	2.63–02	–1.223	B–	2
				133.389	82385.3	832072	5–3	4.16+01	6.65–03	1.46–02	–1.478	B–	*LS*
				133.335	82078.6	832072	3–3	2.50+01	6.66–03	8.77–03	–1.700	C+	*LS*
				133.310	81942.5	832072	1–3	8.33+00	6.66–03	2.92–03	–2.177	C	*LS*
25.		$^3P^o-^3P$		*132.83*	*82234*	*835051*	9–9	6.65+01	1.76–02	6.92–02	–0.801	B	2
				132.845	82385.3	835142	5–5	4.98+01	1.32–02	2.88–02	–1.181	B–	*LS*
				132.821	82078.6	834972	3–3	1.66+01	4.40–03	5.77–03	–1.880	C+	*LS*
				132.875	82385.3	834972	5–3	2.77+01	4.39–03	9.61–03	–1.658	C+	*LS*
				132.845	82078.6	834835	3–1	6.65+01	5.86–03	7.69–03	–1.755	C+	*LS*
				132.791	82078.6	835142	3–5	1.66+01	7.33–03	9.61–03	–1.658	C+	*LS*
				132.797	81942.5	834972	1–3	2.22+01	1.76–02	7.69–03	–1.755	C+	*LS*
26.	$2s2p-$ $2s6d$	$^3P^o-^3D$		*131.79*	*82234*	841036	9–15	5.54+01	2.40–02	9.39–02	–0.665	B	2
				131.813	82385.3	841036	5–7	5.54+01	2.02–02	4.38–02	–0.996	B	*LS*
				131.760	82078.6	841036	3–5	4.16+01	1.80–02	2.35–02	–1.267	B–	*LS*
				131.736	81942.5	841036	1–3	3.08+01	2.41–02	1.04–02	–1.619	B–	*LS*
				131.813	82385.3	841036	5–5	1.38+01	3.61–03	7.83–03	–1.744	C+	*LS*
				131.760	82078.6	841036	3–3	2.31+01	6.01–03	7.83–03	–1.744	C+	*LS*
				131.813	82385.3	841036	5–3	1.54+00	2.40–04	5.22–04	–2.920	D	*LS*
27.	$2s2p-$ $2s7d$	$^3P^o-^3D$		*128.28*	*82234*	*861786*	9–15	3.65+01	1.50–02	5.71–02	–0.869	B	2
				128.302	82385.3	861796	5–7	3.65+01	1.26–02	2.66–02	–1.200	B–	*LS*
				128.255	82078.6	861778	3–5	2.74+01	1.13–02	1.43–02	–1.471	B–	*LS*
				128.232	81942.5	861778	1–3	2.03+01	1.50–02	6.34–03	–1.823	C+	*LS*
				128.305	82385.3	861778	5–5	9.13+00	2.25–03	4.76–03	–1.948	C+	*LS*
				128.255	82078.6	861778	3–3	1.52+01	3.76–03	4.76–03	–1.948	C+	*LS*
				128.305	82385.3	861778	5–3	1.01+00	1.50–04	3.17–04	–3.124	D	*LS*
28.	$2s2p-$ $2s8d$	$^3P^o-^3D$		*126.11*	*82234*	*875173*	9–15	2.46+01	9.77–03	3.65–02	–1.056	B–	2
				126.135	82385.3	875186	5–7	2.46+01	8.20–03	1.70–02	–1.387	B–	*LS*
				126.090	82078.6	875162	3–5	1.84+01	7.33–03	9.13–03	–1.658	C+	*LS*
				126.069	81942.5	875162	1–3	1.37+01	9.77–03	4.06–03	–2.010	C+	*LS*
				126.139	82385.3	875162	5–5	6.14+00	1.46–03	3.04–03	–2.135	C+	*LS*
				126.090	82078.6	875162	3–3	1.02+01	2.44–03	3.04–03	–2.135	C+	*LS*
				126.139	82385.3	875162	5–3	6.82–01	9.77–05	2.03–04	–3.311	D	*LS*
29.	$2s2p-$ $2p5p$	$^3P^o-^3P$		*122.12*	*82234*	*901084*	9–9	3.61+01	8.07–03	2.92–02	–1.139	B–	2
				122.133	82385.3	901165	5–5	2.71+01	6.05–03	1.22–02	–1.519	B–	*LS*
				122.108	82078.6	901023	3–3	9.03+00	2.02–03	2.43–03	–2.218	C	*LS*
				122.154	82385.3	901023	5–3	1.50+01	2.02–03	4.06–03	–1.996	C+	*LS*
				122.133	82078.6	900858	3–1	3.61+01	2.69–03	3.24–03	–2.093	C+	*LS*
				122.087	82078.6	901165	3–5	9.03+00	3.36–03	4.06–03	–1.996	C+	*LS*
				122.088	81942.5	901023	1–3	1.20+01	8.07–03	3.24–03	–2.093	C+	*LS*
30.	$2s2p-$ $2p^2$	$^1P^o-^3P$											
				1815.23	158797.7	213887.0	3–5	1.51–04	1.24–05	2.22–04	–4.429	B	1
				1824.14	158797.7	213618.2	3–3	3.18–06	1.59–07	2.86–06	–6.322	B	1
				1829.33	158797.7	213462.5	3–1	4.42–05	7.39–07	1.34–05	–5.654	B	1
31.		$^1P^o-^1D$		1371.30	158797.7	231721.4	3–5	3.336+00	1.568–01	2.123+00	–0.3277	A+	1
32.		$^1P^o-^1S$		774.518	158797.7	287910.3	3–1	3.804+01	1.140–01	8.724–01	–0.4658	A+	1

O v: Allowed Transitions — Continued

No.	Transition Array	Mult.	λ_{air} (Å)	λ_{vac} (Å) or σ (cm^{-1})*	E_i (cm^{-1})	E_k (cm^{-1})	g_i–g_k	A_{ki} (10^8 s^{-1})	f_{ik}	S (at. u.)	log gf	Acc.	Source
33.	$2s2p$– $2s3s$	$^1P^o$–^1S		248.460	158797.7	561276.4	3–1	5.59+01	1.72–02	4.23–02	–1.287	A	1
34.	$2s2p$– $2s3d$	$^1P^o$–^1D		220.353	158797.7	612615.6	3–5	4.292+02	5.207–01	1.133+00	0.1937	A+	1
35.	$2s2p$– $2p3p$	$^1P^o$–^1P		194.592	158797.7	672693.8	3–3	1.73+02	9.81–02	1.89–01	–0.531	B	2,3
36.		$^1P^o$–^1D		185.745	158797.7	697170.2	3–5	2.13+02	1.83–01	3.36–01	–0.260	B	2,3
37.		$^1P^o$–^1S		182.203	158797.7	707635.5	3–1	5.13+01	8.52–03	1.53–02	–1.593	B–	2,3
38.	$2s2p$– $2s4s$	$^1P^o$–^1S		174.559	158797.7	731670.5	3–1	7.16+01	1.09–02	1.88–02	–1.485	B–	2
39.	$2s2p$– $2s4d$	$^1P^o$–^1D		170.219	158797.7	746274.9	3–5	1.59+02	1.15–01	1.94–01	–0.461	B	2
40.	$2s2p$– $2s5d$	$^1P^o$–^1D		153.952	158797.7	808352.3	3–5	7.75+01	4.59–02	6.98–02	–0.861	B	2
41.	$2s2p$– $2p4p$	$^1P^o$–^1P		149.076	158797.7	829597	3–3	7.28+01	2.42–02	3.57–02	–1.138	B	2
42.		$^1P^o$–^1D		147.263	158797.7	837855	3–5	9.52+01	5.16–02	7.50–02	–0.811	B	2
43.	$2s2p$– $2s6d$	$^1P^o$–^1D		146.351	158797.7	842087	3–5	4.01+01	2.14–02	3.10–02	–1.191	B	2
44.	$2s2p$– $2s7d$	$^1P^o$–^1D		142.122	158797.7	862419	3–5	2.64+01	1.33–02	1.87–02	–1.398	B–	2
45.	$2s2p$– $2p5p$	$^1P^o$–^1P		135.175	158797.7	898580	3–3	3.69+01	1.01–02	1.35–02	–1.518	B–	2
46.	$2s2p$– $2p6p$	$^1P^o$–^1P		128.817	158797.7	935093	3–3	2.12+01	5.27–03	6.70–03	–1.801	D	2
47.	$2p^2$– $2s3p$	^3P–$^3P^o$		*270.91*	*213750*	*582882*	9–9	5.63–01	6.20–04	4.98–03	–2.253	D	2,3
				270.978	213887.0	582920.3	5–5	4.24–01	4.66–04	2.08–03	–2.632	D	3n
				270.838	213618.2	582843.1	3–3	1.38–01	1.52–04	4.06–04	–3.342	D	3n
				271.035	213887.0	582843.1	5–3	2.31–01	1.53–04	6.81–04	–3.117	D	3n
				270.865	213618.2	582806.4	3–1	5.57–01	2.04–04	5.47–04	–3.212	D	3n
				270.781	213618.2	582920.3	3–5	1.44–01	2.65–04	7.08–04	–3.100	D	3n
				270.723	213462.5	582843.1	1–3	1.88–01	6.21–04	5.53–04	–3.207	D	3n
48.	$2p^2$– $2p3s$	^3P–$^3P^o$		*227.53*	*213750*	*653253*	9–9	1.38+02	1.07–01	7.20–01	–0.017	B	2,3
				227.512	213887.0	653423.3	5–5	1.03+02	8.02–02	3.00–01	–0.397	B	3n
				227.551	213618.2	653080.1	3–3	3.42+01	2.65–02	5.97–02	–1.099	B	3n
				227.690	213887.0	653080.1	5–3	5.72+01	2.67–02	9.99–02	–0.875	B	3n
				227.635	213618.2	652918.0	3–1	1.37+02	3.55–02	7.99–02	–0.972	B	3n
				227.373	213618.2	653423.3	3–5	3.46+01	4.47–02	1.00–01	–0.873	B	3n
				227.470	213462.5	653080.1	1–3	4.58+01	1.07–01	7.99–02	–0.972	B	3n

O v: Allowed Transitions — Continued

No.	Transition Array	Mult.	λ_{air} (Å)	λ_{vac} (Å) or σ (cm^{-1})*	E_i (cm^{-1})	E_k (cm^{-1})	g_i-g_k	A_{ki} (10^8 s^{-1})	f_{ik}	S (at. u.)	log gf	Acc.	Source
49.	$2p^2-$ $2p3d$	^{3}P–^{3}D°		*203.86*	*213750*	*704279*	9–15	8.16+02	8.47–01	5.12+00	0.882	B	2,3
				203.890	213887.0	704348	5–7	8.16+02	7.12–01	2.39+00	0.551	B	3*n*
				203.822	213618.2	704242	3–5	6.36+02	6.60–01	1.33+00	0.297	B	3*n*
				203.782	213462.5	704182	1–3	4.71+02	8.80–01	5.90–01	–0.056	B	3*n*
				203.934	213887.0	704242	5–5	1.79+02	1.12–01	3.74–01	–0.254	B	3*n*
				203.847	213618.2	704182	3–3	3.26+02	2.03–01	4.09–01	–0.215	B	3*n*
				203.959	213887.0	704182	5–3	1.83+01	6.84–03	2.30–02	–1.466	B–	3*n*
50.		^{3}P–^{3}P°		*202.30*	*213750*	*708054*	9–9	4.25+02	2.61–01	1.56+00	0.371	B	2,3
				202.391	213887.0	707981.2	5–5	3.43+02	2.11–01	7.02–01	0.022	B	3*n*
				202.222	213618.2	708125.1	3–3	1.21+02	7.39–02	1.48–01	–0.654	B	3*n*
				202.332	213887.0	708125.1	5–3	1.81+02	6.68–02	2.22–01	–0.477	B	3*n*
				202.189	213618.2	708205.0	3–1	4.25+02	8.69–02	1.74–01	–0.584	B	3*n*
				202.281	213618.2	707981.2	3–5	8.18+01	8.37–02	1.67–01	–0.600	B	3*n*
				202.158	213462.5	708125.1	1–3	1.23+02	2.27–01	1.51–01	–0.644	B	3*n*
51.	$2p^2-$ $2s4p$	^{3}P–^{3}P°		*191.50*	*213750*	*735936*	9–9	1.30+01	7.16–03	4.06–02	–1.191	B	2
				191.548	213887.0	735949.1	5–5	9.75+00	5.37–03	1.69–02	–1.571	B–	*LS*
				191.459	213618.2	735922.0	3–3	3.26+00	1.79–03	3.38–03	–2.270	C+	*LS*
				191.558	213887.0	735922.0	5–3	5.42+00	1.79–03	5.64–03	–2.049	C+	*LS*
				191.463	213618.2	735911.6	3–1	1.30+01	2.39–03	4.51–03	–2.145	C+	*LS*
				191.450	213618.2	735949.1	3–5	3.26+00	2.98–03	5.64–03	–2.048	C+	*LS*
				191.402	213462.5	735922.0	1–3	4.34+00	7.16–03	4.51–03	–2.145	C+	*LS*
52.	$2p^2-$ $2s6p$	^{3}P–^{3}P°		*160.10*	*213750*	838375	9–9	2.04+01	7.82–03	3.71–02	–1.152	B	2
				160.131	213887.0	838375	5–5	1.53+01	5.86–03	1.55–02	–1.533	B–	*LS*
				160.062	213618.2	838375	3–3	5.09+00	1.96–03	3.09–03	–2.232	C+	*LS*
				160.131	213887.0	838375	5–3	8.48+00	1.95–03	5.15–03	–2.010	C+	*LS*
				160.062	213618.2	838375	3–1	2.04+01	2.61–03	4.12–03	–2.107	C+	*LS*
				160.062	213618.2	838375	3–5	5.09+00	3.26–03	5.15–03	–2.010	C+	*LS*
				160.022	213462.5	838375	1–3	6.79+00	7.82–03	4.12–03	–2.107	C+	*LS*
53.	$2p^2-$ $2p4d$	^{3}P–^{3}D°		*159.37*	*213750*	*841234*	9–15	2.99+02	1.90–01	8.96–01	0.232	B	2
				159.380	213887.0	841318	5–7	2.99+02	1.59–01	4.18–01	–0.099	B	*LS*
				159.343	213618.2	841195	3–5	2.24+02	1.42–01	2.24–01	–0.370	B	*LS*
				159.327	213462.5	841102	1–3	1.66+02	1.90–01	9.95–02	–0.722	B	*LS*
				159.411	213887.0	841195	5–5	7.46+01	2.84–02	7.46–02	–0.847	B	*LS*
				159.367	213618.2	841102	3–3	1.25+02	4.74–02	7.46–02	–0.847	B	*LS*
				159.435	213887.0	841102	5–3	8.29+00	1.90–03	4.98–03	–2.023	C+	*LS*
54.		^{3}P–^{3}P°		*158.88*	*213750*	*843164*	9–9	1.40+02	5.30–02	2.49–01	–0.322	B	2
				158.926	213887.0	843111	5–5	1.05+02	3.97–02	1.04–01	–0.702	B	*LS*
				158.831	213618.2	843218	3–3	3.50+01	1.33–02	2.08–02	–1.401	B–	*LS*
				158.899	213887.0	843218	5–3	5.83+01	1.32–02	3.46–02	–1.179	B	*LS*
				158.818	213618.2	843270	3–1	1.40+02	1.77–02	2.77–02	–1.276	B–	*LS*
				158.858	213618.2	843111	3–5	3.50+01	2.21–02	3.46–02	–1.179	B	*LS*
				158.792	213462.5	843218	1–3	4.67+01	5.30–02	2.77–02	–1.276	B–	*LS*
55.	$2p^2-$ $2p5d$	^{3}P–^{3}D°		*144.83*	*213750*	*904215*	9–15	1.43+02	7.52–02	3.23–01	–0.170	B	2
				144.837	213887.0	904318	5–7	1.43+02	6.31–02	1.51–01	–0.501	B	*LS*
				144.809	213618.2	904183	3–5	1.08+02	5.64–02	8.06–02	–0.772	B	*LS*
				144.809	213462.5	904027	1–3	7.97+01	7.52–02	3.58–02	–1.124	B	*LS*
				144.865	213887.0	904183	5–5	3.58+01	1.13–02	2.69–02	–1.249	B–	*LS*
				144.842	213618.2	904027	3–3	5.97+01	1.88–02	2.69–02	–1.249	B–	*LS*
				144.898	213887.0	904027	5–3	3.98+00	7.51–04	1.79–03	–2.425	C	*LS*

O v: Allowed Transitions — Continued

No.	Transition Array	Mult.	λ_{air} (Å)	λ_{vac} (Å) or σ (cm^{-1})*	E_i (cm^{-1})	E_k (cm^{-1})	g_i-g_k	A_{ki} (10^8 s^{-1})	f_{ik}	S (at. u.)	log gf	Acc.	Source
56.	$2p^2-$ $2s3p$	^{1}D–^{1}P°		286.448	231721.4	580824.9	5–3	2.28+01	1.68–02	7.94–02	–1.074	B	2,3
57.	$2p^2-$ $2p3s$	^{1}D–^{1}P°		231.073	231721.4	664485.9	5–3	1.12+02	5.37–02	2.04–01	–0.571	B	2,3
58.	$2p^2-$ $2p3d$	^{1}D–^{1}D°		216.019	231721.4	694643.8	5–5	3.07+02	2.15–01	7.64–01	0.031	B	2,3
59.		^{1}D–^{1}F°		207.796	231721.4	712963.5	5–7	8.58+02	7.78–01	2.66+00	0.590	B	2,3
60.		^{1}D–^{1}P°		205.106	231721.4	719274.9	5–3	2.26+01	8.56–03	2.89–02	–1.369	B–	2,3
61.	$2p^2-$ $2s4f$	^{1}D–^{1}F°		193.006	231721.4	749840.5	5–7	6.05+01	4.73–02	1.50–01	–0.626	B	2
62.	$2p^2-$ $2s5f$	^{1}D–^{1}F°		173.252	231721.4	808916.8	5–7	8.05–02	5.07–05	1.45–04	–3.596	D–	5
63.	$2p^2-$ $2p4s$	^{1}D–^{1}P°		168.759	231721.4	824282	5–3	4.36+01	1.12–02	3.10–02	–1.253	B	2
64.	$2p^2-$ $2p4d$	^{1}D–^{1}D°		164.986	231721.4	837833	5–5	1.02+02	4.18–02	1.14–01	–0.680	B	2
65.	$2p^2-$ $2s6f$	^{1}D–^{1}F°		164.177	231721.4	840821.5	5–7	8.35+01	4.73–02	1.28–01	–0.627	B	2
66.	$2p^2-$ $2p4d$	^{1}D–^{1}F°		162.492	231721.4	847136	5–7	2.62+02	1.45–01	3.89–01	–0.138	B	2
67.	$2p^2-$ $2p5d$	^{1}D–^{1}D°		149.038	231721.4	902691	5–5	4.76+01	1.59–02	3.89–02	–1.101	B	2
68.	$2p^2-$ $2s3p$	^{1}S–^{1}P°		341.396	287910.3	580824.9	1–3	1.26+00	6.60–03	7.42–03	–2.180	C+	2,3
69.	$2p^2-$ $2p3s$	^{1}S–^{1}P°		265.551	287910.3	664485.9	1–3	4.39+01	1.39–01	1.22–01	–0.856	B	2,3
70.	$2p^2-$ $2p3d$	^{1}S–^{1}P°		231.822	287910.3	719274.9	1–3	4.36+02	1.05+00	8.05–01	0.023	B	2,3
71.	$2p^2-$ $2s4p$	^{1}S–^{1}P°		222.237	287910.3	737880.8	1–3	6.21+01	1.38–01	1.01–01	–0.860	B	2
72.	$2p^2-$ $2p4s$	^{1}S–^{1}P°		186.438	287910.3	824282	1–3	1.69+01	2.64–02	1.62–02	–1.579	B–	2
73.	$2p^2-$ $2p4d$	^{1}S–^{1}P°		178.715	287910.3	847460	1–3	1.75+02	2.51–01	1.48–01	–0.601	B	2
74.	$2s3s-$ $2s3p$	^{3}S–^{3}P°	*2784.0*	*2784.8*	546972.7	*582882*	3–9	1.40+00	4.88–01	1.34+01	0.165	B	2,3
			2781.01	2781.83	546972.7	582920.3	3–5	1.40+00	2.71–01	7.45+00	–0.089	B	3n
			2786.99	2787.81	546972.7	582843.1	3–3	1.39+00	1.62–01	4.47+00	–0.312	B	3n
			2789.85	2790.67	546972.7	582806.4	3–1	1.38+00	5.38–02	1.48+00	–0.792	B	3n

O V: Allowed Transitions — Continued

No.	Transition Array	Mult.	λ_{air} (Å)	λ_{vac} (Å) or σ (cm^{-1})*	E_i (cm^{-1})	E_k (cm^{-1})	g_i-g_k	A_{ki} (10^8 s^{-1})	f_{ik}	S (at. u.)	log gf	Acc.	Source
75.	2s3s–2s4p	^{3}S–^{3}P°		*529.20*	546972.7	*735936*	3–9	1.27+01	1.60−01	8.37−01	−0.318	B	2
				529.167	546972.7	735949.1	3–5	1.27+01	8.90−02	4.65−01	−0.574	B	*LS*
				529.243	546972.7	735922.0	3–3	1.27+01	5.34−02	2.79−01	−0.796	B	*LS*
				529.272	546972.7	735911.6	3–1	1.27+01	1.78−02	9.30−02	−1.273	B	*LS*
76.	2s3s–2s5p	^{3}S–^{3}P°		*390.78*	546972.7	*802868*	3–9	8.94+00	6.14−02	2.37−01	−0.735	B	2
				390.774	546972.7	802875	3–5	8.94+00	3.41−02	1.32−01	−0.990	B	*LS*
				390.799	546972.7	802859	3–3	8.94+00	2.05−02	7.90−02	−1.212	B	*LS*
				390.799	546972.7	802859	3–1	8.94+00	6.82−03	2.63−02	−1.689	B−	*LS*
77.	2s3s–2s6p	^{3}S–^{3}P°		343.168	546972.7	838375	3–9	5.97+00	3.16−02	1.07−01	−1.023	B	2
78.	2s3s–2s3p	^{1}S–^{1}P°	5114.06	5115.48	561276.4	580824.9	1–3	1.80−01	2.12−01	3.57+00	−0.673	B	2,3
79.	2s3s–2s4p	^{1}S–^{1}P°		566.237	561276.4	737880.8	1–3	1.19+01	1.72−01	3.21−01	−0.764	B	2
80.	2s3s–2s5p	^{1}S–^{1}P°		414.612	561276.4	802466	1–3	8.59+00	6.64−02	9.07−02	−1.178	B	2
81.	2s3p–2s3d	^{1}P°–^{1}D	3144.66	3145.57	580824.9	612615.6	3–5	8.86−01	2.19−01	6.80+00	−0.182	B	2,3
82.	2s3p–2p3p	^{1}P°–^{1}S		788.578	580824.9	707635.5	3–1	1.72+01	5.36−02	4.17−01	−0.794	B	2
83.	2s3p–2s4s	^{1}P°–^{1}S		662.930	580824.9	731670.5	3–1	1.02+01	2.24−02	1.47−01	−1.173	B	2
84.	2s3p–2s4d	^{1}P°–^{1}D		604.412	580824.9	746274.9	3–5	3.58+01	3.26−01	1.95+00	−0.009	B	2
85.	2s3p–2s5d	^{1}P°–^{1}D		439.508	580824.9	808352.3	3–5	2.12+01	1.02−01	4.43−01	−0.514	B	2
86.	2s3p–2s6d	^{1}P°–^{1}D		382.757	580824.9	842087	3–5	1.24+01	4.54−02	1.72−01	−0.866	B	2
87.	2s3p–2s3d	^{3}P°–^{3}D	*5589.9*	*5591.4*	*582882*	*600766*	9–15	1.49−01	1.16−01	1.92+01	0.019	B	2,3
			5597.89	5599.45	582920.3	600779.2	5–7	1.48−01	9.75−02	8.99+00	−0.312	B	3*n*
			5580.12	5581.67	582843.1	600758.9	3–5	1.11−01	8.67−02	4.78+00	−0.585	B	3*n*
			5571.81	5573.36	582806.4	600748.9	1–3	8.33−02	1.16−01	2.13+00	−0.934	B	3*n*
			5604.27	5605.82	582920.3	600758.9	5–5	3.68−02	1.73−02	1.60+00	−1.062	B	3*n*
			5583.23	5584.78	582843.1	600748.9	3–3	6.20−02	2.90−02	1.60+00	−1.061	B	3*n*
			5607.41	5608.97	582920.3	600748.9	5–3	4.08−03	1.15−03	1.07−01	−2.239	B	3*n*
88.	2s3p–2p3p	^{3}P°–^{3}D		*1057.4*	*582882*	*677457*	9–15	4.60+00	1.28−01	4.02+00	0.063	B	2,3
				1055.47	582920.3	677665.0	5–7	4.65+00	1.09−01	1.89+00	−0.265	B	3*n*
				1058.14	582843.1	677348.5	3–5	3.43+00	9.61−02	1.00+00	−0.540	B	3*n*
				1059.95	582806.4	677150.7	1–3	2.53+00	1.28−01	4.46−01	−0.894	B	3*n*
				1059.01	582920.3	677348.5	5–5	1.18+00	1.99−02	3.46−01	−1.003	B	3*n*
				1060.36	582843.1	677150.7	3–3	1.93+00	3.26−02	3.41−01	−1.010	B	3*n*
				1061.23	582920.3	677150.7	5–3	1.26−01	1.28−03	2.24−02	−2.194	C+	*LS*

O v: Allowed Transitions — Continued

No.	Transition Array	Mult.	λ_{air} (Å)	λ_{vac} (Å) or σ (cm^{-1})*	E_i (cm^{-1})	E_k (cm^{-1})	g_i-g_k	A_{ki} (10^8 s^{-1})	f_{ik}	S (at. u.)	log gf	Acc.	Source
89.	$2s3p-2s4s$	$^3P^o-^3S$		*716.34*	*582882*	722480.0	9–3	2.56+01	6.57–02	1.39+00	–0.228	B	2
				716.539	582920.3	722480.0	5–3	1.42+01	6.57–02	7.74–01	–0.484	B	*LS*
				716.143	582843.1	722480.0	3–3	8.54+00	6.57–02	4.65–01	–0.705	B	*LS*
				715.955	582806.4	722480.0	1–3	2.85+00	6.57–02	1.55–01	–1.182	B	*LS*
90.	$2s3p-2s4d$	$^3P^o-^3D$		*627.50*	*582882*	*742245*	9–15	4.50+01	4.42–01	8.22+00	0.600	B	2
				627.635	582920.3	742248.6	5–7	4.49+01	3.71–01	3.84+00	0.269	B	*LS*
				627.349	582843.1	742244.1	3–5	3.37+01	3.32–01	2.06+00	–0.002	B	*LS*
				627.223	582806.4	742239.4	1–3	2.50+01	4.42–01	9.14–01	–0.354	B	*LS*
				627.653	582920.3	742244.1	5–5	1.12+01	6.63–02	6.85–01	–0.479	B	*LS*
				627.367	582843.1	742239.4	3–3	1.87+01	1.11–01	6.85–01	–0.479	B	*LS*
				627.671	582920.3	742239.4	5–3	1.25+00	4.42–03	4.57–02	–1.655	B	*LS*
91.	$2s3p-2s5s$	$^3P^o-^3S$		*469.07*	*582882*	796071	9–3	1.34+01	1.47–02	2.04–01	–0.879	B	2
				469.152	582920.3	796071	5–3	7.42+00	1.47–02	1.13–01	–1.134	B	*LS*
				468.982	582843.1	796071	3–3	4.46+00	1.47–02	6.81–02	–1.356	B	*LS*
				468.901	582806.4	796071	1–3	1.49+00	1.47–02	2.27–02	–1.833	B–	*LS*
92.	$2s3p-2s5d$	$^3P^o-^3D$		*447.29*	*582882*	*806449*	9–15	2.41+01	1.20–01	1.59+00	0.034	B	2
				447.365	582920.3	806451.4	5–7	2.40+01	1.01–01	7.44–01	–0.297	B	*LS*
				447.217	582843.1	806448.4	3–5	1.80+01	9.02–02	3.98–01	–0.568	B	*LS*
				447.147	582806.4	806446.4	1–3	1.34+01	1.20–01	1.77–01	–0.920	B	*LS*
				447.371	582920.3	806448.4	5–5	6.01+00	1.80–02	1.33–01	–1.045	B	*LS*
				447.221	582843.1	806446.4	3–3	1.00+01	3.01–02	1.33–01	–1.045	B	*LS*
				447.375	582920.3	806446.4	5–3	6.68–01	1.20–03	8.85–03	–2.221	C+	*LS*
93.	$2s3d-2s4f$	$^1D-^1F^o$		728.731	612615.6	749840.5	5–7	9.46+01	1.05+00	1.26+01	0.722	B	2
94.	$2s3d-2s5f$	$^1D-^1F^o$		509.421	612615.6	808916.8	5–7	3.13+01	1.71–01	1.43+00	–0.069	B	2
95.	$2s3d-2s6f$	$^1D-^1F^o$		438.201	612615.6	840821.5	5–7	7.52+00	3.03–02	2.19–01	–0.820	B	2
96.	$2p3s-2p3p$	$^3P^o-^3D$	*4130.4*	*4131.6*	*653253*	*677457*	9–15	4.78–01	2.04–01	2.50+01	0.264	B	2,3
			4123.96	4125.12	653423.3	677665.0	5–7	4.81–01	1.72–01	1.17+01	–0.066	B	3*n*
			4119.37	4120.53	653080.1	677348.8	3–5	3.66–01	1.55–01	6.31+00	–0.332	B	3*n*
			4125.49	4126.66	652918.0	677150.7	1–3	2.70–01	2.07–01	2.81+00	–0.685	B	3*n*
			4178.46	4179.64	653423.3	677348.8	5–5	1.12–01	2.93–02	2.02+00	–0.834	B	3*n*
			4153.27	4154.45	653080.1	677150.7	3–3	1.92–01	4.96–02	2.04+00	–0.827	B	3*n*
			4213.35	4214.54	653423.3	677150.7	5–3	1.19–02	1.90–03	1.32–01	–2.022	B	3*n*
97.		$^3P^o-^3S$	*3257.4*	*3258.4*	*653253*	683942.9	9–3	9.19–01	4.87–02	4.70+00	–0.358	B	2,3
			3275.64	3276.58	653423.3	683942.9	5–3	4.76–01	4.60–02	2.48+00	–0.639	B	3*n*
			3239.21	3240.15	653080.1	683942.9	3–3	3.28–01	5.16–02	1.65+00	–0.810	B	3*n*
			3222.29	3223.22	652918.0	683942.9	1–3	1.16–01	5.43–02	5.76–01	–1.265	B	3*n*
98.		$^3P^o-^3P$	*2749.6*	*2750.4*	*653253*	*689611*	9–9	1.82+00	2.07–01	1.68+01	0.269	B	2,3
			2755.13	2755.94	653423.3	689708.5	5–5	1.37+00	1.56–01	7.09+00	–0.107	B	3*n*
			2743.61	2744.42	653080.1	689517.7	3–3	4.38–01	4.95–02	1.34+00	–0.828	B	3*n*
			2769.69	2770.51	653423.3	689517.7	5–3	7.88–01	5.44–02	2.48+00	–0.566	B	3*n*
			2752.23	2753.05	653080.1	689403.5	3–1	1.82+00	6.87–02	1.87+00	–0.686	B	3*n*
			2729.31	2730.12	653080.1	689708.5	3–5	4.52–01	8.40–02	2.27+00	–0.598	B	3*n*
			2731.45	2732.26	652918.0	689517.7	1–3	5.90–01	1.98–01	1.78+00	–0.703	B	3*n*

O v: Allowed Transitions — Continued

No.	Transition Array	Mult.	λ_{air} (Å)	λ_{vac} (Å) or σ (cm^{-1})*	E_i (cm^{-1})	E_k (cm^{-1})	g_i-g_k	A_{ki} (10^8 s^{-1})	f_{ik}	S (at. u.)	log gf	Acc.	Source
99.	$2p3s-$ $2p4p$	$^3P^o-^3D$		*562.16*	*653253*	*831139*	9–15	1.62+01	1.28–01	2.13+00	0.061	B	2
				562.080	653423.3	831334	5–7	1.62+01	1.07–01	9.94–01	–0.270	B	*LS*
				561.959	653080.1	831029	3–5	1.22+01	9.60–02	5.33–01	–0.541	B	*LS*
				561.959	652918.0	830867	1–3	9.01+00	1.28–01	2.37–01	–0.893	B	*LS*
				563.045	653423.3	831029	5–5	4.03+00	1.92–02	1.78–01	–1.019	B	*LS*
				562.471	653080.1	830867	3–3	6.76+00	3.20–02	1.78–01	–1.018	B	*LS*
				563.559	653423.3	830867	5–3	4.47–01	1.28–03	1.18–02	–2.195	B–	*LS*
100.	$2p3s-$ $2p3p$	$^1P^o-^1D$	3058.68	3059.57	664485.9	697170.2	3–5	1.39+00	3.25–01	9.83+00	–0.011	B	2,3
101.	$2p3p-$ $2p3d$	$^1P-^1D^o$	4554.53	4555.81	672693.8	694643.8	3–5	2.41–01	1.25–01	5.61+00	–0.427	B	2,3
102.		$^3D-^3F^o$	*6487.5*	*6489.3*	*677457*	*692867*	15–21	1.11–01	9.82–02	3.14+01	0.168	B	2,3
			6500.24	6502.04	677665.0	693044.8	7–9	1.11–01	9.01–02	1.35+01	–0.200	B	3n
			6466.14	6467.93	677348.5	692809.4	5–7	1.01–01	8.85–02	9.42+00	–0.354	B	3n
			6460.12	6461.91	677150.7	692626.0	3–5	9.37–02	9.77–02	6.23+00	–0.533	B	3n
			6601.28	6603.10	677665.0	692809.4	7–7	1.14–02	7.42–03	1.13+00	–1.284	B	3n
			6543.77	6545.57	677348.5	692626.0	5–5	1.64–02	1.05–02	1.13+00	–1.279	B	3n
			6682.20	6684.05	677665.0	692626.0	7–5	4.19–04	2.00–04	3.09–02	–2.853	B	3n
103.		$^3D-^3D^o$	*3727.1*	*3728.2*	*677457*	*704279*	15–15	1.32–01	2.74–02	5.05+00	–0.385	B	2,3
			3746.64	3747.70	677665.0	704348	7–7	1.18–01	2.48–02	2.14+00	–0.761	B	3n
			3717.31	3718.37	677348.5	704242	5–5	9.63–02	2.00–02	1.22+00	–1.001	B	3n
			3698.36	3699.42	677150.7	704182	3–3	1.03–01	2.10–02	7.68–01	–1.200	B	3n
			3761.58	3762.65	677665.0	704242	7–5	1.61–02	2.44–03	2.11–01	–1.768	B–	3n
			3725.63	3726.68	677348.5	704182	5–3	2.91–02	3.63–03	2.22–01	–1.741	B	3n
			3702.72	3703.77	677348.5	704348	5–7	1.41–02	4.05–03	2.47–01	–1.694	B	3n
			3690.17	3691.22	677150.7	704242	3–5	1.97–02	6.70–03	2.44–01	–1.696	B	3n
104.		$^3D-^3P^o$	*3267.3*	*3268.3*	*677457*	*708054*	15–9	1.51–01	1.45–02	2.34+00	–0.662	B	2,3
			3297.62	3298.57	677665.0	707981.2	7–5	1.30–01	1.51–02	1.15+00	–0.976	B	3n
			3248.28	3249.22	677348.5	708125.1	5–3	1.18–01	1.12–02	6.01–01	–1.250	B	3n
			3219.24	3220.17	677150.7	708205.0	3–1	1.54–01	7.99–03	2.54–01	–1.620	B	3n
			3263.54	3264.49	677348.5	707981.2	5–5	1.86–02	2.97–03	1.59–01	–1.828	B–	3n
			3227.54	3228.47	677150.7	708125.1	3–3	3.38–02	5.28–03	1.68–01	–1.800	B	3n
			3242.61	3243.54	677150.7	707981.2	3–5	1.08–03	2.84–04	9.11–03	–3.069	C+	3n
105.		$^3S-^3P^o$	*4146.3*	*4147.5*	683942.9	*708054*	3–9	3.37–01	2.61–01	1.07+01	–0.107	B	2,3
			4158.86	4160.03	683942.9	707981.2	3–5	3.39–01	1.47–01	6.03+00	–0.356	B	3n
			4134.11	4135.27	683942.9	708125.1	3–3	3.34–01	8.56–02	3.50+00	–0.590	B	3n
			4120.49	4121.65	683942.9	708205.0	3–1	3.33–01	2.83–02	1.15+00	–1.072	B	3n
106.		$^3P-^3D^o$	*6815.5*	*6817.4*	*689611*	*704279*	9–15	7.41–02	8.60–02	1.74+01	–0.111	B	2,3
			6828.95	6830.83	689708.5	704348	5–7	7.35–02	7.19–02	8.08+00	–0.444	B	3n
			6789.62	6791.49	689517.7	704242	3–5	5.79–02	6.67–02	4.47+00	–0.699	B	3n
			6764.72	6766.59	689403.5	704182	1–3	4.37–02	8.99–02	2.00+00	–1.046	B	3n
			6878.76	6880.66	689708.5	704242	5–5	1.65–02	1.17–02	1.32+00	–1.234	B	3n
			6817.40	6819.28	689517.7	704182	3–3	3.00–02	2.09–02	1.41+00	–1.203	B	3n
			6907.27	6909.18	689708.5	704182	5–3	1.70–03	7.28–04	8.28–02	–2.439	B	3n

O V: Allowed Transitions — Continued

No.	Transition Array	Mult.	λ_{air} (Å)	λ_{vac} (Å) or σ (cm^{-1})*	E_i (cm^{-1})	E_k (cm^{-1})	g_i-g_k	A_{ki} (10^8 s^{-1})	f_{ik}	S (at. u.)	log gf	Acc.	Source
107.		^{3}P–^{3}P°	*5420.6*	*5422.1*	*689611*	*708054*	9–9	5.54–02	2.44–02	3.92+00	–0.659	B	2,3
			5471.12	5472.64	689708.5	707981.2	5–5	4.86–02	2.18–02	1.96+00	–0.962	B	3*n*
			5372.71	5374.21	689517.7	708125.1	3–3	1.42–02	6.15–03	3.26–01	–1.734	C+	*LS*
			5428.38	5429.88	689708.5	708125.1	5–3	2.68–02	7.10–03	6.34–01	–1.450	B	3*n*
			5349.74	5351.23	689517.7	708205.0	3–1	7.04–02	1.01–02	5.32–01	–1.520	B–	3*n*
			5414.59	5416.09	689517.7	707981.2	3–5	9.29–03	6.81–03	3.64–01	–1.690	C+	3*n*
			5339.94	5341.42	689403.5	708125.1	1–3	1.85–02	2.37–02	4.16–01	–1.626	B	3*n*
108.	2*p* 3*d*– 2*s* 6*g*	^{3}F°–^{3}G		*678.70*	*692867*	*840208*	21–27	3.94+01	3.50–01	1.64+01	0.866	B	5
				679.132	693044.8	840291.5	9–11	3.93+01	3.32–01	6.69+00	0.476	C+	5
				678.610	692809.4	840169.4	7–9	3.70+01	3.28–01	5.13+00	0.361	C+	5
				677.963	692626.0	840126.7	5–7	3.63+01	3.50–01	3.91+00	0.243	C+	5
				679.696	693044.8	840169.4	9–9	2.45+00	1.70–02	3.42–01	–0.815	C+	*LS*
				678.807	692809.4	840126.7	7–7	3.17+00	2.19–02	3.42–01	–0.815	C+	*LS*
				679.893	693044.8	840126.7	9–7	4.80–02	2.59–04	5.21–03	–2.633	C–	*LS*
109.	2*p* 3*d*– 2*p* 4*f*	^{3}F°–^{3}G		*654.76*	*692867*	*845595*	21–27	4.42+01	3.65–01	1.65+01	0.885	B	2
				655.042	693044.8	845706.7	9–11	4.41+01	3.47–01	6.73+00	0.494	C+	*LS*
				654.719	692809.4	845546.7	7–9	4.14+01	3.42–01	5.16+00	0.379	C+	*LS*
				654.212	692626.0	845481.6	5–7	4.07+01	3.65–01	3.93+00	0.262	C+	*LS*
				655.730	693044.8	845546.7	9–9	2.75+00	1.77–02	3.44–01	–0.797	C+	*LS*
				654.998	692809.4	845481.6	7–7	3.55+00	2.28–02	3.44–01	–0.797	C+	*LS*
				656.010	693044.8	845481.6	9–7	5.38–02	2.70–04	5.25–03	–2.615	C–	*LS*
110.	2*p* 3*p*– 2*p* 3*d*	^{1}D–^{1}F°	6330.05	6331.80	697170.2	712963.5	5–7	1.21–01	1.02–01	1.06+01	–0.294	B	2,3
111.		^{1}D–^{1}P°	4522.66	4523.92	697170.2	719274.9	5–3	1.02–02	1.87–03	1.39–01	–2.029	D–	2,3
112.	2*p* 3*d*– 2*s* 6*g*	^{1}F°–^{1}G		784.796	712963.5	840385.2	7–9	3.07+01	3.65–01	6.60+00	0.407	B	5
113.	2*p* 3*d*– 2*p* 4*f*	^{1}F°–^{1}G		751.998	712963.5	845942.6	7–9	2.76+01	3.01–01	5.21+00	0.323	B	2
114.	2*p* 3*d*– 2*s* 4*d*	^{1}P°–^{1}D	3702.65	3703.70	719274.9	746274.9	3–5	1.98–01	6.77–02	2.47+00	–0.693	B	2
115.	2*s* 4*s*– 2*s* 4*p*	^{3}S–^{3}P°	*7429.6*	*7431.7*	722480.0	*735936*	3–9	2.79–01	6.91–01	5.07+01	0.317	B	2
			7422.36	7424.40	722480.0	735949.1	3–5	2.79–01	3.85–01	2.82+01	0.062	B	*LS*
			7437.32	7439.37	722480.0	735922.0	3–3	2.78–01	2.30–01	1.69+01	–0.161	B	*LS*
			7443.08	7445.13	722480.0	735911.6	3–1	2.77–01	7.67–02	5.64+00	–0.638	B	*LS*
116.	2*s* 4*p*– 2*s* 5*d*	^{3}P°–^{3}D		*1418.2*	*735936*	*806449*	9–15	8.65+00	4.35–01	1.83+01	0.592	B	2
				1418.39	735949.1	806451.4	5–7	8.65+00	3.65–01	8.52+00	0.261	B	*LS*
				1417.91	735922.0	806448.4	3–5	6.49+00	3.26–01	4.57+00	–0.010	B	*LS*
				1417.74	735911.6	806446.4	1–3	4.81+00	4.35–01	2.03+00	–0.362	B	*LS*
				1418.45	735949.1	806448.4	5–5	2.16+00	6.52–02	1.52+00	–0.487	B	*LS*
				1417.95	735922.0	806446.4	3–3	3.61+00	1.09–01	1.52+00	–0.487	B	*LS*
				1418.49	735949.1	806446.4	5–3	2.40–01	4.35–03	1.01–01	–1.663	B	*LS*
117.		^{1}P°–^{1}D		1419.01	737880.8	808352.3	3–5	6.98+00	3.51–01	4.92+00	0.022	B	2

O v: Allowed Transitions — Continued

No.	Transition Array	Mult.	λ_{air} (Å)	λ_{vac} (Å) or σ (cm^{-1})*	E_i (cm^{-1})		E_k (cm^{-1})	g_i-g_k	A_{ki} (10^8 s^{-1})	f_{ik}	S (at. u.)	log gf	Acc.	Source
118.	2s 4d– 2s 5f	^{1}D–^{1}F°		1596.38	746274.9	–	808916.8	5–7	1.51+01	8.05–01	2.12+01	0.605	B	2
119.	2s 4f– 2s 5g	^{1}F°–^{1}G		1708.00	749840.5	–	808388.6	7–9	2.02+01	1.14+00	4.48+01	0.901	B	5
120.	2s 5d– 2s 6f	^{3}D–^{3}F°	*2755.7*	*2756.5*	*806449*	–	*842728*	15–21	3.52+00	5.61–01	7.64+01	0.925	B	2
			2754.70	2755.51	806451.4	–	842742.3	7–9	3.52+00	5.16–01	3.27+01	0.557	C+	*LS*
			2755.91	2756.73	806448.4	–	842723.3	5–7	3.13+00	4.99–01	2.26+01	0.397	C+	*LS*
			2756.98	2757.79	806446.4	–	842707.3	3–5	2.95+00	5.61–01	1.53+01	0.226	C+	*LS*
			2756.14	2756.96	806451.4	–	842723.3	7–7	3.92–01	4.47–02	2.84+00	–0.505	C+	*LS*
			2757.13	2757.94	806448.4	–	842707.3	5–5	5.48–01	6.25–02	2.84+00	–0.505	C+	*LS*
			2757.36	2758.17	806451.4	–	842707.3	7–5	1.55–02	1.26–03	8.00–02	–2.055	C+	*LS*
121.		^{1}D–^{1}F°	3078.95	3079.84	808352.3	–	840821.5	5–7	3.81+00	7.57–01	3.84+01	0.578	B	2
122.	2s 5f– 2s 6g	^{3}F°–^{3}G	*3164.3*	*3165.2*	*808615*	–	*840208*	21–27	4.15+00	8.00–01	1.75+02	1.225	B	5
			3156.11	3157.02	808616.1	–	840291.5	9–11	4.16+00	7.60–01	7.11+01	0.835	C+	5
			3168.10	3169.02	808613.9	–	840169.4	7–9	3.87+00	7.48–01	5.46+01	0.719	C+	5
			3172.31	3173.22	808613.0	–	840126.7	5–7	3.79+00	8.00–01	4.18+01	0.602	C+	5
			3168.32	3169.24	808616.1	–	840169.4	9–9	2.58–01	3.89–02	3.65+00	–0.456	C+	*LS*
			3172.40	3173.31	808613.9	–	840126.7	7–7	3.31–01	4.99–02	3.65+00	–0.457	C+	*LS*
			3172.62	3173.54	808616.1	–	840126.7	9–7	5.04–03	5.91–04	5.56–02	–2.274	C+	*LS*
123.	2s 5f– 2p 4f	^{3}F°–^{3}G	*2703.3*	*2704.1*	*808615*	–	*845595*	21–27	2.55+00	3.60–01	6.72+01	0.878	B	2
			2695.30	2696.10	808616.1	–	845706.7	9–11	2.58+00	3.43–01	2.74+01	0.489	C+	*LS*
			2706.82	2707.62	808613.9	–	845546.7	7–9	2.38+00	3.36–01	2.10+01	0.372	C+	*LS*
			2711.53	2712.34	808613.0	–	845481.6	5–7	2.32+00	3.59–01	1.60+01	0.253	C+	*LS*
			2706.98	2707.78	808616.1	–	845546.7	9–9	1.59–01	1.75–02	1.40+00	–0.804	C+	*LS*
			2711.60	2712.40	808613.9	–	845481.6	7–7	2.03–01	2.24–02	1.40+00	–0.804	C+	*LS*
			2711.76	2712.56	808616.1	–	845481.6	9–7	3.10–03	2.66–04	2.13–02	–2.622	C	*LS*
124.	2s 5f– 2s 6g	^{1}F°–^{1}G	3179.60	3180.52	808916.8	–	840358.2	7–9	4.12+00	8.01–01	5.87+01	0.749	B	5
125.	2s 5f– 2p 4f	^{1}F°–^{1}G	2700.02	2700.82	808916.8	–	845942.6	7–9	2.60+00	3.65–01	2.27+01	0.407	B	2
126.	2s 6f– 2s 7d	^{1}F°–^{1}D	4628.87	4630.17	840821.5	–	862419	7–5	6.59–02	1.51–02	1.61+00	–0.975	B	2
127.	2s 6f– 2s 7g	^{1}F°–^{1}G	4548.48	4549.76	840821.5	–	862800.7	7–9	1.76+00	7.01–01	7.35+01	0.691	B	5

*Wavelengths (Å) are always given unless cm^{-1} is indicated.

O v

Forbidden Transitions

List of tabulated lines

Wavelength (Å)	No.	Wavelength (Å)	No.	Wavenumber (cm^{-1})	No.
in vacuum		1301.15	5	136.1	4
		1303.46	5	155.7	11
431.553	3	1308.69	5	268.8	11
467.537	2	1346.04	13	306.7	4
468.125	2	1350.93	13	424.5	11
486.559	8	1371.30	10	442.8	4
667.651	7	1779.71	14		
668.258	7	1815.23	9		
669.630	7	1824.14	9		
757.894	6				
758.677	6	in air			
760.227	6				
760.446	6	5475.26	12		
762.004	6	5522.35	12		
762.909	6	5605.59	12		
1213.81	1				

We have selected the data of Glass,[1-3] who calculated transition probabilities utilizing the CIV 3 code (see general introduction), i.e., by constructing advanced wavefunctions from configuration interaction expansions. He also included some relativistic corrections. The agreement with earlier, less comprehensive (and somewhat less advanced) calculations by Nussbaumer and Storey[4] is usually within a few percent, except for two weak M1 lines, $^1S_0$ – $^3P_1$ in the $2s^2$–$2p^2$ transition array and $^3P^o_1$ – $^1P^o_1$ in the $2s2p$ configuration, where the discrepancies are large.

For the M1 fine structure transitions in the ^3P terms of the $2s2p$ and $2p^2$ configurations, we have used the line strengths S obtained from general non-relativistic formulas for conditions of pure LS-coupling, given by Pasternak.[5] The A-values published by Glass, converted to S, are in close agreement, while the A-values of Nussbaumer and Storey[4] convert exactly to those of the formulas.

References

[1]R. Glass, Astrophys. Space Sci. **87**, 41 (1982).
[2]R. Glass, Astrophys. Space Sci. **91**, 417 (1983).
[3]R. Glass, Astrophys. Space Sci. **92**, 307 (1983).
[4]H. Nussbaumer and P. J. Storey, Astron. Astrophys. **74**, 244 (1979).
[5]S. Pasternak, Astrophys. J. **92**, 129 (1940).

O v: Forbidden Transitions

No.	Transition Array	Mult.	λ_{air} (Å)	λ_{vac} (Å) or σ (cm^{-1})*	E_i (cm^{-1})	E_k (cm^{-1})	g_i–g_k	Type	A_{ki} (s^{-1})	S (at. u.)	Acc.	Source
1.	$2s^2$–$2s2p$	^1S–$^3P^o$										
				1213.81	0.0	82385.3	1–5	M2	2.17–02	1.91+01	B+	1,4
2.	$2s^2$–$2p^2$	^1S–^3P										
				468.125	0.0	213618.2	1–3	M1	3.60–02	4.11–07	D	2,4
				467.537	0.0	213887.0	1–5	E2	2.99–01	2.99–05	B	3,4
3.		^1S–^1D										
				431.553	0.0	231721.4	1–5	E2	2.92+03	1.95–01	B+	3,4
4.	$2s2p$–$2s2p$	$^3P^o$–$^3P^o$										
				136.1 cm^{-1}	81942.5	82078.6	1–3	M1	4.54–05	2.00+00	A	4
				442.8 cm^{-1}	81942.5	82385.3	1–5	E2	1.34–10	3.52–01	D	3
				306.7 cm^{-1}	82078.6	82385.3	3–5	M1	3.89–04	2.50+00	A	4
				306.7 cm^{-1}	82078.6	82385.3	3–5	E2	4.79–11	7.88–01	D	3

O V: Forbidden Transitions

No.	Transition Array	Mult.	λ_{air} (Å)	λ_{vac} (Å) or σ (cm^{-1})*	E_i (cm^{-1})	E_k (cm^{-1})	g_i-g_k	Type	A_{ki} (s^{-1})	S (at. u.)	Acc.	Source
5.		$^3P^o-^1P^o$										
				1301.15	81942.5	158797.7	1–3	M1	6.73–02	1.65–05	B	2,4
				1303.46	82078.6	158797.7	3–3	M1	1.02–01	2.52–05	E	2,4
				1303.46	82078.6	158797.7	3–3	E2	1.26–03	1.27–05	D	3
				1308.69	82385.3	158797.7	5–3	M1	8.27–02	2.06–05	B	2,4
				1308.69	82385.3	158797.7	5–3	E2	7.55–04	7.76–06	D	3
6.	$2s2p-$ $2p^2$	$^3P^o-^3P$										
				762.909	82385.3	213462.5	5–1	M2	1.39–01	2.42+00	C+	1,4
				760.227	82078.6	213618.2	3–3	M2	1.48–01	7.55+00	C+	1
				762.004	82385.3	213618.2	5–3	M2	1.34–04	6.94–03	C	1
				757.894	81942.5	213887.0	1–5	M2	3.20–02	2.69+00	C+	1
				758.677	82078.6	213887.0	3–5	M2	1.02–05	8.57–04	C	1
				760.446	82385.3	213887.0	5–5	M2	1.88–01	1.60+01	C+	1
7.		$^3P^o-^1D$										
				667.651	81942.5	231721.4	1–5	M2	1.35–01	6.00+00	C+	1
				668.258	82078.6	231721.4	3–5	M2	3.07–01	1.37+01	C+	1
				669.630	82385.3	231721.4	5–5	M2	2.43–01	1.10+01	C+	1
8.		$^3P^o-^1S$										
				486.559	82385.3	287910.3	5–1	M2	9.31–01	1.70+00	C+	1
9.		$^1P^o-^3P$										
				1824.14	158797.7	213618.2	3–3	M2	1.11–03	4.50+00	C+	1
				1815.23	158797.7	213887.0	3–5	M2	2.05–03	1.36+01	C+	1
10.		$^1P^o-^1D$										
				1371.30	158797.7	231721.4	3–5	M2	1.37–04	2.22–01	C	1
11.	$2p^2-$ $2p^2$	$^3P-^3P$										
				155.7 cm^{-1}	213462.5	213618.2	1–3	M1	6.79–05	2.00+00	B+	5
				424.5 cm^{-1}	213462.5	213887.0	1–5	E2	9.64–11	3.12–01	D	3
				268.8 cm^{-1}	213618.2	213887.0	3–5	M1	2.62–04	2.50+00	B+	5
				268.8 cm^{-1}	213618.2	213887.0	3–5	E2	2.10–11	6.68–01	D	3
12.		$^3P-^1D$										
			5475.26	5476.78	213462.5	231721.4	1–5	E2	1.10–07	2.42–06	D	3
			5522.35	5523.89	213618.2	231721.4	3–5	M1	1.21–02	3.77–04	C	2
			5522.35	5523.89	213618.2	231721.4	3–5	E2	4.56–06	1.05–04	D	3
			5605.59	5607.14	213887.0	231721.4	5–5	M1	3.86–02	1.26–03	C	2
			5605.59	5607.14	213887.0	231721.4	5–5	E2	2.94–05	7.27–04	D+	3
13.		$^3P-^1S$										
				1346.04	213618.2	287910.3	3–1	M1	7.75–01	7.00–05	C	2
				1350.93	213887.0	287910.3	5–1	E2	2.28–02	9.17–05	C	3
14.		$^1D-^1S$										
				1779.71	231721.4	287910.3	5–1	E2	7.08+01	1.13+00	B	3,4

*Wavelengths (Å) are always given unless cm^{-1} is indicated.

O VI

Lithium Isoelectronic Sequence

Ground State: $1s^22s\ ^2S_{1/2}$

Ionization Energy: 138.116 eV = 1114010 cm^{-1}

Allowed Transitions

List of tabulated lines

Wavelength (Å)	No.	Wavelength (Å)	No.	Wavelength (Å)	No.	Wavelength (Å)	No.
in vacuum		129.785	13	304.44	47	539.916	86
		129.871	13	318.43	25	539.927	86
91.85	176	129.875	13	332.1	200	541.03	79
93.030	9	130.2	167	333.6	192	555.76	58
93.61	175	130.9	166	345.140	35	588.318	68
93.915	8	132.219	12	345.316	35	588.456	68
94.44	174	132.312	12	345.326	35	588.539	68
95.082	7	132.4	165	348.0	196	597.207	67
95.80	189	140.4	164	353.886	34	597.435	67
95.83	188	146.0	180	354.1	198	600.64	78
96.840	6	147.4	179	354.082	34	601.481	85
98.55	187	150.089	2	355.46	46	601.515	85
98.61	186	150.125	2	357.88	45	601.568	85
99.08	173	151.4	163	445.9	190	603.36	77
99.688	5	159.0	178	447.712	24	664.74	57
102.6	172	170.8	177	447.840	24	710.045	66
102.8	171	172.935	11	449.99	61	710.296	66
103.1	170	173.079	11	450.5	194	710.367	66
103.206	22	173.095	11	468.9	193	727.95	76
103.260	22	183.937	10	471.48	60	728.90	88
103.263	22	184.117	10	473.6	199	729.361	84
103.5	185	229.95	30	474.6	191	729.411	84
103.6	184	235.43	29	480.70	82	729.436	84
104.612	21	242.90	28	495.105	71	731.282	65
104.669	21	249.83	54	495.198	71	731.624	65
104.670	21	250.502	42	495.262	71	734.01	75
104.803	20	250.584	42	498.090	33	792.19	94
104.81	4	250.601	42	498.426	33	839.87	107
104.862	20	254.72	27	498.479	33	861.30	93
106.731	19	256.32	53	502.44	59	898.90	102
106.789	19	258.950	41	504.390	87	914.56	115
106.792	19	259.046	41	504.414	87	917.96	106
107.020	18	259.055	41	504.456	87	970.55	92
107.081	18	260.126	40	505.31	81	986.35	56
109.1	169	260.232	40	506.5	195	1018.2	101
110.157	17	264.81	52	514.0	197	1031.91	1
110.220	17	265.20	51	519.610	44	1036.7	100
110.222	17	272.334	39	519.723	44	1037.1	111
110.656	16	272.432	39	519.748	44	1037.61	1
110.721	16	272.450	39	529.229	70	1038.3	114
111.5	168	274.223	38	529.371	70	1043.1	105
114.2	183	274.341	38	529.408	70	1080.88	64
114.3	182	275.41	26	529.740	43	1081.52	64
115.821	3	278.76	50	529.883	43	1081.63	64
115.830	3	279.35	49	529.919	43	1122.4	74
116.350	15	295.808	37	534.162	69	1126.35	83
116.421	15	295.933	37	534.344	69	1126.46	83
116.422	15	295.945	37	535.955	32	1126.47	83
116.9	181	299.431	36	536.405	32	1147.0	73
117.328	14	299.571	36	539.41	80	1171.56	63
117.401	14	303.39	48	539.889	86	1172.44	63

List of tabulated lines — Continued

Wavelength (Å)	No.	Wavelength (Å)	No.	Wavelength (Å)	No.	Wavenumber (cm^{-1})	No.
1191.3	91	2081.3	135	4773.3	140	129	108
1261.9	99	2082.5	138	5083.6	145	176.2	72
1289.9	110	2102.6	129	5274.0	150	185.5	72
1293.0	113	2108.5	125	5285.7	151	197.6	72
1302.5	104	2114.7	103	5433.4	147	534	161
1303.8	98	2173.2	96	5581.0	143	572	156
1336.1	121	2264.7	142	7015.6	154	694	144
1413.5	131	2403.3	149	7349.5	157	1048	122
1545.3	120	2939.2	141	7662.7	160	1181	153
1607.8	127	3057.5	146	7999.6	158	1776	95
1637.7	136	3071.2	118	9341.75	55	1851	139
1649.8	130	3129.5	152	9397.86	55		
1836.9	90	3176.9	148	10419	162		
1936.3	119	3311.0	124	11744.2	31		
1996.1	97	3423.2	132	11891.6	31		
		3434.7	137	11964.3	31		
in air		3440.3	134	18547.8	89		
		3513.6	128	28190.7	62		
2033.3	126	3615.6	123	28532.7	62		
2065.1	109	3811.35	23	28708.0	62		
2070.1	116	3834.24	23	32778	117		
2074.3	112	4100.7	155				
2078.3	133	4418.4	159				

We have utilized the results of four calculations. For the transitions between lower excited levels we applied the data from the Opacity Project (OP), which is reviewed in the general introduction. For this Li-like ion, the OP calculations were carried out by Peach *et al*.[1]

For the transitions to and between higher levels, we have taken recourse to some less sophisticated calculations.[2,3,5] Since O VI is an ion with only three electrons, with two electrons normally in the closed 1*s* shell, these simpler calculational approaches, the single-configuration self-consistent field approximation applied by Biemont[2] and the semi-empirical Coulomb approximation utilized by Lindgard and Nielsen,[3] also produce fairly accurate results. When there is overlap with OP data, there is indeed very good agreement, typically within ±10%. We therefore have used refs. 2, 3 and 5 for highly excited transitions for which no OP data are available.

A critical compilation[4] performed in 1976 contains data for about 40 multiplets, which were critically selected from the available literature. The OP data normally agree with these earlier results within a few percent. But for some higher *ms*-*np* transitions (m=2–4 and n=4–8), differences up to ±20% are encountered.

Several calculations and measurements have been carried out for transitions between doubly excited quartet levels above the first ionization potential, which are metastable against autoionization. We have selected the results of Laughlin and Fairley[5] which were calculated with the model-potential method. These data are estimated to be quite reliable, which is also indicated by good agreement (in the 10–30% range) with beam-foil lifetime data by Blanke *et al*.[6]

For the quartet transitions, we have always used the calculated wavelengths of Laughlin and Fairley.[5] For some of these transitions, experimental wavelengths from a beam-foil spectroscopy experiment by Blanke *et al*.[7] are available and typically agree with the Laughlin and Fairley data within 0.1%. For consistency and completeness, we have, however, always chosen the Laughlin and Fairley numbers. Energy levels for a few of these transitions have been compiled by Moore.[8] We have not listed these since they are not fully consistent with the above cited sets of wavelength data.

For most multiplets involving high principal quantum numbers, the energy levels for the individual lines within the multiplets coincide, so that we have presented only multiplet data.

References

[1]G. Peach, H. E. Saraph, and M. J. Seaton, J. Phys. B **21**, 3669 (1988).

[2]E. Biémont, Astron. Astrophys., Suppl. Ser. **27**, 489 (1977).

[3]A. Lindgard and S. E. Nielsen, At. Data Nucl. Data Tables **19**, 533 (1977).

[4]G. A. Martin and W. L. Wiese, J. Phys. Chem. Ref. Data **5**, 537 (1976).

[5]C. Laughlin and N. A. Fairley, Z. Phys. D **5**, 91 (1987).

[6]J. H. Blanke, P. H. Heckmann, E. Träbert, Phys. Scr. **32**, 509 (1985).

[7]J. H. Blanke, P. H. Heckmann, E. Träbert, R. Hucke, and H. v. Buttlar, Z. Phys. A **321**, 47 (1985).

[8]C. E. Moore, *Selected Tables of Atomic Spectra*, Natl. Stand. Ref. Data Ser., Natl. Bur. Stand. (U.S.) **3**, Sec. 8 (1979) and *Tables of Spectra of Hydrogen, Carbon, Nitrogen and Oxygen Atoms and Ions*, Edited by J. W. Gallagher, CRC Press, Boca Raton, FL (1993).

O VI: Allowed Transitions

No.	Transition Array	Mult.	λ_{air} (Å)	λ_{vac} (Å) or σ (cm^{-1})*	E_i (cm^{-1})	E_k (cm^{-1})	g_i-g_k	A_{ki} (10^8 s^{-1})	f_{ik}	S (at. u.)	log gf	Acc.	Source
1.	$1s^22s-1s^22p$	$^2S-^2P^o$		*1033.8*	0.0	*96730*	2–6	4.13+00	1.99–01	1.35+00	–0.401	A	1
				1031.91	0.0	96907.5	2–4	4.16+00	1.33–01	9.02–01	–0.576	A	*LS*
				1037.61	0.0	96375.0	2–2	4.09+00	6.60–02	4.51–01	–0.879	A	*LS*
2.	$1s^22s-1s^23p$	$^2S-^2P^o$		*150.10*	0.0	*666218*	2–6	2.62+02	2.65–01	2.62–01	–0.275	A–	1
				150.089	0.0	666269.8	2–4	2.62+02	1.77–01	1.75–01	–0.451	A–	*LS*
				150.125	0.0	666113.2	2–2	2.62+02	8.84–02	8.74–02	–0.752	A–	*LS*
3.	$1s^22s-1s^24p$	$^2S-^2P^o$		*115.82*	0.0	*863376*	2–6	1.23+02	7.41–02	5.65–02	–0.829	B+	1
				115.821	0.0	863397.7	2–4	1.23+02	4.94–02	3.77–02	–1.005	B+	*LS*
				115.830	0.0	863333.8	2–2	1.23+02	2.47–02	1.88–02	–1.306	B+	*LS*
4.	$1s^22s-1s^25p$	$^2S-^2P^o$		104.81	0.0	954080	2–6	6.48+01	3.20–02	2.21–02	–1.193	B+	1
5.	$1s^22s-1s^26p$	$^2S-^2P^o$		99.688	0.0	1003130	2–6	3.78+01	1.69–02	1.11–02	–1.471	B	1
6.	$1s^22s-1s^27p$	$^2S-^2P^o$		96.840	0.0	1032631	2–6	2.38+01	1.00–02	6.40–03	–1.697	B	1
7.	$1s^22s-1s^28p$	$^2S-^2P^o$		95.082	0.0	1051724	2–6	1.59+01	6.48–03	4.06–03	–1.887	B	2
8.	$1s^22s-1s^29p$	$^2S-^2P^o$		93.915	0.0	1064793	2–6	8.55+00	3.39–03	2.10–03	–2.168	B	3
9.	$1s^22s-1s^210p$	$^2S-^2P^o$		93.030	0.0	1074922	2–6	5.82+00	2.27–03	1.39–03	–2.344	B	3
10.	$1s^22p-1s^23s$	$^2P^o-^2S$		*184.06*	*96730*	640039.8	6–2	1.71+02	2.89–02	1.05–01	–0.761	A	1
				184.117	96907.5	640039.8	4–2	1.14+02	2.89–02	7.00–02	–0.937	A	*LS*
				183.937	96375.0	640039.8	2–2	5.70+01	2.89–02	3.50–02	–1.238	A	*LS*
11.	$1s^22p-1s^23d$	$^2P^o-^2D$		*173.03*	*96730*	*674656*	6–10	8.78+02	6.57–01	2.25+00	0.596	A	1
				173.079	96907.5	674676.8	4–6	8.78+02	5.91–01	1.35+00	0.374	A	*LS*
				172.935	96375.0	674625.7	2–4	7.33+02	6.58–01	7.49–01	0.119	A	*LS*
				173.095	96907.5	674625.7	4–4	1.46+02	6.57–02	1.50–01	–0.580	A	*LS*
12.	$1s^22p-1s^24s$	$^2P^o-^2S$		*132.28*	*96730*	852696	6–2	6.52+01	5.70–03	1.49–02	–1.466	B+	1
				132.312	96907.5	852696	4–2	4.34+01	5.70–03	9.93–03	–1.642	B+	*LS*
				132.219	96375.0	852696	2–2	2.18+01	5.71–03	4.97–03	–1.943	B+	*LS*
13.	$1s^22p-1s^24d$	$^2P^o-^2D$		*129.84*	*96730*	*866893*	6–10	2.91+02	1.23–01	3.14–01	–0.133	A	1
				129.871	96907.5	866901.5	4–6	2.91+02	1.10–01	1.89–01	–0.355	A	*LS*
				129.785	96375.0	866880.1	2–4	2.43+02	1.23–01	1.05–01	–0.610	A	*LS*
				129.875	96907.5	866880.1	4–4	4.85+01	1.23–02	2.10–02	–1.310	A	*LS*
14.	$1s^22p-1s^25s$	$^2P^o-^2S$		*117.38*	*96730*	948690	6–2	3.19+01	2.20–03	5.10–03	–1.879	B	1
				117.401	96907.5	948690	4–2	2.13+01	2.20–03	3.40–03	–2.056	B	*LS*
				117.328	96375.0	948690	2–2	1.07+01	2.20–03	1.70–03	–2.356	B	*LS*
15.	$1s^22p-1s^25d$	$^2P^o-^2D$		*116.40*	*96730*	*955856*	6–10	1.35+02	4.56–02	1.05–01	–0.563	A	1
				116.421	96907.5	955860	4–6	1.35+02	4.11–02	6.29–02	–0.785	A	*LS*
				116.350	96375.0	955851	2–4	1.12+02	4.56–02	3.50–02	–1.040	A	*LS*
				116.422	96907.5	955851	4–4	2.24+01	4.56–03	6.99–03	–1.739	A	*LS*

O VI: Allowed Transitions — Continued

No.	Transition Array	Mult.	λ_{air} (Å)	λ_{vac} (Å) or σ (cm^{-1})*	E_i (cm^{-1})	E_k (cm^{-1})	g_i–g_k	A_{ki} (10^8 s^{-1})	f_{ik}	S (at. u.)	log gf	Acc.	Source
16.	$1s^22p$– $1s^26s$	$^2P^o$–^2S		*110.70*	*96730*	1000080	6–2	1.75+01	1.07–03	2.35–03	–2.191	B	2
				110.721	96907.5	1000080	4–2	1.17+01	1.07–03	1.57–03	–2.367	B	*LS*
				110.656	96375.0	1000080	2–2	5.85+00	1.07–03	7.83–04	–2.668	B	*LS*
17.	$1s^22p$– $1s^26d$	$^2P^o$–^2D		*110.20*	*96730*	*1004178*	6–10	7.40+01	2.25–02	4.89–02	–0.870	A–	1
				110.220	96907.5	1004184	4–6	7.40+01	2.02–02	2.93–02	–1.092	A–	*LS*
				110.157	96375.0	1004170	2–4	6.18+01	2.25–02	1.63–02	–1.347	A–	*LS*
				110.222	96907.5	1004170	4–4	1.23+01	2.25–03	3.26–03	–2.047	A–	*LS*
18.	$1s^22p$– $1s^27s$	$^2P^o$–^2S		*107.06*	*96730*	1030780	6–2	1.08+01	6.21–04	1.31–03	–2.429	B	2
				107.081	96907.5	1030780	4–2	7.22+00	6.20–04	8.75–04	–2.605	B	*LS*
				107.020	96375.0	1030780	2–2	3.62+00	6.21–04	4.37–04	–2.906	B	*LS*
19.	$1s^22p$– $1s^27d$	$^2P^o$–^2D		*106.77*	*96730*	*1033324*	6–10	4.53+01	1.29–02	2.72–02	–1.111	B+	1
				106.789	96907.5	1033334	4–6	4.53+01	1.16–02	1.63–02	–1.333	B+	*LS*
				106.731	96375.0	1033310	2–4	3.78+01	1.29–02	9.07–03	–1.588	B+	*LS*
				106.792	96907.5	1033310	4–4	7.54+00	1.29–03	1.81–03	–2.288	B+	*LS*
20.	$1s^22p$– $1s^28s$	$^2P^o$–^2S		*104.84*	*96730*	1050543	6–2	7.16+00	3.93–04	8.15–04	–2.627	B	2
				104.862	96907.5	1050543	4–2	4.77+00	3.93–04	5.43–04	–2.803	B	*LS*
				104.803	96375.0	1050543	2–2	2.39+00	3.93–04	2.72–04	–3.104	B	*LS*
21.	$1s^22p$– $1s^28d$	$^2P^o$–^2D		*104.65*	*96730*	*1052296*	6–10	2.97+01	8.13–03	1.68–02	–1.312	B+	1
				104.669	96907.5	1052301	4–6	2.97+01	7.31–03	1.01–02	–1.534	B+	*LS*
				104.612	96375.0	1052288	2–4	2.48+01	8.13–03	5.60–03	–1.789	B+	*LS*
				104.670	96907.5	1052288	4–4	4.95+00	8.13–04	1.12–03	–2.488	B+	*LS*
22.	$1s^22p$– $1s^29d$	$^2P^o$–^2D		*103.24*	*96730*	*1065327*	6–10	2.07+01	5.53–03	1.13–02	–1.480	B	3
				103.260	96907.5	1065337	4–6	2.07+01	4.97–03	6.76–03	–1.701	B	*LS*
				103.206	96375.0	1065311	2–4	1.73+01	5.53–03	3.76–03	–1.956	B	*LS*
				103.263	96907.5	1065311	4–4	3.46+00	5.52–04	7.51–04	–2.656	B	*LS*
23.	$1s^23s$– $1s^23p$	^2S–$^2P^o$	*3818.9*	*3820.0*	640039.8	*666218*	2–6	5.11–01	3.35–01	8.43+00	–0.174	A	1
			3811.35	3812.43	640039.8	666269.8	2–4	5.14–01	2.24–01	5.62+00	–0.349	A	*LS*
			3834.24	3835.33	640039.8	666113.2	2–2	5.05–01	1.11–01	2.81+00	–0.653	A	*LS*
24.	$1s^23s$– $1s^24p$	^2S–$^2P^o$		*447.75*	640039.8	*863376*	2–6	3.08+01	2.77–01	8.18–01	–0.256	B	1
				447.712	640039.8	863397.7	2–4	3.08+01	1.85–01	5.45–01	–0.432	B	*LS*
				447.840	640039.8	863333.8	2–2	3.07+01	9.25–02	2.73–01	–0.733	B	*LS*
25.	$1s^23s$– $1s^25p$	^2S–$^2P^o$		318.43	640039.8	954080	2–6	1.80+01	8.23–02	1.73–01	–0.784	B	1
26.	$1s^23s$– $1s^26p$	^2S–$^2P^o$		275.41	640039.8	1003130	2–6	1.09+01	3.71–02	6.73–02	–1.129	B	1
27.	$1s^23s$– $1s^27p$	^2S–$^2P^o$		254.72	640039.8	1032631	2–6	6.99+00	2.04–02	3.42–02	–1.390	B	1
28.	$1s^23s$– $1s^28p$	^2S–$^2P^o$		242.90	640039.8	1051724	2–6	4.71+00	1.25–02	2.00–02	–1.602	B	1
29.	$1s^23s$– $1s^29p$	^2S–$^2P^o$		235.43	640039.8	1064793	2–6	2.39+00	5.96–03	9.23–03	–1.924	B	3

O VI: Allowed Transitions — Continued

No.	Transition Array	Mult.	λ_{air} (Å)	λ_{vac} (Å) or σ (cm^{-1})*	E_i (cm^{-1})	E_k (cm^{-1})	g_i-g_k	A_{ki} (10^8 s^{-1})	f_{ik}	S (at. u.)	log gf	Acc.	Source
30.	$1s^23s-1s^210p$	$^2S-^2P^o$		229.95	640039.8	1074922	2–6	1.61+00	3.83−03	5.79−03	−2.116	B	3
31.	$1s^23p-1s^23d$	$^2P^o-^2D$	*11847*	*8439* cm^{-1}	*666218*	*674656*	6–10	1.39−02	4.86−02	1.14+01	−0.535	A	1
			11891.6	8407.0 cm^{-1}	666269.8	674676.8	4–6	1.37−02	4.36−02	6.82+00	−0.759	A	*LS*
			11744.2	8512.5 cm^{-1}	666113.2	674625.7	2–4	1.19−02	4.90−02	3.79+00	−1.009	A	*LS*
			11964.3	8355.9 cm^{-1}	666269.8	674625.7	4–4	2.24−03	4.81−03	7.58−01	−1.716	A	*LS*
32.	$1s^23p-1s^24s$	$^2P^o-^2S$		*536.26*	*666218*	852696	6–2	4.44+01	6.39−02	6.77−01	−0.417	A	1
				536.405	666269.8	852696	4–2	2.96+01	6.39−02	4.51−01	−0.593	A	*LS*
				535.955	666113.2	852696	2–2	1.48+01	6.39−02	2.26−01	−0.893	A	*LS*
33.	$1s^23p-1s^24d$	$^2P^o-^2D$		*498.32*	*666218*	*866893*	6–10	8.98+01	5.57−01	5.48+00	0.524	A	1
				498.426	666269.8	866901.5	4–6	8.97+01	5.01−01	3.29+00	0.302	A	*LS*
				498.090	666113.2	866880.1	2–4	7.49+01	5.57−01	1.83+00	0.047	A	*LS*
				498.479	666269.8	866880.1	4–4	1.50+01	5.57−02	3.66−01	−0.652	A	*LS*
34.	$1s^23p-1s^25s$	$^2P^o-^2S$		*354.02*	*666218*	948690	6–2	2.07+01	1.30−02	9.06−02	−1.109	A−	1
				354.082	666269.8	948690	4–2	1.38+01	1.30−02	6.04−02	−1.286	A−	*LS*
				353.886	666113.2	948690	2–2	6.90+00	1.30−02	3.02−02	−1.586	A−	*LS*
35.	$1s^23p-1s^25d$	$^2P^o-^2D$		*345.26*	*666218*	*955856*	6–10	4.49+01	1.34−01	9.13−01	−0.095	A	1
				345.316	666269.8	955860	4–6	4.49+01	1.20−01	5.48−01	−0.317	A	*LS*
				345.140	666113.2	955851	2–4	3.75+01	1.34−01	3.04−01	−0.572	A	*LS*
				345.326	666269.8	955851	4–4	7.48+00	1.34−02	6.08−02	−1.272	A	*LS*
36.	$1s^23p-1s^26s$	$^2P^o-^2S$		*299.52*	*666218*	1000080	6–2	1.13+01	5.09−03	3.01−02	−1.515	B+	1
				299.571	666269.8	1000080	4–2	7.56+00	5.09−03	2.01−02	−1.691	B+	*LS*
				299.431	666113.2	1000080	2–2	3.79+00	5.09−03	1.00−02	−1.992	B+	*LS*
37.	$1s^23p-1s^26d$	$^2P^o-^2D$		*295.89*	*666218*	*1004178*	6–10	2.52+01	5.52−02	3.23−01	−0.480	A	1
				295.933	666269.8	1004184	4–6	2.52+01	4.97−02	1.94−01	−0.702	A	*LS*
				295.808	666113.2	1004170	2–4	2.11+01	5.52−02	1.08−01	−0.957	A	*LS*
				295.945	666269.8	1004170	4–4	4.21+00	5.52−03	2.15−02	−1.656	A	*LS*
38.	$1s^23p-1s^27s$	$^2P^o-^2S$		*274.30*	*666218*	1030780	6–2	6.92+00	2.60−03	1.41−02	−1.806	B+	1
				274.341	666269.8	1030780	4–2	4.61+00	2.60−03	9.40−03	−1.983	B+	*LS*
				274.223	666113.2	1030780	2–2	2.31+00	2.60−03	4.70−03	−2.283	B+	*LS*
39.	$1s^23p-1s^27d$	$^2P^o-^2D$		*272.40*	*666218*	*1033324*	6–10	1.56+01	2.89−02	1.55−01	−0.761	A	1
				272.432	666269.8	1033334	4–6	1.56+01	2.60−02	9.32−02	−0.983	A	*LS*
				272.334	666113.2	1033310	2–4	1.30+01	2.89−02	5.18−02	−1.238	A	*LS*
				272.450	666269.8	1033310	4–4	2.60+00	2.89−03	1.04−02	−1.937	A	*LS*
40.	$1s^23p-1s^28s$	$^2P^o-^2S$		*260.20*	*666218*	1050543	6–2	4.54+00	1.54−03	7.90−03	−2.035	B	1
				260.232	666269.8	1050543	4–2	3.03+00	1.54−03	5.27−03	−2.211	B	*LS*
				260.126	666113.2	1050543	2–2	1.52+00	1.54−03	2.63−03	−2.512	B	*LS*
41.	$1s^23p-1s^28d$	$^2P^o-^2D$		*259.01*	*666218*	*1052296*	6–10	1.03+01	1.73−02	8.83−02	−0.985	A−	1
				259.046	666269.8	1052301	4–6	1.03+01	1.55−02	5.30−02	−1.207	A−	*LS*
				258.950	666113.2	1052288	2–4	8.58+00	1.73−02	2.94−02	−1.462	A−	*LS*
				259.055	666269.8	1052288	4–4	1.71+00	1.73−03	5.89−03	−2.161	A−	*LS*

O VI: Allowed Transitions — Continued

No.	Transition Array	Mult.	λ_{air} (Å)	λ_{vac} (Å) or σ (cm^{-1})*	E_i (cm^{-1})	E_k (cm^{-1})	g_i–g_k	A_{ki} (10^8 s^{-1})	f_{ik}	S (at. u.)	log gf	Acc.	Source
42.	$1s^23p$–$1s^29d$	$^2P^o$–^2D		*250.56*	*666218*	*1065327*	6–10	7.15+00	1.12–02	5.55–02	–1.172	B	3
				250.584	666269.8	1065337	4–6	7.15+00	1.01–02	3.33–02	–1.394	B	*LS*
				250.502	666113.2	1065311	2–4	5.96+00	1.12–02	1.85–02	–1.649	B	*LS*
				250.601	666269.8	1065311	4–4	1.19+00	1.12–03	3.70–03	–2.348	B	*LS*
43.	$1s^23d$–$1s^24p$	^2D–$^2P^o$		*529.89*	*674656*	*863376*	10–6	6.04+00	1.53–02	2.66–01	–0.817	B+	1
				529.883	674676.8	863397.7	6–4	5.44+00	1.53–02	1.60–01	–1.038	B+	*LS*
				529.919	674625.7	863333.8	4–2	6.04+00	1.27–02	8.87–02	–1.294	B+	*LS*
				529.740	674625.7	863397.7	4–4	6.04–01	2.54–03	1.77–02	–1.993	B+	*LS*
44.	$1s^23d$–$1s^24f$	^2D–$^2F^o$		*519.68*	*674656*	*867083*	10–14	1.79+02	1.02+00	1.74+01	1.007	B	2
				519.723	674676.8	867087.0	6–8	1.79+02	9.68–01	9.93+00	0.764	B	*LS*
				519.610	674625.7	867077.7	4–6	1.67+02	1.02+00	6.95+00	0.609	B	*LS*
				519.748	674676.8	867077.7	6–6	1.19+01	4.84–02	4.97–01	–0.537	B	*LS*
45.	$1s^23d$–$1s^25p$	^2D–$^2P^o$		357.88	674656	954080	10–6	2.58+00	2.97–03	3.50–02	–1.527	B	1
46.	$1s^23d$–$1s^25f$	^2D–$^2F^o$		355.46	674656	955985	10–14	5.91+01	1.57–01	1.83+00	0.195	B	2
47.	$1s^23d$–$1s^26p$	^2D–$^2P^o$		304.44	674656	1003130	10–6	1.35+00	1.13–03	1.13–02	–1.948	B	1
48.	$1s^23d$–$1s^26f$	^2D–$^2F^o$		303.39	674656	1004265	10–14	2.80+01	5.40–02	5.39–01	–0.268	B	2
49.	$1s^23d$–$1s^27p$	^2D–$^2P^o$		279.35	674656	1032631	10–6	8.06–01	5.65–04	5.20–03	–2.248	B–	1
50.	$1s^23d$–$1s^27f$	^2D–$^2F^o$		278.76	674656	1033382	10–14	1.57+01	2.57–02	2.35–01	–0.591	B	2
51.	$1s^23d$–$1s^28p$	^2D–$^2P^o$		265.20	674656	1051724	10–6	5.11–01	3.23–04	2.82–03	–2.490	B	2
52.	$1s^23d$–$1s^28f$	^2D–$^2F^o$		264.81	674656	1052280	10–14	9.82+00	1.45–02	1.26–01	–0.840	B	2
53.	$1s^23d$–$1s^29p$	^2D–$^2P^o$		256.32	674656	1064793	10–6	1.62–01	9.57–05	8.08–04	–3.019	B	3
54.	$1s^23d$–$1s^210p$	^2D–$^2P^o$		249.83	674656	1074922	10–6	9.06–02	5.09–05	4.18–04	–3.294	B	3
55.	$1s^24s$–$1s^24p$	^2S–$^2P^o$	*9360.4*	*9362.9*	852696	*863376*	2–6	1.17–01	4.62–01	2.85+01	–0.034	A	1
			9341.75	9344.31	852696	863397.7	2–4	1.18–01	3.09–01	1.90+01	–0.209	A	*LS*
			9397.86	9400.44	852696	863333.8	2–2	1.16–01	1.53–01	9.49+00	–0.513	A	*LS*
56.	$1s^24s$–$1s^25p$	^2S–$^2P^o$		986.35	852696	954080	2–6	6.86+00	3.00–01	1.95+00	–0.222	B	1
57.	$1s^24s$–$1s^26p$	^2S–$^2P^o$		664.74	852696	1003130	2–6	4.56+00	9.05–02	3.96–01	–0.742	B	1
58.	$1s^24s$–$1s^27p$	^2S–$^2P^o$		555.76	852696	1032631	2–6	3.00+00	4.16–02	1.52–01	–1.080	B	1

O VI: Allowed Transitions — Continued

No.	Transition Array	Mult.	λ_{air} (Å)	λ_{vac} (Å) or σ (cm^{-1})*	E_i (cm^{-1})	E_k (cm^{-1})	g_i-g_k	A_{ki} (10^8 s^{-1})	f_{ik}	S (at. u.)	log gf	Acc.	Source
59.	$1s^24s-1s^28p$	$^2S-^2P^o$		502.44	852696	1051724	2–6	2.05+00	2.32–02	7.69–02	–1.333	B–	1
60.	$1s^24s-1s^29p$	$^2S-^2P^o$		471.48	852696	1064793	2–6	9.72–01	9.72–03	3.02–02	–1.712	B	3
61.	$1s^24s-1s^210p$	$^2S-^2P^o$		449.99	852696	1074922	2–6	6.50–01	5.92–03	1.75–02	–1.927	B	3
62.	$1s^24p-1s^24d$	$^2P^o-^2D$	*28430*	*3517* cm^{-1}	*863376*	*866893*	6–10	4.26–03	8.60–02	4.83+01	–0.287	A	1
			28532.7	3503.8 cm^{-1}	863397.7	866901.5	4–6	4.21–03	7.71–02	2.90+01	–0.511	A	*LS*
			28190.7	3546.3 cm^{-1}	863333.8	866880.1	2–4	3.64–03	8.68–02	1.61+01	–0.761	A	*LS*
			28708.0	3482.4 cm^{-1}	863397.7	866880.1	4–4	6.89–04	8.52–03	3.22+00	–1.468	A	*LS*
63.	$1s^24p-1s^25s$	$^2P^o-^2S$		*1172.1*	*863376*	948690	6–2	1.47+01	1.01–01	2.34+00	–0.217	A	1
				1172.44	863397.7	948690	4–2	9.82+00	1.01–01	1.56+00	–0.393	A	*LS*
				1171.56	863333.8	948690	2–2	4.92+00	1.01–01	7.81–01	–0.694	A	*LS*
64.	$1s^24p-1s^25d$	$^2P^o-^2D$		*1081.3*	*863376*	*955856*	6–10	1.83+01	5.34–01	1.14+01	0.506	A	1
				1081.52	863397.7	955860	4–6	1.83+01	4.81–01	6.85+00	0.284	A	*LS*
				1080.88	863333.8	955851	2–4	1.53+01	5.35–01	3.80+00	0.029	A	*LS*
				1081.63	863397.7	955851	4–4	3.05+00	5.34–02	7.61–01	–0.670	A	*LS*
65.	$1s^24p-1s^26s$	$^2P^o-^2S$		*731.51*	*863376*	1000080	6–2	7.74+00	2.07–02	2.99–01	–0.906	A–	1
				731.624	863397.7	1000080	4–2	5.16+00	2.07–02	1.99–01	–1.082	A–	*LS*
				731.282	863333.8	1000080	2–2	2.58+00	2.07–02	9.97–02	–1.383	A–	*LS*
66.	$1s^24p-1s^26d$	$^2P^o-^2D$		*710.22*	*863376*	*1004178*	6–10	1.10+01	1.39–01	1.95+00	–0.079	A	1
				710.296	863397.7	1004184	4–6	1.10+01	1.25–01	1.17+00	–0.301	A	*LS*
				710.045	863333.8	1004170	2–4	9.18+00	1.39–01	6.49–01	–0.556	A	*LS*
				710.367	863397.7	1004170	4–4	1.83+00	1.39–02	1.30–01	–1.256	A	*LS*
67.	$1s^24p-1s^27s$	$^2P^o-^2S$		*597.36*	*863376*	1030780	6–2	4.61+00	8.22–03	9.70–02	–1.307	A–	1
				597.435	863397.7	1030780	4–2	3.07+00	8.22–03	6.47–02	–1.483	A–	*LS*
				597.207	863333.8	1030780	2–2	1.54+00	8.22–03	3.23–02	–1.784	A–	*LS*
68.	$1s^24p-1s^27d$	$^2P^o-^2D$		*588.42*	*863376*	*1033324*	6–10	6.92+00	5.99–02	6.96–01	–0.444	A–	1
				588.456	863397.7	1033334	4–6	6.92+00	5.39–02	4.18–01	–0.666	A–	*LS*
				588.318	863333.8	1033310	2–4	5.77+00	5.99–02	2.32–01	–0.921	A–	*LS*
				588.539	863397.7	1033310	4–4	1.15+00	5.99–03	4.64–02	–1.621	A–	*LS*
69.	$1s^24p-1s^28s$	$^2P^o-^2S$		*534.28*	*863376*	1050543	6–2	2.98+00	4.25–03	4.48–02	–1.594	B+	1
				534.344	863397.7	1050543	4–2	1.98+00	4.24–03	2.99–02	–1.770	B+	*LS*
				534.162	863333.8	1050543	2–2	9.93–01	4.25–03	1.49–02	–2.071	B+	*LS*
70.	$1s^24p-1s^28d$	$^2P^o-^2D$		*529.33*	*863376*	*1052296*	6–10	4.61+00	3.23–02	3.38–01	–0.713	A–	1
				529.371	863397.7	1052301	4–6	4.61+00	2.91–02	2.03–01	–0.935	A–	*LS*
				529.229	863333.8	1052288	2–4	3.84+00	3.23–02	1.13–01	–1.190	A–	*LS*
				529.408	863397.7	1052288	4–4	7.68–01	3.23–03	2.25–02	–1.889	A–	*LS*
71.	$1s^24p-1s^29d$	$^2P^o-^2D$		*495.17*	*863376*	*1065327*	6–10	3.18+00	1.95–02	1.90–01	–0.932	B	3
				495.198	863397.7	1065337	4–6	3.18+00	1.75–02	1.14–01	–1.154	B	*LS*
				495.105	863333.8	1065311	2–4	2.65+00	1.95–02	6.35–02	–1.410	B	*LS*
				495.262	863397.7	1065311	4–4	5.29–01	1.95–03	1.27–02	–2.109	B	*LS*

O VI: Allowed Transitions — Continued

No.	Transition Array	Mult.	λ_{air} (Å)	λ_{vac} (Å) or σ (cm^{-1})*	E_i (cm^{-1})	E_k (cm^{-1})	g_i–g_k	A_{ki} (10^8 s^{-1})	f_{ik}	S (at. u.)	log gf	Acc.	Source
72.	$1s^24d$–$1s^24f$	^2D–$^2F^o$		*190* cm^{-1}	*866893*	*867083*	10–14	4.09–07	2.37–03	4.11+01	–1.625	B	3
				185.5 cm^{-1}	866901.5	867087.0	6–8	3.80–07	2.21–03	2.35+01	–1.878	B	*LS*
				197.6 cm^{-1}	866880.1	867077.7	4–6	4.28–07	2.47–03	1.64+01	–2.006	B	*LS*
				176.2 cm^{-1}	866901.5	867077.7	6–6	2.17–08	1.05–04	1.17+00	–3.202	B	*LS*
73.	$1s^24d$–$1s^25p$	^2D–$^2P^o$		1147.0	866893	954080	10–6	3.16+00	3.74–02	1.41+00	–0.427	A–	1
74.	$1s^24d$–$1s^25f$	^2D–$^2F^o$		1122.4	866893	955985	10–14	3.36+01	8.88–01	3.28+01	0.948	B	2
75.	$1s^24d$–$1s^26p$	^2D–$^2P^o$		734.01	866893	1003130	10–6	1.56+00	7.58–03	1.83–01	–1.120	B+	1
76.	$1s^24d$–$1s^26f$	^2D–$^2F^o$		727.95	866893	1004265	10–14	1.67+01	1.86–01	4.46+00	0.270	B	2
77.	$1s^24d$–$1s^27p$	^2D–$^2P^o$		603.36	866893	1032631	10–6	8.95–01	2.93–03	5.82–02	–1.533	B	1
78.	$1s^24d$–$1s^27f$	^2D–$^2F^o$		600.64	866893	1033382	10–14	9.55+00	7.23–02	1.43+00	–0.141	B	2
79.	$1s^24d$–$1s^28p$	^2D–$^2P^o$		541.03	866893	1051724	10–6	5.63–01	1.48–03	2.64–02	–1.829	B–	1
80.	$1s^24d$–$1s^28f$	^2D–$^2F^o$		539.41	866893	1052280	10–14	6.00+00	3.67–02	6.51–01	–0.436	B	2
81.	$1s^24d$–$1s^29p$	^2D–$^2P^o$		505.31	866893	1064793	10–6	1.83–01	4.21–04	7.01–03	–2.375	B	3
82.	$1s^24d$–$1s^210p$	^2D–$^2P^o$		480.70	866893	1074922	10–6	1.04–01	2.17–04	3.43–03	–2.664	B	3
83.	$1s^24f$–$1s^25d$	$^2F^o$–^2D		*1126.5*	*867083*	*955856*	14–10	6.63–01	9.01–03	4.68–01	–0.899	B	2
				1126.47	867087.0	955860	8–6	6.31–01	9.01–03	2.67–01	–1.142	B	*LS*
				1126.46	867077.7	955851	6–4	6.63–01	8.41–03	1.87–01	–1.297	B	*LS*
				1126.35	867077.7	955860	6–6	3.16–02	6.01–04	1.34–02	–2.443	B	*LS*
84.	$1s^24f$–$1s^26d$	$^2F^o$–^2D		*729.42*	*867083*	*1004178*	14–10	2.82–01	1.61–03	5.40–02	–1.648	B	2
				729.411	867087.0	1004184	8–6	2.69–01	1.61–03	3.09–02	–1.891	B	*LS*
				729.436	867077.7	1004170	6–4	2.82–01	1.50–03	2.16–02	–2.046	B	*LS*
				729.361	867077.7	1004184	6–6	1.34–02	1.07–04	1.54–03	–3.192	B	*LS*
85.	$1s^24f$–$1s^27d$	$^2F^o$–^2D		*601.53*	*867083*	*1033324*	14–10	1.48–01	5.73–04	1.59–02	–2.096	B	2
				601.515	867087.0	1033334	8–6	1.41–01	5.73–04	9.08–03	–2.339	B	*LS*
				601.568	867077.7	1033310	6–4	1.48–01	5.35–04	6.36–03	–2.494	B	*LS*
				601.481	867077.7	1033334	6–6	7.04–03	3.82–05	4.54–04	–3.640	B	*LS*
86.	$1s^24f$–$1s^28d$	$^2F^o$–^2D		*539.92*	*867083*	*1052296*	14–10	8.83–02	2.76–04	6.86–03	–2.413	B	2
				539.916	867087.0	1052301	8–6	8.41–02	2.76–04	3.92–03	–2.656	B	*LS*
				539.927	867077.7	1052288	6–4	8.83–02	2.57–04	2.75–03	–2.811	B	*LS*
				539.889	867077.7	1052301	6–6	4.21–03	1.84–05	1.96–04	–3.957	B	*LS*

O VI: Allowed Transitions — Continued

No.	Transition Array	Mult.	λ_{air} (Å)	λ_{vac} (Å) or σ (cm^{-1})*	E_i (cm^{-1})	E_k (cm^{-1})	g_i–g_k	A_{ki} (10^8 s^{-1})	f_{ik}	S (at. u.)	log gf	Acc.	Source
87.	$1s^24f$–$1s^29d$	$^2F^o$–^2D		*504.43*	*867083*	– *1065327*	14–10	5.54–02	1.51–04	3.51–03	–2.675	B	3
				504.414	867087.0	– 1065337	8–6	5.27–02	1.51–04	2.00–03	–2.918	B	*LS*
				504.456	867077.7	– 1065311	6–4	5.54–02	1.41–04	1.40–03	–3.073	B	*LS*
				504.390	867077.7	– 1065337	6–6	2.64–03	1.01–05	1.00–04	–4.219	B	*LS*
88.	$1s^24f$–$1s^26g$	$^2F^o$–^2G		728.90	867083	– 1004276	14–18	1.78+01	1.82–01	6.13+00	0.407	B	3
89.	$1s^25s$–$1s^25p$	^2S–$^2P^o$	18547.8	5390 cm^{-1}	948690	– 954080	2–6	3.80–02	5.88–01	7.18+01	0.070	A	1
90.	$1s^25s$–$1s^26p$	^2S–$^2P^o$		1836.9	948690	– 1003130	2–6	2.15+00	3.27–01	3.95+00	–0.185	A–	1
91.	$1s^25s$–$1s^27p$	^2S–$^2P^o$		1191.3	948690	– 1032631	2–6	1.55+00	9.90–02	7.77–01	–0.703	B+	1
92.	$1s^25s$–$1s^28p$	^2S–$^2P^o$		970.55	948690	– 1051724	2–6	1.08+00	4.59–02	2.93–01	–1.037	B	1
93.	$1s^25s$–$1s^29p$	^2S–$^2P^o$		861.30	948690	– 1064793	2–6	4.92–01	1.64–02	9.31–02	–1.484	B	3
94.	$1s^25s$–$1s^210p$	^2S–$^2P^o$		792.19	948690	– 1074922	2–6	3.33–01	9.40–03	4.90–02	–1.726	B	3
95.	$1s^25p$–$1s^25d$	$^2P^o$–^2D		1776 cm^{-1}	954080	– 955856	6–10	1.50–03	1.19–01	1.32+02	–0.148	A	1
96.	$1s^25p$–$1s^26s$	$^2P^o$–^2S	2173.2	2173.9	954080	– 1000080	6–2	5.92+00	1.40–01	5.99+00	–0.077	A	1
97.	$1s^25p$–$1s^26d$	$^2P^o$–^2D		1996.1	954080	– 1004178	6–10	5.41+00	5.39–01	2.12+01	0.510	A	1
98.	$1s^25p$–$1s^27s$	$^2P^o$–^2S		1303.8	954080	– 1030780	6–2	3.38+00	2.87–02	7.39–01	–0.764	A–	1
99.	$1s^25p$–$1s^27d$	$^2P^o$–^2D		1261.9	954080	– 1033324	6–10	3.62+00	1.44–01	3.59+00	–0.063	A	1
100.	$1s^25p$–$1s^28s$	$^2P^o$–^2S		1036.7	954080	– 1050543	6–2	2.13+00	1.14–02	2.34–01	–1.163	B+	1
101.	$1s^25p$–$1s^28d$	$^2P^o$–^2D		1018.2	954080	– 1052296	6–10	2.45+00	6.36–02	1.28+00	–0.419	A–	1
102.	$1s^25p$–$1s^29d$	$^2P^o$–^2D		898.90	954080	– 1065327	6–10	1.73+00	3.49–02	6.19–01	–0.680	B	3
103.	$1s^25d$–$1s^26p$	^2D–$^2P^o$	2114.7	2115.3	955856	– 1003130	10–6	1.58+00	6.34–02	4.41+00	–0.198	A	1
104.	$1s^25d$–$1s^27p$	^2D–$^2P^o$		1302.5	955856	– 1032631	10–6	8.61–01	1.31–02	5.64–01	–0.881	B+	1
105.	$1s^25d$–$1s^28p$	^2D–$^2P^o$		1043.1	955856	– 1051724	10–6	5.25–01	5.14–03	1.77–01	–1.289	B+	1
106.	$1s^25d$–$1s^29p$	^2D–$^2P^o$		917.96	955856	– 1064793	10–6	1.80–01	1.37–03	4.13–02	–1.865	B	3

O VI: Allowed Transitions — Continued

No.	Transition Array	Mult.	λ_{air} (Å)	λ_{vac} (Å) or σ (cm^{-1})*	E_i (cm^{-1})	E_k (cm^{-1})	g_i-g_k	A_{ki} (10^8 s^{-1})	f_{ik}	S (at. u.)	log gf	Acc.	Source
107.	$1s^25d-$ $1s^210p$	$^2D-^2P^o$		839.87	955856	1074922	10–6	1.07–01	6.76–04	1.87–02	–2.170	B	3
108.	$1s^25d-$ $1s^25f$	$^2D-^2F^o$		129 cm^{-1}	955856	955985	10–14	4.44–07	5.61–03	1.43+02	–1.251	B	3
109.	$1s^25d-$ $1s^26f$	$^2D-^2F^o$	2065.1	2065.7	955856	1004265	10–14	9.39+00	8.41–01	5.72+01	0.925	B	2
110.	$1s^25d-$ $1s^27f$	$^2D-^2F^o$		1289.9	955856	1033382	10–14	5.63+00	1.97–01	8.34+00	0.293	B	2
111.	$1s^25d-$ $1s^28f$	$^2D-^2F^o$		1037.1	955856	1052280	10–14	3.58+00	8.08–02	2.76+00	–0.093	B	2
112.	$1s^25f-$ $1s^26d$	$^2F^o-^2D$	2074.3	2075.0	955985	1004178	14–10	5.12–01	2.36–02	2.26+00	–0.481	B	2
113.	$1s^25f-$ $1s^27d$	$^2F^o-^2D$		1293.0	955985	1033324	14–10	2.51–01	4.49–03	2.68–01	–1.201	B	2
114.	$1s^25f-$ $1s^28d$	$^2F^o-^2D$		1038.3	955985	1052296	14–10	1.43–01	1.65–03	7.91–02	–1.636	B	2
115.	$1s^25f-$ $1s^29d$	$^2F^o-^2D$		914.56	955985	1065327	14–10	9.49–02	8.50–04	3.58–02	–1.925	B	3
116.	$1s^25f-$ $1s^26g$	$^2F^o-^2G$	2070.1	2070.8	955985	1004276	14–18	1.43+01	1.19+00	1.13+02	1.220	B	3
117.	$1s^26s-$ $1s^26p$	$^2S-^2P^o$	32778	3050 cm^{-1}	1000080	1003130	2–6	1.45–02	7.02–01	1.51+02	0.147	A	1
118.	$1s^26s-$ $1s^27p$	$^2S-^2P^o$	3071.2	3072.1	1000080	1032631	2–6	8.38–01	3.55–01	7.19+00	–0.148	B	1
119.	$1s^26s-$ $1s^28p$	$^2S-^2P^o$		1936.3	1000080	1051724	2–6	6.38–01	1.08–01	1.37+00	–0.667	B	1
120.	$1s^26s-$ $1s^29p$	$^2S-^2P^o$		1545.3	1000080	1064793	2–6	2.88–01	3.09–02	3.15–01	–1.209	B	3
121.	$1s^26s-$ $1s^210p$	$^2S-^2P^o$		1336.1	1000080	1074922	2–6	1.96–01	1.57–02	1.38–01	–1.502	B	3
122.	$1s^26p-$ $1s^26d$	$^2P^o-^2D$		1048 cm^{-1}	1003130	1004178	6–10	6.74–04	1.53–01	2.89+02	–0.036	A	1
123.	$1s^26p-$ $1s^27s$	$^2P^o-^2S$	3615.6	3616.6	1003130	1030780	6–2	2.74+00	1.79–01	1.28+01	0.031	A	1
124.	$1s^26p-$ $1s^27d$	$^2P^o-^2D$	3311.0	3311.9	1003130	1033324	6–10	2.03+00	5.56–01	3.64+01	0.523	A	1
125.	$1s^26p-$ $1s^28s$	$^2P^o-^2S$	2108.5	2109.1	1003130	1050543	6–2	1.66+00	3.68–02	1.53+00	–0.656	A	1
126.	$1s^26p-$ $1s^28d$	$^2P^o-^2D$	2033.3	2033.9	1003130	1052296	6–10	1.46+00	1.51–01	6.05+00	–0.044	A–	1
127.	$1s^26p-$ $1s^29d$	$^2P^o-^2D$		1607.8	1003130	1065327	6–10	1.04+00	6.71–02	2.13+00	–0.395	B	3

O VI: Allowed Transitions — Continued

No.	Transition Array	Mult.	λ_{air} (Å)	λ_{vac} (Å) or σ (cm^{-1})*	E_i (cm^{-1})	E_k (cm^{-1})	g_i–g_k	A_{ki} (10^8 s^{-1})	f_{ik}	S (at. u.)	log gf	Acc.	Source
128.	$1s^26d$–$1s^27p$	^{2}D–^{2}P°	3513.6	3514.6	1004178	– 1032631	10–6	8.24−01	9.15−02	1.06+01	−0.038	A	1
129.	$1s^26d$–$1s^28p$	^{2}D–^{2}P°	2102.6	2103.2	1004178	– 1051724	10–6	4.85−01	1.93−02	1.33+00	−0.715	B+	1
130.	$1s^26d$–$1s^29p$	^{2}D–^{2}P°		1649.8	1004178	– 1064793	10–6	1.76−01	4.30−03	2.34−01	−1.366	B	3
131.	$1s^26d$–$1s^210p$	^{2}D–^{2}P°		1413.5	1004178	– 1074922	10–6	1.03−01	1.85−03	8.62−02	−1.732	B	3
132.	$1s^26d$–$1s^27f$	^{2}D–^{2}F°	3423.2	3424.2	1004178	– 1033382	10–14	3.37+00	8.29−01	9.34+01	0.918	B	2
133.	$1s^26d$–$1s^28f$	^{2}D–^{2}F°	2078.3	2078.9	1004178	– 1052280	10–14	2.24+00	2.03−01	1.39+01	0.308	B	2
134.	$1s^26f$–$1s^27d$	^{2}F°–^{2}D	3440.3	3441.3	1004265	– 1033324	14–10	3.31−01	4.19−02	6.65+00	−0.231	B	2
135.	$1s^26f$–$1s^28d$	^{2}F°–^{2}D	2081.3	2082.0	1004265	– 1052296	14–10	1.79−01	8.31−03	7.97−01	−0.934	B	2
136.	$1s^26f$–$1s^29d$	^{2}F°–^{2}D		1637.7	1004265	– 1065327	14–10	1.12−01	3.22−03	2.43−01	−1.346	B	3
137.	$1s^26g$–$1s^27f$	^{2}G–^{2}F°	3434.7	3435.7	1004276	– 1033382	18–14	1.47−01	2.02−02	4.11+00	−0.440	B	3
138.	$1s^26g$–$1s^28f$	^{2}G–^{2}F°	2082.5	2083.2	1004276	– 1052280	18–14	6.86−02	3.47−03	4.28−01	−1.204	B	3
139.	$1s^27s$–$1s^27p$	^{2}S–^{2}P°		1851 cm^{-1}	1030780	– 1032631	2–6	6.31−03	8.29−01	2.95+02	0.219	B	2
140.	$1s^27s$–$1s^28p$	^{2}S–^{2}P°	4773.3	4774.6	1030780	– 1051724	2–6	3.74−01	3.84−01	1.21+01	−0.115	B	2
141.	$1s^27s$–$1s^29p$	^{2}S–^{2}P°	2939.2	2940.1	1030780	– 1064793	2–6	1.65−01	6.43−02	1.24+00	−0.891	B	3
142.	$1s^27s$–$1s^210p$	^{2}S–^{2}P°	2264.7	2265.4	1030780	– 1074922	2–6	1.22−01	2.81−02	4.19−01	−1.250	B	3
143.	$1s^27p$–$1s^28s$	^{2}P°–^{2}S	5581.0	5582.5	1032630	– 1050543	6–2	1.37+00	2.13−01	2.34+01	0.106	B	3
144.	$1s^27p$–$1s^27d$	^{2}P°–^{2}D		694 cm^{-1}	1032630	– 1033324	6–10	3.35−04	1.74−01	4.95+02	0.018	B	2
145.	$1s^27p$–$1s^28d$	^{2}P°–^{2}D	5083.6	5085.0	1032630	– 1052296	6–10	8.98−01	5.80−01	5.83+01	0.542	B	2
146.	$1s^27p$–$1s^29d$	^{2}P°–^{2}D	3057.5	3058.4	1032630	– 1065327	6–10	6.62−01	1.55−01	9.35+00	−0.032	B	3
147.	$1s^27d$–$1s^28p$	^{2}D–^{2}P°	5433.4	5434.9	1033324	– 1051724	10–6	4.54−01	1.21−01	2.16+01	0.082	B	2
148.	$1s^27d$–$1s^29p$	^{2}D–^{2}P°	3176.9	3177.8	1033324	– 1064793	10–6	1.53−01	1.39−02	1.45+00	−0.857	B	3

O VI: Allowed Transitions — Continued

No.	Transition Array	Mult.	λ_{air} (Å)	λ_{vac} (Å) or σ (cm^{-1})*	E_i (cm^{-1})	E_k (cm^{-1})	g_i-g_k	A_{ki} (10^8 s^{-1})	f_{ik}	S (at. u.)	log gf	Acc.	Source
149.	$1s^27d-1s^210p$	$^2D-^2P^o$	2403.3	2404.0	1033324	– 1074922	10–6	9.49–02	4.93–03	3.90–01	–1.307	B	3
150.	$1s^27d-1s^28f$	$^2D-^2F^o$	5274.0	5275.5	1033324	– 1052280	10–14	1.43+00	8.34–01	1.45+02	0.921	B	2
151.	$1s^27f-1s^28d$	$^2F^o-^2D$	5285.7	5287.1	1033382	– 1052296	14–10	2.10–01	6.28–02	1.53+01	–0.056	B	2
152.	$1s^27f-1s^29d$	$^2F^o-^2D$	3129.5	3130.4	1033382	– 1065327	14–10	1.22–01	1.28–02	1.85+00	–0.745	B	3
153.	$1s^28s-1s^28p$	$^2S-^2P^o$		1181 cm^{-1}	1050543	– 1051724	2–6	2.94–03	9.49–01	5.29+02	0.278	B	2
154.	$1s^28s-1s^29p$	$^2S-^2P^o$	7015.6	7017.5	1050543	– 1064793	2–6	8.84–02	1.96–01	9.05+00	–0.407	B	3
155.	$1s^28s-1s^210p$	$^2S-^2P^o$	4100.7	4101.9	1050543	– 1074922	2–6	7.69–02	5.82–02	1.57+00	–0.934	B	3
156.	$1s^28p-1s^28d$	$^2P^o-^2D$		572 cm^{-1}	1051724	– 1052296	6–10	2.65–04	2.03–01	7.00+02	0.085	B	2
157.	$1s^28p-1s^29d$	$^2P^o-^2D$	7349.5	7351.5	1051724	– 1065327	6–10	4.34–01	5.86–01	8.51+01	0.546	B	3
158.	$1s^28d-1s^29p$	$^2D-^2P^o$	7999.6	8001.8	1052296	– 1064793	10–6	1.42–01	8.17–02	2.15+01	–0.088	B	3
159.	$1s^28d-1s^210p$	$^2D-^2P^o$	4418.4	4419.7	1052296	– 1074922	10–6	9.22–02	1.62–02	2.36+00	–0.790	B	3
160.	$1s^28f-1s^29d$	$^2F^o-^2D$	7662.7	7664.8	1052280	– 1065327	14–10	1.40–01	8.80–02	3.11+01	0.090	B	3
161.	$1s^29p-1s^29d$	$^2P^o-^2D$		534 cm^{-1}	1064793	– 1065327	10–6	6.45–04	2.04–01	1.26+03	0.309	B	3
162.	$1s^29d-1s^210p$	$^2D-^2P^o$	10419	9595 cm^{-1}	1065327	– 1074922	10–6	7.94–02	7.76–02	2.66+01	–0.110	B	3
163.	$1s2s2p-1s2s3s$	$^4P^o-^4S$		151.4			12–4	3.29+02	3.77–02	2.26–01	–0.344	B	5
164.	$1s2s2p-1s2s3d$	$^4P^o-^4D$		140.4			12–20	1.49+03	7.36–01	4.08+00	0.946	B	5
165.	$1s2s2p-1s2p3p$	$^4P^o-^4D$		132.4			12–20	6.35+01	2.78–02	1.46–01	–0.476	B	5
166.		$^4P^o-^4S$		130.9			12–4	2.77+02	2.37–02	1.22–01	–0.546	B	5
167.		$^4P^o-^4P$		130.2			12–12	3.10+02	7.88–02	4.05–01	–0.024	B	5
168.	$1s2s2p-1s2s4s$	$^4P^o-^4S$		111.5			12–4	8.79+01	5.46–03	2.41–02	–1.184	B	5
169.	$1s2s2p-1s2s4d$	$^4P^o-^4D$		109.1			12–20	4.63+02	1.38–01	5.93–01	0.218	B	5

O VI: Allowed Transitions — Continued

No.	Transition Array	Mult.	λ_{air} (Å)	λ_{vac} (Å) or σ (cm^{-1})*	E_i (cm^{-1})	E_k (cm^{-1})	g_i–g_k	A_{ki} (10^8 s^{-1})	f_{ik}	S (at. u.)	log gf	Acc.	Source
170.	$1s2s2p$–$1s2p4p$	$^4P^o$–^4D		103.1			12–20	9.99+01	2.65−02	1.08−01	−0.497	B	5
171.		$^4P^o$–^4S		102.8			12–4	1.64+02	8.69−03	3.53−02	−0.982	B	5
172.		$^4P^o$–^4P		102.6			12–12	1.51+02	2.38−02	9.64−02	−0.545	B	5
173.	$1s2s2p$–$1s2s5d$	$^4P^o$–^4D		99.08			12–20	2.02+02	4.95−02	1.94−01	−0.226	B	5
174.	$1s2s2p$–$1s2s6d$	$^4P^o$–^4D		94.44			12–20	1.31+02	2.91−02	1.09−01	−0.456	B	5
175.	$1s2s2p$–$1s2p5p$	$^4P^o$–^4P		93.61			12–12	7.96+01	1.05−02	3.87−02	−0.901	B	5
176.	$1s2s2p$–$1s2s7d$	$^4P^o$–^4D		91.85			12–20	6.82+01	1.44−02	5.22−02	−0.763	B	5
177.	$1s2p^2$–$1s2s3p$	^4P–$^4P^o$		170.8			12–12	5.80+00	2.54−03	1.71−02	−1.517	B	5
178.	$1s2p^2$–$1s2p3s$	^4P–$^4P^o$		159.0			12–12	2.44+02	9.24−02	5.81−01	0.045	B	5
179.	$1s2p^2$–$1s2p3d$	^4P–$^4D^o$		147.4			12–20	1.60+03	8.69−01	5.06+00	1.018	B	5
180.		^4P–$^4P^o$		146.0			12–12	8.21+02	2.62−01	1.51+00	0.498	B	5
181.	$1s2p^2$–$1s2p4s$	^4P–$^4P^o$		116.9			12–12	7.17+01	1.47−02	6.79−02	−0.754	B	5
182.	$1s2p^2$–$1s2p4d$	^4P–$^4D^o$		114.3			12–20	5.78+02	1.89−01	8.52−01	0.355	B	5
183.		^4P–$^4P^o$		114.2			12–12	2.96+02	5.78−02	2.60−01	−0.159	B	5
184.	$1s2p^2$–$1s2p5d$	^4P–$^4D^o$		103.6			12–20	2.75+02	7.39−02	3.02−01	−0.052	B	5
185.		^4P–$^4P^o$		103.5			12–12	1.43+02	2.29−02	9.37−02	−0.561	B	5
186.	$1s2p^2$–$1s2p6d$	^4P–$^4D^o$		98.61			12–20	1.53+02	3.72−02	1.45−01	−0.351	B	5
187.		^4P–$^4P^o$		98.55			12–12	6.05+01	8.81−03	3.43−02	−0.976	B	5
188.	$1s2p^2$–$1s2p7d$	^4P–$^4D^o$		95.83			12–20	9.35+01	2.15−02	8.12−02	−0.589	B	5
189.		^4P–$^4P^o$		95.80			12–12	4.96+01	6.82−03	2.58−02	−1.087	B	5
190.	$1s2s3p$–$1s2s4d$	$^4P^o$–^4D		445.9			12–20	9.37+01	4.66−01	8.20+00	0.747	B	5

O VI: Allowed Transitions — Continued

No.	Transition Array	Mult.	λ_{air} (Å)	λ_{vac} (Å) or σ (cm^{-1})*	E_i (cm^{-1})	E_k (cm^{-1})	g_i–g_k	A_{ki} (10^8 s^{-1})	f_{ik}	S (at. u.)	log gf	Acc.	Source
191.	$1s2s3d$–$1s2s4f$	^{4}D–^{4}F°		474.6			20–28	1.89+02	8.92–01	2.79+01	1.251	B	5
192.	$1s2s3d$–$1s2s5f$	^{4}D–^{4}F°		333.6			20–28	6.95+01	1.62–01	3.56+00	0.511	B	5
193.	$1s2p3p$–$1s2p4d$	^{4}P–^{4}D°		468.9			12–20	6.76+01	3.71–01	6.88+00	0.649	B	5
194.		^{4}D–^{4}F°		450.5			20–28	9.42+01	4.01–01	1.19+01	0.904	B	5
195.	$1s2p3d$–$1s2p4f$	^{4}D°–^{4}F		506.5			20–28	1.63+02	8.77–01	2.93+01	1.244	B	5
196.	$1s2p3d$–$1s2p5f$	^{4}D°–^{4}F		348.0			20–28	5.26+01	1.34–01	3.06+00	0.427	B	5
197.	$1s2p3d$–$1s2p4f$	^{4}P°–^{4}D		514.0			12–20	1.45+02	9.54–01	1.94+01	1.059	B	5
198.	$1s2p3d$–$1s2p5f$	^{4}P°–^{4}D		354.1			12–20	5.09+01	1.59–01	2.23+00	0.282	B	5
199.	$1s2p3d$–$1s2p4f$	^{4}F°–^{4}G		473.6			28–36	1.87+02	8.07–01	3.52+01	1.354	B	5
200.	$1s2p3d$–$1s2p5f$	^{4}F°–^{4}G		332.1			28–36	6.71+01	1.43–01	4.37+00	0.601	B	5

*Wavelengths (Å) are always given unless cm^{-1} is indicated.

O VII

Helium Isoelectronic Sequence

Ground State: $1s^2\ ^1S_0$

Ionization Energy: 739.274 eV = 5962800 cm^{-1}

Allowed Transitions

List of tabulated lines

Wavelength (Å)	No.	Wavelength (Å)	No.	Wavelength (Å)	No.	Wavelength (Å)	No.
in vacuum		78.8619	15	81.5947	20	86.1037	13
		78.8656	15	81.9135	6	86.1186	13
18.6270	2	78.8998	15	84.4253	25	86.1200	13
21.6020	1	81.4697	14	86.0748	13	86.2017	19
75.3438	8	81.4737	14	86.0778	13	89.3632	24
77.6946	7	81.5102	14	86.0793	13	91.0780	5

List of tabulated lines — Continued

Wavelength (Å)	No.	Wavelength (Å)	No.	Wavelength (Å)	No.	Wavenumber (cm^{-1})	No.
96.1224	12	262.419	45	821.153	74	220	89
96.1243	12	262.460	45	832.432	77	270	97
96.1280	12	262.467	45	833.264	61	1340	93
96.1289	12	263.380	53	836.960	69	1430	73
96.1751	12	352.386	27	837.240	69	1960	83
96.1788	12	368.080	40	840.760	69	1980	83
96.1797	18	369.235	34	915.081	81		
97.0196	11	369.918	34	924.044	85		
97.0252	11	369.973	34	956.846	91		
97.0770	11	380.764	49	957.029	87		
100.254	23	382.468	59	957.213	87		
120.333	4	383.627	33	959.049	87		
128.177	17	387.252	44	1446.76	80		
128.411	10	387.342	44	1478.63	84		
128.421	10	387.357	44	1521.61	90		
128.500	10	387.417	52	1533.04	92		
128.502	10	392.850	51	1544.88	96		
128.512	10	397.078	72	1554.48	86		
132.767	9	397.141	72	1554.97	86		
132.777	9	397.931	72	1559.82	86		
132.874	9	424.538	58	1623.64	3		
135.820	22	432.788	64	1638.27	3		
137.510	21	435.464	68	1639.88	3		
185.536	31	442.341	76				
194.905	30	442.615	71	in air			
200.807	38	442.693	71				
201.483	43	443.675	71	2384.2	94		
205.120	47	511.797	57	2448.98	16		
205.145	47	523.972	67	2559.40	98		
205.149	47	525.017	63	2575.2	95		
211.457	29	533.960	75	5933.07	26		
218.565	42	538.300	70	8241.76	39		
218.627	37	538.416	70	11399.4	32		
223.534	46	538.851	78	11412.4	32		
223.564	46	539.869	70	11491.1	32		
223.569	46	739.700	82	15970	55		
224.653	54	766.871	88	20000	65		
245.930	28	766.989	88	25630	60		
255.083	41	768.167	88	29400	60		
255.135	36	774.593	56	29750	60		
255.265	36	797.766	66	32460	48		
255.278	36	799.872	62	35300	79		
258.438	35	801.154	62	45860	83		
261.110	50	801.282	62				

We have mainly utilized the results of Cann and Thakkar,[1] who calculated the oscillator strengths of about 100 $n\mathrm{S} - m\mathrm{P}$ and $n\mathrm{P} - m\mathrm{D}$ transitions ($n, m \leq 6$) with variationally determined, highly correlated wavefunctions. This work is an expansion of, and is similar in spirit to, the classic variational calculations on He-like ions by Schiff and Pekeris.[2] Cann and Thakkar improved this variational approach by the use of exponential correlation factors which yield more accurate wavefunctions with a similar number of term expansions. Comparisons between their results, the earlier data by Schiff and Pekeris[2] and similar work by Kono and Hattori[3] in the 1980s show that the differences do not exceed 0.1%.

The presented data are the arithmetic mean values of the ''dipole length'' and ''dipole velocity'' results of Cann and Thakkar.[1] Their numerical values are given to at least three, but most often five or six significant figures, and the differences between the length and velocity data do not exceed seven units in the last digit shown. Thus, these differences, which they consider to be their main indicator of accuracy, are smaller than 0.1% for all but a few transitions.

For some transitions involving higher excited states, which their material does not cover, we have used the results of the Opacity Project (OP), which is reviewed in the general introduction. The calculations for He-like ions were carried out by Fernley *et al.*[4] and also are estimated to be very accurate, as comparisons with the above work of Cann and Thakkar show when there is overlap.

It should be noted that the variational calculations[1] have been done in a non-relativistic framework. This is an excellent approximation for neutral helium, but for highly charged

He-like ions, relativistic effects are expected to become significant. For O VII, some test calculations by Kim and Weiss[5] show that such effects are still very small. They become noticeable only for the line strengths of P-D transitions, where deviations from non-relativistic values may sometimes reach 1%.

References

[1] N. M. Cann and A. J. Thakkar, Phys. Rev. A **46**, 5397 (1992).
[2] B. Schiff and C. L. Pekeris, Phys. Rev. **134**, A638 (1964).
[3] A. Kono and S. Hattori, Phys. Rev. A **30**, 2093 (1984).
[4] J. A. Fernley, K. T. Taylor and M. J. Seaton, J. Phys. B **20**, 6457 (1987).
[5] Y.-K. Kim and A. W. Weiss (private communication), 1993.

O VII: Allowed Transitions

No.	Transition Array	Mult.	λ_{air} (Å)	λ_{vac} (Å) or σ (cm^{-1})*	E_i (cm^{-1})	E_k (cm^{-1})	g_i–g_k	A_{ki} (10^8 s^{-1})	f_{ik}	S (at. u.)	log gf	Acc.	Source
1.	$1s^2$–$1s2p$	^{1}S–^{1}P°		21.6020	0.0	4629201	1–3	3.309+04	6.944–01	4.939–02	–0.1584	AA	1
2.	$1s^2$–$1s3p$	^{1}S–^{1}P°		18.6270	0.0	5368550	1–3	9.365+03	1.461–01	8.962–03	–0.8352	AA	1
3.	$1s2s$–$1s2p$	^{3}S–^{3}P°		*1630.3*	4524640	*4585979*	3–9	7.935–01	9.486–02	1.527+00	–0.5458	AA	1
				1623.64	4524640	4586230	3–5	8.033–01	5.291–02	8.485–01	–0.7993	AA	*LS*
				1638.27	4524640	4585680	3–3	7.820–01	3.146–02	5.091–01	–1.0251	AA	*LS*
				1639.88	4524640	4585620	3–1	7.797–01	1.048–02	1.697–01	–1.5026	AA	*LS*
4.	$1s2s$–$1s3p$	^{3}S–^{3}P°		120.333	4524640	5355670	3–9	5.334+02	3.474–01	4.128–01	0.0179	AA	1
5.	$1s2s$–$1s4p$	^{3}S–^{3}P°		91.0780	4524640	5622600	3–9	2.387+02	8.906–02	8.011–02	–0.5732	AA	1
6.	$1s2s$–$1s5p$	^{3}S–^{3}P°		81.9135	4524640	5745440	3–9	1.237+02	3.732–02	3.019–02	–0.9509	AA	1
7.	$1s2s$–$1s6p$	^{3}S–^{3}P°		77.6946	4524640	5811730	3–9	7.178+01	1.949–02	1.495–02	–1.2331	AA	1
8.	$1s2s$–$1s7p$	^{3}S–^{3}P°		75.3438	4524640	5851890	3–9	4.53+01	1.16–02	8.60–03	–1.460	A	4
9.	$1s2p$–$1s3s$	^{3}P°–^{3}S		*132.83*	*4585979*	5338820	9–3	2.251+02	1.985–02	7.811–02	–0.7480	AA	1
				132.874	4586230	5338820	5–3	1.249+02	1.984–02	4.340–02	–1.0035	AA	*LS*
				132.777	4585680	5338820	3–3	7.512+01	1.986–02	2.604–02	–1.2250	AA	*LS*
				132.767	4585620	5338820	1–3	2.505+01	1.986–02	8.679–03	–1.7021	AA	*LS*
10.	$1s2p$–$1s3d$	^{3}P°–^{3}D		*128.46*	*4585979*	*5364423*	9–15	1.615+03	6.659–01	2.534+00	0.7776	AA	1
				128.500	4586230	5364440	5–7	1.613+03	5.592–01	1.183+00	0.4465	AA	*LS*
				128.411	4585680	5364430	3–5	1.213+03	4.996–01	6.336–01	0.1758	AA	*LS*
				128.411	4585620	5364370	1–3	8.982+02	6.661–01	2.816–01	–0.1764	AA	*LS*
				128.502	4586230	5364430	5–5	4.033+02	9.985–02	2.112–01	–0.3017	AA	*LS*
				128.421	4585680	5364370	3–3	6.735+02	1.665–01	2.112–01	–0.3014	AA	*LS*
				128.512	4586230	5364370	5–3	4.481+01	6.656–03	1.408–02	–1.4778	AA	*LS*
11.	$1s2p$–$1s4s$	^{3}P°–^{3}S		*97.053*	*4585979*	5616340	9–3	8.880+01	4.180–03	1.202–02	–1.4246	AA	1
				97.0770	4586230	5616340	5–3	4.930+01	4.179–03	6.678–03	–1.6800	AA	*LS*
				97.0252	4585680	5616340	3–3	2.963+01	4.181–03	4.007–03	–1.9016	AA	*LS*
				97.0196	4585620	5616340	1–3	9.877+00	4.181–03	1.336–03	–2.3787	AA	*LS*

O VII: Allowed Transitions — Continued

No.	Transition Array	Mult.	λ_{air} (Å)	λ_{vac} (Å) or σ (cm^{-1})*	E_i (cm^{-1})	E_k (cm^{-1})	g_i-g_k	A_{ki} (10^8 s^{-1})	f_{ik}	S (at. u.)	log gf	Acc.	Source
12.	$1s2p-$ $1s4d$	^{3}P°$-^3$D		*96.131*	*4585979*	*5626225*	9–15	5.323+02	1.229−01	3.501−01	0.0438	AA	1
				96.1289	4586230	5626500	5–7	5.323+02	1.032−01	1.634−01	−0.2872	AA	*LS*
				96.1243	4585680	5626000	3–5	3.993+02	9.218−02	8.752−02	−0.5582	AA	*LS*
				96.1224	4585620	5625960	1–3	2.958+02	1.229−01	3.890−02	−0.9104	AA	*LS*
				96.1751	4586230	5626000	5–5	1.329+02	1.843−02	2.917−02	−1.0356	AA	*LS*
				96.1280	4585680	5625960	3–3	2.218+02	3.073−02	2.917−02	−1.0354	AA	*LS*
				96.1788	4586230	5625960	5–3	1.476+01	1.228−03	1.945−03	−2.2117	AA	*LS*
13.	$1s2p-$ $1s5d$	^{3}P°$-^3$D		*86.093*	*4585979*	*5747509*	9–15	2.462+02	4.560−02	1.163−01	−0.3868	AA	1
				86.1037	4586230	5747620	5–7	2.461+02	3.830−02	5.428−02	−0.7178	AA	*LS*
				86.0778	4585680	5747420	3–5	1.848+02	3.421−02	2.908−02	−0.9888	AA	*LS*
				86.0748	4585620	5747400	1–3	1.369+02	4.561−02	1.292−02	−1.3409	AA	*LS*
				86.1186	4586230	5747420	5–5	6.150+01	6.838−03	9.694−03	−1.4661	AA	*LS*
				86.0793	4585680	5747400	3–3	1.026+02	1.140−02	9.694−03	−1.4659	AA	*LS*
				86.1200	4586230	5747400	5–3	6.833+00	4.559−04	6.462−04	−2.6422	AA	*LS*
14.	$1s2p-$ $1s6d$	^{3}P°$-^3$D		*81.494*	*4585979*	5813070	9–15	1.352+02	2.243−02	5.415−02	−0.6950	AA	1
				81.5102	4586230	5813070	5–7	1.351+02	1.883−02	2.527−02	−1.0261	AA	*LS*
				81.4737	4585680	5813070	3–5	1.014+02	1.682−02	1.354−02	−1.2969	AA	*LS*
				81.4697	4585620	5813070	1–3	7.515+01	2.243−02	6.017−03	−1.6491	AA	*LS*
				81.5102	4586230	5813070	5–5	3.377+01	3.363−03	4.513−03	−1.7743	AA	*LS*
				81.4737	4585680	5813070	3–3	5.635+01	5.608−03	4.513−03	−1.7741	AA	*LS*
				81.5102	4586230	5813070	5–3	3.752+00	2.242−04	3.008−04	−2.9503	AA	*LS*
15.	$1s2p-$ $1s7d$	^{3}P°$-^3$D		*78.884*	*4585979*	5853660	9–15	8.28+01	1.29−02	3.01−02	−0.936	A	4
				78.8998	4586230	5853660	5–7	8.28+01	1.08−02	1.40−02	−1.267	A	*LS*
				78.8656	4585680	5853660	3–5	6.22+01	9.66−03	7.53−03	−1.538	A	*LS*
				78.8619	4585620	5853660	1–3	4.61+01	1.29−02	3.34−03	−1.890	A	*LS*
				78.8998	4586230	5853660	5–5	2.07+01	1.93−03	2.51−03	−2.015	A	*LS*
				78.8656	4585680	5853660	3–3	3.45+01	3.22−03	2.51−03	−2.015	A	*LS*
				78.8998	4586230	5853660	5–3	2.30+00	1.29−04	1.67−04	−3.191	A	*LS*
16.	$1s2s-$ $1s2p$	^{1}S$-^1$P°	2448.98	2449.72	4588380	4629201	1–3	2.514−01	6.786−02	5.473−01	−1.1684	AA	1
17.	$1s2s-$ $1s3p$	^{1}S$-^1$P°		128.177	4588380	5368550	1–3	5.055+02	3.735−01	1.576−01	−0.4277	AA	1
18.	$1s2s-$ $1s4p$	^{1}S$-^1$P°		96.1797	4588380	5628100	1–3	2.250+02	9.360−02	2.964−02	−1.0287	AA	1
19.	$1s2s-$ $1s5p$	^{1}S$-^1$P°		86.2017	4588380	5748450	1–3	1.165+02	3.894−02	1.105−02	−1.4096	AA	1
20.	$1s2s-$ $1s6p$	^{1}S$-^1$P°		81.5947	4588380	5813950	1–3	6.770+01	2.027−02	5.445−03	−1.6931	AA	1
21.	$1s2p-$ $1s3s$	^{1}P°$-^1$S		137.510	4629201	5356420	3–1	2.008+02	1.897−02	2.576−02	−1.2448	AA	1
22.	$1s2p-$ $1s3d$	^{1}P°$-^1$D		135.820	4629201	5365470	3–5	1.523+03	7.021−01	9.418−01	0.3235	AA	1
23.	$1s2p-$ $1s4d$	^{1}P°$-^1$D		100.254	4629201	5626670	3–5	4.772+02	1.198−01	1.187−01	−0.4443	AA	1
24.	$1s2p-$ $1s5d$	^{1}P°$-^1$D		89.3632	4629201	5748230	3–5	2.167+02	4.324−02	3.816−02	−0.8870	AA	1

O VII: Allowed Transitions — Continued

No.	Transition Array	Mult.	λ_{air} (Å)	λ_{vac} (Å) or σ (cm^{-1})*	E_i (cm^{-1})	E_k (cm^{-1})	g_i–g_k	A_{ki} (10^8 s^{-1})	f_{ik}	S (at. u.)	log gf	Acc.	Source
25.	1*s*2*p*– 1*s*6*d*	$^1P^o-^1D$		84.4253	4629201	5813680	3–5	1.177+02	2.097–02	1.748–02	–1.2013	AA	1
26.	1*s*3*s*– 1*s*3*p*	$^3S-^3P^o$	5933.1	5934.72	5338820	5355670	3–9	1.002–01	1.588–01	9.308+00	–0.3220	AA	1
27.	1*s*3*s*– 1*s*4*p*	$^3S-^3P^o$		352.386	5338820	5622600	3–9	6.769+01	3.780–01	1.316+00	0.0546	AA	1
28.	1*s*3*s*– 1*s*5*p*	$^3S-^3P^o$		245.930	5338820	5745440	3–9	3.764+01	1.024–01	2.487–01	–0.5127	AA	1
29.	1*s*3*s*– 1*s*6*p*	$^3S-^3P^o$		211.457	5338820	5811730	3–9	2.224+01	4.472–02	9.340–02	–0.8724	AA	1
30.	1*s*3*s*– 1*s*7*p*	$^3S-^3P^o$		194.905	5338820	5851890	3–9	1.41+01	2.42–02	4.65–02	–1.140	A	4
31.	1*s*3*s*– 1*s*8*p*	$^3S-^3P^o$		185.536	5338820	5877800	3–9	4.37+00	6.77–03	1.24–02	–1.692	A	4
32.	1*s*3*p*– 1*s*3*d*	$^3P^o-^3D$	*11422*	*8753* cm^{-1}	5355670	*5364423*	9–15	1.121–02	3.655–02	1.237+01	–0.4829	AA	1
			11399.4	8770 cm^{-1}	5355670	5364440	5–7	1.127–02	3.076–02	5.774+00	–0.8130	AA	*LS*
			11412.4	8760 cm^{-1}	5355670	5364430	3–5	8.426–03	2.743–02	3.093+00	–1.0846	AA	*LS*
			11491.1	8700 cm^{-1}	5355670	5364370	1–3	6.114–03	3.633–02	1.375+00	–1.4397	AA	*LS*
			11412.4	8760 cm^{-1}	5355670	5364430	5–5	2.809–03	5.487–03	1.031+00	–1.5617	AA	*LS*
			11491.1	8700 cm^{-1}	5355670	5364370	3–3	4.585–03	9.082–03	1.031+00	–1.5647	AA	*LS*
			11491.1	8700 cm^{-1}	5355670	5364370	5–3	3.057–04	3.633–04	6.873–02	–2.7408	AA	*LS*
33.	1*s*3*p*– 1*s*4*s*	$^3P^o-^3S$		383.627	5355670	5616340	9–3	6.207+01	4.565–02	5.189–01	–0.3863	AA	1
34.	1*s*3*p*– 1*s*4*d*	$^3P^o-^3D$		*369.61*	5355670	*5626225*	9–15	1.671+02	5.706–01	6.248+00	0.7105	AA	1
				369.235	5355670	5626500	5–7	1.677+02	4.798–01	2.916+00	0.3800	AA	*LS*
				369.918	5355670	5626000	3–5	1.250+02	4.276–01	1.562+00	0.1081	AA	*LS*
				369.973	5355670	5625960	1–3	9.259+01	5.700–01	6.943–01	–0.2441	AA	*LS*
				369.918	5355670	5626000	5–5	4.168+01	8.551–02	5.207–01	–0.3690	AA	*LS*
				369.973	5355670	5625960	3–3	6.944+01	1.425–01	5.207–01	–0.3691	AA	*LS*
				369.973	5355670	5625960	5–3	4.629+00	5.700–03	3.471–02	–1.5452	AA	*LS*
35.	1*s*3*p*– 1*s*5*s*	$^3P^o-^3S$		258.438	5355670	5742610	9–3	2.963+01	9.890–03	7.573–02	–1.0506	AA	1
36.	1*s*3*p*– 1*s*5*d*	$^3P^o-^3D$		*255.21*	5355670	*5747509*	9–15	8.303+01	1.351–01	1.022+00	0.0850	AA	1
				255.135	5355670	5747620	5–7	8.310+01	1.135–01	4.768–01	–0.2459	AA	*LS*
				255.265	5355670	5747420	3–5	6.223+01	1.013–01	2.554–01	–0.5172	AA	*LS*
				255.278	5355670	5747400	1–3	4.609+01	1.351–01	1.135–01	–0.8694	AA	*LS*
				255.265	5355670	5747420	5–5	2.074+01	2.026–02	8.514–02	–0.9943	AA	*LS*
				255.278	5355670	5747400	3–3	3.457+01	3.377–02	8.514–02	–0.9943	AA	*LS*
				255.278	5355670	5747400	5–3	2.304+00	1.351–03	5.676–03	–2.1704	AA	*LS*
37.	1*s*3*p*– 1*s*6*d*	$^3P^o-^3D$		218.627	5355670	5813070	9–15	4.645+01	5.548–02	3.594–01	–0.3016	AA	1
38.	1*s*3*p*– 1*s*7*d*	$^3P^o-^3D$		200.807	5355670	5853660	9–15	2.88+01	2.90–02	1.73–01	–0.583	A	4
39.	1*s*3*s*– 1*s*3*p*	$^1S-^1P^o$	8241.76	8244.02	5356420	5368550	1–3	3.864–02	1.181–01	3.205+00	–0.9278	AA	1

O VII: Allowed Transitions — Continued

No.	Transition Array	Mult.	λ_{air} (Å)	λ_{vac} (Å) or σ (cm^{-1})*	E_i (cm^{-1})	E_k (cm^{-1})	g_i–g_k	A_{ki} (10^8 s^{-1})	f_{ik}	S (at. u.)	log gf	Acc.	Source
40.	$1s3s$–$1s4p$	^{1}S–^{1}P°		368.080	5356420	5628100	1–3	6.671+01	4.065–01	4.926–01	–0.3910	AA	1
41.	$1s3s$–$1s5p$	^{1}S–^{1}P°		255.083	5356420	5748450	1–3	3.684+01	1.078–01	9.053–02	–0.9674	AA	1
42.	$1s3s$–$1s6p$	^{1}S–^{1}P°		218.565	5356420	5813950	1–3	2.179+01	4.681–02	3.368–02	–1.3297	AA	1
43.	$1s3s$–$1s7p$	^{1}S–^{1}P°		201.483	5356420	5852740	1–3	1.38+01	2.52–02	1.67–02	–1.599	A	4
44.	$1s3d$–$1s4p$	^{3}D–^{3}P°		*387.33*	*5364423*	5622600	15–9	1.052+01	1.420–02	2.715–01	–0.6717	AA	1
				387.357	5364440	5622600	7–5	8.835+00	1.420–02	1.267–01	–1.0028	AA	*LS*
				387.342	5364430	5622600	5–3	7.889+00	1.065–02	6.788–02	–1.2738	AA	*LS*
				387.252	5364370	5622600	3–1	1.053+01	7.888–03	3.017–02	–1.6259	AA	*LS*
				387.342	5364430	5622600	5–5	1.578+00	3.549–03	2.263–02	–1.7509	AA	*LS*
				387.252	5364370	5622600	3–3	2.631+00	5.916–03	2.263–02	–1.7508	AA	*LS*
				387.252	5364370	5622600	3–5	1.053–01	3.944–04	1.508–03	–2.9269	AA	*LS*
45.	$1s3d$–$1s5p$	^{3}D–^{3}P°		*262.46*	*5364423*	5745440	15–9	4.492+00	2.783–03	3.607–02	–1.3793	AA	1
				262.467	5364440	5745440	7–5	3.773+00	2.783–03	1.683–02	–1.7103	AA	*LS*
				262.460	5364430	5745440	5–3	3.369+00	2.088–03	9.019–03	–1.9814	AA	*LS*
				262.419	5364370	5745440	3–1	4.494+00	1.547–03	4.008–03	–2.3335	AA	*LS*
				262.460	5364430	5745440	5–5	6.738–01	6.958–04	3.006–03	–2.4585	AA	*LS*
				262.419	5364370	5745440	3–3	1.124+00	1.160–03	3.006–03	–2.4585	AA	*LS*
				262.419	5364370	5745440	3–5	4.494–02	7.733–05	2.004–04	–3.6345	AA	*LS*
46.	$1s3d$–$1s6p$	^{3}D–^{3}P°		*223.56*	*5364423*	5811730	15–9	2.414+00	1.085–03	1.198–02	–1.7883	AA	1
				223.569	5364440	5811730	7–5	2.028+00	1.085–03	5.592–03	–2.1193	AA	*LS*
				223.564	5364430	5811730	5–3	1.811+00	8.140–04	2.996–03	–2.3904	AA	*LS*
				223.534	5364370	5811730	3–1	2.415+00	6.031–04	1.331–03	–2.7425	AA	*LS*
				223.564	5364430	5811730	5–5	3.621–01	2.713–04	9.985–04	–2.8675	AA	*LS*
				223.534	5364370	5811730	3–3	6.038–01	4.523–04	9.985–04	–2.8674	AA	*LS*
				223.534	5364370	5811730	3–5	2.415–02	3.015–05	6.657–05	–4.0435	AA	*LS*
47.	$1s3d$–$1s7p$	^{3}D–^{3}P°		*205.14*	*5364423*	5851890	15–9	1.38+00	5.23–04	5.30–03	–2.105	A	4
				205.149	5364440	5851890	7–5	1.16+00	5.23–04	2.47–03	–2.436	A	*LS*
				205.145	5364430	5851890	5–3	1.04+00	3.92–04	1.33–03	–2.707	A	*LS*
				205.120	5364370	5851890	3–1	1.38+00	2.91–04	5.89–04	–3.059	A	*LS*
				205.145	5364430	5851890	5–5	2.07–01	1.31–04	4.42–04	–3.184	A	*LS*
				205.120	5364370	5851890	3–3	3.46–01	2.18–04	4.42–04	–3.184	A	*LS*
				205.120	5364370	5851890	3–5	1.38–02	1.45–05	2.94–05	–4.360	A	*LS*
48.	$1s3d$–$1s3p$	^{1}D–^{1}P°	32460	3080 cm^{-1}	5365470	5368550	5–3	7.41–04	7.03–03	3.75+00	–1.454	A	1
49.	$1s3d$–$1s4p$	^{1}D–^{1}P°		380.764	5365470	5628100	5–3	7.720+00	1.007–02	6.310–02	–1.2981	AA	1
50.	$1s3d$–$1s5p$	^{1}D–^{1}P°		261.110	5365470	5748450	5–3	3.331+00	2.043–03	8.781–03	–1.9908	AA	1
51.	$1s3p$–$1s4s$	^{1}P°–^{1}S		392.850	5368550	5623100	3–1	5.555+01	4.284–02	1.662–01	–0.8910	AA	1
52.	$1s3p$–$1s4d$	^{1}P°–^{1}D		387.417	5368550	5626670	3–5	1.689+02	6.336–01	2.424+00	0.2789	AA	1

O VII: Allowed Transitions — Continued

No.	Transition Array	Mult.	λ_{air} (Å)	λ_{vac} (Å) or σ (cm^{-1})*	E_i (cm^{-1})		E_k (cm^{-1})	g_i-g_k	A_{ki} (10^8 s^{-1})	f_{ik}	S (at. u.)	log gf	Acc.	Source
53.	1s3p–1s5d	$^{1}P^{o}-^{1}D$		263.380	5368550	–	5748230	3–5	8.087+01	1.402–01	3.646–01	–0.3762	AA	1
54.	1s3p–1s6d	$^{1}P^{o}-^{1}D$		224.653	5368550	–	5813680	3–5	4.452+01	5.614–02	1.246–01	–0.7736	AA	1
55.	1s4s–1s4p	$^{3}S-^{3}P^{o}$	15970	6260 cm^{-1}	5616340	–	5622600	3–9	1.906–02	2.188–01	3.452+01	–0.1828	AA	1
56.	1s4s–1s5p	$^{3}S-^{3}P^{o}$		774.593	5616340	–	5745440	3–9	1.550+01	4.184–01	3.200+00	0.0987	AA	1
57.	1s4s–1s6p	$^{3}S-^{3}P^{o}$		511.797	5616340	–	5811730	3–9	9.775+00	1.152–01	5.821–01	–0.4616	AA	1
58.	1s4s–1s7p	$^{3}S-^{3}P^{o}$		424.538	5616340	–	5851890	3–9	6.30+00	5.11–02	2.14–01	–0.815	A	4
59.	1s4s–1s8p	$^{3}S-^{3}P^{o}$		382.468	5616340	–	5877800	3–9	1.76+00	1.16–02	4.38–02	–1.459	A	4
60.	1s4p–1s4d	$^{3}P^{o}-^{3}D$	*27580*	*3625* cm^{-1}	5622600	–	*5626225*	9–15	3.37–03	6.40–02	5.23+01	–0.240	A	1
			25630	3900 cm^{-1}	5622600	–	5626500	5–7	4.19–03	5.78–02	2.44+01	–0.539	A	*LS*
			29400	3400 cm^{-1}	5622600	–	5626000	3–5	2.08–03	4.50–02	1.31+01	–0.870	A	*LS*
			29750	3360 cm^{-1}	5622600	–	5625960	1–3	1.49–03	5.93–02	5.81+00	–1.227	A	*LS*
			29400	3400 cm^{-1}	5622600	–	5626000	5–5	6.94–04	9.00–03	4.36+00	–1.347	A	*LS*
			29750	3360 cm^{-1}	5622600	–	5625960	3–3	1.12–03	1.48–02	4.36+00	–1.352	A	*LS*
			29750	3360 cm^{-1}	5622600	–	5625960	5–3	7.44–05	5.93–04	2.91–01	–2.528	A	*LS*
61.	1s4p–1s5s	$^{3}P^{o}-^{3}S$		833.264	5622600	–	5742610	9–3	2.123+01	7.366–02	1.819+00	–0.1785	AA	1
62.	1s4p–1s5d	$^{3}P^{o}-^{3}D$		*800.58*	5622600	–	*5747509*	9–15	3.445+01	5.517–01	1.309+01	0.6960	AA	1
				799.872	5622600	–	5747620	5–7	3.454+01	4.639–01	6.107+00	0.3654	AA	*LS*
				801.154	5622600	–	5747420	3–5	2.578+01	4.135–01	3.272+00	0.0936	AA	*LS*
				801.282	5622600	–	5747400	1–3	1.909+01	5.512–01	1.454+00	–0.2587	AA	*LS*
				801.154	5622600	–	5747420	5–5	8.594+00	8.270–02	1.091+00	–0.3835	AA	*LS*
				801.282	5622600	–	5747400	3–3	1.432+01	1.378–01	1.091+00	–0.3836	AA	*LS*
				801.282	5622600	–	5747400	5–3	9.545–01	5.512–03	7.271–02	–1.5597	AA	*LS*
63.	1s4p–1s6d	$^{3}P^{o}-^{3}D$		525.017	5622600	–	5813070	9–15	2.049+01	1.411–01	2.196+00	0.1039	AA	1
64.	1s4p–1s7d	$^{3}P^{o}-^{3}D$		432.788	5622600	–	5853660	9–15	1.30+01	6.09–02	7.80–01	–0.261	A	4
65.	1s4s–1s4p	$^{1}S-^{1}P^{o}$	20000	5000 cm^{-1}	5623100	–	5628100	1–3	9.14–03	1.64–01	1.08+01	–0.784	A	1
66.	1s4s–1s5p	$^{1}S-^{1}P^{o}$		797.766	5623100	–	5748450	1–3	1.573+01	4.502–01	1.182+00	–0.3466	AA	1
67.	1s4s–1s6p	$^{1}S-^{1}P^{o}$		523.972	5623100	–	5813950	1–3	9.837+00	1.215–01	2.095–01	–0.9155	AA	1
68.	1s4s–1s7p	$^{1}S-^{1}P^{o}$		435.464	5623100	–	5852740	1–3	6.281+00	5.357–02	7.680–02	–1.2711	A	4

O VII: Allowed Transitions — Continued

No.	Transition Array	Mult.	λ_{air} (Å)	λ_{vac} (Å) or σ (cm^{-1})*	E_i (cm^{-1})	E_k (cm^{-1})	g_i–g_k	A_{ki} (10^8 s^{-1})	f_{ik}	S (at. u.)	log gf	Acc.	Source
69.	1s4d–1s5p	^{3}D–^{3}P°		*838.82*	*5626225*	5745440	15–9	5.526+00	3.498–02	1.449+00	–0.2801	AA	1
				840.760	5626500	5745440	7–5	4.610+00	3.490–02	6.761–01	–0.6121	AA	*LS*
				837.240	5626000	5745440	5–3	4.168+00	2.628–02	3.622–01	–0.8814	AA	*LS*
				836.960	5625960	5745440	3–1	5.563+00	1.947–02	1.610–01	–1.2334	AA	*LS*
				837.240	5626000	5745440	5–5	8.336–01	8.761–03	1.207–01	–1.3585	AA	*LS*
				836.960	5625960	5745440	3–3	1.391+00	1.461–02	1.207–01	–1.3584	AA	*LS*
				836.960	5625960	5745440	3–5	5.563–02	9.737–04	8.049–03	–2.5344	AA	*LS*
70.	1s4d–1s6p	^{3}D–^{3}P°		*539.07*	*5626225*	5811730	15–9	2.732+00	7.141–03	1.901–01	–0.9701	AA	1
				539.869	5626500	5811730	7–5	2.285+00	7.131–03	8.871–02	–1.3018	AA	*LS*
				538.416	5626000	5811730	5–3	2.056+00	5.362–03	4.753–02	–1.5717	AA	*LS*
				538.300	5625960	5811730	3–1	2.744+00	3.973–03	2.112–02	–1.9238	AA	*LS*
				538.416	5626000	5811730	5–5	4.113–01	1.787–03	1.584–02	–2.0488	AA	*LS*
				538.300	5625960	5811730	3–3	6.859–01	2.980–03	1.584–02	–2.0487	AA	*LS*
				538.300	5625960	5811730	3–5	2.744–02	1.987–04	1.056–03	–3.2248	AA	*LS*
71.	1s4d–1s7p	^{3}D–^{3}P°		*443.13*	*5626225*	5851890	15–9	1.56+00	2.76–03	6.04–02	–1.383	A	4
				443.675	5626500	5851890	7–5	1.31+00	2.76–03	2.82–02	–1.714	A	*LS*
				442.693	5626000	5851890	5–3	1.18+00	2.07–03	1.51–02	–1.985	A	*LS*
				442.615	5625960	5851890	3–1	1.57+00	1.54–03	6.71–03	–2.337	A	*LS*
				442.693	5626000	5851890	5–5	2.35–01	6.91–04	5.03–03	–2.462	A	*LS*
				442.615	5625960	5851890	3–3	3.92–01	1.15–03	5.03–03	–2.462	A	*LS*
				442.615	5625960	5851890	3–5	1.57–02	7.68–05	3.36–04	–3.638	A	*LS*
72.	1s4d–1s8p	^{3}D–^{3}P°		*397.50*	*5626225*	5877800	15–9	3.73–01	5.30–04	1.04–02	–2.100	A	4
				397.931	5626500	5877800	7–5	3.12–01	5.29–04	4.85–03	–2.431	A	*LS*
				397.141	5626000	5877800	5–3	2.80–01	3.98–04	2.60–03	–2.701	A	*LS*
				397.078	5625960	5877800	3–1	3.74–01	2.95–04	1.16–03	–3.054	A	*LS*
				397.141	5626000	5877800	5–5	5.61–02	1.33–04	8.67–04	–3.179	A	*LS*
				397.078	5625960	5877800	3–3	9.35–02	2.21–04	8.67–04	–3.178	A	*LS*
				397.078	5625960	5877800	3–5	3.74–03	1.47–05	5.78–05	–4.355	A	*LS*
73.	1s4d–1s4p	^{1}D–^{1}P°		1430 cm^{-1}	5626670	5628100	5–3	2.697–04	1.186–02	1.365+01	–1.2269	AA	1
74.	1s4d–1s5p	^{1}D–^{1}P°		821.153	5626670	5748450	5–3	4.252+00	2.579–02	3.486–01	–0.8896	AA	1
75.	1s4d–1s6p	^{1}D–^{1}P°		533.960	5626670	5813950	5–3	2.130+00	5.462–03	4.801–02	–1.5636	AA	1
76.	1s4d–1s7p	^{1}D–^{1}P°		442.341	5626670	5852740	5–3	1.21+00	2.13–03	1.55–02	–1.973	A	4
77.	1s4p–1s5d	^{1}P°–^{1}D		832.432	5628100	5748230	3–5	3.625+01	6.276–01	5.160+00	0.2748	AA	1
78.	1s4p–1s6d	^{1}P°–^{1}D		538.851	5628100	5813680	3–5	2.075+01	1.506–01	8.012–01	–0.3452	AA	1
79.	1s5s–1s5p	^{3}S–^{3}P°	35300	2830 cm^{-1}	5742610	5745440	3–9	4.94–03	2.77–01	9.68+01	–0.080	A	1
80.	1s5s–1s6p	^{3}S–^{3}P°		1446.76	5742610	5811730	3–9	4.916+00	4.628–01	6.613+00	0.1425	AA	1
81.	1s5s–1s7p	^{3}S–^{3}P°		915.081	5742610	5851890	3–9	3.37+00	1.27–01	1.15+00	–0.419	A	4

O VII: Allowed Transitions — Continued

No.	Transition Array	Mult.	λ_{air} (Å)	λ_{vac} (Å) or σ (cm^{-1})*	E_i (cm^{-1})	E_k (cm^{-1})	g_i-g_k	A_{ki} (10^8 s^{-1})	f_{ik}	S (at. u.)	log gf	Acc.	Source
82.	1s5s– 1s8p	^{3}S–^{3}P°		739.700	5742610	5877800	3–9	8.00−01	1.97−02	1.44−01	−1.228	A	4
83.	1s5p– 1s5d	^{3}P°–^{3}D	*48320*	*2069* cm^{-1}	5745440	*5747509*	9–15	1.51−03	8.84−02	1.27+02	−0.099	A	1
			45860	2180 cm^{-1}	5745440	5747620	5–7	1.77−03	7.82−02	5.91+01	−0.408	A	*LS*
				1980 cm^{-1}	5745440	5747420	3–5	9.96−04	6.35−02	3.17+01	−0.720	A	*LS*
				1960 cm^{-1}	5745440	5747400	1–3	7.15−04	8.38−02	1.41+01	−1.077	A	*LS*
				1980 cm^{-1}	5745440	5747420	5–5	3.32−04	1.27−02	1.06+01	−1.198	A	*LS*
				1960 cm^{-1}	5745440	5747400	3–3	5.37−04	2.09−02	1.06+01	−1.202	A	*LS*
				1960 cm^{-1}	5745440	5747400	5–3	3.58−05	8.38−04	7.03−01	−2.378	A	*LS*
84.	1s5p– 1s6d	^{3}P°–^{3}D		1478.63	5745440	5813070	9–15	1.023+01	5.590−01	2.449+01	0.7016	AA	1
85.	1s5p– 1s7d	^{3}P°–^{3}D		924.044	5745440	5853660	9–15	6.97+00	1.49−01	4.07+00	0.127	A	4
86.	1s5d– 1s6p	^{3}D–^{3}P°		*1557.1*	*5747509*	5811730	15–9	2.724+00	5.941−02	4.568+00	−0.0500	AA	1
				1559.82	5747620	5811730	7–5	2.276+00	5.931−02	2.132+00	−0.3818	AA	*LS*
				1554.97	5747420	5811730	5–3	2.052+00	4.462−02	1.142+00	−0.6515	AA	*LS*
				1554.48	5747400	5811730	3–1	2.738+00	3.306−02	5.076−01	−1.0035	AA	*LS*
				1554.97	5747420	5811730	5–5	4.103−01	1.487−02	3.807−01	−1.1286	AA	*LS*
				1554.48	5747400	5811730	3–3	6.845−01	2.480−02	3.807−01	−1.1285	AA	*LS*
				1554.48	5747400	5811730	3–5	2.738−02	1.653−03	2.538−02	−2.3046	AA	*LS*
87.	1s5d– 1s7p	^{3}D–^{3}P°		*958.03*	*5747509*	5851890	15–9	1.50+00	1.24−02	5.86−01	−0.731	A	4
				959.049	5747620	5851890	7–5	1.26+00	1.24−02	2.73−01	−1.063	A	*LS*
				957.213	5747420	5851890	5–3	1.13+00	9.29−03	1.46−01	−1.333	A	*LS*
				957.029	5747400	5851890	3–1	1.50+00	6.89−03	6.51−02	−1.685	A	*LS*
				957.213	5747420	5851890	5–5	2.26−01	3.10−03	4.88−02	−1.810	A	*LS*
				957.029	5747400	5851890	3–3	3.76−01	5.16−03	4.88−02	−1.810	A	*LS*
				957.029	5747400	5851890	3–5	1.50−02	3.44−04	3.25−03	−2.986	A	*LS*
88.	1s5d– 1s8p	^{3}D–^{3}P°		*767.51*	*5747509*	5877800	15–9	2.78−01	1.47−03	5.58−02	−1.656	A	4
				768.167	5747620	5877800	7–5	2.33−01	1.47−03	2.60−02	−1.987	A	*LS*
				766.989	5747420	5877800	5–3	2.09−01	1.10−03	1.40−02	−2.258	A	*LS*
				766.871	5747400	5877800	3–1	2.79−01	8.19−04	6.20−03	−2.610	A	*LS*
				766.989	5747420	5877800	5–5	4.18−02	3.68−04	4.65−03	−2.735	A	*LS*
				766.871	5747400	5877800	3–3	6.96−02	6.14−04	4.65−03	−2.735	A	*LS*
				766.871	5747400	5877800	3–5	2.79−03	4.09−05	3.10−04	−3.911	A	*LS*
89.	1s5d– 1s5p	^{1}D–^{1}P°		220 cm^{-1}	5748230	5748450	5–3	8.66−06	1.61−02	1.20+02	−1.094	B	1
90.	1s5d– 1s6p	^{1}D–^{1}P°		1521.61	5748230	5813950	5–3	2.150+00	4.478−02	1.122+00	−0.6500	AA	1
91.	1s5d– 1s7p	^{1}D–^{1}P°		956.846	5748230	5852740	5–3	1.17+00	9.61−03	1.51−01	−1.318	A	4
92.	1s5p– 1s6d	^{1}P°–^{1}D		1533.04	5748450	5813680	3–5	1.099+01	6.454−01	9.772+00	0.2869	AA	1
93.	1s6p– 1s6d	^{3}P°–^{3}D		1340 cm^{-1}	5811730	5813070	9–15	8.01−04	1.11−01	2.46+02	0.001	A	1
94.	1s6p– 1s7d	^{3}P°–^{3}D	2384.2	2384.93	5811730	5853660	9–15	4.20+00	5.96−01	4.21+01	0.730	A	4

O VII: Allowed Transitions — Continued

No.	Transition Array	Mult.	λ_{air} (Å)	λ_{vac} (Å) or σ (cm^{-1})*	E_i (cm^{-1})	E_k (cm^{-1})	g_i–g_k	A_{ki} (10^8 s^{-1})	f_{ik}	S (at. u.)	log gf	Acc.	Source
95.	1s6d– 1s7p	^{3}D–^{3}P°	2575.2	2575.99	5813070	5851890	15–9	1.44+00	8.56–02	1.09+01	0.109	A	4
96.	1s6d– 1s8p	^{3}D–^{3}P°		1544.88	5813070	5877800	15–9	1.73–01	3.71–03	2.83–01	–1.254	A	4
97.	1s6d– 1s6p	^{1}D–^{1}P°		270 cm^{-1}	5813680	5813950	5–3	1.63–05	2.01–02	1.23+02	–0.998	B	1
98.	1s6d– 1s7p	^{1}D–^{1}P°	2559.40	2560.16	5813680	5852740	5–3	1.09+00	6.44–02	2.71+00	–0.492	A	4

*Wavelengths (Å) are always given unless cm^{-1} is indicated.

O VII

Forbidden Transitions

The tabulated data are from calculations which are based on correlated, but nonrelativistic, wavefunctions.[1–4] Drake has used either a nuclear charge expansion calculation[1] with the perturbation coefficient evaluated through ninth order or he has used 50-term variational wavefunctions.[2] For the helium-like carbon and nitrogen spectra, where high precision experimental data for the M1 line are available (see C V and N VI introductions on forbidden lines), the agreement between the experiments and his results is almost perfect. Godefroid and Verhaegen[4] use an extensive multiconfiguration approach and have calculated their data both in the length and velocity form. These two forms usually agree within four significant figures.

References

[1]G. W. F. Drake, Phys. Rev. A **3**, 908 (1971).
[2]G. W. F. Drake, Astrophys. J. **158**, 1199 (1969).
[3]B. Kundu and P. K. Mukherjee, Astrophys J. **298**, 844 (1985).
[4]M. Godefroid and G. Verhaegen, J. Phys. B. **13**, 3081 (1980).

O VII: Forbidden Transitions

No.	Transition Array	Mult.	λ_{air} (Å)	λ_{vac} (Å) or σ (cm^{-1})*	E_i (cm^{-1})	E_k (cm^{-1})	g_i–g_k	Type	A_{ki} (s^{-1})	S (at. u.)	Acc.	Source
1.	1s^2– 1s2s	^{1}S–^{3}S		22.1012	0	4524640	1–3	M1	1.04+03	1.25–06	A+	1
2.	1s^2– 1s2p	^{1}S–^{3}P°		21.8044	0	4586230	1–5	M2	3.31+05	5.47–01	A	2
3.	1s^2– 1s3s	^{1}S–^{3}S		18.7307	0	5338820	1–3	M1	3.57+02	2.61–07	C	3
4.	1s^2– 1s3d	^{1}S–^{1}D		18.6377	0	5365470	1–5	E2	9.656+07	9.695–04	AA	4
5.	1s^2– 1s4s	^{1}S–^{3}S		17.8052	0	5616340	1–3	M1	1.84+02	1.16–07	C	3

O VII: Forbidden Transitions — Continued

No.	Transition Array	Mult.	λ_{air} (Å)	λ_{vac} (Å) or σ (cm^{-1})*	E_i (cm^{-1})		E_k (cm^{-1})	g_i-g_k	Type	A_{ki} (s^{-1})	S (at. u.)	Acc.	Source
6.	$1s2s-$ $1s3d$	$^1S-^1D$											
				128.685	4588380	–	5365470	1–5	E2	4.463+06	7.031−01	AA	4
7.	$1s2p-$ $1s4f$	$^1P°-^1F°$											
				100.237	4629201	–	5626840	3–7	E2	4.841+06	3.062−01	AA	4

*Wavelengths (Å) are always given unless cm^{-1} is indicated.

O VIII

Hydrogen Isoelectronic Sequence

Ground State: $1s\ ^2S_{1/2}$

Ionization Energy: 871.387 eV = 7028394 cm^{-1}

We have tabulated numerical data for hydrogen-like ions only for the spectrum of N VII (nuclear charge $Z = 7$). Data for O VIII ($Z = 8$) may be obtained by scaling the data for the *corresponding transitions* of N VII according to the following relationships:[1]

$$f(\text{O VIII}) = f(\text{N VII}),$$

$$A(\text{O VIII}) = \left(\frac{8}{7}\right)^4 A(\text{N VII}) = 1.7060\,A(\text{N VII}),$$

$$S(\text{O VIII}) = \left(\frac{8}{7}\right)^{-2} S(\text{N VII}) = 0.7656\,S(\text{N VII}).$$

The wavelength and energy level data for O VIII may be obtained from the calculations of Erickson,[2] and Johnson and Soff.[3]

References

[1]A. Corney, Atomic and Laser Spectroscopy, Oxford Univ. Press, Oxford (1977).
[2]G. W. Erickson, J. Phys. Chem. Ref. Data **6**, 831 (1977).
[3]W. R. Johnson and G. Soff, At. Data Nucl. Data Tables **33**, 405 (1985).

American Institute of Physics

500 Sunnyside Boulevard
Woodbury, NY 11797-2999

Roland W. Schmitt	Chair, Governing Board
Marc H. Brodsky	Executive Director and CEO
Roderick M. Grant	Secretary
Arthur T. Bent	Treasurer and CFO
John S. Rigden	Director of Physics Programs
Darlene A. Walters	Vice President, Publishing

The American Institute of Physics was founded in 1931; it is chartered as a membership corporation with leading societies in the fields of physics and astronomy as members. The Institute combines into one operating agency those functions which can best be done by the societies jointly. Its purpose is the advancement and diffusion of the knowledge of physics and its application to human welfare. About 100 000 physicists belong to its ten member societies: The American Physical Society, Optical Society of America, Acoustical Society of America, Society of Rheology, American Association of Physics Teachers, American Crystallographic Association, American Astronomical Society, American Association of Physicists in Medicine, American Vacuum Society, American Geophysical Union. In addition, there are 20 Affiliated Societies, 84 Corporate Associates, and 6500 members of the Society of Physics Students.

For itself and its member societies, AIP publishes 20 archival journals, 17 translated Soviet journals, a general interest magazine, conference proceedings, handbooks, and bulletin-programs, as well as an integrated package of current awareness journals, magnetic tapes, and microfilm. AIP also serves the public by making available to the press and other channels of public information reliable communications on physics and astronomy and their progress; carries on extensive career planning and placement activities; encourages and assists in the documentation and study of the history and philosophy of recent physics; cooperates with local, national, and international organizations devoted to physics and the related sciences; and fosters the relations of the science of physics to other sciences and to the arts and industries.

American Chemical Society

1155 Sixteenth Street, NW
Washington, DC 20036-9976

Paul H. L. Walter	Chairman of the Board
Brian M. Rushton	President
John K Crum	Executive Director
Brian A. Bernstein	Treasurer
D.H. Michael Bowen	Deputy Executive Director and Secretary
Robert J. Massie	Director, Chemical Abstracts Service
Robert H. Marks	Director, Publications Division*
Halley A. Merrell	Director, Membership Division
Sylvia A. Ware	Director, Education Division

The American Chemical Society was founded in 1876 as a scientific and educational society, and in 1937 was granted a national charter by an act of Congress. Its objectives include the advancement of chemistry, the promotion of research in science and industry, improvement of qualifications and usefulness of chemists, increased diffusion of chemical knowledge, and the promotion of scientific interest and inquiry.

ACS publishes 28 magazines, journals, and other scientific periodicals in addition to its participation in publication of the *Journal of Physical and Chemical Reference Data*. It also publishes books and other nonperiodical reports and operates the Chemical Abstracts Service, the key to the world's chemical literature. It pursues its objectives through its services to chemists and chemical engineers and to education, industry, and government.

*Inquiries regarding ACS participation in JPCRD should be directed to this division.

National Institute of Standards and Technology

U.S. Department of Commerce
Gaithersburg, MD 20899-0001

U.S. Department of Commerce	Ronald H. Brown, Secretary
Technology Administration	Mary L. Good, Under Secretary for Technology
National Institute of Standards and Technology	Arati Prabhakar, Director

The National Institute of Standards and Technology (NIST), formerly the National Bureau of Standards, was established by an act of Congress 3 March 1901. NIST works to strengthen U.S. industry's international competitiveness, advance science and engineering, and improve public health, safety, and the environment. With an annual budget of more than $350 million and about 3100 staff members, NIST conducts science and engineering research in commercially important fields such as advanced materials, information systems, biotechnology, optoelectronics, computer-integrated manufacturing, and sensor technology.

NIST's laboratory research is designed to support development of critical emerging technologies and the new measurement methods and standards necessary to make them commercially viable. The ability of U.S.-based industries to exploit these new technologies determines in large part the health of the U.S. economy.

Several new NIST programs are designed to help spur innovation at U.S. businesses through seed money for development of generic technologies; grants to states for support of technology transfer programs; and financial and technical assistance to help small and mid-sized companies adopt more efficient manufacturing methods.

Throughout its history, NIST has helped industry to develop high-quality products and processes by supplying standard reference materials and data; by continuously improving technologies for precision measurements of mass, time, length, voltage, radiation, and other units; and through active participation in voluntary standards-setting bodies.